P9-DTJ-226

Aquatic Insects of California

Aquatic Insects of California

WITH KEYS TO NORTH AMERICAN GENERA AND CALIFORNIA SPECIES

Edited by ROBERT L. USINGER

W. C. BENTINCK
H. G. CHANDLER
W. C. DAY
D. G. DENNING
K. S. HAGEN
S. G. JEWETT, JR.
W. H. LANGE, JR.
IRA LA RIVERS
J. D. LATTIN
H. B. LEECH
A. E. PRITCHARD
D. B. SCOTT, JR.
R. F. SMITH
ALAN STONE
R. L. USINGER
W. W. WIRTH

UNIVERSITY OF CALIFORNIA PRESS
BERKELEY, LOS ANGELES, LONDON

University of California Press
Berkeley and Los Angeles, California

University of California Press, Ltd.
London, England

ISBN: 0-520-01293-3

Manufactured in the United States of America

8 9 0

To

Harry Phylander Chandler

Whose untimely death during the preparation of this work deprived us of an esteemed colleague and close friend.

Preface

Aquatic entomology is a large and diverse science. It partakes of many fields and, in its applied aspects, is pursued by persons with quite unrelated interests and objectives. Specialists in mosquito control, for example, are not directly concerned with problems of stream and lake management for the production of fish, and limnologists generally are inclined to stress physical and chemical studies and plankton investigations rather than trying to deal with numerous and imperfectly known insects.

It is the purpose of this study to point out the central role of insects in aquatic situations, to bring to bear on insects the basic concepts and tools of limnology, and to provide keys and illustrations to aid in the identification of aquatic insects. It is probably this last aspect of the subject that has been the greatest stumbling block to progress in the past. Insects comprise about four-fifths of the Animal Kingdom, and only time and intensive research can overcome the obstacle of such numbers of species. Because of limitations of time, space, and size it has been necessary to restrict the treatment at the species level to California and adjacent regions. However, the California fauna is fairly representative of much of the western United States because of the diverse elements that are included within the geographical boundaries of the state. At the generic level a more comprehensive treatment was feasible, so keys are given to all the genera known from North America.

Most books are the result of the work of many people, and this is particularly true of the present volume. It is a direct outgrowth of University of California Syllabus SS *Biology of Aquatic and Littoral Insects* (Entomology 133, by R. L. Usinger, Ira La Rivers, H. P. Chandler, and W. W. Wirth, University of California Press, 1948), now out of print, which was tested in the laboratory and in the field and was used for several years in classes at various institutions in the West. The syllabus was frankly a compilation of existing knowledge with comparatively little original work. It was meant to be a one-volume working "library." Now, after eight years, it can be said that the objectives of the syllabus have been realized. Extensive collections have been gathered from all parts of the state, and specialists have devoted much time and effort to each group. The results are offered at this time as original contributions, each chapter written by an authority who is most intimately acquainted with our fauna. Illustrations have been added to clarify the text, and a glossary explains the technical terms that are not understandable from the figures. Much remains to be done in every group, but it can now be said that we have a sound foundation on which to build.

Acknowledgments

The editor is personally grateful to the group of distinguished collaborators, each of whom has given generously of his research time to make this volume possible. We are indebted also to all those students who have contributed to the store of knowledge on aquatic insects, a rich heritage from which we have drawn heavily. Detailed acknowledgments are given in each chapter and on nearly every page, but a few persons or institutions have contributed so extensively that they are deserving of special mention here.

For advice and assistance in the introductory sections, thanks are due to Paul R. Needham (general principles, stream and lake management); F. R. Pitelka (ecology); R. F. Peters, J. R. Walker, A. C. Smith, and T. D. Mulhern in the Bureau of Vector Control, State Department of Public Health (mosquito control); E. A. Smith and E. H. Pearl of the Santa Clara County Health Department (mosquito control); W. R. Kellen, Research Entomologist, U.S.P.H.S. grant for study of insects in relation to sewage disposal; and Dana Abell, National Science Foundation Fellow (stream classification). Many of the suggestions of the above-mentioned specialists have been incorporated, but it should be made clear that the author alone is responsible for the final version. Valuable assistance was furnished in the final preparation of the manuscript by the following: J. D. Lattin (photographic copy of illustrations), W. C. Bentinck (glossary), and Jon Herring (index).

For fundamental works in taxonomy, we are indebted to Dr. J. G. Needham and his colleagues and students at Cornell University, and to the late S. A. Forbes and T. H. Frison and to H. H. Ross and others at the Illinois Natural History Survey who have done so much to further our knowledge of the aquatic insects of North America.

Taxonomic work represents the fruition of the labors of countless collectors and curators. In California, field work and collections are recent, as compared with many parts of the world, but are nonetheless impressive. This work is based largely on the collections of the University of California Insect Survey (Berkeley), the University of California collections at Davis, Los Angeles, and Riverside, the California Academy of Sciences (San Francisco), the Los Angeles Museum, the San Diego Museum, and the personal collections of the authors. Thanks are due to the curators in charge of these collections, and especially to E. S. Ross, J. N. Belkin, P. D. Hurd, and A. T. MacClay.

The illustrations may very well prove to be the most useful part of this book. Acknowledgment is given for each borrowed figure by citation in the legend and by listing the original work in the bibliography at the end of each chapter. Original drawings are mostly by Mrs. Celeste Green of the Department of Entomology and Parasitology, University of California, Berkeley (introduction and chapters 4, 10, 11, and 13) and by Arthur Smith, at the British Museum (Natural History) (chapter 7). Photographic work in connection with illustrations was done by J. D. Lattin and W. C. Bentinck. Several chapters have been read and improvements suggested by Jon Herring. For permission to use copyrighted figures we are indebted to the following publishers and authors: University of Toronto Press, E. M. Walker, *The Odonata of Canada and Alaska,* 1953; Pennsylvania Fish Commission, K. F. Lagler, *Freshwater Fishery Biology;* John Wiley & Sons, Inc., G. C. Whipple, *The Microscopy of Drinking Water,* 4th ed., 1927; The Macmillan Company, E. O. Essig, *College Entomology,* 1942; Lane Publishing Company, *Sunset Magazine;* University of California Press, E. S. Ross, *Insects Close Up,* 1953; Institute for Fisheries Research, Michigan Department of Conservation, C. L. Hubbs and R. W. Eschmeyer, *The Improvement of Lakes for Fishing. A Method of Fish Management,* 1938; C. Wesenberg-Lund, *Biologie der Süsswasserinsekten;* The Macmillan Company, E. O. Essig, *Insects of Western North America,* 1926; Entomologica Americana, A. G. Böving and F. C. Craighead, *An Illustrated Synopsis of the Principal Larval Forms of the Order Coleoptera,* 1931; Ohio State University, D. J. Borror and D. M. Delong, *An Introduction to the Study of Insects,* 1954; Ohio State University, Alvah Peterson, *Larvae of Insects;* University of Toronto Press, F. Ruttner, *Fundamentals of Limnology,* 1953; A. S. Barnes and Company, J. Edson Leonard, *Flies,* 1950; Comstock Publishing Company, Inc., J. G. Needham, J. R. Traver, and Yin-Chi Hsu, *The Biology of Mayflies,* 1935; Comstock Publishing Company, Inc., R. Matheson, *Handbook of the Mosquitoes of North America,* 1944; The Ronald Press Company, R. W. Pennak, *Fresh-water Invertebrates of the United States,* copyright. 1953; W. B. Saunders Company, E. P. Odum, *Fundamentals of Ecology,* 1953; American Museum of Natural History, C. H. Curran, *The Families and Genera of North American Diptera;* Ward's Natural Science Establishment, Inc., *How to Make an Insect Collection;* Scientific American, Inc., articles by E. S. Deevey, Jr. and Ralf Eliassen; McGraw-Hill Book Co., Inc., R. E. Snodgrass, *Principles of Insect Morphology,* 1935; McGraw-Hill Book Company, Inc., P. S. Welch, *Limnology,* 1952; Methuen and Company Ltd., N. E. Hickin, *Caddis, a short account of the biology of British Caddis flies with special reference to the immature stages,* 1952; University of California Press, J. G. Needham and M. J. Westfall, Jr., *A Manual of the Dragonflies of North America,* 1955; Charles C. Thomas, Publisher, P. W. Claassen, *Plecoptera Nymphs of America (North of Mexico),* 1931; Charles C. Thomas, Publisher, J. G. Needham and H. B. Heywood, *A Manual of the Dragonflies of North America,* 1929; University of Florida, F. N. Young, *The Water Beetles of Florida,* 1954.

Berkeley, California
June 12, 1956

Robert L. Usinger

Contents

Aquatic Insects of California

Introduction to Aquatic Entomology

A. Principles and Practices

By Robert L. Usinger
University of California, Berkeley

Insects are generally the most conspicuous forms of life in ponds and streams and occur in tremendous numbers in such unlikely places as the bottoms of lakes. No other group of animals shows such diversity in structure and habits. And yet, aquatic insects fall into a pattern, each major group (order, family, or genus) occupying a particular habitat with species represented on each of the continents. Thus a stream in South Africa may resemble a stream in California, each having its own representatives from among the stoneflies, mayflies, caddisflies, and so on. Likewise, a pond in Sumatra and a lake in Sweden will resemble, in a general way, comparable bodies of water in North America. This same phenomenon is noted among the common genera of plankton organisms. However, many of the plankton species are cosmopolitan whereas aquatic insect species are usually limited in their distribution, with some species restricted to local oases in otherwise completely barren deserts. This leads to a multiplicity of insect species and, of course, provides the basis for still greater diversity.

It is thought that aquatic insects were derived from unknown terrestrial types which invaded the water on several occasions during the course of their evolution. The first record is of mayflylike insects of the now extinct order Paleodictyoptera. These appear suddenly in the geological record in rocks of the Upper Carboniferous period. At this remote time, 250 million years ago, the first winged insects had complete mastery of the air because no bird, bat, or flying reptile had yet developed to challenge them. The subsequent history of aquatic insects (and terrestrial forms, as well) was one of increasing diversity. First a wing-folding device rendered all higher insects (Neoptera) more efficient in flight than the dragonfly and mayfly types (Paleoptera). Second, the direct method of development through successive instars with external wing pads (stoneflies, true bugs) was improved by the adoption of a more indirect series of stages with larvae specialized for feeding, pupae for transformation, and adults for reproduction. Wing pads develop internally in such larvae. These advances came before the end of the Paleozoic era (200 million years), and all subsequent evolution was confined to modifications of these basic patterns of structure and development.

At the end of the Paleozoic and during the Mesozoic and Cenozoic eras representatives of many modern groups of insects took to the water. The exact sequence is not known, but probably the stoneflies were among the earliest with dobsonflies, beetles, and true bugs not far behind (Permian, 220 million years). Caddisflies, true flies, and parasitic wasps first appear in the very fragmentary record much later (Jurassic, 160 million years), and the Lepidoptera not until early Tertiary (60 million years). (See Carpenter, 1953, for a summary of information on the geological history of insects.) Furthermore, it seems certain that each of the large orders (Coleoptera, Diptera, etc.) invaded the water not once but several times. As Miall (1895) puts it, "I think we can say with a considerable degree of probability that this change of habitat from terrestrial to aquatic has taken place in the class of insects at least a hundred times quite independently, and the number may be very much higher than a hundred." As a result we have a most amazing variety of insects occupying aquatic habitats, many with a superficial similarity in form owing to the highly selective environment but each group with unique methods for performing essential life functions.

At present, ten orders of insects have truly aquatic forms, and several others may be described as semi-aquatic, at least in part. All these except the beetles and true bugs live on land or in the air as adults and in the water only in their immature stages. In contrast to this, most water bugs and beetles are aquatic throughout their lives but are directly dependent on surface air for respiration as adults. Thus it can be said that despite their great numbers and remarkable diversity, insects are only secondarily and incompletely adapted to aquatic life. This probably accounts for their prevalence in shallow ponds and streams, their scarcity in very large rivers and deep lakes, and their virtual absence from the open waters of the ocean.

Role of Insects in Aquatic Communities

In 1887 Stephan A. Forbes wrote an essay, "The Lake as a Microcosm," showing that a lake is essentially a self-contained or closed community. This concept is also applicable, though to a lesser extent, to ponds and even to larger streams. Each community of plants and animals is more or less attuned to its physical and biotic environment, and the various elements of which it is composed are integrated to form an ecosystem. Insects play an important but not a vital role in such systems. As dominant members of the littoral fauna, together with fishes, they are intermediate in position between the autotrophic or constructive elements (green plants) and such heterotrophic or destructive elements as the bacteria (intro. fig. 1.)

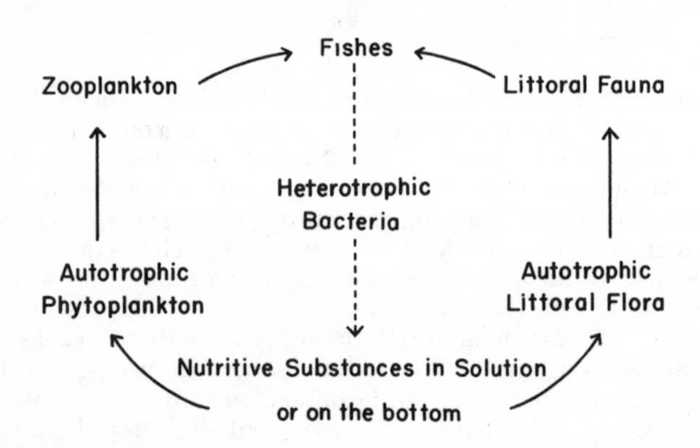

Intro. fig. 1. Simplified diagram of the dynamics of an aquatic community. Solid arrows represent constructive steps; dotted lines, reductive steps.

In the littoral fauna, insects such as mayfly nymphs and midge larvae serve as primary converters of plant materials into animal protoplasm. As pointed out by Elton (1947) such basic herbivores are "key industries" in a community and are usually small in size and large in numbers. Successive links in the food chains are larger and scarcer and are usually carnivorous. Essentially, the food chain concept is simple

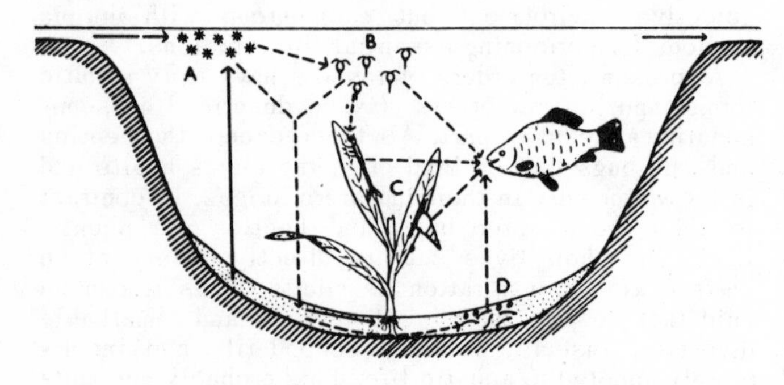

Intro. fig. 2. Diagram of the food chain in a pond. The continuous arrows show the course of inorganic salts and the broken lines indicate their course after they have been built up into living matter. *a*, phytoplankton; *b*, zoöplankton; *c*, weed-dwelling fauna; *d*, bottom fauna (Macan, Mortimer, and Worthington, 1942).

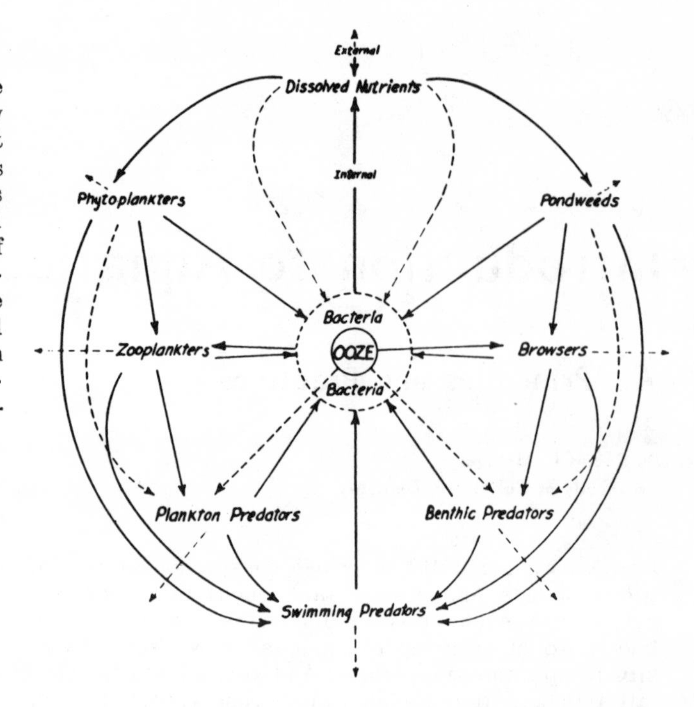

Intro. fig. 3. Food-cycle relationships in a lake (Lindeman, 1941).

(intro. fig. 2) with relatively few links. In nature, however, the situation becomes much more complex, so much so that even in diagrammatic form it has been termed a food web or food cycle (intro. fig. 3). Insects enter into such a cycle as "browsers" and as "swimming predators" on plankton and on benthic organisms.

Probably no two biotic communities are identical in every respect, but there are certain types that are characteristic of particular aquatic situations, and each of these types has a distinctive insect fauna. Three representative types are:

1. Lakes—with Chironomid larvae in the bottom ooze and *Chaoborus* larvae that prey on plankton and perform diurnal migrations from the bottom mud to or near the surface. A varied insect fauna also occurs in the shallow littoral waters of lakes but such forms, with few exceptions, are more characteristic of shallow ponds than of lakes.

2. Ponds—with a large and varied insect fauna including midge larvae in the bottom and mayfly nymphs, caddis larvae, and others as basic herbivores together with numerous predatory beetles, bugs, and odonatan nymphs. Although dependent on the plankton and rooted vegetation, it can truly be said that pond insects are dominant forms of life in their limited environment. Furthermore, they are admirably suited to the uncertain conditions of pond life, with short life histories and ready means of dispersal.

3. Streams—with stonefly nymphs, mayfly nymphs, caddisfly larvae, and various midge larvae as basic herbivores—sieve feeders or grazers—and a host of predaceous forms. Here again, insects predominate and form the staple diet of most fishes.

Adaptations of Aquatic Insects

All living organisms are variously adapted for survival in their respective environments. Many adaptations are so commonplace that they are taken for granted. However, the requirements for existence in aquatic habitats are so rigorous that the adaptations are usually striking. The stream-lined form, as seen in certain fishes and mayfly naiads of swift-flowing streams, and the flattened body with suction discs seen in Psephenid larvae (waterpennies) and certain fly larvae (Blepharoceratidae, Deuterophlebiidae, *Maruina*) that cling to surfaces in rapids, are examples. Other common adaptations, especially among adult aquatics, are reduction in size of antennae which are concealed to reduce water resistance, development of powerful legs with swimming hairs, and presence of hydrofuge hairs or waxy surfaces to prevent wetting. The latter are particularly important at critical periods such as time of hatching of the egg and time of emergence of the adult. Without such adaptations the emergence of a delicate simuliid fly from its pupal case attached to a rock in swift-flowing water would be impossible.

Surface film.—The nature of the surface film is of greatest importance to aquatic insects because of their amphibious existence. To an organism of small size this air-water interface can be an impenetrable barrier, a surface on which to rest, or a ceiling from which to hang suspended. At the surface the water molecules are arranged in such a way that a surface tension is created. This can be demonstrated by a drop of water on a waxy surface (intro. fig. 4). The angle

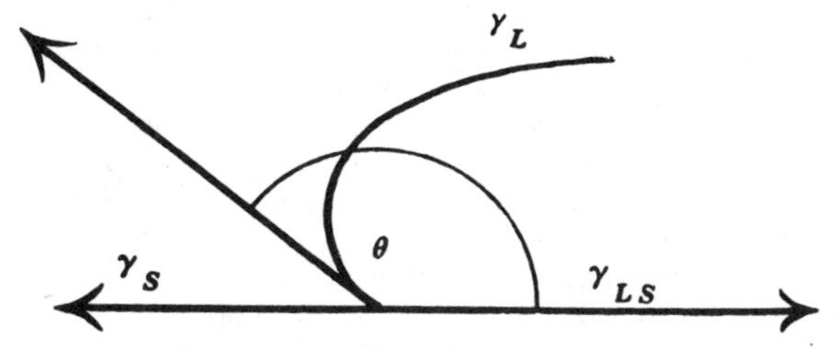

Intro. fig. 4. Diagram to show the angle of contact, made by a drop of fluid on a waxy surface, where γS is the solid-air tension, γLS the liquid-solid tension, and γL the liquid-air tension (Thorpe and Crisp, 1947).

of contact θ under these circumstances is 105°-110°. Addition of soap or some other wetting agent to the water changes the angle of contact so that the bubble spreads across the wax surface.

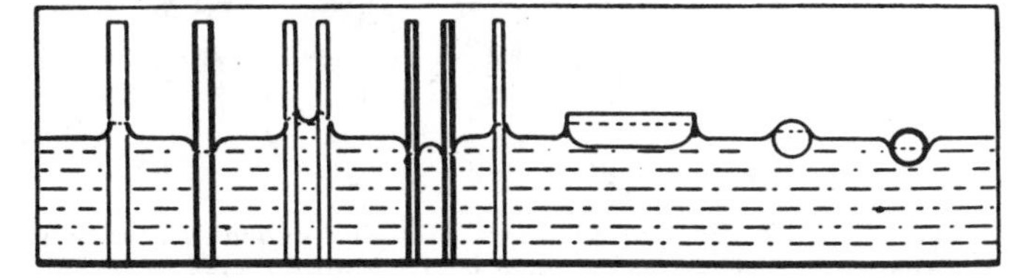

Intro. fig. 5. Diagram showing positive and negative menisci with respect to various emergent and floating objects. Stems that are "wettable" pull the surface about them into upward slopes, or positive menisci, and stems that do not wet readily (with waxy surfaces) bend it downward into negative menisci (Renn, 1943).

In nature the angle of contact usually results in a negative meniscus when the water surface is in contact with the waxy surfaces of green plants with stems or leaves extending above the water (intro. fig. 5). This soon changes to a positive meniscus, however, owing to the accumulation of wettable gelatinous materials or to the death of the plants and consequent loss of wax at the surfaces. The line of intersection between the three interfaces, water-air, water-plant, and plant-air, has been termed the "Intersection Line" by Hess and Hall (1945), and the number of meters of intersection line per square meter of water surface is called the "Intersection Value."

Insects are variously adapted to the intersection line or meniscus. *Anopheles* mosquito larvae, for example, are drawn head first toward a negative meniscus from a distance of 9 mm. by forces independent of their own efforts (intro. fig. 6) (Renn, 1943); the larvae of *Dixa* midges spend most of the time in positive menisci where the water surface meets a wettable surface such as a stone. Other insects, such as water striders, are adapted to life on the surface film where their hydrofuge (non-wettable) tarsi bend but do not break the surface.

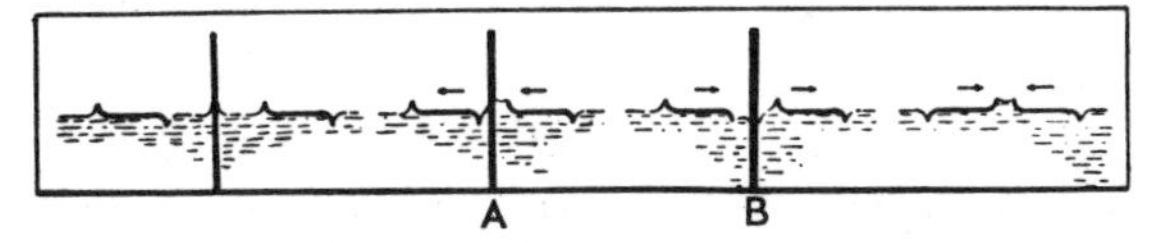

Intro. fig. 6. Diagram showing the pull of positive menisci at a wettable surface (A) on the upward-bent tail of a model *Anopheles* and the reverse action on the downward-bent head. The effects with respect to non-wettable surfaces are shown in B. The pull extends for a distance of 9 mm. (Renn, 1943).

Aquatic respiration.—Possibly because of their origin as terrestrial air breathers, insects have developed the most remarkable adaptations for aquatic respiration. These include: (1) blood gills with hemoglobin (chironomid larvae or bloodworms), (2) cuticular respiration by simple diffusion into the tracheal system (immature stages of most aquatic insects),

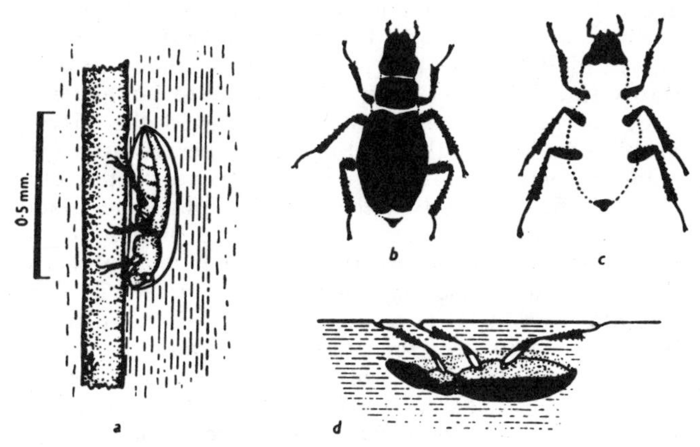

Intro. fig. 7. *a*, *Dryops* freshly submerged, crawling along stem enclosed in its bubble; *b-d*, *Ochthebius*, dorsal, ventral, and lateral views of submerged insect. The extent of the air film is indicated by dotted lines in *b* and *c*, and by stippling in *d*. Wetted areas are black (Thorpe, 1950, in part after Hase).

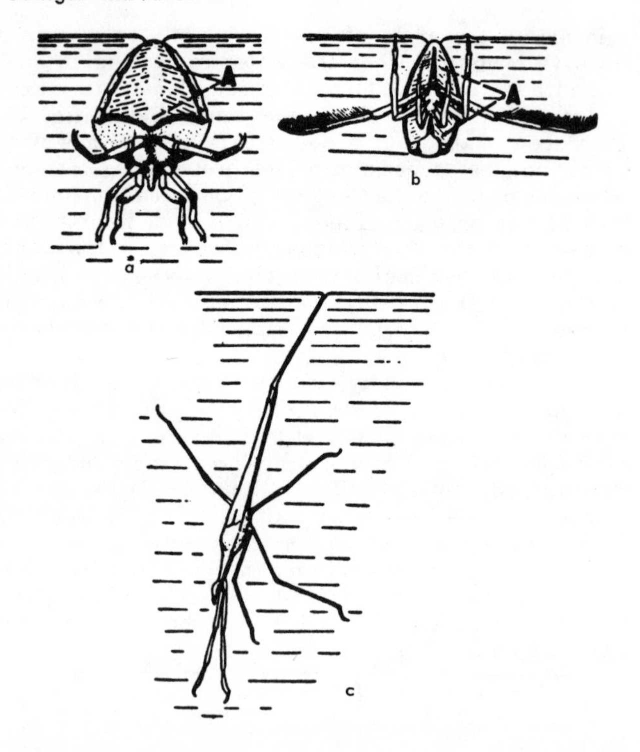

Intro. fig. 8. Surface respiration by water bugs. *a*, fifth instar nymph of *Belostoma flumineum* Say; *b*, fifth instar nymph of *Notonecta undulata* Say; *c*, adult *Ranatra fusca* P. B. (Maloeuf, 1936).

(3) tracheal gills that depend upon diffusion of dissolved oxygen from the water directly into the tracheal system (many aquatic larvae, mayfly naiads, etc.), (4) respiration (intro. fig. 7) by means of an air bubble from which oxygen diffuses into the insect's spiracles and into which oxygen diffuses from the surrounding water (adult bugs and beetles), (5) direct contact with air in plant tissues by inserting tubes into the roots and stems of aquatic plants (beetle larvae of the genus *Donacia*, mosquito larvae of the genus *Mansonia*), and (6) contact with atmospheric air by breaking the surface with hydrofuge hairs or surfaces (intro. fig. 8*a*, *b*) (adult beetles and bugs) or with breathing tubes (intro. fig. 8*c*) (water scorpions, mosquito larvae, etc).

Of all these, the most remarkable is the silvery bubble of adult beetles and bugs that serves as a gill, holding approximately 80 percent N and 20 per cent 0_2 when first formed at the surface. When the insect submerges, the oxygen in the bubble begins to decrease as it is used up, thus lowering the volume of the bubble and reducing the ratio of 0_2 to N. To compensate for this lower oxygen tension, oxygen diffuses into the bubble from the surrounding water, and since the invasion coefficient of oxygen between water and air is three times as great as that of nitrogen, the insect is able to remain submerged much longer (thirteen times as long in one experiment) than if it were dependent on surface oxygen alone (Comstock, 1887). Theoretically it is only when all the nitrogen has diffused outward that the system breaks down and new surface air is required.

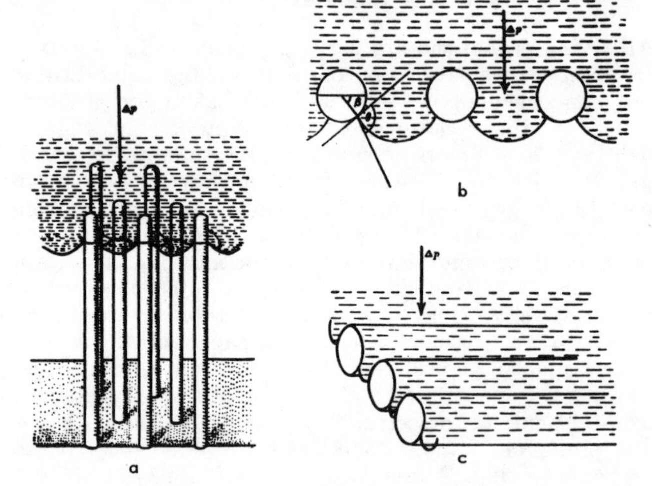

Intro. fig. 9. Diagrams to illustrate the wetting of: *a*, a system of short, stiff, erect hairs; and *b*, *c*, a system of longer hairs bent to form a more or less horizontal and compressible mat (Thorpe and Crisp, 1947).

The bubble of changing volume is held by hydrofuge hairs which lie more or less parallel to the body surface in a compressible mat. The system is illustrated (intro. fig. 9*b* and *c*), showing the angle of contact, of water on the waxy surfaces of the hairs (Thorpe and Crisp, 1947).

A few beetles (Dryopidae) and an old-world water bug *(Aphelocheirus)* have a plastron of fixed volume, maintained by stiff hairs of a density up to 2 million per square millimeter (intro. fig. 9*a*). Unlike the larger bubbles described above, the plastron is virtually incompressible and hence can act as a permanent gill or avenue of diffusion of oxygen from the water to the tracheal system. These are among the very few permanently aquatic insects that do not need to come to the surface at any time during their life cycle (Thorpe, 1950).

Osmoregulation.—The regulation of osmotic pressure of body fluids is another important type of adaptation in aquatic insects. Many marine animals have body fluids that are isotonic with sea water. All fresh-water organisms have some method of regulating the concentration of their body fluids. In the

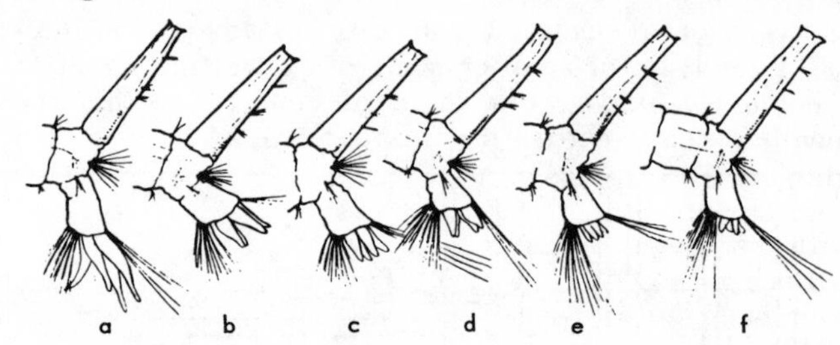

Intro. fig. 10. Terminal segments of *Culex pipiens* L. larvae showing typical appearance of anal papillae when reared in media of increasing salt concentration. *a*, larva reared in distilled water with mean length of papillae 0.82 mm; *b*, tap water (0.006 per cent NaCl)—0.36 mm.; *c*, medium with 0.075 per cent NaCl—0.33 mm.; *d*, medium with 0.34 per cent NaCl—0.22 mm.; *e*, medium with 0.65 per cent NaCl—0.20 mm.; *f*, medium with 0.90 per cent NaCl—0.20 mm. (Wigglesworth, 1938).

simplest cases an impervious body wall protects the internal fluids, and excess water and salts obtained with the food are excreted (most adult beetles and bugs). Wigglesworth (1938) showed that there is a correlation in mosquito larvae between salt concentration of the surrounding medium and degree of development of the anal papillae (intro. fig. 10). Larvae reared in distilled water had well-developed papillae (functional hypertrophy for chloride uptake), whereas larvae reared in a medium with 0.90 per cent NaCl had greatly reduced papillae. This same phenomenon is observed in nature where salt-marsh mosquitoes are able to adapt to varying degrees of salinity.

Aquatic Habitats in California

Probably no area of equal extent in the world can claim a greater variety of aquatic habitats than California. Spanning ten degrees of latitude and nearly 15 thousand feet in altitude, the state offers practically every kind of aquatic situation except the arctic tundra and tropical jungle. The average annual precipitation ranges from 109 inches or more in parts of Del Norte County on the north coast to less than 2 inches in Death Valley, and the climate varies from cool and uniform along the coast to extremes of heat and cold in the interior mountains and deserts. To understand the present climate and topography it is necessary to know something of the geological history of the state.

Evidence from fossils (Camp, 1952) shows that moist climates prevailed throughout most of the Tertiary (70 million years) and that the present period is one of relative aridity. In the Miocene and Pliocene, inland seas occupied such present-day depressions as the Central Valley, the Great Basin, and the southern California deserts. In keeping with this kind of climate redwoods were widely distributed over the western United States, and a broad-leaved deciduous forest occurred in many places. More recently the Pleistocene glacial and pluvial periods (the last as recent as 10 thousand years) resulted in a southward extension of boreal faunas and floras and retreat of the southern biotas. Unlike the great ice sheet of the northeastern states, Sierran glaciers were local, cutting cirques and gouging **U**-shaped valleys with terminal or lateral moraines. These processes, which are still going on to a limited extent, set the stage for the great variety of lakes and streams that are now so characteristic of the Sierra Nevada. Relict glaciers and ice caves now serve as refuges for more northern plants and animals, most of which retreated with the advent of warmer, more arid conditions.

During this same period "pluvial" lakes extended over wide areas in the Southwest (Hubbs and Miller, 1948). Lake Lahontan (including present-day lakes such as Pyramid and Walker in Nevada and arms extending into California and Oregon) and Lake Manly (including Death Valley and parts of Inyo and San Bernardino counties) are examples. These fluctuated from large inland lakes to dry playas. Most of the pluvial lakes are now gone, but a few like Mono Lake still persist and others, like Owens Lake, come and go, depending on surface and ground water fluctuations.

One of the best known of the desert basin lakes is the Salton Sea. It seems certain that at one time the Gulf of California extended northward over most of the Imperial and Coachella valleys (intro. fig. 11*a*). Subsequently the Colorado River built up a silt dam (intro. fig. 11*b*), creating an ancient salt-water lake. This lake had no outlet and, like many other inland waters of the West, eventually dried up. Still later, possibly during a pluvial period, the Colorado River changed its course, emptying into the dry basin rather than into the Gulf, thus creating prehistoric Lake Cahuilla (or Lake LeConte as it is sometimes called) (intro. fig. 11*c*). This may have persisted until the time of the early Cahuilla Indians, judging by a legend handed down to the present (Blake, *fide,* Hubbs and Miller, 1948). In the last stage but one in the story, the Colorado River again shifted its course and Lake Cahuilla dried up. Then, in 1905, the river poured water into the basin for a two-year period, forming the Salton Sea which was 17 by 43 miles in extent, 84 feet in maximum depth, and well below sea level. Since 1907 its area at first slowly decreased by evaporation and its salinity increased (intro. fig. 11*d*). Now irrigation water is reversing the process.

Since the last pluvial period geological processes throughout the state continued as in the past, including upfaulting and subsidence of large areas, erosion, and volcanic action. As a result water courses were formed, dammed at various points by moraines, lava flows, or alluvium, and finally reached the sea or disappeared into the underground water table or were lost through evaporation. This happened repeatedly through time, the processes being continuous, so that our present aquatic habitats and the insects that inhabit them are but a momentary stage in physiographic and biotic evolution.

Stream and Lake Classifications

There have been several attempts by limnologists to classify streams, lakes, and other aquatic habitats for purposes of ecological analysis. Such classifications are doomed from the start because they attempt to fit continuously variable and endlessly diverse situations into stereotyped systems. Nevertheless, the urge to classify runs deep in human nature, and useful generalizations and clearer understanding have resulted from certain broadly based ecological classifications.

In California the streams and lakes are profoundly influenced in form and distribution by the topography of the land. This is of course a truism, but it is especially striking in regions of high relief. For purposes of analysis Meyer (1951) has divided the state into seven major hydrographic areas and numerous drainage basins, subbasins, and stream groups. The biota, too, is determined to a large extent by topography which has largely influenced the migration of floras and faunas. Therefore any stream or lake classification must take into account both physical

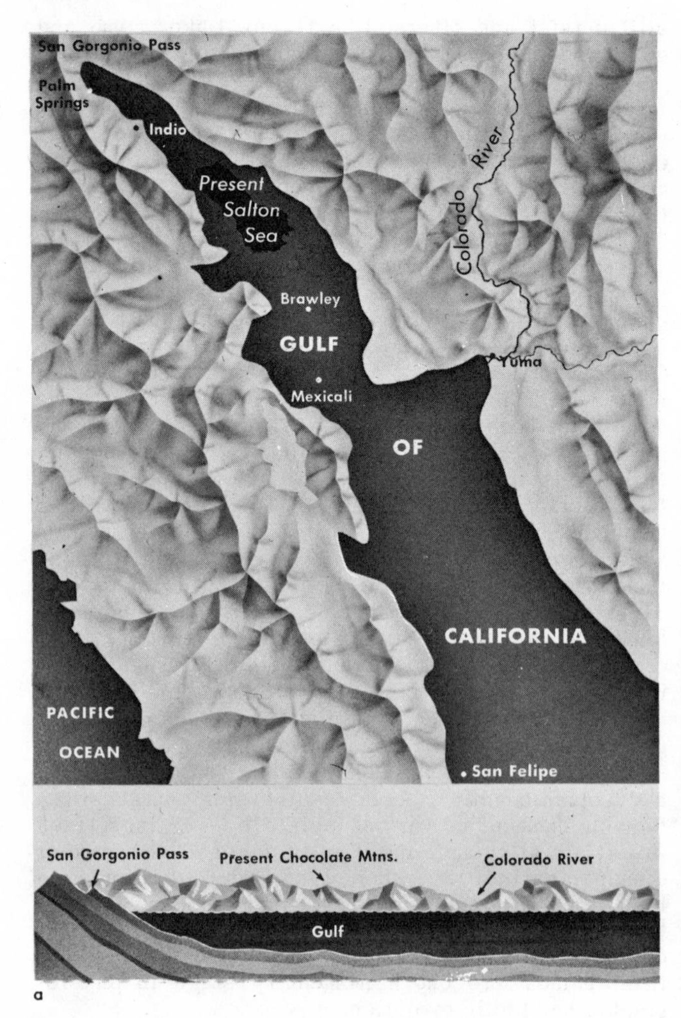

a

b

Intro. fig. 11. Diagrammatic representation (reprinted from Sunset Magazine, February,

and biotic evidence. The principal topographic and aquatic features of the state are shown on the accompanying relief map (intro. fig. 12). Essential features, from the viewpoint of an aquatic biologist, are the high mountain ranges surrounding the Central Valley, with drainage through the Carquinez Straits to San Francisco Bay; the coastal slopes on the west; and the southern California deserts with isolated mountain ranges. Faunistically, the picture is, of course, very complex, but certain broad generalizations have crystallized out of recent analyses of the best-known group of animals—the birds (Miller, 1951) (intro. fig. 13). In general, the pattern derived from birds can be applied directly to aquatic insects, the minor deviations being attributable to the special nature of aquatic habitats.

The dominant fauna of the northern part of California and of the higher mountains in the south is boreal. This includes the barren alpine regions and the evergreen coniferous forests. It is similar to, and more or less continuous with, the widespread coniferous forests of northern North America and of the Rocky Mountains. Minor differences in California are due to the heavy snow pack in the winter and the virtual lack of summer rains with consequent arid conditions, especially in the south.

Intruding from the east are the Great Basin faunal elements, with important sections in Modoc and Lassen counties, along the desert slopes of the Sierra, and in the Owens Valley and Mojave Desert. A striking feature of this region is the relict southern fauna of the hot springs of Death Valley and adjacent parts of Nevada in the Amargosa River system.

Southern elements in the aquatic fauna are now confined (except for the Death Valley relicts) to the Colorado drainage system. The Colorado River itself, with its oxbows, sloughs, and reservoirs, is a rich source of Sonoran and Neotropical types, and streams on the California side are strictly comparable to those of the nearby Gila Mountains in Arizona.

The rest of the state, including the Central Valley, the south and central Coast Range, and the western foothills of the Sierra can best be called Californian. It is a region of winter rains and summer drought. It is not rich in aquatics but, because of its isolation and presumably also its age, is unique in character.

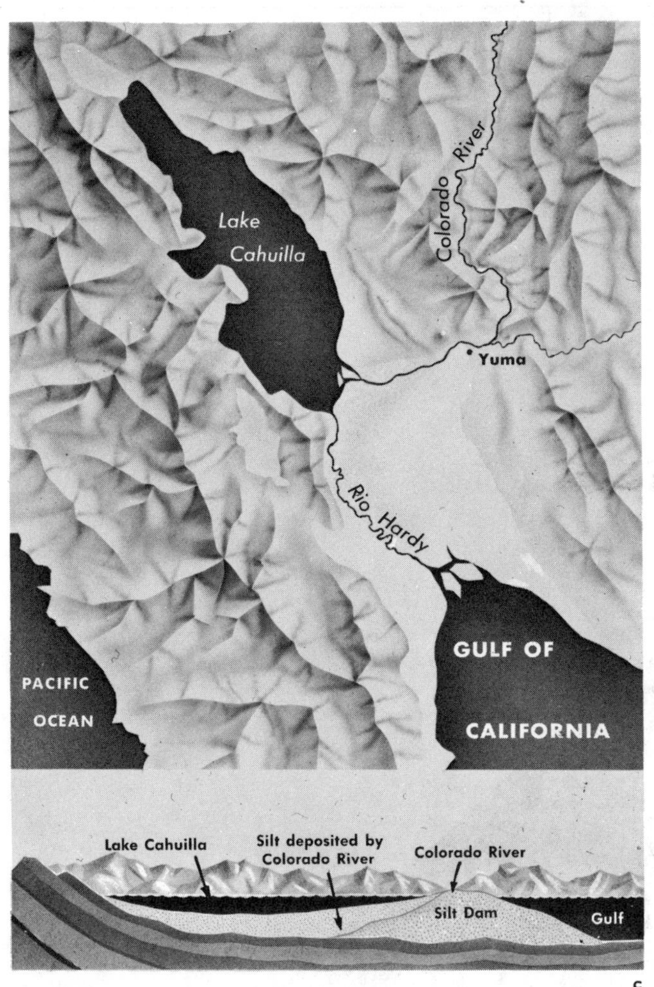

c

d

of the probable history of the Salton Sea 1952, copyright Lane Publishing Company).

The distinctions between these faunas (and floras) are striking to the field naturalist but are only evident to the taxonomist after careful analysis of distributional patterns in particular groups. Such analyses must be based on extensive collections from all parts of the state, and it is unfortunate that such collections of aquatic insects have not yet been made, except in a few groups like the mosquitoes.

Streams

General faunistic relations apply equally to streams, lakes, and other aquatic situations, but many other features are distinctive for each type of water. Streams are especially susceptible to outside influences and hence are infinitely variable. As a result, our knowledge of streams has lagged far behind that of lakes. Nevertheless, certain generalizations have emerged from extensive field investigations, and it may be useful to summarize these and see if they apply to California streams.

Stream habitats.—Ricker (1934) speaks of the "overwhelming variety of habitats presented in streams and rivers." At the same time, certain habitats occur repeatedly within any short section of a stream. For example, there are falls, riffles, runs, and pools in nearly every stream in the world, and bottom habitats including boulders, rubble, gravel, sand, and mud. Also there are endless microhabitats, where each aquatic organism occupies a unique ecological niche with preference for a particular side of a rock, for example, or finds other special conditions different for each stage in its life cycle. The accompanying figure (intro. fig. 14) shows an assemblage of stream organisms as they might appear on a submerged rock in any stream. The species would be different for each part of the world, but the main groups or their ecological equivalents would be the same.

Stream habitats for Yellowstone National Park and other regions of the northwest were classified by Muttkowski (1929) as follows:

1. Permanent habitats, with native (endemic) biota
   a. White water habitats—falls, cascades, white rapids
   b. Clear rapids and stone bottoms—on and under rocks

Intro. fig. 12. Relief map of California showing, in addition to mountain ranges and valleys, the principal lakes, reservoirs, and river systems (original drawing by Celeste Green).

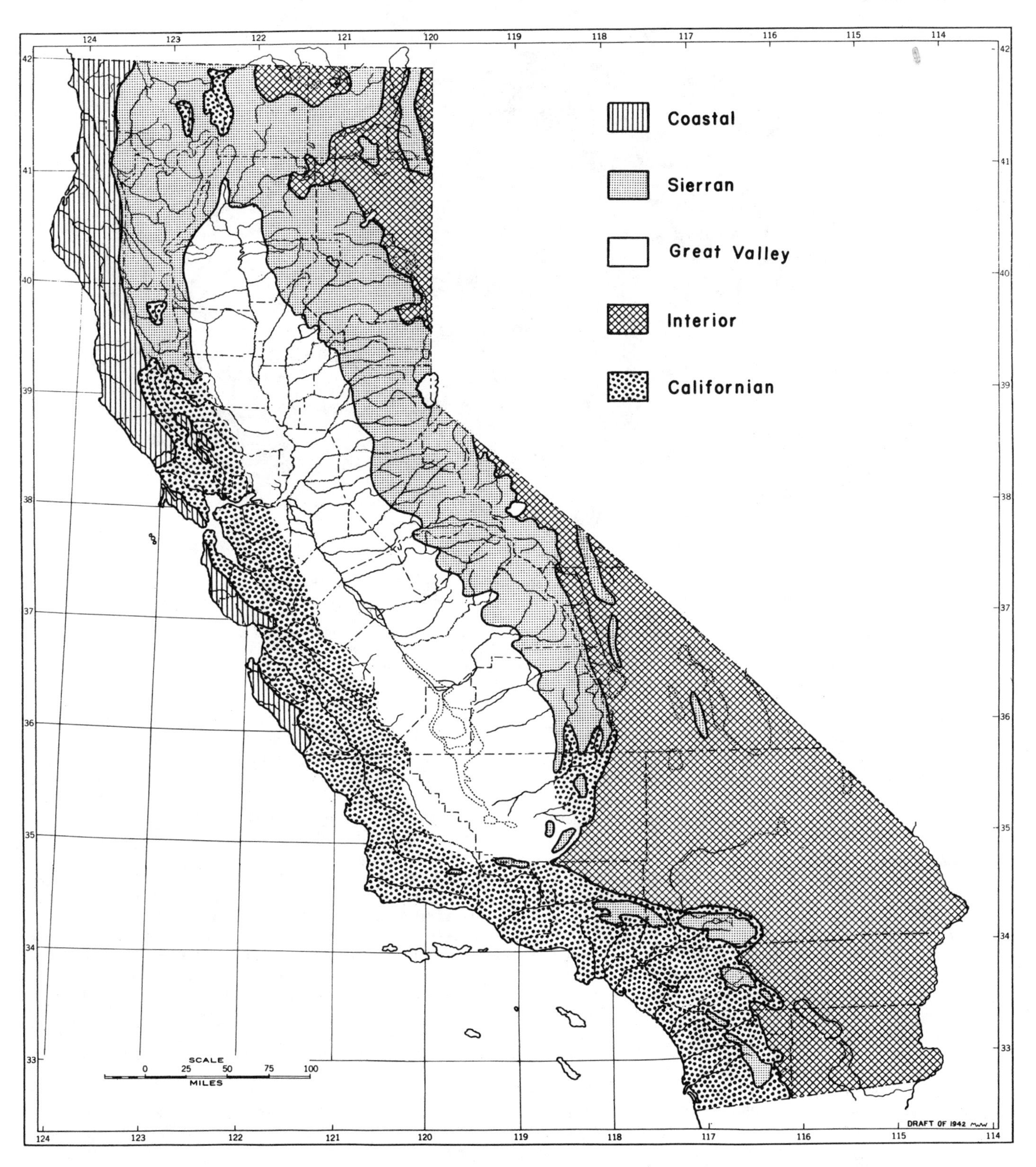

Intro. fig. 13. Biotic provinces of California based on an analysis of the distribution and faunal relationships of the birds (after Miller, 1951).

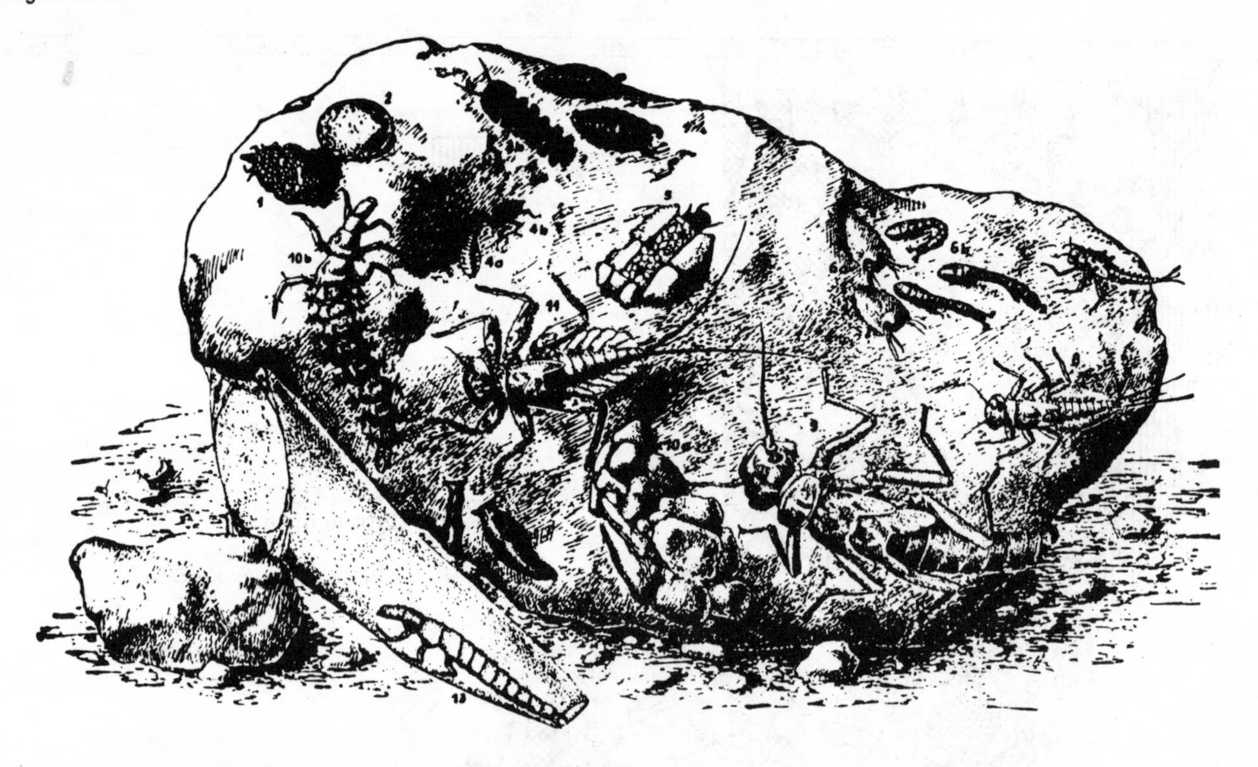

Intro. fig. 14. Stream organisms on a submerged rock—diagrammatic. 1) Rhyacophilidae (Trichoptera), larva in a case of sand grains; 2) *Ancylus* (Gastropoda); 3) Blepharoceratidae (Diptera), *a*, pupae, *b*, larva; 4) Elmidae (Coleoptera), *a*, larva, *b*, adult; 5) Goeridae (Trichoptera), larva; 6) Simuliidae (Diptera), *a*, pupa, *b*, larva; 7) Baetidae (Ephemeroptera), nymph; 8) Heptageniidae (Ephemeroptera), nymph; 9) Perlodidae (Plecoptera), nymph; 10) Rhyacophilidae (Trichoptera), *a*, pupal case, *b*, larva; 11) Heptageniidae (Ephemeroptera), nymph; 12) Planaria (Turbellaria); 13) Philopotamidae (Trichoptera), larva in its catching net (Ruttner, 1953).

c. Placid water habitats—pools and holes
d. Marginal areas—on or under rocks, in soil

2. Interrupted habitats, with native biota
   e. Deposits—on rocks, bottoms, or shores
   f. Splash areas—on rocks
3. Temporary habitats—transient and transitional, with varied biota
   g. Marginal pools
   h. Recession areas

These apply to mountain streams. Obviously other habitats should be mentioned including slow moving rivers of considerable depth with steep banks and mud bottoms. Also, springs are of various types (see below) and offer a variety of special habitats including basins, seepage areas, and the like.

Stream classification.—Different classifications have been proposed for the streams of the European continent (Steinmann, 1907; Thienemann, 1912; Huet, 1948), the British Isles (Carpenter, 1928), Yellowstone National Park (Muttkowski, 1929), Ontario (Ricker, 1934), and other areas. These have been variously based on source of water, size, speed of current, slope, elevation, temperature, substrate, permanence, oxygen and carbon dioxide, pH, hardness of water, productivity, or combinations of several of the above factors. Actually, most of these are interdependent, and it may very well be that no classification can be devised that will reveal in a meaningful way all the complicated interrelations. Therefore, each factor will be discussed separately and will be related, as far as possible, to California conditions.

Source.—The sources of surface waters are glaciers, snow, springs, and surface run-off from rain. The latter results in very temporary storm courses and, in arid regions with little vegetation, in flash floods. Temporary storm courses have no real significance for stream ecology except in rare instances when aquatic organisms may be transported long distances and survive in new regions. Snow is also a temporary source of water but the Sierran snow pack is so great —20 feet or more in many places—that long-flow intermittent streams are numerous and support a special biota of short-lived organisms with adaptations for surviving the periodic dry periods. Glaciers are, of course, a permanent source of water but are so small and so few in number in California that they are a minor factor.

Springs, on the other hand, are of major importance throughout the state and are the only source of perennial streams below snow line and throughout most of the southern part of the state. Muttkowski (1929) said of the springs of Yellowstone Park that "every conceivable type occurs," and this is equally true of California. Muttkowski says further that "One could employ a dozen different criteria for their classification and still not exhaust them. One might

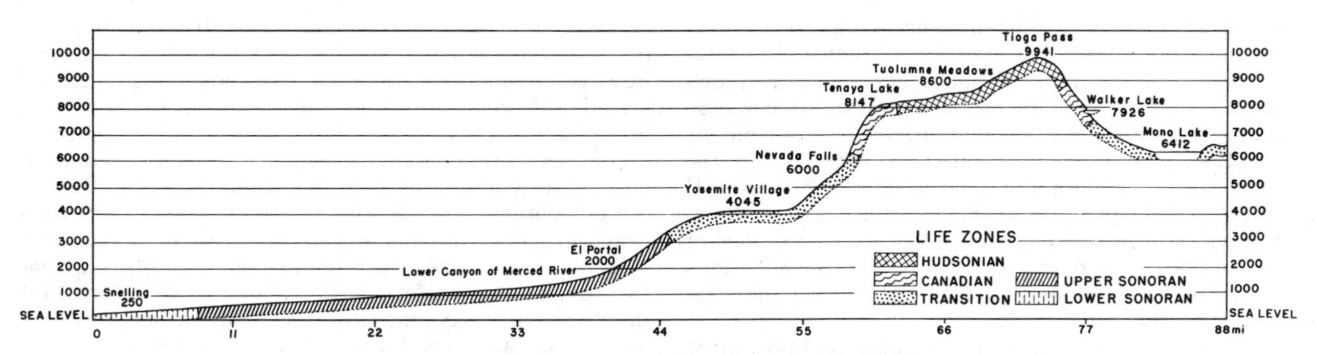

Intro. fig. 15. Stylized profile of the Merced River from the crest of the Sierra to the valley (after Grinnell and Storer, 1924).

classify springs according to topography, whether their waters immediately fall away after leaving the ground or whether they are caught in a pool of varying size (Bornhauser). One could also classify them according to size and temperature; as for the latter, the springs range from icy chillness to boiling point. One might arrange them according to color, which not only differs among different springs, but often in the same spring at different times. Again, one could classify them according to substances in solution, such as carbonates, arsenates, chlorides, silicates, sulphides, etc., or according to substances in suspension, such as sulfur, mud, sand, silt, etc." California examples could be cited for each of the above criteria, from the hottest waters of the Lassen region to the saline springs of Death Valley, but the more normal springs are cool and fresh and form seepage areas and bogs in the humid northwest and feed many streams and lakes throughout the state.

Contingent on the source and also the soil and vegetation in the drainage area, Ricker (1934) recognized hard-water streams with a bicarbonate content greater than 100 parts per million (expressed as $CaCO_3$) and hardness more than 150 parts per million. Comparable figures for soft-water streams were: bicarbonate content less than 25 parts per million and hardness less than 50 parts per million. Hydrogen ion concentration has also been used as a basis for classifying streams, the acid streams that drain bogs being much less productive than streams with a pH of 7.0 or above.

Permanence.—Depending on the source, but also on many other factors, a stream may flow continuously (glacial and most spring-fed streams), may recede to its higher reaches during the dry months of summer and fall, or may dry up completely soon after the rains stop. The first of these is, of course, perennial. The second may be termed a receding intermittent and the third a temporary intermittent. Each of these is characterized by distinctive insects, though the receding intermittent has characteristics of both the perennial and temporary.

Slope, size, and speed of current.—Much has been made of these three interrelated factors in classifying streams. Slope, for example, was used by Huet (1949) as the basis for a classification of the streams of Western Europe. His so-called "slope-rule" states that, "In a given biogeographical area, rivers of like breadth, depth, and comparable slopes, have identical biological properties and specifically similar fish populations." In California, slope is obviously an important factor and follows a stylized profile from the highest mountains to the lowland valleys, deserts, or sea coast (intro. fig. 15).

Size is a factor in its own right, though obviously related to other things. For example, very small streams are largely "extrinsic," being influenced to a great extent by their surroundings. Large rivers, on the other hand, are "intrinsic" in the sense that they determine to a great extent their own course, type of bottom, and productivity and are more or less independent of their immediate environment.

Arbitrary size limits have been set by various authors to define creeks, rivers, and the like. Ricker (1934) chose a volume of flow less than 10 cubic feet per second (on June 1) and a width of less than 10 feet for his "creek" category. Streams with volume and width above these figures he termed "rivers." Huet (1949) found that slope and breadth had a combined influence on fish populations in Western Europe—the smaller the river, the greater the slope tolerated by a given species.

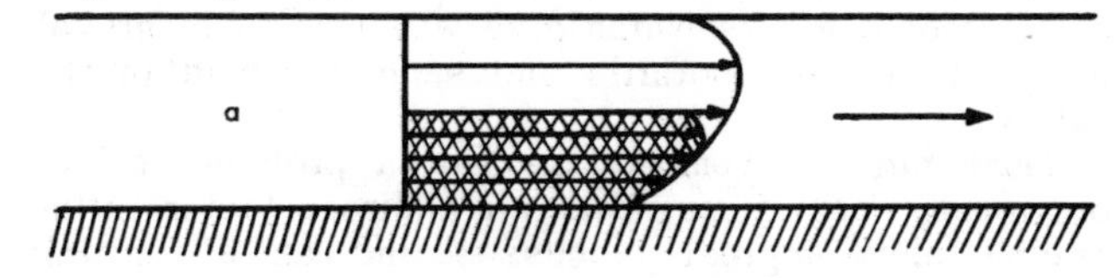

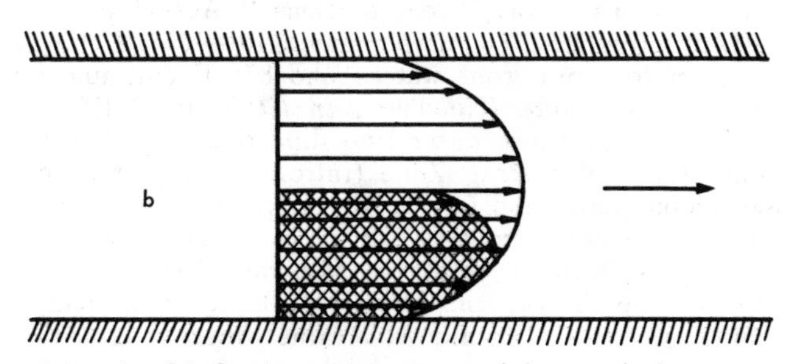

Intro. fig. 16. Graphic representation of the speed of current in a stream. *a*, vertical view; *b*, horizontal view. The dark area represents a stream with half the width and depth (Huet, 1949).

Speed of current has been used as a criterion in classification. Speed, like size, is subject to seasonal fluctuation and also to diurnal fluctuation in snow-melt streams. Readings are not always reliable because of local variations but, in general, as pointed out by Huet (1949), velocity is greatest a little below the surface near the middle of a stream. Also, a stream half as wide and deep as another, given the same slope, will have a slower velocity (intro. fig. 16). Ricker (1934) classified streams arbitrarily as "slow"—less than 1.5 feet per second, and "fast"—more than 1.5 feet per second.

Type of bottom.—There is a positive correlation between speed of current and type of bottom. Tansley worked out the details for streams in the British Isles (quoted by Macan and Worthington, 1951) as shown in table 1.

Table 1 (Introduction)

RELATION OF CURRENT SPEED AND NATURE OF RIVER BED

Velocity of current per second	Nature of bed	Habitat
More than 4 ft. (1.21 m.)	rock	torrential
More than 3 ft. (0.91 m.)	heavy shingle	torrential
More than 2 ft. (0.60 m.)	light shingle	nonsilted
More than 1 ft. (0.30 m.)	gravel	partly silted
More than 8 in. (0.20 m.)	sand	partly silted
More than 5 in. (0.12 m.)	silt	silted
Less than 5 in. (12 cm.)	mud	pondlike

Ricker (1934) recognized several types of bottom materials including bed rock, boulders, stones, gravel, sand, mud, and plant debris. Each of these types occurs in California and, in general, they are correlated with speed of current and hence also with slope, size, and so on. Typically the smallest streams of Sierran and north coast forests have bottoms of gravel and stones with much plant debris; the high mountain streams of great velocity have scoured down to bed rock; and rivers of intermediate elevations have bottoms of boulders which give way in turn to gravel and sand in the foothills and sand and mud in the valleys.

Temperature.—Temperature has a profound effect on aquatic organisms, both as a direct factor influencing physiological processes and as a limiting factor on dissolved oxygen. Ricker (1934) settled on 75° F as the maximum temperature that separates "warm rivers" from "trout streams." Actually, optimum figures may be more meaningful—66° F being optimum for cold trout waters and 85° F optimum for warm bass waters (Needham, *in litt.*). In California the cold-water-warm-water line dips deeply below the generally cool Sierran Zone (intro. fig. 13) or boreal evergreen coniferous forest. Actually, although varying with the season, trout extend down to the lower part of the Californian or Upper Sonoran Zone on each side of the Sierra, and in the northern Coast Range and southern California mountains where vegetation becomes sparse, slopes are gradual, current is slower, and the less protected waters rise in temperature.

Productivity.—Theoretically, productivity should be the best basis for stream classification because it is determined by the interaction of all factors. However, sampling methods are not adequate, and faunal differences would confuse the picture in a large and diverse area like California. In spite of these and other difficulties, productivity has been used in a general way for classifying streams. Patrick (verbal communication), for example, speaks of oligotrophic (poor in nutrients) streams in head-water regions. These are clear and hence entirely exposed to the sun's rays for photosynthesis. They are high in dissolved oxygen, low in temperature, poor in nutrients, and relatively low in productivity. At intermediate elevations temperature rises, velocity decreases, and productivity is greater. Then, rivers pass through the foothills and out onto the alluvial plains, picking up silt and then dropping it onto sand or mud bottoms as velocity decreases. The muddy waters block photosynthesis except near the surface. By this time, algae and other plankton organisms have increased so that the water can be considered eutrophic (rich) but the bottom fauna is very restricted, consisting mainly of Chironomid larvae.

Productivity in terms of bottom fauna has been studied extensively. In general it has been found that riffles are more productive than pools, that rubble and gravel bottoms are more productive than bedrock and sand in that order (Pennak and Van Gerpen, 1947), and that plant beds in streams are very productive (Needham, 1938). Productivity was used by Hazzard (1938) to classify streams. His standards were: Grade I (rich stream)—more than 22 grams of bottom organisms per square meter or 2,152 organisms; Grade II (average)—11-22 grams or 1,076 to 2,152 organisms; Grade III (poor)—less than 11 grams or less than 1,076 organisms. Such figures have been shown (Needham and Usinger, 1955) to be of doubtful significance statistically because of shortcomings of the sampling equipment and techniques. Nevertheless they give an idea of the general levels of productivity and hence are of some value.

Figures for California streams have been given by Needham (1934, 1938, 1939). The maximum was 4,400 organisms in a single square-foot sample taken from the Klamath River near Hornbrook, California. The total wet weight of this sample was more than 105 grams. Seasonal variation was shown by samples taken in Waddell Creek near Santa Cruz. Riffle-dwelling forms varied from a low of 70 pounds per acre (extrapolated from representative square-foot samples) in February to a maximum of 472 pounds in May, 1933. Comparable figures for the Merced River in Yosemite National Park were 103 pounds in February and 85 pounds in August.

All these figures apply only to gross wet weights and total numbers at particular seasons. True productivity should be based on dry weights (and perhaps on nutritive values as fish food) and should be sampled and calculated as a standing crop or as an annual crop, taking into account the emergence of successive generations. Unfortunately, such an ideal study has never been made, but with improved techniques and increased knowledge of the biology of aquatic insects the time may not be far off when work of this kind may become possible.

It does not seem feasible to combine the diverse factors discussed above into a single classification. However, there may be some value in an outline that groups the streams of California under their biotic zones (intro. fig. 13) (which in turn are related to altitude, temperature, precipitation, vegetation, and faunal origins) and, secondarily, according to source and permanence of water.

Classification of the Streams of California

(Based on biotic provinces and source and permanence of water)

A. Alpine (Arctic Alpine Life Zone; above timberline, Sierra and White Mountains)
   1. Glacial "milk" streams (Coness Creek)
   2. Snow-melt intermittent (inlets and outlets of many high Sierra lakes above 9,000 feet)
   3. Spring- and snow-fed permanent (Virginia Creek)

B. Montane (Hudsonian, Canadian, and Transition zones; Boreal coniferous, midelevations, Sierra Nevada, North Coast Range, and southern California Mountains)
   1. Snow-melt intermittent (Angora Creek)
   2. Spring- and snow-fed permanent (Sagehen, Glen Alpine, American, Tuolumne)

C. North Coast Redwood (Transition Life Zone; Moist Sequoia forest)
   1. Spring-fed redwood creeks
   2. Redwood phase of north coastal rivers (Smith R., Eel R., Russian R., etc.)

D. South and Central Coast Range and Coast (Upper Sonoran Life Zone; Open coast, closed-cone pine forest, broad sclerophyll woodland)
   1. Short-flow intermittent
   2. Long-flow, fluctuating intermittent (Temescal Creek)
   3. Permanent streams (Wildcat Creek, Carmel R., Big Sur, etc.)

E. Valley foothill (Upper Sonoran; Digger Pine-Oak Zone Chaparral; moderate gradients)
   1. Short-flow intermittent
   2. Long-flow fluctuating intermittent (Dry Creek, Fresno Co.)
   3. Spring-fed permanent streams
   4. Foothill phase of main rivers (Tuolumne R., Stanislaus R., etc.)

F. Central Valley (Lower Sonoran Life Zone; grassland, agricultural, flat)
   1. Long-flow fluctuating intermittent (Putah Creek)
   2. Permanent plant-choked sloughs (irrigation runoff)
   3. Valley phase of large rivers (Sacramento, San Joaquin rivers, and lower tributaries

G. Desert Foothill (Upper Sonoran Life Zone; Piñon Pine-Juniper, Chaparral)
   1. Short-flow intermittent
   2. Long-flow fluctuating intermittent (lower Walker R.)
   3. Spring-fed permanent (lower Carson R., Hot Creek)
   4. Foothill phase of main rivers (Truckee R.)

H. Desert (Lower Sonoran; Artemisia, Larrea, Mesquite)
   1. Short-flow intermittent (Mojave R.)
   2. Sporadic intermittent (Amargosa R.)
   3. Fluctuating intermittent
   4. Spring-fed permanent (Owens R., Hat Creek, Laurel Creek, Mammoth Creek, Hot Spring runoffs, Death Valley, etc.)
   5. Desert phase of large rivers (Colorado R.)

Lakes

Standing waters such as lakes and ponds differ strikingly from streams not only in superficial appearance but also in physical and biotic characteristics. In general, water movement is less of a dominating influence, and temperature and oxygen supply are more important. Temperature, in particular, exercises a profound effect because of the following inherent thermal properties of water: (1) *specific heat,* which is the greatest of all but a few rare substances; (2) *latent heat of fusion,* which requires that eighty times as much heat be absorbed to transform ice to water at 0°C (and a similar amount given off to reverse the process) as is required to heat or cool water by 1°C; (3) *thermal conductivity,* which is so low that heat transfer in lakes is largely dependent on convection currents and wind action; (4) *density* (intro. fig. 17), which is greatest at 4°C (39.2°F), so that water sinks as it cools to 4°C and then rises as it drops below this, forming ice at the surface rather than at the bottom; (5) *evaporation,* which results in heat loss (cooling); (6) *transparency,* which determines the depth of light penetration and hence also the limits of the photosynthetic zone; and (7) *solvent action,* which is very great and, in the case of oxygen, is greatest at low temperatures so that cold water can hold more dissolved oxygen than warm water.

As a result of these properties lakes are warmed slowly in the summer, cooled in the winter, and stratified at certain times of the year (intro. fig. 18). The period of summer stratification is characterized by an

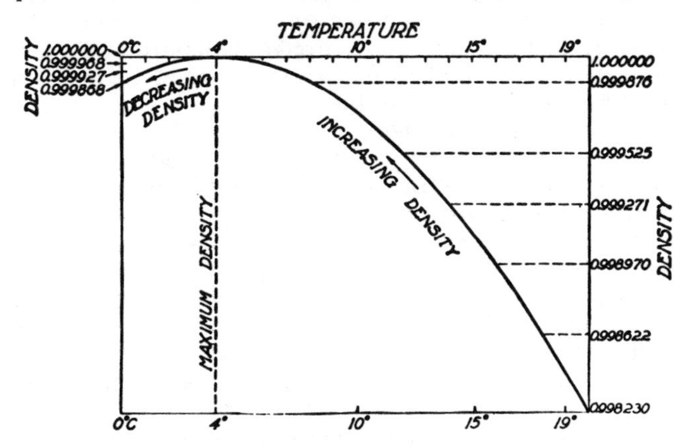

Intro. fig. 17. Graph showing the relation between density and temperature in pure water (Welch, 1952).

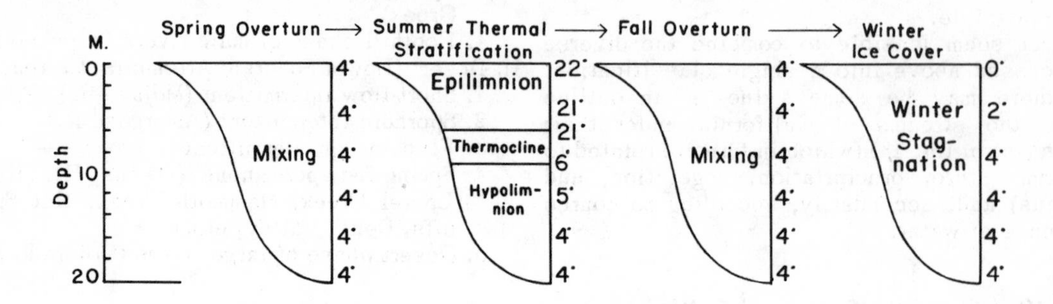

Intro. fig. 18. Diagrams showing thermal stratification and mixing in a temperate lake of the second order during the four principal seasons. Temperatures are given in degrees centigrade (4°C.=39.2°F.; 22°C.=71.6°F.) Figures for the thermocline were chosen arbitrarily to illustrate the general principle (Welch, 1952).

upper zone of summer circulation by wind (epilimnion) with relatively uniform temperatures, an intermediate zone (thermocline) with rapid fall in temperature (Birge Rule—1° C or more per meter), and a lower zone of summer stagnation (hypolimnion). Very deep lakes such as Lake Tahoe and shallow lakes in the tropics never freeze because winter temperatures are not low enough for a sufficiently long period to cool the entire mass of water below 4°C. Ponds may have no thermocline or hypolimnion because wind action mixes the shallow water from top to bottom.

Lake classification.—As in the lotic or running waters, so also in the lentic or standing waters, classification is difficult. Lakes of infinite variety merge into ponds and thence into swamps or bogs, and lines of demarcation between types are necessarily arbitrary. In spite of inherent difficulties, limnologists have devoted far more time and thought to lake classification and to lakes in general than to streams, probably because lakes are more stable and are easier to measure physically and sample quantitatively. Among the criteria used for the classification of lakes, temperature and depth were selected by Forel (1892-1904), productivity was chosen by Thienemann (1926), and source of water, nature of substrate, size, and other criteria have been variously used. None of these has proved to be satisfactory but Forel's thermal classification is theoretically useful, and Thienemann's trophic classification has gained universal acceptance and is used in common parlance without precise limits.

In the thermal classification (intro. fig. 19), lakes are arranged according to "Types," based on surface temperatures, and "Orders," based on bottom temperatures. Three types are recognized: the Polar Type, with surface temperatures never above that of the maximum density of water (4°C, 39.2°F); the Temperate Type, with surface temperatures above 39.2°F in summer and below 39.2°F in winter; and the Tropical Type, with surface temperatures never below 39.2°. Each of these types is then placed as the First Order, deep lakes (approximately 200 feet or more) with bottom temperatures constant near the point of maximum density (39.2°); or the Second Order, intermediate lakes (approximately 25 to 200 feet) with bottom temperatures that fluctuate near 39.2°; or the Third Order, shallow lakes (approximately 25 feet or less) with bottom temperatures at or near those of the surface. This classification, though ill-defined as to depth and misleading in terminology, is based on sound physical principles. In general, California lakes at middle and high elevations are of the temperate type, and lowland lakes are of the tropical type. However, Lake Tahoe, at 6,000 feet elevation, is so deep that it does not freeze. Therefore it would be classed as a tropical lake of the first order. This clearly demonstrates that other criteria must be used if a classification is to be meaningful in a biological sense.

Thienemann's classification is more useful in this sense. It is an established fact that productivity is dependent in a general way on the depth and stage of succession of a lake. Thus in deep lakes of recent origin the bottom is likely to be lacking in organic nutrients, the biota is not rich, and the volume of water below the level of effective penetration of light (heterotrophic or decomposition zone) exceeds that of the producing zone (autotrophic or photosynthetic zone). Such lakes are called Oligotrophic and

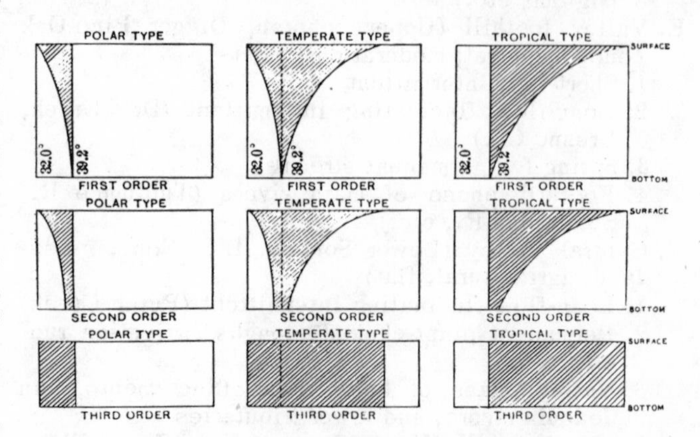

Intro. fig. 19. Thermal classification of lakes arranged according to "types," based on surface temperatures above or below that of the maximum density of water (39.2°F), and "orders," based on bottom temperatures above or below 39.2°. First order lakes are deepest, second order intermediate, and third order shallow (Whipple, 1927).

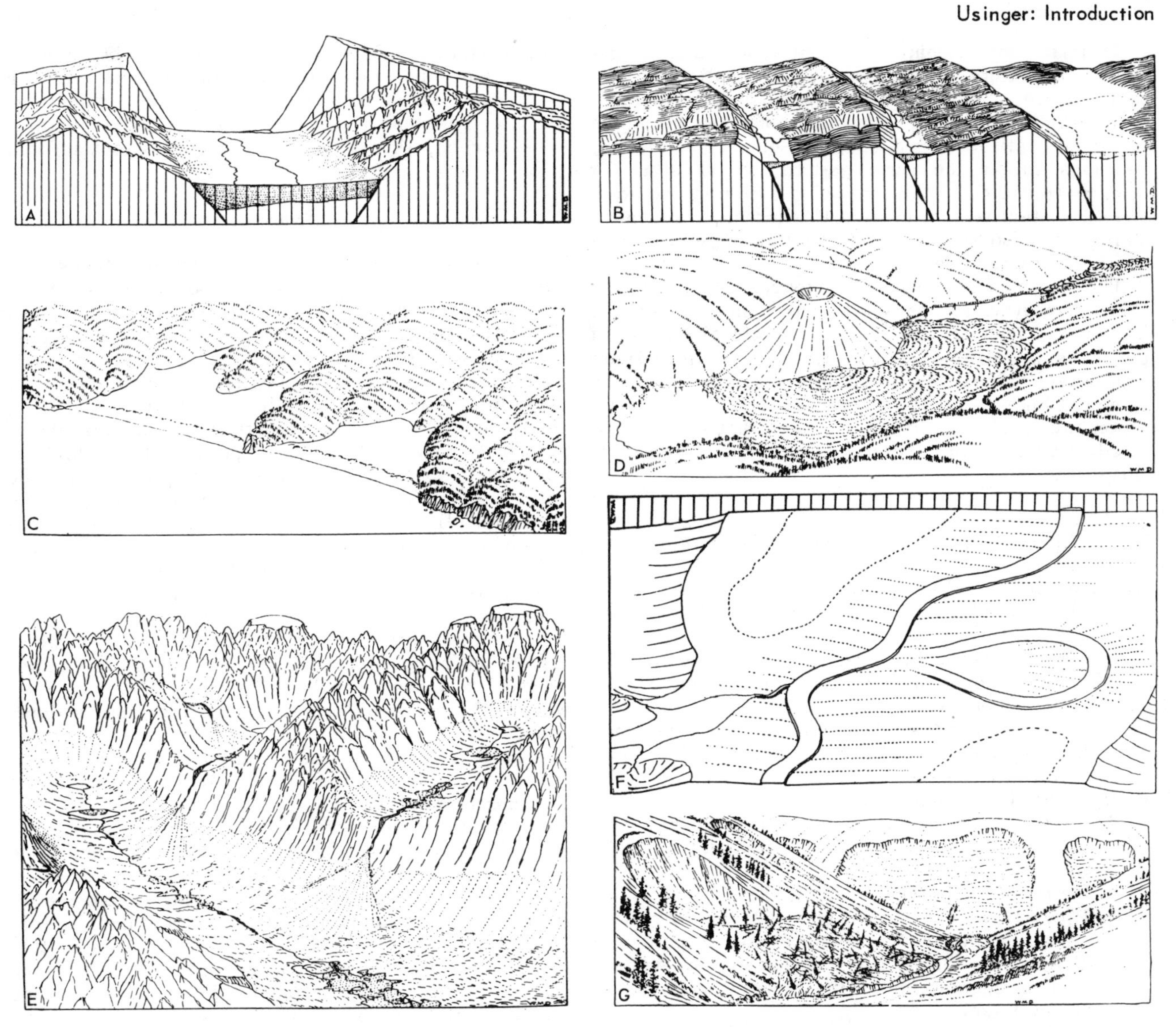

Intro. fig. 20. Diagrams showing the formation of lakes by various geological processes. *a*, a depressed fault block between two upheaved blocks; foreground the same, after a period of erosion and deposition; *b*, great fault blocks of the northern Sierra Nevada, with the plain of Honey Lake in the east; *c*, coastal lagoon formed by wave-and-current-built beach; *d*, lakes formed by volcanic action including crater and impoundments due to lava flows; *e*, glacial cirque lakes with "hanging" troughs; *f*, river-made lakes including an isolated "ox-bow" and sloughs; *g*, lake formed by a landslide (Davis, 1933).

are low in productivity. In contrast to this the so-called Eutrophic lakes are usually old, rich in nutrients, and shallow because of the accumulation of sediments and organic matter. The water in the photosynthetic zone exceeds that in the decomposition zone (or may be entirely in the productive zone), and biological productivity is high. Unfortunately, the precise limits of these zones are difficult to establish because light penetration varies with turbidity, and photosynthesis is influenced by temperature and many other factors. All that can be said is that the Oligotrophic-Eutrophic classification is based on the truism that autotrophic processes exceed hetertrophic processes in surface waters whereas the reverse is true in deeper waters. With this as an axiom, individual lakes can be classified as more or less oligotrophic, and the process of aging or the evolution of a lake into a pond and swamp can be referred to as eutrophication. An alternative sequence involves brown-water bog lakes and peat formation with humic acids hindering organic decomposition. These are called Dystrophic lakes. In California, high elevation lakes are generally oligotrophic (their beds having been scoured out of granite in recent times by glaciers), whereas lakes of the lowlands are at various stages in the process of eutrophication. Dystrophic lakes occur infrequently and mostly at higher elevations in the northern part of the state.

Thermal and trophic classifications are fundamental in nature and world-wide in application. At the local level more detailed classifications become feasible. For example, classifications may be based on faunal affinities or methods of origin. Faunal relations follow the same pattern for lakes as for streams (intro. fig. 13), there being a Sierran and north coastal Boreal region, a Great Basin region, a Central Valley and south coastal California region, and a southern Colorado desert region.

The most conspicuous differences in California lakes, however, are caused by differences in origin. These are used as the basis for a classification of the lakes of California. Illustrations (intro. fig. 20) are from Davis (1933) who lists and discusses 129 of the principal lakes and bays of the state. This, of course, represents but a fraction of the lakes that are encountered, especially in the Sierra Nevada and the northern Coast Range.

Classification of the Lakes of California

A. Lakes of Glacial Origin
1. Ice-scoured highland lakes with shallow granite basins. Desolation Valley Lakes.
2. Cirque lakes. Hanging lakes carved out of granite slopes and dammed with terminal moraines. Upper Angora Lake.
3. Trough-end lakes, with terminal and sometimes lateral moraines. Donner Lake, Fallen Leaf Lake, Convict Lake.

B. Intermont Depression Lakes
4. Fault-block and warped valley lakes dammed by lava flows or landslides. Clear Lake, Lake County.
5. Intermont basin lakes with outlets.
6. Playa lakes with no outlet and therefore saline, sometimes below sea level. These are perennial if the water supply exceeds that lost by evaporation or intermittent, if the reverse. Mono Lake, Honey Lake.

C. Lakes of Volcanic Origin
7. Crater lakes formed in the actual craters or the collapsed cones of craters. Crater Lake (Oregon), Crater Butte Lake, east of Lassen Peak.

D. Limestone Basin Lakes
8. Lakes formed by solution of limestone.

E. River-made Lakes
9. Oxbow lakes representing cut-off river meanders in the delta regions of the Sacramento, San Joaquin, and Colorado rivers. Murphy Lake (Sacramento River), Lake Houghtelin (Colorado River).
10. Flood-plain Lakes. Basins adjacent to the flood plain of the Sacramento River with extensive reed marshes and with 20 to 100 square miles of open water held in check by man-made dykes after winter rains. Sutter Basin, Yolo By-Pass.
11. Lakes barred by Fan Deltas. The saline Salton Sea, a lake of fresh-water origin (in its present form) separated by Colorado River delta-plain from the marine waters of the Gulf of California.
12. Coastal lagoons. Where rivers are blocked by dunes and sand spits to form fresh or brackish water lakes with or without direct connections to the ocean at high tide. Merced Lake, Salinas Lagoon, Abbott's Lagoon, Waddell Creek lagoon.

Ponds

There is no sharp distinction between lakes and ponds. Sooner or later in the course of physiographic evolution and eutrophication all lakes are destined to become ponds unless climate and erosion reverse the process. Generally speaking, ponds are small bodies of water of little depth. Although they vary in turbidity and therefore in penetrability of light, ponds might be defined as entirely within the photosynthetic zone and at least potentially within the zone of rooted aquatic plants. Or ponds might be defined as entirely within the epilimnion, there being no thermocline because all the water is within the zone of wind circulation.

In addition to ponds derived from lakes, there are several types of naturally occurring ponds that arise independent of lakes. Among these may be mentioned (1) pasture ponds of the Central Valley and elsewhere; (2) spring-fed ponds or basins throughout the state; (3) bare rock basins in high Sierran granite and elsewhere; (4) snow-melt pools on rock or forest floor, especially in the Sierra; (5) beaver ponds in Sierran and some coastal streams; and (6) stream-fed pools that become isolated in intermittent streams during the summer and fall.

Many ponds are perennial but pond organisms are generally adapted to life under the rigorous conditions of temporary ponds. Usually they have ready means of dispersal, short life cycles, and resistance to extremes of temperature, desiccation, and so on. Several groups of insects are ideally suited to this type of existence and as a consequence are the dominant forms of life in ponds. Typical pond groups are dragonfly nymphs, mayfly nymphs, true bugs, beetles, and larvae of various groups of Diptera including midges, mosquitoes, and the like.

Lake and pond zones and communities.—Several zones are recognized in standing waters. These are of two main types: those associated with the shore and bottom, and those associated with open water. (Intro. fig. 21 shows the relationships of these zones.) The littoral-benthic series ranges from the shore line to the light compensation level—Littoral Zone, and then to the Benthic Zone. In open water two zones are recognized: the Limnetic Zone above the level of light compensation (=photosynthetic or autotrophic region) and the Profundal Zone (=decomposition or heterotrophic region) below. These terms have been borrowed from marine ecology and are not always used in precisely the same sense by students of fresh-water ecology. Insects occur in each of these zones. For example, shore bugs (Saldidae) and Dixid larvae live at the water's edge; caddisworms and a host of other insects live in the littoral zone near shore; and bloodworms live in the benthic zone. In

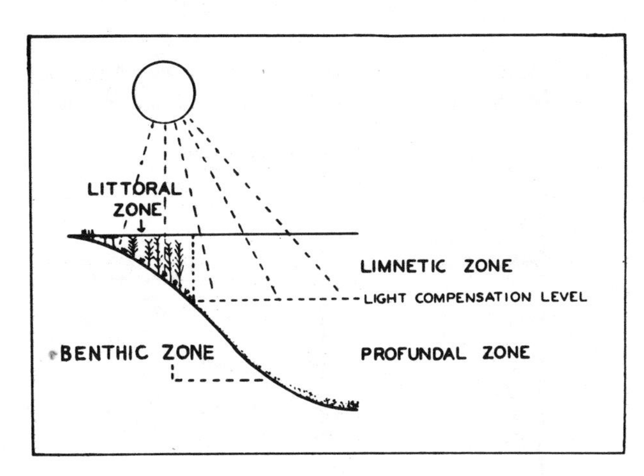

Intro. fig. 21. The major zones of a lake (Odum, 1953).

open water the whirligig beetles (Gyrinidae) and water striders (Gerridae, Veliidae, etc.) live on the surface; nearly all aquatic insects inhabit or traverse the limnetic zone at some time in their life history, and only the *Chaoborus* "phantom larvae" regularly inhabit open water in the profundal zone.

As mentioned earlier, insects are only a part of the aquatic communities in which they live. To understand the dynamics of a lake or pond all aspects of the community must be studied. This can be done only on a large scale, for a lake or pond is in reality a unit. However, within each body of standing water local communities are recognized. These are convenient for study purposes but are always interdependent and should not be permitted to obscure the essential unity of the whole.

The communities associated with the substrate are: the littoral or shore community, the benthos or bottom community, and the periphyton (epilithon) or organisms attached to stems and leaves of rooted plants or other surfaces projecting above the bottom (Odum, 1953). Of these, the littoral is divided into the region above high water and wave action, epilittoral; the region of shore-line fluctuation, eulittoral; and the permanently submerged shore line, sublittoral, above the level of light compensation. Microscopic organisms among the sand grains at the water's edge are referred to as the psammon or psammolittoral. Also a specialized insect fauna including mostly beetles (Heteroceridae) and fly larvae (Psychodidae, Tipulidae, Heleidae, etc.) occurs in mud flats.

The benthos or bottom community contains a rich but specialized insect fauna and plays an important part in the economy of lakes. During the process of eutrophication, bottom materials accumulate. The process has been worked out in detail by Deevey (1942) for Linsley Pond (intro. fig. 22). During an early oligotrophic stage the water was probably clear, and Chironomids of the genus *Tanytarsus* predominated. Later, during the period of maximum production, plankton blooms reduced the transparency, and Chironomids of the genus *Endochironomus* were abundant. Finally, during the present period, the pond has become shallow and relatively stable with *Chironomus* and *Chaoborus* as the dominant benthic insects. This sequence was discovered by boring with a peat sampler and analyzing fragments of microfossils at various levels in the bottom of the lake.

The precise effect of *Chironomid* and *Chaoborus* larvae on the metabolism of a lake is not known. However, it has been demonstrated by Walshe (1951) that *Glyptotendipes* and other larvae live in tubes in the bottom mud. They spin nets (intro. fig. 23) to catch food which is carried in the currents of water that are constantly circulated by undulatory movements of their bodies. As a result of this activity by millions of larvae in the bottom of most lakes, stagnation is reduced at the mud-water interface and nutrients are made available for general circulation. Oxygen, in particular, is at a premium at the bottom of lakes, and Mortimer (quoted in Ruttner, 1953) has

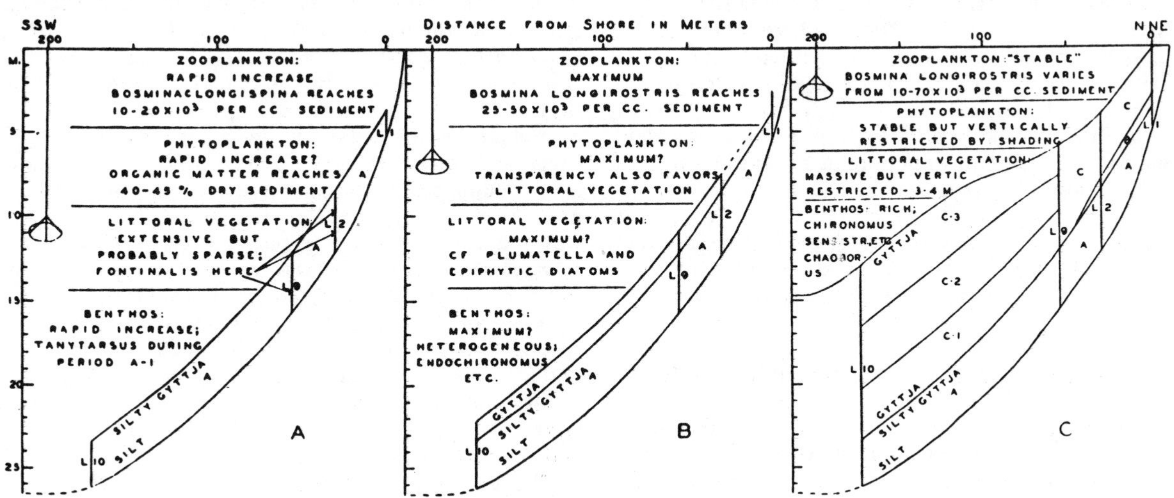

Intro. fig. 22. Schematic bottom profiles and a summary of ecological conditions in Linsley Pond, *a*, during an early oligotrophic period; *b*, at the time of maximum organic production; and c, at present. Estimated transparencies are shown by Secchi discs at the left of each figure (Deevey, 1942).

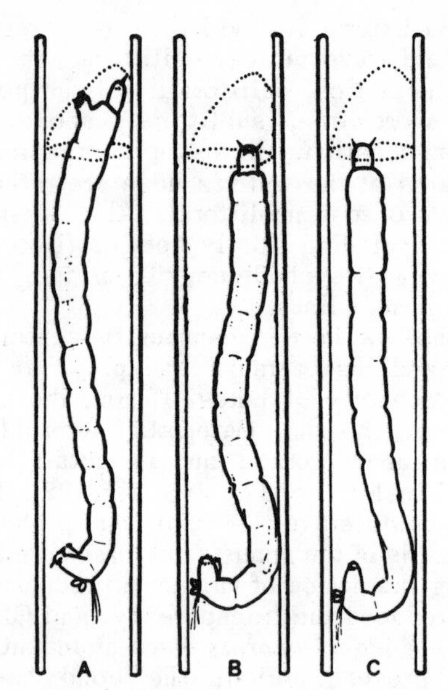

Intro. fig. 23. *Glyptotendipes* larvae (Chironomidae) in tubes showing stages in the spinning of the food-catching net (Walshe, 1951).

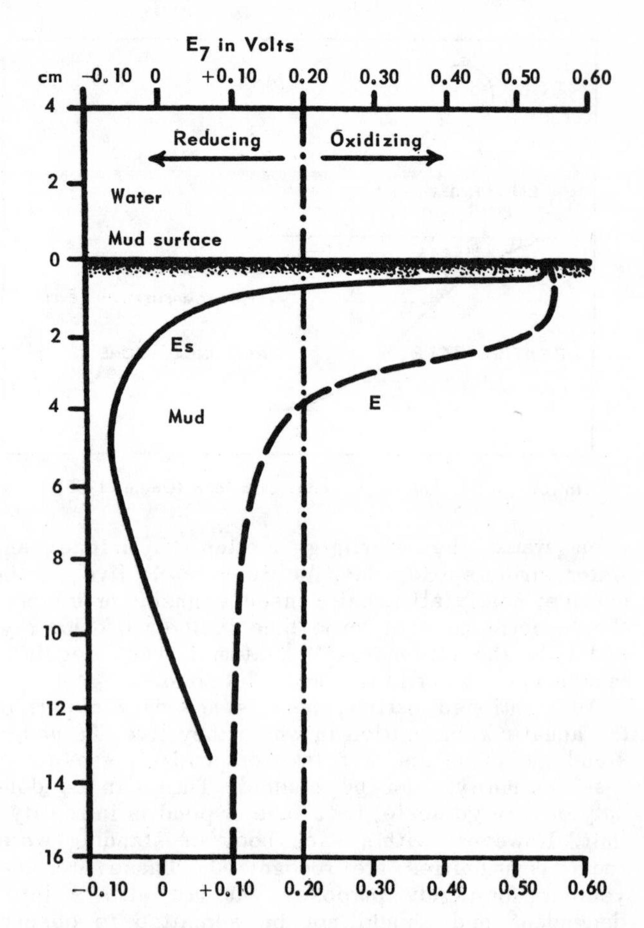

Intro. fig. 24. The redox potential at and below the mud-water interface in two lakes in northern England: E, Ennerdale Water (oligotrophic) and ES, Esthwaite (eutrophic) (after Mortimer in Ruttner, 1953).

shown that "there is a sharp and narrow transition from the more or less oxidized substances at the ooze-water interface to the strongly reduced substances in the deeper layers of the sediments." A measure of the oxidizing or reducing power of solutions has been devised (called the redox potential), and Mortimer discovered that the redox potential, E7 (=E at pH 7), was approximately 0.6 volts in oligotrophic lakes or eutrophic lakes at the time of complete circulation. Within the sediments, however, the redox potential declines very rapidly, reaching a minimum at a depth of about 5 cm. (intro. fig. 24). The effect of Chironomid water circulation on this system is not known but must be considerable.

Open-water communities are the neuston or organisms associated with the surface film (above and below), the plankton or small floating organisms (and other material) whose movements are more or less dependent on currents, and the nekton or larger free-swimming organisms. Insects play a dominant role in the neuston. Their supremacy is challenged in the nekton only by fishes. Periphyton organisms are a chief source of food for "grazing" and "browsing" insects, and plankton forms the staple diet of the sieve feeders or strainers. Others, of course, prey on these primary converters of plant and microscopic animal food.

Marine Marshes, Estuaries, and Intertidal Habitats

The oceans are, of course, the largest aquatic habitat in the world, and it is interesting to speculate as to what would have happened if insects had successfully invaded this habitat. Their failure to do so cannot be attributed to salinity, because a few of them inhabit inland lakes with salts more concentrated than in the ocean. Competition from previous inhabitants may have been a factor, but fishes and crustaceans did not deter them in fresh water. The most plausible explanation is that practically all insects, as mentioned previously, are dependent at some point in their life cycle on surface air, and most are terrestrial or aerial in the adult stage. This effectively limits them to shallow waters not too distant from shore and may account for their absence from the open ocean and their scarcity in the deeper parts of large freshwater lakes. Only one group, the marine water striders *(Halobates),* lives in the open ocean, and they are surface dwellers rather than true aquatics.

However, the situation is quite different along sea coasts and in coastal marshes and estuaries. A small and usually inconspicuous but varied insect fauna exists in such places and includes such notorious pests of mankind as the salt-marsh mosquitoes *(Aedes)* and sand flies *(Culicoides).*

The four principal types of habitat along the California coast are given in the following classification and examples are cited of the insect inhabitants of each.

Classification of Coastal Waters Inhabited by Insects

A. Estuaries and open water bays. Salt or brackish waters. San Francisco Bay, Carquinez Straits. *Trichocorixa reticulata* (Guerin) in shallower pools including brine pools (Leslie Salt Co.).
B. Intertidal rocks. Exposed to the full force of the waves and to tidal fluctuations. Clunionine midges (Chironomidae) with wings often reduced or absent. A crane fly *(Dicranomyia)*. Beetles of the genera *Liparocephalus* (Staphylinidae); *Ochthebius* (Hydraenidae); *Thalassotrechus* (Carabidae); and *Eurystethus* (Eurystethidae). The larvae of flies and larvae and adults of beetles feed on marine algae. Adult beetles are found in cracks in the rocks at low tide. Marine midges emerge, mate, and lay eggs during the relatively brief periods when their habitats are exposed by low tides.
C. Beaches and mud flats. Exposed or partly protected from the waves. Subjected to tidal fluctuation. Aquatic insects scarce. *Deinocerites* mosquito larvae in crab holes. *Ephydra* flies in salt pools. Many essentially terrestrial insects in beach drift.
D. Tidal marshes. Usually protected from wave action. Vegetation of marine algae, salt-marsh grasses and *Salicornia*. Salt-marsh mosquitoes *(Aedes dorsalis* and *squamiger)*. Sand flies *(Culicoides)*.

Miscellaneous Aquatic Habitats

Insects occur in a few unusual situations that deserve special mention.

Snow and ice.—A wingless Tipulid fly, *Chionea nivicola* Doane, and the golden snow flea, *Onychiurus cocklei* (Folson), are found on the surface of the snow in the high Sierra. Eggs of the snow mosquitoes, *Aedes hexodontus, A. communis,* remain beneath the snow all winter and hatch as the snow melts in the spring or early summer. The primitive orthopteran, *Grylloblatta,* lives at the edges of glaciers and snow fields in the high Sierra and in the ice caves of Modoc County. It is not aquatic but is never found far from snow or ice.

Bogs.—Dense sphagnum grows in some parts of California, either as a marginal shore line in bog lakes or as a low dome-shaped carpet in a peat moor (intro. fig. 25) or "hanging bog." The semiaquatic shore bugs (Saldidae) are common under such circumstances, and the small insectivorous sundew plants *(Drosera)* capture a variety of insects on their sticky tentacles. Pitcher plants *(Darlingtonia)* occur in some bogs in the northern part of the state. Insects are trapped in these, but no mosquito larvae *(Wyeomyia)* have been found as in the *Sarracenia* pitcher plants of the southeastern states.

Tree holes.—This inconspicuous aquatic habitat is often overlooked but is so common in deciduous forests *(Quercus, Umbellularia)* at middle elevations in California that the principal insect inhabitant, *Aedes varipalpus* (the tree-hole mosquito), is a major pest. For reasons that are not entirely clear, water in rotten limb holes and other parts of plants gives an acid reaction, is coffee-colored, and is high in organic nitrogen. Under these conditions a unique biota develops with several kinds of insects—a small crustacean, *Cyclops,* a small nematode, several types of rotifers, and the protozoans *Vorticella, Paramecium,* and *Hypotrichia* (Jenkins and Carpenter, 1946). In California only *Aedes varipalpus* and several Heleid larvae have been reported.

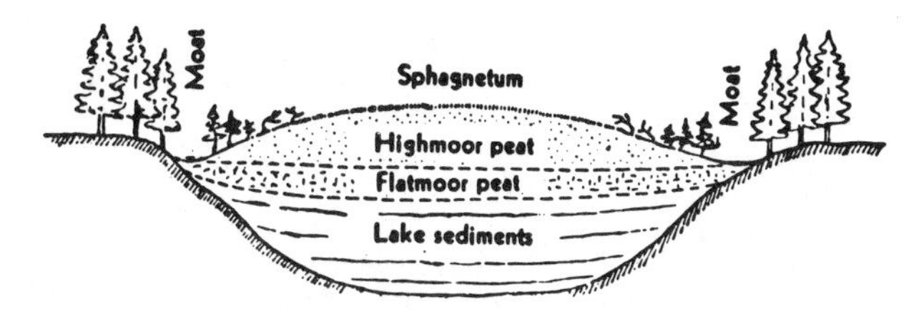

Intro. fig. 25. Diagrammatic cross section through a peat moor that has arisen from a small lake (Ruttner, 1953).

APPLIED AQUATIC ENTOMOLOGY

Man's conquest of the elements and progress toward a civilized state is marked in part by his management of water. Irrigation and pond-fish culture, seen in the remains of prehistoric cultures, were early manifestations of this process. Modern man has perfected these techniques and added measures for navigation, flood control, hydroelectric power, sewage disposal, and storage and transport of drinking water. To these have been added such considerations as recreation and improvement of the public health. It is doubtful if any place in the world exceeds or even equals California in extent of water management because of its semiarid climate, intensive agriculture, and high standard of living. Merely to describe the diverse aspects of water management would take us far from the field of aquatic entomology. For purposes of this Introduction it will suffice to discuss the role of insects and the status of insect problems in relation to water management in the state.

Mosquito and Gnat Control

Mosquito control touches nearly all aspects of aquatic biology and is a primary consideration to the aquatic entomologist. The subject is so large and has been fully dealt with in so many books that it will be treated only briefly here. For fuller information the reader is referred to the following: Herms and Gray (1944), Bates (1949), Horsfall (1955), and Carpenter and La Casse (1955). A summary, *Mosquito Abatement in California,* was issued by the Bureau of Vector Control, State Department of Public Health (Bull. No. VC-1, 1951).

In the half century or more since mosquitoes were found to be vectors of human diseases, control measures have been improved steadily and, at times, spectacularly. To the earliest kerosene spraying (L. O. Howard, 1892) and Paris green dusting (Barber & Hayne, 1921) were added methods for eliminating breeding places (drainage, filling), for altering the habitats of larvae (flooding, stranding), and for controlling larvae by the introduction of top minnows *(Gambusia)*. Then, during World War II, DDT started a revolution in chemical control somewhat comparable to the effect of antibiotics on the practice of medicine. At the present time mosquito control, in spite of complications due to the development of resistance to the newer insecticides, is a highly effective operation. It calls for sound knowledge of the systematics and biology of mosquitoes and the best of modern engineering practices. Furthermore, as a result of public information gained during World War II, and other reasons, mosquito control is more widely demanded and supported than ever before.

In California, certain aspects of mosquito control have been simplified whereas others, owing to climatic and other circumstances, have been accentuated. The dengue-yellow fever vector, *Aedes aegypti*, does not occur here, thus obviating the necessity for routine house-to-house inspection of artificial containers such as tin cans, bottles, flower vases, and so on. Of mosquito-borne diseases, the tropical filariasis is absent. Malaria, on the other hand, was a major problem in times past and remains as a potential threat in periods of mass movement of possible carriers, since the mosquito vector, *Anopheles freeborni*, is as abundant as ever. Historically, California played a pioneering role in malaria control (Herms, 1913). "Thirteen Central Valley counties, with an area of about 20,000 square miles, and harboring three-fifths of all the malaria, had a death rate of 14.2 per 100,000 as late as 1916. These counties comprised an area about half the size of Mississippi, which had a reported death rate of 5.9 per 100,000—and Mississippi was regarded as one of the most malarial states in the country". (*Mosquito Abatement in California*, 1951). Sporadic cases occurred when carriers entered the area during the depression years and when a carrier returned from Korea in 1953 and started a small epidemic, with thirty-two proven cases (Brunetti, *et al.*).

The reasons for the disappearance of endemic malaria are not fully understood in California or elsewhere, but it is thought that widespread control by the public and higher standards of living have contributed to the present freedom from the disease.

Another mosquito-borne disease is encephalitis or sleeping sickness. In California, three strains of virus are known: the Western equine, St. Louis, and California strains. *Culex tarsalis* is the only important vector, according to present knowledge, but thirteen other species have been found harboring the viruses in nature or capable of transmitting them in the laboratory. Information is still not complete (Reeves, 1953) but a diagram of the probable infection chain in Western equine encephalitis is given below (intro. fig. 26), based on studies conducted under the direction of the Hooper Foundation for Medical Research. Great effort is now being made to control the mosquito-borne encephalitides but, unlike malaria, the problem has not yet been solved and cases occur every year, some of which result in damage to brain tissues, paralysis, or death.

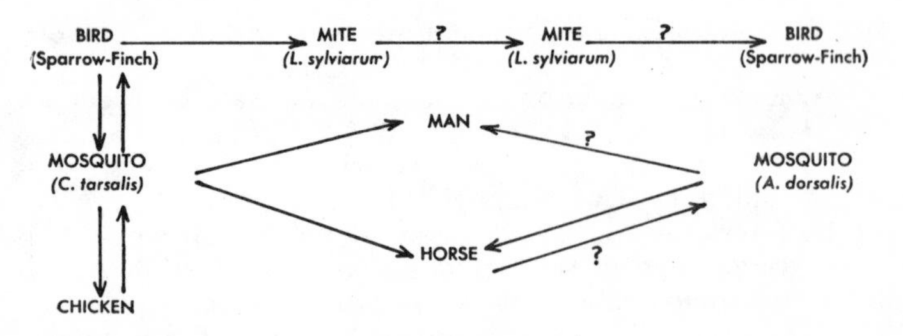

Intro. fig. 26. Infection chain for western equine encephalitis (*Mosquito Abatement in California*, 1951).

Pest mosquitoes.—Long ignored in many parts of the world, pest mosquitoes have received attention mainly near large centers of population (New Jersey, Florida, the Chicago area, the San Francisco Bay area, and elsewhere). In California the most important pest mosquitoes are salt-marsh species *(Aedes dorsalis, Aedes squamiger,* and *Aedes taeniorhynchus);* domestic mosquitoes *(Culex pipiens, Culex quinquefasciatus);* mosquitoes of irrigated areas *(Aedes nigromaculis, Aedes dorsalis, Culex tarsalis);* tree-hole mosquitoes in the deciduous forests of coast range and foothills *(Aedes varipalpus);* and snow mosquitoes *(Aedes communis, Aedes hexodontus).*

The control of pest mosquitoes has lagged in some places because funds are usually earmarked for disease vectors. California is an exception to this, with an efficient and extensive organization for pest-mosquito abatement. This is contributing in no small measure to the high land values, increased enjoyment of recreational areas, and increased efficiency of workers out-of-doors.

Mosquito control.—In 1915 the Mosquito Abatement District Act was passed by the state legislature. In subsequent years this resulted in forty-three local areas (intro. fig. 27) (counties or other areas) organizing for mosquito control and supporting the work with tax rates up to 40 cents on each $100 of assessed valuation. Also nine local health departments are active in mosquito control. Mosquito Abatement Districts are governed by boards of trustees which establish policies and employ a manager to carry out the program of the district. "The District has the power to enter upon and inspect lands for mosquito sources, and to take appropriate measures to abate mosquitoes thereon, whether such lands be within or outside of the district; to acquire land or rights of way for drains or other purposes; to purchase supplies and equipment for the work, etc." The technical staff of a district includes a manager and usually one or more entomologists, inspectors, foremen, and so on. Technical guidance and support is provided to the districts by the Bureau of Vector Control of the State Department of Public Health. The Bureau of Vector Control also conducts investigations of particular problems that are beyond the scope or facilities of the individual districts.

Role of the entomologist in mosquito control programs.—Very early in the development of mosquito control programs it was realized that "shot-gun" methods were expensive and not very effective.

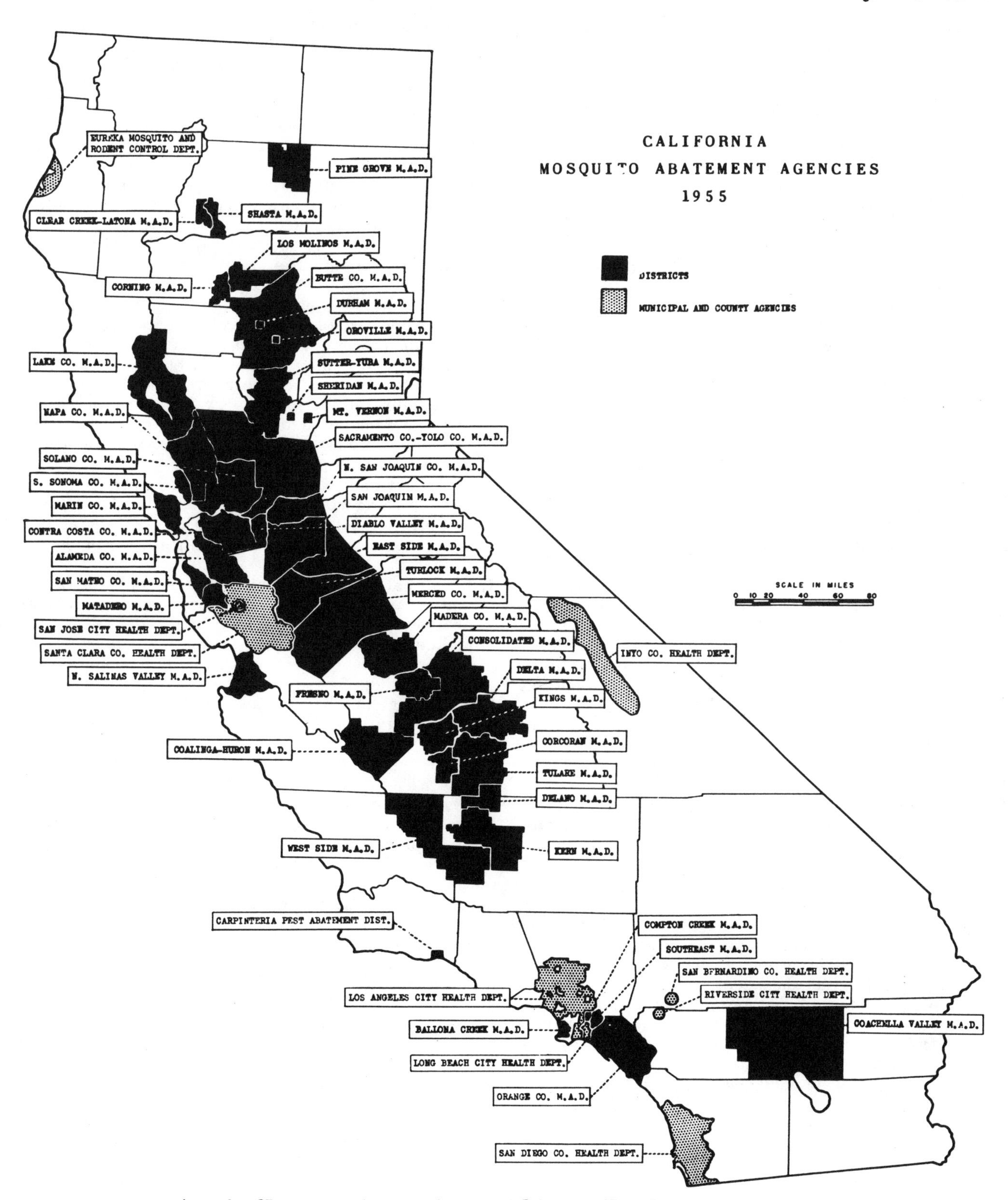

Intro. fig. 27. Mosquito abatement districts of California. (From Bureau of Vector Control, California Dept. of Public Health.)

Instead of spreading larvicides over thousands of acres of water without regard to kinds or numbers of mosquitoes present, the concept of "species sanitation" was adopted. This required that trained entomologists sample potential breeding areas at regular intervals and report on the presence or absence and relative density of the various species encountered. For disease vectors, control measures were undertaken only when counts reached a predetermined level. By this means, *Aedes aegypti* was eliminated from Havana and other Latin American cities, *Anopheles quadrimaculatus* was controlled in the vicinity of war areas in the eastern United States during World War II, and *Anopheles freeborni* was reduced in and around military establishments in California and other western states.

To control mosquitoes effectively and economically, then, requires the services of an entomologist and a crew of trained inspectors. The duties of an entomologist in a mosquito control program are outlined as follows (Bureau of Vector Control Memorandum, April 30, 1948):

1. Accurate appraisal of the existing and potential mosquito sources within and adjacent to the district, with such information systematically recorded on appropriate maps and records. This information to be obtained by:
   a. Surveys of mosquito occurrence such as (1) properly chosen and collected resting stations, light traps, and biting conditions; (2) systematic larval dipping collections; (3) field observations integrated to understand mosquito species ecology.
   b. Accurate identification and systematic recording of species distribution in the district.
   c. Observations of species habitats (aquatic and adult), applied to guide control operations.
2. Evaluation of adequacy and efficiency of control programs by:
   a. Coördination of the routine survey and section survey findings with the over-all control program.
   b. Coördination of the observations of the inspectors with operations of the control crews.
3. Training of organization personnel through:
   a. Casual daily conversations with staff.
   b. Organized training programs to acquaint the staff with mosquito species identification and with the vulnerability of species as determined by knowledge of their ecology.
4. Testing of control methods, materials, and techniques, to determine their reliability in obtaining control and the degree of control. This is accomplished by:
   a. Comparisons of methods, such as aerosol vs. spray; ground vs. aerial approach; adult control vs. larval control.
   b. Testing of materials and determining their capacities one against another, in fresh, foul, salt, brackish, sunlit, and shaded water; also their residual qualities, effects upon mosquito predators, and effects upon agricultural crops, livestock, beneficial insects.
5. Assisting in the planning, preparation, and carrying out of public relations and educational programs, which in the long run serve to document the district's activities by recording its history. This is done by:
   a. Preparation of visual education matter, such as graphs, maps, photographs, exhibits, educational pamphlets.
   b. Preparation of publicity releases for newspaper and magazine publication of the entomological aspects of the district's operations.
   c. Preparation and delivery of talks about mosquitoes, their life histories, and habits, to schools, service clubs, farm organizations, and professional groups.
   d. Personal contacts and professional associations which promote closer appreciation of the entomological problems and the scope and objectives of the entire control program.
6. Immediate application of new methods, materials, and techniques developed by contemporary entomological, chemical, medical, and veterinary workers through:
   a. Review and interpretation of the current literature concerning mosquito control.
   b. Personal contact with research workers at professional meetings and institutions of higher learning.
7. Investigation of encephalitis, encephalomyelitis and malaria cases, including epidemiological analyses of their occurrence by:
   a. Keeping of spot maps.
   b. Correlation of cases with mosquito populations.
   c. Gathering of case histories.
8. Assisting in district administration, particularly in decentralized districts.

Survey methods are of several standard types. Adults may be caught while biting on the bare arms or legs, or counts may be made of the landing rates on trousers or dark cloth. However, counts in natural resting places are preferred because they can be standarized and visited at regular intervals. Favorite resting places include hollow trees (intro. fig. 28), sheds or barns (intro. fig. 29), the shaded undersides of bridges, chicken houses, porches, tank houses, and the like. Artificial resting places may be set up, including cages baited with live animals, or plain boxes (intro. fig. 30). The latter have proved to be effective for sampling *Anopheles* mosquitoes in many places. Another effective method for sampling the populations of some species of mosquitoes is the light trap. A very efficient and standardized type is the so-called "American light trap" (Mulhern, 1953), modified from the "New Jersey" type (intro. fig. 70). Unfortunately, the light-trap method is not uniformly effective for all apecies and varies in efficiency with changes in climate. Likewise, none of the methods mentioned above gives data that can be compared with other samples in the same or different regions because the factors that influence adult mosquitoes are extremely local and elusive. In spite of these difficulties,

Intro. fig. 28. Sampling adult mosquitoes in a hollow trunk of a tree (U.S.P.H.S., C.D.C. photo).

sampling at a single station, properly chosen, can provide valuable data from week to week and year to year and is the best criterion we have for judging the relative abundance of most mosquitoes.

Intro. fig. 29. *Anopheles* mosquitoes resting on the ceiling of a barn. Inset—mosquito in biting position (U.S.P.H.S., C.D.C. photo).

Larval and pupal densities are in some ways easier to determine. A standard, white enamel dipper is commonly used as a sampler (intro. fig. 31). The dipper is placed at the water surface and one edge is tipped so that water flows in, carrying larvae and pupae with it. Specimens can be seen, counted, and collected with a pipette if necessary. Hess (1941) devised a straight-edged screen dipper that samples a standard area of surface, thus giving more meaningful results but this dipper has not come into general use. Dippers are totally inadequate under some circumstances. For example, the water in a weed-choked irrigated pasture is so shallow that other means must be used. As a partial solution to this dilemma Yamaguchi (1949) devised a "sleeve sampler" consisting of a vertical cylinder from which larvae were removed by a hand-operated suction pump (intro. fig. 32). A standard number of dips or samples should be taken

Intro. fig. 30. Red box serving as an artificial resting place for adult Anophelines (U.S.P.H.S., C.D.C. photo).

and the counts recorded so that data will be comparable. Only experience in a particular area can determine at what level of adult or larval counts control measures should be undertaken.

Control measures.—The various methods used in mosquito control are diverse and must be adapted to each situation. For adults, which should be the last line of defense, indoor space sprays are employed—using aerosol "bombs" or even the simple "Flit-gun." Aerosol mists and sprays are designed for quick knockdown of mosquitoes and have been used for this purpose even over extensive outdoor areas with applications made by airplane or specialized equipment on the ground. Most adult sprays contain DDT, pyrethrum, or some other material which has a residual effect when sprayed on walls or other surfaces. Residual sprays are particularly effective in malaria control because engorged females of some Anophelines rest on treated walls and are killed before they are

Intro. fig. 31. Sampling for mosquito larvae with a white enamel dipper (U.S.P.H.S., C.D.C. photo).

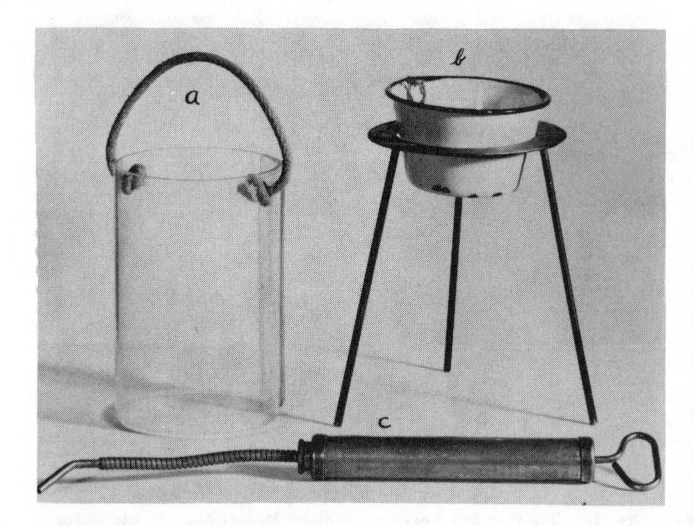

Intro. fig. 32. Sleeve sampler used for shallow water mosquito breeding places in irrigated pastures (Yamaguchi, 1949). Larvae are sucked out of the plastic cylinder by the "pump" and then ejected into the enamel pan for counting.

Intro. fig. 33. Drainage is the most permanent and therefore one of the most effective methods of mosquito control (U.S.P.H.S., C.D.C. photo).

able to incubate the malaria parasites and transmit the disease. Other measures directed at adult mosquitoes include screening, nets of various kinds, and repellents such as "612," dimethylphthalate, and the like.

Most methods of mosquito control are directed against the immature forms—the larvae and pupae. One of the most important of all methods because it is relatively permanent is the elimination of breeding places by drainage (intro. fig. 33). Drainage ditches may be blasted (intro. fig. 34) or dug by dragline (intro. fig. 35) or by hand labor. The cost of maintenance may be greatly reduced if drainage ditches are lined with concrete and sod planted on the upper

Intro. fig. 34. Use of dynamite for construction of a drainage ditch for mosquito control (U.S.P.H.S., C.D.C. photo).

Intro. fig. 35. Dragline ditching at Marysville, California (U.S. Army Signal Corps).

banks (intro. fig. 36*a*, before concrete lining; *b*, after). Underground drainage is commonly used in seepage

Intro. fig. 36. Lining a ditch with concrete and sodding the upper banks to increase effectiveness and reduce maintenance costs. *a*, before; *b*, after (U.S.P.H.S., C.D.C. photo).

areas using tile or buried poles. Breeding places may also be eliminated by filling with a bulldozer or by diking and dewatering with pumps.

In brackish waters such as are found in the salt marshes, drainage ditches may suffice, but more often tidal action requires that other measures be used. Dikes may be constructed with tide gates or automatic siphons to hold back the salt water at high tide and permit drainage outward at low tide.

Biological control has long been a solution to mosquito breeding in local areas. The top minnow, *Gambusia affinis,* is ideal for this purpose and can be introduced into ornamental pools (intro. fig. 37) or even into cisterns and other unlikely bodies of water. Mosquito fish have been distributed so widely that they may be found in almost any body of water or a supply can usually be had by telephoning the nearest mosquito control agency. *Gambusia* are not needed in garden pools if goldfish are present and are not fed excessively.

Mosquito larvicides are considered as emergency measures or temporary means of control, yet larviciding has and probably will continue to occupy a large part of the time of mosquito control crews. Progress in this field has been so rapid during and since World War II that generalizations are likely to be misleading and are certain to be dated.

The principal materials in use at the present time are: oils, either alone or as solvents for organic poisons; chlorinated hydrocarbons such as DDT and related compounds; and organic phosphates. The toxicity of various chemicals was tested with colonized larvae of *Culex quinquefasciatus* Say by Isaak (1952) with results as shown in table 2.

Intro. fig. 37. Stocking a garden pool with mosquito fish, *Gambusia affinis* (U.S.P.H.S., C.D.C. photo).

Of the larvicides tested, EPN was the most effective followed by Parathion, Aldrin, colloidal Aldrin, Heptachlor, Dieldrin, DDD, DDT, Q-137, chlordane, lindane, and toxaphene in that order. Field applications are made at concentrations that seem ridiculously low by prewar standards. Formerly, 20 gallons of fuel oil or 20 pounds of Paris green dust were applied per acre of water surface, at great expense and effort.

TABLE 2 (Introduction)

Toxicity Range and LD-50 of Various Insecticides Against Colonized *Culex quinquefasciatus* Say Larvae (4th Instar)

Larvicide	1	.2	.1	.04	.03	.02	.015	.01	.008	.005	.004	.003	.0025	.002	LD-50[a]
EPN										93	74	44	25	17	.0031
Parathion								100	99	63	52	15	6		.0039
Aldrin				100		92		72		39			4		.0060
Colloidal aldrin				100		90	80	73	44	36	31	16	0		.0062
Heptachlor				100		99		76		17			0		.0070
Dieldrin			100	95		79		49		22					.01
DDD		100	95	82		40	28	10							.023
DDT		100	84	79		16		0							.028
Q-137			100	58		37	28		4						.029
Chlordane		100	98	50		9	2								.040
Lindane		100	98	38		10									.0454
Toxaphene	100	89	75	27		4									.055

Now DDT may be applied at as low a figure as half a gallon (½ pound) per acre.

With such small amounts of material the method of application becomes very important. In general, materials are applied as dusts, usually wettable, as sprays in solution, or as finely divided mists (aerosols) or smoke particles. Applications may be made by knapsack sprayer (intro. fig. 38), hand duster (intro. fig. 39), or by various means from a boat, jeep (intro. fig. 40), motorcycle (intro. fig. 41), or airplane (intro. fig. 42). Special methods include the addition of larvicides to irrigation water.

In spite of the spectacular progress with new larvicides, control has been complicated and made difficult by the development of resistant strains of several of our most important mosquitoes. Gjullin and Peters (1952) reported on recent studies of mosquito resistance as follows: "*Aedes nigromaculis* larvae from the treated areas were found to be from three to seven times as resistant to DDT as larvae from the untreated areas (in the San Joaquin Valley). Resistance to toxaphene was less than twice that of larvae from the untreated area . . . Little or no resistance to lindane and aldrin was shown . . . In a few tests with *A. dorsalis,* resistance to DDT was 3 to 12 times that of larvae from the untreated area, but no resistance to the other insecticides was indicated. *Culex tarsalis* larvae from a duck club in the Kern District where applications of toxaphene and aldrin were failing were found to be 10, 33, 11, 215, and 1300 times as resistant respectively, to DDT, toxaphene, lindane, aldrin, and heptachlor as larvae from an untreated area."

Just what the answer to the resistance problem will be is difficult to say. Certainly, it reëmphasizes the importance of permanent control measures, and it also suggests that reliance cannot be placed on a single panacea but rather on a sequence of larvicides including, no doubt, materials that have not as yet been formulated.

Finally, a word should be said as to the timing of control programs. Salt-marsh mosquitoes occur as larvae in the winter and early spring. Therefore drainage and larviciding should be completed before mid-March. Flood water species such as *Aedes vexans* must be treated as soon as pools begin to form when the water recedes in spring or early summer. Control in irrigated pastures must be adjusted to water schedules during the spring and summer months. Tree-hole mosquitoes can be treated with wettable dusts during the breeding season, and filling with sand and asphalt or concrete may be done during the winter months.

Intro. fig. 38. Knapsack sprayer for distributing mosquito larvicides (U.S.P.H.S., C.D.C. photo).

Intro. fig. 39. Hand duster for distributing mosquito larvicides (U.S.P.H.S., C.D.C. photo).

Intro. fig. 40. Jeep equipped for power larviciding (Consolidated Mosquito Abatement District).

Intro. fig. 42. Application of larvicides by means of an airplane (Kern County Mosquito Abatement District).

Snow-pool mosquitoes may be treated in the spring when the snow is melting and the eggs are hatching or in the fall in anticipation of the spring hatch. For domestic mosquitoes such as *Culex pipiens* control measures should be continued throughout the year.

Unfortunately, with the increase in potency of mosquito larvicides greater hazards have been introduced to fish and other aquatic organisms. Ordinarily this problem does not arise because it is considered bad mosquito control to introduce larvicides in areas where fish occur. Nevertheless, the danger exists and evidence is not yet entirely clear as to the limits of tolerance of various species under diverse conditions. However, certain generalizations can be made from the work of Tarzwell (1950) and his associates. Field

Intro. fig. 41. Hand sprayer mounted on a motorcycle for treating catch basins with larvicide (U.S.P.H.S., C.D.C. photo).

experiments were conducted at weekly intervals by airplane using a standard rate of 0.1 pound per acre of DDT mosquito larvicide applied as a spray or aerosol. "Studies on the effect of DDT (dichloro-diphenyl-trichloroethane) and certain other new insecticides indicate that they are all toxic to fishes if used in large doses. With DDT the type of pond or water in which it is used greatly influences the onset and severity of toxic action on fishes. Vegetation, organic material, type of water, and silt or turbidity are all factors influencing this action. Crabs, crayfish, amphipods, isopods, and *Palaemonetes* are very sensitive to DDT, being considerably more so than fishes. Among the fishes, some of the Centrarchidae are the first to be affected, especially the bluegill . . . Although top minnows were among the first fish to be killed, they continued to be present during the period of treatment and were in evidence when most other fish had been eliminated. A few frogs and snakes were killed by routine dosages of 0.1 and 0.05 pound DDT per acre. At routine dosages of 0.1 pound per acre, DDD (dichloro-diphenyl-dichloroethane), chlordane, and DDT are toxic to fish and will significantly reduce the population of ponds. At dosages of 0.05 pound per acre, DDT appears to be somewhat more toxic than chlordane or DDD. Studies carried on in 1947 indicated that DDD was considerably less toxic to fish than DDT. These three insecticides appear to have no significant effect on the fish population at dosages of 0.025 pound per acre. Toxaphene was found to be very toxic to fishes, giving complete kills at 0.2 and 0.1 pound per acre after two and three applications in deep ponds. Kills were obtained at dosages of less than 1 part in 27 million, indicating that this material is as toxic or more toxic to fish than rotenone and may be useful as a substitute for it in fish management work." Doudoroff, Katz, and Tarzwell (1953) added data on other insecticides, stating that, "Aldrin appears to be less toxic to goldfish than toxaphene, but much more toxic than DDT and BHC (benzene hexachloride)."

As regards invertebrates, Tarzwell (1947) reported that DDT is less toxic applied as a dust than in oil. He found that treatment at the rate of 1 to 2 pounds per acre in fuel oil killed Hemiptera, Coleoptera, Odonata, Ephemeroptera, and Chironomids. At 0.025 pound per acre in fuel oil Dytiscids, Gyrinids, Hydrophilids, and Corixids were killed. Seasonal effects after periodic treatment were: an increase in the number of Oligochaetes, nematodes and copepods; a decrease in the numbers of Chironomids, Hemiptera, Coleoptera, and Ephemeroptera. Insects as a group decreased, with the greatest effect of the treatment on the Chironomids.

Repopulation after treatment with DDT was studied by Hoffmann, Townes, Sailer, and Swift (1946). As might be expected, the insect fauna of ponds, which is characterized by short life histories, came back to normal within a few weeks. Streams, on the other hand, required a year or more.

Gnat control.—Gnat control is a peripheral activity of mosquito abatement districts in a few places. Chironomid gnats are the principal pests in Lake Elsinore (southern California) (Miller, 1951) and in Klamath Lake and nearby waters along the Oregon-California border. By far the worst pest, however, is the "Clear Lake gnat," *Chaoborus astictopus* D. and S. In 1940 it was estimated that the total seasonal emergence in the upper part of Clear Lake (44 sq. mi.) was 712 billion gnats or 356 tons (Lindquist and Deonier, 1942), and the "phantom larvae" were estimated on the basis of adequate samples to number 800 billion. One light trap captured 88½ pounds of gnats in two hours. Eggs at a density of 10 million per square foot occurred near shore in drifts 20 feet wide and several miles in length. Fork-tail catfish, square-tail catfish, and split-tail were important feeders on all stages of the gnat; one 9-inch fish was found with more than 1,000 larvae in its stomach.

In former years control seemed to be impossible but the advent of chlorinated hydrocarbon larvicides opened up new possibilities. Experiments had shown that DDD would kill larvae at a dilution of one part to 75 million parts of water whereas fish and other aquatic life in the lake were not killed unless the concentration was increased to 1 in 45 million. With this margin of safety, and after a preliminary trial in nearby Blue Lakes, Lake County organized a mosquito abatement district and, with state and federal help, treated the entire 41,600 acres of lake surface on September 15 and 16, 1949 (Lindquist, Roth, and Walker, 1950). The lake is eutrophic and relatively shallow (27 to 50 feet) without a thermocline so the wettable insecticide was thoroughly mixed by the wind. Control was complete and no gnats were found in the lake for several years, though they gradually increased until 1954 when, on September 25 and 26, a second treatment was carried out. It was a remarkable fact that the removal of so much fish food had no apparent effect on the over-all economy of the lake, probably because *Chaoborus* larvae are carnivorous and are an intermediate link in the food chain and hence were bypassed.

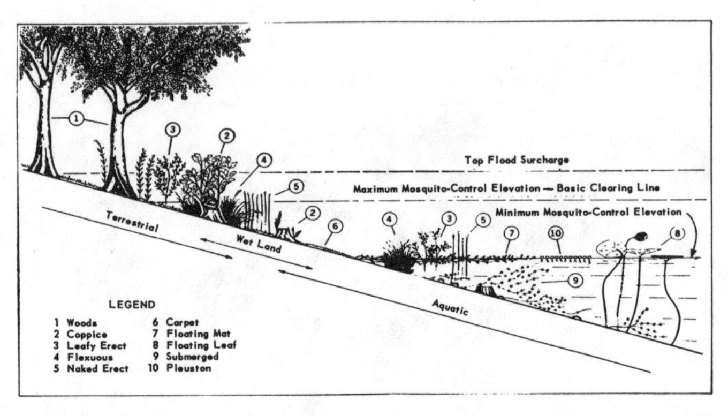

Intro. fig. 43. Classification of plant types along the shore line of a reservoir in relation to water level management (Hess and Hall, 1945).

Man-Made Impoundments

Literally hundreds of dams have been built or are planned for California. The resulting impoundments vary in size from small stock ponds to local reservoirs and enormous multipurpose lakes. They do not differ fundamentally from naturally dammed lakes and ponds, but economic considerations are more likely to arise because of the effects on erosion, mosquito breeding, fish production, and recreation. Therefore, limnological studies have now become an important part of reservoir planning.

Reservoirs.—Some important considerations in a preimpoundage entomological survey (*Malaria Control on Impounded Water,* 1947) are:

1. Location in relation to known pest or disease-bearing insects is a critical factor. For example, a mosquito survey should be made to determine the species present in the locality and their relative abundance, the present and potential breeding areas, and the flight range of potential pests. Periodic density observations should be made throughout the season and during a period of several seasons in order to provide a basis for comparing mosquito production before and after impoundage.

2. Soil and vegetation have a significant bearing on aquatic life. Therefore a reconnaissance survey should include: a study of the timber in the basin

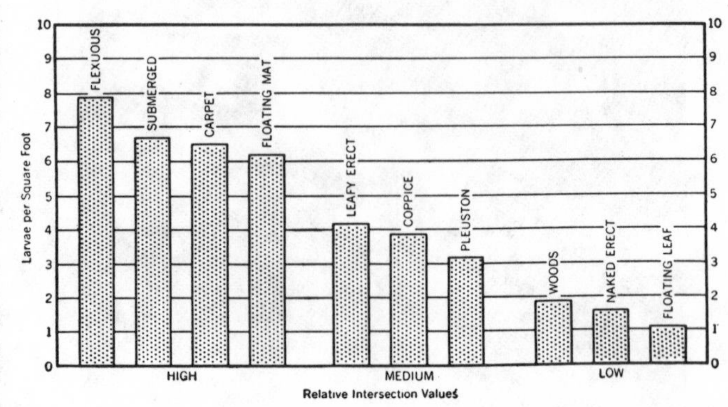

Intro. fig. 44. *Anopheles quadrimaculatus* production potentials of plant types (Hess and Hall, 1945).

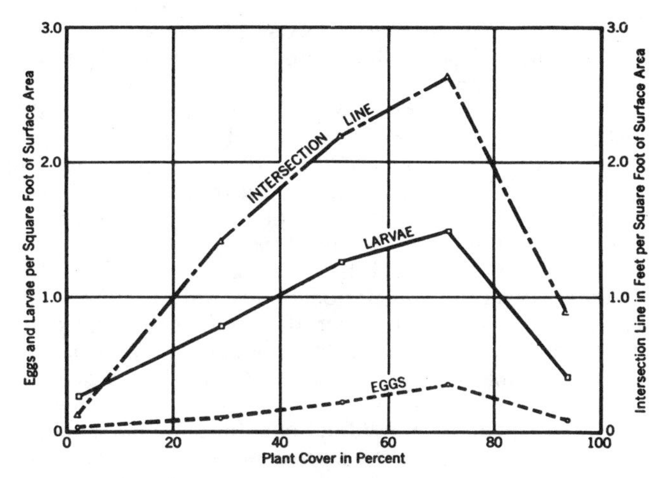

Intro. fig. 45. Relationship between intersection line and production of eggs and larvae of *Anopheles quadrimaculatus* (*Malaria Control on Impounded Water*, 1947).

(acreage to be cleared, density, predominating species, and tolerance to flooding); soil conditions and types; and the existence, location, and extent of marginal and aquatic plants. Shore-line plants may be classified into ecological types (as was done by Hess and Hall, 1945, for the southeastern United States) (intro. fig. 43) and then rated in terms of intersection line (air-water-plant-interface) values (Hess and Hall, 1943) (p. 5, intro. fig. 5). An accompanying figure (intro. fig. 44) shows the relative intersection values of each of the ecological types (woods, coppice, etc.) and gives the production in terms of eastern *Anopheles quadrimaculatus* larvae per square foot. The correlation between intersection line and numbers of eggs and larvae is shown (intro. fig. 45).

From the above, it is evident that marginal vegetation is of primary importance to *Anopheles* mosquito production. It also influences other aquatic insects. Therefore, shore-line filling (intro. fig. 46) and clearing (intro. fig. 47) are essential operations both during the preparation of a reservoir and after the water has been impounded. On surveys it is useful to try to estimate the cost of such operations by dividing the reservoir into areas classified according to type of shore line.

3. Water level schedules are of great importance in reservoir management. They not only affect the

Intro. fig. 46. Filling operations used in preparing reservoirs of the Tennessee Valley Authority (*Malaria Control on Impounded Water*, 1947).

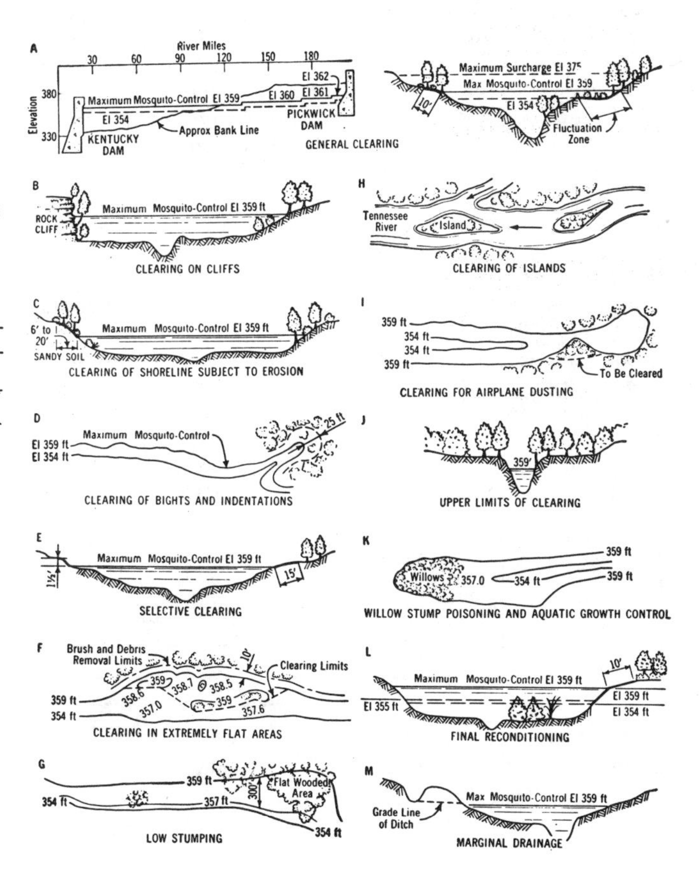

Intro. fig. 47. Clearing operations as used in preparing reservoirs of the Tennessee Valley Authority (*Malaria Control on Impounded Water*, 1947).

production of aquatic insects but also influence the growth of marginal vegetation and consequently the cost of maintenance. In planning a water level schedule for a reservoir it is necessary to consider:

a. The primary purpose or purposes of the project—whether for flood control, power, navigation, water supply, irrigation, recreation, wildlife, or a combination of these.
b. The stream flow and volume of storage in the fluctuation zone—the probability of filling each year, amount of fluctuation possible, and the seasonal recession are all functions of these two items.
c. The design of the dam and the water level control facilities—the maximum pool level, flood surcharge, and rate and extent of recession, all depend upon the type of design, elevation, and capacity of the control facilities.
d. The topography and vegetation in the fluctuation zone—a steep, rugged shore line exposed to wave action will require much less precise water level management than a shore line of extensive, flat, shallow areas; and water level fluctuation is sometimes ineffective for mosquito control where the shore line or margin is colonized with certain types of marginal or aquatic vegetation.

The simplest schedule for high storage reservoirs with steep banks and little or no mosquito production is direct seasonal recession. For large reservoirs at low elevations a more complicated schedule may be required. The classical example of this is the combination of seasonal recession and cyclical fluctuation (intro. fig. 48) used on twenty-four of the large Tennessee Valley Authority impoundments involving 735,000 acres of water surface and 10,000 miles of shore line. This schedule, which is not directly applicable to California conditions, calls for maximum elevation for a short time before April 1 to strand the winter accumulation of drift and floatage. Then there follows a constant pool level during the spring growth period (April 1 to May 15). This prevents the invasion of marginal vegetation into the zone of fluctuation and delays the germination of annual plants, thus decreasing the cost of annual shore line conditioning. During the period of moderate mosquito production (May 15 to July 1) the water level is raised and lowered one foot at weekly or ten-day intervals. This alternately strands larvae and eggs on the shore or flushes them out of protective vegetation and exposes them to predators. During the period of heaviest mosquito breeding (July 1 to October 1) cyclical fluctuation is combined with an over-all recession of about 0.1 foot per week. Finally (after October 1) during the period of low rainfall the water level is lowered to or near the minimum required for navigation and power. This exposes broad expanses of shore line for the annual job of shore-line conditioning.

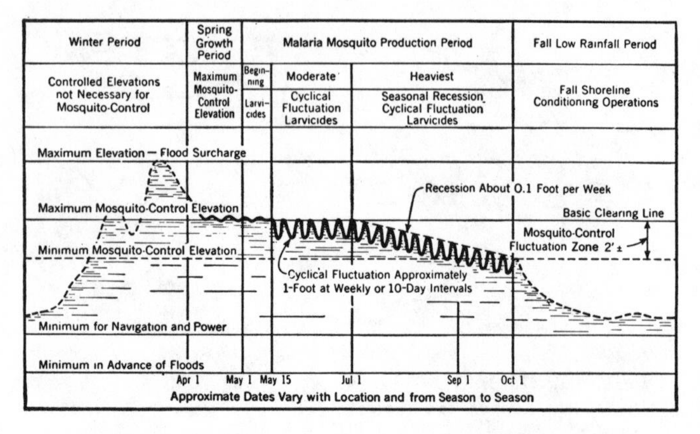

Intro. fig. 48. Schedule for water level management on main river reservoirs of the Tennessee Valley Authority, combining cyclical fluctuation and seasonal recession (after Hess and Kiker, 1944).

4. Preimpoundage studies should also be made of the productivity of lakes and reservoirs in the vicinity as an indication of the probable adequacy of fish-food organisms. In this connection a fish-stocking plan may be developed to ensure the best utilization of fish-food organisms in the reservoir.

5. The productivity of the stream below the proposed dam should be determined as accurately as possible in order to predict the effects of periodic flooding or drying and to recommend the optimum flow for maintenance of adequate bottom food organisms for fish. Such figures are often used as a basis for legal action when adverse effects are noted after impoundage.

Duck ponds.—The impoundment of water by clubs for duck hunting has become a common practice in parts of California. The subject was difficult in the past because the interests of sportsmen and mosquito control agencies appeared to be in conflict. More recently it has become evident that good practices for duck clubs are also good for mosquito control. After considerable study the Wildlife Committee of the California Mosquito Control Association made the following recommendations pertaining to the management of duck ponds:

1. If it is desired to hold ducks in a hunting area, a certain proportion of the area should be prepared as good sharp banked and properly maintained ponds, and permanently flooded rather than flooding the whole area early in the summer to achieve this purpose
2. An attempt should be made to adjust the duck hunting season by smaller regions allowing for a later season in the Central Valley. This is concurred in by many hunters; however, the setting of the season is done by the United States Fish and Wildlife Service and direct recommendations will have to be made by them.
3. Efforts should be made to control or eliminate cattails, tules, and other emergent vegetation, especially in permanently flooded ponds.
4. Where seepage areas occur outside of ponds a ditch should be dug around the pond to cut off this water and the water in the drain ditch should be disposed of by draining into an existing drainage system or pumping back into the pond.
5. Ponds should be drained immediately after the close of the season.
6. The planting of food grains in the ponds should be done in such a manner that no mosquito problems can be caused.

7. When pond areas are used as cattle pasture between seasons the land should be so prepared that good and proper pasture irrigation practices can be used.

Ponds created or influenced by faulty engineering practices.—Roadside ditches are found in many places where highways and railroad beds interrupt normal drainage and impound water because culverts are improperly placed. Borrow pits and quarries almost always hold water and provide breeding places for mosquitoes and other aquatic insects. Such conditions can be corrected only at considerable expense.

Swimming pools.—Outdoor swimming pools are becoming a feature of suburban life. Most of them are easy to clean and are free from insect pests. However, biting insects such as backswimmers *(Notonecta)* and "toe-biters" (Belostomatidae) sometimes fly in and annoy swimmers, usually in the heat of the summer when large flights of insects are attracted to nearby electric lights. The most practical method of control is to drain the pool and destroy the insects when they are concentrated at the deepest point. The mosquitoes, *Culex pipiens* and *quinquefasciatus,* occasionally breed in swimming pools. They may be controlled with emulsifiable pyrethrum or kerosene sprays.

Irrigation

In California most of the precipitation occurs in the north and only during the winter months. Therefore elaborate irrigation systems have been developed for transporting water to the arid south, and the underground water supply is tapped by thousands of irrigation pumps. According to Henderson (1951) California leads the nation in area under irrigation with 6 million acres. With the huge volume and extensive surface of water it is not surprising that insects intrude into the picture at several points. Caddisworms, for example, have been reported obstructing water in irrigation tunnels in southern California (Simmons, Barnes, Fisher, and Kaloostian, 1942) and many irrigation ditches and canals are inhabited by Simuliid larvae and other stream insects.

In irrigated fields the insect fauna is determined largely by the nature of the crop and by the water schedule. Any irrigated crop can produce mosquitoes but in general, row crop irrigation is less troublesome than sheet irrigation. Rice, for example, is constantly flooded over wide areas and presents ideal conditions for the development of aquatic insects. In 1954 more than 453,000 acres of rice were harvested in California, and most of this area was under water from April or May into September. Leaf- and stem-boring aquatic flies *(Cricotopus, Hydrellia)* are pests in the Sacramento Valley, and a host of pond-dwelling insects commonly invade the fields. *Hydrellia griseola* var. *scapularis* Loew, in particular, built up to epidemic proportions in 1953, causing an estimated loss of 10 to 20 per cent of the crop (Lange, Ingebretsen, and Davis, 1953). The rice leaf miner, as it is called, belongs to the family Ephydridae. The eggs are laid on leaves lying prone in the water and hatch in about four days. The larvae feed in the leaves. Control was achieved by lowering the water to a depth of about two inches and spraying with dieldrin or heptachlor. After forty-eight hours the water level was raised and the checks were blocked off so that no water was spilled from the fields for two weeks.

Mosquitoes have long been a problem in rice fields. *Culex tarsalis* is the most important of these. *Anopheles freeborni* is the malaria vector that caused the epidemics of former years (Herms, 1949) and was responsible for the recent outbreak initiated by a malaria carrier just returned from Korea (Brunetti, Fritz and Hollister, 1954; Fontaine, Gray and Aarons, 1954). *A. freeborni* is mainly a breeder in rice fields and outlying areas and can be reduced in numbers by good irrigation practices. *Aedes dorsalis* breeds in the rice fields early in the season in response to initial flooding (Portman and Williams, 1952). The eggs of this mosquito overwinter in the soil and may remain viable for several years. Soon after flooding, larvae appear in great numbers and produce myriads of adults unless controlled by application of insecticides such as wettable DDT powder mixed with the seed rice at the rate of one and one-half to two pounds of 50 per cent powder to the acre.

Other aquatic pests in the rice fields (Portman and Williams, 1952) include fairy shrimps, *Apus* spp., which undergo a life history somewhat like *Aedes dorsalis* and chew off the tender leaves and dislodge the soil around the roots of the seedlings, and giant scavenger beetles, *Hydrous triangularis* (Say), the larvae of which dig in the bottom mud and uproot entire plants.

Alfalfa fields and irrigated pastures provide extensive aquatic habitats, with more than a million acres of each in California.

Because of the increased availability of water in the past two decades, the practice of irrigating pasture lands has spread over much of the Central Valley. Coincident with this, *Aedes nigromaculis* and *Culex tarsalis* have extended their ranges and the former has now become the number one pest mosquito over a wide area. *Culex tarsalis* is a known vector of encephalitis. In irrigated pastures a succession of generations is produced through the season, the larvae breeding in shallow water choked with grass and weeds at low points in the fields.

Another method of using irrigation water that is quite common in California is the spreading of water for percolation purposes. In such cases individual pumps are used to replenish the water supply. The use of percolation beds creates mosquito problems. Also, of course, mosquitoes breed in irrigation structures of all kinds including canals, ditches, standpipes, and the like.

From the above discussion it is obvious that mosquito breeding is intimately related to irrigation. The solution to the problem should therefore be sought in better irrigation practices. Henderson (1951) refers to the problem as "conservation irrigation" or the use of "irrigated soils and irrigation water in a way that will insure high production without the waste of either water or soil . . . Generally, the direct economic benefits of conservation irrigation materially exceed the cost. Under proper technical guidance, resultant freedom from nuisance mosquitoes and encephalitis hazard are dividends obtained at negligible cost to society."

Water Pollution

The disposal of domestic and industrial wastes is a major problem of modern civilization. The pollution of surface and underground waters threatens the 15 billion gallons of water used by cities in the United States each day. It also endangers recreational fishing, boating, and swimming and may lower land values. The enormity of the problem is difficult to grasp by mere citing of figures. For example, the City of Sacramento produces 45 million gallons of raw sewage per day, and a single river such as the Delaware (Eliassen, 1952) is estimated to receive each day 500 million gallons of domestic sewage and hundreds of millions of gallons of industrial wastes.

Legal basis.—To control the undesirable consequences of water pollution (intro. fig. 49) California operated for many years under a statute of the State Health Department. In 1949 this was changed and the Water Pollution Control Act was passed. This law sets up a state board and nine regional boards. The regional boundaries, shown on the accompanying map (intro. fig. 50), are based on the major watersheds of the state. The regional boards have four principal duties: prescribing regulations for waste discharges; obtaining coördinated action in controlling pollution; enforcing orders for correcting pollution by means of administrative hearings, followed, if necessary, by court action; and formulating and adopting long-range plans and policies for water pollution control (Water Pollution Control Board Publication No. 5, 1952).

According to California law there are two basic aims of water pollution control: protection of public health and conservation of water quality for various beneficial uses. The first aim considers only the health aspects of waste treatment and disposal. The second considers the economic aspects and requires that the cost of waste disposal be balanced against the beneficial uses of the receiving waters. These two aspects are recognized and defined in California statutes as follows: *Contamination* is impairment of water quality by sewage or industrial waste causing an actual hazard to the public health. Primary responsibility for reducing contamination rests with the State Department of Public Health and the local health agencies, although the State Water Pollution Control Board may legally assume final responsibility in the case of an uncorrected contamination. *Pollution*

Intro. fig. 49. Through water pollution control domestic and industrial wastes are treated and clean water is maintained (Water Poll. Contr. Board Publ. No. 5, 1952).

Intro. fig. 50. Water pollution control regions and location of board offices (Water Poll. Contr. Board Publ. No. 5, 1952).

adversely and unreasonably impairs the beneficial use of water even though no actual health hazard is involved. The regional and state water pollution control boards are the agencies primarily concerned with reducing pollution. (The regional boards also are responsible for controlling *nuisance,* such as odors or unsightliness caused by unreasonable waste disposal practices.)

Since insects have little to do with the contamination of natural waters, the present discussion will be limited to pollution. Insects do play an important role at various stages in the treatment of polluted waters and are used as indicators of pollution.

Types of wastes.—Wastes are of three principal types: physical, including silt and other erosive agents; chemical, including toxic materials from industry, and agricultural chemicals such as insecticides and weed killers; and organic, including domestic sewage, industrial wastes from canneries, and fertilizers.

Effects of physical wastes.—The effects of physical agents are diverse and far-reaching. Silt, for example, may be abrasive and injure the gills of aquatic organisms; it may coat the gills and interfere with respiration; or it may settle out and cover natural habitats. It may also reduce light penetration and thus restrict the photosynthetic zone with consequent loss of oxygen and food supply for higher organisms.

Excessive silting is usually caused by faulty agricultural practices and by the deep cuts and exposed banks now so characteristic of our countryside along superhighways. Such uncontrolled erosion loads streams, fills lakes, and eliminates some of the best fish-food organisms (stonefly and mayfly nymphs, etc.). The solution is well-known but too often neglected—roadside plantings to stabilize the soil and, in agricultural areas, contour plowing, and other soil conservation methods.

Effects of chemical wastes.—Chemicals usually affect aquatic organisms by their direct toxic action, though there may be secondary effects such as extremes of pH or changes in osmotic pressure. Attempts have been made to establish precise tolerances for various chemical substances, based largely on their toxicity to fish. However, such figures have been shown to be extremely misleading (Doudoroff and Katz, 1950, 1953) because "the minimal harmful concentration of a toxic substance may vary greatly, depending on the duration of the test, the species and age of the test animals, the dissolved mineral content of the water used as a solvent or diluent, the concentration of other waste components having a pronounced synergetic or antagonistic effect, the temperature, and other factors." Under the circumstances, safe concentration limits of toxic wastes were not definitely prescribed.

Doudoroff and Katz did conduct extensive tests, however, using the method of "bioassay" or exposure of known concentrations to fish. Their conclusions are too voluminous to repeat here but a few points may be listed: (1) pH values above 5.0 and ranging upward to 9.0, at least, are not lethal for most freshwater fishes; (2) none of the strong alkalies which are important as industrial wastes (NaOH, $CA(OH)_2$, and KOH) has been clearly shown to be lethal to fully developed fish when its concentration is insufficient to raise the pH well above 9.0; (3) solutions of ammonia, ammonium hydroxide, or ammonium salts can be very toxic to fish even when the pH is not very high (that is, below pH 9.0); (4) the common strong mineral acids (that is, H_2SO_4, HCl, and HNO_3) and also phosphoric acid (H_3PO_4) and some moderately weak organic acids apparently can be directly lethal to fully developed fish only when the pH is reduced thereby to about 5.0 or lower; (5) a number of weak inorganic acids (such as carbonic, tannic, etc.) can cause pronounced toxicity without lowering the pH as low as 5.0; (6) the susceptibility to free carbon dioxide varies greatly; for example, sensitive species may succumb rapidly at free CO_2 concentrations between 100 and 200 ppm in the presence of much dissolved oxygen; (7) solutions of hydrogen sulfide, free chlorine, cyanogen chloride, carbon monoxide, and ozone all are extremely toxic and have been reported as lethal to sensitive fish in concentrations near 1 ppm or less; (8) all metal cations can be toxic in rather dilute (less than 0.05 m) physiologically unbalanced solutions of single metal salts; (9) sodium, calcium, strontium, and magnesium ions are among the least harmful of the metallic cations; (10) silver, mercury, copper, lead, cadmium, aluminum, zinc, nickel, and trivalent chromium, and perhaps also tin and iron, can be classed as metals of high toxicity; (11) cupric, mercuric, and silver salts are extremely toxic.

These are, of course, generalizations and would differ not only for each species of fish but also for other kinds of aquatic organisms including insects.

In some cases insecticides are regarded as pollutants or toxic agents from the viewpoint of fish production. Either as mosquito larvicides applied directly to the water or as applications on agricultural crops that run off or are washed into streams or lakes, these materials are a threat to aquatic resources. Fortunately, the conflicting interests of agriculturists, mosquito control agencies, and sport fishermen can be resolved in most cases by judicious choice of materials, careful timing of applications, and adjustment of concentrations to fit the tolerances of fish and fish-food organisms. (See discussion under mosquito control.)

Tolerances to chemicals (except insecticides) have not been studied for many insects but it has been observed that the copper sulfate treatment for reduction of algae in swimming pools has no apparent effect on insects. On the other hand, chlorine at the concentrations used for purification of drinking water may produce a residual that is toxic to aquatic insects and fish; hence treated water cannot be used safely for rearing insects in the laboratory unless it is detoxified by boiling, filtering, or by holding in a container for twenty-four hours.

Effects of organic wastes.—The effects of organic pollutants are more complex than those of mechanical or chemical agents. In general, the action of bacteria on organic material causes a deficiency of oxygen. Patrick (1953) has described the process as follows: "These organisms use the complex wastes . . . as a source of energy in their metabolism. In so doing they break down the wastes into substances that can be used as a source of food by other organisms. These processes, which are often referred to as decay or decomposition, occur most rapidly when the bacterial population is of optimum size. When the bacteria become too numerous the processes are slowed down. The protozoa and other small invertebrates which feed on bacteria are instrumental in keeping the bacterial populations in check. The algae are also at the base of the food chain. They are able to utilize inorganic substances to make proteins and carbohydrates, which are used as a source of food by other organisms. Indeed, algae have often been referred to as the grasses of the sea. Upon them not only the many different invertebrates, but also some fish and other vertebrates, feed directly. Besides their value as a source of food they also replenish the oxygen supply . . . by photosynthesis."

Insects fit into this complex picture in diverse ways. When exposed to organic pollution in a stream they follow a typical pattern (intro. fig. 51). Immediately below a sewage outfall a septic zone develops with turbid water and noxious odors. The bottom is coated with zoögloea and inhabited by tubificid worms. Oxygen is low or absent. The bacterial count is high and plankton organisms include *Oscillatoria* and *Sphaerotilus*. Fish are absent. The insect fauna is limited to larvae which breathe surface air through tubes—*Tubifera (Eristalis), Culex*. Below the septic zone is the zone of recovery with clearer water, cleaner bottom, and more dissolved oxygen. Here

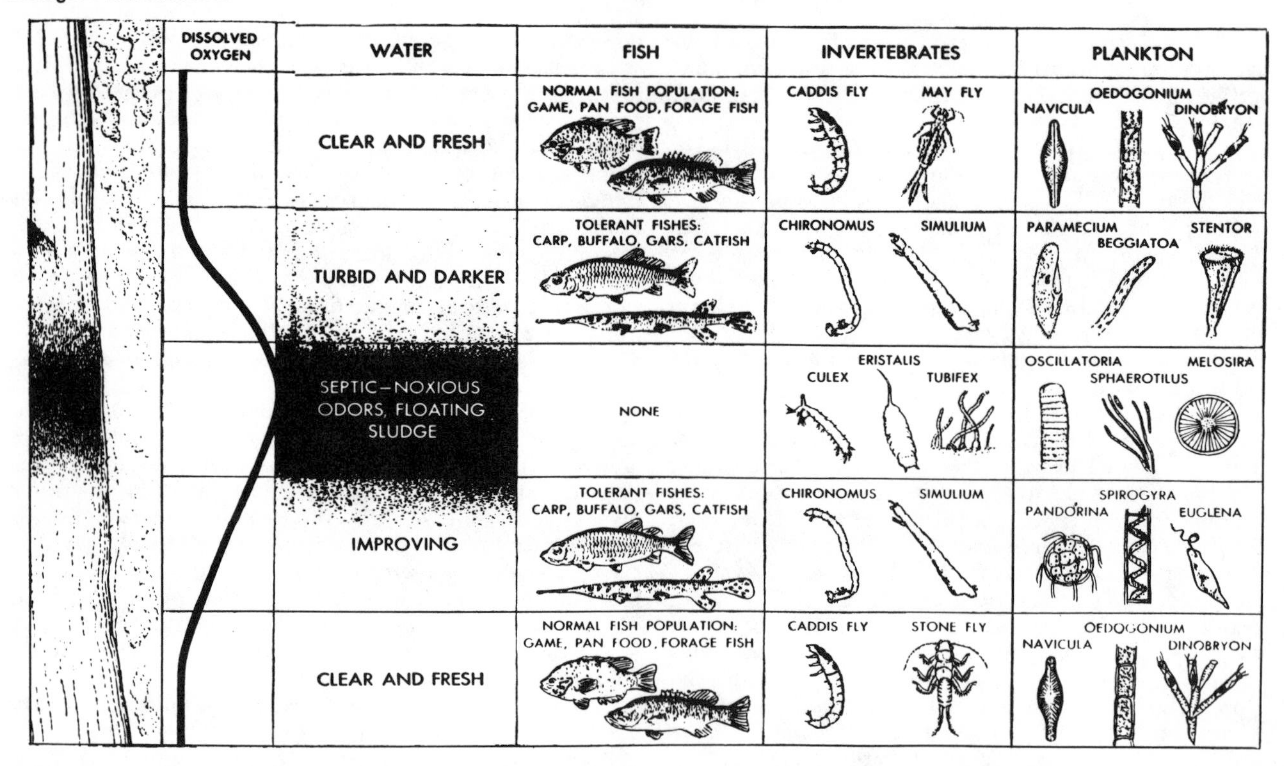

Intro. fig. 51. Pollution and recovery in a stream. As the oxygen dissolved in the water decreases (curve at left), so do certain microörganisms and, in turn, the insects and fishes that depend upon them for food (Eliassen, 1952).

the oxidation of organic matter by bacteria is hastened because sunlight can penetrate the clearer water and produce oxygen through the agency of the more numerous algae. *Spirogyra* and *Euglena* are often abundant in the plankton, and Chironomid larvae occur in enormous numbers on the bottom. Finally, and often within a surprisingly short distance, the stream returns to normal with clear water, clean bottom, abundant oxygen, and with the wide variety of organisms described elsewhere.

A similar pattern is seen when organic matter is dumped into lakes, but the effect of water movement is less evident and in smaller bodies of water the chief agents for recovery of oxygen are the algae.

Trickling filters.—The ability of natural waters to purify themselves is limited. For example, a stream loses its ability to absorb organic pollution if its microörganisms use up the dissolved oxygen faster than it can be replenished (Eliassen, 1952). The rate at which the micropopulation uses up oxygen depends primarily on the amount of organic matter in the water. This establishes what is known as the biochemical oxygen demand (B.O.D.). Various methods have been devised to reduce the B.O.D. to a level that a stream can absorb. Perhaps the commonest method is the trickling filter (intro. fig. 52). The various steps in the trickling filter process are as follows: After preliminary grease removal and chemical treatment, if necessary, the sewage is run through a primary settling tank. There, solids are taken out and pumped to a sludge digestion tank where anaerobic bacteria act to reduce the material and methane gas is produced. After treatment the dried sludge is used as fertilizer. Meanwhile the fluid sewage is sprayed onto filter beds where it trickles between rocks of 1- to 3-inch diameter to a depth of 3 or 4 feet. The surfaces of the rocks soon become coated with zoögloea, an organic film containing millions of bacteria, algae, and other microscopic organisms. Oxidation (and hence treatment) of the sewage takes place as it passes over the zoögloea. For efficient operation the filter must be rich in zoögloea but still open enough so that the liquid can pass through at a high rate. Psychodid larvae play a decisive role in this process. They occur in enormous numbers in filter beds where they feed on the zoögloea. Too many larvae scour the rocks excessively and reduce the effectiveness of the treatment. On the other hand, too few larvae result in excessive growth of the zoögloea which clogs the filter and produces "ponds" of standing water that may halt the entire process. Larvae have been maintained at optimal levels in recent years by careful application of DDT in wettable form at dilutions of 5 parts per million. After treatment in the filter the effluent may be run through a secondary settling tank and chlorinated to kill pathogenic bacteria. Finally it is discharged into a stream or other body of water which has been found capable of carrying the reduced but nevertheless considerable organic load.

Oxidation ponds or sewage treatment lagoons.—Another method for the treatment of domestic and

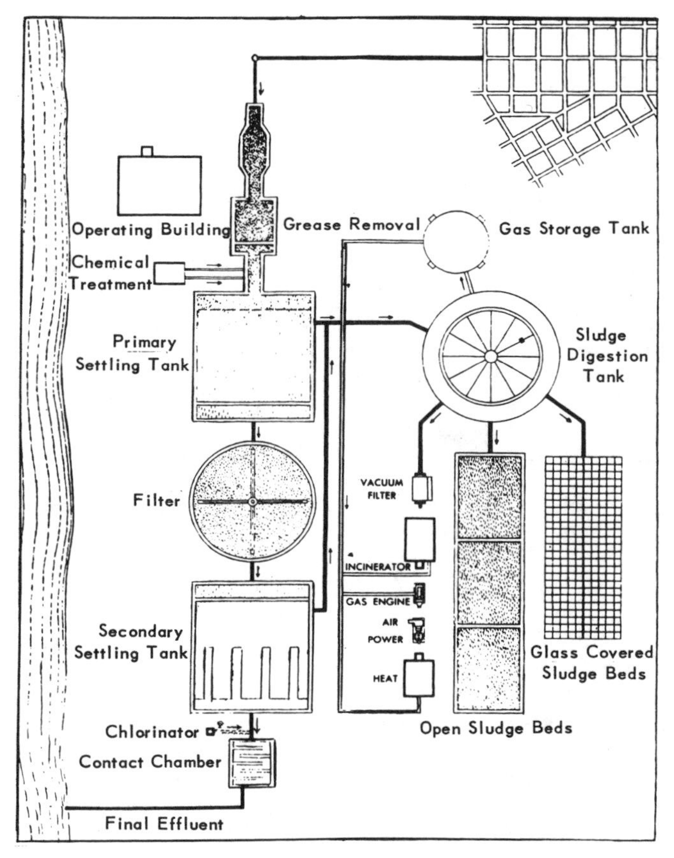

Intro. fig. 52. Model sewage treatment plant showing the various processes including settling, filtration, and sludge digestion. The arms on the filter rotate slowly, spraying fluid which trickles over the zoögloeal surfaces and through the interstices of the rocks (Eliassen, 1952).

industrial wastes is by oxidation in shallow ponds. This is the procedure used in many suburban and rural parts of California. Sewage is settled and strained to eliminate solid matter and then the fluid is run through a series of shallow ponds several acres in extent. Organic matter is acted on by bacteria, first under virtually anaerobic conditions and later in the presence of oxygen supplied in large part through photosynthesis by symbiotic algae. Insects play an important part in this system. Only surface breathing *Tubifera* larvae and other Diptera with respiratory tubes can live in the milky, anaerobic water of the first ponds. Later, however, a rich insect fauna develops with many beetles, true bugs, dragonfly and mayfly nymphs, and mosquito larvae.

By far the most numerous forms, however, are Chironomid larvae (*Glyptotendipes*, etc.). Counts of 1,500 or more per square foot of bottom surface have been made, and the total dry weight of the annual crop of larvae in a 4½ acre pond at Concord has been calculated at 575 pounds (Kellen, 1955). The algal blooms utilize end products of bacterial metabolism and form a high percentage of the final effluent. However, the larvae extract algae by filter feeding; also, they burrow at the mud-water interface, extending oxygen deeper and promoting the exchange of oxygen and nutrients between mud and water. In the biodynamics of oxygen pond treatment insects represent one of the end points of energy transfer—unless fish are introduced to eat the insects. Also insects are the only agents that actually remove part of the energy and organic load from the system as they emerge and fly away. The rest of the unsettled material is transformed into dead algal cells with a B.O.D. that may not be very different from the original sewage and hence creates a considerable load when discharged into a stream.

Detection of pollution.—Various tests have been devised to recognize the presence of pollution and to detect the effects of past exposure to wastes. The simplest methods and therefore the ones most generally used are physical and chemical. It is relatively easy to take water samples from a stream, for example, and test for dissolved oxygen (DO), biochemical oxygen demand (B.O.D.), pH, turbidity, and the presence of toxic chemicals. However, such tests are not very revealing because pollutants are seldom discharged continuously and therefore their presence may be missed by sampling at the wrong time.

What is needed is a method of determining the effects of pollution on the resident biota, and this is precisely where insects fit into the picture. "Specifically (as stated by Gaufin and Tarzwell, 1953), the degree and extent of pollution in a stream can be determined accurately by reference to the macroinvertebrate fauna, particularly that found in the riffles. A biological analysis of the pollutional status of a stream can be obtained in the field through recognition of the orders, families, or genera in the invertebrate associations encountered. This type of biological inventory is superior to chemical data, [because] the complex of such organisms which develops in a given area is . . . indicative of present, as well as past, environmental conditions . . . Bottom organisms are more fixed in their habitat than are fish or plankton and cannot move to more favorable surroundings when pollutional conditions are most critical."

Because of the importance of the bottom fauna and more particularly in streams, the insect fauna, attempts have been made to set up criteria of abundance or to fix upon the presence or absence of certain indicator organisms as evidences of pollution. Unfortunately, these efforts have failed because, as pointed out by Patrick (1953), it is the total spectrum of all groups of organisms that provides the best criterion for judging the "health" of a river. In general a normal stream will be rich in species with no single group predominating, whereas a polluted stream is poor in number or variety of species but often rich in individuals. Therefore it has been suggested (Needham and Usinger, 1954) that biological sampling for detection of pollution be done by means of bottom samples (Surber or drag-type samplers). The organisms in the samples should be sorted out according to orders or other major taxonomic groups. In this way the spectrum of organisms can be determined with only two or three square-foot samples giving reliable evidence as to the presence or absence of main groups.

By way of summary it should be emphasized that wastes are not intrinsically bad. In many cases they are legitimate by-products of man's civilization. Too often industrialists try to ignore pollution problems by "looking the other way" and conservationists would simply legislate them out of existence. The result is an impasse and the solution, of course, lies not in either extreme but requires that issues be faced squarely and dealt with on a continuing basis. Actually, substantial progress has been made in recent years by recovering useful chemical wastes, improving erosion control, and by increased knowledge of the processes involved in reduction of organic wastes by natural waters. It is now realized that organic wastes simply speed the process of eutrophication in a lake or stream. In limited or controlled amounts they increase the productivity of natural waters. It is only when an excessive load is dumped at a time or place which affects the interests of other people that trouble starts. At such times the limnologist is usually called upon to provide evidence for or against pollution.

Pond Fish Culture

Aquiculture has been practiced since the dawn of history and has been a source of protein in China and India for centuries. Strictly from the viewpoint of productivity in pounds per acre, carp are unsurpassed, yielding up to 3,000 pounds per annum in the Orient. However, in the United States carp are not considered desirable for human consumption, and fish ponds are not intended solely for food production. They may be purely aesthetic garden ponds, primarily practical farm ponds for watering stock, or recreational pools for fishermen. In California there has been increased interest in the multipurpose farm pond and in commercial "trout farm" ponds where fishermen are guaranteed results and pay for their catches by the linear inch or pound.

Experiments in Alabama (Swingle and Smith, 1941) have shown that each pond has a normal carrying capacity for a particular species of fish, regardless of depth (18 to 54 inches) and regardless of number of fish stocked. If a pond is overstocked, the fish will be small, if understocked, they will be large; in either condition the pounds per acre will be the same. Phytoplankton feeders such as goldfish yielded up to 1,000 pounds per acre; insect feeders such as bluegill, 600 pounds per acre; and fish feeders such as large-mouthed black bass, 200 pounds per acre. Ideally, a pond should be stocked with a ratio of carnivorous and forage fish of approximately 1:10 or 1:15.

The total productivity of a pond can be increased by addition of fertilizer. The following mixture is recommended: 40 pounds of sulfate of ammonia, 60 pounds of superphosphate (16 per cent), 5 pounds of muriate of potash, and 15 pounds of finely ground limestone per acre. The materials should be mixed before applying. Fertilizer was applied in Alabama (Swingle and Smith, 1951) beginning in April or May and continuing every four weeks until September or October. Thus eight to fourteen applications were made at an annual cost per acre of $11.00 to $20.00 or 3 to 6 cents per pound of fish. The stocking policy of the U. S. Fish and Wildlife Service for new farm ponds in the southeastern states is: 50 bass and 500 bluegills per acre in unfertilized waters, and 100 bass and 1,000 bluegills per acre in fertilized waters (Holloway, 1951).

Weeds are generally considered to be undesirable in fish ponds. They supply protection for mosquito larvae, hinder bass from their essential role in preventing an overpopulation of plankton and insect-feeding fish, utilize fertilizer without greatly increasing food for fish, and interfere with sport fishing. To prevent the rooting of weeds along shore lines, the edges of ponds should be deepened. Periodic clearing also may be necessary to remove volunteer plants before they become heavily rooted and spread (Davison and Johnson, 1943).

The role of insects in farm ponds was studied intensively by Wilson (1923) in Iowa. It was found, as might be expected, that insects are an important element in the food of pond fishes and that they are, in turn, dependent on phytoplankton and other organic matter for their existence. However, a special situation exists for predatory insects in farm ponds. Immediately after stocking, all the fish are small and hence are at the mercy of the larger insects, with no fish yet large enough to eat them. The beetle genera *Dytiscus, Hydrous,* and *Cybister,* Belostomatids, and large nymphs of Odonata are especially troublesome in ponds with fish fry but seldom bother fish more than one year old and, in fact, are a valuable source of food for the larger fish. Therefore it is recommended to: screen small ponds for fish fry in areas where large predatory insects occur in abundance; remove strong lights from the immediate vicinity of ponds since the large predators are attracted to lights; remove fry from infested ponds and stock with larger fish; and remove fry and drain infested ponds, thus exposing large predators so that they may be destroyed by hand. Chemical control has been recommended by Meehean (1937), using oil film at the surface (1 part cod-liver oil to 3 parts kerosene, or straight kerosene at 10-12 gallons per acre), but this has no effect on the gill-breathing immature insects which take oxygen beneath the surface.

A purely negative approach is not adequate for proper management of insects in pond fish culture. By far the majority of aquatic insects are desirable and should be maintained at high population levels. Fortunately, insects need not be stocked because they migrate readily from one pond to another. Wilson (1923) found eight species of beetles in a new pond twenty-four hours after it was filled with water and five more species invaded the pond after three days. In ponds where predation by fish is so intensive that insects cannot maintain themselves, fish-free side ponds or troughs may be used. In this way stocks of insects can be developed and washed into fish ponds periodically as needed.

Stream and Lake Management

Sport fishing has long been a means of recreation and, in recent years, has achieved the status of big business. In addition to the large investments for the manufacture and sale of fishing gear, there is enormous expenditure for hatchery production and stocking of fish. More recently attention has been directed to the improvement of natural waters in an attempt to provide more fish at less cost than has been possible with hatchery methods. This is a complicated subject with ramifications that have no place in a book on aquatic insects; therefore the present discussion will be limited to the entomological aspects of stream and lake management.

Analysis of food grades and preferences.—Hatchery-bred fish have been stocked in California waters for more than half a century. Much of this was done without regard to the natural food supply and hence with no knowledge of the carrying capacity of the stream or lake, but in recent years this situation has changed. Pioneer studies by Embody (1927) provided the basis for a stocking policy, and more recently investigations have been made of the fish-food organisms of many of our lakes and streams; stocking recommendations have been based on them.

Unfortunately, most surveys of stream-bottom organisms have been made with a Surber Square Foot Sampler. Although this is undoubtedly the most practical sampler thus far devised for shallow riffles, Needham and Usinger (1955) showed that it does not give statistically significant data for total weights and numbers of organisms, even in a single relatively uniform riffle. It was found that 194 and 73 samples, respectively, would be required to give reliable figures at the 95 per cent level of significance. Hence all existing data on stream-bottom organisms is statistically inadequate, but it is the best we have and may be the best that can be obtained because of the extreme variability of stream habitats. Examples of this type of data are found in the unpublished reports of Needham and Hanson (1935), Smith and Needham (1935), and Taft and Shapovalov (1935) for the Klamath, Shasta, Sierra, Mono, and Inyo national forests. Some idea of average and extreme numbers of bottom food organisms for various streams (Surber samples) and lakes (Ekman samples) in California are given in table 3.

TABLE 3 (Introduction)

AVERAGE NUMBERS OF BOTTOM FOOD ORGANISMS TAKEN BY SQUARE FOOT SURBER SAMPLERS IN STREAMS AND BY 1/4 SQUARE FOOT EKMAN SAMPLERS IN LAKES OF CALIFORNIA IN 1934

National forests	No. of stream organisms	No. of lake organisms
Klamath and Shasta	380	150
Mono and Inyo	370	66
Sierra	143	26

Numbers of organisms can be quite misleading because of the great disparity in size of bottom-dwelling organisms. Therefore the total volume or "wet weight" may be more satisfactory. Wet weights are commonly taken after drying the samples for one minute on blotting paper. In a quantitative test using 100 samples taken from a single riffle in Prosser Creek near Truckee, California, Needham and Usinger (1954) found a minimum of 2 and a maximum of 198 organisms per square foot with wet weights from 0.15 grams to 2.31 grams. For comparative purposes wet weights are sometimes calculated in pounds per acre. Table 4 shows such figures for Waddell Creek near Santa Cruz, the Merced River in Yosemite National Park, and Convict Creek at the head of the Owens River (Needham, 1938, 1939; Maciolek and Needham, 1952).

TABLE 4 (Introduction)

STANDING CROPS OF BOTTOM ORGANISMS IN POUNDS PER ACRE AT VARIOUS SEASONS IN WADDELL CREEK, CONVICT CREEK, AND THE MERCED RIVER.

Location	Month	Standing crop lbs. per acre
Waddell Creek	February, 1933	70
	May, 1933	472
Convict Creek	*Low*	
	May to Sept., 1938	68
	High	
	May to Sept., 1942	197
Merced River	February	103
	August	85

Attempts have been made to improve the techniques for sampling fish-food organisms in order to obtain more meaningful results. Such efforts are of greatest importance to the aquatic entomologist who has found himself too often in the past expending great effort in the field to obtain data which are virtually meaningless.

One attempt at increasing precision is the "forage ratio" (Hess and Swartz, 1941) proposed to aid in interpreting the results of stomach contents investigations. It was pointed out that the organisms found in the stomach of a fish reveal nothing as to food preferences unless it is also known what organisms occurred in the immediate environment of the fish at the time it was feeding. The forage ratio (FR) is obtained by dividing the percentage of a given kind of organism in the stomachs by its percentage in the environment. The formula may be expressed as follows:

$$FR = \frac{\frac{n}{N}}{\frac{n'}{N'}}$$

where n = the number of any organism in the stomachs, N = the total number of organisms in the stomachs, n' = the number of the same organism in the environment inhabited by the fish, and N' = the total number of food organisms in the environment. Weight or volume may be substituted for number if desired.

A forage ratio of 1 indicates that an organism is being taken at random according to its relative abundance in the environment; a forage ratio of more than

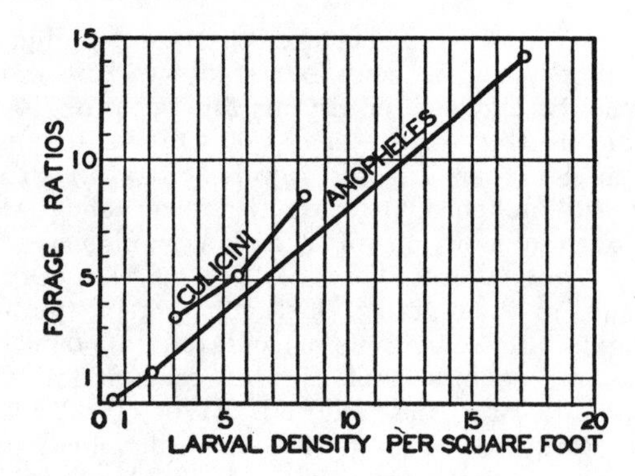

Intro. fig. 53. Correlation between forage ratio and density of mosquito larvae (Hess and Tarzwell, 1942).

1 indicates that an organism is either being selected in preference to other organisms, or that it is more available than others; and a forage ratio of less than 1 indicates that an organism is either less preferable or less available. In a test using top minnows *(Gambusia affinis)* and anopheline and culicine larvae, Hess and Tarzwell (1942) found that there was a direct correlation between forage ratio and population density (intro. fig. 53). "This would seem to indicate that as *Anopheles* became more abundant, *Gambusia* became more accustomed to feeding upon them and began to select them in preference to other food organisms present."

A knowledge of the biology of aquatic insects should be used in conjunction with forage ratio data in order to interpret the results correctly. For example, during one period of their development different species of food organisms may be of great value to the fish and at another period be of no value. Hess and Swartz cite emerging caddisfly pupae as a case in point. Trichoptera larvae may be relatively unavailable to the fish owing to their stone cases; the pupal stage is quiescent and spent under a stone case, but when the pupae emerge and make their way to the surface of the water, they are free of their cases and are readily available to all fish.

The forage ratio may also be used to make more meaningful the calculation of food grades of streams (Hess and Swartz, 1941). Food grades have commonly been determined without regard to the value of the various organisms as food for the fish. The following method was proposed to correct this: The forage ratio for a given group multiplied by the mean density of the group per square foot will give a measure of the "effective food grade" in terms of that group, in relation to the species of fish for which the forage ratio was determined. Only those organisms which

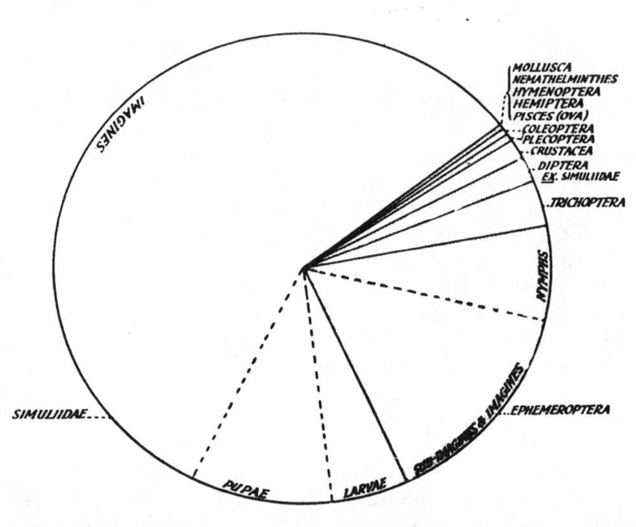

Intro. fig. 55. Composition of food material from the stomachs of trout feeding in the same area at the same time as Intro. fig. 54 (Neill, 1938).

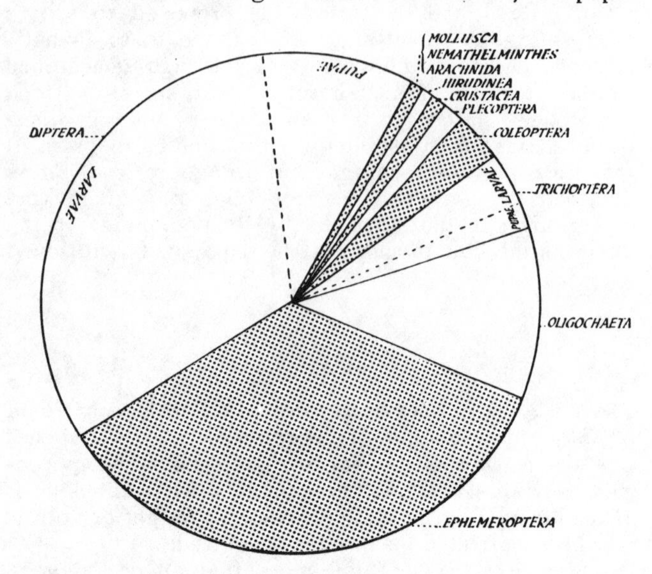

Intro. fig. 54. Composition of the bottom fauna in a limited area of the River Don (Scotland) during spring and summer (Neill, 1938).

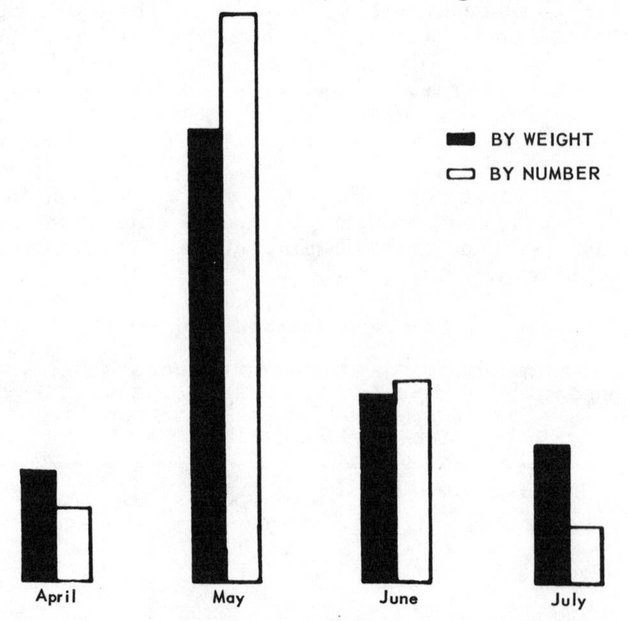

Intro. fig. 56. Seasonal fluctuations in amount of food taken by trout (based on the amount of Ephemeroptera, Trichoptera, and Simuliidae found in the stomachs) (Neill, 1938).

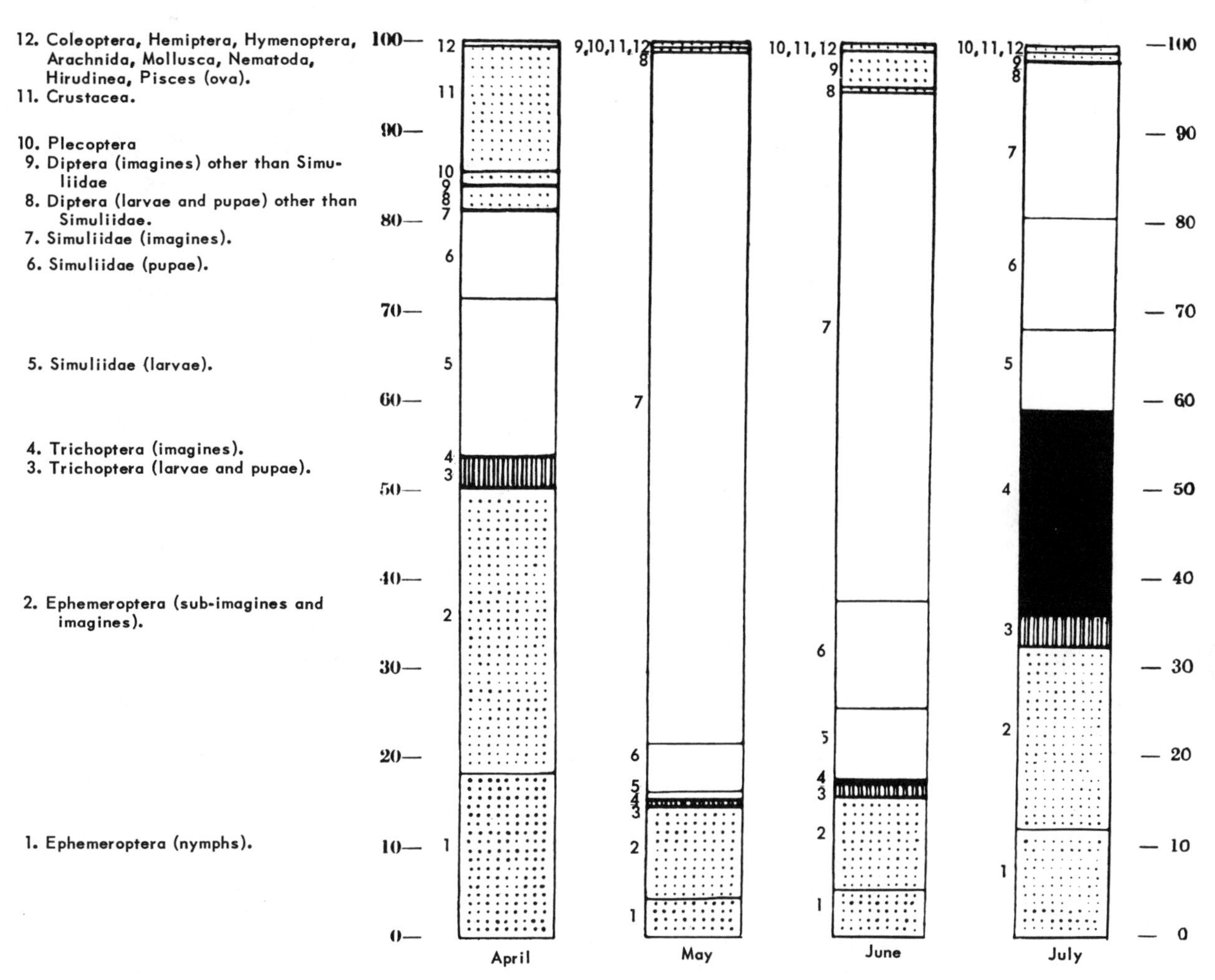

Intro. fig. 57. Composition of food material month by month. The sections of each column represent, superimposed in the order shown on the left, the percentage of the month's food material formed by the different groups drawn on (Neill, 1938).

make up 1 per cent or more of the fish's diet will be considered as food organisms. Sufficient numbers of both bottom samples and fishes' stomachs shall be taken to keep the standard error within 10 per cent of the mean.

Another factor to consider in food preference studies is the rate of digestion of different organisms eaten by fish. Hess and Rainwater (1939) found that soft-bodied organisms, such as many dipterous larvae, are digested and passed through the alimentary tract much more rapidly than heavily chitinized forms such as stonefly nymphs. Differences in rates of digestion of various organisms and differences in time of exposure to digestion in different stomachs (owing to delays in dissecting stomachs, gathering fish from traps or gill nets, etc.) therefore should be considered in interpreting food preference data.

It is probably too early in the development of the science of aquatic entomology to outline an ideal procedure for the analysis of fish-food organisms because too many of the basic techniques are inadequate and too little information is available on insect life histories. However, Neill (1938) has conducted a comprehensive study on the River Don in Scotland that might well serve as a model for future investigations: 1. The general characteristics of the site were recorded on a sketch map and summarized, including area, gradient, current velocity, depth, nature of bottom, nature of banks, influence of pollution, and general nature of the biota, plant and animal, including fish, and, as far as possible, synchronous field observations were made both in regard to day and hour (noon); 2. Daily observations were made of air temperature, barometric pressure, light intensity, rainfall, wind, water temperature, depth, current speed, water samples for pH, alkali reserve, carbon dioxide content, oxygen content, bottom samples (using a cylindrical sampler somewhat like the Hess modification of the Surber sampler), plankton samples, surface samples, and fish captures, and preservation of stomachs; 3. The bottom fauna for the limited area during spring and summer (intro. fig. 54) was compared with the food material (intro. fig. 55) taken from the stomachs of fish feeding in the same area over the same

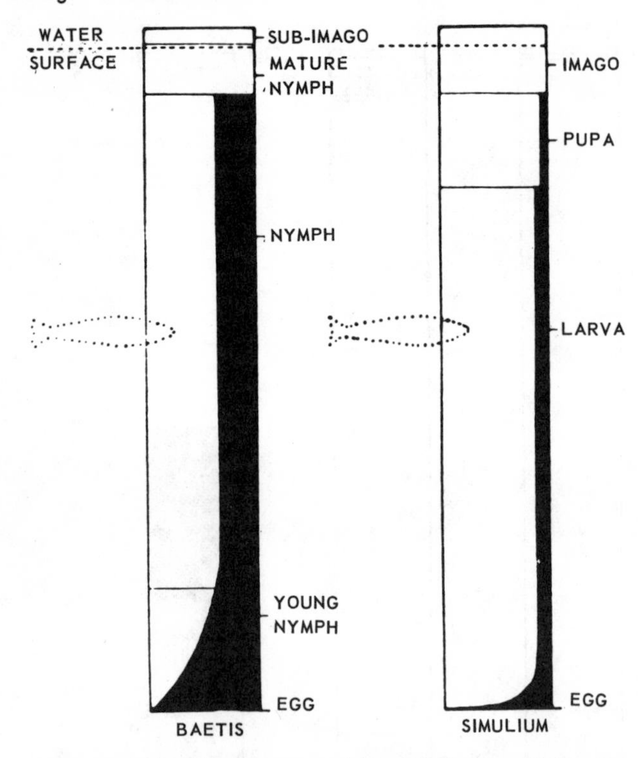

Intro. fig. 58. Relative accessibility of *Baetis* and *Simulium* during the aquatic stages of the life cycle. Degree of shelter—black. Degree of exposure—white (Neill, 1938).

period of time; 4. Seasonal variations in amount of food (intro. fig. 56) and composition of food (intro. fig. 57) were recorded; 5. The effects of current, vegetation, type of bottom, and chemical character of the water were related to densities of food organisms; 6. Fish were taken at a time to correspond to the end of a given day's feeding, thus ensuring adequate fresh material; they were captured mostly by angling rather than by seining, (but no significant differences were found in amount and kinds of food in the stomachs of 496 fish whether taken with nets or by angling, Dimick and Mote, 1934); 7. All organisms were identified to the species level with notations as to developmental stages; 8. For each group the life cycles of the species present were described, so that the times of greatest abundance could be associated with the presence of various organisms in the food; 9. The relative accessibility (intro. fig. 58) of each of the principal food organisms was worked out by grouping the whole fauna into five categories, to each of which a conservatively estimated "coefficient of accessibility" was assigned. All available bionomic data were taken into account in assigning these figures which ranged from 1.0 for freely exposed groups (e.g., *Simulium*) to 0.0 for completely secure groups (Oligochaeta) (table 5).

Total numbers in the bottom samples were multiplied by the appropriate coefficient to bring out the accessible fauna. In spite of the somewhat arbitrary method of ranking, a high degree of correspondence was found when figures for the "accessible" fauna were compared with those for stomach contents. This led Neill to the conclusion that "the predatory relationship between the brown trout and its organic environment therefore is that the trout feeds on the whole range of animals present in whatever type of habitat it finds itself to an extent dependent on the degree of accessibility and the extent of their representation in the fauna. This is sufficient to account for the nature of its stomach contents without invoking discrimination on the part of the fish."

Summary of the methods of assessment of the food of fresh-water fish (after Hynes, 1950).—Most workers studying the food of fresh-water fish have based their conclusions on study of the contents of the stomach or, more rarely, of the entire gut of captured fish. Digestion is less advanced in the stomach, and thus identification of the contents is usually more satisfactory. There has been great variety in the methods of analyzing and presenting the data. Some authors have merely listed the food organisms found in each fish, but this, without analysis, gives no indication of the relative importance of each type of food organism, and most workers have analyzed their data by one or more of the following methods.

(a) *The occurrence method.* The number of fish in which each food item occurs is listed as a percentage of the total number of fish examined. Often the number of occurrences of all items is summed and scaled down to a percentage basis to show the percentage composition of the diet.

(b) *The number method.* The total numbers of individuals of each food item are given, and are also usually expressed as percentages of the total number of organisms found in all fish examined.

(c) *The dominance method.* The number of fish in which each food item occurs as the dominant foodstuff is expressed in one of the two ways used in the occurrence method.

TABLE 5 (Introduction)

"COEFFICIENT OF ACCESSIBILITY" OF COMMON GROUPS OF AQUATIC ORGANISMS

Freely exposed	Partly protected or hidden		Largely secure	Secure
1.0	0.75	0.5	0.25	0.0
Simulium (all stages)	Ephemerid nymphs	Trichoptera larvae and pupae	Coleoptera larvae	Oligochaeta
Coleoptera (adults)	Plecoptera nymphs	Hirudinea	Diptera larvae	Hydracarina
		Nematoda		
Diptera pupae				
Crustacea				
Mollusca				

(d) *The volume and weight methods.* The volume or weight of each food item, or of the total food of each fish, is given, and is usually expressed as a percentage of the total weight of the fish. Some authors calculate the weight of food eaten from the known average weight of each individual of each food item. Most workers use this method only to supplement some other method, and it is often used to show seasonal variation in food intake. Ricker, however, by counting and "weighting" each type of food organism according to its known average weight, has evolved a method similar to the points method (below).

(e) *The fullness method.* Some workers wishing to demonstrate seasonal variation in food intake have used an arbitrary estimate of the fullness of the stomachs. This is only a special extension of the total volume method.

(f) *The points method.* Swynnerton and Worthington used a method in which the food items in each fish were listed as common, frequent, etc., on the basis of rough counts and judgment by eye, due regard being taken of the size of the organisms as well as of their abundance (i.e., one large organism counted as much as a large number of small ones). Each category was then allotted a number of points and all the points gained by each food item were summed and scaled down to percentages, to give percentage composition of the food of all the fish examined. This is essentially a volumetric method, and is similar in principle to that of Ricker above. However, until dietetic values of food species are known, volume would appear to be a satisfactory basis for assessment, and this method has the advantage of being rapid. The points system of Swynnerton and Worthington has been modified slightly by taking into account also the degree of fullness of the stomach.

Consideration of the various methods indicates that the points method is the most satisfactory. Facts in its favor are that it is rapid and easy, requires no special apparatus for measurement, is not influenced by frequent occurrence of a small organism in small numbers, nor of heavy bodies, like snail shells and caddis cases, and does not involve trying to count large numbers of small and broken organisms. It also does not give the spurious impression of accuracy which is given by some other methods.

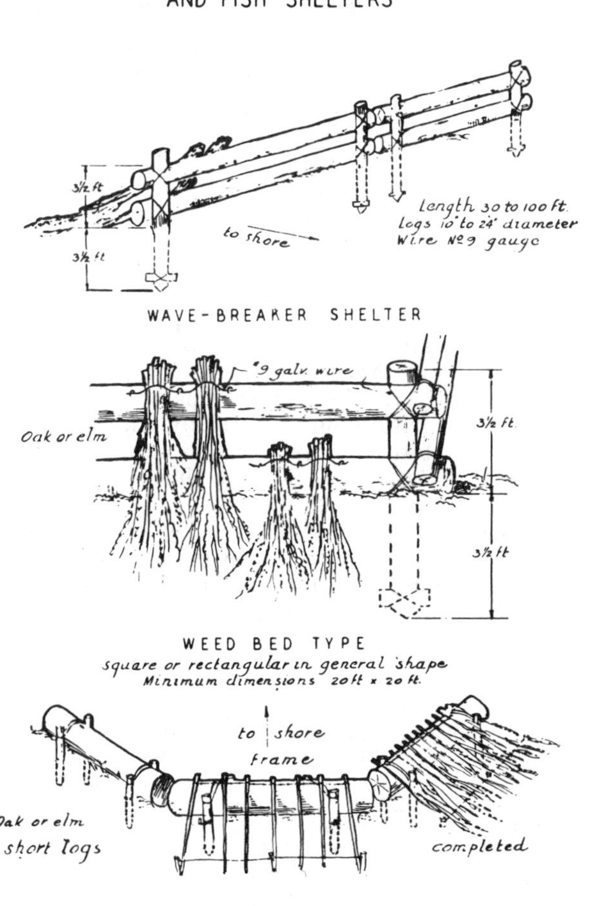

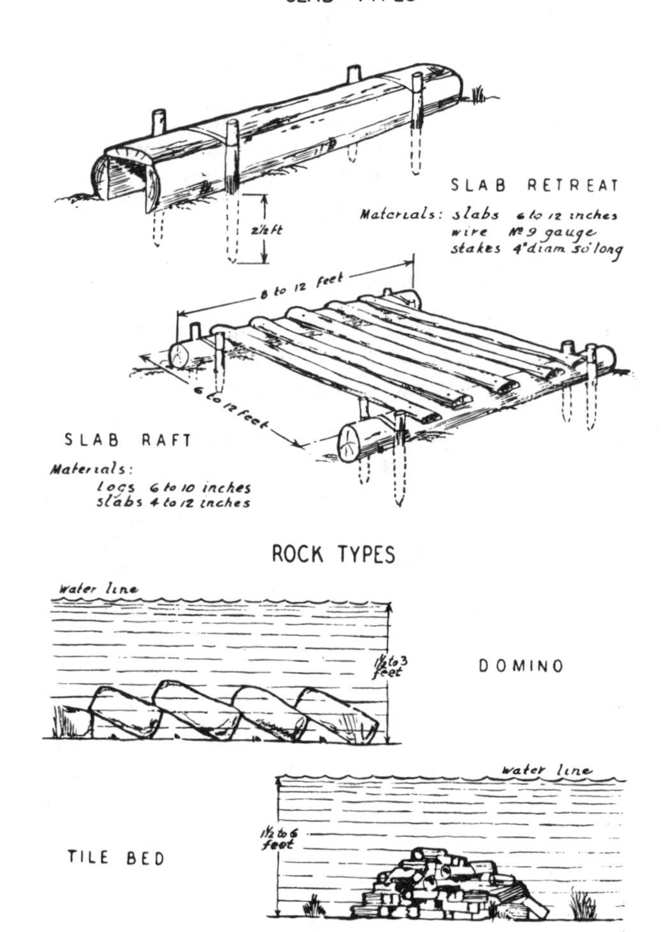

Intro. fig. 59. Methods of improving shelter for aquatic organisms and fish in lakes (Hubbs and Eschmeyer, 1938).

Improvement of streams and lakes.—Stream and lake management for the production of fish-food organisms was first developed in England (Mosely, 1926) where private trout streams have been maintained for centuries and close attention could be given to methods for the production of fish-food organisms. Many of the techniques advocated for British streams, such as collecting eggs of mayflies to introduce into a barren stream, are totally impractical here where miles of unimproved waters exist. However, some practical techniques have been developed in this country (Hubbs and Eschmeyer, 1938) and abroad, mainly applying to the fish alone but in some cases with definite advantages from the standpoint of fish-food production.

Certain practices have been found to be effective in increasing fish food in lakes. These include: (1) management of water level by dams so as to maintain maximum littoral productivity and, in small lakes, to prevent drying up with consequent mortality; (2) stabilizing banks by plantings to avoid excessive silting; (3) regulating the abundance of fish and balance of kinds of fish so that fish populations can maintain themselves at optional densities in relation to food supplies; (4) improving the shelter for both aquatic organisms and fish (intro. fig. 59); and (5) managing the growth of aquatic plants. Other measures such as fertilizing and dredging or filling have been done in small artificial impoundments but are not practical or even desirable in most natural lakes of the state.

In streams fish need pools for protection and riffles for food production and spawning. Therefore stream improvement seeks to provide both types of habitat in the proper proportions, usually in a ratio of about 1:1. Riffles are maintained by clearing trash and logs and by fencing to prevent trampling by stock. Pools are created by small dams or by deflectors strategically placed (intro. fig. 60). Banks are stabilized by plantings to prevent erosion. In burned-over areas shade is restored by reforestation so that water temperatures in trout streams can be maintained below 70°-75° F.

It is very difficult to measure the results of stream improvement because of the lack of precise sampling methods. Tarzwell (1937) attempted an evaluation in Michigan by determining the average production for various types of bottom and then calculating the potential volume of food production on the basis of the area of each bottom type before and after improvement (table 6). The results showed a threefold increase after improvement.

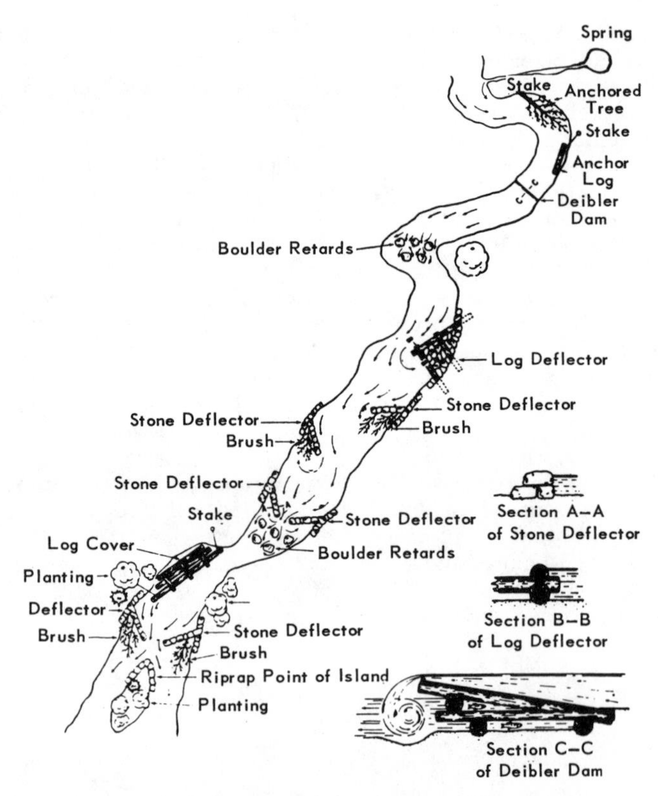

Intro. fig. 60. Summary of some stream improvement techniques (Lagler, 1952).

More general methods for improving the health of streams include the following: 1. Pollution is prevented whenever possible. 2. Check dams are installed in large drainage areas primarily for flood control but also to maintain some flow throughout the year in otherwise intermittent streams (flow maintenance dams, Cronemiller, 1955) thus permitting the survival of organisms with long life cycles and creating a perennial population of fish food organisms; 3. Stream flow is equalized below large dams by judicious management to avoid the scouring effects of flash floods and the mortality caused by drying parts of the bed; 4. Finally, efforts are being made in a few places to control the ratio of predators to bottom-feeding fish so that optimum productivity of fish food

TABLE 6 (Introduction)

VOLUME OF FOOD PRODUCTION IN A STREAM BEFORE AND AFTER IMPROVEMENT, CALCULATED ON THE BASIS OF AVERAGE PRODUCTION OF VARIOUS BOTTOM TYPES AND INCREASE IN AREA OF EACH BOTTOM TYPE

Bottom type	Average production on 4 sq. ft. (in cc.)	Before improvement		After improvement	
		Area of each bottom type in sq. ft.	Total calculated production (in cc.)	Area of each bottom type in sq. ft.	Total calculated production (in cc.)
Sand	0.27	76,105	5,137	48,995	3,307
Muck	3.99	4,942	23,397	23,397	23,397
Gravel	22.76	17,791	12,276	14,719	10,156
Gravel riffle	12.48	......	.....	7,142	22,283
Plant beds	5.32	......	......	4,585	6,098
Total		98,838	22,355	98,838	65,241

is permitted. Ultimately it is hoped that streams can be managed in such a way that they will produce maximum numbers of game fishes and at the same time be restored as nearly as possible to their natural condition.

Artificial flies and the fisherman's entomology.— Whether entomology has a place in the art and lore of angling is a moot question. Leonard (1950) takes a positive position, stating that, "the fisherman with a knowledge of aquatic insects and the important relationship they bear to the fish he wants to catch is better equipped with a single fly than is the man who knows nothing of such things though he sports a jacket full of fly-boxes stuffed with crisp, unmouthed flies of every description. The man with an understanding of aquatic life knows how and where to place his casts, fishing those places his knowledge tells him suit the lure and the fish, whereas the other fellow will cast at random, forever changing flies, fondly hoping that eventually he will discover a fly of some sort that will catch a fish . . . The fly-dresser in particular is obliged to know as much as he possibly can about the life cycles of the insects his flies are designed to represent. The more he knows about their aquatic and aerial stages, the more intelligently he will design, balance, and dress the copies."

That Leonard is not alone in his position is indicated by a literature that runs into hundreds of titles. One of the earliest accounts (*ca.* 200 A.D.) is by Aelianus in *De Animalium Natura* (1611) so the Macedonians are credited with the first use of artificial flies in the river Astraeus. In England the subject came into its own with such classics as *A Treatyse of Fysshynge With an Angle* by Dame Juliana Berners (1496). Later classics include: Izaak Walton's *The Compleat Angler, or the contemplative man's recreation, being a discourse of rivers, fishponds, fish and fishing* (1653) (5th ed., 1676, with Cotton's *Instructions how to angle for a Trout or Grayling in a clear stream,* containing many entomological notes); Ronalds' *The Fly-Fisher's Entomology* (1836); Halford's *Dry-fly Entomology* (1897); Mosely's *The Dry-fly Fisherman's Entomology* (1921); and Harris' *An angler's Entomology* (1952). A noteworthy American title of recent date is J. Edson Leonard's *Flies, their origin, natural history, tying, hooks, patterns and selections of dry and wet flies, nymphs,* etc. (1950).

"Fundamentally [as stated by Leonard, 1950], artificial flies are made according to two schools of thought. The first, the Impressionistic, believes that approximate size, general appearance and color are sufficient to lure a trout under all conditions, while the second, the Realistic, demands precise duplication of an insect."

In the books mentioned above and in literally hundreds of others, we find an astonishing amount of fact and fiction, of novelty and tradition, of superficiality and meticulous care. As an illustration of the latter the following quote from "Piscator" in the Preface to the sixth edition of Ronalds (1862) is typical, stating that, "he has been induced to paint both the natural and artificial fly from nature, to etch them with his own hand, and to colour, or superintend the colouring of each particular impression."

Scientific competence was added to the empirical observations of former times by the late Martin Mosely, deputy keeper of entomology in charge of the principal groups of aquatic insects at the British Museum (Natural History). Mosely was a close associate of Halford and after his death controlled a length of the River Test at and below Mottisfont (England), for a joint period of about eighteen years. Mosely's contribution to dry-fly entomological literature was to supply at Halford's request "a series of plates, . . . based on modern scientific ideas, and illustrating in color the insects which are of main importance to the dry-fly fisherman." In setting out to accomplish this Mosely stated that "I am . . . inclined to regard such a task as this with the eye of an entomologist rather than that of the fly-fisherman; and throwing aside such considerations as whether this fly is acceptable to the trout, whether that fly has a bitter taste and is allowed to float away unnoticed, I have attempted to describe the flies which I myself have found in plenty, and which I think my brother anglers will also meet with by the river's bank."

"Dry flies" are meant to imitate insects that fall onto the water surface and float without wetting. Actually, such "driftfood" includes all sorts of terrestrial insects that fall onto the water from overhanging vegetation. In practice, however, most dry flies are made to imitate adults of the aquatic groups —mayflies, stoneflies, caddisflies, and true flies. "Wet flies" are those that are fished under water, including adult insects that have become wet and immature nymphs and larvae of various kinds. Because of the skills required to select and cast flies properly the dry-fly fisherman enjoys a higher status than those who employ other techniques and lures.

In England, a special fisherman's nomenclature has been developed for the commonest species and higher groups of aquatic insects. Since our own literature and culture has drawn so heavily on British sources many of these names have been carried over, at times inappropriately, to our fauna. The resulting confusion is probably of little consequence to the fisherman but

TABLE 7 (Introduction)

SCIENTIFIC AND ANGLER'S NAMES FOR SOME COMMON GROUPS OF AQUATIC INSECTS

Scientific name	Angler's names
Ephemeroptera	Mayfly, drake, quill, Brown, Cahill Developmental stages: nymph, dun (subimago), spinner (imago), spent-wing (wet imago)
Plecoptera	stonefly, perlid, willow, sally
Trichoptera	caddisfly, caddicefly, sedge, grannom, fish moth, stickworm, caseworm
Diptera	crane fly, midge, punkie, gnat mosquito, bloodworm
Megaloptera	
Corydalidae	Dobsonfly, hellgrammite
Sialidae	Alderfly
Odonata	Dragonfly, damselfly, snake feeder, devil's darning needle, mosquito hawk

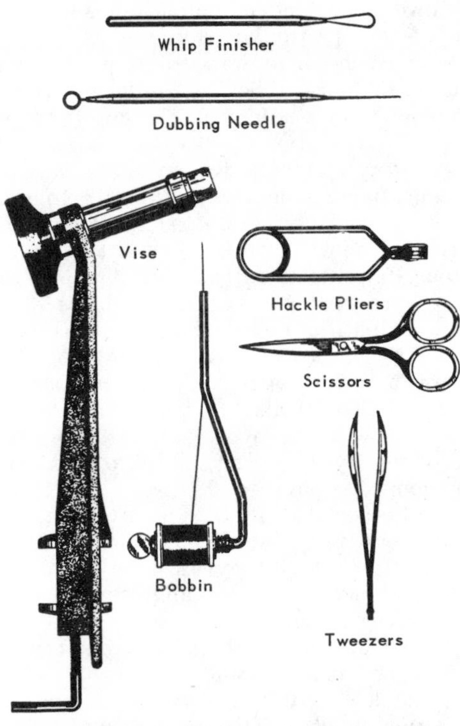

Intro. fig. 61. Tools used for tying flies (J. Edson Leonard, 1950).

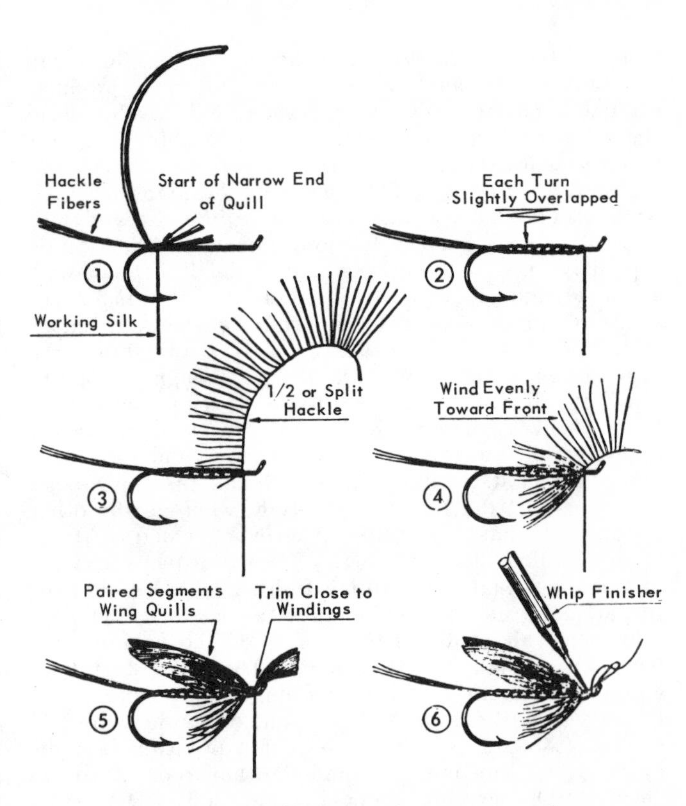

Intro. fig. 63. Technique for tying a wet fly (J. Edson Leonard, 1950).

is confusing to the entomologist. Hence the preceding list is given, not to explain common names of British species, but to show the common names of the principal groups of aquatic insects here and abroad with equivalent scientific name (table 7).

Much has been made of the technique of fly-tying and of the skill required to achieve a close approximation to nature. The subject is too extensive for detailed treatment here but a few key illustrations are reproduced from the excellent book by Leonard (1950) with permission of the author and the publishers. These figures are largely self-explanatory but the original work should of course be consulted for details. The tools illustrated (intro. fig. 61) may be obtained from dealers in sporting goods. Materials such as hooks, feathers, silk, wax, and so on, may

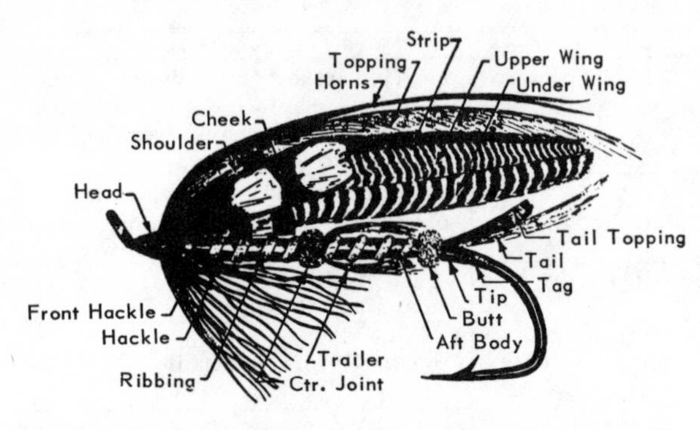

Intro. fig. 62. Terminology applied to the various parts of a fly (J. Edson Leonard, 1950).

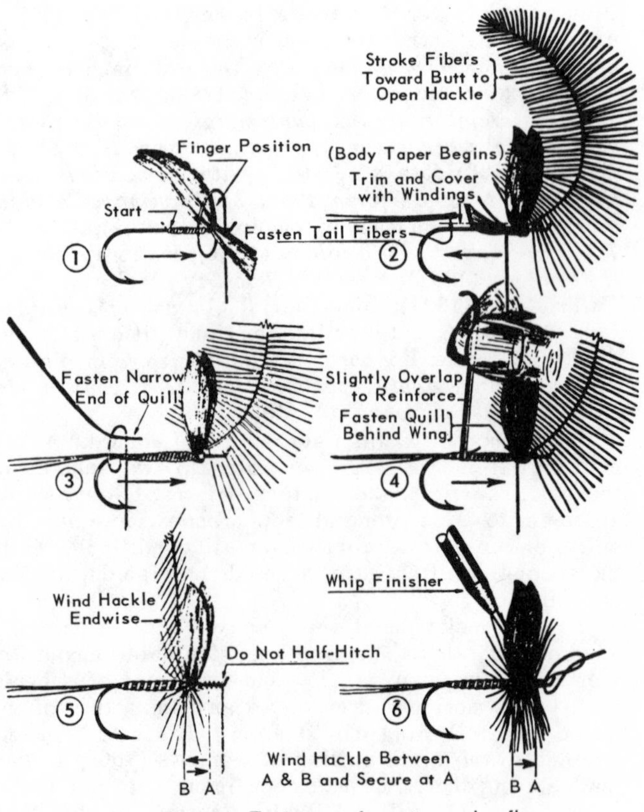

Intro. fig. 64. Technique for tying a dry fly (J. Edson Leonard, 1950).

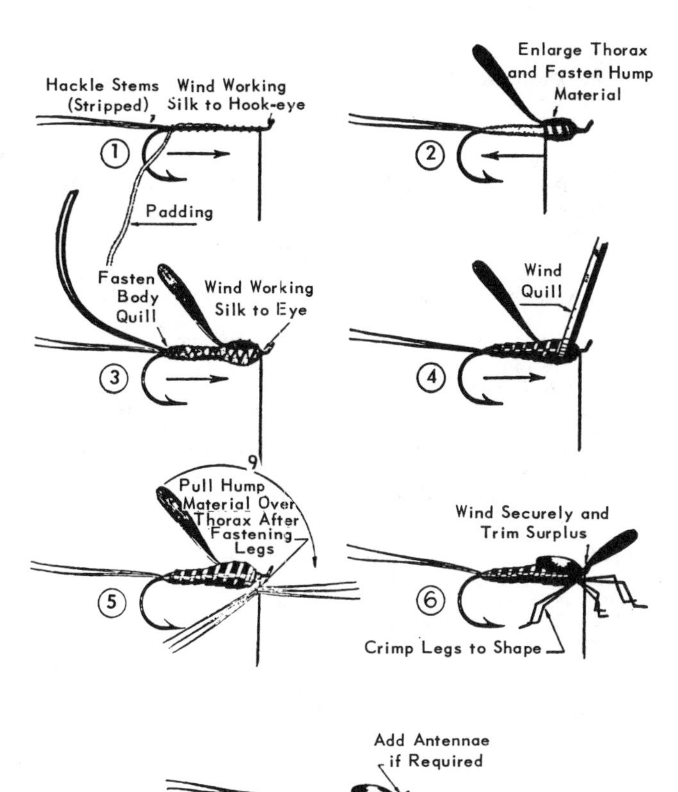

Intro fig. 65. Technique for tying a nymph (J. Edson Leonard, 1950).

also be obtained commercially but many of the body materials are commonly available as quills, wool, kapok, tinsel, chenille, and chicken feathers. Special terminology (intro. fig. 62) is used for the various parts of a "fly." Techniques are shown (intro. figs. 63-65) for tying a wet fly, a dry fly, and a nymph, respectively.

With sufficient skill it is possible to copy such standard fly patterns as the "Royal Coachman" or to imitate insects found flying at the stream side or discovered by dissecting fish stomachs. However, as pointed out by Neill (1938), it is not known whether dry flies and wet flies "are invariably accepted by trout for what they are intended to represent." In fact, evidence has been obtained indicating that fish take food strictly according to its accessibility without exercising fine discrimination or food preferences.

* * *

From the preceding discussion it is evident that aquatic entomology is a large and complex subject. Cutting across many specialized fields, it includes much of general entomology and limnology and certain aspects of public health and wildlife management. The aquatic entomologist should be well grounded in these diverse fields and in addition should be able to use the technical keys and descriptions given or referred to in the following chapters and thus make the study of insects an integral part of his work.

REFERENCES

AELIANUS, CLAUDIUS
1611. De animalium natura libri XVII. Lyons. Apud. I. Tornaesium. pp. 4 + 1018 + 94.
BARBER, M. A., and T. B. HAYNE
1921. Arsenic as a larvicide for anopheline larvae. Publ. Hlth. Reps., 36:3027-3034.
BERNERS, DAME JULIANA
1496. The treatyse of fysshinge with an angle. pp. 37-48. *in* Book of hawking, hunting, and heraldry. Westminster. Wynkyn de Worde. 74 pp. 2d ed. (1st ed. 1486. "The boke of Saint Albans." Lacks the "Treatyse".)
BRUNETTI, ROSEMARY, R. F. FRITZ, and A. C. HOLLISTER, JR.
1954. An outbreak of malaria in California, 1952-1953. Amer. Jour. Trop. Med. Hyg., 3:779-788.
CAMP, C. L.
1952. Earth song, A prologue to history. University of California Press. pp. 1-127.
CARPENTER, F. M.
1953. The geological history and evolution of insects. Amer. Sci., 41:256-270.
CARPENTER, K. E.
1928. Life in inland waters, with especial reference to animals. London; Sidgwick and Jackson, Ltd. pp. xviii + 267.
CARPENTER, S. J., and W. J. LACASSE
1955. Mosquitoes of North America. Berkeley: University of California Press. pp. vii + 360, 127 plates.
COMSTOCK, J. H.
1887. Note on the respiration of aquatic bugs. Amer. Nat., 21:577-578.
CRONEMILLER, F. P.
1955. Making new trout streams in the Sierra Nevada. *in* Water, the Yearbook of Agriculture for 1955. U.S.D.A. pp. xiii + 751 (583-586).
DAVIS, W. M.
1933. The Lakes of California. Calif. Jour. Mines, 29:175-236.
DAVISON, V. E., and J. A. JOHNSON
1943. Fish for food from farm ponds. U.S.D.A., Fmrs' Bull. 140. pp. 1-22.
DEEVEY, E. S., JR.
1942. Studies on Connecticut lake sediments. III. The biostratonomy of Linsley Pond. Amer. Jour. Sci., 240:233-264.
DIMICK, R. E., and D. C. MOTE
1934. A preliminary survey of the food of Oregon trout. Bull. Oregon Agric. Expt. Sta., 323:1-23.
DOUDOROFF, P., and M. KATZ
1950. Critical review of literature on the toxicity of industrial wastes and their components to fish. I. Alkalies, acids, and inorganic gases. Sewage and Ind. Wastes, 22:1432-1458.
1953. II. The metals, as salts. Sewage and Ind. Wastes, 25:802-839.
DOUDOROFF, P., M. KATZ, and C. M. TARZWELL
1953. Toxicity of some organic insecticides to fish. Sewage and Ind. Wastes, 25:840-844.
ELIASSEN, R.
1952. Stream pollution. Sci. Amer., 186:17-21.
ELTON, CHARLES
1947. Animal ecology. 3d Impression. London; Sidgwick and Jackson, Ltd. pp. xx + 209.
EMBODY, G. C.
1927. An outline of stream study and the development of a stocking policy. Contr. Aquiculture Lab. Cornell Univ. p. 1-21. (Privately printed.)
FONTAINE, R. E., H. F. GRAY, and T. AARONS
1954. Malaria control at Lake Vera, California, in 1952-53. Amer. Jour. Trop. Med. Hyg., 3:789-792.
FORBES, S. A.
1887. The lake as a microcosm. Reprinted, Bull. Illinois Nat. Hist. Surv., 15:537-550, 1925.
FOREL, F. A.
1892-1904. Le Léman. Monographie limnologique. Lausanne, 3 vols.

GAUFIN, A. R., and C. M. TARZWELL
1953. Discussion *after* Patrick, 1953. Sewage and Ind. Wastes, 25:214-217.
GJULLIN, C. M., and R. F. PETERS
1952. Abstract of recent studies of mosquito resistance to insecticides in California. Proc. Pap. 20th Ann. Conf. Calif. Mosq. Contr. Ass'n., pp. 44-45.
GRINNELL, J., and T. I. STORER
1924. Animal life in the Yosemite. University of California Press. pp. xviii + 741.
HALFORD, F. M.
1897. Dry-fly entomology. London. Vinton & Co., Ltd. pp. xii + 323.
HARRIS, J. R.
1952. An angler's entomology. The new naturalist library. London; Collins. pp. xv+268.
HAZZARD, A. S.
1938. *in* Davis, H. S. Instructions for conducting stream and lake surveys. U.S. Bureau of Fisheries, Circ. 26, pp. 1-55.
HENDERSON, J. M.
1951. Irrigation and mosquito problems. Proc. Pap. 19th Ann. Conf. Calif. Mosq. Contr. Ass'n., pp. 49-51.
HERMS, W. B.
1949. Looking back half a century for guidance in planning and conducting mosquito control operations. Proc. Pap. 17th Ann. Conf. Calif. Mosq. Contr. Ass'n., pp. 89-92.
HESS, A. D.
1941. New limnological sampling equipment. Limnol. Soc. Amer. Spec. Publ. No. 6.
HESS, A. D., and T. F. HALL
1943. The intersection line as a factor in anopheline ecology. Jour. Nat. Malaria Soc., 2:93-98.
1945. The relation of plants to malaria control on impounded waters with a suggested classification. Jour. Nat. Malaria Soc., 4:20-46.
HESS, A. D., and C. C. KIKER
1944. Water level management for malaria control on impounded waters. Jour. Nat. Malaria Soc., 3:181-196.
HESS, A. D., and J. H. RAINWATER
1939. A method for measuring the food preference of trout. Copeia, No. 3, pp. 154-157.
HESS, A. D., and A. SWARTZ
1941. The forage ratio and its use in determining the food grade of streams. Trans. 5th N. Am. Wildlife Conf., 1940:162-164.
HESS, A. D., and C. M. TARZWELL
1942. The feeding habits of *Gambusia affinis affinis,* with special reference to the malaria mosquito, *Anopheles quadrimaculatus*. Amer. Jour. Hyg., 35:142-151.
HOFFMANN, C. H., H. K. TOWNES, R. I. SAILER, and H. H. SWIFT
1946. Field studies on the effect of DDT on aquatic insects. U.S.D.A. Bur. Ent. Pl. Quar., E-702:1-20.
HOLLOWAY, A. D.
1951. An evaluation of fish pond stocking policy and success in the southeastern states. The Progr. Fish-Culturist, 13:171-180.
HORSFALL, W. R.
1955. Mosquitoes, their bionomics and relation to disease. New York: Ronald, 723 pp.
HOWARD, L. O.
1892. An experiment against mosquitoes. Insect Life, 5:12-14.
HUBBS, C. L., and R. W. ESCHMEYER
1938. The Improvement of lakes for fishing. A method of fish management. Bull. Inst. Fisheries Res. No. 2, Ann Arbor: University of Michigan, pp. 1-233.
HUBBS, C. L., and R. R. MILLER
1948. The Great Basin, with emphasis on glacial and postglacial times. II. The zoological evidence. Bull. Univ. Utah, 38:18-166.
HUET, M.
1949. Aperçu des relations entre la pente et les populations piscicoles des eaux courantes. Rev. Suisse d' Hydrologie, 11:332-351.
HYNES, H. B. N.
1950. The food of fresh-water sticklebacks (Gasterosteus aculeatus and Pygosteus pungitius), with a review of methods used in studies of the food of fishes. Jour. Anim. Ecol., 19:36-58.
ISAAK, L. W.
1952. Progress Report CMCA Operational Investigations Project—Insecticide investigations. Proc. Pap. 20th Ann. Conf. Calif. Mosq. Contr. Ass'n., pp. 31-36.
JENKINS, D. W., and S. J. CARPENTER
1946. Ecology of the tree hole breeding mosquitoes of nearctic North America. Ecol. Monog., 16:31-48.
LAGLER, K. F.
1952. Freshwater fishery biology. Dubuque, Iowa: W. C. Brown. pp. x + 360.
LEONARD, J. EDSON
1950. Flies, their origin, natural history, tying, hooks, patterns and selections of dry and wet flies, nymphs, streamers, salmon flies for fresh and salt water in North America and the British Isles, including a dictionary of 2200 patterns. New York: H. S. Barnes. pp. xii + 340.
LINDEMAN, R. L.
1941. Seasonal food-cycle dynamics in a senescent lake. Amer. Midl. Nat., 26:636-673.
LINDQUIST, A. W., and C. C. DEONIER
1942. Emergence habits of the Clear Lake Gnat. Jour. Kansas Ent. Soc., 15:109-120.
LINDQUIST, A. W., A. R. ROTH, and J. R. WALKER
1950. Report on the control of the Clear Lake Gnat, *Chaoborus astictopus* Dyar and Shannon, in Clear Lake, California. Unpubl. Rep. U.S.D.A. Bur. Ent. Pl. Quar.
Malaria Control on Impounded Water
1947. pp. xiii + 422. U.S. Public Health Service and Tennessee Valley Authority. Washington, D.C.
MACAN, T. T., C. H. MORTIMER, and E. B. WORTHINGTON
1942. The production of freshwater fish for food. Freshw. Biol. Assn. Brit. Emp. Sci. Publ. No. 6, pp. 1-36.
MACAN, T. T., and E. B. WORTHINGTON
1951. Life in lakes and rivers. The New Naturalist Series. London: Collins. pp. xvi + 272.
MALOEUF, N. S. R.
1936. Quantitative studies on the respiration of aquatic arthropods and on the permeability of their outer integument to gases. Jour. Exp. Zool., 74:323-351.
MEEHEAN, O. L.
1937. Control of predaceous insects and larvae in ponds. Progr. Fish-Culturist, No. 33, pp. 15-16.
MEYER, C. B.
1951. Water resources of California. State Water Resources Board, Bull. 1, pp. 1-648.
MIALL, L. C.
1895. The natural history of aquatic insects. London: Macmillan. pp. ix + 395.
MILLER, A. H.
1951. An analysis of the distribution of the birds of California. Univ. Calif. Publ. Zoöl., 50:531-644.
MILLER, LEONARD
1951. The gnat problem of Elsinore. Proc. Pap. 19th Ann. Conf. Calif. Mosq. Contr. Ass'n., pp. 96-97.
MOSELY, M. E.
1921. The dry fly fisherman's entomology. London. pp. xx + 109.
1926. Insect life and the management of a trout fishery. London. pp. 1-112.
Mosquito abatement in California
1951. State of California, Dept. Pub. Health, Bull. VC-1, pp. 1-47.
MULHERN, T. D.
1953. Better results with mosquito light traps through standardizing mechanical performance. Mosquito News, 13:130-133.
MUTTKOWSKI, R. A.
1929. The ecology of trout streams in Yellowstone National Park. Bull. N.Y. St. Coll. For. Roosevelt Wild Life Annals, Vol. 2, No. 2 , pp. 151-240.

NEEDHAM, P. R.
1934. Quantitative studies of stream bottom foods. Trans. Amer. Fish. Soc., 64:239-247.
1938. Trout streams. Ithaca, New York: Comstock. pp. 233.
1939. Quantitative and qualitative observations on fish foods in Waddell Creek Lagoon. Trans. Amer. Fish. Soc., 69:178-186.
NEEDHAM, P. R., and R. L. USINGER
1956. Variability in the macrofauna of a single riffle in Prosser Creek, California, as indicated by the Suber sampler. Hilgardia, Vol. 24, No. 14, pp. 383-410.
NEILL, R. M.
1938. The food and feeding of the brown trout (*Salmo trutta* L.) in relation to the organic environment. Trans. Roy. Soc. Edin., 59:481-520.
ODUM, E. P.
1953. Fundamentals of ecology. Philadelphia: W. B. Saunders. pp. xii + 384.
PATRICK, R.
1953. Aquatic organisms as an aid in solving waste disposal problems. Sewage and Ind. Wastes, 25:210-214.
PENNAK, R. W., and E. D. VAN GERPEN
1947. Bottom fauna production and physical nature of the substrate in a northern Colorado trout stream. Ecology, 28:42-48.
PORTMAN, R. F., and A. H. WILLIAMS
1952. The control of *Aedes dorsalis* and other aquatic pests in rice fields. Proc. Pap. 20th Ann. Conf. Calif. Mosq. Contr. Ass'n., pp. 88-89.
REEVES, W. C.
1953. The knowns and the unknowns in the natural history of encephalitis. Proc. Pap. 21st Ann. Conf. Calif. Mosq. Contr. Ass'n., pp. 53-55.
REIMERS, N., J. A. MACIOLEK, and E. P. PISTER
1955. Limnological study of the lakes in Convict Creek Basin, Mono County, California. U.S.D.I., Fish and Wildlife Service, Fishery Bull. 103, pp. 437-503.
RENN, C. E.
1943. Emergent vegetation, mechanical properties of the water surface, and distribution of Anopheles larvae. Jour. Nat. Malaria Soc., 2:47-52.
RICKER, W. E.
1934. An ecological classification of certain Ontario streams. Univ. Toronto Studies, Biol. Ser. No. 37, pp. 1-114.
RONALDS, A.
1836. The fly-fisher's entomology. London. pp. viii + 115.
RUTTNER, F.
1953. Fundamentals of limnology. Transl. by D. G. Frey and F. E. J. Fry. University of Toronto Press. pp. xi + 242.
SIMMONS, P., D. F. BARNES, C. K. FISHER, and G. H. KALOOSTIAN
1942. Caddisfly larvae fouling a water tunnel. Jour. Econ. Ent., 35:77-79 (Simmons, *in litt.,* 1955).
STEINMANN, PAUL
1907. Die Tierwelt der Gebirgsbäche. Eine faunistisch-biologisch Studie. Ann. Biol. Lac., 2:1-137.
SWINGLE, H. S., and E. V. SMITH
1941. The management of ponds for the production of game and pan fish. *In* "A Symposium on Hydrobiology". University Wisconsin Press. pp. ix + 405 (218-226).
TARZWELL, C. M.
1937. Experimental evidence on the value of trout stream improvement in Michigan. Trans. Amer. Fish. Soc., 66:177-187.
1947. Effects of DDT mosquito larviciding on wildlife. I. The effects on surface organisms of the routine hand application of DDT larvicides for mosquito control. Public Health Rep., 62:525-554.
1950. Effects of DDT mosquito larviciding on wildlife. V. Effects on fishes of the routine manual and airplane application of DDT and of the mosquito larvicides. Public Health Rep., 65:231-255.
THIENEMANN, A.
1912. Der Bergbach des Sauerlandes. Int. Rev. ges. Hydrobiol. Hydrogr., Biol. Suppl., 4, pp. 1-125.
1926. Die Binnengewässer Mitteleuropas. Die Binnengewässer, Bd. 1. 225 pp.
THORPE, W. H.
1950. Plastron respiration in aquatic insects. Biol. Rev., 25:344-390.
THORPE, W. H., and D. J. CRISP
1947. Studies on plastron respiration. I. The biology of *Aphelocheirus* and the mechanism of plastron retention. Jour. Exp. Biol., 24:227-269.
USINGER, R. L., and W. R. KELLEN
1955. The role of insects in sewage disposal beds. Hilgardia, 23:263-321.
USINGER, R. L., and P. R. NEEDHAM
1954. A plan for the biological phases of the periodic stream sampling program. State Water Pollution Control Board, Unpublished Report, 59 pp.
1956. A drag-type riffle-bottom sampler. The Progr. Fish-Culturist, 18:42-44.
WALSHE, B. M.
1951. The feeding habits of certain chironomid larvae (subfamily Tendipedinae). Proc. Zool. Soc. London, 121:63-79.
WALTON, IZAAK
1653. The compleat angler or the comtemplative man's recreation. Being a discourse of fish and fishing, not unworthy the perusal of most anglers. London; Maxey. 246 pp. (1676, 5th ed. by Charles Cotton, Instructions how to angle for a trout or grayling in a clear stream, London.)
Water Pollution Control Progress Report for 1950 through 1952. Water Pollution Control Board Publication No. 5, Sacramento, pp. 1-56.
WELCH, P. S.
1952. Limnology. 2d ed. McGraw-Hill. pp. ix + 538.
WHIPPLE, G. C.
1927. The microscopy of drinking water. 4th ed. revised by G. M. Fair and M. C. Whipple. New York. 586 pp.
WIGGLESWORTH, V. B.
1938. The regulation of osmotic pressure and chloride concentration in the haemolymph of mosquito larvae. Jour. Exp. Biol., 15:235-247.
WILSON, C. B.
1923. Water beetles in relation to pond fish culture, with life histories of those found in fishponds at Fairport, Iowa. Bull. U.S. Bur. Fisheries, 39:231-345.

B. Equipment and Technique

By John D. Lattin
Oregon State College, Corvallis

Field work in aquatic entomology is carried out by many persons with diverse points of view. Some approach the subject as taxonomists specializing in one or another of the groups of aquatic insects; others are primarily interested in life histories or ecological relations. And there is an increasing number of professional workers whose primary concern is quantitative sampling of populations for mosquito control or stream and lake management. It is the object of this chapter to describe the equipment and methods best suited to each of these varied purposes. A fuller discussion of entomological equipment is given in Peterson (1953), culture methods are described in the compendium edited by J. G. Needham (1937), and general limnological methods are treated in detail in Welch (1948).

GENERAL COLLECTING

Terrestrial

Most aquatic insects are terrestrial or aerial as adults and can be identified to the species level only at this stage. Therefore it is important to collect generally in the vicinity of water where adult insects are most likely to occur. Stones, boards, and piles of debris should be turned over and carefully examined. All types of vegetation growing near the water should be examined and the insects dislodged by beating or "sweeping" into a net. Swarms of gnats and other aerial insects should be swept with the net.

Nets.– Although many different types of nets are used by collectors, the basic design is the same (intro. fig. 66). A light, strong handle is used with a ring of stout, spring-steel wire to which is fastened a bag of nylon netting. Aerial nets, for catching insects in flight, should be light enough to handle easily. The sweeping net, on the other hand, is of

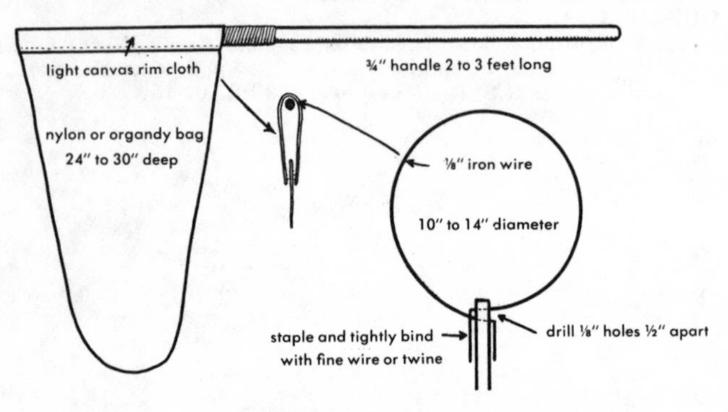

Intro. fig. 66. Specifications for construction of insect net (Ross, 1953).

Intro. fig. 67. Sweeping net in action (Ross, 1953).

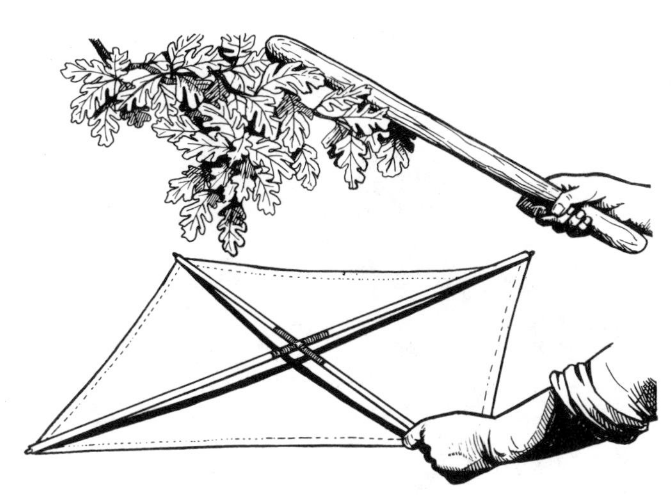

Intro. fig. 68. The correct use of the beating sheet (Ross, 1953).

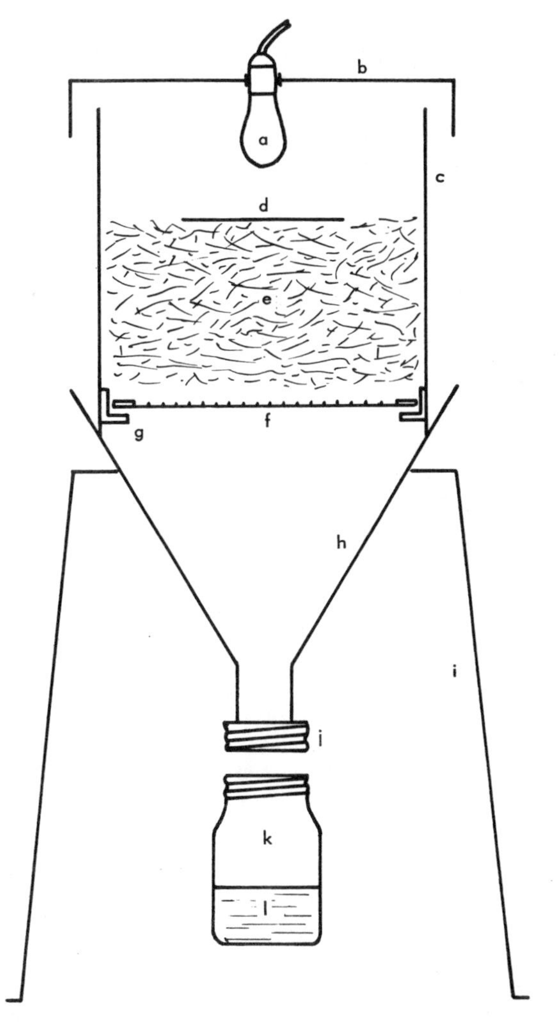

Intro. fig. 69. Details of the Berlese Funnel; *a*, light bulb; *b*, metal cover; *c*, metal cylinder to hold debris; *d*, sheet of aluminum foil; *e*, debris; *f*, circular piece of 1/2" hardware cloth with metal rim; *g*, one of three metal flanges fastened to inside of cylinder to support screen; *h*, metal funnel; *i*, supporting stand; *j*, lid of mason jar; *k*, mason jar; *l*, 70 per cent ethyl alcohol (original).

heavier material and may be made of heavy-duty muslin with canvas reinforcement along the leading edge. It is swept through heavy vegetation, jarring loose and capturing many small insects that would otherwise be overlooked (intro. fig. 67). Nets and other equipment may be homemade or obtained from biological supply houses such as Ward's Natural Science Establishment, Rochester, N.Y.; Turtox (General Biological Supply House, Chicago, Ill.), and others. Specialized limnological equipment may be ordered from the equipment shop of the California Academy of Sciences, San Francisco.

Beating sheet.—This device consists of a sheet of canvas stretched over a wooden frame (intro. fig. 68). It is held under vegetation from which insects are jarred loose with a stick. On a warm, sunny day considerable dexterity is needed to capture quick-flying insects before they leave the sheet. A hook-handled umbrella, used upside down, is a useful "beating sheet" because the handle can be slipped around the neck, leaving both hands free to collect the insects.

Berlese funnel.—The Berlese funnel is useful for collecting small insects living in duff or debris at or near the edge of the water. It consists of a large funnel with a screen inserted to hold the material (intro. fig. 69). The debris is brought back to the laboratory in sacks and dumped into the funnel. A strong light is placed above, and the heat and light drive the insects down the funnel, through the screen, and into a jar of 70 per cent alcohol. A sheet of aluminum foil placed on the debris increases the effectiveness of the funnel and prevents the possibility of fire. The Berlese funnel is particularly useful for collecting springtails and many small beetles and bugs which are seldom seen or collected by any other method.

Light traps.—Many different types of light traps have been designed. Briefly, they consist of a light source that attracts insects, and some method of

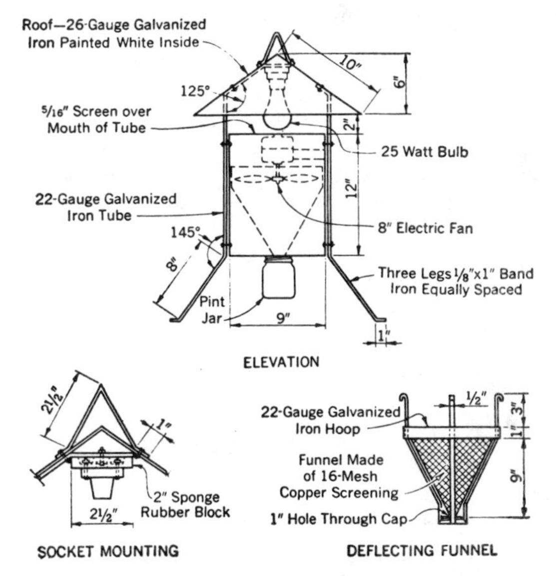

Intro. fig. 70. Structural details of New Jersey light trap (Malaria Contr. Imp. Waters).

retaining and killing the specimens. Intro. figure 70 shows a common type, known as the New Jersey light trap, which can be constructed with comparative ease. It is desirable to have an electric outlet close to the trap although battery operated traps can be set up. A Coleman lantern on a white sheet makes an efficient "light trap" and frequently is the best way to collect many of the adult stages of aquatic insects. Collecting at any light source near water usually yields many adult aquatic insects, and neon lights (especially blue) in towns are sometimes productive. In general, insects fly to lights in greatest numbers on warm sultry nights when there is little or no wind and the moon is not too bright.

Miscellaneous methods.—The *aspirator* (intro. fig. 71) is a device used to collect insects from the net, from resting places under bridges and on vegetation, and from microhabitats at the water's edge. The insects are sucked into a tube and are later transferred to a killing bottle. The insect fauna of sand or mud may be collected by *splashing* water on the bank and washing out individual specimens. The reverse can likewise be done; that is, sand can be thrown into the water to float off shore dwelling forms. This type of collecting will produce larvae and adults of such beetle families as Omophronidae, Staphylinidae, Heteroceridae, and Carabidae and certain dipterous larvae. Rocks in the intertidal zone of the seashore are the special habitat of members of several families of beetles (Eurystethidae, Staphylinidae). A *crow bar* is useful to split such rocks and expose beetles that have retreated deep into cracks. Some flying insects, including large dragonflies, are practically inaccessible during the heat of the day but may be picked from resting places on vegetation in the early morning. Strong fliers which are otherwise unobtainable may be "felled" at close range with fine dust shot from a smooth-bore .22 caliber gun. A good collector does not rely solely on specialized equipment but examines every possible microhabitat—under stones, boards, grass, and mats of vegetation or debris.

AQUATIC COLLECTING

The same intensive approach is necessary in aquatic as in terrestrial collecting. Close attention should be paid to the seemingly endless microhabitats including surfaces of stones, aquatic vegetation, sunken logs, and accumulations of debris. If trash and plant materials are removed from the water and spread out

Intro. fig. 72. Method of using insect net for stream collecting (Ross, 1953).

in the sun to dry, dozens or even hundreds of small insects will crawl out. Such trash may be raked from the bottom of a pond or stream, gathered in a net, or collected from snags after a flash flood. Small Dryopids and other beetles may continue to emerge from drying trash for one-half hour or more.

Nets and screens.—Water nets should be sturdier than aerial nets. Nylon is desirable because of its strength; also it dries quickly and hence can be used for terrestrial collecting as well. The size mesh should be 24 to 32 strands per inch with mesh as open as is practicable to hold insects of the desired size. Mesh of too fine gauge drags the net and hinders the capture of quick-moving insects. In streams, the net is held close to the bottom, and stones and trash are disturbed as the collector moves upstream (intro. fig. 72). Insects are dislodged and carried downstream into the net by the current. A piece of window screen fastened to two strips of lath is particularly useful for this type of collecting (intro. fig. 73*c*). The screen is stretched across a narrow section of stream, or held with both hands, while the collector backs upstream scuffing the bottom with his boots. In a rich stream this method will yield enough specimens to occupy the collector for a half hour or more, picking from the drying screen in direct sunlight. Small dip

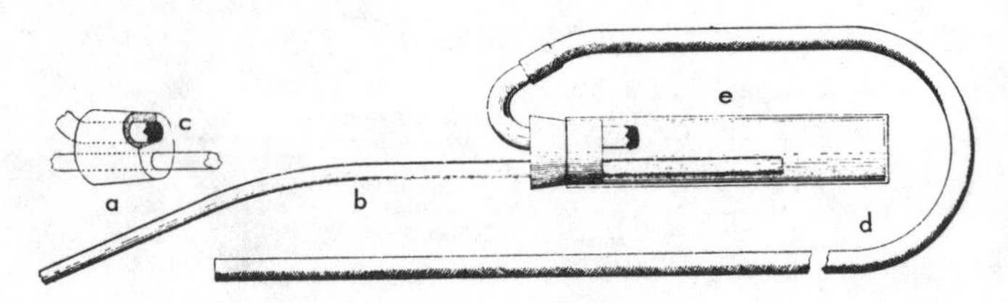

Intro. fig. 71. Aspirator; *a*, two-hole rubber stopper; *b*, glass or copper tubing; c, fine copper screen; *d*, rubber tubing; e, glass or plastic vial (Oman and Cushman, 1946).

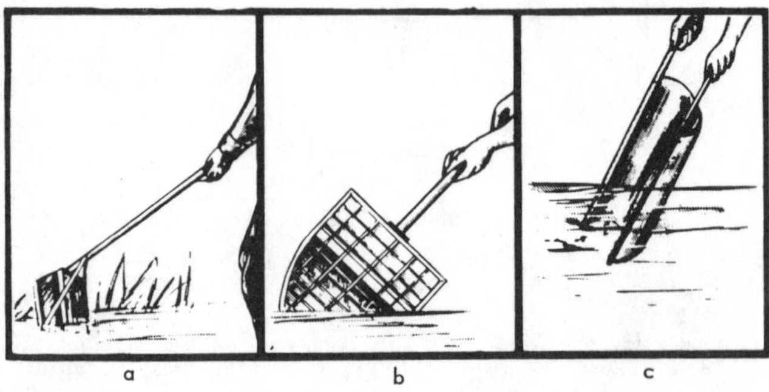

Intro. fig. 73. Specialized stream collecting equipment: *a*, Needham scraper; *b*, Needham net; c, Lath screen collector (Traver, 1940).

nets and even kitchen strainers make good collecting tools in streams and ponds.

Intro. figure 73*b* shows a Needham net that can be used in areas where weeds are thick. The coarse screen on top permits the insects to enter but keeps out most of the vegetation that would foul the net. The Needham scraper (intro. fig. 73*a*) is used to sample the bottom fauna. It is extended and dragged back to shore along the bottom, and helps to increase the collecting range beyond the depth possible in hip boots. Although less efficient, the common garden rake can be used to drag in large masses of weeds that often contain many insects.

Miscellaneous methods.—There are many unusual or highly specialized methods of collecting for particular purposes. In bogs, for example, sphagnum is pressed down, and the insects are floated out into open water where they may be netted with ease. Tree holes that hold water may be collected by siphoning with a short length of hose. Mosquito larvae may be readily collected from just under the surface film with a hand dipper. As the edge of the dipper is tipped below the surface, water enters and with it the larvae; specimens may then be taken up with a pipette or eye dropper. Mosquito larvae and, in fact, most aquatic insects may be washed into a shallow enamel pan from nets, trash, dippers, and so on, and picked at leisure with forceps or a pipette. During the sorting process, whether from the drying surface of net, screen, or trash, or from a pan, individual specimens are placed in vials containing 70 per cent ethyl alcohol. If the collector is a specialist he may sort specimens to species or higher groups by placing them in separate vials. The general collector usually segregates only by locality or habitat and sorts his material later.

Intro fig. 74. Ekman dredge, note metal messenger at top of picture.

QUANTITATIVE SAMPLING METHODS

One of the greatest needs in aquatic biology today is improved methods for sampling populations of insects. All existing techniques are inadequate or impractical for purposes of statistical analysis but in spite of obvious shortcomings, they are the best we have. Therefore standard sampling devices and some specialized equipment and techniques are described below.

Bottom sampling devices.—The *Ekman dredge* (intro. fig. 74) is the commonest type of bottom sampler used in lake and pond surveys. It is a six-inch (or larger) square brass box equipped with closing jaws. The jaws are cocked at the surface, and the dredge is lowered on a rope either by hand or by a winch. When it reaches the bottom a metal messenger is sent down the rope to trip the jaws shut. Then the sampler is hauled to the surface and the bottom material is collected in a bucket and later washed through a sieve or a series of graded screens (intro. fig. 75), the coarsest screen uppermost. After the muck and silt have been removed, the specimens are picked from the sieve and preserved in vials of alcohol for future study. Ekman samples should be taken at several different localities around a pond or lake, wherever the bottom fauna might vary. Transects are frequently run in several directions across a body of water or a random pattern may be followed. The Ekman dredge is usually employed in soft-bottomed lakes and ponds or slow-moving rivers, since the jaws do not bite well into hard clay or rough-stone bottoms.

Intro. fig. 75. Graded screens used with bottom samplers.

Intro. fig. 76. Surber square-foot bottom sampler being used in shallow stream.

Heavy-duty dredging under all types of bottom conditions is done with a *Peterson dredge* in which the mechanism holding the jaws open on the way down is released when it touches bottom. This dredge is much heavier than the Ekman, often weighing up to 70 pounds, so a sturdy winch is required to operate it.

The *Surber square-foot bottom sampler* (intro. fig. 76) is designed for use in riffles of shallow streams. It consists of two square frames hinged together. One bears the net that extends downstream and the other delimits the area of bottom to be sampled. The frame is placed in position on the bottom and the enclosed pebbles or other material are carefully gone over with the hands to dislodge the specimens. The insects are carried back into the net by the current. This sampler has the disadvantage of permitting some specimens to escape around the edges, especially if a fine mesh net is used. Side wings are used to reduce this hazard. One great advantage of the Surber sampler is that the frames fold flat and can be easily carried in the field.

The *Hess sampler* (intro. fig. 77) is a variation of the Surber sampler designed to correct its disadvantage of allowing specimens to escape. It is a cylinder of one-half-inch hardware cloth on a steel frame. The hardware cloth is covered except on the front to keep the smaller organisms from escaping. A square hole is cut in the back and a net attached. The sampler is placed on the bottom and is turned by the handles until it is firmly imbedded in the bottom. The technique of collecting is the same as for the Surber sampler. When using either of these samplers in rapid water, the operator should stand downstream from the sampler, bracing it with his legs, to prevent its movement by the current. The standard Surber sampler and the improved Hess sampler are limited in usefulness to depths of less than arm's length.

A more versatile sampler was devised by Usinger and Needham (1956) (intro. fig. 78). This *drag-type sampler* consists of a rectangular iron box with bars on the open face to exclude large rocks and trash. The two leading edges are provided with tines that dig into the rock and gravel of the bottom. The curved tines serve the same purpose as the fingers of the operator of the shallow-water type samplers. The current sweeps the organisms into the net, just as in those samplers. The net bag is removable and can be opened by means of a zipper to gather the organisms. A canvas sleeve is provided for the protection of the net. In operation the sampler is lowered into a riffle and pulled. The area of bottom to be sampled is determined by the length of bottom over which it is pulled.

The *square-foot tray* ("basket method," Wene and Wickliff, 1940) may be used to determine the number and type of organisms per unit area on the bottom of a stream or lake. The tray is made with wooden sides and a one-fourth-inch mesh screen bottom. It is filled with clean bottom material. It is then placed flush with the surrounding bottom of the stream or lake and should be left long enough for the fauna to become stabilized. A collar is placed around the tray at the time of removal to avoid disturbing the organisms. The trays are lifted out at intervals and all the organisms collected and counted.

Traps.—Several traps have been devised which

Intro. fig. 77. Hess sampler, collecting technique.

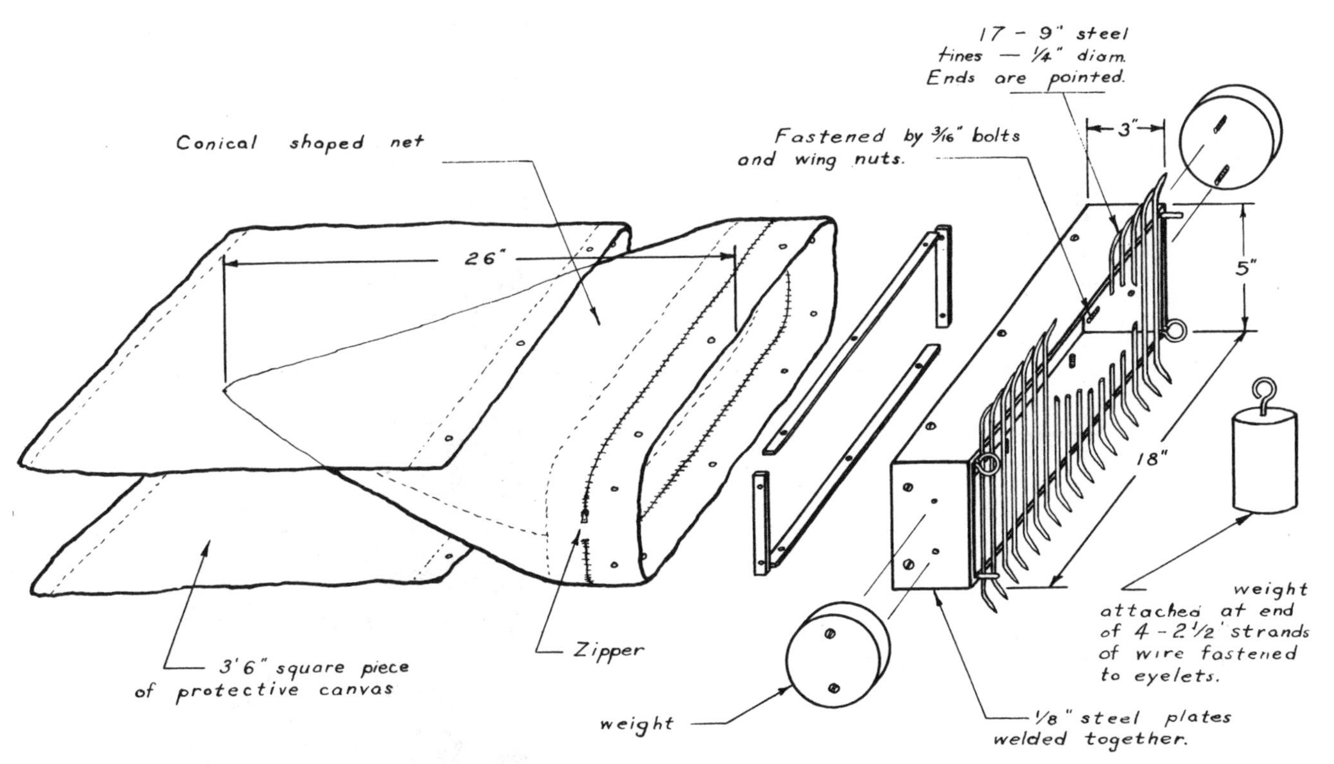

Intro. fig. 78. Drag-type sampler showing structural details (Usinger and Needham, 1956).

Intro. fig. 79. Submerged trap, used to sample free-swimming fauna in ponds (W. R. Kellen).

are useful in sampling known areas of stream or lake bottoms, particularly for Trichoptera, Ephemeroptera, Plecoptera, Odonata, and Diptera. One type is the *tent trap* (Needham, 1908; Sprules, 1947). This is a cage that is set on the bottom in the shallow water of a stream, for example, with a large part projecting above the surface. Adult insects emerge and fly or crawl up onto the screen, and are removed from time to time through a door on the side. In streams, care should be taken to keep the upstream side of the cage clean of debris that might slow down the current and thus alter the habitat. On lakes, *floating tent traps* have been used by Adamstone and Harkness (1923) and others (intro. fig. 81).

Inverted cone emergence traps or "funnel traps" have been used to sample insects as they emerge from the bottom of ponds or lakes. These cones are submerged to various depths (Brundin, 1949) and held by anchors or wires where they catch the midges or other organisms which emerge over a given area (intro. fig. 80). The underwater cones catch pupae which may go on to molt in the air trapped in the tube at the small end of the cone. Emergence traps are particularly useful in studies of the turnover or the annual crop of bottom organisms in contrast to the standing crop measured by other samplers. A comparison of such sampling techniques is given by Guyer and Hutson (1955).

The *submerged "fly" trap* (intro. fig. 79) is constructed along the lines of a standard fly trap; that is, it consists of a square screen cage with an inverted

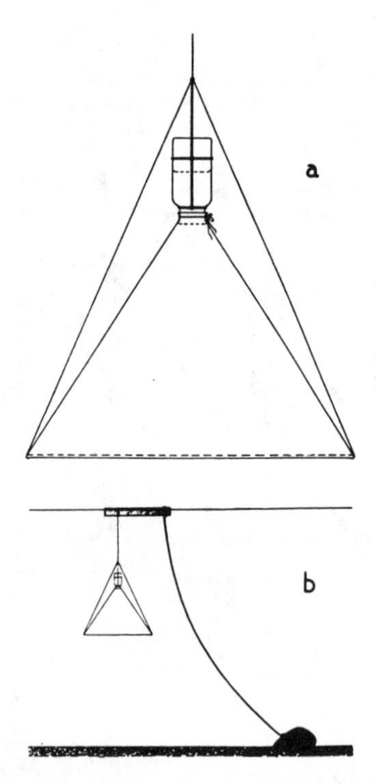

Intro. fig. 80. Submerged, inverted cone emergence trap used to sample Chironomidae (Brundin, 1949).

cone-shaped end. The trap is anchored to the bottom by ropes and weights, and can be lowered or raised to any depth in order to sample different levels. The apex of the cone is open and permits the insects to swim into the cage but prevents them from escaping once they are inside. These traps have been used with some success to sample the beetle and bug populations in oxidation ponds (Usinger and Kellen, 1955). Certain quantitative information can be drawn from the results, since the trap is left in the water for definite periods of time. However, as it is not known what area or volume of water outside the trap is sampled the results must be used with caution.

Intro. fig. 81. Floating inverted cone emergence trap 8 (W. R. Kellen).

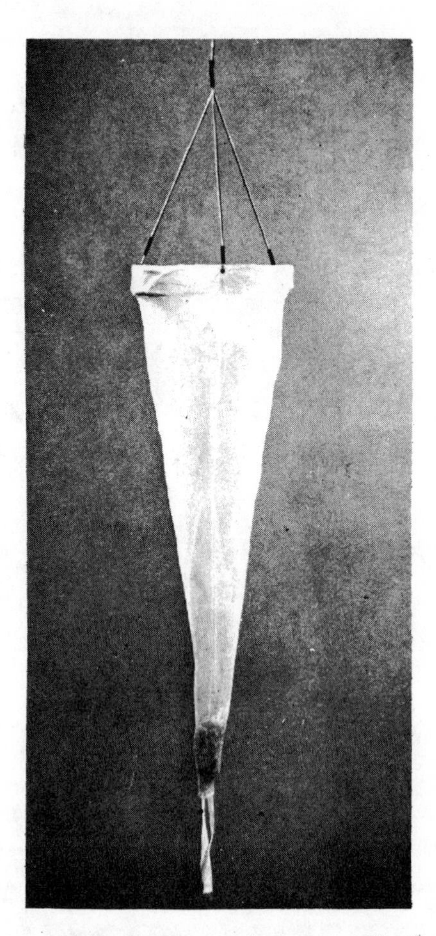

Intro. fig. 82. Wisconsin type plankton net (Ward's Nat. Sci. Est., 1949).

An *underwater light trap* has been developed by J. G. Needham (1924) and others. The light source can be either a large flashlight or a light bulb and reflector connected by an insulated wire to some source of electricity on shore. A surprising variety of aquatic insects has been collected in this manner.

Plankton nets.—Plankton forms the chief source of food for some fish and many invertebrates; hence it may be an important part of a comprehensive insect survey. For more detailed treatment of plankton, the reader is referred to Welch (1935, 1948). A plankton net (intro. fig. 82) is used to sample the small-sized floating organisms of open water. The particular net shown is called the Wisconsin plankton net. It consists of a fine silk cone that may have a special filter trap at the tip. The net is attached to a tow rope and dragged through the water for a known period of time. The plankters accumulate on the sides of the net and are washed down into the filter trap at the bottom. The trap is removed and the contents flushed in a vial of 5 per cent formalin solution for preservation. Later this material is identified in the laboratory. More accurate results as to the amount of water sampled can be obtained if a known volume of water is poured or pumped through the net. By using samples of known quantity and a special counting cell slide, relative numbers or densities of various forms can be calculated. Nets similar to a plankton net may be used to sample the free-swimming insect fauna of the open water, particularly those forms such as *Chaoborus* that migrate between the bottom and the surface of the water.

Miscellaneous techniques for quantitative sampling.—The *Berlese funnel* is useful in obtaining small insects, often overlooked, from the bottom debris of streams and lakes. The trash is collected and blotted to remove the excess water and then placed in the funnel. The recovery of the specimens is made in the manner previously mentioned. *Drift nets* (P. R. Needham, 1928) may be stretched across a stream at surface level to sample floating organisms, especially terrestrial forms that fall onto the water and provide an important part of the food of fish. Unfortunately, such nets are quickly fouled and hence must be cleaned frequently throughout the day. Smaller nets may be used to sample a limited area of surface but here again, fouling occurs and back eddies carry organisms around and past the net. *Pumping and draining* parts of a stream, after damming and by-passing the water, will reveal the total population of conspicuous insects over a given area. *Sialis* larvae, for example, are revealed as very abundant in some streams when pumped and drained but are seldom taken in riffle samples.

The microscopic biota associated with the surface of rocks and other objects in the water (periphyton) is an important source of food for insects. It may be sampled through the use of *collodion* (Margalef, 1949; Wenzl, 1940, 1941). The rocks to be examined are collected in 5 per cent formalin solution. They are next stained with Delafield's haematoxylin, washed in distilled water, and run through a series of alcohols, ending in absolute alcohol mixed with ether. The process up to this time takes about four hours. The stones are picked out of the solution and immediately covered with a solution of collodion dissolved in alcohol and ether. They are then allowed to dry completely. This process takes only a few minutes. After the solvents have dried, the collodion film is peeled off with a pair of sharp forceps, and a microscopic examination made of the contents. Quantitative counts may be made through the use of calibrated ocular grids on the microscope. Permanent slides can be made of the film by mounting it in Canada balsam and covering with a glass cover slip. Margalef ran repeat tests on "used" stones but found that the original treatment removed all the microörganisms from the surface of the rock.

Most sampling methods (Surber, Hess, drag-type, and the like) require time-consuming *hand picking* of individual organisms. This essential step may require so many man hours that adequate sampling becomes impractical. To obviate this, a *calcium chloride flotation method* has been employed with some success (Welch, 1948). The basis of the method is to alter the specific gravity of the liquid so that insects will float to the surface while rocks and trash remain at the bottom. A $CaCl_2$ solution with a specific gravity of 1.1 is placed in a bucket. Unsorted trash from a Surber sampler, for example, is emptied into the bucket, and the net is rinsed to dislodge clinging organisms. The trash is then mixed and swirled to help loosen insects from crevices. All but the heavy case-encumbered caddisworms and mollusks may be floated in this way.

Whether or not $CaCl_2$ is used, a circular motion of the hand will swirl the water in the bucket and float organisms free of the heavier debris. Then, while the water is still turning rapidly, it is poured through a 24-mesh sieve to concentrate the organisms. The solution should be poured through a rather small, single area on the sieve bottom to avoid spreading the organisms. The reason for this is that in the next step, when transfer of the organisms to a preservation jar is made, the solution is gently back-poured through the sieve while tilting it over the opening of a wide-mouth jar. If the organisms are concentrated close to the edge of the brass frame of the sieve in a small area, it is much easier to decant them into a jar than if they are spread all over the 8-inch sieve bottom. If they do become widely spread over the bottom, it is easy to concentrate them again by holding the sieve in an inch or so of solution and gently swirling it in a circular movement. A small, "pouring" bottle should be carried to flush the fluid back through the sieve. As a final precaution, the person sampling should carefully examine the remaining trash in the bucket to make sure that it is clean and does not contain any organisms. It is important that no sand, gravel, or small stones be decanted from the bucket into the sieve or from there into the preservation bottle because such materials may grind and break up the tender, soft-bodied insects and other invertebrates (P. R. Needham, unpubl.).

Once the sample has been transferred to a preservation jar, 70 per cent alcohol should be added and the sample properly labeled. Care should be taken to prevent dilution of the preservative with water already poured into the jar with the organisms. To avoid excessive dilution, once the organisms have settled to the bottom the surplus water may be decanted, leaving only about a quarter of an inch in the bottom. Then if the bottle is filled to the top with 70 per cent alcohol, dilution, though still occurring, is reduced to a minimum.

Later, in the laboratory, samples may be drained and dried on blotting paper for one minute and then weighed (wet weight). Final counting (total numbers) and sorting for identification may be done in shallow trays, and individuals of a given species may then be stored in vials as described below.

KILLING, PREPARING, AND STORING MATERIAL

It frequently happens that insects are collected with great enthusiasm and care and then left to deteriorate in a museum box or collecting bottle. This kind of neglect is wasteful of time and money and retards the development of science. The proper care of specimens is not a simple or invariable matter. Each group of organisms requires special handling, and collections are dealt with in different ways, depending on the purposes which they are expected to serve. For example, mosquito larvae collected on a malaria survey might be stored in the original field-collecting vials in 70 per cent ethyl alcohol. If rubber stoppers are used, such material may be held for several years in case further checking becomes necessary. More permanent preservation for museum purposes would require that the rubber stopper be replaced by a cotton plug, then the vial (or vials) can be inverted in a larger museum jar half filled with alcohol. This facilitates storage in an arranged collection and provides a ready means of checking against evaporation. Finally, such material (or representatives of each species) might be mounted in one of several media on microscope slides.

Slide mounting techniques.—Although many methods have been used for mounting small insects on slides, only a few will be described. Intro. figures 83 and 84 show a common method of preparation: 1) clear specimen in 10 per cent potassium or sodium hydroxide, either cold overnight or hot for 5-10 minutes (wet dried specimens in detergent solution before clearing); 2) transfer to small dish and remove body contents under microscope by means of small needles (intro. fig. 83); 3) place in acetic acid to neutralize alkali; 4) transfer to oil of cloves until specimen becomes transparent (15-60 minutes) (intro. fig. 84*a*); 5) rinse in xylol to remove oil (intro. fig. 84*b*); 6) transfer to drop of Canada balsam on glass slide (intro. fig. 84*c*); 7) arrange on slide with legs and antennae extended (intro. fig. 84*d*); 8) put on glass cover slip avoiding air bubbles (intro. fig. 84*e*); and 9) label with necessary information (intro. fig. 84*f*) and place slide in drying oven for 3-7 days.

The following method may be used for mounting larvae of large size from either fresh or alcohol preserved material: 1) slit specimens to ensure adequate penetration of reagents; 2) place in hot 10 per cent potassium hydroxide for 15 minutes; 3) transfer to water and remove body contents; 4) place in 70 per cent alcohol for 10 minutes, 80 per cent for 15 minutes, 95 per cent for 15 minutes; 5) stain in Gage's stain

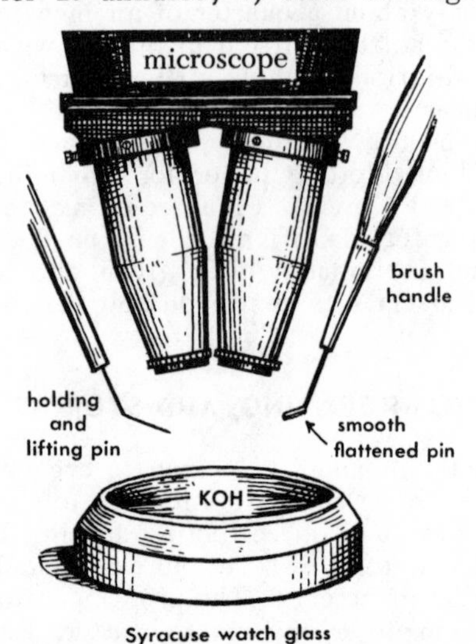

Intro. fig. 83. Equipment used in preparing small insects for slide mounting (Ross, 1953).

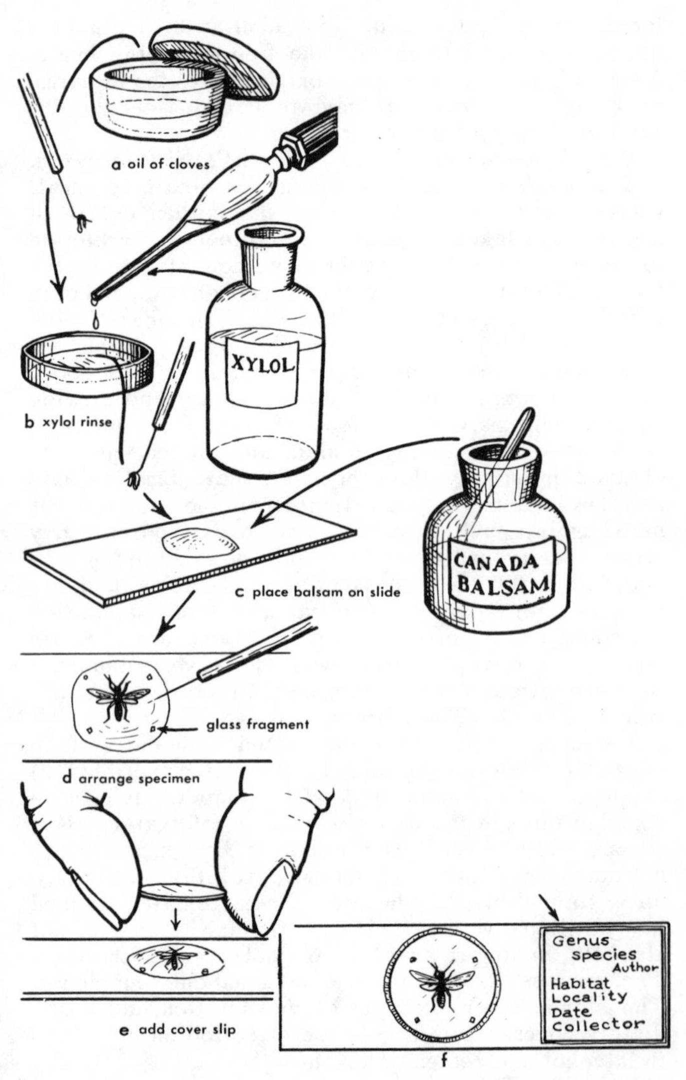

Intro. fig. 84. Steps used in mounting small insects on slides: *a*, clove oil for clearing; *b*, rinse in xylol; *c*, transfer to balsam on slide; *d*, orient specimen on slide; *e*, put cover slip over mount; *f*, label slide with necessary information (Ross, 1953).

for 3-5 minutes; 6) destain in 95 per cent alcohol; 7) rinse in Euparal solvent; 8) mount in Euparal (Verde); and 9) put on cover slip and place in drying oven for 3 days.

Carpenter and LaCasse (1955) recommend the following technique for mounting mosquito larvae: 1) remove from storage in 70-80 per cent ethyl alcohol; 2) slit body; 3) place in cellosolve (ethylene glycol monoethyl ether) for 10 minutes; and 4) mount in Canada balsam. The specimen should be oriented on the slide and the tip of the abdomen partly severed to enable the siphon to lie flat. Male genitalia of mosquitoes are mounted by means of the first method given.

Dried material is likewise susceptible to various levels of care and handling, depending on the purposes of the collector or collection. For an ecological study dry mounting should be limited to such forms as moths, mosquitoes, and a few other types of insects which are altered beyond recognition in alcohol, or which, by tradition, have always been mounted on pins or points. When collecting for an entomological museum an entirely different approach is necessary. In many groups, such as the moth flies (Psychodidae), adults should be collected and preserved both wet and dry, in split series. Parallel collections of this kind make it possible to study color pattern and pubescence on dry specimens and structural details on specimens in alcohol. Dried material is very fragile and requires constant care to guard against breakage and damage from museum pests. Collectors and administrators of collections should be prepared to care for scientific collections or should see that they are placed in a museum where they will be preserved and made available to others.

Alcohol vials.—The aquatic stages of nearly all insects are best collected and preserved in vials of 70 or 80 per cent ethyl alcohol. This fluid penetrates and preserves all but the most fleshy larvae and has none of the undesirable characteristics of formalin. As a convenience in handling and in storage only two or three sizes of vials should be used. Two-dram homeopathic (lipped) vials are very satisfactory for general collecting. Smaller vials of a diameter similar to dental procaine tubes came into general use during World War II and are now used extensively for collecting and storing very small insects. Stoppers should be of synthetic rubber, rather than cork, so as to prevent drying, shrinking, and evaporation. Alcohol should be changed within a few days after collecting because it becomes diluted with water and body fluids of the specimens. Large beetles, bugs, and Megaloptera larvae should not be placed in the same vials with delicate specimens. White, fleshy crane fly larvae and caterpillars should be killed quickly in boiling water before preserving in alcohol. This fixes the body proteins and prevents decomposition and blackening of specimens.

If a collecting vial contains more than one species, the material should be sorted in the laboratory to facilitate future arrangement in a collection. Sorting may be done in a white enamel pan, petri dish, or some other convenient flat container. After sorting, each vial should be labeled with full data as to locality, date, collector, and field notes or notebook number. Labels should be written in India ink on good grade bond paper.

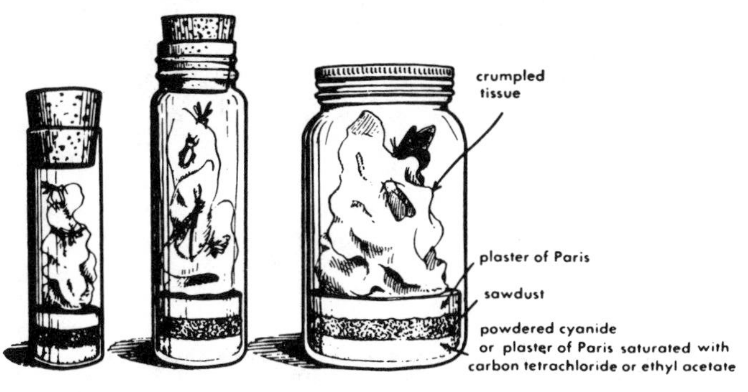

Intro. fig. 85. Details of cyanide bottles (Ross, 1953).

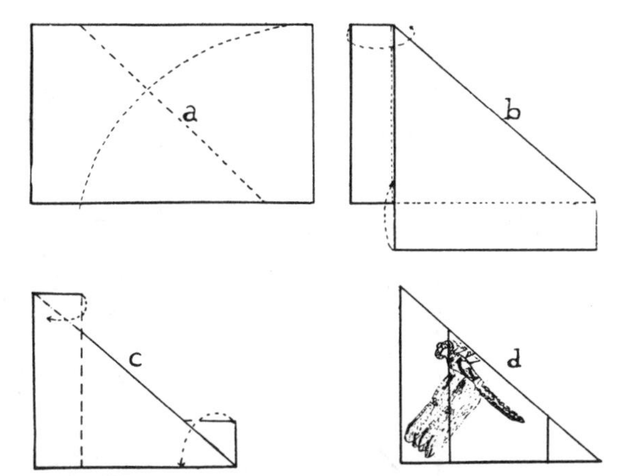

Intro. fig. 86. Technique of folding paper triangles for insect storage (DeLong and Davidson, 1946).

Killing bottles.—Most adult insects collected on land or in the air should be killed in a cyanide bottle or other dry container. Exceptions are the stoneflies, caddisflies, and mayflies which are better preserved in alcohol. A cyanide bottle (intro. fig. 85) contains crystals of calcium, sodium, or potassium cyanide covered by sawdust and held in place by a layer of plaster of Paris. A small amount of moisture is sometimes necessary to activate the cyanide, but usually the problem is too much rather than too little water. Absorbent tissue should be placed in the bottle to remove excess moisture and to prevent fragile insects from damaging each other. Cork stoppers should be used rather than screw tops because the latter are inconvenient or difficult to manipulate while collecting. The lips and bottoms of cyanide bottles should be taped to reduce the chances of breakage or air-tight plastic bottles may be used. *Cyanide is, of course, a deadly poison and must be used with great care.* If a bottle is broken it should

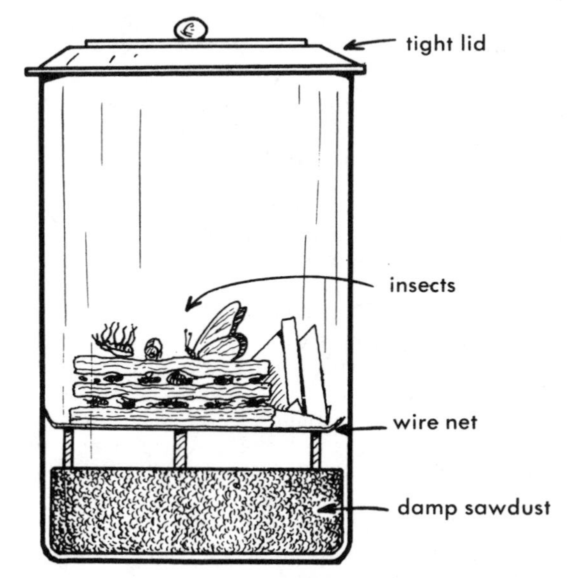

Intro. fig. 87. Relaxing jar used to soften insects for mounting (Ross, 1953).

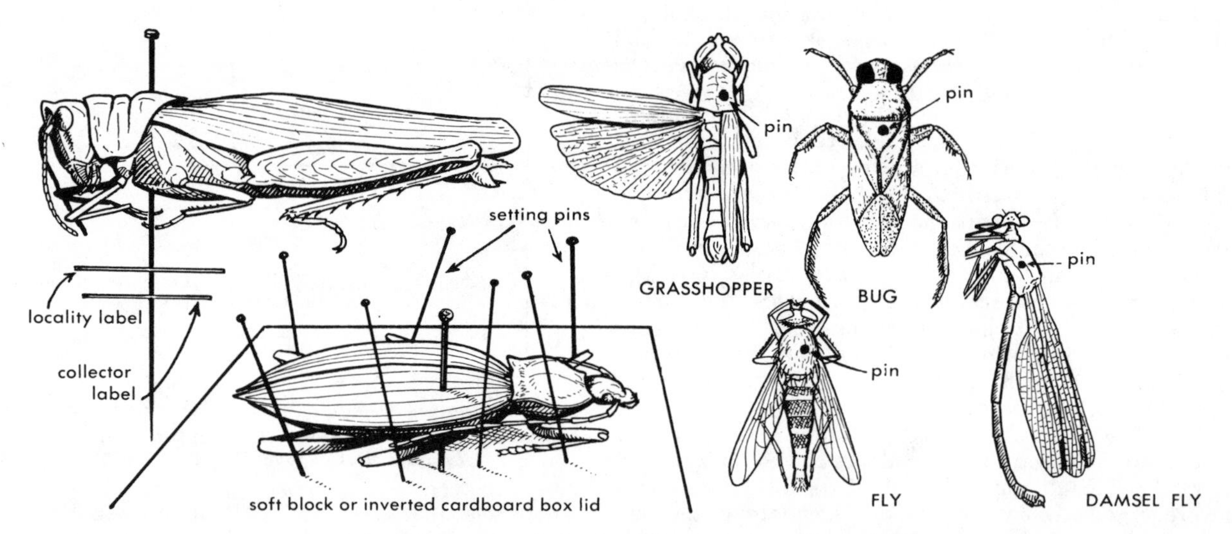

Intro. fig. 88. Position of pins for different insects (Ross, 1953).

be buried deep in the soil and away from human habitations.

Safer killing bottles can be made by using ethyl acetate, carbon tetrachloride, or chloroform as the toxic agent. These materials may be poured onto a matrix of plaster of Paris or tightly packed paper and the bottle may be recharged as needed.

Specimens should be removed from killing bottles as soon as possible and not later than the end of the day of collection. Delicate insects such as mosquitoes should be transferred to small pill boxes while still in the field. Larger insects such as beetles or scaly insects such as moths should never be placed in the same bottles with fragile forms.

Mounting and storage of material.—Insects killed in the field should be mounted on pins or points while still fresh and soft. However, this is sometimes impractical on extended field trips. If specimens are to be held unmounted for more than a few hours they should be placed between layers of cellucotton in cardboard or wooden boxes (cigar boxes are satisfactory) or between cellucotton or soft tissue paper in pill boxes. Dragonflies are best handled in paper triangles (intro. fig. 86). Crystals of paradichlorobenzene or naphthalene may be added to repel dermestid beetles and other museum pests. In the tropics, silica gel or some other dehydrating agent should be used in airtight containers to prevent formation of mold.

In the laboratory, dried material is softened in a "relaxing jar" (intro. fig. 87) before mounting. This jar should have a wide mouth to facilitate handling of specimens. A layer of moist sand is placed on the bottom, and a small amount of carbolic acid is added to prevent the growth of fungus. A piece of hardware cloth is bent down at the sides and placed on the sand to prevent the specimens from coming in direct contact with water. The specimens are placed on the screen and the top is closed tightly. Within twenty-four hours the specimens will be soft.

Fresh or "relaxed" specimens are usually mounted on pins that are especially designed for the purpose. These may be obtained from most biological supply houses. They are longer and thinner than common pins and are black with gold heads. No. 2 pins are

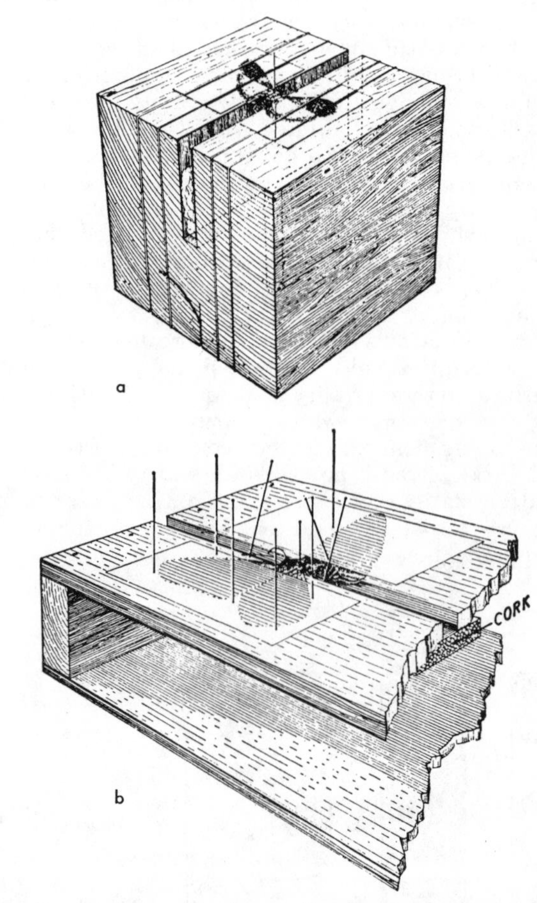

Intro. fig. 89. Two types of spreading boards showing technique of preparation; *a*, block for very small insects; *b*, typical board for larger Lepidoptera and Odonata (Oman and Cushman, 1946).

used for small insects, but larger insects may be mounted on the stiffer No. 3 pins. As a general rule, beetles are pinned through the right wing cover, true bugs through the scutellum, and all other insects through the main part of the thorax slightly off-center (intro. fig. 88).

Lepidoptera and some other insects with important venational characters should be mounted on a spreading board (intro. fig. 89). After pinning through the thorax, the relaxed wings are spread out and held in place by strips of paper until setting is complete.

Specimens which are too small for No. 2 pins should be mounted on *Minuten nadeln* pins (intro. fig. 90) or glued to paper points (intro. fig. 91). These points are small paper triangles punched or cut from thin cardboard or stiff drawing paper. The tip of the point is bent downward to match the side of the thorax of a particular insect. A drop of glue or other adhesive (shellac, colorless nail polish, or the like) is placed at the tip and the specimen is then fastened at its side. Lepage's iron glue is very satisfactory since it is water soluble, permitting the specimen to be removed easily for closer examination.

Locality labels are printed by hand or press and impaled on the pin below the insect. The precise locality, county when necessary, and state or country are given on the label, and also the date of collection and name of collector. Additional information as to habitat, host plants, and so on is desirable but may be given by means of a date or numbering cross index system to field notebooks.

Insect-proof boxes or trays, obtainable at a few of the larger biological supply houses, should be used for storage and for arrangement of the systematic collections (intro. fig. 92). Individual units of cardboard facilitate handling, sorting, and arranging large collections. A good grade of cork bottom is desirable as a pinning surface in all boxes and trays, to prevent corrosion of the pins. However, pressed wood-fiber boards may be used and even flat strips of balsa wood. Insect collections should be inspected from time to time for possible damage by museum pests. Paradichlorobenzene is a satisfactory repellent.

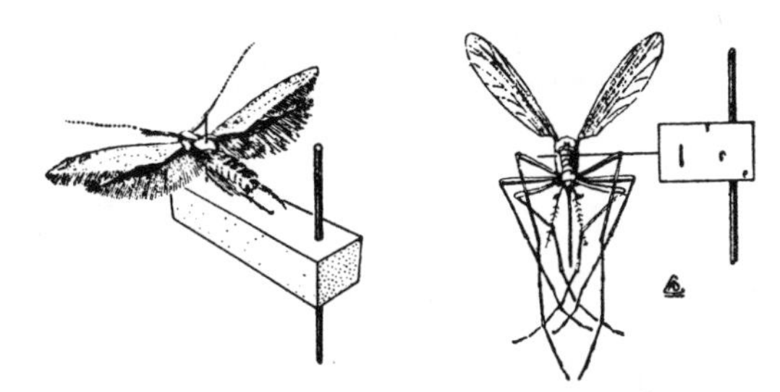

Intro. fig. 90. Details of mounting small insects on *minuten nadeln* pins (Oman and Cushman, 1946).

REARING METHODS

Since most aquatic insects are in the immature stage, they must be reared in order to make positive identification. Usually it is sufficient merely to carry last instar larvae or nymphs through to the adult stage, saving cast skins to associate stages. In some cases, however, it is desirable to work out complete life histories. The techniques and equipment for life history studies vary for each group of organisms and even for each species so only the most generally used methods are described here.

Rearing of pupae in the field.—The pupae of black flies and some other aquatic insects may be reared by carefully removing them from objects in the water and placing them on moist cotton or blotting paper in a cotton-stoppered vial. Excessive water and high temperatures should be avoided.

Transporting live specimens.—The greatest mortality occurs when live specimens are carried from the field to the laboratory. This is especially true of stream forms. A bucket of water including some bottom material may suffice for hardier specimens, such as dragonfly nymphs. However, it usually happens that water splashes in an automobile, temperatures rise if the car is closed and stands in the direct sun, and the supply of dissolved oxygen is depleted. These problems may be solved by carrying aquatic insects not in water but in moist moss and other aquatic vegetation. Temperatures are reduced by evaporation

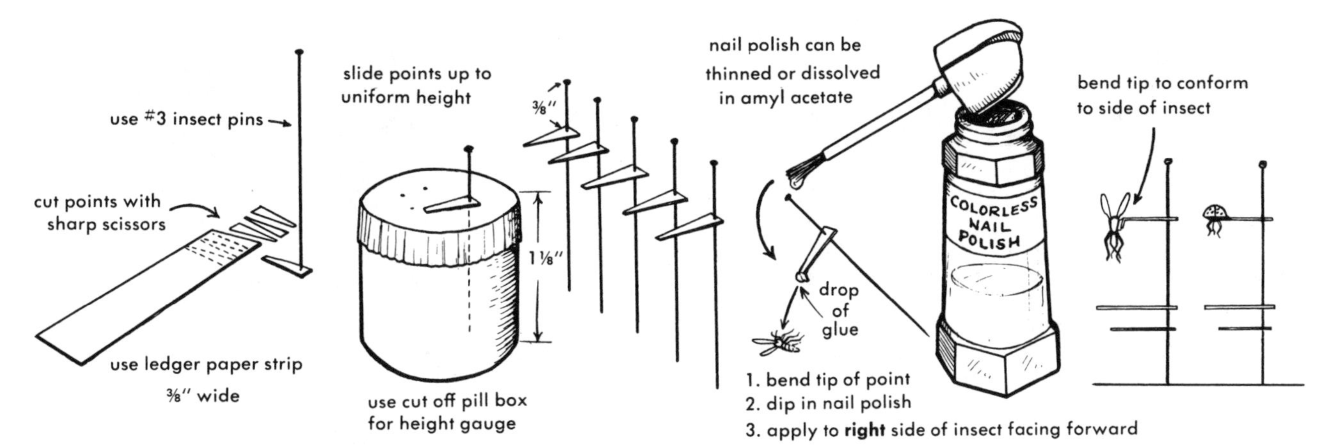

Intro. fig. 91. Paper point mounting of small insects (Ross, 1953).

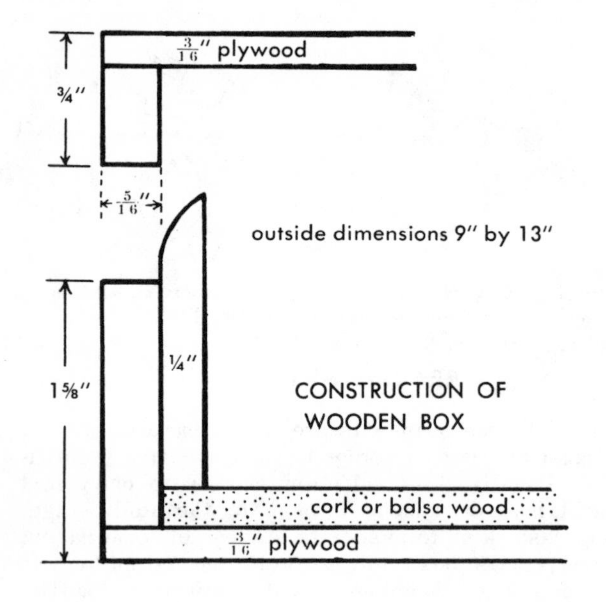

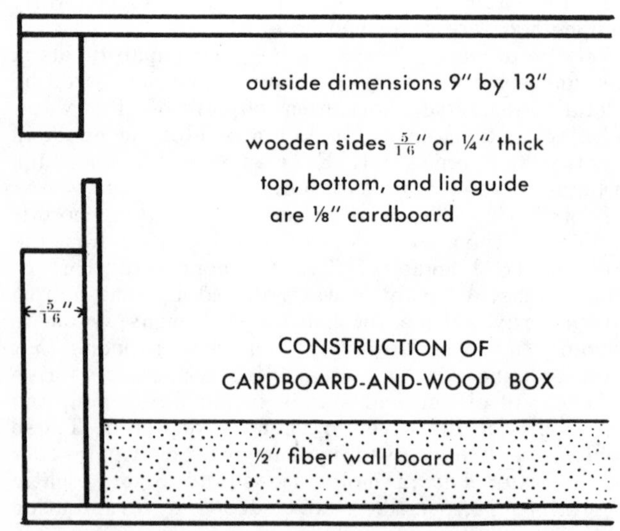

Intro. fig. 92. Construction details of boxes for insect storage (Ross, 1953).

or may be held very low by dry ice or by various coolers that are now available for motorists, or travel may be arranged during the cool hours of the night and early morning.

Field cages.—Various cages have been designed for rearing insects in the field. The simplest of these is the "pillow cage" (intro. fig. 93) made of screen and anchored in a stream with the lower part below water and the upper part dry. Specimens are placed in these cages to develop under conditions that are in some respects identical to the natural habitat. Nymphs of stoneflies, dragonflies, mayflies, and caddis larvae have been reared successfully in this way. Another type of cage consists of a cylinder of window screen with the bottom made of the same material. A cloth sleeve is attached to the top. The sleeve is closed with a rubber band or piece of string, but is readily opened to remove specimens that have emerged. These cages are fastened to stakes driven into the bottom of streams or ponds, and a few clean stones are placed in the cage to simulate natural conditions. Though there is no size limit to this type, small cages are particularly useful for rearing small numbers of specimens of a particular species. Many such cages can be carried in the trunk of a car.

Cages for rearing adults from immature stages may be designed like the "emergence traps" described earlier.

Laboratory rearing.—Most insects are difficult to rear when removed from their natural environment. Success can be attained only by paying close attention to the temperature of the water, dissolved oxygen, toxic substances, and seemingly endless details. The size and shape of the aquarium may be important. The kind of substrate should be as natural as possible, and the water should be from a natural source, or, if from a domestic water supply, should be dechlorinated by filtering through activated charcoal or by allowing it to stand for several days. The temperature should be maintained as nearly as possible like that of the natural habitat for most temperate zone insects. This means cooling by means of a cold-water bath in which the aquarium is set or by means of a pumping system with the water line passing through a refrigerator or cold water bath until the desired temperature is attained.

Aeration is necessary for many insects, including all stream forms. Fortunately various air pumps are available for this purpose and may be obtained at low cost from aquarium supply stores. Air is pumped through a block of pumice and forms bubbles that oxygenate the water and also agitate it to a certain extent.

Most stream insects require running water to complete their development. Running water troughs, as found in fish hatcheries, may be used for this purpose but most laboratories are not equipped with such facilities. For more limited operations a simple circulating system can be set up. Water is brought

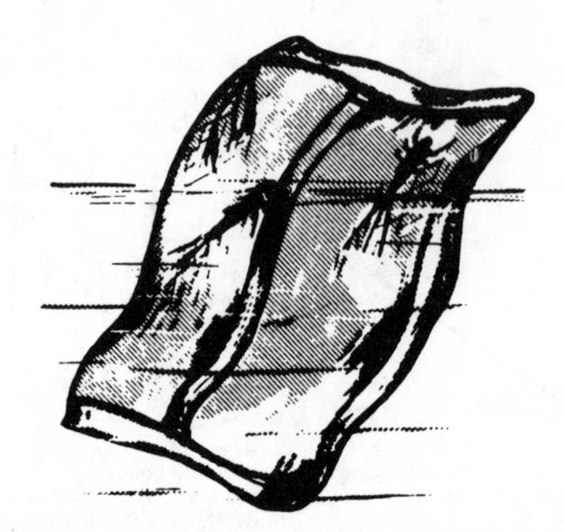

Intro. fig. 93. Simple "pillow" cage used for rearing insects in the field (Ward's Nat. Sci. Est., 1940).

in to a large aquarium, preferably from a natural source, and is circulated by means of a small pump. A pump with stainless steel and plastic fittings is needed, and plastic tubing should be used to avoid possible toxic effects from metal surfaces. The current can be adjusted by the speed of the pump, and microhabitats are created by stones and other bottom materials from the natural habitat or by plastic baffles placed at strategic points.

Food is of course an important consideration in rearing aquatic insects. The readily available aquarium plants may be used for plant feeders and will contribute to the conditioning of the water or may even result in a "balanced aquarium." Predatory insects must be supplied with particular kinds of living food. This can become a major problem but here again the aquatic entomologist is indebted to suppliers of equipment and materials for home aquaria. Brine shrimp may be purchased at small cost or may be reared from eggs that are obtainable commercially. Other live food material that can be maintained in culture includes mosquito larvae (especially *Aedes aegypti* in areas where it occurs naturally), house flies, *Drosophila* flies, meal worms, and water fleas (*Daphnia*). Techniques for rearing these are relatively simple (see Needham, *et al.*, 1937). On a more limited scale, nonpredaceous insects such as mosquito larvae and some mayfly nymphs can be collected in the field and added to the aquarium as needed for food.

The running water habitat poses special problems in food supply. Most stream insects are grazers on the periphyton that coats rocks and other objects in the water. Small rocks should be moved from the natural habitat to supply this type of food and fresh periphyton-covered surfaces may have to be supplied at intervals.

FISH COLLECTING AND EXAMINATION OF STOMACH CONTENTS

When conducting an aquatic survey, it is desirable to know what the fish are eating and to correlate the percentage of the various insects in the stomachs of the fish with the percentage in the habitat (Hess and Swartz, 1941). Frequently one insect species may be in great abundance, but the fish stomachs reveal that some other species is the dominant one selected by the fish. To obtain such information, fish are collected and the stomachs removed. As soon as possible after collecting, the stomachs are placed in vials of alcohol or 5 per cent formalin. Each specimen should be labeled, giving the species of fish, locality, date, collector, and any other pertinent information. The stomachs are usually examined later in the laboratory, where they can be opened with care and where the contents can be sorted under a dissecting microscope. Identification is difficult and sometimes impossible because many of the specimens are represented only by a few fragments. The work is considerably easier when an adequate collection has been made of the insects in the environment. This permits a correlation to be made between the fragments in the stomachs and the intact specimens collected in the field.

Fish collecting methods.—One of the commonest methods of "collecting" fish is, of course, by hook and line. Students of fish foods sometimes find this a convenient way to combine business and pleasure. Other methods which are more likely to produce quantities of fish for scientific study include seining, poisoning, shocking, and draining. These mass collecting methods are illegal for game fish in most states but special permits may be obtained for scientific studies.

Seining is a common method of collecting fish for scientific purposes. The nets used vary in size and mesh, depending on the type of fish being collected. *Gill nets* are used when it is possible to remain in an area for several hours or overnight. They are designed to allow the fish to pass through the mesh as far as the gills. Once in the net, the fish tries to back out and becomes entangled. A net of several sized meshes is used to ensure catching most types of fish present. Fish caught in gill nets are less satisfactory than those collected by other methods because the stomach contents are more completely digested and hence more difficult to identify. The *fish poison, rotenone,* is useful for collecting in small ponds, slow streams, and protected bays in larger bodies of water. The poison constricts the capillaries in the gills, preventing the flow of blood in this region and thus greatly reducing the exchange of oxygen and carbon dioxide. This causes the fish to rise to the surface to get more air where they are collected by means of a hand net. The *electric shocker* is used to collect and tag fish in streams and small ponds. A battery or generator supplies the power. Shocking temporarily stuns the fish, permitting them to be collected. The *pump and drain method,* when feasible, is perhaps the best method of determining the exact number of fish in a given area of stream. A section of the stream is blocked off and the water diverted around it. The section to be sampled is then pumped out and fish are collected, counted, weighed, and utilized for other scientific purposes, after which they may be returned to the restored stream bed. Smaller fish, usually overlooked or missed by other methods, are taken in great numbers by pumping and draining. Small ponds can also be pumped dry.

MEASUREMENT OF PHYSICAL AND CHEMICAL FACTORS

In addition to biological collections, it is often necessary to measure and record certain physical and chemical features of the aquatic environment. The following is a brief summary of the equipment and techniques used to obtain the minimum data that should be taken on any survey. This section is taken largely from Davis (1938).

Physical Factors

Temperature.—A simple laboratory thermometer, properly calibrated, will serve to record air and water surface temperatures. If a series of temperatures is required from various depths in a pond or lake, then some other means must be used. The *reversing the, mometer* may be used in recording temperature series in a lake. The thermometer is lowered to the desired depth and allowed to remain there one or two minutes to permit the instrument to adjust to the temperature of that level. Then a metal messenger is sent down the line; it trips the mechanism and the thermometer swings upside down, breaking the column of mercury at the recorded temperature. The instrument is pulled to the surface, read, reset, and returned to the next level. A *maximum-minimum thermometer* may also be used at various levels in lakes. The instrument is set, lowered to the desired depth, and allowed to remain in position five minutes; then it is pulled back to the surface and read. Still another method is to read temperatures from the thermometer that is built into the Kemmerer water sampler. Unfortunately, these methods require considerable time in order to take a complete series of temperatures. The *electric-resistance thermometer* is more efficient and saves a great deal of time. It is based on the principle that different metals have different resistance at different temperatures. It consists of a box containing batteries, galvanometer, Wheatstone bridge dial, and a switch. A cable is calibrated in feet and has a terminal bulb containing temperature sensitive ends. The device is lowered into the water, the switch closed, and the temperature read on the Wheatstone bridge dial in the box. The advantages of this type of recorder are that the adjustment interval for each level is very short, and it is not necessary to bring the instrument back to the surface after each temperature has been read. Although the initial expense is higher than with the other two types mentioned, the increased efficiency and greater accuracy offset the cost.

Relative humidity.—The *sling psychrometer* is used to obtain relative humidity readings. The wet bulb is moistened with water and the device swung in circles for three to five minutes. The two temperatures are read, one from the dry bulb and the other from the wet bulb. A calibrated chart will then give the relative humidity.

Current velocity.—The speed of current is determined most efficiently by means of a *current meter*. This consists of a propeller mounted on a finlike or rudderlike metal base that serves for orientation in the stream. An electric sounding device records a click for each revolution of the propeller, and the number of clicks per minute as detected with earphones is recorded. Measurements should be made at various depths and in a transect across a stream because the current speed is greatest toward the center and at intermediate depths.

Stream velocity may also be determined in a rough way by means of a *surface float*. The float is placed in the water and timed as it is carried over a known distance. By this means the velocity is determined in feet per second. Since the surface speed is generally faster than the average channel speed, it is necessary to divide the above mentioned figure by 1.05 for channels less than 2 feet deep and by 1.33 for channels 10 or more feet in depth. A *subsurface float* can be made that will give a reasonably accurate recording of the channel speed. The float is made of a small plastic bottle with a screw top that tapers to a small hole. By squeezing the bottle, air is forced out and water can be drawn in. With a little practice, the proper proportions of air and water will be attained so that the bottle will float beneath the surface. The smaller the bottle, the less resistance is offered. Plastic bottles will not break and the narrow-holed top permits small amounts of water or air to be added. The volume of flow is given in feet per second or, in small streams, in gallons per minute. The flow is obtained by multiplying the average velocity in feet per second by the cross section area of the stream in square feet. The cross section is obtained by multiplying the width of the stream by the average depth. The average depth is obtained by measuring the depth of water at uniform intervals across the stream and dividing the sum of these depths by the number of measurements plus one.

Light.—On land, the intensity of light is measured by a *photoelectric cell* or light meter. For measuring the light beneath the water, the same instrument can be used except that it must be connected to a galvanometer by means of a long wire. A simple device, known as a *Secchi disc,* is often used to measure the relative penetration of light. The Secchi disc is a white, round plate eight inches in diameter divided into four alternating white and black quarters. The plate is lowered until it disappears, dropped a few feet further, and raised until it appears again. The average is taken of the two depths. Although giving only relative results, the Secchi disc is a convenient instrument and has been used by limnologists for years. It serves in a general way to indicate the limits of effective light penetration for photosynthesis.

Chemical Factors

Water sampling.—Water for chemical analysis is usually collected from a stream in 500 cc. glass-stoppered bottles. Care should be taken when filling the bottles that the water is not disturbed too much since that might alter the gaseous content. Water to be analyzed for carbon dioxide should be collected in a *low-form Nessler tube* (200 × 32 mm.). Water samples taken from ponds and lakes are collected by means of a *Kemmerer water sampler* (intro. fig. 94). The samples are taken at the surface, at or near the thermocline, and at the bottom. In deep lakes, it may be necessary to take a longer series of water samples in order to have a complete picture of the distribution of the various gases through the water levels. The open sampler is lowered to the desired depth on a graduated rope, then a metal messenger is sent down to trip the closing mechanism, shutting the upper and lower valves. Two or three minutes are allowed for the thermometer to adjust to the water temperature. Then the instrument is pulled to the surface, the thermometer read, and the water drained into a water sample bottle by means of the rubber hose on the bottom valve. Water should be permitted to overflow the sample bottle until the contents have been changed twice.

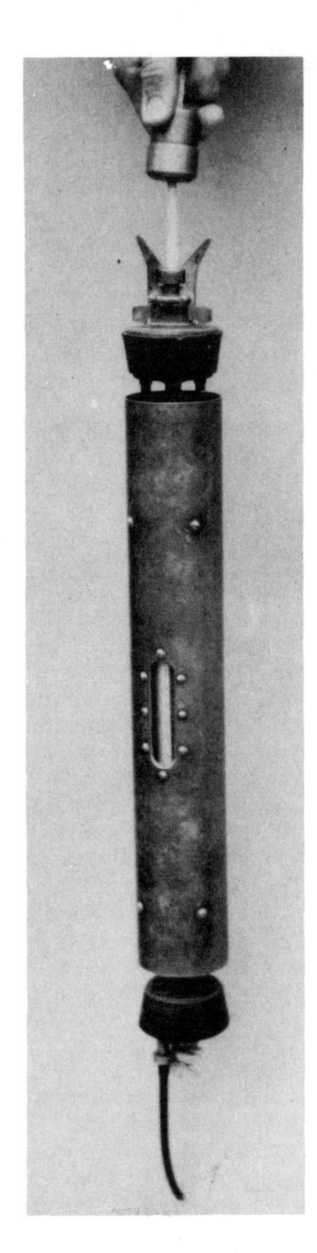

Intro. fig. 94. Kemmerer water sampler. Note metal messenger at top of photo.

Dissolved oxygen determination.—1) To the sample collected in the 250–300 cc. ground-glass stoppered bottle, add 1 cc. of manganous sulfate and 3 cc. of alkaline potassium iodide beneath the surface of the liquid. 2) Replace the stopper, at the same time preventing the entrapping of air bubbles. Shake vigorously for fifteen seconds and allow the precipitate to settle until it is contained in the bottom half of the bottle. If this does not occur quite rapidly the shaking should be repeated. 3) When the precipitate has settled, add 2 cc. of concentrated sulfuric acid above the water level, holding the tip of the pipette against the neck of the bottle. Replace the stopper with precautions against trapping air bubbles and shake the bottle vigorously to mix the contents. Up to this point the procedure must be carried out in the field. The final titration should be made as soon as possible thereafter, but an elapse of a few hours is permissible. 4) Transfer 200 cc. of the treated sample, measured with a 200 ml. volumetric flask, into a 500 cc. Erlenmeyer flask. This is conveniently accomplished by carrying into the field a 200 cc. flask cut off a few millimeters above the mark and inserted into a pierced stopper which fits the neck of the larger flask. By inserting the mouth of the smaller flask into the larger flask and reversing the two with a slight rotary motion the sample will transfer rapidly. 5) Titrate the sample with 0.025 N sodium thiosulfate. One or two cc. of starch solution should be added only when the color has become a faint yellow after the addition of thiosulfate. Titrate over a white background until the blue color disappears. Dissolved oxygen is reported in parts per million by weight. If a 200 cc. sample is used the dissolved oxygen in parts per million is equal to the number of cc. of 0.025 N thiosulfate required. No corrections are necessary except for work of unusual precision. Since the 0.025 N solution of thiosulfate is unstable, this solution should be restandardized or replaced occasionally. Starch solutions, even if preservatives such as chloroform or zinc chloride are added, deteriorate quite rapidly, especially in warm weather. A satisfactory method is to use sterilized solutions in small bottles which are opened as required.

Carbon dioxide determination.—Collect 100 cc. of the sample in a low-form Nessler tube (200 × 32 mm.) according to one of the methods mentioned above. Add 10 drops of phenolphthalein and titrate rapidly with N/44 sodium hydroxide until a faint but permanent pink is produced (3 minutes). The sample may be mixed by swinging with a circular motion of the wrist. Any agitation of the surface of the liquid tends to change the gaseous content and should therefore be avoided. The free carbon dioxide is equal to ten times the number of cubic centimeters of N/44 sodium hydroxide used. Since N/44 hydroxide or stronger solutions deteriorate once the bottle is opened, this solution should be restandardized, or a fresh one obtained at frequent intervals. It is convenient to have this reagent supplied in small, tightly corked containers.

Alkalinity.—The alkalinity of natural waters represents its content of carbonates, bicarbonates, hydroxides, and occasionally borates, silicates, and phosphates. It is determined by titration with a standard solution of strong acid to certain datum points or hydrogen-ion concentrations. Indicators are selected which show definite color changes at these points. Since dilute bicarbonate solutions have a hydrogen-ion concentration of about pH 8.0 and dilute carbonic acid solutions a hydrogen-ion concentration of about pH 4.0, these are chosen as datum points, and indicators should be selected which show definite color changes at about these two points. The amount of standard acid required to bring the water to the first point measures the hydroxides plus one-half the normal carbonates (phenolphthalein), and the amount to bring it to the second point corresponds to the total akalinity (methyl orange). The sample may be collected by any method since there is no great danger of changing the alkalinity. Both end points may be determined on the same sample by using phenolphthalein first and continuing with methyl orange after the first end point has been determined. 1) *Phenolphthalein alkalinity:* add four drops of phenolphthalein indicator to 50 or 100 ml. of the sample in a white porcelain casserole or an Erlenmeyer flask over a white surface. If the solution becomes colored, hydroxide or normal carbonate is present. Add 0.02 N sulfuric acid from a burette until the coloration disappears. The phenolphthalein alkalinity in parts per million of calcium carbonate is equal to the number of milliliters of 0.02 N acid multiplied by 20 if 50 ml. of sample was used, or by 10 if 100 ml. was used. 2) *Methyl orange alkalinity:* add two drops of methyl orange indicator to 50 or 100 ml. of the sample or to the solution to which phenolphthalein has been added, in a white porcelain casserole or an Erlenmeyer flask over a white surface. If the solution becomes yellow, hydroxide, normal carbonate, or bicarbonate is present. Add 0.02 N sulfuric acid until the faintest pink coloration appears, that is, until the color of the solution is no longer pure yellow. The methyl orange alkalinity in parts per million of calcium carbonate is equal to the total number of milliliters of 0.02 N sulfuric acid used multiplied by 20 if 50 ml. of sample was used, or by 10 if 100 ml. was used.

Hydrogen-ion concentration.—The pH, or hydrogen-ion concentration, is a measure of the degree of acidity or alkalinity of a solution. A pH from 1-7 is acid, 7 is neutral, 7-14 is alkaline. The pH of a water sample is taken by adding a few drops of an indicator to the water and comparing the color to a standard color chart of pH values. One method is to use 0.5 ml. of "Universal pH Indicator" in a 10 ml. sample of water in a glass tube 16 mm. in diameter. The color is checked against a Harleco color chart and the pH read off. It is also possible to use indicators of known pH range. Here the color is checked for exact pH against a set of color tubes of known pH color value. The following are indicators with their pH ranges given:

Indicator	pH	Color Change
Bromphenol blue	3.0-4.6	yellow-blue
Bromcresol green	4.0-5.6	yellow-blue
Chlorphenol red	5.2-6.8	yellow-red
Bromthymol blue	6.0-7.6	yellow-blue
Phenol red	6.8-8.4	yellow-red
Thymol blue	8.0-9.6	yellow-blue

MAPPING

Choose two points, A and B, as widely separated as is convenient, 100 to 300 yards, near the shore, and from which most of the lake or stream section can be seen. Measure the straight line distance between these points. Set up the plane table over one end of the base line with its surface level and one side oriented on a north-south line.

Mark a convenient point on the paper to represent the point over which the paper is set and label this point A. Using point A as a center, sight with an alidade or rule to the other end of the base line and draw a fine line toward point B. With the known distance between points A and B, determine a suitable scale, preferably one listed below, and mark off on the AB line the distance, according to scale, between A and B.

From point A, sight around the shore line from left to right, bringing the alidade to bear on each major point and draw a line from A toward each point. Label these lines 1, 2, 3, and so on, in order, and record to what conspicuous object on the shore line each number refers. Complete the circuit of the body of water in this manner and then move the plane table to point B. Reorient the table at point B and check by taking a backsight from point B to point A, making sure that the line between A and B corresponds to the edge of the alidade when point A is viewed from B.

From point B sight around the shore as before to the various objects which were viewed from point A, draw a line toward each from point B and number each line to correspond with the lines from point A. The point at which the lines intersect, when extended, is the location on the map of the points selected.

If the lake or stream is small and the shore line entirely visible from the two points, two stations will be enough. If part of the area is not visible, it will be necessary to proceed to some point already located by the intersection of two lines and orient the plane table to correspond, and then set up another base line to map the area that was not visible from the first two stations.

After all the shore line points have been located by intersections, the details of the shore line between the points may be filled in by careful freehand sketching.

Determination of area.—To determine the area of a mapped lake, lay over the completed map a semi-transparent sheet of coördinate paper ruled to one inch squares. Count the number of whole squares within the limits of the lake and carefully estimate the inner area of the squares cut by the shore line. Add the two together and record as square inches of lake area. The following table gives the number of acres per square inch of map area for various scales.

Acres per square inch of map area

Scale (feet per inch)	Acres (per square inch)
50	0.057
100	.230
200	.92
350	2.81
500	5.74

REFERENCES

ADAMSTONE, F. B., and W. J. K. HARKNESS
1923. The bottom organisms of lake Nipigon. Univ. Toronto Stud., Biol. Ser., 22:123-170, illus.
BRUNDIN, LARS
1949. Chironomiden und andere bodentiere der südschwedischen urgebirgsseen. Institute of Freshwater Research, Drottningholm. Report No. 30, Fishery Board of Sweden. v + 914 pp., illus.
CARPENTER, STANLEY J., and WALTER J. LACASSE
1955. Mosquitoes of North America. University of California Press. vi + 360 pp. + 127 pls.
DAVIS, H. S.
1938. Instructions for conducting stream and lake surveys. U.S. Dept. Comm., Bur. Fisheries, Fishery Cir. No. 26, 1-55, illus.
DELONG, DWIGHT M., and RALPH H. DAVIDSON
1936. Methods of collecting and preserving insects. Ohio State University Press. 20 pp., illus.
GUYER, G., and R. HUTSON
1955. A comparison of sampling techniques utilized in an ecological study of aquatic insects. Jour. Econ. Ent., 48:662-665.
HESS, A. D.
1941. New limnological sampling equipment. Limnol. Soc. Amer., Spec. Publ. No. 6:1-5, illus.
HESS, A. D., and ALBERT SWARTZ
1941. The forage ratio and its use in determining the food grade of streams. Trans. Fifth N. A. Wildlife Conference, 1940. pp. 162-164.
IDE, F. P.
1940. Quantitative determination of the insect fauna of rapid water. Univ. Toronto Stud., Biol. Ser., 47:1-20, 4 pls.
LAGLER, KARL F.
1952. Freshwater fishery biology. Dubuque, Iowa: W. C. Brown Co. x + 360 pp., illus.
Malaria control on impounded water.
1947. Supt. Doc., Washington, D.C. xiii + 590 pp., illus.
MARGALEF, R.
1949. A new limnological method for the investigation of thin-layered epilithic communities. Hydrobiologia, 1:215-216.
NEEDHAM, JAMES G.
1908. Report of the entomological field station conducted at Old Forge, N.Y., in the summer of 1905. N.Y. State Mus. Bull. No. 124:167-168.
1924. Observations of the life of the ponds at the head of Laguna Canyon. Pomona Coll. Jour. Ent. Zool., 16:1-12, illus.
1937. (ed.) Culture methods for invertebrate animals. Ithaca, N.Y.: Comstock Publ. Co. xxxii + 590 pp., illus.
NEEDHAM, P. R.
1928. A net for the capture of stream drift organisms. Ecology, 9:339-342.
OMAN, P. W., and ARTHUR D. CUSHMAN
1946. Collection and preservation of insects. U.S.D.A. Misc. Publ. No. 601, Washington, D.C. 42 pp., illus.
PETERSON, ALVAH
1953. A manual of entomological techniques. Ann Arbor, Michigan. v + 367 pp., illus.
ROSS, EDWARD S.
1953. Insects close up. University of California Press. 81 pp., illus.
SPRULES, WM. M.
1947. An ecological investigation of stream insects in Algonquin Park, Ontario. Univ. Toronto Stud., Biol. Ser. No. 56, Publications of the Ontario Fisheries Research Laboratory, No. 69. vi + 81 pp., illus.
TRAVER, JAY R.
1940. Compendium of entomological methods. Part I, Collecting mayflies (Ephemeroptera). Rochester, N.Y.: Ward's Natural Science Establishment. 7 pp., illus.
USINGER, R. L., IRA LA RIVERS, H. P. CHANDLER, and W. W. WIRTH
1948. Biology of aquatic and littoral insects. Univ. Calif. Syllabus Ser., 1-244.
USINGER, R. L., and W. R. KELLEN
1955. The role of insects in sewage disposal beds. Hilgardia, 23:263-321, illus.
USINGER, R. L., and P. R. NEEDHAM
1956. A drag-type riffle-bottom sampler. Progr. Fish-Cult., 18:42-44.
Ward's Natural Science Establishment, Rochester, N.Y.
1940. How to make an insect collection. 33 pp., illus.
1949. Equipment and supplies. Biological, Entomological. Catalog 495. 64 pp., illus.
WELCH, PAUL S.
1935. Limnology. New York: McGraw-Hill. xiv + 471 pp., illus.
1948. Limnological Methods. Philadelphia, Pa.: Blakiston. xviii + 381 pp., illus.
WENE, G. and E. L. WICKLIFF
1940. Modification of a stream bottom and its effect on the insect fauna. Canad. Ent., 72:131-135.
WENZL, H.
1940. Die Untersuchung epiphytische Pilze nach dem Abdruckverfahren. Chronica Botanica, 6:103.
1941. Die Bestimmung des Spaltöffnungszustandes mittels Zelloidinabdrücken. Chronica Botanica, 6:250.

CHAPTER 1

Structure and Classification

By W. C. Bentinck
University of California, Berkeley

Arthropods differ from other animals by a combination of characters. They have a hardened or sclerotized exoskeleton secreted by the epidermis and molted at intervals. The body is segmented externally in varying degree, and typically each somite bears a pair of jointed appendages though these have been lost from some somites of various forms. The circulatory system is open with a dorsal vessel or heart, distributing blood to organs and tissues, whence it returns through body spaces to the heart. Respiration is accomplished by means of tracheae, gills, lung books, or by diffusion directly through the integument. The nervous system is characterized by the paired ventral nerve cord and dorsal brain. Reproduction is usually bisexual, and most arthropods are oviparous.

Insects may be distinguished from the other classes of Arthropoda by the following characteristics of the adults. The body is divided into three distinct regions: the head, thorax, and abdomen. The head bears one pair of antennae and mouth parts for chewing, sucking, or lapping. The thorax typically consists of three somites, each bearing a pair of jointed legs, and the posterior two somites usually each bearing also a pair of wings. The abdomen is comprised of eleven somites or less with the terminal parts modified as genitalia.

The basic insect pattern has been modified in an almost endless variety of ways throughout the course of evolution. Hence there are numerous exceptions to some of the generalizations mentioned above, especially among immature stages.

For the benefit of the beginning student, there follows a brief discussion of the basic features of insect anatomy. Most of the structures and regions to be mentioned are illustrated in figures 1:1, 1:2, 1:3, and 1:4.

EXTERNAL ANATOMY

The exoskeleton is made up of numerous hardened or sclerotized plates or regions called sclerites. These are joined together by pliable membranes which lend flexibility to the total structure. The seamlike furrows or narrow membranous areas separating sclerites are called sutures. A typical unmodified segment consists of four main regions; a dorsal region or tergum, a ventral region or sternum, and on each side a lateral region or pleuron. In each of these regions the cuticle may be differentiated into separate sclerites called tergites, sternites, and pleurites, respectively.

The head bears compound eyes on each side with the antennae situated anterior to them and the three simple eyes or ocelli arranged in a triangle between them. Anterior to the median ocellus is a transverse suture, the epistomal suture, which separates the clypeus from the frons or region between the compound eyes.

The mouth parts are shown in detail in figure 1:2. The labrum or upper lip is hinged to the front margin of the clypeus and behind it is a pair of powerful mandibles. Back of the mandibles are the maxillae, each with three appendages; a lateral five-segmented maxillary palpus; a somewhat fleshy appendage, the galea; and an inner appendage, the lacinia, which somewhat resembles the mandibles. The labium forms the posterior wall of the mouth and bears laterally the three-segmented labial palpi. The hypopharynx is a soft tonguelike structure situated centrally between the mouth parts.

The thorax is comprised of three segments: the prothorax, mesothorax, and metathorax, each bearing a pair of legs and the latter two segments usually each bearing a pair of wings. Generally there are two pairs of thoracic tracheal openings or spiracles.

The legs typically consist of five parts: the coxa, a short basal segment which attaches the leg to the body; the trochanter; the femur; the tibia; and the tarsus which bears a pair of claws apically. The tarsus may have one to five segments or, rarely, be entirely absent.

The adult plecopteron of figure 1:1 represents a fully winged or alate insect of generalized type. Some insects have shortened, stublike wings and are termed brachypterous whereas others have none

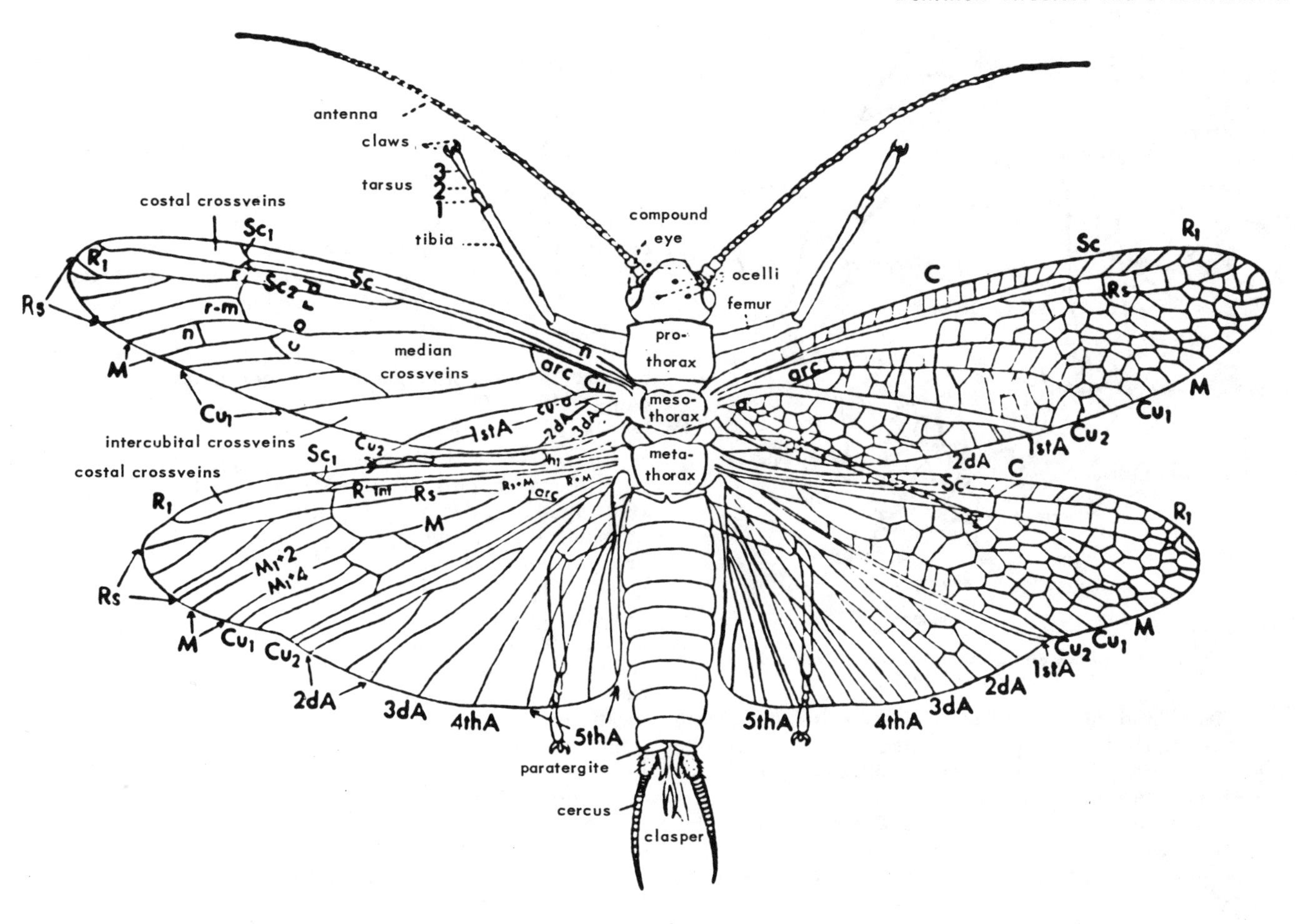

Fig. 1:1. An adult Plecoptera, *Pteronarcys californica* Newport, illustrating general insect anatomy (Essig, 1942).

at all or only tiny vestiges thereof and are referred to as apterous. The main wing veins develop along the course of tracheae present in the developing wing of the immature insect. The wing venation varies greatly from order to order but certain similarities in basic pattern lend support to the theory that all the diverse patterns have been derived from a single type of primitive venation. In figure 1:4 is shown the venation pattern which students generally regard as the archetype pattern from which the venation of modern insects has been derived. The names of the veins are indicated with the standard abreviations used for the Comstock-Needham system of wing-vein nomenclature.

In wings of modern insects the precosta (Pc) is lacking, and the anterior fork of the media (MA) is usually absent. The remaining veins may undergo modification by fusion of adjacent veins or by partial or complete suppression of one or more veins. In some orders the pattern is greatly complicated by the addition of secondary veins, and in other orders there are forms in which practically all venation has been lost.

The abdomen typically consists of eleven segments the last of which bears the cerci. The Collembolla are a notable exception in possessing never more than six abdominal segments, either in the embryo or the adult. Various numbers of the terminal segments are involved in the formation of the external genital apparatus and ovipositor or egg-laying structure. Typically there is a pair of spiracles on each of the first eight abdominal segments, but some or all of these may be absent in some insect. Likewise the number of abdominal segments may be actually or apparently reduced or the first segment may be incorporated into the thoracic region.

INTERNAL ANATOMY

Just within the body wall and attached to it are the muscles. The alimentary canal, divisible into fore-, mid-, and hind-gut, occupies the center of the body and extends from one end to the other. The fore-gut is further divided into the pharynx, esophagus, crop, and proventriculus. The mid-gut represents the stom-

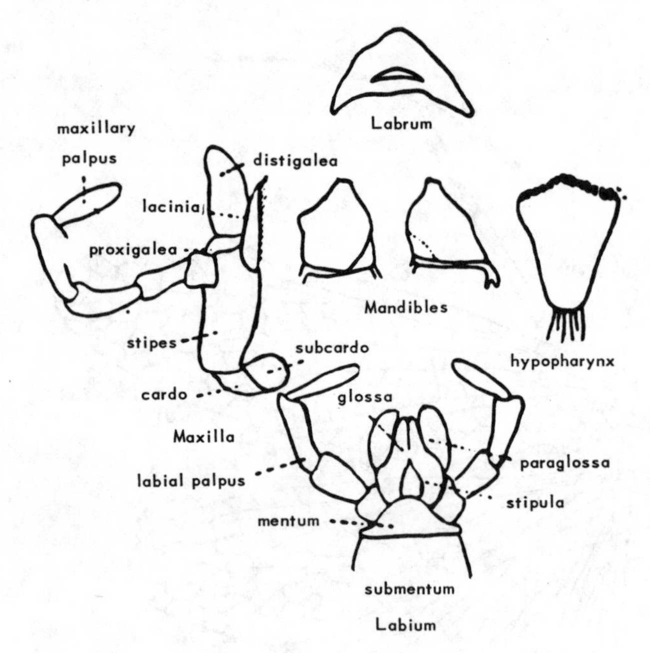

Fig. 1:2. Mouth parts of *Pteronarcys californica* Newport (Essig, 1942).

ach. The hind-gut is differentiated into the small intestine with the Malpighian tubules, or excretory organs, arising from its anterior extremity, and the rectum terminated by the anus.

The heart is an open tube lying dorsally above the alimentary canal.

The paired salivary glands lie anteriorly beneath the intestine and open via a common duct on the hypopharynx.

The reproductive system consists of a pair of gonads usually connected by a common duct leading to the exterior. Accessory structures in the female include one or more spermathecae or sacs for storage of seminal fluid, and colleterial glands for secretion of cement, gelatin, or other substance for attaching or protecting the eggs. The external structures consist of appendages of the caudal segments variously modified as an ovipositor in the female or as a clasping organ in the male.

The central nervous system consists of a paired ventral nerve cord connecting segmentally arranged ganglia. The first ganglionic mass which constitutes the brain lies in the head above the esophagus. It is connected with the second mass or subesophageal ganglion by a pair of connectives. In the more primitive insects the postcephalic ganglia tend to be segmentally arranged, but in the more specialized groups there is a tendency for consolidation of these ganglia into a single mass in the thorax.

Respiration is accomplished in most insects by means of the tracheal system, a network of fine tubes by means of which air is brought directly to the tissue cells. Typically the tracheal system consists of two longitudinal tracheal trunks running the length of the body, one on each side. Smaller branches lead to the external openings or spiracles, two pairs of which are found in the thorax and up to eight pairs in the abdomen. The lateral trunks give rise to numerous branches which divide many times and penetrate the tissues. This system may be developed to various degrees; it may function in conjunction with gills in aquatic forms, or it may be entirely lacking. In the latter case respiration may proceed by diffusion through the cuticle.

The fat body or adipose tissue occurs throughout the body and often completely surrounds the alimentary canal and other organs. Its primary function appears to be storage of nutrients.

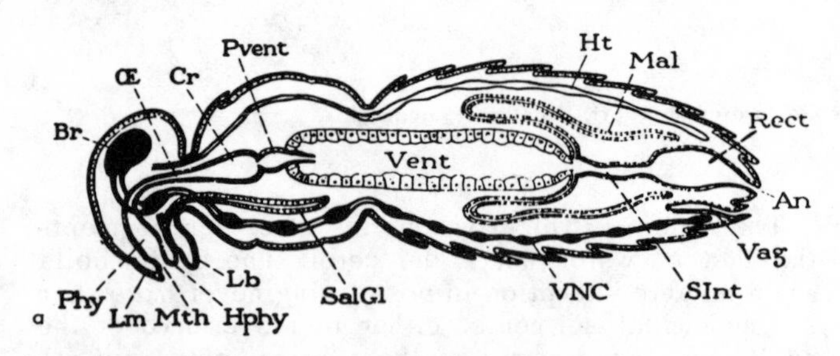

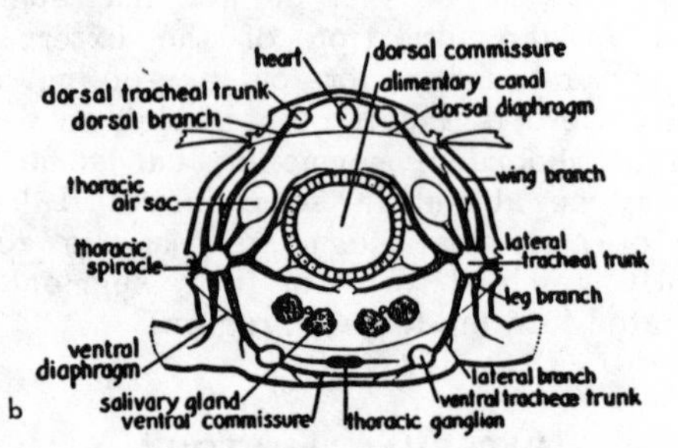

Fig. 1:3. Diagramatic sections through an insect body showing arrangement of principal internal organs. *a*, sagittal section; *b*, cross section through thorax. An, anus; Br, brain; Cr, crop; Hphy, hypopharynx; Ht, heart; Lb, labium; Lm, labrum; Mal, Malpighian tubules; Mth, mouth; Oe, esophagus; Pvent, proventriculus; Rect, rectum; SalGl, salivary gland; SInt, small intestine; Soe Gng, subesophageal ganglion; Vag; vagina; Vent, ventriculus or stomach; VNC, ventral nerve cord (*a*, Snodgrass, 1925; *b*, Essig, 1942).

GROWTH AND DEVELOPMENT

From the time of hatching until adulthood the individual passes through a period of growth and change. The insect integument has little capacity to stretch, so to accommodate increase in size the insect periodically sheds or molts its skin and replaces it with a larger one. The process of molting is sometimes called ecdysis, and the cast-off skins are called exuviae. Most insects molt three to six times during normal development but in some cases thirty or more molts occur. The period between any two consecutive molts is called a stadium or instar. The period between hatching and the first ecdysis is the first stadium, and any individual in this stage of development is a first instar.

The adult or imago is the stage having functional reproductive organs and associated mating or egg-laying structures. It is the stage that usually bears fully developed or functional wings.

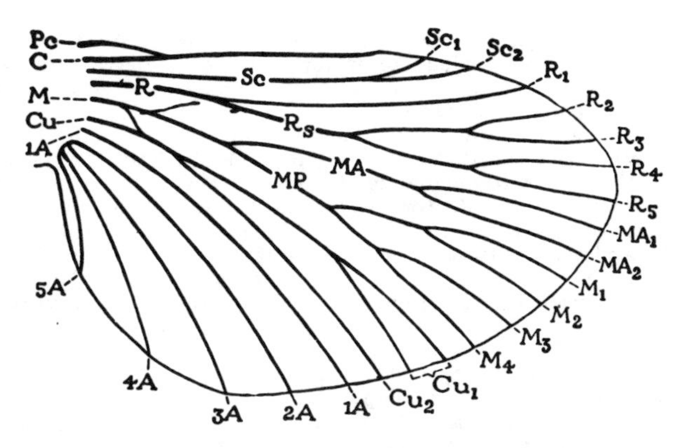

Fig. 1:4. Diagram of archetype wing venation with veins named according to Comstock-Needham system. A, anal veins; C, costa; Cu, cubitus; M, media; MA, media anterior; MP, media posterior; Pc, precosta; Sc, subcosta; R, radius; R_s, radial sector (Snodgrass, 1935).

In most insects there is considerable difference between the first instar and the adult. The process of change which occurs between these stages is called metamorphosis. The types of insect metamorphosis can be classified in three principal categories according to the degree of change which takes place.

1. *Ametabolous metamorphosis.*—Some insects, such as the Collembola, emerge from the egg resembling miniature adults, and change of form from instar to instar is very slight, being limited primarily to increase in size and development of functional reproductive organs. Such insects are called Ametabola, and others exhibiting distinct metamorphosis are termed Metabola. The latter may be further divided into the Hemimetabola and the Holometabola.

2. *Hemimetabolous metamorphosis.*—Insects in this category undergo simple metamorphosis, often termed direct or incomplete. The immature stages greatly resemble the imago in general structure, body form, and mode of life and are called nymphs. Change of form consists primarily of gradual development of wings as external pads and of the reproductive organs. The Hemiptera typify this type of metamorphosis.

In the Plecoptera, Ephemeroptera, and Odonata the nymphs are aquatic and the adults aerial. Since the nymphs of these three orders possess special respiratory organs and other adaptive modifications for life in the water, the final transformation into the adult involves more profound morphological change than among other Hemimetabola. Plecoptera nymphs have only relatively minor structural adaptations for aquatic life; the most conspicuous are the gills possessed by some species, the vestiges of which are carried over to the adults. In the Ephemeroptera and Odonata the nymphs have well-developed lateral or anal gills and other special structures not carried over to the adult. They are so entirely unlike the adult in general appearance that the relationship is not at all self-apparent. The nymphs of these orders are called naiads by some writers.

3. *Holometabolous metamorphosis.*—In this type of development, also called complex or indirect, there are three distinct growth forms: the larva or feeding and growing stage, the pupa or resting and transformation stage, and the adult. The larva is quite different in appearance from the adult and frequently bears no resemblance whatsoever to it. It is often wormlike, without legs, and the eyes and antennae may be small or aborted. They do not have external wing pads as do hemimetabolous nymphs, but the wings develop as internal pads and are called imaginal buds or histoblasts. When the larva transforms to a pupa these are everted and are then external wing pads. Frequently the larva spins a cocoon, builds a protective case, or excavates a chamber in which to pass the pupal stadium.

Among the many types of larvae there are a few so common as to merit distinctive names. The more important of these encountered among the aquatic insects are the following:

Campodeiform (fig. 13:10*n*).—In this type the mandibles are well developed, the body is elongate, flattened, with three pairs of thoracic legs and usually with cerci or caudal filaments. Typical examples are larvae of Dytiscid beetles.

Eruciform (fig. 11:1*c*).—Not common among aquatic insects, this type is represented by larvae of Lepidoptera. The body is cylindrical, the thoracic legs short, and the abdomen is furnished with prolegs on some segments.

Vermiform (fig. 14:63*d, e, k, m*).—Larvae which are more or less wormlike in form and without legs are termed vermiform or apodous. Characteristic of this type are the larvae of most Diptera, many Hymenoptera, and the weevils among the Coleoptera.

The pupa is a nonfeeding, quiescent stage during which transformation from larva to adult occurs. At this stage the wings and appendages of the future adult become evident externally. There are three principal types of pupae.

Exarate (fig. 13:39).—In this type all appendages are free of any secondary attachments to the body. Pupae of Trichoptera and most Coleoptera are characteristic of this type.

Obtect.—Obtect pupae have the appendages tightly appressed or fused to the body. Most Lepidoptera are of this type.

Coarctate (fig. 14:63*c, h, j, s*).—In the Diptera Cyclorrhapha, the last larval exuvium persists and becomes transformed into a hardened barrellike capsule or puparium that protects the exarate pupa within. A pupa so encased is said to be coarctate.

The adult is the final instar and possesses functional reproductive organs and, in most cases, wings. Its primary function is reproduction.

Classification may be defined as the systematic arrangement of organisms in groups or categories according to some definite plan or sequence. The principal categories used herein are, in order of descending rank: class, order, family, genus, and species. The criteria by which these categories are defined may be purely arbitrary, in which case the resulting classification would be an artificial one; or an attempt may be made to select criteria which elucidate the natural relationships and evolutionary history of the various forms involved. In the latter case the desired result would be a natural or phylogenetic classification with organisms grouped in categories according to true relationship based on common ancestry.

The classification of insects has passed through many changes proceeding from the earlier artificial classifications toward the more natural system of today which is based essentially upon characteristics of the mouth parts, wings, and metamorphosis. Although no single classification of insects is at present universally accepted, the major differences of opinion are concerned with the rank assigned various categories rather than with the unity of the categories themselves.

The most efficient method of identifying insects involves the use of keys. There are various types of keys but all are essentially a tabulation of diagnostic characters of species, genera, and so on in dichotomous couplets for the purpose of facilitating rapid identification. Determinations arrived at through the use of keys alone are not definitive, however, and for maximum certainty specimens should be compared with illustrations, original or subsequent descriptions, and, if possible, with types or correctly identified specimens.

The diagnostic characters utilized in keys are merely attributes by which an organism (or group of organisms) differs from other organisms belonging to different taxonomic categories or resembles organisms belonging to the same category. Good diagnostic characters have a minimum of variability within the category to which they apply and are easily visible. They do not necessarily refer to attributes of particular biological significance or to basic differences between categories. Their primary function is utilitarian.

The first step in the identification of insects is to place them in the proper order. This would be a simple matter if every specimen were typical and if only the adult stages were to be considered. Actually numerous atypical forms are encountered, especially among the immature insects found inhabiting inland waters of California. For this reason the key may appear to be more complicated than the subject matter would warrant. In using this key one should keep in mind the general facies or appearance of typical representatives of each order. Most of these are so distinctive that, with a little experience, they will become recognizable at a glance.

Key to the Orders of Aquatic Insects

1. Wings well developed 2
— Wings absent or vestigial 14
2. Wings of mesothorax horny, leathery, or parchmentlike at least basally; metathoracic wings entirely membranous 3
— Wings of mesothorax entirely membranous 5
3. Mouth parts in the form of a jointed beak, fitted for piercing and sucking. True bugs. (in part) HEMIPTERA
— Mouth parts mandibulate, fitted for chewing 4
4. Front wings without veins, horny; hind wings folded both lengthwise and crosswise in repose. Beetles .... (in part) COLEOPTERA
— Front wings usually with distinct veins; hind wings folded lengthwise in repose. Grasshoppers .... (in part) ORTHOPTERA
5. Only 1 pair of wings, the mesothoracic or front wings, present; hind wings often represented by small clublike halteres 6
— Two pairs of wings present 7
6. Abdomen with long caudal filaments. Mouth parts greatly reduced. Mayflies . (in part) EPHEMEROPTERA
— Abdomen without caudal filaments. Mouth parts usually well developed, forming a proboscis. Flies, midges, mosquitoes ....(in part) DIPTERA
7. Abdomen petiolate; wing venation reduced; parasitic on aquatic insects .... (in part) HYMENOPTERA
— Abdomen broadly joined to thorax; wings with numerous veins 8
8. Tarsi 3-segmented 9
— Tarsi 4- or 5-segmented 10
9. Wings differing in shape and size, the hind wings with a large longitudinal folded area; antennae long, many segmented .... (in part) PLECOPTERA
— Wings similar in shape and size; antennae short and bristlelike, with 3-7 segments. Dragonflies, damselflies .... (in part) ODONATA
10. Wings covered with scales or hairs, cross veins few 11
— Wings at most with hairs only on the veins, cross veins numerous 12
11. Wings with scales; mandibles absent; feeding apparatus a coiled sucking tube derived from galeae .... (in part) LEPIDOPTERA
— Wings with hairs; mandibles usually reduced; feeding apparatus not a coiled sucking tube .... (in part) TRICHOPTERA
12. Hind wings much smaller (less than ½) than the front wings; antennae short and bristlelike; cerci long and prominent. Mayflies ......(in part) EPHEMEROPTERA
— Hind wings not greatly smaller or larger than front wings; antennae long, not bristlelike; cerci absent 13
13. Hind wings not longitudinally folded. Spongilla-flies .... (in part) NEUROPTERA
— Hind wings longitudinally folded. Dobsonflies .... (in part) MEGALOPTERA
14. Abdomen with a springing apparatus or furcula. Springtails .... COLLEMBOLA
— Abdomen without a springing apparatus 15
15. Wing rudiments usually present as external flaplike appendages; body form not wormlike, mummylike, or with appendages fused to the body covering. Nymphs, naiads, and wingless adults of Hemimetabola 16
— Wing rudiments usually invisible but never present as external flaplike appendages (larvae, Holometabola) or, if externally visible (pupae of most Holometabola), the general form is mummylike with appendages encased in sheaths which may be free or fused to the body covering 20
16. Mouth parts fitted for piercing and sucking, forming a slender proboscis. True bugs... (in part) HEMIPTERA
— Mouth parts fitted for chewing 17
17. Labium, when extended, long and scooplike, and, when folded, serving as a mask covering the other mouth parts. Dragonfly and damselfly naiads .... (in part) ODONATA
— Labium neither scooplike nor hinged and extensile . 18
18. Abdomen without long, taillike cerci or tracheal gills; hind femora enlarged for jumping. Grasshoppers .... (in part) ORTHOPTERA
— Abdomen with 2 or 3 long taillike cerci; thorax and abdomen usually with plate-, feather-, tassel-, or fingerlike tracheal gills; hind femora not enlarged for jumping 19
19. Tarsi with one claw; gills plate-, feather-, or tassellike and present on one or more of first 7 abdominal segments except absent in first instars. Mayfly naiads .... (in part) EPHEMEROPTERA
— Tarsi with 2 claws; gills usually present and fingerlike and may occur on abdomen, thorax, or labium. Stonefly naiads ....(in part) PLECOPTERA
20. Body wormlike, elongate, and cylindrical to short and

obese, or flattened and oval, with or without appendages and distinct sclerotized head capsule. Larvae 21
— Body completely enclosed in a hard, capsulelike case or puparium, or mummylike with appendages free or fused to the body covering. Pupae 29
21. Thorax with 3 pairs of legs 22
— Thorax without legs 26
22. Underside of abdomen with at least 2 pairs of fleshy prolegs tipped with many small hooks. Caterpillars .... (in part) LEPIDOPTERA
— Abdomen without prolegs 23
23. Anal segment with a pair of hook-bearing appendages; antennae inconspicuous, 1-segmented; eyes consisting of a single facet; tracheal gills, if present, seldom confined to lateral margins; usually living in silken cases which may have vegetable or mineral matter incorporated in their construction ....(in part) TRICHOPTERA
— Anal segment without hooks; or if hooks are present, then antennae consist of more than 1 segment, eyes are of more than 1 facet, and tracheal gills are lateral 24
24. Mandibles and maxillae united to form long straight or slightly recurved suctorial tubes; abdomen with segmented gill appendages folded beneath; size small, 10 mm. or less; feed on fresh-water sponges .... (in part) NEUROPTERA
— Maxillae not united with the mandibles; mandibles if suctorial are strongly curved; gill appendages, if present, seldom segmented and never folded beneath abdomen 25
25. Abdomen with 7 or 8 pairs of segmentally arranged lateral filaments or gills; ninth segment with a pair of hooked anal feet or with a single long filament. Hellgramites .... (in part) MEGALOPTERA
— Abdomen usually without such lateral filaments; but, when present, with hooked anal feet absent, or located on tenth segment and 4 in number, or with no median filament. Beetle larvae ....(in part) COLEOPTERA
26. Head capsule distinct, at least anteriorly 27
— Without a distinct, sclerotized head capsule 28
27. Abdomen with gills or a breathing tube at posterior end. Mosquito larvae, etc. .... (in part) DIPTERA
— Abdomen without terminal gills or a breathing tube. Weevil larvae .... (in part) COLEOPTERA
28. Larvae elongate, more or less cylindrical, or spindle-shaped, or peg-shaped with anterior end pointed and posterior end blunt. Maggots, crane fly larvae, etc. .... (in part) DIPTERA
— Larvae not as above; less than 2 mm. in length; parasitic .... (in part) HYMENOPTERA
29. All appendages invisible, the pupa enclosed in a hardened, barrel-shaped capsule with segmental annulations (Coarctate pupae) .... (in part) DIPTERA
— Appendages visible, either free or fused to pupal case (if enclosed in a silken or hardened cocoon or case, this should be removed to see the pupal characters) 30
30. Appendages free, though sometimes held in a fixed position, mummylike (Exarate type) 31
— Appendages fused to the body wall, forming a continuous covering (Obtect type) 36
31. Abdomen constricted at base, narrowly joined to thorax .... (in part) HYMENOPTERA
— Abdomen broadly joined to the thorax 32
32. Thorax with only 1 pair of wing pads .... (in part) DIPTERA
— Thorax with 2 pairs of wing pads 33
33. Antennae usually 11-segmented, pads of front wings thickened .... (in part) COLEOPTERA
— Antennae always with 12 or more segments; pads of front wings not appearing like leathery wing covers . 34
34. Mandibles curved, base stout, usually projecting forward and crossing each other; usually in cases or webs constructed by the larva.....(in part) TRICHOPTERA
— Mandibles broad and stout, never overlapping or crossing each other 35
35. Size small, 10 mm. or less; cocoon spun in protected spots near stream above ground ....(in part) NEUROPTERA
— Size larger, 12 mm. or more; pupae in burrows in ground or wood (in part) MEGALOPTERA
36. A pair of short breathing tubes present anterodorsally and a pair of leaflike gills at tip of abdomen. Mosquito pupae .... (in part) DIPTERA
— Without anterodorsal or posterior breathing tube and without leaflike gills........(in part) LEPIDOPTERA

REFERENCES

BRUES, C. T., A. L. MELANDER, and F. M. CARPENTER
1954. Classification of Insects. Bull. Mus. Comp. Zool., 73: v + 917, 1219 figs.

ESSIG, E. O.
1942. College Entomology. New York: Macmillan. vii + 900 pp., 308 figs.

CHAPTER 2

Aquatic Collembola

By D. B. Scott, Jr.
Salinas, California

Collembola, more commonly referred to as springtails, are generally placed with the primitive insects of the subclass Apterygota. However, they are not closely related to other apterygotans, such as silverfish, campodeids, and proturids. With a few exceptions, Collembola are very small, generally less than three mm. in length. A few species, however, may reach ten mm. The body may be elongate (distinctly segmented) or globular (indefinite segmentation) and clothed with either scales or hairs. Coloration consists of brilliant pigmentation in hair-clothed species, bright irridescence in the scaled types, or no color at all in subterranean species. The mouth parts are adapted for sucking in some species and for chewing in others, and the head may be either prognathous or hypognathous.

The antennae, often a key characteristic, vary greatly in structure and length. The joints may be simple, or the terminal one may be annulated as in the family Smynthuridae. Eyes are present in most Collembolans; they occur as simple ocelli in patches behind the antennae. The eyes are never compound, and the ocelli never exceed eight to a patch. In many species pseudocelli occur on various parts of the dorsal surface (figs. 2:2*a*; 2:4*e*).

A specialized sensory organ, the postantennal organ is a distinct and unique structure in many groups. It exhibits a variety of designs and arrangements from simple, elliptical depressions to elaborate rosettes or fernlike protuberances (fig. 2:3*a-e*). Sensory organs are found behind the base of each antenna and sometimes special microscope techniques are required to discern them.

The organ by which the order is distinguished, the ventral tube, arises on the ventral side of the first abdominal segment. This is a bilobed appendage with filaments capable of extrusion. Although its purpose is not fully understood, this tube is presumably for adhesion or for some respiratory function.

The "spring" or furca (furcula), when present, is composed of a pair of whiplike appendages, the basal parts (the manubrium) of which are fused together and attached to the fourth, or occasionally the fifth, abdominal segment. A pair of dentes arises from the manubrium and accounts for a major part of the spring. Each dens is equipped with a hooklike, chitinous tip of many variations (the mucro) by which certain genera are separated. When retracted under the body,

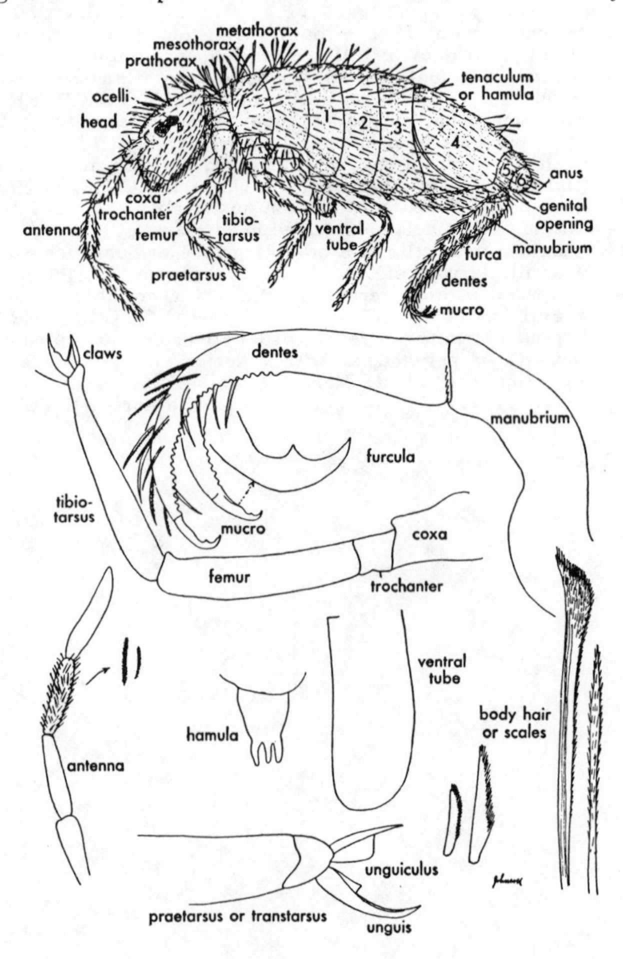

Fig. 2:1. *Entomobrya laguna* Bacon and anatomical features (Essig, 1942.)

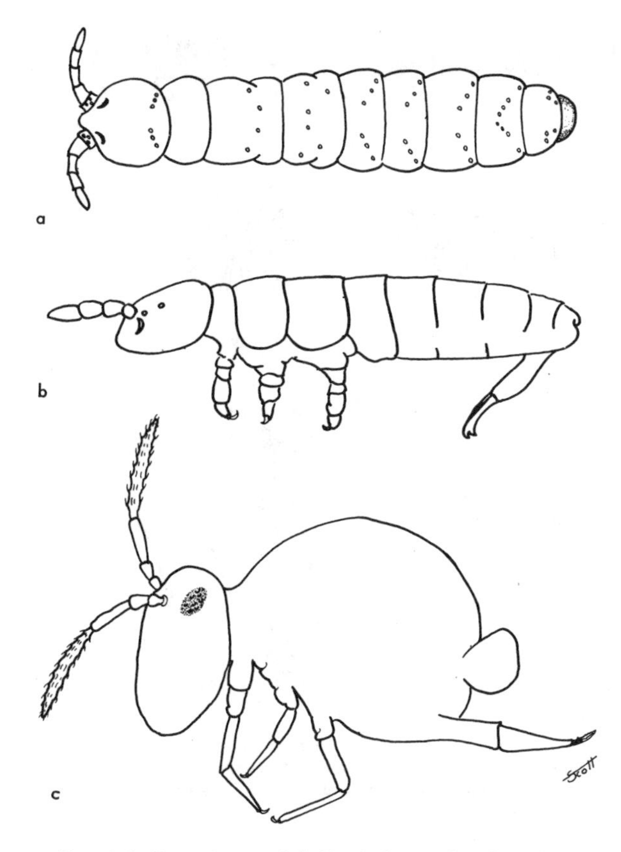

Fig. 2:2. Body forms of Collembola. *a*, Onychiuridae; *b*, Isotomidae; *c*, Smynthuridae.

the spring is held by the retinaculum or catch on the third abdominal segment (fig. 2:4*a-d*).

The legs of most Collembola are fitted with elaborately toothed claws, consisting usually of two parts; the unguis or large claw and, when present, the unguiculus or small claw opposing the unguis. In conjunction with the claw, special guard setae known as tenent hairs may be present. The tenent hairs are often clavate or knobbed (fig. 2:3*f-g*).

With the possible exception of three genera, respiration is cutaneous over the entire body. Womersley (1939) reports that the genera *Smynthurus, Smynthurides,* and *Actaletes* are provided with true tracheae.

So far as is known, there are few external sexual differences between individuals. Metamorphosis is lacking; the only external changes between newly emerged young and fully grown adults are progressive intensification of pigmentation and further differentiation of segmented appendages.

The food of aquatic Collembola consists of decaying animal or vegetable matter which usually occurs in abundance in stagnant water, intertidal zones, rain pools, or on melting snow.

In spite of their relative obscurity, Collembola are believed to be the most widely distributed of all insects in temperate and arctic regions. Many species are cosmopolitan, whereas others, such as those of the Australian fauna, are endemic and typical only of specialized environments. The aquatic or semiaquatic types with which we are concerned here require special environmental conditions. Standing water generally is a preferred habitat though not always a necessary condition. Collembola are not "aquatic" in the true sense of the word, that is, they may feed and live primarily on the water, but they do not lay their eggs in the water.

COLLECTION AND PREPARATION OF COLLEMBOLA

Collecting.—Although we are concerned here with aquatic types, it is appropriate to mention also methods used in collecting from moss, leaves, under wood and bark, and from fungi.

A flat, white enameled pan into which moss and trash can be shaken will permit rapid collecting of some species, especially if a small amount of water is first put in the pan. The insects can then be picked from the surface with a fine brush. If it is inconvenient to carry a pan in the field, samples of moss, leaves, fungi, or rotten wood may be collected in ice-cream cartons and brought to the laboratory for separation. A few drops of xylol on the surface of the water will prevent most of the jumping forms from escaping. For complete coverage of soil, leaves, or moss, the Berlese funnel can be used most effectively. A vial of 90 per cent alcohol is placed beneath the funnel to catch the specimens as they fall out.

For aquatic collecting, a sieve frame lined with fine porous nylon netting is recommended. The insects can be picked out with a suction tube and transferred to alcohol vials. To prevent distortion from shrinkage and to preserve color patterns, specimens are best preserved in 90 to 95 per cent alcohol. However, color records should be made as soon as possible since color retention in alcohol is, at best, limited.

Preparation.—Because of the delicate structure of the Collembola, only the mild solutions suggested below should be used for clearing, staining, or fixing. Carbo-xylol, KOH solutions, and pure xylol are not recommended.

Direct clearing.—Many species need only be passed from alcohol through clove oil before mounting. Clearing in clove oil requires one to several hours. Xylol added at the rate of 1 cc. xylol to 15 cc. clove oil assists the process in some instances.

Clearing and staining.—Many Collembola require clearing with caustic and subsequent staining to differentiate key characters used in identification. This applies particularly to those possessing intricate postantennal organs or other sensory apparatus. Treatment in 5 per cent NaOH for periods ranging from a few minutes to several hours will remove pigmentation. Specimens are then transferred to 10 per cent acetic acid solution to neutralize the caustic. They can then be stained in an alcohol-base basic fuchsin to the desired intensity. Overstaining can be remedied to some degree by returning to a water wash. After staining, specimens must be dehydrated in 95 per cent ethyl alcohol before final clearing in clove oil.

Mounting.—After clearing in clove oil, specimens may be transferred directly to the slide-mounting medium. Euparal and hyrax have proven very satisfactory; Canada balsam darkens with age whereas the other two do not. Since it may be necessary to examine specimens under the oil immersion objective, no. 1 cover-glass circles should be used, and only mild heat, if any, should be used to cure the slides. The medium will usually set to durable strength in a month or six weeks at room temperatures.

Key to the California families of Aquatic Collembola

1. Body elongate, thorax and abdomen distinctly separated, abdominal segmentation distinct except for posterior fusion in some species (fig. 2:2*a,b*). suborder ARTHROPLEONA 2
— Body conspicuously subglobular, not distinctly segmented except for 5th and 6th abdominal segments which are usually distinct (fig. 2:2*c*) suborder SYMPHYPLEONA .... SMYNTHURIDAE
2. Prothorax well developed, visible dorsally, and similar to other segments, furnished with hairs; cuticle granular. superfamily PODUROIDEA 3
— Prothorax not distinct, usually hidden, not hairy; cuticle smooth, hairy, or scaled. superfamily ENTOMOBRYOIDEA 5
3. Pseudocelli absent on thorax or abdomen 4
— Pseudocelli present on thorax and abdomen (fig. 2:2*a*) .... ONYCHIURIDAE
4. Head hypognathous; ocelli on back part of head; dentes annulated distally, reaching beyond ventral tube when retracted (*Podura aquatica* Linneaus 1758).PODURIDAE
— Head prognathous; ocelli, if present, on fore part of head; dentes not annulated, usually not reaching ventral tube when retracted .... HYPOGASTRURIDAE
5. Hind coxae with specialized area of fine hairs; inner edge of claw with a basal groove .... ENTOMOBRYIDAE
— Hind coxae without specialized area of fine hairs; inner edge of claw without groove .... ISOTOMIDAE

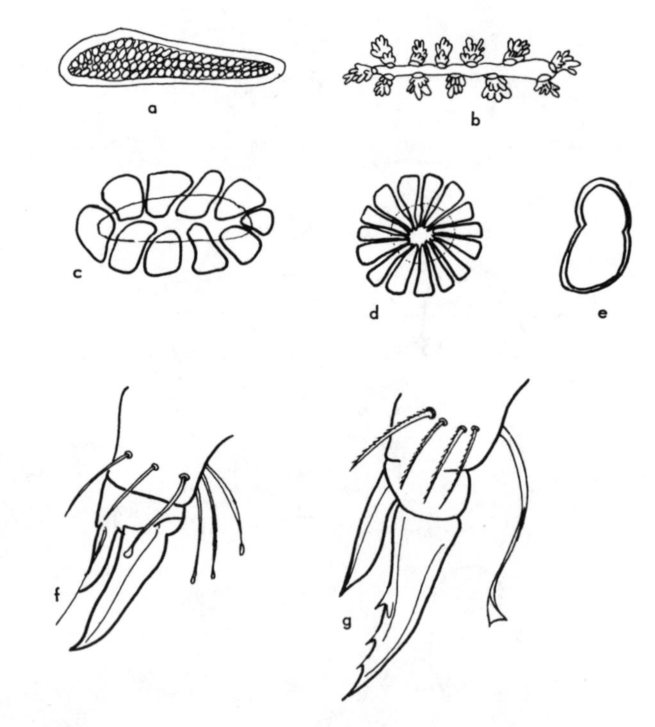

Fig. 2:3. a-e, postantennal organ; f, g, apex of tibiotarsus; a, *Tullbergia* sp.; b, *Onychiurus fimetarius*; c, *Pseudachorutes* sp.; d, *Anurida granaria*; e, *Isotoma* sp.; f, *Smynthurinus aureus*; g, *Entomobrya* sp.

Family ONYCHIURIDAE

Key to the Genera

1. Two sensory clubs of 3rd antennal segment curved together .... *Tullbergia* Lubbock 1876
— Two sensory clubs of 3rd antennal segment not curved toward each other .... *Onychiurus* Gervais 1841

Genus *Onychiurus* Gervais 1841

1. Anal spines absent; tubercles of postantennal organs compound, with many branched tubercles (fig. 2:3*b*); color white; pseudocelli present on all segments; length 1.8-2.1 mm.; cosmopolitan, stagnant water, fish ponds, and rain pools ..... *fimetarius* (Linneaus) 1767
— Anal spines present; tubercles of postantennal organ simple 2
2. Postantennal organs with 18-44 tubercles, closely set but not crowded together; 3 pseudocelli at base of each antenna arranged in triangles; unguiculus as long as unguis; color white; cosmopolitan, rain pools, and stagnant water .... *armatus* (Tullberg) 1869
— Postantennal organs with 8-14 separated tubercles; 1 pseudocellus at base of each antenna; unguiculus two-thirds as long as unguis; California, Oregon, Washington, on snow or water at 6,000 feet or higher .... *cocklei* Folsom

Family PODURIDAE

This family is represented by only a single genus and species, *Podura aquatica* Linneaus 1758. The head is hypognathous, claws long and slender, color red brown to black. The mucrones are trilamellate and granular. This is a cosmopolitan species found on stagnant water.

Family HYPOGASTRURIDAE

Key to the Genera

1. Mouth parts adapted for chewing; mandibles with well-developed molar plate 2
— Mouth parts adapted for piercing and sucking; mandibles without molar plate 3
2. Postantennal organ present, eyes 8 on each side. Furcula present ....*Achorutes* Templeton 1835
— Postantennal organ absent, eyes 5 on each side ....*Xenylla* Tullberg 1869
3. Anal spines present............*Odontella* Schffr. 1897
— Anal spines absent 4
4. Ocelli 5 on each side; body elongate .... *Anurida* Laboulbene 1865
Ocelli 8 on each side; body stout; furca present, but stout and short; not aquatic .... *Pseudachorutes* Tullberg 1871

Genus *Xenylla* Tullberg 1869

1. Dentes and mucrones fused, not readily differentiated

at junction; length 1.5 mm.; cosmopolitan, found occasionally on rain pools and stagnant water. . . . . *maritima* Tullberg 1869
— Dentes and mucrones clearly articulated at junction . 2
2. Anal spines large; California, rain pools, standing water . . . . *baconae* Folsom 1916
— Anal spines small 3
3. Mucrones with very broad lamellae or wings; unguiculus present as small knob; color gray or violet, mottled; cosmopolitan, stagnant water . . . . . *welchi* Folsom 1916
— Mucrones with narrow lamellae; unguiculus completely absent; color dark blue; cosmopolitan, marshlands bordering ponds . . . . *humicola* Tullberg 1876

Genus *Achorutes* Templeton 1835

1. One long tenent hair on tibiotarsus, not knobbed; anal spines equal to length of hind claws (fig. 2:5); cosmopolitan, stagnant water, fishponds, irrigation ditches . . . . *armatus* (Nicolet) 1841
— Tenent hairs usually 3 to each tibiotarsus, clavate or knobbed; anal spines about 1/5 as long as hind claws; cosmopolitan, stagnant water . . . . *viaticus* Tullberg 1872

Genus *Anurida* Laboulbene 1865

1. Color dark blue, eyes 5 on each side 2
— Color white, eyes absent; cosmopolitan, seashore . . . . *granaria* (Nicolet) 1847
2. Postantennal organs with 6-10 peripheral tubercles in a circle; 1 tenent hair on tibiotarsus; cosmopolitan, along seashore in intertidal zone . . . . *maritima* Guerin 1836
— Postantennal organs with 17-30 peripheral tubercles in an irregular ellipse; no knobbed tenent hairs on tibiotarsus; cosmopolitan, along seashores on rocks and in tidepools; inland in fish ponds and lake margins . . . . *tullbergi* Schött 1891

Family ISOTOMIDAE

Key to the Genera

1. Anus ventral, not terminal . . . . . *Folsomia* Willem 1902
— Anus terminal 2
2. Body with bothriotricha 3
— Body without bothriotricha . . . . *Proisotoma* Börner 1901
3. Hind femur with a thornlike process; manubrium without ventral setae; mucrones tridentate, nonlamellate; tibiotarsus with a distal subsegment . . . . *Archisotoma* Linnaniemi 1912
— Hind femur without a thornlike process; manubrium with many ventral setae; mucrones quadridentate, lamellate; (fig. 2:4*c*) tibiotarsus without a distal subsegment . . . . *Isotomurus* Börner 1903

Genus *Folsomia* Willem 1902

1. Eyes absent; dentes with many dorsal crenulations; length 2 mm.; fish ponds, rain pools . . . . *fimetaria* Linneaus 1758
— Eyes present, 8 on each side; dentes with few dorsal crenulations; on seashore under rocks in intertidal zone . . . . *elongata* (MacGillivary) 1896

Genus *Proisotoma* Börner 1901

Two species have been reported from California, *P. aquae* Bacon in stagnant water and rain pools, and *P. laguna* Folsom in the intertidal zone along the coast.

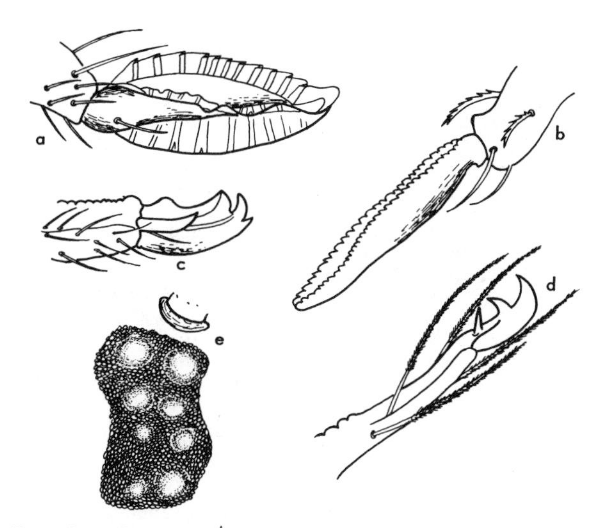

Fig. 2:4. Diagnostic characters of Collembola. *a*, mucro of *Smynthurides aquaticus*; *b*, mucro of *Ptenothrix maculosa*; *c*, mucro of *Isotomurus palustris*; *d*, mucro of *Entomobrya* sp.; *e*, eye patch and ocelli of *Entomobrya* sp.

Genus *Isotomurus* Börner 1903

The only species found in California is *I. palustris* (Muller) 1776 (fig. 2:4*c*). It is a cosmopolitan species found on stagnant water.

Genus *Isotoma* Bourlet 1839

The one representative of this genus (fig. 2:3*e*) found in California is *I. besselsii* Packard. It is found along the seashore in the intertidal zone under rocks.

Family ENTOMOBRYIDAE

Key to the Genera

1. Basal tooth of claw winged; clavate tenent hairs present but weak . . . . *Sinella* Brook 1882
— Basal tooth of claw not winged; clavate tenent hairs well developed and strong (figs. 2:3*g*; 2:4*d,e*) . . . . *Entomobrya* Rondani 1861

Genus *Sinella* Brook 1882

A single cosmopolitan species, *S. hoftii* Schaeffer, found on stagnant water, represents this genus.

Genus *Entomobrya* Rondani 1861

1. Color mottled dark brown with dark blue legs (fig. 2:1); California, seashore in intertidal zone . . . . *laguna* Bacon 1914
— Color orange with a distinct dark blue saddle; legs pale; inland species on stagnant water, ponds . . . . *atrocincta f. clitellaria* Guthrie 1903

Family SMYNTHURIDAE

Key to the Genera

1. Exserted filaments of ventral tube not warty 2

— Exserted filaments of ventral tube warty 3
2. Anal segments fused, with 2 sensory setae on each side; fourth antennal segment simple or subsegmented .... *Smynthurides* Börner 1900
— Anal segments separate, with 1 sensory seta on each side; fourth antennal segment not subdivided .... *Smynthurinus* Börner 1901
3. Antennae bent between segments 3 and 4, fourth longer than third, third segment undivided; furca without dorsal sensory papillae 4
— Antennae bent between segments 2 and 3, fourth shorter than the third; third and fourth segments sometimes subdivided; furca with dorsal papillae 5
4. Two or 3 clavate tenent hairs on each tibiotarsus ... . *Bourletiella* Banks 1899
— Tibiotarsus without multiple tenent hairs, if present then 1 only and prominent ... *Smynthurus* Latreille 1804
5. Third and fourth antennal segments not markedly subdivided .... *Dicyrtoma* Bourlet 1843
— Third antennal segment and usually the fourth distinctly subdivided .... *Ptenothrix* Börner 1906

Genus *Smynthurides* Börner 1900

1. Tibiotarsal organ present on third pair of legs composed of 2 sacs and enlarged spine 2
— Tibiotarsal organ absent on third pair of legs; fish ponds...................... *sexpinnatus* Denis 1931
2. Mucrones half as long as broad; (fig. 2:4*a*) dorsal segmentation of abdomen clearly evident; cosmopolitan, quiet streams, salt ponds .... *aquaticus f. levanderi* (Reuter) 1891
— Mucrones one-third as long as broad; dorsal segmentation of abdomen indistinct; cosmopolitan, freshwater pools.... *malmgreni f. palustris* Folsom and Mills 1938

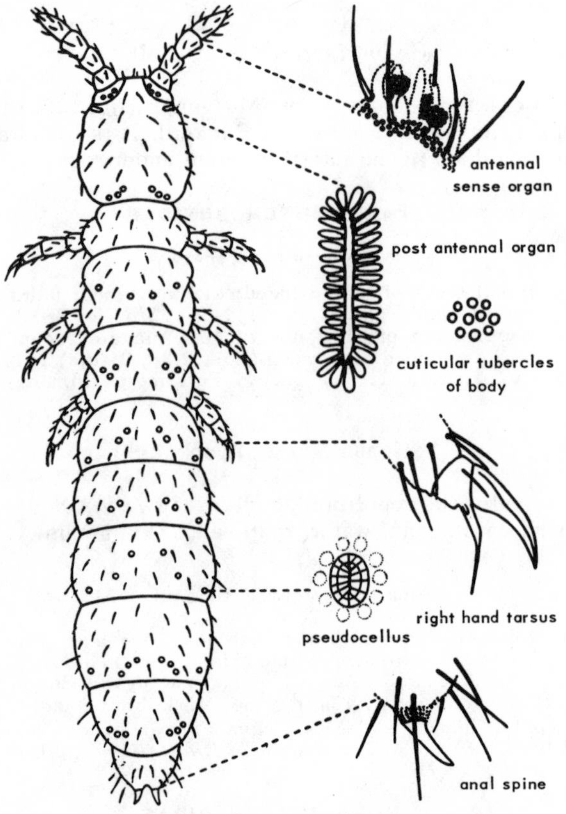

Fig. 2:5. *Achorutes armatus* Nicolet and important anatomical features (Essig, 1942).

Genus *Smynthurinus* Börner 1901

1. Tenent hairs on tibiotarsi usually 2; stagnant water . ... *remotus* Folsom 1896
— Tenent hairs on tibiotarsi 3-8 (fig. 2:3*f*) 2
2. Both edges of mucrones serrate; unguis with 3 inner teeth; cosmopolitan, stagnant water, rain pools .... *quadrimaculatus* (Ryder) 1879
— Mucrones with outer edge entire, inner edge serrate; unguis usually untoothed (fig. 2:3*f*) 3
3. Tenent hairs 4-8 on tibiotarsi (fig. 2:3*f*); color orange with black eye spots; cosmopolitan, stagnant water, fish ponds .... *aureus* (Lubbock) 1862
— Tenent hairs 3 or 4 on tibiotarsi; color pale yellow or green with black lateral stripes; cosmopolitan, stagnant water .... *elegans* (Fitch) 1863

Genus *Bourletiella* Banks 1899

Only a single species, *B. hortensis* (Fitch) 1863, is found in California. It is a cosmopolitan species found on stagnant water and on rain pools.

Genus *Ptenothrix* Börner 1906

A single species, *P. maculosa* (Schött) 1891 (fig. 2:4*b*) is recorded from California and is found on rain pools and stagnant water.

REFERENCES

BACON, G.
1914. Distribution of Collembola in the Claremont-Laguna region of California. Pomona Coll. Jour. Ent. Zool., 6:137-184, 5 figs.
DAVENPORT, C. B.
1903. Collembola of Cold Spring Beach. Brooklyn Institute of Arts and Sciences. Cold Spr. Harb. Monog. II, 33 pp.
FOLSOM, J. W.
1934. Redescriptions of North America Smynthuridae. Iowa St. Coll. Jour. Sci., 8:461-511. figs.
1938. Contributions to the knowledge of the genus *Smynthurides*. Bull. Mus. Comp. Zool., 82:229-274.
LABOULBENE, A.
1864. Recherces sur l'*Anurida maritima,* insecte Thysanura de la famille des Podurides. Ann. Soc. Ent. Fr., 4:705-720. figs.
MAYNARD, E. A.
1951. Monograph of the Collembola or Springtail Insects of New York State. Ithaca, N.Y.: Comstock Publ. Co. xxiv-339 pp.
MILLS, H. B.
1934. Collembola of Iowa. Ames, Iowa: Collegiate Press. 143 pp.
SCHÖTT, HARALD
1896. North American Apterygogenae. Proc. Calif. Acad. Sci., 6:169-196.
SCOTT, D. B., JR.
1942. Collembola Records for the Pacific Coast. Pan-Pac. Ent., 18:177-186.
WOMERSLEY, H.
1939. Primitive Insects of South Australia. Govt. Publ., Adelaide, pp. 79-278. figs.

CHAPTER 3

Ephemeroptera

By W. C. Day
Oakland, California

Mayflies are the familiar insects with fragile bodies and slender tails which swarm in the vicinity of ponds and streams. They occupy an important place in writing and folklore because of their ephemeral existence in the adult stage. The angler's fly box contains many feather and fur imitations of the mayflies, and his literature has been filled for five hundred years with descriptions of "duns" and "spinners" as the winged forms are called.

The nymphs of mayflies occupy an important place in the economy of aquatic communities. They are the "cattle" or "rabbits" of the aquatic environment, transforming plant tissue into animal. They are practically defenseless, making up in numbers for their vulnerability. Thus they fulfill all the requirements of a basic herbivore. Since they are so readily available, mayfly nymphs are preyed upon by practically every aquatic predator.

Adult mayflies also play their role in the economy of nature; they are eaten by birds, dragonflies, and fish, to mention only a few of their more important enemies.

Mayfly nymphs respire primarily by means of articulated gills which are situated at the sides of the abdominal segments. Of the ten abdominal segments, from four to seven bear gills at their posterolateral corners. The various nymphs are well adapted, each to particular types of aquatic habitats, and Needham, Traver, and Hsu (1935) proposed the following classification based upon body form and other adaptations which show a close correlation with particular habitats.

A. Still water forms
   - Climbers amid vegetation *(Siphlonurus, Callibaetis)*
   - Sprawlers on the bottom *(Choroterpes, Tricorythodes)*
   - Burrowers in the bottom *(Hexagenia)*

B. Rapid water forms
   - Agile, free-ranging, streamlined forms *(Isonychia, Ameletus, Baetis)*
   - Close-clinging, limpet forms found under stones *(Heptagenia, Ironodes)*
   - Stiff-legged trash-, silt-, and moss-inhabiting forms (some *Ephemerella*)

Although the nymphs of some species mature within a few months, and other species require a period of two years, the great majority of mayfly nymphs live in the water for one year. Each species has its own time for emergence—from February through November in California. The nymphs of a small number of species leave the water by climbing onto rocks or stems of plants. Most of them swallow air, rise to the surface film, break the nymphal skin along the back and, clambering from it, dry their wings for a moment and fly into the surrounding vegetation.

When they first emerge from the water, mayflies are dull in appearance and more or less pubescent and at this stage are known as "subimagoes." In a period of from a few minutes to forty-eight hours (about twenty-four hours for most species) the subimago molts to become the final adult or imago. No other order of insects undergoes such a second stage of development between nymph or naiad and final adult.

The true adult of the mayfly is ordinarily smooth and shining, usually with glassy but perhaps lightly tinted wings, and with longer tails and legs than the subimago. Adults live only for a few hours or, at the most, for a few days, and the mouth parts are vestigial and nonfunctional. The sole function of the adults is reproduction, and to this end they expend most of their energy.

Mating swarms consist of males with an occasional female which flies in for mating. Swarms of many millions of these insects occur in the Sacramento Valley, but large flights are rarely encountered elsewhere in California; most swarms consist of just a few hundred. Gravid females deposit eggs on the water surface, and the eggs sink to the bottom. An adult mayfly at rest, *Ameletus amador* Mayo, is shown in figure 3:1.

COLLECTING MAYFLIES

Nymphs.—The one piece of essential equipment for collecting mayfly nymphs, effective in taking undamaged specimens in most lotic or moving California waters, is the simple hand screen. A good screen is made from a three-foot length of sixteen-mesh copper

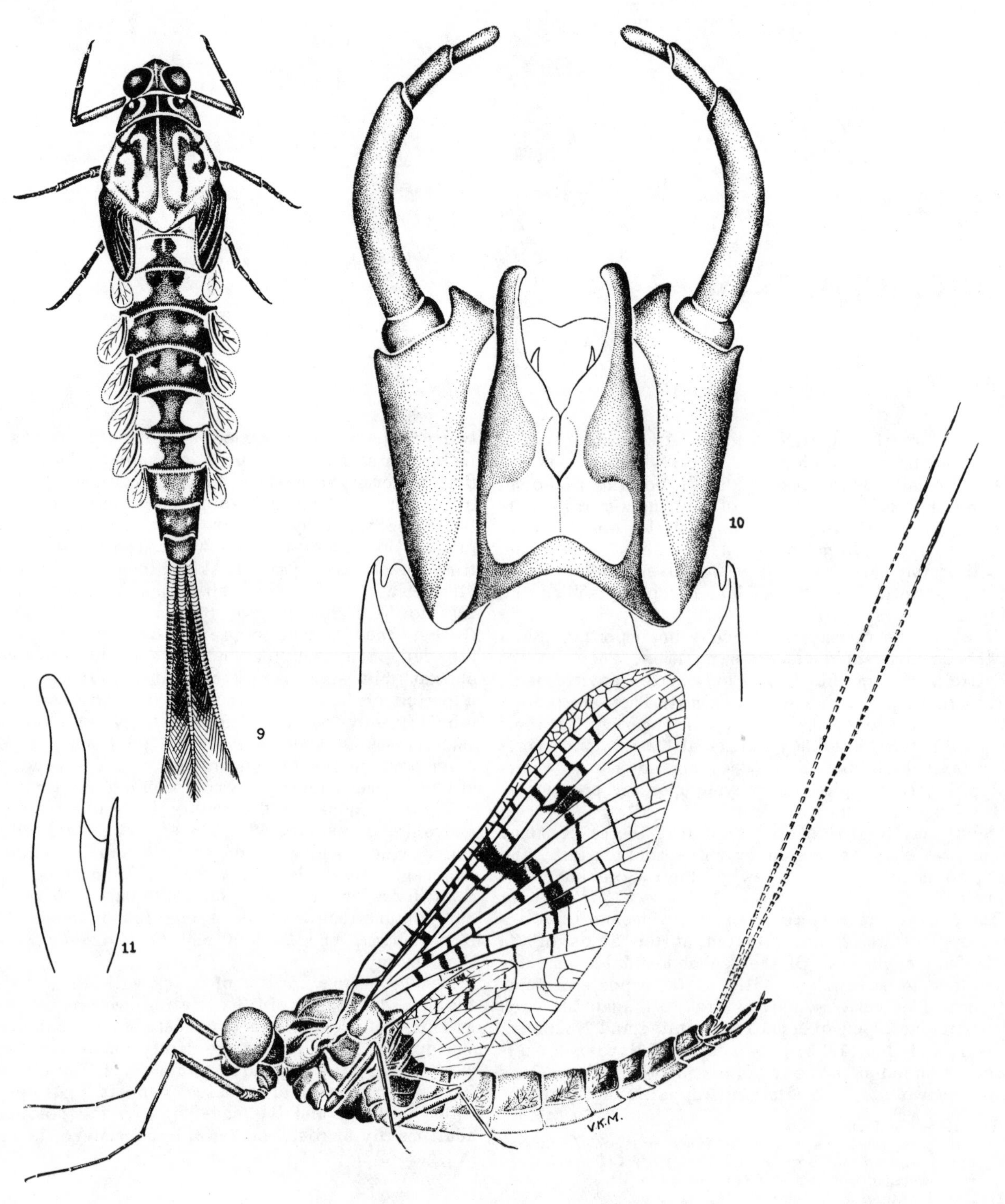

Fig. 3:1. *Ameletus amador*, adult. (V. K. Mayo, 1939.)

window screen of two-foot width, and a one and one-quarter inch square strip of hardwood twenty-seven inches in length fastened at each twenty-four-inch end, the strips extending about one and one-half inches beyond the edges of the screen.

Wading into the water with or without boots, the collector faces upstream, rests one three-foot edge of the screen on the bottom and rests the other three-foot edge across his knees, permitting the center part of the screen to sag more or less downstream. The collector, leaning over the screen and reaching upstream, lifts rocks from the bottom to inspect for clinging specimens; at the same time, the removal of rocks from the bottom dislodges the nonclingers which are washed into the screen. In gravel or sand, the upstream bottom is carefully stirred with hands or garden fork, or the hands are used to "paddle" a current of water across the bottom and into the screen.

To minimize injury to the nymphs, frequent trips should be made to the bank to remove them from the screen. Nymphs are removed from the screen with forceps or brush and placed in fresh water in a white enameled pan about three-by-six inches in size and two inches in depth. Ordinarily a low-power glass is adequate for separating the different kinds of nymphs, some of which usually go into 70 per cent alcohol and others into rearing cages.

In ponds and other quiet waters many specimens can be taken by simply holding the hand screen close to the bottom and "walking" it through the water. In most instances it is necessary to rest the screen on the bottom and across the knees, then paddle currents of water across the bottom and through the screen, the bottom-dwelling nymphs coming to the screen with the current. If weeds are present, they are broken and paddled into the screen and taken to the bank for inspection.

Where weeds are dense, or in collecting fossorial nymphs that lie an inch or two below the mud or silt surface of quiet or slow-moving water, the Ward's Scraper Net is a great convenience. With scraper net, shovel, or bucket, a scoop of mud, silt, or weed is taken from the bottom and placed in a sixteen-by-twenty-inch white enameled pan of six- or eight-inch depth that is half filled with water. Gentle stirring and further addition of water to the debris in the pan soon brings to light any nymphs that may have been captured.

With some dampening of the collector, the hand screen may be well employed in waters up to perhaps four feet in depth, but the Ekman Dredge should be used for deeper waters. Bottom samples brought up by the Ekman are handled in the large white enameled pan.

The single disadvantage found in using the all-purpose hand screen is that a relatively coarse sixteen-mesh wire cloth is necessarily used and very small specimens may pass through the screen. Finer mesh screens soon clog so that water currents wash around the screen rather than through it. At each collecting location, a few screenings should be made in likely places with a two-foot square hand screen made of twenty-four-mesh screen, this being brushed clean after each immersion.

Great variation in ecological conditions is ordinarily present even in a very short stretch of water. Many mayfly nymphs are highly selective in their habitat, and each possibility should be investigated by the collector: fast and slow riffles both deep and shoal; rock surfaces together with areas behind, in front of, and between them; quiet water; trash beds and leafy deposits—all should be carefully worked, with particular attention given to shoals right at the edge of the stream. In working new waters, the collector of mayfly nymphs should work the whole bottom, always ready to expect the unexpected.

If nymphs are to be added to the collection as such, they should be taken from the collecting screen and placed immediately in four-dram vials of 70 per cent alcohol, this fluid being drained and replaced after a few days.

Winged stages.—It is frequently possible to capture mayflies in their first winged stage as they rise from the surface of the water or as they rest on stones for a moment after emergence. These subimagoes, together with those often found in shaded parts of nearby shrubs and trees, can be easily reared to the final adult, or imago, form.

Subimagoes must be handled with extreme care as the slightest injury will prevent transformation into the desired imago. The ordinary aerial insect net, wielded slowly and gently, can be used to capture one flying specimen at a time, the specimen being shaken individually into the open mouth of a paper bag. Subimagoes at rest should be "urged" into the paper bag by the slightest touch on wings or tails.

A standard eight-inch Kraft paper bag with a six-inch branch of leaves at the bottom and the open end folded over provides a good housing for a half-dozen subimagoes. The subimagoes will transform into imagoes in from twelve to forty-eight hours and should then be permitted to rest for another six or eight hours after transformation in order to develop full coloration. Then they should be placed in four-dram vials of 70 per cent alcohol.

Mayfly imagoes are most easily collected by means of the insect net when engaged in their nuptial flight. Different species swarm at different hours and may be found over the water, near it, or more than a mile from it. Most mayflies are poor fliers and usually swarm in areas protected from the wind; some prefer to swarm in bright sunlight whereas others are always found in shade. If females are found depositing eggs on the water, the direction of their approach may be noted and followed to the swarm; collecting activities of birds sometimes indicate the presence of swarming mayflies. Flights are most easily detected when looking directly toward the sun.

Collected imagoes are taken from the aerial net and placed directly into four-dram vials of 70 per cent alcohol, save for a few that go into the ethyl acetate killing jar for color determinations.

Rearing.—From the taxonomic standpoint, the rearing of nymphs through the adult winged stages, with consequent positive association of the life forms, must be considered as a superior procedure in collecting.

Field rearing is accomplished by placing a dozen perfect and mature nymphs of a single species in a sixteen-mesh copper wire screen cylinder about six inches in diameter and eighteen inches long, one end of which is permanently closed with copper screen. With a stone in the bottom, or closed end, the cylinder is placed upright in the stream in eight or nine inches depth of water. A square of cheesecloth is fastened over the open end with a rubber band, and a few leafy branches are laid over the cage to give light shade. In rain areas, a thin wooden cover is placed loosely over the cheesecloth.

The nymphs emerge and climb to the top of the cage where they transform from subimago to imago.

Every effort should be made to avoid any possibility of confusion in associating nymphs and adults of a single species; therefore the nymphal skins and imagoes of the same species are finally added to the same vial of 70 per cent alcohol wherein a half-dozen similar nymphs had been placed originally.

Mature nymphs of certain species may be transported over quite long distances for rearing in the aquarium. At the point of collection, several dozen mature nymphs may be placed in a single six-by-eighteen-inch aluminum screen cylinder. During the day's collecting, this cylinder is half immersed in the stream as for field rearing. At the end of the day, the screens are taken from the stream and placed in three-gallon aluminum milk cans having about six and one-half inch mouths, the cans being two-thirds filled with fresh and well-aerated stream waters. The cans are well wrapped with thick cotton cloths which, when thoroughly wetted, provide quite cool temperatures through evaporation. By traveling during the cool evening hours, with an occasional stop for re-wetting cloths, even many fast-water species may be safely carried more than two hundred miles and reared in the aquarium.

A successful mayfly rearing aquarium in use for several years is eighteen-by-thirty inches in size with a depth of nine inches, constructed of glass and stainless steel. A strong current of water is produced in the tank by Samuel S. Gelber Company's stainless steel, rubber centrifugal pump No. 205, with sheets of heavy plastic arranged in the aquarium to create areas of quiet water, slow current, and rapid current. A single brick stands on end for the use of nymphs which clamber out of the water for emergence.

The aquarium must be kept clean, and the nymphs reared in water brought from the stream where they were collected. The oxygen level of the water in the aquarium should be maintained through the use of a Venturi tube at the discharge end of the Gelber pump, or by continuous operation of a Marco air pump, or equivalent. A small amount of sand and coarse gravel should be scattered over the bottom of the aquarium, and a half-dozen rocks from the collecting stream disposed about the tank. With water coils or floating heater, water temperatures can be held within a few degrees of stream temperatures.

Any number of well-known genera may be reared together in one tank since the adults are easily separated; however, nothing but confusion comes forth from a mixture of the nymphs of closely allied species.

CLASSIFICATION

In the taxonomy of the adult mayflies, the arrangement and form of the wing veins is of utmost importance. The following keys are based on those prepared by Dr. J. R. Traver for *Biology of Mayflies,* Needham, Traver, and Hsu (1935). The system of wing venation used herein is that of Dr. James G. Needham, and is shown in figure 3:2*a*. In all keys Cu_1 is used interchangeably with CuA and Cu_2 with CuP. The keys which follow are for the families and genera of mayflies of the United States and Canada, and for the California species.

Key to the Families of the Order Ephemeroptera

Adults

1. Veins M and Cu_1 of fore wing strongly divergent at base, with M_2 strongly bent toward Cu_1 basally; outer fork (Of) in hind wing wanting (fig. 3:2*b*); hind tarsi 4-jointed.......................... EPHEMERIDAE
— Veins M and Cu_1 little divergent at base and fork of M more nearly symmetrical; outer fork (Of) of Rs in hind wing present or absent; hind tarsus 4- or 5-jointed 2
2. Hind tarsus with 5 freely movable joints; cubital intercalaries in 2 parallel pairs, long and short alternately; venation never greatly reduced (fig. 3:2*c*); eyes of the male simple .... HEPTAGENIIDAE
— Hind tarsi with 3 or 4 freely movable joints; cubital intercalaries not as above (except in Metretopinae); venation sometimes greatly reduced (fig. 3:2*d*); eyes of male often divided .... BAETIDAE

Nymphs

1. Mandible with an external tusk projecting forward and visible from above the head (except in Neoephemerinae) (fig. 3:2*e*) .... EPHEMERIDAE
— Mandible with no such tusk, except in some species of *Paraleptophlebia* 2
2. Head strongly depressed; eyes dorsal[1] (fig. 3:2*f*) ... . HEPTAGENIIDAE
— Head not strongly depressed; eyes lateral (fig. 3:2*g*) .... BAETIDAE

Family EPHEMERIDAE

Key to the Genera

Adults

1. Costal cross veins normally developed 2
— Basal costal cross veins vestigial or wanting .... *Neoephemera* McDunnough
2. Marginal veinlets absent 3
— Marginal veinlets present 4
3. Middle and hind legs reduced to functionless, membranous vestiges, but with all leg parts discernible ....*Tortopus* Needham & Murphy
— Middle and hind legs aborted beyond the trochanters .... *Campsurus* Eaton
4. Cubital intercalaries simple ...... *Ephoron* Williamson
— Cubital intercalaries widely forking 5
5. Cubital intercalaries 1 or 2, long and deeply forked .... *Potamanthus* Pictet
— Cubital intercalaries shorter and with shallow forks, more like marginal veinlets, decurrent from vein Cu_1 6
6. Cross veins somewhat crowded at and below bulla; wings with a variable pattern of dark spots .... *Ephemera* Linnaeus
— Cross veins not crowded at bulla; wings with no pattern of dark spots 7
7. Penes tubelike, tips not incurved.... *Pentagenia* Walsh
— Penes broader at base than apex, tips more or less incurved (fig. 3:3*d*) .... *Hexagenia* Walsh

---

[1]Certain Leptophlebiinae, of family Baetidae, which may seem to fall here, have gills of each pair filiform or lamelliform, whereas in Heptageniidae the upper member of each gill pair is platelike the lower member fibrillate.

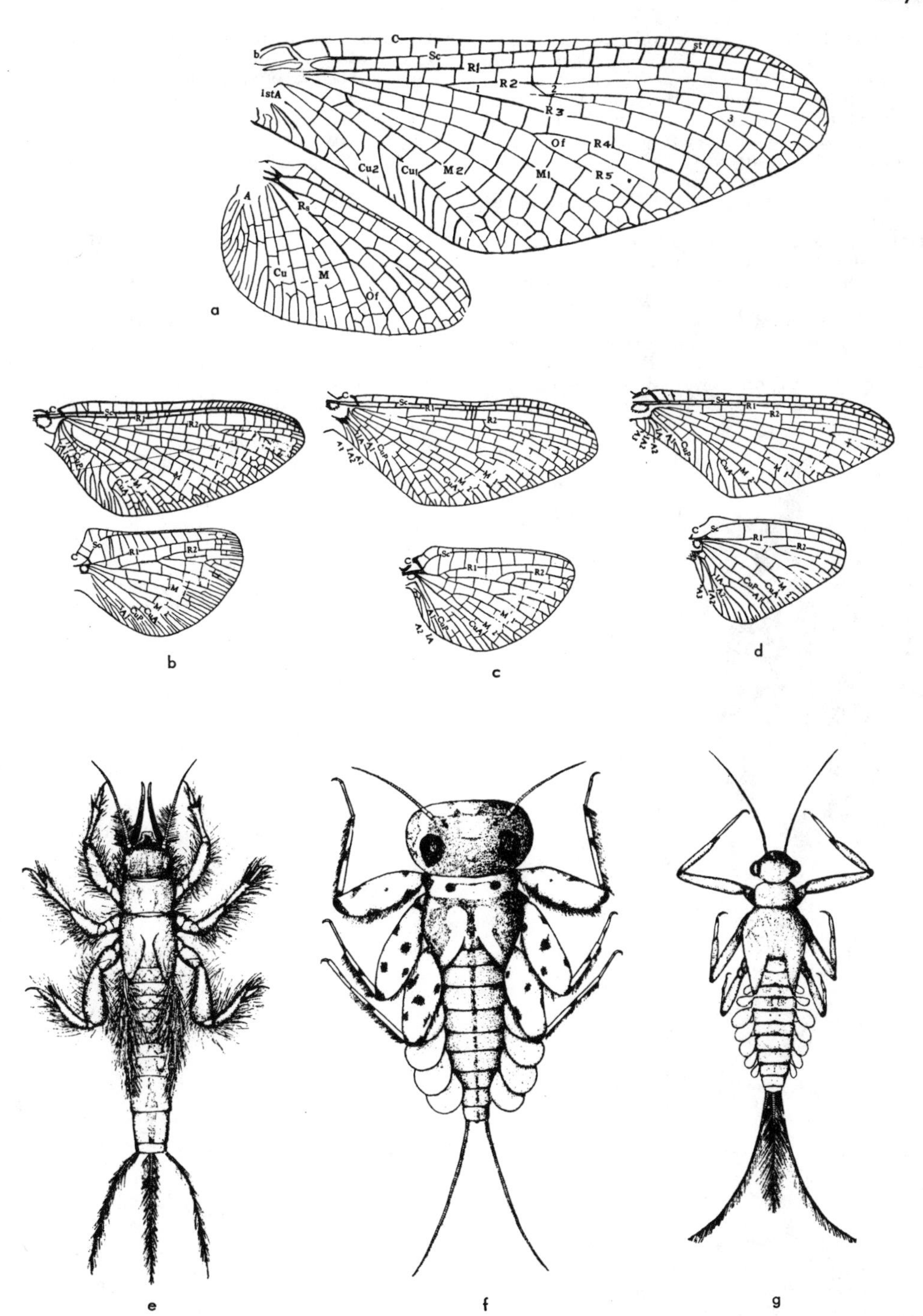

Fig. 3:2. *a*, Wings of *Siphlonurus berenice* illustrating Needham system of wing venation notation; *b*, wings of Ephemeridae; c, wings of Heptageniidae; *d*, wings of Baetidae; e, Ephemeridae, nymph; *f*, Heptageniidae, nymph; *g*, Baetidae, nymph (*a*, Needham, Traver and Hsu, 1935; *b-d*, adapted from Spieth, 1933; e-*g*, Murphy, 1922).

Nymphs

1. Gills present on segments 1-6 only, rudimentary on 1, operculate on 2, so that gills on 3-6 are under opercula; mandibular tusk absent..... *Neoephemera* McDunnough

— Gills present on segments 1-7, rudimentary on 1; mandibular tusk present 2

2. Gills lateral in position, outspread at sides of abdomen; fore tibiae long and slender, not flattened and widened .... *Potamanthus* Pictet

— Gills dorsal in position, curving up over abdomen; fore tibiae more or less flattened and widened (burrowing form) 3

3. Front of head rounded, lacking frontal process ...... *Campsurus* Eaton
— Head with a conspicuous frontal process 4
4. Gills broad, laminate, tapering only near tips, with many prominent lateral tracheal branches; marginal fringe of hairs short, inconspicuous; mandibular tusks curve downward apically in lateral view .... *Ephoron* Williamson
— Gills narrow, tapering, with no distinct lateral tracheal branches, margined on each side by a fringe of hairs at least twice as long as width of gill; mandibular tusks upcurved apically, in lateral view 5
5. Frontal process of head entire; truncate, conical, or rounded (fig. 3:2*e*; 3:3*a*,*b*) .... *Hexagenia* Walsh
— Frontal process of head bifid 6
6. Mandibular tusks crenate on outer (upper) margin; labial palp 2-jointed .... *Pentagenia* Walsh
— Mandibular tusks smooth on margins; labial palp 3-jointed .... *Ephemera* Linnaeus

Fig. 3:3. *a*,*b*, posterior and anterior third gill lamellae of *Hexagenia* spp.; *c*, *Hexagenia limbata californica*, male genitalia; *d*, *Hexagenia* sp.; male genitalia; *e*, *Heptagenia* sp.; male genitalia (*a*,*b*,*c*,*e*, Spieth, 1933; *c*, W. C. Day).

Family EPHEMERIDAE

Only one species of the entire family Ephemeridae has been found in California, namely *Hexagenia limbata californica* (Upholt) 1937. This species is characterized as follows.

Male adult.—Large, yellow, tinged with red on abdomen; extensive dark color pattern. Fore tarsal joints red brown. Costal area of fore wing light brown .Cross veins of fore wing from costa to M_1 and in hind wing disk, heavily marginated with black. Entire rear margin of hind wing widely and heavily infuscated. Penes hook-shaped. Length fore wing 17-18 mm. (fig. 3:3*c*).

Nymph.—Fossorial, yellowish white, dorsum marked as in adult; with low, rounded frontal process of head. Length of mature nymph, 20-27 mm.

Distribution: Sacramento and San Joaquin river systems.

Family HEPTAGENIIDAE

Key to the Genera

Adults

1. Outer fork (Of) of Rs of hind wing wanting; male forceps 5-jointed .... *Arthroplea* Bengtsson
— Outer fork of Rs of hind wing present; male forceps 4-jointed 2
2. Fore tarsus of male not more than 3/4 as long as tibia; apical margin of forceps base deeply excavated .... *Anepeorus* McDunnough
— Fore tarsus of male distinctly longer than tibia; apical margin of forceps base not as above 3
3. Basal joint of fore tarsus of male 1/2 or less than 1/2 as long as second joint 4
— Basal joint of fore tarsus of male at least 2/3 as long as second joint; may equal or slightly exceed it in length 6
4. Stigmatic cross veins more or less anastomosed; basal joint of fore tarsus of male 1/6 to 1/4 as long as second (fig. 3:4*a*) .... *Rhithrogena* Eaton
— Stigmatic cross veins not anastomosed; basal joint of fore tarsus of male variable 5
5. Basal joint of fore tarsus of male 1/3 to 1/2 as long as second; penes more or less distinctly L-shaped .... *Stenonema* Traver
— Basal joint of fore tarsus of male 1/6 to 1/3 as long as second; penes not distinctly L-shaped (fig. 3:3*e*) .... *Heptagenia* Walsh
6. Basal joint of fore tarsus of male 2/3 to 3/4 as long as second joint 7
— Basal joint of fore tarsus of male equal to or slightly longer than second joint 8
7. Stigmatic area of forewing divided by a fine line into an upper and lower series of cellules; basal costal cross veins very weak (fig. 3:4*b*)...... *Cinygma* Eaton
— Stigmatic area of fore wing not divided as above; basal costal cross veins usually well developed (fig. 3:5*a*) .... *Cinygmula* McDunnough

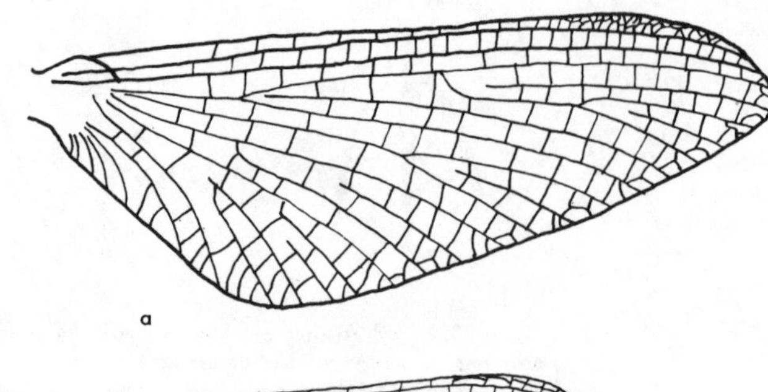

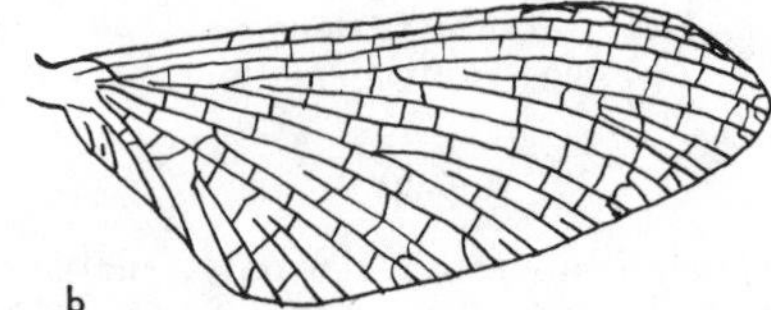

Fig. 3:4. Fore wings. *a*, *Rhithrogena flavianula*; *b*, *Cinygma* sp. (W. C. Day).

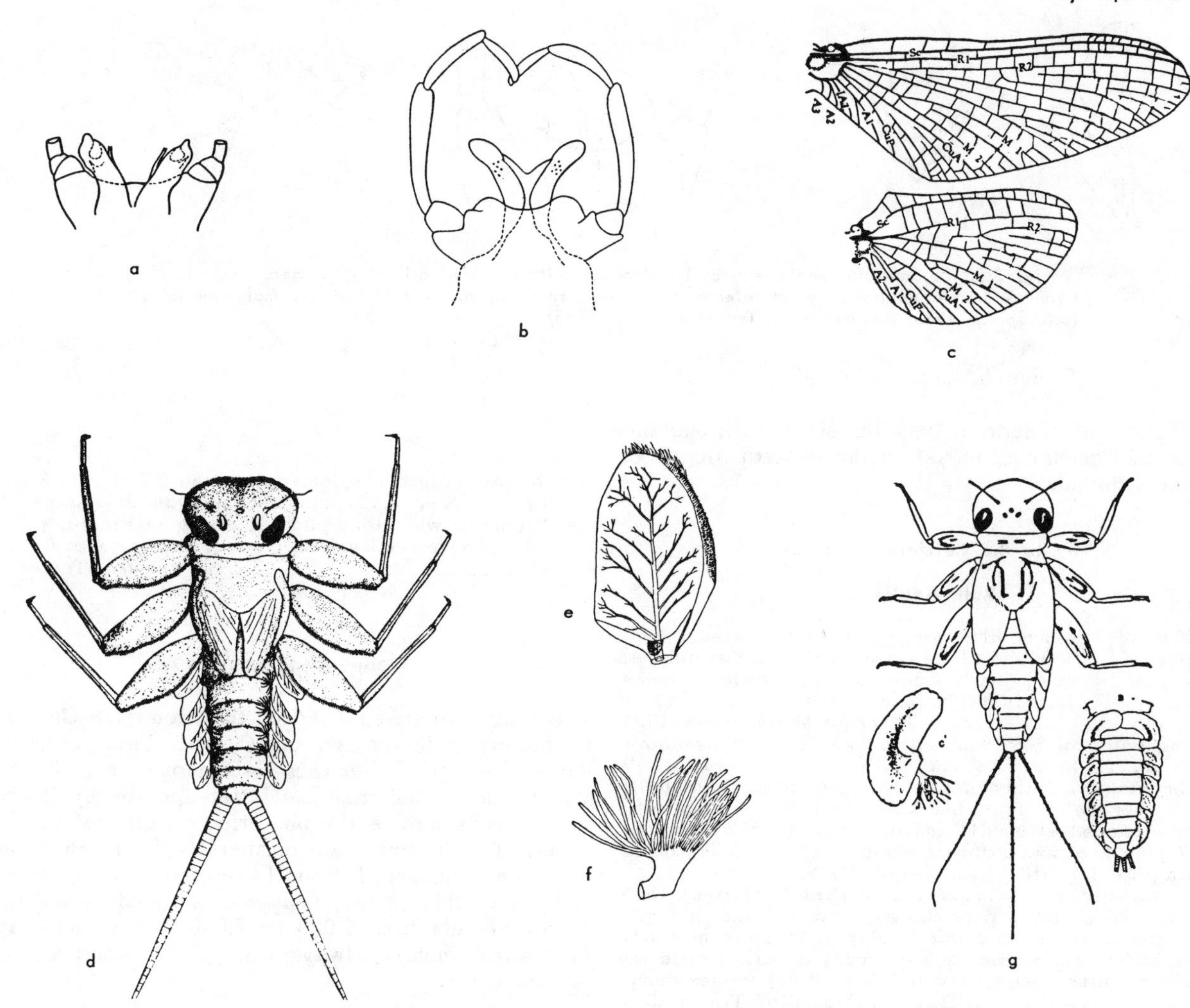

Fig. 3:5. *a, Cinygmula uniformis*, male genitalia; *b, Ironopsis grandis*, male genitalia; *c, Iron* sp., wings; *d, Ironodes* sp., nymph; *e,f*, anterior and posterior lamellae of third nymphal gill of *Iron* sp.; *g, Rhithrogena* sp., nymph, with ventral aspect of abdomen and first gill (*a,b,d,g*, Needham, Traver and Hsu, 1935; *c,e,f*, Spieth, 1933).

8. Stigmatic cross veins of fore wing more or less strongly anastomosed, sometimes forming 2 series of cellules; basal costal cross veins weak (fig. 3:5*b*) .... *Ironopsis* Traver
— Stigmatic cross veins of fore wing not anastomosed; basal costal cross veins variable 9
9. Basal costal cross veins of fore wing weak, stigmatic cross veins not slanting (fig. 3:5*c*) .... *Iron* Eaton
— Basal costal cross veins of fore wing strong; stigmatic cross veins distinctly slanting ...... *Ironodes* Traver

Nymphs

1. Fibrillar part of each pair of gills wanting; second joint of maxillary palp at least 4 times as long as galealacinia .... *Arthroplea* Bengtsson
— Fibrillar part of each pair of gills present; second joint of maxillary palp much shorter 2
2. Tails 2 3
— Tails 3 5
3. A triad of stout spines present at tip of galea-lacinia; no submedian spines on tergites 4
— No such triad of spines on galea-lacinia, a pair of submedian spines present on tergites (fig. 3:5*d*) .... *Ironodes* Traver
4. Median row of long hairs on abdominal tergites; front of head slightly emarginate; first gills meet beneath the body .... *Ironopsis* Traver
— No such row of median hairs on tergites; front of head not emarginate (fig. 3:5*e,f*) .... *Iron* Eaton
5. Gills of seventh pair reduced to a single slender tapered filament or spine; tracheae, if present in this filament, without lateral branches .... *Stenonema* Traver
— Gills of seventh pair flat and platelike, quite similar to preceding pairs; tracheae with lateral branches always present in seventh pair 6
6. Fibrillar part of gills reduced to a few tiny threads; front of head distinctly emarginate .... *Cinygmula* McDunnough
— Fibrillar part of gills well developed on segments 1-6, may be wanting on 7; front of head entire or very slightly emarginate 7
7. Gills of first and last pairs much enlarged, converging beneath body (fig. 3:5*g*) .... *Rhithrogena* Eaton
— Gills of first and last pairs not as large as middle pairs; all gills directed laterally, not converging beneath body 8
8. Labrum narrow, about 1/3 the width of anterior margin of head; a chitinized area present on each mandible below molar region .... *Cinygma* Eaton
— Labrum broad, 2/3 to 3/4 the width of anterior margin of head; no such chitinized area on mandible; (fig. 3:6*a,b*) .... *Heptagenia* Walsh

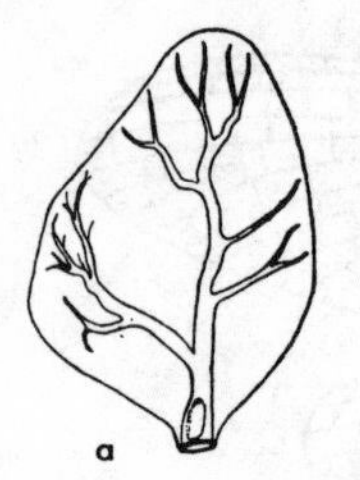

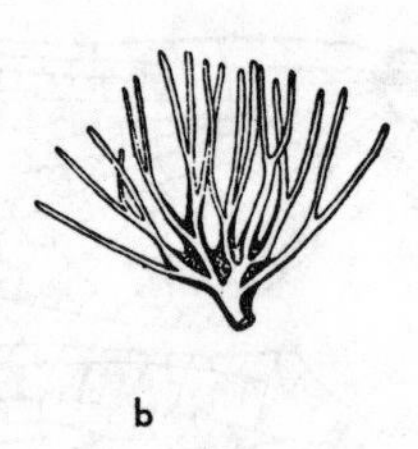

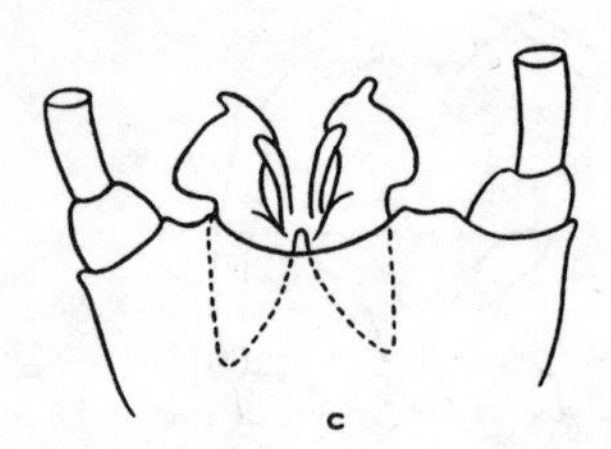

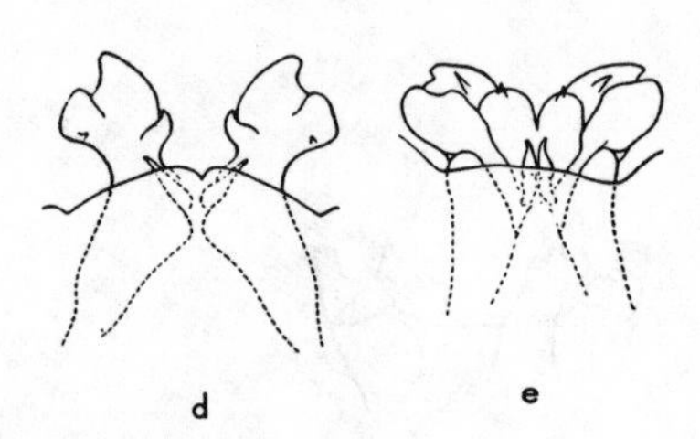

Fig. 3:6. *a,b,* anterior and posterior lamellae of third nymphal gill of *Heptagenia* sp.; *c, H. kennedyi,* male genitalia; *d, H. elegantula,* male genitalia; *e, H. rubroventris,* male genitalia (*a,b,* Spieth, 1933; c-e, Needham, Traver and Hsu, 1935).

Genus *Heptagenia* Walsh

Distribution is general in smaller streams throughout California from the seacoast to the eastern slopes of the Sierra Nevada.

Key to the California Species

Male Adults

1. Veins of fore and hind wings entirely colorless; legs pale, fore femora and tibiae tinged with black at apices; tergites 2-9 pale, faintly smoky along posterior margins (fig. 3:6*c*) (Alameda and Napa counties)............ .... *kennedyi* McDunnough 1924
— Long veins of fore wing dark; fore legs not as above 2
2. Subcosta and radius of fore wing amber in basal half; all other veins fine, blackish; fore leg light brown, femur banded at middle and darkened at apex; tergites 2-7 pale hyaline, posterior margins smoky; 8-10 tinged with pink (fig. 3:6*d*) (San Joaquin Valley) .... *elegantula* (Eaton) 1885
— Long veins fine, dark brown; cross veins entirely colorless; fore femur dark ruddy with apex narrowly black; fore tibia pale basally, ruddy apically; abdomen widely marked with dark red (fig. 3:6*e*) (widespread) .... *rubroventris* Traver 1935

Although flattened and apparently well adapted for swift water habitation, the two following nymphs prefer slow moving currents. They are found under rocks in midstream or in rocky situations at the water's edge.

Known nymphs

— Pronotum widest at front; claws not pectinate; postero-lateral abdominal spines on segments 6, 7, and 8 absent ....*elegantula* (Eaton) 1885
— Pronotum widest at middle; claws pectinate, postero-lateral abdominal spines present on segments 6, 7, and 8......................... *rubroventris* Traver 1935

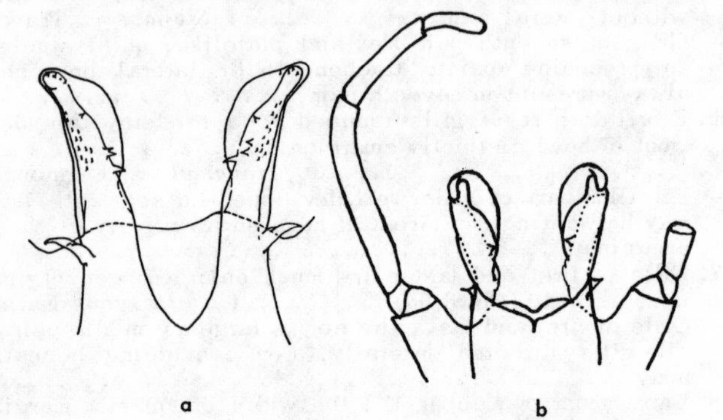

Fig. 3:7. Male genitalia of *Rhithrogena. a, flavianula; b, doddsi* (*a,* Day, 1954; *b,* McDunnough, 1934).

Genus *Cinygma* Eaton

The single representative of this genus in California is herewith designated as *C.* sp? This species is quite close to *C. dimicki* McDunnough but the male may be separated from the latter species by its wide black bands across the posterior margins of the sternites. The nymph has comparatively much smaller gills on segment 1 than have been found in other nymphs of this genus. *Cinygma* is found in the lower Sierra Nevada from 5,000 to 7,000 feet in moderately fast water, nearly always clinging to wood and bark of dark color.

Genus *Rhithrogena* Eaton

With the exception of three species, this genus has been found in California only above 6,000 feet in the Sierra Nevada. *R. petulans* may extend its range from southern California into the redwood belt of northern California. *R. morrisoni* has been found on both sides of the Sierra Nevada, and an undescribed related species occurs in the Redwood Belt and Coast Range. The nymphs are limpetlike forms with expanded gills forming perfect suction cups for attachment to stones in fast currents.

Key to the California Species

Male Adults

1. Fore wing 8 mm. or less in length (fig. 3:27*b*); Shasta County ....*decora* Day 1954
— Fore wing 10 mm. or longer 2
2. Lateral spine near base of each long, slender division of penes, 5 or 6 dorsal spines in central part of each division; larger species, wing 15 mm. in length (fig. 3:7*a*); Alpine County ...*flavianula* (McDunnough) 1924
— Basal lateral spines of penes as above; ventral spines present on each division of penes in central part; wing 13 mm. or less in length 3

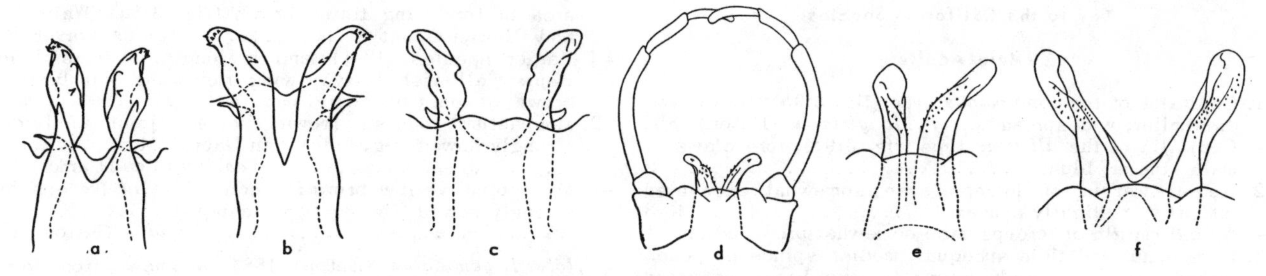

Fig. 3:8. Male genitalia. *a, Rhithrogena petulans; b, R. morrisoni; c, R. brunnea; d, Ironodes lepidus; e, I. californicus; f, I. nitidus* (Needham, Traver, and Hsu, 1935).

3. Divisions of penes only slightly divergent apically 4
— Divisions of penes strongly divergent apically 5
4. Each division of penes broad at apex; lateroapical spines minute (fig. 3:7*b*); Sierra Nevada .... *doddsi* McDunnough 1926
— Each division of penes narrowed at apex; lateral-apical spines better developed (fig. 3:8*a*); San Bernardino County .... *petulans* Seeman 1927
5. Each division of penes narrowed at apex; 5-6 small apical ventral spines (fig. 3:8*b*); widespread .... *morrisoni* (Banks) 1924
— Each division of penes broader at apex; mid-ventral spine strong, apical spines less well developed (fig. 3:8*c*); Sierra Nevada .... *brunnea* (Hagen) 1875

Known Nymphs

1. Body 7 mm. or less in length (fig. 3:27*p*) .... *decora* Day 1954
— Body 9 mm. or longer 2
2. Body green brown; gills sometimes rose tinted, legs pale with darker femora marked with a pale V .... *doddsi* McDunnough 1926
— Body red brown; tergites 8-10 with pale submedian spots; gills sometimes rose tinted; legs pale, outer surface of femora brown with large median dark brown spot and with black apices ..... *morrisoni* (Banks) 1924

Genus *Cinygmula* McDunnough

The nymphs are often found in water only one to two inches deep at the foot of riffles, in crevices, and on lower surfaces of small stones in the small streams of the Redwood Belt, Coast Range, and Sierra Nevada.

Key to California Species

Male Adults

1. Each division of penes with ventral spines 2
— Penes without ventral spines 3
2. Ventral spines long, distinct; lateral in position and near base of penes; Madera County.... *par* (Eaton) 1885
— Ventral spines small; subapical in position (fig. 3:5*a*); central California coastal .. *uniformis* McDunnough 1934
3. Wings usually suffused with amber, strongly so at base; margin of styliger plate straight; no tracheations on abdominal tergites; widespread (fig. 3:9*a*) .... *mimus* (Eaton) 1885
— Wings evenly and faintly amber tinted; styliger plate slightly produced; prominent dark tracheations on abdominal tergites; Sierra Nevada .... *tioga* Mayo 1952

Nymphs

— Femora ochreous, with 2 broad brown bands......... .... *uniformis* McDunnough 1934
— Femora ochreous, with pale L-shaped, wide longitudinal mark in center of basal half and another wide, pale area along dorsal margin in basal half .... *mimus* (Eaton) 1885

Genus *Iron* Eaton

Widely distributed at all elevations in rocky streams of good current and moderate to cold temperatures, this typically flattened fast-water nymph is found in the strongest currents of rapids clinging to the undersides of medium to large boulders. They are removed only by individual picking.

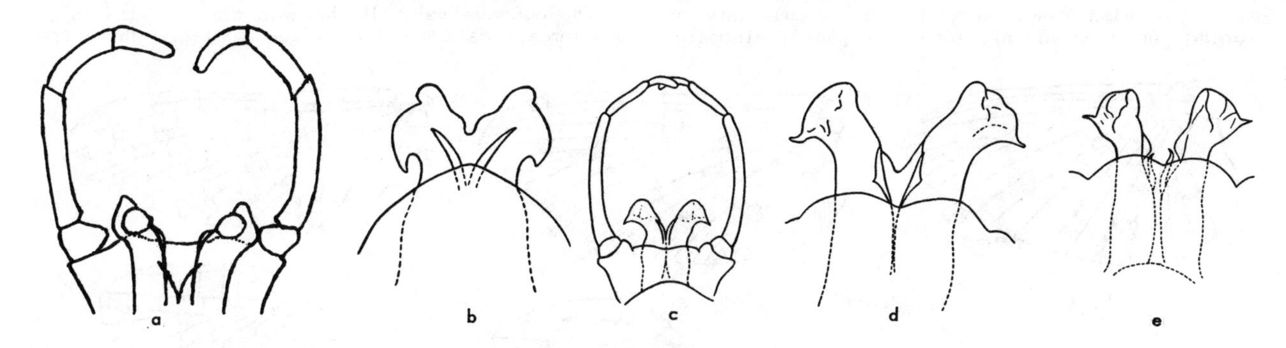

Fig. 3:9. Male genitalia. *a, Cinygmula mimus; b, Iron longimanus; c, I. albertae, d, I. sanctagabriel; e, I. dulciana* (Needham, Traver, and Hsu, 1935.)

Key to the California Species

Male Adults

1. Genitalia of the *longimanus* type (fig. 3:9*b*); fore claws dissimilar; widespread .... *longimanus* (Eaton) 1881
— Genitalia of the *albertae* type (fig. 3:9*c*); fore claws of male similar, blunt 2
2. Apical margin of forceps base somewhat excavated laterally or slightly convex 3
— Apical margin of forceps base somewhat produced ... 4
3. Fore femur and tibia subequal; median spines on penes rather stout, relatively short; forceps base excavated laterally (fig. 3:9*c*); central California coastal .... *albertae* McDunnough 1924
— Fore femur shorter than tibia; median spines on penes longer and more slender; apical margin forceps base convex; Marin County .... *lagunitas* Traver 1935
4. Venation brown, all cross veins distinct (fig. 3:9*d*); San Gabriel Mountains .... *sancta-gabriel* Traver 1935
— Venation pale, cross veins mostly indistinct (fig. 3:9*e*); east slope Sierra Nevada... *dulciana* McDunnough 1935

Known Nymphs

1. Outer posterolateral spines on middle abdominal segments long and slender, about 1/2 the length of segment on which found .... *albertae* McDunnough 1924
— Outer posterolateral spines on middle abdominal segments short and blunt, about 1/4 the length of segment on which found 2
2. First pair of lamellate gills meet beneath the body .... *longimanus* (Eaton) 1881
— First pair of lamellate gills definitely not meeting beneath the body .... *sancta-gabriel* Traver 1935

Genus *Ironodes* Traver

Found at higher elevations in the southern California mountains, this genus has also been taken in northern California in the Coast Range and the Cascades at elevations less than 5,000 feet. The nymphs are ordinarily found in small tributary streams of moderately fast current clinging to rocks and wood close to the surface.

Key to the California Species

Male Adults

1. Small species, 8-9 1/2 mm. in length; abdominal segments yellowish brown except for narrow pale anterior margin; small subapical spines on penes; stigmatic area of fore wing tinted brown (fig. 3:8*d*); White Mts. and Plumas County .... *lepidus* Traver 1935
— Larger species, 12-14 mm. in length; abdominal segments yellowish brown; wing membrane faintly tinted brown, at least on costal margin 2
2. Mesonotum reddish brown; second joint of forceps strongly bowed (fig. 3:8*e*); San Gabriel Mts. .... *californicus* (Banks) 1910
— Mesonotum yellow brown; second joint of forceps less strongly bowed (fig. 3:8*f*); widespread .... *nitidus* (Eaton) 1885

Note: I. geminatus (Eaton) 1885 is known from female adult only; 8-9 1/2 mm. in length; yellow tinted wings; tergites yellowish.

Nymphs

1. Small species, 9-10 mm. in length.. *lepidus* Traver 1935
— Larger species, 12-15 mm. in length 2
2. Tracheae of gills distinct; no distinct dark lateral streaks on abdominal sternites .... *californicus* (Banks) 1910
— Tracheae of gills indistinct; distinct dark lateral streaks on basal and middle sternites .... *nitidus* (Eaton) 1885

Genus *Ironopsis* Traver

The single California species reported is *I. grandis* (McDunnough) 1924.

Male.—The male is 13-16 mm. in length; mesonotum light brown; fore leg blackish brown, paler at base of femur, with other legs light olive to red brown; mesonotum light red brown; small spines on ventral surface of penes few in number (fig. 3:5*b*).

Nymph.—In addition to characters given in keys to the Heptageniidae herein, the nymph has a fringe of long hairs on the femur, tibia, and tarsus, and three lateral pectinations near apex of each claw; gills on seventh segment meet beneath the body.

Distribution.—Found in California in the Sierra Nevada, above 6,000 feet elevation.

Family BAETIDAE

Key to the Subfamilies

Adults

1. The concave veins lie beneath the convex veins, and intercalaries are mostly absent, so that there appear

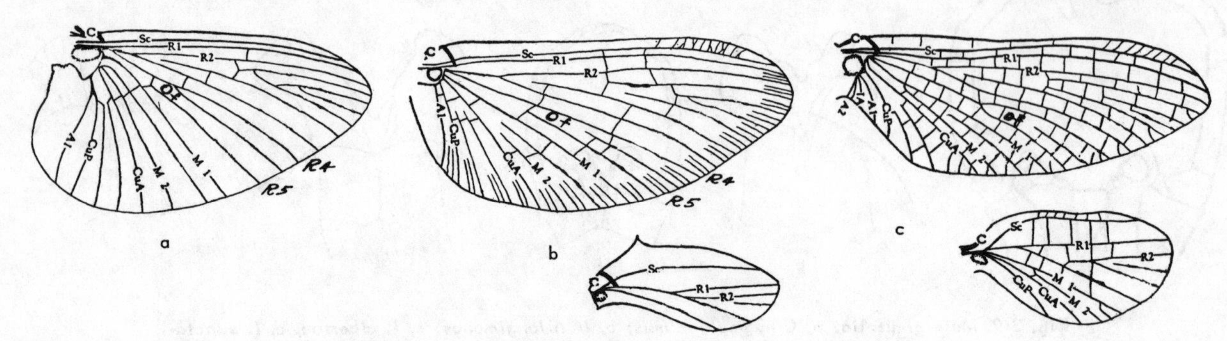

Fig. 3:10. Wings. *a*, *Caenis* sp., fore wing only; *b*, *Baetis* sp.; *c*. *Paraleptophlebia* sp. (adapted from Spieth, 1933).

to be 8 or less longitudinal veins; cross venation greatly reduced .... OLIGONEURIINAE
— Fore wing with more extensive venation than above 2
2. First anal vein of fore wing ends in outer margin; no cubital intercalaries .... BAETISCINAE
— First anal vein ends in hind margin; cubital intercalaries present 3
3. Hind wings present or absent; outer fork of Rs (*Of*) of fore wing and posterior branch of M detached basally from their respective stems (fig. 3:10*b*); eyes of male turbinate .... BAETINAE
— Hind wings present or absent; *Of* of fore wing attached basally; eyes of male not as above 4
4. Hind wings absent in imago; posterior branch of M in fore wing may be attached or detached basally from anterior branch (fig. 3:10*a*) .... CAENINAE
— Hind wings present; posterior branch of M in fore wing normal, attached basally to anterior branch 5
5. *Of* of hind wing (R_4 and R_5) present; tails 2 6
— *Of* of hind wing absent; tails 3 7
6. Cubital intercalaries of fore wing consist of a series of forking or sinuate veinlets attaching Cu_1 to hind margin (fig. 3:2*a*) .... SIPHLONURINAE
— Cubital intercalaries of fore wing 2 to 4 in number, free basally, not as above .... METRETOPINAE
7. Two pairs of cubital intercalaries, the anterior pair being the longer; first anal vein of fore wing attached to hind margin by a series of veinlets .... AMETROPINAE
— Two cubital intercalaries only; first anal vein not attached to hind margin as above 8
8. Two short intercalaries in fore wing between median intercalary and posterior branch of M, also between the latter vein and Cu_1; male forceps with a single short terminal joint (fig. 3:11*a*) ... EPHEMERELLINAE
— No true intercalaries in fore wing in positions indicated above; male forceps with 2 (occasionally 3) short terminal joints (fig. 3:10*c*) .... LEPTOPHLEBIINAE

Nymphs

1. Gills of the first abdominal segment stem from the ventral surface; a tuft of gills at the base of each maxillae, but no gill tufts at base of fore coxae .... OLIGONEURIINAE
— Gills of the first abdominal segment not as above; gill tufts at base of maxillae may be present when accompanied by gill tufts at the base of the fore coxae 2
2. Gills completely concealed beneath an enormously enlarged thoracic shield or carapace ... BAETISCINAE
— Gills exposed; thoracic notum normal 3
3. Outer tails fringed[2] on both sides 4
— Outer tails with heavy fringes on inner side only (may have a few short hairs on outer side) 6
4. Gills present on abdominal segments 1-7 .... LEPTOPHLEBIINAE
— Gills absent from one or more of segments 1-7 5
5. Gills present on segments 3-7 or 4-7; in latter case, may be elytroid on segment 4; a rudimentary gill often present on segment 1 .... EPHEMERELLINAE
— Gills present on segments 1-6 or 2-6; if present, rudimentary on segment 1; gill on segment 2 elytroid, covering all those behind it .... CAENINAE[3]
6. Claw of fore leg usually differs from others in structure but, if similar in structure, margins of gills deeply fissured; claws of middle and hind legs long and slender 7
— Claws on all legs similar, sharp pointed, and usually much shorter than tibiae; margins of gills not fissured 8
7. Claw of fore leg bifid .... METRETOPINAE
— Claw of fore leg sometimes slender, curved, bearing several long spines; when fore claw without such spines, gills deeply fissured .... AMETROPINAE
8. Posterolateral angles of the apical abdominal segments prolonged into thin, flat lateral spines; frontal margin of labrum usually with not more than a shallow, broad emargination...................... SIPHLONURINAE[4]
— Posterolateral angles of apical abdominal segments hardly more than acute, not prolonged into flat, thin, lateral spines; a well-defined notch at center of frontal margin of labrum .... BAETINAE[5]

[2]*Baetodes* of subfamily Baetinae with tails almost bare.

[3]*Neoephemera* of the family Ephemeridae has similar characters but differs in having buds and tiny gills on segment 1; in the Caeninae, hind wing buds are usually lacking but, if present, gills are lacking entirely on segment 1.

Subfamily SIPHLONURINAE

Key to the Genera

Adults

1. Claws on each tarsus dissimilar; costal angulation of the hind wing acute (fig. 3:11*b*)........ *Ameletus* Eaton
— Claws on each tarsus similar; costal angulation of hind wing obtuse or wanting 2
2. Hind tarsus longer than tibia (except in *Siphlonurus marshalli*); all claws sharp pointed (fig. 3:2*a*) 3
— Hind tarsus shorter or subequal to tibia; claws of middle and hind legs sharp pointed, variable on fore leg of male 4
3. Fore leg as long or longer than body; posterior margin of styliger plate of male produced, straight, or slightly concave .... *Siphlonurus* Eaton
— Fore leg two-thirds the length of the body; posterior margin of styliger plate of male with deep, median V-shaped cleft (fig. 3:27*a*) .... *Edmundsius* Day
4. Abdominal segments 5-9 dilated laterally into flat, broad expansions; no costal angulation on hind wing .... *Siphlonisca* Needham
— Abdominal segments 5-9 not dilated as above; costal angulation of hind wing present 5
5. Media of hind wing simple, not forked; all claws sharp pointed .... *Parameletus* Bengtsson
— Media of hind wing forked near outer margin (fig. 3:2*d*); fore claws of male similar, blunt, in most species ... . *Isonychia* Eaton

Nymphs

1. Fore legs conspicuously fringed with long hairs (fig. 3:11*f*); gill tufts present on bases of maxillae and fore coxae .... *Isonychia* Eaton
— Fore legs without such conspicuous fringes; no maxillary or coxal gill tufts 2
2. Gill lamellae double on at least segments 1 and 2 .. 3
— Gill lamellae single on all segments 4
3. Gills oval, the posterior lamellae on segments 1 and 2 about two-thirds as large as the anterior lamellae (fig. 3:27*q*); meso- and metathoracic claws almost twice as long as the prothoracic claws........ *Edmundsius* Day
— Gills with retuse apical margins, the posterior lamellae on segments 1 and 2 as large as the anterior lamellae (fig. 3:11*c,d*); meso- and metathoracic claws not noticeably longer than the prothoracic claws .... *Siphlonurus* Eaton

[4]In *Parameletus* and some species of *Ameletus*, these spines are very short and weakly developed. The pincerlike process at the tip of the labial palp in *Parameletus* and the series of pectinate spines on the crown of the maxilla in *Ameletus* easily separate these 2 genera from all of the Baetinae. In *Edmundsius* the frontal margin of the labrum is broadly and rather deeply excavated.

[5]In *Apobaetis*, frontal margin of labrum without center notch.

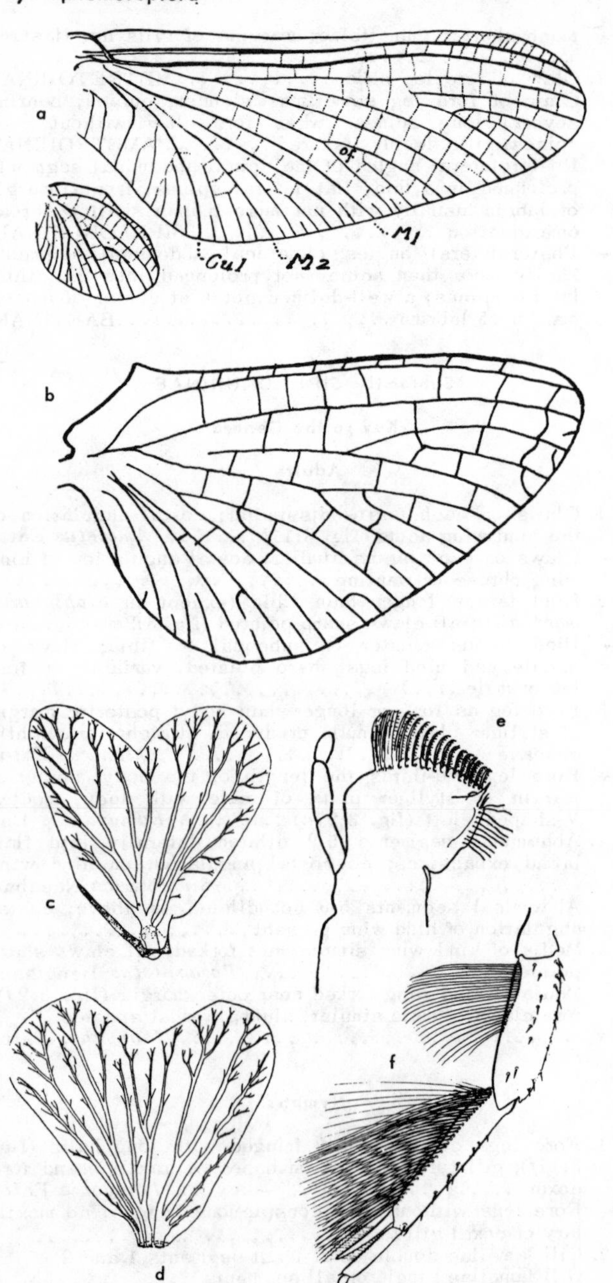

Fig. 3:11. *a, Ephemerella* sp., wings; *b, Ameletus dissitus,* hind wing; *c,d,* anterior and posterior lamellae of first nymphal gill of *Siphlonorus* sp.; *e, Ameletus* sp., maxilla of nymph; *f, Isonychia velma,* fore leg of nymph (*a,e,* Needham, Traver and Hsu, 1935; *b,f,* Day; *c,d,* Spieth, 1933).

4. Abdominal segments 5-9 greatly expanded laterally .... *Siphlonisca* Needham
— Abdominal segments 5-9 normal, not greatly expanded laterally 5
5. A pincerlike process near tip of labial palp; no pectinate spines on crown of maxilla .... *Parameletus* Bengtsson
— No such pincerlike process on labial palp; a fringe of pectinate spines on upper margin of maxilla (fig. 3:11*e*) .... *Ameletus* Eaton

Genus *Ameletus* Eaton

This genus is widely distributed in California at all elevations. Although found in many fast streams the nymph often seeks the quieter parts. It is a fine swimmer. Because of insufficient data a key to the nymphs of California species is impractical.

Key to the California Species

Male Adults

1. Several cross veins in fore wing widely margined with brown, so that wing appears speckled (fig. 3:15*a*); Amador County .... *amador* Mayo 1939
No widely margined cross veins; wings not speckled 2
2. Ganglionic areas of abdominal sternites strongly marked with black or brown 3
— Ganglionic areas of sternites not strongly marked with black or brown 4
3. A single long slender spine attached to inner margin of each division of penes; abdominal tergites yellow (fig. 3:12*a*); Alameda County, Napa County, and "San Geronimo" .... *dissitus* Eaton 1885
— Penes without such spines; abdominal tergites yellow-brown (fig. 3:13*a*); southern California, montane .... *velox* Dodds 1923
4. Wings strongly brown tinted 5
— Wings clear, very faintly brown, or milky 6
5. Apical ends of penes rounded, stimuli absent (fig. 3:13*b*); Redwood Belt and Alpine County .... *shepherdi* Traver 1934
— Apical ends of penes fine pointed, directed outward; short, straight, fine stimuli present (fig. 3:14*a*); montane .... *validus* McDunnough 1923
6. Wings faintly milky; ganglionic areas of sternites faintly dark; stimuli of penes straight; second joint of forceps bent strongly inward at base (fig. 3:12*b*); coastal and Sierra .... *imbellis* Day 1952
— Wings vitreous or faintly brown; ganglionic areas of sternites opaque white; stimuli of penes bent outward at right angles at tips (fig. 3:12*c*); central California, coastal .... *facilis* Day 1952

Genus *Siphlonurus* Eaton

S. spectabilis is a very common species of general distribution, but collection data indicates it is restricted to the lower elevations. *S. columbianus* has been found only at elevations above 4,000 feet. The perfectly streamlined nymphs of this genus attain lengths

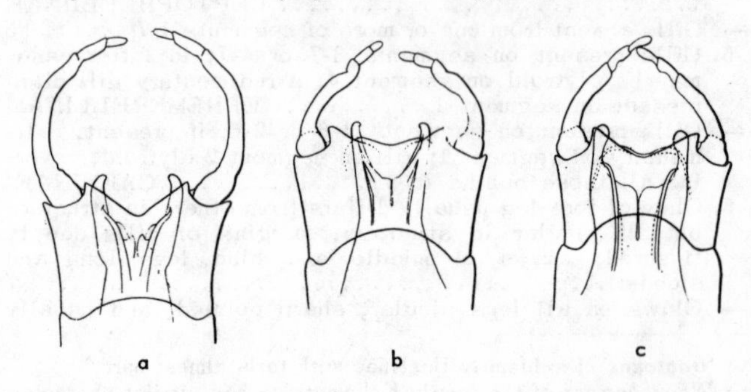

Fig. 3:12. Male genitalia of *Ameletus. a, dissitus; b, imbellis; c, facilis* (Day, 1952).

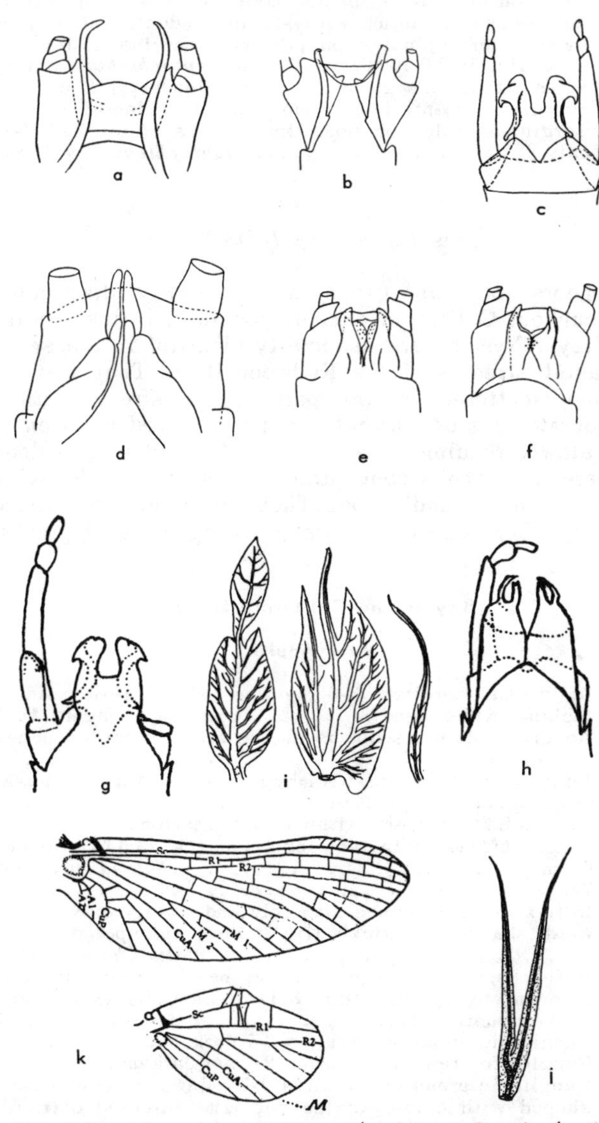

Fig. 3:13. *a-h*, male genitalia; *a*, *Ameletus velox*; *b*, *A. shepherdi*; *c*. *Paraleptophlebia* sp.; *d*, *Siphlonurus spectabilis*; *e*, *S. columbianus*; *f*, *S. occidentalis*; *g*, *Paraleptophlebia debilis*; *h*, *P. pallipes*; *i*, *Choroterpes* sp., anterior and posterior lamellae of third gill and first gill; *j*, *Paraleptophlebia* sp., third nymphal gill; *k*, *Choroterpes* sp., wings (*a,b,d,e,f*, Needham, Traver, and Hsu; *c,i-k*, Spieth, 1933; *g,h*, McDunnough, 1926).

up to 18 mm. They prefer quiet water and are often found in warm, poorly aerated, detached pools.[6]

Key to the California Species

Adults

1. Hind wing of male often orange tinted; fore and hind wings of both sexes with a large dark brown blotch in the radial space (fig. 3:13*d*); widespread below 5,000 ft. elevation .... *spectabilis* Traver 1934
— Fore and hind wings of both sexes hyaline; prominent dark U-shaped marks on abdominal sternites 2
2. U-marks on sternites 8 and 9 diffuse and poorly defined; a pale ruddy spot on the rear of the mesonotum (fig. 3:13*e*); Sierra Nevada. . . .*columbianus* McDunnough 1925
— U-marks clearly defined on sternites 8 and 9; no such ruddy mark on the mesonotum (fig. 3:13*f*); widespread .... *occidentalis* Eaton 1885

Key to the California Species[7]

Nymphs, female

1. Posterior margin of styliger plate with long, narrow medial projection, sides of which are smoothly concave from base to acute, pointed tip ....*spectabilis* Traver 1935
— Posterior margin of styliger plate not as above 2
2. Posterior margins of styliger plate with short, wide medial projection, sides of which are straight and form a right angle at tip ....*occidentalis* Eaton 1885
— Posterior margin of styliger plate with low, rounded medial projection ......*columbianus* McDunnough 1925

Genus *Edmundsius* Day

The genus *Edmundsius* was described from a single species, *Edmundsius agilis* Day 1953. This genus has been reported from California only, and both nymph and adult resemble, in a general way, the same life forms of the genus *Siphlonurus*.

The nymphs of *E. agilis* have been taken in Madera County on the western slope of the Sierra Nevada, at elevations of from 5,000 to 8,000 feet. The species frequent large, shallow, but well-aerated pools in a typical small Sierra stream of high gradient. Nothing is known of the adult life of this species except that maturity is reached at the above locality during July and early August.

Adults.—The large male, about 16-17 mm. in length, has wings with typical venation of the Siphlonurinae (fig. 3:2*a*). Eyes contiguous, not divided, with lower part darker; genitalia as in figure 3:27*a*.

Nymph.—The gray brown nymph, up to 17 mm. in length, is distinguished by long claws on the middle and hind legs which are relatively twice as long as the claws of any nymph of the closely related *Siphlonurus*. Tergites 2-8 distinctively marked with prominent white spots, one on each side. Hind leg as in figure 3:27*m*.

Genus *Isonychia* Eaton

The genus *Isonychia* is represented in California by but one species, *I. velma* Needham 1932. The nymph has been found only in very fast riffles in the larger rivers and streams of northern California and is a large dark species up to 20 mm. in body length.

Adults.—Forceps base of male and subanal plate of female deeply excavated apically; venation strong, dark brown (see fig. 3:2*d* for wing venation).

Nymph.—Body darkest yellow-brown; head almost black; tergites with faint, narrow pale median stripe in anterior halves paralleled by a submedian pale stripe on each side.

[6] *S. maria* Mayo, 1939, is here regarded as a synonym of *S. spectabilis*.

[7] This key is useful when applied to series of mature female nymphs.

Gills pale gray brown with chitinous ridge along lower edges and chitinous band 1/3 distance below dorsal edge; a spur almost as long as tarsus projects from apical end of fore tibia (fig. 3:11*f*).

Subfamily LEPTOPHLEBIINAE

Key to the Genera

Adults

1. Hind wing much reduced in size, venation rather scanty; a distinct costal angulation halfway to apex 2
— Hind wing somewhat larger, venation less reduced; no costal angulation, costal margin slightly concave at center 6
2. Costal angulation of hind wing obtuse; no sag in *Of* of forewing 3
— Costal angulation of hind wing acute; a distinct sag usually evident in *Of* of forewing 4
3. M of hind wing forked; fore tarsus of male shorter than tibia .... *Thraulodes* Ulmer
— M of hind wing not forked (fig. 3:13*k*); male fore tarsus subequal to tibia .... *Choroterpes* Eaton
4. Forceps base of male entire; apical margin of subanal plate of female only slightly cleft .... *Traverella* Edmunds
— Forceps base of male rather deeply cleft, as is the apical margin of subanal plate of female 5
5. In hind wing subcosta ends in margin at outer side of costal angulation; 2 end joints of forceps short, together not 1/4 length of preceding joint.................. .... *Habrophlebiodes* Ulmer
— In hind wing subcosta extends almost to apex of wing; 2 end joints of forceps together equal in length the preceding joint .... *Habrophlebia* Eaton
6. Penes of male separated almost to base; a long flaplike appendage is attached near apex of each and extends inward and downward between lobes of penes .... *Leptophlebia* Westwood
— Penes usually not as above; usually 2 sets of appendages, an upper short pair extending laterally and a lower longer pair (the reflex spurs) extending backward, often obliquely (figs. 3:10*c*; 3:13*c*) .... *Paraleptophlebia* Lestage

Nymphs

1. Gills of first pair differing in form from those of the following pairs 2
— Gills of first pair similar in form to those of the following pair 4
2. Gills filamentous; each gill, on segments 2-7, consists of 2 clusters of slender filaments ...*Habrophlebia* Eaton
— Gills lamelliform on segments 2-7, not as above; each gill of the pairs on middle segments with long terminal extension 3
3. Gill on segment 1 single, unbranched; terminal extension of upper gill of each middle pair (sometimes of lower gill also) rather broad and spatulate (fig. 3:13*i*) .... *Choroterpes* Eaton
— Gill on segment 1 definitely bifid except at base; terminal extension of each gill on middle segments very slender .... *Leptophlebia* Westwood
4. All gills double; gill 1 largest, diminishing in size to gill 7; gills 1-5 bilamellate, each gill with fimbriate margin; posterior member of each pair about 2/3 as large as anterior member; posterior gill of segment 6, and both pairs on 7, fibrilliform; labrum as broad as head ....*Traverella* Edmunds
— Gills lanceolate; labrum narrower than above 5
5. Lateral spines present on abdominal segments 2-9; gills diminishing in size to rearward ......*Thraulodes* Ulmer
— Lateral spines present on abdominal segments 8 and 9 only; gills not noticeably diminishing in size to rearward 6
6. Spine on ninth segment not more than 1/4 the length of that segment; labrum with shallow indentation only on fore border; spinules on posterior margins of tergites 1-10 (fig. 3:13*j*) .... *Paraleptophlebia* Lestage
— Spine on ninth segment long and slender, 1/2 as long as that segment; labrum more deeply indented on fore margin; spinules on posterior margins of tergites 7-10 only .... *Habrophlebiodes* Ulmer

Genus *Paraleptophlebia* Lestage

Members of this genus are generally distributed throughout California except, possibly, in the central valleys. Nymphs are especially plentiful in the small coastal streams of the Redwood Belt. They may be found scattered in any part of the stream with a moderate or slow current but the preferred location is in slowly running shoal water 1/4 to 12 inches deep where the nymphs concentrate in deposits of decaying leaves, bark, and wood. They are also often found at the very edge of the water among roots of plants.

Key to the California Species

Male Adults

1. Abdominal tergites 3-6 hyaline white; reflex spurs lacking; widespread .... *pallipes* (Hagen) 1875
— Abdominal tergites not as above; reflex spurs on penes 2
2. Long joint of forceps with large rounded superior basal dilation 3
— Long joint of forceps without such dilation 5
3. Wings suffused with deep red brown (fig. 3:14*d*); Sonoma County........................... *helena* Day 1952
— Wings clear 4
4. Reflex spurs of penes smoothly and evenly curved outward, shaped as a sickle (fig. 3:13*g*); widespread ... . *debilis* (Walker) 1853
— Reflex spurs of penes wide at base, nearly straight, tapering to a point (fig. 3:14*b*); Santa Cruz and San Mateo counties .... *zayante* Day 1952
5. Ganglionic areas of sternites 2-7 darkened 6
— Ganglionic areas of sternites 2-7 not darkened 7
6. Ganglionic areas of sternites 2-7 red brown; penes lyre-shaped with apices of the two arms directed outward; reflex spurs widely separated (fig. 3:14*f*); Amador County.......................... *placeri* Mayo 1939
— Ganglionic areas of sternites 2-7 usually faintly brown tinted; penes with small rounded opening between arms; reflex spurs straight, sharp pointed, directed slightly outward (fig. 3:27*g*); Yolo County .... *cachea* Day 1953
7. Ganglionic areas of sternites 2-7 vaguely marked with hyaline or opaque white; sternites 2-4 with faint, short oblique dashes based on centers of anterior margins 8
— Ganglionic areas of sternites 2-7 unmarked 10
8. Ganglionic areas of sternites 2-7 faintly white, sternites 2-4 with oblique sternal dashes hyaline; reflex spurs of uniform width from base to apex, somewhat rounded at tip, curving slightly outward; apical notch between the penes almost closed at top (fig. 3:14*g*); coastal and east slope Sierra Nevada .... *associata* (McDunnough) 1924
— Ganglionic areas of sternites 2-7 hyaline; sternites 2-4 with oblique sternal dashes faintly dark; reflex spurs tapering from base to apex, blunt or sharp pointed; apical notch between penes widely open 9
9. Apical notch between penes deep and wide, U-shaped; reflex spurs as long as depth of apical notch, straight and rather sharp pointed (fig. 3:14*h*); central California coastal .... *californica* Traver 1934

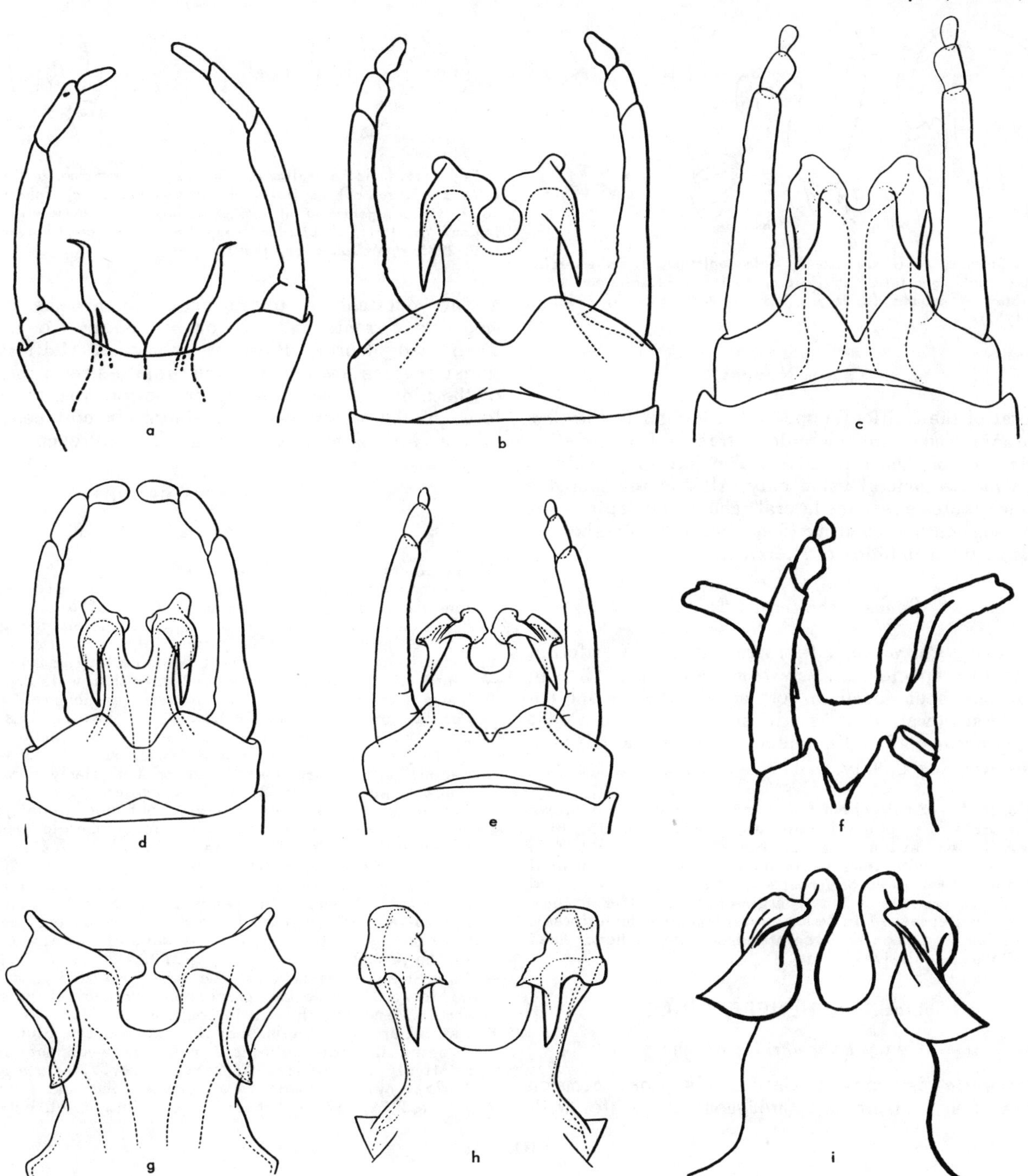

Fig. 3:14. Male genitalia. *a*, *Ameletus validus*; *b*, *Paraleptophlebia zayante*; *c*, *P. clara*; *d*, *P. helena*; *e*, *P. quisquilia*; *f*, *P. placeri*; *g*, *P. associata*; *h*, *P. californica*; *i*, *P. gregalis* (*a-e*, *g-i*, Day, 1952, 1954; *f*, Mayo, 1939).

— Apical notch between penes very shallow, semicircular; reflex spurs much longer than depth of apical notch; these spurs very wide at base, sometimes sinuate, tapering to a blunt point (fig. 3:14*c*); central California coastal .... *clara* (McDunnough) 1924

10. Medium size species, 7½ to 9 mm. in length; apical notch between penes relatively narrow and deep, nearly closed at the top; reflex spurs strongly tapering, curved evenly outward to point (fig. 3:14*i*); sternites 2-7 with paler anterior margins; coastal... *gregalis* (Eaton) 1884

— Small species, 6 to 6½ mm. in length; apical notch between penes almost round, all but closed at top; reflex spurs straight on inner margins and sinuate on outer margins, tapering to sharp point directed downward (fig. 3:14*e*); sternites 1-7 with oblique submedian hyaline dashes and 2 hyaline dots on each; Lake County .... *quisquilia* Day 1952

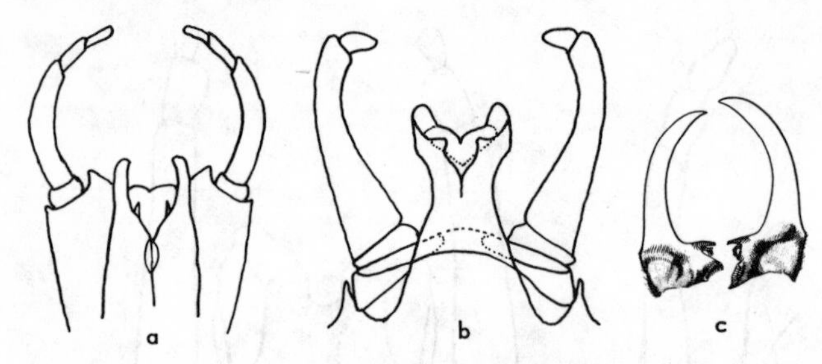

Fig. 3:15. *a, Ameletus amador,* male genitalia; *b, Ephemerella hecuba,* male genitalia; *c, Paraleptophlebia helena,* tusks on mandibles of nymphs (*a,* Mayo, 1939; *b,* McDunnough, 1935; *c,* Day, 1952).

Known Nymphs

Several of the California species of the genus *Paraleptophlebia* show considerable intraspecific variation of size, color, and maculation. The following table is given for its general value only. All species included in the table possess lateral abdominal spines on both segments 8 and 9. See figure 3:15*c* showing tusks of the mandibles of *P. helena.*

Genus *Choroterpes* Eaton

The genus *Choroterpes* is represented in California by but one species, *C. terratoma* Seeman 1927. The nymph has been found in northern California only in lotic, warm waters in small streams of the Coast Range, north of San Francisco. The type locality is in Los Angeles County.

Male adult.—Small, about 6 to 7 mm. in length; gray brown and blackish in general coloration, with sternite nine entirely brown, and a dark ganglionic mark on eight; fore wings hyaline with long veins dark brown except in anal area; long, first joint of forceps bent strongly inward and with bulbous process on inner margins near base (fig. 3:16*a*).

Nymph.—Depressed in form; eyes of male nymph dorsal; eyes of female nymph at dorsolateral margins of head; head wide, flat, and relatively large.

Subfamily EPHEMERELLINAE

Genus *Ephemerella* Walsh

Ephemerella is one of California's most common genera, found in streams throughout the state. With

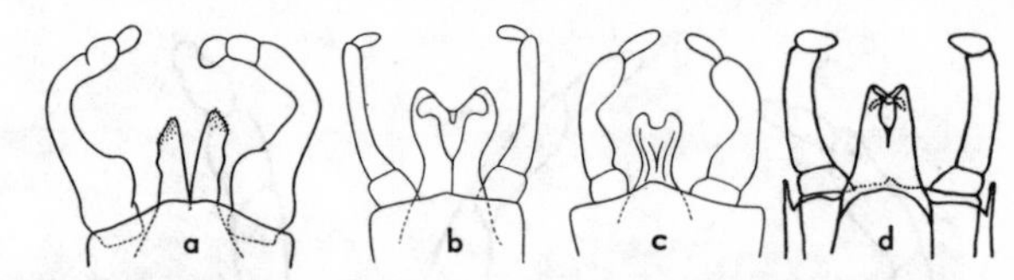

Fig. 3:16. Male genitalia. *a, Choroterpes terratoma; b, Ephemerella* sp ?, typical genitalia of *simplex* group; *c, Ephemerella* sp ?, typical genitalia of *walkeri* group; *d, Ephemerella* sp ?, typical genitalia of *bicolor* group (*a-c,* Needham, Traver, and Hsu, 1935; *d,* McDunnough, 1930).

a few outstanding exceptions, the nymphs prefer moderate currents and live on silt bottoms or among trash and debris. However, some of California's finest species are found in the granite-strewn torrents of the Sierra Nevada, taking protection from the main force of the water behind boulders, in crevices, and in quiet situations near the edge of the stream.

Key to the SPECIES GROUPS

Male Adults

1. On pleural fold of segments 4-7, small fingerlike projections directed backward, those on 5-7 fully half as long as their respective segments (fig. 3:15*b*) .... *hecuba* group
— No such projections along pleural fold 2
2. Third joint of forceps at least twice as long as wide . 3
— Third joint of forceps scarcely longer than wide 4
3. Third joint of forceps 4 to 5 times as long as wide; second joint not strongly bowed, but swollen basally, slender apically; brownish species ..... *attenuata* group
— Third joint of forceps at least twice as long as wide; second either strongly bowed or irregularly swollen; blackish species (fig. 3:16*c*) ... *walkeri (fuscata)* group
4. Penes united, swollen basally, but not apically; no spines present on penes; a distinct tubercle present between them at the base (fig. 3:16*d*) ..... *bicolor* group
— Penes united, somewhat swollen apically; spines usually present on penes; with or without a tubercle between the bases of the penes 5
5. Lateral apical margins of the penes projecting as distinct processes forward; usually several to many spines on the penes 6
— Lateral apical margins of the penes rounded, and with no forward projecting processes; spines, if present, barblike and dorsolateral in position 7
6. Lateral apical processes of the penes short, blunt, and directed inward; the apical notch between the penes relatively shallow (fig. 3:17*a*) .... *invaria* group
— Lateral apical processes of penes long and slender, directed forward or outward; the apical notch between

TABLE 3:1

SUMMARY OF CHARACTERS FOR NYMPHS OF PARALEPTOPHLEBIA

Nymphs	Head marked with	Tusks extended forward beyond head	Lat. branchlets on gill tracheae	Mid. abd. tergites pale spotted	Fore tibiae marked with
associata	White spot	None	Short, fine	Yes	Unmarked
cachea	White areas	None	Very few	No	Unmarked
californica	2 white spots	None	Sparse, fine	Yes	Dark area
gregalis	White spot	None	None	Few	Dark margin
debilis	Dark vertex	None	None	Many	Dark bands
helena	Black V	Present	None	Strongly	Brown spot
placeri	Pale spots	None	None	Yes	Dark bands
quisquilia	Dark brown	None	None	Small	Light brown
pallipes	Pale occiput	None	Yes	Yes	Unmarked
zayante	Black wash	Present	None	Yes	Brown bands

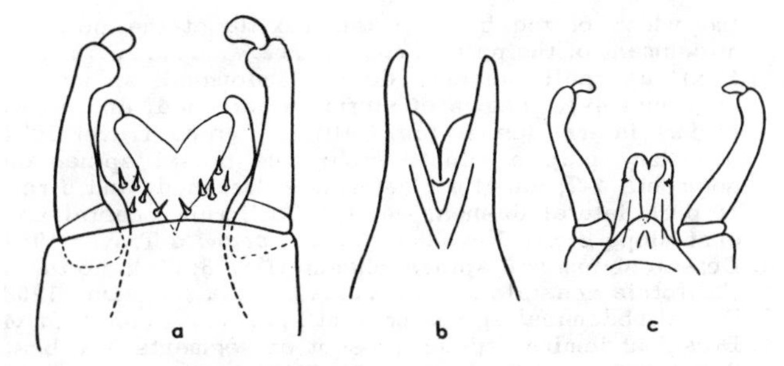

Fig. 3:17. Male genitalia of *Ephemerella* spp. *a*, typical genitalia of *invaria* group; *b*, typical genitalia of *needhami* group; *c*, typical genitalia of *serrata* group (Needham, Traver, and Hsu, 1935).

the penes deeper (fig. 3:17*b*) .... *needhami* group
7. Penes broadened more at apex than at base; minute papillae borne near apex on each side (fig. 3:16*b*) .... *simplex* group
— Penes broadened as much or more at base than at apex; no such minute papillae near apex; usually with barblike lateral dorsal spines, 1 on each side (fig. 3:17*c*) .... *serrata* group

Nymphs

1. Gills present on abdominal segments 3-7 2
— Gills absent on segment 3, present on 4-7 5
2. Teeth or spines generally present on the anterior margin of fore femur; when absent, head tubercles present and spines of abdomen very large (fig. 3:18*a,b*) .... *walkeri (fuscata)* group
— Teeth or spines wanting on anterior margin of the fore femur 3
3. A whorl of spines at each tail joining; maxillary palp weak or wanting .... *serrata* group
— Tails fringed with long hairs in the apical part, may have whorls of spines at the base; maxillary palp normally developed 4
4. Rather long spines along the posterior margins of the femora, and on the upper surface of the fore femur, where they are arranged in an irregular transverse band near apical end; (fig. 3:20*c*); tails without whorls of spines at the base .... *invaria* group

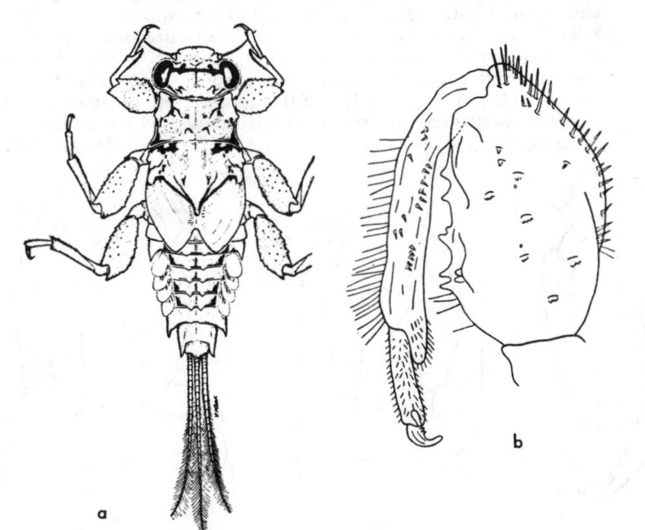

Fig. 3:18. *Ephemerella* sp ? of *walkeri* group. *a*, nymph; *b*, fore leg of nymph showing spines on femur (Needham, Traver, and Hsu, 1935).

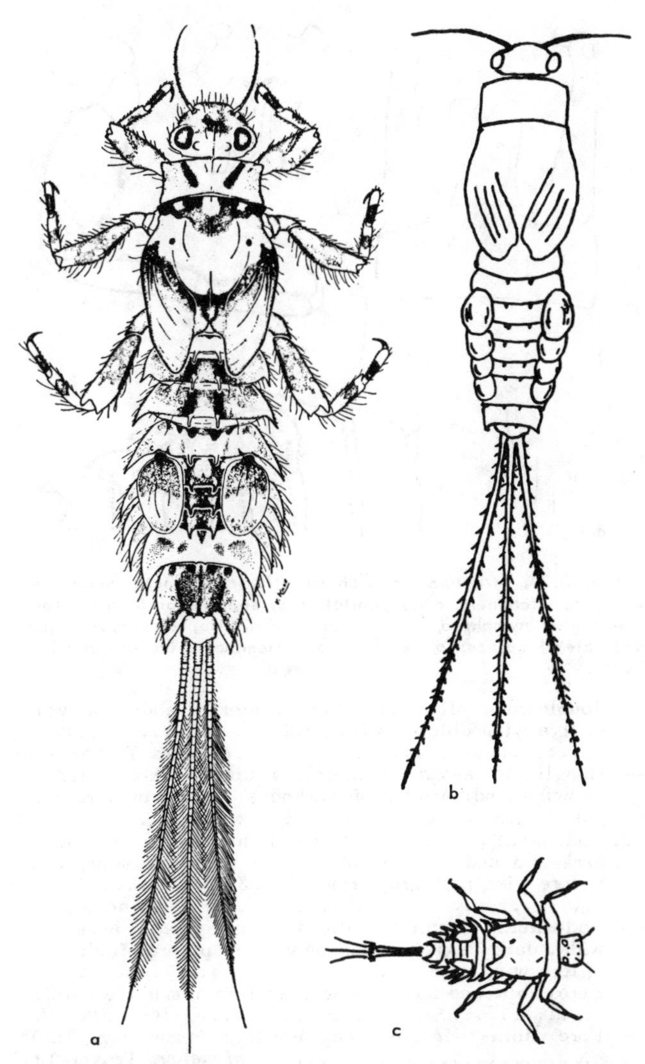

Fig. 3:19. *Ephemerella* nymphs. *a*, species of *bicolor* group; *b*, *tibialis*; *c*, *hecuba* (*a*, Needham, Traver, and Hsu, 1935; *b*, Walley, 1930; *c*, Day).

— Only short and inconspicuous spines along the margins of the femora; usually no spines on upper surface of fore femur; but if present, generally distributed and not in a transverse band; tails may have whorls of spines in the basal half .... *needhami* group
5. Knee spines present on all femora (fig. 3:19*c*) .... *hecuba* group
— Knee spines absent 6
6. Gill on segment 4 operculate; maxillary palp wanting; segment 9 longer than segment 8 (fig. 3:19*a*) .... *bicolor* group
— Gill on segment 4 usually semioperculate; maxillary palp present; segment 9 not longer than segment 8 (fig. 3:20*e*) .... *simplex* group

Genus *Ephemerella* Walsh: *Serrata* Group

Key to the California Species

Adults

1. Ganglionic areas of sternites white or hyaline; dark,

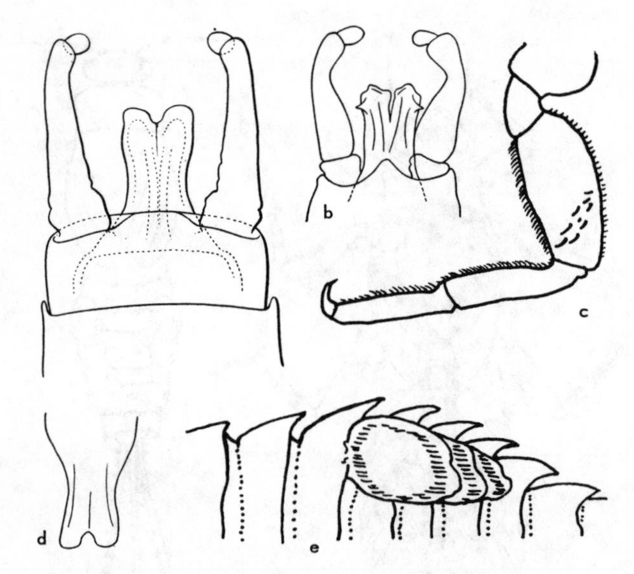

Fig. 3:20. Structures of *Ephemerella. a, cognata,* male genitalia; *b, micheneri,* male genitalia; *c,* species of *invaria* group, fore leg of nymph; *d, tibialis,* male genitalia; *e, simplex,* gills and lateral spines (*a,c,e,* Day; *b,d,* Needham, Traver, and Hsu, 1935).

double submedian stripes on venter of abdomen; veins of fore wing colorless (fig. 3:20*a*) ....*cognata* Traver 1934
— Ganglionic areas of sternites either wholly dark or showing indications of darkness; veins of fore wing pale or gray brown 2
2. Abdominal sternites with ganglionic areas only faintly darkened and with no other dark marks present; veins of fore wing pale gray brown (fig. 3:20*d*) .... *tibialis* McDunnough 1924
— Abdominal sternites with dark ganglionic areas and wide, dark lateral streak near each pleural fold; veins of fore wing pale 3
3. Fore tibia one-half longer than fore femur (fig. 3:27*h*) ....*levis* Day 1953
— Fore tibia twice as long as fore femur (fig. 3:20*b*) .... *micheneri* Traver 1934

Nymphs

1. Maxillary palp minute or absent 2
— Maxillary palp better developed, 3 jointed and normal in form, the total length of the palp being almost equal to the width of the body of the maxilla at the point of attachment of the palp 3
2. Maxillary palp absent; dorsal abdominal spines on segments 4-7; ganglia of sternites darkened, and a row of dark lateral dashes (So. Calif.)...*teresa* Traver 1934
— Maxillary palp a minute stub; true dorsal spines on segments 4-7; ganglia of sternites darkened, and a row of dark lateral dashes; central California coastal and east slope Sierra Nevada .... *cognata* Traver 1934
3. Dorsal abdominal spines absent (fig. 3:27*o*); northern California coastal .... *levis* Day 1953
— Dorsal abdominal spines present 4
4. Dorsal abdominal spines present on segments 3-8, best developed on 4-7; dark markings present on venter of abdomen; widespread .... *micheneri* Traver 1934
— Dorsal abdominal spines present on segments 2-9; no dark markings on venter of abdomen (fig. 3:19*b*); widespread .... *tibialis* McDunnough 1924

Genus *Ephemerella* Walsh: *Walkeri (Fuscata)* Group

Key to the California Species

Adults

1. Second joint of male forceps strongly bowed, with a deep constriction near the middle (fig. 3:21*a*); fore wing length 14 mm.; venation dark brown; abdominal segments brown banded .... *doddsi* Needham 1927
— Second joint of forceps not strongly bowed 2
2. Second joint of male forceps with a sharp constriction near the apex, the tip expanded and bent inward (fig. 3:21*b*); fore wing length 8 mm., venation pale .... *flavilinea* McDunnough 1926
— Second joint of forceps straight or curved slightly inward; venation dark brown 3
3. Dark brown species; abdomen uniform dark brown except that margins of segments are pale, as is pleural fold; fore wing length, 12 mm. (fig. 3:21*c*) ....*coloradensis* Dodds 1923
— Reddish or rose-colored species 4
4. Species with head and thorax dark yellow-brown and abdomen deep ruddy, tergites somewhat lighter in tone; fore wing length 18 mm.; wide, creamy line along pleural fold; wide, creamy margins of each segment gives abdomen a strongly marked annulate appearance (fig. 3:21*d*) ....*glacialis carsona* Day 1952
— Species with pale yellow head, thorax yellow-brown and abdomen bright purple-rose; fore wing length, 15 mm.; wide, white stripe full length of pleural fold; tergites 4-10 and all sternites with wide, white posterior margins (fig. 3:21*e*) .... *spinifera* Needham 1927

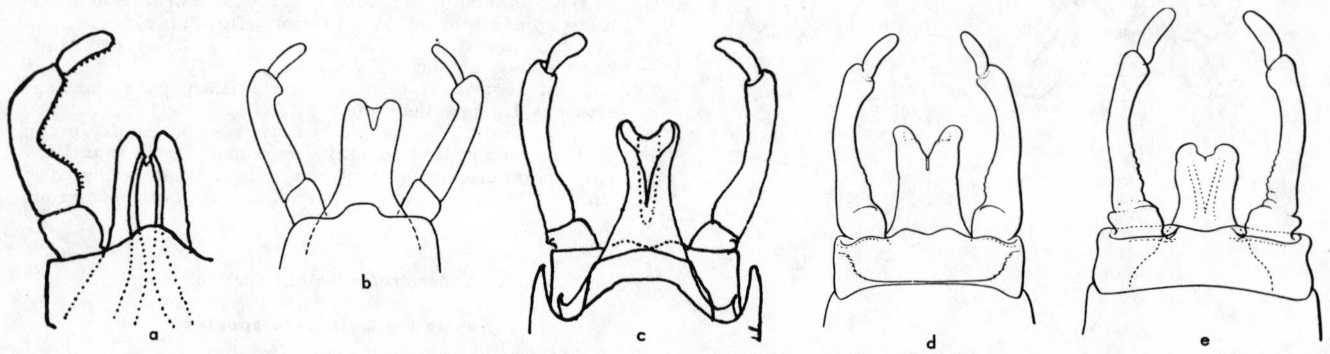

Fig. 3:21. Male genitalia of *Ephemerella. a, doddsi; b, flavilinea; c, coloradensis; d, glacialis carsona; e, spinifera* (*a,* Dodds, 1923; *b,* Needham, Traver and Hsu, 1935; *c,* McDunnough, 1929; *d,e,* Day, 1952).

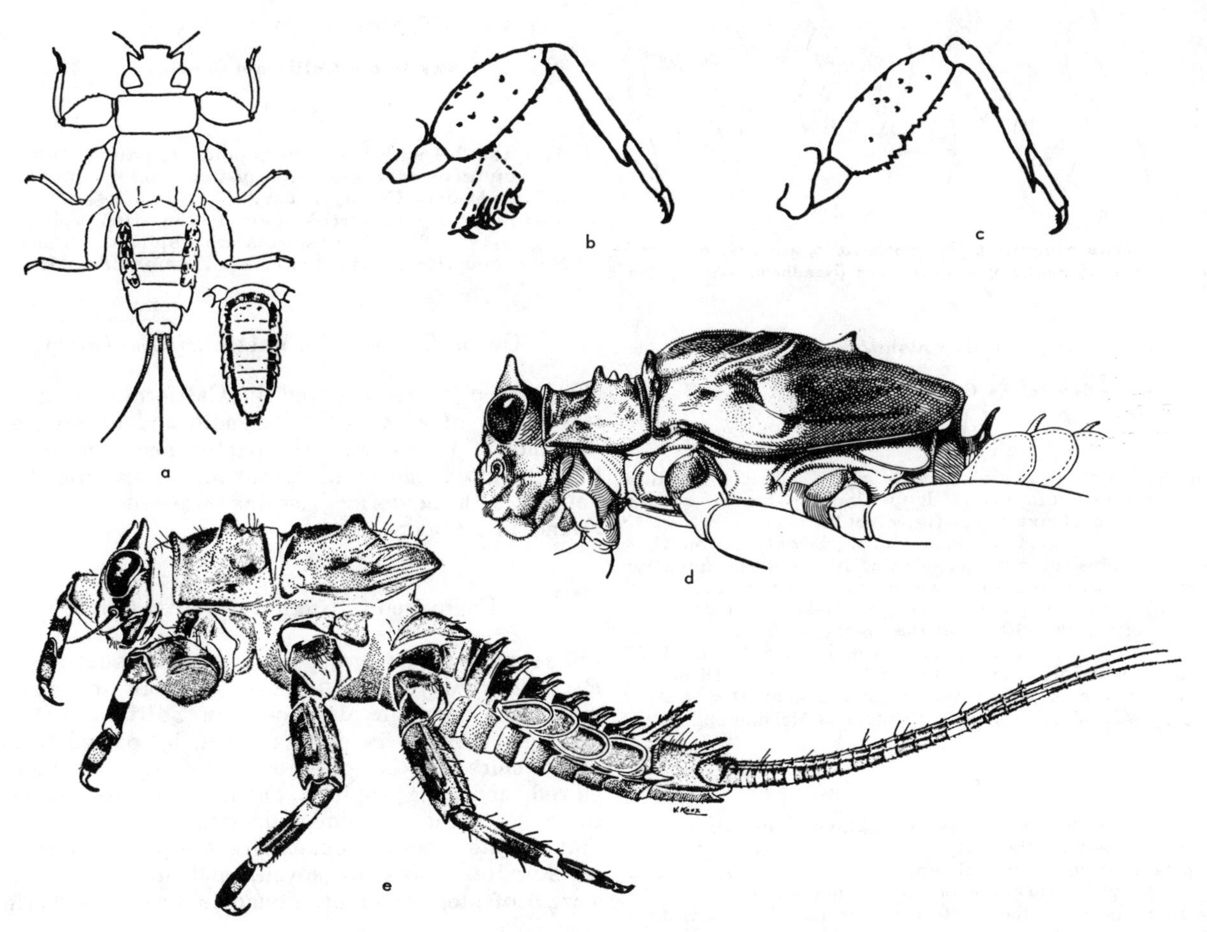

Fig. 3:22. *Ephemerella* nymphs. *a, doddsi,* in dorsal view with ventral view of abdomen showing suction disc; *b, flavilinea,* fore leg; *c, coloradensis,* fore leg; *d, glacialis* carsona, head and thorax; *e, proserpina* (*a,e,* Needham, Traver, and Hsu, 1935; *b,c,* Walley, 1937; *d,* Day, 1952).

Nymphs

1. Teeth or spines present on the anterior margin of the fore femur 2
— No teeth or spines on the anterior margin of fore femur; large occipital tubercles present 4
2. Occipital tubercles absent; a complete adhesive disc on venter of abdomen (fig. 3:22*a*); widespread, montane ....*doddsi* Needham 1927
— Occipital tubercles present 3
3. "Thumb" at distal end of fore tibia long and sharp (fig. 3:22*c*); abdominal spines moderate; widespread, montane .... *coloradensis* Dodds 1923
— "Thumb" at distal end of fore tibia short and blunt (fig. 3:22*b*); abdominal spines very short; widespread .... *flavilinea* McDunnough 1926
4. Dorsal abdominal spines forming a rather regular series; submedian tubercles of prothorax 3 on each side, the anterior high and blunt, the posterior pair small; moderate, well-defined tubercles arise from submedian ridges of the mesonotum; dorsal spines very prominent on segments 2-9, those on 8 and 9 about 40 per cent longer and heavier than those of segments 4-7 (fig. 3:22*d*); east slope Sierra Nevada ....*glacialis carsona* Day 1952
— Dorsal abdominal spines on tergites 8 and 9 very much enlarged, being from 2 to 4 times as large as preceding pairs 5
5. Submedian spines of pronotum heavy and blunt, 3 on each side, with the anterior spine much larger and higher than the posterior pair (fig. 3:22*e*); widespread, montane .... *proserpina* Traver 1934
— Submedian spines of pronotum rather slender, sharp pointed, 2 on each side; anterior spine sometimes higher than posterior spine; widespread, montane .... *spinifera* Needham 1927

Genus *Ephemerella* Walsh: *Needhami* Group

Key to the California Species

Adults

1. Wing 10-11 mm. in length; fore leg yellow; venation red-brown; tails purple-brown (fig. 3:23*a*) .... *euterpe* Traver 1934
— Wing 8 mm. in length; fore leg brown; venation dark purplish brown; tails dark brown (fig. 3:23*b*) .... *maculata* Traver 1934

Nymphs

1. Legs banded; venter of abdomen marked with reddish lateral triangles; length 10-11 mm.; tails dark brown (central California coastal) ....*euterpe* Traver 1934
— Legs not banded; ventral markings consist of dark bands on the anterior margins and lateral extensions of these; length 8-9 mm.; tails yellowish (coastal and San Gabriel Mts.) .... *maculata* Traver 1934

[8] *E. yosemite* Traver 1934, reported from California, is here regarded as a synonym of *E. proserpina* Traver 1934. On the basis of description given, *E. sierra* Mayo 1952 cannot be separated from the type specimens of *E. spinifera*.

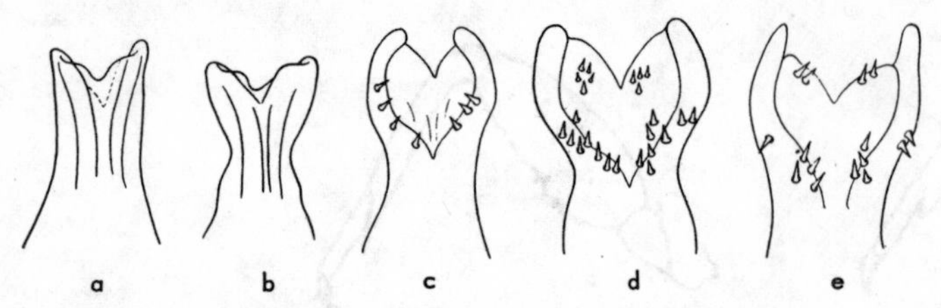

Fig. 3:23. Male genitalia of *Ephemerella*. *a*, *euterpe*; *b*, *maculata*; *c*, *inermis*; *d*, *mollitia*; *e*, *infrequens* (Needham, Traver, and Hsu, 1935).

Genus *Ephemerella* Walsh: *Invaria* Group

Key to the California Species

Adults

1. Second joint of male forceps not distinctly swollen apically; thoracic notum deep blackish brown; a dark spot at base of fore wing (fig. 3:23*c*) .... *inermis* Eaton 1884
— Second joint of forceps distinctly swollen apically; thoracic notum reddish brown or yellow-orange 2
2. Thorax deep yellow and orange; 24-26 spines on the penes, of which 8-10 are at the apex (fig. 3:23*d*) .... *mollitia* Seemann 1927
— Thorax deep olive brown to red-brown; 14 to 16 spines on the penes of which 4-5 are at the apex (fig. 3:23*e*) .... *infrequens* McDunnough 1924

Nymphs

1. Pronotum with distinct pale markings; pale submedian spots present on the posterior margins of the abdominal tergites; dorsal spines absent 2
— No such pale markings on the pronotum; no pale submedian spots on the posterior margins of the tergites; dorsal spines absent; widespread .... *infrequens* McDunnough 1924
2. Lateral extensions of the abdominal segments marked with a dark median transverse band; tails with several median and apical dark bands; widespread.......... .... *inermis* Eaton 1884
— Lateral extensions of abdominal segments with no such dark median band; tails banded only at the base; widespread .... *mollitia* Seemann 1927

Genus *Ephemerella* Walsh: *Hecuba* Group

The group consists of one unique species, *E. hecuba* Eaton 1884, which is found generally distributed at all elevations in northern California.

Adult.—On pleural fold of segments 4 to 7 are small fingerlike projections directed backward and slightly downward. The length of the fore wing is 14 mm., head and thorax black-brown. Dorsally, the abdomen is dark yellow-brown and, ventrally, whitish (fig. 3:15*b*).

Nymph.—Body greatly flattened and hairy; a short tubercle on the dorsal surface of the middle coxa; all femora with large, single acute spine at apex; gill on segment 4 operculate (fig. 3:19*c*).

Genus *Ephemerella* Walsh: *Simplex* Group

Key to the California Species

Nymphs

1. Tergites 2-9 with fine, sharp pointed, paired submedian spines; segments 3-9 with posterolateral spines (fig. 3:27*n*); Madera County .... *soquele* Day 1953
— Tergites 4-8 with wartlike paired submedian projections; segments 5-8 with posterolateral spines; Placer and Mono counties .... *delantala* Mayo 1952

Genus *Ephemerella* Walsh: *Bicolor* Group

This group is represented in California by *E. lodi* Mayo 1952 of which only the male and female adults are known. Wings are amber-tinted and venation yellow. Tergites one to three are brown and four to ten yellow with posterior margins washed with brown. Found in Amador County.

Ungrouped California Species: Part I

Unassigned to any group is a very unusual species, *E. pelosa* Mayo 1951, known only in the nymphal stage and found to date only in California. A very hairy species, it is distinguished by paired tufts of short hairs in the position ordinarily occupied by paired abdominal spines on the tergites; sternites three to eight are covered with long, white hairs that radiate from the centers of the posterior margins of sternites three to seven, and from the anterior margin of sternite eight. Found in Fresno and Alpine counties.

Ungrouped California Species: Part II

Ephemerella heterocaudata McDunnough 1929, was assigned in "Biology of Mayflies" to the *serrata* group. In the description of *E. spinosa* Mayo 1951, this species was referred to the *needhami* group by Dr. Mayo.

E. spinosa Mayo is here regarded as a synonym of *E. hystrix* Traver, and these two species should probably be placed in a new group of *Ephemerella* because they have similarities to each other that seem more positive than their attachments to present groups.

Descriptions have been given for both male adult and nymph of *E. heterocaudata* but only the nymphal form is known for *E. hystrix*. Both species have been collected in California and can be quickly separated from all other California species of the genus *Ephemerella* since the length of the center tail of adult and nymph considerably exceeds that of the outer tails; tails of the nymph have whorls of spines at each joining.

E. heterocaudata McDunnough 1929, male: Middle tail more than twice as long as the outer ones; length of fore wing, 7 mm.; thorax and abdominal tergites olive brown, the thorax darker with dark shadings and markings; legs dark olive with fore femur and tibia tinged with black. Wings hyaline, long veins smoky and cross veins pale.

E. heterocaudata McDunnough 1929, nymph: Middle tail three or more times length of outer ones; gills on tergites three to seven; general color dark brown with head blackish and legs without pale bands; dorsal spines large and strong, incurved, becoming progressively longer and further apart from tergites two to seven, but shorter and closer together on eight to nine. Found in Alpine and Plumas counties.

E. hystrix Traver 1927, nymph: Middle tail one-half again as long as outer ones; length of body, 7 mm.; length of middle tail, 11 mm.; color dark brown, with legs yellow-brown; very prominent spines on tergites two to nine, those on two close together and straight and those on three farther apart and directed outward; on four to seven spines are widely divergent, increasing in size to rearward; they are directed laterally and posteriorly and widely curved, each spine beset with coarse spinules from base to tip; on segments eight to nine spines are stout but shorter and straighter. Found in Inyo, Nevada, Sierra, Mono, and Shasta counties.

Subfamily CAENINAE

Key to the Genera

Adults

1. Posterior branch of M in fore wing and intercalary between branches of this vein much shorter than anterior branch, not extending to base of wing; male forceps 3-jointed 2
— Posterior branch of M in fore wing and intercalary between branches as long as anterior branch, extending to base of wing; male forceps 1-jointed 3
2. Tails 2; fore wing elongate and narrow, broadest in the center .... *Leptohyphes* Eaton
— Tails 3; fore wing broad, widest in anal region (fig. 3:24*a*) .... *Tricorythodes* Ulmer
3. Prosternum twice as wide as long, fore coxae therefore widely separated; second joint of antennae 3 times as long as first ....*Brachycercus* Curtis
— Prosternum 2 to 3 times longer than wide, coxae much closer together; second joint on antennae not more than twice as long as first (fig. 3:10*a*) ....*Caenis* Stephens

Nymphs

1. First abdominal segment without gills; gills of second segment well separated on middorsal line, operculate, covering all following gills 2
— First segment bearing tiny single gills; gills of second meet or overlap on middorsal line, operculate, covering all following gills 3
2. Operculate gills triangular (fig. 3:24*b,c*) .... *Tricorythodes* Ulmer
— Operculate gills elongate and oval. . *Leptohyphes* Eaton
3. Three prominent tubercles on head; maxillary and labial palps 2-jointed...........*Brachycercus* Curtis
— No tubercles on head; maxillary and labial palps 3-jointed (fig. 3:24*d,e,f*) .... *Caenis* Stephens

Genus *Caenis* Stephens

This tiny nymph is from 2 mm. to 4 mm. in length, a lotic form living in the silt of the bottom.

Besides *C. tardata,* found in San Diego County, one unnamed species of this tiny, two-winged genus has been collected in Napa County; this undescribed species is rather closely related to *C. tardata* McDunnough 1931, the male of the latter being described as follows:

Forceps with one joint only, short (fig. 3:24*k*); head blackish; mesonotum deep brown and abdomen dorsally black shaded on segments one to six, entirely pale on seven and eight, and faintly dark shaded on nine and ten. All sternites are wholly pale yellowish white. The wings are 3 mm. in length.

Genus *Tricorythodes* Ulmer

This is another genus with hind wings absent; the species are quite small in size, the wing length being from 3 mm. to 7 mm. *Tricorythodes* is represented in California by the single species *T. fallax* Traver 1935 which is very widespread in the northern coastal region, the nymph being found in quiet water on the bottom of the stream. Notes on *T. fallax* are as follows:

Adult.—Length of fore wing 5-6 mm.; a blackish or brownish black species. Entire head, thorax, and abdomen blackish brown, with tergites one, two, eight, and nine black, and blackish markings numerous. All tibiae and tarsi white except the apical tarsal joint which is faintly dark; apex of tibiae narrowly black; tails gray with hyaline joinings (fig. 3:24*l*).

Nymph.—Length of body 3½-5 mm.; pale whitish species with black surfacing on head, pronotum and abdominal tergites: a black spot at femorotibial knees; tibiae smoky brown, with all other leg joints white. The large, triangular operculate gill on segment two is dark in color.

Subfamily BAETINAE

Key to the Genera

Male Adults

1. Hind wings present 2
— Hind wings absent............................ 4
2. Fore wings usually with numerous costal cross veins before the bulla; hind wings with a moderate number of cross veins, at least in the costal region (fig. 3:24*g*) .... *Callibaetis* Eaton
— Fore wings without costal cross veins before the bulla; hind wings with no cross veins, or with very few of them 3
3. Marginal intercalaries of fore wing occur singly (fig. 3:24*h*) .... *Centroptilum* Eaton
— Marginal intercalaries of fore wing occur in pairs (fig. 3:10*b*) .... *Baetis* Leach
4. Marginal intercalaries of fore wing occur singly 5
— Marginal intercalaries of fore wing occur in pairs ... 6
5. Basal joint of hind tarsus very long, equal in length to the 3 other joints together; second joint of forceps short and broad, its inner margin with a prominent, wide extension .... *Neocloeon* Traver
— Basal joint of hind tarsus only moderately long, equal in length to the third and fourth joints together; second joint of forceps longer and narrower, conical, tapering apically, with no extension on its inner margin .... *Cloeon* Leach
6. Fore tibia 1½ to 1⅘ times as long as femur ....*Pseudocloeon* Klapalek, *Paracloeodes* Day, *Apobaetis* Day
— Fore tibia more than twice as long as femur .... *Baetodes* Needham & Murphy

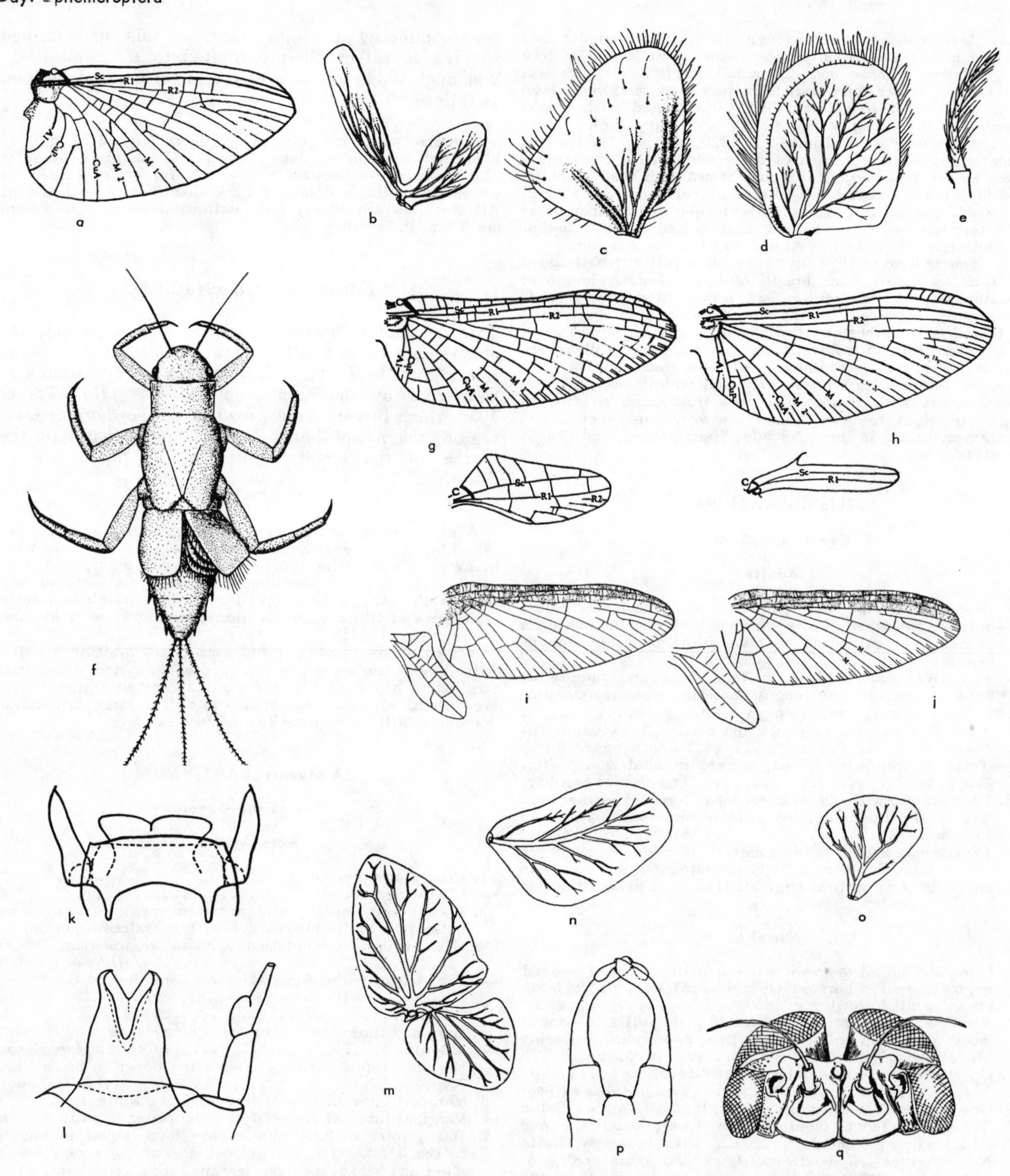

Fig. 3:24. *a*, *Tricorythodes* sp ?, fore wing; *b,c*, *Tricorythodes* sp ?, posterior and anterior lamellae of first nymphal gill; *d,e*, *Caenis* sp ?, second and first gills of nymph; *f*, *Caenis* sp ?, nymph; *g*, *Callibaetis* sp ?, wings; *h*, *Centroptilum* sp ?, wings; *i*, *Callibaetis californicus*, wings of male; *j*, *Callibaetis pacificus*, wings of male; *k*, *Caenis tardata*, male genitalia; *l*, *Tricorythodes explicatus*, male genitalia; *m*, *Callibaetis* sp ?, fourth gill of nymph; *n*, *Baetis* sp ?, third gill of nymph; *o*, *Centroptilum* sp ?, first gill of nymph; *p*, *Callibaetis* sp ?, male genitalia; *q*, *Callibaetis* sp ?, head of male showing divided eyes (*a-e,g,h,m-p*, Spieth, 1933; *f,i-l,q*, Needham, Traver, and Hsu, 1935).

Nymphs

1. Gills on segments 1-5 only; lateral tails bare, or at the most with only a few inconspicuous setae .... *Baetodes* Needham and Murphy
— Gills on segments 1-7 2
2. Gill lamellae double (often with a mere recurved flap on ventral or dorsal surface) on abdominal segments 1-6 or 1-7 3
— Gill lamellae single on all abdominal segments 5
3. Tracheae of gill lamellae palmately branched; second pair of wing buds absent .... *Cloeon* Leach
— Tracheae of gill lamellae pinnately branched; second pair of wing buds present 4
4. Double part of gill a flap on ventral surface (fig. 3:24*m*); maxillary palp 2-jointed .... *Callibaetis* Eaton
— Double part of gill a flap on dorsal surface; gills double on 1-7; maxillary palp 3-jointed (fig. 3:24*o*) .... *Centroptilum* Eaton
5. Tails 2 6
— Tails 3 7
6. Second pair of wing buds absent; labium as in figure 3:28*a* .... *Pseudocloeon* Klapalek
— Second pair of wing buds present (*bicaudatus*, and the *propinquus* group) (fig. 3:24*n*) .... *Baetis* Leach
7. Middle tail shorter and weaker than outer ones .... *Baetis* Leach
— Middle tail practically as long and stout as outer ones 8
8. Second pair of wing buds present... *Centroptilum* Eaton
— Second pair of wing buds absent 9
9. Labial palp 3-jointed, form of distal segment as in *Centroptilum* .... *Neocloeon* Traver
— Labial palp 2-jointed 10
10. Labrum with median notch on anterior margin; distal margin of apical segment of labial palp pointed on outer side and rounded on inner side (fig. 3:28*b*) ... . *Paracloeodes* Day
— Labrum without median notch on anterior margin; distal margin of apical segment of labial palp squarely truncate with deep V on inner margin (fig. 3:28*c*) ... . *Apobaetis* Day

Genus *Callibaetis* Eaton

Distributed in all parts of California in lakes, streams, and ponds, the nymph is strictly a lotic water form, dainty and gracefully streamlined and an excellent swimmer. For some general features of adult *Callibaetis* see figure 3:24*p, q*.

Key to the California Species

Adults

1. Cross veins of fore wing behind the vitta (or behind the fourth longitudinal vein) relatively few in number—only about 15-25; usually none very near the hind margin; usually a single irregular row of them across the wing 2
— Cross veins of fore wing behind the vitta more numerous—about 35-60 of them; usually several are close to the hind margin; 2 or more very irregular rows of them across the wing 6
2. Intercalaries of middle area of outer margin of fore wing usually occur in pairs 3
— Intercalaries of middle area of outer margin usually occur singly 5
3. Wing 5½ to 7 mm. in length; in wing of female, several brownish clouds on veins behind vitta, also on hind margin; cross veins brown in male, locality given as "California" .... *pictus* (Eaton) 1871
— Wing 7-9 mm. in length; no such distinct brown clouds behind vitta in wing of female; cross veins in wing of male pale or brownish 4

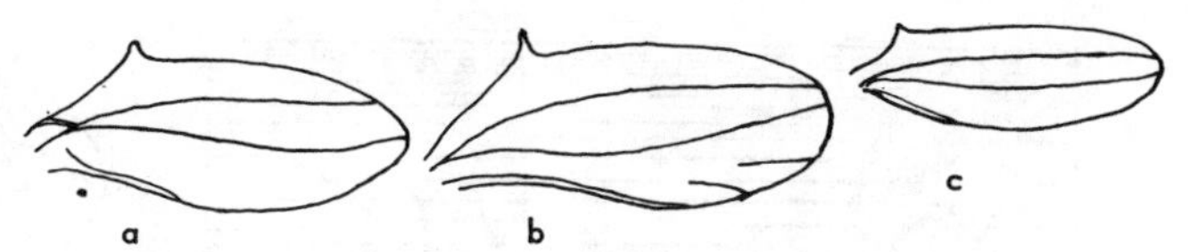

Fig. 3:25. Hind wings of *Baetis*. *a*, *bicaudatus*; *b*, *tricaudatus*; *c*, *intermedius* (W. C. Day).

4. Abdominal tergites 5, 6, and 9 of male with large pale areas; in female, tergites 1, 4, 7, and 8 almost wholly dark brown, others pale with dark markings; cross veins brown in male wing (fig. 3:24*j*); widespread .... *pacificus* Seemann 1927
— Abdominal tergites 7 and 9 of male with large pale areas; in female all tergites pale brown; cross veins in wing of male pale; Redlands ... *signatus* Banks 1918
5. Tails dark brown at the joinings (fig. 3:24*i*); widespread .... *californicus* Banks 1900
— Tails wholly pale or pale yellowish only at joinings; Sacramento Valley............ *montanus* (Eaton) 1885
6. Intercalaries of the middle area of the fore wing usually occur singly 7
— Intercalaries of the middle area of the fore wing usually occur in pairs 8
7. Clouds present in disc and along outer margin of wing of male; tails blackish at joinings; San Bernardino Mts. and Alameda County .... *carolus* Traver 1935
— Wings of male not pigmented; tails faintly yellowish at joinings; San Joaquin and Sacramento valleys .... *traverae* Upholt 1937
8. Vitta in wing of female dark brown, and many clouds along veins and outer margin; male fore wing 12 mm. in length; San Joaquin Valley and San Francisco .... *hageni* (Eaton) 1885
— Vitta in wing of female irregularly mottled with reddish brown; no clouds on veins or outer margin; male fore wing 9½ mm. in length; Marin and Alameda counties .... *hebes* Upholt 1936

Insufficient data are available to make possible a key to the nymphs of the California *Callibaetis* that would be of practical value in the separation of species. For the general appearance of the *Callibaetis* nymph see figure 3:26*a*. The distinctive labium is shown in figure 3:26*b*. The following nymphs have been reported from California:

C. californicus Banks
C. carolus Traver
C. hebes Upholt
C. montanus Eaton
C. pacificus Seeman

Genus *Baetis* Leach

Members of this genus occur almost wherever running water is found in California. The small nymphs, not longer than 8 mm., are very adaptable, living in the open waters of streams in currents ranging from slow to very swift.

Key to the California Species

Male Adults

1. Hind wing with strong, curved costal projection, much as in *Centroptilum* (fig. 3:27*e, j*); central California coastal .... *leechi* Day 1954
— Hind wing without strong, curved costal projection . 2
2. Hind wing with 3 longitudinal veins of about equal length 3

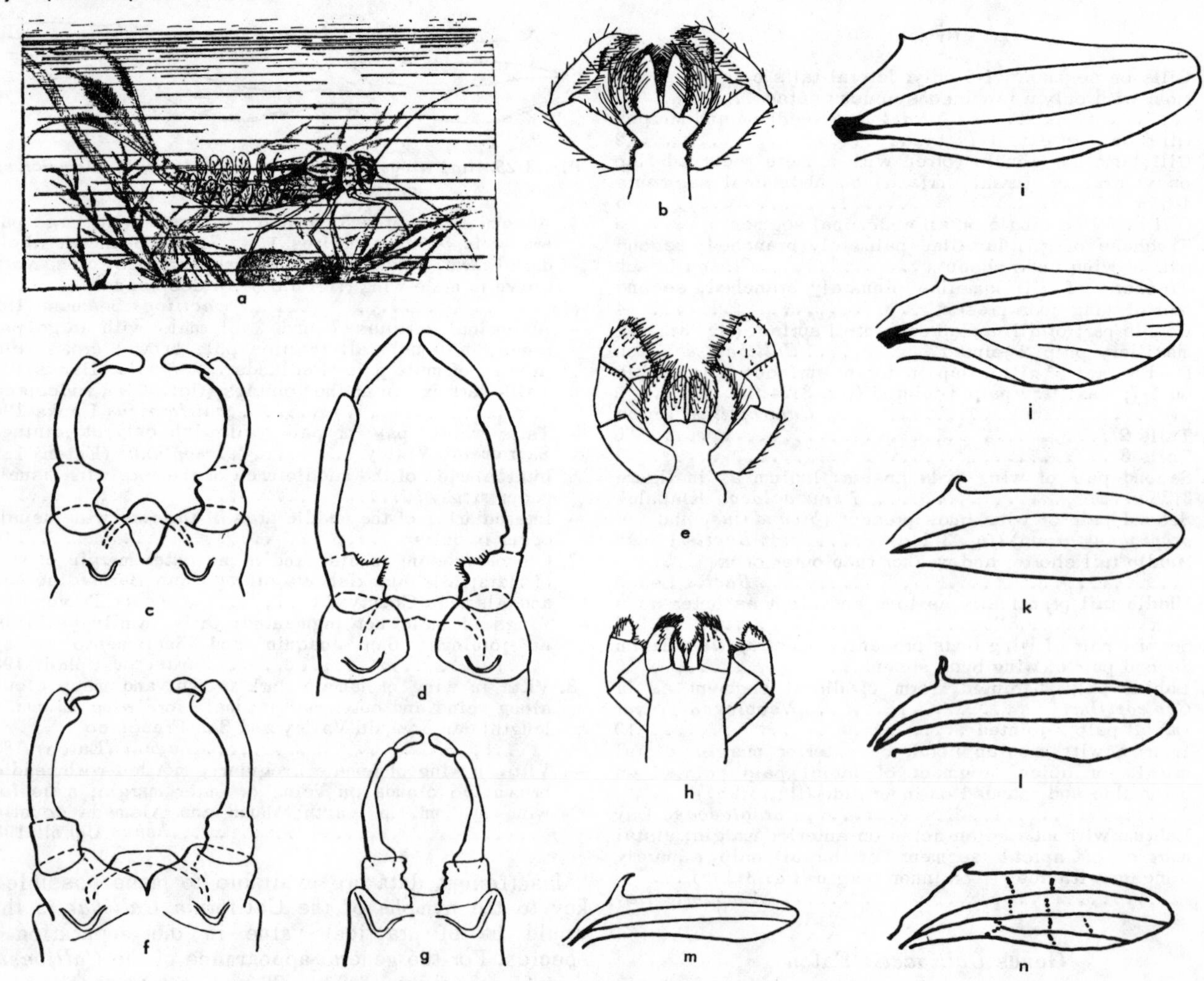

Fig. 3:26, *a, Callibaetis* sp., nymph; *b, Callibaetis* sp., labium of nymph; *c, Centroptilum conturbatum*, male genitalia; *d, Centroptilum asperatum*, male genitalia; *e, Centroptilum* sp., labium of nymph; *f, Centroptilum venosum*, male genitalia; *g, Baetis devinctus*, male genitalia; *h, Baetis* sp., labium of nymph; *i, Baetis thermophilos*, hind wing; *j, Baetis devinctus*, hind wing; *k, Centroptilum asperatum*, hind wing; *l, Centroptilum conturbatum*, hind wing; *m, Centroptilum convexum*, hind wing; *n, Centroptilum venosum*, hind wing (*a,c,d,f,g,i-m*, Needham, Traver and Hsu, 1935; *b,e,h*, Spieth, 1933).

— Hind wing with 2 or 3 longitudinal veins, of which the first and second are about equal in length; the third, if present, is much shorter, ending near the middle of the hind margin 4

3. Fore wing 4 mm. in length; fourth segment of forceps 1/2 as long as third segment (fig. 3:27*d,i*); central California coastal .... *sulfurosus* Day 1954

— Fore wing 5 mm. in length; fourth segment of forceps nearly as long as third segment (fig. 3:26*i*); Eel River .... *thermophilos* McDunnough 1926

4. Hind wing with but 2 longitudinal veins, the third wholly wanting 5

— Third vein present in hind wing, although it may be weakly developed 6

5. Costal projection of hind wing present (fig. 3:25*a*); mountains .... *bicaudatus* Dodds 1923

— Costal projection of hind wing absent; Klamath River .... *insignificans* McDunnough 1926

6. Second longitudinal vein of hind wing forked 7

— Second longitudinal vein of hind wing not forked ... 8

7. Second longitudinal vein of hind wing forked about 3/5 the distance from base to distal margin; genitalia of *moffati* type (fig. 3:27*f,k*); central California coastal .... *diablus* Day 1954

— Second longitudinal vein of hind wing forked about 1/3 the distance from base to distal margin; genitalia of *intercalaris* type (fig. 3:26*g,j*); Santa Cruz and Napa counties .... *devinctus* Traver 1935

8. Two strong, well-developed intercalaries between veins 2 and 3 of hind wing; veins 1 and 2 convergent (fig. 3:25*b*); Santa Clara County .... *tricaudatus* Dodds 1923

— Not more than one well-developed intercalary, sometimes none, between veins 2 and 3 of hind wing; other faint intercalaries may be present 9

9. Fore wing about 5 mm. in length 10

Fore wing 6-8 mm. in length; genitalia of *moffati* type 11

10. Fore wing 5½ mm. long; genitalia of *intercalaris* type (fig. 3:27*c,l*); central California coastal .... *alius* Day 1954

— Fore wing 5 mm. in length; genitalia of *moffati* type; San Gabriel Mts. and Sonoma County .... *adonis* Traver 1935

11. Fore wing 8 mm. in length; no intercalaries in hind wing; Inyo County .... *palisadi* Mayo 1952

— Fore wing 6-6½ mm. in length 12

12. Vein 3 of hind wing faint; no intercalary between veins 1 and 2 of hind wing, but 1 short intercalary between

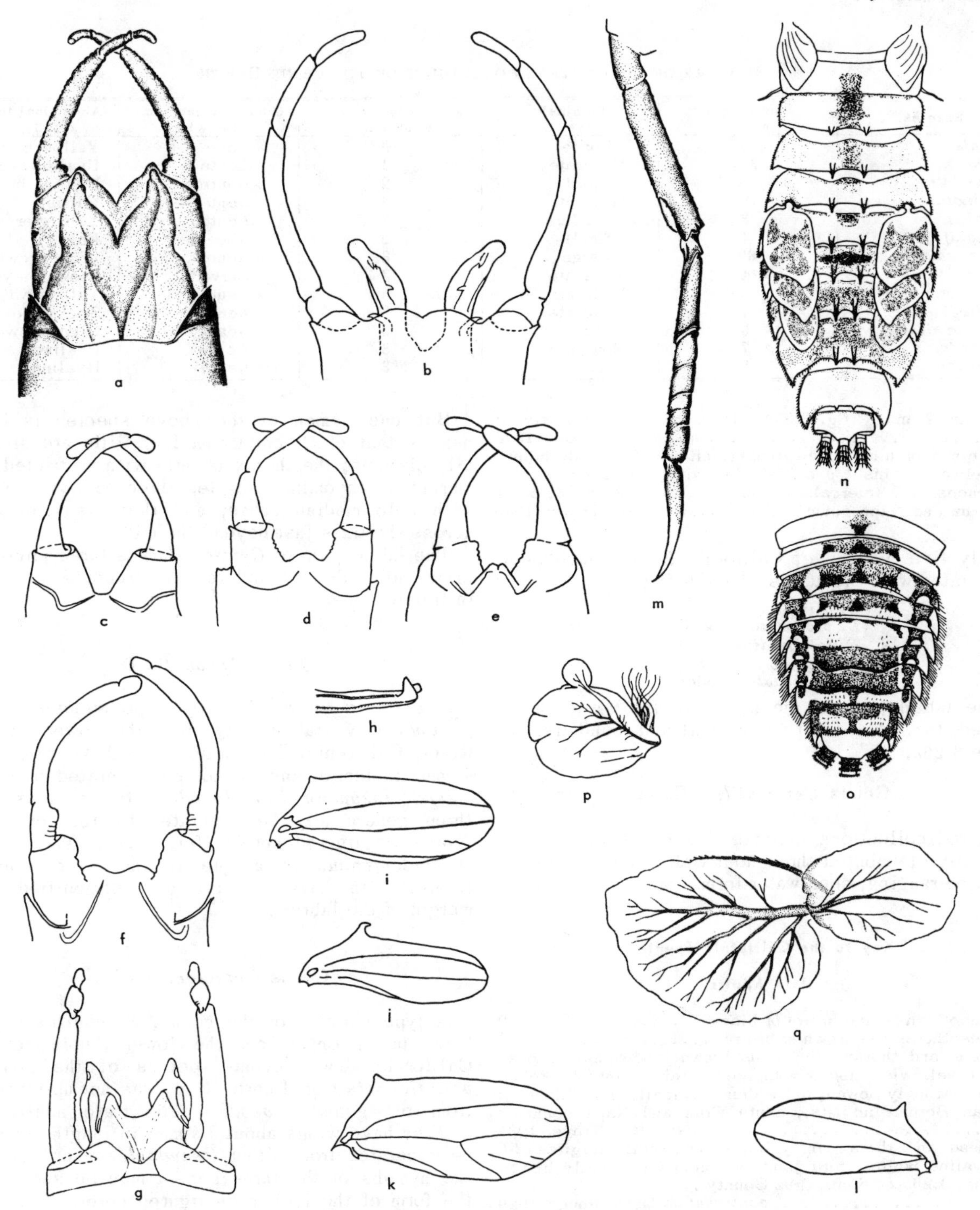

Fig. 3:27. *a*, *Edmundsius agilis*, male genitalia; *b*, *Rhithrogena decora*, male genitalia; *c*, *Baetis alius*, male genitalia; *d*, *Baetis sulfurosus*, male genitalia; *e*, *Baetis leechi*, male genitalia; *f*, *Baetis diablus*, male genitalia; *g*, *Paraleptophlebia cachea*, male genitalia; *h*, *Ephemerella levis*, male penes in lateral aspect; *i*, *Baetis sulfurosus*, male hind wing; *j*, *Baetis leechi*, male hind wing; *k*, *Baetis diablus*, hind wing, *l*, *Baetis alius*, male hind wing; *m*, *Edmundsius agilis*, nymphal hind leg; *n*, *Ephemerella soquele*, nymphal tergites; *o*, *Ephemerella levis*, nymphal tergites; *p*, *Rhithrogena decora*; nymphal gill five; *q*, *Edmundsius agilis*, nymphal gill one (*a,m,q*, Day, 1953; *b-l*, *n-p*, Day, 1954).

TABLE 3:2

SUMMARY OF CHARACTERS FOR ADULTS OF THE GENUS BAETIS

Species	Wing length, mm. ♂	Turbinate eye, ♂	Veins in hind wing	Costal projection of hind wing	Abdominal tergites, ♂ 2-6
adonis	5	Large	3	Acute	Yellow brown
alius	5½	Moderate	3	Acute	Hyaline brown
bicaudatus	6½	Small	2	Acute	Hyaline brown
devinctus	5	Moderate	3	Acute	Whitish
diablus	8	Med. large	3	Acute	Yellow brown
insignificans	4	Small	2	Absent	Pale
intermedius	6½	Large	3	Acute	Light brown
leechi	4	Moderate	3	Curved	Brownish yellow
piscatoris	6	Moderate	3	Acute	Yellow white
sulfurosus	4	Moderate	3	Weak	Smoky brown
thermophilos	5	?	3	Weak	Olive brown
tricaudatus	7	Moderate	3	Acute	Hyaline
palisadi	8	Large	3	Acute	Hyaline

veins 2 and 3 (fig. 3:25*c*); Truckee and Feather rivers .... *intermedius* Dodds 1923

— Vein 3 of hind wing stronger; often a faint intercalary between veins 1 and 2 of hind wing, and usually faint traces of 2 intercalaries between veins 2 and 3; Santa Cruz and Napa counties........ *piscatoris* Traver 1935

Only five nymphs of California *Baetis* have been associated with the adults, these being:

B. alius Day
B. bicaudatus Dodds
B. intermedius Dodds
B. piscatoris Traver
B. tricaudatus Dodds

The labial palp of the nymph of *Baetis* is three-jointed, the distal joint being short and rounded as in figure 3:26*h*.

Genus *Centroptilum* Eaton

Scattered collections indicate this genus is probably generally distributed throughout California; the nymph is a free-ranging, open water form.

Key to the California Species

Male Adults

1. Mesothorax olive brown or black 2
— Mesothorax red brown or brown 3
2. Head and thorax olivaceous brown; abdominal tergites 2-6 yellowish tinged with brown; third joint of forceps not strongly bowed, but widened apically (fig. 3:26*d,k*); San Bernardino Mts., Santa Cruz and Napa counties .... *asperatum* Traver 1935
— Head and thorax shiny black; abdominal tergites 2-6 hyaline white; third joint of forceps distinctly bowed (fig. 3:26*c,l*); Santa Cruz County .... *conturbatum* McDunnough 1929
3. Hind wing long and narrow, about 6 times as long as broad, with two longitudinal veins but no intercalaries or cross veins; abdominal tergites 2-6 hyaline white, sometimes faintly pink tinged (fig. 3:26*m*); Sonoma and Contra Costa counties .... *convexum* Ide 1930
— Hind wing broader than above, about 4 times as long as broad, with a distinct intercalary below the second of the two longitudinal veins, and 7 to 11 distinct cross veins present; abdominal tergites 2-6 yellowish white, semitranslucent, with posterior margins and lateral areas reddish or orange brown (fig. 3:26*f,n*); San Gabriel Mts........................... *venosum* Traver 1935

But one nymph of the above species is known, namely that of *C. convexum* Ide. Gills are single on all segments; the thorax is pale with restricted brown markings; abdominal tergites three to five are brown with pale median areas; and there is a dark band across the tails just beyond the middle.

The labial palp of *Centroptilum* is three-jointed, the distal joint dilated and roundly truncate apically as in figure 3:26*e*.

Genus *Apobaetis* Day

Apobaetis is known from the type species *A. indeprensus* Day, taken only from the lower Tuolumne River, California. The male adults have wings about 4 mm. in length and cannot be separated from either *Pseudocloeon* or *Paracloeodes;* the nymphs of the three genera may be separated by the form of the labium as figured herein (fig. 3:28*c*). The nymph of *A. indeprensus* is unique in the entire subfamily Baetinae in having a straight, unmodified frontal margin of the labrum.

Genus *Paracloeodes* Day

The type species of the genus, *P. abditus* Day, has been taken only from the lower Tuolumne River, California. The second species of the genus, *P. portoricensis* from Puerto Rico, was set up by removal from the genus *Cloeodes*. The male adults of *P. abditus* have wings about 3½ mm. in length and cannot be separated from either *Pseudocloeon* or *Apobaetis;* the nymphs of the three genera may be separated by the form of the labium as figured herein (fig. 3:28*b*).

Subfamily AMETROPINAE

In the Nearctic this subfamily consists of two genera, *Ametropus* Albarda and *Metreturus* Burks. Neither genus is represented in California.

The adult of *Ametropus* is characterized by the fore tarsus of male adults being very long in proportion to the short tibia, and by claws dissimilar on all

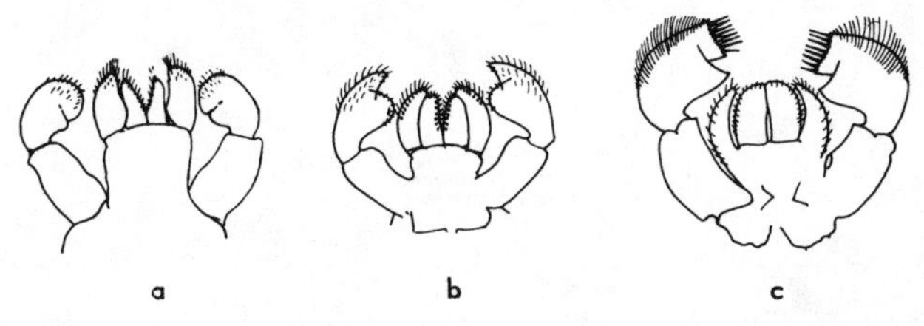

Fig. 3:28. *a, Pseudocloeon sp ?,* labium of nymph; *b, Paracloeodes abditus,* labium of nymph; *c, Apobaetis indeprensus,* labium of nymph. (W. C. Day).

tarsi. In the hind wing the costal angulation is acute and the *Of* is not forked. Forceps are four-jointed.

Key to the Genera

Nymphs

1. Claw of fore leg bears several long spines; a fleshy appendage is attached to inner margin of fore coxae; margins of gills smoothly oval..... *Ametropus* Albarda
— Claw of fore leg without spines; margins of gills fissured .... *Metreturus* Burks

Subfamily METRETOPINAE

This subfamily is not represented in California.

Key to the Genera

Adults

1. A single pair of cubital intercalaries in fore wing; costal angulation of hind wing acute .... *Metretopus* Eaton
— Two pairs of cubital intercalaries in fore wing; costal angulation of hind wing obtuse 2
2. Fore tarsus of male fully 3 times the length of the tibia; basal joint of fore tarsus of male slightly longer than the second; subanal plate of female obtuse .... *Siphloplecton* Clemens
— Fore tarsus of male only about twice as long as tibia; basal joint of male fore tarsus three-fourths of second; subanal plate of female excavated apically .... *Pseudiron* McDunnough

Known Nymphs

1. Tarsi longer than tibiae; maxillary palpi 2-jointed .... *Metretopus* Eaton
— Tarsi subequal to tibiae; maxillary palpi 3-jointed .... *Siphloplecton* Clemens

Subfamily OLIGONURIINAE

This subfamily is represented by two genera in the Nearctic region, *Oligoneuria* Pictet and *Lachlania* Hagen, neither represented in California. The subfamily is separated in the keys to the subfamilies of the Baetidae.

Subfamily BAETISCINAE

This subfamily consists of the single Nearctic genus *Baetisca* Walsh, which is separated by the characters in the keys to the subfamilies of Baetidae.

Baetisca obesa was reported from California by Eaton in Revised Monograph, 1885, p. 228, but has not subsequently been taken in this state.

REFERENCES

BURKS, B. D.
1953. The mayflies, or Ephemeroptera, of Illinois. Bull. Illinois Lab. Nat. Hist., 26:1-216, figs. 1-395.
DAY, W. C.
1952. New species and notes on California mayflies. Pan-Pac. Ent., 28:17-39, 47 figs.
1953. A new mayfly genus from California. Pan-Pac. Ent., 29:19-24, 13 figs.
1954. New species and notes on California mayflies II. Pan-Pac. Ent., 30:15-29, 14 figs.
1954. New species of California mayflies in the genus *Baetis.* Pan-Pac. Ent., 30:29-34, 8 figs.
1955. New genera of mayflies from California. Pan-Pac. Ent., 31:121-137, 27 figs.
EDMUNDS, G. F., JR.
1948. A new genus of mayflies from western N. A. (Leptophlebiinae). Proc. Biol. Soc. Wash., 61:141-148, 16 figs.
1950. New records of the mayfly genus *Baetodes,* with notes on the genus. Ent. News., 61:203-205.
IDE, F. P.
1941. Mayflies of two Neotropical genera, *Lachlania* and *Campsurus,* from Canada. Canad. Ent., 73:153-156, 3 figs.
MAYO, VELMA K.
1939. New western Ephemeroptera. Pan-Pac. Ent., 15: 145-154, 21 figs.
1951. New western Ephemeroptera II. Pan-Pac. Ent., 27:121-125, 10 figs.
1952. New western Ephemeroptera III. Pan-Pac. Ent., 28:93-103, 11 figs.
1952. New western Ephemeroptera IV. Pan-Pac. Ent., 28:179-186, 11 figs.
NEEDHAM, J. G., J. R. TRAVER, and Y. HSU.
1935. The biology of mayflies. Ithaca, N.Y.: Comstock.
SPIETH, HERMAN T.
1933. The phylogeny of some mayfly genera. Jour. N.Y. Ent. Soc., 41:55-86; 327-390, pls. 16-24.
1937. An Oligoneurid from North America. Jour. N.Y. Ent. Soc., 65:139-145, 11 figs,
1941. Taxonomic studies on the Ephemeroptera II. The genus *Hexagenia.* Amer. Midl. Nat., 26:233-280, 62 figs.
UPHOLT, W. M.
1936. A new species of mayfly from California. Pan-Pac. Ent., 12:120-122, 3 figs.
1937. Two new mayflies from the Pacific Coast. Pan-Pac. Ent., 13:85-88.

CHAPTER 4

Odonata

By Ray F. Smith and A. Earl Pritchard
University of California, Berkeley

The dragonflies and damselflies are usually the most conspicuous insect group near any body of inland water. Their beauty and wonderful powers of flight attract the attention of all. The order is distinguished from other insects by having two subequal pairs of net-veined wings, three-segmented tarsi, short setiform antennae, strong chewing mouth parts, and large compound eyes.

The large size of some dragonflies (our largest has a wing spread of about five inches) and their activities have given rise to a large number of superstitions. This is reflected in common names such as "devil's darning needle," "snake feeder," "snake doctor," and "horse stinger." They, of course, do not sting, but the larger forms can give a small, harmless bite. The common name "mosquito hawk" refers to their value in destroying mosquitoes, gnats, and other pests. On the other side of the ledger certain forms are pests about bee yards, and some of the larger naiads occasionally take trout fry. On the whole, however, they are a very beneficial and attractive group of insects.

The damselflies (members of the suborder Zygoptera) hold the wings together above the body or tilted upward when at rest. They also have narrowly transverse heads with eyes separated by more than their own width. The dragonflies (members of the suborder Anisoptera) hold the wings horizontally in repose, sometimes tilted up or down, and the eyes are never separated by more than their own width.

The dragonflies and damselflies are handsomely constructed. In certain features they are unique among the insects. It is not possible to cover this subject here (for details see Tillyard, 1917; Needham and Heywood, 1929; Walker, 1953; and Needham and Westfall, 1955). Special attention should be drawn

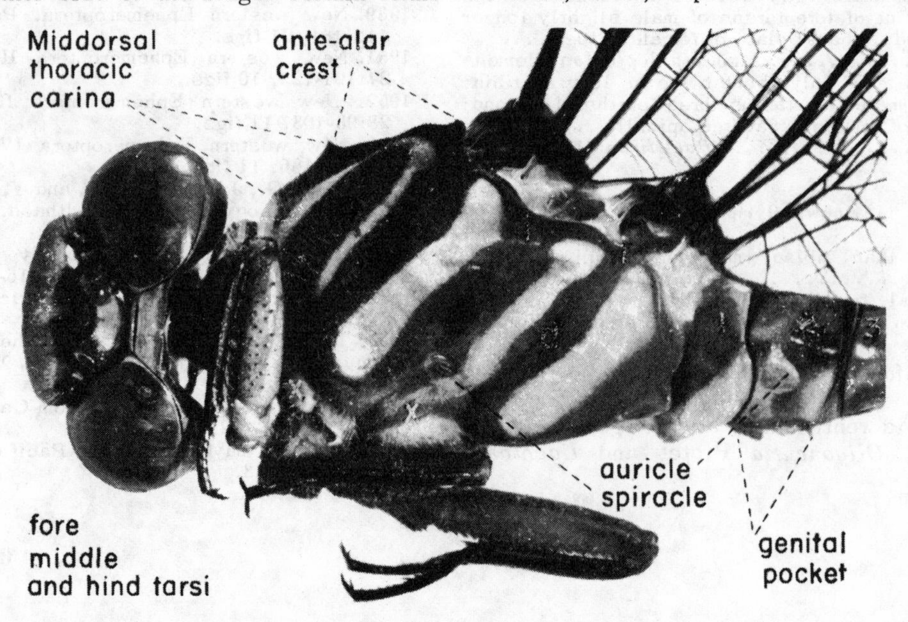

Fig. 4:1. Head (dorsal view) and thorax (lateral view) of *Aphylla protracta* (Needham and Westfall, 1955).

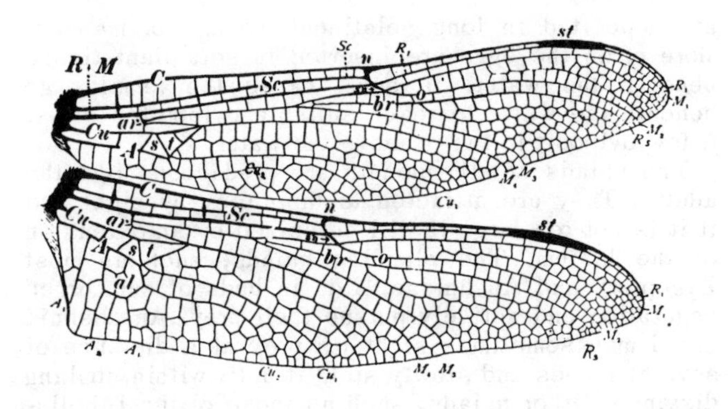

Fig. 4:2. Main wing veins of anisopterous wings. A, anal vein; al, anal loop, ar, arculus; br, bridge; C, costa; Cu, cubitus; M, media; n, nodus; o, oblique vein; R, radius; s, subtriangle; Sc, subcosta, sn, subnodus; st, stigma; t, triangle (Needham and Heywood, 1929).

to the modifications of the thorax (fig. 4:1) and the special copulatory mechanism. The meso- and metathorax are greatly enlarged and fused to form a pterothorax. This pterothorax contains the large flight muscles. The legs are crowded together and moved forward on the thorax. They are not used for walking, but they are adapted for perching and for catching and handling the prey. In flight the legs act as sort of a basket to catch prey. The front legs are also used to hold the prey, either at rest or in flight, while it is being chewed by the mandibles. The abdomen is ten-segmented and carries the genital appendages at the distal end. The males of all Odonata have a pair of movable, unsegmented dorsolateral appendages

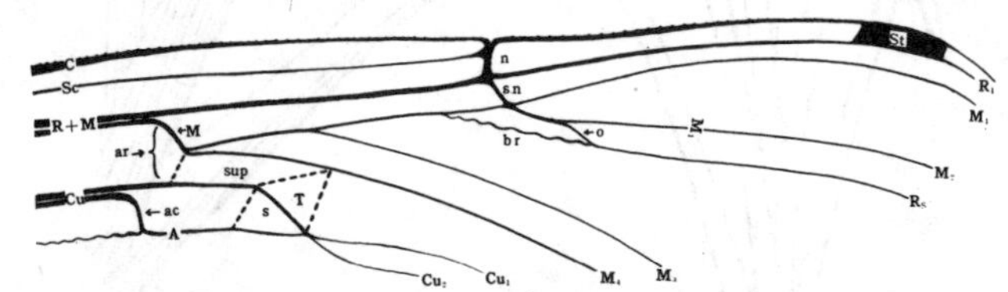

Fig. 4:3. Diagram of wing illustrating principal veins and their connections. sup, supratriangle; ac, anal crossing. Other lettering as in fig. 4:2 (Needham and Westfall, 1955).

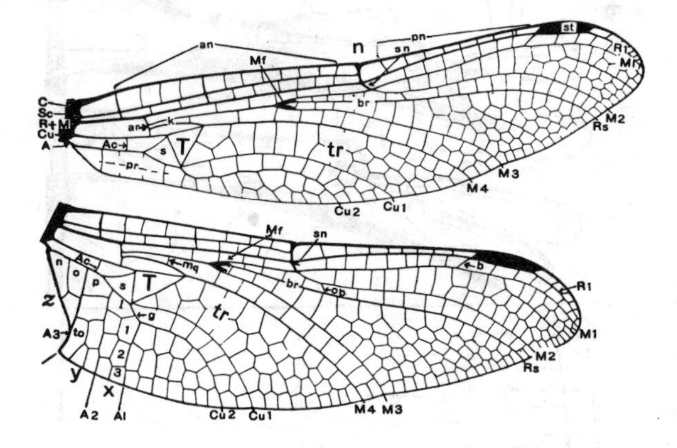

Fig. 4:4. Venational characters in anisopterous wing (*Gomphus cavillaris*). Middle fork (MF) thickened for easier recognition; Ac, anal crossing; an, antenodal cross veins; b, brace vein; pn, postnodal cross veins. Other lettering as in fig. 4:2 (Needham and Westfall, 1955).

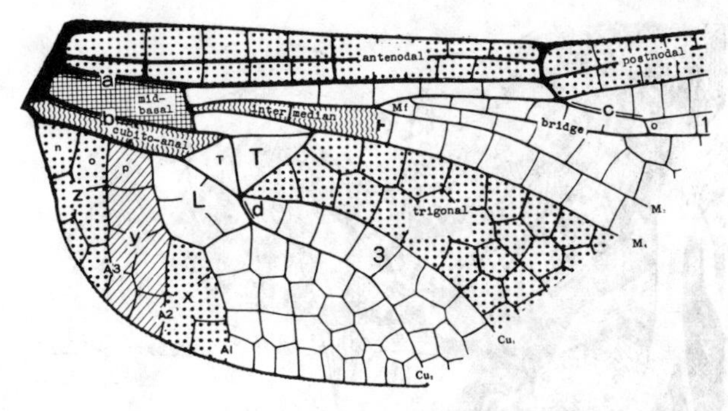

Fig. 4:5. Diagram showing paired veins, fused veins, and principle interspaces in venation of dragonfly wing (Needham and Westfall, 1955).

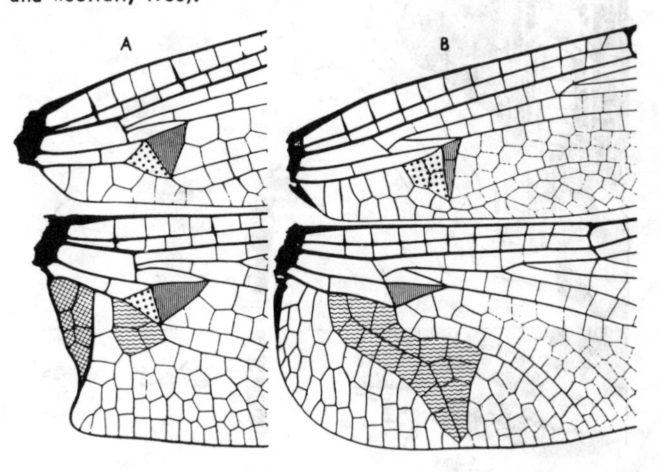

Fig. 4:6. Comparative diagrams of the bases of anisopterous wings. *a*, family Gomphidae (*Ophiogomphus carolus*); *b*, family Libellulidae (*Erythemis simplicicollis*). Triangles striated, subtriangles dotted, anal loops wavy-lined, basal triangle cross-hatched (Needham and Westfall, 1955).

immediately behind the tenth segment. In the suborder Anisoptera there is an additional median inferior appendage (fig. 4:23), whereas in the Zygoptera there are two inferior appendages (fig. 4:68). These male appendages are adapted to grasp the head or thorax of the female in copulation. In addition, the males possess a penis on the venter of the second and third abdominal segment. The female Odonata lack this penis on the second and third abdominal segments; however, the superior anal appendages are usually simpler and reduced or vestigial. In all Zygoptera and in the Aeshnidae and Petaluridae of the Anisoptera, a well-developed ovipositor consisting of three pairs of ventral processes is present near the tip of the abdomen. In other families it is reduced or absent. The well-developed male naiads show the developing genitalia on the venter of abdominal segment two, and in the female the ovipositor may be seen on the venter of segments eight and nine.

The male transfers sperm capsules from the genital aperture on the ninth segment to the penis vesicle on the second segment. In mating the male grasps the female either by the head (Anisoptera) or the prothorax (Zygoptera) with the terminal abdominal appendages. The female then curls her abdomen forward to reach

Fig. 4:7. Generalized habitat of *Cordulegaster dorsalis:* Below, naiad with protective coat of algae on stream bottom; above, exuvium and female ovipositing in stream bed (Kennedy, 1917).

the intromittent organ on the second abdominal segment of the male and to receive the sperm (fig. 4:9). The male frequently accompanies the female while she oviposits (fig. 4:8).

The naiads (fig. 4:36) are strikingly different in appearance from the adults. They are short and compact as compared to the adults and have smaller heads. They are cryptically colored and breathe by means of gills. In the Zygoptera these consist of three caudal lamellae; the Anisoptera have rectal gills. The most striking feature of the naiad is the development of an extensile labium. It is long and jointed so that it can be extended quickly to capture prey.

Most Odonata develop in permanent fresh water. A few are semiaquatic in that they occur in bogs; others occur in saline water, and some with short life cycles can develop in temporary waters. They are found in ponds, lakes, streams, tanks, rivers, and canals of all sizes and descriptions. Although the naiads are aquatic the adults sometimes range many miles from water.

The eggs are laid in various ways (fig. 4:7). They may be simply dropped into the water, or they may be attached to objects in the water. In a few cases they are deposited in long gelatinous strings or masses. More often the eggs are inserted in soft plant tissue beneath the water. In some cases the females go beneath the water to reach suitable oviposition sites. A few oviposit in twigs above the water.

The naiads of all Odonata are predaceous like the adults. They are attracted to their prey by sight and if it is not too large it will be seized by an extension of the labium. The climbing naiads, such as most Zygoptera and Aeshnidae, hide in beds of submerged vegetation and actively pursue their prey. An aeshnid naiad may sometimes sight its prey at a distance of several inches and slowly stalk it until within striking distance. Other naiads, such as those of the Libellulidae that sprawl on the bottom, are very sluggish and do not strike until the prey comes within reach of the labium. These sprawlers are frequently covered with a camouflage of algae or a layer of silt. The naiads of the Gomphidae burrow into the bottom sand or mud so that only the tip of the abdomen reaches the water. After about ten to fifteen instars and a period from less than one to five years the naiad is fully grown. It then crawls out of the water and attaches itself to some suitable object. The adult then

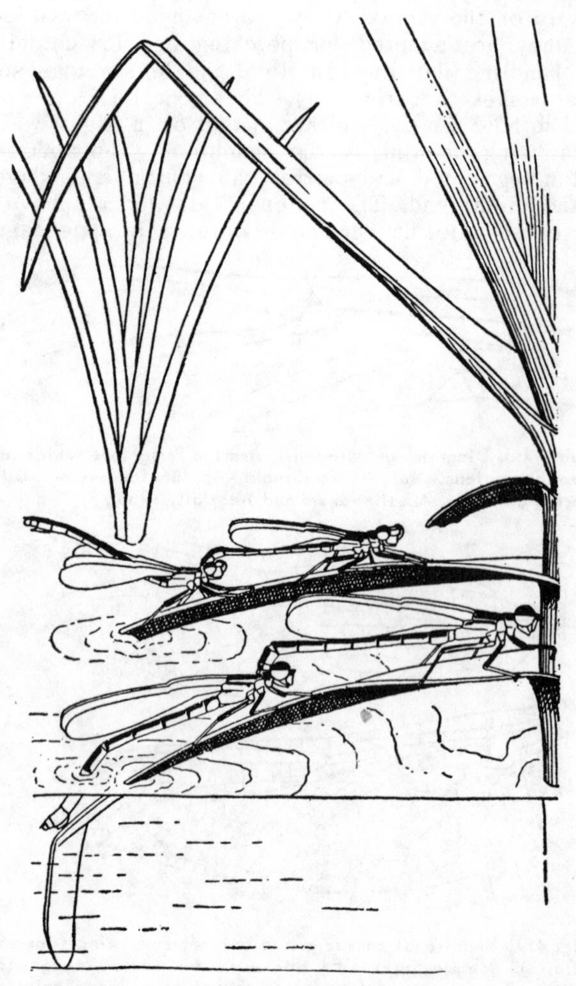

Fig. 4:8. Damselfly pairs *(Ischnura denticollis)* ovipositing (below) and in resting position (above) (Kennedy, 1917).

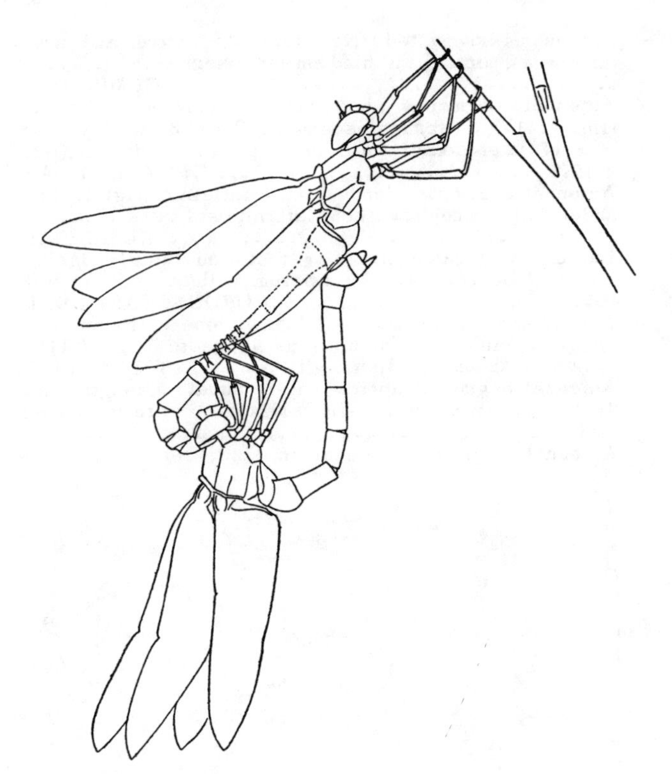

Fig. 4:9. Copulation of *Macromia magnifica* (Kennedy, 1915).

emerges through a slit in the dorsum of the head and thorax.

The newly emerged adults are pale and without the darker markings. Within a few hours the pattern appears, but the full color does not develop until later. Blues and reds are especially slow to develop. Different kinds of Odonata have characteristic flight and habits. Details are given below under the special discussions.

In general, the dragonflies have very few natural enemies. The teneral adults are especially vulnerable owing to their weak powers of flight. At this time and later they may be attacked by birds, lizards, frogs, spiders, and other Odonata. The naiads are eaten by fish, birds, frogs, and other aquatic insects. Of these the most important are the fish. A few hymenopterous egg parasites have been reported on some of the species that lay their eggs in plant tissue. The parasite goes beneath the water to oviposit in the eggs. Certain water mites are also occasionally abundant on the naiads.

Adult dragonflies are usually collected with a standard insect net. The net opening should be fifteen inches or more in diameter and the handle at least three feet long. In netting a dragonfly, a quick following stroke from below and behind is better than a head-on stroke. Many species have regular habits of flight and their movements can be anticipated. In some cases sweeping vegetation in the early morning or at night will be very profitable. For damselflies a large fly swatter is helpful and for large, high-flying forms the finest dust shot in a .22 target pistol will yield a fair proportion of good specimens. Rearing the adults from the naiads frequently gives the best specimens and, of course, associates the two stages.

The adults should not be killed in cyanide but papered and allowed to die slowly. When handled in this manner the contents of the gut are excreted and the colors of the preserved specimens are at their best. Artificial drying is sometimes necessary to fix the colors properly. The adults may then be pinned, or better, placed with wings folded in transparent envelopes, or they may also be placed directly into 80 per cent alcohol.

Naiads may be collected by any of the usual aquatic insect collecting methods. They should be preserved in 80 per cent alcohol.

Key to Nearctic Families

Adults

1. Fore and hind wings dissimilar in size and shape, the proximal part of hind wing broader than that of fore wing; supratriangle and triangle present (fig. 4:2); male with 3 caudal appendages—2 superior and 1 inferior (fig. 4:32) .... ANISOPTERA 2
— Fore and hind wings similar in size and shape, the proximal part of both fore and hind wings of about equal width (fig. 4:74); quadrangle present; male with 4 caudal appendages—2 superior and 2 inferior (fig. 4:60) .... ZYGOPTERA 6
2. Triangles of fore and hind wings about equally distant from arculus and similarly shaped (fig. 4:6*a*) 3
— Triangle more distant from the arculus in fore wing than in hind wing and with its long axis at right angle to costa (fig. 4:6*b*) .... LIBELLULIDAE
3. Stigma with a brace vein at its inner end (fig. 4:4) 4
— Stigma without a brace vein at inner end (fig. 4:2) .... CORDULEGASTRIDAE
4. Eyes widely separated on top of head 5
— Eyes meeting on top of head or nearly so .... AESHNIDAE
5. Stigma linear, not widened medially (fig. 4:15) .... PETALURIDAE
— Stigma rhomboid, widened medially (fig. 4:4) .... GOMPHIDAE
6. Wings distinctly petiolate, with 2 to 4 antenodal cross veins (figs. 4:61*b,c*; 4:74) 7
— Wings not distinctly petiolate, with 5 or more antenodal cross veins (fig. 4:61*a*) .... AGRIONIDAE
7. Wings with vein M_3 arising nearer to arculus than to nodus; short intercalary veins present between M_3 and principal adjacent veins and running to wing margin (fig. 4:61*b*) .... LESTIDAE
— Wings with vein M_3 arising nearer the nodus than to the arculus; no such intercalary veins present (fig. 4:61*c*) .... COENAGRIONIDAE

Naiads

1. Anus surrounded by 3 stiff pointed valves (fig. 4:13); head not markedly wider than thorax and abdomen (fig. 4:10*p*) .... ANISOPTERA 2
— Three external gills present at caudal end of abdomen; head wider than thorax and abdomen (figs. 4:14; 4:72) .... ZYGOPTERA 6
2. Mentum of labium (including lateral lobes) flat, or nearly so, without stout setae 3
— Mentum of labium (including lateral lobes) spoon-shaped, covering face to base of antennae (fig. 4:10*q*), armed with stout setae (fig. 4:11*l*) 5
3. Antenna 6- or 7-segmented (fig. 4:11*a,b*); tarsi 3-segmented, the fore tarsus 2-segmented in *Gomphaeschna* 4

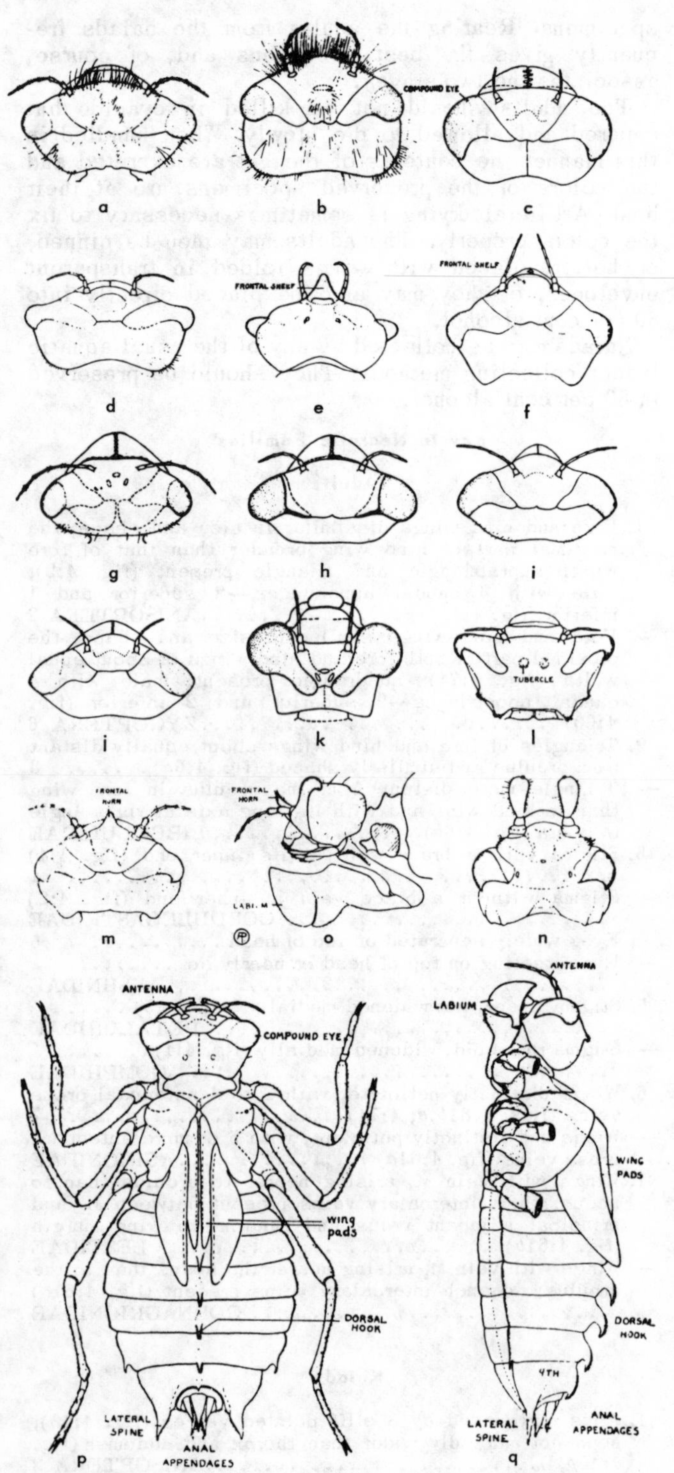

Fig. 4:10. Anisopterous naiads. *a*, *Libellula*; *b*, *Plathemis*; *c*, *Paltothemis*; *d*, *Somatochlora*; *e*, *Platycordulia*; *f*, *Neurocordulia*; *g*, *Tramea*; *h*, *Pachydiplax*; *i*, *Leucorrhinia*, *j*. *Celithemis*; *k*, *Basiaeschna*; *l*, *Nasiaeschna*; *m*, *o*, *Macromia*; *n*, *Octogomphus*; *p*, *q*, *Epicordulia*, dorsal and lateral views; *a-n*, dorsal view of head; *o*, lateral view of head (Wright and Peterson, 1944).

— Antenna 4-segmented (fig. 4:11*c*,*d*,*e*); fore and mid-tarsi 2-segmented, the hind tarsus 3-segmented .... GOMPHIDAE

4. Antennal segments short, thick, heavily setiferous (fig. 4:11*b*); abdominal segments 2 (or 3) to 9 with a pair of laterodorsal tufts of long black bristles (fig. 4:12*i*) .... PETALURIDAE

— Antennal segments slender and bristlelike (fig. 4:11*a*); abdominal segments without laterodorsal tufts .... AESHNIDAE

5. Labium with large irregular teeth on distal edge of lateral lobe (fig. 4:11*h*); mentum with a median cleft .... CORDULEGASTRIDAE

— Labium with distal edge of lateral lobe entire or with small, even-sized crenulations or teeth (fig. 4:11*i*); mentum without a median cleft .... LIBELLULIDAE

6. Antennal segments approximately equal in length (fig. 4:14*a*); mentum entire (fig. 4:14*f*) or with a closed median cleft 7

— Antennal segments unequal in size, 1st segment as

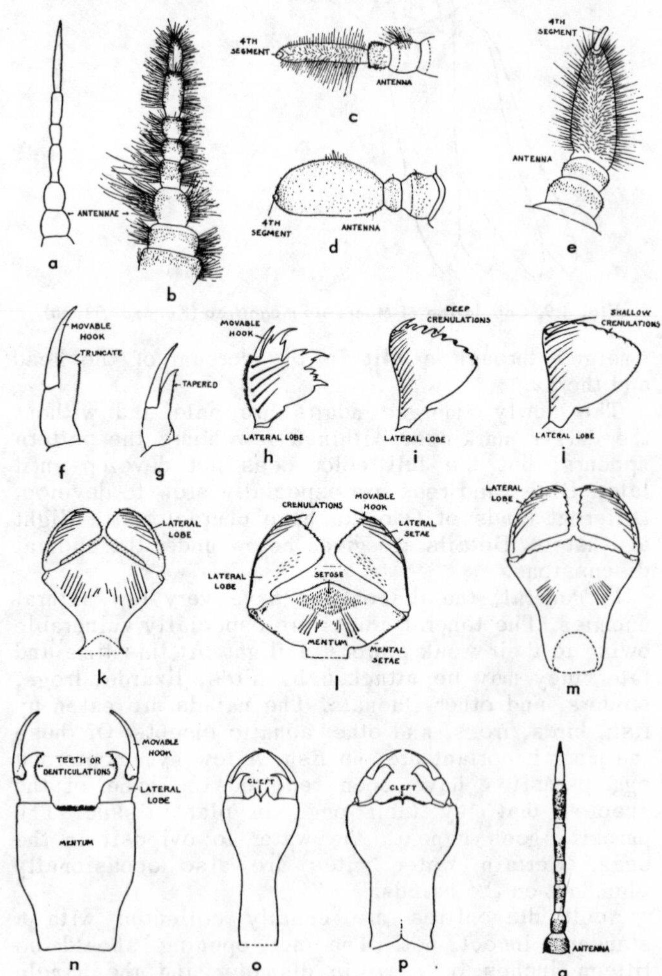

Fig. 4:11. Anisopterous naiads, antennae and parts of labium. *a*, *p*, *Aeshna*; *b*, *Tachopteryx*; *c*, *n*, *Dromogomphus*, *d*, *Octogomphus*; *e*, *Progomphus*; *f*, *Boyeria*; *g*, *Basiaeschna*; *h*, *Cordulegaster*; *i*, *Pantala*; *j*, *Tramea*; *k*, *Libellula*; *l*, *Epicordulia*; *m*, *Plathemis*; *o*, *Coryphaeschna*; *q*, *Pachydiplax*; *a-e* and *q*, antennae; *f-j*, lateral lobe of labium; *k-p*, inner surface of mentum and lateral lobes of labium (Wright and Peterson, 1944).

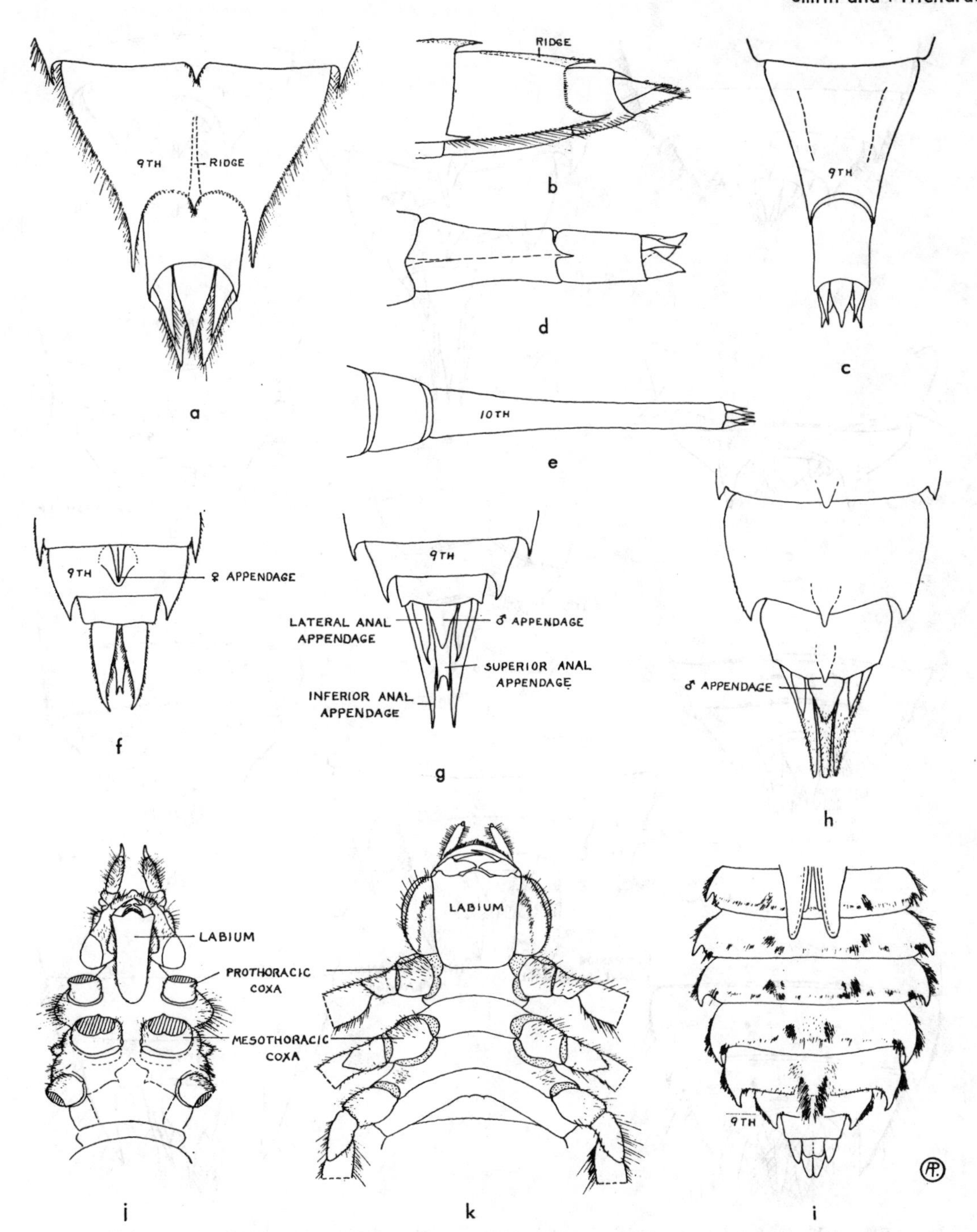

Fig. 4:12. Anisopterous naiads. *a,b, Dromogomphus; c,d,k, Gomphus; e, Aphylla; f,g, Basiaeschna; h, Nasiaeschna; i, Tachopteryx; j, Progomphus; a-i,* caudal segments of abdomen; *j,k,* ventral view of thorax (Wright and Peterson, 1944).

long as 6 following together (fig. 4:14*e*); mentum with a deep, open, median cleft (fig. 4:14*d*) ....AGRIONIDAE

7. Basal half of labium greatly narrowed like a stalk; lateral lobes with distal margin deeply cut by 2 or 3 incisions; median cleft present, closed (fig. 4:14*b*) .... LESTIDAE

— Basal half of labium not greatly narrowed; lateral lobes not deeply cleft; median cleft lacking (fig. 4:14*c,f*) .... COENAGRIONIDAE

ANISOPTERA

Family PETALURIDAE

The petalurids comprise the most generalized group of living Anisoptera. A single species, *Tachopteryx thoreyi* Selys, occurs in the eastern United States, whereas the genus *Tanypteryx* is found in western North America and in Japan. The sluggish naiads

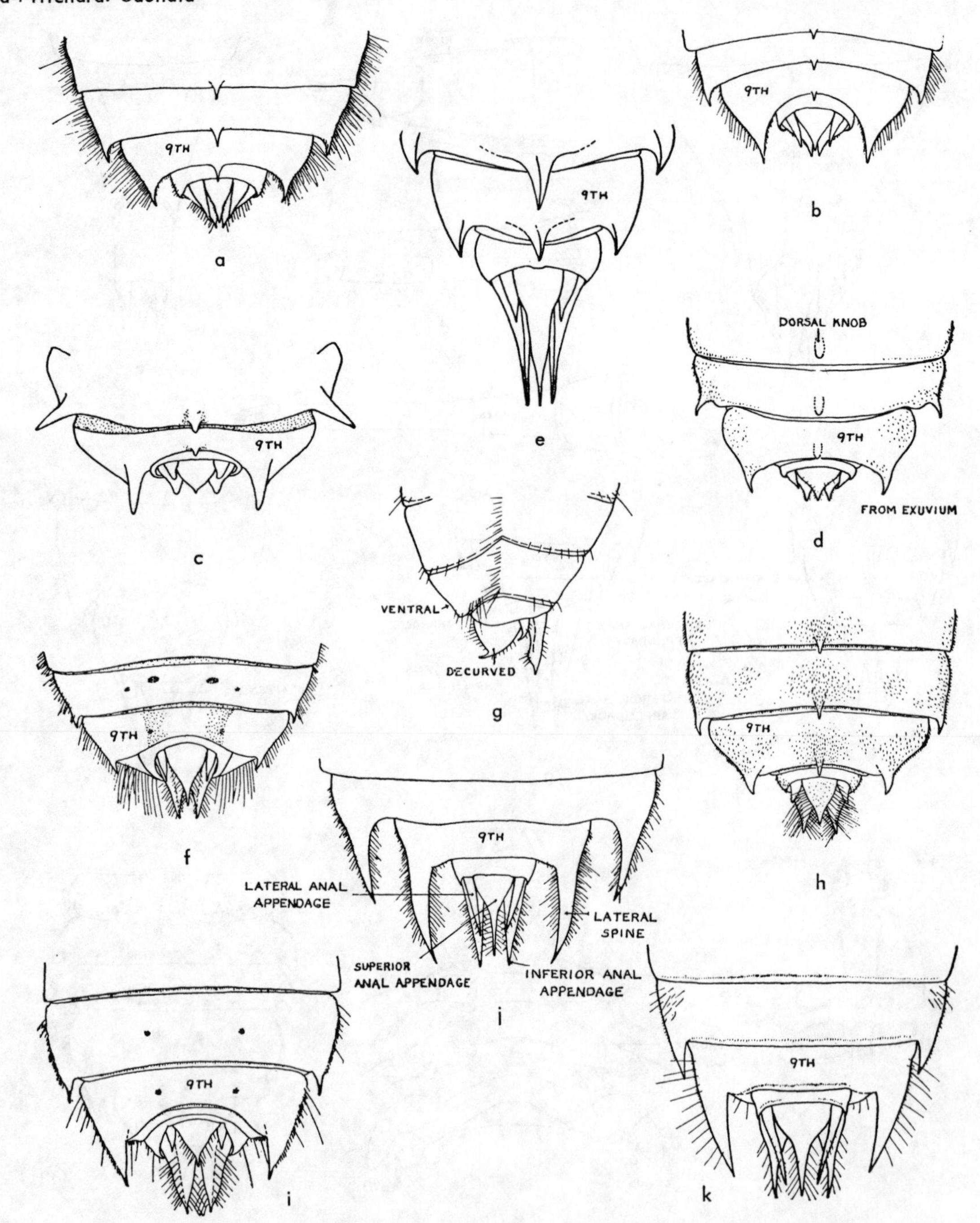

Fig. 4:13. *Anisopterous naiads, caudal segments of abdomen, all dorsal views except g, (lateral). a, Macromia; b, Didymops; c,d, Neurocordulia; e, Cannacria; f, Sympetrum; g, Erythemis; h, Perithemis; i, Pachydiplax; j, Tramea; k, Pantala* (Wright and Peterson, 1944).

occur in the muck of seepage waters and in spring bogs.

Key to Nearctic Genera

Adults

1. Hind wing with triangle having a cross vein; thorax grayish, striped with black .... *Tachopteryx* Selys
   Hind wing with triangle lacking a cross vein (fig. 4:15); thorax black, spotted with yellow .... *Tanypteryx* Kennedy

Naiads

1. Antenna with 7 segments (fig. 4:11*b*); no extra hooks present .... *Tachopteryx* Selys
   Antenna with 6 segments; lateral lobe of labium with minute extra hook on upper rim of base of movable hook .... *Tanypteryx* Kennedy

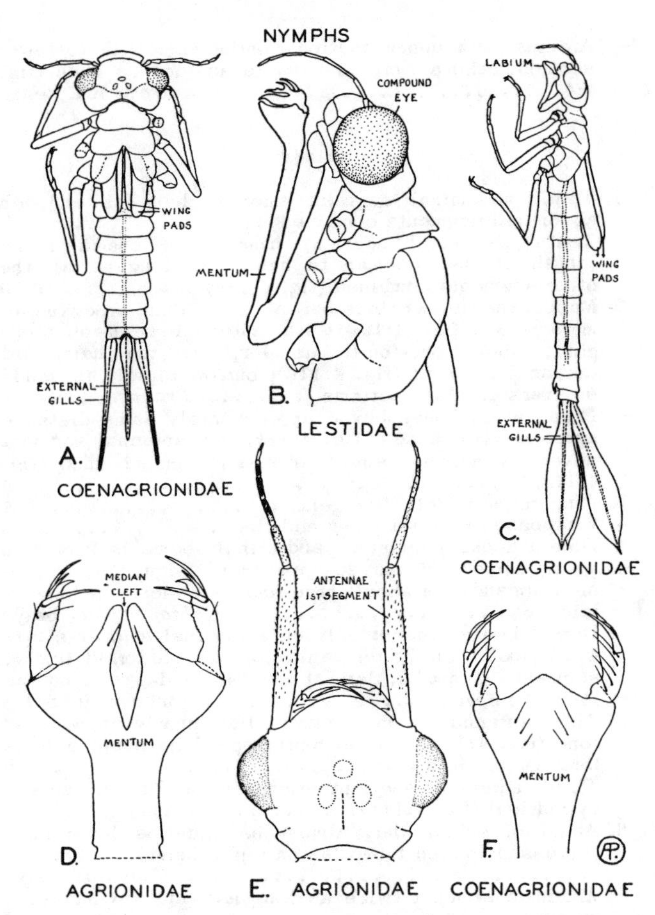

Fig. 4:14. Zygopterous naiads. *a,c,f,* Coenagrionidae; *b,* Lestidae; *d, e,* Agrionidae (Wright and Peterson, 1944).

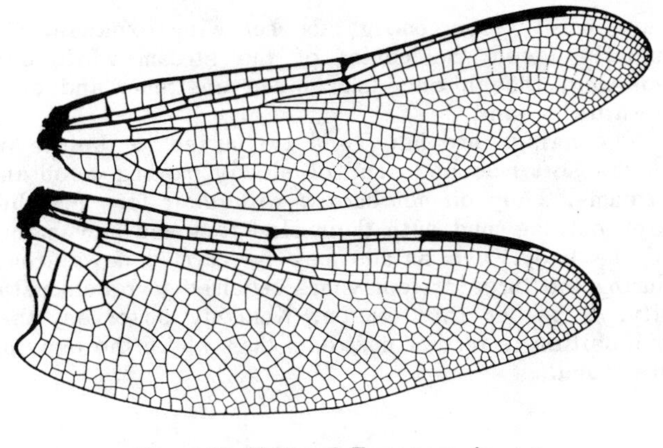

Fig. 4:15. Wings of *Tanypteryx hageni* (Needham and Westfall, 1955).

Genus *Tanypteryx* Kennedy, 1917

Adults of *Tanypteryx* have been taken in alpine meadows, from California to British Columbia, but very few specimens are known. The only naiads referable to this genus in North America were collected by Vincent Roth in a mountain bog in Oregon. His field notes state that he collected two naiads "on a moss covered bank at Parker Creek, about 3000 feet on Marys Peak, Benton County, Oregon. The bank is essentially a rock formation over which water seeps throughout the year. The moss forms a thin cover on the rocks, seldom being over an inch or two in depth. The specimens were collected in wet moss at the junction of one rock and another where a slight niche was formed." The habits of the adults have been described by Whitney (1947).

A single species, *Tanypteryx hageni* (Selys) 1879, has been described from western North America.

Family CORDULEGASTRIDAE

The family Cordulegastridae contains a single genus in North America, although some workers split off *Taeniogaster* Selys and *Zoraena* Kirby.

Genus *Cordulegaster* Leach, 1815

A single species, *Cordulegaster dorsalis* Hagen 1858, is widespread in western North America, being known from Alaska to southern California along the Pacific Coast, and eastward to Wyoming and Utah. Another species, *C. diadema* Selys 1868, is known from Utah, the Southwest, and northern Mexico, but it has not been found in California.

Adults of *Cordulegaster dorsalis* are very large and are strong fliers. Kennedy (1917, p. 517) observes that, "In the steep and narrow mountain gorges where the rushing torrents pour down through the shade of the redwoods and alders, this dragonfly adds a note of mystery to the scene, for the individuals with their strange ophidian coloration glide noiselessly upstream or down, never showing that curiosity toward strangers or unusual surroundings which is exhibited by the libellulines of the sunny valleys, but always moving straight ahead as though drawn irresistably onward. Only males are common on the streams, the females seldom resorting to the water except to oviposit. The males, as indicated above, fly on the longest beats I have observed for any dragonfly, for they fly continuously upstream or down until they come to the head of the stream or to the slow water below, or until some unusual obstruction turns them aside, when they face about and fly as steadily in the opposite direction. The course is usually a foot or two above the surface of the stream and goes through dense shade and any loose brush or foliage which may hang over the water. Because of this habit of flying in long beats this dragonfly is not easily taken, as the collector has but a single chance at each individual" (figs. 4:7).

Regarding oviposition, Kennedy writes, "The female flew hurriedly up the creek and every few yards stopped, and with a sudden backing or downward stroke, while hovering with the body in a perpendicular position, stabbed her large ovipositor into the coarse sand beneath. Four to ten such perpendicular thrusts were made at each stop. Some stops were along the open beaches, but more were in quiet nooks between large rocks where she would have barely room enough for her wing expanse. She usually faced the center of the stream while ovipositing, though once she faced upstream and once toward the bank."

The naiads are hairy and lie buried to their eyes in the soft mud and sand of slowly moving woodland streams. They do not burrow with their fore feet but kick out the mud with their hind feet and movements of the body. The naiads may be carried downstream during the three or four years required to reach maturity. It is for this reason, Kennedy observes, that oviposition is farther upstream than where the exuviae are found.

Family GOMPHIDAE

Adult gomphids are clear winged, with yellow and brown or black bodies, and the caudal end of the male abdomen is more or less enlarged. They spend much of their time perched on the ground or low objects, making short flights.

The naiads are found buried shallowly in the sandy or muddy beds of streams or ponds. They do not usually climb stems of plants for transformation but move to the adjacent shore or objects on the shore.

Key to Nearctic Genera

Adults

1. Fore wing with nodus located beyond middle (fig. 4:16); basal subcostal cross vein usually present 2
— Fore wing with nodus at middle (fig. 4:4); basal subcostal cross vein absent 5
2. Anal loop well defined, with 2 or 3 cells 3
— Anal loop indistinct 4
3. Anal loop with 3 or more cells; male with inferior appendage well developed and deeply forked .... *Gomphoides* Selys
— Anal loop with 2 cells; male with inferior appendage very small or not evident .... *Phyllocycla* Calvert
4. Supratriangle with cross veins; triangle 3-sided ... . *Aphylla* Selys
— Supratriangle without cross veins; triangle 4-sided (fig. 4:16)........................ *Progomphus* Selys
5. Triangle with a cross vein; anal loop with 4 cells .... *Hagenius* Selys
— Triangle without a cross vein; anal loop with 3 cells or less, if present (figs. 4:18; 4:6) 6
6. Anal loop distinct, with 2 or 3 cells (fig. 4:18)..... *Ophiogomphus* Selys
— Anal loop indistinct or absent (figs. 4:4; 4:21) 7
7. Hind femur with many short spines and a row of prominent long spines .... *Dromogomphus* Selys
— Hind femur with many short spines only (fig. 4:26*i*) 8
8. Hind wing with cells below subtriangle twice as long as wide (fig. 4:21); thorax with middorsal stripe yellow and contrasting .... *Octogomphus* Selys
— Hind wing with cells below subtriangle little longer than wide; thorax with middorsal stripe dark 9
9. Stigma distinctly wider than subtended cells (fig. 4:22); male with forks of inferior appendage contiguous .... *Erpetogomphus* Selys
— Stigma approximately as wide as subtended cells (fig. 4:4) 10
10. Arculus with upper section much shorter than lower section; stigma about twice as long as wide .... *Lanthus* Needham
— Arculus with upper section similar in length to lower section; stigma more than twice as long as wide (fig. 4:4) .... *Gomphus* Leach

Naiads

1. Tenth abdominal segment shorter than 8th and 9th abdominal segments combined 2
— Tenth abdominal segment from about one-third the length of the abdomen to nearly as long as all the other segments combined (fig. 4:12*e*) 10
2. Mesocoxae closer together at base than procoxae or metacoxae (fig. 4:12*j*); 4th antennal segment elongate, about one-fourth as long as the hairy 3rd antennal segment (fig. 4:11*e*); burrow rapidly in sands of rivers and lake bottoms .... *Progomphus* Selys
— Procoxae and mesocoxae approximately same distance apart at their bases (fig. 4:12*k*); 4th antennal segment never as above, usually a small rounded knob (fig. 4:11*c*,*d*) 3
3. Wing cases widely divergent 4
— Wing cases parallel along mid-line 5
4. Dorsal hooks present on abdominal segments 2 or 3 to 9, hooklike and curved on caudal segments; lateral anal appendages about three-fourths or less as long as inferiors .... *Ophiogomphus* Selys
— Dorsal hooks present only on abdominal segments 2 to 4, at most a slight thickening on the middorsal line of segments 8 and 9; lateral anal appendages about as long as inferiors .... *Erpetogomphus* Selys
5. Third antennal segment ovate, flat, nearly as wide as long (fig. 4:11*d*); lateral anal appendage about half as long as inferiors 6
— Third antennal segment elongate or linear, usually cylindrical (fig. 4:11*c*)........................ 8
6. Abdomen subcircular, almost as wide as long; body depressed; paired tubercles on top of head .... *Hagenius* Selys
— Abdomen at least twice as long as wide; no tubercles on head 7
7. Abdominal segments 7 to 9 with short lateral spines .... *Octogomphus* Selys
— Abdominal segments 8 and 9 with short lateral spines .... *Lanthus* Needham
8. Abdominal segment 9 rounded dorsally and without a sharp dorsal hook, or, if dorsal hook present on abdominal segment 9, then the segment longer than wide at its base (fig. 4:27) .... *Gomphus* Leach
— Abdominal segment 9 with an acute middorsal ridge with a dorsal hook at its apex, this segment never as long as wide at its base 9
9. Mentum with median lobe moderately produced in a low rounded curved spinulose border; abdominal segment 10 a little longer than 9 .... *Gomphoides* Selys
— Mentum with straight front border; abdominal segment 10 shorter than 9 .... *Dromogomphus* Selys
10. Abdomen with sharp lateral spines on segments 6 or 7 to 9; labium with inner margin of lateral lobe entirely smooth, 3 teeth before end hook of lateral lobe .... *Phyllocycla* Calvert
— Abdomen without lateral spines; labium with inner margin of lateral lobe armed with large sharp pointed recurved teeth, 4 or 5 teeth before end hook of lateral lobe .... *Aphylla* Selys

Genus *Progomphus* Selys, 1854

The genus *Progomphus* (considered by Muttkowski, 1910, to be *Gomphoides*) is primarily Neotropical in distribution. Two species are wide ranging in the United States. One species, *P. obscurus* (Rambur) 1842, is a greenish, yellowish species striped with brown (fig. 4:16). It is common in the eastern United

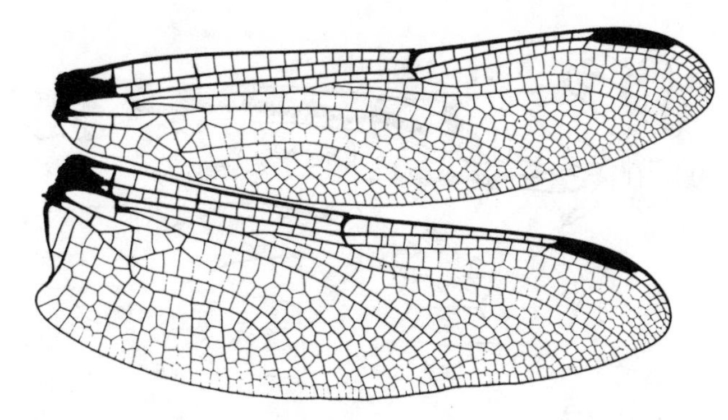

Fig. 4:16. Wings of *Progomphus obscurus* (Needham and Westfall, 1955).

States.[1] Another species, *P. borealis* MacLachlan 1873, is widespread in the western United States and Mexico, having been recorded from Arizona, California, Colorado, New Mexico, Oklahoma, Oregon, Texas, and Utah. It is a grayish brown and dull yellow species.

The naiads of *Progomphus* live in the sandy beds of permanent streams and lakes. They are adept burrowers, and their colors match the sand in which they burrow. *P. borealis* is found in the sandy shallows of permanent desert streams and in sandy areas of intermittent streams in the Sierra Nevada foothills. Adults rest on the banks of the streams or on snags protruding from the water.

Genus *Ophiogomphus* Selys, 1854

Ophiogomphus naiads are found in the gravelly beds of mountain lakes and streams. Adults rest on the gravelly shores when they are not in sallies of flight.

Three of the five species recognized from the western United States are subject to considerable variation in the development of the thoracic stripes (fig. 4:17). Names based on such differences in coloration are here considered to be synonyms, although subspeciation may be indicated.

Key to California Species

Males

1. Inferior appendage broad near base and narrowing distally 2
— Inferior appendage slender proximally and enlarging distally, the dorsal margin concave (fig. 4:17*g*) (British Columbia to Utah and California) (= *phaleratus* Needham 1902; = *occidentis californicus* Kennedy 1917) .... *occidentis* Hagen 1882
2. Superior appendages with ventral angulation near distal end bearing tiny denticulations; tibiae usually with outer face pale 3
— Superior appendages slender, without ventral angulation and with large denticulations; tibiae black (fig. 4:17*a*) (California, Nevada) (= *sequoiarum* Butler 1914) ... . *bison* Selys 1873
3. Posterior hamule acutely pointed (California, Oregon, Nevada) (fig. 4:17*b,c*) (= *morrisoni nevadensis* Kennedy 1917) .... *morrisoni* Selys 1879
— Posterior hamule spatulate distally (western U.S. and Canada) (fig. 4:17*d,e*) (= *montanus* (Selys) 1878) ....*severus* Hagen 1874

Females

1. Occipital spurs present 2
— Occipital spurs absent 3
2. Head with a pair of postoccipital spurs .... *occidentis* Hagen
— Head without postoccipital spurs .... *bison* Selys
3. Humeral stripe usually double .... *morrisoni* Selys
— Humeral stripe with anterior part reduced to an oval spot or absent .... *severus* Hagen

Naiads

1. Lateral spines on abdominal segments 6-9 (fig. 4:19*a*) 2
— Lateral spines on abdominal segments 7-9 (fig. 4:19*b*) 3
2. Lateral anal appendages about nine-tenths inferiors; lateral spines on abdominal segments 7 and 8 subequal (fig. 4:19*a*) .... *bison* Selys
— Lateral anal appendages about seven-tenths inferiors; lateral spine on abdominal segment 8 longer than spine on segment 7 (fig. 4:19*e*)........... *occidentis* Hagen
3. Dorsal hooks on abdominal segments 8 and 9 weak, slender, flattened; tips of hooks on segments 2 and 3 tapered, erect (fig. 4:19*c*) .... *severus* Hagen
— Dorsal hooks on abdominal segments 8 and 9 stout, erect; tips of hooks on segments 2 and 3 very blunt (fig. 4:19*b*) .... *morrisoni* Selys

Genus *Octogomphus* Selys, 1873

The genus *Octogomphus* is based on a single species, *O. specularis* (Hagen) 1859. It is found along the Pacific Coast, from Mexico to British Columbia.

Adults of *Octogomphus specularis* (figs. 4:20; 4:21) are found primarily along the upper reaches of densely shaded streams in the coastal mountains. The males perch on low objects in sunlit openings of the stream, but the females are seldom found near the water except for oviposition.

Kennedy (1917) writes regarding an ovipositing female: "She came volplaning down through an opening in the canopy of alders and, while going through evolutions involving several figures, 8's and S's, she touched the surface of the pool lightly with the tip of her abdomen at intervals of two to six feet. After twenty seconds of this she airily spiraled up and out into the sunshine, where she alighted on a bush on the hillside above the creek."

The naiads live in the loose trash on the bottom of pools and eddies. Kennedy estimated that the naiads spend three years in the water, emergence occurring throughout the spring and summer.

Genus *Erpetogomphus* Selys, 1858

Members of the genus *Erpetogomphus* are found along sandy streams in the western United States, Mexico, and Central America. Two species are known to occur in California.

[1] Reports from the far West are undoubtedly misidentification or records before recognition of *P. borealis* as a distinct species.

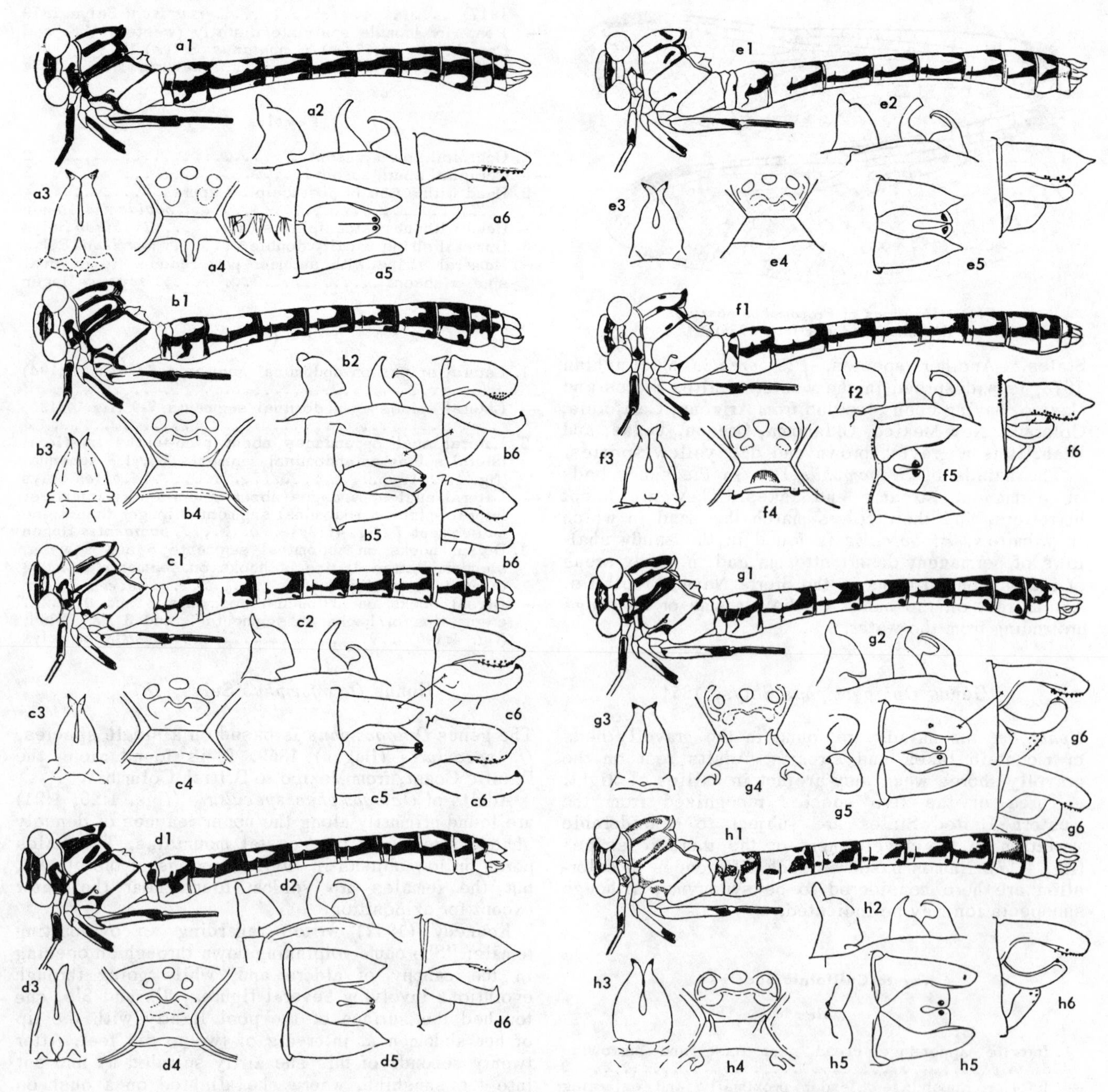

Fig. 4:17. Adult characters in the genus *Ophiogomphus*. *a*, *O. bison*; *b*, *O. morrisoni*; *c*, *O. morrisoni nevadensis*; *d*, *O. severus montanus*; *e*, *O. severus*; *f*, *O. arizonicus*; *g*, *O. occidentis*; *h*, *O. occidentis californicus*; 1, color pattern; 2, hamules; 3, valva; 4, occiput of female; 5, dorsal view of male abdominal appendages; 6, lateral view of male abdominal appendages (Kennedy, 1917).

Key to California Species

Males

1. Superior appendages with a dorsal angulation (California to Texas and Oklahoma) . . . *lampropeltis* Kennedy 1918
— Superior appendages without a dorsal angulation (Oregon and California to Wyoming and Texas) . . . . *compositus* Hagen 1858

Females

1. Occiput with caudal margin emarginate medially . . . . *lampropeltis* Kennedy
— Occiput with caudal margin trilobed . . . . *compositus* Hagen

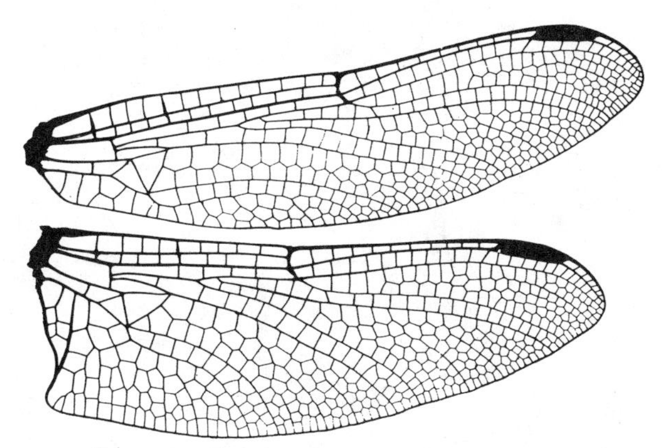

Fig. 4:18. Wings of *Ophiogomphus carolus* (Needham and Westfall, 1955).

Naiads

1. Dorsal hooks rudimentary or absent on abdominal segments from 4 to 9; lateral spines on 6 and 9 smaller than on 7 and 8 .... *lampropeltis* Kennedy
— Dorsal hooks vestigial on 7 to 9; lateral spines about equal in size on 6 to 9 .... *compositus* Hagen

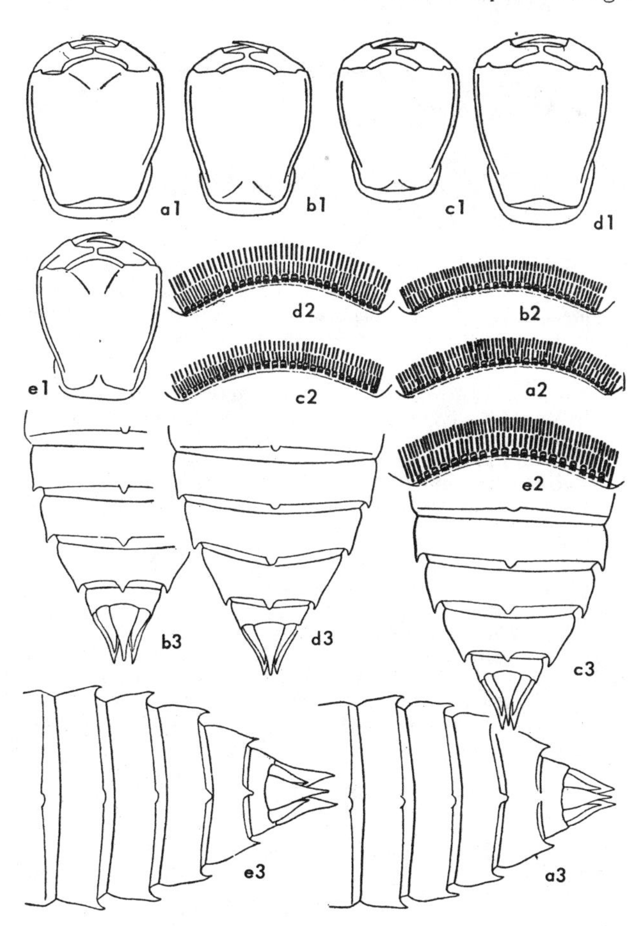

Fig. 4:19. Naiad characters in the genus *Ophiogomphus*. *a*, *O. bison*; *b*, *O. morrisoni*; *c*, *O. severus*; *d*, *O. morrisoni nevadensis*; *e*, *O. occidentis*; 1, mentum; 2, teeth and setae on middle lobe of mentum; 3, dorsal view of abdominal segments 6-10 (Kennedy, 1917).

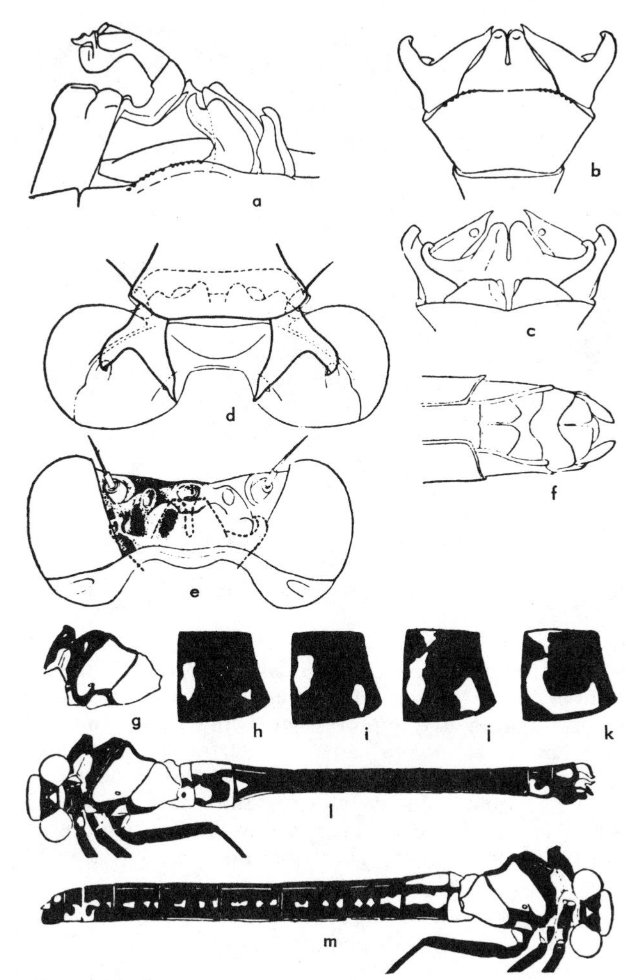

Fig. 4:20. *Octogomphus specularis*. *a*, second abdominal segment of male; *b*, dorsal view of male abdominal appendages; *c*, ventral view of male abdominal appendages; *d-e*, male abdominal appendages applied to head of female; *f*, ventral view of female abdominal segments 9-10; *g*, female thoracic color pattern variation; *h-k*, color variation in male abdominal segment 9; *l*, male color pattern; *m*, female color pattern (Kennedy, 1917).

Genus *Gomphus* Leach, 1815

Gomphus is a very large Holarctic genus, the members of which breed in diverse aquatic environments. Although many species occur in the eastern United States, only three are known from California. The naiads have special hooks for burrowing in silt.

Of the California species, *Gomphus intricatus* and *G. olivaceous* develop in warm and muddy streams or sometimes ponds. *G. confraternus confraternus*[2] Selys 1873 (=*sobrinus* Selys 1873) is also found in sluggish streams of the valleys. However, *G. confraternus donneri* is found in clear mountain lakes. This subspecies differs from the nominate subspecies in that the yellow spot on the ninth abdominal segment is smaller.

[2]Gloyd (1941) showed *Gomphus confraternus* to be a synonym of *G. kurilis* Hagen (1857). We prefer to use the better known name.

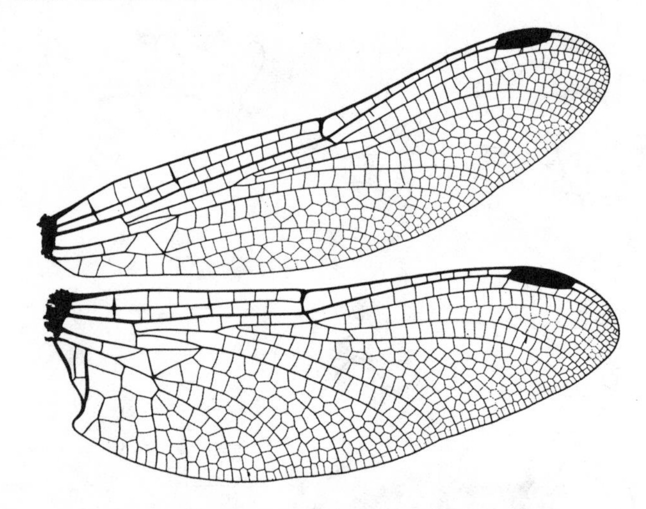

Fig. 4:21. Wings of *Octogomphus specularis* (Needham and Westfall, 1955).

Kennedy (1917) writes with regard to *Gomphus intricatus:* "As with most species of *Gomphus* this species spends much of its time seated on some bush or piece of driftwood, rarely alighting on the ground. However, when it is on the wing it is very energetic, and the males fly rapidly back and forth in short beats, about six inches above the surface of the water. The females oviposit while flying in the same quick, nervous manner. In copulation the male picks the female up either from over the water or from some bush, and after a very short nuptial flight settles for a very long period of copulation."

Key to California Species

Males

1. Posterior hamule stout, sharply bent forward near distal end (Washington to California) (figs. 4:23; 4:24) .... *confraternus* Selys 1876
— Posterior hamule slender, tapering distally 2
2. Tibia yellow externally; club of abdomen bright orange (British Columbia to California, east to Nebraska and Texas) (figs. 4:25*a*; 4:26) .... *intricatus* Hagen 1858
— Tibia black; club of abdomen pale yellowish and largely black dorsally (British Columbia to Utah and California) (= *olivaceous nevadensis* Kennedy 1917) (fig. 4:25*d,f*) .... *olivaceous* Selys 1873

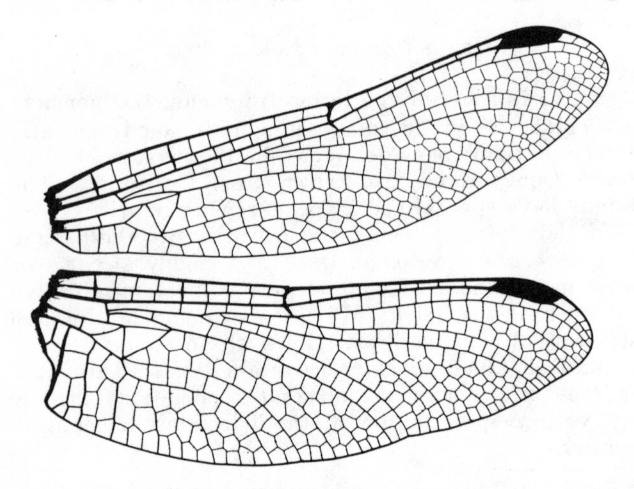

Fig. 4:22. Wings of *Erpetogomphus coluber* (Needham and Westfall, 1955).

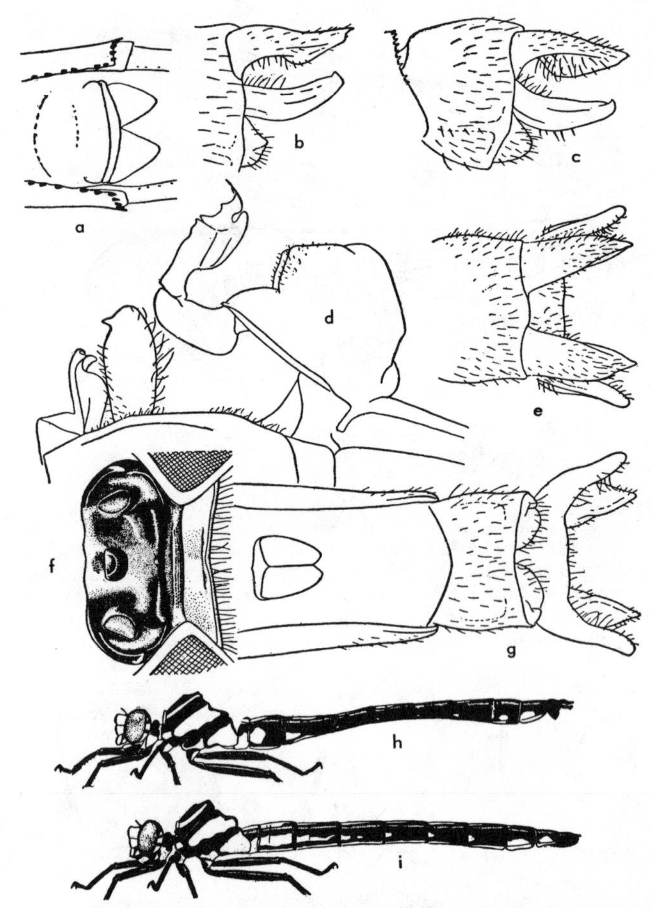

Fig. 4:23. *Gomphus confraternus donneri*. *a*, vulva; *b-c*, lateral view of male abdominal appendages; *d*, second abdominal segment of male; *e*, dorsal view of male abdominal appendages; *f*, female occiput; *g*, ventral view of male abdominal appendages; *h*, color pattern of male; *i*, color pattern of female (Kennedy, 1917).

Females

1. Tibiae black 2
— Tibiae with outer face yellow .... *intricatus* Hagen
2. Thorax with dark lateral stripe (figs. 4:23; 4:24)..... .... *confraternus* Selys
— Thorax without lateral stripe (fig. 4:25*e,g*) .... *olivaceous* Selys

Naiads

1. Dorsal groove present on segments 3-7; concave inner edge of lateral lobe of mentum with 1-3 rounded teeth (fig. 4:27*a-b*) 2
— Dorsal groove absent, but apex of segments 2-7 with low rounded middorsal tubercle; inner margin of lateral lobe of mentum with 6 or more low rounded teeth followed by smaller ones (fig. 4:27*c-d*) ..... *confraternus* Selys
2. Segment 9 with no median spine; anal appendages

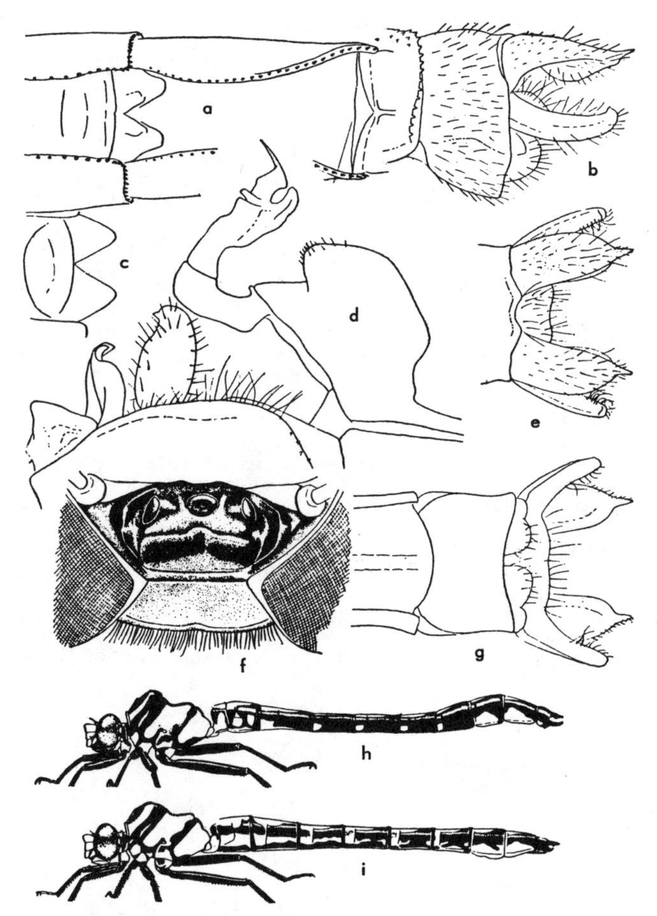

Fig. 4:24. *Gomphus confraternus confraternus. a,c,* vulva; *b,* lateral view of male abdominal appendages; *d,* second abdominal segment of male; *e,* dorsal view of male abdominal appendages; *f,* female occiput; *g,* ventral view of male abdominal appendages; *k,* color pattern of male; *i,* color pattern of female (Kennedy, 1917).

twice the length of segment 10 (fig. 4:27*b*) .... *intricatus* Hagen

— Segment 9 with a minute middorsal spine at apex; anal appendages equal to segment 10 (fig. 4:27*a*) .... *olivaceous* Selys

Family AESHNIDAE

Adults belonging to the family Aeshnidae are large, robust, and strong flying dragonflies. They are commonly seen over ponds or lakes, feeding on swarming insects, and many species roam far from water, even into cities and buildings. Most of the group are diurnal, but a few species are crepuscular in flight.

The female has a well-developed ovipositor that she inserts into succulent stems of aquatic plants for oviposition. Most species breed in the quiet water of lakes or along the margins of slow flowing streams.

The naiads possess a smooth, elongate body with long thin legs. They are active and clamber over the aquatic vegetation and bottom trash, sometimes pausing to stalk their prey. Almost any living animal that they can capture is eaten, even small fish, and they are also notoriously cannibalistic.

Many of the genera found in North America are indigenous to the eastern United States or else they are primarily tropical in distribution. Only four genera are known to occur in California; a fifth genus, *Oplonaeschna,* occurs in the neighboring state of Arizona.

Key to Nearctic Genera

Adults

1. Arculus with upper sector shorter than lower sector (fig. 4:28) .... *Anax* Leach

— Arculus with upper sector as long as or longer than lower sector (figs. 4:29; 4:33) 2

2. Basal space with 2 or more cross veins .... *Boyeria* MacLachlan

— Basal space with a single cross vein or cross veins lacking 3

3. Radial sector simple 4

— Radial sector forked; stigma surmounting 3 or more cross veins not counting brace vein (fig. 4:29) 6

4. Stigma surmounting 1 cross vein not counting brace vein; supratriangle without cross veins .... *Gomphaeschna* Selys

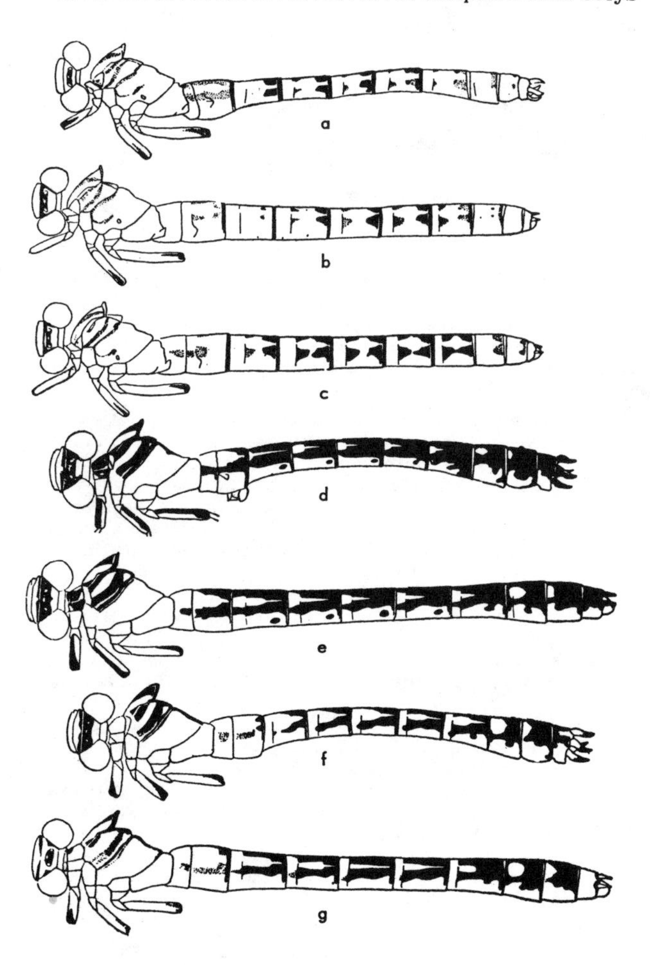

Fig. 4:25. Color patterns in the genus *Gomphus. a-c, G. intricatus; d-g, G. olivaceous* (Kennedy, 1917).

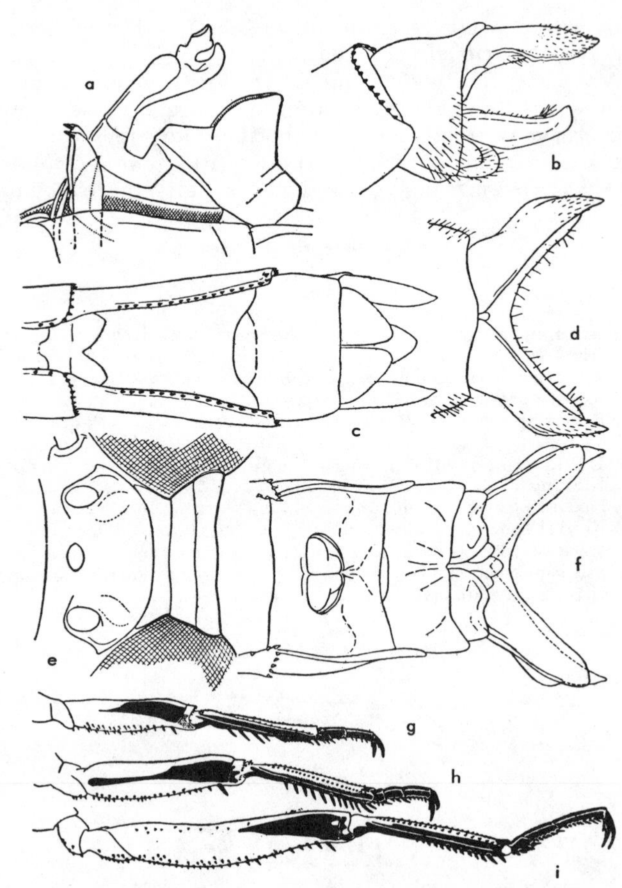

Fig. 4:26. *Gomphus intricatus*. *a*, second abdominal segment of male; *b*, lateral view of male abdominal appendages; c, ventral view of female abdominal segments 9-10; *d*, dorsal view of male abdominal appendages; *e*, female occiput; *f*, ventral view of terminal segments of male abdomen; *g-i*, color pattern on male legs (Kennedy, 1917).

— Stigma surmounting 2 or more cross veins not counting brace vein; supratriangle with cross veins 5
5. Base of wings with large brown spot; 1 row of cells between Cu_1 and Cu_2 .... *Basiaeschna* Selys
— Base of wings hyaline; 2 rows of cells between Cu_1 and Cu_2 .... *Oplonaeschna* Selys
6. Radial sector with fork symmetrical, and with not more than 2 rows of cells between it and its planate 7
— Radial sector with fork unsymmetrical, and with 3 or more rows of cells between it and its planate (fig. 4:33) 8
7. Frons strongly projecting, the dorsal margin acute; radial planate subtending 1 row of cells .... *Nasiaeschna* Selys
— Frons not projecting; radial planate subtending 2 rows of cells .... *Epiaeschna* Hagen
8. Radial sector forked under the stigma............. .... *Coryphaeschna* Williamson
— Radial sector forked proximad to the stigma 9
9. Supratriangle no longer than midbasal space (fig. 4:29) .... *Aeshna* Fabricius
— Supratriangle distinctly longer than midbasal space . 10
10. Hind wing with 2 rows of cells between M_1 and M_2 beginning distad to base of stigma; female with trifid process on venter of 10th abdominal segment (fig. 4:34) .... *Triacanthagyna* Selys
— Hind wing with 2 rows of cells between M_1 and M_2 beginning at base or proximal to base of stigma; female with bifid process on venter of 10th abdominal segment (fig. 4:33) .... *Gynacantha* Rambur

Naiads

1. Lateral lobes of labium with stout raptorial setae .. 2
— Lateral lobes of labium lacking raptorial setae (fig. 4:11*o-p*) 3
2. Lateral setae of labium nearly uniform in length .... *Triacanthagyna* Selys
— Lateral setae of labium very unequal in length, diminishing to very small ones at proximal end of row ... . *Gynacantha* Rambur
3. Distal border of mentum with a pair of sharp, parallel spines, 1 on each side of the mental cleft; head flattened and elongate-rectangular seen from above (fig. 4:11*o*) .... *Coryphaeschna* Williamson
— Distal border of mentum not so armed (fig. 4:11*p*) ... 4
4. Lateral spines present on abdominal segments 7-9 .. 5
— Lateral spines present on abdominal segments 4, 5, or 6-9 7
5. Superior anal appendage but slightly shorter than inferior 6
— Superior anal appendage about three-fourths inferiors; inferiors about subequal in length to middorsum of segments 9 and 10; male appendage triangular with a blunt, rounded apex .... *Aeshna* Fabricius
6. Inferior anal appendage about one and one-half times as long as middorsal length of segment 9 and 10; superior anal appendage cleft at apex ..... *Anax* Leach
— Inferior anal appendage less than middorsal length of segment 9 and 10; superior anal appendage bluntly rounded at apex .... *Gomphaeschna* Selys

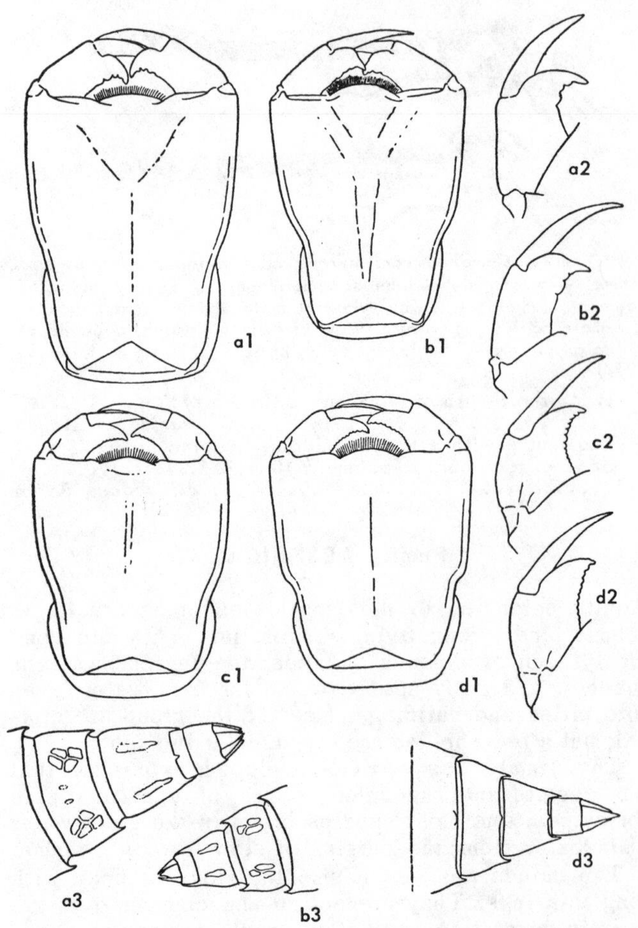

Fig. 4:27. Naiad characters in the genus *Gomphus*. *a*, *G. olivaceous*; *b*, *G. intricatus*; *c*, *G. confraternus donneri*; *d*, *G. confraternus confraternus*; 1, mentum, 2, detail of mental lobe; 3, terminal abdominal segments. (Kennedy, 1917).

7. Caudolateral margin of head from dorsal view with 2 large, well-developed tubercles (fig. 4:10*l*); eyes small, occupying only one-third of the lateral margin of the head 8
— Caudolateral margin of head from dorsal view never with 2 tubercles as described above; eyes large, occupying about half of the lateral margin of the head (fig. 4:10*k*) 9
8. Low but distinct dorsal hooks (best seen from lateral view) present on abdominal segments 7 to 9; apex of lateral lobe broadly rounded; lateral anal appendages less than half the length of the superior (fig. 4:12*h*) . . . . *Nasiaeschna* Selys
— Dorsal hooks absent on all abdominal segments; apex of lateral lobe truncate; lateral anal appendages more than half the length of the superior . *Epiaeschna* Hagen
9. Lateral spines present on abdominal segments 6-9 . . . . *Aeshna* Fabricius
— Lateral spines present on abdominal segments 3, 4, or 5-9 10
10. Hind angles of head strongly angulate (fig. 4:10*k*) . . 11
— Hind angles of head rounded or slightly angulate . . . 12
11. Lateral lobe of labium obtuse or subtruncate at tip; median border of lateral lobe with distinct, more or less square-cut teeth (fig. 4:11*f*) . . . . . . *Boyeria* MacLachlan
— Lateral lobe of labium with a taper-pointed tip; median border of lateral lobe with indistinct denticulation (fig. 4:11*g*) . . . . *Basiaeschna* Selys
12. Superior anal appendage nine-tenths length of inferiors; inferiors strongly incurved at tips . . . . *Oplonaeschna* Selys
— Superior anal appendage about three-fourths as long as inferiors . . . . *Aeshna* Fabricius

Genus *Anax* Leach, 1815

Four species of the cosmopolitan genus *Anax* occur in the United States. One of these, *A. junius* (fig. 4:28), is widespread and common in North America, but the others are more southern in distribution.

Adults are commonly known as green darners. They are large and very strong fliers. *Anax walsinghami* is the largest of our North American dragonflies, with a wing length up to five inches and a length up to nearly ten inches.

Eggs are inserted beneath the water into the water-soaked stems of reeds or other plants or floating sticks. The slender, green and brown naiads are active climbers on submerged pond vegetation and also move by ejection of water from the respiratory chamber. They are notoriously cannibalistic.

Key to California Species

Males

1. Superior appendages not bifid; abdomen 47-58 mm. long (North America, Central America, West Indies) . . . . *junius* (Drury) 1773
— Superior appendages bifid; abdomen 100-116 mm. long (California, Utah, and Texas to Central America) . . . . *walsinghami* MacLachlan 1882

Females

1. Occiput with 2 blunt teeth on hind margin . . . . *junius* (Drury)
— Occiput without teeth. . . . . *walsinghami* MacLachlan

Naiads

1. Lateral lobes of labium tapering to a hooked point; no teeth on mentum on either side of median cleft . . . . *junius* (Drury)
— Lateral lobes of labium squarely truncate, a little rounded on the superior angle; mentum with small teeth on either side of the median cleft . . . . *walsinghami* MacLachlan

Genus *Aeshna* Fabricius, 1775

The original spelling of *Aeshna,* rather than *Aeschna,* as used by many workers, is retained because of a ruling by the International Commission on Zoological Nomenclature.

Adults are strong fliers, usually blue and brownish in color, and they are often commonly called blue darners. Features of the coloration are often indistinct in specimens that are not dried rapidly.

They are found most abundantly near waters in which they breed, but they may wander far inland, particularly shortly after maturity. Generally, the imagoes follow no regular course but fly up and down over marshes and sluggish streams, shallow lakes, ponds, or bays containing vegetation. In hot weather the adults have a tendency to hang in the shade from the underside of leaves of trees. Two species of *Aeshna* have been recorded as swarming.

Females oviposit soon after becoming fully mature and continue to do so from time to time throughout

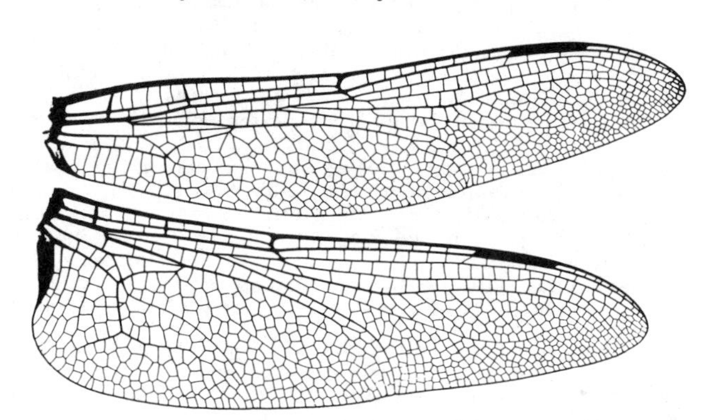

Fig. 4:28. Wings of *Anax junius* (Needham and Westfall, 1955).

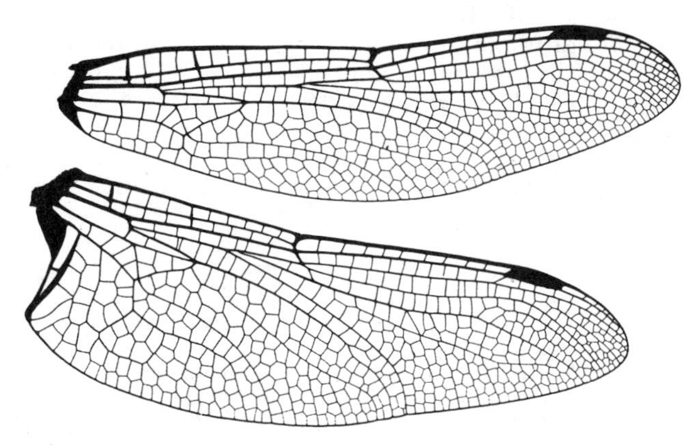

Fig. 4:29. Wings of *Aeshna juncea* (Needham and Westfall, 1955).

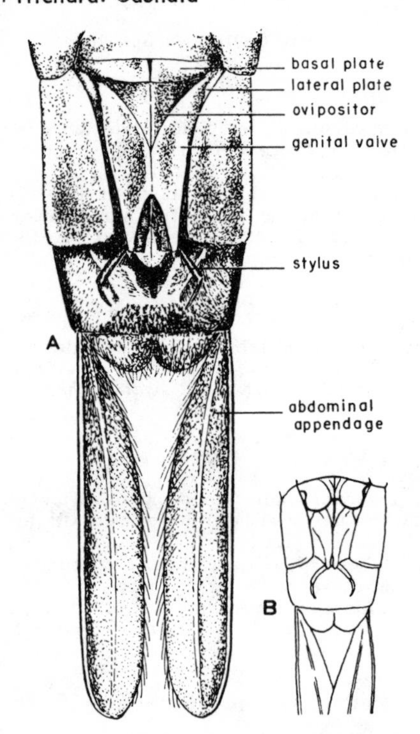

Fig. 4:30. Ventral view of terminal abdominal segments of female *Aeshna*. *a*, *A. clepsydra;* *b*, *A. subarctica* (Needham and Westfall, 1955).

the rest of their life. The eggs may be inserted in the stem of an aquatic plant every two or three seconds until the female is entirely submerged. Rarely, an immersed log or even the sand or mud may be used for egg deposition.

Three years appear to represent the normal term of life of an *Aeshna* naiad, representing seven immature stages. Most naiads are found in fresh water that is rather shallow and harboring aquatic vegetation. However, *A. californica* has been recorded from brackish water. Naiads of smaller dragonflies, water beetles, mayflies, and aquatic bugs, as well as leeches, tadpoles, and small fish, all serve as food for the voracious naiads.

Walker (1912) presented an excellent monograph of the genus *Aeshna,* but the later papers of Walker must also be consulted.

The California species fall into four distinct species groups as far as the male terminalia are concerned, but it is rather difficult to determine some of the females with certainty.

The following species are related to those recorded from California, but they are not yet known from this state:

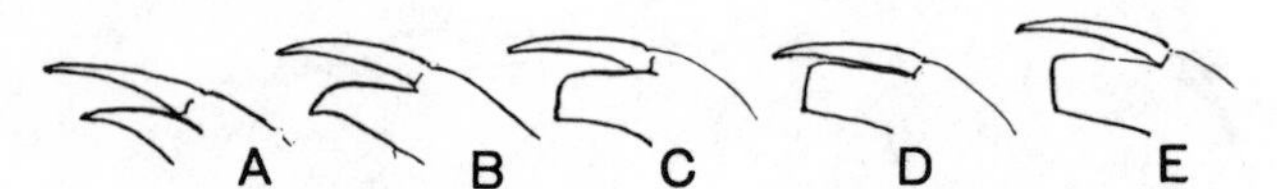

Fig. 4:31. Types of lateral lobes of naiads in the genus *Aeshna*. *a*, *constricta;* *b*, *canadensis;* *c*, *californica*, *multicolor* and *verticalis;* *d*, *walkeri* and *interrupta;* *e*, *palmata* and *umbrosa* (Needham and Westfall, 1955).

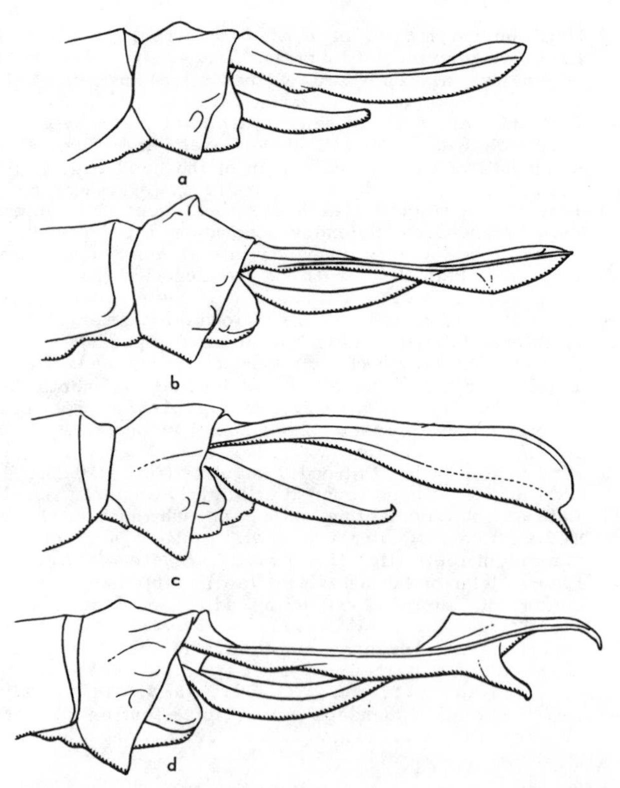

Fig. 4:32. Lateral views of male abdominal appendages in the genus *Aeshna*. *a*, *californica;* *b*, *interrupta;* *c*, *palmata;* *d*, *multicolor* (Celeste Green).

Aeshna dugesii Calvert 1908, Baja California, Mexico, and Texas (related to *A. multicolor*).

Aeshna manni Williamson 1930, Baja California (related to *A. californica* and misidentified by Calvert 1908, as *A. cornigera* Brauer 1865).

Aeshna arida Kennedy 1918, Arizona and New Mexico (related to *A. palmata*).

Key to California Species

Males

1. Abdominal segment 1 with a ventral tubercle 2
— Abdominal segment 1 without a ventral tubercle 3
2. Superior appendage with a sharp ventral angulation or spine near distal end (fig. 4:32*d*) (British Columbia to Wyoming, south to Texas and California and Panama) .... *multicolor* Hagen 1861
— Superior appendage without a ventral projection near distal end (fig. 4:32*a*) (British Columbia and Idaho to California and Arizona)...... *californica* Calvert 1895
3. Anal triangle with 2 cells; superior appendage without a distoventral tooth (fig. 4:32*b*) 4
— Anal triangle with 3 cells; superior appendage with a sharp, distoventral tooth (fig. 4:32*c*) 5
4. Mesothorax with dorsal pale stripes reduced to small, isolated spots or narrow, incomplete lines (Canada, western U.S.) .... *interrupta*[3] Walker 1908
— Mesothorax with dorsal pale stripes complete, expanded at upper ends (U.S.)............ *verticalis* Hagen 1861

[3]*A. interrupta interna* Walker 1908 (British Columbia, Oregon to Colorado and New Mexico), a montane subspecies.

A. interrupta nevadensis Walker 1908 (British Columbia, western Nevada, and California), a valley subspecies.

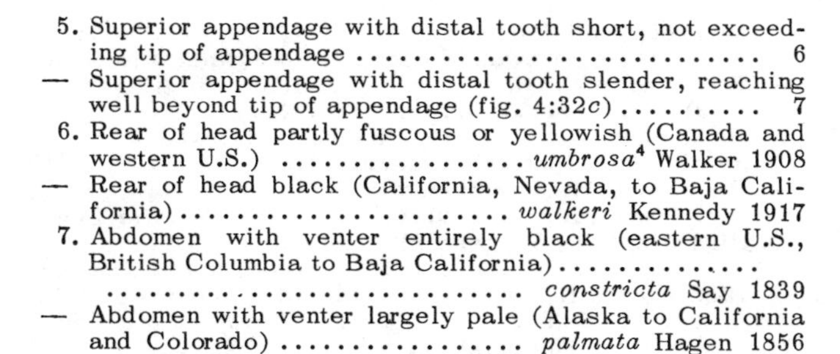

5. Superior appendage with distal tooth short, not exceeding tip of appendage 6
— Superior appendage with distal tooth slender, reaching well beyond tip of appendage (fig. 4:32*c*) 7
6. Rear of head partly fuscous or yellowish (Canada and western U.S.) .... *umbrosa*[4] Walker 1908
— Rear of head black (California, Nevada, to Baja California) .... *walkeri* Kennedy 1917
7. Abdomen with venter entirely black (eastern U.S., British Columbia to Baja California) .... *constricta* Say 1839
— Abdomen with venter largely pale (Alaska to California and Colorado) .... *palmata* Hagen 1856

Females

1. Abdomen with a tubercle on the venter of segment 1 2
— Abdomen without a tubercle on the venter of segment 1 3
2. Face with a brown or black line on frontoclypeal suture .... *californica* Calvert
— Face without a black line on frontoclypeal suture .... *multicolor* Hagen
3. Styli as long as the dorsum of abdominal segment 10; appendages broadest before middle (fig. 4:30) .... *constricta* Say
— Styli much shorter than dorsum of abdominal segment 10; appendages broadest beyond middle 4
4. Genital stylus with a tiny pencil of hairs; genital valves with apices not elevated 5
— Genital stylus without a tiny pencil of hairs; genital valves with apices elevated 7
5. Mesothorax with dorsal pale stripes absent or represented by a small spot; face with a black line on frontoclypeal suture .... *interrupta* Walker
— Mesothorax with dorsal pale stripes present; face without a black line on frontoclypeal suture 6
6. Mesothorax with first lateral pale stripe having anterior margin sinuate.................... *verticalis* Hagen
— Mesothorax with first lateral pale stripe having anterior margin straight .... *walkeri* Kennedy
7. Rear of head pale; face without a black line on frontoclypeal suture .... *umbrosa* Walker
— Rear of head black; face with a black line on frontoclypeal suture .... *palmata* Hagen

Naiads

1. Blade of lateral lobe of labium wider than in figure 4:31*a* or *b* 2
— Blade of lateral lobe of labium shaped as in figure 4:31*a*; femora concolorous .... *constricta* Say
2. Blade about like figure 4:31*c*; mentum of labium about 1¼ times as long as greatest width 3
— Blade wider than figure 4:31*c*; mentum of labium about 1.3-1.6 times as long as greatest width 5
3. Blade with minute triangular tooth on innermost angle; femora concolorous .... *verticalis* Hagen
— Blade merely sharply angulate at innermost angle; femora striped 4
4. Lateral spines of abdominal segment 8 shorter than 9, 6 rudimentary .... *multicolor* Hagen
— Lateral spines of abdominal segment 8 longer than 9, 6 well developed................ *californica* Calvert
5. Blade of lateral lobe of labium shaped as shown in figure 4:31*d* 6
— Blade of lateral lobe of labium shaped as shown in figure 4:31*e* 7
6. Blade of lateral lobe of labium with minute tooth on innermost angle .... *interrupta* Walker

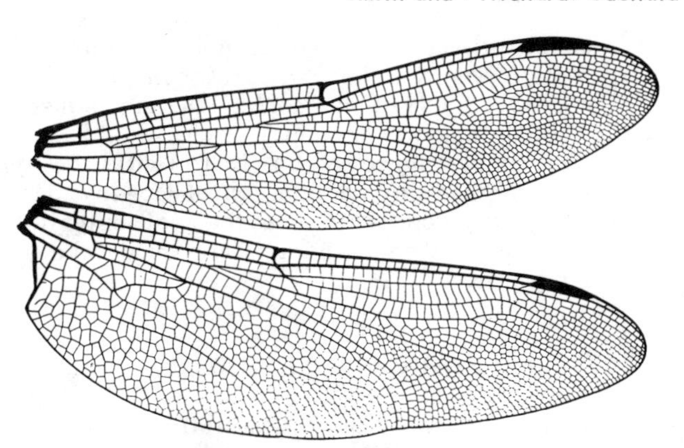

Fig. 4:33. Wings of *Gynacantha nervosa* (Needham and Westfall, 1955).

— Blade of lateral lobe of labium sharply angulate at innermost angle .... *walkeri* Kennedy
7. Mentum about 1.3 times as long as greatest width; ovipositor of female about one and one-third times as long as segment 9 .... *palmata* Hagen
— Mentum about 1.6 times as long as greatest width; ovipositor of female about one and one-tenth times as long as segment 9 .... *umbrosa* Walker

Genus *Gynacantha* Rambur, 1842

Gynacantha represents a group of aeshnine genera of dragonflies that are found in the tropics. The only species found in the United States, *G. nervosa* Rambur, 1842 (fig. 4:33), ranges widely over Central and South America, and it has been found in southern Florida and southern California. This is a large, dusky brown species with green markings, particularly at the base of the abdomen. Adults fly only for a short period in the evening and early morning. Females lay eggs in soil near water.

The type of the genus *Gynacantha* was first designated by Kirby 1890, as *G. trifida*. Previously, however, Selys, 1857, made *G. trifida* the type of his genus *Triacanthagyna,* because the female is generically recognizable by the triacanthagyne characteristics.

It is true that *Gynacantha,* with seven originally included species, was without a type species until Kirby's designation, which should be accepted under

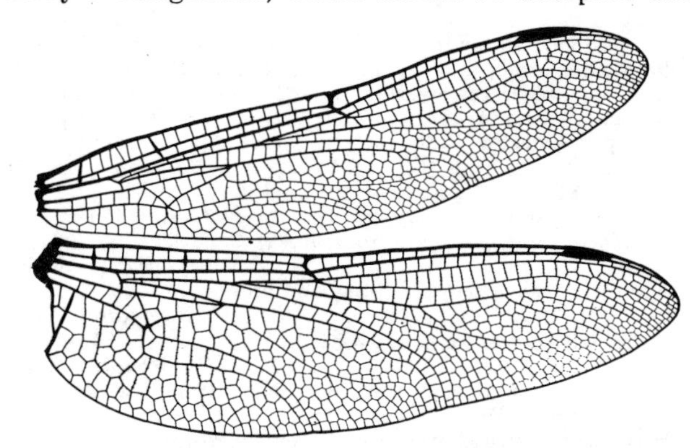

Fig. 4:34. Wings of *Triacanthagyna trifida* (Needham and Westfall, 1955).

[4] *A. umbrosa umbrosa* Walker 1908, a Canadian subspecies. *A. umbrosa occidentalis* Walker 1912, occurring from Alaska to Utah, Nevada and California.

the present rules of zoölogical nomenclature. This interpretation would make *Triacanthagyna* a synonym of *Gynacantha,* and *Gynacantha,* in the sense of having biacanthagyne characteristics, would be replaced by *Acanthogyna* Kirby 1890 (with the type *Gynacantha nervosa* by designation of Cowley, 1934).

However, we propose to use the well-established interpretations of these genera rather than make such a confusing switch.

Family LIBELLULIDAE

The members of the family Libellulidae are mostly showy dragonflies commonly seen hovering over the surface of still water. They vary greatly in size; in our fauna the smallest is about an inch in length, and the largest about four inches. The bodies of these dragonflies are in general stouter and less elongated than the Aeshnidae and Gomphidae.

The females do not possess a well-developed ovipositor but merely drop their eggs into the water or place them about plants at the surface of the water.

The naiads are usually protectively colored and sprawl on the bottom in shallow water or clamber over fallen plant debris. They possess a characteristic spoon-shaped, masklike labium.

Many authors split this group into two or three parts, each of family rank. Because of difficulties involved in characterizing the naiads at the family level, they are considered here as subfamilies.

Key to Nearctic Genera

Adults

1. Anal loop much longer than wide, its cells divided into 2 rows; hind wing with triangle close to arculus (fig. 4:40) 2
— Anal loop little longer than wide, its cells not arranged in 2 rows; hind wing with triangle remote from arculus (fig. 4:35) .... MACROMIINAE 3
2. Anal loop not boot-shaped, distal angles similarly developed (figs. 4:37; 4:38; 4:39); posterior margin of eye emarginate; male with auricles on second abdominal segment and with anal lobe of hind wing notched .... CORDULIINAE 4
— Anal loop boot-shaped, the toe rarely absent; posterior margin of eye rounded (fig. 4:40); male without auricles on second abdominal segment and with anal lobe of hind wing smooth .... LIBELLULINAE 11
3. Occiput larger than vertex; space above bridge with 4 to 6 cross veins .... *Didymops* Rambur
— Occiput smaller than vertex; space above bridge with 2 or 3 cross veins (fig. 4:35) .... *Macromia* Rambur
4. Fore wing with M_4 and Cu_1 divergent 5
— Fore wing with M_4 and Cu_1 convergent (fig. 4:37) ... 6
5. Fore wing with triangle equilateral .... *Neurocordulia* Selys
— Fore wing with triangle long and narrow .... *Williamsonia* Davis
6. Wing with large maculations at the base, nodus, and distal end .... *Epicordulia* Selys
— Wing without distal marking and without a separate nodal maculation 7
7. Hind wing with 2 cubito-anal cross veins; bisector of anal loop closer to A_2 than to A_1; wings without fuscous spots (fig. 4:37) .... *Somatochlora* Selys
— Hind wing with 1 cubito-anal cross vein; bisector of anal loop equidistant from A_1 and A_2 (fig. 4:38)) 8
8. Fore wing with triangle open ... *Dorocordulia* Needham
— Fore wing with triangle having a cross vein (figs. 4:38; 4:39) 9
9. Hind wing with proximal infuscation (except in one species from Florida) 10
— Hind wing without infuscation (northern U.S., Canada) (fig. 4:38) .... *Cordulia* Leach
10. Hind wing with 6 antenodal cross veins .... *Helocordulia* Needham
— Hind wing with 4 or 5 antenodal cross veins (fig. 4:39) .... *Tetragoneuria* Hagen
11. Stigma with ends parallel (figs. 4:43; 4:45) 12
— Stigma trapezoidal (figs. 4:41; 4:55; 4:57) 33
12. Anal loop complete, more or less foot-shaped 13
— Anal loop open at wing margin .... *Nannothemis* Brauer
13. Pronotum with caudal lobe erect, as wide as pronotum and bilobulate 14
— Pronotum with caudal lobe directed caudad, narrower than the pronotum and usually entire 23
14. Bisector of anal loop nearly straight 15
— Bisector of anal loop distinctly angulate beyond middle (fig. 4:44) 16
15. Fore wing with triangle as broad as long .... *Perithemis* Hagen
— Fore wing with triangle longer than broad .... *Celithemis* Hagen
16. Hind wing with Cu_1 arising from outer face of triangle (fig. 4:40) 17
— Hind wing with Cu_1 arising from posterior angle of triangle (fig. 4:48) 20
17. Fore wing with distal antenodal matched (fig. 4:53) 18
— Fore wing with distal antenodal unmatched (fig. 4:52) 19
18. Wing with 2 or more cross veins under stigma .... *Leucorrhinia* Brittinger
— Wing with only 1 cross vein under stigma .... *Pachydiplax* Brauer
19. Radial planate subtending 1 row of cells (fig. 4:52); tibial spines about as long as intervals .... *Erythemis* Hagen
— Radial planate subtending 2 rows of cells, tibial spines much longer than intervals .... *Lepthemis* Hagen
20. Wing with 2 or more cross veins under stigma.......... 21
— Wing with not more than 1 cross vein under stigma (figs. 4:48; 4:51) 22
21. Median planate subtending 2 rows of cells; hind wing with 7 or 8 antenodal cross veins .... *Cannacria* Kirby
— Median planate subtending 1 row of cells; hind wing with 6 antenodal cross veins .... *Brachymesia* Kirby
22. Abdomen with an extra transverse carina on 4th segment .... *Tarnetrum* Needham and Fisher
— Abdomen without an extra transverse carina on 4th segment .... *Sympetrum* Newman
23. Hind wing narrow at base, with 2 cubito-anal cross veins, and with vein Cu_1 arising from outer side of triangle (Texas and West Indies) .... *Cannaphila* Kirby
— Hind wing wider at base; veins not as above 24
24. Fore wing with distal antenodal matched 25
— Fore wing with distal antenodal unmatched (fig. 4:54) 30
25. Hind wing with Cu_1 arising from hind angle of triangle 26
— Hind wing with Cu_1 arising from outer face of triangle .... *Erythrodiplax* Brauer
26. Wing with 1 bridge cross vein (fig. 4:43) .... *Orthemis* Hagen
— Wing with 2 or more bridge cross veins (fig. 4:44) 27
27. Fore wing triangle with 3 or more cells (fig. 4:44; 4:45) 28
— Fore wing triangle with only 2 cells .. *Ladona* Needham
28. Arculus near middle of distance between first and second antenodal cross vein (fig. 4:45); male with a

pair of stout processes on venter of first abdominal segment; fore wing triangle distinctly convex on outer side .... *Plathemis* Hagen
— Arculus at or very close to second antenodal cross vein (fig. 4:44); male without ventral processes of first abdominal segment; fore wing triangle straight or very slightly convex on outer side 29
29. Wings marked with reddish .... *Belonia* Kirby
— Wings marked with black, white, or hyaline .... *Libellula* Linnaeus
30. Wing with anal crossing directly opposite A_2 31
— Wing with anal crossing proximal to origin of A_2 32
31. Fore wing subtriangle with 2 cells .... *Macrothemis* Hagen
— Fore wing subtriangle with 3 cells (fig. 4:54) .... *Brechmorhoga* Kirby
32. Hind wing triangle with a cross vein (fig. 4:42) .... *Pseudoleon* Kirby
— Hind wing triangle without a cross vein .... *Micrathyria* Kirby
33. Radial planate subtending 1 row of cells........... 34
— Radial planate subtending 2 rows of cells 36
34. Vein M_2 undulate .... *Macrodiplax* Brauer
— Vein M_2 smoothly curved......................... 35
35. Wing with one cross vein beneath stigma......... .... *Miathyria* Kirby
— Wing with 2 cross veins under stigma...*Tauriphila* Kirby
36. Abdominal segment 4 with a median transverse carina 37
— Abdominal segment 4 with only the transverse carina near caudal end of segment 38
37. Vein M_2 strongly undulate (fig. 4:57)....*Pantala* Hagen
— Vein M_2 evenly curved (fig. 4:56) *Tramea* Hagen
38. Hind wing with bisector of anal loop about twice as far from A_1 as from A_2 at base of anal loop (fig. 4:58) .... *Dythemis* Hagen
— Hind wing with bisector of anal loop at least 3 times as far from A_1 as from A_2 (fig. 4:55) .... *Paltothemis* Karsch

Naiads[5]

1. Head with a prominent upturned frontal horn between the bases of the antennae (figs. 4:10*m,o*; 4:36); legs very long .... MACROMIINAE 2
— Head without a prominent upturned frontal horn (fig. 4:10*d-j*) .... CORDULIINAE and LIBELLULINAE 3
2. Lateral spines of 9th abdominal segment reaching to apex of anal appendages (fig. 4:13*b*); lateral setae of labium 3 or 5; dorsal hook on segment 10 absent .... *Didymops*
— Lateral spines of 9th abdominal segment reaching less than halfway to apex of anal appendages (fig. 4:13*a*); lateral setae of labium 6; small dorsal hook on segment 10 ....*Macromia*
3. Abdomen with dorsal hooks or knobs present on 1 or more segments (best seen in lateral views) 4
— Abdomen with dorsal hooks or knobs absent on all segments, sometimes vestigial tubercles or tufts of hair present 27
4. Dorsal hook or knob present on abdominal segment 9 5
— Dorsal hook or knob absent on abdominal segment 9 16
5. Lateral spines of abdominal segment 9 reaching to or beyond the tips of the anal appendages (figs. 4:10; 4:13*c*) 6
— Lateral spines of abdominal segment 9 shorter, not reaching to tip of anal appendages (fig. 4:13*e-h*) ... 8
6. Frontal shelf present between bases of the antennae, lateral setae 5 to 6 (fig. 4:10*f*) ....*Neurocordulia*
— Frontal shelf not present between bases of the antennae, low and rounded, not produced into a flat triangle (fig. 4:10*p*) 7
7. Distal half of dorsal surface of mentum heavily setose; lateral setae 4 to 5 (fig. 4:11*l*) ....*Epicordulia*
— Distal half of dorsal surface of mentum with few or generally no setae; lateral setae 6 to 8 ..*Tetragoneuria*
8. Lateral anal appendages nearly as long as the superior appendage 9
— Lateral anal appendages one-half or less the length of superior appendage 10
9. Dorsal hooks absent on segments 3 and 4, crenulations of the lateral lobes deep (fig. 4:11*i*) .....*Helocordulia*
— Dorsal hooks present on segments 3 and 4; crenulations of the lateral lobes shallow ....*Somatochlora* (in part)
10. Dorsal hooks cultriform, the series in lateral view like teeth of a circular saw; lateral setae of labium 5 or 6 (fig. 4:13*h*) .... *Perithemis*
— Dorsal hooks more spinelike or low and blunt; lateral setae of labium 6 to 10 11
11. Dorsal hooks long and laterally flattened 12
— Dorsal hooks short and thick 14
12. Abdomen broadly depressed, little longer than wide; lateral setae of labium 8 ....*Tauriphila*
— Abdomen about twice as long as wide 13
13. Superior anal appendage, seen from above, slightly less than half as long as its basal width ......*Brachymesia*
— Superior anal appendage about twice as long as its basal width; tip of hind wing case extends posteriorly about halfway across abdominal segment 6 (fig. 4:13*e*) ....*Cannacria*
14. Teeth on opposed edges of lateral lobes of labium large 15
— Teeth on opposed edges of lateral lobes of labium obsolete; lateral setae 6-7; mental setae 9 ...*Dythemis*
15. Lateral setae of labium 6; mental setae 9-10 ....*Macrothemis*
— Lateral setae of labium 7 to 9; mental setae 14-15 ....*Brechmorhoga*
16. Postocular distance less than the length of the eye when viewed from above; eyes large and prominent on an usually triangular-shaped head (fig. 4:10*g-j*) 17
— Postocular distance equal to or greater than the length of the eye when viewed from above; eyes usually small and not very prominent (fig. 4:10*a,b,c*) 23
17. Lateral anal appendages nearly as long as superior appendage; lateral setae 6 or 7 18
— Lateral anal appendages usually about half the length of the superior appendage; lateral setae 7 to 14..... 19
18. Abdomen with lateral spines on segment 9 one-third as long as that segment; dorsal hooks present on segments 5 to 8, those on 7 and 8 not prominent ....*Dorocordulia*
— Abdomen with lateral spines on segment 9 one and one-half times as long as that segment; dorsal hooks present on segments 3 or 4 to 8 increasing in size posteriorly ....*Miathyria*
19. Dorsal hook present on segment 3; inferior anal appendages subequal in length to superior appendage 20
— Dorsal hook absent on segment 3; inferior anal appendages markedly longer than superior appendage (except in *Sympetrum costiferum*) 22
20. Dorsal hook present on segments 2-6; lateral setae 7 to 9 .... *Paltothemis*
— Dorsal hook absent on segment 2 21
21. Mental setae 10-15; lateral spine on segment 9 less than middorsal length of segment 9; lateral setae 9-12 ....*Leucorrhinia*
— Mental setae 16-17; lateral spine on segment 9 greater than middorsal length of segment 9; lateral setae 10; dorsal hooks present on abdominal segments 7 and 8 .... *Macrodiplax*
22. Lateral spines of abdomen long and straight, those of segment 9 extending to or beyond the tips of the inferior anal appendages ....*Celithemis*
— Lateral spines of abdomen short, not reaching tips of anal appendages, curved toward meson (fig. 4:13*f*) .... *Sympetrum* (in part)
23. Inferior and superior anal appendages subequal in length 24

[5]Naiads of *Williamsonia* and *Cannaphila* are unknown.

— Inferior anal appendages noticeably longer than the superior .... *Sympetrum* (in part)
24. Mental setae 0 to 4 .... *Ladona*
— Mental setae 8 to 15 (fig. 4:11*k*) 25
25. Margin of median lobe of labium crenulate on its distal margin; abdominal segments 7 to 9 with dark, shining middorsal ridges .... *Plathemis*
— Margin of median lobe of labium evenly contoured; abdominal segments 7 to 9 without such ridges 26
26. Abdomen with lateral spines present on segments 8 and 9; dorsal hooks normally present on abdominal segments 3 to 8, those on 7 and 8 rudimentary and hidden among scurfy hairs .... *Libellula* (in part)
— Abdomen with lateral spines vestigial on segments 8 and 9; dorsal hooks absent .... *Belonia*
27. Apical third of inferior and lateral anal appendages strongly decurved (fig. 4:13*g*) 28
— Apical third of all anal appendages straight, not decurved (fig. 4:13*i*) 29
28. Minute lateral spine on abdominal segment 9; lateral setae of labium 11 or 12 .... *Lepthemis*
— No lateral spines on abdomen; lateral setae of labium 7 to 9 .... *Erythemis*
29. Postocular distance equal to or greater than the length of the eye when viewed from above; eyes usually small and not very prominent (fig. 4:10*a,b,d*) 30
— Postocular distance less than the length of the eye when viewed from above; eyes large and prominent on a somewhat triangularly shaped head (fig. 4:10*g-i*) . 36
30. Lateral anal appendages nearly as long as the superior 31
— Lateral anal appendages one-third to two-thirds as long as the superior; crenulations of lateral lobes shallow or absent (fig. 4:11*j*) 33
31. Crenulations of the distal margin of the lateral lobe obsolete, merely indicated by about 15 single spinules; abdomen abruptly rounded to tip; caudal appendages protruding but little beyond the ventral margin of segment 9 .... *Pseudoleon*
— Crenulations of the distal margin of the lateral lobe shallow to deep, with groups of 2-7 spinules on each tooth 32
32. Crenulations on lateral lobes deep and separated by rather wide notches; thorax unicolored .... *Somatochlora* (in part)
— Crenulations on lateral lobes shallow, teeth low; thorax with a dorsal longitudinal dark stripe .... *Cordulia* (in part)
33. Lateral spines present on abdominal segments 8 and 9 34
— A single small lateral spine present only on abdominal segment 9 or none 35
34. Mentum with distal margin crenulate; abdominal segments 4 to 7 with dorsal tufts of long hair (fig. 4:11*m*) .... *Orthemis*
— Mentum with distal margin entire, evenly contoured; abdominal segments 4 to 7 not as above (fig. 4:11*k*) .... *Libellula* (in part)
35. Mental setae 5-11, lateral setae 8-10. . .*Belonia* (in part)
— Mental setae 12-16, lateral setae 10-12 .... *Tarnetrum* (in part)
36. Anal appendages long, slender and needle-pointed; lateral spines of segments 8 and 9 long and curved toward meson, those on segment 8 at least as long as middorsal length of segment 9 (fig. 4:13*j-k*) 37
— Anal appendages short and heavy, not projected into a long needlepoint; lateral spines on segments 8 and 9 flat and straight, those on 8 not as long as middorsal length of segment 9 38
37. Lateral spines of abdominal segment 8 but slightly shorter than those of segment 9 (fig. 4:13*j*); lateral spines of segment 9 reaching tips of anal appendages, crenulations of distal margin of lateral lobe shallow (fig. 4:11*j*) .... *Tramea*
— Lateral spines of abdominal segment 8 only one-third size of those of segment 9; lateral spines of segment 9 not reaching tips of anal appendages (fig. 4:13*k*); crenulations of distal margin of lateral lobe deep (fig. 4:11*i*) .... *Pantala*
38. Lateral setae 6 or 7; lateral anal appendages more than half as long as the inferiors 39
— Lateral setae usually 9 to 16, sometimes as few as 6; lateral anal appendages half or less than half as long as inferiors 40
39. Lateral setae 6; mental setae 9 to 11; inferior and superior anal appendages subequal in length .... *Nannothemis*
— Lateral setae 7; mental setae about 14; inferior anal appendages longer than superior appendage .... *Cordulia* (in part)
40. Lateral spines absent or vestigial on abdominal segment 8 .... *Tarnetrum* (in part)
— Lateral spines present on abdominal segment 8..... 41
41. Superior anal appendages nearly as long as the inferiors (mainly northern North America) .... *Leucorrhinia* (in part)
— Superior anal appendages usually much shorter than the inferiors (mainly southern North America) 42
42. Lateral spines of abdominal segments 8 and 9 subequal in length 43
— Lateral spines of abdominal segment 8 about half as long as those of 9; lateral spine of 9 equal to or greater than the middorsal length of segment 9 (figs. 4:11*q*; 4:13*i*) .... *Pachydiplax*
43. Lateral spines of abdominal segment 8 approximately half as long as segment 9 middorsally; some species with prominent bunches of setae present on the dorsum of abdominal segments 4 to 9 .... *Erythrodiplax*
— Lateral spines of abdominal segment 8 nearly as long as segment 9 middorsally; without prominent bunches of setae as described above .... *Micrathyria*

Subfamily MACROMIINAE

This is a small group containing large, actively flying, brown or blackish dragonflies marked with yellow. They fly high, forage widely, and are very difficult to capture. The naiads have a short, flat, almost circular abdomen, an erect horn on the front of the head, and long legs with long simple claws (fig. 4:36). They live sprawled in the silt of bare areas awaiting their prey. Only two genera, *Macromia* and *Didymops*, occur in the United States; the latter is restricted to the East.

Genus *Macromia* Rambur, 1862

Most of the species of this group in the United States occur in the area east of the Mississippi River. However, two are found along the Pacific Coast. *Macromia magnifica* McLachlan 1874, occurs from British Columbia south to California and Arizona (fig. 4:35). Kennedy (1915, pp. 313-322) studied this species in Washington. He states: "the male *Macromias* were usually found patrolling the larger pools or sometimes a patrol would include two or three of the shorter pools. Seldom were more than three or four males seen at any one time, and each male's beat was rarely over three hundred feet long. The flight was very swift, ordinarily about two feet above the surface of the water and straight down the middle of the pools or, on the broader pools, up one side and down the other. For speed few dragonflies can

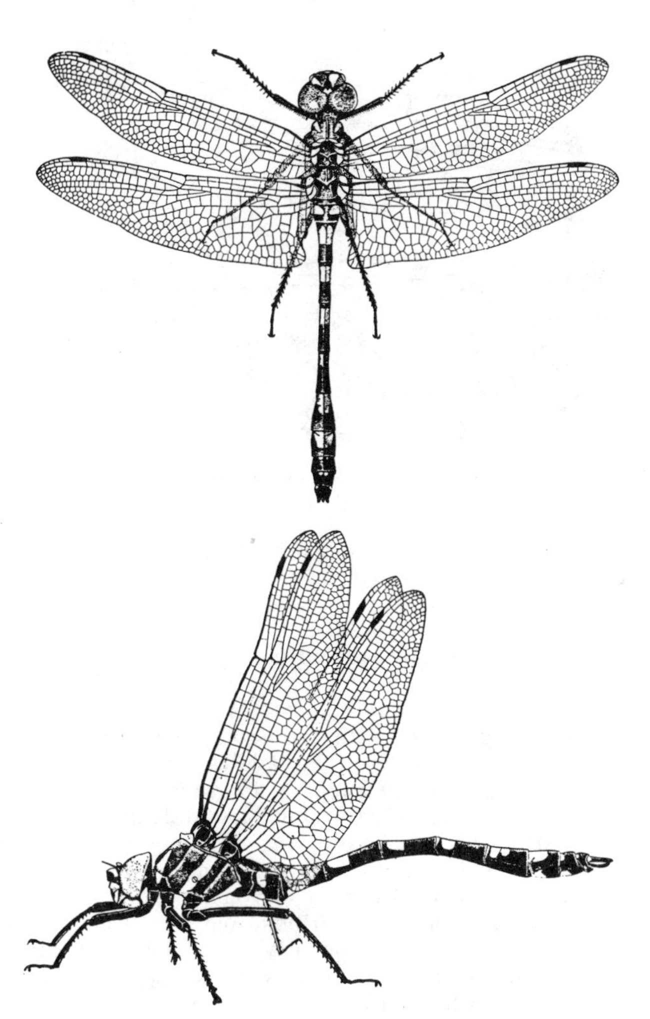

Fig. 4:35. Male of *Macromia magnifica* (Kennedy, 1915).

equal it . . . This species was found most commonly over the water on calm days between the morning hours of seven and ten. Few were found in the afternoons or on windy days. The flight over the water appeared to be controlled by the ovipositing females, who resorted to the water to oviposit early in the day in calm weather, where they were sought by the males. As the females oviposited by striking the end of the abdomen on the surface of the largest pools only, this could not be done except when the surface was smooth. At other times even until late twilight individuals of both sexes might be found patrolling glades and barnyards as much as a half mile from water. Here the flight varied from close over the ground to as high as the trees . . . In ovipositing the female would fly several times back and forth over a short beat of forty or fifty feet, striking her abdomen on the surface of the water at three to five foot intervals. This beating back and forth generally lasted until a male discovered her, when she would be taken away in copulation. At such times the male swooped and grasped the female's head with his feet, then bending the abdomen forward and grasping the female's head with the abdominal appendages he would free his feet and she would bend her abdomen forward and copulate. The copulatory flight was ordinarily away from water over the surrounding trees, but ended in a long period of copulation while resting on some bush or tree. One pair, observed resting in copulation for fifteen minutes, on being disturbed flew away still in copulation."

Kennedy (1915, pp. 318-319) also figures the naiads (fig. 4:36) which he found in a mass of fibrous alder roots in a pool about three feet deep.

M. pacifica is largely restricted to the Midwest, and the California records of this species are doubtful. *M. magnifica* can easily be separated from *M. pacifica* by the yellow markings which broadly cover the upper surface of the vertex in *M. magnifica* but are restricted to the summit in *M. pacifica*.

Subfamily CORDULIINAE

This subfamily contains large, strong flying dragonflies often brilliantly colored with metallic green, blue, or purple. The hairy, dark-colored naiads sprawl on the bottom or climb through the bottom vegetation. Of the nine genera in the United States only three reach California.

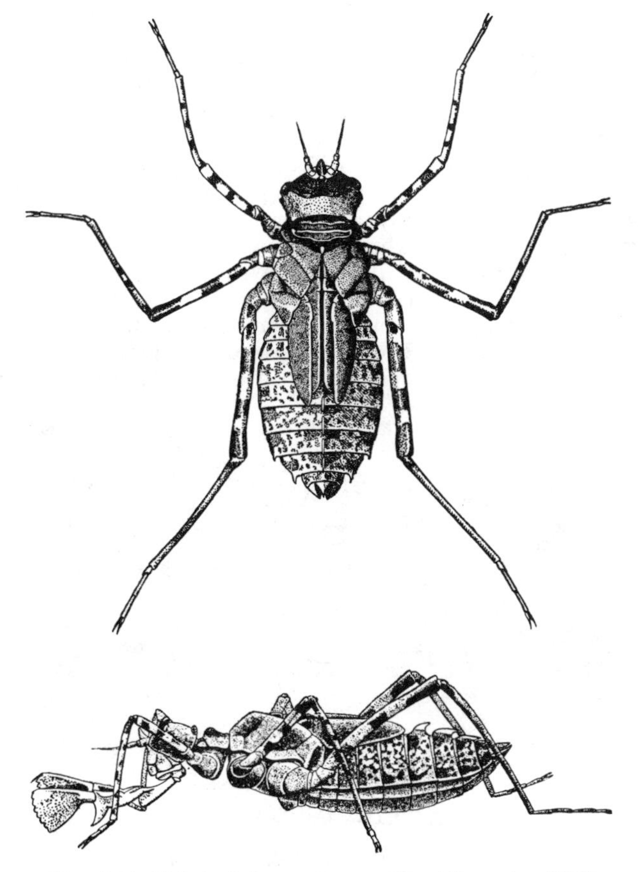

Fig. 4:36. Naiad of *Macromia magnifica* (Kennedy, 1915).

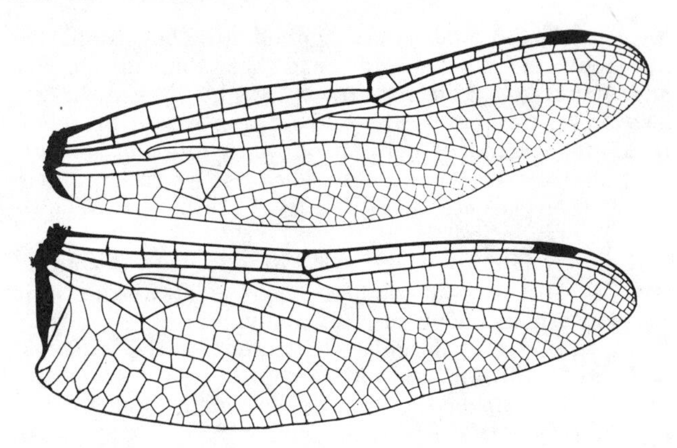

Fig. 4:37. Wings of *Somatochlora tenebrosa* (Needham and Westfall, 1955).

Genus *Cordulia* Leach, 1815

This genus contains one species in North America, *C. shurtleffi* Scudder 1866 (fig. 4:38). It is a bog-loving species with a northern distribution. It occurs in Canada and northern United States, but ranges as far south as Utah and California in the West. The naiads are thick set and hairy and occur in shaded trashy areas at margins of ponds or bogs.

Genus *Somatochlora* Selys, 1871

Somatochlora is a large circumpolar genus inhabitating the northern parts of the Palaearctic and Nearctic regions (fig. 4:37). One species, *S. semicircularis* (Selys) 1871, ranges from western Alaska south to the high mountains of California, Utah, and Colorado. The adults fly in sunny openings in wooded mountain slopes or river valleys. The immature stages occur in swamps or spring bogs. While feeding they fly at heights of thirty to fifty feet or more. In the breeding areas they fly back and forth low over the bog. Eggs are laid in masses on the surface of the water in the more open pools. These masses disintegrate and fall to the bottom. For more details on the habits of this and other *Somatochlora* see Walker (1925) and Kennedy (1913).

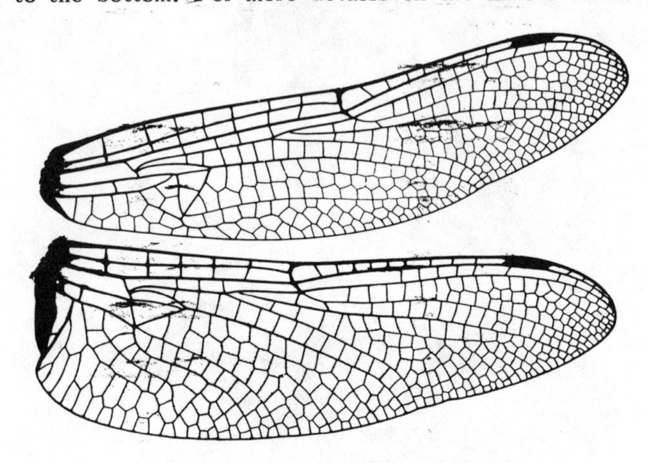

Fig. 4:38. Wings of *Cordulia shurtleffi* (Needham and Westfall, 1955).

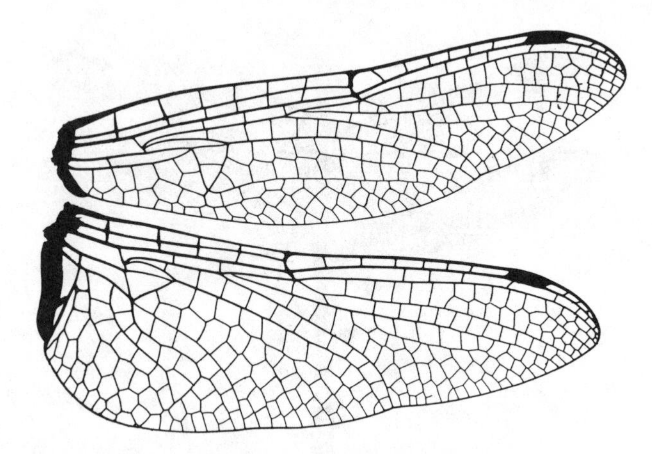

Fig. 4:39. Wings of *Tetragoneuria sepia* (Needham and Westfall, 1955).

Genus *Tetragoneuria* Hagen, 1861

The genus *Tetragoneuria* contains brownish, non-metallic dragonflies with the thorax heavily clothed with hairs (fig. 4:39). The adults are of a roving habit and most of the species are widely distributed. They are sometimes very abundant at long distances from water where in a clearing or along a road they may be seen patrolling a few feet above the ground. Two wide ranging species occur on the Pacific Coast. The naiads are smooth, hairless, and with a depressed abdomen. They are frequently very abundant along the edges of ponds and streams where they crawl over the bottom and loose trash.

Key to California Species

Males

1. Dorsal appendages each with a strong dorsal tooth, the medioventral tooth on inside short and blunt (Canada, northern U.S., Washington to California)............ .... *canis* MacLachlan 1886
— Dorsal appendages each without a dorsal tooth, and the medioventral tooth on inside long and slender (Canada, northern U.S., Washington to California)............ ....*spinigera* Selys 1871

Females

1. Frons (as seen from above) with anterior margin pale, abdominal appendages 2.3-2.7 mm. in length ..... *canis*
— Frons (as seen from above) with anterior margin black, abdominal appendages 3.5 mm. in length......*spinigera*

Naiads

1. Lateral spines of the ninth abdominal segment very slightly or not at all divergent, six-tenths middorsal length of segment .... *canis*
— Lateral spines of the ninth abdominal segment strongly divergent longer than middorsal length of segment ....*spinigera*

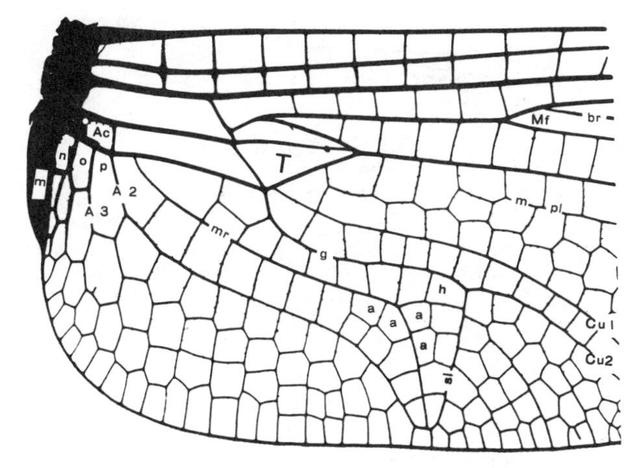

Fig. 4:40. Base of hind wing of *Lepthemis vesiculosa*, male. g, gaff; h, heel cell; a, interpolated ankle cells; sl, sole; n,o, paranals; m, membranule; Ac, anal crossing; Mf, middle fork; mr, midrib; br, bridge; m-pl, median planate; ob, oblique cross vein; r, reverse cross vein (Needham and Westfall, 1955).

Subfamily LIBELLULINAE

This group contains the commonest and best known of the Odonata. The adults are common about every pond, ditch, and roadside. They are generally non-metallic but often of brilliant coloration. In some cases the color becomes obscured by pruinosity in old age. The sexes frequently differ in color and markings. This cosmopolitan subfamily is the largest and most dominant member of the order comprising about one-fourth of all known species.

The majority of the species breed in still water, and adults rarely stray far from water.

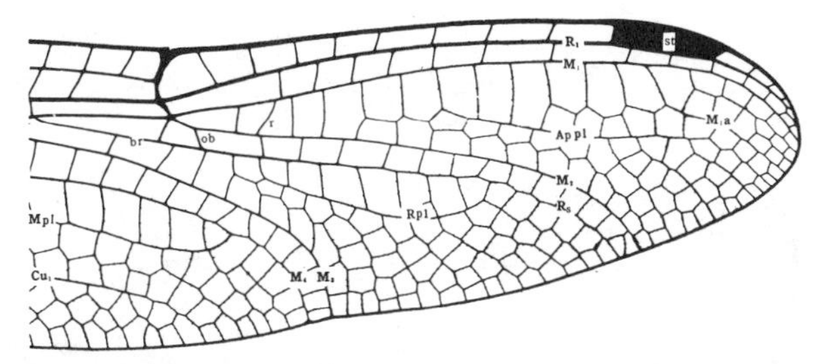

Fig. 4:41. Venation of *Tauriphila australis*, fore wing, showing planates in relation to principal veins: Ap pl, apical planate; R pl, radial planate; M pl, median planate.

The naiads have squarish heads when viewed from above. They are usually active and climb through the aquatic vegetation.

Genus *Pseudoleon* Kirby, 1889

A single ornate species, *P. superbus* (Hagen) 1861, is found in the Southwest and south to Guatemala. The adult is easily recognized by the heavy pattern of brown on the wings (fig. 4:42) and dull yellow oblique markings on the abdomen. The naiads occur in cooler parts of southwestern streams. The females prefer ovipositing in and about algae and other debris in slow moving pools.

Genus *Pachydiplax* Brauer, 1868

This genus contains the single species, *P. longipennis* (Burmeister) 1839 (fig. 4:53). It is wide ranging, occurring in southern Canada, the United States, West Indies, and northern Mexico.

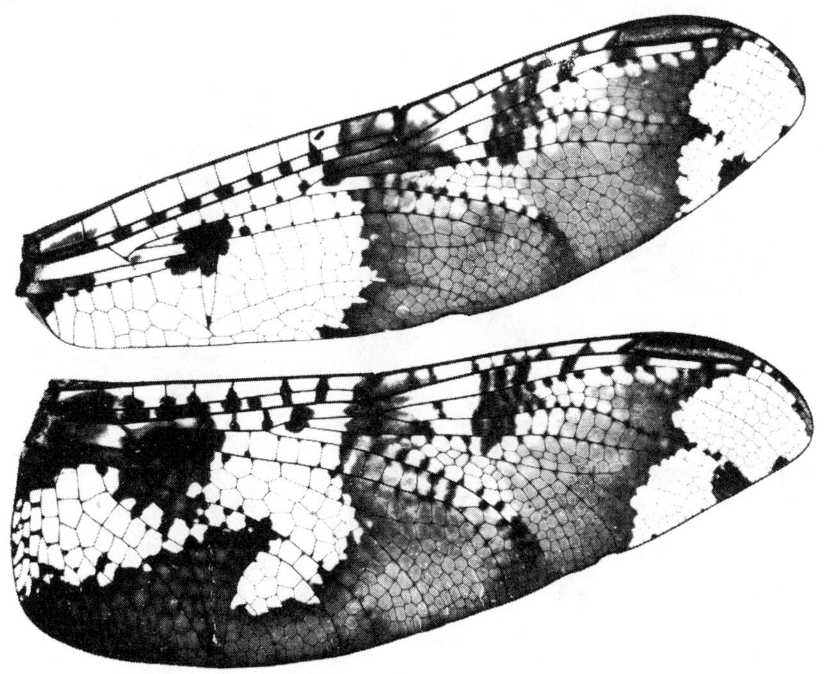

Fig. 4:42. Wings of *Pseudoleon superbus* (Needham and Westfall, 1955).

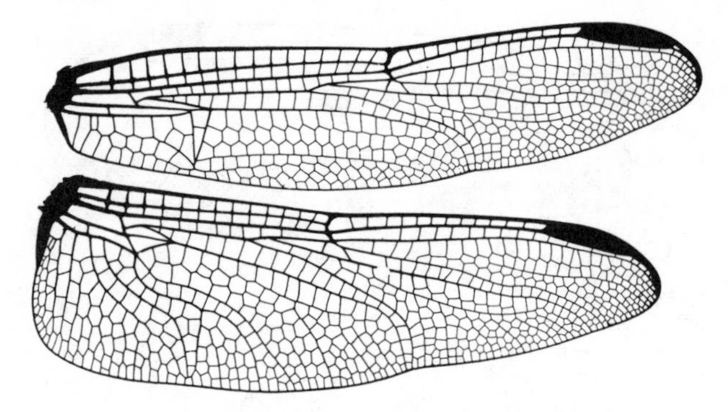

Fig. 4:43. Wings of *Orthemis ferruginea* (Needham and Westfall, 1955).

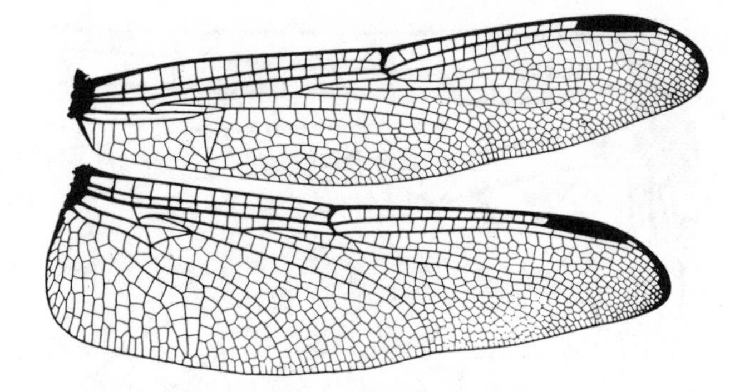

Fig. 4:44. Wings of *Libellula incesta* (Needham and Westfall, 1955).

The males are conspicuous and swift in flight. They hover near the surface of water or rest briefly on projecting twigs. They challenge all newcomers; when two males meet they face each other, then dart upward together to great heights.

The females are less in evidence as they rest back from shore except when foraging or ovipositing. When ovipositing over open water they fly horizontally close to the surface and occasionally move the abdomen down to the water. Among vegetation, they fly up and down as do most Libellulidae. They are most common in the neighborhood of bushes and small trees at the edges of woods.

The naiads crawl about the trash in the bottom of ponds and transform close to the margin of the water. They occur in static water with mud bottoms such as ponds, borrow pits, or creeks.

Genus *Erythemis* Hagen, 1861

These are pond species of moderate size. Characteristically, the adults rest on the ground, floating logs, or other low objects. They wait for the appearance of suitable food and then dart out to take it. The female, unattended by the male, oviposits by touching her abdomen to the surface of the water at widely scattered points. As the adults age, they change in color. For example, *Erythemis simplicicollis* Say (fig. 4:52) is bright green with black when young and then becomes a pruinose blue gray in old age. In some places near the margins of ponds large numbers of these dragonflies may be found congregating.

The thick-bodied naiads have bulging green eyes and usually occur in static water with a mud bottom. Bick (1941) gives details of the life history of *E. simplicicollis*.

Key to California Species

Adults

1. Face green; abdominal appendages yellow (Canada,

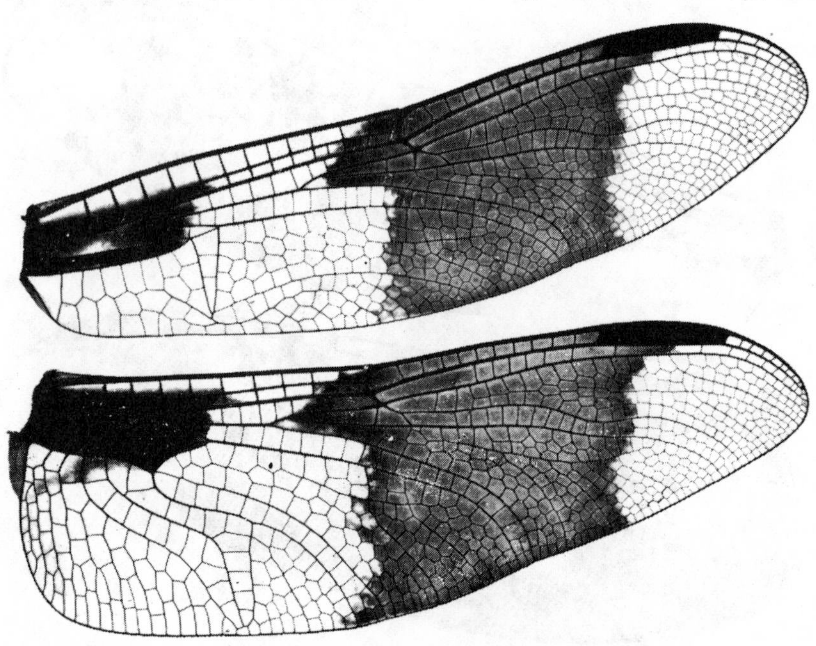

Fig. 4:45. Wings of *Plathemis lydia* (Needham and Westfall, 1955).

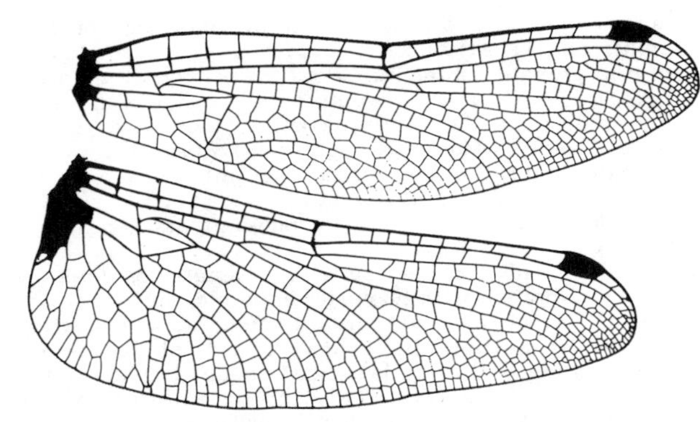

Fig. 4:46. Wings of *Leucorrhinia intacta* (Needham and Westfall, 1955).

U.S. and Mexico, West Indies) ....
.... *simplicicollis* (Say) 1839
— Face black across frons; abdominal appendages black (British Columbia to Utah, California and Mexico) .... *collocata* (Hagen) 1861

Naiads

1. Lateral anal appendages not more than one-half as long as superior appendage .... *simplicicollis*
— Lateral anal appendages two-thirds as long as superior appendage .... *collocata*

Genus *Leucorrhinia* Brittinger, 1850

This is a northern group of small, white-faced dragonflies. They have red or tawny bodies with hyaline wings (fig. 4:46). The males are similar to the females except that they are somewhat larger.

Of the seven species in the United States, three reach California.

The green and brown slender naiads are climbers among green vegetation. The naiads and adults occur in and about sphagnum pools and boggy places. The adults fly low near shore and are fond of bright sunlight.

Key to California Species

Adults

1. Abdomen with middorsal pale triangles on segments 4 to 7 or at least 6 and 7 (Canada, northern U.S., Oregon, California, Nevada, Utah) ...... *hudsonica* (Selys) 1850
— Abdomen without pale spots on segments 4 to 6 2
2. Abdomen with a pair of pale spots on segment 7 (Canada, northern U.S., British Columbia to California) .... *intacta* Hagen 1861
— Abdomen with segment 7 entirely black (Canada, northern U.S., British Columbia to California) .... *glacialis* Hagen 1890

Naiads

1. Dorsal hooks present on abdominal segments 7 and 8; no distinct band on under side of abdomen .... *intacta*
— Dorsal hooks absent on abdominal segments 7 and 8; 3 wide longitudinal dark bands on underside of abdomen 2
2. Lateral spines of abdominal segment 9 pointing straight to rearward, with axes parallel .... *hudsonica*
— Lateral spines of abdominal segment 9 distinctly convergent .... *glacialis*

Genus *Tarnetrum* Needham and Fisher, 1936

Two species of this genus, closely related to *Sympetrum,* occur in the United States. These two species are probably the commonest dragonflies in our area. The adults can be found along roadsides, in fields, and near ditches (fig. 4:51).

Key to California Species

Adults

1. Vertex with tip emarginate; wings with orange suffusion proximally and along costa; legs yellowish (British Columbia and Wyoming, south to California, Mexico, Argentina) .... *illotum* (Hagen) 1861
— Vertex with tip truncate; wings hyaline; legs black, (North America to British Honduras; Asia) .... *corruptum* (Hagen) 1861

Naiads

1. Lateral anal appendages two-thirds as long as inferiors; lateral setae 9; mental setae 13 .... *illotum*
— Lateral anal appendages half as long as inferiors; lateral setae 13-14; mental setae 17 .... *corruptum*

Genus *Sympetrum* Newman, 1833

The species of this genus are mostly reddish in color and are most abundant in the autumn (fig. 4:48). They occur in or near ponds and wet meadows. They are poor fliers and easy to catch. The beautiful red colors fade in preserved specimens.

Adults frequently congregate in large numbers in sunny locations. The slender naiads crawl over the bottom trash and vegetation in ponds.

Key to California Species

Males

1. Tibia pale, at least on outer face 2

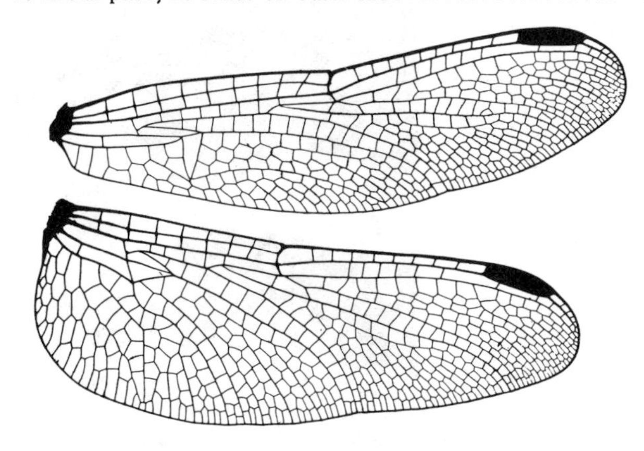

Fig. 4:47. Wings of *Erythrodiplax berenice* (Needham and Westfall, 1955).

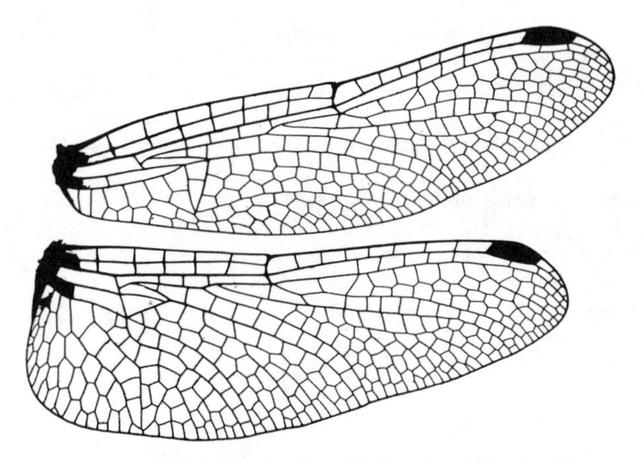

Fig. 4:48. Wings of *Sympetrum vicinum* (Needham and Westfall, 1955).

— Tibia black 3

2. Hamule broadly bifid on distal fifth (fig. 4:50*g*); dorsal abdominal appendages convex on distoventral margin (fig. 4:49*g*) (British Columbia to California and east to Montana and Colorado) .... *pallipes* Hagen 1874

— Hamule with anterior branch ending well before end of posterior branch (fig. 4:50*b*); dorsal abdominal appendages concave on distoventral margin (fig. 4:49*b*) (British Columbia to California east to Maine and Wyoming) .... *costiferum* Hagen 1861

3. Radial planate subtending 1 cell row; dorsal abdominal appendages with a ventral tooth at distal end of denticulations (fig. 4:49*c,d,f*) 4

— Radial planate subtending 2 cell rows; dorsal abdominal appendages without a ventral tooth at distal end of denticulations (fig. 4:49*e*) (British Columbia to California, east to Montana) .... *madidum* (Hagen) 1861

4. Hamule bifid for less than one-half its length (fig. 4:50*d,f,h*) 5

— Hamule bifid for more than one-half its length (fig. 4:50d) 7

5. Hamule bifid for one-fourth its length or less, the posterior branch with posterior edge bicuspidate 6

— Hamule bifid for one-third its length, the posterior branch semielliptical, concave, and shell-like (fig. 4:50*h*); face yellowish brown, (Canada, Nevada east to New York) .... *rubicundulum* (Say) 1839

6. Hamule bifid for one-fourth its length (fig. 4:50*d*); face cherry red (Alaska to California east to Main) (=*decisum auct.*) .... *internum* Montgomery 1943

— Hamule bifid for one-fifth its length (fig. 4:50*f*); face white (Alaska to California east to Maine (=*decisum* Hagen) .... *obtrusum* Hagen 1874

7. Hamule with anterior branch nearly as wide as posterior branch; wings with only proximal spot flavescent (Alaska to California east to New York)(=*scoticum* Kennedy) .... *danae* Sulzer 1776

— Hamule with anterior branch much narrower and much shorter than posterior branch 8

8. Dorsal abdominal appendages in mature specimens pale (California, Nevada) .... *occidentale californicum* Walker 1951

— Dorsal abdominal appendages in mature specimens dark, (California and Oregon to Colorado) ....*atripes* (Hagen) 1873

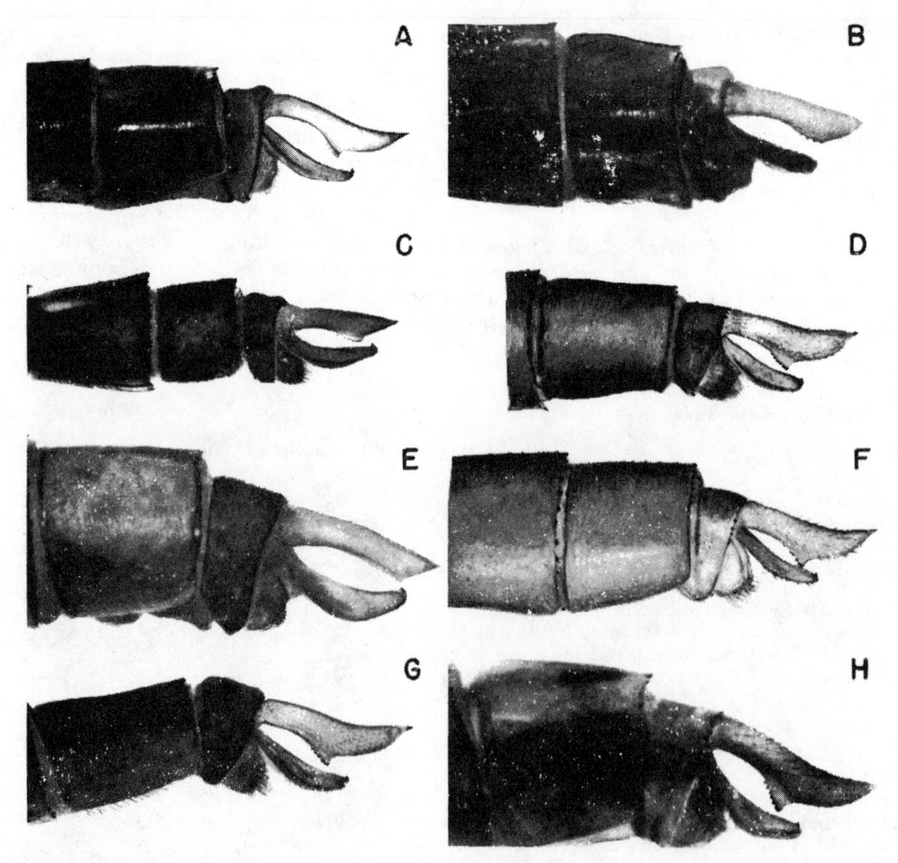

Fig. 4:49. Lateral view of posterior end of male abdomen in the genus *Sympetrum*. *a*, *ambiguum*; *b*, *costiferum*; *c*, *danae*; *d*, *internum*; *e*, *madidum*; *f*, *obtrusum*; *g*, *pallipes*; *h*, *rubicundulum* (Needham and Westfall, 1955).

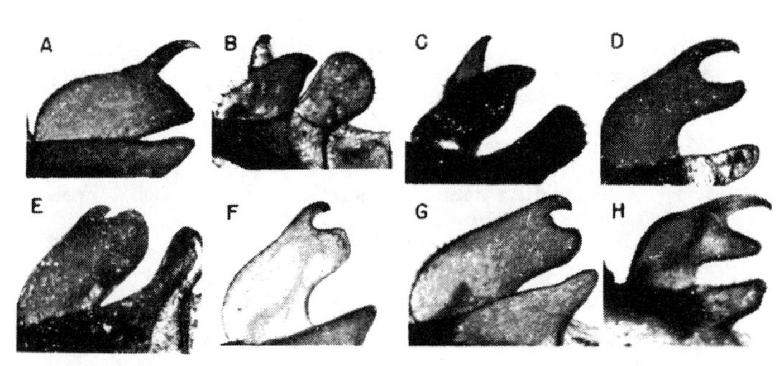

Fig. 4:50. Hamules of *Sympetrum*. *a, ambiguum; b, costiferum; c, danae, d, internum; e, madidum; f, obtrusum; g, pallipes; h, rubicundulum* (Needham and Westfall, 1955).

Females

1. Genital plate entire or shallowly emarginate 2
— Genital plate deeply emarginate and divided 4
2. Tibiae entirely black 3
— Tibiae pale, at least on outer face.......... *costiferum*
3. Genital plate broadly angulate mediodistally ..... *danae*
— Genital plate broadly emarginate mediodistally .... *occidentale californicum* and *atripes*
4. Tibiae entirely black 5
— Tibiae pale........................... *pallipes*
5. Radial planate subtending 1 cell row 6
— Radial planate subtending 2 cell rows........ *madidum*
6. Vulvar lamina with lobes widely divaricate 7
— Vulvar lamina with lobes contiguous......... *obtrusum*
7. Vulvar lamina inflated with lobes directed dorsad .... *rubicundulum*
— Vulvar lamina not inflated, with lobes not directed dorsad .... *internum*

Naiads[6]

1. Dorsal hooks on segments 6-8 as long as the segments which bear them; lateral appendages half as long as the inferiors, the latter not acuminate; lateral setae 9; mental setae 12 .... *occidentale californicum*
— Dorsal hooks shorter than the segments which bear them; lateral appendages not as above 2
2. Lateral spine of segment 9 equal to one-half lateral margin of segment 9 including spine; lateral setae 12; mental setae 12 .... *madidum*
— Lateral spine of segment 9 not more than one-third lateral margin of segment 9 including spine; lateral setae usually less than 12, typically 10-11; mental setae usually more than 12 3
3. Lateral spines of segment 9 one-third lateral margin of segment 9 including spine; dorsal hooks present on segments 4 to 8, well developed; mental setae 13-14 4
— Lateral spines of segment 9 not more than one-fourth lateral margin of segment 9 including spine; dorsal hooks usually absent from segments 4 or 8 5
4. Base of mentum of labium distinctly broader than middle coxae and somewhat more than one-fourth the greatest width; inhabitant of shallow marshy bogs and lagoons .... *costiferum*
— Base of mentum scarcely, if at all, broader than middle coxae and about one-fourth greatest width; in small, semipermanent ponds .... *pallipes*
5. Dorsal hook present, though small, on abdominal segment 8 6
— Dorsal hooks absent on segments 4 and 8 .... *obtrusum* and *rubicundulum*
6. Lateral spine of abdominal segment 9 about one-sixth lateral margin; lateral setae 10-11 .... *internum*

[6]The naiad of *atripes* has not been described.

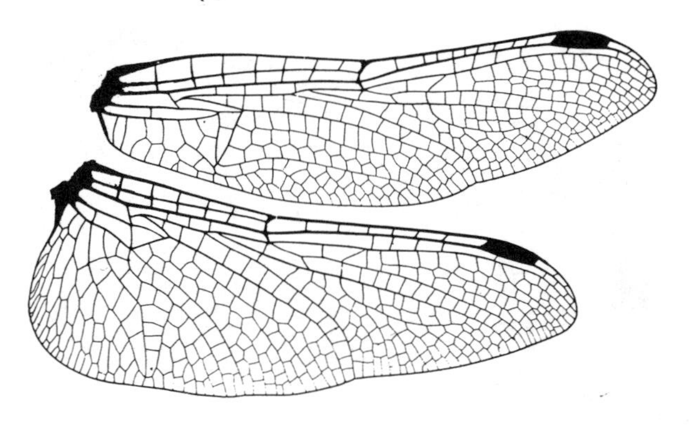

Fig. 4:51. Wings of *Tarnetrum corruptum* (Needham and Westfall, 1955).

— Lateral spine of abdominal segment 9 about one-fourth lateral margin; lateral setae 11-12 .... *danae*

Genus *Erythrodiplax* Brauer, 1868

This is a large genus of Neotropical dragonflies which enters the southern margin of the United States (Boror, 1942) (fig. 4:47). Only one species, *Erythrodiplax funerea* (Hagen) 1861, has been recorded from southern California. It is a wide-ranging form occurring from California and Texas south to Ecuador. The naiads live in tangled submerged vegetation.

Genus *Orthemis* Hagen, 1861

A Neotropical genus with one brown and red species, *Orthemis ferruginea* (Fabricius) 1775, reaching the southern border of the United States from California and Utah to Florida (fig. 4:43). The nymph is elongate, with a subcylindric and slowly tapering abdomen.

Genus *Belonia* Kirby, 1889

This is a western group of dragonflies with dark red or reddish wings closely related to *Libellula*. A big red species, *Belonia saturata* (Uhler) 1857, is the only one in our territory. This very common form ranges throughout much of the territory west of the

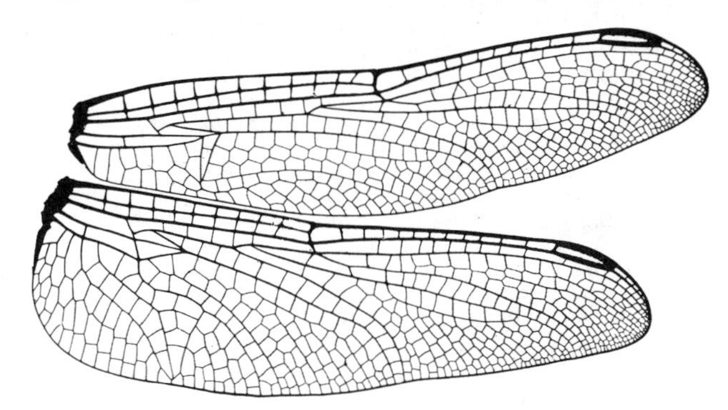

Fig. 4:52. Wings of *Erythemis simplicicollis* (Needham and Westfall, 1955).

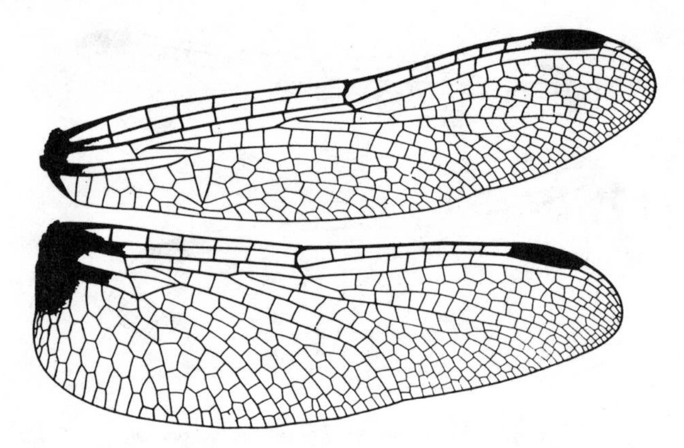

Fig. 4:53. Wings of *Pachydiplax longipennis* (Needham and Westfall, 1955).

Mississippi River and is found about ponds and slowly moving streams. The hairy nymphs lie in the ooze at the bottom of stagnant ponds.

Genus *Libellula* Linnaeus, 1758

These showy insects are among our best known dragonflies. They are very common and hover about ponds, tanks, marshes, canals, and sluggish rivers or streams (fig. 4:44). The naiads are hairy and sprawl in the bottom silt and trash. For the most part they are fresh-water forms. The genus is Holarctic with most of the species occurring in the United States.

Key to the California Species

Males

1. Wing with a proximal dark band covering wing and reaching nodus .... *luctuosa* Burmeister 1839
— Wing with proximal marking much smaller or absent 2
2. Pterostigma bicolored (half brown, half yellow) (California and Wyoming south to Mexico) .... *comanche* Calvert 1907
— Pterostigma uniformly colored 3
3. Fore wing without a proximal dark brown or black streak; costa pale 4
— Fore wing with proximal brown streak; costa dark... 5
4. Hind wing with proximal dark spot having dark veins (Wyoming to California and Arizona) .... *composita* Hagen 1873
— Hind wing with proximal dark spot triangular and with yellowish veins (Holarctic) .... *quadrimaculata* Linnaeus 1758
5. Wing with a large maculation in apical third (America north of Mexico) .... *pulchella* Drury 1773
— Wing without a maculation in apical third 6
6. Wing with spot in middle third covering most of breadth of wing (British Columbia to Montana south to California) .... *forensis* Hagen 1861
— Wing with median spot confined to small area around nodus (Wyoming to California and south to Columbia and Venezuela) .... *nodisticta* Hagen 1861

Females

1. Stigma bicolored .... *comanche*
— Stigma uniformly colored 2
2. Wing with a broad dark band proximally, covering wing and extending nearly to nodus .... *luctuosa*
— Wing with proximal marking much smaller or absent 3
3. Wing with maculation in middle third traversing over one-half of wing 4
— Wing with median maculation restricted to small area around nodus or absent 5
4. Wing with maculation beyond stigma .... *pulchella*
— Wing hyaline distally ....*forensis*
5. Fore wing with dark brown or black maculation proximal to nodus ....*nodisticta*
— Fore wing without dark proximal maculation 6
6. Hind wing with proximal spot triangular with yellowish veins; face yellowish ....*quadrimaculatus*
— Hind wing with proximal spot with dark veins; face white ....*composita*

Naiads[7]

1. Head squarish seen from above, not narrowed behind eyes; 6 deeply cut, sharply serrate teeth on lateral lobes of labium; single raptorial seta on each side of mental lobe; lateral setae 7-8 ....*composita* Hagen
— Head narrowed behind eyes; 10 or 11 shallow, broadly truncated teeth on lateral lobes of labium; mental setae 9 or more 2
2. Dorsal hooks present on some middle abdominal segments 3
— Dorsal hooks absent on abdomen; lateral setae of labium 6........................*comanche* Calvert
3. Dorsal hooks lacking on abdominal segment 8; lateral setae 5-9 4
— Dorsal hooks present on segments 4 to 8; lateral setae 7 or more; mental setae 10 5
4. Lateral setae of labium 5-7; mental setae 9-11 ....*forensis* Hagen
— Lateral setae of labium 8-9; mental setae 13 .... *pulchella* Drury
5. Lateral abdominal appendages about half as long as inferiors; dorsal hooks present on segments 4-8 ....*luctuosa* Burmeister
— Lateral abdominal appendages about seven-tenths as long as inferiors; dorsal hooks present on segments 3-8 ....*quadrimaculata* Linnaeus

Genus *Plathemis* Hagen, 1861

This genus, which is closely related to *Libellula,* contains two common and widespread species (fig. 4:45). They exhibit marked sexual dimorphism. They occur near ponds and ditches and frequently are seen hovering over the water or resting on plants near shore. The females work in and out of sheltered nooks nearby to lay their eggs. They drop twenty-five to fifty eggs each time their abdomen strikes the surface of the water.

Key to California Species

Adults

1. Male with ventral process on abdominal segment 1 deeply bifid, wing with median crossband uniformly brown (fig. 4:45); female with tip of wing black (widespread in U.S. and Canada) .... *lydia* Drury 1773
— Male with ventral process on abdominal segment 1 shallowly and broadly emarginate, wing with median crossband subdivided by a paler band; female with

[7] The naiad of *nodisticta* Hagen is unknown.

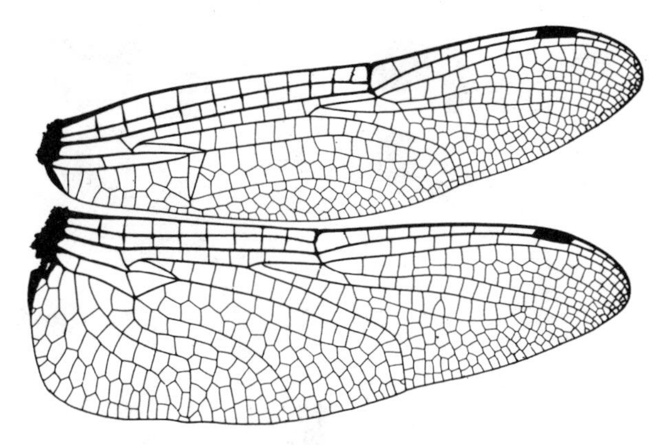

Fig. 4:54. Wings of *Brechmorhoga mendax* (Needham and Westfall, 1955).

wing tip hyaline, (British Columbia and Nebraska south to California and Mexico) ..... *subornata* (Hagen) 1861

Naiads

1. Dorsal hooks on abdominal segments 2 to 6, highest on 5; all sharp, thornlike .... *lydia*
— Dorsal hooks on abdominal segments 2 to 6, highest on 4; all blunt and hairy .... *subornata*

Genus *Brechmorhoga* Kirby, 1894

Only one species, *Brechmorhoga mendax* (Hagen) 1861, of this Neotropical genus enters the United States (fig. 4:54). This large grayish species is found in desert regions from California to Texas and Oklahoma and south into Mexico. The naiads are found in sand and gravel beds of pools in torrential streams.

Kennedy (1917, p. 605) states that this very graceful species takes short beats over streams. He noted that they "fly with a swinging may-flylike motion" and "in the heat of the day they floated around among the tree tops".

Genus *Dythemis* Hagen, 1861

This is a Neotropical genus with three species entering the southern border of the United States. *Dythemis velox* Hagen 1861 occurs as far south as Argentina and is found in Alabama, California, Mississippi, New Mexico, Oklahoma, and Texas (fig. 4:58). It is a slender blackish species marked with yellow and brown. It is a very swift flier. The adults perch on tall dry stems with the tip of the abdomen lifted, and they frequently return to the same perch. The naiads are active stream dwellers. Their body colors match the sand over which they run.

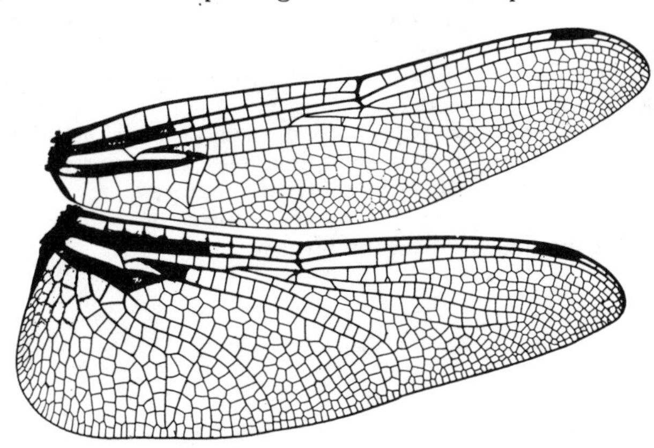

Fig. 4:55. Wings of *Paltothemis lineatipes* (Needham and Westfall, 1955).

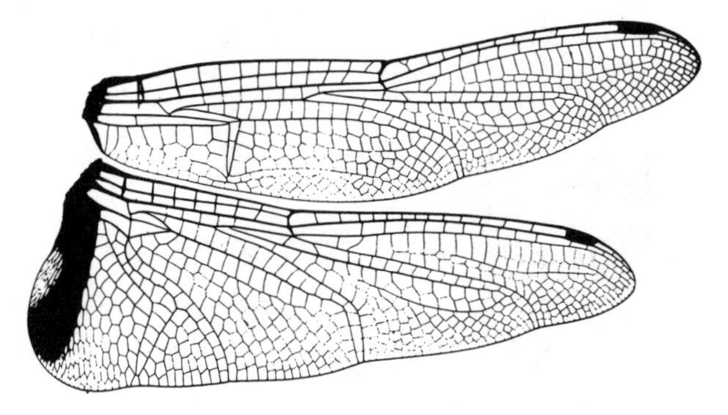

Fig. 4:56. Wings of *Tramea abdominalis* (Needham and Westfall, 1955).

Genus *Paltothemis* Karsch, 1890

This genus contains the single species, *P. lineatipes* Karsch 1890 (=*russata* Calvert 1895) (fig. 4:55). It occurs from Oklahoma and Texas west to California and south to Brazil. It is a fine large species, rusty red in the male and hoary gray in the female.

Genus *Tramea* Hagen, 1861

These are large, wide-ranging dragonflies, conspicuously marked with bands of brown across the base of the hind wing (fig. 4:56). The active naiads are green marked with brown. They crawl among trash, silt, and vegetation near the shores of warm quiet ponds and lakes. We prefer to use the well-known name of *Tramea* rather than the little-known name of *Trapezostigma* applied to this group by Crowley (1935).

Key to California Species

Adults

1. Basal wing band reddish; top of head red (Canada and U.S. south to Panama and West Indies) .... *onusta* Hagen 1861
— Basal wing band blackish; top of head black (Canada, U.S., and northern Mexico) .... *lacerata* Hagen 1861

Naiads

1. Fourth segment of antennae two-thirds length of the third; lateral anal appendages about nine-tenths as long as superiors .... *onusta*
— Fourth segment of antennae half as long as the third;

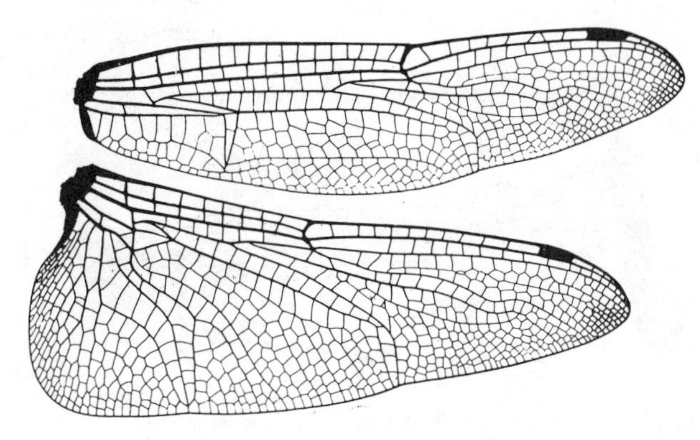

Fig. 4:57. Wings of *Pantala flavescens* (Needham and Westfall, 1955).

lateral anal appendage about four-fifths as long as superiors . . . . *lacerata*

Genus *Pantala* Hagen, 1861

These strong fliers are our most wide-ranging dragonflies. *P. flavescens* (Fabr.) (fig. 4:57) is cosmopolitan and has frequently been reported migrating in large numbers. It is one of the first species to move into new waters, especially where there is a rich growth of green algae. They prefer sunny open spaces where they fly tirelessly, usually at almost the same level (about five feet from ground).

Key to California Species

Adults

1. Hind wing with a distinct brown spot at anal angle (Canada to Chile) . . . . *hymenaea* (Say) 1839
— Hind wing with only flavescence proximally (fig. 4:57) (world-wide) . . . . *flavescens* (Fabricius) 1798

Naiads

1. Body pattern of brown conspicuous; movable hook of lateral lobe of labium twice as long as the crenulations of the distal edge of the lobe; lateral setae 15; mental setae 17-18 . . . . *hymenaea*
— Body pattern pale; movable hook of lateral lobe of labium less than twice as long as the crenulations of the distal edge of the lobe; lateral setae 12-14; mental setae 15 . . . . *flavescens*

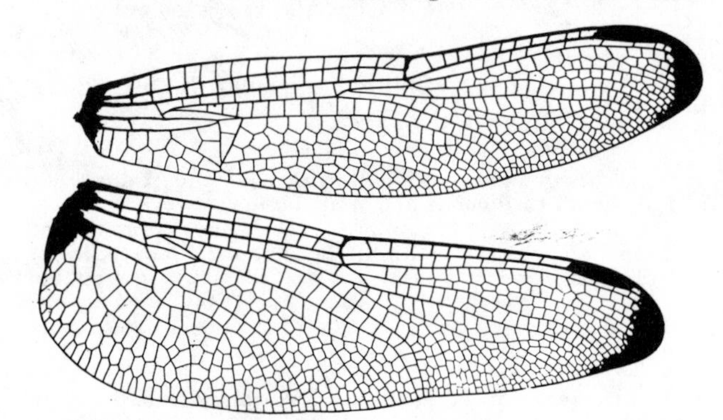

Fig. 4:58. Wings of *Dythemis velox* (Needham and Westfall, 1955).

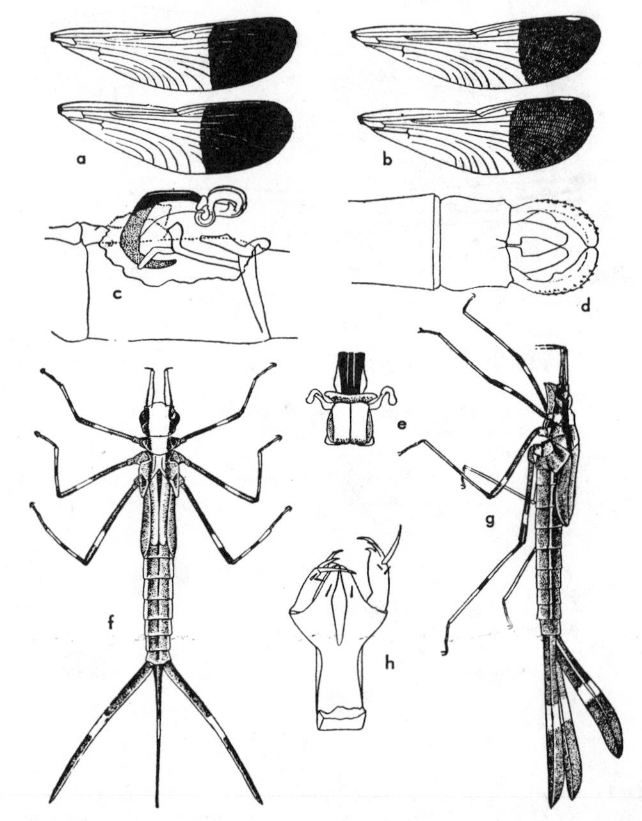

Fig. 4:59. *Agrion aequabile*. *a*, wings of male; *b*, wings of female; *c*, lateral view of second abdominal segment of male; *d*, dorsal view of male abdominal appendages; *e*, penis; *f-g*, naiad; *h*, labium of naiad (Kennedy, 1915).

ZYGOPTERA

Family AGRIONIDAE (=CALOPTERYGIDAE)

The family Agrionidae has been referred to as the Agriidae and the Calopterygidae in certain of the more recent papers on North American damselflies. We accept the designation of Latreille, 1810, of *Libellula virgo* Linnaeus as the type of genus *Agrion;* and *Calopteryx* Leach, 1815, with the same type species, is, therefore, a synonym.

These insects have broad wings with metallic bodies, and they flutter conspicuously along permanent streams. Naiads are protectively colored, and they awkwardly cling to roots and stems of submerged vegetation and trash in the stream.

Key to Nearctic Genera

Adults

1. Wing with cross veins before arculus (fig. 4:61*a*) . . . . *Hetaerina*
— Wing without cross veins before arculus (fig. 4:59*a,b*) . . . . *Agrion*

Naiads

1. Middle lobe of labium cleft only to base of lateral lobes .... *Hetaerina*
— Middle lobe of labium cleft far below base of lateral lobes (figs. 4:14*d*, 4:59*h*) .... *Agrion*

Genus *Agrion* Fabricius, 1775

The members of this genus are among our largest damselflies. They are easily recognized by the wings conspicuously marked with black or entirely black, and by the metallic green body (fig. 4:59). The young adults fly near bushes in the sunlight or rest on the foliage; the older adults fly along the banks of small streams, with a dancing flight much like that of satyrid butterflies. Where there are alternate rapids and smooth stretches, they tend to congregate in the rapids. Walker (1953) describes the mating habits as follows: "We observed a male hovering before a female, which was resting on foliage over a stream. The four wings were spread apart like a cross and vibrating rapidly. The male was within a few inches of the female and, in less than a minute, the male seized the female *per collum* and copulation ensued. The pair remained on the leaf, the abdomen of the female bending downward, but about half a minute passed before the female finally bent the abdomen forward to engage the genitalia of the male."

Oviposition takes place without the accompaniment of the male; however, the male may be close at hand to ward off other males. The eggs are deposited in almost any kind of plant tissue just beneath the surface of the water.

The long-legged, awkward naiads cling to roots and stems in the current of medium-sized streams. *Agrion aequabile* occurs in slightly larger streams than does *A. maculatum*.

Key to California Species

Males

1. Wings entirely blackish (eastern U.S. and Canada; Nevada and California in the West) .... *maculatum* Beauvais 1805
— Wings with only distal part blackish (Washington to California) (fig. 4:59) .... *aequabile californicum* Kennedy 1917

Females

1. Occiput with a prominent tubercle on each side; pterostigma with many cross veins .... *maculatum*
— Occiput without a prominent tubercle on each side; pterostigma without cross veins except near ends ... . *aequabile californicum*

Naiads

1. Antenna with proximal segment usually shorter than head .... *maculatum*
— Antenna with proximal segment usually longer than head (fig. 4:59*f*) .... *aequabile*

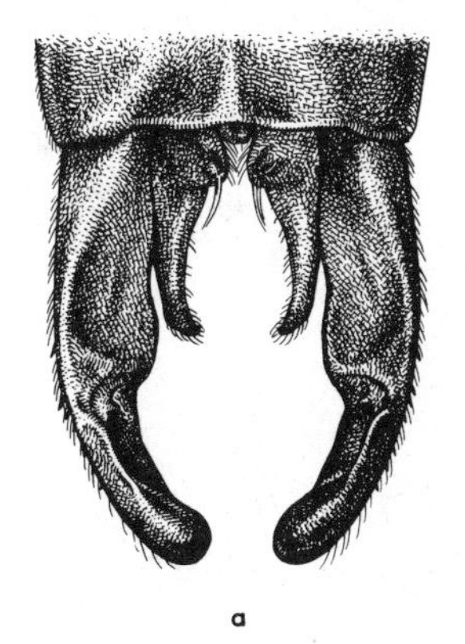

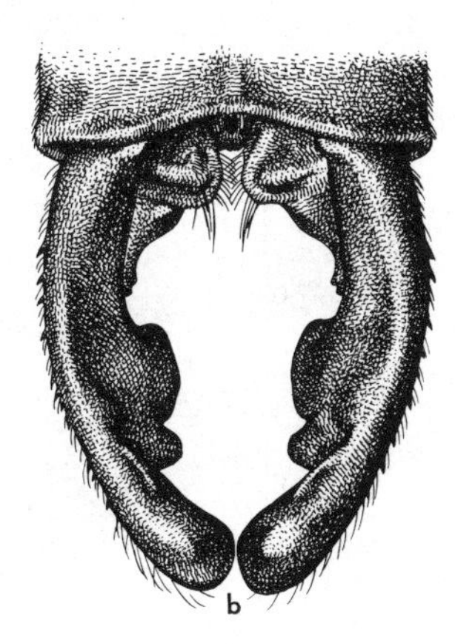

Fig. 4:60. Dorsal view of male abdominal appendages of *Hetaerina*. *a*. *vulnerata*; *b*, *americana* (Celeste Green).

Genus *Hetaerina* Hagen, 1854

These are beautiful, slender, bronzy-brown damselflies (fig. 4:61*a*). The males are conspicuously marked with a ruby or carmine basal wing spot, but in the females this spot is reddish or amber. The genus is most abundant in the tropics, but is represented in the United States by four species. Only the widespread *Hetaerina americana* is recorded from California, but *H. vulnerata* is also included in the key as it is likely to occur in southern California.

The adults frequent the edges of slowly but noticeably moving streams where the banks are overhung with willows or other vegetation. They also flit low over the rapids but do not fly far from water. The females rest at the water's edge on logs or other partly submerged objects and oviposit by thrusting the abdomen beneath the water and inserting the eggs into soft tissue.

The naiads cling to trash, plants, and rocks at the edge of the current of slow streams. Transformation takes place a few inches above the surface of the water. The naiad of *Hetaerina vulnerata* is unknown.

Key to Species

Males

1. Dorsal abdominal appendages (as seen dorsally) with a large, bilobulate expansion inside (fig. 4:60*b*) (southern Quebec and Ontario south to Florida, California, and Guatemala) .... *americana* (Fabricius) 1798
— Dorsal abdominal appendages (as seen dorsally) with an entire and slight expansion inside (fig. 4:60*a*) (Arizona and Utah to Colombia and Brazil) .... *vulnerata* Hagen 1853

Females

1. Mesothorax with middorsal stripes almost as wide as the dorsum .... *americana*
— Mesothorax with middorsal stripes about one-half as wide as the dorsum, at least at anterior end .. *vulnerata*

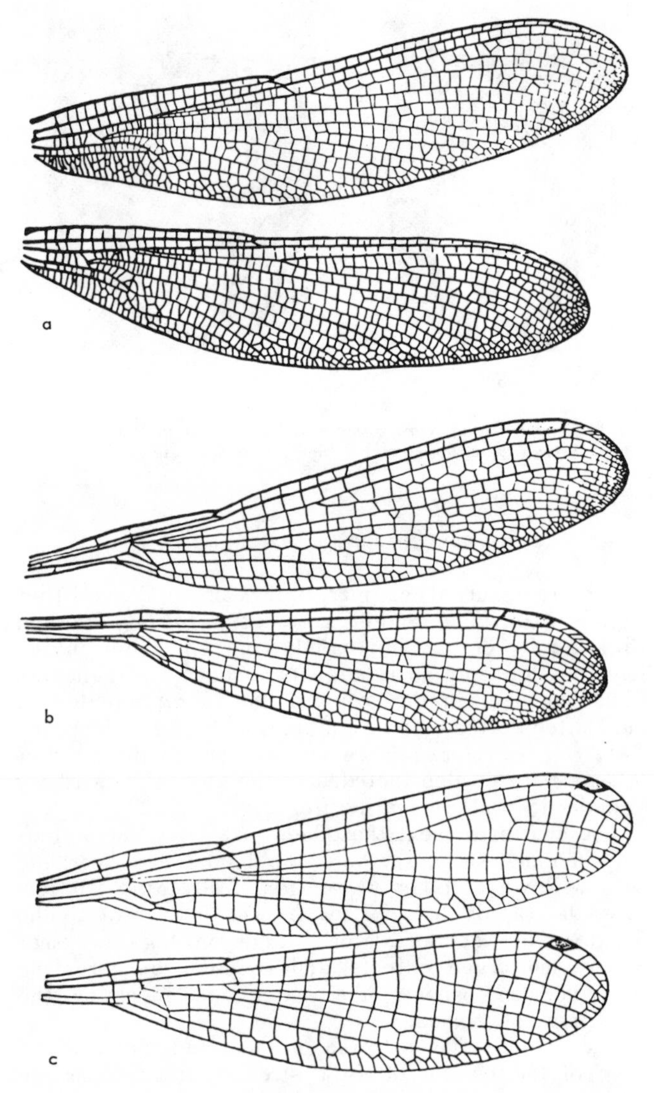

Fig. 4:61. Wings of Zygoptera. *a, Hetaerina americana; b,* Lestes *vigilax; c, Argia apicalis* (Walker, 1953).

Family LESTIDAE

The members of this family are large elongate, clear-winged damselflies. Their flight is not swift, and they are easily captured. The naiads are slender, climbing forms in still water (fig. 4:65).

Key to Nearctic Genera

Adults

1. Vein M_2 originating about 1 cell beyond nodus (fig. 4:63) .... *Archilestes*
— Vein M_2 originating several cells beyond nodus (fig. 4:61*b*) .... *Lestes*

Naiads

1. Trifid lateral lobe of labium with upper notch simple (fig. 4:65*b*) .... *Archilestes*

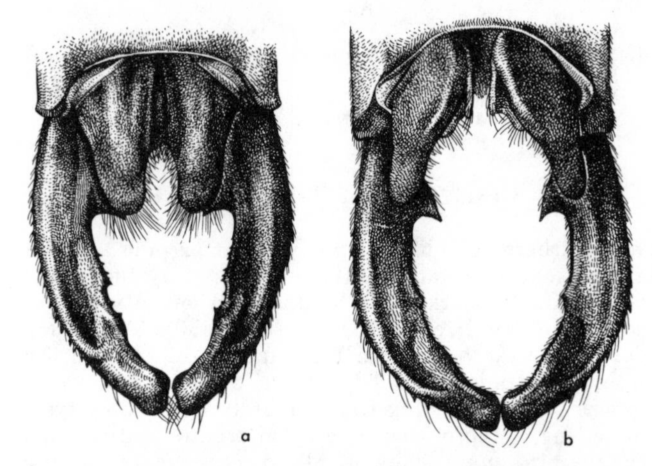

Fig. 4:62. Ventral view of male abdominal appendages of *Archilestes. a, californica; b, grandis* (Celeste Green).

— Trifid lateral lobe of labium with serrated border within upper notch (fig. 4:69*b*) .... *Lestes*

Genus *Archilestes* Selys, 1862

These are large damselflies with rather stout bodies and broad wings. *Archilestes grandis* (Rambur) is our largest damselfly (fig. 4:63). The naiads crawl over vegetation in still water (fig. 4:65). *Archilestes* is a fall form and becomes abundant as the common summer species are disappearing.

Adults commonly hang on the leaves and stems on the sunny sides of willows and alders. They fly about six to ten feet from such resting spots to take prey, and then return to the same spot. They do not fly more than twenty to thirty feet at a time, but when disturbed they dart into the densest part of the bush.

The eggs are laid in woody stems of willows and alders high above the water (fig. 4:64). Kennedy (1915) has described the mating and oviposition of *Archilestes californica* MacLachlan as follows:

"In capturing the female, the male flies toward

Fig. 4:63. *Archilestes grandis,* male (Kennedy, 1915).

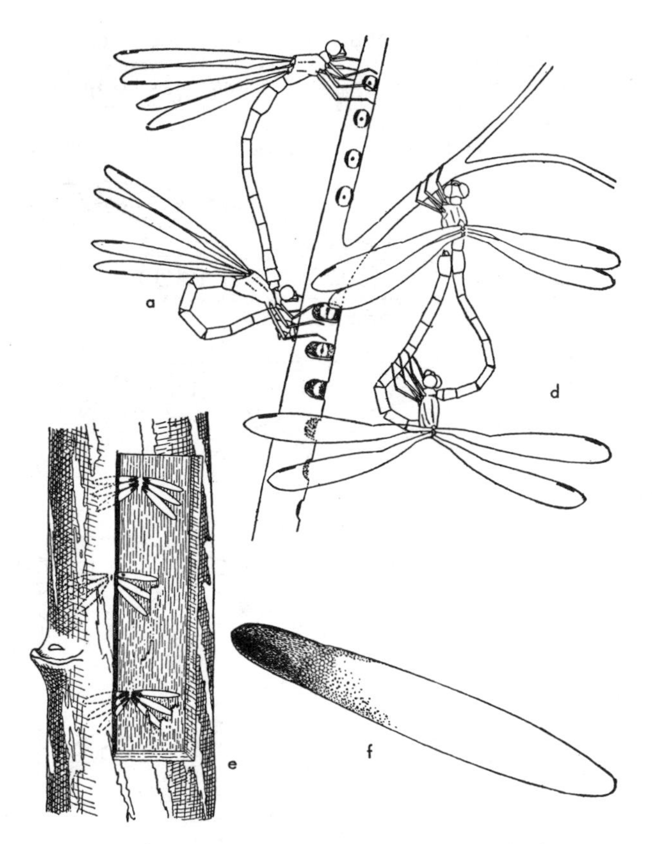

Fig. 4:64. *Archilestes californica. a,* ovipositing; *b,* oviposition scars one year old; *c,* oviposition scars two years old; *d,* copulation; *e,* bark cut away showing eggs in cambium; *f,* egg (Kennedy, 1915).

her while she is on the wing, or if she is alighted, as is the usual case, she flies up to meet him, when he first seizes her head with his feet, then bending his abdomen. She usually copulates at once, which is a lengthy process, the pair in copulation restlessly wandering from place to place.

"After many minutes in copulation they settle down on a vertical willow twig from one-fourth to one-half an inch in diameter overhanging some pool, or which may be even three feet back from the water and at a distance of from two to ten feet above the surface of the water, and begin the tedious process of oviposition. The male holds the female during oviposition, . . . " The female lays six eggs in each puncture and may lay from seventy to one hundred eighty eggs. In some cases the twigs may be girdled by egg punctures. The eggs probably pass the winter in the cambium and hatch in the spring.

Key to California Species

Males

1. Ventral abdominal appendages strongly divergent (fig. 4:62*b*); hind wing more than 35 mm. long (midwestern and western U.S. to Colombia) . . *grandis* (Rambur) 1842
— Ventral abdominal appendages parallel-sided; hind wing less than 30 mm. long (fig. 4:62*a*) (Washington to Baja Californica . . . . *californica* MacLachlan 1895

Naiads

1. Abdomen pale yellowish or greenish brown with a double row of brownish clouds on each side of the abdomen, 2 little transverse marks on dorsum of segments 3 to 7; length when mature, 40 mm. including gills . . . . *grandis*
— Abdomen pale with a narrow white stripe above each lateral keel, and a black and a white spot on side of each segment just above white stripe; length 28-31 mm. . . . . *californica*

Genus *Lestes* Leach, 1815

The members of this genus are elongate damselflies of rather large size. The adults rarely fly out over open water but frequent the margins of ponds, bogs, marshes, or other areas where there is emergent vegetation. They fly low over the water and rest frequently on plants, with wings half spread. They are difficult to see when at rest, but because of their slow flight they are easy to capture.

They oviposit above the water line in standing aquatic plants such as *Typha, Scirpus, Sparganium, Eleocharis,* and occasionally willow and grass. They generally fly in tandem while the female oviposits.

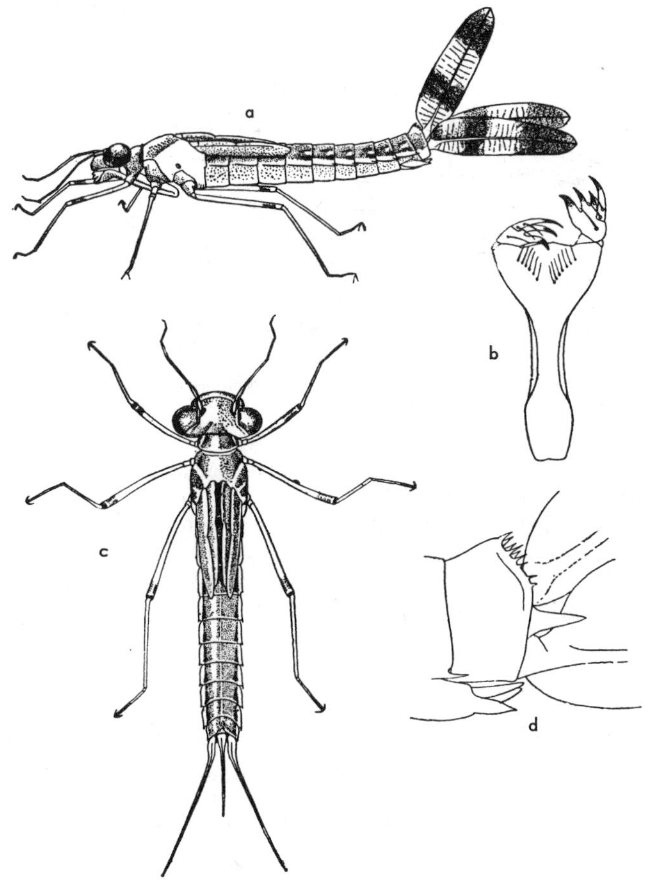

Fig. 4:65. *Archilestes californica* naiad. *a,* lateral view; *b,* labium; *c,* dorsal view; *d,* abdominal segment 10 (Kennedy, 1915).

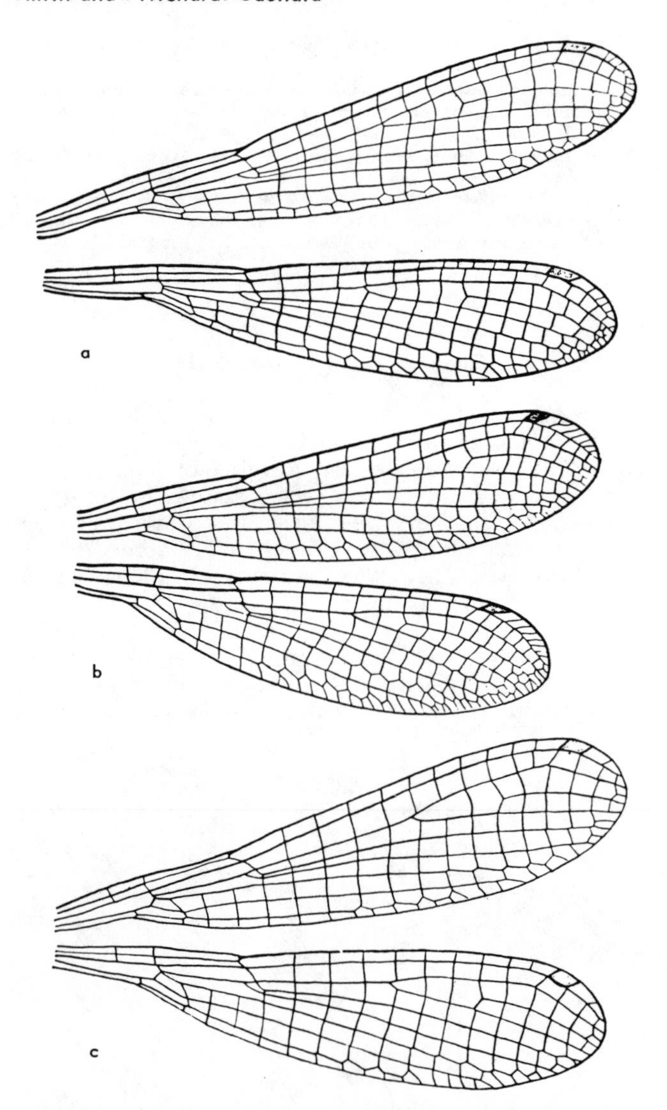

Fig. 4:66. Wings of Zygoptera. *a*, *Chromagrion conditum; b*, *Amphiagrion saucium; c*, *Nehalennia irene* (Walker, 1953).

The eggs aestivate through the summer and hatch when the pools are filled in the fall.

The naiads move quickly with sweeps of the broad gills. Some of the species can develop in temporary water.

Key to California Species

Males

1. Ventral abdominal appendages reaching end of median expansion of dorsal abdominal appendages 2
— Ventral abdominal appendages falling well short of end of median expansion of dorsal abdominal appendages (fig. 4:68*c*) (Canada and northern U.S. to Texas and California) .... *congener* Hagen 1861
2. Ventral abdominal appendages expanded distally (fig. 4:68*d,e*) 3
— Ventral abdominal appendages with distal part slender (fig. 4:68*a,f*) 4
3. Body metallic green; abdomen 26-28 mm. long (Alaska, Canada, northern U.S. to Colorado and California (=*uncatus* Kirby 1890)...............*dryas* Kirby 1890
— Body black; abdomen 30-34 mm. long (California) .... *stultus* Hagen 1861
4. Ventral abdominal appendages (as seen from above) sigmoid, the tip directed outward (fig. 4:68*f*) (Canada

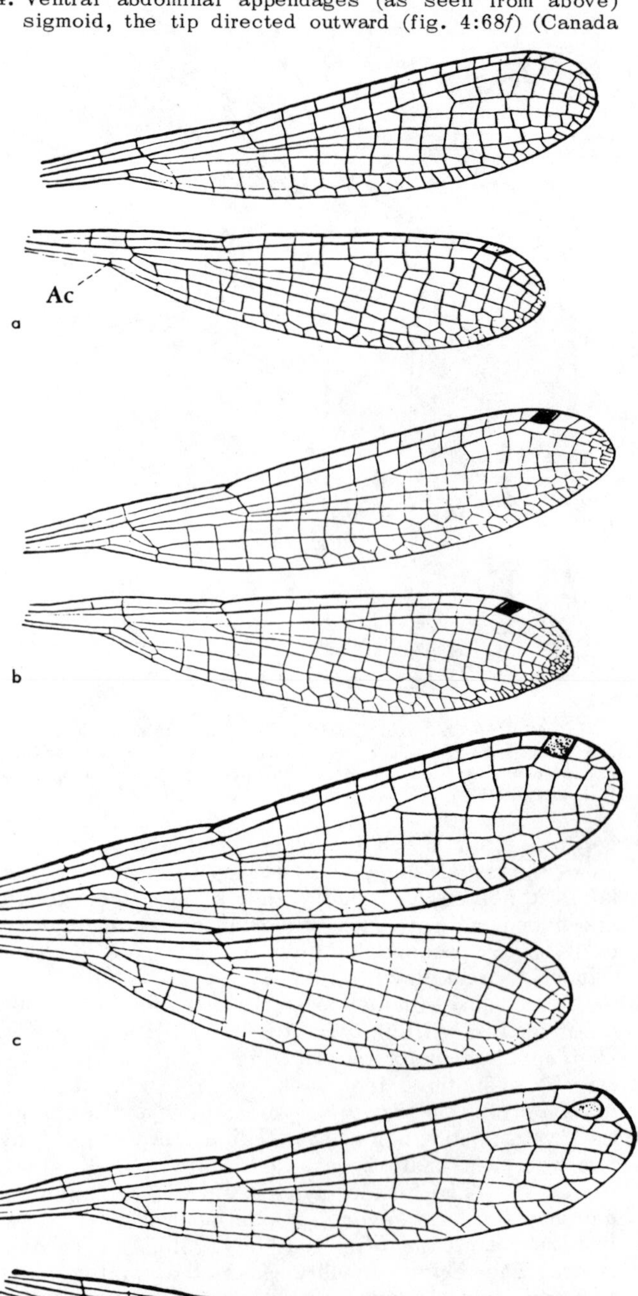

Fig. 4:67. Wings of Zygoptera. *a*, *Coenagrion resolutum* (ac, anal crossing); *b*, *Enallagma boreale; c*, *Ischnura perparva; d*, *Anomalagrion hastatum* (Walker, 1953).

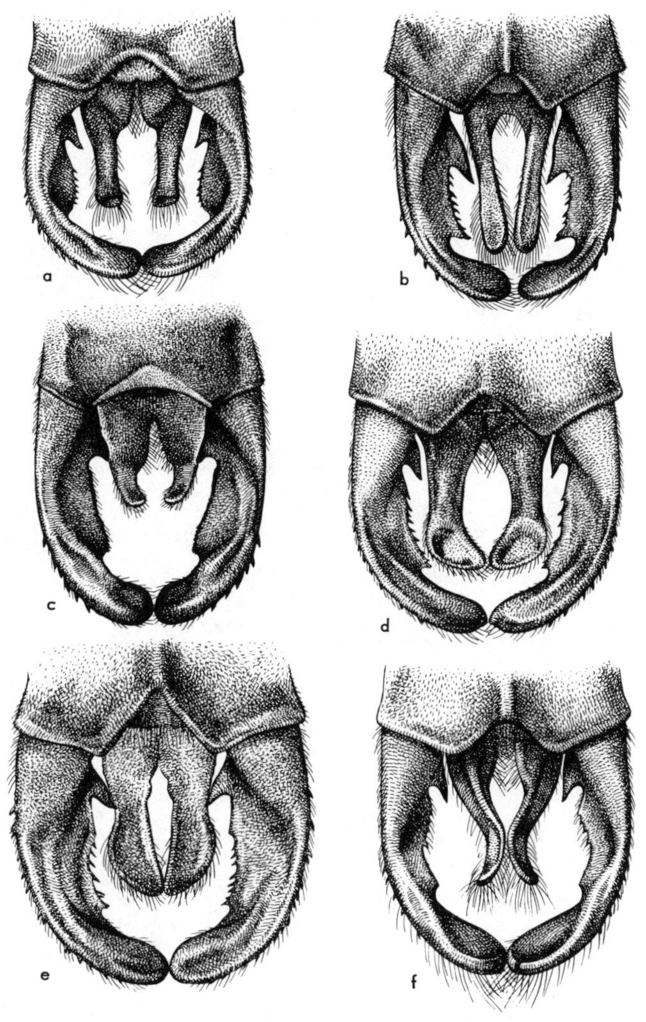

Fig. 4:68. Dorsal views of male abdominal appendages of Lestes. *a, alacer; b, disjunctus; c, congener; d, dryas; e, stultus; f, unguiculatus* (Celeste Green).

and northern U.S.; British Columbia to Utah and California in West) .... *unguiculatus* Hagen 1861

— Ventral abdominal appendages (as seen from above) straight distally 5

5. Dorsal abdominal appendages with a differentiated tooth at distal end of median expansion (fig. 4:68*b*) (Alaska, northern U.S. to northern Mexico, and California) .... *disjunctus* Selys 1862[8]

— Dorsal abdominal appendages without a differentiated tooth at distal end of median expansion (fig. 4:68*a*) (northern Mexico, and California south to Guatemala) .... *alacer* Hagen 1861

Females

1. Thorax with dorsum metallic greenish .... *dryas*
— Thorax with dorsum brown or black 2
2. Occiput largely yellowish; abdomen with a greenish tint .... *unguiculatus*
— Occiput blackish except around foramen; abdomen not greenish 3
3. Metepimeron with a black spot anteroventrally and ventrally 4

[8]California record of *L. forcipatus* prob. = *L. disjunctus* (see Walker, 1952).

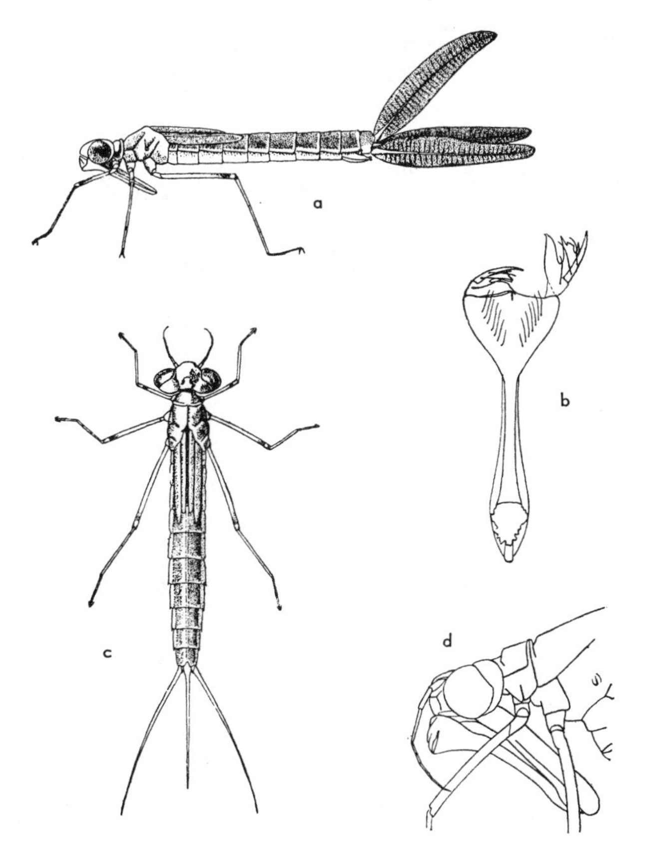

Fig. 4:69. *Lestes dryas* naiad. *a,* lateral view; *b,* labium; *c,* dorsal view; *d,* head (Kennedy, 1915).

— Metepimeron with anteroventral black spot indistinct or absent and ventral spot absent 5
4. Abdominal segments 4 to 7 with a separate posteroventral stripe on each side .... *stultus*
— Abdominal segments 4 to 7 without a posteroventral stripe on each side .... *congener*
5. Abdomen mostly black above .... *disjunctus*
— Abdomen largely pale .... *alacer*

Naiads[9]

1. Lateral setae 4-5, 3 or 4 of these on movable hook and one before its base; mentum of labium broad at its proximal end, its narrowest width equal to about one-third its width at the bases of the lateral lobes .... *congener*
— Lateral setae normally 3; mentum of labium narrow at its proximal end, its narrowest width equal to about one-fifth to one-eighth its width at the bases of the lateral lobes 2
2. Slender proximal part of labial mask at least twice as long as expanded distal part (fig. 4:69*b*); ovipositor extending beyond apical margin of segment 10 3
— Slender proximal part of labial mask not more than 1.5 times as long as distal part; ovipositor just reaching apical margin of segment 10.. *unguiculatus* and *stultus*
3. Mentum of labium 4.3-5.1 mm. long, reaching apex of hind coxae or slightly beyond (fig. 4:69*d*); ovipositor 3-3.5 mm., extending beyond basal joint of gills .. *dryas*
— Mentum of labium 4.75-5.5 mm., reaching beyond apex of hind trochanters; ovipositor 2 mm. long, extending very little beyond apical margin of segment 10 .... *disjunctus*

[9]Naiad of *alacer* Hagen is undescribed.

Family COENAGRIONIDAE (=AGRIONIDAE *auct.*)

These are small, mostly brightly colored damselflies with clear wings. The colors run to black and blue, but red, orange, green, brown, and yellow also occur. The females are often dichromatic, the homeochromatic form being like the male in color and often also in markings, whereas the heterochromatic form differs from the male not only in color but often also in the pattern. The naiads are green or brown and are usually climbing forms in still water.

Key to Nearctic Genera

Adults

1. Vein Cu_1 only 3 cells long; Cu_2 rudimentary .... *Neoneura*
— Veins Cu_1 and Cu_2 both well developed (fig. 4:74) 2
2. Tibiae with spines much longer than intervals between them 3
— Tibiae with spines about as long as intervals between them 5
3. Wing with 2 distinct rows of cells behind Cu_2 .... *Hyponeura*
— Wing with 1 row of cells behind Cu_2 (sometimes a few scattered double cells) (fig. 4:61*c*) 4
4. Vein Cu_2 4 or 5 cells long .... *Argiallagma*
— Vein Cu_2 10 or more cells long (figs. 4:61*c*; 4:71) .... *Argia*
5. Occiput almost entirely pale or with pale postocular spots or markings 6
— Occiput dark without pale postocular spots or markings 12
6. Male with stigma of fore and hind wings similar (fig. 4:67*a,b*) 7
— Male with stigma of fore and hind wings different as to size or shape (fig. 4:67*c,d*) 11
7. Wings stalked to the anal crossing .... *Teleallagma*
— Wings not stalked as far as anal crossing 8
8. Stigma with costal margin shorter than "radial" margin .... *Hesperagrion*
— Stigma with costal margin not shorter than "radial" margin (figs. 4:67; 4:78) 9
9. Male with a bifurcate process dorsally on eighth abdominal segment; female with large ventral spine on 8 (California) (fig. 4:77).............. *Zoniagrion*
— Male without a bifurcate process on eighth abdominal segment (figs. 4:80; 4:84) 10
10. Female without a ventral spine on eighth abdominal segment; penis with long membranous processes distally or subdistally (fig. 4:80)........... *Coenagrion*
— Female with a spine on eight abdominal segment; penis without long distal or subdistal processes (figs. 4:84; 4:85) ....*Enallagma*
11. Male with stigma of fore wing removed from costa; female with hind tibia entirely pale (fig. 4:67*d*) .... *Anomalagrion*
— Male with stigma of fore wing adjacent to costa; female hind tibia with outer face black (figs. 4:67*c*; 4:88) .... *Ischnura*
12. Hind wing with costal side of stigma shorter than radial side .... *Hesperagrion*
— Hind wing with costal side of stigma as long as or longer than radial side (fig. 4:66) 13
13. Body black or tan and red 14
— Body black and blue or green 15
14. Male with dorsal appendages bifid; female with ventral spine on abdominal segment 8; body short and thick set .... *Amphiagrion*
— Male with dorsal appendages entire (fig. 4:76*a*); female without ventral spine on eighth abdominal segment; body slender .... *Telebasis*
15. Body metallic green; frons angular in profile .... *Nehalennia*
— Body blue and black; frons rounded in profile .... *Chromagrion*

Naiads[10]

1. Gills not more than twice as long as broad (fig. 4:72); no mental setae 2
— Gills usually 4 or more times as long as broad, rarely only 2½ times as long as broad; mental setae present (figs. 4:79; 4:82) 3
2. One to 4 lateral setae....................... *Argia*
— Lateral setae absent, or occasionally 1 weak seta just before base of hook ....*Hyponeura*
3. Posterolateral margins of head angulate 4
— Posterolateral margins of head rounded (figs. 4:79; 4:81) 5
4. Gills ovate-lanceolate, widest near middle, one-third as broad as long, with little or no pigmentation (fig. 4:82) .... *Amphiagrion*
— Gills oblanceolate, widest toward the distal end, 1/6 as broad as long, darkly pigmented except on apical margin .... *Chromagrion*
5. Labium with 1 or 2 mental setae 6
— Labium with 3 or more mental setae, usually 5 or 6 (fig. 4:79*f,g*) 8
6. Mental setae 2 7
— Mental setae 1 and a small rudiment; lateral setae 5-6; main gill tracheae proximally sparse, distally close together (fig. 4:82) .... *Nehalennia*
7. Lateral setae 6-7 (southern California to Texas); dark colored .... *Telebasis*
— Lateral setae usually 4-5, sometimes 6, generally pale .... a few *Enallagma*
8. Tips of median gill broadly rounded or, if acute, at most angulate or slightly pointed (fig. 4:81; 4:83) .. 9
— Tips of median gill ending in slender, tapering taillike tips—tips much longer than wide (fig. 4:79)........ 13
9. Mental setae 5-6, or if less, then lateral setae 6-7 (fig. 4:81) 10
— Mental setae 3-4; lateral setae 4-5; marginal setae of mentum 3-9 .... many *Enallagma*
10. Gills broadly rounded at tip; median gill 2½ times as long as greatest width; Arizona .... *Hesperagrion*
— Gills acutely tipped; median gill 3-5 times as long as greatest width (fig. 4:81) 11
11. Third antennal segment 3 times as long as first segment; gills with pigment in tracheae only; mental setae 3; length 18 mm. (Atlantic Coast)......... *Teleallagma*
— Third antennal segment usually not more than 2½ times as long as first segment; if longer, then gills have more than tracheal pigment 12
12. Antennae 6-segmented; mental setae 0-5, often 3 .... many *Enallagma*
— Antennae 7-segmented; mental setae 4-5 ... *Coenagrion*
13. Gills with 5 or 6 black crossbars; lateral keels of abdomen edged with a row of spicules and numerous long hairs; lateral setae 5 (California) (fig. 4:79) .... *Zoniagrion*
— Gills with one to three crossbars or none; lateral keels of abdomen with at most slight indication of setae; lateral setae 5 or 6 (fig. 4:82).................... 14
14. Gills without pigment except in tracheae (fig. 4:82); small, body length 9.5 mm. when mature, gills 4.5 mm.; lateral setae 5; mental setae 4 (eastern U.S.)...... .... *Anomalagrion*
— Gills usually with cuticular pigment (fig. 4:82); larger, body length 10 mm. or more when mature, gills 5 mm. or more; lateral setae 5-6 .... *Ischnura*

[10]Naiads of *Neoneura* and *Argiallagma* are unknown.

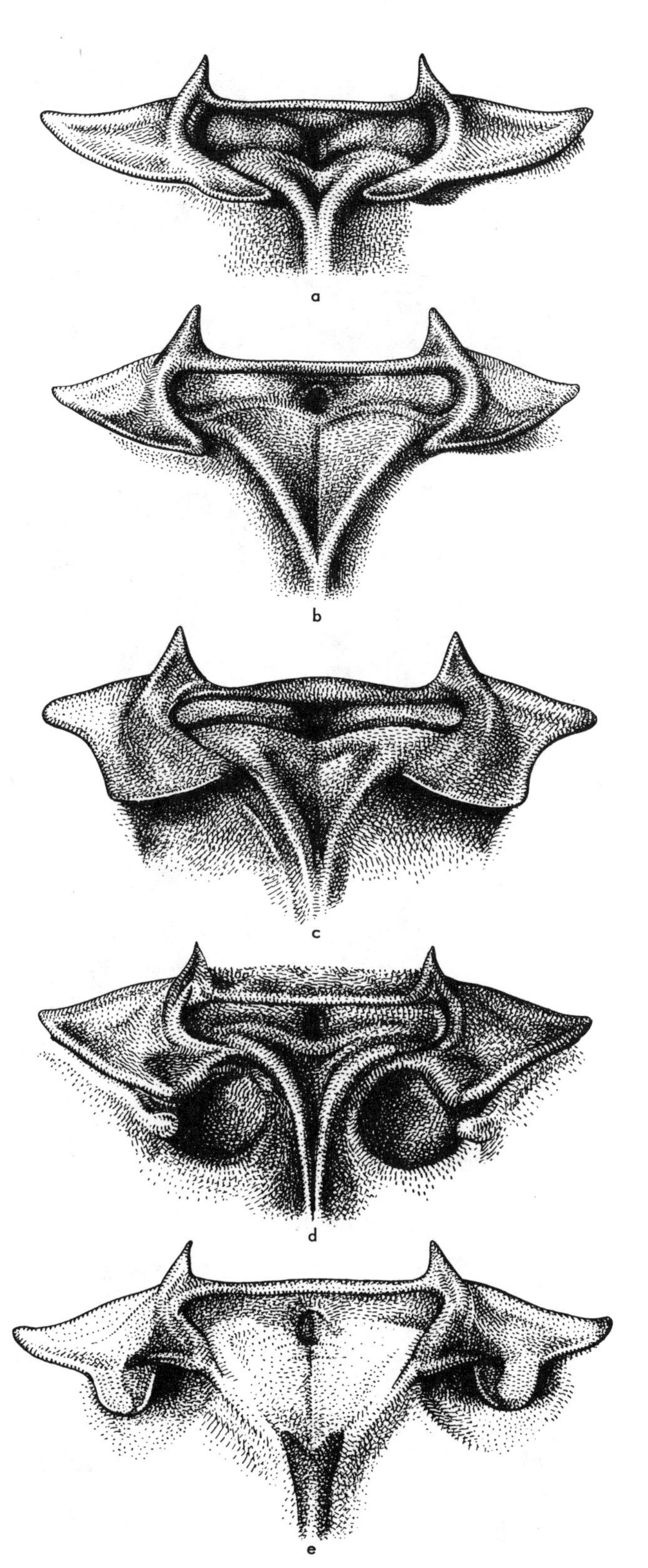

Fig. 4:70. Mesostigmal laminae of females of *Argia*: *a*, *nahuana*; *b*, *alberta*; *c*, *vivida*; *d*, *emma*; *e*, *moesta* (Celeste Green).

Genus *Hyponeura* Selys, 1854

This is a Sonoran genus that occurs from California, Utah, and western Oklahoma south into Mexico. One species, *Hyponeura lugens* Hagen 1861, enters our Southwest. These rather large, dull-colored, heavily built, clear-winged damselflies frequent permanent streams in the desert regions. The dark-colored naiad has very thick gills.

Genus *Argia* Rambur, 1842

Damselflies of the genus *Argia* are common throughout North America, particularly around larger bodies of water of streams (figs. 4:61; 4:71). Males are blue or violet and black, and the females are tan. Adults differ from most other damselflies in that they prefer to alight in open spaces, on logs, stones, and bare banks rather than on vegetation. However, they tend to be nervous, seldom resting for long. Frequently they travel some distance from water.

Eggs are deposited in surface mats of algae, or they may be laid in water-soaked logs, roots, or other types of wood. The male usually accompanies the female when ovipositing, even while submerged, but males sometimes forsake their mates at this time. In oviposition, some females are submerged for as long as an hour, but others merely oviposit from floating objects.

The naiads are stocky and have short, dark-colored gills (fig. 4:72). They are found predominantly in rather still water, but some live in the swift riffles of clear streams (Kennedy, 1915).

Our treatment of the genus *Argia* in California is based on determinations by Mrs. Leonora K. Gloyd. The presence of *A. nahuana* Calvert in California is based on her identifications (our specimens of *A. agrioides* Calvert 1895 are all from Baja California). *A. tonto* Calvert 1902 is known from Arizona and Mexico; *A. solita* Kennedy 1918 and *A. rita* Kennedy 1919 are known from Arizona. We have not determined

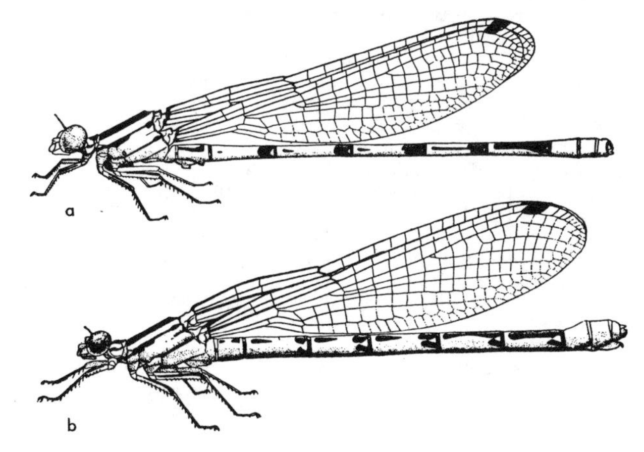

Fig. 4:71. Lateral views of *Argia vivida*. *a*, male; *b*, female (Kennedy, 1915).

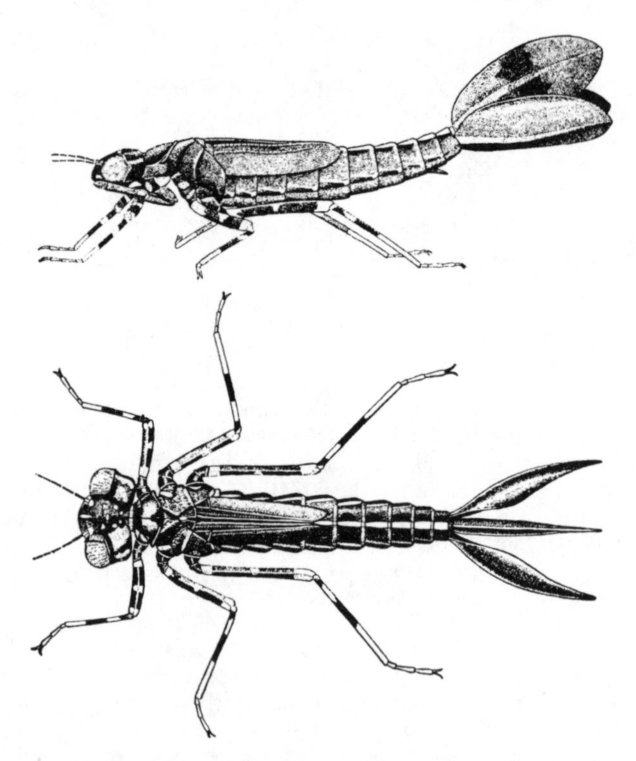

Fig. 4:72. *Argia emma*, naiad (Kennedy, 1915).

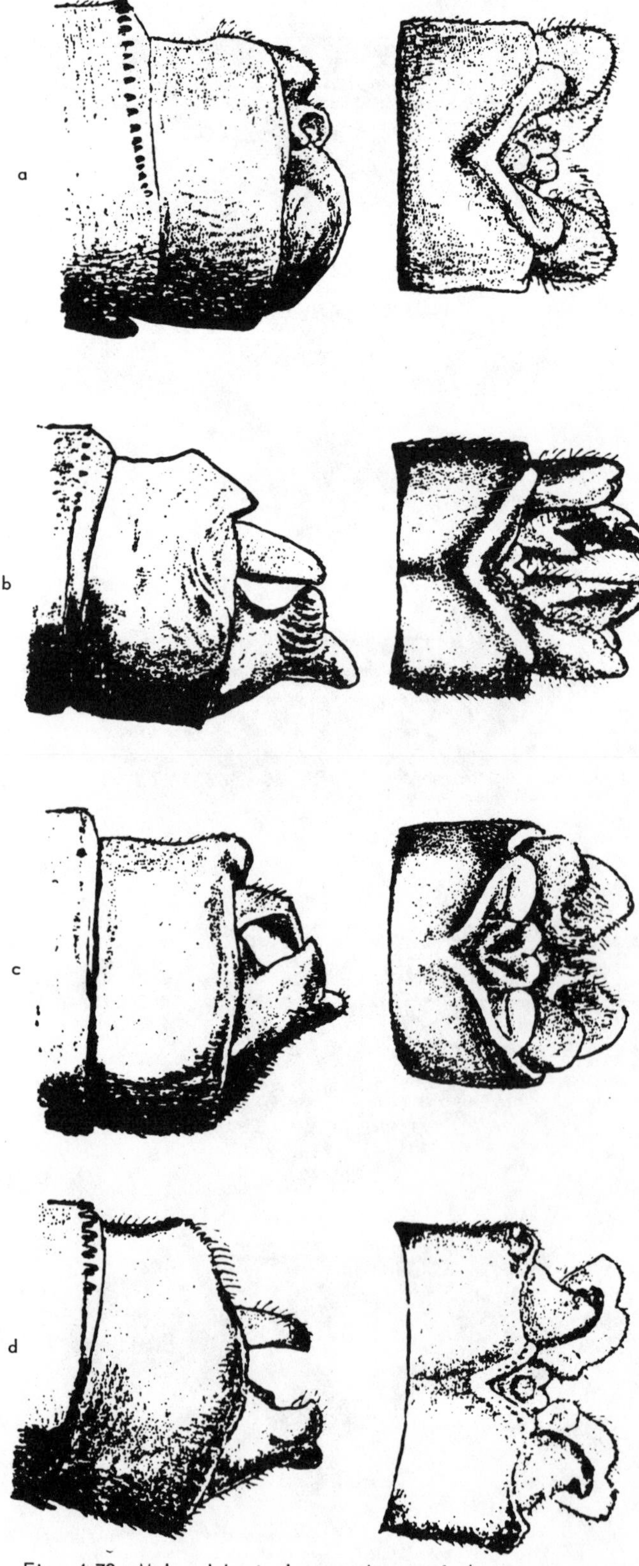

Fig. 4:73. Male abdominal appendages of *Argia*. *a*, *moesta*; *b*, *sedula*; *c*, *vivida*; *d*, *emma*. (left) lateral view; (right) dorsal view (Walker, 1953).

the status of *A. salaasi* Valle 1942 described from California. Although Gloyd (1941) showed *A. kurilis* Hagen 1865 to be a prior name for *A. vivida*, we have adhered to the well-known latter name.

Key to California Species

Males

1. Mesothorax with black middorsal stripe a very thin line (British Columbia and Utah to California) (fig. 4:73*d*) .... *emma* Kennedy 1915
— Mesothorax with black middorsal stripe wide 2
2. Mesothorax with humeral stripe not forked, wide on proximal part and with small posterior end (midwestern U.S., British Columbia to Mexico) (fig. 4:73*c*) .... *vivida* Hagen 1865
— Mesothorax with humeral stripe forked or else broad mostly throughout 3
3. Stigma of wing surmounting one cross vein (fig. 4:71*a*) 4
— Stigma of wing surmounting more than one cross vein (Texas to California and Mexico) (fig. 4:73*a*) .... *moesta moesta* (Hagen) 1861
4. Abdominal segments 4 to 6 mostly black dorsally ... 5
— Abdominal segments 4 to 6 mostly pale dorsally 6
5. Inferior appendages bifid (eastern U.S. to southern California and Mexico) (fig. 4:73*b*) .... *sedula* (Hagen) 1861
— Inferior appendages not bifid (Colorado to California) .... *alberta* Kennedy 1918
6. Inferior appendages distinctly bifid (Arizona, California) .... *hinei* Kennedy 1918
— Inferior appendages not bifid (California, Mexico) .... *nahuana* Calvert 1901

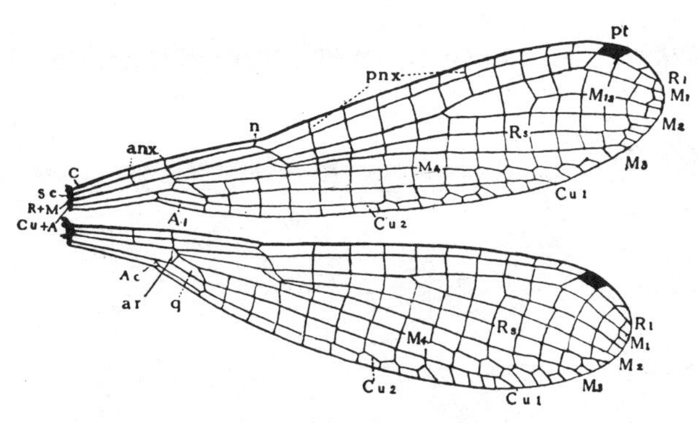

Fig. 4:74. Wings of *Argia* illustrating venation of the Zygoptera. A, anal vein; Ac, anal crossing; anx, antenodal cross veins; ar, arculus; C, costal vein; Cu, cubital vein; M. median vein; n, nodus; pnx, postnodal cross veins; pt, pterostigma; q, quadrangle; R, radial vein; Rs, radial sector; Sc, subcostal vein (Walker, 1953).

Females

1. Mesothorax with a pair of conspicuous, black pits on either side of the anterior fork of the mediodorsal carina (fig. 4:70*c-e*) 2
— Mesothorax without more than a tiny, transverse depression on either side of the anterior fork of the mediodorsal carina (fig. 4:70*a,b*) 5
2. Mesothoracic pits large, circular and open (fig. 4:70*d*) .... *emma*
— Mesothoracic pits partly covered by mesostigmal laminae (fig. 4:70*c,e*) 3
3. Mesostigmal lamina with flange on either side constricted, no broader than long (fig. 4:70*e*) 4
— Mesostigmal lamina with flange on either side broadly angulate, obviously wider than long (fig. 4:70*c*) .... *vivida*
4. Wing with stigma surmounting more than one cell (fig. 4:71) .... *moesta*
— Wing with stigma surmounting one cell or less ... *hinei*
5. Mesothorax with a mere black line along mediodorsal carina .... *sedula*
— Mesothorax with mediodorsal stripe well developed . 6
6. Abdominal segments 4 to 6 mostly black dorsally; with a slight median depression between anterior fork of mediodorsal lamina (fig. 4:70*b*) .... *alberta*
— Abdominal segments 4 to 6 mostly pale dorsally; with a deep depression between anterior fork of mediodorsal lamina (fig. 4:70*a*) .... *nahuana*

Naiads[11]

1. Gills with a marginal fringe of stiff bristles mixed with fine long hairs toward apex 2
— Gills without stiff marginal setae, or with only a few near the base, uniformally dark except along apical margin which is paler; labium with 1 lateral seta .... *moesta*
2. Gills broadest at or before the middle, rounded at base (fig. 4:72) 3
— Gills broadest beyond the middle, narrowed toward the base; labium with 3 lateral setae; gills with coarse dark blotches on a pale background and a dark crossband immediately beyond middle .... *sedula*
3. Antennae shorter than head; labium with 4 lateral setae; gills acute at apex .... *vivida*
— Antennae slightly longer than head; labium with 1 lateral seta; gills with round-angulate apices and a fine filament at tip (fig. 4:72) .... *emma*

[11]The naiads of *alberta, hinei,* and *nahuana* are not described.

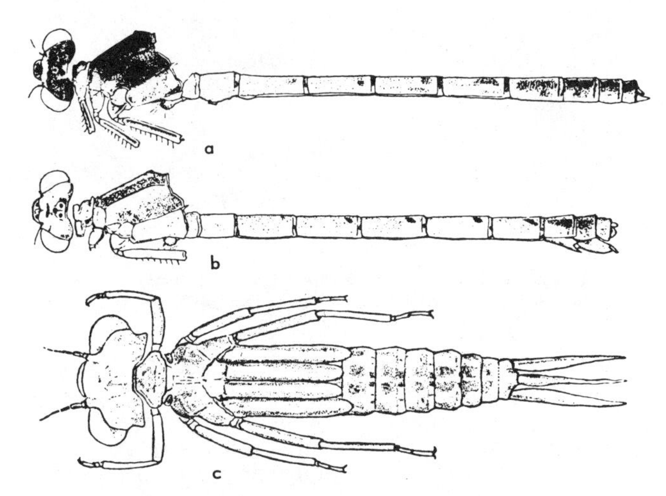

Fig. 4:75. *Amphiagron. a,* male of *abreviatum; b,* female of *abreviatum; c,* naiad of *saucium* (Walker, 1953).

Genus *Amphiagrion* Selys, 1876

These are small, stout-bodied damselflies, the males of which are red and black in color and the females of which are nearly tan (fig. 4:75).

The genus is widespread in North America, *Amphiagrion saucium* (Burmeister) 1839 being the eastern representative and *A. abbreviatum* (Selys) 1876 being the species found from Oklahoma and Kansas to British Columbia and California.

The naiad is thickset with prominent hind angles of the head. It breeds in shallow seepage pools, the spring-fed water of swales, and the small shallow beds of rivulets, all of which are densely filled with aquatic vegetation such as sedges and reeds.

Genus *Telebasis* Selys, 1865

The genus *Telebasis* is comprised of dainty, slender, red damselflies that are mostly tropical American in distribution. The one known species in our fauna, *T. salva* (Hagen) 1861 (fig. 4:76), occurs throughout the southwestern United States, from Oklahoma to California and south to Panama. One of the closely related species known from Mexico, *T. incolumis* Williamson 1930, occurs from Baja California southward to Guatemala.

Telebasis salva is found in shallow waters containing palustral vegetation. They fly low in and about the grasses. Females oviposit in algal mats or under floating sticks. The naiad has the peculiar habit of flapping the lateral gills against the middle gill about once a second.

Genus *Zoniagrion* Kennedy, 1917

The genus *Zoniagrion* is based on a single, slender, blue and black damselfly that is closely allied to the genus *Enallagma.*

Zoniagrion exclamationis (Selys) 1876 (figs. 4:77;

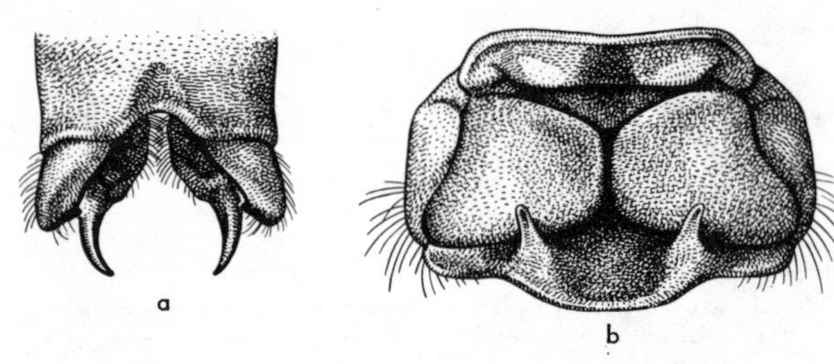

Fig. 4:76. *Telebasis salva. a,* dorsal view of male abdominal appendages; *b,* mesostigmal lamina of female (Celeste Green).

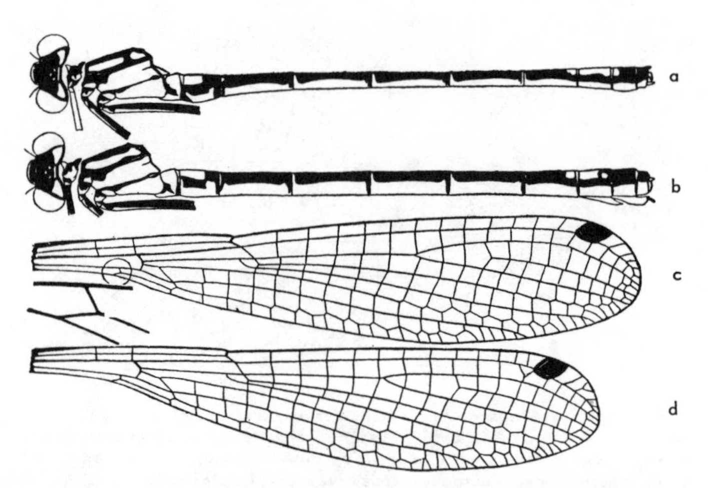

Fig. 4:78. *Zoniagrion exclamationis. a,* color pattern of male; *b,* color pattern of female; *c,* fore wing; *d,* hind wing (Kennedy, 1917).

4:78; 4:79) is known only from California. This species occurs in permanent pools of mud-banked streams.

The female, unassisted by the male, inserts her eggs into the leaf blades of *Sparganium* that hang over into the water. In oviposition the female first backs down into the water (in contradistinction to most damselflies), and lays several or more eggs as she climbs upward. A zigzag row of incisions is left behind as oviposition progresses.

The naiads are found in the roots of the *Sparganium* clumps.

Genus *Coenagrion* Kirby, 1890

The genus is primarily Palaearctic, and it is closely allied to *Enallagma*. The fact that only the vulvar spine on segment eight of the female is absent and that more or less distal membranous appendages are present in the male indicates how closely the genera are related. Adult males are similarly blue and black in coloration, and females are more brownish.

Three species of *Coenagrion* are found in boreal North America of which one, *C. resolutum* (Hagen) 1876 (figs. 4:80; 4:81; 4:82) ranges across Canada and the northern United States, south to Colorado and California in the West. This species is found along the reedy margins of streams and rivers.

Genus *Enallagma* Charpentier, 1840

The genus *Enallagma* contains the bright blue and black damselflies that are commonly seen by every American naturalist (figs. 4:83; 4:84; 4:85; 4:86). Females are less conspicuous, because they are less obviously colored. Adults abound in diverse aquatic habitats, but they usually are found where there is still, shallow, fresh water with abundant vegetation. Some species are found over brackish water or desert alkaline pools.

Adults fly low over the surface of the water or through grasses along the shore. Rarely, they are found some distance from water. Floating debris or emergent vegetation commonly serves as a support on which to rest. Females do not usually submerge for oviposition.

Our key to naiads indicates that Walker (1954) and others are correct in showing that there are significant differences among the various groups of species of which the genus *Enallagma* is now comprised. Kennedy (1920) proposed a generic division on the basis of the structure of the intromittent organ, but he did not continue these studies on a comprehensive basis.

Fig. 4:77. *Zoniagrion exclamationis,* lateral view of male abdominal appendages (Celeste Green).

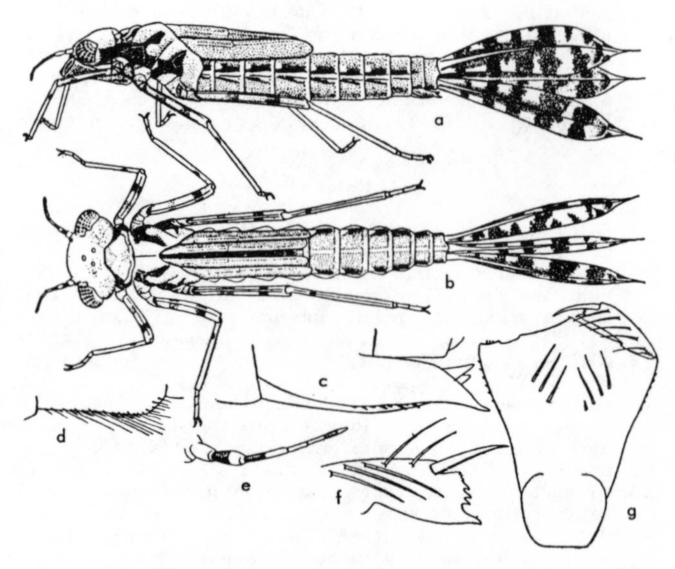

Fig. 4:79. *Zoniagrion exclamationis,* naiad. *a,* lateral view; *b,* dorsal view; *c,* segment 9 of female; *d,* lateral carina; *e,* antenna; *f,g,* mentum (Kennedy, 1917).

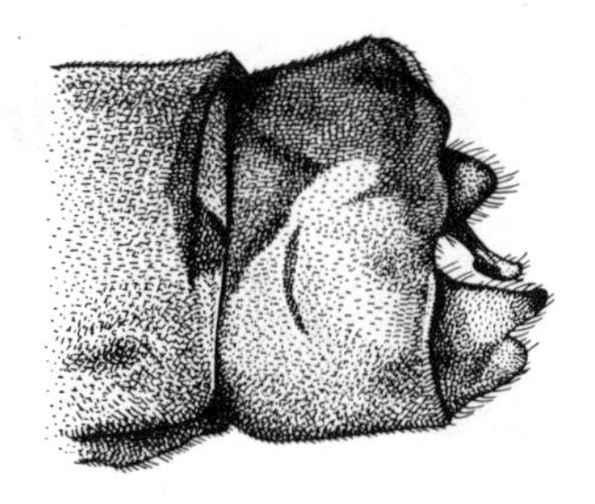

Fig. 4:80. *Coenagrion resolutum*, lateral view of male abdominal appendages (Celeste Green).

Key to California Species

Males

1. Superior appendage forked distally and the fork filled by a large, pale nodule (fig. 4:84*c,f*) 2
— Superior appendage without a large, pale nodule filling a distal fork (fig. 4:84*a,b,d,e*) 3
2. Superior appendage with fork shallow and the branches even (Canada and northern U.S., south to Colorado and Baja California) (fig. 4:84*f*) .. *caruncu latum* Morse 1895
— Superior appendage with the upper branch of the fork larger and longer than the lower branch (Canada and U.S., south to Colombia and West Indies) (fig. 4:84*c*) .... *civile* (Hagen) 1861
3. Superior appendage declivate or short and globular, shorter than the ventral appendage (fig. 4:84*a,b,e*) .. 4
— Superior appendage with a slender dorsal branch and a ventral projection, as long as or longer than the ventral appendage (figs. 4:84*d*; 4:85) 6
4. Superior appendage (as seen laterally) subglobular, with tip broadly rounded (Alaska, Canada, and northern U.S., south to New Mexico and California) (fig. 4:84*e*) .... *boreale* (Selys) 1875
— Superior appendage declivate with narrow distal end (as seen laterally) 5
5. Superior appendage with tip turned up (Holarctic: Alaska, Canada, U.S.) (fig. 4:84*b*)................ .... *cyathigerum* (Charpentier) 1840
— Superior appendage with tip directed caudad (Great Basin: western Canada and U.S., south to Kansas and eastern California) (fig. 4:84*a*) ... *clausum* Morse 1895
6. Superior appendage with tip of lower branch directed caudad and without a tubercle (southern U.S. from Louisiana to Utah and California, south to Guatemala) (fig. 4:84*d*) .... *praevarum* (Hagen) 1861
— Superior appendage with tip of lower branch directed ventrad and bearing a large caudally directed tubercle (western U.S., Wyoming and Utah to California) (fig. 4:85) .... *anna* Williamson 1900

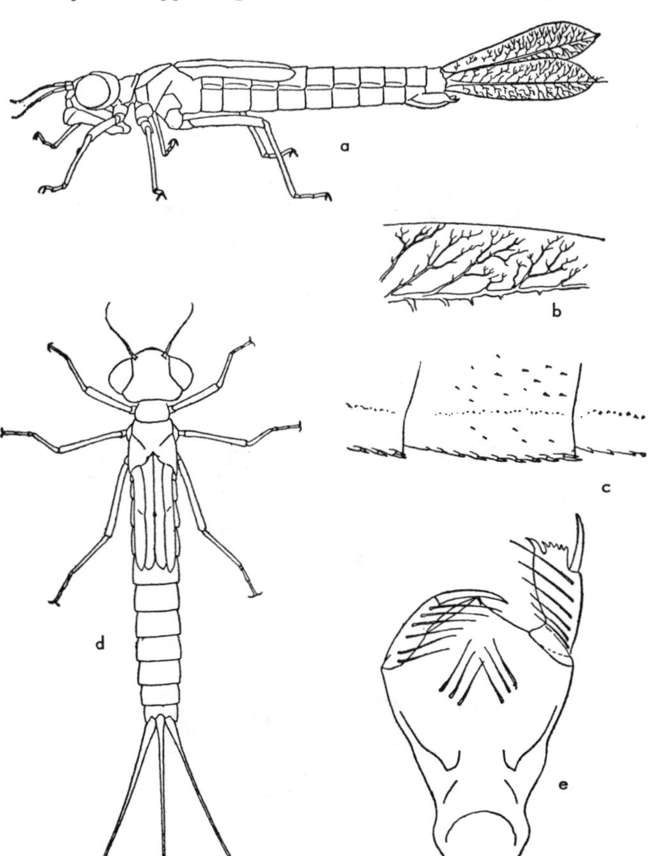

Fig. 4:81. *Coenagrion resolutum*, naiad. *a*, lateral view; *b*, detail of caudal gill; *c*, lateral fold of abdomen showing the spinulose edge; *d*, dorsal view; *e*, labium (Kennedy, 1915).

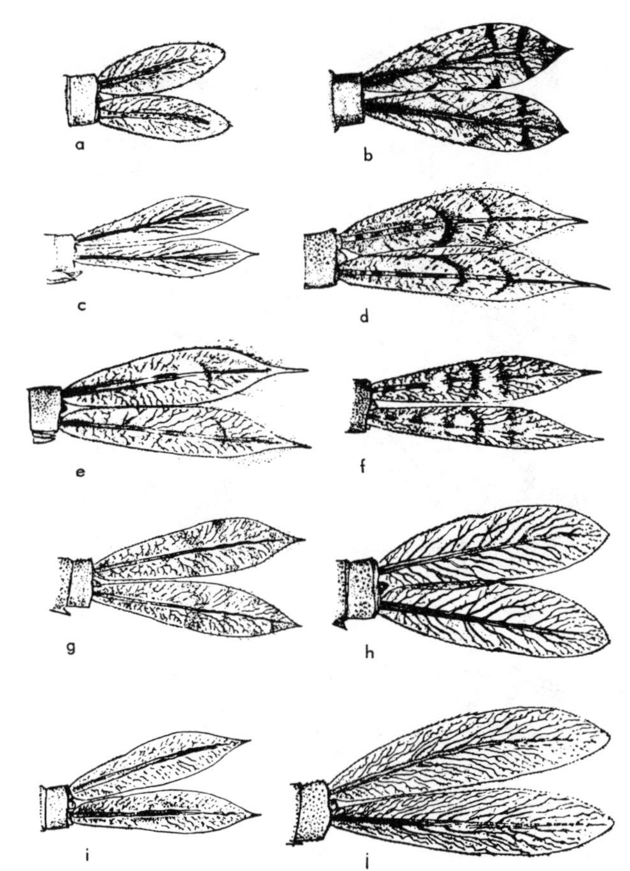

Fig. 4:82. Caudal gills of naiads. *a*, *Amphiagrion saucium*; *b*, *Nehallenia irene*; *c*, *N. gracilis*; *d*, *Ischnura cervula*; *e*, *I. damula*; *f*, *I. verticalis*; *g*, *I. posita*; *h*, *Coenagrion angulatum*; *i*, *C. resolutum*; *j*, *Anomalagrion hastatum* (Walker, 1953).

Females

1. Prothorax with hind margin convex throughout 2
— Prothorax with hind margin concave medially, slightly concave or truncate on either side .... *praevarum* and *anna*

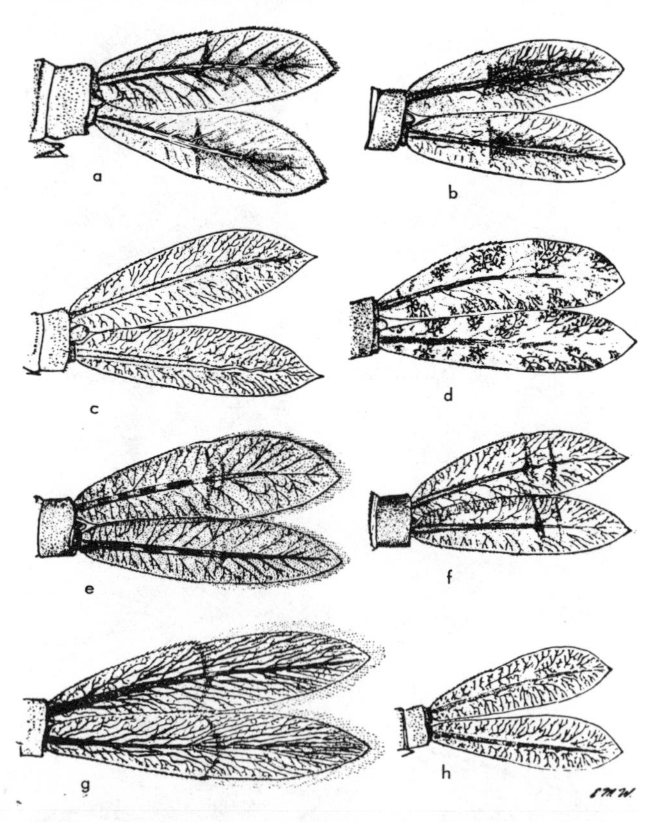

Fig. 4:83. Caudal gills of *Enallagma* naiads. *a, clausum; b, carunculatum; c, civile; d, ebrium; e, boreale; f, cyathigerum; g, vernale; h, geminatum* (Walker, 1953).

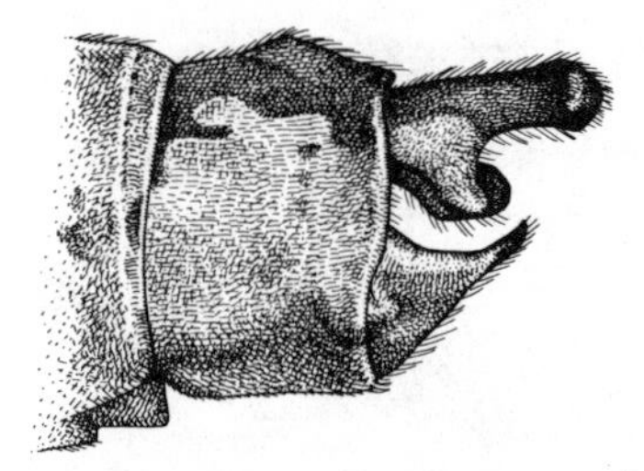

Fig. 4:85. Lateral view of male abdominal appendages of *Enallagma anna* (Celeste Green).

2. Mesostigmal lamina with a broadly scooped-out hollow diagonally across each lateral lobe (fig. 4:86*a,c*) . . . 3
— Mesostigmal lamina with the diagonal ridge very narrowly interrupted at most 5
3. Mesostigmal lamina with a deep hollow at the inner-anterior end of each lateral lobe (fig. 4:86*c*) . . . . *cyathigerum*
— Mesostigmal lamina without an obvious hollow on the lateral lobe (fig. 4:86*a*) 4
4. Abdominal segment 8 with black markings; humeral stripe without an abruptly narrowed section (fig. 4:86*a*) . . . . *boreale*
— Abdominal segment 8 entirely blue; humeral stripe abruptly narrowed on caudal part . . . . *clausum*
5. Mesostigmal lamina with posterolateral margin sharply elevated and lobelike . . . . *carunculatum*
— Mesostigmal lamina with posterolateral expansion nearly flat . . . . *civile*

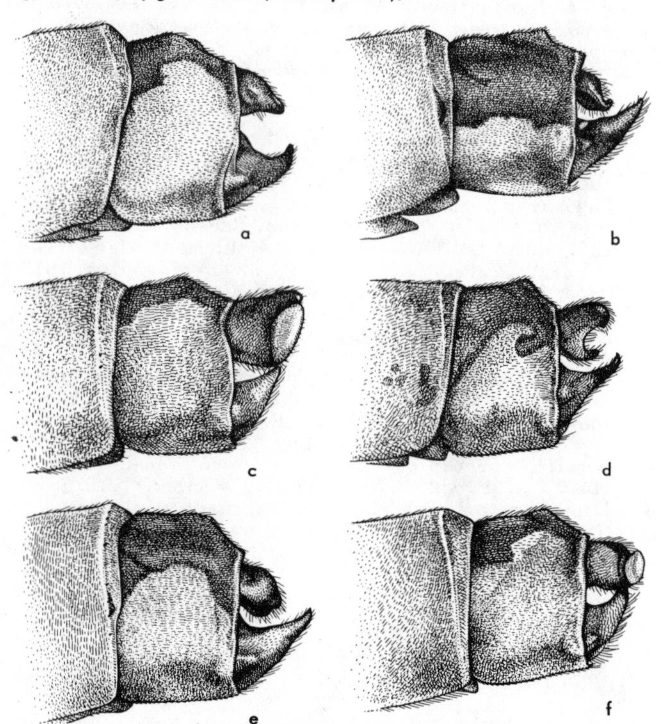

Fig. 4:84. Lateral views of male abdominal appendages of *Enallagma*. *a, clausum; b, cyathigerum; c, civile; d, praevarum; e, boreale; f, carunculatum* (Celeste Green).

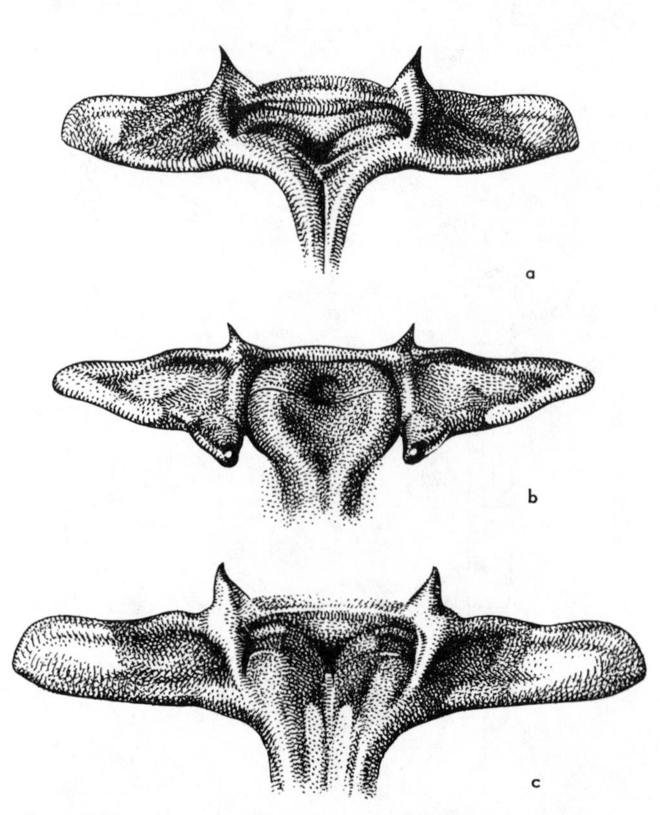

Fig. 4:86. Mesostigmal laminae of females. *a, Enallagma boreale; b, Coenagrion resolutum; c, Enallagma cyathigerum* (Celeste Green).

Naiads[12]

1. Dorsal setigerous margin of the median gill extending beyond the middle of the gill 2
— Dorsal setigerous margin of the median gill not extending beyond the middle of the gill (fig. 4:83*a*) 4
2. Mental setae 3-5; lateral setae 6-7; marginal setae 5-9 3
— Mental setae 3; lateral setae 5; marginal setae 3 .... *praevarum*
3. Greatest width of middle gill about 1/4 of its length; marginal setae 5-6 (fig. 4:83*e*) .... *boreale*
— Greatest width of middle gill about 1/3 of its length; marginal setae 6-9 (fig. 4:83*f*) .... *cyathigerum*
4. Tracheae of caudal gills with numerous branches which tend to cŭrve; dorsal setigerous margin of middle gill not extending to middle (fig. 4:83*b*) 5
— Tracheae of caudal gills with fewer and straighter branches; dorsal setigerous margin of middle gill extending to about middle (fig. 4:82*a*) ....*clausum*
5. Lateral carina of segment 1 with a few spinose setae; dorsal setigerous margin of middle gill extending about three-sevenths of length; gills diffusely pigmented (fig. 4:83*c*) .... *carunculatum*
— Lateral carina of segment 1 without spinose setae; dorsal setigerous margin of middle gill extending about 1/3 of length; gills with pigment only in small tracheae (fig. 4:83*c*) ....*civile*

Genus *Ischnura* Charpentier, 1840

Ischnura males are black with greenish markings on the thorax and blue markings caudally on the abdomen (figs. 4:90*a*; 4:92).

Females exhibit marked dichromatism (figs. 4:90; 4:91; 4:92). The homeochromatic form exhibits color markings similar to that of the male, but it becomes dull pruinose in later life. The heterochromatic female is often the more common form of this sex, and it is black with a more extensive, orange coloration. However, this form also becomes pruinose with age, the pale markings becoming indistinct.

Adults are found where there is an abundance of aquatic vegetation in still water, but they may stray into fields. Copulation usually takes place while resting on plants, and the males usually leave the females before oviposition. Eggs are laid singly in the stems and leaves of aquatic plants. Naiads (fig. 4:93) occur among the vegetation of quiet waters of lakes, ponds, swamps, and streams.

Kennedy (1917, p. 500) writes concerning *Ischnura denticollis:* "The habits of this species are in general ischnuran but indicate greater feebleness. Early in the morning it is found in the sedges and grasses bordering the water but during the heat of the day it spends the greater part of its time over the surface of the water, usually seated on trash or aquatic vegetation.

"The females resorted to the little drain ditches to oviposit; there the males in great numbers awaited their coming. After a considerable time in copulation seated on some grass stem, the female, still accompanied by the male, would fly to the surface of the stream, preferably a quiet lateral pool, and commence ovipositing.

[12]The naiad of *anna* is not described.

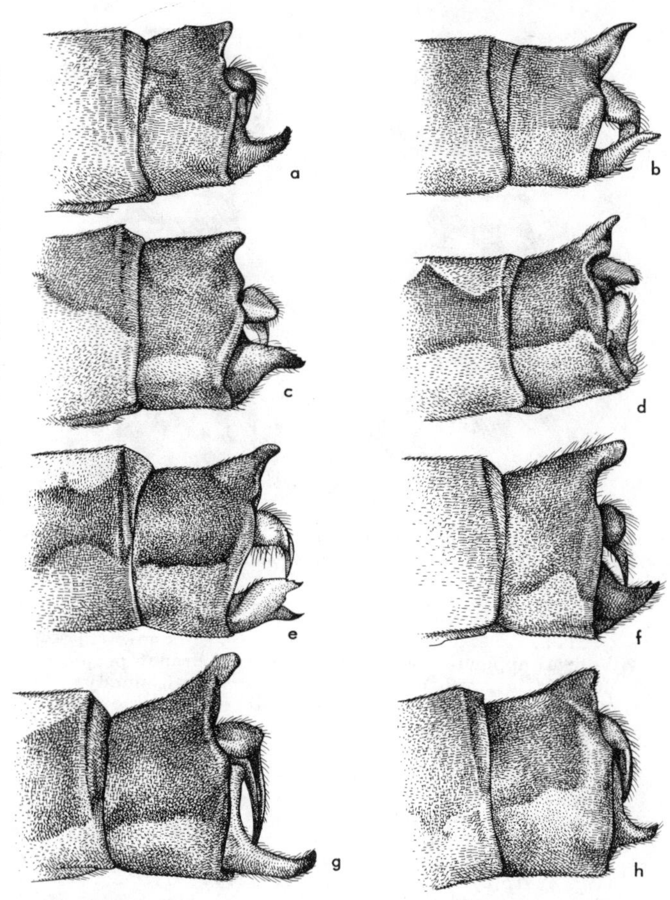

Fig. 4:87. Lateral views of male abdominal appendages of *Ischnura. a, barberi; b, demorsa; c, ramburii; d, gemina; e, perparva; f, cervula; g, erratica; h, denticollis* (Celeste Green).

"In ovipositing the male held the female by the head. The pair would alight on floating vegetation, in a horizontal position, and the female would bend her abdomen slightly and make usually one or two incisions, after which she would raise the end of her abdomen considerably above the horizontal and wait in this position several seconds, when the pair would fly to another straw and repeat the one or two thrusts followed by the wait with the tip of the female's abdomen in the air. This was kept up, by a pair under observation, for twenty minutes. In no place did they make more than one or two thrusts, . . ."

Kennedy (1917) proposed the genus *Celaenura* for *Ischnura denticollis* and *I. gemina,* and the generic distinction is probably valid. However, later workers have withheld the use of this name pending more revisionary studies.

Key to California Species

Males

1. Mesothorax with dorsum solid black (fig. 4:92) 2
— Mesothorax with pale spots or stripes on dorsum 3
2. Ventral appendage slender and pointed, the dorsal branch not projecting (fig. 4:87*h*) (Utah to California,

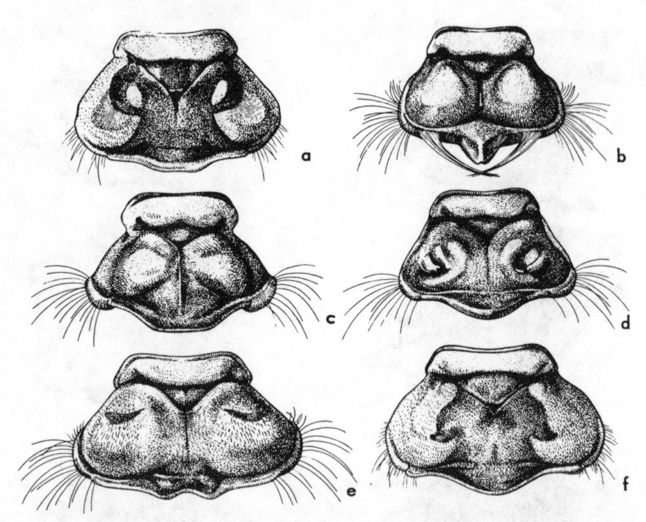

Fig. 4:88. Prothoraces of female *Ischnura. a, barberi; b, cervula; c, perparva; d, denticollis; e, demorsa; f, ramburii* (Celeste Green).

south to Mexico) (=*extriata* Calvert 1895) ....
.... *denticollis* (Burmeister) 1839

— Ventral appendage rounded, the dorsal branch projecting dorsad (fig. 4:87*d*) (California) ..*gemina* (Kennedy) 1917

3. Mesothorax with the antehumeral pale stripes represented only by widely separated spots (fig. 4:89) (western U.S. and southwestern Canada, south to Baja California and new Mexico) .... *cervula* Selys 1876

— Mesothorax with antehumeral pale stripes complete 4

4. Ventral appendage widened and with 2-4 teeth distally (fig. 4:87*b,c*) 5

— Ventral appendage with a single, slender branch (fig. 4:87*a,b,g*) 6

5. Prothorax medially emarginate on caudal margin; abdominal segment 9 usually entirely blue (fig. 4:87*b*) (Montana and Colorado, south to Mexico City) .... *demorsa* (Hagen) 1861

— Prothorax with hind margin convex medially; abdominal segment 9 with a lateral black stripe (fig. 4:87*e*) (British Columbia and western U.S., from Montana and Colorado to Texas and California).............. .... *perparva* Selys 1876

6. Ventral appendage with dorsal margin forming a very large right angle (fig. 4:87*g*); abdominal segment 10 with a pronounced dorsal development (British Columbia to California)................ *erratica* Calvert 1895

— Ventral appendage with dorsal margin forming at most a small angulation (fig. 4:87*a*); abdominal segment 10 with a small dorsomedian development 7

7. Ventral appendage with an obvious emargination dorsally, the tip directed dorsad (fig. 4:87*a*) (Utah, Colorado New Mexico, Arizona) (=*utahensis* Muttkowski 1910) .... *barberi* Currie 1903

— Ventral appendage without a dorsal emargination and with the tip direct caudad (fig. 4:87*c*) (California, Baja California to Central America, West Indies and Florida) (=*defixum* Hagen 1861)...*ramburii credula* (Hagen) 1861

Females

1. Prothorax with a dorsomedian angulation on posterior margin, accompanied by pencils of incurved hairs from each mediolateral lobe (figs. 4:88*b*; 4:89).......*cervula*

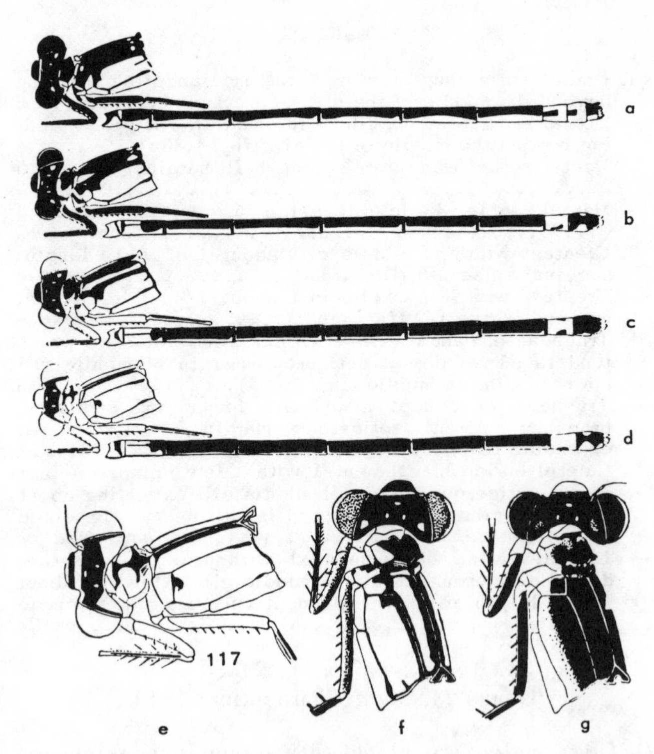

Fig. 4:89. Coloration of *Ischnura cervula. a,* mature male; *b,* dark female; *c,f,* teneral dark female; *d,e,* teneral light female; *g,* mature female (Kennedy, 1915).

— Prothorax without a dorsomedian projection on caudal margin 2

2. Prothorax with a dorsal projection on each mediolateral lobe (fig. 4:88*d*)........................ 3

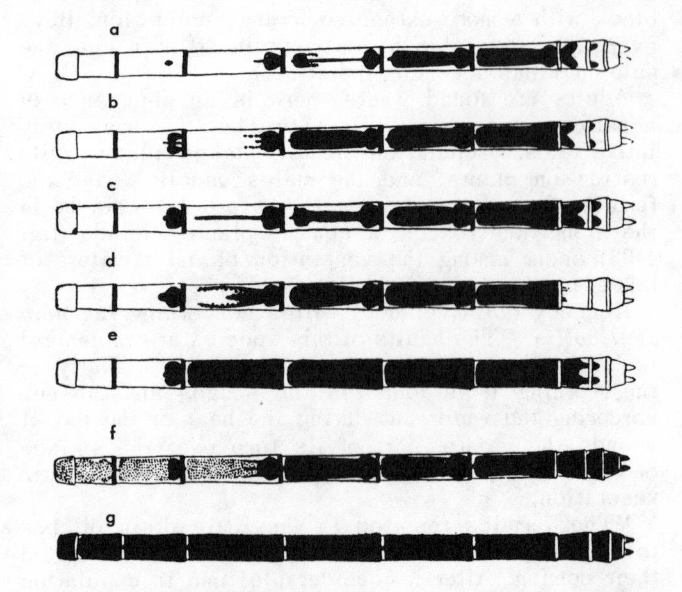

Fig. 4:90. Coloration of *Ischnura perparva,* female. *a-e,* stages in teneral coloration. *f,* intermediate stage between teneral and adult coloration; *g,* adult coloration (Kennedy, 1915).

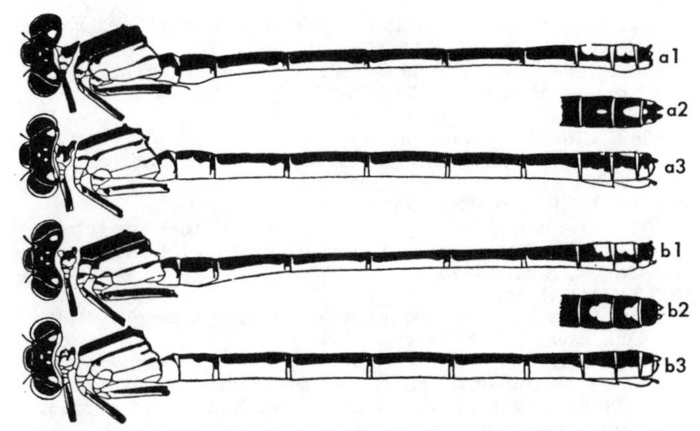

Fig. 4:91. Coloration of *Ischnura*. *a*, *gemina*; *b*, *denticollis*; 1, lateral view of male; 2, dorsal view of posterior end of female abdomen; 3, lateral view of female (Kennedy, 1917).

— Prothorax without dorsal projections 4
3. Abdomen with dorsum of segment 10 entirely black (figs. 4:88*d*; 4:91*b*2) .... *denticollis*
— Abdomen with blue spots on dorsum of segment 10 (figs. 4:89; 4:91*a*2) .... *gemina*
4. Prothorax with dorsoposterior margin evenly convex 5
— Prothorax with small, dorsomedian emargination on caudal margin (fig. 4:88*e*) .... *demorsa*
5. Abdominal segment 8 with a well-developed medioventral spine (fig. 4:88*a*) .... *barberi*
— Abdominal segment 8 without a medioventral spine or with this spine rudimentary 6
6. Abdomen more than 21 mm. long 7
— Abdomen less than 20 mm. long (figs. 4:88*c*; 4:90) .... *perparva*
7. Hind wing more than 19 mm. long (fig. 4:88*f*) .... *ramburii credula*
— Hind wing less than 14 mm. long .... *erratica*

Naiads[13]

1. Dorsal setigerous margin of the median gill extending more than 1/2 length of gill; mental setae 3-4, sometimes a vestigal fifth; lateral setae 5 .... *perparva*
— Dorsal setigerous margin of median gill extending 1/2 length of gill or less 2
2. Dorsal setigerous margin extending 1/2 length of gill; marginal setae of median lobe 4-5; lateral setae 6; femora with preapical rings of brown .... *ramburii credula*
— Dorsal setigerous margin not extending beyond 1/3 length of gill 3
3. Body length, not including gills, 15 mm. ...... *barberi*
— Body length, not including gills, 12 mm. or less .... *cervula* and *denticollis*

[13]Naiads of *gemina*, *demorsa*, and *erratica* are undescribed.

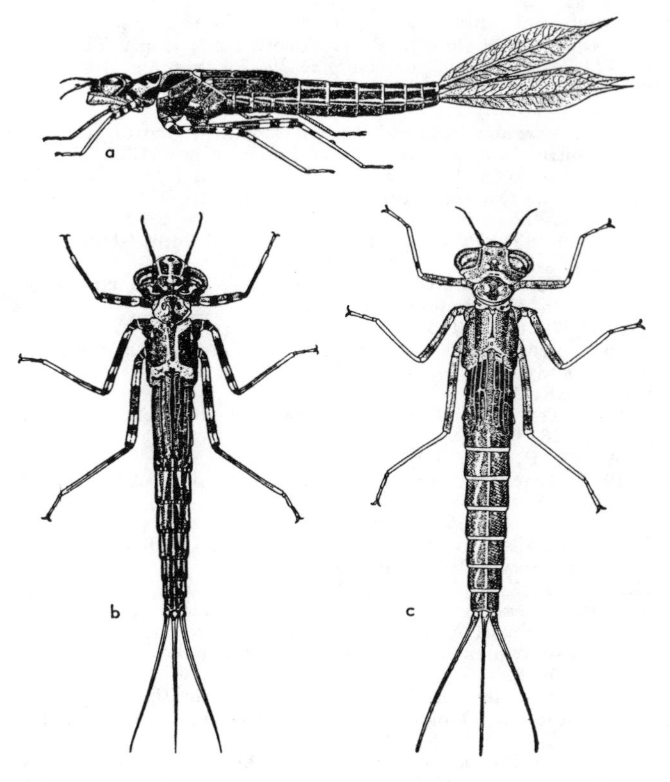

Fig. 4:92. *Ischnura* naiads. *a,b*, *cervula*; *c*, *perparva* (Kennedy, 1915).

REFERENCES

AHRENS, C.
1938. A list of dragonflies taken during the summer of 1936 in western United States (Odonata). Ent. News, 49:9-16.
BICK, GEORGE H.
1941. Life-history of the dragonfly, *Erythemis simplicollis* (Say). Ann. Ent. Soc. Amer., 34:215-230.
1953. The nymph of *Miathyria marcella* (Selys). Proc. Ent. Soc. Wash., 55:30-36.
BORROR, D. J.
1934. Ecological studies of *Argia moesta* Hagen (Odonata: Coenagrionidae) by means of marking. Ohio Jour. Sci., 34:97-108.
1942. A revision of the libelluline genus *Erythrodiplax*. Ohio State Univ. Grad. Sch. Stud., Biol. Series, Contrib. Zool. Ent., 4:XV 1-286.
1945. A key to the new world genera of Libellulidae. Ann. Ent. Soc. Amer., 38:168-194.
BYERS, C. F.
1927*a*. Key to the North American species of *Enallagma*, with a description of a new species (Odonata: Zygoptera). Trans. Amer. Ent. Soc., 53:249-260.
1927*b*. *Enallagma* and *Telagrion* from Western Florida, with a description of a new species. Ann. Ent. Soc. Amer., 20:385-392.
1927*c*. Notes on some American dragonfly nymphs (Odonata, Anisoptera). Jour. N.Y. Ent. Soc., 35:65-74.
1927*d*. The nymph of *Libellula incesta* and a key for the separation of the known nymphs of the genus *Libellula*. Ent. News, 38:113-115.
1930. A contribution to the knowledge of Florida Odonata. Univ. Florida Publ. Biol. Sci. Series, 1:1-327.
1936. The immature form of *Brachymesia gravida*, with notes on the taxonomy of the group (Libellulidae). Ent. News, 47:35-37.
1937. A review of the dragonflies of the genera *Neurocordulia* and *Platycordulia*. Univ. Mich. Misc. Publ., 36, 36 pp.
1939. A study of the dragonflies of the genus *Progomphus (Gomphoides)* with a description of a new species. Proc. Fla. Acad. Sci., 4:19-86.
CALVERT, P. P.
1901-1908. Biologia Centrali-Americana. Insecta. Odonata. 1-VIII, 17-420.
1928. Report on Odonata, including notes on some internal organs of the larvae collected by the Barbados-Antigua Expedition from the University of Iowa in 1918. Univ. Iowa Stud. Nat. Hist., 12:1-54.
1934. The rates of growth, larval development and

seasonal distribution of dragonflies of the genus *Anax* (Aeshnidae). Proc. Amer. Phil. Soc., 73:1-70.
1941. *Aeshna (Coryphaeschna) luteipennis* and its subspecies (Odonata: Aeshnidae). Ann. Ent. Soc. Amer., 34:389-396.
1942. Increase in knowledge of Odonate fauna of Mexico, Central America and West Indies since 1908. Proc. Eighth Amer. Sci. Congr., Biol. Sci. Zool., pp. 323-331.
1947. The Odonate collections of the California Academy of Sciences from Baja California and Tepic, Mexico, of 1889-1894. Proc. Calif. Acad. Sci.,[4] 23:603-609.
COWLEY, J.
1935. Nomenclature of Odonata: Three generic names of Hagen. Entomologist, 68:283-284.
DAVIS, WILLIAM T.
1933. Dragonflies of the genus *Tetragoneuria*. Bull. Brooklyn Ent. Soc., 28:87-104.
GARDNER, A. E.
1952. The life history of *Lestes dryas* Kirby. Ent. Gaz., 3:4-26.
GARMAN, P.
1917. Zygoptera or damselflies of Illinois. Bull. Illinois Lab. Nat. Hist., 12:411-587.
1927. The Odonata or dragonflies of Connecticut. Guide to the Insects of Connecticut Part V. Conn. Geol. Nat. Hist. Survey Bull., 39:1-331.
GLOYD, L. K.
1938. Notes on some dragonflies (Odonata) from Admiralty Island, Alaska. Ent. News, 49:198-200.
1939. A synopsis of the Odonata of Alaska. Ent. News, 50:11-16.
1941. The identity of three geographically misplaced species of Odonata. Bull. Chicago Acad. Sci., 6:130-132.
1943. *Enallagma vernale,* a new species of Odonata from Michigan, Occ. Pap. Mus. Zool., Univ. Mich., 479:1-8.
GRIEVE, E. D.
1937. Studies on the biology of the damselfly, *Ischnura vesticalis* Say, with notes on certain parasites. Ent. Amer. (n.s.), 17:121-152.
HAGEN, H. A.
1889. Synopsis of the Odonata of North America, No. 1. Psyche, 5:241-250.
KENNEDY, C. H.
1913. Notes on the Odonata, or dragonflies, of Bumping Lake, Washington. Proc. U.S. Nat. Mus., 46(2017): 111-126.
1915. Notes on the life history and ecology of the dragonflies (Odonata) of Washington and Oregon. Proc. U.S. Nat. Mus., 49:259-345.
1917. Notes on the life history and ecology of the dragonflies (Odonata) of Central California and Nevada. Proc. U.S. Nat. Mus., 52:483-635.
1918. New species of Odonata from the Southwestern United States. Canad. Ent., 50:256-261, 297-300.
1919. A new species of *Argia* (Odonata). Canad. Ent., 51:17-18.
1920*a*. The phylogeny of the zygopterous dragonflies as based on the evidence of the penes. Ohio Jour. Sci., 21:19-32.
1920*b*. Forty-two hitherto unrecognized genera and subgenera of zygoptera. Ohio Jour. Sci., 21:83-88.
KLOTS, E. B.
1932. Insects of Porto Rico and the Virgin Islands. Odonata or Dragonflies. Scien. Surv. Porto Rico and the Virgin Islands. N.Y. Acad. Sci., 16:1-107.
KRULL, WENDELL H.
1929. The rearing of dragonflies from eggs. Ann. Ent. Soc. Amer., 22:651-658.
LA RIVERS, IRA
1938. An annotated list of the *Libelluloidea* (Odonata) of Southern Nevada, Pomona Coll. Jour. Ent. Zool., 73-85.
1940*a*. A preliminary synopsis of the dragonflies of Nevada. Pan-Pac. Ent., 16:111-122.
1940*b*. Some dragonfly notes from Northern Nevada. Pomona Coll. Jour. Ent. Zool., 32:61-68.
1941. Additions to the list of Nevada dragonflies. Ent. News, 52:126-130, 155-157.
1946. Some dragonfly observations in alkaline areas in Nevada. Ent. News, 57:209-217.
MARTIN, ROSEMARY D. C.
1939. Life histories of *Agrion aequabile* and *Agrion maculatum* (Agridae: Odonata). Ann. Ent. Soc. Amer., 35:601-619.
MONTGOMERY, B. ELWOOD
1943. *Sympetrum internum,* new name for *Sympetrum decisum* Auct., nec Hagen (Odonata, Libellulidae). Canad. Ent., 57-58.
MUTTKOWSKI, R. A.
1910. Catalogue of Odonata of North America. Bull. Pub. Mus. Milwaukee, 1:1-207.
MUNZ, P. A.
1919. A venational study of the suborder Zygoptera (Odonata) with keys for the identification of genera. Mem. Amer. Ent. Soc., 3:1-78.
NEEDHAM, J. G.
1903. Aquatic insects in New York State. Life histories of Odonata. Suborder Zygoptera. Damselflies. N.Y. State Mus. Bull., 68:218-279.
1904. New dragonfly nymphs in the United States National Museum. Proc. U.S. Nat. Mus., 27:685-720.
1924. Observations of the life of the ponds at the head of Laguna Canyon. Pomona Coll. Jour. Ent. Zool., 16:1-12.
1937. The nymph of *Pseudoleon superbus* Hagen. Pomona Coll. Jour. Ent. Zool., 29:107-109.
1941*a*. Life history studies on *Progomphus* and its nearest allies (Odonata: Aeschnidae). Trans. Amer. Ent. Soc., 67:221-245.
1941*b*. Life history notes on some West Indian Coenagrionine Dragonflies (Odonata). Jour. Agric. Univ. Puerto Rico, 25:1-18.
1943. Life history notes on *Micrathyria* (Odonata). Ann. Ent. Soc. Amer., 35:185-189.
1944. Further studies on Neotropical gomphine dragonflies (Odonata). Trans. Amer. Ent. Soc., 69:171-224.
1948. Studies on the North American species of the genus *Gomphus* (Odonata). Trans. Amer. Ent. Soc., 73:307-339.
1951. Prodrome for a manual of the dragonflies of North America, with extended comments on wing venation systems. Trans. Amer. Ent. Soc., 77:21-62.
NEEDHAM, J. G., and C. BETTEN
1901. Aquatic insects in the Adirondacks. N.Y. State Mus., 47:383-599.
NEEDHAM, J. G., and E. BROUGHTON
1927. The venation of the Libellulinae (Odonata). Trans. Amer. Ent. Soc., 53:157-190.
NEEDHAM, J. G., and T. D. A. COCKERELL
1903. Some hitherto unknown nymphs of Odonata from New Mexico. Psyche, 10:134-139.
NEEDHAM, J. G., and E. FISHER
1936. The nymphs of North American libelluline dragonflies. Trans. Amer. Ent. Soc., 62:107-116.
NEEDHAM, J. G., and H. B. HEYWOOD
1927. Guide to the study of fresh water biology. New York: Amer. Viewpoint Soc., 88 pp.
NEEDHAM, J. G., and M. J. WESTFALL, Jr.
1955. A manual of the dragonflies of North America including the Greater Antilles and the provinces of the Mexican Border. University of California Press, 615 pp.
NEVIN, F. R.
1929. Larval development of *Sympetrum vicinum.* Trans. Amer. Ent. Soc., 55:79-102.
RIS, F.
1930. A revision of the libelluline genus *Perithemis* (Odonata). Univ. Mich. Mus. Zool. Misc. Publ., 21:1-50.
SEEMAN, M. T.
1927. Dragonflies, mayflies and stoneflies of southern California. Jour. Ent. Zool., 19:1-69.

SNODGRASS, R. E.
1954. The dragonfly larva. Smithsonian Misc. Coll. 123:1-38.
TILLYARD, R. J.
1917. The biology of dragonflies (Odonata or Paraneuroptera). Cambridge University Press. xii, 396 pp.
VALLE, K. T.
1942. A small list of Odonata from U.S.A. collected summer 1928 by Prof. U. Saalas and Mrs. Anna-Liisa Saalas. Ann. Ent. Fenn., 8:163-166.
WALKER, E. M.
1912. The North American dragonflies of the genus *Aeshna*. Univ. Toronto Stud., Biol. Ser. II:1-213.
1913. New nymphs of Canadian Odonata. Canad. Ent., 45:161-170.
1914. The known nymphs of the Canadian species of *Lestes*. Canad. Ent., 46:189-200, 349-50.
1914. New and little known nymphs of Canadian Odonata. Canad. Ent., 46:349-56, 369-377.
1916*a*. The nymphs of *Enallagma cyathigerum* and *E. calverti*. Canad. Ent., 48:192-196.
1916*b*. The nymphs of the North American species of *Leucorrhinia*. Canad. Ent., 48:414-422.
1917. The known nymphs of the North American species of *Sympetrum* (Odonata). Canad. Ent., 49:409-418.
1925. The North American dragonflies of the genus *Somatochlora*. Univ. Toronto Stud., Biol. Ser., 26:1-202.
1933. The nymphs of the Canadian species of *Ophiogomphus*. Canad. Ent., 65:217-229.
1934. The nymphs of *Aeschna juncea* L. and *subarctica* Wlk. Canad. Ent., 66:267-274.
1937. A new *Macromia* from British Columbia (Odonata: Corduliidae) Canad. Ent., 69:5-13.
1940*a*. A preliminary list of the Odonata of Saskatchewan. Canad. Ent., 72:26-35.
1940*b*. Odonata from the Patricia portion of the Kenora district of Ontario with description of a new species of *Leucorrhinia*. Canad. Ent., 72:4-15.
1941. The nymph of *Aeschna verticallis* Hagen. Canad. Ent., 73:229.
1944. The nymphs of *Enallagma clausum* Morse and *E. boreale* Selys. Canad. Ent., 76:233-237.
1951. *Sympetrum semicinctum* (Say) and its nearest allies (Odonata). Ent. News, 62:153-163.
1952. The *Lestes disjunctus* and *forcipatus* complex (Odonata: Lestidae) Trans. Amer. Ent. Soc., 78:59-74.
1953. The Odonata of Canada and Alaska Vol. I., Part I, General Part II, The Zygoptera-damselflies. University of Toronto Press, 292 pp.
WESTFALL, M. J.
1953. The nymph of *Miathyria marcella* Selys (Odonata). Fla. Ent., 36:21-25.
WHEDON, A. D.
1927. The structure and transformation of the labium of *Anax junius*. Biol. Bull. Woods Hole, 53:286-300.
WHITEHOUSE, F. C.
1941. British Columbia dragonflies (Odonata), with notes on distribution and habits. Amer. Midl. Nat., 26:488-557.
1948. Catalogue of the Odonata of Canada, Newfoundland and Alaska. Trans. Royal Can. Inst., 27:3-56.
WHITNEY, R. C.
1947. Notes on *Tanypteryx hageni*. Ent. News, 58:103.
WILLIAMS, F. X.
1937. Notes on the biology of *Gynacantha nervosa* Rambur, a crepuscular dragonfly in Guatemala. Pan-Pac. Ent., 13:1-8.
WILLIAMSON, E. B.
1900. Notes on a few Wyoming dragonflies. Ent. News, 11:455-458.
1909. The North American dragonflies (Odonata) of the genus *Macromia*. Proc. U.S. Nat. Mus., 37(1710):369-398.
1933. The status of *Sympetrum assimilatum* (Uhler) and *Sympetrum decisum* (Hagen). (Odonata: Libellulinae). Occ. Pap. Mus. Zool. Univ. Mich., 11(264):1-7.
WILLIAMSON, E. B., and J. H. WILLIAMSON
1930. Five new Mexican dragonflies (Odonata). Occ. Pap. Mus. Zool. Univ. Mich., 216:1-34.
WRIGHT, MIKE, and A. PETERSON
1944. A key to the genera of anisopterous dragonfly nymphs of the United States and Canada. (Odonata, suborder Anisoptera). Ohio Jour. Sci., 44:151-166.

CHAPTER 5

Aquatic Orthoptera

By Ira La Rivers
University of Nevada, Reno

Grasshoppers and crickets are not usually thought of as aquatic animals; however, one group is intimately associated with water—the Tridactylidae or pygmy molecrickets. These peculiar and unfamiliar little insects are fossorial, burrowing in the loose, saturated sand bordering water, and are able to swim by means of specially modified natatory lamellae ("calcaria") or swimming plates borne on the ends of the hind tibiae. However, the ability of these animals to handle themselves competently about water does not require such remarkable adaptations, since most saltatorial Orthoptera are so constructed that locomotion through water comes rather naturally. In contrast to most strictly terrestrial insects, even the desert grasshoppers swim readily and strike out efficiently for shore when they accidentally fall into water.

The grouse or pygmy locusts (Acrydiidae), in particular, exhibit a proficiency about water though lacking such definite structural modifications as the tridactylid natatory lamellae, and may dive beneath the surface of the water when disturbed. Grouse locusts are not treated here because they are primarily terrestrial or certainly no more than semiaquatic.

Little has been published on the biology of tridactylids. Within the past thirty years, only a few authors have made even passing mention of the aquatic associations of our species. Blatchley (1920) mentions their swimming plates or "calcaria," and notes that they burrow in moist sand and can leap about on the surface of the water. Hebard's extensive 1934 treatment of the Orthoptera and Dermaptera of Illinois suggests only that they occur "about lakes and watercourses"; Urquhart (1937) takes some notice of them; and Chopard (1938), repeating Urquhart's observations in a more comprehensive survey, has something to say about their habitat (p. 84), their hibernation (p. 260), and their family associations (p. 506).

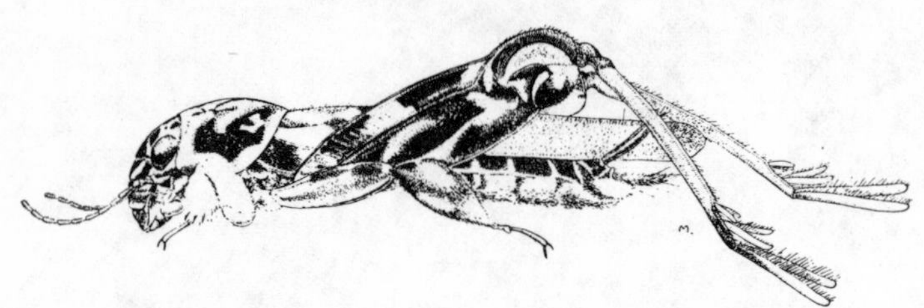

Fig. 5:1. Adult male of the pygmy molecricket, *Tridactylus minutus* Scudder (courtesy Illinois Natural History Survey.)

Family TRIDACTYLIDAE

Genus *Tridactylus* Olivier 1789
(*Ellipes* Scudder 1902)

Key to California Species (from Hebard 1934)

1. Hind tarsus present, 1-segmented; pronotum with a weak transverse furrow; shining brown in color, little or not at all maculate; Santa Ana R., near Ontario, Riverside Co., Calif. .... *apicalis* Say 1825

— Hind tarsus entirely absent; pronotum lacking furrow; blackish-brown in color, usually strikingly maculate with a buffy color; Calaveras, Humboldt, Los Angeles, Monterey, Riverside, San Diego counties .... *minutus* Scudder 1862

REFERENCES

BLATCHLEY, WILLIS LINN
1920. The Orthoptera of Northeastern America. Indianapolis: Nature Publ. Co., 784 pp.
CHOPARD, L.
1938. La Biologie des Orthopteres. Encycl. Entom., 20:1-541, 4 pls.
HEBARD, MORGAN
1934. The Dermaptera and Orthoptera of Illinois. Illinois Nat. Hist. Surv. Bull., 20:125-279.
URQUHART, F. A.
1937. Some notes on the sand cricket (*Tridactylus apicalis* Say). Canad. Field Nat., 51:28.

CHAPTER 6

Plecoptera

By Stanley G. Jewett, Jr.
U. S. Fish and Wildlife Service, Portland, Oregon

INTRODUCTION

This is a relatively small order of aquatic insects with a world fauna of approximately twelve hundred species. Stoneflies require moving water for the development of the nymphs, and hence the adults are usually found near streams. In some northern regions the early life history is passed in cold lakes where the shore area is composed of gravel, but in most areas the immature stages are passed in creeks and rivers. There is a marked seasonal succession in the emergence of stoneflies, particularly in the northern hemisphere; adult stoneflies can be collected every month of the year in California if the proper locality is visited.

Adult members of the genus *Brachyptera* are occasionally destructive to soft fruit crops in the Pacific Northwest, where they are reported to feed on the tender buds of these plants (Newcomer, 1918). The principal economic importance of the majority of species, however, lies in their value as food for fish. Dimick and Mote (1934) rate stoneflies as the second most important order of insects in the diet of Oregon rainbow trout in streams.

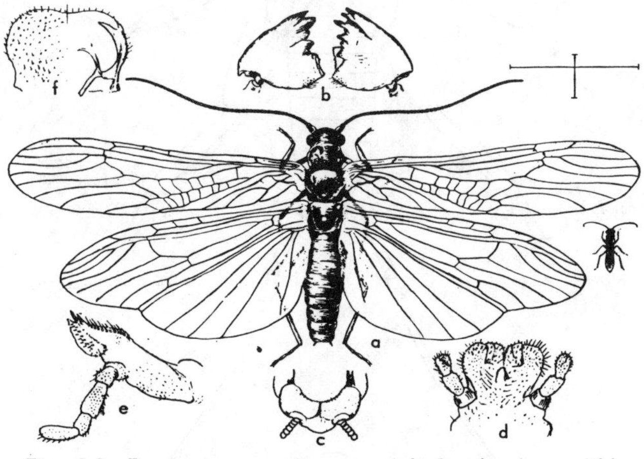

Fig. 6:1. *Brachyptera pacifica. a*, adult female; *b*, mandibles in ventral view; *c*, female terminal abdominal segments; *d*, labium; *e*, maxilla; *f*, labrum, dorsally at left and ventrally at right (Newcomer, 1918).

The order is divided into two suborders: the Filipalpia, in which the nymphs and adults of many genera are primarily vegetarians, and the Setipalpia, in which the nymphs are usually carnivores and the adults are nonfeeding in most genera. The world fauna includes nine families (Ricker, 1951), three of which are primarily Notogaean. Six occur in North America, and all of these are represented in California. Of the approximately four hundred described North American species, ninety-three are known to occur in California. As the fauna of this state becomes better known, the list of species will probably be increased by at least twenty.

The system of classification adopted here is that proposed by Ricker (1950, 1952). The keys include the North American fauna to subgenera and the known California fauna to species. Distributional notes are given for genera which occur only within one general region of the continent. Specific names are included for all monotypic genera and subgenera. Synonyms established since the Needham and Claassen monograph (1925) are indicated parenthetically in the keys.

Adult.—The adult stonefly (fig. 6:1) is readily distinguished from other insects with which it might be confused, such as male Embioptera and certain Neuroptera, by its relatively primitive venation and mouth

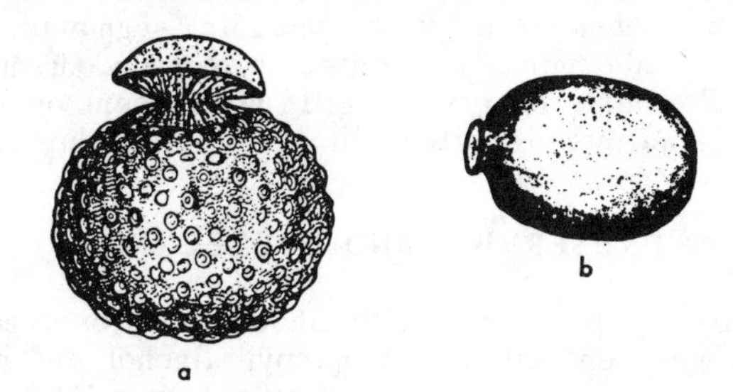

Fig. 6:2. Plecoptera eggs. *a*, *Pteronarcys dorsata*; *b*, *Isogenus frontalis colubrinus*. (Needham and Claassen, 1925).

parts and by the fact that stoneflies always have three tarsal segments and two or three ocelli. In size stoneflies vary in body length from 4-5 mm. (small *Capnia* and *Nemoura*) to 40-50 mm. (large *Pteronarcys*). Most stoneflies are normally winged, but a number are known to be brachypterous; one little-known western species, recorded only from New Mexico, is apparently wingless in the male (*Capnia fibula* Claassen); this is true also of another capniid found in the Midwest, *Allocapnia vivipara* (Claassen). Filipalpia are primarily diurnal; many Setipalpia are crepuscular or nocturnal and are attracted at night to artificial lights.

Egg.—The eggs of stoneflies (fig. 6:2) are most frequently deposited in flight over water, but some species, notably among the Filipalpia, crawl to the water's edge for egg deposition. Egg laying may occur only once or several times. The total number of eggs deposited is known to exceed a thousand in some species studied. Egg shape usually differs in the two suborders: those of the Filipalpia are spherical in general shape and have a sticky coating when moistened, an adaptation which enables them to adhere to the substrate; those of the Setipalpia are usually longer than wide but variously shaped and sculptured, without an adhesive coating but with an anchor plate.

Nymph.—The nymphs of European stoneflies and those found in the midwestern United States (fig. 6:3) are rather well known, but those of a great many North American species have not yet been described; this is particularly true of our western species. Many stonefly nymphs occur in waters with a gravel bottom, but some species occur where the substrate is mostly detritus. Most species of Filipalpia feed primarily on plant tissue, but some are omnivorous; most Setipalpia are carnivorous. Generally, the Filipalpia are found most abundantly in cooler waters, and the Setipalpia, most commonly in warmer waters. Many exceptions occur, and in some tropical regions only Setipalpia are found. For the stoneflies which have been reared from eggs, the number of instars has varied from twenty-two to thirty-three (Claassen, 1931, pp. 7-8). Development of the nymph may occur gradually, but for the majority of species there is apparently a period shortly after hatching when growth virtually ceases (Brinck, 1949, pp. 131-140). The nymphal stage lasts about a year in most species, but two or three years are required in some. Nymphs may or may not have external gills located on the mentum, submentum, neck, thoracic segments, the first few abdominal segments, or extruded from the anus. Remnants of nymphal gills are present on some adults and are important in classifying the order.

PRESERVING AND COLLECTING

Stonefly nymphs and adults should be preserved in 70-75 per cent ethyl or isopropyl alcohol and preferably placed in this liquid as they are collected or very soon afterward. Stoneflies are soft-bodied insects and particularly during warm weather will soon dry and shrivel if left exposed to the air after death. A convenient sized vial for temporary preservation is one of three-dram capacity measuring 65 by 17 mm. Cork or rubber stoppers may be used but the latter give better protection from evaporation. Care should be exercised not to place so many specimens in a vial they they will be improperly preserved. Specimens in vials of somewhat smaller diameter with cotton plugs are best stored permanently in large jars filled with 70 per cent alcohol.

Adult stoneflies may be collected in several ways. From late autumn to early spring, when most Capniids and many Nemourids emerge, concrete bridges over streams are excellent sources of specimens which can be easily collected with a pair of forceps. During the rest of the year the most productive method of collecting is by sweeping vegetation along streams with an insect net. The foliage of conifers is a favorite resting place for stoneflies where such trees border streams. Nocturnal species may be taken at artificial lights, from under large stones bordering streams, and particularly from under loose bark of logs which extend into the water. Cast nymphal skins found on rocks, tree trunks, and so on near the water should be preserved.

Nymphs are easily collected by overturning rocks and stirring gravel in stream beds upstream from the

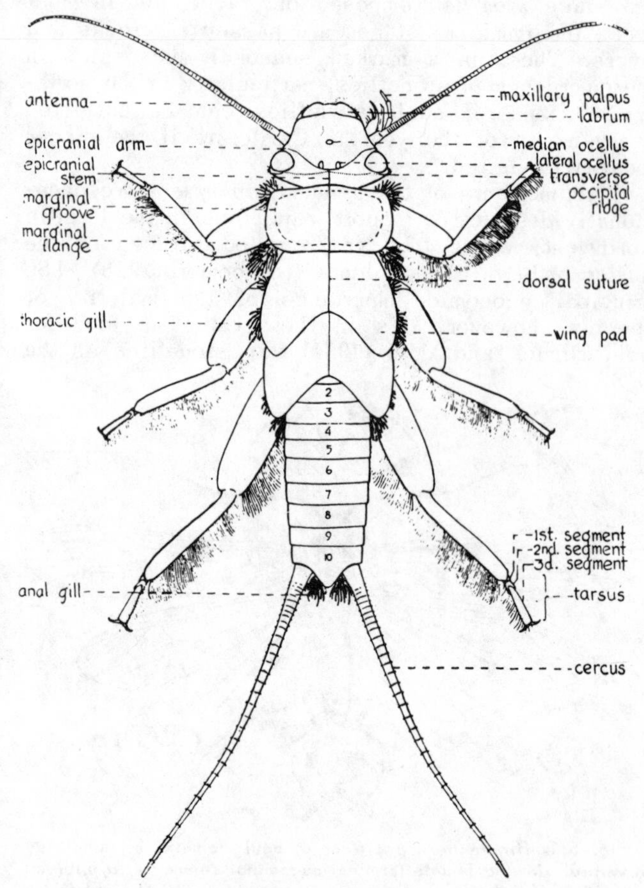

Fig. 6:3. Dorsal view of a constructed stonefly nymph (Frison, 1935).

opening of an aquatic net or fine-meshed sieve. An excellent source of nymphs is among the debris which collects at grills at water diversion structures. Nymphs may be reared by placing them in small wire cages fixed to a floating raft in streams in such a manner that the upper part of each cage is above the water level (Frison, 1935*a*, pp. 305–307).

TAXONOMIC CHARACTERS

As indicated in the following keys, wing venation, gills, number of ocelli, and male genitalia are the principal morphological features used in classifying stoneflies above the species level. The genitalia of both sexes, but particularly those of the male, are used primarily for differentiating species. The dorsal color patterns of the head, thorax, and abdomen are very useful in classifying the Setipalpia and are specifically distinct for many species. The mouth parts and chaetotaxy of nymphs seem to be distinctive for most species, but the association of nymphs with adults is best accomplished through rearing. Where external gills occur, nymphs may sometimes be associated with adults of the same species, since remnants of nymphal gills persist in the adult stage.

Usually clearing of genitalia in adults is unnecessary in studying stoneflies if the specimens have been preserved correctly in alcohol. Exceptions occur with some of the smaller species, with dry or poorly preserved alcoholic specimens, and with genera where the aedeagus is useful or necessary for specific identification. The mouth parts of nymphs should be cleared. Clearing is easily accomplished by placing the mouth parts or the terminal third of the abdomen in 10–15 per cent KOH. Softening of the nonsclerotized parts is usually accomplished within twenty-four hours, and these can be removed; heating the KOH will greatly speed the process. After several baths in distilled water, the structure may be placed in a small shell vial which may be kept in the larger vial containing the rest of the specimen.

Key to the Families and Genera of North American Plecoptera Adults [1]

1. Paraglossae and glossae of about equal length (fig. 6:4*a*) .... suborder FILIPALPIA 2
— Paraglossae much longer than the glossae (fig. 6:4*b*) .... suborder SETIPALPIA 16
2. Abdomen without branched gills on the ventral side; anal area of fore wing without cross veins or with only 1 row of them 3
— Branched gills on the ventral side of abdominal segments 1 and 2; anal area of fore wing with 2 or more full rows of cross veins ........PTERONARCIDAE 15
3. Form cockroachlike (fig. 6:5); ocelli 2; at least 10 costal cross veins in fore wing .... PELTOPERLIDAE *Peltoperla* Needham 1905
— Form typical; ocelli 3; less than 10 costal cross veins in fore wing except in *Isocapnia* which may have 10 or more ....NEMOURIDAE 4
4. Second tarsal segment much shorter than the first (fig. 6:6*a*) 5

[1] Adapted largely from Ricker (1943, 1952).

Fig. 6:4. Ventral view of nymphal labium. *a, Taeniopteryx maura; b, Isoperla patricia* (*a*, Frison, 1935; *b*, Frison, 1942*b*).

— Second tarsal segment at least as long as the first (fig. 6:6*b*) ................TAENIOPTERYGINAE 14
5. Wings lying nearly flat when at rest; second anal vein of fore wing forked (fig. 6:7*b*); cerci 1-segmented .... NEMOURINAE *Nemoura* Pictet 1841
— Wings either rolled around the body at rest, or wings flat when at rest and with second anal vein of the fore wing simple; cerci either 1-segmented or with more than 4 segments 6
6. Wings rolled around the body; intercubital cross veins of the fore wing usually more than 5 (fig. 6:7*a*); second anal vein of the fore wing forked; cerci 1-segmented ....LEUCTRINAE 7
— Wings flat; 1, or rarely 2, intercubital cross veins in the fore wing (fig. 6:7*e,f-h*); 2nd anal vein of the fore wing simple; cerci with at least 4 segments ....CAPNIINAE 9
7. Veins Rs and M in the fore wing with a common origin on R (fig. 6:7*d*); male 9th sternite greatly produced, sharply upturned and fingerlike at the tip; 7th

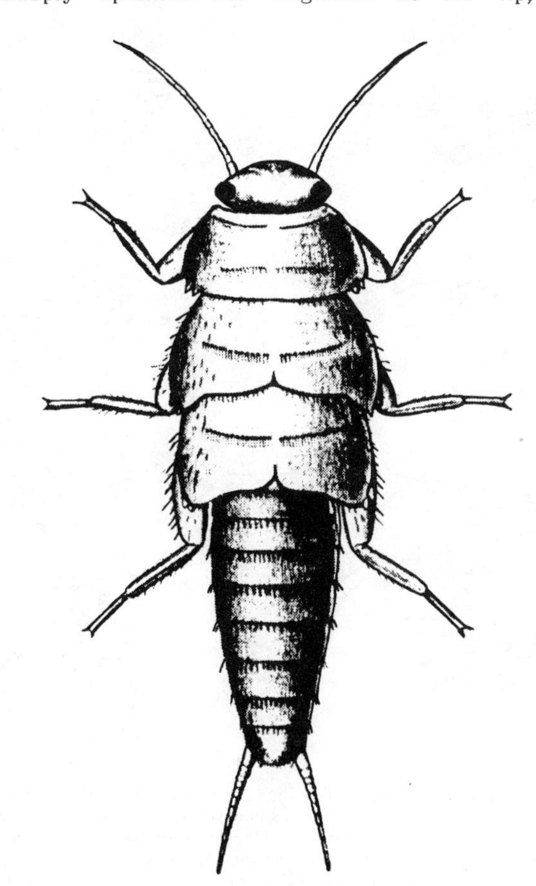

Fig. 6:5. Nymph of *Peltoperla brevis* (Frison, 1942*b*).

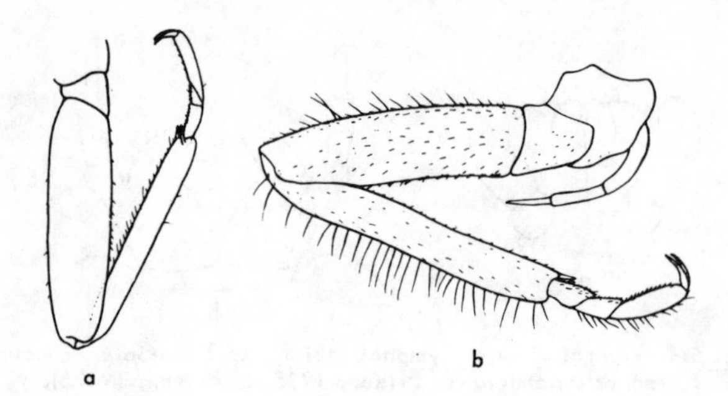

Fig. 6:6. Hind leg of: *a, Leuctra claasseni; b, Taeniopteryx maura* (Frison, 1935).

sternite of female produced over the 8th and with a canal opening beneath it; western North America .... *Perlomyia* Banks 1906

— Vein Rs in the fore wing arising from R beyond the origin of M (fig. 6:7*a,m*); male 9th sternite somewhat produced and usually broadly rounded, never sharply upturned; 7th sternite of female not produced 8

8. Anal area of hind wing with 6 long veins (fig. 6:7*m*); female with 8th or 9th sternites greatly elongated and laterally compressed (fig. 6:25*b*); male with a long recurved process developed from the 10th tergite near the base of each cercus (fig. 6:25*a*) .... *Megaleuctra* Neave 1934

— Anal area of hind wing with only 3 long veins; genitalia not as above .... *Leuctra* Stephens 1835

9. R_1 of fore wing bent abruptly upward at its origin; first anal vein of fore wing bent abruptly caudad at its

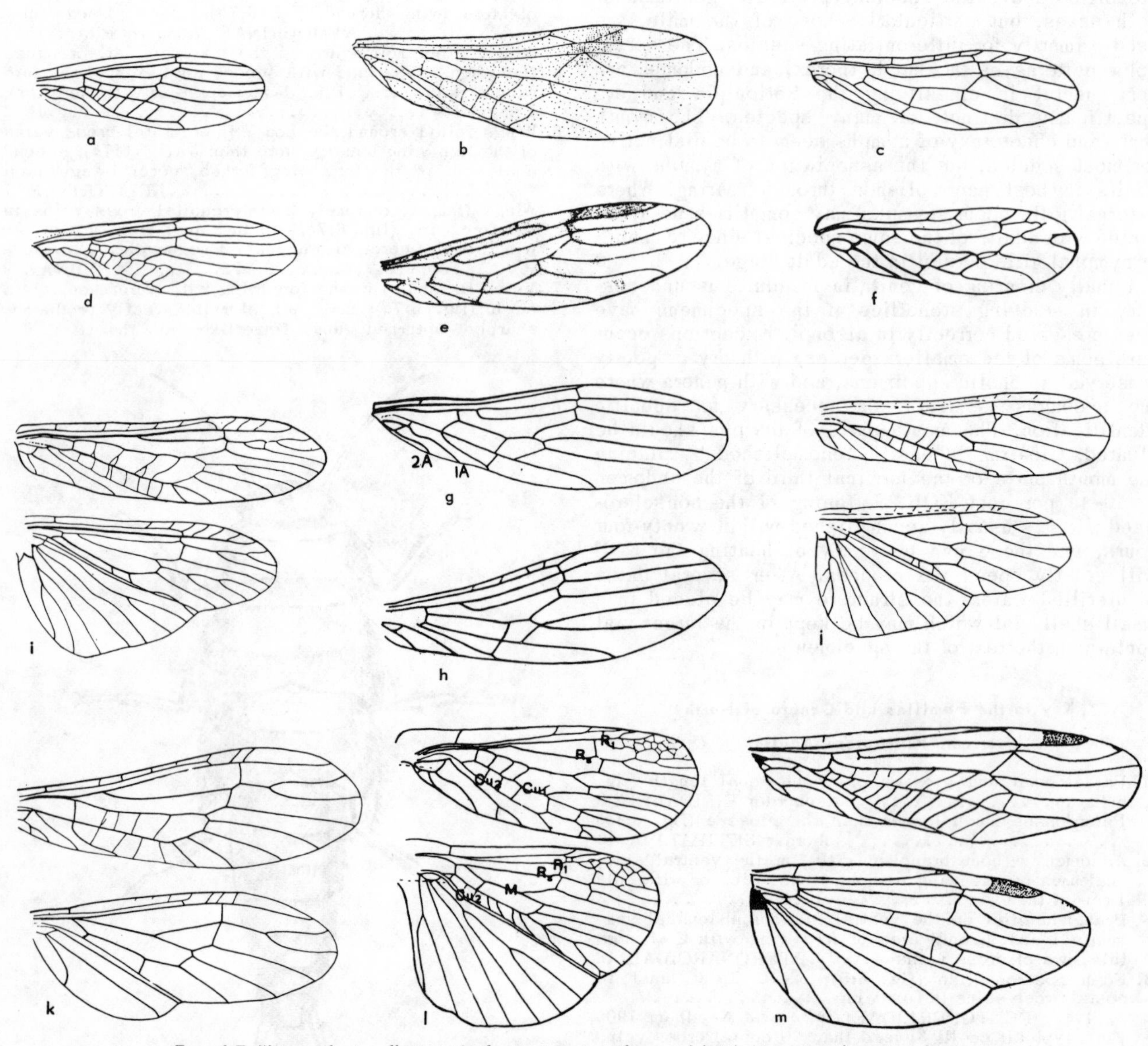

Fig. 6:7. Wings of stoneflies. *a-h*, fore wing; *i-m*, fore and hind wings. *a, Leuctra claasseni; b, Nemoura besametsa; c, Isocapnia* sp.; *d, Perlomyia utahensis; e, Capnia projecta; f, Allocapnia* sp.; *g, Nemocapnia* sp.; *h, Eucapnopsis brevicauda; i, Isogenus frontalis colubrinus; j, Perlesta placida; k, Alloperla borealis; l, Arcynopteryx aurea; m, Megaleuctra spectabilis* (*a*, Frison, 1935; *b,c*, Ricker, 1943; *d,i,j,k,l*, Needham and Claassen, 1925; *e,h*, Frison, 1937; *f,g*, Hanson, 1946; *m*, Neave, 1934).

junction with cu-a and then curved outwardly again (fig. 6:7*e*) .... *Capnia* Pictet 1841
— R_1 of fore wing not bent abruptly upward at its origin; first anal vein of fore wing without an abrupt bend at its junction with cu-a (fig. 6:7*c,f-h*) 10
10. Sc ending much before the cord (fig. 6:7*f*); supra-anal process of male genitalia double; east of Rocky Mountains .... *Allocapnia* Claassen 1928
— Sc ending at or beyond the cord (fig. 6:7*c,g,h*); supra-anal process of male genitalia never double 11
11. First anal vein of fore wing slightly bent beyond cu-a (fig. 6:7*g*); base of male 9th sternite without a lobe; east of Rocky Mountains ...... *Nemocapnia* Banks 1938
— First anal vein of fore wing without a bend beyond cu-a (fig. 6:7*c,h*) 12
12. Base of male 9th sternite without a lobe; eastern North America .... *Paracapnia* Hanson 1946
— Base of male 9th sternite with a lobe 13
13. Cerci with fewer than 11 segments; 1 cross vein in costal area beyond the cord (fig. 6:7*h*); western North America .... *Eucapnopsis* Okamoto 1922
— Cerci with more than 11 segments; 2 or more cross veins in costal area beyond the cord (fig. 6:7*c*); western North America .... *Isocapnia* Banks 1938
14. Coxae without membranous areas on ventral surfaces; male cerci with at least 3 segments; female 9th sternite with a long projection .... *Brachyptera* Newport 1851
— Coxae with small round membranous areas on ventral surfaces; male cerci 1-segmented; female 9th sternite without a long projection..... *Taeniopteryx* Pictet 1841
15. Ventral abdominal gills on segment 3; western North America................. *Pteronarcella* Banks 1900
— No ventral abdominal gills on segment 3 ....
.... *Pteronarcys* Newman 1838
16. Branched gills absent from the thorax; cubitoanal cross vein, if present, usually distant from the anal cell by more than its own length (figs. 6:7*i,k,l;* 6:9*a-c*) 17
— Profusely branched gills at the lower angles of the thorax; cubitoanal cross vein of fore wing usually either in the anal cell or distant from it by no more than its own length (figs. 6:7*j;* 6:11*a-c*) ....
.... PERLIDAE 28
17. Pronotum nearly rectangular, the corners acute or narrowly rounded; fork of 2nd anal vein of the fore wing included in the anal cell so that its 2 branches leave the cells separately (fig. 6:7*i,l*) ....
..............................PERLODIDAE 18
— Pronotum ellipsoidal; 2nd anal vein of the fore wing either not forked or forked beyond the anal cell except in *Kathroperla* which has the fork at the margin of the cell or included in it (figs. 6:7*k;* 6:9*a-c*) ....
.... CHLOROPERLIDAE 23
18. Male 10th tergite completely cleft; supra-anal process of male genitalia broadly and irregularly U-shaped in side view, attached to the anterior dorsal end of the segment, surrounded by a fleshy cowl which is slit dorsally, and usually with 2 lateral stylets inserted on it .... ISOGENINAE 19
— Male 10th tergite entire, or at most slightly notched; supra-anal process of male simple 20
19. Wings with 4 to many cross veins beyond the cord, and these usually arranged in an irregular network (fig. 6:7*l*); 7th abdominal sternite of male without a lobe
.... *Arcynopteryx* Klapalek 1904
— Wings normally with no more than 2 cross veins beyond the cord (fig. 6:7*i*); 7th abdominal sternite of male usually with a lobe behind .... *Isogenus* Newman 1833
20. Male subanal lobes produced inward and backward, meeting along their inner face (fig. 6:40*a*); no lobe on the 8th sternite of male ....
................PERLODINAE *Diura* Billberg 1820
— Male subanal lobe not as above—either little modified or produced into erect or recurved hooks (fig. 6:8*b*); usually a lobe at the hind margin of the male 8th sternite (sometimes obsolescent) (fig. 6:8*a*) ....
.... ISOPERLINAE 21
21. Subanal lobes of male not formed into hooks; male 10th

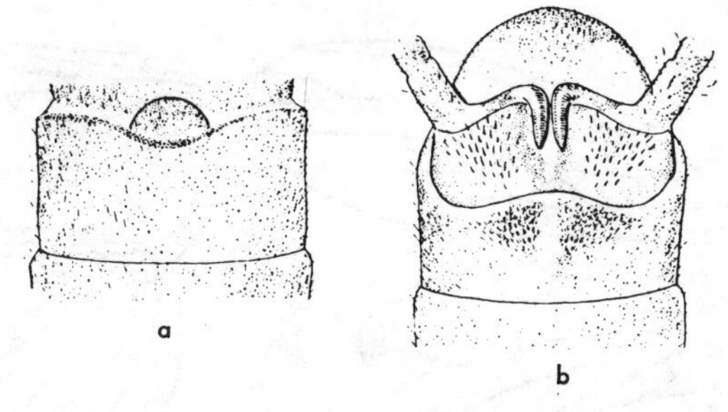

Fig. 6:8. Male terminalia of *Isoperla patricia*. *a*, ventral view; *b*, dorsal (Frison, 1942*b*).

tergite slightly notched dorsally; Oregon and California
.... *Calliperla* Banks 1947
— Subanal lobes of male produced into erect or recurved hooks (fig. 6:8*b*); male 10th tergite entire 22
22. Male with lobe on hind margin of 8th sternite ....
.... *Isoperla* Banks 1906
— Male without lobe on hind margin of 8th sternite but a lobe on the 7th sternite; Oregon ... *Rickera* Jewett 1954
23. Eyes set far forward except in *Utaperla;* body narrow and elongate; anal area of hind wing of normal size; anal veins with 5 to 7 branches reaching the margin of the wing (fig. 6:9*b*) .... PARAPERLINAE 24
— Eyes normally situated; body less elongate; anal area of hind wing often reduced or even absent; anal veins with 4 branches at most (figs. 6:7*k*; 6:9*a,c*) ....
.... CHLOROPERLINAE 26
24. Head longer than wide; western North America .....
.........................*Kathroperla* Banks 1920
— Head about as wide as long 25
25. Ocellar area dark; at least 3 cross veins in costal area beyond the subcosta; posterior margin of the male 7th sternite not raised and hairy; western North America
............................*Paraperla* Banks 1906
— Ocellar area light; fewer than 3 costal cross veins beyond the subcosta; posterior margin of male 7th sternite raised and hairy; western North America
............................*Utaperla* Ricker 1952
26. Anal area of hind wing absent (fig. 6:9*c*) ....
.... *Hastaperla* Ricker 1935
— Anal area of hind wing present (figs. 6:7*k;* 6:9*a*) ... 27
27. Third anal vein of fore wing with basal part fused with 2nd anal vein so that 2nd anal vein appears branched (fig. 6:7*k*) .... *Alloperla* Banks 1906
— Third anal vein of fore wing not present, or if present, not fused with 2nd anal vein so that 2nd anal vein appears branched (fig. 6:9*a*); east of Rocky Mountains
.... *Chloroperla* Newman 1836
28. Males with subanal lobes produced inward and upward, sharply pointed or hooked (fig. 6:10*a*); middle of the hind margin of the 10th tergite not cleft; a definitely raised knob or "hammer" present on the 9th sternite, except in *Perlesta;* dorsal prolongations of the hind margin of the 10th tergite usually present in *Claassenia* where they are developed from the lateral angles rather than from the sides of a median cleft ....
.... ACRONEURINAE 29
— Subanal lobes of male not modified as above; hind margin of the 10th tergite deeply cleft at the middle and with dorsal prolongations developed from the sides of the cleft and adjacent parts of the hind margin (fig. 6:10*c*); no hammer on the 9th sternite ... PERLINAE 34
29. With 2 ocelli 30
— With 3 ocelli 31
30. Several costal cross veins before the end of the subcosta; course of the 2nd anal vein sinuate (fig. 6:11*a*); south of U. S. border; one record from Texas ......
.... *Anacroneuria* Klapalek 1909
— Usually no costal cross veins before the end of the subcosta; course of 2nd anal vein not sinuate (fig.

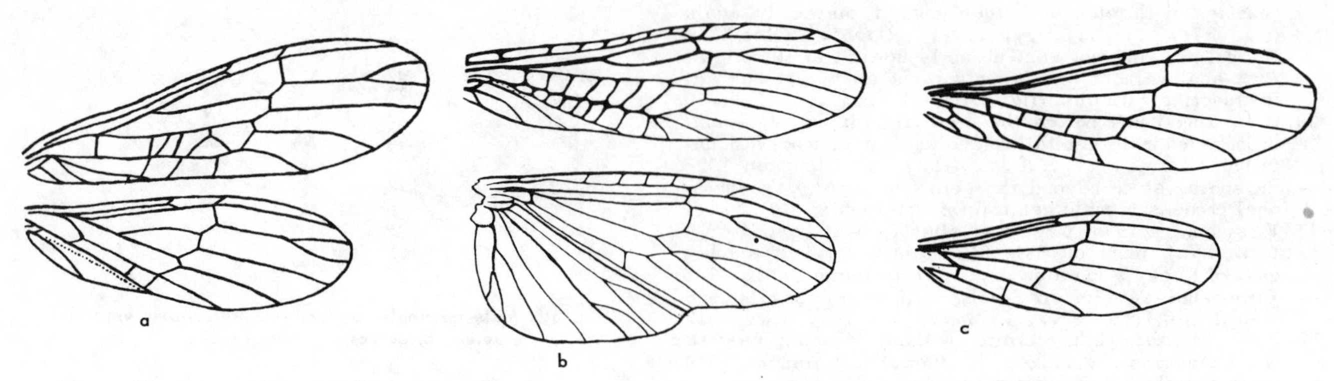

Fig. 6:9. Wings of: *a*, *Chloroperla terna*; *b*, *Kathroperla perdita*; *c*, *Hastaperla brevis* (*a*, Frison, 1942*b*; *b,c*, Needham and Claassen, 1925).

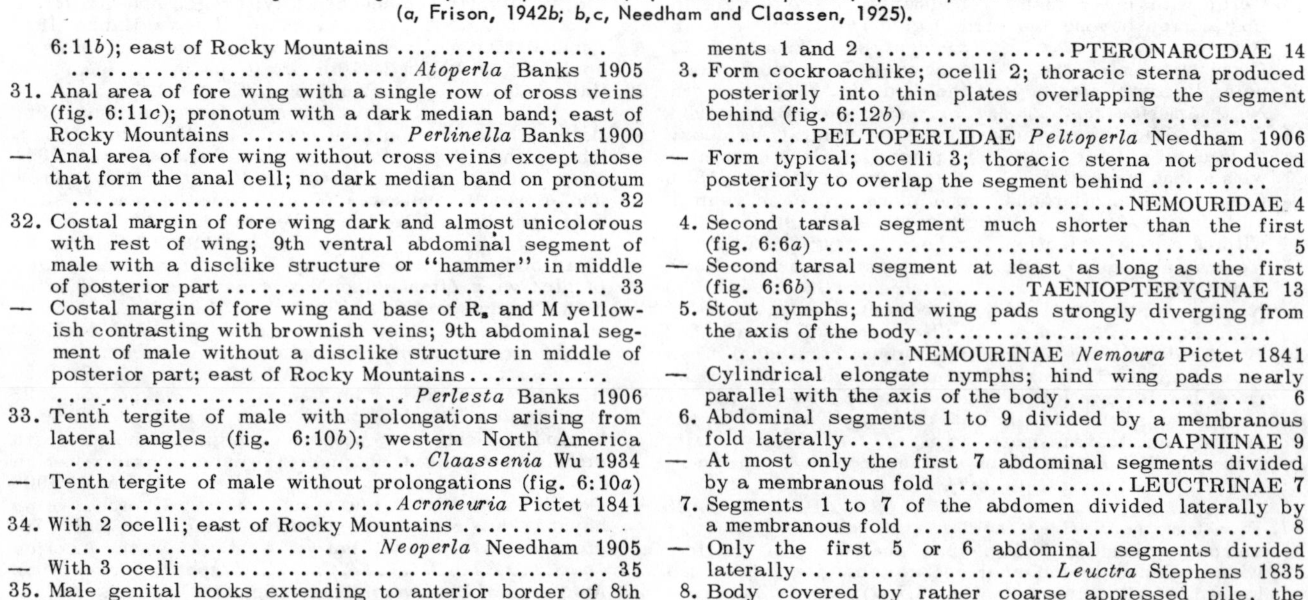

6:11*b*); east of Rocky Mountains ....
.... *Atoperla* Banks 1905

31. Anal area of fore wing with a single row of cross veins (fig. 6:11*c*); pronotum with a dark median band; east of Rocky Mountains .... *Perlinella* Banks 1900
— Anal area of fore wing without cross veins except those that form the anal cell; no dark median band on pronotum 32
32. Costal margin of fore wing dark and almost unicolorous with rest of wing; 9th ventral abdominal segment of male with a disclike structure or "hammer" in middle of posterior part 33
— Costal margin of fore wing and base of R_s and M yellowish contrasting with brownish veins; 9th abdominal segment of male without a disclike structure in middle of posterior part; east of Rocky Mountains ....
.... *Perlesta* Banks 1906
33. Tenth tergite of male with prolongations arising from lateral angles (fig. 6:10*b*); western North America .... *Claassenia* Wu 1934
— Tenth tergite of male without prolongations (fig. 6:10*a*) .... *Acroneuria* Pictet 1841
34. With 2 ocelli; east of Rocky Mountains ....
.... *Neoperla* Needham 1905
— With 3 ocelli 35
35. Male genital hooks extending to anterior border of 8th tergite; east of Rocky Mountains ....
.... *Neophasganophora* Lestage 1922
— Male genital hooks extending at most to posterior margin of 9th tergite (fig. 6:10*c*); east of Rocky Mountains .... *Paragnetina* Klapalek 1907

Key to the Families and Genera of North American Plecoptera Nymphs

1. Paraglossae and glossae of about equal length (fig. 6:4*a*) .... suborder FILIPALPIA 2
— Paraglossae much longer than the glossae (fig. 6:4*b*) .... suborder SETIPALPIA 15
2. Abdomen without branched gills on the ventral side . 3
— Branched gills on the ventral side of abdominal segments 1 and 2 .... PTERONARCIDAE 14
3. Form cockroachlike; ocelli 2; thoracic sterna produced posteriorly into thin plates overlapping the segment behind (fig. 6:12*b*) ....
.......PELTOPERLIDAE *Peltoperla* Needham 1906
— Form typical; ocelli 3; thoracic sterna not produced posteriorly to overlap the segment behind ....
.... NEMOURIDAE 4
4. Second tarsal segment much shorter than the first (fig. 6:6*a*) 5
— Second tarsal segment at least as long as the first (fig. 6:6*b*) .... TAENIOPTERYGINAE 13
5. Stout nymphs; hind wing pads strongly diverging from the axis of the body ....
...............NEMOURINAE *Nemoura* Pictet 1841
— Cylindrical elongate nymphs; hind wing pads nearly parallel with the axis of the body 6
6. Abdominal segments 1 to 9 divided by a membranous fold laterally .... CAPNIINAE 9
— At most only the first 7 abdominal segments divided by a membranous fold .... LEUCTRINAE 7
7. Segments 1 to 7 of the abdomen divided laterally by a membranous fold 8
— Only the first 5 or 6 abdominal segments divided laterally .... *Leuctra* Stephens 1835
8. Body covered by rather coarse appressed pile, the individual hairs of which are about 1/5 as long as a middle abdominal segment; galea exceeding lacinia .... *Megaleuctra* Neave 1933
— Body with extremely fine pile, appearing naked; galea not quite reaching the end of the lacinia ....
.... *Perlomyia* Banks 1906
9. Cerci with mesal and lateral fringes of long silky hairs, several to a segment; western North America .... *Isocapnia* Banks 1938
— Cerci without silky fringes, though sometimes with 1 or 2 rather long bristles on each segment 10
10. Body and appendages densely covered with bristles including several on the subanal lobes; east of Rocky Mountains .... *Paracapnia* Hanson 1946

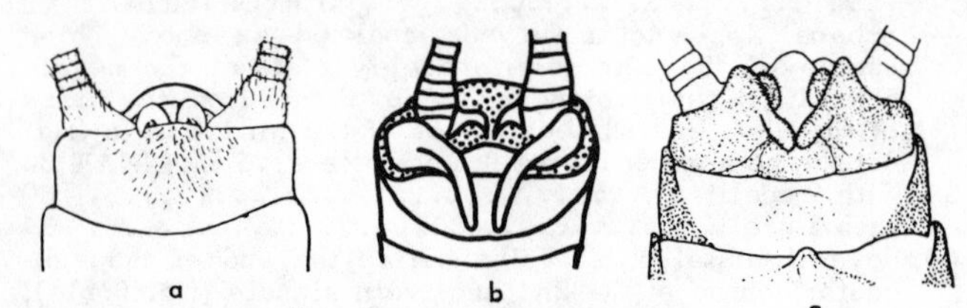

Fig. 6:10. Dorsal view of male terminalia of: *a*, *Acroneuria theodora*; *b*, *Claassenia sabulosa*; *c*, *Paragnetina fumosa* (*a*, Needham and Claassen, 1925; *b*, Ricker, 1938; *c*, Ricker, 1949).

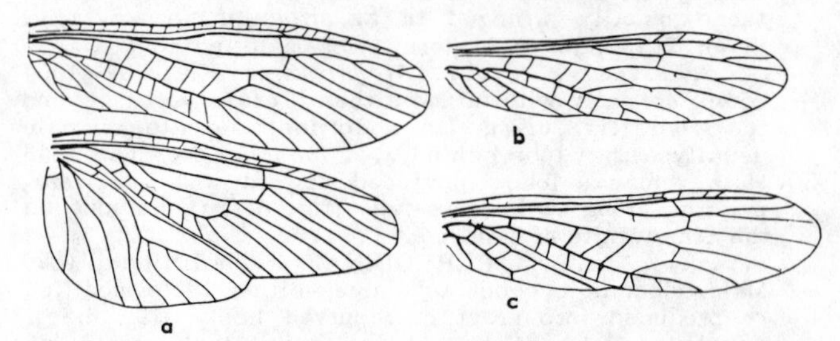

Fig. 6:11. *a*, Wings of *Anacroneuria naomi*; *b*, fore wing of *Atoperla ephyre*; *c*, fore wing of *Perlinella drymo* (*a*, Needham and Broughton, 1927; *b,c*, Frison, 1935).

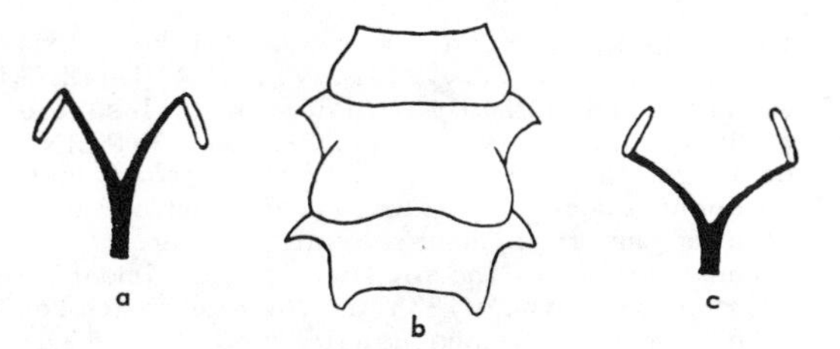

Fig. 6:12. *a,c,* diagram of mesosternal ridge pattern; *b,* sternal plates of *Peltoperla brevis; a, Arcynopteryx* (in part); *c, Isogenus* (in part). (*a,c,* Ricker, 1952; *b,* Ricker, 1943).

— Body with few or no bristles, none on the subanal lobes 11
11. Abdominal segments with a few slender bristles; western North America .... *Eucapnopsis* Okamoto 1922
— Abdominal segments with no bristles 12
12. Inner margin of the hind wing pad with a notch about half way from base to tip; *Nemocapnia* has been recorded only from North Carolina, Virginia, Arkansas, Indiana, and Illinois where *Capnia* does not occur .... *Capnia* Pictet 1841 and *Nemocapnia* Banks 1938
— Inner margin of the hind wing pad notched very close to the tip if at all; east of Rocky Mountains .... *Allocapnia* Claassen 1928
13. Coxae without gills .... *Brachyptera* Newport 1848
— A single gill on each coxa .... *Taeniopteryx* Pictet 1841
14. Ventral abdominal gills on segment 3; western North America .... *Pteronarcella* Banks 1900
— No ventral abdominal gills on segment 3 .... *Pteronarcys* Newman 1838
15. Profusely branched gills at the lower angles of the thorax .... PERLIDAE 24
— Branched gills absent from the thorax 16
16. Body usually pigmented in a distinct pattern; cerci usually at least as long as the abdomen; pads of the hind wing in nearly mature nymphs set at an angle so that their central axis diverges considerably from the axis of the body; gills absent or simple gills present on submentum, thorax, or abdomen ... PERLODIDAE 17
— Body almost concolorous, without a pattern; cerci not

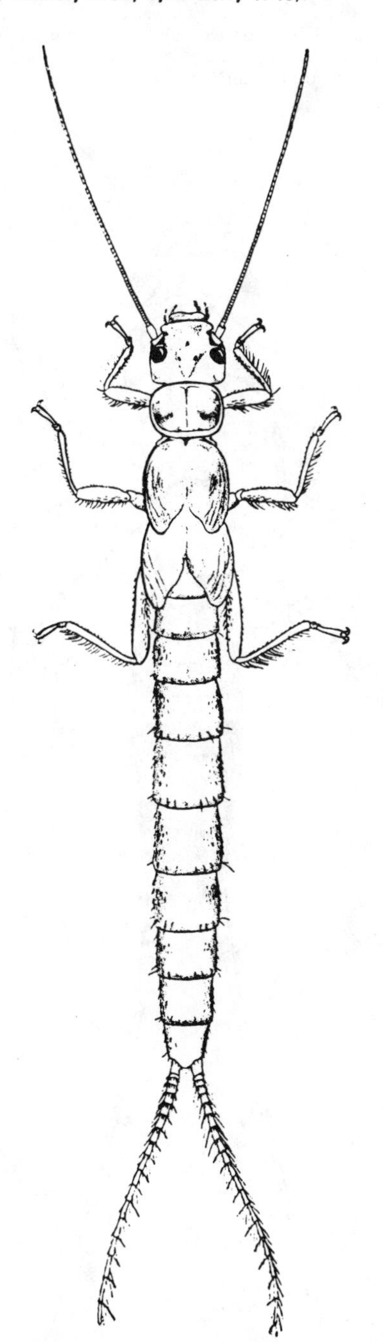

Fig. 6:13. Nymph of *Paraperla frontalis* (Claassen, 1931).

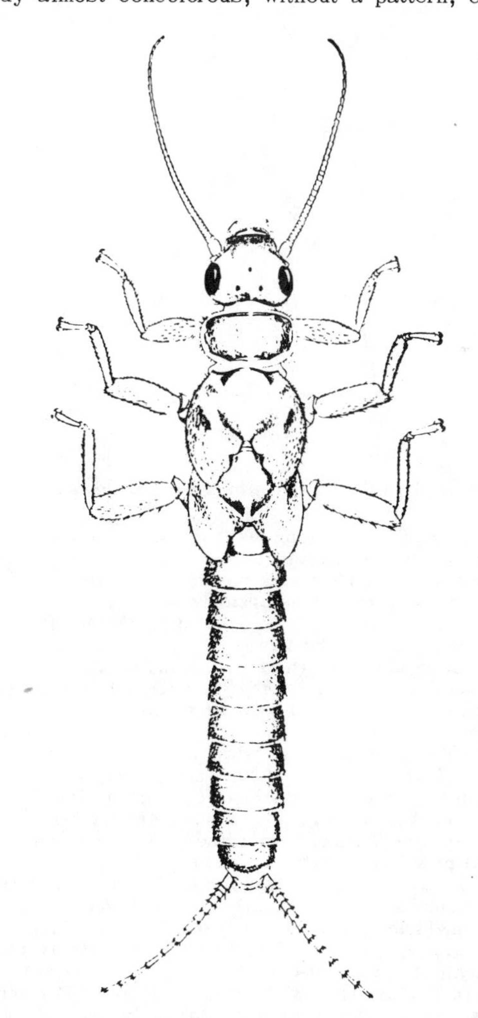

Fig. 6:14. Nymph of *Alloperla borealis* (Claassen, 1931).

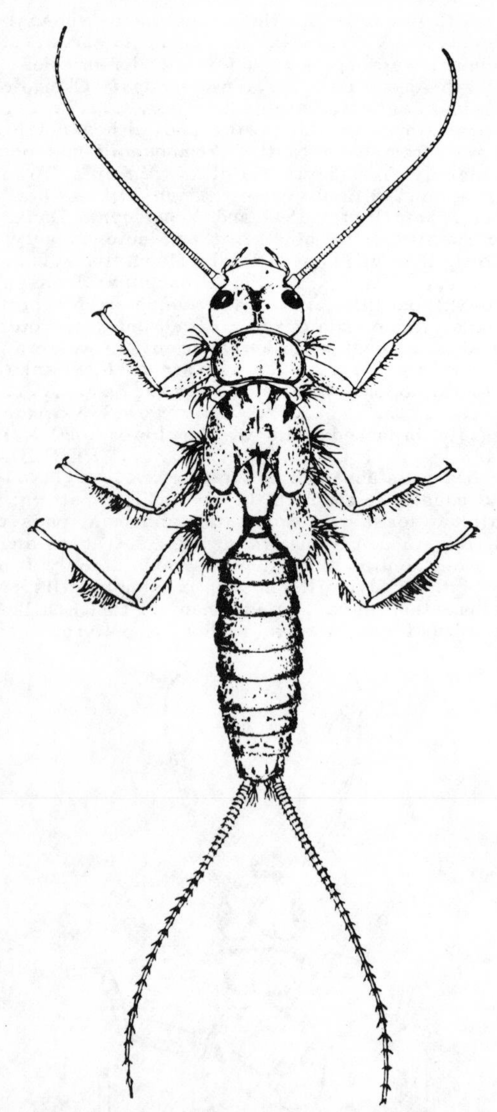

Fig. 6:15. Nymph of *Perlinella drymo.* (Claassen, 1931).

more than 3/4 as long as the abdomen; pads of hind wing nearly parallel to the axis of the body except in mature *Kathroperla* which has the head elongated behind the eyes; external gills entirely lacking ....CHLOROPERLIDAE 21

17. Gills absent from the thorax 18
— Gills present on the thorax ....*Arcynopteryx* Klapalek 1904 (in part)
18. Submental gills at least twice as long as their greatest width 19
— Submental gills about as wide as long, or absent ... 20
19. Arms of the Y-ridge of the mesosternum meet or approach the anterior corners of the furcal pits (fig. 6:12*a*) ....*Arcynopteryx* (in part)
— Arms of the Y-ridges meet the posterior corners of the furcal pits (fig. 6:12*c*) ....*Isogenus* Newman 1833 (in part)
20. Abdominal segments usually with a longitudinal striped pattern. (The nymph of *Calliperla* is unknown.) ....ISOPERLINAE *Isoperla* Banks 1906
— Abdominal segments without a longitudinal striped pattern .....*Isogenus* (in part) and *Diura* Billberg 1820
21. Eyes small and set far forward; body narrow and elongated (fig. 6:13). (The nymph of *Utaperla* Ricker 1952 has not been described and might not kev out here.) ..............................PARAPERLINAE 22
— Eyes large and normally situated; body less elongate (fig. 6:14)..................CHLOROPERLINAE 23
22. Head longer than wide .... *Kathroperla* Banks 1920
— Head about as long as wide..... *Paraperla* Banks 1906
23. Length of mature nymphs, excluding cerci, 5-7 mm., the inner margins of the hind wing pads almost straight ..*Chloroperla* Newman 1836 and *Hastaperla* Ricker 1935
— Length of mature nymphs usually more than 6 mm.; the inner margins of the hind wing pads curved or notched ....*Alloperla* Banks 1906
24. Eyes situated much anterior to the hind margin of the head (fig. 6:15) 25
— Eyes situated close to the hind margin of the head (fig. 6:16) 26
25. Two ocelli; body uniformly colored; east of Rocky

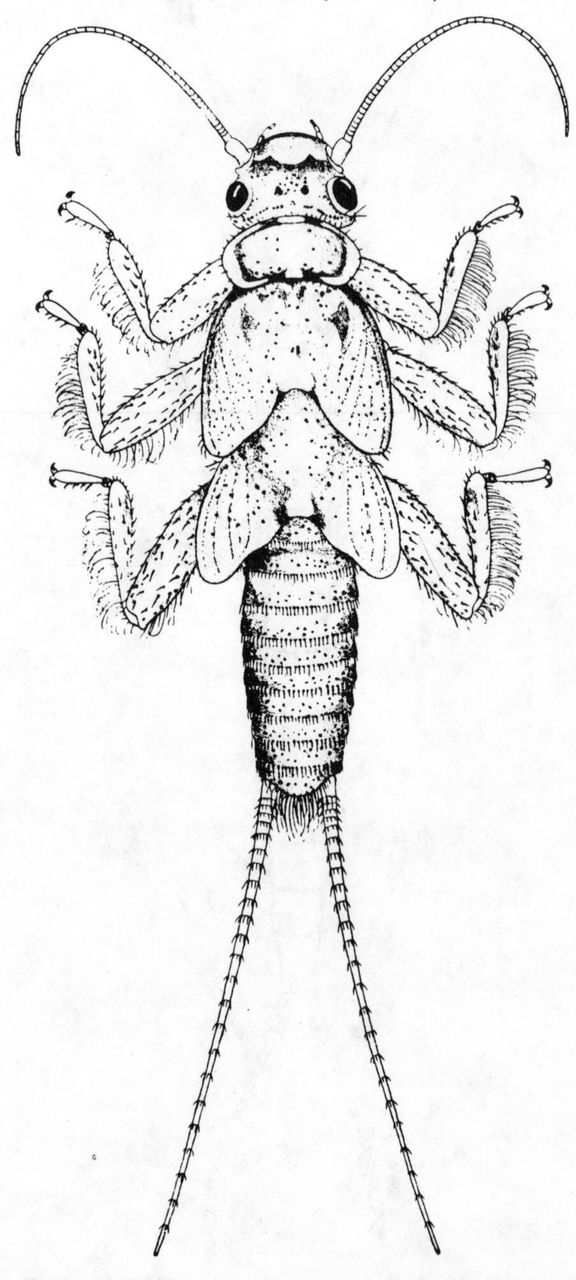

Fig. 6:16. Nymph of *Perlesta placida* (Claassen, 1931).

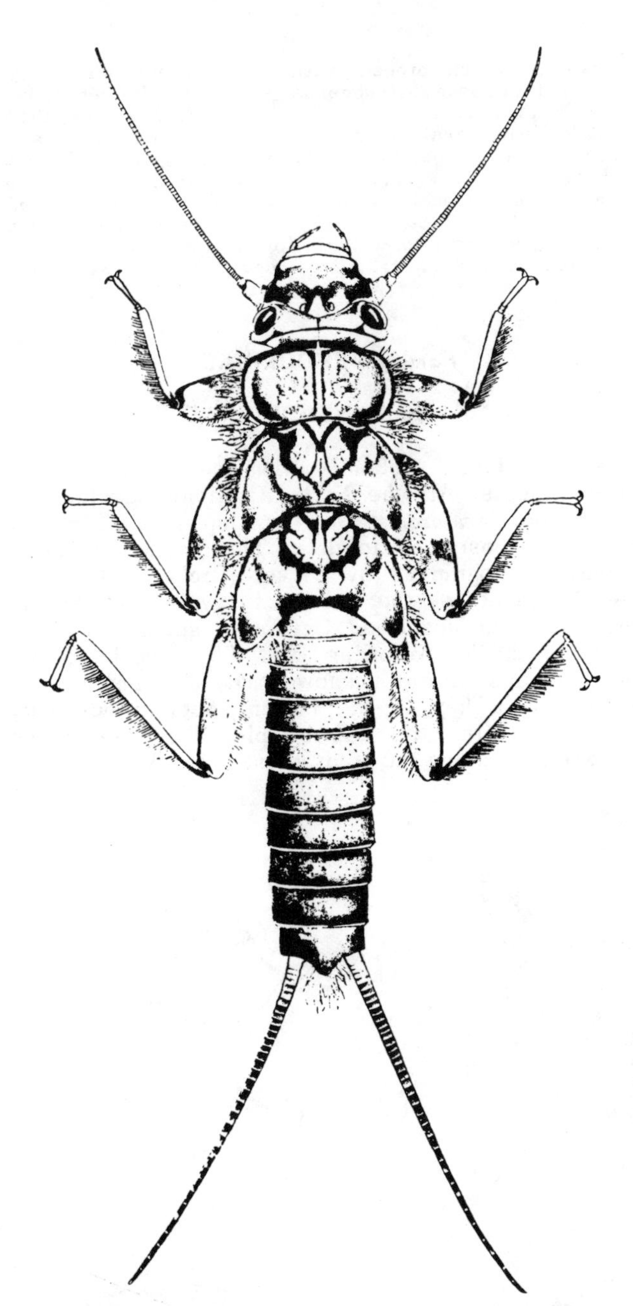

Fig. 6:17. Nymph of *Claassenia sabulosa* (Claassen, 1931).

Mountains .... *Atoperla* Banks 1905
— Three ocelli, the anterior ocellus inconspicuous in immature nymphs; body patterned; east of Rocky Mountains .... *Perlinella* Banks 1900
26. A closely set row of spinules across the back of the head, usually set on a low ridge (fig. 6:16) 27
— Occipital ridge absent; spinules absent from back of head except at the sides 28
27. Three ocelli 29
— Two ocelli set close together; east of Rocky Mountains .... *Neoperla* Needham 1905
28. Two ocelli; south of U. S. border; one record from Texas .... *Anacroneuria* Klapalek 1909
— Three ocelli........... *Acroneuria* Pictet 1841 (in part)
29. Abdomen without frecklelike spots 30
— Abdomen covered with freckles; east of Rocky Mountains .... *Perlesta* Banks 1906
30. Dorsal surface patterned 31
— Dorsal surface almost uniformly brown; eastern North America .... *Acroneuria ruralis* Hagen (1861)
31. Anal gills present (fig. 6:17) 32
— Anal gills absent; east of Rocky Mountains .... *Paragnetina* Klapalek 1907
32. Abdominal segments yellow, broadly bordered with black; east of Rocky Mountains .... *Neophasganophora* Lestage 1922
— Abdominal segments almost wholly brown; western North America .... *Claassenia* Wu 1934

Suborder FILIPALPIA

Family PELTOPERLIDAE

Members of this family are found in the Americas and in eastern Asia and bordering islands. Their roachlike body form is quite unlike that of any other group of stoneflies. The North American fauna consists of eleven described species which were placed in four subgenera of *Peltoperla* by Ricker (Ricker, 1952, pp. 152-157); one of these has been further divided by Jewett (1954*b*). Three subgenera occur in California, the species of which can be separated by the following key. Nymphs are not keyed because those of only a few species are known. The gill characters in the adult key can be used to place specimens of nymphs to subgenus.

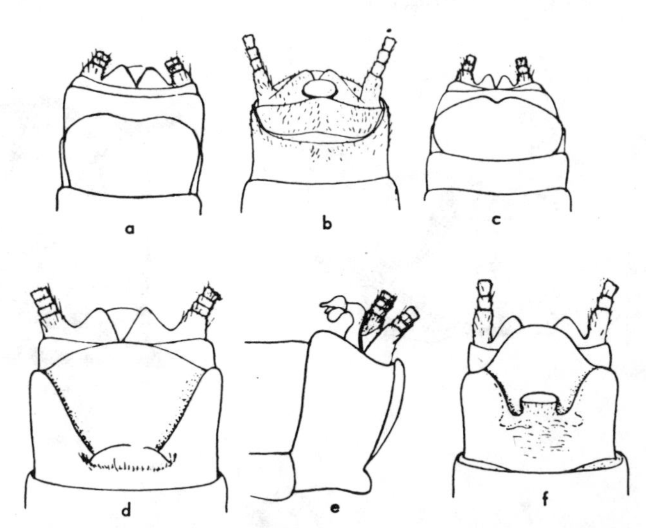

Fig. 6:18. Terminalia of *Peltoperla*. *a*, *cora*, ventral view of female; *b*, *brevis*, dorsal view of male; *c*, *brevis*, ventral view of female; *d*, *cora*, ventral view of male; *e*, *thyra*, lateral view of male; *f*, *brevis*, ventral view of male (Needham and Claassen, 1925).

Key to the North American Subgenera and Adults of California Species of Peltoperla

1. One pair of cervical gills 2
— No cervical gills 4
2. Two pairs of gills on the side of all 3 thoracic segments (subgenus *Yoraperla* Ricker 1952) 3

Fig. 6:19. *Nemoura producta*, basal abdominal segments of male in lateral view (Frison, 1937).

— A pair of gills on each side of the pro- and mesothorax, a single gill on the sides of the metathorax (fig. 6:18*a,d*); California and Nevada......(subgenus *Sierraperla* Jewett 1954).......*cora* Needham and Smith 1916

3. Length to tip of wings of male, 9-11 mm., of female, 10-13 mm. (fig. 6:18*b,c,f*); Pacific Coast and Alta .. ..(synom: *nigrisoma* Banks) *brevis* Banks 1907

Length to tip of wings of male, 11 mm., of female, 14-15 mm.; British Columbia to Oregon.............. ....*mariana* Ricker 1943

4. Two pairs of gills on the side of the mesothorax and metathorax set just above the coxae; one pair near the posterior edge of the metasternum projecting from under the produced sternal plate; one pair projecting from under the subanal lobes; eastern North America ....(subgenus *Peltoperla* Needham 1905)

— One pair of gills each on the meso- and metathorax; no other external gills 5

5. Male supra-anal process with a slender tip; known only from Tennessee...(subgenus *Viehoperla* Ricker 1952) ....*zipha* Frison 1942

— Male supra-anal process strongly sclerotized and complex in shape; Pacific Coast...(subgenus *Soliperla* Ricker 1952) 6

6. Aedeagus of male with a large bilobed sclerotized process (fig. 6:18*e*).....*thyra* Needham and Smith 1916

— Aedeagus with 4 sclerotized spines (fig. 6:53*a*)..... ....*quadrispinula* Jewett 1954

Family NEMOURIDAE

Subfamily NEMOURINAE

This is a large subfamily with representatives on all continents. All the North American species are placed in *Nemoura,* twenty-four species of which have been described to date. This fauna has been recently monographed by Ricker (Ricker, 1952, pp. 10-61) who places the species in twelve subgenera. The adults of described California species can be placed with the following key. The nymphs of so many species remain unknown that it is impractical to present a key to the known species; however, specimens with gills can be placed in the proper subgenera by using the adult keys.

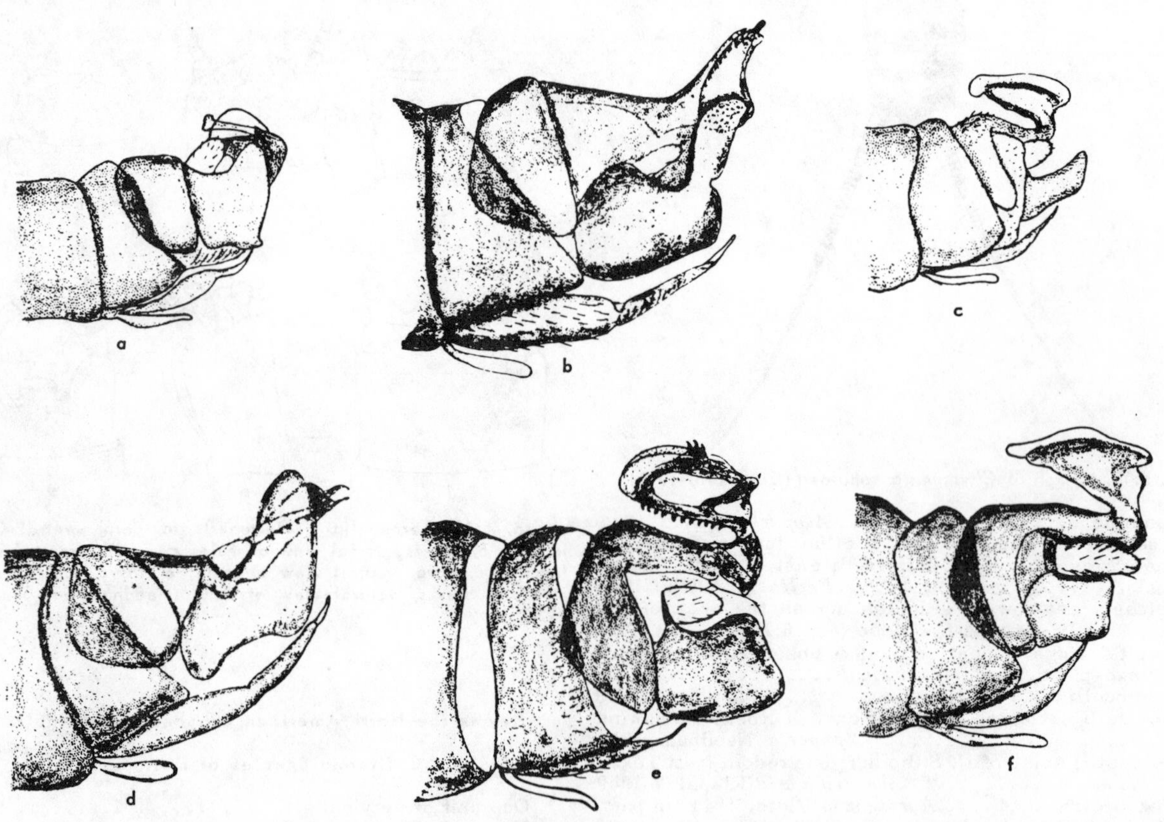

Fig. 6:20. Male terminalia of *Nemoura*, in lateral view. *a, columbiana; b, nevadensis interrupta; c, frigida; d, nevadensis nevadensis; e, oregonensis, f, cinctipes* (Needham and Claassen, 1925).

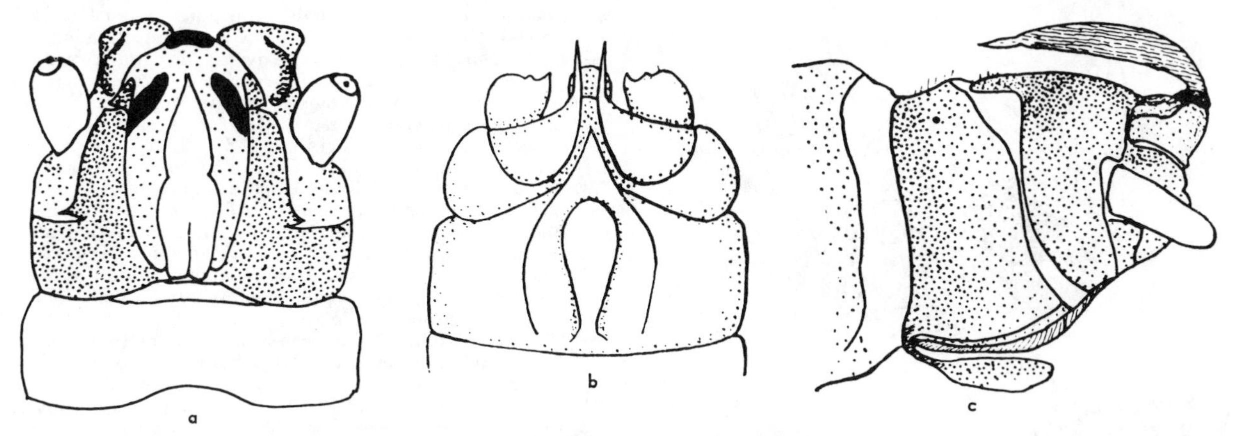

Fig. 6:21. Male terminalia of *Nemoura*. *a*, *haysi*, dorsal view; *b*, *cataractae*, ventral view; *c*, *besametsa*, lateral view (*a,c*, Ricker, 1952; *b*, Neave, 1933).

Key to the Males of the North American Subgenera and California Species of Nemoura[3]

1. Cerci elongated and heavily sclerotized to the tip, on at least the outer surfaces; the tip sharp or with 2 or more processes 2
— Cerci membranous or weakly sclerotized (except for a distinct mesobasal process, often sclerotized, in some species of *Malenka*); the tip blunt, without spines or processes 5
2. Supra-anal process simple, slender, subacute; subanal lobes broad with a long slender spine mesally; with or without gills 3
— Supra-anal process complex, broad; subanal lobes not as above; gills absent 4
3. Two-branched gills inserted at the sides of the mentum (fig. 6:21*b*); Alberta and British Columbia to California ....(subgenus *Visoka* Ricker 1952)...*cataractae* Neave 1933
— No external gills; central California (fig. 6:53*b*) (not assigned to subgenus)..........*spiniloba* Jewett 1954
4. Cerci only a little elongated, membranous on the inner surface; abdominal segments all completely sclerotized and of approximately equal width; northern and northeastern North America ..... *Nemoura, s.s.* Pictet 1841
— Cerci long and slender, completely sclerotized; 9th and 10th abdominal segments sclerotized and much wider than the weakly sclerotized remaining segments; eastern North America and British Columbia to Oregon .... (subgenus *Ostrocerca* Ricker 1952)
5. No lobe on the 9th sternite 6
— Lobe present at the base of the 9th sternite 7
6. Terminal slanting costal cross vein between Sc and C; 10th tergite not elevated or armed; eastern North America .... (subgenus *Paranemoura* Needham and Claassen 1925)
— Terminal slanting cross vein between R_1 and C; 10th tergite elevated, heavily sclerotized, and armed with short spines; known only from Montana ....(subgenus *Lednia* Ricker 1952) *tumana* Ricker 1952
7. Veins A_1 and A_2 of the fore wing united a little before their outer end; subanal lobes long, upcurved, and flattened in side view .... (subgenus *Soyedina* Ricker 1952) 8
— Veins A_1 and A_2 separate at the tip; subanal lobes not as above 10
8. Tergites 2-4 with elevated processes (fig. 6:19); British Columbia to Oregon ....(synonym: *N. tuberculata* Frison 1937) *producta* Claassen 1923
— Tergites 2-4 without processes 9
9. Subanal lobes tapering regularly to their broad tips, no notch on the mesal surface when viewed from the side or from behind though there is a broad excavation near the base (fig. 6:20*d*); California and Nevada .... *nevadensis nevadensis* Claassen 1923
— Mesal margin of subanal lobes notched; their tips rather narrow (fig. 6:20*b*); British Columbia to Oregon .... (synonym: *N. pseudoproducta* Frison 1942) *nevadensis interrupta* Claassen 1923
10. Cervical gills present 11
— Cervical gills absent 21
11. Gills unbranched except in *cinctipes* which has gills commonly 5-branched and wings with contrasting clear and pigmented bands ....(subgenus *Zapada* Ricker 1952) 12
— Gills with 6 or more branches; wings without any clear transverse bands 16
12. Wings uniformly dark; gills 12-15 times as long as wide (fig. 6:20*c*); southern Alaska to California, Colorado, Wyoming .... *frigida* Claassen 1923
— Wings conspicuously banded or mostly clear; gills less than 10 times as long as wide 13
13. Gills branched one to several times (very rarely unbranched); tip of the wings clear (fig. 6:10*f*); Alaska to California and Alberta to Colorado and Nevada; east to Manitoba and South Dakota ..... *cinctipes* Banks 1897
— Gills simple.................................. 14
14. Gills constricted at the base and one or more times (usually twice) beyond the base (fig. 6:20*a*); southern Alaska to California and Utah .... *columbiana* Claassen 1923
— Gills constricted at the base only, if at all (a slight subterminal constriction occasionally in *haysi*) 15
15. Subanal lobes quadrangular, the inner terminal angle rather sharp, without any sclerotized knob on the inner membrane (fig. 6:20*e*); Yukon to California, Utah, Wyoming, and Colorado .....*oregonensis* Claassen 1923
— Subanal lobes with the inner terminal angle obtuse and rounded so that the lobe appears almost triangular, and bearing a sclerotized knob on the distal lateral corner of the inner membrane (fig. 6:21*a*); Alaska to California, Wyoming, and Colorado ....*haysi* Ricker 1952
16. Subanal lobes divided almost to the base into 2 parts, one or both of them spinulose; cerci without basal processes; North America generally except on the plains and Pacific Coast ..(subgenus *Amphinemoura* Ris 1902)
— Subanal lobes variously modified, never spinulose, cleft, if at all, for less than half their length; cerci with a membranous or sclerotized process mesally near their base (this appears to be distinct from the cercus in some species—more like a process of the 10th tergite) .... (subgenus *Malenka* Ricker 1952) 17
17. Mesobasal lobe of the cercus sclerotized and sharply

[3]Adapted from Ricker (1952).

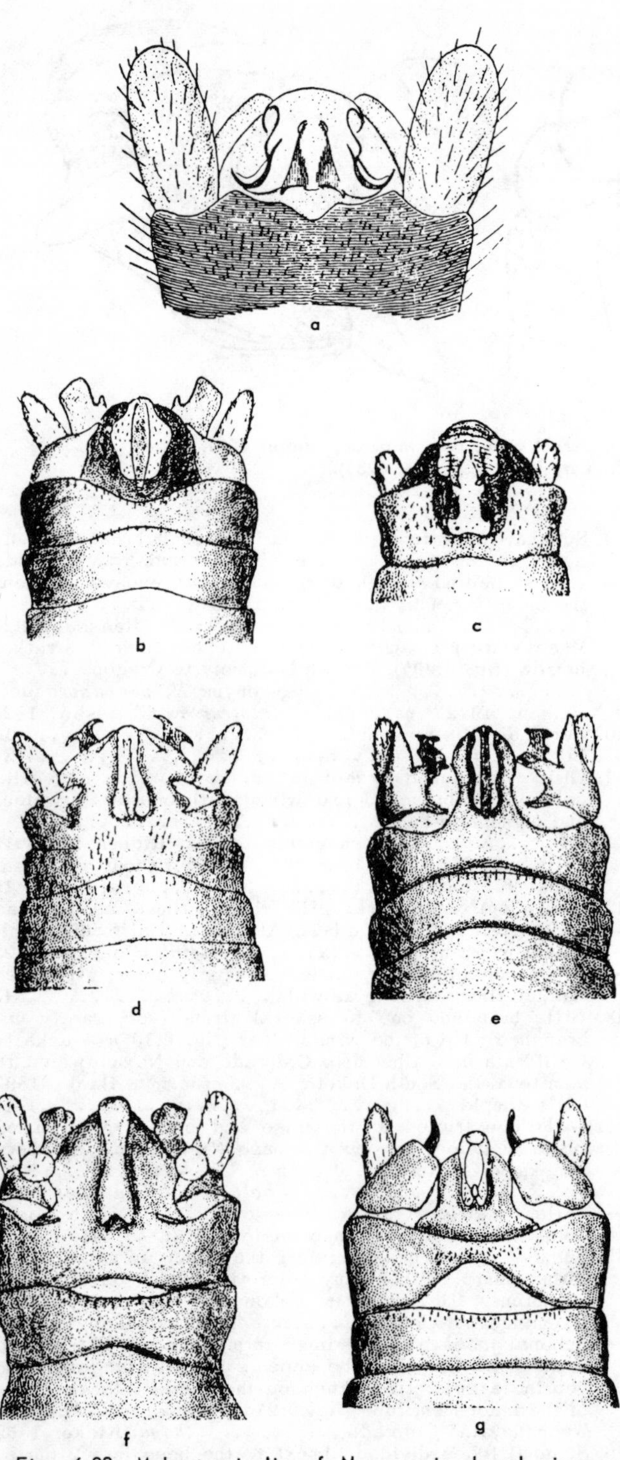

Fig. 6:22. Male terminalia of *Nemoura* in dorsal view. *a*, *decepta*; *b*, *californica*; *c*, *delicatula*; *d*, *biloba*; *e*, *cornuta*; *f*, *coloradensis*; *g*, *depressa* (*a*, Frison, 1942a; *b-g*, Needham and Claassen, 1925).

pointed and appearing to be a part of the 10th tergite 18
— Mesobasal lobe of the cercus rounded and usually membranous 19
18. Mesobasal lobe directed straight back; outer part of the subanal lobes narrow and twice constricted (fig. 6:22*e*); British Columbia to California...*cornuta* Claassen 1923
— Mesobasal lobe directed inward and backward; outer part of the subanal lobes broad and not constricted but with a notch on the mesal margin (fig. 6:22*b*); Alberta and British Columbia to California and New Mexico (synonym: *N. lobata* Frison 1936) .... *californica* Claassen 1923
19. Tip of the subanal lobes simple 20
— Tip of the subanal lobes divided into 2 divergent and acute prongs (fig. 6:22*d*); southwestern California .... *biloba* Claassen 1923
20. Mesobasal lobe of the cercus about twice as broad as the cercus proper and moderately sclerotized (fig. 6:22*g*); Oregon and California..... *depressa* Banks 1898
— Mesobasal lobe of the cercus no broader than the cercus itself and unsclerotized (fig. 6:22*f*); Colorado and Utah to New Mexico ....*coloradensis* Banks 1897
21. Sides of the 10th tergite produced into erect spiny incurved processes; wings with alternate clear and pigmented bands; Alaska to Ontario and Maryland .... (subgenus *Shipsa* Ricker 1952) *rotunda* Claassen 1923
— Sides of the 10th tergite not as above; wings banded or uniformly colored 22
22. Supra-anal process slender, completely recurved along the 10th and 9th tergites (fig. 6:21*c*); southern British Columbia to California and Colorado .... (subgenus *Prostoia* Ricker 1952) *besametsa* Ricker 1952
— Supra-anal process rather short, thick, complex in structure, and only slightly bent forward .... (subgenus *Podmosta* Ricker 1952) 23
23. Supra-anal process with sharp, curved, laterally directed horns near the tip; 10th abdominal tergite only slightly excavated behind (fig. 6:22*a*); British Columbia to Wyoming ....*decepta* Frison 1942
— Supra-anal process without such horns; 10th abdominal tergite deeply depressed and excavated behind (fig. 6:22*c*); central British Columbia to California and Colorado .... *delicatula* Claassen 1923

Key to the Females of North American Subgenera and California Species of Nemoura[4]

1. Gills present under the neck or head 2
— Gills absent 13
2. One branched gill present on each side of the mentum; wings not banded; 7th sternite moderately produced (fig. 6:23*a*); Alberta and British Columbia to California .... (subgenus *Visoka* Ricker 1952) *cataractae* Neave 1933
— Two gills present on each side of the neck, branched or simple; wings usually banded, spotted, or completely infuscated 3
3. Gills simple filaments (except in *cinctipes*); 7th sternite produced completely over the 8th which is very weak .... (subgenus *Zapada* Ricker 1952) 4
— Gills branched; 7th sternite moderately or little produced, not covering all of the 8th; 8th sternite bearing a distinct terminal or subterminal notch 8
4. Wings uniformly dark; gills 12-15 times as long as broad; southern Alaska to California, Colorado, and Wyoming .... *frigida* Claassen 1923
— Wings conspicuously banded or mostly clear; gills less than 10 times as long as wide 5
5. Gills branched 1 to several times (very rarely unbranched); tip of the wing clear; Alaska to California and Alberta to Colorado and Nevada; east to Manitoba and South Dakota .... *cinctipes* Banks 1897
— Gills simple 6
6. Gills constricted at the base and one or more times (usually twice) beyond the base; southern Alaska to California and Utah ....*columbiana* Claassen 1923

[4]Adapted from Ricker (1952).

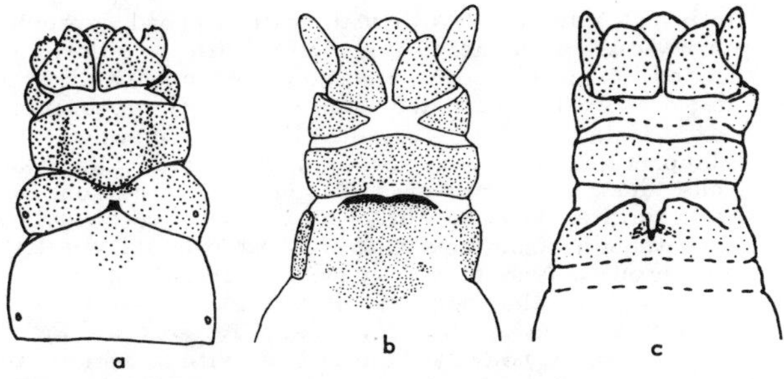

Fig. 6:23. Female terminalia of *Nemoura* in ventral view. *a, cataractae; b, besametsa; c, cornuta* (Ricker, 1943).

— Gills constricted only at the base if at all (a slight subterminal constriction occasionally in *haysi*) 7

7. Produced part of the 7th sternite usually light-colored and narrowly rounded, almost semicircular; Yukon to California, Utah, Wyoming, and Colorado .... *oregonensis* Claassen 1923

— Produced part of the 7th sternite dark colored, at least along the broadly rounded or nearly straight hind margin; Alaska to California, Utah, Wyoming, and Colorado .... *haysi* Ricker 1952

8. Notch of the 8th sternite in a sclerotized band set before the hind margin of the segment (except in the eastern *wui*); North America generally except the plains and Pacific Coast .. (subgenus *Amphinemura* Ris 1902)

— Notch of the 8th sternite terminal .... (subgenus *Malenka* Ricker 1952) 9

9. Seventh sternite produced and rounded but lacking a nipple; 8th sternite greatly swollen, considerably produced, and often very irregular in shape at either side of the notch; the anterior third of the notch bordered by a square patch of heavy sclerotization (fig. 6:23*c*); British Columbia to California .... *cornuta* Claassen 1923

— Produced part of the 7th sternite bearing a distinct nipple, the base of which is often anterior to the hind margin of the sternite (best seen in side view; sometimes obscure in *depressa*); 8th sternite not as above 10

10. Median notch not completely bisecting 8th sternite 11

— Median notch completely bisecting 8th sternite; southwestern California .... *biloba* Claassen 1923

11. Notch of the 8th sternite extending about halfway through the sternite, nowhere margined by extra sclerotization 12

— Notch of the 8th sternite extending almost to the anterior margin near which it is narrowly margined by heavy sclerotization (fig. 6:24*b*); Utah to New Mexico .... *coloradensis* Banks 1897

12. Nipple of the 7th sternite low, inconspicuous, often scarcely recognizable; its base confluent with the hind margin of the sternite (fig. 6:24*a*); Oregon and California .... *depressa* Banks 1898

— Nipple of the 7th sternite erect, easily distinguished in side view at least, its base set somewhat anterior to the hind margin of the sternite (fig. 6:24*c*); Alberta and British Columbia to California and New Mexico ..... (synonym: *lobata* Frison 1936) *californica* Claassen 1923

13. Veins A_1 and A_2 united near the margin of the wing; 7th sternite produced over the full length of the 8th ....(subgenus *Soyedina* Ricker 1952) 14

— Veins A_1 and A_2 not united (sometimes united in *obscura*, a small species occurring in Washington and Oregon in which the 7th sternite does not cover the 8th) 15

14. Subgenital plate of the 7th sternite more broadly rounded; California and Nevada...*nevadensis* Claassen 1923

— Subgenital plate less broadly rounded; British Columbia to Oregon ....(synonym: *N. tuberculata* Frison 1937) *producta* Claassen 1923

15. Wings mostly dark with a clear or relatively clear band across the middle of the outer field 16

— Wings entirely clear or with some or all veins margined with brown or with a marginal or central dark spot at the level of the cord 19

16. Terminal costal cross vein running between Sc and C, proximad of the cord; eastern North America .... (subgenus *Paranemoura* Needham and Claassen 1925)

— Terminal costal cross vein running between R_1 and C beyond the cord 17

17. Seventh sternite sclerotized and produced over the full length of the 8th, its hind margin straight to broadly rounded; northern and northeastern North America .... *Nemoura, s.s.* Pictet 1841 (in part)

— Seventh sternite only slightly or not at all produced over the 8th (the normally rounded margin of the subgenital plate of the 8th sternite in *Shipsa* may be mistaken for the 7th sternite) 18

18. Eighth sternite with a subgenital plate terminating anterior to the well-developed hind margin and bearing contrasting dark and light bands; Alaska east to Ontario and Maryland.................................... ..(subgenus *Shipsa* Ricker 1952) *rotunda* Claassen 1923

— Eighth sternite with a terminal subgenital plate not separated from the hind margin of the sternite and uniformly colored (fig. 6:23*b*); southern British Columbia to California and Colorado .... (subgenus *Prostoia* Ricker 1952) *besametsa* Ricker 1952

19. Eighth sternite very weak and unsclerotized medially; 7th sternite produced over the full length of the 8th and not laterally excavated; northern and northeastern North America ..... *Nemoura, s.s.* Pictet 1841 (in part)

— Eighth sternite sclerotized medially at least near the posterior margin; 7th sternite only slightly or not at all produced (except in *Ostrocerca prolongata*, found in northeastern North America, where it is excavated laterally) 20

20. Eighth sternite produced and rounded medially, not notched; 2 squarish patches of sclerotization on the hind margin, separated from the median produced part; known only from Montana ....(subgenus *Lednia* Ricker 1952) *tumana* Ricker 1952

— Eighth sternite produced and notched or not produced; its lateral sclerotization, if present, continuous with that of the median area 21

21. Eighth sternite with a narrow median sclerotized band contrasting with the unsclerotized field at either side ....(subgenus *Podmosta* Ricker 1952) 22

— Eighth sternite without any median stripe, usually uniformly sclerotized; eastern North America and British Columbia to Oregon .... (subgenus *Ostrocerca* Ricker 1952)

22. Median sclerotized stripe of the 8th sternite 3-4 times as long as its greatest breadth, of nearly uniform width throughout; central British Columbia to California and Colorado ....*delicatula* Claassen 1923

— Median sclerotized stripe of the 8th sternite not more than twice as long as its greatest breadth, frequently rather obscure; British Columbia to Wyoming ....*decepta* Frison 1942

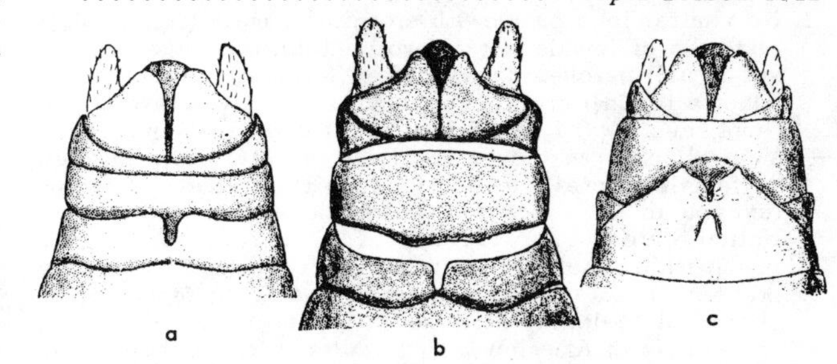

Fig. 6:24. Female terminalia of *Nemoura* in ventral view. *a, depressa; b, coloradensis; c, californica* (Needham and Claassen, 1925).

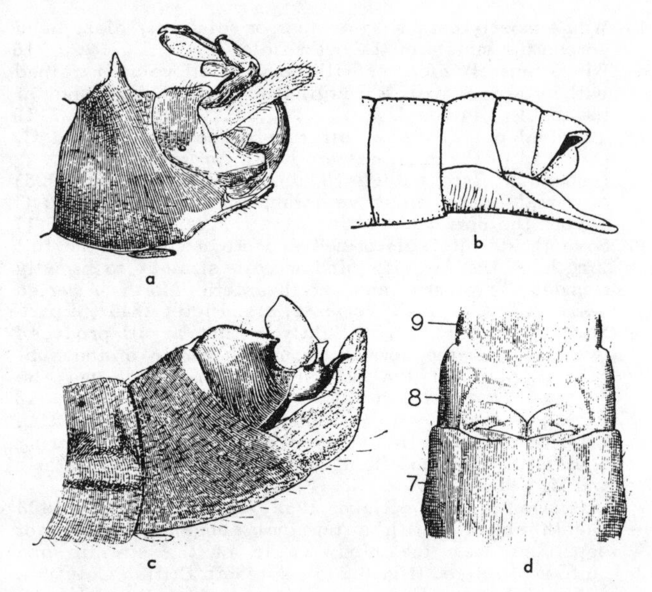

Fig. 6:25. *a, Megaleuctra kincaidi; b, M. spectabilis; c,d, Pelomyia collaris; a,c,* lateral view of male terminalia; *b,* lateral view and *d,* ventral view of female terminalia (*a,* Frison 1942a; *b,* Neave, 1934; *c,d,* Frison, 1936).

Subfamily LEUCTRINAE

Members of this subfamily are distributed throughout the Holarctic region, South Africa, and Tierra del Fuego. The North American fauna includes about twenty-eight species in three genera. The genus *Megaleuctra* Neave 1934 (figs. 6:7*m;* 6:25*a,b*) has been rarely collected; no California specimens have been recorded, but. two species occur in Oregon. *Perlomyia* Banks 1926 is confined to western North America, and *P. collaris* Banks 1906 (synonyms: *P. solitaria* and *P. sobrina* Frison 1936) (fig. 6:25*c,d*) is known to occur in California. Adults of the preceding two genera may be placed by employing the key on pages 157 to 160. The genus *Leuctra* contains five recognized subgenera, three of which occur in California. The California species are keyed below.

Key to Adults of North American Subgenera and California Species of Leuctra[5]

1. No ventral lobe on the 9th sternite of male (fig. 6:26*b*); abdomen of female not sclerotized dorsally and only in very small patches on the sides (fig. 6:26*d*); Alaska and Alberta to California. . .(synonym: *glabra* Claassen 1923) (subgenus *Despaxia* Ricker 1943) *augusta* Banks 1907

— Ventral lobe present on the male 9th sternite; abdomen of female normally sclerotized on the sides and below (except in *L. claasseni* which occurs in midwestern United States) 2

2. In the male, backward-pointing processes nearly always present on one or more of tergites 6 to 8; female subgenital plate deeply cleft medially; cerci always simple; eastern North America .... *Leuctra, s.s.* Stephens 1835

— In the male, backward-pointing processes absent from the tergites; female subgenital plate not simply cleft except usually in *Paraleuctra* 3

3. Ninth and 10th tergites of male with a broad depressed membranous area dorsally; cerci with a sclerotized rounded hump near the apex on the inner and upper side; dorsum of female abdomen completely membranous; West Virginia; Ohio; Indiana; Illinois; Missouri, and Oklahoma .... (subgenus *Zealeuctra* Ricker 1952) *claasseni* Frison 1929

— Ninth tergite of male fully sclerotized; cerci not as above though sometimes with 2 or more heavily sclerotized prongs; female with a median dorsal sclerotized stripe on the abdomen on either side of which is a membranous area 4

4. Cerci of male heavily sclerotized with terminal and lateral pointed projections; body color usually black; subgenital plate of female with lateral projections and a median notch ....(subgenus *Paraleuctra* Hanson 1941) 5

— Cerci of male large, membranous or weakly sclerotized, without sharp angles (fig. 6:26*a*); subgenital plate of female with a median projection below and anterior to the 2 lateral projections (fig. 6:27*c*); body color brown with a whitish stripe along the costal space of the fore wing; British Columbia to California .... (subgenus *Moselia* Ricker 1943) *infuscata* Claassen 1923

5. Titillator of the male with a large membranous bulb at the tip; the longer prong of the bifurcate cercus is

[5]Adapted from Ricker (1943, 1952).

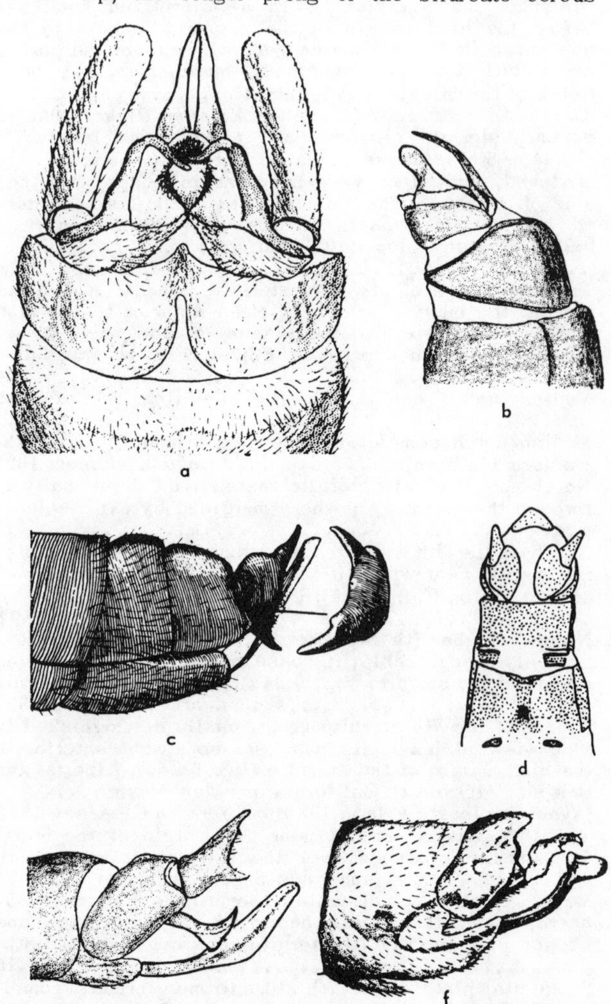

Fig. 6:26. Terminalia of *Leuctra. a, infuscata,* dorsal view of male; *b, augusta,* lateral view of male; *c, forcipata,* lateral view of male; *d, augusta,* ventral view of female; *e, sara,* lateral view of male; *f, occidentalis,* lateral view of male, (*a,b,e,f,* Needham and Claassen, 1925; *c,* Frison, 1937; *d,* Ricker, 1943)

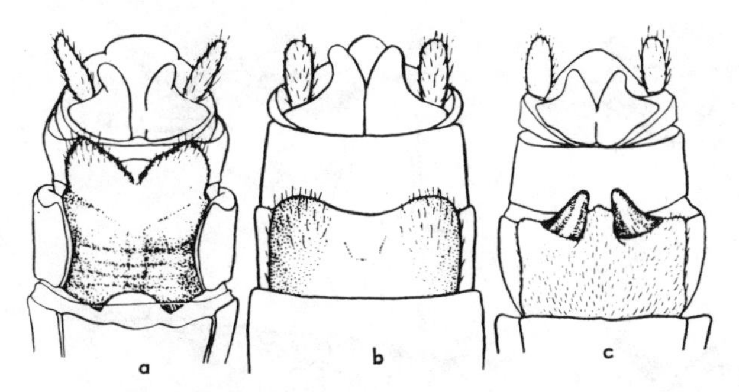

Fig. 6:27. Female terminalia of *Leuctra*, in ventral view. *a, sara; b, occidentalis; c, infuscata* (Needham and Claassen, 1925).

hooked (fig. 6:26*f*); female subgenital plate extending little beyond the 8th sternite (fig. 6:27*b*); southern British Columbia to California and Utah..........
......................(synonyms: *bradleyi* Claassen 1923; *projecta* Frison 1942)...*occidentalis* Banks 1907

— Titillator of male without bulbous tip; female subgenital plate extending about its own length beyond the 8th sternite 6

6. The upper prong of the male cercus longer than the lower and with a small tooth on the inner margin (figs. 6:26*e*; 6:27*a*); southern British Columbia to California and Colorado .... *sara* Claassen 1937

— The two prongs of the male cercus widely spaced and of about equal length (fig. 6:26*c*); Washington and Montana to California .... *forcipata* Frison 1937

Key to the Known Nymphs of California Species of Leuctra[6]

1. Body nearly naked (legs moderately hairy); femora of the hind legs unusually large ....
.... *L. (Despaxia) augusta* Banks 1907

— Body at least moderately hairy; hind femora normal 2

2. Body moderately hairy; abdominal tergites with a band of unusually long hairs on either side, about a third of the way from the anterior margin ....
............*L. (Paraleuctra) occidentalis* Banks 1907

— Body very hairy; longest dorsal abdominal hairs near the hind margin of each segment ....
...............*L. (Moselia) infuscata* Claassen 1923

Subfamily CAPNIINAE

This family is Holarctic in distribution and is represented in North America by six genera, three of which occur in California. The genus *Eucapnopsis* Okamoto 1922 is represented in California by *brevicauda* (Claassen) 1924 (fig. 6:28*a,b*) which ranges from British Columbia and Montana to California. Two described species of the genus *Isocapnia* Banks 1938 occur in the state. *I. grandis* (Banks) 1908 (synonyms: *C. fumigata* Claassen 1937 and *I. fumosa* Banks 1938) is a large species with a body length of more than 10 mm. *I. abbreviata* Frison 1942 is a much smaller species with a body length of less than 8 mm., and the supra-anal process of the male (fig. 6:29*a*) is very much shorter than the long one of *I. grandis* (fig. 6:29*b*). More than thirty species of *Capnia* Pictet 1841, have

[6]From Ricker (1943).

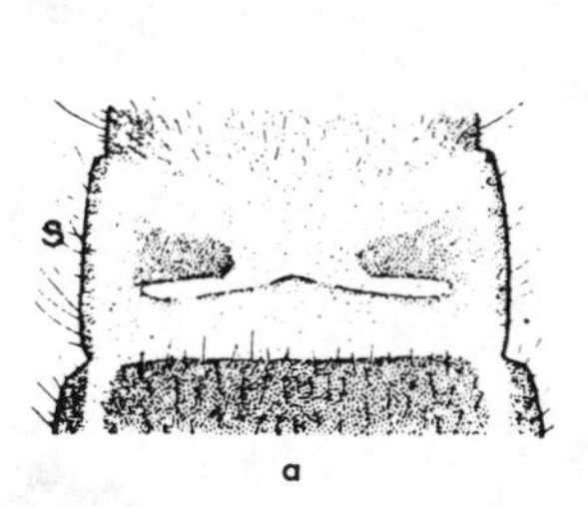

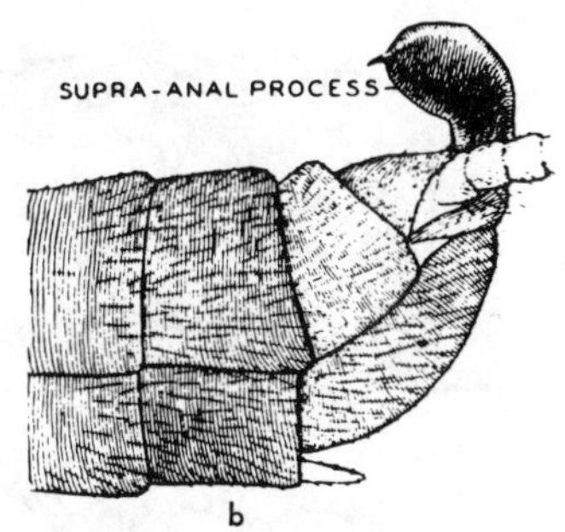

Fig. 6:28. *Eucapnopsis brevicauda, a,* ventral view of female terminalia; *b,* lateral view of male terminalia (Frison, 1937).

been described from North America; the males of those known to occur in California are keyed below. Lack of knowledge concerning the females and nymphs precludes the construction of keys for these.

Key to Males of California Species of Capnia (Male of *C. bakeri* (Banks) 1918 unknown)[7]

1. Ninth sternite with a ventral appendage; wings spotted; central California (fig. 6:53*c*) ....*maculata* Jewett 1954

— Ninth sternite without a ventral appendage; wings not spotted but may be banded 2

2. Supra-anal process slender, completely divided into a dorsal and a ventral part 3

— Supra-anal process not both slender and divided 4

3. A pair of spinous processes present on tergites 8 and 9 (fig. 6:30*h*); southern California ....
..........................*spinulosa* Claassen 1937

— No process on tergites 8 and 9 (fig. 6:30*a*); British Columbia to Oregon ....*columbiana* Claassen 1924

4. Supra-anal process rather broad and divided into a longer ventral process and a shorter dorsal process (fig. 6:30*b*); Plumas County, California............
.... *barberi* Claassen 1924

— Supra-anal process not divided, often slender 5

5. Supra-anal process much expanded at the middle, about half as wide as long 6

— Supra-anal process slender and tapered, or if expanded, the greatest width less than a third of the total length 9

6. Ninth tergite with raised knobs on either side of the mid-line 7

— Ninth tergite without knobs 8

7. Seventh tergite with a median tubercle overhanging the 8th (fig. 6:30*c*); southern California; Yosemite National

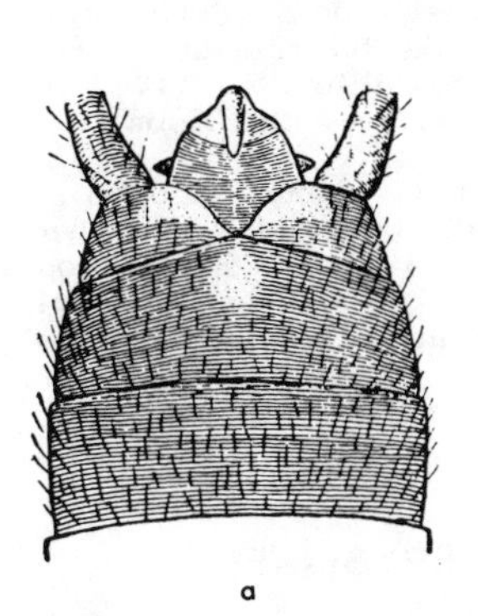

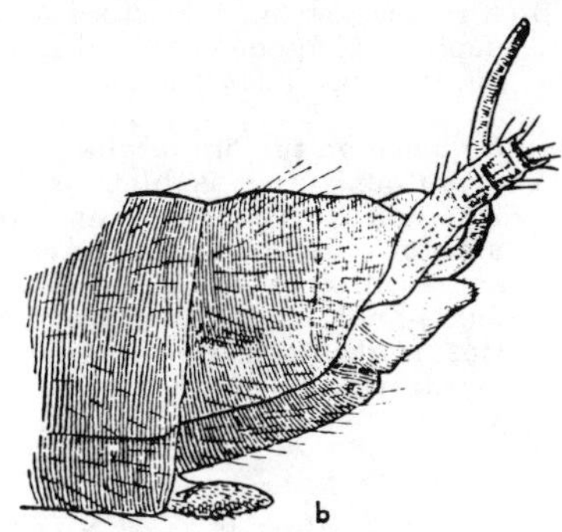

Fig. 6:29. Male terminalia of *Isocapnia*. *a, abreviata,* dorsal view; *b, grandis,* lateral view (Frison, 1942*a*).

[7]Adapted from Ricker (1943).

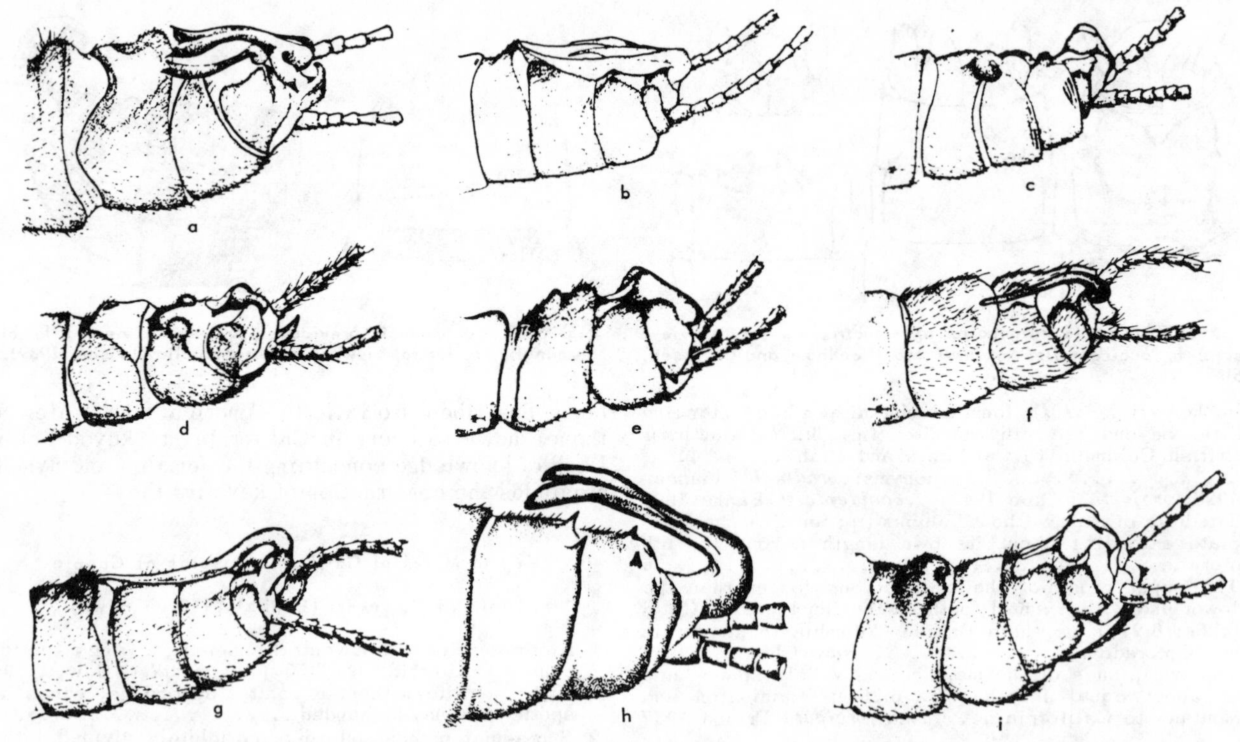

Fig. 6:30. Male terminalia of *Capnia* in lateral view. *a, columbiana; b, barberi; c, teresa; d, californica; e, excavata; f, glabra; g, elongata; h, spinulosa; i, tumida* (*a-g,i*, Needham and Claassen, 1925; *h*, Claassen, 1937*a*).

Park .... *teresa* Claassen 1924

— Seventh tergite without a tubercle (fig. 6:30*d*); Santa Clara and Sonoma counties, California .... *californica* Claassen 1924

8. Eighth tergite with a median elevation which is notched at the tip in side view; no knob on the 7th tergite; supra-anal process lacking long bristles (fig. 6:30*e*); British Columbia to California .... *excavata* Claassen 1924

— Eighth tergite largely membranous; 7th tergite with a knob; supra-anal process fringed dorsally with long forward-pointing bristles (fig. 6:30*i*); British Columbia to California .... *tumida* Claassen 1924

9. A conspicuous hump or process present on the 7th tergite; supra-anal process reaching to the 7th tergite, concave upward (fig. 6:30*g*); British Columbia to California .... *elongata* Claassen 1924

— No hump on the 7th tergite 10

10. A rather low hump or process present on the 8th tergite; supra-anal process reaching to the tubercle of the 8th tergite (fig. 6:31*b*); Washington to California .... *promota* Frison 1937

— No hump on the 8th tergite 11

11. Supra-anal process blunt or merely pointed at the tip; 9th tergite with longitudinal raised tubercles on either side of a median membranous area (fig. 6:30*f*); Oregon and California .... *glabra* Claassen 1924

— Supra-anal process with a definite acute spine at the tip, marked off from the process in dorsal and side views (fig. 6:31*a*); British Columbia to California ... . *projecta* Frison 1937

Subfamily TAENIOPTERYGINAE

This is another group of Nemourids with a Holarctic distribution. The North American representatives are placed in the genera *Brachyptera* Newport (1851) and *Taeniopteryx* Pictet (1841). The key on page 159 will readily separate the two genera, and the key below will separate the North American subgenera and the males of the California species of *Brachyptera*. Females and nymphs of western species in this genus are too little known at present to be keyed. *Taeniopteryx maura* (Pictet) 1842 (synonym: *T. nivalis* Fitch 1847) (fig. 6:32) is a widespread North American species which occurs in California.

Key to North American Subgenera and California Species of Brachyptera[8]

Males

1. Subgenital plate of the male sharply recurved upwards;

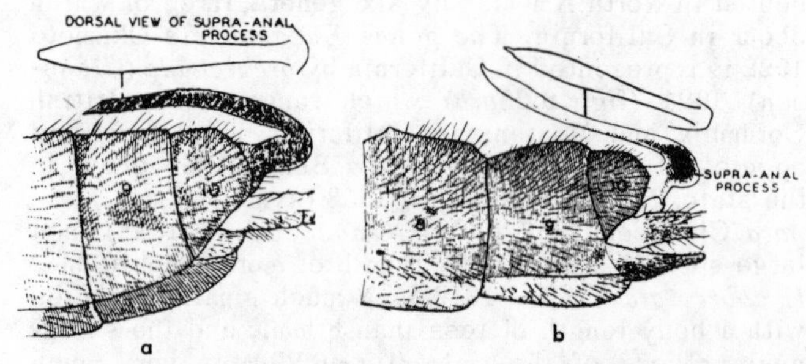

Fig. 6:31. Male terminalia of *Capnia* in lateral view. *a, projecta; b, promota* (Frison, 1937).

[8]Adapted in part from Needham and Claassen (1925) and Ricker (1943).

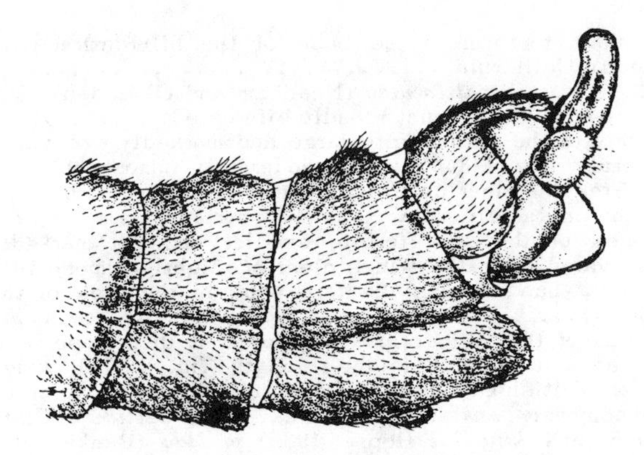

Fig. 6:32. Male terminalia of *Taeniopteryx maura* in lateral view (Needham and Claassen, 1925).

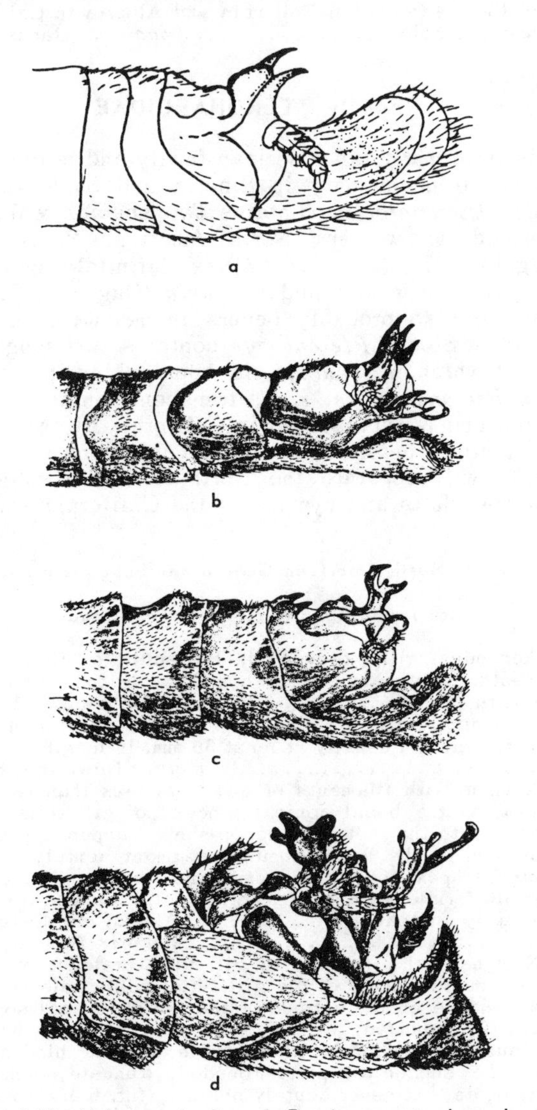

Fig. 6:33. Male terminalia of *Brachyptera* in lateral view. *a, grinnelli; b, nigripennis; c, californica; d, occidentalis* (Needham and Claassen, 1925).

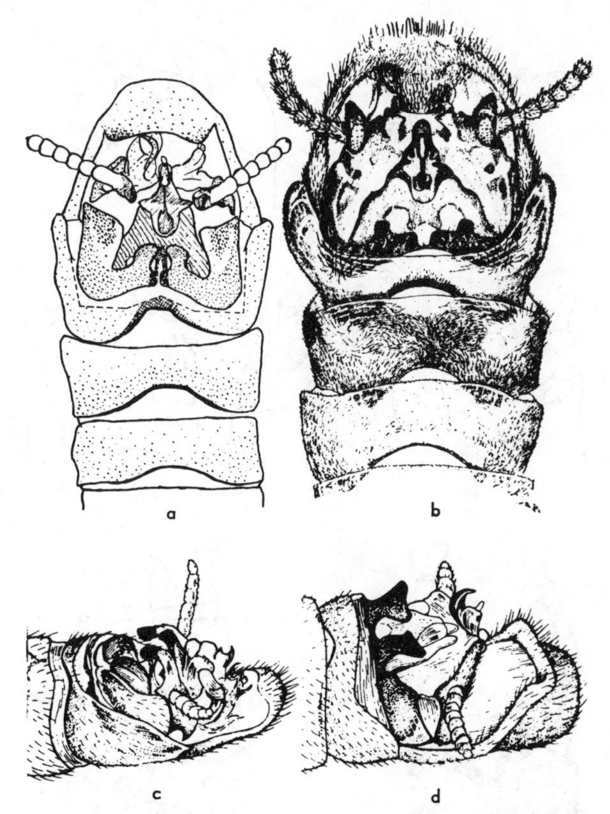

Fig. 6:34. Male terminalia of *Brachyptera. a, pallida,* dorsal view; *b, pacifica,* dorsal view; *c, vanduzeei,* dorsolateral view; *d, raynoria,* dorsolateral view (*a,* Ricker, 1943; *b,* Needham and Claassen, 1925; *c,d,* Claassen, 1937*b*).

male 9th tergite often produced into 2 rounded membranous knobs; 9th sternite of female with a narrow rearward extension the sides of which are broadly concave; eastern North America ....(subgenus *Strophopteryx* Frison 1929)

— Subgenital plate of male curved upwards gradually, if at all; hind margin of the male 9th tergite without lobes, though often sinuate; 9th sternite of female produced into a broad extension the sides of which are convex, at least beyond the middle, or in one case notched (*B. raynoria*).. 2

2. Subanal lobes of male in 2 parts, the upper part a sclerotized prong pointing mostly inward, the lower mostly membranous and asymmetrical; costal field of fore wing with no cross veins between the basal one and the end of Sc in C, nor beyond the cord; 10th tergite of male not produced into lobes; northeastern North America and Utah ....(subgenus *Oemopteryx* Klapalek 1902)

— Subanal lobes consisting of mostly membranous and asymmetrical parts only; costal field with 1 to several cross veins before the cord, and 1 or more beyond it; 10th tergite usually produced into lobes or processes 3

3. A median keel present near the end of the subgenital plate; 10th tergite with 2 processes, each very narrow at the base, broad and with 2 sharp corners at the tip (fig. 6:33*d*); Cu_1 in the fore wing with 4 or 5 branches; British Columbia to Oregon, Montana, and Colorado ....(subgenus *Doddsia* Needham and Claassen 1925) *occidentalis* (Banks) 1900

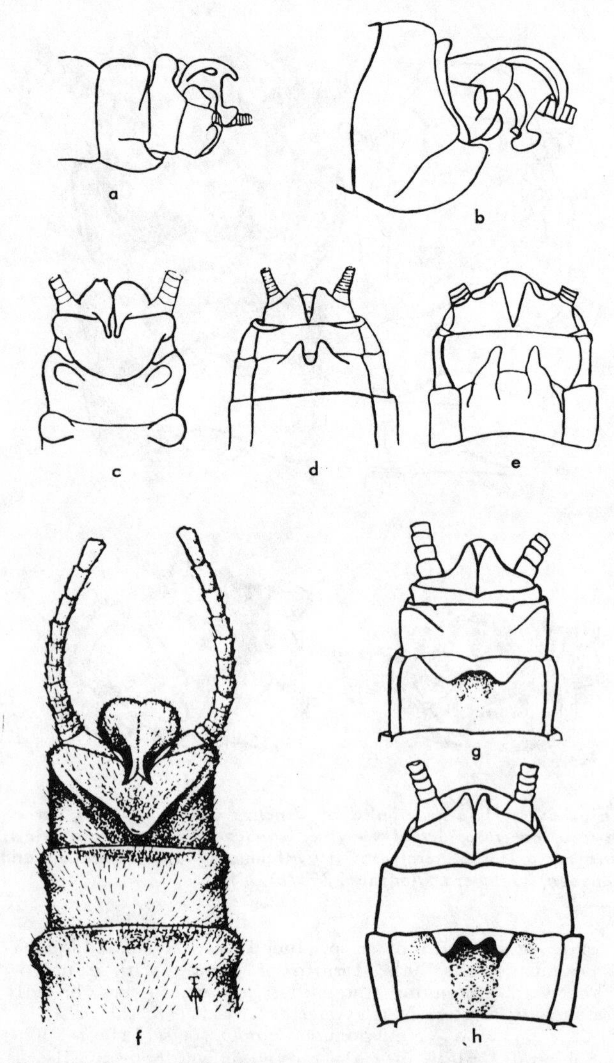

Fig. 6:35. *a,d, Pteronarcys californica; b,e, Pteronarcys princeps; c,g, Pteronarcella badia; f,h, Pteronarcella regularis; a,b, lateral view of male terminalia; c,f, dorsal view of male terminalia; d,e,g,h, ventral view of female terminalia* (Needham and Claassen, 1925).

— Ninth sternite without a keel; processes of the 10th tergite rounded if present; vein Cu_1 in the fore wing with usually only 2 branches .... (subgenus *Taenionema* Banks 1905).... 4

4. Tenth abdominal tergite without 2 raised, rearward-pointing appendages 5

— Tenth abdominal tergite with 2 raised rearward-pointing appendages 7

5. Lobes at base of cerci large, rounded, and directed forward (fig. 6:34*c*); Alpine Creek, Tahoe, California .... *vanduzeei* Claassen 1937

— Lobes at base of cerci spinelike and directed backward 6

6. Supra-anal process gradually tapering to a point; wings usually heavily infuscated (fig. 6:33*b*); British Columbia and Alberta to California and Colorado .... *nigripennis* (Banks) 1918

— Supra-anal process enlarged toward tip which is broadly rounded; wings lightly infuscated with a clear band across the cord and the tip (fig. 6:33*a*) southern California .... *grinnelli* (Banks) 1918

7. Supra-anal process broadly bifurcate in side view with a short tooth near the base of the bifurcation (fig. 6:33*c*); California .... *californica* (Needham and Claassen) 1925

— Supra-anal process not broadly bifurcate 8

8. Lobes of the 10th tergite large and markedly excavated laterally (fig. 6:34*d*); wings moderately fumose but with an irregular clear band across the fore wing in the area of the cord; California .... *raynoria* (Claassen) 1937

— Lobes of the 10th tergite smaller and not markedly excavated laterally; wings clear or slightly fumose with a clear band across the fore wing in the area of the cord 9

9. Lobes of the 10th tergite separated at their base by a distance no more than their length (fig. 6:34*a*); wings clear; British Columbia to California; Utah and Colorado (synonyms: *banksii* N. & Clsn. 1925, *pallidura* Clsn. 1936, and *kincaidi* Hoppe 1938) *pallida* (Banks) 1902

— Lobes of the 10th tergite separated at their base by a distance 2 or 3 times their length (fig. 6:34*b*); wings clear or slightly fumose with a clear band in the area of the cord; British Columbia and Alberta to California and Colorado .... *pacifica* (Banks) 1900

Family PTERONARCIDAE

This is primarily an American family and is otherwise known to occur only in Siberia and Sakhalin. The genus *Pteronarcella* contains two species which are confined to western North America. *P. regularis* (Hagen) 1874 (fig. 6:35*f,h*) is definitely known to occur in California, and *P. badia* (Hagen) 1874 (fig. 6:35*c,g*) also probably occurs in mountains east of the main Sierra. *Pteronarcys* contains two subgenera, one of which, *Allonarcys*, is Appalachian in distribution. *Pteronarcys, s. s.*, is transcontinental in distribution and is represented in California by two widely distributed western species. The following key to the family will separate the North American subgenera and the adults and nymphs of the California species.

Key to North American Genera and Subgenera and California Species of Peteronarcidae[9]

1. Abdominal gills, reduced in the adult, on first 3 segments; adults and mature nymphs less than 30 mm. in length .... *Pteronarcella* Banks (1900) 2

— Abdominal gills on first 2 segments only; adult and mature nymph usually at least 35 mm. in length..... ....*Pteronarcys* Newman (1838) 3

2. Nymphs with filaments of gill tufts less than twice as long as the basal conical process of gill tufts; adult male with the recurved, scoop-shaped appendage on the dorsum of the 9th abdominal segment acutely pointed at the apex and its side margins straight (fig. 6:35*f*); adult female with hind margin of the subgenital plate acutely notched (fig. 6:35*h*); Alaska to California .... *regularis* (Hagen) 1873

— Nymphs with filaments of gill tufts at least twice as long as the bases of the gill tufts; adult male with the appendage on the dorsum of the 9th abdominal segment broadly rounded at the apex and its side margins sinuous (fig. 6:35*c*); adult female with the hind margin of the subgenital plate rounded, truncate, somewhat trilobate, but never acutely notched (fig. 6:35*g*); Alberta to Nevada and Colorado ....*badia* (Hagen) 1873

3. Nymph with paired lateral projection on the abdomen;

[9]Adapted from Needham and Claassen (1925) and Claassen (1931).

adult male with a slender, erect supra-anal process; Appalachian ....
...(subgenus *Allonarcys* Needham and Claassen 1925)[10]
— Nymphs without paired lateral projections on the abdomen; adult male with a massive decurved supra-anal process .... *Pteronarcys, s.s.* Newman (1838) 4
4. Nymph with lateral thoracic processes long, slender, directed outward and wing pads pointed; erect lobes of the divided 10th tergite of male rather broadly rounded (fig. 6:35*a*); processes on the apex of the female subgenital plate are somewhat equilateral triangles approximated at the base (fig. 6:35*d*); British Columbia to California .... *californica* Newport 1851
— Nymph with lateral thoracic processes short, not markedly directed outward and wing pads rounded; erect lobes of the divided 10th tergite of the male rather narrow, much higher than wide (fig. 6:35*b*); processes at the apex of the female subgenital plate are very elongate triangles, twice as high as broad and more widely separated at the base (fig. 6:35*e*); British Columbia to California ....
.... *princeps* Banks 1907

Suborder SETIPALPIA

Family PERLODIDAE

This family, confined to the Holarctic region and North America, is very rich in species, about eighty of which have been described to date. These have been placed in three subfamilies by Ricker in his recent monograph of the world fauna of this family (Ricker, 1952, pp. 62–145). Ricker's classification of this family is adopted here. The male genitalia must be relied upon for placing many species to subfamily. Keys to Ricker's subgenera and to California species of the family are included below.

Subfamily ISOGENINAE

Members of this subfamily are placed in two genera, *Arcynopteryx* Klapalek 1904 and *Isogenus* Newman 1833. The former includes seven subgenera in North America, four of which definitely occur in California. One of these, *Oroperla,* occurs in streams of the Sierra Nevada, but adults have not yet been taken; their discovery will be of great interest. *Isogenus* is divided into twelve North American subgenera, four of which are known to occur in California. An additional species found in the state, *I. sorptus* (Needham and Claassen) 1925 (fig. 6:36*c*) has not been placed in a subgenus as only the female is described. The subgenus *Isogenoides* has not been recorded in California but the species *Isogenoides frontalis colubrinus* Hagen 1874 (figs. 6:7*i*; 6:36*a,d*) very probably occurs in some of the larger coastal streams of the northern part of the state.

[10] *P. (Allonarcys) proteus* Newman is recorded by Needham and Claassen (1925) from California; this is regarded as a probable error.

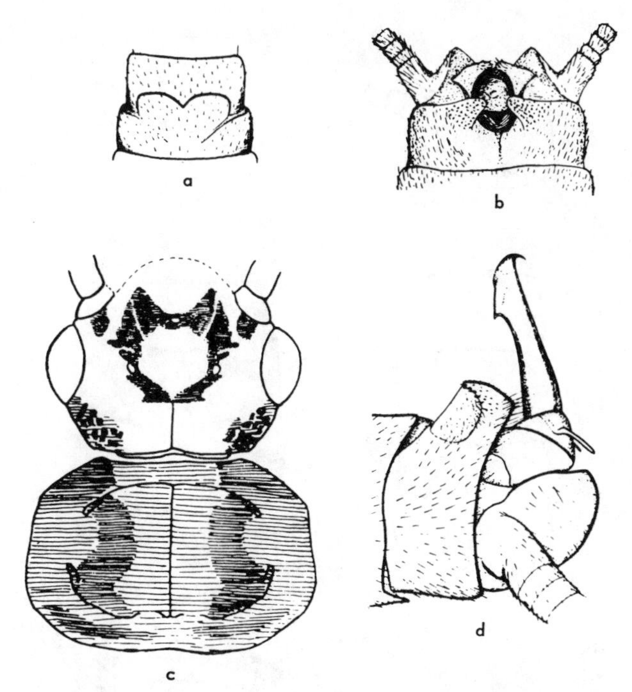

Fig. 6:36. *a, Isogenus frontalis colubrinus,* ventral view of female terminalia; *b, Isogenus alameda,* dorsal view of male terminalia; *c, Isogenus sorptus,* head and pronotum of female; *d, Isogenus f. colubrinus,* lateral view of male terminalia (*a,d,* Hanson; 1943; *b,* Needham and Claassen, 1925; *c,* Ricker, 1952).

Key to the North American Subgenera and California Species of Arcynopteryx[11]

1. Lateral abdominal gills present (adult unknown); California....................................(subgenus *Oroperla* Needham 1933) *barbara* (Needham) 1933
— Abdominal gills absent 2
2. Three pairs of thoracic gills and 1 pair of cervical gills present (fig. 6:37*d,g*); Washington to California ....
......(synonym: *A. vagans* Smith 1917) (subgenus *Perlinodes* Needham and Claassen 1925) *aurea* Smith 1917
— Cervical gills absent 3
3. Gills present on meso- and metathorax, absent from prothorax; Cordilleran but not known in California.
.... (subgenus *Setvena* Ricker 1952)
— Gills present on all 3 thoracic segments, or entirely absent from the thorax 4
4. Gills present on all 3 thoracic segments (fig. 6:37*b*); Washington to California in Alpine Zone ....
.............................(subgenus *Megarcys* Klapalek 1912) *yosemite* (Needham and Claassen) 1925
— Gills absent from the thorax 5
5. Transverse mesosternal ridge present in adult; nymphal lacinia tapered to a short terminal spine which is equal to 1/4 of the total lacinial length (fig. 6:37*a,c*); British Columbia to Oregon ....
.... (synonym: *A. walkeri* Ricker 1943) (subgenus *Frisonia* Ricker 1943) *picticeps* Hanson 1942
— Transverse mesosternal ridge absent; terminal spine of nymphal lacinia equal to 1/3 or more of the total length 6
6. Male supra-anal process very long and needlelike;

[11] Adapted from Ricker (1952).

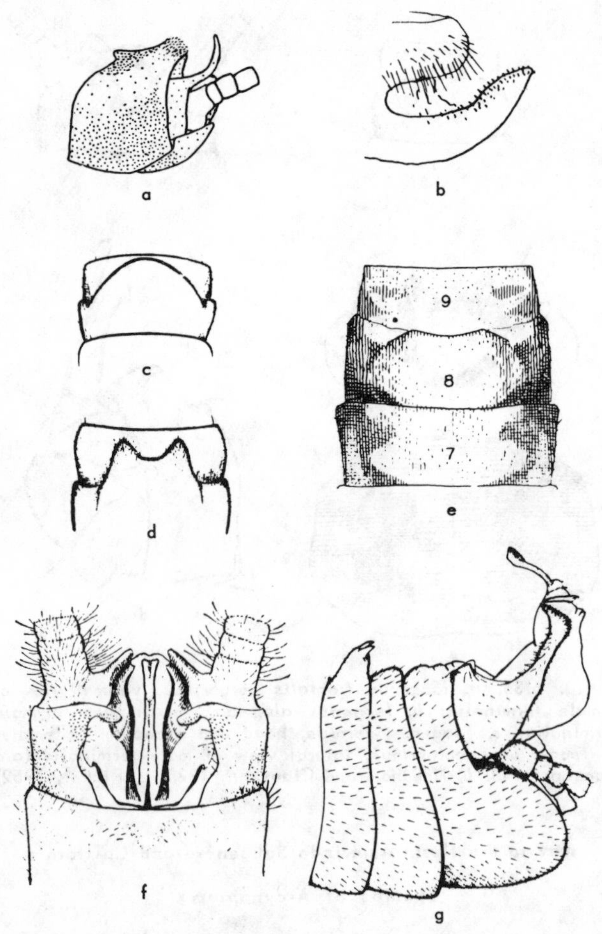

Fig. 6:37. Terminal abdominal structures of *Arcynopteryx*. *a, picticeps*, lateral view of male terminalia; *b, yosemite*, hook on tenth tergite of male; *c, picticeps*, ventral view of female terminalia; *d, aurea*, ventral view of female terminalia; *e, parallela*, ventral view of female terminalia; *f*, dorsal view of male terminalia; *g, aurea*, lateral view of male terminalia (*a*, Ricker, 1943; *b*, Needham and Claassen, 1925; *c,d,g*, Hanson, 1942; *e*, Frison, 1937; *f*, Frison, 1936).

denticles absent from the cusps of the nymphal mandible, or a very few present on the outer left cusp only; Holarctic; northern North America, Colorado, New Hampshire .... *Arcynopteryx*, *s.s.* Klapalek 1904

— Male supra-anal process blunt, not unusually long (fig. 6:37*f*); denticles numerous along both sides of the outer cusps of both mandibles; British Columbia to Oregon and California; Montana, and Utah ....
(subgenus *Skwala* Ricker 1943) *parallela* (Frison) 1936

Key to the North American Subgenera and California Species of Isogenus[12]

Males

1. Arms of the mesosternal Y-ridge meet the anterior corners of the furcal pits (indistinct in males of *Chernokrilus*) 2

— Arms of the mesosternal Y-ridge meet or approach the posterior corners of the furcal pits 4

2. Wings almost clear; male lateral stylets short with 2 or 3 terminal spinules (figs. 6:38*e*; ♀ 6:39*a*); Washington to California.................................
(subgenus *Osobenus* Ricker 1952) *yakimae* (Hoppe) 1938

— Wings dark brown; male lateral stylets long, smoothly rounded ....
.....(subgenus *Chernokrilus* Ricker 1952) (in part) 3

3. Supra-anal process bluntly hooked at the tip (fig. 6:38*h*); Fieldbrook, California .... (synonym: *Perla venosa* Needham and Claassen 1925) *erratus* (Claassen) 1936

— Supra-anal process straight at the tip (fig. 6:38*i*; ♀, 6:38*d*); Oregon ....(synonym: *Perla obscura*

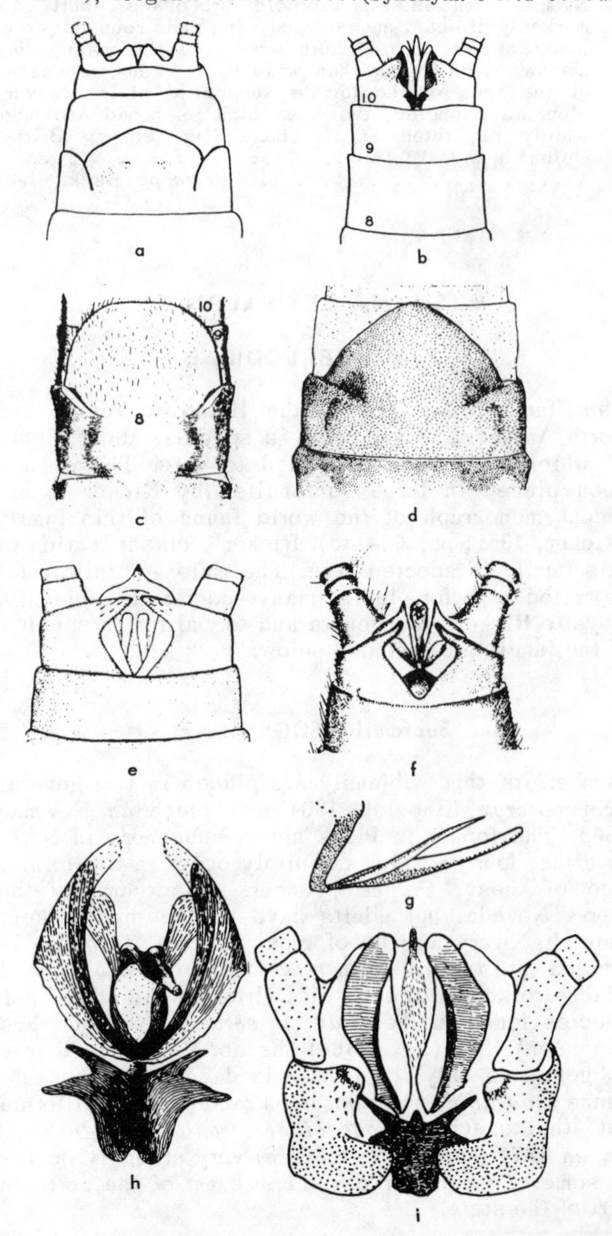

Fig. 6:38. Terminal abdominal structures of *Isogenus*. *a,f, nonus*; *b,c, pilatus*; *d,i, misnomus*; *e, yakimae*; *g, tostonus*; *h, erratus*; *a,c,d*, ventral view of female terminalia; *b,e,f,h,i*, dorsal view of male terminalia; *g*, supra-anal process of male, lateral view; (*a*, Needham and Claassen, 1925; *b,c*, Frison, 1942*b*; *d,h,i*, Ricker, 1952; *e*, Hoppe, 1938; *f*, Claassen, 1937*b*; *g*, Ricker, 1943).

[12] Adapted from Ricker (1952). See text for *Isogenus sorptus* (Needham and Claassen).

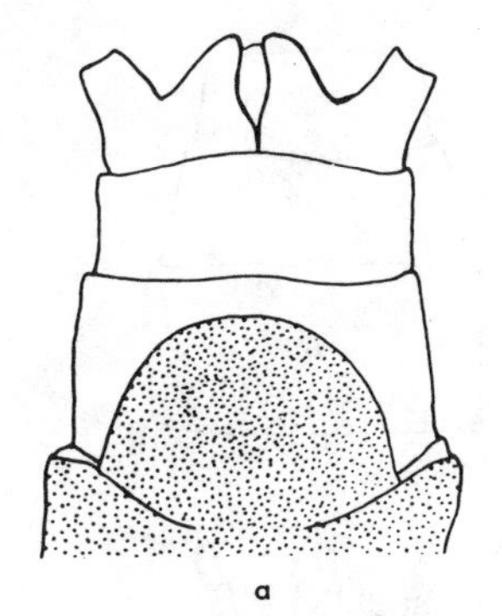

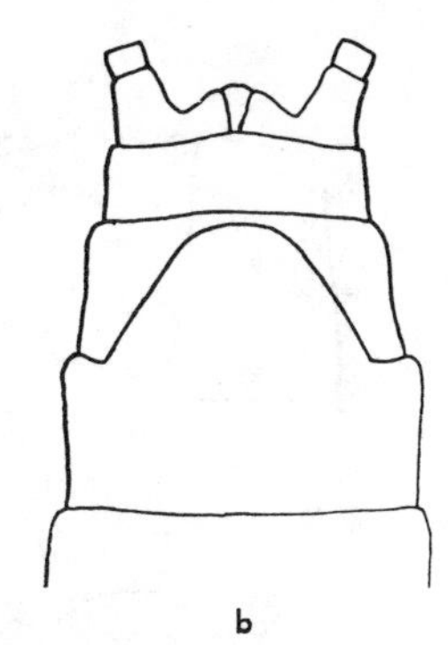

Fig. 6:39. Ventral view of female terminalia of *Isogenus*. *a*, *yakimae*; *b*, *tostonus* (Ricker, 1952).

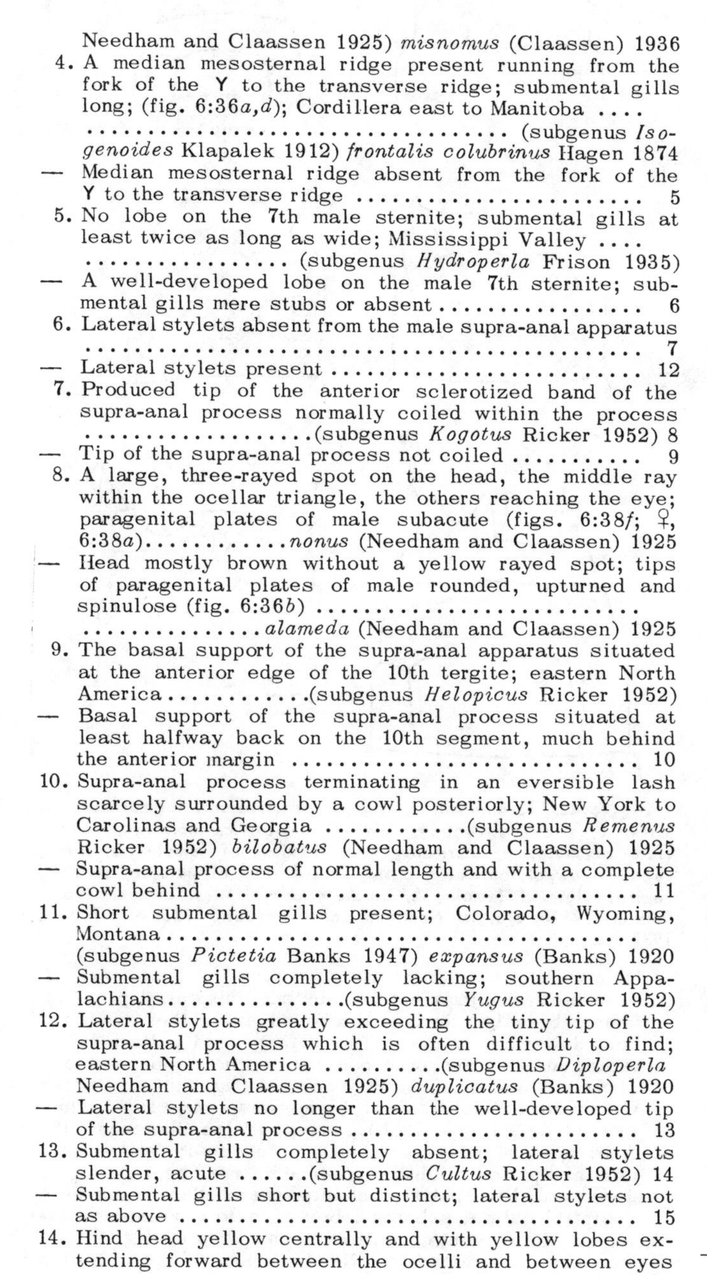

Needham and Claassen 1925) *misnomus* (Claassen) 1936

4. A median mesosternal ridge present running from the fork of the Y to the transverse ridge; submental gills long; (fig. 6:36*a,d*); Cordillera east to Manitoba .... (subgenus *Isogenoides* Klapalek 1912) *frontalis colubrinus* Hagen 1874
— Median mesosternal ridge absent from the fork of the Y to the transverse ridge 5
5. No lobe on the 7th male sternite; submental gills at least twice as long as wide; Mississippi Valley .... (subgenus *Hydroperla* Frison 1935)
— A well-developed lobe on the male 7th sternite; submental gills mere stubs or absent 6
6. Lateral stylets absent from the male supra-anal apparatus 7
— Lateral stylets present 12
7. Produced tip of the anterior sclerotized band of the supra-anal process normally coiled within the process ....(subgenus *Kogotus* Ricker 1952) 8
— Tip of the supra-anal process not coiled 9
8. A large, three-rayed spot on the head, the middle ray within the ocellar triangle, the others reaching the eye; paragenital plates of male subacute (figs. 6:38*f*; ♀, 6:38*a*)...........*nonus* (Needham and Claassen) 1925
— Head mostly brown without a yellow rayed spot; tips of paragenital plates of male rounded, upturned and spinulose (fig. 6:36*b*) ....*alameda* (Needham and Claassen) 1925
9. The basal support of the supra-anal apparatus situated at the anterior edge of the 10th tergite; eastern North America...........(subgenus *Helopicus* Ricker 1952)
— Basal support of the supra-anal process situated at least halfway back on the 10th segment, much behind the anterior margin 10
10. Supra-anal process terminating in an eversible lash scarcely surrounded by a cowl posteriorly; New York to Carolinas and Georgia ....(subgenus *Remenus* Ricker 1952) *bilobatus* (Needham and Claassen) 1925
— Supra-anal process of normal length and with a complete cowl behind 11
11. Short submental gills present; Colorado, Wyoming, Montana .... (subgenus *Pictetia* Banks 1947) *expansus* (Banks) 1920
— Submental gills completely lacking; southern Appalachians..............(subgenus *Yugus* Ricker 1952)
12. Lateral stylets greatly exceeding the tiny tip of the supra-anal process which is often difficult to find; eastern North America ....(subgenus *Diploperla* Needham and Claassen 1925) *duplicatus* (Banks) 1920
— Lateral stylets no longer than the well-developed tip of the supra-anal process 13
13. Submental gills completely absent; lateral stylets slender, acute ......(subgenus *Cultus* Ricker 1952) 14
— Submental gills short but distinct; lateral stylets not as above 15
14. Hind head yellow centrally and with yellow lobes extending forward between the ocelli and between eyes and ocelli, 3 in all (fig. 6:38*b,c*); British Columbia to Oregon ....*pilatus* (Frison) 1942
— Head mostly yellow, the only important dark marking being the bands which join the anterior of the lateral ocelli (figs. 6:38*g*; 6:39*b*); British Columbia to California, Montana, and Wyoming .... *tostonus* Ricker 1952
15. Submental gills conical; lateral stylets hooked at the tip; eastern North America.................... (subgenus *Malirekus* Ricker 1952) *hastatus* (Banks) 1920
— Submental gills rounded; lateral stylets rounded at the tip (see also couplet 3) ....(subgenus *Chernokrilus* (in part)

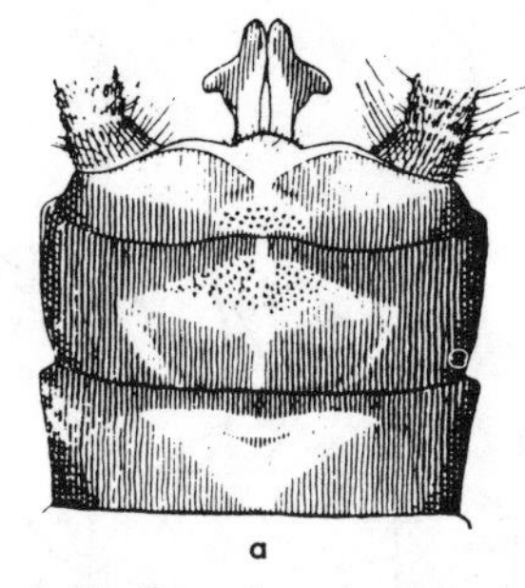

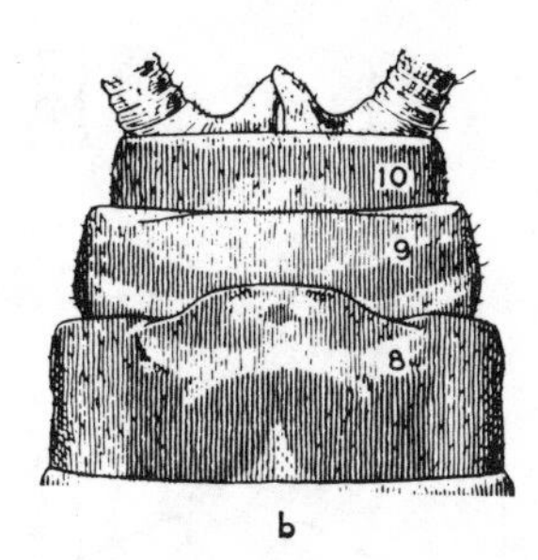

Fig. 6:40. *Diura knowltoni*, terminalia. *a*, dorsal view of male; *b*, ventral view of female (*a*, Frison, 1937; *b*, Frison, 1942*b*).

Subfamily PERLODINAE

This subfamily is represented in North America by the genus *Diura* Billberg 1820, which comprises two subgenera, *Diura, s. s.*, confined to the Appalachian region, and the monotypic *Dolkrila* Ricker 1952 which is cordilleran. The latter has not been recorded from California, but *D. (Dolkrila) knowltoni* (Frison) 1937 (fig. 6:40*a,b*) occurs to the north in Oregon.

Subfamily ISOPERLINAE

This group contains the genera *Calliperla* Banks 1948, *Isoperla* Banks 1906, and *Rickera* Jewett 1954. *Calliperla* has but one species, *luctuosa* (Banks) 1906 (fig. 6:41*a,b*) which occurs in California and Oregon. Several species of *Isoperla* occur in the Palaearctic region, but the greatest development of the genus occurs in North America from which about forty species have been described. The genus contains several complexes but has not been broken into subgenera. The following key and accompanying illustrations may be used to classify the described California species. Much work needs to be done on this genus, particularly on our western species. Nymphs of only a few California species are known.

Key to Adults of California Species of Isoperla[13]

1. Tip of male 10th tergite with 2 small recurved processes; no lobe on male 8th sternite (fig. 6:42*a*); female 8th sternite not produced except for a small median process (fig. 6:42*b*); head with a complete median dark stripe, darkest between the ocelli; pronotum with broad median and lateral light stripes; Washington to California ....*trictura* (Hoppe) 1938

[13] Adapted in part from Ricker (1943).

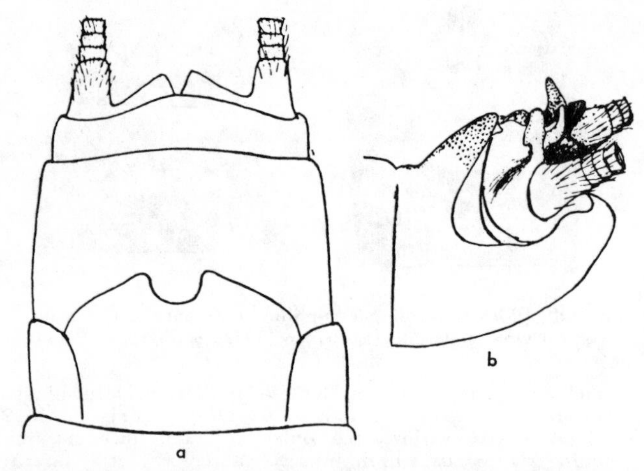

Fig. 6:41. ***Calliperla luctuosa,*** **terminalia.** ***a,*** **ventral view of female;** ***b,*** **lateral view of male (Needham and Claassen, 1925).**

— Tip of male 10th tergite without process; male 8th sternite with a lobe (obsolescent in *ebria*); female 8th sternite not as above; color not as above 2

2. Lobe at the tip of the male 8th sternite nearly square with subacute angles; spinules present on the male 9th tergite (fig. 6:42*c*); female subgenital plate very little produced (fig. 6:42*d*); a small yellowish species; British Columbia to California and Montana to Wyoming and Arizona ....
(synonym: *I. insipida* Hoppe 1938) *mormona* Banks 1920

— Lobe on the male 8th sternite broadly rounded behind; male 9th tergite without spinules, though with short stout hairs in *patricia;* female subgenital plates vary; medium to large species 3

3. Pronotum with a checkered pattern of black on yellow; female subgenital plate moderately produced, slightly excavated at the middle (fig. 6:42*e*); male subanal lobes flat, recurved, acute, bent outward near the tips (fig. 6:42*f*); British Columbia to California and Wyoming .... *pinta* Frison 1937

— Pronotum striped or reticulately marked; female subgenital plates varied; male subanal lobes not bent outward close to their tips 4

4. Female subgenital plate long, its sides parallel or nearly so near the base (fig. 6:43*a*); hind margin of the male 8th sternite with a hairless yellow area, but scarcely a lobe (fig. 6:43*e*); head with a sharp pattern of brown or yellow, including an ocellar and a preocellar yellow spot; British Columbia to California .... *ebria* (Hagen) 1875

— Female subgenital plate shorter, usually its sides turned inward making an angle of at least 135° with the side margin of the segment; male with a definite lobe on the 8th sternite 5

5. Female subgenital plate only slightly produced (fig. 6:42*g*); male subanal lobes acutely bent upward, long, thin, pointed (fig. 6:42*h*); Washington to California .... *sordida* (Banks) 1906

— Female subgenital plate distinctly produced; male subanal lobes not long, thin, and acutely bent 6

6. Subgenital plate of female with a wide, deep excavation at the tip (fig. 6:43*b*); male with patches of short stout hairs on the posterior margin of the 9th tergite (fig. 6:43*c*); a yellowish species; British Columbia to California; South Dakota to Colorado ....
.... *patricia* Frison 1942[14]

— Subgenital plate of female without a median excavation; male without patches of short stout hairs on the posterior margin of the 9th tergite 7

7. Anal area of wings not infuscated; subgenital plate of female broadly rounded, thickened and rather suddenly bent downward near tip (fig. 6:43*f*); head pattern interrupted at the transverse occipital suture; subanal lobes of male stout, tips abruptly pointed (fig. 6:43*d*); British Columbia to California...(synonyms: *chrysannula* and *cascadensis* Hoppe 1938)......... *fulva* Claassen 1937

— Anal area of wings distinctly infuscated; female subgenital plate slightly produced, not bent abruptly downward; head pattern not interrupted at the transverse occipital suture; subanal lobes of male gradually tapering to sharp points; California and Oregon (fig. 6:53*e*) .... *marmorata* (Needham and Claassen) 1925

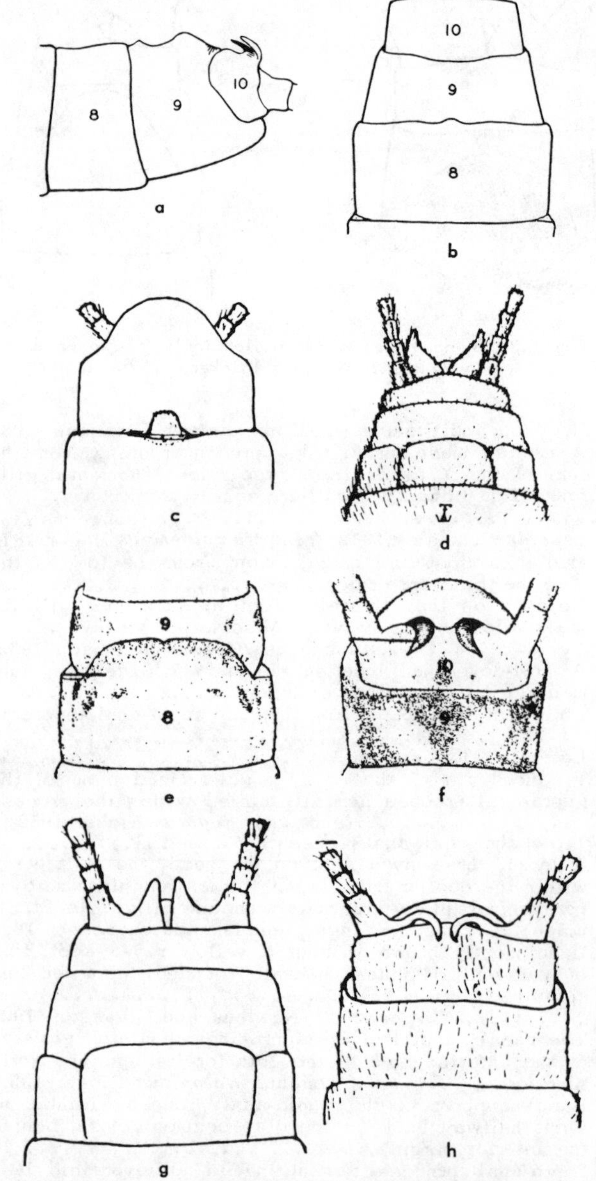

Fig. 6:42. Terminalia of ***Isoperla. a, trictura,*** **lateral view of male;** ***b, trictura,*** **ventral view of female;** ***c, mormona,*** **ventral view of male;** ***d, mormona,*** **ventral view of female;** ***e, pinta,*** **ventral view of female;** ***f, pinta,*** **dorsal view of male;** ***g, sordida,*** **ventral view of female;** ***h, sordida,*** **dorsal view of male (*a,b,* Frison, 1942*b*; *c,d,g,h,* Needham and Claassen, 1925; *e,f,* Frison, 1937).**

[14] *I. 5-punctata* (Banks) 1902, recorded for California by Needham and Claassen (1925, p. 152) and Seemann (1927, p. 57) may key here; it is said to have greenish wings.

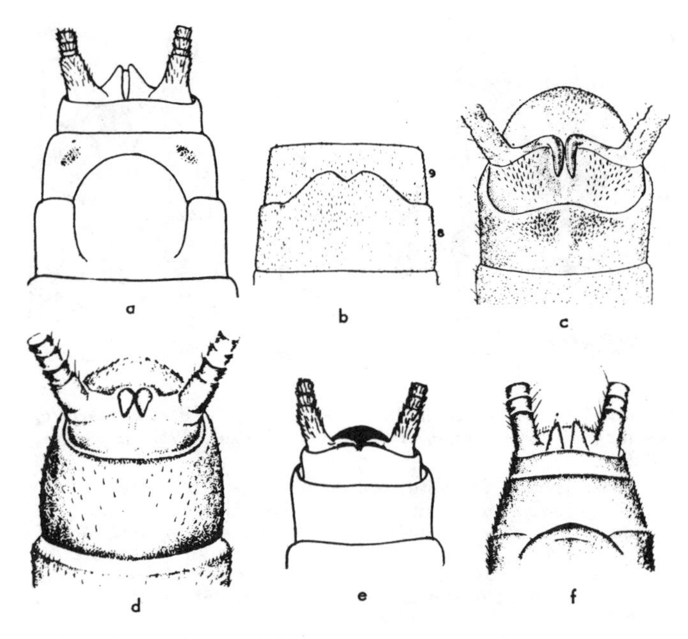

Fig. 6:43. Terminalia of *Isoperla*, *a,e*, *ebria*; *b,c*, *patricia*; *d,f*, *fulva*; *a,b,f*, ventral view of female; *c,d,e*, dorsal view of male (*a,e*, Needham and Claassen, 1925; *b,c*, Frison, 1942*b*; *d,f*, Claassen, 1937*a*).

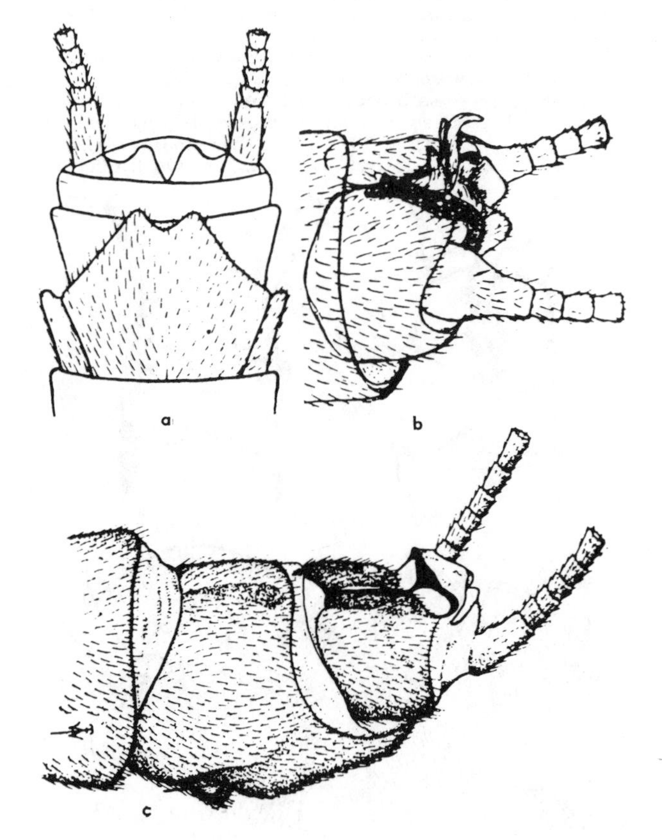

Fig. 6:44. *a*, *Kathroperla perdita*, ventral view of female terminalia; *b*, *Paraperla frontalis*, lateral view of male terminalia; *c*, *K. perdita*, dorsolateral view of male terminalia (Needham and Claassen, 1925).

Family CHLOROPERLIDAE

These stoneflies are Holarctic in distribution and include two subfamilies, the Nearctic Paraperlinae and the Holarctic Chloroperlinae. The Paraperlinae contain three monotypic genera, two of which occur in California: *Kathroperla* Banks 1920, species *perdita* Banks 1920 (figs. 6:9*b*; 6:44*a,c*); and *Paraperla* Banks 1906, species *frontalis* (Banks) 1902 (figs. 6:44*b*; 6:45*b*). The recently described *Utaperla* is recorded from Yukon Territory and Utah (Ricker, 1952, p. 176). The Chloroperlinae contains many North American species which are placed in three genera. *Chloroperla* Newman 1836 is represented in North America by the subgenus *Rasvena* Ricker 1952 which contains a single species found in the eastern United States. *Hastaperla* Ricker 1935 is represented by three species, one of which, *H. chilnualna* Ricker 1952, is known from California. Another widespread species, *H. brevis* (Banks) 1895, may occur in the state but positive records are lacking. It differs in the male from *H. chilnualna* principally in that the supra-anal process in lateral view is not hooked and is broader in dorsal view. About forty species of *Alloperla* Banks 1906 in five subgenera occur in North America. The following key separates the subgenera, and keys the males of species known to occur in California. The nymphs of *Alloperla* resemble one another very closely and few have been described.

Key to the North American Subgenera and California Species of Alloperla[15]

Males

1. Basal segment of the cerci greatly elongated, deeply concave inward, with a sharp spine inside near the base and with 4 blunt knobs on the inside toward the tip; 9th tergite with its hind border produced backward, from which comes a stout forward-pointing decurved hook; 10th tergite with erect processes from the hind margin pointing upward and inward; (figs. 6:47*b*; ♀, 6:50*a*); Alberta, British Columbia, and Washington .. .. (subgenus *Neaviperla* Ricker 1943) *forcipata* Neave 1929
— Cerci normal; 9th tergite without a median forward-projecting hook; 10th tergite without dorsal processes on the hind margin 2
2. A fingerlike process pointing inward from the basal segment of each cercus; supra-anal body a membranous lobe with a very small hairy process at its tip (figs. 6:48*d*; ♀, 6:51*h*); British Columbia to California and Alberta to Wyoming...(synonym: *A. dubia* Frison 1935) .... (subgenus *Suwallia* Ricker 1942) *pallidula* (Banks) 1904
— No process at the base of the cercus; supra-anal body elongate, its terminal process usually larger 3
3. Color usually green in life; no dark abdominal stripe; no process on the 9th tergite; ....(subgenus *Alloperla, s.s.* Banks 1906) 4
— Color mostly yellow in life; a dark dorsal stripe present on the abdomen; with or without a process on the 9th tergite 6
4. Process at the tip of the supra-anal body dumbbell-shaped, longer than broad, rounded in front and behind 5
— Process at the tip of the supra-anal body short, its

[15]Adapted in part from Ricker (1943, 1952).

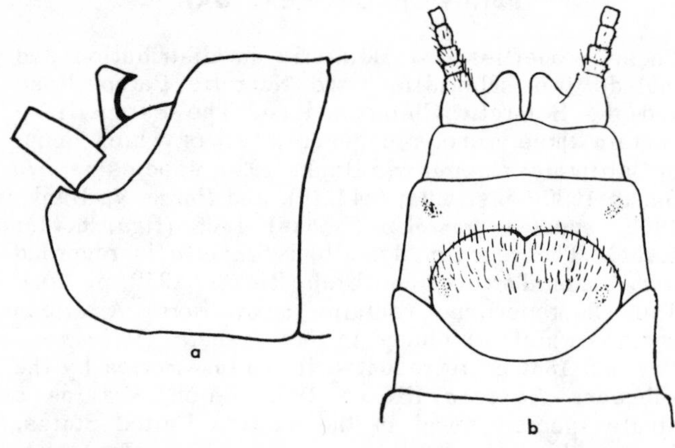

Fig. 6:45. *a, Hastaperla chilnualna,* lateral view of male terminalia; *b, Paraperla frontalis,* ventral view of female terminalia (*a,* Ricker, 1952; *b,* Needham and Claassen, 1925).

anterior corners produced into acute angles (fig. 6:46*a*) British Columbia to Oregon...... *delicata* Frison 1935

5. In lateral view, depth of the tip of the supra-anal process about equal to half the length; central California (fig. 6:53*f*)............. *chandleri* Jewett 1954

— In lateral view, depth of the tip of the supra-anal process less than 1/3 the length; British Columbia to Oregon...(synonyms: *elevata* Frison 1935; *thalia* Ricker 1952) (fig. 6:48*c*).................... *severa* (Hagen) 1861

6. Body of the supra-anal apparatus obscure, short, and lying along the surface of and fused with the 10th tergite, usually in a slight depression but never in a deep groove; supra-anal process short, sharply recurved .... (subgenus *Triznaka* Ricker 1952) 7

— Body of the supra-anal apparatus lying in a deep groove of the 10th tergite and attached to its sides, bearing at its posterior end a terminal part which is well marked off from the rest of the apparatus ....(subgenus *Sweltsa* Ricker 1943) 8

7. Head unmarked except for the ocellar rings; pronotum yellow except for dusky lateral margins; no lobe on the male 7th sternite (fig. 6:46*b*); British Columbia to California .... *diversa* Frison 1935

— Head and pronotum with conspicuous median markings of black on yellow; anterior dark mark on head about twice as wide as long; a small posterior lobe on the male 7th sternite; Washington to California; Colorado to North Dakota and South Dakota............... .... *pintada* Ricker 1952

8. A definite elevated transverse and usually notched

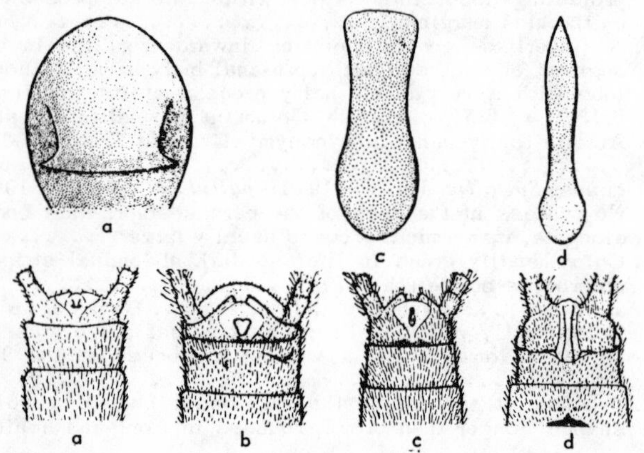

Fig. 6:46. Male terminalia of *Alloperla,* in dorsal view with enlarged dorsal view of tip of supra-anal process for *a,c* and *d. a, delicata; b, diversa; c, fraterna; d, exquisita* (Frison, 1935).

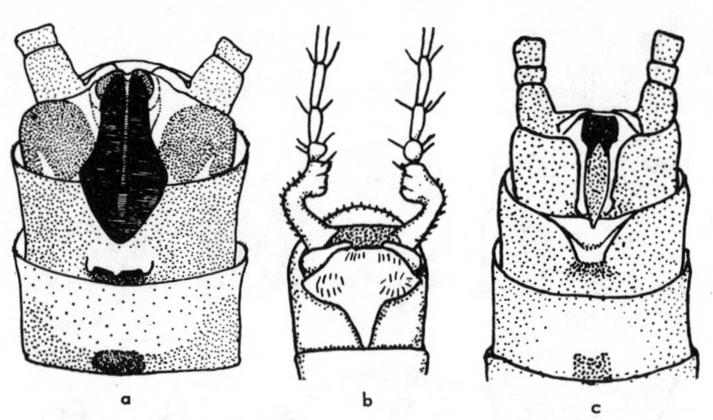

Fig. 6:47. Male terminalia of *Alloperla,* in dorsal view. *a, townesi; b, forcipata; c, tamalpa* (*a,c,* Ricker, 1952; *b,* Neave, 1929).

process near the anterior border of the 9th tergite ... 9

— No process on the 9th tergite (figs. 6:46*c*; ♀, 6:51*e*); no dark U-marks on meso- and metanota; margins of the pronotum only faintly darkened; British Columbia to California .... *fraterna* Frison 1935

9. A bifurcate transverse process on the 8th tergite (figs. 6:49*a*; ♀, 6:51*f*); Montana to California............ .... (synonym: *A. spatulata* Needham and Claassen 1925) *pacifica* (Banks) 1895

— No process on the 8th tergite 10

10. Disc of the pronotum clear yellow; supra-anal process terete, slender, bluntly pointed (figs. 6:46*d*; ♀, 6:51*c*); British Columbia to Oregon..... *exquisita* Frison 1935

— Disc of the pronotum with black reticulate markings 11

11. Supra-anal process slender, somewhat expanded in dorsal view toward the tip 12

— Supra-anal process broadly flattened 15

12. Supra-anal process extending barely beyond the 10th tergite (fig. 6:47*c*; ♀, 6:50*b*); Mt. Tamalpais, California .... *tamalpa* Ricker 1952

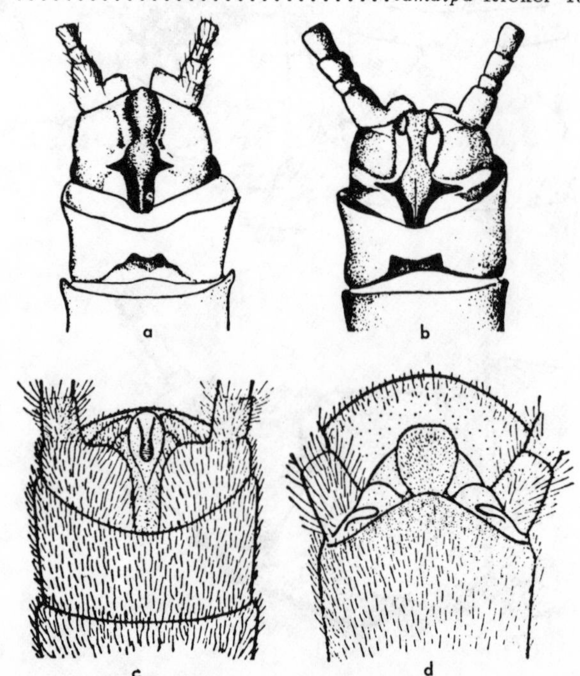

Fig. 6:48. Male terminalia of *Alloperla,* in dorsal view. *a, continua; b, fidelis; c, severa; d, pallidula* (*a,b,* Needham and Claassen, 1925; *c,d,* Frison, 1935).

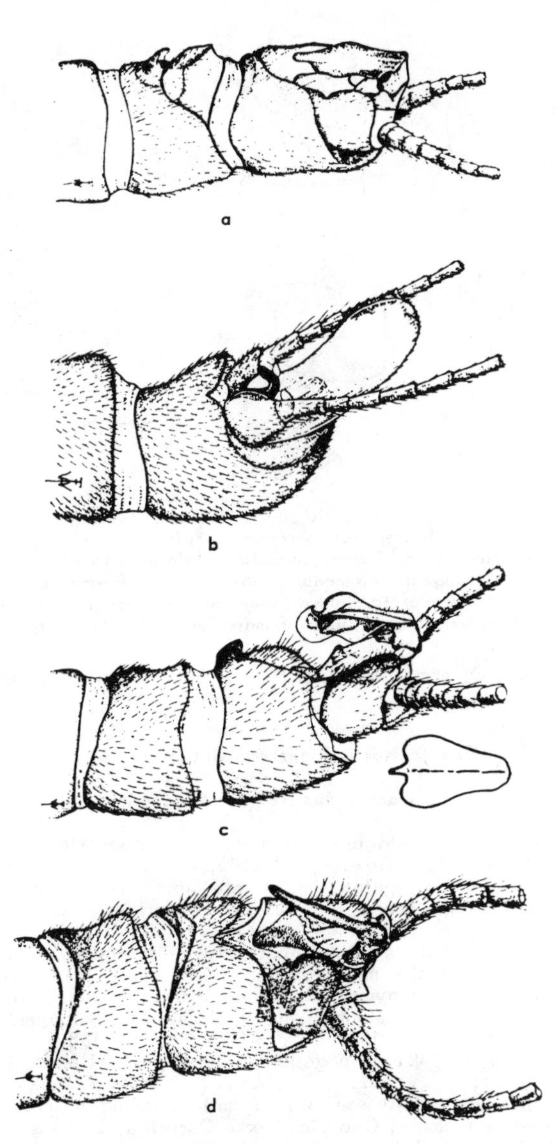

Fig. 6:49. Male terminalia of *Alloperla*, *a,b,* and *d* in dorsolateral view and c in lateral view. *a, pacifica; b, signata; c, coloradensis; d, borealis* (Needham and Claassen, 1925).

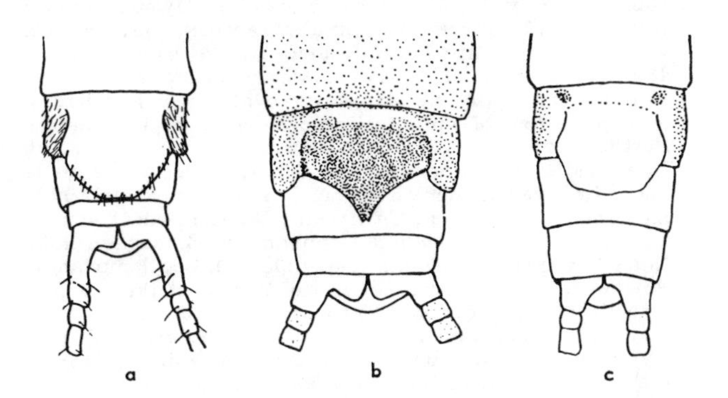

Fig. 6:50. Female terminalia of *Alloperla*, in ventral view. *a, forcipata; b, tamalpa; c, townesi* (*a*, Ricker, 1943; *b,c*, Ricker, 1952).

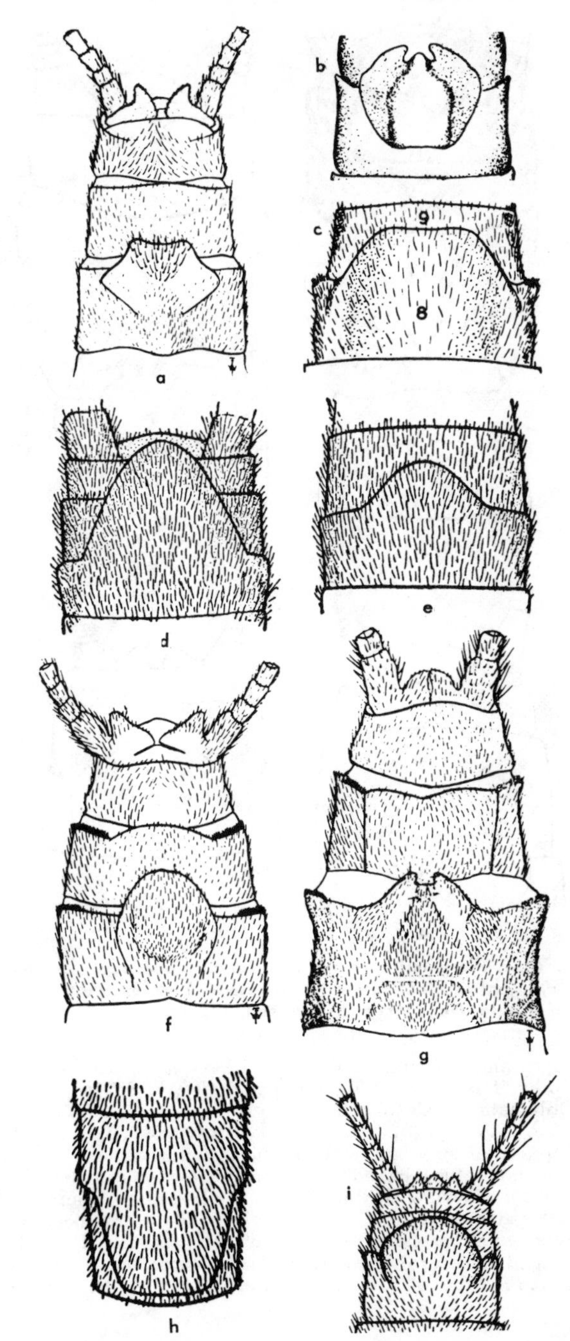

Fig. 6:51. Female terminalia of *Alloperla*, in ventral view. *a, coloradensis; b, fidelis; c, exquisita; d, severa; e, fraterna; f, pacifica; g, borealis; h, pallidula; i, signata* (*a,b,f,g,i*, Needham and Claassen, 1925; *c*, Frison, 1937; *d,e,h*, Frison, 1935).

— Supra-anal process extending well beyond the 10th tergite 13

13. Disc of the pronotum with a dusky central stripe (fig. 6:48*a*); southern California ...... *continua* Banks 1911

— Disc of the pronotum with reticulate dark markings (occasionally clear in *fidelis*) 14

14. Supra-anal process about 1.7 times as broad near the tip as it is near the base (figs. 6:48*b*; ♀, 6:51*b*); British Columbia to California; Idaho to Wyoming .... *fidelis* Banks 1920

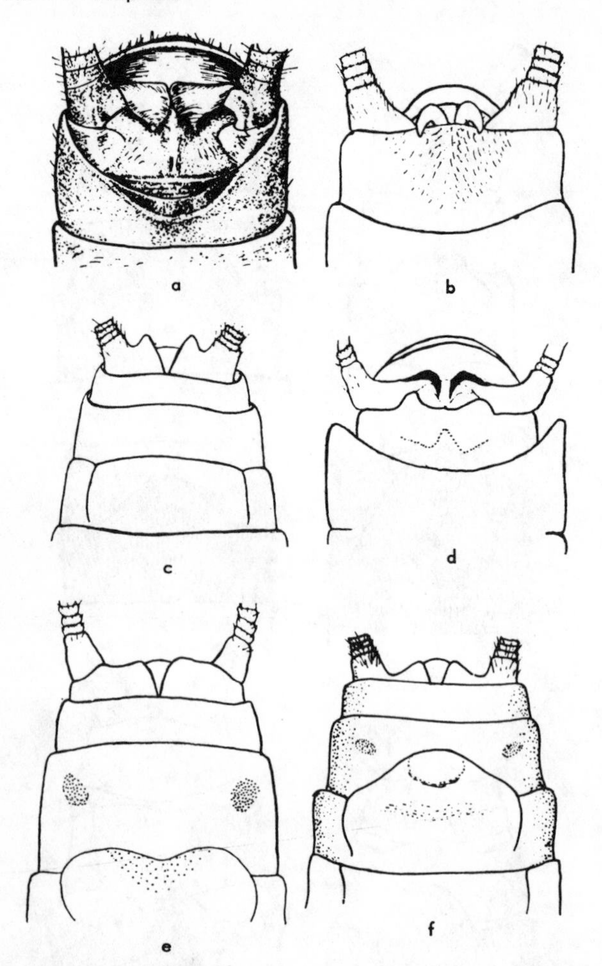

Fig. 6:52. Terminalia of *Acroneuria*. *a,f, pacifica*; *b,c, theodora*; *d,e, californica*; *a,b,d*, dorsal view of male; *c,e,f*, ventral view of female (Needham and Claassen, 1925).

— Supra-anal process about 1.2 times as broad near the tip as it is near the base (figs. 6:49*d*; ♀, 6:51*g*); British Columbia to California; Colorado .... *borealis* (Banks) 1895

15. Supra-anal process with a short, upturned hook at the tip, not club-shaped in lateral view (figs. 6:49*c*; ♀, 6:51*a*); British Columbia to California; Montana to Colorado .... *coloradensis* (Banks) 1898

— Supra-anal process without a hook at the tip; club-shaped in lateral view (figs. 6:47*a*; ♀, 6:50*c*); California ....*townesi* Ricker 1952

Family PERLIDAE

This family is divisible into two well-characterized subfamilies, the Perlinae of the Old World tropics and temperate regions of Africa, Eurasia, and eastern North America, and the Acroneurinae of the Americas and eastern and southeastern Asia. The keys on pages 159-160 and 162-163 will separate the subfamilies and genera found in North America, and the following key includes the recognized American subgenera and the three California species of *Acroneuria* Pictet 1841.

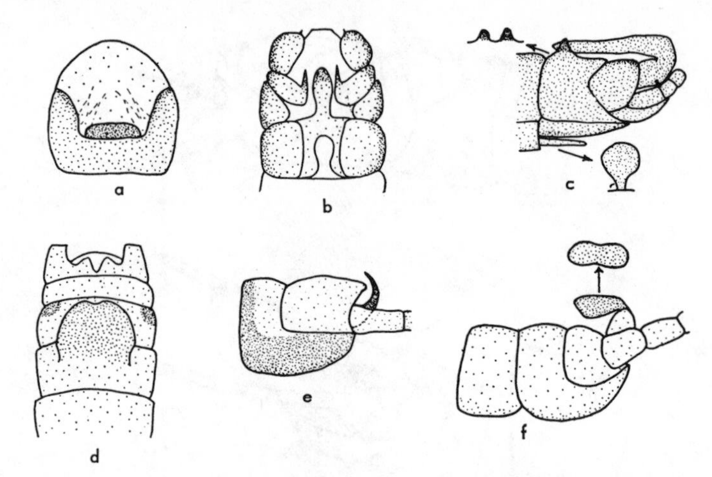

Fig. 6:53. *a*, *Peltoperla quadrispinula*, aedeagal structure through the ninth sternite; *b*, *Nemoura spiniloba*, ventral view of male terminalia; *c*, *Capnia maculata*, lateral view of male terminalia; *d*, *Isogenus alameda*, ventral view of female terminalia; *e*, *Isoperla marmorata*, lateral view of male terminalia; *f*, *Alloperla chandleri*, lateral view of male terminalia (Jewett, 1954*b*).

Key to North American Subgenera and California Species of Acroneuria

1. Fore wings suddenly widened beyond the origin of Rs; color wholly yellow; Appalachian .... (subgenus *Eccoptura* Klapalek 1921) *xanthenes* (Newman) 1828

— Fore wings gradually widened to stigma; not wholly yellow 2

2. Grooves of the mesosternum widely divergent 3

— Grooves of the mesosternum short and nearly parallel (fig. 6:52*a,f*); nymph with anal gills; British Columbia to California; Montana to Colorado...(synonyms: *A. pumila* Banks 1906, *H. obscura* Banks 1938, *A. delta* Claassen 1937) (subgenus *Hesperoperla* Banks 1938) ....*pacifica* Banks 1900

3. Ninth abdominal sternite of male bearing a triangular disc or hammer; Georgia, North Carolina, Tennessee (subgenus *Beloneuria* Needham and Claassen 1925) ....*georgiana* (Banks) 1914

— Ninth abdominal sternite of male bearing an oval or longitudinally rectangular disc or hammer 4

4. Hammer of male elongate-rectangular; nymph with a fairly complete row of spinules across the occiput; western North America...(subgenus *Calineuria* Ricker 1954) 5

— Hammer of male transversely oval; nymph with or without a row of spinules across the occiput; eastern North America 6

5. Head and thorax blackish (fig. 6:52*b,c*); ocellar triangle dark in nymph, and head and thorax not strikingly patterned; Oregon and California, Montana and Wyoming ....*theodora* Needham and Claassen 1922

— Head and thorax brownish (fig. 6:52*d,e*); ocellar triangle yellow in nymph, and head and thorax strikingly patterned; British Columbia to California............. ....*californica* (Banks) 1905

6. Nymph with a wavy but complete row of spinules across the occiput...(subgenus *Attaneuria* Ricker 1954) ....*ruralis* (Hagen) 1861

— Nymph without a row of spinules across the occiput .... (subgenus *Acroneuria*, *s.s.*)

REFERENCES

The literature listed below, with the exception of Newcomer's paper, has all been published since 1925, when Needham and Claassen's monograph of North American Plecoptera north of Mexico was published. Full references for citations in the preceding text before 1939 may be found in Claassen's catalogue (1940). For the period covered, this bibliography is not complete for the United States but does include all references cited in the preceding text, all papers which include treatment of California stonefly material or taxonomic treatment of species known to occur in the state, and a few other outstanding papers. Brinck's recent contribution on Swedish stoneflies, written in English, contains outstanding contributions concerning the biology of the order.

BANKS, NATHAN
1938. A new genus of Perlidae. Psyche, 45:136-137.
1947. Some characters in the Perlidae. Psyche, 54:266-291.
1948. Notes on Perlidae. Psyche, 55:113-130.
BRINCK, PER
1949. Studies on Swedish stoneflies (Plecoptera). Opuscula Entomologica, Supplementum XI, pp. 1-250, Lund.
CLAASSEN, P. W.
1931. Plecoptera nymphs of America (north of Mexico). Thos. Say Foundation of the Ent. Soc. Amer., Publ. 3, 199 pp. Springfield, Ill. Chas. C. Thomas.
1937*a*. New species of stoneflies (Plecoptera). Canad. Ent., 69:79-82.
1937*b*. New species of stoneflies (Plecoptera). Jour. Kans. Ent. Soc., 10:42-51.
1940. A catalogue of the Plecoptera of the world. Mem. Cornell Univ. Agric. Exp. Sta., 232:1-235.
DIMICK, R. E., and DON C. MOTE
1934. A preliminary survey of the food of Oregon trout. Oregon Agric. Exp. Sta. Bull., 323 pp. 1-23.
ESSIG, E. O.
1942. Plecoptera, *In* College Entomology. New York: Macmillan. pp. 148-158.
FRISON, T. H.
1929. Fall and winter stoneflies, or Plecoptera, of Illinois. Bull. Ill. Nat. Hist. Surv., 18:345-409.
1935*a*. The stoneflies, or Plecoptera, of Illinois. Bull. Ill. Nat. Hist. Surv., 20:281-471.
1935*b*. New North American species of the genus *Alloperla* (Plecoptera: Chloroperlidae). Trans. Amer. Ent. Soc., 61:331-344.
1936. Some new species of stoneflies from Oregon (Plecoptera). Ann. Ent. Soc. Amer., 29:256-265.
1937. Descriptions of Plecoptera, with special reference to the Illinois species. Bull. Illinois Nat. Hist. Surv., 21:78-99.
1942*a*. Descriptions, records, and systematic notes concerning western North American stoneflies (Plecoptera). Pan-Pac. Ent., 18:9-16, 61-73.
1942*b*. Studies of North American Plecoptera, with special reference to the fauna of Illinois. Bull. Illinois Nat. Hist. Surv., 22(2):235-355.
HANSON, JOHN F.
1942. Records and descriptions of North American Plecoptera. Part II. Notes on North American Perlodidae. Amer. Midl. Nat., 28:389-407.
1943. Records and descriptions of North American Plecoptera. Part III. Notes on *Isogenoides*. Amer. Midl. Nat., 29:657-669.
1946. Comparative morphology and taxonomy of the Capniidae. (Plecoptera). Amer. Midl. Nat., 35:193-249.
HANSON, JOHN F., and JACQUES AUBERT
1952. First supplement of the Claassen catalogue of the Plecoptera of the world. pp. 1-23.
HOPPE, GERTRUDE N.
1938. Plecoptera of Washington. Univ. Wash. Publ. Biol., 4:139-174.
JEWETT, S. G., JR.
1954*a*. New stoneflies (Plecoptera) from western North America, Jour. Fish. Res. Bd. Canad., 11(5):543-549.
1954*b*. New stoneflies from California and Oregon (Plecoptera), Pan-Pac. Ent., 30:167-179.
NEAVE, FERRIS
1929. Reports of the Jasper Park lakes investigations. II. Plecoptera. Contr. to Canad. Biol. and Fisheries, N. S. 4:159-168.
1933. Some new stoneflies from western Canada. Canad. Ent., 65:235-238.
1934. Stoneflies from the Purcell Range, B. C. Canad. Ent., 66:1-6.
NEEDHAM, J. G.
1933. A stonefly nymph with paired lateral abdominal appendages. Jour. Ent. Zool., 25:17-19.
NEEDHAM, J. G., and ELSIE BROUGHTON
1927. Central American stoneflies, with descriptions of new species (Plecoptera). Jour N.Y. Ent. Soc., 35:109-120.
NEEDHAM, J. G., and P. W. CLAASSEN
1925. A monograph of the Plecoptera or stoneflies of America north of Mexico. Lafayette, Ind.: Thos. Say Foundation of the Ent. Soc. Amer., publ. 2, pp. 1-397.
NEWCOMER, E. G.
1918. Some stoneflies injurious to vegetation. Jour. Agric. Res., 13:37-41.
RICKER, W. E.
1935. Description of three new Canadian perlids. Canad. Ent., 67:197-201.
1938. Notes on specimens of American Plecoptera in European collections. Trans. Roy. Canad. Inst., 22:129-156.
1943. Stoneflies of southwestern British Columbia. Indiana Univ. Publ. Sci. Ser. No. 12, pp. 1-145.
1949. The North American species of *Paragnetina*. (Plecoptera, Perlidae). Ann. Ent. Soc. Amer., 42:279-288.
1950. Some evolutionary trends in Plecoptera. Proc. Indiana Acad. Sci., 59:197-209.
1952. Systematic studies in Plecoptera. Indiana Univ. Publ. Sci. Ser. No. 18, pp. 1-200.
1954. Nomenclatorial notes on Plecoptera, Proc. Ent. Soc. B.C., 51:37-39.
SEEMANN, THERESA MARIAN
1927. Plecoptera. *In* Dragonflies, mayflies, and stoneflies of southern California. Jour. Ent. Zool., 19:51-59.

CHAPTER 7

Aquatic Hemiptera

By R. L. Usinger
University of California, Berkeley

Sixteen families of Hemiptera occur in, on, or near the water. These include the water boatmen (Corixidae), back swimmers (Notonectidae), water scorpions (Nepidae), giant electric light bugs (Belostomatidae), toad bugs (Gelastocoridae), shore bugs (Saldidae), the several families of surface striders, and a few others. All agree in having sucking mouth parts usually in the form of a slender, three- or four-segmented beak; antennae of not more than five segments, usually of four segments but variously reduced and concealed in the true aquatics; tarsi of not more than three segments; and the fore tarsi sometimes greatly reduced or absent. The wings, when present, have relatively few veins. Scent gland openings are present at the sides of the thorax in some forms, including Corixidae. A single opening occurs at the middle of the metasternum in Gerridae and Hydrometridae, and dorsal abdominal scent glands occur in the nymphs of Corixidae and Dipsocoridae (3), Saldidae, Mesoveliidae, and Macroveliidae (1), and Naucoridae (1 pair).

The aquatic Hemiptera are not homogeneous. Each family is distinctive in structure and habits, and the group as a whole is certainly polyphyletic. China (1955*a*, 1955*b*) summarized knowledge of hemipterous phylogeny and gave his views on the probable evolutionary lines of surface bugs and under water bugs in the form of diagrams (figs. 7:1; 7:2).

Acknowledgments

Dr. H. B. Hungerford of the University of Kansas gave a brief history of studies on aquatic Hemiptera in the introduction to his classical work, *The Biology and Ecology of Aquatic and Semiaquatic Hemiptera* (1920). This pioneer contribution is quoted on nearly every page of the present work. In addition, Dr. Hungerford has been an inspiration and help to me and to generations of students in the years since 1920. He is personally responsible for most of the great advances in this field, and he was good enough to read much of the present work in manuscript form. Dr. R. I. Sailer also read nearly all the manuscript and made several valuable suggestions. J. L. Herring made substantial contributions to the manuscript during the later stages of its preparation.

Others who have contributed indirectly or directly to this work are P. R. Uhler (1884), G. W. Kirkaldy, whose *Guide to the Study of British Waterbugs* (1898-1908) served as an introduction to the subject; the late J. R. de la Torre-Bueno, whose collection is now at the University of Kansas; the late E. P. Van Duzee, whose *Catalogue* (1917) is still the main source of information on literature and whose collection at the California Academy of Sciences was a valuable source of records; C. J. Drake and H. M. Harris, whose work at Ames, Iowa, has provided indispensable monographs of several groups, and a host of graduate students at the University of Kansas including: Kuitert, Cummings, Sailer, Hidalgo, McKinstry, Hodgden, Martin, Porter, Gould, Wiley, Evans, Truxal, Bare, Hoffmann, and Todd. Important contributions by others include O. Lundblad and T. Jaczewski (Corixidae), G. E. Hutchinson (Corixidae and Notonectidae), José de Carlo (Belostomatidae, Nepidae, and Naucoridae) and Ira La Rivers (Naucoridae). Valuable help in collecting has been given by the late Harry P. Chandler and J. D. Lattin. Some general works on aquatic Hemiptera of other parts of the world are: Blatchley (1926), Butler (1923), Abbott and Torre-Bueno (1923), Uhler (1884), Macan (1941), Herring (1950-1951), Zimmerman (1948).

The most recent account of the aquatic Hemiptera of California is by Usinger (*in* Usinger, La Rivers, Chandler, and Wirth, 1948). For the present study the 1948 keys have been completely revised, synonymy has been rechecked and revised, and many new records have been added.

For the illustrations of whole insects I am indebted to Arthur Smith, British Museum (Natural History) and for funds, to the Research Committee, University of California. Mrs. Frieda Abernathy is responsible for figures 7:7, 7:26, 7:36, and 7:37, prepared originally for Zimmerman's *Insects of Hawaii*, and figure 7:12. The anatomical drawings of Corixidae (figs. 7:3; 7:9) and *Microvelia* (fig. 7:31) were made by Mrs. Celeste Green.

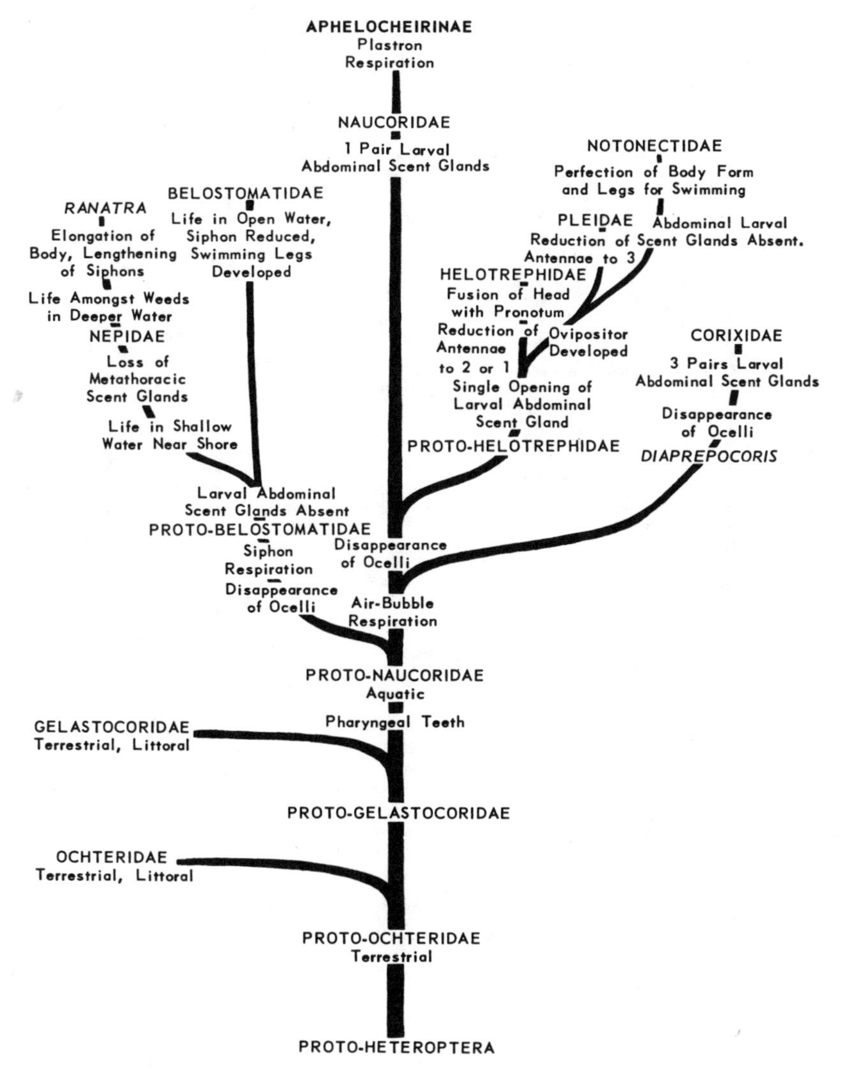

Fig. 7:1. Probable evolutionary lines of water bugs (China, 1955).

DISTRIBUTION

California has a diverse hemipterous fauna. On the north coast and in the mountains (Transition Zone and above) are found boreal elements common to the northern United States and Canada. On the east slope of the Sierra, including parts of Lassen County and the Owens Valley, Great Basin species are found. Then, along the Colorado River bordering Arizona occur species typical of the Sonoran region. Of particular interest are the relict forms found in isolated springs in the midst of the southern deserts. These springs are relict habitats, mere vestiges of more extensive lakes and streams of former times. That the aquatic Hemiptera were not very different during the recent past, however, is indicated by species in the La Brea and McKittrick tar pits (Pierce, 1948) (Belostomatidae, Notonectidae, and Nepidae) which differ only slightly, if at all, from present-day forms in the same area. The age of these fossils is presumably about 10,000 years.

TABLE 7:1

COMPARISON OF NUMBER AND PERCENTAGES OF AQUATIC HEMIPTERA OF CALIFORNIA AND THE BRITISH ISLES

Hemiptera	California		British Isles	
	Species	Per cent	Species	Per cent
Corixidae	29	25.6	35	41.2
Notonectidae	11	9.7	4	4.8
Pleidae	0	0	1	1.2
Naucoridae	8	7	2	2.3
Belostomatidae	5	4.5	0	0
Nepidae	3	2.6	2	2.3
Gelastocoridae	4	3.5	0	0
Ochteridae	1	1	0	0
Gerridae	11	9.7	10	11.7
Veliidae	8	7	4	4.8
Hydrometridae	1	1	2	2.3
Macroveliidae	1	1	0	0
Hebridae	4	3.5	2	2.3
Mesoveliidae	2	1.7	1	1.2
Saldidae	24	21 2	20	23.5
Aepophilidae	0	0	1	1.2
Dipsocoridae	1	1	1	1.2
Total	113	100	85	100

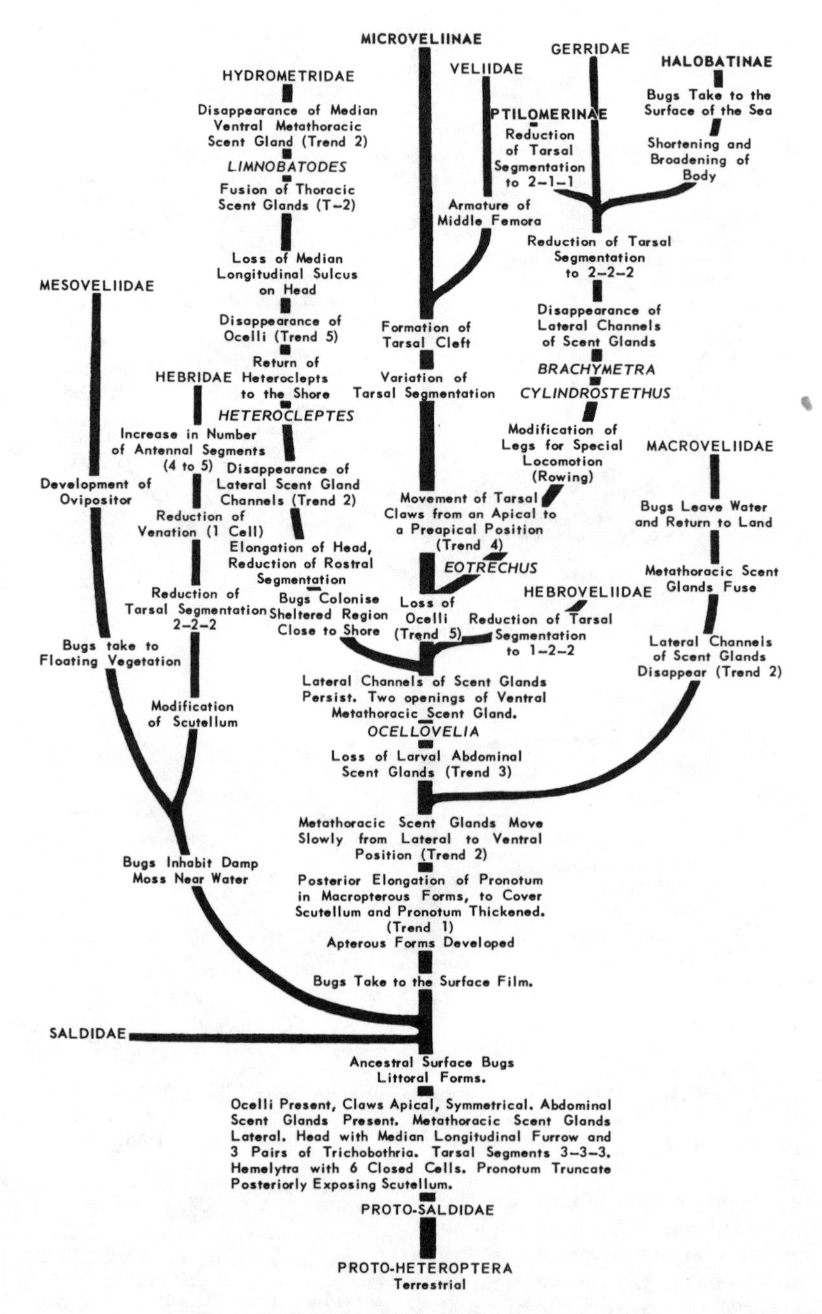

Fig. 7:2. Probable evolutionary lines of surface bugs (China, 1955).

The California fauna is roughly comparable in number of species and percentage representation of families to other parts of the world. Table 7:1 shows a detailed comparison with the best known and most recently catalogued fauna, that of the British Isles (Kloet and Hincks, 1945). California has roughly one-third more species than Britain. The percentage composition, however, is quite different, the Corixidae being significantly more numerous in species in Britain and the Notonectidae, Naucoridae, and Veliidae being less numerous. One species each of Pleidae and Aepophilidae occur in Britain but are lacking in California, and the families Belostomatidae, Gelastocoridae, Ochteridae, and Macroveliidae, present in California, do not occur in the British Isles.

Jaczewski (1937) has carried this type of analysis still further, and from his tables it is evident that the California fauna compares favorably with those of Poland, Germany, Sweden, Holland, France, and New York, being larger than any of these both in number of species and in area. However, the percentage of Corixidae is about half that of each of the above faunas. The fauna of the Malay Archipelago is larger than that of California whereas the Australian fauna is poorer. In the latter cases, particularly, more collecting may alter the figures.

ECOLOGY

The aquatic and semiaquatic Hemiptera occupy a wide variety of habitats from salt-water pools to mountain lakes and from hot springs to large rivers. In general, they fill the role of predators at intermediate stages in the food chains of their respective communities. Some, such as the surface striders, appear to be complete masters of their environment, whereas others, such as the relatively defenseless Corixidae, are generally preyed upon. Corixids are partly responsible for the primary conversion of plant material into animal food, but it now seems clear that they cannot subsist on the "flocculent ooze" alone but must take animal food, such as small mosquito larvae, as a part of their diet.

Although each genus of aquatic Hemiptera occupies a characteristic habitat and exhibits distinctive behavior patterns, it is difficult to organize such information in a useful way. Hungerford (1920) gave a "Habitat Key" which has been modified below, adapted to the California fauna, and enlarged with new information gleaned from the literature and experience of the last thirty-five years.

Key to California Aquatic and Semiaquatic Hemiptera

Based on Habits and Habitats

1. True aquatics, living beneath the water 2
— Surface or shore bugs 10
2. Respiration by breaking surface film with pronotum 3
— Respiration via tip of abdomen or abdominal appendages 4
3. Salt or brackish waters ..... CORIXIDAE—*Trichocorixa*
— Fresh waters .... other CORIXIDAE
4. Respiration by breaking surface film with tip of abdomen 5
— Respiration by long, slender "tube" or short, retractile flaps 8
5. Back swimmers 6
— Swimming with dorsum uppermost 7
6. Limnetic, resting poised in open water ....NOTONECTIDAE—*Buenoa*
— Resting on submerged objects or at surface film ....NOTONECTIDAE—*Notonecta*
7. Living amidst rocks in streams and lake margins ... .NAUCORIDAE—*Ambrysus*
— Living amidst aquatic vegetation in ponds ....NAUCORIDAE—*Pelocoris*
8. Awkward swimmers, living in trash and tangled plants ....NEPIDAE
— Strong swimmers, hiding in protected places 9
9. Eggs laid on stems of rushes and the like .... BELOSTOMATIDAE—*Lethocerus*
— Eggs laid on backs of males ....BELOSTOMATIDAE—*Abedus, Belostoma*
10. Surface bugs 11
— Shore bugs 17
11. Living on open water surface 12
— Living at water's edge or on floating vegetation, running on open water when disturbed 15
12. Marine .... GERRIDAE—*Halobates*
— Fresh waters 13
13. Living on ponds, lakes, and quiet waters of streams .... GERRIDAE—*Gerris, Trepobates*
— Living on swiftly flowing streams, gregarious 14
14. Living on riffles, usually of small streams and rivers ....VELIIDAE—*Rhagovelia*
— Living on open water, usually of large rivers .... GERRIDAE—*Metrobates*
15. Delicate "treaders" on long, stiltlike legs in protected places .... HYDROMETRIDAE
— Rapid "skaters" on shorter legs in open places 16
16. Resting on surface film near shore, gregarious .... VELIIDAE—*Microvelia*
— Resting on floating leaves, and the like, more solitary ..............MESOVELIIDAE—*Mesovelia mulsanti*
17. Walking or running forms 18
— Jumping or flying forms 22
18. Swift runners, beneath stones at water's edge .... DIPSOCORIDAE
— Slow walkers at water's edge, stones, moss, etc. ... 19
19. Capable of skating on surface film when disturbed ................MESOVELIIDAE—*Mesovelia amoena*
— Not capable of skating on surface film 20
20. On dry land near water...........MACROVELIIDAE
— At water's edge 21
21. Under stones or on moss ....HEBRIDAE—*Hebrus*
— On floating mats of algae...... HEBRIDAE—*Merragata*
22. Jumping forms, concealed by coloration of sand or mud 23
— Jumping and/or flying forms 24
23. Exposed on sand or mud; eggs in sand ....GELASTOCORIDAE—*Gelastocoris*
— Hiding in mud under rocks; eggs brooded in mud cell .... GELASTOCORIDAE—*Nerthra*
24. Moving by short, quick jump-flights ....SALDIDAE
— Moving by longer flights or hiding near water ....OCHTERIDAE

BIOLOGY.

In general, water bugs overwinter as adults, lay eggs in the spring, develop during the summer, and then repeat their yearly cycle. Only a few exceptions to this are noted in the following pages. There are five nymphal instars in all but a few species of *Microvelia*. Eggs are laid in a wide variety of places, each usually characteristic for a particular species. Unlike many aquatic insects no water bug is known to deposit its eggs freely to float on the surface or in the water. The eggs are glued to various objects including the backs of males (*Belostoma* and *Abedus*). Bug eggs can usually be recognized by the relatively tough, hexagonally reticulate chorion and by buttonlike or peglike micropylar processes. They are usually oval, occasionally spindle-shaped, and sometimes stalked. A key to eggs and egg-laying habits is given below with some diffidence because information on this subject is so inadequate.

Key to the Eggs and Egg-Laying Habits of California

Aquatic and Semiaquatic Hemiptera

1. Eggs stalked and attached by the end opposite the micropyle 2
— Eggs not stalked, inserted in plant tissue or glued to the substrate 3
2. About 6 times as long as greatest diameter; brown in color with longitudinal fluting at middle, the tapered end reticulate .................HYDROMETRIDAE

— Suboval in shape, white, with the micropylar end briefly produced as a small nubbin (fig. 7:4) ..... CORIXIDAE
3. Two threadlike filaments at micropylar end, longer than the egg itself; egg inserted in plant tissue with only the filaments projecting .... NEPIDAE—*Ranatra*
— No threadlike filaments 4
4. Micropylar end with a small, cylindrical, bent projection 5
— Micropylar end without such a bent projection 6
5. A smooth elongate oval area on anterior half of upper surface. Eggs inserted in plant tissue, the smooth area exposed .... NOTONECTIDAE—*Buenoa*
— Without such a smooth area, eggs inserted or glued to the substrate .... NOTONECTIDAE—*Notonecta*
6. Micropylar end narrowed and bent as a curved neck terminating in a flat surface which is exposed, the rest of the egg being inserted in plant tissue .... MESOVELIIDAE
— Form oval or subcylindrical, not narrowed as above, and not inserted in plant tissue 7
7. Attached vertically by the end opposite micropyle; color brown; laid in clusters of 100 or more, the individual eggs contiguous 8
— Attached horizontally, glued to leaves of aquatic or shore plants, floating or submerged objects, or laid free 9
8. Laid in a clump on the folded hemelytra of the male .... BELOSTOMATIDAE—*Abedus, Belostoma*
— Laid in a mass above the water on cattail stalks or other objects .... BELOSTOMATIDAE—*Lethocerus*
9. Oval, about twice or less as long as wide 10
— Elongate-oval, about 3 or more times as long as wide 13
10. Twice as long as wide; glued to objects under water .... NAUCORIDAE
— Less than twice as long as wide, laid in sand or mud or at the bases of clumps of grass, out of water 11
11. Laid in clumps of grass and roots ...... OCHTERIDAE
— Laid in sand or mud 12
12. Laid amidst sand grains .... GELASTOCORIDAE—*Gelastocoris*
— Laid in small holes in mud several feet from shore, beneath stones, the female "guarding" the eggs .... GELASTOCORIDAE—*Nerthra*
13. Laid just beneath the surface film on floating or emergent objects 14
— Laid out of water at the bases of clumps of grass or moss 15
14. Length usually more than 1 mm.; often laid in rows, the individual eggs not contiguous, laid side by side .... GERRIDAE
— Length less than 1 mm.; laid individually or in irregular clusters .... VELIIDAE
15. Shape elongate-oval .... HEBRIDAE
— Slightly tapering and bent toward micropylar end .... SALDIDAE

Flight.—Only the marine *Halobates,* among California Hemiptera, have lost their wings completely. Others are capable of flying at times, and there is no doubt that this ability has real survival value in an area where rainfall is low and seasonal, and many ponds and streams dry up for a part of each year. Macan (1939) recorded the recolonizing of a fountain at King's College, Cambridge, England, throughout a period of six years, and found twenty-six species of aquatic insects including seven corixids and one notonectid. No study of this kind has been made in California, but corixids *(Corisella)* and belostomatids *(Lethocerus)* are frequently attracted to lights, sometimes in large numbers, during the summer months in the Sacramento and San Joaquin valleys.

Wing polymorphism is typical of members of the superfamily Gerroidea and is dealt with in the discussion of each of the families in which it occurs. Despite the studies by Poisson (1924) and others on wing polymorphism, a full understanding of the genetic factors and environmental influences is lacking. Curiously, some of the macropterous Gerridae (Halobatinae) break off the wings, presumably during mating.

Stridulation.—A great deal has been written about sound production in water bugs, but firsthand records of noises that have been heard are rare indeed. Hungerford (1924) heard a *Buenoa* stridulate, and it appears certain that the structures seen in certain corixids are for this purpose. Details are given under each family.

Respiration.—Each of the truly aquatic families has a distinctive type of respiration which will be discussed later in detail. In general, water bugs are dependent on surface air obtained through tubes (Nepidae), flaps (Belostomatidae), the tip of the abdomen (Naucoridae, Notonectidae), or the pronotum (Corixidae). Only certain exotic naucorids *(Aphelocheirus)* can live exclusively on dissolved oxygen, obtained through a submicroscopic plastron. Corixids, notonectids, naucorids, and belostomatids gain a large part of their oxygen by diffusion into the air "bubble" or film held by hydrofuge hairs on their ventral surfaces. This type of respiration was first described by Comstock (1887) and later elaborated by Thorpe (1950).

Collecting methods and rearing techniques.—Water bugs are obtainable without special equipment. A nylon net will serve for all forms in or on the water, and a screen or dredge may be used in dense aquatic vegetation. For shore bugs and species that live at the water's edge painstaking search must be made beneath rocks and in moss and other marginal plants. Bogs may be collected by pressing the vegetation down and floating specimens out. Swift-moving water striders are best collected by a quick downward stroke of the net, followed by a twist as it splashes on the water. Stream forms such as *Ambrysus* may be taken by holding the net downstream and dislodging the bugs by moving rocks above the net. Saldids and other jumping forms can be made to jump into the net by holding it still with one hand while approaching the specimen from the opposite side with the free hand. Light traps are useful for a very few species that fly regularly, but a light at the water's edge has been found to attract various species including *Ambrysus.* Berlese funnels are useful for small shore dwellers such as hebrids, dipsocorids, and saldids that regularly inhabit trash and vegetation at the water's edge.

Any water bug may be preserved in alcohol, but most study collections consist of pinned or pointed specimens. In most cases it is useful to have some specimens dried and others in alcohol. The latter are ideal for dissection and for careful study under the higher powers of the dissecting microscope. Of course dried specimens may be softened in KOH and dissected or studied intact, but this usually results in damaged or destroyed specimens.

Water bugs, like other aquatic insects, should be transported alive in containers with moist water plants rather than in jars of sloshing water. They do well in home aquaria if cared for daily. However, the mortality is usually quite high during successive molts, and very few species have been reared through many generations. For the true aquatics a constant supply of fresh air is desirable and can be had by bubbling through a line from an aquarium aerator. Shore bugs do well in small jars containing a substrate of plaster of Paris darkened with lamp black. Excessive moisture must be avoided, lest the small nymphs become entangled in the surface film.

Food can be provided by sweeping terrestrial vegetation, by maintaining a culture bottle of *Drosophila,* by supplying freshly killed or immobilized house flies, or by keeping a culture of mosquito larvae or brine shrimp.

Bugs must be kept in individual containers to avoid cannibalism and to keep track of cast skins.

SYSTEMATICS

Characters.—Figure 7:3 shows the external structures of a corixid. Other bugs vary in details but these drawings will suffice for a general understanding of the group. It should be noted that the first visible abdominal segment as seen from below is really the second segment. The genital capsule or pygophore is usually concealed and is in the ninth segment. Genitalia are used for the separation of species in most groups of aquatic Hemiptera. The genital segment may be removed in dried specimens by softening the tip of the abdomen in a solution of phenol (1/3), water (1/3), and alcohol (1/3). A dissecting needle with its tip bent should be inserted at the side of the eighth segment and then twisted and drawn out, pulling the genital segment. For specimens preserved in alcohol the operation is simpler, because the parts are soft enough to dissect without special relaxing fluid. Genitalia, including the claspers and aedeagus, are best studied in fluid, but in a few cases the structures are hard enough to observe in a dried condition. The latter can be glued to a card beneath the specimen, but the genitalia of most water bugs should be mounted in balsam on microscope slides or between cover slips on small cards or points beneath the specimen. Perhaps the commonest method of preserving genitalia is in micro vials containing glycerine. These are stoppered with long cylindrical corks and are held at an angle on the pin below the insect, care being taken to keep the glycerine from permanent contact with the cork.

Measurements are made with an eye-piece micrometer. Total length is the over-all length of the insect in one plane. Antennal segments are measured ignoring the small, intercalary, ringlike segments seen in some forms.

Key to the Families of Aquatic and Semiaquatic Hemiptera

Adults and Nymphs.

1. Antennae shorter than head, inserted beneath eyes, not visible from above (fig. 7:3*a,b*) except in Ochteridae (fig. 7:23). True aquatics and a few shore bugs. Series Cryptocerata 2
— Antennae longer than head, inserted in front of eyes and plainly visible from above (fig. 7:24). Shore bugs and surface bugs. Series Gymnocerata 9
2. Rostrum very short, broad, scarcely distinguishable from the broad apex of head, not distinctly segmented (fig. 7:3*b*); front tarsi developed as comblike palae (fig. 7:9); base of head overlapping front margin of pronotum (fig. 7:6); nymphs with 3 dorsal abdominal scent gland openings .... CORIXIDAE
— Rostrum cylindrical or cone-shaped and distinctly 3- or 4-segmented; front tarsi not developed as comblike palae; base of head inserted in pronotum; nymphs with scent gland openings lacking or present on only one segment 3
3. Abdomen with a pair of long, slender posterior appendages, forming a respiratory siphon; hind coxae short, free, rotary .... NEPIDAE
— Abdomen without a pair of long, slender posterior appendages; hind coxae broadly joined to thoracic pleura 4
4. Ocelli absent; middle and hind legs provided with more or less extensive fringes of long swimming hairs. Water bugs 5
— Ocelli present; middle and hind legs without fringes of swimming hairs. Shore bugs 8
5. Front legs not chelate; body strongly convex above 6
— Front legs chelate, the femora enlarged and tibiae curved and articulating against femora; body subflattened above 7
6. Form elongate; adults large, more than 5 mm.; hind legs long, oarlike, without distinct claws (fig. 7:10) .... NOTONECTIDAE
— Form oval, adults small, less than 3 mm.; hind legs not long and oarlike, with 2 distinct claws...... PLEIDAE
7. Membrane of hemelytra with reticulate veins, abdomen with a pair of short, flat, retractile posterior appendages (fig. 7:16); nymphs without dorsal abdominal scent gland openings .... BELOSTOMATIDAE
— Membrane of hemelytra without veins; abdomen without posterior, straplike appendages (fig. 7:13); nymphs with a pair of widely separated scent gland openings between 2nd and 3rd abdominal tergites (fig. 7:12).......... .... NAUCORIDAE
8. Front legs raptorial, the femora very broad and grooved along inner edges nearest the curved tibiae; antennae concealed in grooves beneath strongly protuberant eyes (fig. 7:21) .... GELASTOCORIDAE
— Front legs similar to middle pair, fitted for running; antennae exposed (fig. 7:23) .... OCHTERIDAE
9. Hind coxae short, freely movable, rotatory. Shore and surface bugs 10
— Hind coxae long, broadly joined to thoracic pleura. Shore bugs 15
10. Claws of at least front tarsi inserted before apex ... 11
— Claws all inserted at tips of tarsi 12
11. Hind femora very long, greatly exceeding apex of abdomen (fig. 7:24); adults with a median metasternal scent gland opening .... GERRIDAE
— Hind femora scarcely, if at all, surpassing tip of abdomen (fig. 7:29); adults with lateral metathoracic scent gland openings .... VELIIDAE
12. Head as long as the entire thorax; body long and cylindrical (fig. 7:32); nymphs without dorsal abdominal scent gland openings .... HYDROMETRIDAE

— Head shorter, not exceeding the combined length of pronotum and scutellum; nymphs with a single dorsal abdominal scent gland opening 13
13. Tarsi 2-segmented in adults; under surface of head deeply grooved to form a rostral sulcus (figs. 7:35; 7:36) .... HEBRIDAE
— Tarsi 3-segmented in adults; under surface of head without a rostral groove, the base of rostrum clearly visible 14
14. Legs with scattered stiff black bristles; winged forms with scutellum exposed and double, and ocelli present; wingless adults with scutellum and ocelli absent (figs. 7:37; 7:38)........................ MESOVELIIDAE
— Legs without scattered, stiff black bristles; adults fully winged or brachypterous, with scutellum concealed by backwardly projecting plate of pronotum; ocelli distinct (fig. 7:34)................ MACROVELIIDAE
15. Ocelli close together, nearer to each other than to eyes; antennal segments all about equally thick or apical segments thicker (figs. 7:40; 7:41); nymphs with 1 dorsal abdominal scent gland opening .... SALDIDAE
— Ocelli widely separated, nearer to eyes than to each other; first 2 antennal segments very stout, the apical 2 slender; nymphs with 3 or 4 dorsal abdominal scent gland openings (fig. 7:43) .... DIPSOCORIDAE

REFERENCES

(Sometimes cited but not repeated in later sections)

ABBOTT, F. A., and J. R. de la TORRE-BUENO
1923. *In* W. E. Britton, Guide to the insects of Connecticut. Part IV, The Hemiptera or sucking insects of Connecticut, pp. 1-807, 20 pls.
BLATCHLEY, W. S.
1926. Heteroptera or true bugs of eastern North America. Indianapolis: Nature Publ. Co., pp. 1-1116.
BUTLER, E. A.
1923. A biology of the British Hemiptera—Heteroptera. London: H. F. and G. Witherby. pp. viii + 682.
CHINA, W. E.
1955*a*. The evolution of the water bugs. National Institute of Sciences of India, Bull. 7:91-103.
1955*b*. A reconsideration of the systematic position of the family Joppeicidae Reuter, with notes on the phylogeny of the suborder. Ann. Mag. Nat. Hist., (12)8:353-370.
COMSTOCK, J. H.
1887. Note on respiration of aquatic bugs. Amer. Nat., 21:577-578.
HERRING, JON L.
1950-1951. The aquatic and semiaquatic Hemiptera of northern Florida. Part I: Gerridae. Florida Ent., 33:23-32; Part II: Veliidae and Mesoveliidae. Florida Ent., 33:145-150; Part III: Nepidae, Belostomatidae, Notonectidae, Pleidae and Corixidae. Florida Ent., 34:17-29; Part IV: Classification of habitats and keys to the species. Florida Ent., 34:146-161.
HUNGERFORD, H. B.
1920. The biology and ecology of aquatic and semiaquatic Hemiptera. Kansas Univ. Sci. Bull. Vol. II, 341 pp., 31 pls., 3 color pls. (Whole Series, Vol. 21, No. 17).
1924. Stridulation of *Buenoa limnocastoris* Hungerford and systematic notes on the *Buenoa* of the Douglas Lake region of Michigan, with the description of a new form. Ann. Ent. Soc. Amer., 17:223-227.
JACZEWSKI, T.
1937. Allgemeine Züge der geographischen Verbreitungen der Wasserhemipteren. Arch. für Hydrobiol., 31:565-591.
KIRKALDY, G. W.
1898-1908. A guide to the study of British waterbugs. Entomologist, 31:177-180, 203-206; 32:3-8, 108-115, 151-154, 200-204, 296-300; 33:148-152; 38:173-178, 231-236; 39:60-64, 79-83, 154-157; 41:37.
KLOET, G. S., and W. D. HINCKS
1945. A check list of British insects. Stockport. 483 pp.
MACAN, T. T.
1939. Notes on the migration of some aquatic insects. Jour. Soc. Brit. Ent., 2:1-6.
1941. A key to the British water bugs, with notes on their ecology. Freshwater Biol. Assn. Brit. Emp., Sci. Publ. No. 4, pp. 1-36.
PIERCE, W. D.
1948. Fossil arthropods of California. 15. Some Hemiptera from the McKittrick asphalt field. Bull. So. Calif. Acad. Sci., 47:21-33.
POISSON, R.
1924. Contribution a l'étude des Hémipteres aquatiques. Bull. Biol. France et Belgique., 38:49-305, 35 figs. pls. 1-13.
SINGH-PRUTHI, HEM
1925. The morphology of the male genitalia in Rhynchota. Trans. Ent. Soc. London, pp. 127-267, 32 pls.
THORPE, W. H.
1950. Plastron respiration in aquatic insects. Biol. Reviews, 25:344-390.
UHLER, P. R.
1884. Order IV. Hemiptera. *in* Standard Natural History, Vol. II, pp. 204-296.
VAN DUZEE, E. P.
1917. Catalogue of the Hemiptera of America north of Mexico. Univ. Calif. Publ. Ent., 2:1-902.
ZIMMERMAN, E. C.
1948. Insects of Hawaii. Vol. 3. Heteroptera. University of Hawaii Press. pp. 1-255.

Family CORIXIDAE

Water Boatmen

Water boatmen are the most numerous of all aquatic Hemiptera, both in species and in individuals. They occur from below sea level (Death Valley, California) to 15,000 feet elevation (Himalaya Mts.). They seem equally adapted to the cold waters of the subarctic and to tropical waters.

Corixids play an important role in aquatic communities because they are primary converters of plant material (see below under feeding habits) and also serve as an early link in the animal food chain by scooping up and ingesting small benthic organisms. Other food consists of small midge and mosquito larvae, and to this extent corixids act as typical predators.

Analyses of fish stomachs show that corixids are a preferred food in many instances (Forbes, 1888). The experiments of Popham (1941) show clearly that the resemblance of corixids to the immediate background colors of their environment is owing to selective predation by fish. But to what extent the bugs are rendered unpalatable by their nymphal and adult scent glands is not known.

Stridulation was first recorded in the Corixidae more than one hundred years ago. Several persons have heard a chirping sound while observing the insects in aquaria. According to Von Mitis (Hungerford, 1948), "Only those species in which the males are equipped with a field of pegs on the base of the front femur are capable of producing sound, and the sound is produced by rubbing this peg field over the

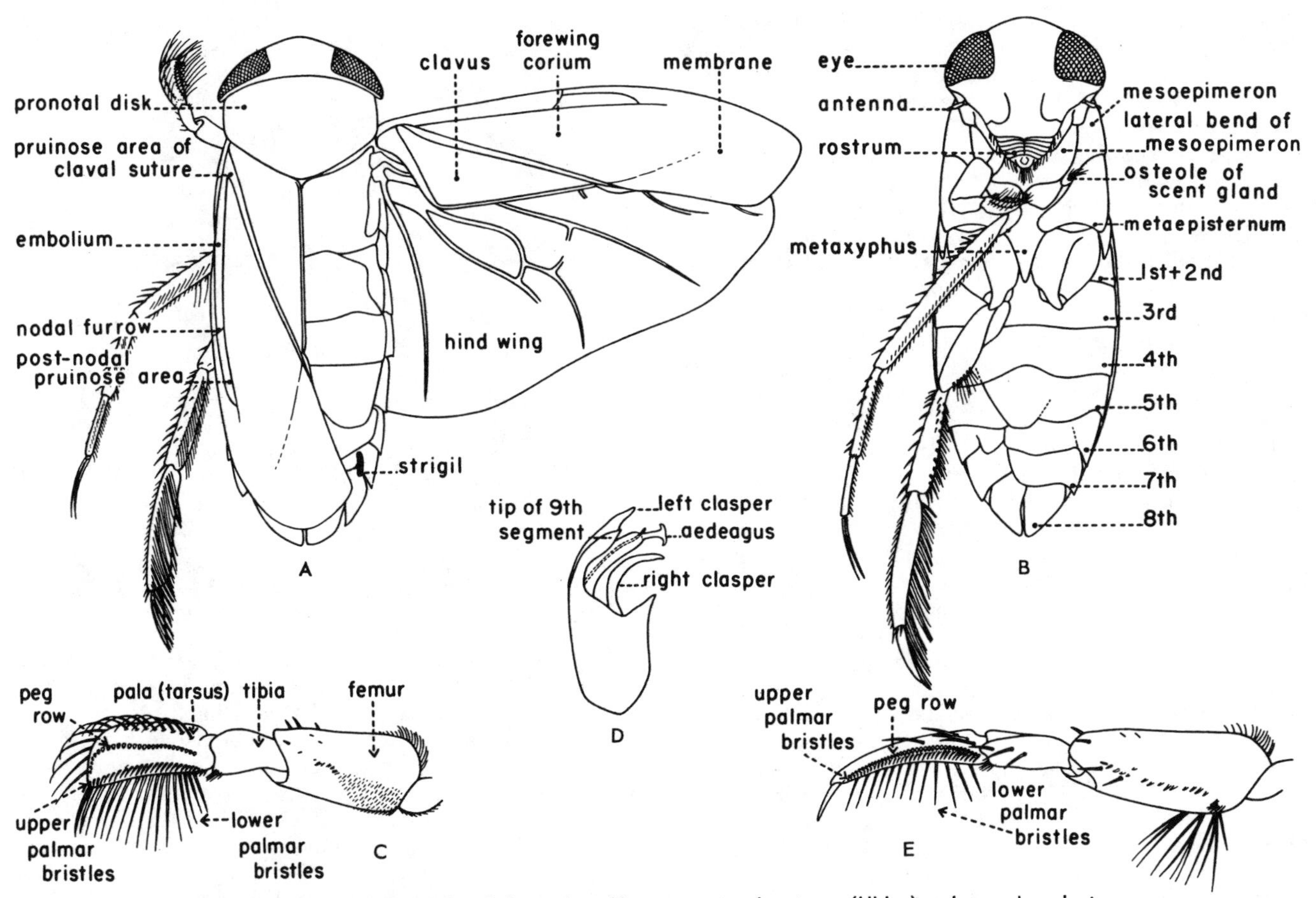

Fig. 7:3. Structural details of Corixidae. *Hesperocorixa laevigata* (Uhler) male: *a*, dorsal view; *b*, ventral view; *c*, front leg of male; *d*, genital capsule of male; *e*, *Graptocorixa californica* Hungerford, front leg of male.

sharp lateral cephalic margin of the head." According to Hungerford members of the following California genera should be capable of sound production: *Corisella, Hesperocorixa,* and *Cenocorixa.*

Corixids have entered into commercial trade channels- for at least three hundred years and probably for many centuries. According to Hungerford (1948) *Corisella mercenaria* (Say) "is often found in Mexican markets mixed with *C. edulis, C. tarsalis* (Fieb.) . . . *Krizousacorixa femorata* (Guer.), *K. azteca* Jacz., and *Notonecta unifasciata* Guer. Tons of these dried insects are also shipped abroad for bird or fish food, and their eggs are gathered for human food from Lake Texcoco, Mexico." The adult bugs are called "Mosco" and the eggs "ahuatle." They are for sale over the counter throughout Mexico, and the adults are packaged in cellophane envelopes and sold as food for birds and pet turtles in pet stores in the United States. The eggs are laid in enormous numbers on submerged objects, each egg attached by a short petiole. In Mexico eggs are collected (Ancona, 1933) by placing reeds in the water and returning at a later date to harvest them.

An interesting oviposition relationship between a corixid and a crayfish was first noted by Forbes (1878). According to Griffith (1945) the eggs of *Ramphocorixa acuminata* are preferably deposited upon crayfish, usually *Cambarus immunis* and *C. simulans.* These species are typically associated with the corixid in water holes. The oviposition habit is not obligatory, amounting to a fixation of choice in the behavior pattern of the insect, in which preference extends to areas affected by branchial current. Protection of eggs from drought and enemies is a possible benefit accruing from the association; aeration by branchial currents may be a factor in the choice of special areas; while the convenience of the crayfish in long association with the ovipositing water bugs serves largely to explain the relationship.

Flight.—Some corixids are attracted to lights in enormous numbers. The most striking example of this in California is *Corisella decolor.* On the day after a large flight dead specimens of this species are seen in heaps on the ground beneath lights. This has been observed at Davis and elsewhere in the Central Valley.

Habitats.—The various species of Corixidae show marked preferences for particular habitats and thus may serve as indicators of local conditions. In California *Trichocorixa reticulata* and *T. verticalis* are the halobionts, living exclusively in brackish or saline waters. The former species has been found in the ocean (Hutchinson, 1931) and occurs in the brine pools (Leslie Salt Company) on the south shore of San Francisco Bay together with the brine shrimp,

Artemia salina, and the brine fly, *Ephydra gracilis.* The bugs also occur in the saline waters of former lake beds in the California deserts. They are the dominant insect of the extensive inland Salton Sea and thrive at Badwater in Death Valley.

Corisella decolor and *C. inscripta* likewise show a preference, or at least a tolerance, for saline waters. Although absent from the situations described for *Trichocorixa,* they occur in enormous numbers in Little Borax Lake, near Clear Lake, and in the septic waters of the sewage oxidation ponds at Concord, Contra Costa County.

It is interesting to note that both *Trichocorixa* and *Corisella* are commonly found in pairs, two closely related but perfectly distinct species living together under what appear to be identical conditions.

The remaining genera of California corixids comprise typically fresh-water species. These are found under varied conditions such as quiet pools in streams, ponds, and lakes. Many of these species are widely distributed, from south to north and from the coast to the Sierra Nevada.

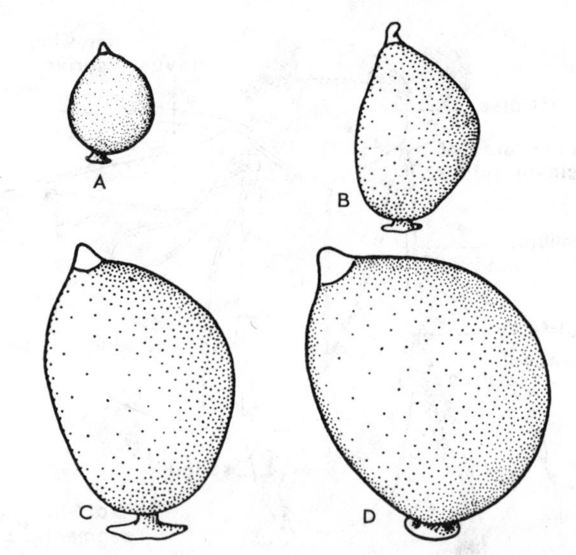

Fig. 7:4. Eggs of Corixidae. *a, Sigara alternata* (Say); *b, Cenocorixa bifida* (Hungerford); *c, Corisella edulis* (Champion); *d, Graptocorixa abdominalis* (Say) (Hungerford, 1948*a*).

Feeding.—The feeding mechanism of Corixidae is unique among Hemiptera. The rostrum is completely altered, forming a broad, cone-shaped apex of the head. The stylets are used more for piercing and rasping than for sucking and lack the well-defined food and salivary channels that are so characteristic of other Hemiptera. The mouth opening is on the face of the broad "beak." It permits entry of entire organisms of small size as well as the cell contents of pierced strands of algae. Diatoms, rotifers, and other whole organisms are ground up by the "masticator," a structure of the buccal and pharyngeal region which is generally lacking in other Hemiptera that ingest only liquids. Griffith (1945) summarizes his studies of the corixid diet with the statement that "the presence of algal, protozoan, and rotiferan remains in adult stomachs indicates for *Ramphocorixa acuminata* a diet neither wholly animal, nor vegetable. The scooping movements of the forelegs seem designed to winnow out of the ooze a nutritious salad from both kingdoms, mixed in one digestible mass in the pharyngeal grinder." In addition to this "salad" and the habit of evacuating the cell contents of filamentous algae, corixids capture and feed upon whole chironomid and mosquito larvae.

Respiration.—Corixids differ from other water bugs in the manner in which they renew the air in their plastron. Notonectids, naucorids, and dytiscids break the surface film with the posterior end of the body. Corixids break the surface with the head and pronotum, so quickly that it is difficult to observe. Because of the extensive plastron surface corixids are able to remain below the surface for comparatively long periods of time, using oxygen which diffuses into the bubble from the water.

Life History.—Hungerford (1920) has worked out the life history of one of our commonest corixids, *Sigara alternata* (Say). The winter is passed in the adult stage, "the adults exhibiting considerable activity even in waters covered by a layer of ice." Eggs are laid in the spring and are attached to stems and leaves of various water plants, sticks, boards, and even the shells of living snails. The incubation period was one to two weeks at Ithaca, New York. The five nymphal instars required about one week each except the last stage which occupied a few days longer. Approximately the same sequence was observed for *Corisella edulis* by Griffith (1945).

The various nymphal instars may be recognized apart from the sequence of sizes, by the progressive development of the hemelytral pads. First instars show no prolongation of the posterolateral margins of the mesothorax; in second instars the hind margin is sinuate sublaterally; in third instars the hemelytral pads are half as long as the entire thorax; in fourth instars the pads attain the level of the first abdominal segment; and in fifth instars they reach the middle of the third abdominal segment.

Eggs.—Hungerford (1948) describes eggs of thirty species of Corixidae, six of which occur in California. All the California genera are represented. In general the eggs are ovoid but not symmetrical, being more convex on one side than the other and the surface is hexagonally reticulate. The egg is fastened on a pedestal and has a nipple at the apical or micropylar end (fig. 7:4).

Relationships.—The family Corixidae is placed in the suborder Cryptocerata by most recent authors because of the reduced antennae. However, this is a superficial criterion. Others, recognizing the unique mouth parts, have proposed the series Sandaliorrhyncha for this group of water bugs alone. Some doubt was cast on this extreme view by the studies of Singh-Pruthi, who found that the male genitalia are not inconsistent with the basic pattern found among other aquatic Hemiptera. At present, then, the true relationships of the family are not clear, but the three pairs of nymphal scent gland openings and the peculiar mouth parts set the family apart from all other water bugs. An Australasian genus *Diaprepocoris,* which is presumably primitive, has well-developed ocelli, and *Diaprepocoris, Micronecta,* and *Tenagobia* have the scutellum exposed. All other corixids have the scutellum almost or entirely covered by the pronotum.

Taxonomic characters.—The taxonomy of the Corixidae has long been based on the chaetotaxy of the highly specialized front tarsi (palae) of the males. The position and arrangement of the pegs in the so-called peg row are characteristic for the members of each genus and are usually distinctive for each species. Therefore diagrams are given of the palae of the California species (fig. 7:9).

The male genital characters are also definitive in most cases. The abdominal segments are asymmetrical, and this asymmetry may be dextral (the genitalia directed toward the right) or sinistral. This proves to be a very conservative and useful character, consistent for all the species in a given genus (at least in our fauna). However, curious cases of reversed symmetry have been recorded. For example, Hungerford (1948) reports two sinistral specimens in a series of six hundred males of *Hesperocorixa laevigata* (Uhl.). A so-called strigil (which apparently has nothing to do with stridulation) is located on the sixth abdominal tergite in all our genera but *Callicorixa.* It is made up of rows of comblike teeth and is said to grip against the female venter in copulation. It is on the left side in sinistral asymmetry and on the right side in dextral species.

Characters of the internal genitalia are easy to observe if the tip of the abdomen is softened and the capsule is teased out with a needle. The shape of the right clasper is usually distinctive for each species.

Simplified Key to Nearctic Genera of Corixidae

Based largely on characters of males

1. Rostrum without transverse sulcations. Cymatiinae; Holarctic, northern North America ... *Cymatia* Flor 1860
— Rostrum with distinct transverse sulcations. Corixinae 2
2. Eyes protuberant with inner anterior angles broadly rounded. Glaenocorisini 3
— Eyes not protuberant and inner angles normal 4
3. Male pala dorsally expanded at base; northern Canada, Europe .... *Glaenocorisa* Thomson 1869
— Male pala not dorsally expanded at base; northern North America .... *Dasycorixa* Hungerford 1948
4. Palae similar in both sexes, peg row adjacent to upper palmar bristles. Graptocorixini 5
— Palae dissimilar, males with peg row separated from upper palmar bristles. Corixini 6
5. Male abdomen sinistral. Strigil absent; southwestern United States, Mexico ..... *Neocorixa* Hungerford 1925
— Male abdomen dextral. Strigil present; western United States, Central America .. *Graptocorixa* Hungerford 1930
6. Male abdomen sinistral; North, Central, and South America .... *Trichocorixa* Kirkaldy 1908
— Male abdomen dextral 7
7. Strigil absent 8
— Strigil present 10
8. One row of pegs above upper palmar bristles; southwestern United States, Mexico .... *Morphocorixa* Jaczewski 1931
— Two rows of pegs above upper palmar bristles 9
9. Front tibiae triangularly produced and flattened apically; Sonoran, Neotropical ...... *Centrocorisa* Lundblad 1928
— Front tibiae not triangularly produced and flattened apically; northern North America .... *Callicorixa* White 1873
10. Vertex of male with a distinct longitudinal carina anteriorly; pala deeply impressed dorsally; eastern and southern United States; Mexico .... *Ramphocorixa* Abbott 1912
Vertex of male not carinate; pala not deeply impressed dorsally 11
11. Seventh abdominal tergite with a small dextral hook; eastern North America .... *Palmacorixa* Abbott 1912
— Seventh abdominal tergite without a dextral hook 12
12. Pala with 2 rows of pegs, one in row of upper palmar bristles; southwestern United States to Guatemala .... *Pseudocorixa* Jaczewski 1931
— Pala without a row of pegs in upper palmar bristles 13
13. Pala with 2 distinct rows of pegs above upper palmar bristles (fig. 7:9*e-h*); United States and Mexico .... *Corisella* Lundblad 1928
— Pala with 1 variously curved, sinuate, or interrupted row of pegs 14
14. Prothoracic lobe quadrate or trapezoidal; Holarctic, northern North America.... *Hesperocorixa* Kirkaldy 1908
— Prothoracic lobe narrower than long 15
15. Hemelytral pattern not reticulate; hemelytra and face not hairy (fig. 7:5*a*); cosmopolitan .... *Sigara* Fabricius 1775
— Hemelytral pattern reticulate; hemelytra and face hairy 16
16. Median carina well defined for entire length of pronotal disc; Holarctic, northern North America .... *Arctocorixa* Wallengren 1894
— Median pronotal carina poorly defined except on anterior third (fig. 7:5*c*); United States and Canada .... *Cenocorixa* Hungerford 1948

Key to California Genera of Corixidae

Males

1. Pala with peg row adjacent to upper palmar bristles (fig. 7:3*e*); face usually with dense mat of hairs (fig. 7:6) .... *Graptocorixa*
— Pala with peg row separated from upper palmar bristles (fig. 7:3*c*); face sometimes with sparse hairs but not with a dense mat of hairs 2
2. Asymmetry of abdomen sinistral (directed toward the left with strigil on left side) .... *Trichocorixa*
— Asymmetry of abdomen dextral 3
3. Strigil absent .... *Callicorixa*
— Strigil present 4
4. Pala with 2 rows of pegs, the upper row sometimes reduced to 2 or even 1 spine; apex of front tibia with a small fleshy lobe or pad (fig. 7:9*e-h*); base of clavus with brown bands broken or effaced (fig. 7:5*b*) .... *Corisella*
— Pala with 1 frequently curved, sinuate, or interrupted row of pegs; apex of front tibia without a small fleshy lobe; base of clavus with brown bands not conspicuously broken or effaced 5
5. Lateral lobes of prothorax trapezoidal, nearly as broad as long; strigil long and narrow, oriented with longitudinal axis of body; size large, 9-11 mm.... *Hesperocorixa*
— Lateral lobes of prothorax distinctly narrower than long; strigil oval or transverse; size smaller, less than 8 mm. 6
6. Hemelytra rugose but not strongly rastrate; size relatively large, 6.8-8 mm. (fig. 7:5*c*)........ *Cenocorixa*
— Hemelytra strongly rastrate; size smaller, less than 6.8 mm. (fig. 7:5*a*)........ *Sigara*

Key to California Genera of Corixidae

Females

1. Front femora with 3 or 4 tufts of long bristles or a close set row of bristles subbasally on inner margin; face

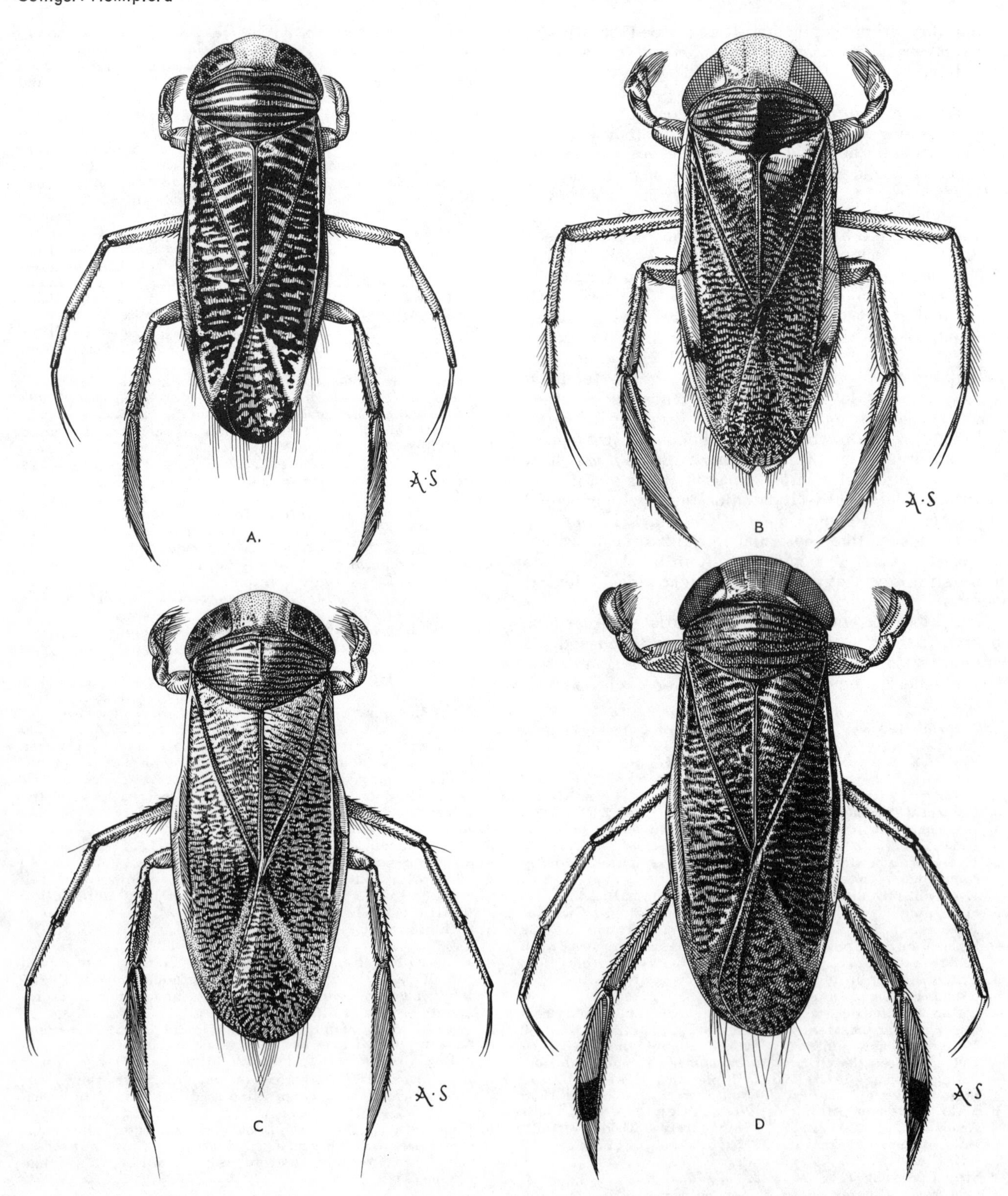

Fig. 7:5. *a, Sigara mckinstryi* Hungerford, male, Moss Beach, San Mateo County, California, November 4, 1946; *b, Corisella decolor* (Uhler), male, Manzanita Lake, Lassen National Park, California, September 18, 1946 (H. P. Chandler); *c, Cenocorixa kuiterti* Hungerford, male, Chilkoot Lake, Madera County, California, July 23, 1946 (R. L. Usinger); *d, Callicorixa vulnerata* (Uhler), male, Moss Beach, San Mateo County, California, July 4, 1929 (R. L. Usinger).

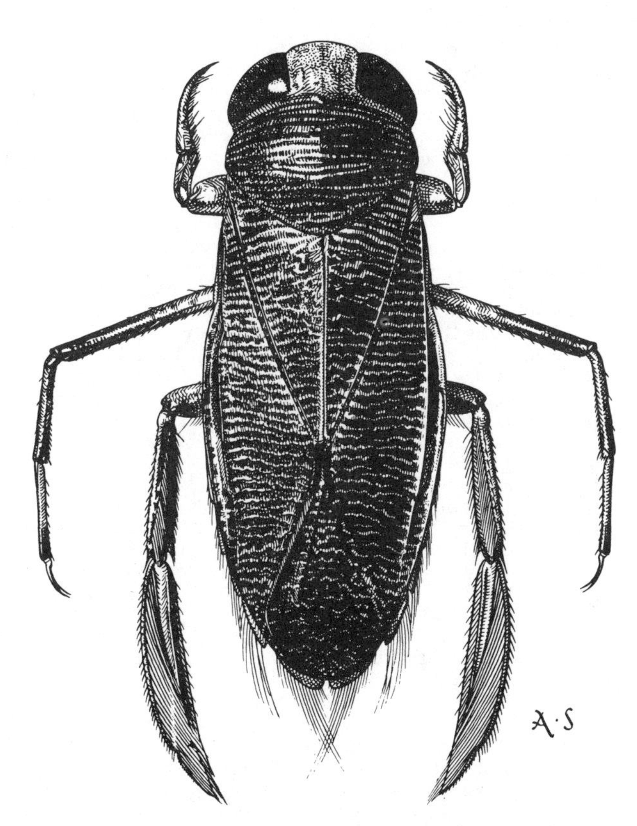

Fig. 7:6. *Graptocorixa californica* Hungerford, male, Mokelumne Hill, Calaveras County, California, May 29, 1931 (R. L. Usinger).

usually with a dense mat of hairs........*Graptocorixa*
— Front femora without prominent tufts or a row of bristles on inner margin; face sometimes with sparse hairs but not with a dense mat of hairs 2
2. Apices of clavi not exceeding a line drawn through costal margins of hemelytra at nodal furrows...... .... *Trichocorixa*
— Apices of clavi exceeding a line drawn through costal margins at nodal furrows 3
3. Base of clavus with brown bands broken or effaced (fig. 7:5*b*) .... *Corisella*
— Base of clavus without broken or effaced brown bands 4
4. Lateral lobes of prothorax trapezoidal, nearly as broad as long; size large, 9-11 mm .... *Hesperocorixa*
— Lateral lobes of prothorax distinctly narrower than long; size smaller, less than 8 mm. 5
5. Size small, less than 6.8 mm. .... *Sigara*
— Size larger, 6.8 mm. or more 6
6. Corium rugose but not strongly rastrate; apex of hind tibia and usually also base of first tarsal segment infuscated in *vulnerata* but not in *audeni* ... *Callicorixa*
— Corium strongly rastrate; hind tibia and tarsus not infuscated as above .... *Cenocorixa*

Preliminary Key to Genera of California Corixidae

Based on Fifth Instar Nymphs

(Mesonotal pads reaching to middle of third abdominal segment)

1. Scent gland openings very small and widely separated, the middle pair separated by about 10 times the diameter of 1 opening. (Middle claws less than two-thirds as long as tarsus; face with a dense mat of hairs; 3 tufts of long bristles at base of front femur) (*californica* Hung.) .... *Graptocorixa*
— Scent gland openings larger, the middle pair less than 5 times as far apart as diameter of a single opening. (Middle claws subequal or longer than tarsus; face with scattered hairs but not with a dense mat of hairs; front femora without 3 tufts of long bristles near base) ... 2
2. Mesonotum entirely covered with long shaggy hairs; middle claws subequal in length to tarsus (*laevigata* Uhl.) .... *Hesperocorixa*
— Mesonotum bare on either side of middle posteriorly; middle claws longer than tarsus 3
3. Metanotum at middle, and abdomen clothed with short, sparse, bristles backward directed (*inscripta* Uhl. *decolor* Uhl.) .... *Corisella*
— Metanotum at middle and abdominal tergites bare 4
4. Abdomen with 6 longitudinal stripes above (*mckinstryi* Hung.) .... *Sigara*
— Abdomen without longitudinal stripes 5
5. Scent gland openings large, the space between middle pair about twice the diameter of a single opening; middle of mesonotum bare except at anterior third and narrowly along a median-longitudinal carina (*reticulata* Guer.) .... *Trichocorixa*
— Scent gland openings small, the middle pair about 5 times as far apart as diameter of a single opening; middle of mesonotum broadly pubescent, the lateral bare areas confined to hind margin near inner angles of hemelytral pads (*wileyae* Hung.).......... *Cenocorixa*

Key to California Species of Graptocorixa

1. Prothoracic lateral lobe elongate; small, length less than 7 mm. (fig. 7:9*x*); Oregon, Nevada, Arizona .... *serrulata* (Uhler) 1879
— Prothoracic lateral lobe quadrate, if longer than broad, then of uniform width on species with a white spot on

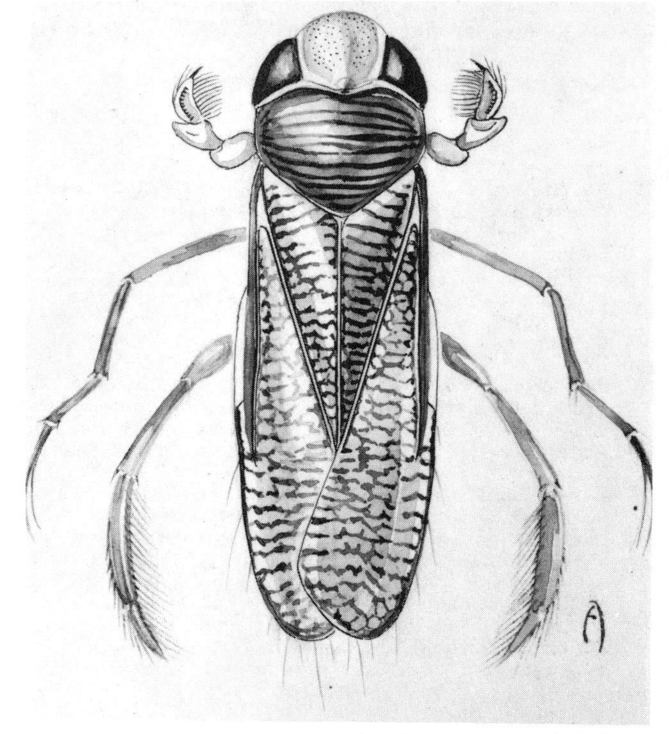

Fig. 7:7. *Trichocorixa reticulata* (Guerin-Meneville), Oahu, T. H., 1936 (R. L. Usinger).

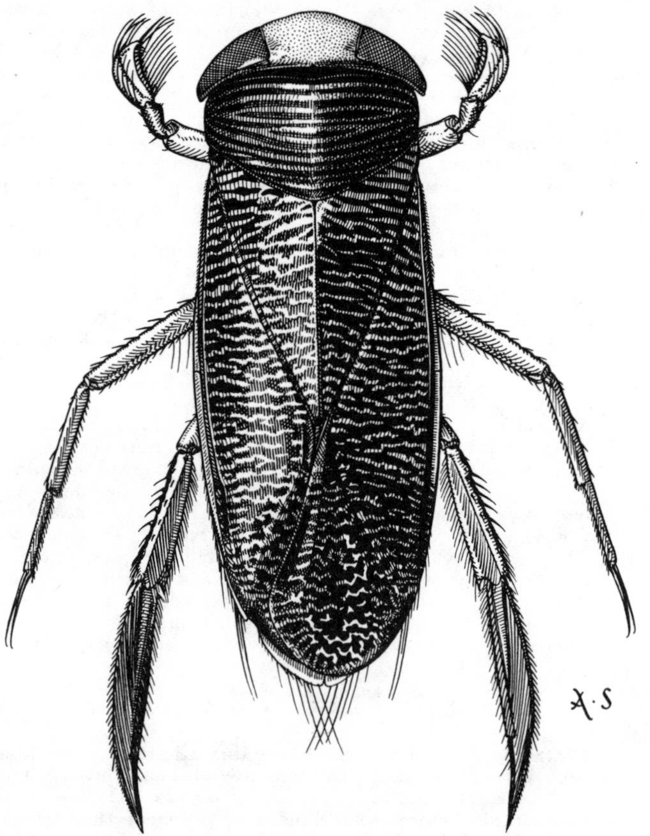

Fig. 7:8. *Hesperocorixa vulgaris* (Hungerford), Lake Britton, Shasta County, California, September 17, 1946 (H. P. Chandler).

distal angle of corium; size larger, length 8 mm. or more 2

2. Infraocular width of genae at level of hypoöcular suture as great as diameter of middle femur; anterior femur of male conspicuously produced at inner base (fig. 7:9*v*); Nevada, Arizona, Texas, Mexico, southern California desert . . . . *abdominalis* (Say) 1832
— Infraocular width of genae at level of hypoöcular suture narrower than diameter of middle femur; anterior femur of male not conspicuously produced at inner base . . . 3
3. Prothoracic lateral lobe about equal in width to width of middle femur, its rear margin not conspicuously shorter than distal margin; anal lobes of female shallowly notched on inner ventral margin (fig. 7:9*u*); southern California . . . . *uhleri* (Hungerford) 1925
— Prothoracic lateral lobe plainly broader than width of middle femur, its rear margin conspicuously shorter than distal margin; anal lobes of female deeply notched on inner ventral margin 4
4. Lower basal angle of front femur roundly but not greatly produced; male right genital clasper broadest at basal third and pointed at tip (fig. 7:9*w*); southern and central California . . . . *uhleroidea* (Hungerford) 1938
— Lower basal angle of front femur not produced; male right genital clasper broadest at distal third and rounded at tip (figs. 7:6; 7:9*t*); widely distributed in fresh water streams, California and Oregon . . . . *californica* (Hungerford) 1925

Key to California Species and Subspecies of Trichocorixa

1. Width of interocular space at narrowest point distinctly exceeding width of an eye along hind margin as seen from above; nodal furrow located at apex of pruinose area of embolar groove in females (fig. 7:7); widely distributed, Neotropical region, central and southern California in brackish and saline waters (fig. 7:9*b*). . . . . *reticulata* (Guerin) 1857
— Width of interocular space subequal to or less than width of an eye along hind margin; nodal furrow dividing pruinose area of embolar groove subapically in females 2
2. Male strigil long, slender and curved, appearing as a heavy dark line along margin of left lateral lobe of sixth abdominal tergite, at point of greatest width less than one-sixth as wide as long; females with costal margins broadly, shallowly indented in front of middle; eastern and southern United States, Imperial County, California (fig. 7:9*d*) . . . . *calva* (Say) 1832
— Male strigil not appearing as a heavy dark line; at point of greatest width at least one-sixth as wide as long; females with costal margins abruptly dilated at or behind middle 3
3. Male strigil arcuate; female costal margins abruptly dilated at about middle; Colorado River near Needles, California (fig. 7:9*c*) . . . . *uhleri* Sailer 1948
— Male strigil elongate-oval, straight; female costal margins abruptly dilated beyond middle; North America and West Indies, brackish and saline waters (*verticalis* Fieber) 1851 (fig. 7:9*a*). 4
4. Length of head as seen from above greater than length of tarsus of middle leg in males; prenodal polished area of costal margin not exceeding length of tarsus of middle leg in female; saline waters, Salton Sea, Death Valley . . . . subspecies *saltoni* Sailer 1948
— Length of head as seen from above less than length of tarsus of middle leg in males; prenodal polished area of costal margin definitely exceeding length of tarsus of middle leg in females; coastal brackish waters . . . . subspecies *californica* Sailer 1948

Key to California Species of Corisella

1. Male pala with only 1 or 2 conspicuous pegs in upper row position (fig. 7:9*f*); both segments of hind tarsus brown; generally distributed, United States and Mexico (=*tumida* Uhler) . . . . *tarsalis* (Fieber) 1851
— Male pala with a row of 8 or more pegs near upper margin; first segment of hind tarsus, at least, yellow, not brown 2
2. Pronotum with a median longitudinal carina on anterior fourth, disc crossed by 6-8 dark bands, none of which is partly or completely interrupted (figs. 7:5*b*; 7:9*g*); size small, 4-5.8 mm.; widely distributed in western United States alkali waters (=*dispersa* Uhl.) . . . . *decolor* Uhler 1871
— Pronotum without a median longitudinal carina, crossed by 10 or more partly interrupted bands; size larger, over 6.3 mm. 3
3. Male vertex only briefly, roundly produced; male pala broader than tibia, its dorsal margin projecting above tibia (fig. 7:9*e*); rear margin of last ventral abdominal segment of female broadly, medianly produced; western United States and Mexico, southern and central California . . . . *inscripta* (Uhler) 1894
— Male vertex strongly, broadly produced; male pala not broader than tibia, or at least the dorsal part not projecting above tibia (fig. 7:9*h*); rear margin of last ventral abdominal segment of female nearly straight; Mexico, southern and western United States, Oregon, Nevada, Arizona . . . . *edulis* (Champion) 1901

Key to California Species of Hesperocorixa

1. Pronotum nonrastrate, shining; clavus and corium somewhat rastrate to rugulose (fig. 7:9*q*); widely distributed, North America . . . . *laevigata* (Uhler) 1893

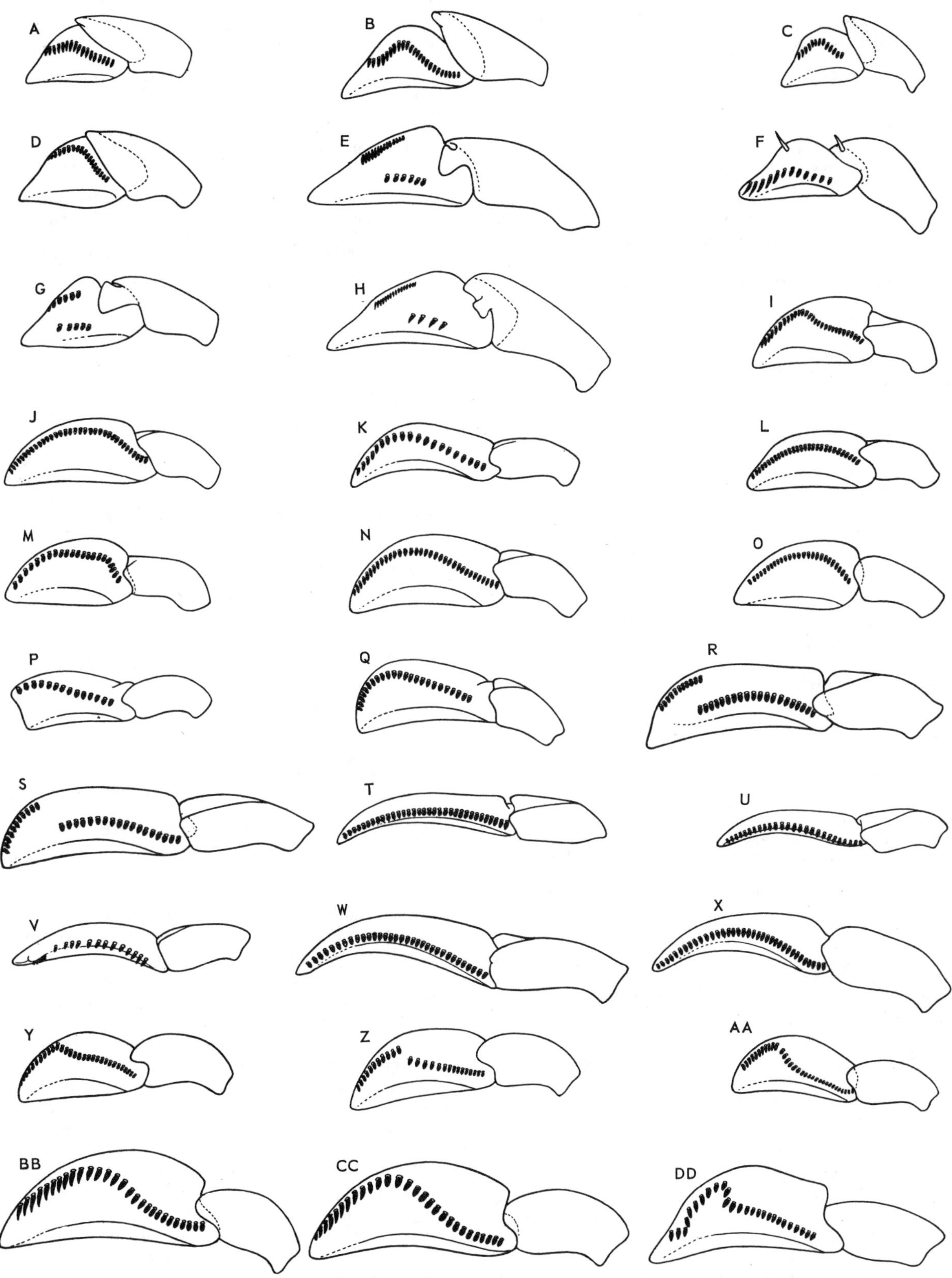

Fig. 7:9. Palae of California species of Corixidae, diagrammatic; showing shape and also position and number of pegs. See figure 7:3, c and e, for characteristic arrangement of bristles. *a*, *Trichocorixa verticalis* (Fieber); *b*, *T. reticulata* (Guerin); c, *T. uhleri* (Sailer); *d*, *T. calva* (Say); e, *Corisella inscripta* (Uhler); *f*, *C. tarsalis* (Fieber); *g*, *C. decolor* Uhler; *h*, *C. edulis* (Champion); *i*, *Sigara nevadensis* (Walley); *j*, *S. mckinstryi* Hungerford; *k*, *S. omani* (Hungerford); *l*, *S. grossolineata* Hungerford; *m*, *S. vandykei* Hungerford; *n*, *S. alternata* (Say); o, *S. washingtonensis* Hungerford; *p*, *Hesperocorixa vulgaris* (Hungerford); *q*, *H. laevigata* (Uhler); *r*, *Callicorixa audeni* Hungerford; *s*, *C. vulnerata* (Uhler); *t*, *Graptocorixa californica* (Hungerford); *u*, *G. uhleri* (Hungerford); *v*, *G. abdominalis* (Say); *w*, *G. uhleroidea* (Hungerford); *x*, *G. serrulata* (Uhler); *y*, *Cenocorixa andersoni* Hungerford; *z*, *C, bifida* (Hungerford); *aa*, *C. utahensis* (Hungerford); *bb*, *C. blaisdelli* (Hungerford); cc, *C. kuiterti* Hungerford; *dd*, *C. wileyae* (Hungerford).

— Pronotum and hemelytra heavily rastrate (fig. 7:9*p*); widely distributed, North America ....*vulgaris* (Hungerford) 1925

Key to California Species of Callicorixa

1. First tarsal segment of hind leg infuscated for at least a part of its length (figs. 7:5*d*; 7:9*s*); western North America, Alaska to California ... *vulnerata* (Uhler) 1861
— First tarsal segment of hind leg concolorous throughout (fig. 7:9*r*); northern United States, Alaska and Canada. Sierra Nevada, California...... *audeni* Hungerford 1928

Key to California Species of Sigara

(Subgenus Vermicorixa Walton)

1. Metaxyphus short, plainly broader than long 2
— Metaxyphus about as broad as long 6
2. Ostiole almost in lateral bend of mesoepimeron, and connected to metasternum by the broad, often dark post-coxal piece (fig. 7:9*l*); widespread, northern United States and Canada. Mammoth Lakes, California .... *grossolineata* Hungerford 1948
— Ostiole not in lateral bend of mesoepimeron, closer to tip of mesoepimeron 3
3. Anterolateral third of clavus ridged and projecting laterally over pruinose area along claval suture (fig. 7:9*m*); Washington and California, mouth of Van Duzen River .... *vandykei* Hungerford 1948
— Anterolateral margin of clavus not as above 4
4. Mesoepimeron of equal width from ostiole to lateral bend; ostiole posterior to distolateral angle of mesosternum; North America, widely distributed, California, Yuba County (fig. 7:9*n*)...........*alternata* (Say) 1825
— Mesoepimeron broader at lateral bend than at level of scent gland ostiole; ostiole on a level with distolateral angle of mesosternum 5
5. Right clasper of male genital capsule roundly elbowed, slightly widened apically, abruptly rounded, and minutely produced at tip; pronotal disc of female more than half as long as broad (figs. 7:5*a*; 7:9*j*); California coast .... *mckinstryi* Hungerford 1948
— Right clasper of male genital capsule angulately elbowed, tapering apically with the tip subacute and bent; pronotal disc of female less than half as long as broad; western United States, widespread in California ....*washingtonensis* Hungerford 1948
6. Male pala broadest near its base, an oblique ridge across palar face (fig. 7:9*i*); lateral margin of pronotal disc acute; Nevada, Utah, Wyoming .... *nevadensis* (Walley) 1936
— Male pala broadest beyond middle, no ridge across palar face; lateral margin of pronotal disc rounded; western United States, northern and central California (fig. 7:9*k*) .... *omani* (Hungerford) 1930

Key to California Species of Cenocorixa

1. Male pala with peg row divided (fig. 7:9*z*); western and northern United States, Sierran California ....*bifida* (Hungerford) 1926
— Male pala with peg row variously curved but not divided 2
2. Last segment of hind tarsus deeply embrowned for its entire length (fig. 7:9*cc*); California and Utah, Sierra and Wasatch Mountains ....*kuiterti* Hungerford 1948
— Hind tarsi concolorous throughout, or with only the tip of last segment embrowned 3
3. Male pala with peg row doubly curved (fig. 7:9*dd*); strigil very small, of about 5 regular combs; western United States............ *wileyae* (Hungerford) 1926
— Male pala with peg row not doubly curved; strigil larger, with 10 or more combs 4
4. Right genital clasper of male bifurcate at tip; interocular space greater than width of an eye (fig. 7:9*aa*); western and central United States, Sierran California .... *utahensis* (Hungerford) 1925
— Right genital clasper of male not bifurcate at tip; interocular space equal to or slightly less than width of an eye 5
5. Interocular space slightly narrower than an eye; right genital clasper of male narrowed beyond subapical bend (fig. 7:9*y*); Oregon, Washington .... *andersoni* Hungerford 1948
— Interocular space equal to width of an eye; right genital clasper of male widened beyond subapical bend and then slightly narrowed to broad apex (fig. 7:9*bb*); California coastal and Modoc County .... *blaisdelli* (Hungerford) 1930

REFERENCES

ANCONA, L.
1933. El Ahuatle de Texcoco. Ann. Inst. Biol., 4:51-69.
FORBES, S. A.
1878. Breeding habits of *Corixa*. Amer. Nat., 12:820.
1888. On the food relations of freshwater fishes: a summary and discussion. Bull. Illinois Lab. Nat. Hist., 2:475-538.
GRIFFITH, M. E.
1945. The environment, life history and structure of the water boatman, *Ramphocorixa acuminata* (Uhler). Univ. Kansas Sci. Bull., 30:241-365.
HUNGERFORD, H. B.
1948*a*. The Corixidae of the Western Hemisphere. Univ. Kansas Sci. Bull., 32:1-827.
1948*b*. The eggs of Corixidae. Jour. Kansas Ent. Soc., 21:141-146.
HUTCHINSON, G. E.
1931. On the occurrence of *Trichocorixa* Kirkaldy in salt water and its zoo-geographical significance. Amer. Nat., 65:573-574.
POPHAM, E. J.
1941. The variation in the colour of certain species of *Arctocorisa* and its significance. Proc. Zool. Soc., Lond. (A), 111:135-172.

Family NOTONECTIDAE

Back Swimmers

Back swimmers are a characteristic feature of pond life in most parts of the world. They are fierce predators, attacking any invertebrate of appropriate size and even relatively larger fish fry. The common name is derived from the unique habit of rowing with the ventral (in a morphological sense) side uppermost. In body form they are long and slender, deep bodied, and convex dorsally. Ocelli are absent. The antennae are short, partly concealed, and of three or four segments. The rostrum is short, stout, and four-segmented. The front and middle legs are adapted for grasping prey or for holding on to objects in the water. The hind legs are very long, oarlike, and fringed with long swimming hairs. The adult tarsi in our species are two-segmented, with two claws present but inconspicuous on hind legs. The abdominal venter is longitudinally keeled at the middle with long hairs which, together with the inwardly directed hairs on the sides of the venter, close over two troughs which form air chambers.

Relationships.—Notonectids are commonly grouped

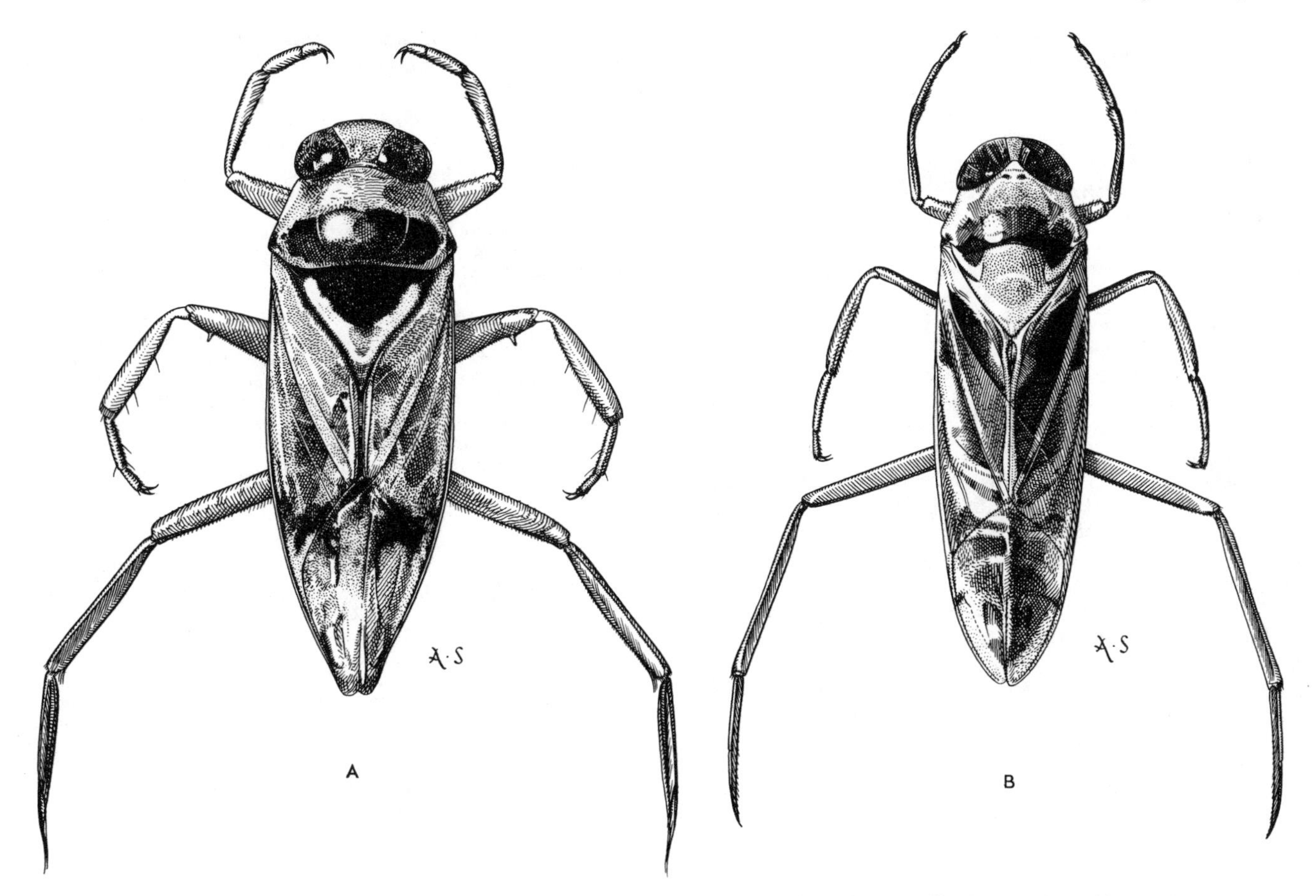

Fig. 7:10. *a, Notonecta unifasciata* Guerin, male, Davis, Yolo County, California, June 12, 1932 (R. L. Usinger); *b, Buenoa scimitra* Bare, male, Walnut Creek, Contra Costa County, California, July 9, 1929 (R. L. Usinger).

with the other aquatic families, except Corixidae, in the Notonectoidea. They appear to be somewhat isolated from the naucoroid type by the elongate, deep-bodied form, absence of dorsal abdominal scent glands in the nymphs, and back-swimming habit. Yet in male genitalia and in egg type, they are closest to the naucoroid families and are somewhat distinct from the belostomatids, nepids, and corixids. The Pleidae are now generally given family rank but are closely related to notonectids. They are very small, suboval bugs with hind legs that are not oarlike, though possessing swimming hairs. The abdominal venter is not narrowly keeled but rather is elevated into variously formed carinae at the middle of each segment.

Respiration.—Like so many water bugs, notonectids come to the surface at intervals to replenish their supply of air. The surface film is broken by the tip of the abdomen with the bug suspended, head downward at an angle and swimming legs outstretched. Air enters the abdominal troughs where it comes in contact with the ventral spiracles and diffuses forward to the thoracic spiracles and to the subelytral air spaces.

In *Buenoa* and its close old-world ally, *Anisops,* a red color is clearly visible through the thin body wall. Hungerford (1922) first reported that the color is due to hemoglobin, and Bare (1928) found that it occurs in large cells that surround the tracheae in abdominal segments three to seven. In some cross sections nearly one-third of the abdominal space is occupied by hemoglobin cells. The function of hemoglobin in these bugs is not known, but it would be logical to assume that it is in some way associated with respiration or hydrostatic equilibrium.

Life history.—Most notonectids overwinter in the adult stage, either actively or in hibernation, depending on the climate. Hungerford (1933) reports that *Notonecta* has been observed swimming beneath the ice in midwinter. Eggs are laid in the spring and summer, there being one or more overlapping broods in a season. However, Rice (1954) reported that *Notonecta borealis* passes the winter in the egg stage and suggested that two species of *Buenoa* may do likewise.

Eggs are laid in plant tissue or on plant or rock surfaces. Rice (1954) found that eggs of *Plea, Buenoa,* and two species of *Notonecta* from Douglas Lake region (Michigan) were inserted in stems and those of three other *Notonecta* species were not inserted. All eggs that were glued to the surface of stems had characteristic hexagonal reticulations, and those which were inserted had characteristic nodules in addition to hexagonal ridges. *Buenoa* eggs differ from those of *Notonecta* and *Plea* in having a distinct anterodorsal cap. The incubation period varies with the season and species. There are five nymphal instars in all species studied.

Habits and distribution.—Back swimmers occur in a wide variety of fresh and stagnant pools and in the quiet water of streams and lakes. In California they are found from sea level to high elevations in the Sierra Nevada. Food consists of small Crustacea, chironomids, mosquito larvae, fish fry, and in fact anything that they can overpower.

Adults fly readily and disperse over considerable distances. Hungerford (1933) reports swarms in flight and says that they are attracted to lights. Walton (1935), in a series of experiments with *Notonecta maculata* (Fabr.) in Britain, carried specimens away from water for distances ranging from one and a half to thirty yards. After drying on the ground 200 bugs took wing and of these 45 per cent were seen to drop into water.

Pierce (1948) described a new species, *Notonecta badia,* from the McKittrick asphalt field, Kern County, and reported that *Notonecta* was also present in the Los Angeles La Brea deposits. According to Pierce, the age "is unquestionably Pleistocene, because all of the recovered fragments come from the skull cavities of saber-tooth cats."

Stridulation.—Definite chirping sounds have been recorded for *Buenoa* by Hungerford (1924). Several structures are described by Bare (1928) as possible stridulatory organs. The most obvious of these are the stridulatory comb at the base of the male tibia and the rostral prong with filelike teeth that stands in opposition when the fore limbs are brought up to the head. Also reportedly of use in sound production are the fine sclerotized ridges on the inner face of the fore femur.

The habits of *Buenoa* and *Notonecta* are strikingly different as stated by Truxal (1953). "The *Buenoa* swim gracefully on their backs in almost perfect equilibrium with the water." Thus, in a sense, *Buenoa* could be described as limnetic. *Notonecta,* on the other hand, are jerky swimmers, working constantly to maintain their position in the water when they are not clinging to some object or hanging from the surface film.

In distributional pattern, the notonectoid genera present an unusual picture. *Notonecta* is world-wide but *Buenoa* is restricted to the Western Hemisphere and Hawaii. The Pleidae are most abundant in the tropics but extend into temperate regions including the southern and eastern United States. They have not yet been found in the Pacific states.

Taxonomic characters.—The principal characters used in classification include the male genitalia, which differ in most species but must be dissected out and therefore are not used in the present keys. Stridulatory ridges on the fore femora of *Buenoa* can be seen only if the front legs are bent outward or detached and mounted on a card. Measurements, including the synthlipsis (narrowest point of interocular space at hind margins of eyes) and vertex (anterior interocular width), should be made with the transverse and longitudinal axes of the insect horizontal. The length of the rostral prong in *Buenoa* is measured by holding the insect so that the surface of the prong is horizontal.

Key to the Nearctic Genera of Notonectidae

Adults

1. Hemelytral commissure with a pit at anterior end, just behind tip of scutellum; antennae 3-segmented, the last segment longer than penultimate segment; relatively small, slender, 5 to 8 mm. (fig. 7:10*b*)............ ....*Buenoa* Kirkaldy 1904
— Hemelytral commissure without a pit; antennae 4-segmented, the last segment much shorter than penultimate segment, robust, more than 10 mm. (fig. 7:10*a*) ....*Notonecta* Linnaeus 1758

Nymphs

1. Abdominal spiracles large, oval, about 1/4 as long as segments, located mesad of ventral hair fringe; hind legs without rows of conspicuous, stout spines ....*Buenoa*
— Abdominal spiracles small, round, less than 1/10 as long as segments; hind legs with rows of short, stout, black spines ....*Notonecta*

Key to the California Species of Notonecta[1]

1. Keel of 4th abdominal sternite bare, hairs confined to sides 2
— Keel of 4th abdominal sternite not bare 3
2. Head broad, 5/6 as wide as pronotum; eyes large, rear width of one eye as great as length of lateral margin of pronotum; scutellum 1/2 again as broad as long; southern California .... *hoffmanni* Hungerford 1925
— Head narrower, 2/3 as wide as pronotum; eyes smaller, rear width distinctly less than length of lateral margin of pronotum; scutellum less than 1/4 broader than long; widespread ....*kirbyi* Hungerford 1925
3. Pronotum distinctly longer than scutellum; males with a stout tubercle at angle of fore trochanter and a very stout broad hook; dichromatic, one form white, the other black, marked with white; widespread ....*shooterii* Uhler 1894
— Pronotum distinctly shorter than scutellum; males without a tubercle or hook on fore trochanter 4
4. Trochanter of middle legs rounded or nearly so 5
— Trochanter of middle legs distinctly angulate or produced into a tooth or stout, spinelike process 6
5. Lateral margins of pronotum distinctly though shallowly concave; length 10-11 mm., last abdominal sternite of female scarcely or shallowly notched at tip, the notch wider than deep; Imperial Valley..... *indica* Linn. 1771
— Lateral margins of pronotum straight; length usually larger, 11-12 mm.; last abdominal sternite of female with an apical notch which is deeper than wide; widespread .... *undulata* Say 1832
6. Middle trochanter angulate or produced into a short tooth at inner posterior angle (fig. 7:10*a*); widespread .... *unifasciata* Guerin 1857[2]

[1]Excluded from key: *N. ochrothoe* Kirk. 1897, a South American species not clearly distinguishable from *N. shooterii,* with a single doubtful California record, San Diego, one specimen.

[2]Hungerford (1933) divided *unifasciata* into several subspecies, based on the shape of the vertex and male genital clasper. California specimens pertain to the subspecies *andersoni* which

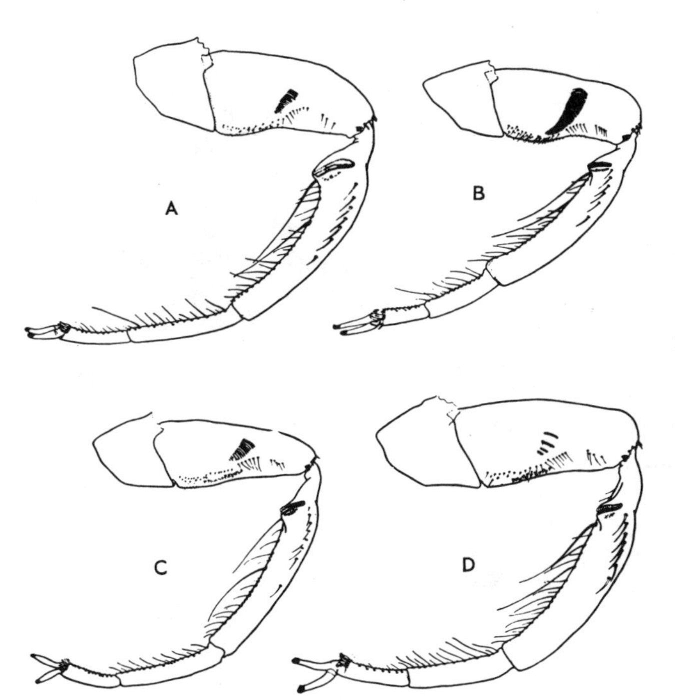

Fig. 7:11. Front legs of males. *a*, *Buenoa uhleri* Truxal; *b*, *B. scimitra* Bare; *c*, *B. margaritacea* Bueno; *d*, *B. omani* Truxal (Truxal, 1953).

— Middle trochanter produced into a long, spinose process; northern California and Mono County ....*spinosa* Hungerford 1930

Key to Males of Buenoa

1. Synthlipsis narrow, less than 1/2 the anterior width of vertex; fore femur with long, conspicuous, sword-shaped stridulatory area consisting of about 60 extremely fine, sclerotized ridges (figs. 7:10*b*; 7:11*b*); widespread in the southern United States .... *scimitra* Bare 1925
— Synthlipsis wide, 1/2 or more the anterior width of vertex, fore femur with smaller stridulatory area consisting of less than 25 sclerotized ridges 2
2. Fore femur with only 4 sclerotized ridges in stridulatory area (fig. 7:11*d*); southern California and Mexico .... *omani* Truxal 1953
— Fore femur with 15 to 25 sclerotized ridges in stridulatory area 3
3. Rostral prong longer than 3rd rostral segment, fore femur with 19-24 sclerotized ridges in stridulatory area (fig. 7:11*a*); southern California, Texas, and Mexico .... *uhleri* Truxal 1953
— Rostral prong subequal to 3rd rostral segment; fore femur with 15-18 sclerotized ridges (fig. 7:11*c*); widely distributed .... *margaritacea* Bueno 1908

REFERENCES

BARE, C. O.
1928. Haemoglobin cells and other studies of the genus *Buenoa*. Univ. Kansas Sci. Bull., 18:265-349, 14 pls.
HUNGERFORD, H. B.
1922. Oxyhaemoglobin present in backswimmer, *Buenoa margaritacea*. Canad. Ent., 54:263.

has a long slender clasper. Typical *unifasciata* is Mexican but specimens identified by R. I. Sailer have recently been collected on Santa Catalina Island by Lee, Ryckman, and Christianson (1955).

1924. Stridulation of *Buenoa limnocastoris* Hungerford and systematic notes on the *Buenoa* of the Douglas Lake region of Michigan, with the description of a new form. Ann. Ent. Soc. Amer., 17:223-227.
1933. The genus *Notonecta* of the world. Univ. Kansas Sci. Bull., 21:5-195, 17 pls.
LEE R. D., R. E. RYCKMAN, and C. P. CHRISTIANSON
1955. A note on several species of aquatic Hemiptera from Santa Catalina Island, California. Bull. So. Calif. Acad. Sci., 54:20-21.
PIERCE, W. D.
1948. Fossil arthropods of California. 15. Some Hemiptera from the McKittrick asphalt field. Bull. So. Calif. Acad. Sci., 47:21-33.
RICE, L. A.
1954. Observations on the biology of ten notonectoid species found in the Douglas Lake, Michigan, region. Amer. Midl. Nat., 51:105-132, 3 pls.
TRUXAL, F. S.
1953. A revision of the genus *Buenoa*. Univ. Kansas Sci. Bull., 35:1351-1523, 17 pls.
WALTON, G. A.
1935. Field experiments on the flight of *Notonecta maculata* Fabr. Trans. Soc. Brit. Ent., 2:137-143.

Family NAUCORIDAE

Creeping Water Bugs

Naucorids are among the least known but widely distributed groups of aquatic Hemiptera. In size, body form, and in their role in the aquatic community, they are comparable to the Dytiscidae among the beetles. Although fiercely predaceous, they are quite defenseless when molting and fall easy prey to damselfly nymphs and other predators at this time.

Relationships.—Among the Cryptocerata the naucorids are perhaps closest in structure to the backswimming Notonectidae and more particularly to an obscure group of Notonectoidea known only from tropical waters, that is, the Helotrephidae. In general, the naucorids are characterized by their oval form, subflattened body, raptorial front legs, and middle and hind legs modified for swimming.

Respiration.—Most naucorids breathe through the cuticle as nymphs and through spiracles in contact with the air bubble as adults. Adults break the surface film with the tip of the abdomen to replenish the air in their subelytral air space and in the air bubble which appears as a silvery sheen on the pubescent venter. When at rest, naucorids "row" with their hind legs, thus maintaining a current of fresh water over the surface of the bubble.

A few naucorids, *Aphelocheirus* of the Old World and *Cryphocricos* of the New World, are dimorphic and lack conspicuous pubescence on the venter. The common form in each of the above genera is brachypterous, and thus lacks the subelytral air store. Thorpe (1950) has studied the special problems of respiration and orientation encountered in *Aphelocheirus*. The spiracles open as small tubular "rosettes" and the lumen of the rosette tubes and the entire surface of the body are covered with an ultramicroscopic hair pile. By means of the electron microscope it was determined that there are approximately 2 million hairs per square millimeter. This

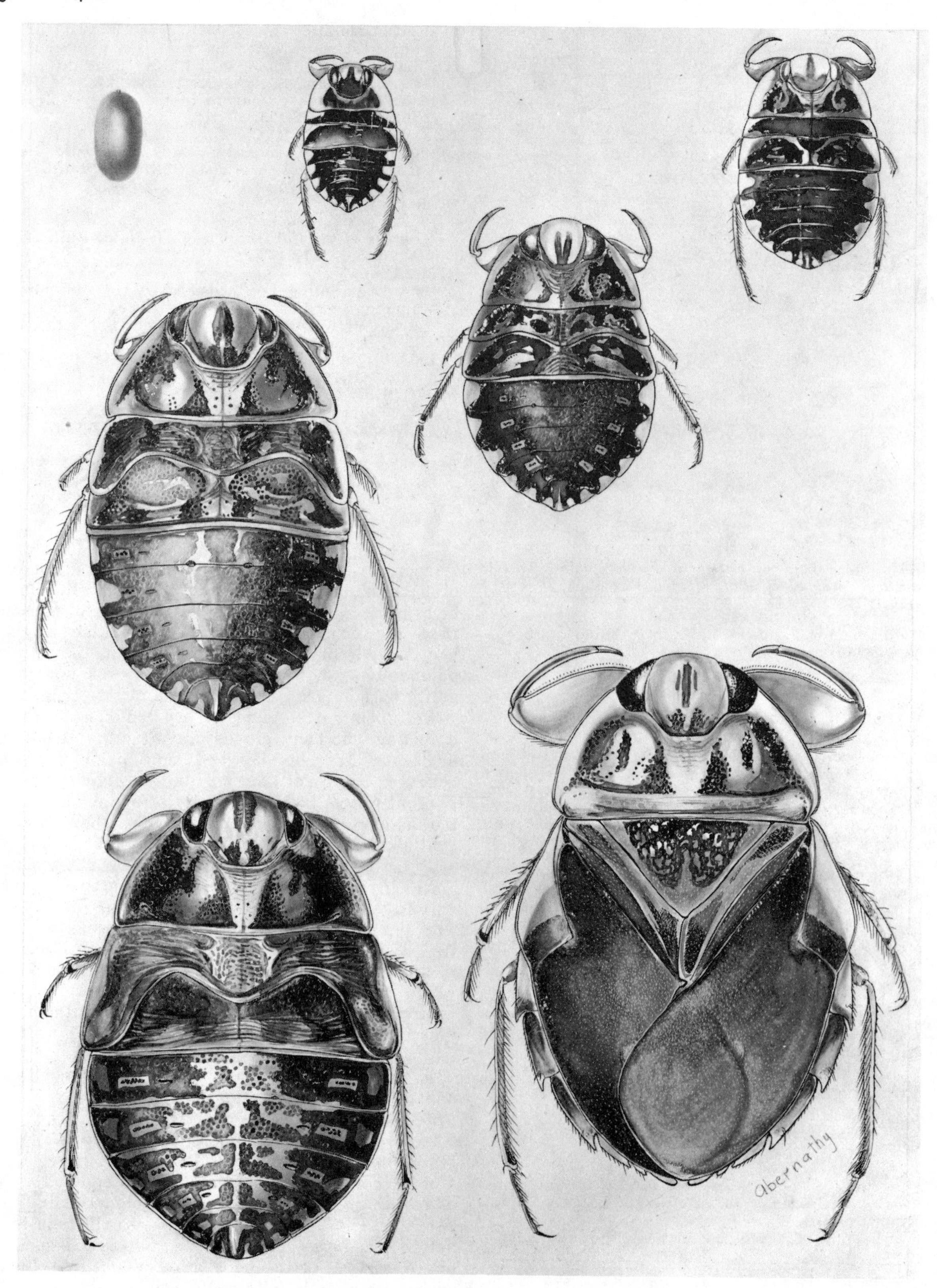

Fig. 7:12. *Ambrysus mormon* Montandon, egg, five nymphal instars, and adult. Davis, Yolo County, California, 1932 (R. L. Usinger).

hair pile provides a very thin but virtually incompressible air store or "plastron" which enables the insect to remain below the surface in well-oxygenated waters for an indefinite period, extracting dissolved oxygen from the water. Lacking a subelytral air store, *Aphelocheirus* has no "hydrostatic" mechanism. Therefore, special pressure receptors are developed at the base of the abdomen to aid in orientation.

Life history (Usinger, 1946).—The life history of the common genera of Naucoridae is relatively simple (fig. 7:12). The winter is passed in the adult stage. Eggs are laid in the spring, the eggs in our species being glued to stones in shallow waters of lakes and streams. The eggs are suboval with a small, buttonlike micropyle at the anterior end. When first laid, they are cream-colored, but they change to gray with reddish eyespots in a week or so. Hatching occurred in about four weeks in the Sacramento Valley *(Ambrysus mormon),* the nymph emerging through a crescent-shaped tear at the micropylar end of the egg. The first four nymphal instars each required about a week during May, June, and July, whereas the fifth and last instar required three weeks. Adults fly by day, but this must be a rare phenomenon because I have observed it only once. They do not fly to lights at night but are attracted to a light held near the water.

Nymphal instars may be determined by the following table (7:2), based upon *Ambrysus mormon* Montd. In the fifth nymphal instar the lateral margins of the mesonotum (hemelytral pads) completely cover the metanotal pads.

Habitat and distribution.—Naucorids occur in a wide variety of habitats including ponds, lakes, rivers, hot springs, and saline waters. In the United States *Cryphocricos* is a Neotropical group which extends only into southern Texas where it inhabits swift-flowing streams. *Usingerina* is the northernmost representative of the Neotropical subfamily Limnocorinae.

Pelocoris is the only genus of Naucoridae found in the eastern United States. In the West, it has been found only in two hot springs in Nevada and at Saratoga Springs, Death Valley, California. *Pelocoris* occurs in quiet waters and is commonly found amidst aquatic vegetation.

Ambrysus is the dominant naucorid genus throughout the western United States. It ranges from the hot springs of Yellowstone to Argentina. In general, *Ambrysus* species are found in clear flowing, well-oxygenated waters, especially those with pebbly or rocky bottoms. Although the preferred habitat seems to be streams, *Ambrysus mormon,* at least, invades the margins of lakes within its range. Thus Clear

TABLE 7:2

DEVELOPMENT OF HEMELYTRAL PADS IN SUCCESSIVE INSTARS OF AMBRYSUS MORMON (MONTD.)

	1st instar	2d instar	3d instar	4th instar	5th instar
Ratio of lateral margin of mesonotum to lateral margin of metanotum	1:2	2:3	1:1	2:1	1:0

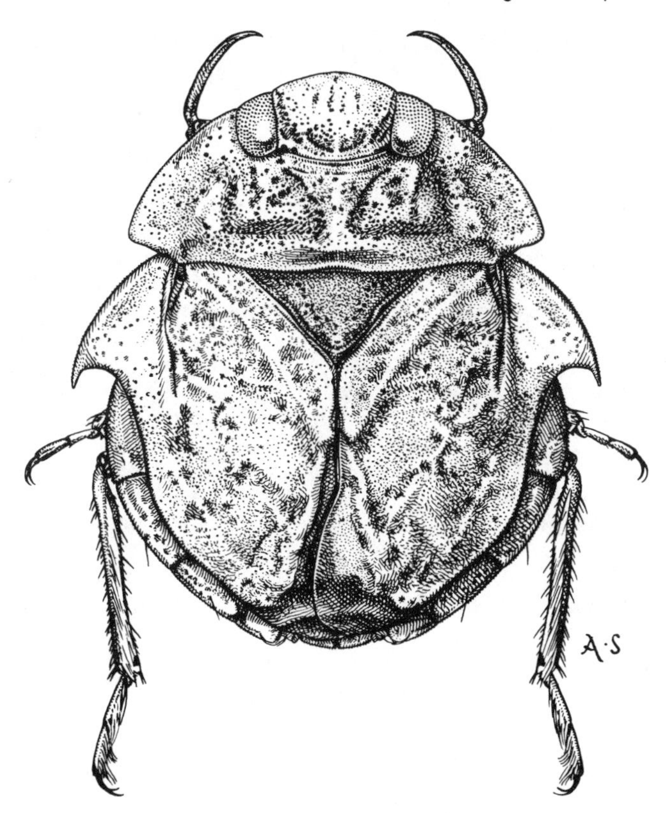

Fig. 7:13. *Usingerina moapensis* La Rivers, female, Warm Springs, Clark County, Nevada, December 26-27, 1948 (La Rivers).

Lake and Eagle Lake are inhabited by this species, and the specimens from Clear Lake appear to be slightly smaller and darker than usual, whereas the opposite is true for Eagle Lake in Lassen County. *Ambrysus mormon* has been taken also in shallow water at the north end of Lake Tahoe. It is most abundant, however, in the saline waters of Pyramid Lake, Nevada, where it is the dominant insect.

In spite of interesting local occurrences in lakes, *Ambrysus mormon* is typically an inhabitant of streams. It occurs in all the streams draining into the Sacramento and San Joaquin valleys from the Sierra Nevada and from the Coast Ranges. It is abundant in north coastal streams such as the Eel River. In southern California it is replaced by *Ambrysus occidentalis* (=*Ambrysus signoreti* of authors, not Stål 1862). Coincident in range with both of these, at least in coastal streams, and sometimes found in the same streams is *Ambrysus californicus,* a smaller pale green species. The typical form is found in southern California, whereas a form with spined connexival angles occurs in northern California.

The largest *Ambrysus* known from California is *A. puncticollis*. This species occurs in streams draining into the Colorado River and has been taken near Parker Dam on both the California and Arizona sides of the river.

By far the most local species in California are *Ambrysus funebris* and *A. amargosus*. These little species have thus far been taken only in the water flowing from hot springs in the Amargosa River system in southern Nevada and Death Valley, California. The characteristics of the water are as follows (La Rivers, 1953): temperature 32.2°C (90°F); pH—7.3; CO_2—20 ppm.; CO_3—lacking; HCO_3—256 ppm; DO—2.7 ppm.

Taxonomic characters.—Aside from size, color, and contour, which are distinctive for each species but difficult to delimit, the chief distinguishing characters of naucorids are the form and extent of the propleura and the genital process of the male and genital plate of the female (fig. 7:15). The male genital segments move as a unit and are fixed to the basal abdominal segments only at the middle. In mating they are bent downward and turned slightly to expose a prominent hook (absent in *A. funebris* and *amargosus*) on the caudal margin of the fifth tergite on the right side. This process is very distinctive in size and shape for each species and may be examined by softening the specimen and bending the genital segments downward. The female subgenital plate is the large median plate at the apex of the ventral surface of the abdomen. The form of the apical margin of this plate is distinctive for our species.

Key to Nearctic Genera of Naucoridae

Adults

1. Anterior margin of pronotum straight or scarcely concave behind interocular space 2
— Anterior margin of pronotum deeply concave behind interocular space 3
2. Inner margins of eyes anteriorly divergent; meso- and metasterna bearing prominent longitudinal carinae which are broad and foveate along middle; body broadly oval, subflattened, the embolium produced outward and backward as an acute spine. Limnocorinae (fig. 7:13) ....*Usingerina* La Rivers 1950
— Inner margins of eyes anteriorly convergent; meso- and metasterna without longitudinal carinae at middle; body strongly convex above, the embolium rounded. Naucorinae (fig. 7:14)...........*Pelocoris* Stål 1876
3. Prosternum completely exposed, separated from the flattened pleura by simple sutures; abdominal venter naked and with a disclike area near each spiracle; dimorphic, the brachypterous forms with hemelytra truncate at apices, about half as long as abdomen. Cryphocricinae ....*Cryphocricos* Signoret 1850
— Propleura produced platelike over posterior part of prosternum, subcontiguous at middle and completely covering this part of prosternum; abdominal venter densely pubescent, interrupted by small holes at spiracular openings and by a transverse row of small holes behind each spiracle; macropterous (fig. 7:12) Ambrysinae .... *Ambrysus* Stål 1862

Last Instar Nymphs

1. Abdominal venter naked; front femora long and slender, with 1 ventral and 2 posterior longitudinal ridges of granules ....*Cryphocricos* Signoret 1850
— Abdominal venter pubescent; front femora roundly inflated, without ventral and posterior longitudinal ridges 2
2. Inner margins of eyes anteriorly divergent; mesosternum bearing a longitudinal carina which is foveate along the middle .... *Usingerina* La Rivers 1950
— Inner margins of eyes convergent anteriorly and also to a certain extent posteriorly; mesosternum without a foveate carina 3
3. Anterior margin of pronotum scarcely concave behind interocular space ....*Pelocoris* Stål 1876
— Anterior margin of pronotum deeply concave behind interocular space...............*Ambrysus* Stål 1862

Usingerina moapensis La Rivers 1950 (fig. 7:13) is known only from Warm Springs, Clark County, Nevada, where it was taken in the gravel bed of a stream. The temperature of the water was 75° to 89° F.

Pelocoris shoshone La Rivers 1948 (fig. 7:14) is known from the White River system in southeastern Nevada and from the Amargosa River system in Nevada and Death Valley, California (La Rivers, 1953). Two nymphs in the collection of the University of Kansas bear the label, "Nipomo, Santa Barbara County, California."

Key to the California Species of Ambrysus

Adults

1. Connexival margins of first and second visible segments smooth, those of segments 3-5 minutely serrate (best seen from below); size small, 6-6.5 mm. (fig. 7:15*b*); Hot Springs, Ash Meadows, Nye County, Nevada ....*amargosus* La Rivers 1953
— Connexival margins smooth; size usually larger ... 2
2. Propleura not contiguous, separated by a small but distinct gap at middle over prosternum; hemelytra greenish- or yellowish-brown with ill-defined paler markings at base of embolium. *californicus* subspp. 3
— Propleura contiguous over middle of prosternum; hemelytra black with distinct pale markings at least at base of embolium and usually at inner apex of embolium and at middle of apical margin of corium 4
3. Connexival angles produced into small but distinct

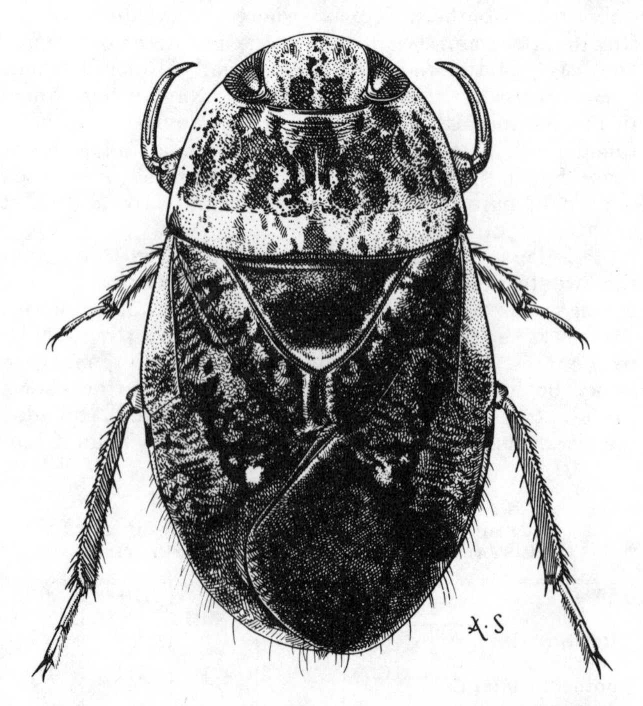

Fig. 7:14. *Pelocoris shoshone* La Rivers, female, Warm Springs, Clark County, Nevada, December 26-27, 1948 (La Rivers).

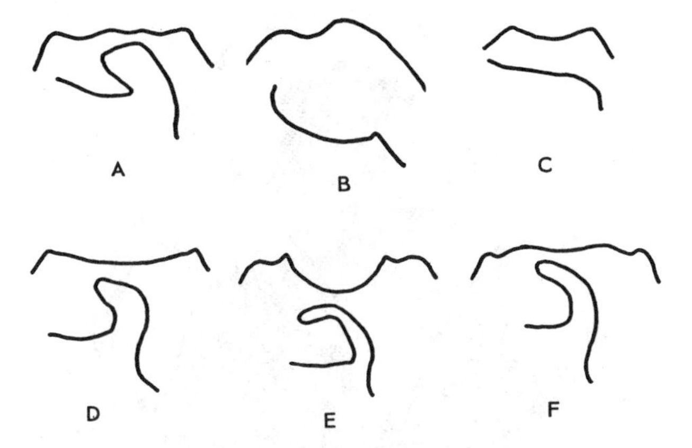

Fig. 7:15. *Ambrysus* genital segments, the top outline in each case representing the hind margin of the female subgenital plate, and the outline beneath it the dorsal process of the hind margin of the fifth visible tergite in the male. *a, Ambrysus occidentalis* La Rivers; *b, A. amargosus* La Rivers; *c, A. funebris* La Rivers; *d, A. puncticollis* Stal; *e, A. mormon* Montandon; *f, A. californicus* Montandon.

spines; northern California; streams in the coast Ranges .... *bohartorum* Usinger 1946

— Connexival angles scarcely produced, little more than right angles (fig. 7:15*f*); southern California ....*californicus* Montandon 1897

4. Size small, 6-6½ mm.; no dorsal genital process in the male (fig. 7:15*c*); Death Valley, stream from a hot spring, 36° C ....*funebris* La Rivers 1949

— Size larger, 8-15½ mm.; genital process of fifth abdominal tergite of male well developed 5

5. Corium unspotted, unicolorous; size large, 13 mm. or more; female genital plate unisinuate (simply, shallowly concave) at apex (fig. 7:15*d*); stream near Parker Dam, Colorado River .... *puncticollis* Stål 1876

— Corium variously marked with pale areas; size smaller, 8-12½ mm.; female genital plate trisinuate at apex 6

6. Head more than half as wide, eyes included, as greatest width of pronotum, in ratio of 7:13, corium with a large bilobed spot at inner apex of embolium and a large crescent-shaped spot at middle of apical margin; male genital process as broad as long; female genital plate only feebly concave at middle of apex (fig. 7:15*a*); southern California streams .... *occidentalis* La Rivers 1951

— Head slightly less than 1/2 as wide as pronotum; corium with a small spot at inner apex of embolium and at middle of apical margin or rarely without spots (fig. 7:12); male genital process much longer than wide; female genital plate deeply concave at middle of apex (fig. 7:15*e*); northern and central California, streams and shallow margins of lakes .... *mormon* Montandon 1909

Key to Last Instar Nymphs of Ambrysus

1. Lateral margins of abdominal segments entire, smooth; size large, 12 mm.; color rather uniform dark brown on meso- and metanota and abdomen .....*puncticollis* Stål

— Lateral margins of abdominal segments minutely but distinctly serrate or crenulate; size less than 10 mm.; ground color ochraceous with a regular pattern of brown markings on thorax and abdomen 2

2. Head approximately 1/2 as wide across eyes as greatest width of pronotum (0.49 to 0.52) 3

— Head much more than 1/2 as wide across eyes as greatest width of pronotum 4

3. Head relatively small, 1/2 again as wide across eyes as long (51:35) .... *occidentalis* La Rivers

— Head broader, more than 1/2 again as wide across eyes as long (57:35) (fig. 7:12)..... *mormon* Montandon

4. Pronotum relatively narrow, 1¾ times as wide as head; pronotum distinctly spotted laterally .... *funebris* La Rivers

— Pronotum broader, 1⅔ as wide as head, the lateral areas unspotted ....*californicus* Montandon

REFERENCES

LA RIVERS, IRA
1949. A new species of *Pelocoris* from Nevada, with notes on the genus in the United States. Ann. Ent. Soc. Amer., 41:371-376.
1950. A new Naucorid genus and species from Nevada. Ann. Ent. Soc. Amer., 43:368-373.
1951. A revision of the genus *Ambrysus* in the United States. Univ. Calif. Publ. Entom., 8:277-338.
1953. New Gelastocorid and Naucorid records and miscellaneous notes, with a description of the new species, *Ambrysus amargosus*. Wasmann Jour. Biol., 11:83-96.

THORPE, W. H.
1950. Plastron respiration in aquatic insects. Biol. Rev., 25:344-390.

USINGER, R. L.
1941. Key to the subfamilies of Naucoridae with a generic synopsis of the new subfamily Ambrysinae. Ann. Ent. Soc. Amer., 34:5-16.
1946. Notes and descriptions of *Ambrysus* Stål, with an account of the life history of *Ambrysus mormon* Montd. Univ. Kansas Sci. Bull., 31:185-210.
1947. Classification of the Cryphocricinae. Ann. Ent. Soc. Amer., 40:329-343.

Family BELOSTOMATIDAE

Giant Water Bugs

Giant water bugs have attracted the attention of naturalists in America for more than one hundred years by their large size, their habit of flying to lights, and the curious characteristic exhibited by some genera of laying eggs on the backs of the males. Other common names are "fish killers," "electric-light bugs" and "toe-biters." Belostomatids are large, dorsally subflattened, brown bugs with the rostrum short and stout, antennae concealed in ventral grooves behind the eyes, front legs raptorial, and apex of abdomen with a pair of short, straplike, retractile appendages. The hind legs are flattened and ciliated for swimming. The tarsi, in the adults of our species, are two-segmented with one claw on the front legs and two claws on the middle and hind legs. The tarsi are one-segmented in the nymphs, and there are two prominent claws on the middle and hind tarsi of nymphs and two equally developed claws on the front tarsi of nymphs of *Lethocerus* and *Benacus*. The outer claw of the front tarsus of *Belostoma* and *Abedus* nymphs, on the other hand, is reduced and is totally absent in later instars (Bueno, 1906).

Students of animal behavior have studied the curious "death-feigning" habit which is reported for *Belostoma* and *Abedus*. When disturbed by removal from water or by contact with the dorsal and ventral surfaces of the body, the bugs assume a characteristic rigid position and remain in this so-called "death-

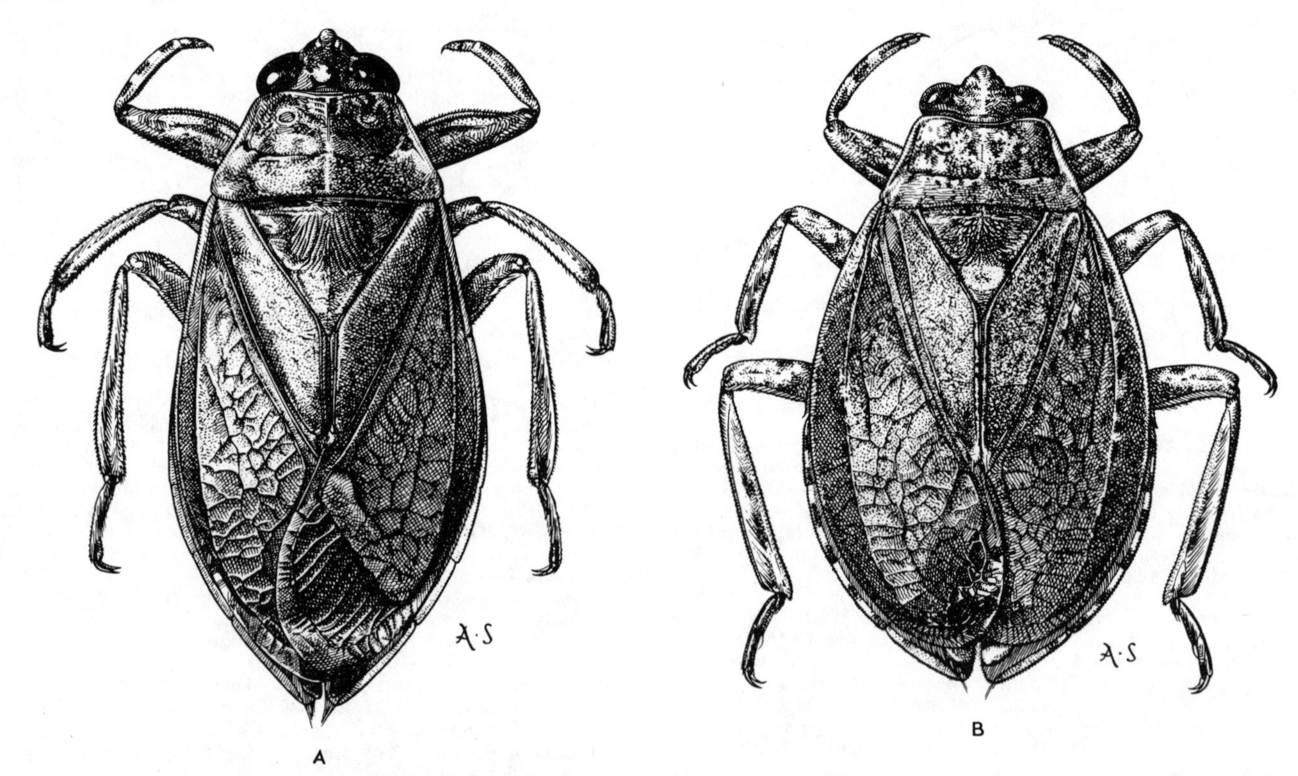

Fig .7: 16. *a, Belostoma bakeri* Montd., Davis, Yolo County, California, July 9, 1932 (R. L. Usinger); *b, Abedus indentatus* (Hald.), Walnut Creek, Contra Costa County, California (R. L. Usinger).

feint" for varying periods which averaged 16.4 minutes in five successive experiments involving six belostomas (Severin and Severin, 1911).

Another curious habit, namely, the forceful ejection of fluid from the anus, has been noted when the bugs have been held captive out of water (Harvey, 1907). This fluid may be responsible for the odor noted by some observers.

Lethocerus indicus (L. & S.) is boiled in salt water and eaten by the Cantonese and other Asiatic peoples. Specimens may be purchased in food shops in China and in Chinatown in San Francisco.

Stridulation has not been verified in this family but Hungerford (1925) noted "a wheezing noise" made by *Benacus griseus,* and Harvey (1907) reported a "soft chirping noise" for *Abedus "macronyx."*

Flight.—Hungerford (1925) documents the well-known propensity of *Benacus* and *Lethocerus* to fly to electric lights. At Lawrence, Kansas, 44 males and 57 females were taken on May 5, 1920; 9 males and 7 females on May 8; 22 and 29 on May 9; 9 and 12 on May 21; and 29 and 16 on May 22. All these figures apply to *Benacus griseus* but 10 males and 1 female of *Lethocerus americanus* were taken at the same time.

Habitat.—Ponds and quiet pools in streams are the favorite haunts of belostomatids. In spite of their large size, these bugs are not conspicuous because they commonly rest at the surface in trash or mats of vegetation. They rest with the body extending obliquely downward and the tip of the abdomen slightly above the surface film.

Feeding.—Belostomatids are clearly the masters of their immediate environment. They capture and suck dry a variety of insects and other aquatic arthropods as well as tadpoles and fish. Large bugs have been known to destroy fish several times their size (Dimmock, 1886). Some attention has been given to control methods in pond fish culture in the United States and in China.

Respiration.—Belostomatids obtain air by breaking the surface film with the short, straplike appendages at the tip of the abdomen. Each of these flaps has a spiracular opening near the base which connects with a main longitudinal tracheal trunk. Air is admitted to the space between the hemelytra and the dorsal surface of the abdomen and appears as a large silvery bubble when the bug dives below the surface.

Life history.—Like most water bugs, belostomatids overwinter as adults. Eggs are laid in the spring and early summer. *Benacus* and *Lethocerus* lay masses of one hundred or more eggs on supports above the water, for example, on cattail stalks. The eggs are glued at the posterior ends and project outward. They are arranged in rows and are contiguous. They increase in size as they develop, the dimensions for *Benacus griseus* being 4.5 mm. long and 2.25 mm. in diameter when laid and 6.57 mm. long twelve days later. The egg surface is hexagonally reticulate, and the color becomes progressively darker and longitudinally striped with brown as development proceeds. The incubation period has been reported as one to two weeks. Hatching takes place by a breaking open of the shell at the head end. *Lethocerus* eggs from California measure 5.25 mm. by 2.5 mm., tapering basally to 2 mm. The color is grayish to dark brown with darker brown stripes at the apical end and a distinct white ring marking the edges of the lid which is pushed upward at hatching. The cap remains hinged to the eggshell after hatching.

The eggs of *Belostoma* and *Abedus* are glued to the backs of the males and are carried in this position until hatching. This is a remarkable adaptation, presumably serving as protection for the eggs. The female may add to the batch of eggs on successive days until a single male may carry more than one hundred eggs. Flight is, of course, impossible during this period, though the whole mass of eggs is readily dislodged as a unit when dry. *Abedus* eggs have been described as 5 mm. long by 1 mm. thick. They are brown with a small, paler elevation at the micropylar end. Harvey (1907) gives the incubation period for *Abedus "macronyx"* as 10-12 days. *Belostoma flumineum* eggs from California measure 1.78 by 0.9 mm.

Nymphal development was found by Bueno to require 43 to 54 days in *Belostoma flumineum* Say. The approximate length of each of the five successive instars is given by Hungerford (1920) as 4.6, 5.5, 8, 11.5, and 17 days. Several overlapping broods have been observed during the year.

Distribution.—Belostomatidae are found in most tropical and temperate regions but are lacking in the northern Palaearctic region. *Lethocerus* is widespread, but *Belostoma* is found exclusively in the Western Hemisphere as are *Abedus* and *Benacus*. *Benacus* is confined to the eastern United States and *Abedus* is exclusively western, extending from California to Panama, with the exception of one species confined to Florida and Georgia.

A specimen of *Lethocerus* was taken by E. S. Ross in the La Brea asphalt pits. Since the head and hind tibiae are missing it cannot be placed positively to species.

Relationships.—The family Belostomatidae is quite distinctive in general appearance but is usually placed near the Nepidae in phylogenetic diagrams. The two families have been compared on the basis of terminal respiratory appendages, though the appendages are entirely different in gross appearance and degree of development. The eggs of the two families are entirely different and the rotatory coxae would require placement in a separate "series" according to one classification of the Hemiptera. The male genitalia are also quite distinct so it is fair to say that the family Belostomatidae has no really close relatives.

Taxonomic characters.—The first monographic works on Belostomatidae were by Mayr early in 1863 and Dufour late in 1863. Excellent though these works were, they were the source of much subsequent confusion because of insufficient knowledge of the limits of variation of North American species. De Carlo (1938) produced a modern revision, but this and his more recent work on *Abedus* (1948) suffer from lack of knowledge of the limits of variation of North American species. Hidalgo (1935) also revised *Abedus* on the basis of the large collections at the University of Kansas, and Cummings (1933) straightened out the taxonomy of the giant water bugs of the genera *Lethocerus* and *Benacus*. The present keys are based largely on the above quoted works with new characters introduced for separating the genera in nymphal stages and with corrections in synonymy and distributional records based on study of long series of California species.

Generic characters as indicated in the keys are reasonably adequate and distinctive, though difficulties are encountered when Neotropical forms are considered. Specific characters are much less distinctive—so much so that existing keys have proved difficult to use in identifying our species of *Abedus* and *Belostoma*.

In *Abedus* the number of antennal segments (three or four) and the shape of the prolongations of the second and third segments have been found to vary [Hidalgo, 1935, and Hussey, *in litt.*, *Abedus signoreti* Mayr and *Abedus immaculatus* (Say)]. Size (31 to 35 mm.) likewise appears to be of doubtful significance. Only one species, *Abedus indentatus* (Haldeman), is found in California. *Abedus mayri* De Carlo, *hungerfordi* De Carlo, *macronyx* (Mayr) and *brachonyx* (Mayr) all described from California, fall within the range of variation of *indentatus* and are reduced to synonymy. Records of *dilatatus* (Say) from the state are the result of misidentifications by Uhler, and the species is only known from Mexico.

Two species of *Lethocerus* are recognized from California. Kirkaldy and Bueno (1908) record *annulipes* (H. S.) from the state, but I have not seen specimens of this.

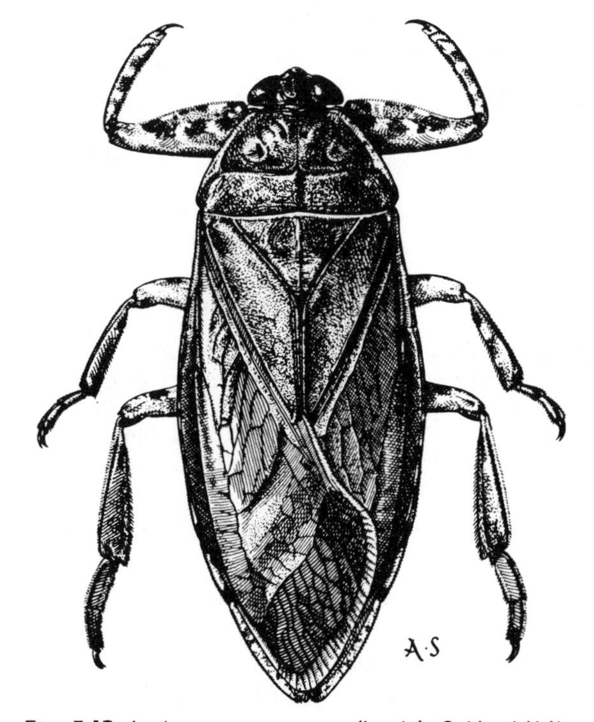

Fig. 7:17. *Lethocerus americanus* (Leidy), Oakland Hills, Alameda Co., Calif., Feb. 22, 1927 (R. L. Usinger).

Five species of *Belostoma* have been recorded from California but three of these, *boscii* L. & S., *fusciventre* (Dufour), and *apache* Kirkaldy, remain unverified. *Fusciventre* was described from Mexico, and Mexican specimens which agree with the description have distinct interocular depressions and are much darker than anything I have seen from California. *Apache* was proposed as a new name for *minor* Dufour, a Brazilian species, dubiously recorded also from California by Kirkaldy (1908). *Boscii* appears also to have been recorded first from California in the Kirkaldy-Bueno catalogue, though the early records and misidentifications of Uhler (1870-1894) may have been the source.

Key to the Genera of Belostomatidae

Adults

1. Basal segment of rostrum much thicker and shorter than second; size large, more than 40 mm.; eggs not laid on backs of males 2
— Basal segment of rostrum not conspicuously thickened, subequal to second segment; smaller, 36 mm. or less; eggs laid on backs of males 3
2. Anterior femora sulcate on inner face opposite tibiae (fig. 7:17) .... *Lethocerus* Mayr 1853
— Anterior femora not sulcate on inner face, one species, *B. griseus* (Say, 1832), eastern United States .... *Benacus* Stål 1861
3. Membrane of hemelytra reduced (fig. 7:16*b*); length more than 27 mm.; except *A. immaculatus* (Say) from Florida and Georgia .... *Abedus* Stål 1862
— Membrane of hemelytra not reduced (fig. 7:16*a*); length less than 22 mm. .... *Belostoma* Latr. 1807

Last Instar Nymphs

1. Front legs with 2 equally developed claws 2
— Front legs with only 1 well-developed claw 3
2. Front femora sulcate .... *Lethocerus* Mayr
— Front femora not sulcate .... *Benacus* Stål
3. Inner anterior angles of eyes separated by a smooth area and then indented in the form of a right angle with tapering anteocular part of head .... *Abedus* Stål
— Inner anterior angles of eyes continuous with gradually tapering anteocular part of head ...... *Belostoma* Latr.

The California Species of Abedus

Abedus indentatus (Haldeman) 1854 (fig. 7:16*b*) appears to be the only species in California. It is restricted to California and northern Baja California. It occurs as far north as the 39th parallel and is abundant in many southern California streams.

Key to California Species of Lethocerus

1. Inner apical angle of under side of hind tibia produced into a sharp point (fig. 7:18); interocular space distinctly narrower than width of an eye; Mexico and Saratoga Springs, Death Valley, California .... *angustipes* Mayr 1871
— Inner apical angle of under side of hind tibia rounded; interocular space not narrower than width of an eye (fig. 7:17); widely distributed over the United States .... *americanus* Leidy 1847

Key to California Species of Belostoma

1. First rostral segment longer than second; vertex along inner margins of eyes without conspicuous pubescence; fore femora typically with 3 spots on outer and 3 spots on inner surface (fig. 7:16*a*); Sacramento, San Joaquin, and Imperial valleys .... *flumineum* Say 1832
— First rostral segment shorter than second; vertex along inner margins of eyes with conspicuous gray pubescence; fore femora not spotted as above; widespread .... *bakeri* Montd. 1913

REFERENCES

CUMMINGS, CARL
1933. The giant water bugs. Bull. Univ. Kansas, 34:197-219.
DeCARLO, J. A.
1938. Los Belostómidos Americanos. An. Mus. Argent. Cien. Nat. Buenos Aires, 39:189-252, Lam. I-VIII.

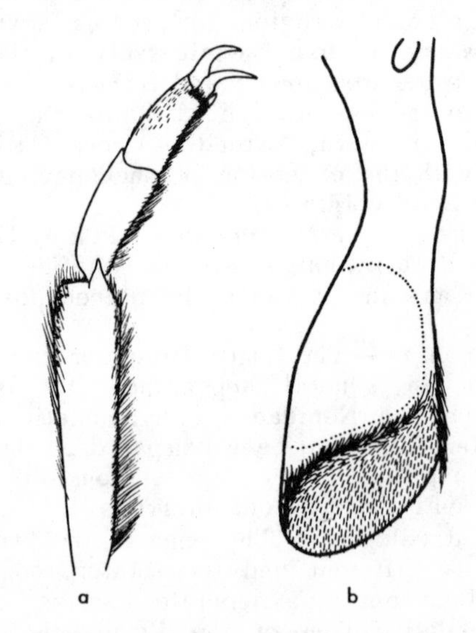

Fig. 7:18. *a*, *Lethocerus angustipes* (Mayr), under side of hind leg (Cummings, 1933).
b. *Abedus indentatus* (Hald.), caudal filament.

1948. Revision del Genero "*Abedus*" Stål. Com. Mus. Argent. Cien. Nat. (Zool.). No. 5, pp. 1-24, Tab. I-III.
DIMMOCK, GEORGE
1886. Belostomidae and some other fish-destroying bugs. Ann. Rpt., Fish and Game Comm., Mass. Pub. Doc. 25 (D.), pp. 67-74.
DUFOUR, L.
1863. Essai Monographique sur les Bélostomides. Ann. Soc. Ent. France, 3:373-400.
HARVEY, G. W.
1907. A ferocious water-bug. Canad. Ent., 39:17-21.
HIDALGO, JOSE
1935. The genus *Abedus*. Bull. Univ. Kansas, 36:493-519.
HUNGERFORD, H. B.
1920. The biology and ecology of aquatic and semiaquatic Hemiptera. Kansas Univ. Sci. Bull., 21:1-328, 30 pls.
1925. Notes on the giant water bugs. Psyche, 32:88-91, 3 figs.
KIRKALDY, G. W.
1908. *In* Kirkaldy and Bueno, A catalogue of American aquatic and semiaquatic Hemiptera. Proc. Ent. Soc. Wash., 10:173-215.
MAYR, G. L.
1863. Hemipterologische studien. Die Belostomiden. Verh. Zool.-Bot. Ges. Wien, 13:339-364, 10 figs.
1871. Die Belostomiden. Verh. Zool.-Bot. Ges. Wien, 21:399-438.
SEVERIN, H. H. P., and H. C. SEVERIN
1911. An experimental study on the death-feigning of *Belostoma* (=*Zaitha* Aucct.) *flumineum* Say and *Nepa apiculata* Uhler. Behavior Monogr. Henry Holt and Co., Vol. 1, No. 3, pp. 1-44, 1 pl.
TORRE-BUENO, J. R. de la
1906. Life-histories of North-American Water-bugs. I. Life-history of *Belostoma fluminea,* Say . Canad. Ent., 38:189-197.

Family NEPIDAE

Water Scorpions

Members of the family Nepidae are distinguished from all other water bugs by their long, slender, respiratory filaments and by the three pairs of small, oval disclike "static sense organs" at the sides of the second, third, and fourth visible ventral segments. The anterior legs are raptorial and the remaining legs are slender and thus better suited to progress among tangled mats of vegetation than to swimming in open water. Nevertheless, an ungainly *Ranatra* may be seen occasionally in open water with its legs moving alternately.

Life history.—The life history of *Ranatra* has been studied by Torre-Bueno (1906) in the eastern United States. Winter is passed in the adult stage. Egg-laying takes place during the growing season, the eggs being inserted in plant tissue with two slender filaments protruding (fig. 7:20*c,d*). These filaments are longer than the egg, and the incubation period varies from two weeks to a month. The five nymphal instars each required a week or more according to Torre-Bueno.

Feeding habits.—Water scorpions are voracious feeders on practically any organism of suitable size in their environment. Small Crustacea, mayfly nymphs, and mosquito larvae have been mentioned, among others. Water scorpions do not pursue their prey, but lie in wait and capture organisms that chance to come close to their place of concealment. Ranatras are especially well concealed by trash or tangled plant

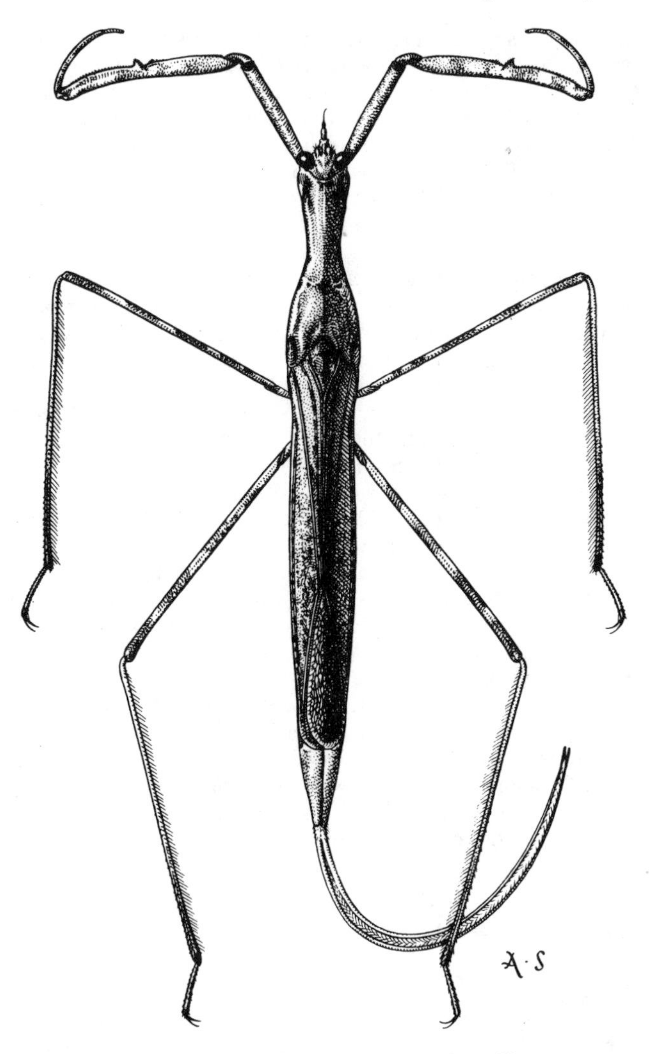

Fig. 7:19. *Ranatra brevicollis* Montandon, male, Walnut Creek, Contra Costa Co., Calif., Aug. 9, 1927 (R. L. Usinger.)

growth because of their sticklike appearance. Brooke and Proske (1946) showed by precipitin tests that a *Ranatra fusca* adult had recently fed upon immature mosquitoes. Omerod (1889) records the European *R. linearis* as attacking small fish.

Distribution.—Only three genera of water scorpions are known from North America. One of these, *Nepa,* is strictly eastern, the single species, *N. apiculata* Uhl., being confined to the eastern states with its nearest ally, *N. cinerea* Linn., occurring in Europe. *Curicta* is a Neotropical genus with only two species extending into the southern part of the United States. *Ranatra* is by far the dominant genus in North America and throughout most of the rest of the world as well. It is the only genus that occurs in California. *R. brevicollis* Montd. (fig. 7:19) is the commonest California species and is apparently confined to the state. Two additional species can now be recorded, *fusca* P. B. from Shasta County in northern California and *quadridentata* Stål from the Colorado River.

Pierce (1948) described two fossil species, *Ranatra*

bessomi and *R. asphalti* from the McKittrick asphalt pits in what is "now very dry country around McKittrick in the western foothills of Kern County, Calif."

Stridulation.—A squeaking noise is produced by *Ranatra* and is made by rubbing a thickened area at the base of the front coxa along the crenulate margins of the flaring coxal cavity (fig. 7:20*e,f*).

Respiration.— Water scorpions are surface breathers, obtaining air by means of their caudal appendages. In the nymphs the respiratory tube consists of a single prolongation of the abdomen with a longitudinal groove which is open along the ventral surface. The adults have two caudal appendages which, when held together, form a closed tube. The exact role of the "static sense organs" is not known, but the theory has been advanced that they serve for orientation when the bugs descend into deep water. Most water bugs have air stores which help in maintaining their position in relation to the water surface. Without some such "hydrostatic organ" a water bug is heavier than water and hence is in danger of sinking and thus losing access to the surface air which is essential to its continued existence. Water scorpions, it is said, utilize their "static sense organs" for orientation in lieu of a hydrostatic air store.

Taxonomic characters.—The species of *Ranatra* are distinguished mainly by differences in proportions of pronotum, front femora, lateral prolongations of antennal segments, and extension of the metaxyphus between the hind coxae. The presence or absence and degree of development of subapical spines on the front femora provide reliable characters for some species, but the spines are variable in our species. Male genitalia have been figured by Hungerford (1922) but the differences are so slight for our species that they are of little or no value in classification.

The principal work on the taxonomy of Nepidae of North America is by Hungerford (1922).

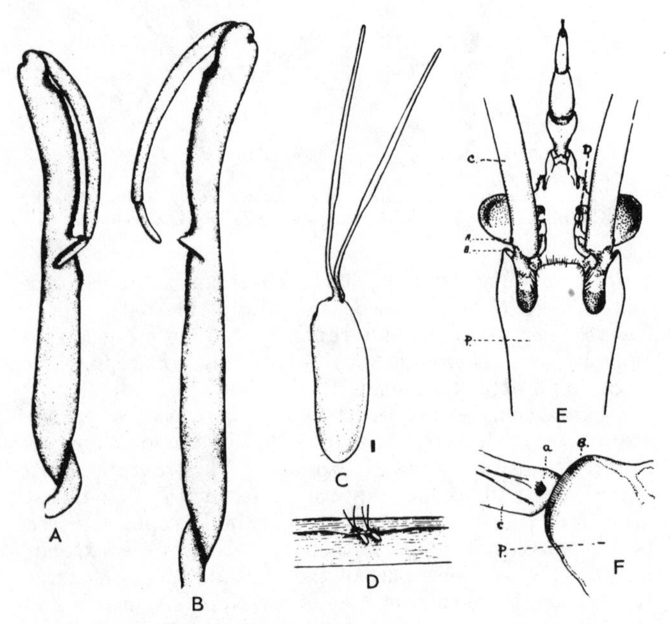

Fig. 7:20. Front legs. *a, Ranatra brevicollis* Montd.; *b, R. fusca* P. B.; *c*, egg of *R. fusca* P. B. dissected from a water-soaked dead cattail blade; *d*, eggs of *R. fusca* P. B. in cattail leaf with a part of the leaf tissue removed to expose the eggs; *e*, stridulation mechanism in *Ranatra*, showing ventral view of head and part of prothorax. The rough patch (A) on the base of coxa (C) is rubbed against the file (B) on the inner edge of the anterolateral margin of prothorax (P), producing a chirping or squeeking noise. Antennae (D) are partly concealed by coxae. *f*, parts of stridulatory mechanism enlarged. (Hungerford, 1922.)

Key to Genera of Nepidae

1. Body slender, subcylindrical; hind coxae approximate, the intercoxal distance less than width of a coxa (fig. 7:19) .... *Ranatra* Fabr. 1790
— Body oval or elongate oval, subflattened; hind coxae widely separated, the intercoxal distance much greater than width of a coxa 2
2. Broadly oval, more than 1/3 as wide as long; propleura widely separated, leaving the prosternum broadly exposed posteriorly; fore femora with a longitudinal trough for the reception of the tibia, the trough wider on basal half than the tibia; eastern United States .... *Nepa* Linn. 1758
— Elongate oval, less than 1/4 as wide as long; propleura contiguous over prosternum posteriorly; fore femora with longitudinal groove much narrower than tibia throughout its length; southern states and Mexico .... *Curicta* Stål 1861

Key to California Species of Ranatra

1. Pronotum relatively short and broad, less than 2½ times as long, measured on the side, as width across humeri, the anterior dilation of pronotum more than 90 per cent as wide as head across eyes; posterior lobe much more than 2/3 as long, measured on the side, as anterior part, 85:100; front femur relatively broad, the proximal edge of tooth located midway between apex of trochanter and apex of femur (figs. 7:19; 7:20*a*); metaxphus reaching nearly to base of abdomen; California, south and central .... *brevicollis* Montd. 1910
— Pronotum longer and more slender, nearly or quite 3 times as long as wide across humeri; the anterior dilation of pronotum about 85 per cent as wide as head across eyes; posterior lobe about 2/3 as long as anterior part; front femur relatively slender, the proximal edge of tooth located beyond the mid-point between apex of trochanter and apex of femur; metaxyphus reaching only 2/3 the distance to base of abdomen 2
2. Anterior lobe of pronotum 1¼ times as long as posterior lobe, measured on the side; hind femora reaching only to basal third of penultimate abdominal segment (fig. 7:20*b*); Lake Britton, Shasta County, California .... *fusca* P. B. 1805
— Anterior lobe of pronotum relatively longer, nearly 1½ times as long as posterior lobe; hind femora reaching beyond middle of penultimate abdominal segment; Colorado River .... *quadridentata* Stål 1861

REFERENCES

BROOKE, M. M., and H. O. PROSKE
1946. Precipitin test for determining natural insect predators of immature mosquitoes. Jour. Nat. Malaria Soc., 5:45-56.
HUNGERFORD, H. B.
1922. The Nepidae in North America north of Mexico. Kansas Univ. Sci. Bull., 14:425-469, 8 pls.
ORMEROD, E. A.
1889. *Ranatra linearis* attacking small fish. Entomologist, 11:95, 119-120.
PIERCE, W. D.
1948. Fossil arthropods of California. 15. Some Hemiptera from the McKittrick asphalt field. Bull. S. Calif. Acad. Sci., 47:21-33.

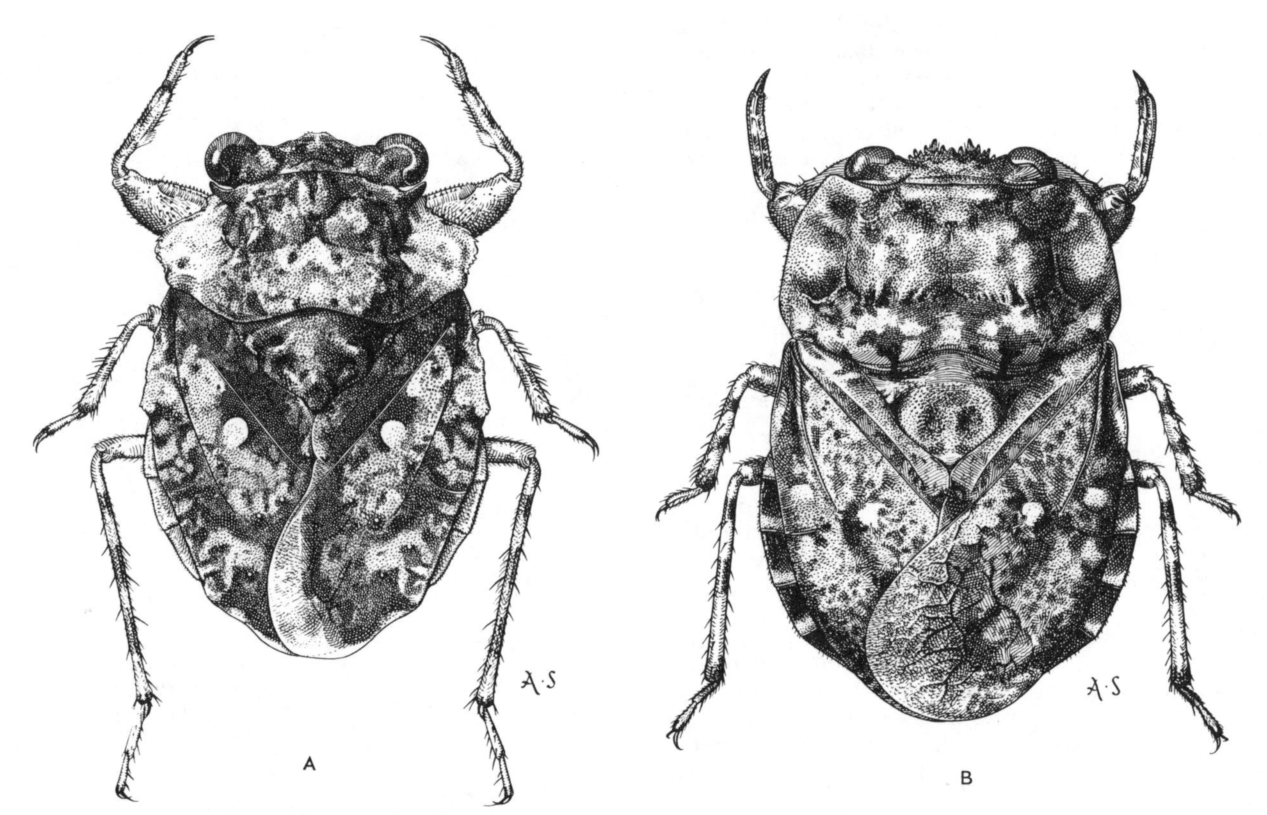

Fig. 7:21. *a, Gelastocoris oculatus* (Fabr.) Mokelumne Hill, Calaveras County, California, May 27, 1931, (R. L. Usinger); *b, Nerthra martini* Todd, Frenchman Flats, Tehachapi Mountains, California, April 8, 1951, (R. L. Usinger.)

TORRE-BUENO, J. R. de la
1906. Life history of *Ranatra quadridentata*. Canad. Ent., 38:242-252.

Family GELASTOCORIDAE

Toad Bugs

Toad bugs are squat, compact, mud and sand inhabitants with prominent eyes, raptorial front legs, ocelli more or less distinct, and hemelytra usually separate and fully developed but sometimes with the membrane reduced and rarely with the hemelytra fused. The nymphs lack dorsal abdominal scent gland openings.

Life history.—From the meager records at hand *Gelastocoris* overwinters as an adult. Eggs are laid in the sand. They are broadly oval with roughly granular surface, 1.25 mm. long and .91 mm. in diameter. The incubation period in Kansas (Hungerford, 1922) was found to be 12 days. The average length of each of the five nymphal instars was: 15 days, 16, 15, 15½, 22. No member of the genus *Nerthra* has been reared through a complete life history, but Kevan (1942) found that the last nymphal instar of *N. nepaeformis* (Fabr.) lasts for at least two weeks. Egg laying was observed in *N. martini* Todd at Frenchman Flats, Tehachapi Mountains, California, in June. The eggs are similar in form to *Gelastocoris* but are longer (1.35 mm.) and smaller in diameter (.85 mm.). The *Nerthra* eggs were laid in small holes in the mud beneath stones several feet from the water's edge. Most remarkable was the fact that in both instances when eggs were found, the female *Nerthra* was in a position on top of the egg cluster in the hole. Since the eggs were in an advanced stage of development as indicated by conspicuous eyespots, it is suggested that females may remain with (guard?) the eggs throughout the incubation period.

Feeding habits.—Hungerford (1922) fed his *Gelastocoris oculatus* on a variety of insects taken by sweeping grass. In the natural habitat they no doubt pounce upon any helpless creatures of small size. Kevan (1942) reports that *N. nepaeformis* (Fabr.) took termites and a soil-dwelling earwig (*Euborella* sp.) in Trinidad.

Habitat.—As mentioned above, *Gelastocoris* is found on sandy and muddy shores of ponds and streams. The mottled color pattern renders them inconspicuous against their normal background. *Gelastocoris oculatus* is an extremely variable species and superficial observations would suggest that this variation is correlated with the color of the substrate. Hungerford tried to test this "by placing them in pans containing sand on one side and sandy loam on the other" but reports that his results were inconclusive.

Nerthra occurs in muddy situations and often at considerable distance from the water. This was noted by Kevan (1942) in Trinidad and confirmed independently by me in California.

Distribution.—*Nerthra* is a world-wide genus, though confined to tropical and warmer temperate regions. In California *N. martini* is known only from the south. It also occurs in adjacent parts of Nevada and extends southward to the tip of Lower California. *N. usingeri* is known only from the type locality, a small wash which empties into the Colorado River near Parker Dam.

The genus *Gelastocoris* is exclusively American and our commonest species, *oculatus,* formerly known in the West as *G. variegatus,* is distributed throughout the entire United States and southward as well. *G. rotundatus* is a Central American species which extends northward into Arizona and California along the Colorado River drainage (Adams and MacNeill, 1951).

Taxonomy.—The taxonomy of the genus *Gelastocoris* has been hopelessly confused because of the unusual variation in color and, to some extent, in size and in other details of external structure. Melin (1928) added to this confusion by describing several species on the basis of superficial characters. His *G. californiensis* cannot be recognized as distinct from *oculatus*. Martin (1928) made a sound contribution to the taxonomy of the group, evaluating the various characters and illustrating the male genitalia. Unfortunately, no key was given. At present, then, two species are known from California and these are readily distinguished by the shape of the lateral margins of the pronotum.

The name *Nerthra* was proposed by Say one year earlier than the better known *Mononyx* of Laporte (Todd, 1952, personal communication). The hemelytra are sometimes undifferentiated into corium and membrane in this group and rarely are fused along the mid-line. Two species are known from California, both possessing very distinctive male genital claspers. The commonest species in southern California, *Nerthra martini* Todd, was known in all previous literature as *Mononyx fuscipes* Guer. Todd (1954) has shown that the latter is a more southern species which does not occur in the United States.

Key to the Genera of Gelastocoridae

1. Fore tarsi 1-segmented with 2 claws in nymphs and adults; fore femora only moderately enlarged at base, twice as long as wide across base, not subtriangular, the inner side broad, with 2 rows of widely separated short spines in apposition to tibiae; rostrum arising from rounded anterior margin of head, directed backward (fig. 7:21*a*) .... *Gelastocoris* Kirkaldy 1897
— Fore legs without a tarsal segment, the single claw (adults) or 2 claws (nymphs) inserted at apex of tibia; fore femora very wide at base, about as wide as long, subtriangular, the inner side extending platelike to form a tuberculate edge in apposition to the tibia; rostrum arising beneath notched anteroventral margin of head, directed backward and then downward (fig. 7:21*b*) .... *Nerthra* Say 1832

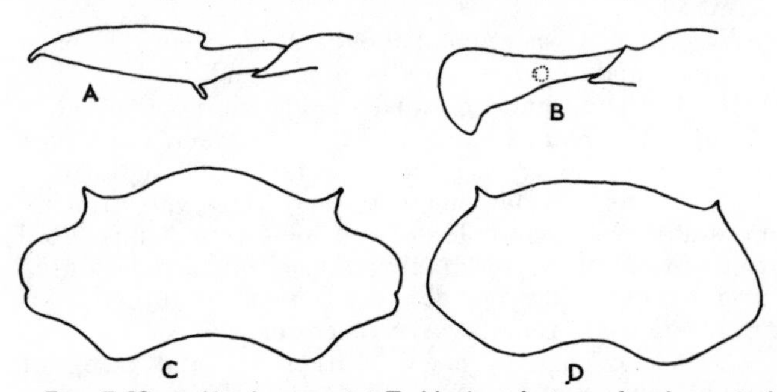

Fig. 7:22. *a, Nerthra martini* Todd, dorsal view of male genital clasper; *b, N. usingeri* Todd, same; *c, Gelastocoris oculatus* (Fabr.), outline of pronotum; *d, G. rotundatus* Champion, same.

Key to the Species of *Gelastocoris*

1. Lateral margins of pronotum strongly sinuate, deeply indented and strongly flaring platelike behind (figs. 7:21*a;* 7:22*c*); generally distributed .... *oculatus* (Fabricius) 1798
— Lateral margins of pronotum feebly sinuate, only slightly concave anteriorly and bent downward rather than flaring behind (fig. 7:22*d*); near Parker Dam, San Bernardino County, California .... *rotundatus* Champion 1901

Key to the Species of Nerthra

1. Length less than 7 mm.; male genital clasper broad, spatulate, at apex with inner side angulate (fig. 7:22*b*); female with last ventral abdominal segment broadly, roundly emarginate on posterior margin; near Parker Dam, San Bernardino County ...... *usingeri* Todd 1954
— Length more than 7 mm.; male genital clasper abruptly thickened at middle, roundly tapering to subacute tubercle at apex, with a membranous fingerlike projection on inner surface at middle (fig. 7:22*a*); female with posterior margin of last ventral abdominal segment more or less angulately emarginate (fig. 7:21*b*)..... .... *martini* Todd 1954

REFERENCES

ADAMS, P. A., and C. D. MacNEILL
1951. *Gelastocoris rotundatus* Champion in California. Pan-Pac. Ent., 27:71.
HUNGERFORD, H. B.
1922. The life history of the Toad Bug. *Gelastocoris oculatus* Fabr. Kansas Univ. Sci. Bull., 14:145-167, 2 pls.
KEVAN, D. K. McE.
1942. Some observations on *Mononyx nepaeformis* (Fabr., 1775), a toad bug. Proc. Roy. Ent. Soc. Lond. (A), 17:109-110.
MARTIN, C. H.
1928. An exploratory survey of characters of specific value in the genus *Gelastocoris* Kirkaldy, and some new species. Kansas Univ. Sci. Bull., 18:351-369, 2 pls.
MELIN, D.
1928. Hemiptera from South and Central America. I. Zool. Bidrag från Uppsala, 12:151-198.
TODD, E. L.
1954. New species of *Nerthra* from California. Pan-Pac. Ent., 30:113-117.
1955. A taxonomic revision of the family Gelastocoridae. Kansas Univ. Sci. Bull. (In press).

Family OCHTERIDAE

Ochterids are small, shore-dwelling bugs related to the toad bugs. They are oval in form and dark in color, with yellowish spots. The vertex has two well-

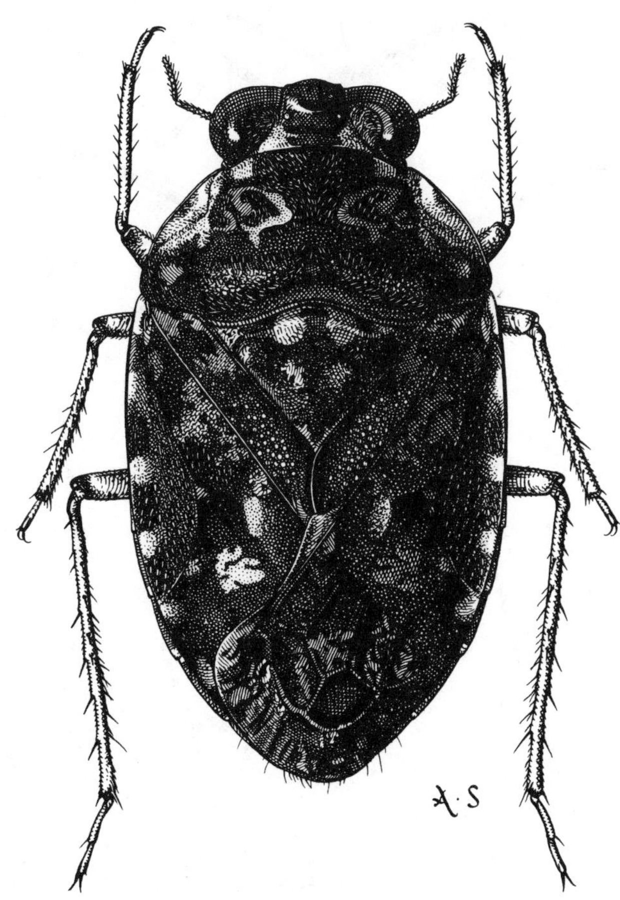

Fig. 7:23. *Ochterus barberi* Schell, Lake Houghtelin, near Bard, Calif., Nov. 13, 1951 (R. L. Usinger).

developed ocelli. The antennae are four-segmented and, though inserted beneath the eyes, are not concealed in grooves and hence do not fit precisely into the Cryptocerata in terms of key characters. Nevertheless, Singh-Pruthi (1925) found that the male genitalia are closely related to those of the Gelastocoridae. The rostrum is four-segmented, with the first segment broadly joined to the head, the second segment small, and the third segment very long and tapering. The legs are slender and are fitted for running rather than for swimming, jumping, or grasping as in related families. The front and middle tarsi are two-segmented and the hind tarsi are three-segmented, the first segment in each case being very short.

The genus *Ochterus* Latr. is widely distributed in tropical and temperate regions of the Old and New Worlds. Fifteen species are recorded from the Western Hemisphere by Schell in her revision (1943), but only one of these, *Ochterus barberi* Schell (fig. 7:23), occurs in California and this only along the Colorado River in Imperial County.

Life history data are available for only one species, *Ochterus banksi* Barber, which was reared in Virginia (Bobb, 1951). According to this account, the eggs are white and broadly oval without a cap. They are deposited singly in clumps of grass and roots and on plant debris. The incubation period was found to be 15 to 22 days. The five nymphal instars lasted as follows: first, 11 to 17 days; second, 8 to 21 days; third, 22 to 41 days; fourth, 181 to 229 days; fifth, 18 to 28 days. The winter is passed in the fourth nymphal instar, hibernating under leaves and debris along the shores of ponds and streams. The nymphs are said to construct small individual cells in the moist sand in which they molt. All nymphal stages carry sand grains on their bodies, presumably for the purpose of concealment.

REFERENCES

BOBB, M. L.
1951. Life history of *Ochterus banksi* Barber. Bull. Brooklyn Ent. Soc., 46:92-100, 1 pl.
SCHELL, D. V.
1943. The Ochteridae of the Western Hemisphere. Jour. Kansas Ent. Soc., 16:29-47, 1 pl.

Family GERRIDAE

Water Striders, Pond Skaters, Wherrymen

Gerrids are the most familiar inhabitants of the surface film of ponds, lake margins, and pools in streams. More specialized forms, such as *Metrobates,* prefer

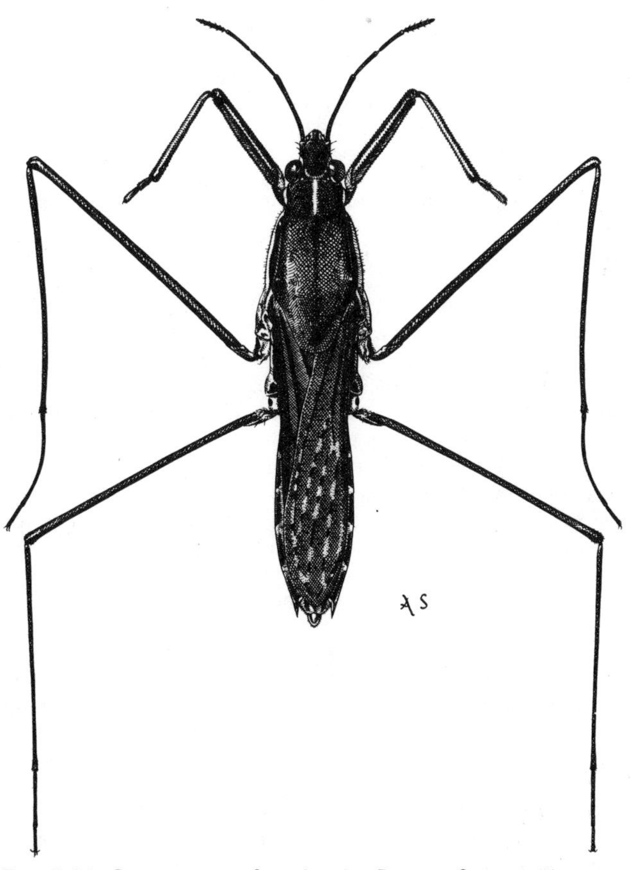

Fig. 7:24. *Gerris remigis* Say, female, Paraiso Springs, Monterey Co., Calif., May 27, 1924 (L. S. Slevin).

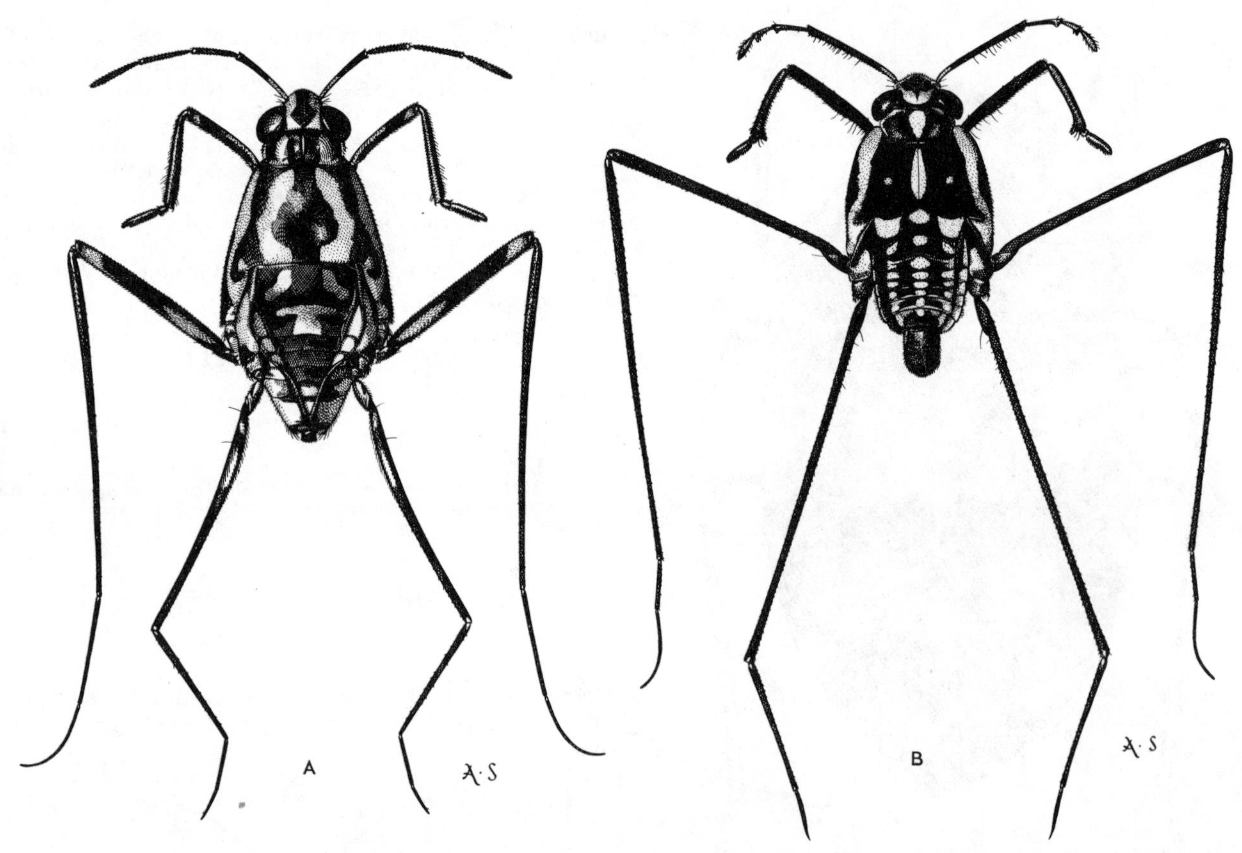

Fig. 7:25. *a, Trepobates becki* D. & H., female, Lake Houghtelin, near Bard, California, November 13, 1951, (R. L. Usinger); *b, Metrobates trux infuscatus* Usinger, male, Davis, Yolo County, California, October 24, 1942 (R. L. Usinger).

the swiftest flowing parts of larger streams and rivers, and *Halobates* are the only insects that have successfully occupied the surface of the open ocean.

Gerrids are characterized by a velvety hydrofuge hair pile, long legs with preapical claws, and a single scent gland opening at the middle of the metasternum (omphalium). The claws are inserted before the apices of the tarsi and thus do not break through the surface film. The film is simply bent by the tarsi, and this phenomenon may be observed in clear, shallow water by the shadows cast on the bottom.

Wing polymorphism is a characteristic of all members of the family except the permanently apterous marine forms (Parshley, 1919). In a single colony of *Gerris,* specimens may be found with fully developed wings (macropterous), partly developed wings (brachypterous), and with no trace of wings (apterous). The body changes that coincide with the various degrees of wing development are considerable, involving a totally different appearing thorax and even a difference in length of the abdominal spines. Wing polymorphism is controlled by a combination of genetic and environmental factors but has not been fully investigated as yet. In macropterous specimens of *Metrobates* and other Halobatinae the wings are often broken apically, possibly as a result of mating (Torre-Bueno, 1908).

Gerrids have been observed only rarely in flight. Riley (1920) reports flights at dusk and on moonlight nights and presumes that they fly on darker nights but are unobserved. Macropterous forms have an obvious advantage where pools dry up (Parshley, 1922), and completely apterous forms are at no disadvantage in the permanent marine habitat.

Most species of Gerridae are gregarious, and this is especially noticed in *Metrobates* and *Halobates*. The apparent immunity of these bugs from attack by fish may be owing to their scent gland secretions.

Feeding habits.—Gerrids are predaceous on a wide variety of organisms, including aquatic forms that come to the surface and terrestrial species that fall on the water. Water striders are very agile and do not hesitate to seize one of their own kind. This is particularly true in confined spaces such as a drying pool or an aquarium where the bugs are crowded together. Individuals are particularly vulnerable when they are molting.

Life history.—Hungerford (1920) has summarized life history studies for *Gerris remigis* and *G. marginatus*. The bugs overwinter in the adult stage in protected places near the water. Eggs are laid throughout the spring and summer. They are glued to objects at the water's edge, usually to floating objects which, of course, remain at approximately the level of the water surface throughout the season. The eggs are laid in rows, parallel to one another, the shape of individual eggs being long and cylindrical. Egg parasites, *Patasson gerrisophaga* (Doutt), have been reared from *Gerris* in California. Hungerford found that the incubation period for *G. remigis* was two weeks in Kansas. There were five instars, each lasting about one week.

Distribution.—Most of our *Gerris* species are widely distributed. *G. remigis* occurs throughout North America and *marginatus* is nearly as widespread. *Nyctalis* is very closely related to *remigis* and has not been recognized with certainty from California since 1928 when Drake and Harris reported it from Fresno. The other large species, *notabilis,* was formerly considered to be the Holarctic *Limnoporus rufoscutellatus,* but the latter is now regarded as distinct, occurring in Europe and in Alaska, and two other North American species are recognized, *dissortis* in the East and *notabilis* in the West. The remaining species of *Gerris* are all small and all northern in distribution.

Metrobates is strictly an American genus with its center of distribution in the tropics. Although widespread and common in the eastern United States it has been only recently recorded from California (Usinger, 1953). The northern California form (*M. trux infuscatus* Usinger) is darker than typical *M. trux* of the southwestern United States and the Colorado River region of southern California.

Trepobates is also a tropical American genus which is widespread in the eastern United States but is here recorded for the first time from the southern border of California at the Colorado River.

Halobates is tropicopolitan, occurring in the open ocean and in protected reefs of tropical islands. *H. sericeus* has been taken in the warm (16.22° C) waters about two hundred miles off the Monterey Coast (34° 54.5 N latitude; 124° 29 W longitude) (H. R. Attebery, Scripps Inst. Oceanography, *in litt.*)

Taxonomic characters.—The most distinctive characters of gerrid species are the male genitalia. Fortunately the apical connexival spines of females are also distinctive in many instances, although the latter vary to some extent with wing polymorphism.

Key to Genera of Gerridae

1. Inner margins of eyes sinuate or concave behind the middle, body comparatively long and narrow. Gerrinae 2
— Inner margins of eyes convexly rounded; body comparatively short and broad. Halobatinae 3
2. Pronotum sericeous, dull; basal segment of front tarsi but little shorter than apical segment (fig. 7:24); widespread .... *Gerris* Fabr. 1794
— Pronotum glabrous, shining; basal segment of front tarsi much shorter than apical segment; eastern and southern United States and tropicopolitan .... *Limnogonus* Stal 1868
3. Tibia and first tarsal segment of middle leg with a fringe of long hairs; always apterous, the meso- and metanota fused, without trace of a dividing suture (fig. 7:26); marine; widely distributed .... *Halobates* Esch. 1822
— Tibia and first tarsal segment of middle leg without a fringe of long hairs; dimorphic, the meso- and metanota of apterous forms with a distinct dividing suture situated at the bases of the posterior (dorsal) acetabula 4
4. First antennal segment subequal to the remaining 3 together; body relatively short, broad and subflattened, about half as wide as long from clypeus to apex of abdomen (fig. 7:25*b*); swift-flowing streams; widely distributed in the Western Hemisphere .... *Metrobates* Uhl. 1878
— First antennal segment much shorter than the remaining 3 together; body relatively longer and narrower and more convex, about 1/3 as wide as long 5
5. Third antennal segment with several stiff bristles; males often with strikingly incrassate or curved antennae and hind femora; eastern and southern United States .... *Rheumatobates* Bergr. 1892
— Third antennal segment with fine pubescence only; male antennae and femora not modified (fig. 7:25*a*); eastern and southern United States to southern California .... *Trepobates* Uhler 1894

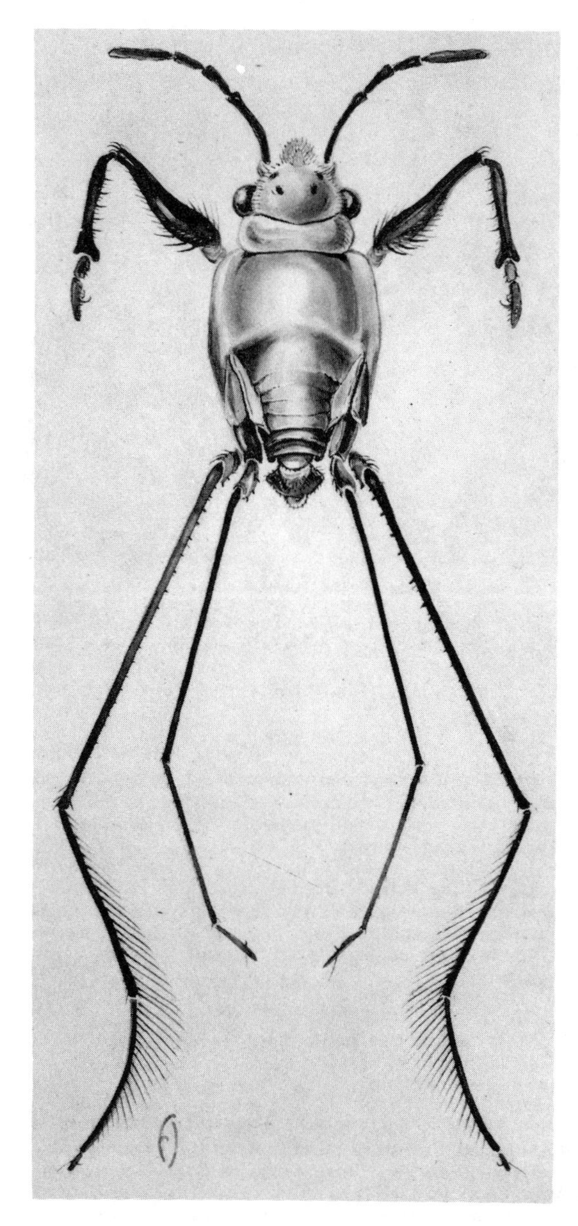

Fig. 7:26. *Halobates sericeus* Esch., male, Kailua, Oahu, T. H. (V. Fosberg) (Zimmerman, 1948).

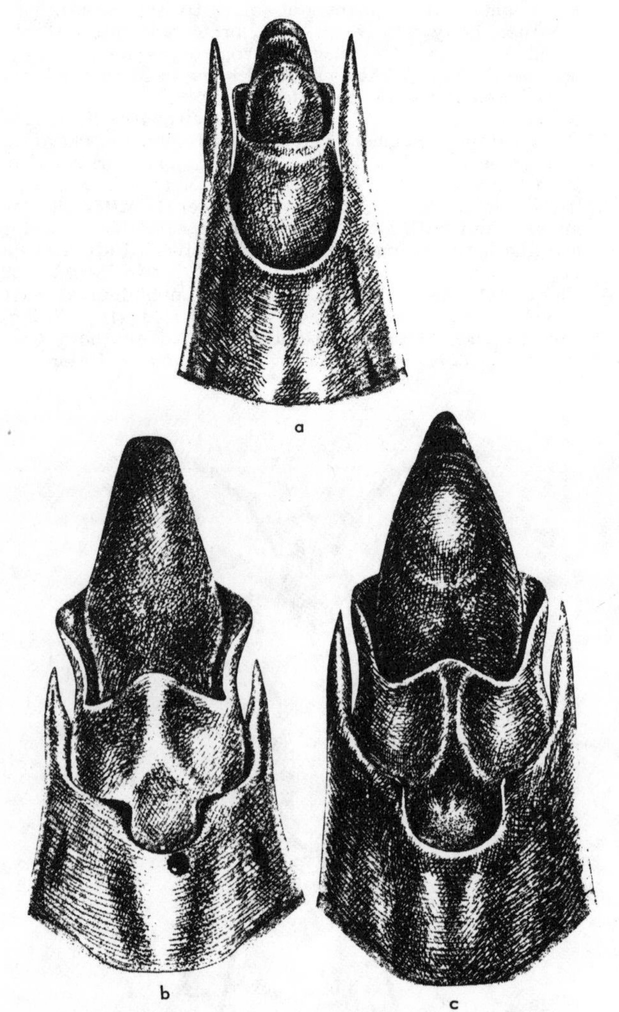

Fig. 7:27. Male genital segments of *Gerris* in ventral view. *a, notabilis; b, nyctalis; c, remigis* (Drake and Harris, 1934).

Key to California Genera of Gerridae

Nymphs

1. Inner margins of eyes sinuate behind the middle . *Gerris*
— Inner margins of eyes convexly rounded 2
2. Tibia and first tarsal segment of middle leg with a fringe of long hairs .... *Halobates*
— Tibia and first tarsal segment of middle leg without a fringe of long hairs 3
3. First antennal segment long, subequal to the remaining 3 together .... *Metrobates*
— First antennal segment much shorter than the remaining 3 together .... *Trepobates*

Key to California Species of Gerris

1. Antennae very long, more than 1/2 as long as body; body length 15-20 mm.; reddish-brown; males with venter roundly emarginate at apex (fig. 7:27*a*); widely distributed, western North America (=*rufoscutellatus* Auctt.). ....*notabilis* Drake & Hottes 1925
— Antennae shorter, less than 1/2 as long as body; body length less than 16 mm.; color dark brown to black; males with venter doubly emarginate at apex 2
2. Large, robust species (11 mm. or more) 3
— Smaller species (less than 11 mm. long) 4
3. Color deep black; average size smaller, ranging from 11-13 mm., with relatively short legs and antennae; male genitalia as in fig. 7:27*b*, the eighth segment seen from below dilated subbasally and the spines of 6th segment reaching only halfway from base to apex of inner lateral margin of 7th segment; Fresno, California, C. J. Drake ....*nyctalis* Drake & Hottes 1925
— Color brownish-black or even reddish-brown; average size larger, 11½-16 mm., with relatively longer legs and antennae (fig. 7:24). Male genitalia, fig. 7:27*c*, 8th segment evenly rounded subbasally, spines of 6th segment long; widely distributed, North America, common throughout California ....*remigis* Say 1832
4. Anterolateral margins of pronotum without a pale stripe; males with omphalium strongly elevated 5
— Anterolateral margins of pronotum with a pale stripe; males with omphalium scarcely elevated 6
5. Notch at hind margin of 6th ventral segment of males relatively shallow, the apices of 6th segment short, reaching about to level of base of 8th (2nd genital) segment as seen from below (fig. 7:28*e*); female connexival spines abruptly incurved, not reaching level of apex of 7th tergite; northwestern United States, British Columbia .... *incurvatus* Drake & Hottes 1925
— Notch at apex of 6th ventral segment in males deeper, the apices of 6th segment longer and more attenuate, surpassing level of base of 8th (2nd genital) segment as seen from below (fig. 7:28*c*); female connexival spines not abruptly incurved, reaching to apex of 7th (1st genital) segment; widespread, North America, scarce in California .... *marginatus* Say 1832
6. First genital segment of males as broad as long, the notch in middle at apex of 6th ventral segment subrectangular (fig. 7:28*d*); size small, 7-8.2 mm.; widespread, North America ....*buenoi* Kirkaldy 1911
— First genital segment of males distinctly longer than broad, narrowed posteriorly, the median apical notch of venter rounded; size relatively larger, 8½-10½ mm. 7
7. First genital segment of male with conspicuous long silvery hairs on either side of ventral surface (fig. 7:28*b*); northern and western North America......... .... *incognitus* Drake & Hottes 1925
— First genital segment of male without conspicuous long

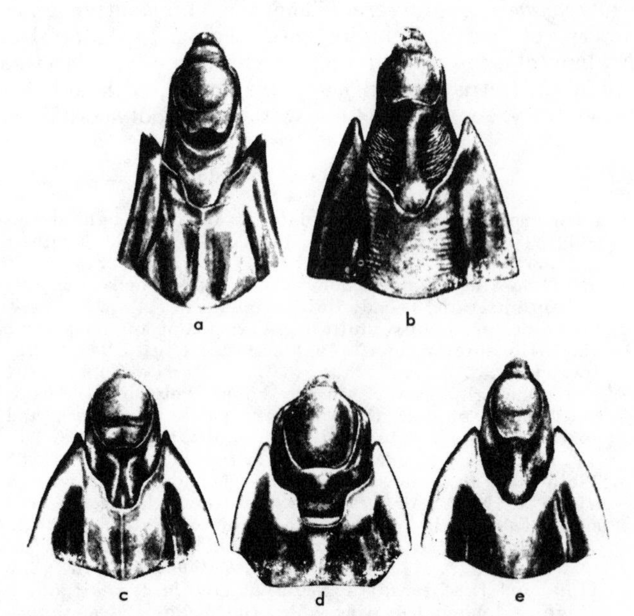

Fig. 7:28. Male genital segments of *Gerris* in ventral view. *a, gillettei; b, incognitus; c, marginatus; d, buenoi; e, incurvatus* (Drake and Harris, 1934).

silvery hairs on either side of ventral surface (fig. 7:28*a*); western United States .... *gillettei* Lethierry & Severin 1896

Halobates sericeus Eschscholtz 1822

California coast, 200 miles west (H. R. Attebery).

Trepobates becki Drake and Harris (fig. 7:25a)

Colorado River, Imperial County (R. L. Usinger).

Key to Subspecies of Metrobates trux (Bueno) 1921

1. Second antennal segment pale, at least basally; apterous forms predominantly pale; Colorado River .... subspecies *trux* (Bueno) 1921
— Second antennal segment uniformly dark; apterous forms with dark markings of body predominant (fig. 7:25*b*); streams of northern California .... subspecies *infuscatus* Usinger 1953

REFERENCES

DRAKE, C. J., and H. M. HARRIS
1922. A note on the migration of certain water-striders with descriptions of three new species. Ohio Jour. Sci., 28:269-276.
1934. The Gerrinae of the Western Hemisphere. Ann. Carnegie Mus., 23:179-240.
KUITERT, L. C.
1942. Gerrinae in the University of Kansas Collections. Univ. Kansas Sci. Bull., 28, pt. 1, pp. 113-143.
PARSHLEY, H. M.
1922. A note on the migration of certain water-striders. Bull. Brooklyn Ent. Soc., 17:136-137.
1929. Observations on *Metrobates hesperius* Uhler. I. Pterygopolymorphism. Bull. Brooklyn Ent. Soc., 24:157-160.
RILEY, C. F. C.
1920. Migratory responses of water-striders during severe droughts. Bull. Brooklyn Ent. Soc., 15:1-10.
TORRE-BUENO, J. R. de la
1908. The broken hemelytra of certain Halobatinae. Ohio Naturalist, 9:389-392.
USINGER, R. L.
1953. Notes on the genus *Metrobates* in California with description of a new subspecies. Pan-Pac. Ent., 29:178-179.

Family VELIIDAE

Small Water Striders, Riffle Bugs

Veliids, like their relatives the gerrids, have preapical claws and thus are able to move about without breaking the surface film. In general, veliids inhabit more protected or secluded places than gerrids, even the marine forms (*Trochopus, Halovelia,* and *Husseyella*) occurring in the tropics within protected reefs near the shore or in mangrove swamps rather than on the open ocean. Veliids are characterized by a velvety hydrofuge pile, relatively short legs, and scent gland canals which extend laterally from the middle of the metasternum to the outer sides of the hind acetabula, with a tuft of hairs extending outward on either side and visible from above. In contrast to this the Gerridae have a median scent gland opening (omphalium) on the metasternum which lacks lateral canals. Annectent types are discussed in terms of the world-wide classification of the Veliidae by China and Usinger (1949).

As in the Gerridae, wing polymorphism is characteristic of all but the marine forms. Frick (1949) bred alate males of *Microvelia capitata* Guerin in Panama. He found that the percentage of alates decreased in confined breeding containers but the factors which determine wing dimorphism in nature are not known.

The two basic types of Veliidae in California are the pond bugs of the genus *Microvelia* and the riffle bugs of the genus *Rhagovelia.* Microvelias are abundant on almost any body of standing water. They abound in ponds and quiet pools of streams. Rhagovelias, on the other hand, occur in large numbers in the swiftest riffles of streams. They are especially fitted for this habitat by virtue of a remarkable plume of hairs which is inserted in a deep cleft in the tarsus of the middle or swimming legs (see fig. 7:30).

Feeding habits.—Veliids prey on almost any small organisms that occur in their habitat. Frick (1949) fed them through four generations on anopheline mosquito larvae.

Life history.—Bueno (1910, 1917), Hungerford (1920), and Hoffmann (1925) have reared North American species of *Microvelia.* In general, they hibernate as adults. Eggs are laid singly or in clusters on floating or stationary objects at the water's edge. They are white and oval and are attached by a gelatinous material. The incubation period is about one to three weeks. Hatching is accomplished by means of a black shiny "egg burster" which is shed with the postnatal molt (Hungerford, 1920). Hoffmann (1925) found that *Microvelia borealis* and *M. buenoi* have only four nymphal instars whereas *M. americana* (Uhler), *M. hinei,* and *M. albonotata* Champion have the normal number of five. Whether or not this is constant for each species remains to be seen. Frick (1949) found that the normal number of instars for *M. capitata* Guerin in Panama was five but that "Eight apterous males and one apterous female had only four nymphal stages," out of a total of 110 complete rearings, 76 of which were apterous. The total nymphal period ranges from two to five weeks in the United States.

Distribution.—Both *Microvelia* and *Rhagovelia* are world-wide in distribution. *Velia* is also widespread but has not been found along the Pacific Coast of the United States. *Trochopus* is the marine equivalent of *Rhagovelia.* It occurs in the Caribbean area and in Central American waters. *Husseyella* is known from Florida and Central and South America.

Taxonomic characters —As in other groups of water striders, the male genitalia are the most distinctive features of the Veliidae. However, no monograph of the western American species utilizing these characters has been attempted. *Microvelia,* in particular,

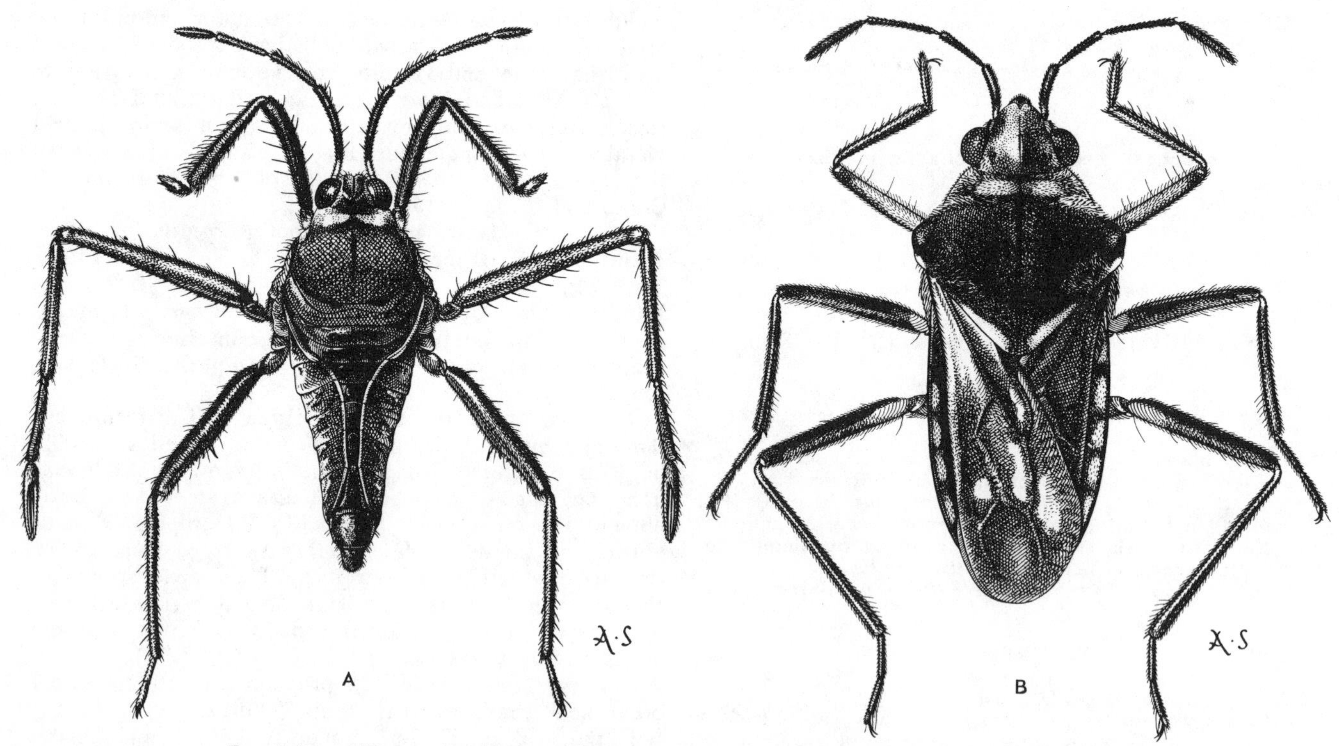

Fig. 7:29. *a, Rhagovelia distincta* Champion, female, Davis, Yolo County, California, October 24, 1942, (R. L. Usinger); *b, Microvelia californiensis* McKinstry, female, Mokelumne Hill, Calaveras County, California, May 29, 1931 (R. L. Usinger).

needs careful study to correlate winged and wingless forms and to establish the identity or differences between our species and various poorly known but earlier described species from Central America and the West Indies. Drake and Hussey (1955) have made a start on this but their treatment of *incerta* (Kirby) as a subspecies of *pulchella* Westw. is not in accord with the current definition of subspecies.

Rhagovelia has received more attention (Gould, 1931), but the geographical variation in southern and western species was treated under several "varieties" of *R. distincta*. Either these "varieties" are geographical subspecies (in which case several more populations could be differentiated from California) or *distincta* is a single extremely variable species. A third possibility is that each slightly different population represents a distinct local species but there seems to be no evidence for this at the moment. Dimorphism is even more striking in *Rhagovelia* than in *Microvelia* because of the great differences between the sexes (see figs. 7:29, 7:30). In the macropterous females the pronotum is often produced backward and upward into a long, curved, blunt spine.

For the present, keys are offered only for apterous males. This is because the apterous forms are most commonly encountered and are the best known. Also the thoracic sclerites present the most obvious characters and therefore have been used in the best existing keys (Parshley, 1921; Bueno, 1924). I am indebted to A. P. McKinstry for making available his unpublished key to *Microvelia,* parts of which have been used in the present key.

Key to the Nearctic Genera of Veliidae

1. Middle tarsi deeply cleft, with leaflike claws and plumose hairs arising from base of cleft. Subfamily Rhagoveliinae 2
— Middle tarsi not deeply cleft and without plumose hairs arising from base of cleft 3
2. Hind tarsi 2-segmented, the basal segment very short; apterous, marine, tropical American in distribution .... *Trochopus* Carpenter 1898
— Hind tarsi 3-segmented, the basal segment very short; apterous or macropterous (fig. 7:29*a*). Cosmopolitan, riffles of rivers .... *Rhagovelia* Mayr 1865
3. Tarsal formula 3:3:3. First antennal segment distinctly longer than the others. Subfamily Veliinae .... *Velia* Latreille 1807
— Tarsal formula 1:2:2. First antennal segment subequal to others (fig. 7:29*b*) . . Subfamily Microveliinae ... 4
4. Middle legs nearly equidistant from other two, tarsal claws normal .... *Microvelia* Westwood 1834

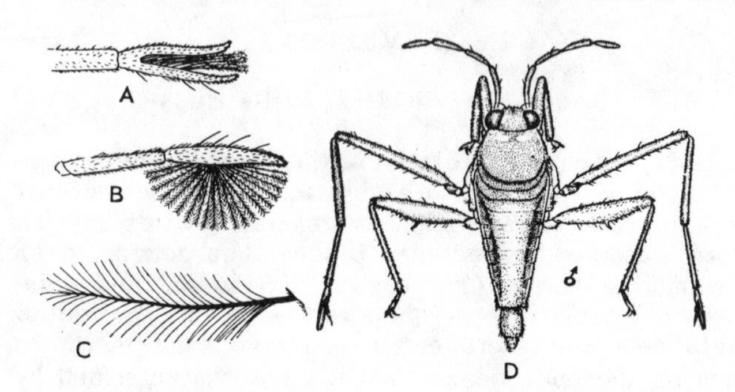

Fig. 7:30. *Rhagovelia obesa* Uhler. *a,* cleft tarsus of middle leg; *b,* same, with fringed hairs spread; *c,* one of the fringed hairs enlarged; *d,* apterous male (Hungerford, 1920).

— Middle legs closely approximated to hind pair and far removed from front pair, tarsal claws and arolia of middle legs modified into four broad, membranous plates; tropical America, Marine .... *Husseyella* Herring 1955

Key to California Species of Rhagovelia

Apterous Specimens

1. Size relatively small, 3½ to 4 mm.; male with pronotum covering all the mesonotum; female with connexiva contiguous or nearly so along mid-line (fig. 7:30); Blythe, California .... *obesa* Uhler 1872
— Size larger, distinctly more than 4 mm. long; male with pronotum not quite covering all the mesonotum, the hind margin narrowly but distinctly exposed; female with connexiva separated by a distinct gap along mid-line (fig. 7:29*a*); generally distributed in California .... *distincta* Champion 1898

Key to California Species of Microvelia

Apterous Males[3]

1. Pronotum extending backward over mesonotum to level of sublateral pits (fig. 7:31*a,b*) 2
— Pronotum not produced over mesonotum, shorter than mesonotum at middle with a distinct sinuate suture at hind margin which clearly reaches lateral margins (fig. 7:31*c,d*) 5
2. Pronotum produced over both lobes of the mesonotum leaving only the lateral triangles of the metanotum showing .... *cerifera* McKinstry 1937
— Pronotum produced over the anterior lobe of the mesonotum only 3
3. Tarsal segment of the front legs not noticeably longer than segments of middle and hind legs .... *signata* Uhler 1894
— Tarsal segment of the front legs noticeably longer than segments of middle and hind legs 4
4. Pronotum about 3 times as long at middle as mesonotum, the hind margin of pronotum straight at middle; abdomen with conspicuous tufts of silvery pubescence; color black, the anterior lobe of pronotum with a pale spot; Corvallis, Oregon .... *buenoi* Drake 1920
— Pronotum less than twice as long as mesonotum at middle, the hind margin broadly concave at middle; abdomen without conspicuous tufts of silvery pubescence; color dark brown with numerous paler areas (fig. 7:31*c-d*); southern California .... *hinei* Drake 1920
5. Hind tibiae curved, length less than 2 mm. Corvallis, Oregon (= *incerta* Kirby, = *borealis* Bueno) .... *pulchella* Westwood 1834
— Hind tibiae straight, length more than 2 mm. 6
6. Hind femora inflated, 5 times as long (from apices of trochanters) as greatest thickness, with several prominent spines; hind tibiae with long hairs on apical half, the individual hairs longer than thickness of tibiae; apical segment of hind tarsi distinctly longer than basal segment; widely distributed, western United States and Central America.. *paludicola* Champion 1898
— Hind femora more slender, about 7 times as long as greatest thickness, with spines greatly reduced or obsolete; hind tibiae with only short hairs, the individual hairs shorter than thickness of tibia; hind tarsal segments subequal 7
7. Front femur swollen, black in the middle region and flattened on anterior or inner surface; western United States .... *gerhardi* Hussey 1924
— Front femur neither swollen, black nor flattened ... 8
8. Body relatively short and broad, 2½ times as long as broad, the last abdominal tergite transverse (fig. 7:29*b*); California .... *californiensis* McKinstry 1937
— Body longer and more slender, about 3 times as long as broad, the last abdominal tergite longer than average width, subquadrate; California, Arizona, Texas, Mexico .... *beameri* McKinstry 1937

[3]Nymphs may be eliminated since all nymphs have 1-segmented tarsi on all legs. Adults have 2 segments on middle and hind legs.

REFERENCES

CHINA, W. E., and R. L. USINGER
1949. Classification of the Veliidae (Hemiptera) with a new genus from South Africa. Ann. Mag. Nat. Hist., (12)2:343-354.
DRAKE, C. J. and R. F. HUSSEY
1955. Concerning the genus *Microvelia* Westwood, with descriptions of two new species and a check-list of the American forms. Florida Ent., 38:95-115.
FRICK, K. E.
1949. Biology of *Microvelia capitata* Guerin, 1857, in the Panama Canal Zone and its role as a predator on Anopheline larvae. Ann. Ent. Soc. Amer., 42:77-100.
GOULD, G. E.
1931. The *Rhagovelia* of the Western Hemisphere, with notes on world distribution. Univ. Kansas Sci. Bull., 20:1-61.
HOFFMANN, W. E.
1925. Some aquatic Hemiptera having only four nymphal stages. Bull. Brooklyn Ent. Soc., 20:93-94.
HUNGERFORD, H. B.
1920. The biology and ecology of aquatic and semi-aquatic Hemiptera. Kansas Univ. Sci. Bull., 11:1-341.
PARSHLEY, H. M.
1921. On the genus *Microvelia* Westwood. Bull. Brooklyn Ent. Soc., 16:87-93.
TORRE-BUENO, J. R. de la
1910. Life histories of North American water bugs. III. *Microvelia americana* Uhler. Canad. Ent., 42:176-186.
1917. Life history of the northern *Microvelia—Microvelia borealis* Bueno. Ent. News, 28:354-359.
1924. A preliminary survey of the species of *Microvelia* Westwood of the Western world, with description of a new species from the southern United States. Bull. Brooklyn Ent. Soc., 19:186-194.

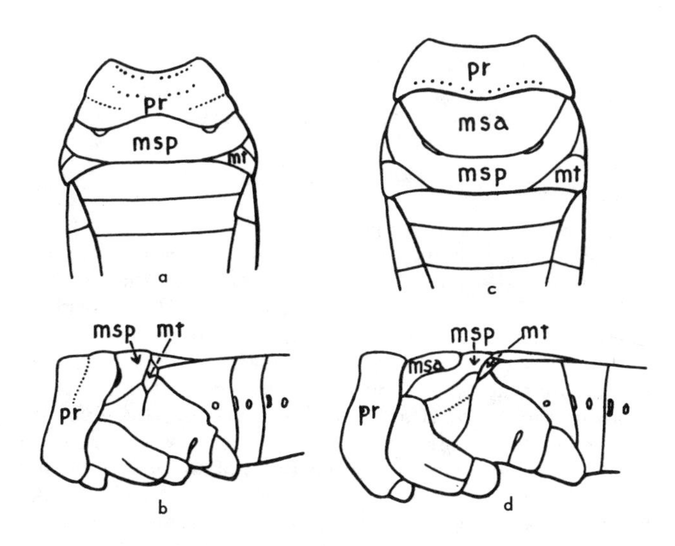

Fig. 7:31. *Microvelia hinei* Drake: male, *a*, dorsal view; *b*, lateral view. *Microvelia californiensis* McKinstry: male, *c*, dorsal view; *d*, lateral view.

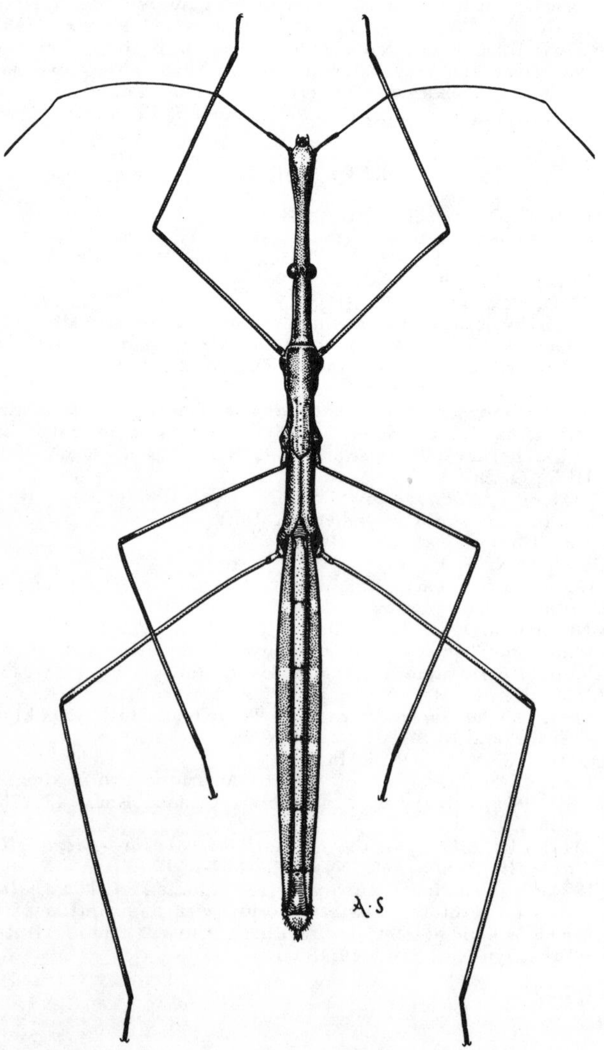

Fig. 7:32. *Hydrometra australis* Say, female, Imperial Dam, Colorado River, Calif., Nov. 14, 1951 (R. L. Usinger).

Family HYDROMETRIDAE

Marsh Treaders

Hydrometrids are perhaps the most fragile of the surface water bugs, with very slender bodies and threadlike legs (fig. 7:32). In our species the head is subequal in length to the abdomen. The swollen anterior part of the head above has two pairs of small brown pits in which are inserted long, erect bristles or trichobothria, and another pair is present at the base of the head. Brachyptery is common, the hemelytra sometimes being reduced to virtually unrecognizable stubs or rounded elevations. The body surface is more or less granular and variously pitted. Unlike most surface dwellers, the minute claws of *Hydrometra* are terminal, and this no doubt renders them less efficient in open water than such masters of the surface film as the Gerridae and Veliidae.

Life history.—Martin (1900) and Hungerford (1920) have given excellent accounts of the biology of a common North American species, *Hydrometra martini* Kirkaldy. Adults overwinter in the adult stage. Eggs are laid, at least in the eastern and midwestern United States, in the spring. The eggs are spindle-shaped and are attached by a slender stalk to solid objects at or just above the surface of the water. According to Hungerford, the average durations of the various stages are: egg, 7 days; 5 nymphal instars, 2 days each.

Habitat and distribution.—*Hydrometra* is an inhabitant of quiet waters where it dwells in the protection of various kinds of plants. Emergent grasses, duck weed, and cattails provide support and protection. In the Colorado River drainage system it was found most commonly in small seepage pools choked with vegetation and without fish. Whether or not its scarcity in larger bodies of water is due to the predatory pressure of fish is not known, but it is a fact that *Hydrometra* lacks the thoracic scent glands which are so characteristic a feature of other surface bugs.

The geographical distribution of the genus in California is unusual. *H. lillianis* has never been reported since the original collection at Santa Barbara, in spite of the fact that a diligent search was made for it in habitats that appeared to be ideal. *H. australis* has been found only along the Colorado River under what could best be described as subtropical conditions. This would not be remarkable but for the fact that the only other record on the Pacific Coast is Corvallis, Oregon (*H. martini* Kirkaldy?) Negative evidence for the intervening territory is, of course, not conclusive, but might be owing to the arid conditions which prevail during the summer months in most of California. Careful comparison of a Corvallis specimen loaned by V. Roth with Colorado River specimens shows no significant differences. Elsewhere *H. martini* occurs over most of northern North America and *australis* is Southern.

Feeding habits.—Hungerford (1920) reports that *Hydrometra* will feed upon insects which fall onto the water surface, but that the normal inhabitants of the surface film are much preferred. He cites ostracods and mosquito larvae and pupae as favorite foods. The role of hydrometrids as predators of *Anopheles* larvae, which spend most of their time at the under side of the surface film, led F. X. Williams to consider introducing them to areas where they do not normally occur as a possible means of biological control.

Taxonomic characters.—*Hydrometra* species are distinguished by the relative proportions of antennae, rostrum, and legs, by the ratio of anteocular to postocular parts of the head, by average size, by the shape of the clypeus, by the arrangement and number of pits on the acetabula, and especially by the shape of the genital segments in both sexes, and by the presence or absence and form of a pair of projections or carinae near the base of the sixth ventral segment in the male. The principal taxonomic works on Hydrometridae are by Torre-Bueno (1926) and Hungerford and Evans (1934).

with descriptions of three new species. Ohio Jour. Sci., 28:269-276.

HUNGERFORD, H. B., and N. E. EVANS
1934. The Hydrometridae of the Hungarian National Museum and other studies in the family. Ann. Mus. Nat. Hungarici, 28:31-112, 12 pls.

MARTIN, J. O.
1900. A study of *Hydrometra lineata*. Canad. Ent., 32:70-76, pl. 3, figs. 7-8.

TORRE-BUENO, J. R. de la
1926. The family Hydrometridae in the Western Hemisphere. Entomologica Americana VII (n.s.), pp. 83-128.

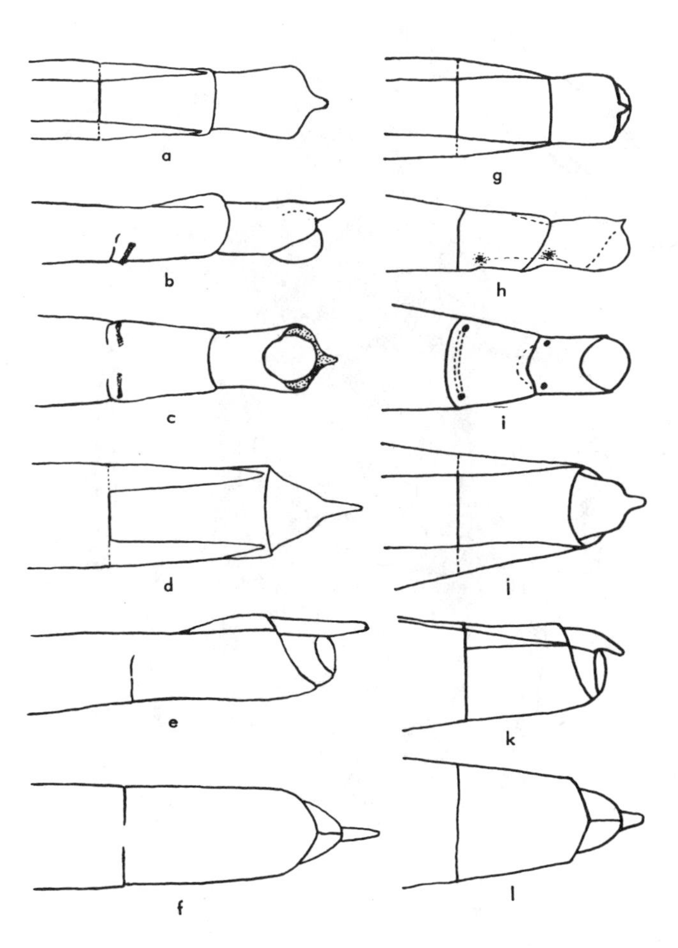

Fig. 7:33. *Hydrometra australis* Say, tip of abdomen. *a*, dorsal view of male; *b*, side view of male; *c*, ventral view of male; *d*, dorsal view of female; *e*, side view of female; *f*, ventral view of female. *Hydrometra lillianis* Bueno. *g* to *l*, views comparable to above.

Key to California Species of Hydrometra

1. Numerous pits on all 3 pairs of acetabula, scarcely evident on posterior pair; body surface velvety; clypeus truncate anteriorly; 2nd antennal segment only about 1/3 longer than 1st. segment (13:10, 15:11); male abdominal processes widely separated, blunt, inclined outward (fig. 7:33*g*-*l*); Santa Barbara, California .... *lillianis* Bueno 1926

— Two pits only on anterior and middle acetabula, posterior acetabula unpitted; body surface not conspicuously velvety; clypeus rounded anteriorly; 2nd antennal segment more than twice as long as 1st segment (56:25); male abdominal processes slightly arched, transverse, separated by a distance equal to transverse length of a process (figs. 7:32; 7:33*a*-*f*); Colorado River drainage, Yuma to Needles, Death Valley... *australis* Say 1832

REFERENCES

DRAKE, C. J., and H. M. HARRIS
1928. Concerning some North American water striders

Family MACROVELIIDAE

Macrovelia is a monotypic genus which shows certain affinities to the South African *Ocellovelia* China and Usinger (1949). Whether these two annectent types belong in a single subfamily (or family) or as separate groups does not concern us here. Certainly *Macrovelia* is a unique type in the North American fauna, the six-celled hemelytra being without parallel here. The well-developed ocelli and apical claws and the single dorsal abdominal scent gland of the nymphs

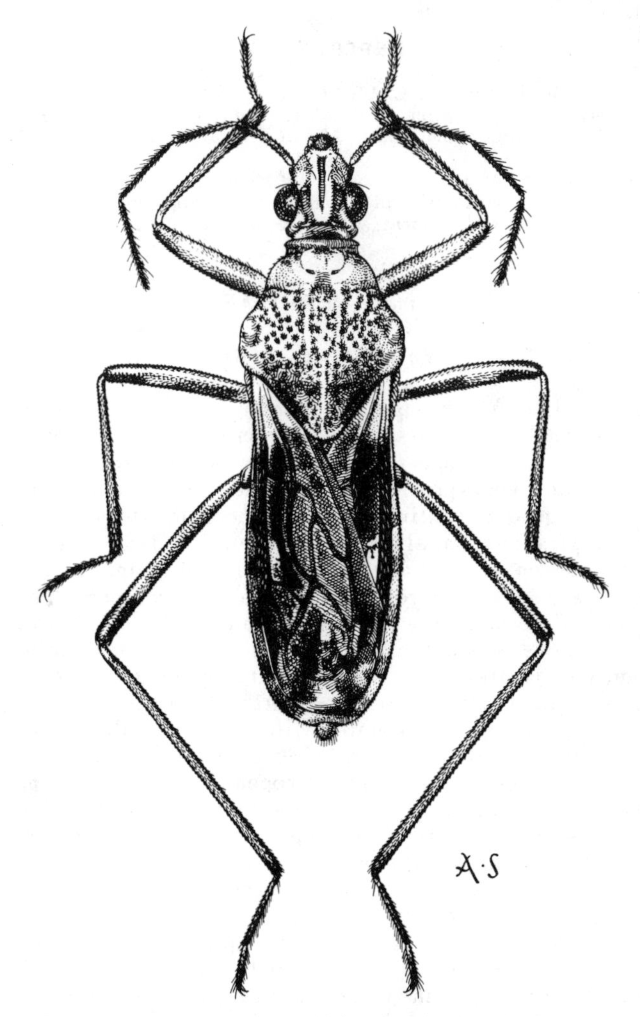

Fig. 7:34. *Macrovelia hornii* Uhler, female, Mokelumne Hill, Calaveras Co., Calif., May 21, 1931 (R. L. Usinger).

indicate a relationship with the Hebridae and Mesoveliidae whereas the concealment of the mesonotal scutella by the backward extension of the pronotum is a typically veliid character. Brachypterous specimens are known in which the pronotum is smaller and the wings are reduced to short pads.

Macrovelia hornii Uhler (fig. 7:34) was described (as a veliid) eighty years ago from specimens collected in New Mexico, Arizona, and California. McKinstry (1942) first called attention to its anomalous position and recorded it from the Dakotas and from Colorado. California records span the entire state from Siskiyou County to Palm Springs. It occurs at low elevations in the Coast Range and in the foothills of the Sierra. Typical habitats are in the vicinity of permanent streams or springs. Nymphs and adults are found in moss at the water's edge, usually in protected niches behind rocks or logs or among debris. They are never found in or on the water, being incapable of walking on the surface film because of their terminal claws. On the other hand, they are never more than a few feet away from water. Detailed life history studies have not been made.

REFERENCES

CHINA, W. E., and R. L. USINGER
1949. Classification of the Veliidae (Hemiptera) with a new genus from South Africa. Ann. Mag. Nat. Hist., (12)2:343-354.

McKINSTRY, A. P.
1942. A new family of Hemiptera-Heteroptera proposed for *Macrovelia hornii* Uhler. Pan-Pac. Ent., 18:90-96.

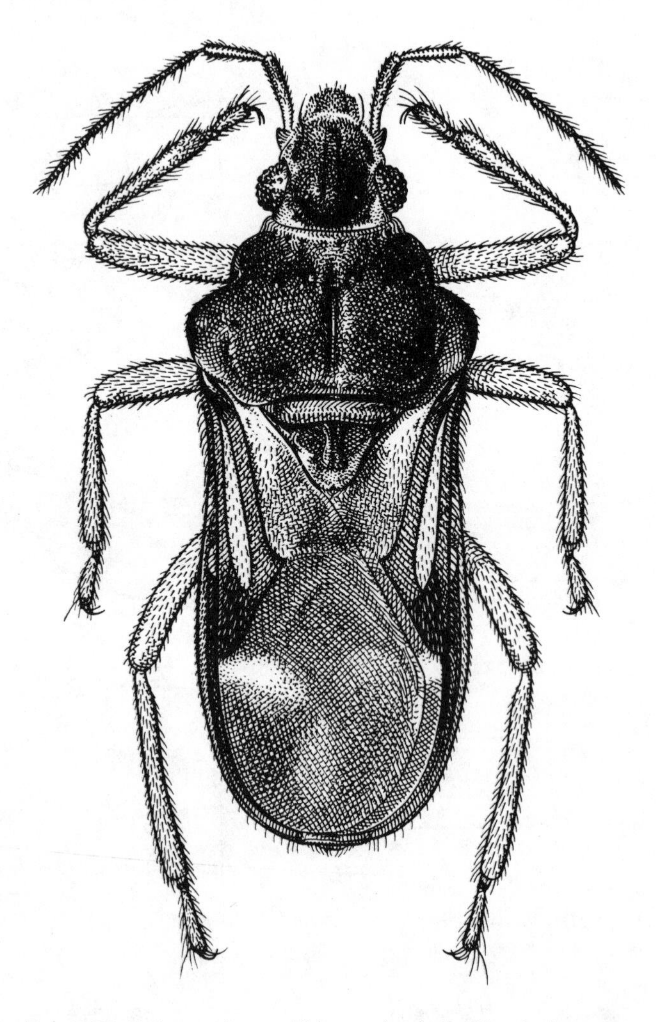

Fig. 7:35. *Hebrus sobrinus* Uhler, male, Lake Houghtelin, near Bard, Calif. Nov. 13, 1951 (R. L. Usinger).

Family HEBRIDAE

Velvet Water Bugs

The velvet water bugs are inconspicuous because of their small size, less than 2½ mm. They are stocky with well-developed ocelli, four or five-segmented antennae, very prominent bucculae, and a trough or rostral groove continuing along thoracic sterna. The hemelytra have a single closed cell, and the clavus and corium are continuous and without veins in our species. Brachypterous forms are not common. The body is clothed with a velvety pile of hydrofuge hairs. The claws are apical and are provided with minute, padlike arolia. The nymphs have a single median scent gland opening on the fourth abdominal tergite.

Life history.—No detailed life history studies have been published for our species of Hebridae. Kulgatz (1911) suggests that the European *Hebrus ruficeps* overwinters as an adult. Hungerford (1920) observed ovipositing bugs in June near Ithaca, New York (*H. concinnus* Uhl.), the eggs being laid partly concealed between the leaves of sprigs of moss. The eggs are elongate-oval with rounded ends.

Habitat.—*Hebrus* lives in moist places at the edge of the water. It is found most commonly beneath stones or in trash along the margins of streams. Hungerford (1920) fed his specimens on plant lice, midges, and mosquitoes, but they doubtless feed on any small insects that they can overpower in their secluded habitat.

Merragata is an inhabitant of floating plants and is found most commonly on mats of *Spirogyra* at the surface of the water. Drake (1917) describes them as "aquatic pedestrians." He states further that, "It is not uncommon to find them on the under side of floating leaves, or even among the roots of floating water plants. When submerged in the water, the insects are surrounded by a film of air which enables them to stay beneath the surface film for a considerable period of time." I found enormous numbers of *M. hebroides* in flight in the Sacramento Valley near Williams in the summer of 1932.

Distribution.—Both *Hebrus* and *Merragata* are widely distributed throughout the world. *Hebrus* is the larger genus with twenty-two species known from the Western Hemisphere. *Hebrus sobrinus* Uhl., *H. concinnus* Uhl., and *H. hubbardi* Porter (1952) have been reported from California, but I have seen only *sobrinus* and *hubbardi*. The latter was described from Palm Springs. *Sobrinus* is apparently rather widely distributed. Records at hand include the Pinnacles National Monument, San

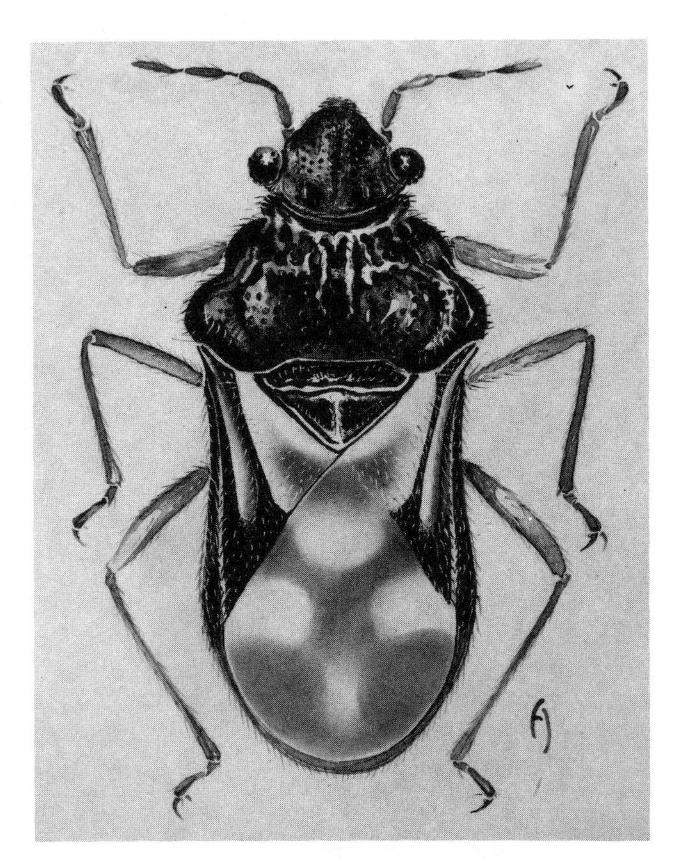

Fig. 7:36. *Merragata hebroides* White, Oahu, T. H. (R. L. Usinger) (Zimmerman, 1948).

Benito County, May 3, 1946 (H. P. Chandler); Frenchman's Flat, Tehachapi Mountains, California; Pine Canyon, Mt. Diablo, May 23, 1937 (R. L. Usinger); Navarro River, Mendocino County, June 15, 1950 (H. B. Leech); and Lake Houghtelin near Bard, Imperial County (R. L. Usinger).

Merragata hebroides White is very common in California, being found in low elevations throughout the state. It was originally described from the Hawaiian Islands but was almost certainly introduced there from California. *M. brevis* Champion has been reported from California (Drake and Harris, 1943), but I know it only from Mexico.

Taxonomy.—The two principal genera of Hebridae are differentiated by a rather superficial character, that is, the number of antennal segments. No other characters appear to support this division so it may very well be that the present classification will prove to be one of convenience only. At the species level the form and proportions of antennal segments and sculpturing of the head and pronotum are the most useful external characters. In some species of *Hebrus* long setae project from the margins of the male genital segment, and these provide useful characters for differentiating species. The most comprehensive treatment of the taxonomy of our species is by Drake and Harris (1943) with additional new species described by Porter (1952).

Key to the Genera of Hebridae

1. Antennae 5-segmented (fig. 7:35) ... *Hebrus* Curtis 1833
— Antennae 4-segmented (fig. 7:36) .... *Merragata* White 1877

Key to Species of Hebrus

1. Male genital capsule with long hairs at sides, plainly visible in ventral view (fig. 7:35) .. *sobrinus* Uhler 1877
— Male genital capsule without long hairs projecting at sides; Palm Springs .... *hubbardi* Porter 1952

Key to the Species of Merragata

1. Third and 4th antennal segments relatively long and slender, 1/2 again as long as 1st and 2nd segments; Los Angeles County, Drake and Harris, 1943 .... *brevis* Champion 1898
— First 3 antennal segments subequal or 3rd slightly shorter than 2nd, 4th a little longer, fusiform (fig. 7:36); California, widely distributed, also Hawaii .... *hebroides* White 1877

REFERENCES

DRAKE, C. J.
1917. A survey of the North American species of *Merragata*. Ohio Jour. Sci., 17:101-105.

DRAKE, C. J., and H. M. HARRIS
1943. Notas sobre Hebridae del Hemisferio Occidental. Notas del Museo de la Plata, 8:41-58.

HUNGERFORD, H. B.
1920. The biology and ecology of aquatic and semi-aquatic Hemiptera. Kansas Univ. Sci. Bull., 11:1-327, 2 color pls. plus 30 pl.

KULGATZ, TH.
1911. Die aquatilen Rhynchoten Westpreussens. Bericht. Westpreuss. Bot. Zool. Ver., 33:175.

PORTER, T. WAYNE
1952. Three new species of Hebridae (Hemiptera) from the Western Hemisphere. Jour. Kansas Ent. Soc., 25:9-12. 6 figs.

Family MESOVELIIDAE

Water Treaders

Members of the genus *Mesovelia* are small, delicate, surface and shore dwellers. Apterous specimens show no sign of hemelytral pads, and the thorax is reduced to simple segments separated by simple sutures. Macropterous specimens, which are less common than the wingless forms, have an enlarged pronotum and both a mesonotal and a metanotal scutellum. The hemelytra are long and have two or three closed cells on the outer half. The membranous apices of the hemelytra are frequently broken, presumably during mating. *Mesovelia* is related to *Macrovelia* and to the hebrids, both of which share the following characteristics: ocelli, apical claws, and a single median scent gland opening on the fourth abdominal tergite in the nymphs.

Life history.—The biology of the American species of *Mesovelia* was summarized by Hoffmann (1932). It is thought that the winter is passed in the adult stage,

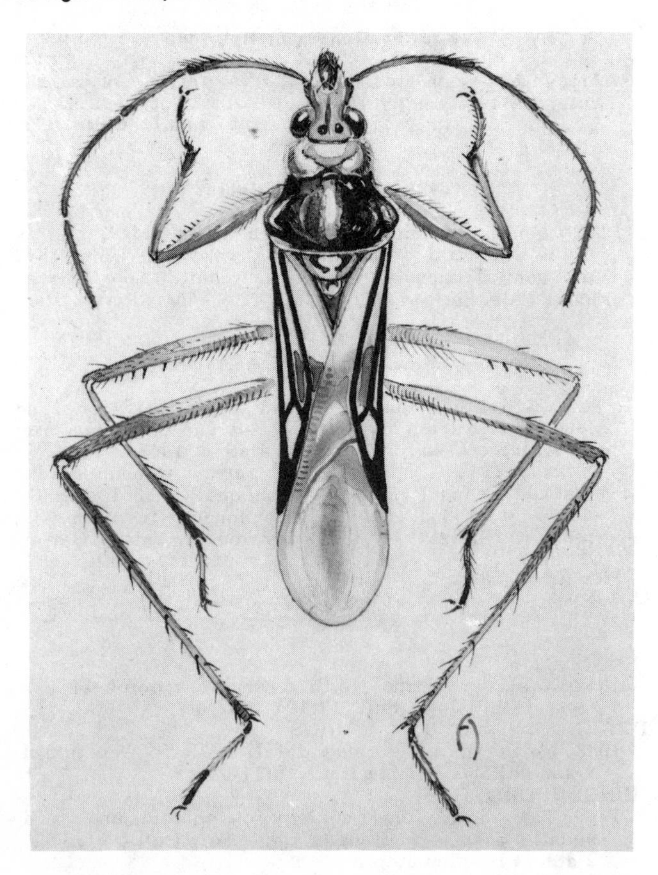

Fig. 7:37. *Mesovelia mulsanti* White, macropterous male. Oahu, T. H. (R. L. Usinger) (Zimmerman, 1948).

but eggs embedded in plant tissue have been observed to withstand winter temperatures, so it is possible that this may be the normal method of overwintering. Eggs are inserted in plant tissue at the water's edge by means of the ovipositor. They are "elongate-oval with a curved neck terminating in a flat surface which marks the exposed end of the egg as it lies *in situ* in the stem of some plant" (Hungerford, 1920). The incubation period for *M. mulsanti* was recorded by Hungerford as 7 to 9 days, for *douglasensis* (= *amoena*?) 12 days (Hoffmann). The average length of each of the five nymphal instars for *mulsanti* was 3.4, 3.2, 3.5, 4.4, and 5.4. Comparable figures for *douglasensis* were 4.3, 4.4, 5.4, and 7.1 days. Both species undergo several generations in a season.

Habitat.—*Mesovelia mulsanti* lives on the surface of ponds and other bodies of standing water. Bog lakes and ponds with much surface vegetation are preferred. Typically, they forage on the floating leaves of pond lillies, *Typha,* and the like, and run out over the open water with remarkable agility, considering that their claws are inserted at the apices of the tarsi. Hungerford (1917) found that they feed readily on freshly killed insects on the surface of the water, and he expressed the opinion that they also spear small Crustacea that are associated with algae and floating *Typha.*

Mesovelia amoena is found in the same general area as *mulsanti* but in a quite different microhabitat. These small bugs have been found in protected crevices at the edges of ponds along the Colorado River near Bard and also on moss-covered rocks in a small hot-spring cave in Death Valley. In Colorado River ponds they took to the open water only if forced to do so, whereas *mulsanti* was running freely a few inches away.

Distribution.—*Mesovelia mulsanti* was described from Brazil and *M. amoena* from the West Indies. As pointed out by Jaczewski (1930), topotypical males of both species must be studied before their taxonomic status can be settled. Jaczewski proposed three subspecies of *mulsanti,* basing them on the form of the male genital claspers. His southern subspecies, *meridionalis,* was recorded from as far south as Argentina, *caraiba* was described from Central America and the West Indies, and *bisignata* Uhler was considered to be the North American subspecies. The differences described by Jaczewski were not particularly striking, and subsequent investigations have not substantiated the classification into subspecies.

In North America *M. mulsanti* is widely distributed,

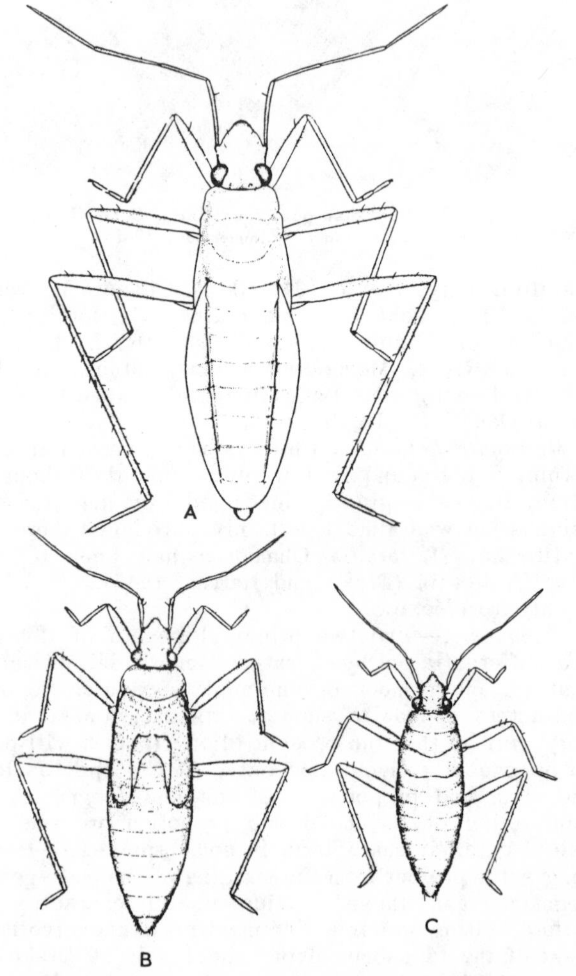

Fig. 7:38. *Mesovelia mulsanti* White. *a,* apterous female; *b,* fifth (last) nymphal instar of macropterous form; *c,* fourth nymphal instar of apterous form (Hungerford, 1920).

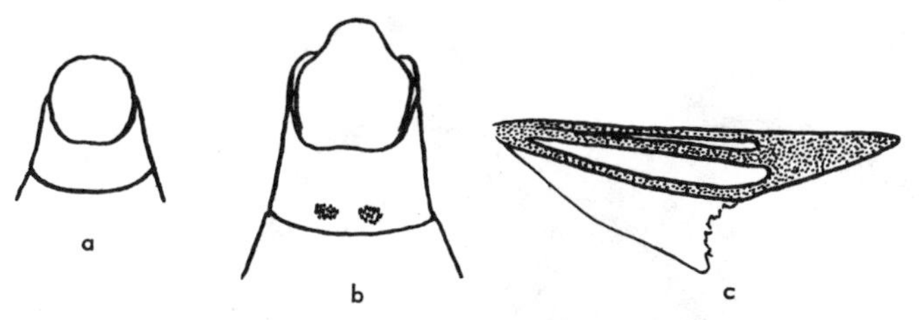

Fig. 7:39. *a, Mesovelia amoena* Uhler, male genital segments, ventral view. *b, M. mulsanti* White, same; *c, M. amoena* Uhler, hemelytra with only two closed cells, the membranous area broken off.

at least from southern Canada (British Columbia) throughout most, if not all, of the United States and Mexico. Scarcity of records from the Pacific Coast is owing in part to lack of collecting and in part to a peculiarly discontinuous distribution. Usinger (1942) first recorded the species from California, taking it in a bog lake in the central part of the Sierra Nevada (Swamp Lake, Yosemite National Park). Although common in this one lake, it has not been found in other places in the Sierra in ten years of collecting. Elsewhere in California *mulsanti* has been found commonly in scattered localities from Tule Lake in the north to the Colorado River in the south.

The taxonomy of *M. amoena* is even less clear, but if, as suggested by Jaczewski, *M. douglasensis* Hungerford is a synonym of *amoena,* then the published distribution of the species is: Michigan, Florida, Louisiana, Mississippi, and Texas; Puerto Rico and Grenada in the West Indies; and Brazil and Panama. I can add Jamaica and three localities in California—Death Valley, the Colorado River near Bard, and Borego Palm Canyon.

Taxonomy.—The family Mesoveliidae is small, less than two dozen species having been described in four genera for the world. Three of the genera are monotypic. The species of *Mesovelia* are easily distinguished by characters of the penultimate abdominal segment (first genital segment) in the male. The presence or absence and number and arrangement of tufts of hair on the ventral surface of this segment are diagnostic for most species. In addition to this, the usual proportions of appendages are used and, in *amoena,* the venation of the hemelytra.

Key to the California Species of Mesovelia

1. Length more than 3 mm. Front and middle femora armed beneath with a row of black spines. Males with 2 black tufts of minute spines on the ventral side of the 8th abdominal segment. Macropterous specimens with 3 closed cells in the hemelytra. Rostrum extending only to hind margins of middle coxae (figs. 7:37, 7:38, 7:39*b*) .... *mulsanti* White 1879

— Length about 2 mm. Front and middle femora without a row of black spines beneath. Males without black tufts on the ventral side of 8th abdominal segment. Macropterous specimens lacking the small closed cell at apex of corium. Rostrum extending to apices of hind coxae (fig. 7:39*a,c*) .... *amoena* Uhler 1894

REFERENCES

HOFFMANN, C. H.
1932. The biology of three North American species of *Mesovelia.* Canad. Ent., 64:88-133. 1 pl.

HUNGERFORD, H. B.
1917. The life history of *Mesovelia mulsanti* White. Psyche, 24:73-81.

JACZEWSKI, T.
1930. Notes on the American species of the genus *Mesovelia* Muls. Ann. Musei Zool. Polonici, 9:3-12. 3 pls.

USINGER, R. L.
1942. Notes on the variation and distribution of *Mesovelia mulsanti* White. Bull. Brooklyn Ent. Soc., 37:177-178.

Family SALDIDAE

Shore Bugs

Saldids are intermediate in structure and in habits between the terrestrial bugs and the true water bugs. They have exposed antennae like land bugs but are able to survive in or on the water when they accidentally land on the surface after a jump or when they are submerged on beaches at high tide.

Saldids have four-segmented antennae, a long, slender, three-segmented beak, and large eyes with the inner margins notched behind like many water bugs. There are two well-developed ocelli and pairs of erect trichobothria on the vertex. The legs are relatively slender but permit rapid movements including leaps which, when combined with short flights, provide a ready means of escape. The nymphs are oval in outline like the adults but have only two-segmented tarsi instead of three (the first being very small in both stages). There is only a single dorsal abdominal scent gland opening in the nymphs. The rostrum is long, slender, three-segmented, and tapering. The body is clothed with one or more types of pubescence which serves to protect them from wetting. Drake and Hottes (1951) have noted peglike or spinelike "stridulatory" structures anterolaterally on the second visible connexival segments of males.

Life history.—Hungerford (1920) and Wiley (1922) have given accounts of the biology of some of our species. In general, it appears that they overwinter as adults, lay eggs in the spring and early summer, and develop through five instars in about a month. The incubation period of the egg for *Saldula major* (Prov.) in Kansas was found by Wiley to be 12 days, and the length of nymphal instars was: first, 4 days; second, 3 days; third, 3 days; fourth, 2 or 3 days; fifth, 4 days. *Saldula pallipes* Fabr. followed essentially the same pattern. The eggs are elongate, larger and more broadly rounded at one end, tapering and smaller at the other end, with the dorsal part arched. They are laid at the bases of clumps or between the leaflets of moss or other marginal vegetation.

Habitat and distribution.—Saldids are typical inhabitants of the beaches and shore lines of lakes, streams, and ocean. They also inhabit springs, bogs, and salt marshes. In a sense, they are not aquatic

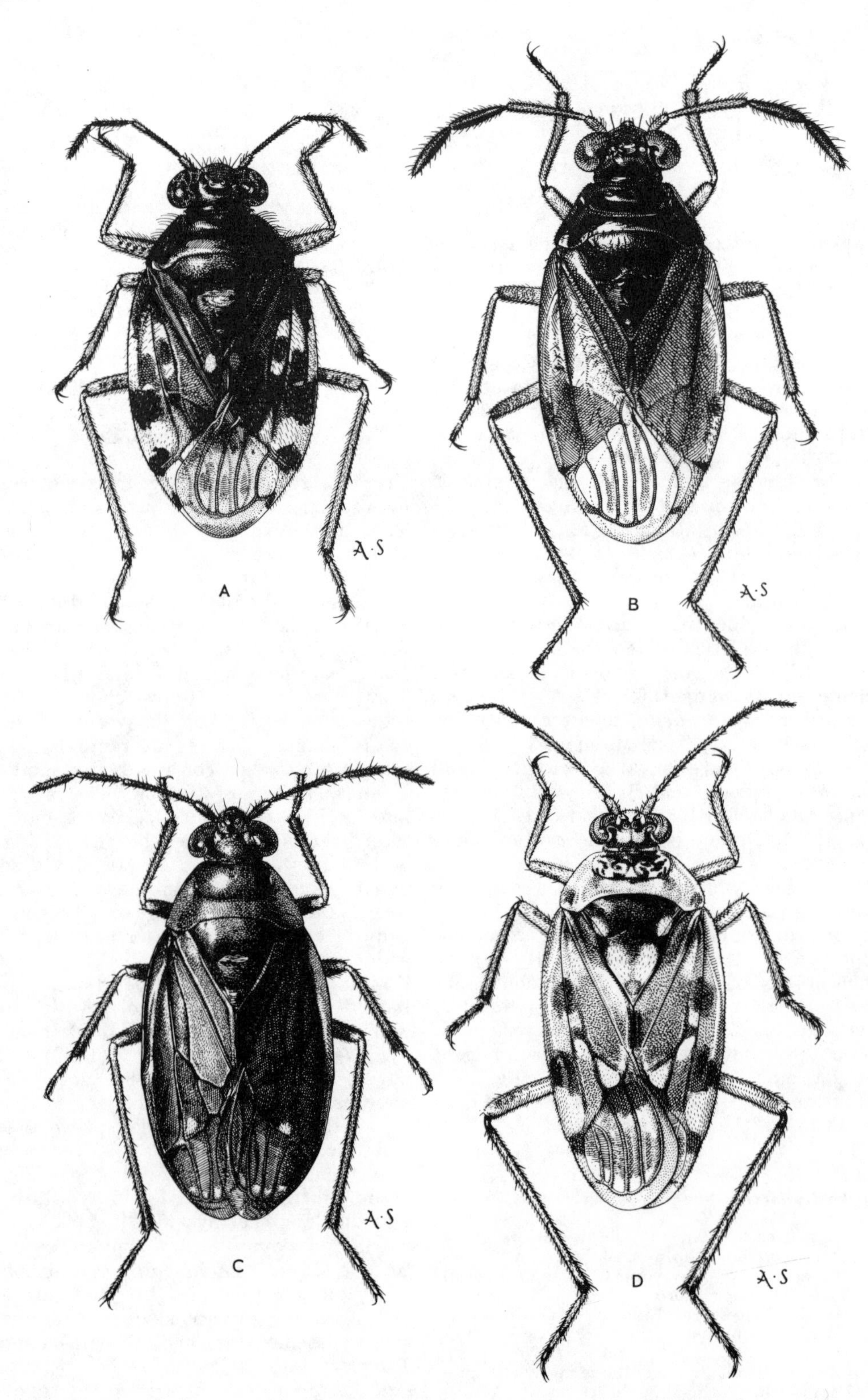

Fig. 7:40. *a, Saldula comatula* (Parshley), female, Mokelumne Hill, Calaveras County, California, May 27, 1931 (R. L. Usinger); *b, Ioscytus politus* Uhler, female, Lone Pine, Inyo County, California, June 10, 1929 (R. L. Usinger); *c, Salda buenoi* McD., female, Steen Mountains, Harney County, Oregon, June 24, 1922 (E. C. Van Dyke); *d, Pentacora signoreti* (Guerin), female, Salton Sea, California, May 25, 1940 (R. L. Usinger).

insects, but in Europe species have been found consistently below the high-tide line and a close relative, *Aepophilus,* lives under sea water most of the time along the coasts of France and Britain. Saldids are world-wide in distribution, and one species, *Chiloxanthus stellatus* Curtis, occurs at Pt. Barrow, the northernmost point on the North American mainland.

In California shore bugs may be found in all kinds of aquatic habitats. *Pentacora signoreti* prefers salt-water shores along the coast and at the Salton Sea. *Saldula luctuosa* is a salt-marsh form known from the San Francisco and Monterey bays. *Ioscytus nasti* is a bog inhabitant, and *I. politus* has been taken in the vicinity of hot springs. However, most of our species occur along the shores of fresh-water streams and lakes.

Feeding habits.—Wiley (1922) fed her colonies on small mirids and cicadellids, living or recently killed. I reared two species in Hawaii on freshly killed katydids. In nature the bugs are seen to forage among grasses and rocks, seeking any small insect life. Saldids share their microhabitat with other predators such as *Gelastocoris, Ochterus,* and the small but very numerous carabids, *Bembidion.* However, they are more agile and fly more readily than any of these.

Taxonomic characters.—The pioneer work on North American Saldidae was by Uhler (1877). Unfortunately, no key was given by Uhler, so the taxonomy of the group remained obscure for more than half a century. Recently, C. J. Drake has published a dozen or more papers in which new species are described from California, and North American species are, in some cases, synonymized under older European names. Here again, no keys are given, so the present keys to California species are entirely original, though based on adequate series including paratypes supplied by Drake. Dr. Drake also checked the keys and offered valuable suggestions. The *Catalogue of Genera and Species of Saldidae* by Drake and Hoberlandt (1950) was useful in the course of this work. On a trip to Washington during the preparation of the keys, the Uhler types were studied.

The following California records listed in the Van Duzee *Catalogue* (1917) are now considered to be misidentifications or synonyms of other species: *Salda littoralis* (Linn.), *Saldula dispersa* (Uhl.), *S. interstitialis* (Say), *S. xanthochila* (Fieber), and *Micracanthia humilus* (Say).

The most reliable specific characters are seen in the type and arrangement of pubescence. In addition to subappressed hairs, one group of species has stiff, erect hairs. These can be seen only with good light under a binocular dissecting microscope with magnification of 18 diameters or more. The bristles are more easily seen in profile than from above. Although size and color pattern vary within a species, these characters are useful within limits, as has been demonstrated in the much better known fauna of Europe (fig. 7:42). Male genitalia have not been found useful thus far either at the species level or for higher groups.

The generic classification is taken directly from the classical work of Reuter (1912), though the keys have been simplified to cover only the North American genera. Although Reuter's genera have stood the test of time, the genera *Teloleuca* and *Micracanthia* can scarcely be separated from the large and diverse genus *Saldula.* Also *Ioscytus* may not be a natural group since the addition of *franciscanus* (Drake) and *nasti* Drake and Hottes. However, it was felt that no changes should be made in the generic classification until a world-wide study is undertaken.

Key to Nearctic Genera of Saldidae

1. Pronotum with 2 prominent, erect tubercles on front lobe, occupying nearly all of the disc .... *Saldoida* Osborn 1901
— Pronotum without conical tubercles on front lobe 2
2. Membrane with 5 closed cells; always macropterous 3
— Membrane with 4 closed cells in macropterous forms, variously reduced in brachypterous forms 4
3. Sublateral cell of membrane short, not reaching apices of the cells on either side; northern North America .... *Chiloxanthus* Reuter 1891
— Sublateral cell of membrane reaching apices of cells on either side (fig. 7:40*d*) .... *Pentacora* Reuter 1912
4. Lateral margins of pronotum concave, the humeral angles acutely produced ... *Lampracanthia* Reuter 1912
— Lateral margins of pronotum convex, the humeral angles rounded 5
5. First or inner cell of membrane produced forward 2/5 or 1/2 its length beyond base of 2nd 6
— Inner cell of membrane produced forward only slightly or not more than 1/3 its length beyond base of 2nd 7
6. Second antennal segment less than twice as long as 1st; upper surface with more or less extensive pale areas; the lateral margins of pronotum pale .... *Calacanthia* Reuter 1891
— Second antennal segment twice as long as 1st or more; upper surface usually entirely dark brown or black, rarely with a few small pale spots on corium and membrane (fig. 7:40*c*) .... *Salda* Fabricius 1803
7. Antennae relatively thick, the 3rd and 4th segments thicker than apex of second segment (fig. 7:40*b*) .... *Ioscytus* Reuter 1912
— Antennae relatively slender, the 3rd and 4th segments not thicker than apex of second segment 8
8. Inner cell of membrane reaching only 4/5 of the distance to apex of adjacent cell; 2nd segment of hind tarsi nearly 1/2 again as long as 3rd segment .... *Teloleuca* Reuter 1912
— Inner cell of membrane usually reaching almost to apex of adjacent cell; 2nd segment of hind tarsi subequal or slightly longer than 3rd segment 9
9. Veins of corium obsolete; size small, usually less than 3 mm. (fig. 7:41) .... *Micracanthia* Reuter 1912
— Veins of corium more or less distinct, size larger, more than 3 mm. (fig. 7:40*a*) .... *Saldula* Van Duzee 1914

Key to California Species of Ioscytus

1. Most or all of corium pale or red 2
— Corium entirely black or with a pale spot near outer apex 3
2. Costal margins broadly infuscated at least apically (fig. 7:40*b*); western United States .... typical *politus* (Uhler) 1877
— Costal margins entirely pale .... variety *flavicosta* Reuter 1912
3. Third and 4th antennal segments very stout, about 1/4

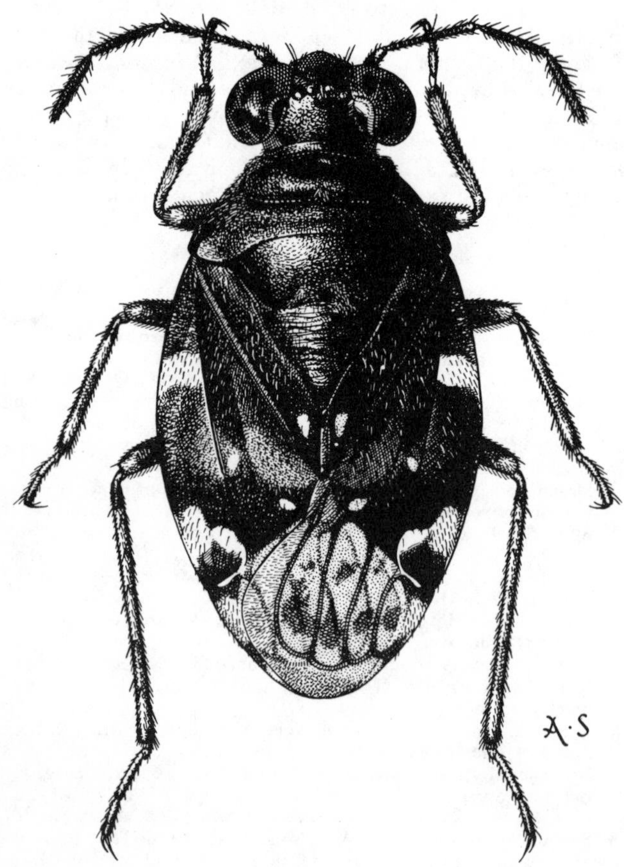

Fig. 7:41. *Micracanthia quadrimaculata* (Champion), female, Mokelumne Hill, Calaveras County, California, May 27, 1931 (R. L. Usinger).

as thick as long. Corium entirely black or dark brown apically; Sierra and Nevada counties .... *nasti* Drake and Hottes 1955
— Third and 4th antennal segments more slender, about 1/6 as thick as long. Corium with a pale spot near outer apex; Marin and Mendocino counties .... *franciscanus* (Drake) 1949*b*

Micracanthia quadrimaculata (Champion) 1900

This very small, oval species (fig. 7:41) is widely distributed in the western United States and Mexico. *M. pusilla* Van D. (1941) described from San Diego, is a synonym (see Drake and Hottes, 1950).

Pentacora signoreti (Guerin) 1856

Our largest saldid, this species is easily recognized by the five-celled membrane (fig. 7:40*d*). It occurs along the southern California coast and Salton Sea.

Salda buenoi (McDunnough) 1925

This is a dull black species with broad but abbreviated hemelytra (fig. 7:40*c*). It has been taken in Lassen County and Inyo County, California, and occurs northward into Oregon and Canada (see Drake and Hottes, 1950).

Key to California Species of Saldula

1. Upper surface with long, erect hairs and/or short, appressed pubescence 2
— Upper surface with pubescence all short and subappressed 8
2. Eyes with a few short, erect and short appressed hairs 3
— Eyes without short, erect hairs 4
3. Body subrounded, 4½ mm. long; membrane greatly reduced, only 2/3 as long beyond level of apices of coria as length in front of this; hemelytra with extensive pruinose areas (central California) (=*orbiculata* of Van D. not Uhler) .... *villosa* (Hodgden) 1949
— Body elongate-oval, 3½ mm. in length; membrane well developed, longer beyond level of apices of coria than in front of this; hemelytra with pale spots but no pruinose areas; Los Angeles (=*severini* Harris, see Drake and Hottes, 1954) .... *orbiculata* (Uhler) 1877
4. Second antennal segment relatively long, distinctly longer than narrowest width of vertex and one eye, equaling 1/2 the width of 2nd eye as well; Utah, Colorado, California .... *andrei* Drake 1949
— Second antennal segment subequal to narrowest width of vertex and 1 eye 5
5. Pronotum less than 1½ times as wide as head, including eyes 6
— Pronotum more than 1½ times as wide as head, including eyes 7
6. Corium with broad subbasal and subapical spots and with an inner subapical spot pale; central California, widely distributed .... typical *hirsuta* (Reuter) 1888
— Corium with subbasal and subapical spots much reduced, the inner subapical spot wanting .... variety *pexa* Drake 1950
7. Color black with a more or less distinct pale spot near apex of corium; clypeus and labrum black; membrane largely brown; pubescence rather fine, relatively short on lateral margins of pronotum; length about 4 mm.; San Francisco Bay, Carmel, see Drake, 1949*a* .... *luctuosa* (Stål) 1858
— Color paler posteriorly, the apical half of corium, clypeus, labrum, and membrane more or less broadly pale; pubescence relatively long and stiff, especially noticeable on lateral margins of pronotum; length 4½ to 5 mm. (fig. 7:40*a*); widely distributed, western United States .... *comatula* Parshley 1922
8. Second antennal segment relatively long, distinctly longer than narrowest width of vertex and one eye; costal margins usually entirely black; widely distributed, northern California .... *nigrita* Parshley 1922
— Second antennal segment shorter, subequal to or less than width of vertex and 1 eye; costal margins usually more or less pale 9
9. Callosities of anterior lobe of pronotum large, about twice as long as pronotum behind; body roundly narrowed posteriorly, the membrane relatively small; front tibiae with an ill-defined dark area at middle 10
— Callosities smaller, less than twice as long as pronotum behind; body broad posteriorly, the membrane relatively large 11
10. Pale markings of outer part of corium near middle fused in the shape of a "C"; Santa Cruz, Palo Alto (see Drake and Hottes, 1950) .... *c-album* Fieber 1859
— Pale markings of outer part of corium near middle not fused in the shape of a "C"; Siskiyou County .... *saltatoria* (Linn.) 1758
11. Head relatively large, the pronotum 1⅓ times as wide as head across eyes; size small, less than 3 mm.; Riverside, San Francisco ..... *bassingeri* Drake 1949*b*
— Head relatively smaller, the pronotum about 1½ times

as wide as head including eyes; size larger, more than 3 mm. 12

12. Labrum entirely black; size small, 3¼ mm.; Inyo County .... *ourayi* D. & H. 1949
— Labrum pale, at least on basal half; size larger, 3½ mm. or more 13
13. Corium broadly clothed with subappressed golden pubescence on basal half (as seen with light from in front of the specimen); size approximately 5 mm.; Oregon, British Columbia, Alaska .... *fernaldi* Drake 1949*b*
— Corium with golden pubescence lacking, or inconspicuous and scattered or confined to extreme base of corium 14
14. Extreme lateral margins of corium entirely dark; size relatively large, about 5 mm.; Lake Tahoe and Argus Mountains, California ....*explanata* (Uhler) 1893
— Lateral margins of corium more or less interrupted with pale; size smaller, 4 to 4¾ mm. 15
15. Pubescence of upper surface short but dense and suberect, shaggy on head; central California .... *notalis* Drake 1950
— Pubescence of upper surface even, subappressed ... 16
16. Lateral margins of pronotum entirely black, relatively strongly arcuate 17
— Lateral margins of pronotum pale, comparatively feebly arcuate 18
17. Corium glabrous, with fine golden yellow pubescence. Hemelytra varying from predominately pale to black, the intermediate forms with anterior transverse fascia interrupted or obscured (fig. 7:42); widespread, Europe and North America............. *pallipes* (Fabr.) 1794
— Corium dull, with gray pubescence. Hemelytra varying from predominately pale to black, the intermediate forms with anterior transverse fascia clear or scarcely interrupted (fig. 7:42); widespread .... *arenicola* (Scholtz) 1846
18. Pronotum broadly pale laterally and posteriorly, except at middle behind black callosities; size relatively small, about 3½ mm.; Arizona........*balli* Drake 1950
— Lateral margins of pronotum narrowly pale; size larger, more than 4 mm. 19
19. Pale color of pronotal margins not reaching anterior angles and scarcely reaching humeral angles; length approximately 4 mm.; southern and central California (see Drake and Hottes, 1950) ......*coxalis* (Stål) 1953
— Pale color of pronotal margins reaching both anterior and humeral angles; length nearly 5 mm.; El Dorado, Mono, and Inyo counties ....*opiparia* D. & H. 1955

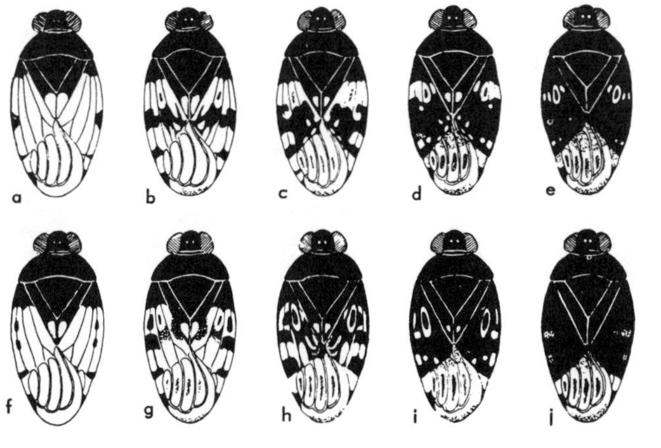

Fig. 7:42. *a* to *e*, *Saldula arenicola* Sz. showing types with different degrees of darkening of hemelytra. *f–j*, *S. pallipes* Fabr., same (Wagner, 1950).

REFERENCES

DRAKE, C. J.
1949*a*. Concerning North American Saldidae. Arkiv för Zool., 42B (3):1-4.
1949*b*. Some American Saldidae. Psyche, 56:187-193, 1 pl.
1950. Concerning North American Saldidae. Bull. Brooklyn Ent. Soc., 45:1-7.

DRAKE, C. J., and LUDVIK HOBERLANDT
1950. Catalogue of genera and species of Saldidae. Acta Ent. Mus. Nat. Pragae, 26(376):1-12.

DRAKE, C. J., and F. C. HOTTES
1949. Two new species of Saldidae from western United States. Proc. Biol. Soc. Wash., 62:177-184.
1950. Saldidae of the Americas. Great Basin Naturalist, 10:51-61.
1951. Stridulatory organs in Saldidae. Great Basin Naturalist, 11:43-46.
1954. Synonymic data and description of a new Saldid. Occ. Paps. Mus. Zool. Univ. Michigan, 553:1-5.
1955. Concerning Saldidae of the Western Hemisphere. Bol. Ent. Venezolana, 11:1-12.

HODGDEN, B. B.
1949. New Saldidae from the Western Hemisphere. Jour. Kansas Ent. Soc., 22:149-165.

PARSHLEY, H. M.
1922. A report on some Hemiptera from British Columbia. Proc. B.C. Ent. Soc., 18:13-24, 1921.

REUTER, O. M.
1912. Zur generischen Teilung der paläarktischen und nearktischen Acanthiaden. Öfv. Finska Vet.-Soc. Förh., 54A(12):1-24.

SLATER, J. A.
1955. The macropterous form of *Lampracanthia crassicornis* (Uhler). Jour. Kansas Ent. Soc., 28:107-109.

STICHEL, W.
1934. Illustr. Bestimmungstabellen Deutsch. Wanzen, Lief., 10:301-302.

UHLER, P. R.
1877. Report upon the insects collected by P. R. Uhler during the explorations of 1875, including monographs of the families Cydnidae and Saldae, and the Hemioptera collected by A. S. Packard, Jr., M.D. Bull. U.S. Geol. Geog. Surv. Terr., 3:355-475. 2 pls.

WAGNER, E.
1950. Notes on Saldidae. I. The Saldula pallipes-group. Acta Ent. Mus. Nat. Pragae, 26(371):1-4.

WILEY, G. O.
1922. Life history notes on two species of Saldidae. Kansas Univ. Sci. Bull., 14:301-308, 2 pls.

Family DIPSOCORIDAE

Only the nominate genus *Cryptostemma* (H. S., 1835) (=*Dipsocoris* Haliday 1855) is semiaquatic, other dipsocorids living in leaf mold, rotten logs, and the like. The family is characterized (McAtee and Malloch, 1925) by three-segmented tarsi, three-segmented beak, four-segmented antennae with the first two segments short and stout, the apical two long, slender, and pilose. The ocelli are distinct in macropterous specimens but reduced or absent in short-winged forms. The hemelytra are usually long and slender with several closed cells, the costal margin thickened and broken by a distinct fracture. There are no metathoracic scent gland openings in adults, and the nymphs have three or four dorsal abdominal scent gland openings. The body is covered by a fine pubescence imparting a velvety appearance and serving as a protection from wetting.

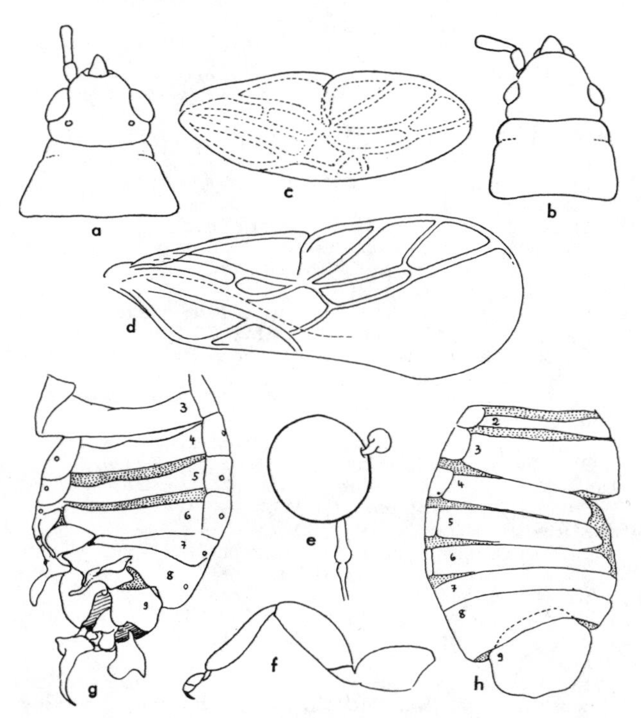

Fig. 7:43. *Cryptostemma usingeri* Wygodzinsky, Hot Creek, Inyo Co., Calif., July 17, 1953 (R. L. Usinger, J. D. Lattin). *a*, head and pronotum of macropterous male; *b*, same, brachypterous female; *c*, fore wing of brachypterous male; *d*, fore wing of macropterous male; *e*, spermatheca of female; *f*, fore leg of macropterous male; *g*, abdomen of male, dorsal view; *h*, same, ventral view (Wygodzinsky, 1955).

Life history.—Butler (1923) summarized information on the biology of the British *Cryptostemma alienum* H.S. Adults were found from March to October, and nymphs were found with the adults throughout much of this period. Therefore it is surmised that a continuous succession of broods occurs. No life history data are available on California species, but possibly the occurrence in the vicinity of hot springs would permit year-round development.

Habitat and distribution.—It has long been known that the European *C. alienum* is an inhabitant of the shore line of streams, running about beneath stones at the water's edge. Haliday (1855) described it as "gliding among the wet gravel, its silky down protecting it from the wet." A similar habit has been described for American species in Georgia and Puerto Rico (Usinger, 1945), and the California species, *C. usingeri* Wygodzinsky 1955, likewise occurs at the water's edge. In the latter case, the species was found under rocks along Hot Creek, in Inyo County. A single female of a much larger undescribed species was taken at a hot spring in Death Valley, California.

REFERENCES

BUTLER, E. A.
1923. A biology of the British Hemiptera—Heteroptera. London: H. F. & G. Witherby, pp. 1-682 (pp. 308-309).

McATEE, W. L., and J. R. MALLOCH
1925. Revision of bugs of the family Cryptostemmatidae in the collection of the United States National Museum. Proc. U.S. Nat. Mus., 67, Art. 13:1-42, pl. 1-4.

USINGER, R. L.
1945. Notes on the genus *Cryptostemma* with a new record for Georgia and a new species from Puerto Rico. Ent. News, 56:238-241.

WYGODZINSKY, P.
1955. Description of a new *Cryptostemma* from North America. Pan-Pac. Ent., 31:199-202.

CHAPTER 8

Megaloptera

By H. P. Chandler
California Department of Fish and Game

Alderflies, Dobsonflies, Fishflies

Adult dobsonflies are medium- to large-sized insects with complete metamorphosis. The large, broad, gray to black wings are held rooflike over the body and are similar in shape, but the hind wings have a small folded area. The head is broad and flat, the mouth parts are of the chewing type, and the antennae are slender and multisegmented. Adults may be seen making short, slightly awkward flights along the margins of streams and lakes. The aquatic larvae (fig. 8:1*b*) are called hellgrammites and the larger species are highly prized as fish bait. They are fiercely predaceous and have strong chewing mouth parts. The abdomen bears seven or eight pairs of lateral gills.

The Megaloptera are among the most primitive of the winged insects, being close to the ancestral stock of the Neuroptera, Mecoptera, and Hymenoptera. Almost every feature of their wings can be duplicated in the wings of the Paleozoic fossil insects. The branching of the accessory veins from the forward side of the radial sector found in several fossils can be seen in *Sialis*. The order contains two families, Sialidae and Corydalidae, with a total of about twenty genera. Both families and seven genera are represented in the Nearctic region.

The adults are not aquatic and do not contact the water even to deposit their eggs and thus have no special need for adaptations for aquatic respiration. The larvae have long, slender, tapering lateral processes, or tufts of filaments at the bases of these processes that serve as gills.

Key to the Families

Adults

1. Without ocelli; fourth tarsal segment bilobed; wing expanse less than 40 mm. .... SIALIDAE
— With 3 ocelli; fourth tarsal segment simple, not bilobed; wing expanse 40 to 160 mm........... CORYDALIDAE

Larvae

1. Last abdominal segment produced as a long median filament; lateral gill processes on abdominal segments 1-7 .... SIALIDAE
— Last abdominal segment without median filament, but with a pair of lateral hooks; lateral filaments on abdominal segments 1-8 .... CORYDALIDAE

Family SIALIDAE

Alderflies

These slow, awkward flying, smoky to black insects vary little in size, ranging from 10-15 mm. in length. They are usually seen making flights along the margins of streams. The larvae (fig. 8:1*b*) are easily recognized by their lateral abdominal gills, the prolongation of the last abdominal segment, and their pale testaceous to white color.

Relationships.—The adults are characterized by the absence of ocelli. The head is broad, with long, multisegmented antennae. The pronotum is large and rectangular, two and one-half times as wide as long. *Sialis* (fig. 8:1*a*), the only Nearctic genus, contains about twenty species. Ross (1937) places these in seven phylogenetic groups, based largely on the structure of the male genitalia. The "*californica* group," his largest, contains seven species including three of the four species known from California. Three groups contain but one species, one of which is *S. nevadensis,* the fourth known California species.

Respiration.—The larvae have five-segmented lateral gill processes on the first seven abdominal segments. The terminal process contains tracheae but is not segmented.

Life history.—The adults are probably short-lived as is indicated by the soft mouth parts which appear unsuited for extensive feeding. The eggs are deposited in rows, forming large masses on objects overhanging the water. The eggs are cylindrical, rounded at the top with a curved process. They hatch during the

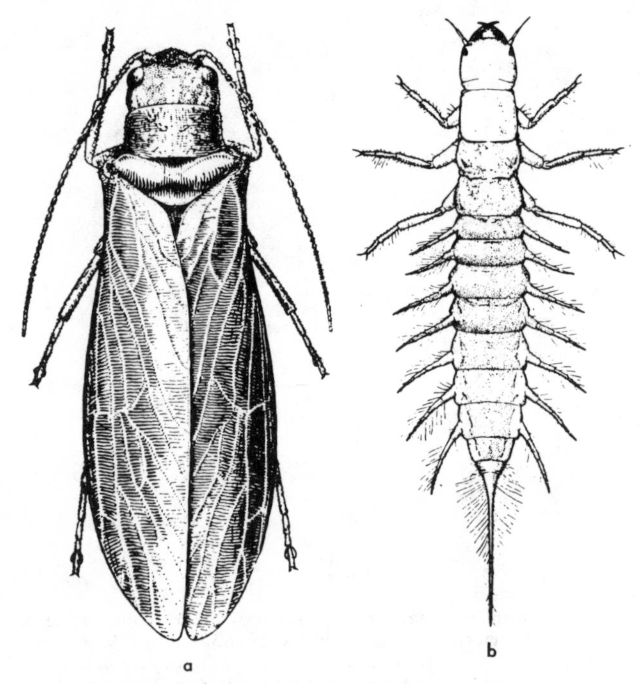

Fig. 8:1. *a*, adult of *Sialis mohri*; *b*, larva of *Sialis* sp. (Ross and Frison, 1937).

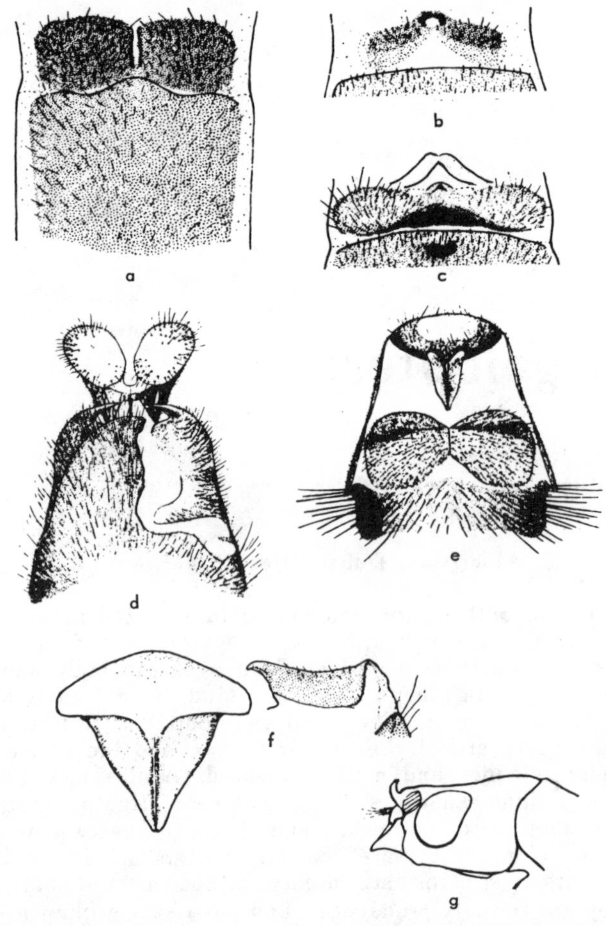

Fig. 8:2. *a-c*, terminal abdominal sternites of *Sialis*; *a*, *nevadensis*; *b*, *rotunda*; *c*, *californica*; *d-e*, ventral aspect of genitalia and ninth segment of male *Sialis*; *d*, *nevadensis*; *e*, *californica*; *f*, male genitalia of *S. occidens* in ventral and lateral views; *g*, lateral view of head of male *S. rotunda* (Ross and Frison, 1937).

night in about nine or ten days. The larvae drop into the water where their development takes place. They are predaceous, seeking their prey on the bottom in the mud and under stones. When the larvae are mature they seek the banks above the water and burrow into the ground to form an earthen cell. In this cell the larval skin is shed and pupation takes place.

Habitat and distribution.—The larvae are found in well-aerated standing or running water. They are not as characteristic of the riffle areas in California as are the Corydalidae but may be frequently found in numbers in the back eddies of streams where leaves and other debris accumulate. The adults are diurnal and are most active during the middle of the day. When disturbed, they will often run rather than fly. *Sialis* is found throughout the United States and well into Canada, with the greatest number of species found in the eastern part of the continent.

Taxonomic characters.—The male genitalia offer the principal means of distinguishing the species. Although the wings sometimes present characters, in general they vary too much to be of any significant value, particularly since the venation often differs on opposite sides of a single specimen. The sculpturing of the head and pronotum is used in a few cases to separate species.

Key to the California Species of Sialis

Adults

1. Females 2
— Males 4
2. Eighth abdominal sternite large with a long narrow cleft on the apical margin (fig. 8:2*a*) ....*nevadensis* Davis 1903
— Eighth abdominal sternite without cleft 3
3. Posterior margin of eighth sternite with a large, deep, median depression (fig. 8:2*b*) ..... *rotunda* Banks 1920
— Posterior margin of eighth sternite with a median prominence (fig. 8:2*c*)........................ ....*californica* Banks 1920 and *occidens* Ross 1937
4. Front between the eyes with prominent hornlike process (fig. 8:2*g*) .... *rotunda* Banks 1920
— Front without hornlike process 5
5. Ninth abdominal sternite produced into a flap covering most of the genitalia (fig. 8:2*d*); northern California south to Mariposa County...... *nevadensis* Davis 1903
— Ninth sternite not produced into a flap 6
6. Basal part of genital plate neither enlarged nor projecting over terminal plates; 9th sternite with fairly short lateral setae; San Joaquin Valley and foothills ....*arvalis* Ross 1937
— Basal part of genital plates projecting above apex of terminal plate; 9th sternite with lateral setae long and whiskerlike 7
7. Genital plate with base narrow and produced into two bulbous elevations (fig. 8:2*e*); British Columbia south to northern and western California ....*californica* Banks 1920
— Genital plate with base wider and not produced into two bulbous knobs (fig. 8:2*f*); Plumas to Mariposa County .... *occidens* Ross 1937

The female of *S. arvalis* Ross, described from Mokelumne Hill, California is unknown.

Family CORYDALIDAE

Dobsonflies, Fishflies

This family contains the largest of the Megaloptera; some species have a wing span of more than six inches. However, these large insects are weak flyers. *Corydalis* is probably the most conspicuous representative of the family because of the enormously elongate mandibles found on the males of some species. Many of the genera have morphological characters that are considered to be quite primitive.

Relationships.—The genitalia of *Dysmicohermes ingens* Chandler, with four pairs of paired appendages are undoubtedly very primitive. The wings of *Nigronia* and *Chauliodes* with the unbranched forks of the radial sector and the simple anal veins and cells are also primitive. *Neohermes* and *Protochauliodes* with R_3 and R_4 fused basally, and A_2 fused with A_1, are less primitive in this respect. *Corydalis* seems more advanced in most characteristics than the other genera.

The adults are characterized by three ocelli, long antennae which range in shape from filiform to pectinate and five-segmented tarsi with the fourth segment cylindrical, not bilobed, and the claws toothed.

Respiration.—All the larvae have long lateral projections from the sides of abdominal segments one to eight with a small pair on the prolegs. These projections serve as gills in most of the genera. However, in *Corydalis* they are less modified for this purpose, and there is a tuft of filamentous gills at the base of the lateral projection except in the first instar. *Nigronia,* which is found in quiet or stagnant water, has the last pair of abdominal spiracles situated at the end of a pair of projections for surface breathing.

Life history.—The life history of *Chauliodes* was described by Davis (1903). Eggs are deposited on stones or on branches or under surfaces of bridges that overhang the water. In general, hatching occurs at night five or six days after the eggs were deposited. The larva emerges from the egg near the micropylar projection and drops into the water. The first instar larvae differ from the later instars by having the lateral filaments longer and the head large. The fiercely predaceous larvae feed chiefly in the dark and will eat almost anything that they can subdue, including their own kind. *Corydalis* larvae have been observed to swim either forward or backward with a snakelike motion but are more prone to crawl. The length of the larval period is not known but may be as long as three years. When mature, they seek the bank of the stream above the water level and burrow into the ground or a rotten log and construct a pupal chamber. In this cavity, the pupa forms and the larval skin is shed. The pupal period is about two weeks. *D. ingens* sheds its pupal skin after emerging from the pupal cell. This cell may be situated twenty to thirty feet from the margin of the stream.

Maddux (1954) studied the life history of *Protochauliodes aridus* in the foothills of the Sierra Nevada and found numerous points of difference from the observations of previous authors. The larvae were found in streams running only during the winter and spring rainy seasons. When the streams dried up in late spring, the mature larvae would burrow beneath large stones in the stream bed and construct pupal chambers. The mandibles, and to a slight extent the legs of the pupae, are functional so that when two pupae are placed together in the same container, usually only one survives. After emergence, the adults deposit their eggs on rocks in the dry stream bed. Upon hatching, the larvae burrow into the loose gravel of the stream bed. This part of the cycle occurs before the onset of summer during which time the ground surface temperatures may exceed 125° F. Specimens kept in dry soil in the laboratory did not become dormant but fed actively on termites and other small insects supplied them. Their abundance and the development of the mouth parts and digestive tract indicate that it is doubtful that adults live more than a few days. *Neohermes* also pupates in dry stream beds.

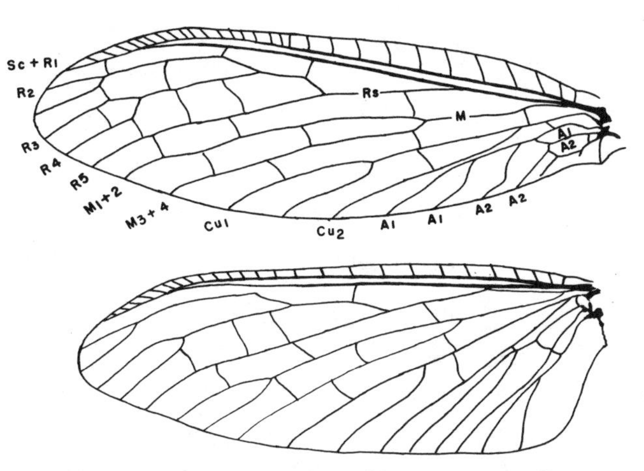

Fig. 8:3. Wings of *Neohermes* (Chandler, original).

Habitat and distribution.—The larvae prefer coarse or rubble bottoms where they can move about freely below the surface of the stream bed, but occasionally they are found on mud bottoms. *Nigronia* is found in quiet or stagnant water. They are taken in many habitats such as streams flowing only four months of the year, moderately large rivers, and small streams cascading down the sides of canyons. The frequency of occurrence of the larvae in stream samples would indicate a much larger adult population than is evidenced by the small amount of material to be found in insect collections. This is no doubt owing to the short life and retiring nature of the adults, whose flying activities are usually restricted to early evening. Some, like *P. infuscatus,* are active during the middle of the day but fly at large over the countryside so that they are not seen in any numbers along the streams.

The family is found on all continents in temperate and tropical regions. In North America, members extend into eastern Canada, across the southern United States, and into Canada again along the Pacific Coast. Apparently they are not found in the Rocky Mountains or in most of the Great Basin area.

Taxonomic characters.—The length and type of

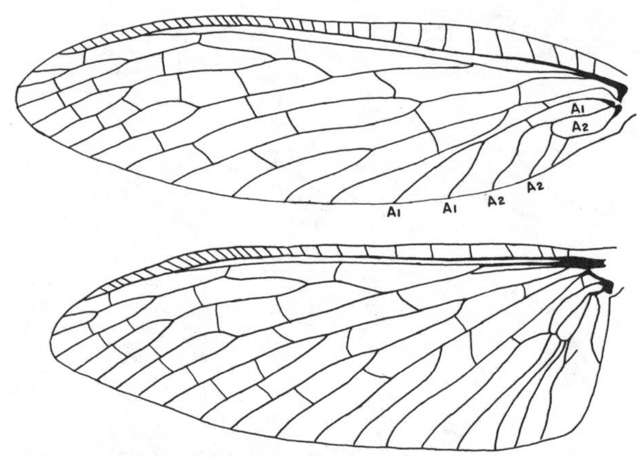

Fig. 8:4. Wings of *Dysmicohermes ingens* (Chandler; original).

antennae, size and placement of the ocelli, venation of the wings, and the genitalia are the most useful taxonomic characters. Except for those with black wings, the coloration is very difficult to use or to describe. This is also true of the micro- and macrosculpture of the pronotum and head.

Key to the Nearctic Genera of Corydalidae

Adults

1. Mandibles large, not concealed by the labrum; cells of the wings with white dots, radial sector usually with 6 or more forks; eastern and southern United States ....*Corydalus* Latreille 1802
— Mandibles smaller, when closed mostly concealed by labrum; cells of wing without white dots, radial sector usually with 5 or less forks 2
2. Cell A_1 (fig. 8:3) closed distally by anterior fork of vein A_2 which fuses with vein A_1 for a short distance so that vein A_1 appears 3-branched; radial sector with only 3 forks, R_3 and R_4 fused for nearly 1/2 their length 3
— Cell A_1 (fig. 8:4), closed distally by a cross vein from A_1 to A_2; A_1 2-branched; radial sector with 4-5 forks 4
3. Antennae of male moniliform, as long as body with numerous erect hairs in a band around each segment; R_3 and R_4 with cross vein beyond fork (fig. 8:3); western and southern United States.... *Neohermes*[1] Banks 1908
— Antennae of both sexes filiform, 1/2 to nearly as long as body, without erect hairs; R_3 and R_4 without cross vein beyond fork; western United States and Canada .... *Protochauliodes* Weele 1910
4. Second anal cell (fig. 8:5) not reaching the fork of second anal vein, R_5 never forked; antennae either serrate or pectinate 5
— Second anal cell (fig. 8:4) with end closed by posterior branch of second anal vein, R_5 forked near middle; antennae filiform; western United States and Canada .... *Dysmicohermes* Monroe 1953
5. Anterior ocellus small, round, more than twice its diameter from the posterior ocelli; costal cells about as long as broad, fork of second anal vein (fig. 8:5) well beyond cross vein between first and second anal vein; black species; eastern United States .... *Nigronia* Banks 1908
— Anterior ocellus large, not round, not twice its long diameter from the others; cross vein between first and second anal at or beyond fork of second anal vein; gray, mottled species; eastern United States............. .... *Chauliodes* Latreille 1796

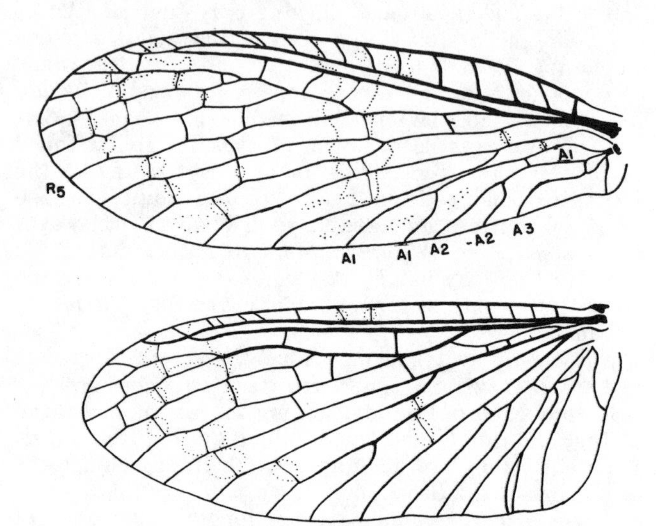

Fig. 8:5. Wings of *Nigronia* sp. (Chandler, original).

Larvae

1. First 7 abdominal segments with a large tuft of filamentous gills at the base of each lateral gill process; patch of hydrofuge pile on ventral side of 9th and 10th segments; eastern to southwestern United States ... . *Corydalus* Latreille 1802
— Without filamentous gills at base of lateral gill processes; no hydrofuge pile on 9th and 10th segments 2
2. Eighth abdominal segment with spiracles at end of long contractile respiratory tubes which extend well past the end of the terminal claws; eastern United States ....*Chauliodes* Latreille 1796
— If 8th abdominal spiracles situated at end of respiratory tubes, they are short and do not reach middle of 9th segment 3
3. Spiracles of 8th abdominal segment small, anteromesad to lateral gill processes; lateral wings of submentum pointed at apex; western United States ....*Dysmicohermes crepusculus* Chandler 1954
— Spiracles of 8th abdominal segment larger than those of the anterior segments and posteromesad to lateral gill processes; lateral wings of submentum with apex emarginate 4
4. Lateral gill processes short, 1/2 as wide as abdominal segments (longer in early instars); spiracle of 8th abdominal segment large, situated on short respiratory tube, about as long as wide; western United States .... *Dysmicohermes* Munroe 1953
— Lateral gill processes longer, at least as long as width of abdominal segments; spiracles of 8th segment smaller 5
5. Spiracles of 8th abdominal segment at the end of a short, tapered respiratory tube, about 1½ times as long as wide; clypeus narrower than base of vertex where joined; eastern United States ..... *Nigronia* Banks 1908
— Spiracles of 8th abdominal segment on posterior edge of 8th segment without respiratory tube; clypeus as wide as base of vertex; western United States ..... *Neohermes* Banks 1908 and *Protochauliodes* Weele 1909

The genus *Corydalis* is represented in California by only a single species, *C. cognata* Hagen 1861. This species is rare in California but has been taken at Sacramento and the Sequoia National Forest.

[1]It appears that the genus *Neohermes* is represented in California by the single species, *N. californicus* (Walker) 1869 and that *N. nigrinus* Van Dyke is a synonym of *Protochauliodes infuscatus* Caudell (Editor).

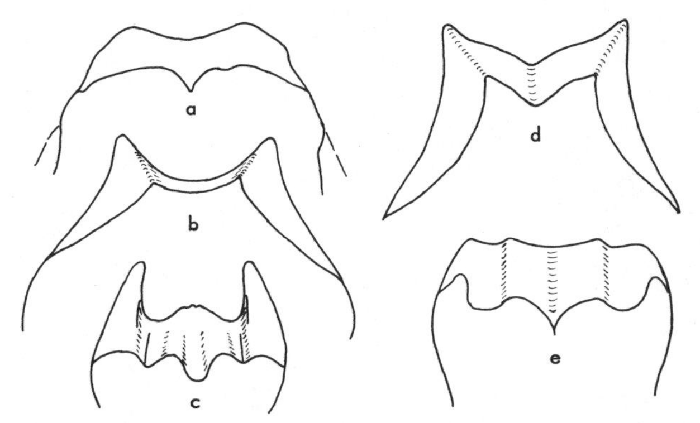

Fig. 8:6. Aedeagus of *Protochauliodes* species. *a, montivagus; b, simplus; c, minimus; d, aridus; e, infuscatus* (Chandler, original).

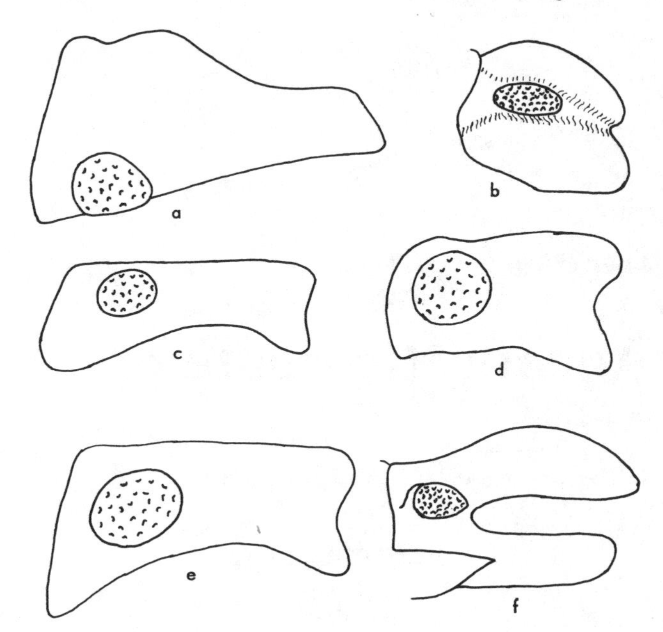

Fig. 8:7. Cerci in dorsolateral view. *a, Protochauliodes aridus; b, Dysmicohermes crepusculus; c, P. minimus; d, P. montivagus; e, P. simplus; f, D. ingens* (Chandler, original).

Key to the California Species of Protochauliodes Weele 1909

Adults

1. Wings of male and hind wings of female black, fore wings of female mottled, black areas predominating; truncate end of aedeagus (fig. 8:6*e*) divided into thirds by 2 ridges; wing spread 45-70 mm.; foothills of lower Sacramento Valley .... *infuscatus* Caudell 1933
— Wings mottled with dark and light, dark not predominating 2
2. Cerci short (fig. 8:7*d*), twice as long as width of apex, base and apex about equal in width, upper and lower apical lobes equal; aedeagus blunt (fig. 8:6*a*), roundly emarginate at middle, wing spread 66-88 mm.; Sierra and Madera counties above 4,000 feet .... *montivagus* Chandler 1954
— Length of cerci much more than twice the width of apex 3
3. Base of cercus with a large dorsomedial tuberocity (fig. 8:7*a*), apex less than 1/2 the maximum width of the base, apical lobes reduced; apex of aedeagus with acute lateral corners (fig. 8:6*d*), emarginate middle part a broad V-shape, wing spread 58-70 mm.; Butte to Mariposa counties .... *aridus* Maddux 1954
— Base of cerci without dorsomedial tuberocities, apex more than 1/2 the maximum width of base 4
4. Cercus with lower lobe even with upper but narrower (fig. 8:7*e*), definitely emarginate between lobes; aedeagus flat and simple (fig. 8:6*b*), broad at middle, apex 1/2 as wide, acutely pointed at sides, roundly emarginate medially; wing spread 76–84 mm.; Los Angeles County .... *simplus* Chandler 1954
— Cercus with lower lobe longer (inverted footlike) to short (fig. 8:7*c*) and equal to upper lobe, emargination between lobes nearly even with upper lobe; aedeagus more or less bent upward (fig. 8:6*c*), with prominent lateral earlike processes, medial part bluntly produced 5
5. Wing spread 42–50 mm.; color dark smoky with maculations rather indistinct; central California .... *minimus* Davis 1903
— Wing spread 65–86 mm.; membrane transparent and maculations distinct; Alameda County to British Columbia .... *spenceri* Munroe 1953

Key to the California Species of Dysmicohermes Munroe 1953

Adults

1. Spots on the wings relatively small, seldom completely crossing a cell except at base of wing, hump at base of first anal vein prominent and black, media of hind wing (fig. 8:4) 4-branched; male gentalia (fig. 8:7*f*) with cerci slightly divergent, upper and lower lobes much longer than wide; wing spread 120-160 mm.; Mariposa County .... *ingens* Chandler 1954
— Spots larger, occasionally completely crossing the cells in the distal half of the wing, hump at base of first anal vein less noticeable, seldom entirely black, media of hind wing 3-branched (rarely 4-branched); male genitalia (fig. 8:7*b*) with cerci short, upper and lower lobes indicated by a horizontal groove around the end of the cerci; wing spread 76–104 mm.; Shasta County to Mariposa County .... *crepusculus* Chandler 1954

REFERENCES

CHANDLER, H. P.
1954. Four new species of dobsonflies from California (Megaloptera: Corydalidae). Pan-Pac. Ent., 30:105-111.
DAVIS, K. C.
1903. Sialidae of North and South America. Bull. 68 N.Y. St. Mus. Ent., 18:442-486.
MADDUX, D. E.
1954. A new species of dobsonfly from California (Megaloptera: Corydalidae). Pan-Pac. Ent., 30:70-71.
MUNROE, E. G.
1953. *Chauliodes disjunctus* Walker: a correction, with the descriptions of a new species and a new genus (Megaloptera: Corydalidae). Canad. Ent., 85:190-192, 4 figs.
ROSS, H. H.
1937. Nearctic alder flies of the genus *Sialis* (Megaloptera: Sialidae). Bull. Illinois Nat. Hist. Surv., 21:57-78.
WEELE, H. W.
1910. Megaloptera. *In* Coll. Zool. Selys Longchamps, Brussels, fasc. 5, no. 115:1-95.

CHAPTER 9

Aquatic Neuroptera

By H. P. Chandler
California Department of Fish and Game

Family SISYRIDAE

Spongilla-Flies

The only strictly aquatic family of Neuroptera is the Sisyridae. Its larvae occur as parasites on fresh-water sponges belonging to the family Spongillidae. They creep about in and on the sponge, inserting their styletlike mouth parts into the living cells for nourishment. When mature they emerge from the water and pupate in a double-walled cocoon. The adults are known as spongilla-flies. They are small typical neuropterans with large eyes, long filamentous antennae, and large membranous brown or spotted wings folded rooflike over the body.

Relationships.—This family is regarded as most closely related to the Osmylidae, which have semi-aquatic larvae and are found throughout most of the world except North America. The larvae of Osmylidae do not have gills but hide in moist places near the margins of streams and feed upon dipterous larvae. The adult Sisyridae differ from the Osmylidae in that the cross veins are fewer and unbranched, the larvae have gills, the one-segmented tarsi are single clawed, and the styletlike mandibles and maxillae are curved downward. The family consists of six genera and about thirty-two species distributed throughout the world. Only two genera, *Sisyra* Burm. 1839 and *Climacia* McLach. 1869, are known from North America.

Respiration.—Respiration of the second and third larval instars takes place by means of ventral jointed tracheal appendages, folded medially and invisible from above, on abdominal segments one to seven. These gill appendages are vibrated rapidly, creating a current across the gill surfaces which aids in respiration. Since they are not present in the first instar, they are not believed to be remnants of the abdominal appendages found in primitive insects. Apparently respiration is cuticular in the first instar.

Life history.—The eggs are oval with a micropylar knob at one end. They are usually placed in a crevice or depression on some object overhanging the water such as a branch, rock, pier, or bridge. Two to five eggs are covered by a blanket composed of silken threads applied in a Z-pattern by spinnerets situated at the end of the ninth sternite. Hatching occurs in about eight days. The larva escapes from the egg by means of an egg-burster situated on the tip of the amnionic membrane which it soon sheds. It then drops upon the surface of the water and, often with considerable difficulty, forces its way through the surface film. It may crawl on the under side of the

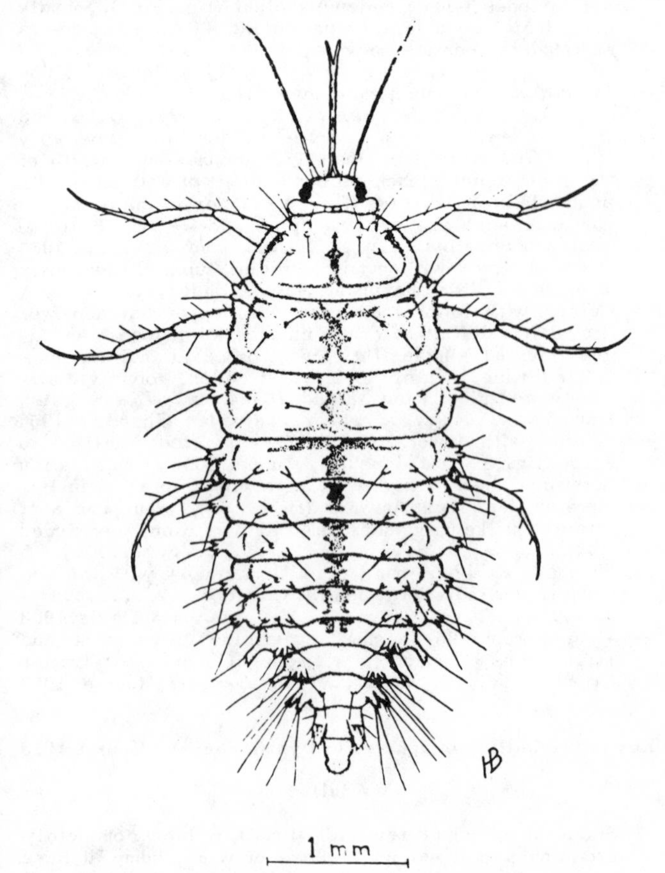

Fig. 9:1. Third instar larva of *Climacia* (Brown, 1952).

surface film, swim by curling the abdomen ventrally and snapping it out straight, float or sink at will, or wander about on the bottom. The sponge host is found in one of these ways. Once settled upon a sponge, the larvae do not readily leave it unless the sponge dies. They frequently enter the ostioles of the sponge. The first instar larva commonly feeds for one-half to two minutes, rests or wanders about for one to five minutes, then reinserts its mouth parts for another meal. The sucking mouth parts consist of the mandibles and maxillae on each side joined so as to make a tube. The right and left pairs can also be joined. Color varies from green to yellow-brown depending somewhat on the color of the sponge. The mature third instar larvae (fig. 9:1) crawl or swim to an object upon which they can emerge from the water. Most larvae emerge from one to five hours after dark. They may travel as far as fifty feet before selecting a site to spin a cocoon, usually in a somewhat protected location. The larvae take a firm stance and do not move their feet while spinning the outer net. The tenth abdominal segment is modified into a spinneret with the silk exuding from the anus. The spinning is accomplished by bending the abdomen this way and that with some telescopic movement. The resulting hemispherical outer net has a radius about the same length as that of the abdomen (fig. 9:2). In *Climacia* this outer net has a delicate hexagonal mesh whereas in *Sisyra* the fibers are merely crisscrossed. The cocoon is about half the diameter of the outer net and is composed of crisscrossed fibers in both genera. The cast pupal skin is wadded into the end of the cocoon. The sex of the pupa can be determined by examining the genitalia. The mandibles are heavily sclerotized and used by the pupa to chew its way out of the cocoon and net. The appendages are immovable until just before emergence, which usually occurs after sunset. Within two hours the insect assumes its adult form and color (fig. 9:3). Shortly thereafter a fecal pellet is deposited which probably contains the accumulated waste solids from the entire previous existence, since in the larvae the alimentary tract beyond the stomach is atrophied or modified into silk glands. Little is known of the food habits of the adults; fecal pellets of captured adults were found to be largely composed of pollen grains. Eggs were obtained from laboratory-reared specimens fed only on a gum drop and water. Adults live for two weeks or more and mating may occur at any time during this period, usually from dusk until midnight. Oviposition takes place two to six hours after sunset, and forty-five or more eggs are laid by each female. Two or three complete cycles may occur each year. Winter is probably passed in the larval stage.

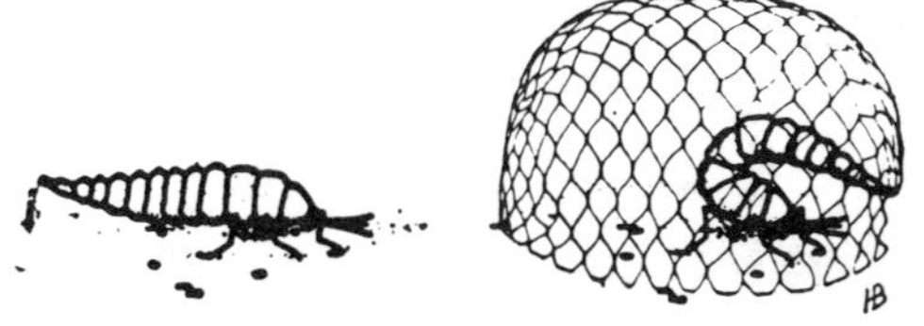

Fig. 9:2. Larva of *Climacia* spinning a net (Brown, 1952).

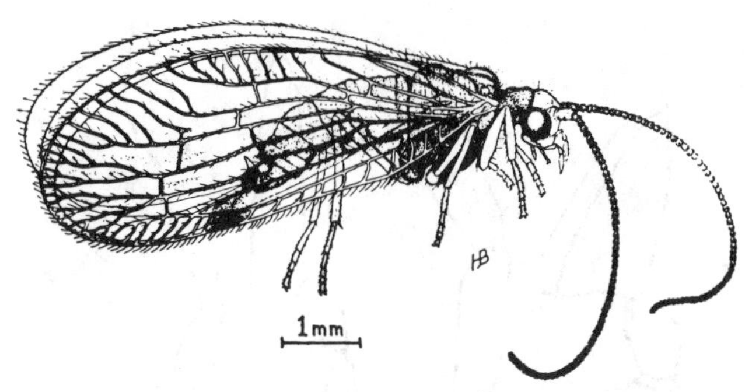

Fig. 9:3. Adult female *Climacia* (Brown, 1952).

Members of this family are exposed to the usual physical hazards of the environment and to predators both in the water and upon the land. *Climacia* at least serves as a host for *Sisyridivora cavigena* Gahan (Hymenoptera, Chalcidoidea), a small black parasitic wasp. By means of a long ovipositor, this wasp is able to pierce the net and cocoon, sting the larva or pupa, and lay its eggs on it. *Sisyridivora cavigena* was described from Ohio on *Climacia areolaris*. The same species, or a nearly related species, was found in pupal cases of *C. californica* Chandler at Clear Lake, California.

Habitat and distribution.—Spongilla-flies are likely to occur wherever the proper species of sponge is to be found. Most larval records are from *Spongilla fragilis,* but other species of *Spongilla* and *Ephydatia* have been listed as hosts. Some genera of fresh-water sponges are notably free of these parasites even when nearby *Spongilla* are heavily infested. Brown (1951) noted that in his study area, *Spongilla fragilis* occurred in the cool, clean lake and in a relatively warm, polluted pond; but the former yielded only *Climacia* and the latter *Sisyra*. Larvae may occasionally be found feeding on Bryozoa or Algae, but sponges are always present nearby. Adults are usually taken at lights or by sweeping bushes near the place of emergence.

Sisyra contains nineteen species scattered throughout the world except Australia. Only one species,

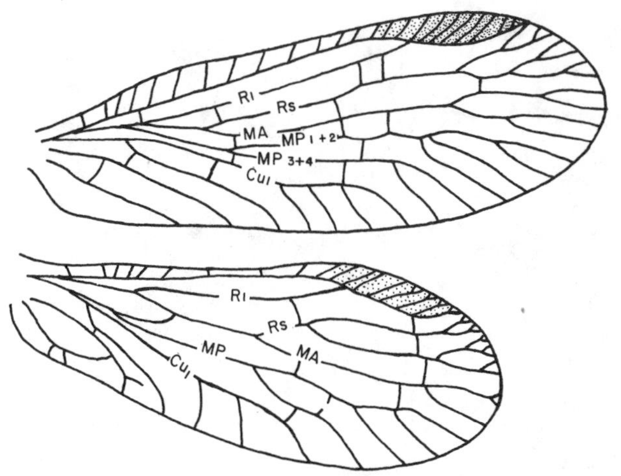

Fig. 9:4. Wing of *Climacia californica* (H. P. Chandler, original).

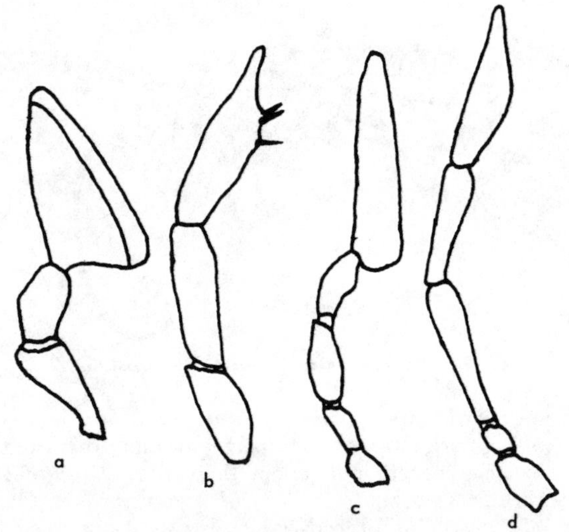

Fig. 9:5. *Sisyra; a,* labial palp; *c,* maxillary palp; *Climacia: b,* labial palp; *d,* maxillary palp (H. P. Chandler, original).

S. vicaria Walker 1853, is described from the Nearctic region. It occurs across the United States and Canada from Nova Scotia and Florida to British Columbia and Oregon. There are no records from California, but it probably occurs in the northern part of the state.

Climacia is found only in North and South America and contains five described species. *C. areolaris* (Hagen) 1861 and *C. dictyona* Needham 1901 are found in the eastern United States, and *C. californica* Chandler 1953 has been taken at Clear Lake, Lake County, California, and Deschutes River, Oregon.

Key to Nearctic Genera

Adults

1. Rs of fore wing (fig. 9:4) with 1 fork, which is below pterostigma; fore wing with brown markings; ultimate segment of palpus (fig. 9:5*b,d*) swollen, widest near middle but only slightly wider than other segments .... *Climacia* McLachlan 1869

— Rs with 2 main forks, both proximal to the pterostigma; fore wing uniformly brown; ultimate segment of palpus (fig. 9:5*a,c*) much wider at base, triangular in shape .... *Sisyra* Burm. 1839

Pupae

1. Outer net (fig. 9:2) woven with hexagonal meshlike bobbinet .... *Climacia* McLachlan 1869

— Outer net of crisscrossed threads .... *Sisyra* Burm. 1839

Larvae

1. Dorsum (fig. 9:1) with conspicuous tubercles bearing 2-3 setae each .... *Climacia* McLachlan 1869

— Dorsum with setae present but without conspicuous tubercles .... *Sisyra* Burm. 1839

REFERENCES

ANTHONY, MAUD H.
1902. The metamorphosis of *Sisyra*. Amer. Nat., 36:615-631.

BROWN, HARLEY P.
1952. The life history of *Climacia areolaris* (Hagen), a Neuropterous "parasite" of fresh water sponges. Amer. Midl. Nat., 47:130-160.

CARPENTER, F. M.
1940. A revision of nearctic Hemerobiidae, Berothidae, Sisyridae, Polystoechotidae, and Dilaridae. Amer. Acad. Arts and Sci. Proc., 74:193-280.

CHANDLER, H. P.
1953. A new species of *Climacia* from California (Sisyridae, Neuroptera). Jour. Wash. Acad. Sci., 43:182-184.

KILLINGTON, F. J.
1936. A monograph of the British Neuroptera Ray Soc., 1:1-269.

NAVAS, LONGINOS
1935. Monografia de lo Familia de los Sisiridos. Mem. Acad. Cienc. Zaragoza, 4:1-83.

NEEDHAM, JAMES G.
1901. Family Hemerobiidae, pp. 550-560, in Needham and Betten, Aquatic insects in the Adirondacks. N.Y. State Mus. Bull., 47:383-612.

PARFIN, S. I., and A. B. GURNEY
1956. The spongilla-flies, with special reference to those of the Western Hemisphere. Proc. U. S. Nat. Mus. 105:421-529.

CHAPTER 10

Trichoptera

By D. G. Denning
Walnut Creek, California

At present seventeen families, one hundred thirty-five genera, and nine hundred eighty recognizable species of caddisflies are known to occur in North America north of Mexico. In California the comparable figures are fourteen families, fifty-eight genera, and one hundred and sixty-eight species. Thus 43 per cent of the North American genera are represented but only 17 per cent of the species. The only North American families not yet recorded from the state are Beraeidae, Goeridae, and Molannidae. The family Goeridae is known to occur in Oregon and probably will be collected in northern California.

Adult caddisflies (or caddiceflies) are generally small to moderate size, somber colored, and secretive. The adults (fig. 10:1) may be distinguished from other insects as follows: head with long, threadlike antennae; mandibles vestigial; maxillary and labial palpi well developed; two pairs of wings (reduced in a few females) held rooflike over the body and covered with hairlike setae (in some genera with a few scattered scales), fore wings slightly coriacious, hind wings shorter but usually broader; and venation simple.

The classification of adult Trichoptera is based largely on the male genitalia. In most cases, species identification is possible only by reference to these characters, so species keys are given only to males. In a few large genera such as *Rhyacophila,* species keys are not given because monographic treatment is needed but in such cases it is nevertheless possible to determine California species by referring to the illustrations. Because of their importance, figures of genitalia have been reproduced for every species, including a few original drawings of species not illustrated with the original description.

The eggs of caddisflies are deposited in or near the water, either as strings or in a mass. The females of such families as Rhyacophilidae, Philopotamidae, Psychomyiidae, Hydropsychidae, and Hydroptilidae enter the water to attach their eggs to submerged objects. Others (Limnephilidae) may deposit their egg masses above the water on various objects.

The larvae are little known, only about 15 per cent of the known species having been associated with the adults. The terminology commonly used in the description of the larvae and pupae is illustrated in figure 10:6. One of the best accounts of the various larval types is given by Ross (1944) and is quoted here.

"*Free-Living Forms.* —The larvae of the genus *Rhyacophila* are completely free living, having no case or shelter; they lay a thread trail and have many modifications for free life in flowing water, including widely spaced, strong legs and large, strong anal hooks. For pupation they form a stone case or cocoon.

"Also free living are the early instars of many Hydroptilidae.

"*Net-Spinning Forms.*—Larvae of Hydropsychidae, Philopotamidae and Psychomyiidae spin a fixed abode which is fastened to plants or other supports in the water, sometimes in still water but more frequently in running water. Three common types of these structures are found, all of them spun from silk and forming some sort of net; when taken out of water they collapse into a shapeless string. There is always an escape exit at the end of the tube.

"1.—Finger nets. These are long, narrow pockets of fine mesh, with the front end anchored upstream, the remainder trailing behind with the current. They are built by the Philopotamidae (fig. 10:2*e*).

"2.—Trumpet nets. In this type the opening of the net is funnel-shaped, and the end is fastened in such a way that the water movement distends the net into a trumpet-shaped structure. This type of net is used extensively by the Psychomyiidae.

"3.—Hydropsychid net. Peculiar to the family Hydropsychidae is the habit of erecting a net directly in front of a tubelike retreat concealed in a crevice or camouflaged by bits of wood, leaves, or similar material. These nets may be erected between two supports in the open, as in the case of *Potamyia,* or the net may be constructed as one side of an antechamber, as in the case of many species of *Hydropsyche.*

"In all these types the caddisfly larva cleans the food and debris off the net, ingesting anything edible swept into it by the current. Normally the larva spends

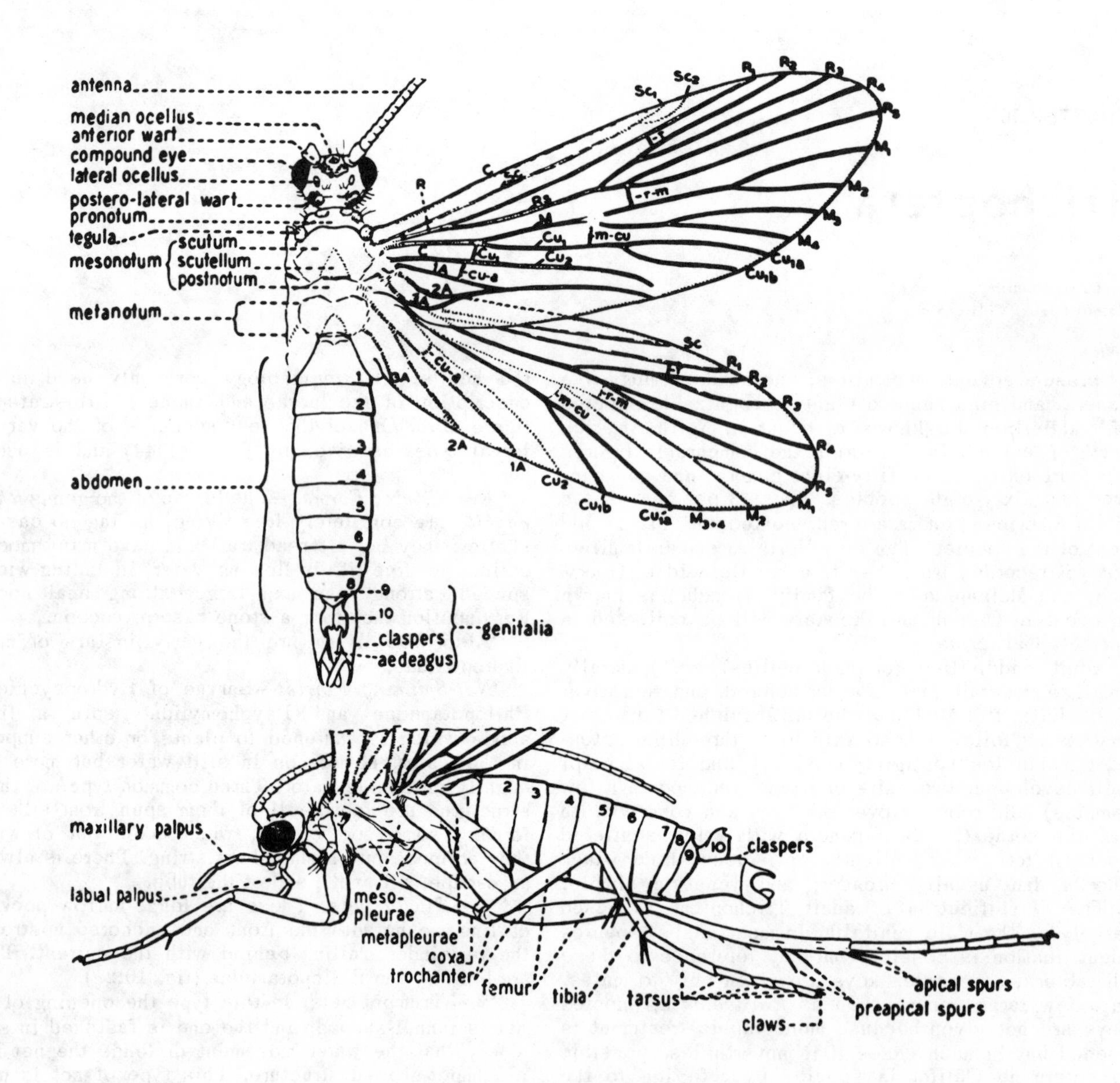

Fig. 10:1. *Rhyacophila lobifera*, illustrating terminology of structures of adult Trichoptera (Ross, 1944).

most of its time with its head near the net ready to pounce on any prey. When disturbed, it backs out of the net or retreats with great agility. The flexible body structure enables the larva to move backward rapidly, but it can move forward only slowly.

"*Tube-Making Forms.*—Some psychomyiid larvae, notably of the genus *Phylocentropus,* burrow into sand at the bottom of streams, cementing the walls of the burrow into a fairly rigid structure which may be dug out intact. The mechanics of food gathering in this group are not well understood.

"In both the net-spinning and tube-making forms, pupation takes place in the end of the tube or retreat. The larva constructs a cocoon of leaf fragments, stones or whatever other material is available, lining it with silk. The pupa is formed here.

"*Saddle-Case Makers.*—Larvae of the rhyacophilid subfamily Glossosomatinae make a portable case

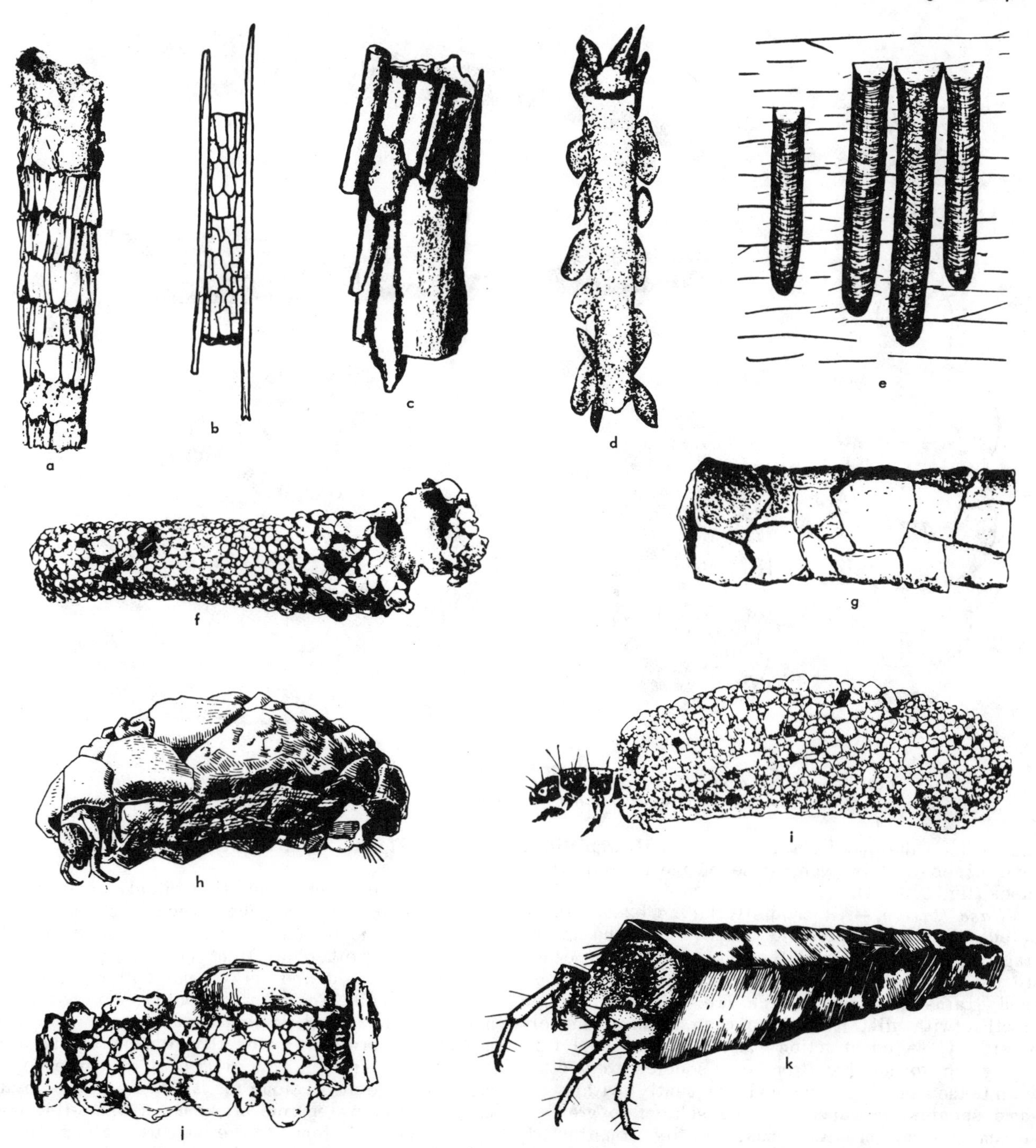

Fig. 10:2. Representative Trichoptera cases. *a, Phryganea; b, Mystacides; c, Limnephilus; d, Mystacides longicornis; e, Chimarra; f, Hesperophylax designatus; g, Limnephilus; h, Glossosoma; i, Ochrotrichia; j, Neophylax; k, Lepidostoma* (*a,f,j,* Betten, 1934; *b,c,e,g,* Pennak, 1953; *d,* Denning, 1937; *h,i,* Ross, 1944; *k,* Hickin, 1952).

which consists of an oval top made of stones and a ventral strap made of the same material. The larva proceeds with its head and legs projecting down in front of the strap and the anal hooks projecting down at the back of the strap. For pupation, the strap is cut away and the oval dome is cemented to a support, the pupa being formed in the stone cell thus made (fig. 10:2*h*).

"*Purse-Case Makers.*—Following exactly the same principle as the above are many cases of the Hydroptilidae. In general appearance they resemble a purse. The larva occupies the case with the head and legs projecting out of a slit in the front margin while the anal hooks project out of a slit in the posterior margin. For pupation, however, the case is cemented along one side to a support and the slits are cemented

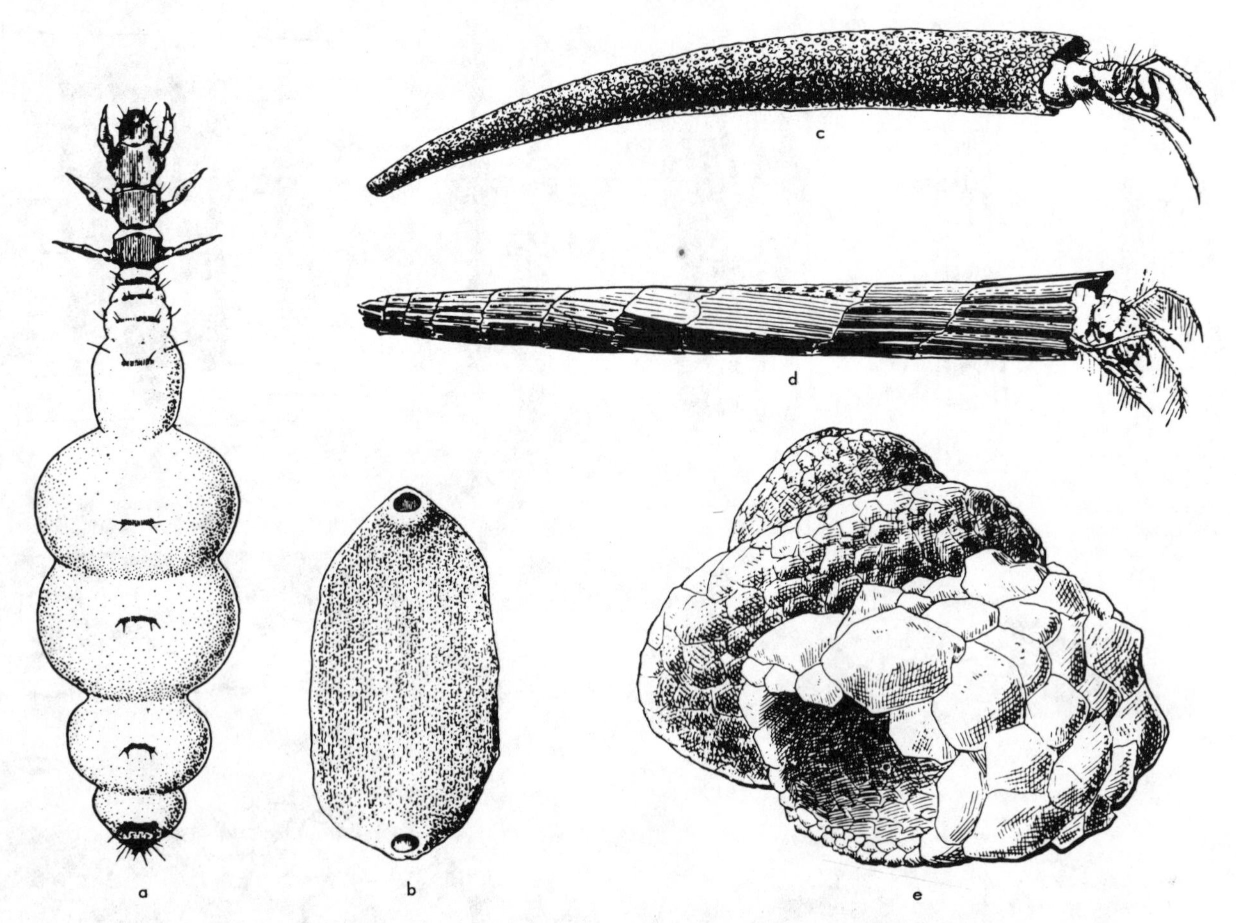

Fig. 10:3. Representative Trichoptera cases and larvae. *a,b, Leucotrichia pictipes; c, Leptocella albida; d, Triaenodes tarda; e, Helicopsyche borealis* (Ross, 1944).

shut to form the pupal chamber. Not all Hydroptilidae have cases of this type, some of them having true cases (fig. 10:3*a,b*).

"*Case Makers.*—All caddisfly larvae except those listed above make portable cases which the larvae drag with them in their daily movements. These cases are usually made of pieces of leaves, bits of twigs, sand grains or stones which are cemented or tied together with silk. Rarely the case is made entirely of silk. Case construction varies a great deal from one group to another, from one species to another within the same genus, and frequently within the same species. In general, cases subject to greatest stream current are the most solidly constructed, whereas those in small ponds where there is scarcely any water movement are the most loosely constructed.

"For pupation the case is anchored to a support and a top added to the case; the pupa is formed inside this shelter and no additional cocoon is made."

The cases constructed by the larvae are of many kinds and of such materials as sand, sticks, leaves, and shells. The case-building larvae are phytophagous whereas the free-living larvae are carnivorous. Representative larval cases are illustrated in figures 10:2; 10:3.

SUGGESTIONS FOR COLLECTING CADDISFLIES

Adult caddisflies are secretive, highly excitable, and wary. During the day they seek cool, dark, damp situations, and seldom fly actively until dusk. Their movement, whether in flight or running about on a surface, is rapid and jerky, and collecting is often rather difficult. For most productive collecting it is necessary to find their daytime resting places. These are wooden bridges, especially in mountainous areas, when the underside is dark and cool; concrete highway bridges; vegetation along the water's edge; vegetation hanging over water; and branches of shore-line coniferous trees. A lone coniferous tree along an open stretch of the bank will often yield an abundant variety of adults. Caddisflies may be also found in crevices of bark, under ledges, or in protected parts of buildings—especially if near water. The adults may be collected resting on vegetation, stones, or exposed sticks in the water. Sweeping or beating is the best method for dislodging adults from vegetation, and a cautious approach with an open killing bottle is effective for specimens resting on an open surface.

On warm, cloudy nights with little or no wind, large numbers of adults may be taken in flight. They are strongly attracted to lights, especially to blue neon lights. Yellow and red lights are usually the least attractive. Conventional type light traps will yield large numbers, especially if placed near a productive source of water. If electricity is not available, automobile lights, gasoline lanterns, or a portable mercury glow-tube light may be used to advantage.

Adult caddisflies should always be preserved in 70 to 80 per cent alcohol. The only exceptions are *Leptocella*—slender, white species, which should be kept dry.

Larvae and pupae.—Immature stages of Trichoptera are collected by turning over stones, logs, and other objects in the water. In streams many larvae wedge themselves into narrow crevices under stones; a few species live exposed on sand or gravel, and some burrow beneath the surface. In lakes or ponds, especially if marshy, larvae often are found around emergent plants. Drift, weeds, or debris often contain larvae. In general, caddis larvae are restricted to particular habitats, each species showing a preference for certain types of water and kinds of bottom. Pupation usually takes place in the larval case, or in pupal cells fastened to rocks (fig. 10:2*j*).

Larvae and pupae should be preserved in 70 to 80 per cent alcohol. Some may be removed from their cases, but larval and pupal cases or pupal cocoons should also be collected with their contents intact because pupae can be identified if the male genitalia are well developed. This makes possible the association of adults with larval cases and sometimes also with sclerites of the cast larval skins. Vials containing larvae and pupae, especially the larger forms, should be decanted and refilled with alcohol sometime after the original collection is made. This is advisable because water in the cases and body fluids dilutes the alcohol and permits decomposition to occur.

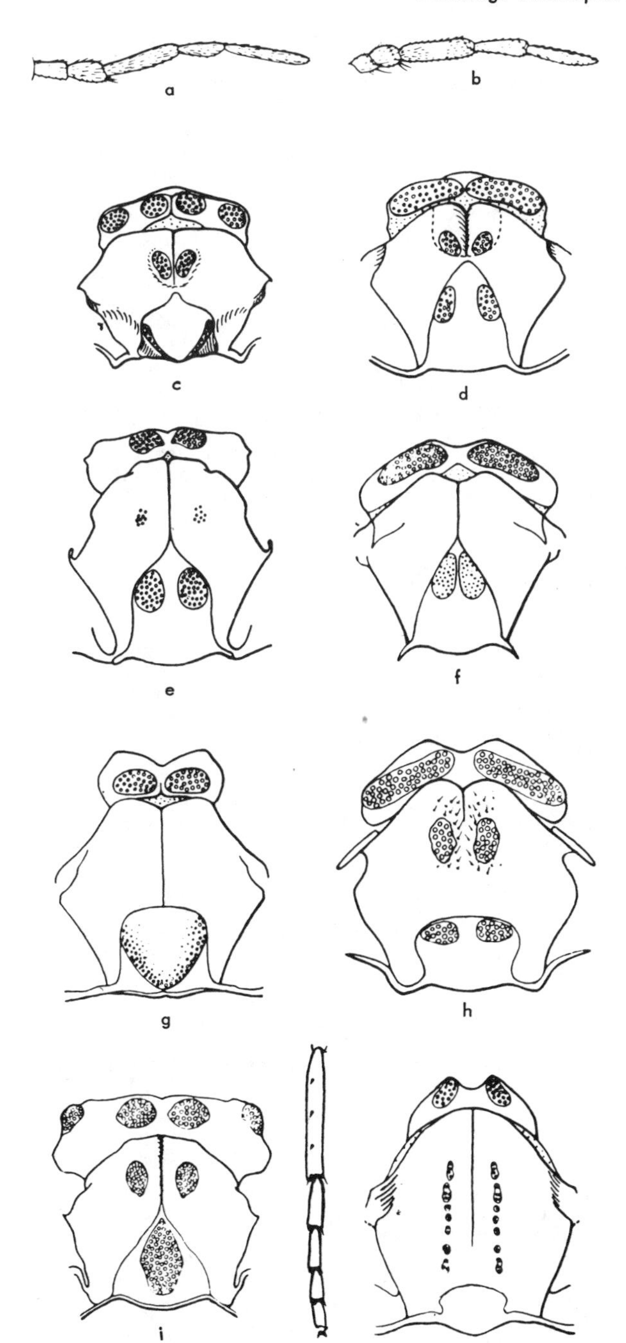

Fig. 10:4. Generic characters of Trichoptera. *a,b*, maxillary palpus; *c-i* and *k*, dorsal view of adult pronotum and mesonotum; *j*, tarsus; *a*, *Wormaldia*; *b*, *Ryacophila*; *c*, *Psychomyia*; *d*, *Sericostoma*; *e*, *Brachycentrus*; *f*, *Beraea*; *g*, *Hydropsyche*; *h*, *Helicopsyche*; *i*, *Goera*; *j*, *Beraea*; *k*, *Ganonema* (Ross, 1944).

Key to Families[1]

Adults

1. Size small, not more than 6 mm. long; wings long, narrowed, fringes very long, often those of hind wing longer than breadth of wings; antennae shorter than fore wings, front tibia with only one spur .... HYDROPTILIDAE
— Larger insects, size ranges from 5 to 40 mm.; wing fringes not longer than breadth of wings; antennae seldom shorter than wings 2
2. Ocelli present 3
— Ocelli absent................................ 8
3. Maxillary palpi 3-segmented .... LIMNEPHILIDAE
— Maxillary palpi 4- or 5-segmented 4
4. Maxillary palpi of males 4-segmented .... PHRYGANEIDAE
— Maxillary palpi 5-segmented 5
5. Maxillary palpi with 5th segment 2-3 times length of 4th segment (fig. 10:4*a*) .... PHILOPOTAMIDAE
— Maxillary palpi with 5th segment only slightly longer than 4th segment (fig. 10:4*b*) 6
6. First 2 segments of maxillary palpi short, thick, subequal (fig. 10:4*b*) .... RHYACOPHILIDAE
— Second segment of maxillary palpi much longer than first 7
7. Anterior tibiae with 1 spur, middle tibiae with 2 or 3 spurs .... LIMNEPHILIDAE
— Anterior tibiae with 2 or more spurs, middle tibiae with 4 spurs .... PHRYGANEIDAE
8. Maxillary palpi with 5 or more segments 9
— Maxillary palpi with less than 5 segments 11

[1]Adapted from Ross (1944).

9. Anterior tibiae with preapical spur .... PSYCHOMYIIDAE (in part)
— Anterior tibiae without preapical spur 10
10. Mesoscutum without warts (fig. 10:4*g*), hind wings with radius about normal and the stem distinct from subcosta .... HYDROPSYCHIDAE
— Mesoscutum with a pair of warts (fig. 10:4*c*), hind wings with the radius reduced and the stem absent or fused with the subcosta .... PSYCHOMYIIDAE
11. Middle pair of tibiae without preapical spurs 12
— Middle pair of tibiae with preapical spurs 16
12. Antennae long, slender, about twice the length of the fore wings; mesonotum with a long, irregular line of setate spots .... LEPTOCERIDAE
— Antennae stouter, equal to or slightly longer than fore wings; mesonotum either with scutal warts small (fig. 10:4*h*), or with none (fig. 10:4*f*).............. 13
13. Anterior tibiae with 1 spur; hamuli present on hind wings .... HELICOPSYCHIDAE
— Anterior tibiae with 2 spurs; hamuli absent 14
14. Middle and hind tarsi with a group of 4 black spines at apex of each segment (fig 10:4*j*) ...... BERAEIDAE
— Middle and hind tarsi with apical spines separated, not grouped together as above 15
15. Scutal warts of mesoscutum near meson (fig. 10:4*d*); front wings with a long cross vein between R_1 and R_2 .... SERICOSTOMATIDAE
— Scutal warts some distance from meson (fig. 10:4*e*); front wings without cross vein between R_1 and R_2 .... BRACHYCENTRIDAE
16. Middle femora with a row of 6-10 black spines .... MOLANNIDAE
— Middle femora with 0-2 black spines 17
17. Mesonotum with scutal warts arranged as a lineal area of small setate spots extending the full length of the scutum (fig. 10:4*k*) ...... CALAMOCERATIDAE
— Mesonotum with scutal warts oval or lanceolate and short (fig. 10:4*e*) 18
18. Mesoscutellum with single large wart extending its entire length (fig. 10:4*i*) 19
— Mesoscutellum with 2 warts (fig. 10:4*e*) 20
19. Maxillary palpi of males 5-segmented; tibial spurs not hairy .... ODONTOCERIDAE
— Maxillary palpi of males 3-segmented; tibial spurs hairy (fig. 10:4*i*) .... GOERIDAE
20. Middle tibiae with spines; preapical spurs of tibiae bare .... BRACHYCENTRIDAE
— Middle tibiae without spines; preapical spurs of tibiae hairy .... LEPIDOSTOMATIDAE

Larvae[2]

1. Larvae campodeiform. Terminal abdominal prolegs not fused to form an apparent 10th segment; tubercles absent on 1st abdominal segment; prosternal horn absent 2
— Larvae eruciform or suberuciform. Basal segments of terminal abdominal prolegs fused to form an apparent 10th segment; tubercles present on 1st abdominal segment; prosternal horn present (fig. 10:6)........ 6
2. Abdomen swollen, much wider than thorax, without gills; small larvae living in silken cases (fig. 10:3*a*, *b*) .... HYDROPTILIDAE
— Abdomen only slightly wider than thorax, cases when present not entirely made of silk but utilizing bits of debris, most larvae free living.................. 3
3. Sclerotized shield present on dorsum of 9th abdominal segment (fig. 10:5*b*) .... RHYACOPHILIDAE
— Sclerotized shield absent, dorsum of 9th abdominal segment entirely membranous 4
4. Tracheal gills branched, abundant, on venter of abdomen .... HYDROPSYCHIDAE
— Tracheal gills absent on abdomen................ 5

[2]Adapted from Ross (1944) and Betten (1934).

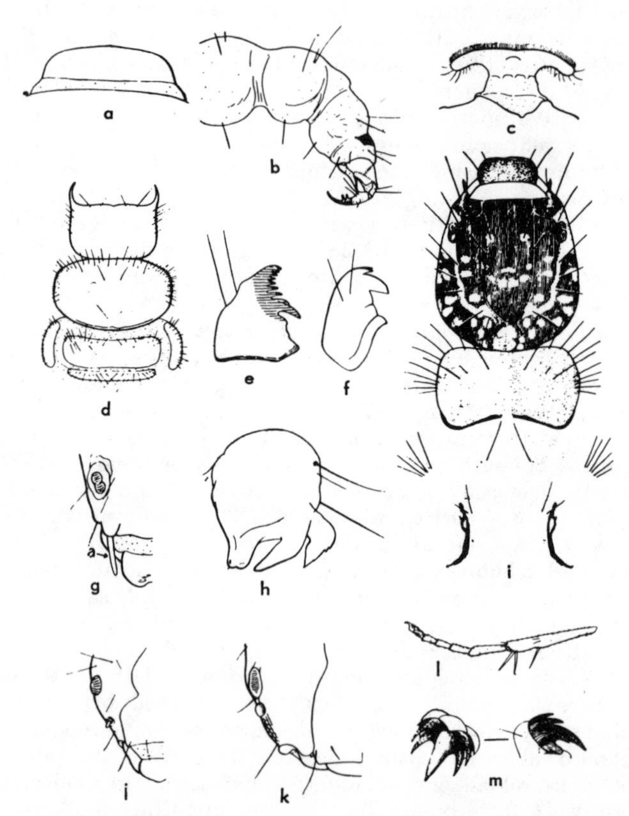

Fig. 10:5. Larval structures. *a, Polycentropus*, labrum; *b, Rhyacophila*, apex of abdomen; *c, Chimarra*, labrum; *d, Psilotreta*, thorax; *e, Helicopsyche*, anal hooks; *f, Brachycentrus*, anal hooks; *g, Leptocerus*, larval antenna; *h, Sericostoma*, anal hooks; *i, Athripsodes alagmus*, head and pro- and mesonota; *j, Limnephilus*, larval antenna; *k, Lepidostoma*, larval antenna; *l, Rhyacophila*, adult front tibia; *m, Protoptila*, anal hooks (Ross, 1944).

5. Labrum expanded distally, entirely membranous (fig. 10:5*c*)........................... PHILOPOTAMIDAE
— Labrum shorter, entirely sclerotized (fig. 10:5*a*) .... PSYCHOMYIIDAE
6. Claws of hind legs very small, those of front and middle legs large .... MOLANNIDAE
— Claws of hind legs long, subequal to those of middle legs 7
7. Antennae long, arising at base of mandibles (fig. 10:5*g*) .... LEPTOCERIDAE
— Antennae short, often very inconspicuous; arising from various locations (fig. 10:5*j*,*k*) 8
8. Mesonotum submembranous except for a pair of parenthesislike sclerotized bars (fig. 10:5*i*) *(Athripsodes)* .... LEPTOCERIDAE
— Mesonotum without such sclerotized bars or with plates of different shape than above.................. 9
9. Meso- and metanotum entirely membranous, each with lateral tuft of long setae; lateral gills covered with hair .... PHRYGANEIDAE
— Meso- and usually metanotum with at least some conspicuous sclerotized plates 10
10. Labrum with distinct transverse row of 20 or more stout setae across middle of dorsal surface .... CALAMOCERATIDAE
— Labrum lacking such a row of setae, 6-8 long setae at most 11
11. Teeth on anal hooks comblike (fig. 10:5*e*); trachael gills very small; case shaped like a snail shell (fig. 10:3*e*) .... HELICOPSYCHIDAE

— Teeth on anal hooks not comblike (fig. 10:5*f*); larval case not snaillike 12
12. Labrum much longer than broad; pronotum and mesonotum sclerotized, metanotum with wide anterior plate and a pair of oblong lateral plates (fig. 10:5*d*) .... ODONTOCERIDAE
— Labrum broader than long; pronotum sclerotized, mesonotum may or may not be sclerotized, metanotum with 1 or 2 small round or poorly defined sclerites...... 13
13. Anal hooks formed of 2 or 3 long, stout teeth (fig. 10:5*h*); mesonotum with small sclerotized plates .... SERICOSTOMATIDAE
— Anal hooks formed of a single large tooth with 1 or more small teeth on dorsal edge (fig. 10:5*f*)....... 14
14. Pronotum with deep transverse impression across the sclerite; antennae small, closer to mandibles than to eyes .... BRACHYCENTRIDAE
— Pronotum without such a deep transverse impression or with only a slight concave depression across sclerite 15
15. Tarsal claws of hind legs very long, as long as tibia .... BERAEIDAE
— Tarsal claws of hind legs much shorter than tibia.. 16
16. Mesonotum divided into 2 or 3 pairs of plates and metanotum with 3 or 4 pairs of plates; tracheal gills in groups of 3 or 4.................... GOERIDAE
— Mesonotum consisting of a single large sclerite, not divided into several pairs of plates 17
17. Antennae located close to the eye (fig. 10:5*k*); first abdominal segment lacking a dorsal spacing tubercle .... LEPIDOSTOMATIDAE
— Antennae usually located midway between the eyes and base of the mandibles, if not, then closer to mandibles than to the eyes (fig. 10:5*j*); first abdominal segment with a dorsal tubercle (fig. 10:6) .... LIMNEPHILIDAE

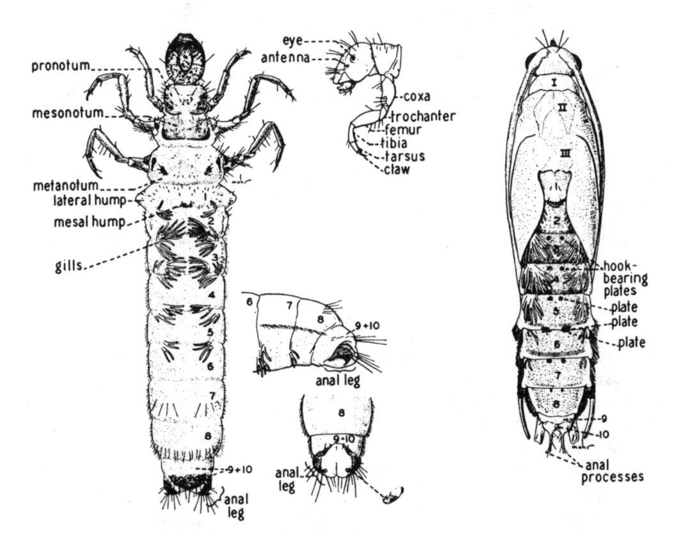

Fig. 10:6. *Limnephilus submonilifer*, larva and pupa, illustrating terminology of structures (Ross, 1944).

Family RHYACOPHILIDAE

About one hundred fifty Nearctic species representing seven genera have been described to date. Six genera and forty-three species have thus far been recorded from California. The free-living larvae of the Rhyacophilinae and the larvae of *Glossosoma* are almost always found in clear, cold, fast-flowing streams. The mountain streams of the West are particularly

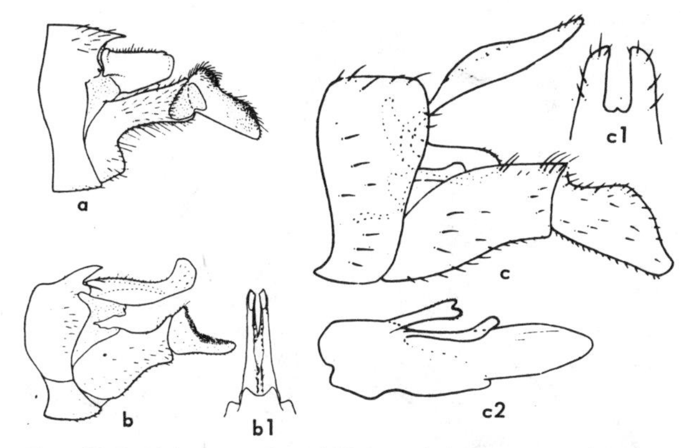

Fig. 10:7. Male genitalia of *Rhyacophila*. *a*, *acropedes*, lateral view; *b*, *grandis*, lateral; b_1, tenth tergite; *c*, *angelita*, lateral; c_1, tenth tergite; c_2, aedeagus (*a,b*, Ross, 1938; *c*, Denning, 1948*b*).

productive, and a high percentage of the known species are from that area.

Key to Genera[3]

Known Larvae

1. Anal hooks long, large, (fig. 10:5*b*); larvae free-living .... *Rhyacophila* Pictet 1834
— Anal hooks very small; larvae living in cases...... 2
2. Front legs attached at anterolateral angle of pronotum .... *Glossosoma* Curtis 1834
— Front legs attached at mid-point of lateral margin of pronotum 3
3. Anal hooks divided into many teeth (fig. 10:5*m*) .... *Protoptila* Banks 1904
— Anal hooks divided into 1 large and 1 small tooth ..... *Agapetus* Curtis 1834 and *Anagapetus* Ross 1938

Adults

1. Front tibiae lacking apical spurs .... *Protoptila* Banks 1904
— Front tibiae with 2 prominent apical spurs......... 2
2. Front tibiae with preapical spur (fig. 10:5*l*) .... *Rhyacophila* Pictet 1834
— Front tibiae without preapical spur................ 3
3. Mesal pair of warts on pronotum nearly touching .... *Paleagapetus* Ulmer 1912
— Mesal pair of warts on pronotum well separated 4
4. Hind wing with radial sector 2-branched; male with lateral plate of 5th abdominal segment highly specialized .... *Agapetus* Curtis 1834
— Hind wing with radial sector 3- or 4-branched....... 5
5. Hind wing with radial sector 3-branched, male claspers large, 2-segmented........... *Atopsyche* Banks 1905
— Hind wing with radial sector 4-branched 6
6. Front wing with callosity; apical spur of hind tibiae modified; pair of small warts on mesoscutellum..... .... *Glossosoma* Curtis 1834
— Front wing without a callosity; apical spur of hind tibiae simple; single transverse wart on mesoscutellum .... *Anagapetus* Ross 1938

Genus *Rhyacophila* Pictet 1834

Nearly ninety species of *Rhyacophila* are known from

[3]Modified from Ross (1944).

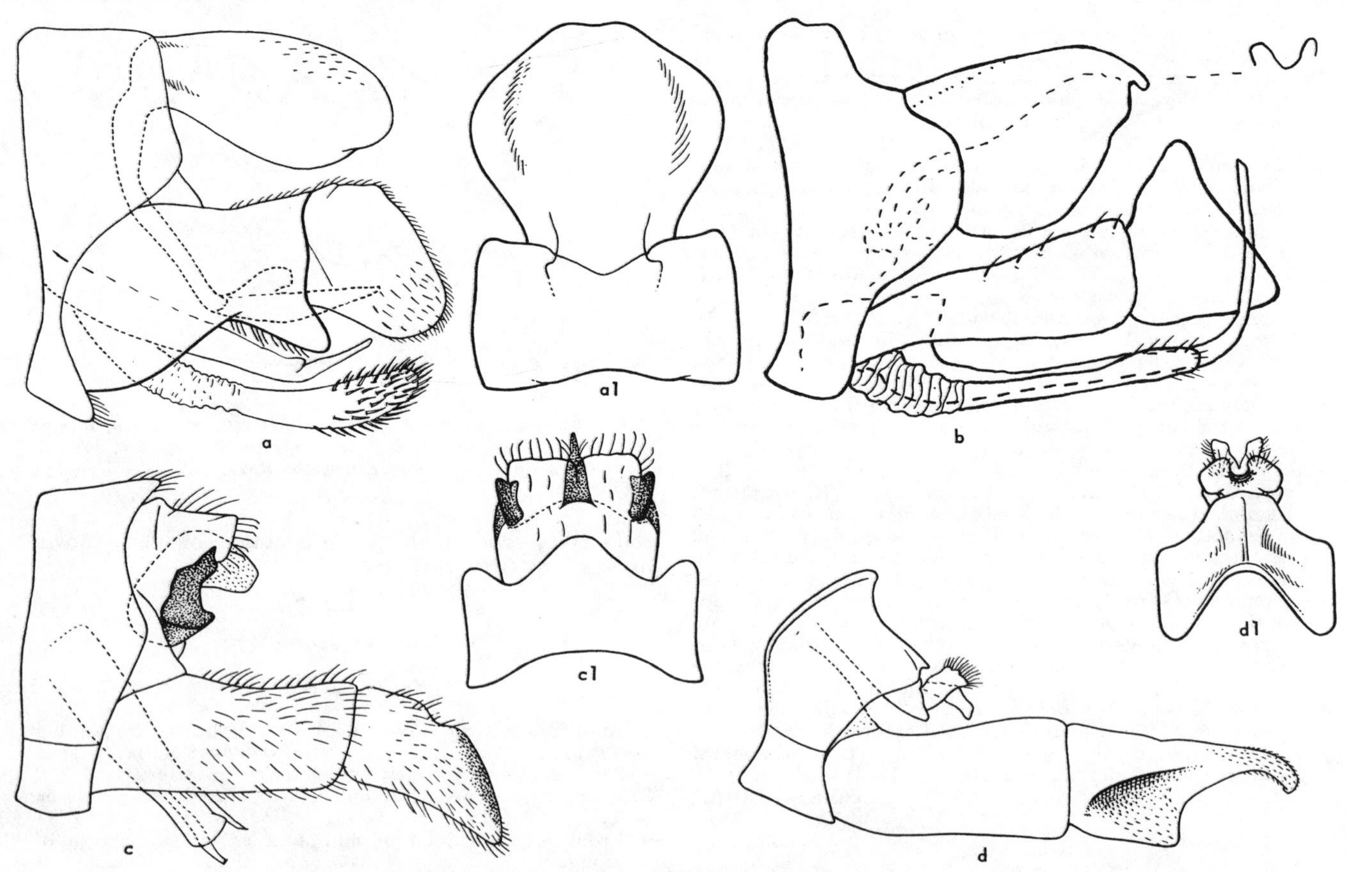

Fig. 10:8. Male genitalia of *Rhyacophila*. *a*, *rotunda*, lateral view; a_1, tenth tergite, dorsal view; *b*, *basalis*, lateral; *c*, *nevadensis*, lateral; c_1, tenth tergite, dorsal; *d*, *betteni*, lateral; d_1, tenth tergite, dorsal (Celeste Green, 1955).

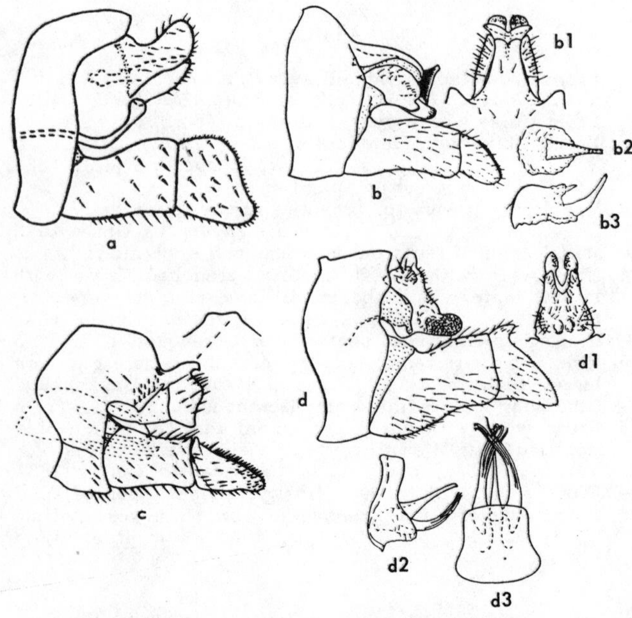

Fig. 10:9. Male genitalia of *Rhyacophila*. *a*, *bifila*, lateral view; *b*, ecosa, lateral; b_1, tenth tergite, dorsal; b_2 and b_3, aedeagus, ventral and lateral views, respectively; *c*, *karila*, lateral; *d*, *harmstoni*, lateral; d_1, tenth tergite, dorsal view; d_2 and d_3, aedeagus, lateral and ventral views, respectively (*a*,*c*, Denning, 1948*b*; *b*, Ross, 1941; *d*, Ross, 1944).

the Nearctic region. Of these, twenty-nine are known from California. The California species may be distinguished by the accompanying figures of male genitalia.

acropedes Banks 1914 (fig. 10:7*a*) — Shasta County
angelita Banks 1911 (fig. 10:7*c*) — Widespread
basalis Banks 1911 (fig. 10:8*b*) — Los Angeles County
betteni Ling 1938 (fig. 10:8*d*) — Marin to Shasta
bifila Banks 1914 (fig. 10:9*a*) — Sierra, Merced, Inyo
chandleri Denning 1955 (fig. 10:36*a*) — Siskiyou County
ecosa Ross 1941*b* (fig. 10:9*b*) — Shasta County
grandis Banks 1911 (fig. 10:7*b*) — Shasta, Nevada, Marin
harmstoni Ross 1944 (fig. 10:9*d*) — Lassen County
karila Denning 1948*b* (fig. 10:9*c*) — Humboldt County
kernada Ross 1950*a* (fig. 10:10*a*) — Tulare County
lineata Denning 1955 (fig. 10:36*b*) — Shasta County
neograndis Denning 1948*b* (fig. 10:10*g*) — Alameda, Sonoma, Fresno
nevadensis Banks 1924 (fig. 10:8*c*) — Lassen, Shasta, Butte
norcuta Ross 1938*d* (fig. 10:10*c*) — Santa Cruz County
oreta Ross 1941*b* (fig. 10:10*d*) — Shasta, San Bernardino
pellisa Ross 1938*d* (fig. 10:10*b*) — Sierra, Lassen, Mono, Inyo, San Bernardino
phryganea Ross 1941*b* (fig. 10:10*f*) — Shasta County
rayneri Ross 1951*b* (fig. 10:11*a*) — Riverside, Los Angeles
rotunda Banks 1924 (fig. 10:8*a*) — Shasta to Los Angeles

This species is similar to *norcuta* (fig. 10:10*c*), differing in that the mesal part of the 10th tergite is not prolonged into a prominent lobe.

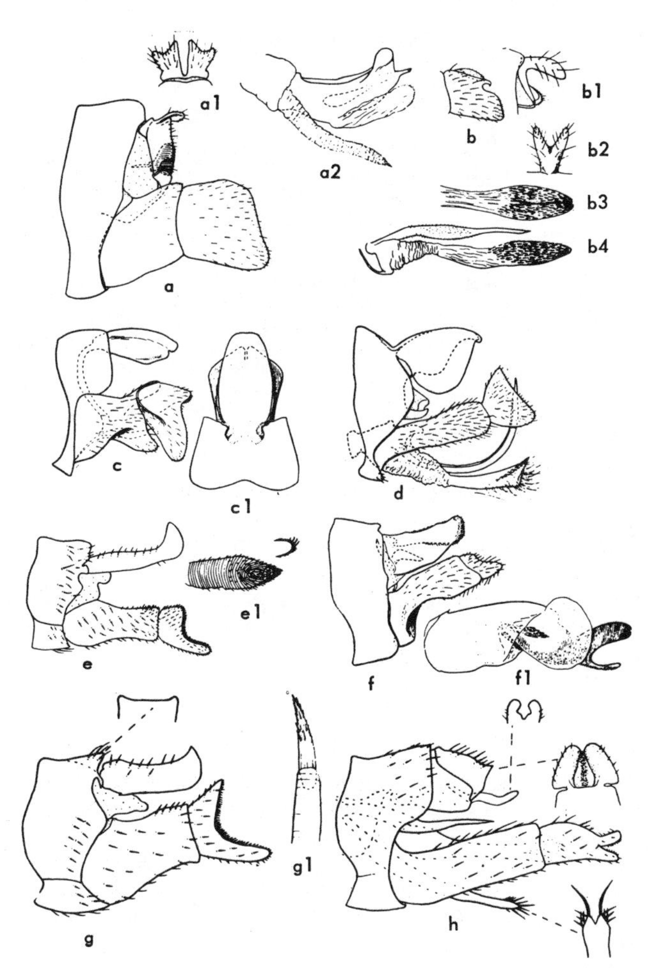

Fig. 10:10. Male genitalia of *Rhyacophila*. *a*, *kernada*, lateral view; a_1, tenth tergite; a_2, aedeagus, lateral; *b*, *pellisa*, apical segment of clasper; b_1, tenth tergite, lateral view; b_2, tenth tergite, dorsal; b_3 and b_4, aedeagus, ventral and lateral views, respectively; *c*, *norcuta*, lateral view; *c*, tenth tergite, dorsal view; *d*, *oreta*, lateral view; *e*, *sequoia*, lateral view; *e*, apex, lateral arm of aedeagus; *f*, *phryganea*, lateral view; *f*, aedeagus, lateral; *g*, *neograndis*, lateral view; *g*, apex, lateral arm of aedeagus; *h*, *vaccua*, lateral view (*a*, Ross, 1950*a*; *b*, Ross, 1938*d*; *d*,*f*, Ross, 1941*b*; *e*, Denning, 1950*c*; *g*, Denning 1948*c*; *h*, Denning, 1948*b*).

sequoia Denning 1950*b* (fig. 10:10*e*) — Tulare, Shasta, Sierra
sonoma Denning 1948*b* (fig. 10:11*b*) — Sonoma, Tulare
vaccua Milne 1936 (fig. 10:10*h*) — Shasta County
valuma Milne 1936 — Placer, Lassen
This species may be separated from *pellisa* (fig. 10:10*b*), by the clavate apical brush of the lateral arms of the aedeagus.
vedra Milne 1936 (fig. 10:11*d*) — Widespread
velora Denning 1954*a* (fig. 10:12*a*) — Shasta County
vepulsa Milne 1936 (fig. 10:11*c*) — Santa Clara, Shasta
verrula Milne 1936 (fig. 10:11*f*) — Shasta County
vocala Milne 1936 (fig. 10:11*e*) — Butte County

Genus *Glossosoma* Curtis 1834

All but two of the twenty-one described species in this genus are found only in the western United States

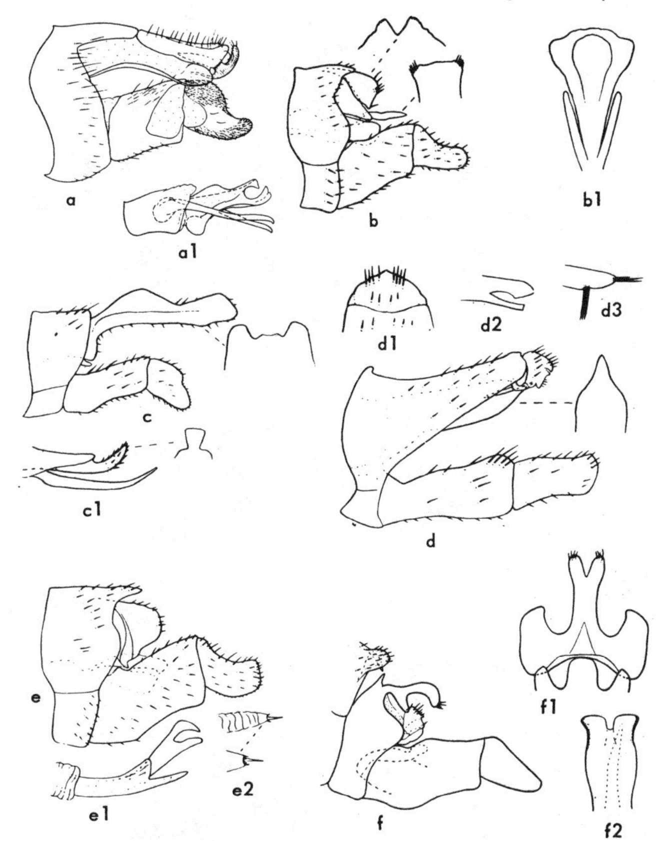

Fig. 10:11. Male genitalia of *Rhyacophila*. *a*, *rayneri*, lateral; a_1, aedeagus, lateral; *b*, *sonoma*, lateral; b_1, aedeagus, ventral sheath; *c*, *vepulsa*, lateral; c_1, aedeagus, lateral; *d*, *vedra*, lateral; d_1, tenth tergite; d_2, apex of carinate process; d_3, apex of ventral process of aedeagus; *e*, *vocala*, lateral; e_1 and e_2, lateral view of aedeagus; *f*, *verrula*, lateral; f_1, ninth and tenth tergites, dorsal view; f_2, ventral view of aedeagus sheath (*a*, Ross, 1951*b*; *b*-*f*, Denning, 1948*b*).

and Canada. Only five species are recorded from California, but doubtless several more will be found. Nearly all the species are small, seldom exceeding a quarter of an inch in length, and are generally

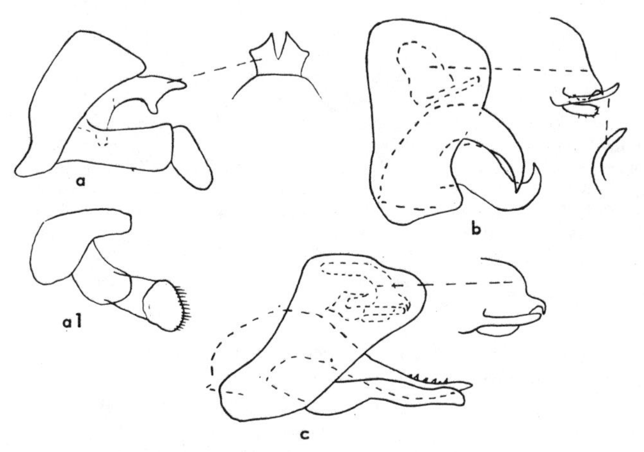

Fig. 10:12. Male genitalia. *a*, *Rhyacophila velora*, lateral; a_1, aedeagus, lateral; *b*, *Glossosoma bruna*, lateral; *c*, *G. oregonense*, lateral (Denning, 1954*a*).

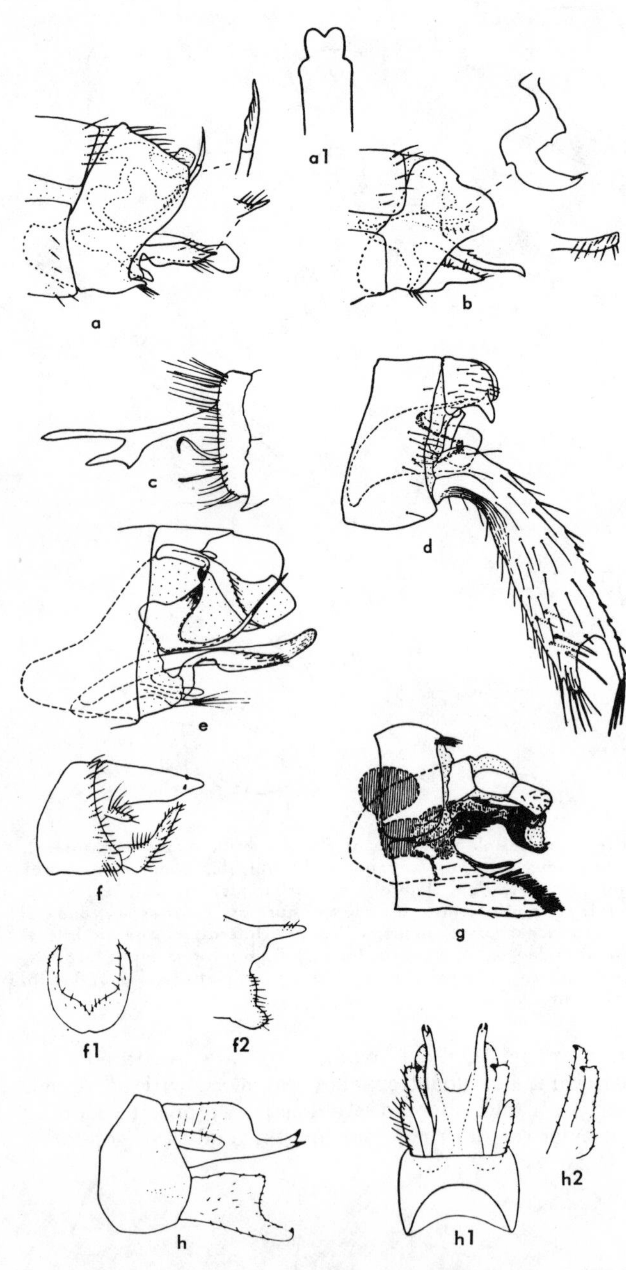

Fig. 10:13. Male and female genitalia. *a*, *Glossosoma califica*, lateral; a_1, aedeagus, ventral; *b*, *G. mereca*, lateral; *c*, *Palaeagapetus nearcticus*, lateral; *d*, *Anagapetus chandleri*, lateral; *e*, *Glossosoma pterna*; *f*, *Agapetus arcita*, lateral; f_1, claspers, ventral; f_2, female genitalia, lateral; *g*, *Protoptila coloma*, lateral; *h*, *Agapetus celatus*, lateral; h_1, tenth tergite, dorsal; h_2, claspers, dorsal (*a*,*b*, Denning, 1948*b*; *c*, Banks, 1936; *d*, Ross, 1951*a*; *e*, Ross, 1947; *f*, Denning, 1951; *g*, Ross, 1941; *h*, Kimmins and Denning, 1951).

dark brown or blackish in color. Distinguishing characters are found in the male genitalia.

The larvae construct dome-shaped cases of small stones (fig. 10:2*h*); at pupation the case is cemented to the side of a stone or other submerged object and occasionally may be found in large numbers in streams.

Key to Adult Males

1. Cerci elongate, apically a long curved rod 2
— Cerci short, curved, spinelike; or slender and fingerlike 3
2. Apex of 10th tergite acute, apex of clasper blunt (fig. 10:13*e*); Santa Cruz County.... *pterna* Ross 1947
— Apex of 10th tergite enlarged, apex of clasper attenuated (fig. 10:13*a*); Inyo, Sierra, Placer, and Mono counties .... *califica* Denning 1948
3. Cerci slender, fingerlike; apex of 10th tergite hooked (fig. 10:12*c*); Siskiyou to Mariposa counties .... *oregonense* Ling 1938
— Cerci short, curved, spinelike 4
4. Clasper prominent, apex acute and curved (fig. 10:12*b*); Shasta County .... *bruna* Denning 1954
— Clasper with apical part narrowed, directed caudad (fig. 10:13*b*); Tuolumne County.. *mereca* Denning 1948

Genus *Anagapetus* Ross 1938

Only four species have been placed in this genus and all are restricted to the West. At present only a single species, *Anagapetus chandleri* Ross 1951 (Mariposa County), is recorded from California (fig. 10:13*d*).

Genus *Palaeagapetus* Ulmer 1912

Three species are known from the Nearctic region, one species being eastern and the other two western in distribution. The members of this genus are uniformly blackish and about 4 mm. in length. The only species found in California is *Palaeagapetus nearcticus* Banks 1936 (Tuolumne County to Tulare County) (fig. 10:13*c*).

Genus *Protoptila* Banks 1904

All members of this genus are small and dark-colored; they resemble the Hydroptilidae in general appearance and action. The long fringe of hairs on the anal margin of the wing is a character also found in the family Hydroptilidae. Although about a dozen Nearctic species are known, and several of these are western in distribution, only a single species, *Protoptila coloma* Ross 1941 (Siskiyou County to Mono County), has thus far been recorded from California (fig. 10:13*g*). Doubtless several additional species will be collected.

Genus *Agapetus* Curtis 1834

Of the twenty-three Nearctic species, six have been taken in California. The adults of all the species resemble each other in small size and general appearance; distinguishing characteristics seem to be largely restricted to the male genitalia.

Key to Adult Males

1. Clasper directed dorsad, apex acute (fig. 10:13*f*); Riverside County .... *arcita* Denning 1951

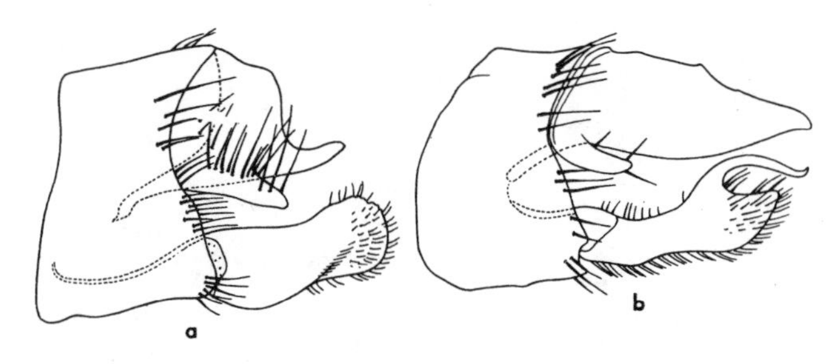

Fig. 10:14. Male genitalia of *Agapetus*. *a*, *malleatus*, lateral; *b*, *marlo*, lateral (Celeste Green, 1955).

— Clasper, at most, directed only slightly dorsad 2
2. Apex of clasper decidedly emarginate, forming a distinct projection 3
— Apex of clasper with no projection 4
3. Apical emargination of clasper forming a ventral, fingerlike projection (fig. 10:13*h*); southern California .... *celatus* McLachlan 1871
— Apical emargination of clasper forming a long, curved, acute dorsal projection (fig. 10:14*b*); Santa Clara County to Monterey County .... *marlo* Milne 1936
4. Clasper longer than 10th tergite, gradually tapering toward apex 5

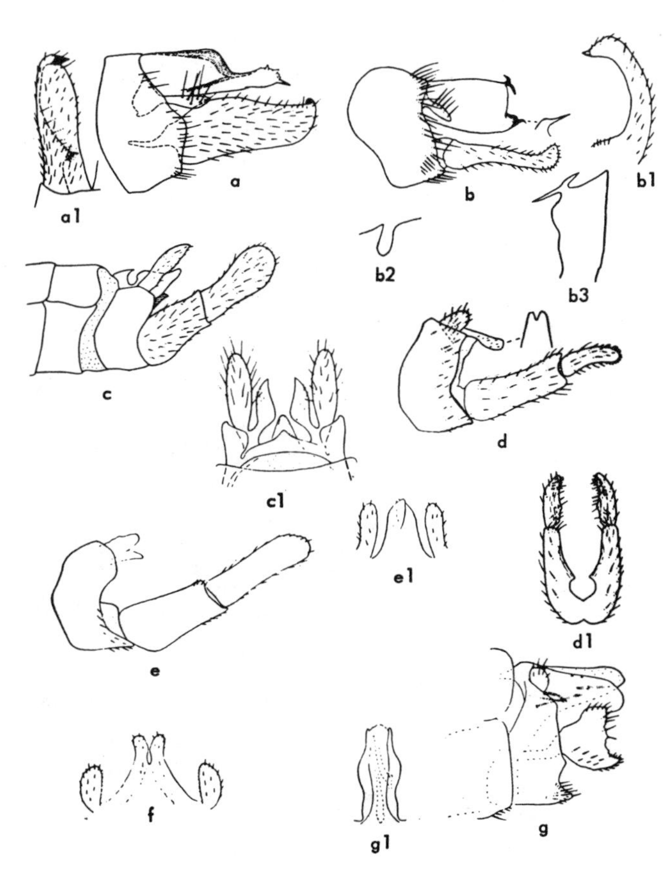

Fig. 10:15. Male genitalia. *a*, *Agapetus taho*, lateral; a_1, clasper, ventral; *b*, *A. orosus*, lateral; b_1, clasper, ventral; b_2, mesal process of sixth sternite; b_3, tenth tergite, dorsal; *c*, *Dolophilodes dorcus*, lateral; c_1, tenth tergite, dorsal; *d*, *D. novusamericanus*, lateral; d_1, claspers, ventral; *e*, *D. pallidipes*, lateral; e_1, tenth tergite, dorsal; *f*, *D. aequalis*, tenth tergite, dorsal; *g*, *Chimmara siva*, lateral; g_1, tenth tergite, dorsal (*a*, Ross, 1947; *b*, Denning, 1950*b*; *c*, Ross, 1938*a*; *d-g*, Denning, 1949c).

— Clasper only slightly longer than 10th tergite, apical half widened, apex truncate, (fig. 10:14*a*); Mariposa County to Los Angeles County .. *malleatus* Banks 1914
5. Clasper long, slender, mesally curved, dorsodistal corner of 10th tergite bearing a pair of spines (fig. 10:15*b*); Los Angeles County.... *orosus* Denning 1950
— Clasper wide throughout, apex of 10th tergite with a long ventral projection (fig. 10:15*a*); Placer and Modoc counties .... *taho* Ross 1947

Family PHILOPOTAMIDAE

The Philopotamidae is a small family of world-wide distribution. The thirty-two known Nearctic species are arranged in five genera. Most of the known larvae are found in rapid streams where they construct long, narrow silken nets (fig. 10:2*e*). Pupation occurs in cocoons of small stones and debris. The adults are rather small, blackish, and secretive. Three of the five genera are found in California and contain a total of thirteen species.

Key to Genera[4]

Known Larvae

1. Apex of frons asymmetrical ....*Chimarra* Stephens 1829
— Apex of frons only slightly or not at all asymmetrical 2
2. Apex of frons slightly asymmetrical .... *Sortosa* Navas 1918
— Apex of frons perfectly symmetrical .... *Wormaldia* McLachlan 1865

Adults

1. Wings reduced to stubs .... *Sortosa* Navas 1918 (some)
— Wings normal, reaching beyond apex of abdomen 2
2. Front tibiae with 1 apical spur or none, clasper 1-segmented .... *Chimarra* Stephens 1829
— Front tibiae with 2 apical spurs, clasper 2-segmented. 3
3. Front wing with Sc_2 deeply bowed near apex, R_1 sinuate, the two nearly touching or touching .... *Gatlinia* Ross 1948
— Front wing having Sc and R_1 nearly straight, not close to each other 4
4. R_{2+3} either not branched, or branching at or near radial cross veins .... *Wormaldia* McLachlan 1865
— R_{2+3} branched near margin of wing .... *Sortosa* Navas 1918

Genus *Dolophilodes* Ulmer 1909

There are five species described from the Nearctic region and four have been recorded from California. *Sortosa distincta* (Walker), occurs in the eastern half of the United States and is of interest in that adults occur throughout the year and most of the females are apterous.

Key to Adult Males

1. Apical segment of clasper considerably narrowed,

[4]Adapted from Ross (1944).

curved (fig. 10:15*d*); Shasta to Santa Cruz counties. .... *novusamericanus* (Ling) 1938
— Apical segment of clasper not much narrower than basal segment................................ 2
2. Tenth tergite with lateral portions widely separated on meson (fig. 10:15*c*); Del Norte County .... *dorcus* (Ross) 1938
— Tenth tergite with lateral portions confluent most of the distance 3
3. Cerci, in dorsal aspect, long, slender; apex of 10th tergite subacute, slightly overlapping, slightly exceeding cerci (fig. 10:15*e*); Tuolumne County...... .... *pallidipes* (Banks) 1936
— Cerci, in dorsal aspect, somewhat orbicular; apex of 10th tergite blunt, divergent, greatly exceeding cerci (fig. 10:15*f*); Placer to San Bernardino counties.... .... *aequalis* (Banks) 1924

Genus *Chimarra* Stephens 1829

Only two of the fifteen Nearctic species of *Chimarra* are known to occur in California. It is possible that *angustipennis* Banks and *elia* Ross eventually will be collected here.

The adults are generally small, 6 to 8 mm. in length, dark brown to black in color, and are very secretive. The quick flying adults are usually found in humid areas alongside streams.

The larvae are found in rapid and clear streams, and one species, *feria* Ross, has been found in streams which become dry in the summer and fall. Larvae of our western species are unknown.

Key to Adult Males

1. Apicodorsal margin of clasper with no mesal projection (fig. 10:15*g*); Humboldt County to Yolo County .. .. *siva* Denning 1949
— Apicodorsal margin of clasper with a distinct mesal projection (fig. 10:16*b*); Shasta County to Santa Barbara County .... *utahensis* (Ross) 1938

Genus *Wormaldia* McLachlan 1865

Seven of the thirteen Nearctic species have been taken in California. The larvae are associated with clear, rapid streams. The adults, about 7 to 12 mm. in length, are usually dark brown to black.

Key to Adult Males[5]

1. Tergum of 8th segment normal.................. 2
— Tergum of 8th segment elongated, extended dorsocaudad beyond 9th segment (fig. 10:16*d*); Trinity County.................... *hamata* Denning 1951
2. Seventh sternite with a long, narrow mesal process; 8th sternite with a short process or none (fig. 10:16*c*); Siskiyou County to Los Angeles County .... *gabriella* Banks 1930
— Seventh sternite with a short triangular process or none 3
3. Clasper elongate and slender......................... 4
— Clasper short and stout............................ 5
4. Clasper with apical segment slender and shorter than basal segment (fig. 10:16*h*); Tulare County .... *sisko* Ross 1949
— Clasper with apical segment considerably longer than basal segment (fig. 10:37); Nevada County .... *pachita* Denning 1955
5. Seventh sternite with a triangular apicomesal process; apical segment of clasper shorter than basal segment (fig. 10:16*g*); Glenn County ...... *occidea* (Ross) 1938
— Seventh sternite without a triangular apicomesal process 5
6. Apical segment of clasper parallel-sided (fig. 10:16*e*); Santa Cruz County .... *cruzensis* (Ling) 1938
— Apical segment of clasper markedly constricted (fig. 10:16*a*); Shasta County .... *anilla* (Ross) 1941

[5]Adapted from Ross (1944).

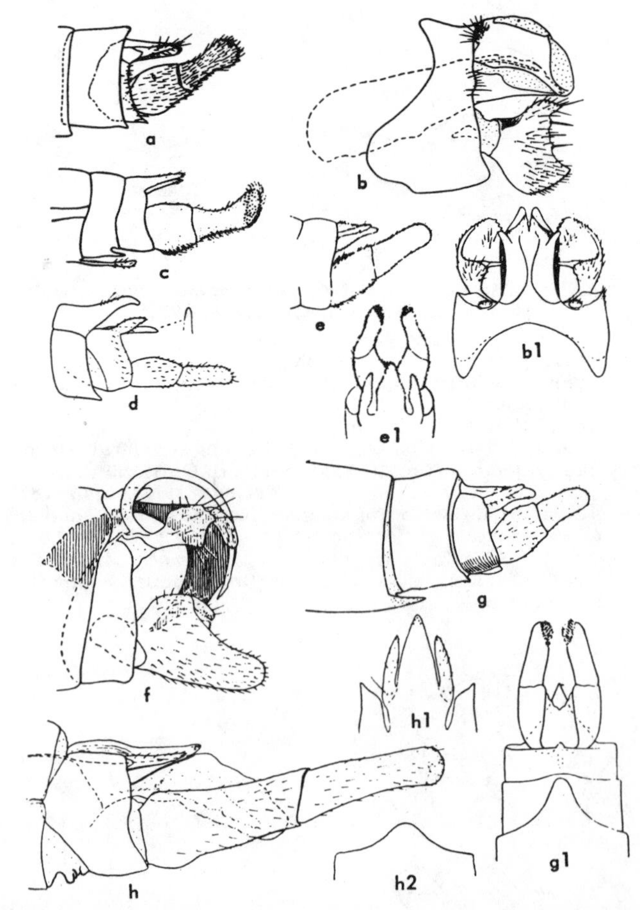

Fig. 10:16. Male genitalia. *a*, *Wormaldia anilla*, lateral; *b*, *Chimmara utahensis*, lateral; b_1, ninth and tenth segments, dorsal; *c*, *Wormaldia gabriella*, lateral; *d*, *W. hamata*, lateral; *e*. *W. cruzensis*, lateral; e_1, tenth tergite, dorsal; *f*, *Polycentropus flavus*, lateral; *g*, *Wormaldia occidea*, lateral; g_1, ventral view; *h*, *Wormaldia sisko*, lateral; h_1, tenth tergite; h_2, process of sixth sternite (*a*, Ross, 1941*b*; *b*,*g*, Ross, 1938*a*; *c*,*e*, Ling, 1938, unpublished figs.; *d*, Denning, 1951; *f*, Ross, 1944; *h*, Ross, 1949*b*).

Family PSYCHOMYIIDAE

This is a large family of world-wide distribution and contains nine Nearctic genera with sixty-two species. Some of the species are Holarctic or transcontinental in range and are locally very abundant. The larvae spin a long silken net and are found in water varying from lakes to rapid rivers. In California three genera and nine species have thus far been collected.

Key to Genera[6]

Adults

1. Front tibia with a preapical spur 2
— Front tibia without a preapical spur 6
2. Both pairs of wings with R_2 present, R_2 branches from R_3 at radial cross vein ....*Phylocentropus* Banks 1907
— Both pairs of wings with R_2 absent, or branching from R_3 near margin of wing 3
3. Hind wings with media 3-branched ....*Neureclipsis* McLachlan 1864
— Hind wings with media 2-branched 4
4. R_2 present in both wings or in either the front or the hind wings ....*Polycentropus* Curtis 1835
— R_2 lacking in both pairs wings 5
5. Maxillary palpi with 3rd segment only slightly longer than second ....*Cyrnellus* Banks 1913
— Maxillary palpi with 3rd segment 3 times as long as second ....*Nyctiophylax* Brauer 1865
6. Maxillary palpi with 2nd segment having apex developed into a protuberance ......*Cernotina* Ross 1938
— Maxillary palpi with 2nd segment uniformly and evenly cylindrical 7
7. Maxillary palpi with 3rd segment longer than 2nd; middle legs of female not dilated ....*Tinodes* Stephens 1829
— Maxillary palpi with 3rd segment not longer than 2nd; middle legs of female dilated 8
8. Apex of fore and hind wings somewhat rounded ....*Lype* McLachlan 1879
— Apex of fore and hind wings somewhat pointed ....*Psychomyia* Pictet 1834

Key to the Known Larvae[7]

1. Anal hooks with a row of 4 or 5 long teeth along inner ventral margin; 10th segment short, with scarcely any ventral margin; mentum forming a pair of distinct sclerotized plates ....*Psychomyia* Pictet 1834
— Anal hooks with at most very short, inner teeth; 10th segment longer and tubular; mentum not divided into 2 sclerotized plates 2
2. Mandibles short and triangular, each with a large, thick brush on the mesal side ....*Phylocentropus* Banks 1907
— Mandibles longer, with only a thin brush on left mandible, none on the right 3
3. Basal segment of anal appendages (10th segment) without hair; left mandible with basal tooth small and with a linear brush on mesal face near base ....*Neureclipsis* McLachlan 1864
— Basal segment of anal appendages (10th segment) with long hair; left mandible with basal tooth large, subequal to one above and with brush small........ ....*Polycentropus* Curtis 1835

Genus *Polycentropus* Curtis 1835

This is the dominant genus in the family and contains about one-half of the described species. Three species are known to occur in California. The adults are moderate in size, generally 8 to 11 mm. in length, and are usually dark gray in color. Adults are frequently collected along the shores of lakes.

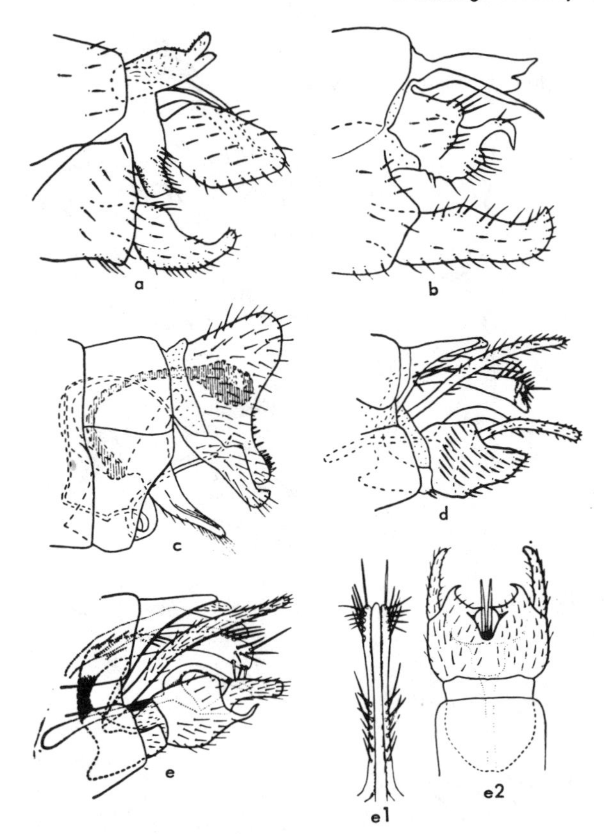

Fig. 10:17. Male genitalia. *a, Polycentropus halidus*, lateral; *b, Polycentropus variegatus*, lateral; c, *Psychomyia flavida*, lateral; *d, Tinodes belisa*, lateral; e, *Tinodes consueta*, lateral; e_1, aedeagus and sheaths, dorsal; e_2, ventral view of genitalia (*a,b*, Denning, 1948*a*; c, Ross, 1938; *d*, Denning, 1950*b*; e, Kimmins and Denning, 1951).

Key to Adult Males

1. Clasper narrow, much longer than wide 2
— Clasper with basal half wide, length only slightly more than the width (fig. 10:16*f*); Plumas County ....*flavus* (Banks) 1908
2. Clasper long, slender, about the same width throughout (fig. 10:17*b*); Shasta County to Alameda and Fresno counties ....*variegatus* Banks 1900
— Clasper short, apical portion narrowed, upturned (fig. 10:17*a*); Santa Clara County to Los Angeles County ....*halidus* Milne 1936

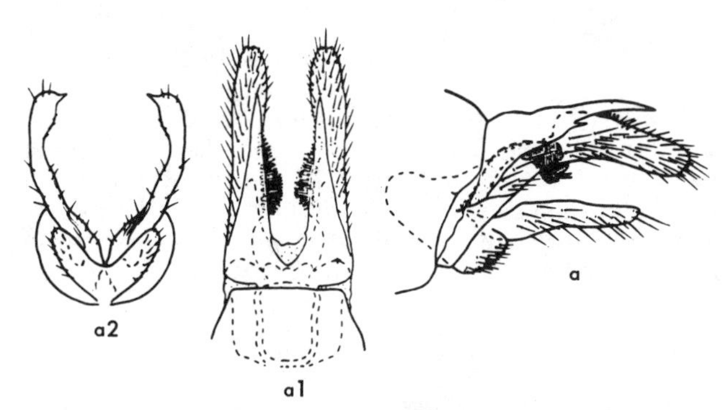

Fig. 10:18. Male genitalia of *Psychomyia lumina*. *a*, lateral view; a_1, tenth tergite, dorsal; a_2, claspers, ventral (Ross, 1938*a*).

[6]Slightly modified from Ross (1944).

[7]Slightly modified from Ross (1944).

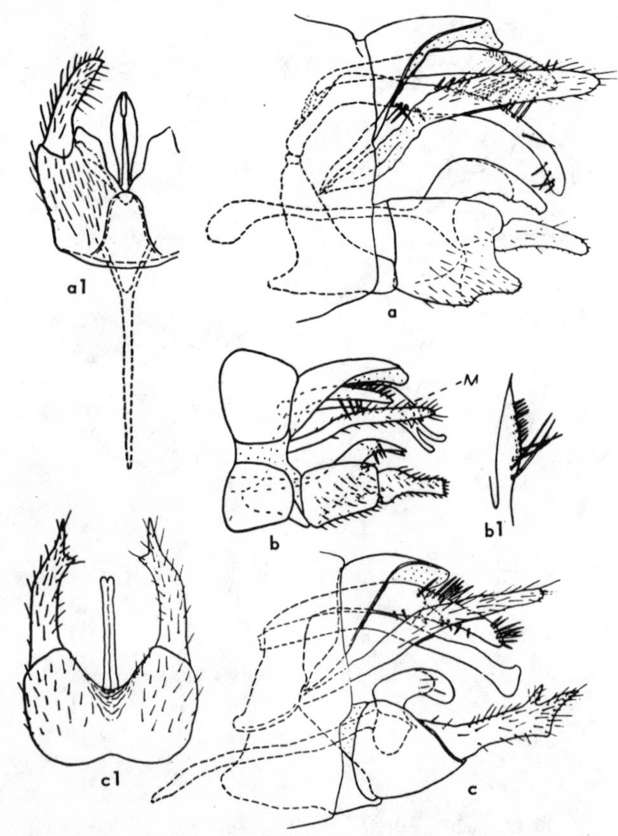

Fig. 10:19. Male genitalia of *Tinodes*. *a*, *provo*, lateral; a_1, ventral view; *b*, *parvula*, lateral; b_1, sheaths, dorsal; *c*, *sigodana*, lateral; c_1, claspers, ventral (*a,c*, Ross and Merkley, 1950; *b*, Denning 1950*b*).

Genus *Psychomyia* Pictet 1834

The genus contains only three North American species. One species, *P. nomada* (Ross), is eastern and one species, *P. lumina* (Ross) (fig. 10:18), is known only from Oregon but probably will be found in California. The only species known from California is *P. flavida* Hagen 1861 (fig. 10:17*c*) (Siskiyou and Shasta counties). Because of the small size of the adults, *Psychomyia* are frequently confused with the Hydroptilidae.

Genus *Tinodes* Stephens 1829

All six of the Nearctic species occur only in the western states. Thus far, five of the six species have been taken in California. None of the larvae has been described. Adults of this typically western genus are frequently found along warm, slowly flowing streams and are often found in abundance.

Key to the Adult Males

1. Apical segment of the clasper pointed and oblique at apex with a dorsal spur near apex (fig. 10:19*c*); Los Angeles County .... *sigodana* Ross 1950
— Apical segment of clasper rounded or truncate at apex 2
2. Apical segment of clasper rounded 3
— Apical segment of clasper truncate (fig. 10:19*b*); southern California .... *parvula* Denning 1950
3. Apical segment of clasper elongate, sinuous, as long or longer than basal segment (fig. 10:17*d*); Shasta County to Sonoma County .... *belisa* Denning 1950
— Apical segment not any longer than basal segment, somewhat sausage-shaped 4
4. Mesal process of basal segment of clasper short and wide from lateral aspect (fig. 10:19*a*); widespread .... *provo* Ross & Merkley 1950
— Mesal process of basal segment long and narrow (fig 10:17*e*); widespread .... *consueta* McLachlan 1871

Family HYDROPSYCHIDAE

This is world-wide in distribution and is one of the larger families in the order. There are eleven Nearctic genera containing about one hundred and fourteen described species. The known California fauna consists of seven genera and twenty-one species. Most of the species are found along streams, and some members of the two dominant genera, *Hydropsyche* and *Cheumatopsyche*, become very abundant.

The larvae are very active and possess a bushy gill along each abdominal segment. They prefer fast-flowing streams where they build a net under or on logs, stones, or other submerged objects. Near these nets the larvae spin ovoid cocoons for pupation.

Key to the Genera[8]

Adults

1. Head with anterior warts large; posterior warts much smaller .... *Macronemum* Burmeister 1839
— Head with anterior warts small or indistinct; posterior warts large 2
2. Front wings with R_4 and R_5 very close together at base, forming a long, narrow V .... *Smicridea* McLachlan 1871
— Front wings with R_4 and R_5 widely divergent at base, not running close together 3
3. Hind wing with Sc and R_1 bowed deeply at apex 4
— Hind wing with Sc and R_1 not bowed, or both wings with apical margin incised........................ 5
4. Hind wing with apex round, Sc and R_1 bowed deeply at apex................. *Diplectrona* Westwood 1840
— Hind wing with Sc joining R_1 at base of stigmal region, then separating and diverging again .... *Homoplectra* Ross 1938
5. Second segment of maxillary palpi distinctly shorter than 3rd.. 6
— Second segment of maxillary palpi as long as or longer than 3rd 7
6. Eighth sternite forming a short wide projection extending under the genital capsule, 5th segment of maxillary palpi longer than the other 4 .... *Parapsyche* Betten 1934
— Eighth sternite not projecting under genital capsule, 5th segment of maxillary palpi shorter than other 4 .... *Arctopsyche* McLachlan 1868
7. Front tibiae without apical spurs .... *Potamyia* Banks 1900
— Front tibiae with well-developed apical spurs 8

[8] Adapted from Ross (1944).

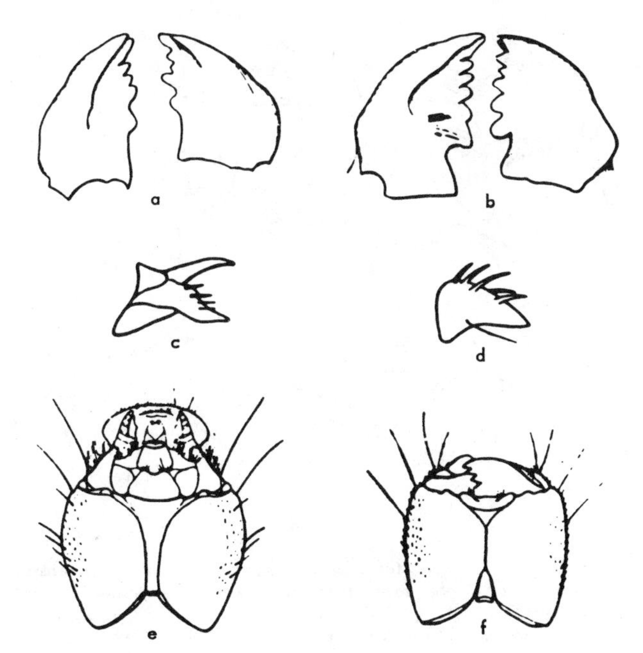

Fig. 10:20. Larval structures of Hydropsychidae. *a, Hydropsyche*, mandibles; *b, Cheumatopsyche*, mandibles; *c, Hydropsyche cheilonis*, larval stridulator of front leg; *d, Smicridea fasciatella*, larval stridulator of front leg; *e, Arctopsyche*, head of larva, ventral; *f, Diplectrona*; head of larva, ventral (Ross, 1944).

8. Front and hind wings similar in shape and size, broad and short 9
— Front and hind wings considerably different, hind wing margin arcuate 10
9. Apical margin of both wings incised ....*Oropsyche* Ross 1941
— Apical margin of both wings evenly rounded .... *Aphropsyche* Ross 1941
10. Base of aedeagus cylindrical; cell R_2 of hind wings present ....*Hydropsyche* Pictet 1834
— Base of aedeagus bulbous; cell R_2 of hind wings absent ....*Cheumatopsyche* Wallengren 1891

Key to the Known Larvae[9]

1. Head with a broad, flat dorsal area set off by an extensive arcuate carina....*Macronemum* Burmeister 1839
— Head without a dorsal area set off by a carina...... 2
2. Left mandible with a high thumblike, dorsolateral projection .... Genus A
— Left mandible without a dorsolateral projection, at most with a carina 3
3. Stridulator of front leg forked (fig. 10:20*c*) 4
— Stridulator of front leg not forked (fig. 10:20*d*) 5
4. Prosternal plate with a pair of sclerites; posterior basal tooth of mandibles single (fig. 10:20*a*) ....*Hydropsyche* Pictet 1834
— Prosternal plate without a pair of sclerites posterior to it; basal tooth of mandibles double (fig. 10:20*b*) ....*Cheumatopsyche* Wallengren 1891
5. Gula rectangular and long (fig. 10:20*e*), separating genae completely; each branched gill with all its branches arising at top of basal stalk 6
— Gula triangular and short (fig. 19:20*f*), genae therefore fused for most of their length; each branched gill with branches arising from both sides and top of basal stalk 7
6. Gula rectangular and of even width; abdomen with stout, short, black, scalelike hairs arranged in tufts along dorsum near sides, frequently with broad scales scattered between them...... *Parapsyche* Betten 1934
— Gula narrowed posteriorly; abdomen without distinct setal tufts, with coarse hairs of varying lengths, some of them scalelike but narrow and long (fig. 10:20*e*) ....*Arctopsyche* McLachlan 1868
7. Mandibles with winglike dorsolateral flanges along basal half; submentum cleft.....*Potamyia* Banks 1900
— Mandibles without distinct dorsolateral flanges; submentum subconical, not cleft..................... 8
8. Frons expanded laterad, its lateral extensions sharp .... *Diplectrona* Westwood 1840
— Frons not expanded laterad, its lateral extensions scarcely produced.........*Smicridea* McLachlan 1871

[9]Slightly modified from Ross (1944).

Genus *Arctopsyche* McLachlan 1868

This is a small genus containing five Nearctic species. Three species are western and two are found in eastern Canada and United States. The species are large and generally dark grayish.

Key to Adult Males

1. Tenth tergite with a pair of long slender prongs 2
— Tenth tergite with a pair of short dorsal and ventral prongs (fig. 10:21*a*); Shasta County.............. ....*californica* Ling 1938
2. Prongs of 10th tergite convergent, sharply acuminate in lateral aspect (fig. 10:21*d*); Santa Catalina Mts. ....*inermis* Banks 1943
— Prongs of 10th tergite not especially convergent, generally wide throughout except extreme tip (fig. 10:21*b*); Trinity County to El Dorado County ....*grandis* (Banks) 1900

Genus *Diplectrona* Westwood 1840

This is a small genus represented by only three species, two eastern and one, *D. californica* Banks 1914, from California. The latter was originally described from a specimen collected at Claremont, Los Angeles County, and at present cannot be identified.

Genus *Homoplectra* Ross 1938

The genus is represented by five western species. All species except *H. alseae* Ross (from Oregon) have been found in California.

Key to Adult Males

1. Armature of the aedeagus with a short, dorsal spine arising at the base of the structure, the spine not much more than half as long as rest of structure (fig. 10:21*e*) 2
— Armature of aedeagus with a long, slender, dorsal spine, which is nearly as long as rest of structure (fig. 10:21*h*) 4
2. Dorsal spine at base of aedeagus curved dorsad, reaching considerably above rest of structure (fig. 10:21*c*); Marin County .... *nigripennis* (Banks) 1911

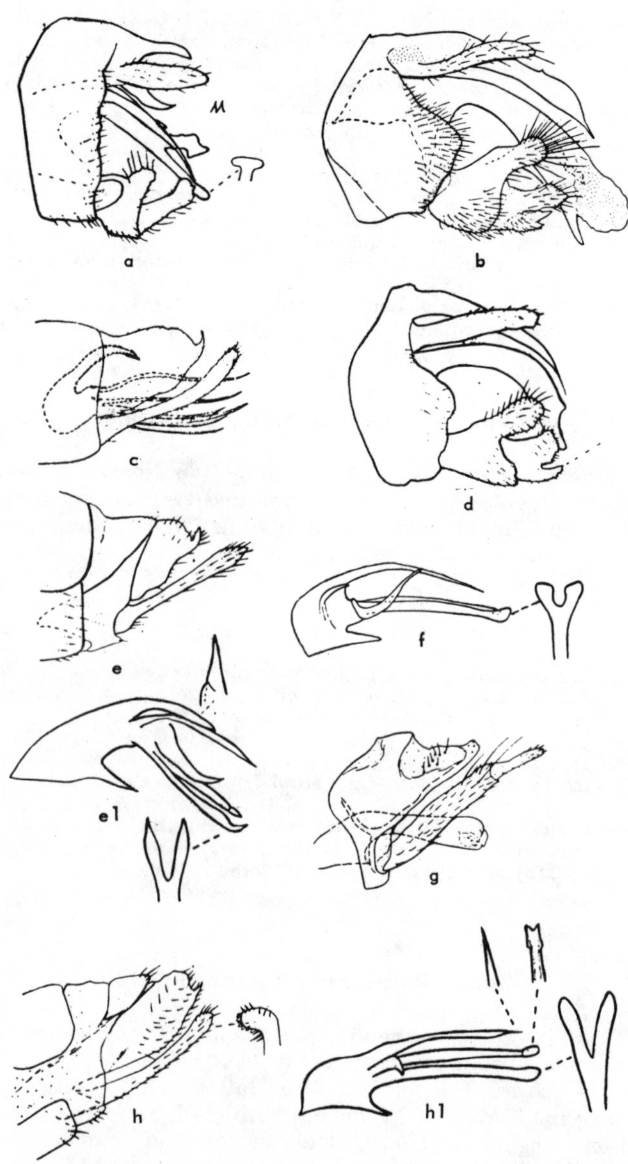

Fig. 10:21. Male genitalia. *a, Arctopsyche californica,* lateral; *b, Arctopsyche grandis,* lateral; *c, Homoplectra nigripennis,* lateral; *d, Arctopsyche inermis,* lateral; *e, Homoplectra shasta,* lateral; e_1, aedeagus, lateral; *f, Homoplectra oaklandensis,* aedeagus, lateral; *g, Smicridea utico,* lateral; *h, Homoplectra spora,* lateral; h_1, aedeagus, lateral (*a,d,* Denning, 1950*a*; *b,* Ross, 1938*c*; *c,* Ling, 1938, unpublished fig.; *e,* Denning, 1949*c*; *f,h,* Denning, 1952; *g,* Ross, 1947).

— Dorsal spine at base of aedeagus in a nearly horizontal position, not much above remainder of structure 3
3. Dorsal spine subdivided at apex..... *alseae* Ross 1938
— Dorsal spine furcate near the base (fig. 10:21*e*); Shasta County .... *shasta* Denning 1949
4. Lateral spur of armature of aedeagus directed ventrad (fig. 10:21*h*); Santa Cruz and San Mateo counties ... . *spora* Denning 1952
— Lateral spur of armature of aedeagus directed caudodorsad (fig. 10:21*f*); Alameda County to Santa Cruz County .... *oaklandensis* (Ling) 1938

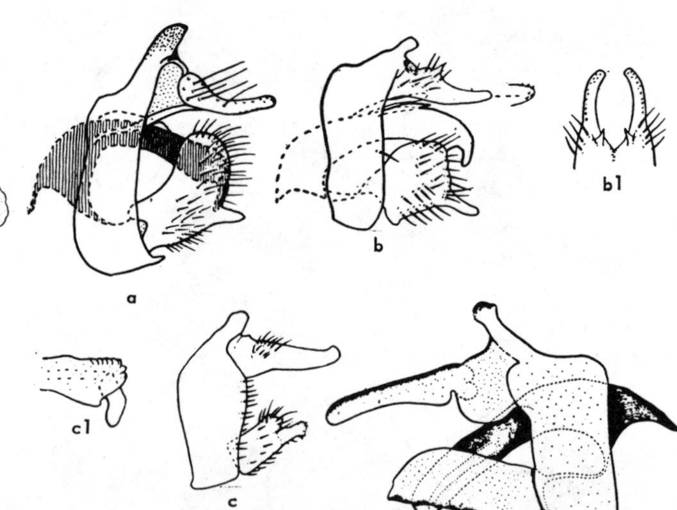

Fig. 10:22. Male genitalia of *Parapsyche. a, almota,* lateral; *b, spinata,* lateral; b_1, tenth tergite, dorsal; *c, extensa,* lateral; c_1, aedeagus, lateral; *d, elsis,* lateral (*a,* Ross, 1938; *b,* Denning, 1949; *c,* Denning, 1949*a*; *d,* Milne and Milne, 1938).

Genus *Parapsyche* Betten 1934

This is a small genus represented by six Nearctic species. Two species are restricted to the East and four are found in the West.

Key to the Adult Males

1. Clasper long, slender, somewhat sinuate, extending caudad almost as far as apex of 10th tergite (fig. 10:22*d*); Shasta County to Fresno County .... *elsis* Milne 1936
— Clasper short, either quadrate or quadrangular...... 2
2. Clasper with the main body quadrate, bearing a short distoventral digitate process (fig. 10:22*b*); Tulare County .... *spinata* Denning 1949
— Clasper with the main body short, about the same width throughout 3
3. Clasper with short acute thumblike process, dorsodistal corner rounded (fig. 10:22*a*); Shasta County to San Bernardino County .... *almota* Ross 1938
— Clasper with truncate thumblike process almost half length of structure, dorsodistal corner truncate (fig. 10:22*c*); Shasta County......... *extensa* Denning 1949

Genus *Smicridea* McLachlan 1871

Only one of the three species in this genus is known in California where occasionally it occurs in abundance. The single California species is *Smicridea utico* Ross 1947 (fig. 10:21*g*) (Imperial County).

Genus *Cheumatopsyche* Wallengren 1891

This is the second largest genus in the family and contains about thirty Nearctic species. At present

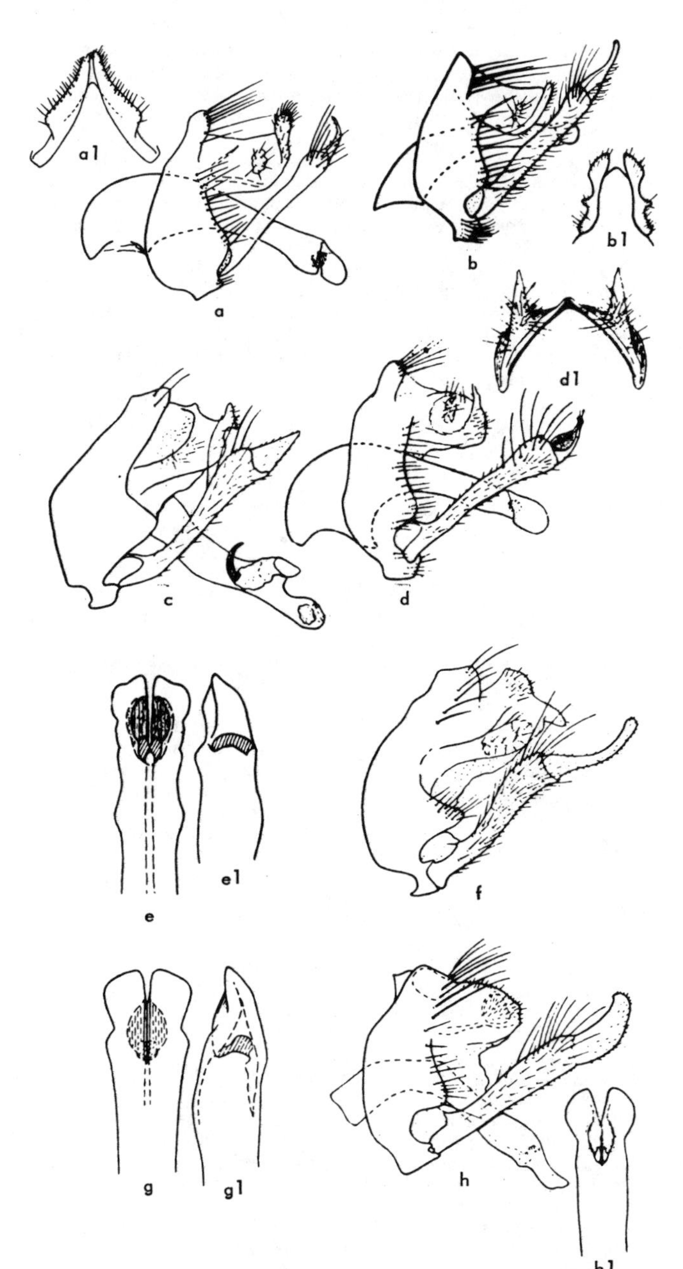

Fig. 10:23. Male genitalia. *a*, *Cheumatopsyche campyla*, lateral; a_1, tenth tergite; *b*, *Cheumatopsyche mickeli*, lateral; b_1, tenth tergite; *c*, *Hydropsyche cockerelli*, lateral; *d*, *Cheumatopsyche mollala*, lateral; d_1, tenth tergite; *e*, *Hydropsyche californica*, aedeagus, ventral; e_1, aedeagus, lateral; *f*, *Hydropsyche oslari*, lateral; *g*, *Hydropsyche occidentalis*, aedeagus, ventral; g_1, aedeagus, lateral; *h*, *Hydropsyche philo*, lateral (*a*, Ross, 1944, *b*, Denning, 1942; *c,e,f,g*, Ross, 1938; *d,h*, Ross, 1941).

we know of only three species in California. Generally the larvae are found under the same ecological conditions as *Hydropsyche*, and similarly built nets are constructed which strain food from the flowing water.

Key to the Adult Males

1. Apical segment of clasper long, slender; apical lobes of 10th tergite long, narrow (fig. 10:23*a*) 2
— Apical segment of clasper short, stout; apical lobes of 10th tergite short, acutely triangular (fig. 10:23*d*); Del Norte County .... *mollala* Ross 1941
2. Lateral margin of apical lobes of 10th tergite with no lateral expansion (fig. 10:23*a*); Siskiyou County .. .. *campyla* Ross 1938
— Lateral margin of apical lobes of 10th tergite with distinct lateral expansions (fig. 10:23*b*); Sierra to Los Angeles counties .... *mickeli* Denning 1942

Genus *Hydropsyche* Pictet 1834

Hydropsyche, with about fifty-five Nearctic species, is the largest genus in the family. Thus far only five species have been recorded from California. Locally some species are very abundant. Larvae are free-living and spin webs on rocks in swift-flowing streams.

Key to Adult Males

1. Aedeagus with membranous lateral processes bearing a long curved spine (fig. 10:23*c*); Siskiyou and Sierra counties .... *cockerelli* Banks 1905
— Aedeagus without membranous lateral processes 2
2. Apical segment of clasper rather long, curved 3
— Apical segment of clasper short, approximately quadrate 4
3. Apical segment of clasper attenuated, long, slender and curved (fig. 10:23*f*); Trinity County to Riverside County .... *oslari* Banks 1905
— Apical segment of clasper flattened, saber-shaped (fig. 10:23*h*); Monterey County to San Bernardino County .... *philo* Ross 1941
4. Aedeagus, ventral aspect, with lateral margin irregular, a sharp angulation along margin a short distance before apex (fig. 10:23*e*); Shasta County to Santa Clara and Fresno counties .... *californica* Banks 1899
— Aedeagus, ventral aspect, with lateral margin not irregular, no angulation before apex, a triangular notch along margin near apical margin (fig. 10:23*g*); widespread .... *occidentalis* Banks 1900

Family HYDROPTILIDAE

This is the family of microcaddisflies, which generally average less than 6 mm. in length. The small adults are very hairy and often have a dark mottled appearance. They are found along ponds, large lakes, or streams and often occur in abundance.

Twelve genera and about one hundred and thirty species occur in the United States and Canada, but many others probably will be taken because this family is usually poorly represented in collections. In California only five genera and ten species are recorded.

Key to the Genera[10]

Adults

1. Ocelli absent 2
— Ocelli present 4
2. Front tibiae with apical spur .... *Dibusa* Ross 1939

[10] Adapted from Ross (1944).

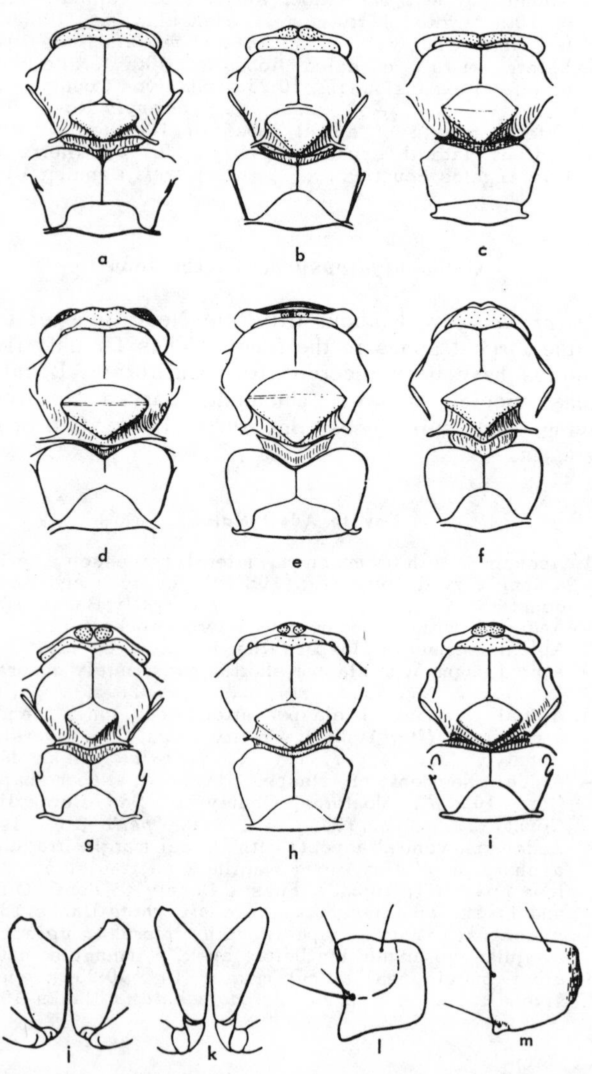

Fig. 10:24. Adult and larval structures of Hydroptilidae. *a, Orthotrichia americana,* thorax; *b, Hydroptila hamata,* thorax; *c, Stactobiella brustia,* thorax; *d, Metrichia nigritta,* thorax; *e, Leucotrichia pictipes,* thorax; *f, Ochrotrichia tarsalis,* thorax; *g, Agraylea multipunctata; h, Oxyethira pallida,* thorax; *i, Ithytrichia clavata,* thorax; *j, Leucotrichia pictipes,* front coxae; *k, Metrichia nigritta,* front coxae; *l, Ochrotrichia tarsalis,* larval metanotum; *m, Hydroptila waubesiana,* larval metanotum (Ross, 1944).

— Front tibiae without apical spur 3
3. Metascutellum almost rectangular (fig. 10:24*a*) .... *Orthotrichia* Eaton 1873
— Metascutellum pentagonal to triangular (fig. 10:24*b*) .... *Hydroptila* Dalman 1819
4. Metascutellum wide, short, and rectangular (fig. 10:24*c*) .... *Stactobiella* Martynov 1924
— Metascutellum either triangular (fig. 10:24*d*) or narrower than scutum (fig. 10:24*e*)....................... 5
5. Front tibiae with an apical spur 6
— Front tibiae without an apical spur 7
6. Front coxae wide (fig. 10:24*j*) .... *Leucotrichia* Mosely 1934
— Front coxae narrow (fig. 10:24*k*)...*Metrichia* Ross 1939
7. Hind tibiae with 1 preapical spur .... *Neotrichia* Morton 1905
— Hind tibiae with 2 preapical spurs 8
8. Middle tibiae without a preapical spur .... *Mayatrichia* Mosely 1937
— Middle tibiae with a preapical spur 9
9. Mesoscutellum with a linelike fracture dividing the sclerite almost equally (fig. 10:24*f*)............... .... *Ochrotrichia* Mosely 1934
— Mesoscutellum without such a transverse groove 10
10. Mesoscutellum narrow and diamond-shaped, posterodorsal edge widely separated from posterior margin (fig. 10:24*g*) .... *Agraylea* Curtis 1834
— Mesoscutellum wider, posterodorsal edge close to or touching posterior margin 11
11. Posterodorsal edge of mesoscutellum touching posterior margin on meson (fig. 10:24*h*)... *Oxyethira* Eaton 1873
— Posterodorsal edge of mesoscutellum separated from posterior margin by a narrow strip (fig. 10:24*i*) .... *Ithytrichia* Eaton 1873

Known Larvae[11]

1. Abdomen enlarged, at least some part of it much thicker than thorax (fig. 10:3*a*); living in case (later instars) 2
— Abdomen slender, not appreciably thicker than thorax; free living, not with case (early instars) ..... not keyed
2. Each segment of abdomen with a dark, sclerotized dorsal area 3
— Abdomen with at least segments 2-7 without dark, sclerotized dorsal area, at most with a small, delicate ring.. 4
3. Abdomen with dorsal sclerites solid; segments 1 and 2 small, 3-6 greatly expanded. Case translucent, ovoid (fig. 10:3*b*).............. *Leucotrichia* Mosely 1934
— Abdomen with dorsal sclerites membranous across middle; segments 1-6 evenly expanded .... *Ochrotrichia* Mosely 1934
4. Abdominal segments with lateral projections .... *Ithytrichia* Eaton 1873
— Abdominal segments without lateral projections..... 5
5. Middle and hind legs almost 3 times as long as the front legs .... *Oxyethira* Eaton 1873
— Middle and hind legs not more than 1½ times as long as front legs 6
6. Middle and hind legs with very long, slender tarsal claws which are much longer than tarsi; case purselike .... *Agraylea* Curtis 1834
— Middle and hind legs with tarsal claws shorter or stouter; cases of various types 7
7. Anal legs distinctly projecting from body mass; 8th abdominal tergite with a brush of setae 8
— Anal legs apparently combined with body mass and only the claws projecting; 8th abdominal tergite with only 1 or 2 pairs of weak setae 9
8. Thoracic tergites clothed with long, slender, erect, inconspicuous setae; case of sand grains, evenly tapered, and without posterior slit .... *Neotrichia* Morton 1905
— Thoracic tergites clothed with shorter, stout, black setae which are conspicuous and appressed to the surface of the body; case translucent, evenly tapered, and with dorsal side either ringed or fluted with raised ridges .... *Mayatrichia* Mosely 1937
9. Tarsal claws with long, stout inner tooth; case purselike, robust............ *Stactobiella* Martynov 1924
— Tarsal claws without prominent inner tooth; case either purselike or cylindrical.................... 10
10. Middle and hind legs with tibiae cylindrical and long; case long, smooth and round in cross section, tapered at each end and with an indented slit at both ends .... *Orthotrichia* Eaton 1873
— Middle and hind legs with tibiae stout and widened at apex; case of purse type 11

[11]Adapted from Ross (1944).

11. Metanotum with a distinct, widened ventrolateral area (fig. 10:24*l*), case as in fig. 10:2*i* .... *Ochrotrichia* Mosely 1934
— Metanotum without a widened, lateral area (fig. 10:24*m*) .... *Hydroptila* Dalman 1819

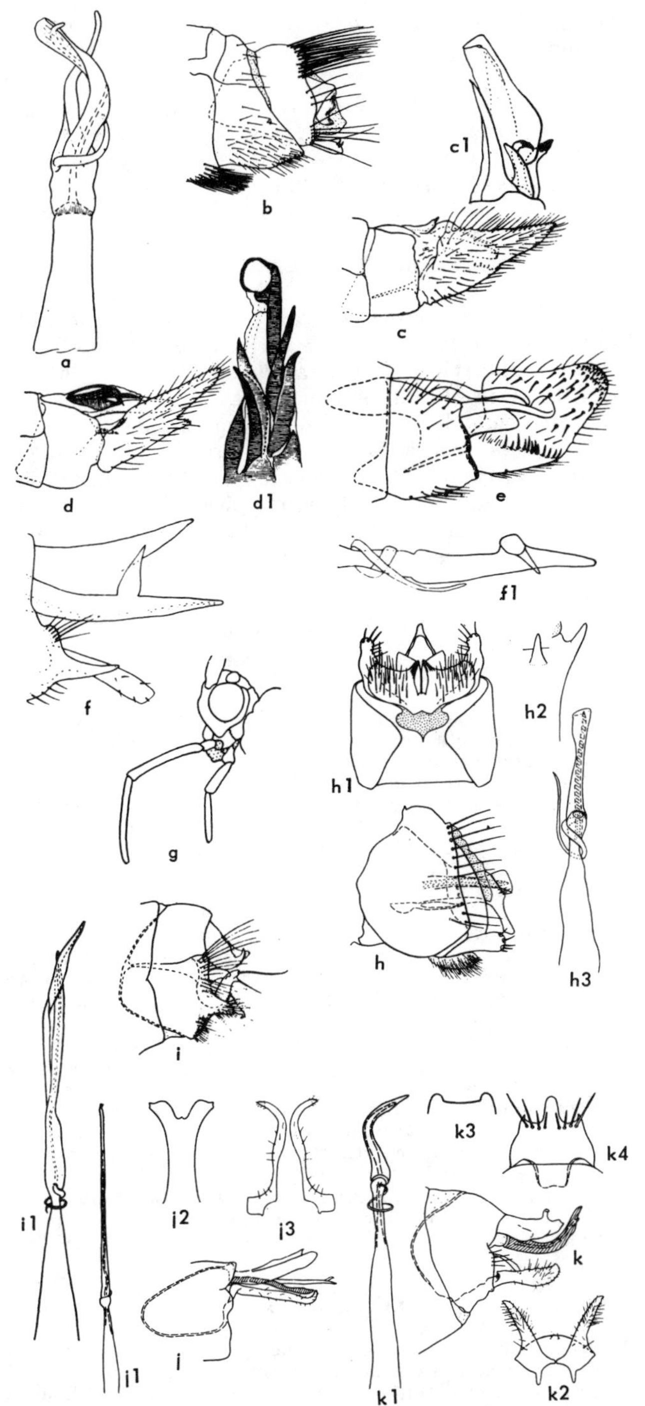

Fig. 10:25. Distinguishing structures of Trichoptera. *a*, *Oxyethira pallida*, aedeagus; *b*, *Leucotrichia pictipes*, male genitalia, lateral; *c*, *Ochrotrichia mono*, male genitalia, lateral; c_1, tenth tergite; *d*, *Ochrotrichia lometa*, male genitalia, lateral; d_1, tenth tergite; *e*, *Ochrotrichia trapoiza*, male genitalia, lateral; *f*, *Hydroptila acoma*, male genitalia, lateral; f_1, aedeagus; *g*, *Chyranda centralis*, head of male; *h*, *Agraylea saltesea*, male genitalia, lateral; h_1, ventral view; h_2, style of seventh sternite; h_3, aedeagus; *i*, *Hydroptila rono*, male genitalia, lateral; i_1, aedeagus; *j*, *Hydroptila xera*, male genitalia, lateral; j_1, aedeagus; j_2, tenth tergite; j_3, claspers, ventral; *k*, *Hydroptila argosa*, male genitalia, lateral; k_1, aedeagus; k_2, claspers, ventral; k_3, eighth tergite of female; k_4, eighth sternite of female (*a,b,g*, Ross, 1944; *c,d,i*, Ross, 1941; *e*, Ross, 1947; *f*, Denning, 1947*a*; *h,j,k*, Ross, 1938).

Genus *Oxyethira* Eaton 1873

Although about twenty-four Nearctic species occur in this genus, only one, *O. pallida* (Banks) 1904 (fig. 10:25*a*) (Riverside County), is known to occur in California. It is certain that additional species will be collected in the state.

Genus *Leucotrichia* Mosely 1934

Four species have been assigned to this genus. One species, *L. pictipes* (Banks) 1911 (fig. 10:25*b*), is transcontinental in distribution and is recorded from California (Placer and Tulare counties). The other three species occur in the Rocky Mountain region and Texas.

Genus *Ochrotrichia* Mosely 1934

Of the twenty-two species in this genus, only three have been collected in California. Generally the species are found in clear, cold, rapidly flowing streams. They may be identified by reference to figures of the male genitalia.

lometa (Ross) 1941 (fig. 10:25*d*)	Mono County
mono (Ross) 1941 (fig. 10:25*c*)	Mono County
trapoiza Ross 1947 (fig. 10:25*e*)	Mono County

Genus *Agraylea* Curtis 1834

Three Nearctic species belong to the genus. In the Midwest and certain of the Mountain States members of this genus are often abundant along lakes or other permanent bodies of water. The only species known from California is *A. saltesea* Ross 1938 (fig. 10:25*h*) (Tuolumne and Madera counties).

Denning (1937) has described the egg-laying habits of a closely related species, *A. multipunctata*. One female is recorded as remaining submerged for one hour and eight minutes.

Genus *Hydroptila* Dalman 1819

This is the largest genus in the family, containing about fifty species. Four species have thus far been collected in California. Since the genus is so abundant and widespread it is probable that many more species will eventually be collected in the state.

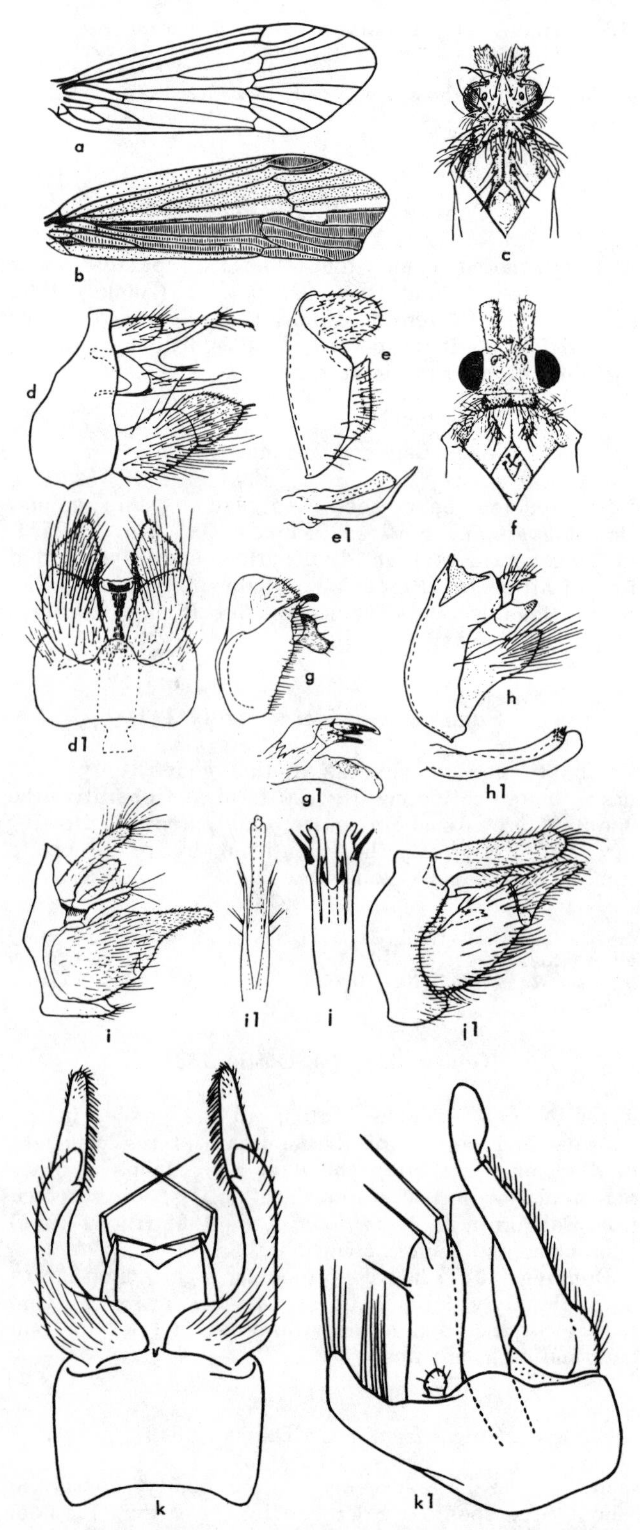

Fig. 10:26. Distinguishing structures of adult Trichoptera. *a, Apatania stigmatella*, front wing; *b, Psychoglypha avigo*, front wing; *C, Limnephilus submonilifer*, head and thorax; *d, Apatania sorex*, male genitalia, lateral; d_1, male genitalia, ventral; *e, Chyranda centralis*, male genitalia, lateral; e_1, aedeagus; *f, Pycnopsyche subfasciata*, head and thorax; *g, Pedomoecus sierra*, male genitalia, lateral; g_1, aedeagus; *h, Clostoeca disjunctus*, male genitalia, lateral; h_1, aedeagus; *i, Dicosmoecus atripes*, male genitalia, lateral; i_1, aedeagus, *j, Dicosmoecus unicolor*, male genitalia, lateral; j_1, aedeagus; *k, Agrypnia glacialis*, male gentalia, lateral; k_1, dorsal view (a-c,f, Ross, 1944; d,g, Ross, 1941; e,h-j, Ross, 1938; k, Celeste Green, 1955).

The male genitalia are distinctive for each of the species.

acoma Denning 1947 (fig. 10:25*f*)	Santa Clara County
argosa Ross 1938 (fig. 10:25*k*)	Siskiyou County
rono Ross 1941 (fig. 10:25*i*)	Humboldt to Tuolumne
xera Ross 1938 (fig. 10:25*j*)	Mono to Shasta

Family PHRYGANEIDAE

Caddisflies belonging to this family are generally large, seldom less than 12-14 mm. in length. Seven genera containing twenty-nine species are known from the United States and Canada. The larvae are generally found among submerged plants in lakes and marshes, but a few species are found in slowly flowing streams. The larval cases (fig. 10:2*a*) are generally long and built in a spiral. The family is very poorly represented in California, there being only three genera and three species recorded to date.

Key to the Genera[12]

Adult males

1. Ninth sternite produced caudad as a shelf beyond the base of the claspers .... *Ptilostomis* Kolenati 1859
— Ninth sternite not shelflike 2
2. Clasper short, biscuit-shaped, produced caudad into a short, slightly upturned point .... *Phryganea* Linneaus 1758
— Claspers produced dorsad into a prominent process 3
3. Tenth tergite consisting of 2 long, black sclerotized rods .... *Eubasilissa* Martynov 1930
— Tenth tergite not rodlike....................... 4
4. Hind wings mostly black except for yellow reticulations near apex .... *Oligostomis* Kolenati 1848
— Hind wings with only black spots on a gray or clear ground color 5
5. Ninth tergite forming a transverse, somewhat hood-shaped area, and bearing a brush or pair of brushes of long setae ....*Agrypnia* Curtis 1835
— Ninth tergite continuous with outline of 10th tergite, usually not bearing long setae 6
6. Middle tibiae thorny, the black spines very prominent; wings shiny with black markings .... *Banksiola* Martynov 1924
— Middle tibiae not thorny, the black spines not prominent, wings dull, tawny brown .....*Fabria* Milne 1934

Known Larvae[13]

1. Frons with a median black line 2
— Frons without a median black line 3
2. Pronotum with anterior margin black, and without a diagonal black line......... *Phryganea* Linneaus 1758
— Pronotum with a diagonal black line, anterior margin mostly yellow .... *Banksiola* Martynov 1924

[12]Adapted from Ross (1944).

[13]After Ross (1944).

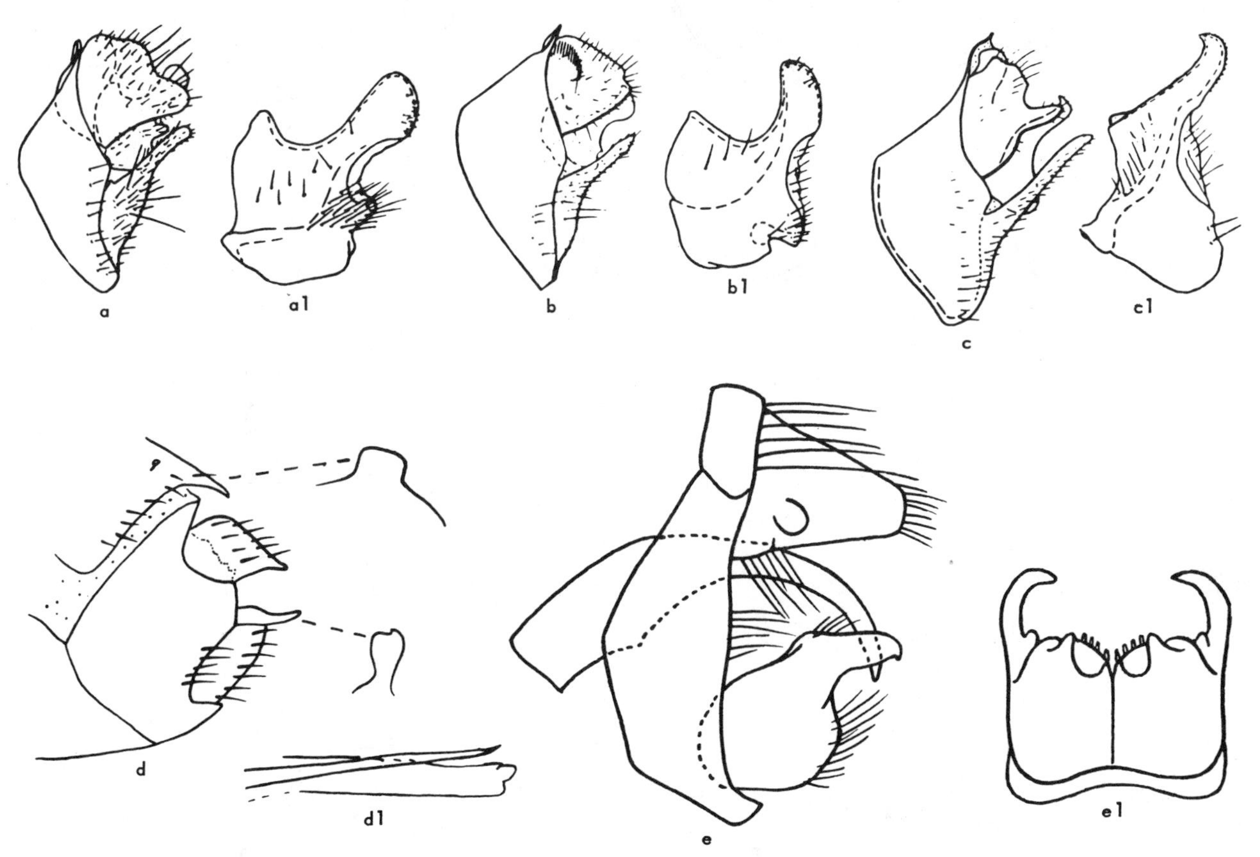

Fig. 10:27. Male genitalia. *a*, *Hesperophylax incisus*, lateral; a_1, tenth tergite, lateral; *b*, *H. occidentalis*, lateral; b_1, tenth tergite, lateral; *c*, *H. magnus*, lateral; c_1, tenth tergite, lateral; *d*, *Desmona bethula*, male genitalia, lateral; d_1, aedeagus; *e*, *Banksiola crotchi*, male genitalia, lateral; *e*, *claspers*, *ventral* (*a-c*, Ross, 1944; *d*, Denning, 1954*a*; *e*, Celeste Green, 1955).

3. Mesonotum with a pair of small sclerites near anterior margin 4
— Mesonotum without a pair of sclerites, sometimes with a very small sclerotized area around the base of 1 seta 5
4. Mature larvae attaining length of 30 mm. .... *Eubasilissa* Martynov 1930
— Mature larvae attaining length of only 20 mm. .... *Oligostomis* Kolenati 1848
5. Pronotum with anterior margin black, and without a diagonal black line ..... *Agrypnia vestita* (Walker) 1852
— Pronotum with a diagonal black line, anterior margin mostly yellow 6
6. Diagonal marks on pronotum meeting at posterior margin to form a V-mark .... *Phryganeid Genus A*
— Diagonal marks on pronotum not reaching posterior margin but joining each other on meson to form an arcuate mark .... *Ptilostomis* Kolenati 1859

Genus *Agrypnia* Curtis 1835

Nine species are known from the Nearctic region but only a single species, *A. glacialis* Hagen 1864 (fig. 10:26*k*) (Shasta County), has been reported from California.

Genus *Yphria* Milne 1934

Three species of this genus are known from the Nearctic region and a single species, *Y. californica* Banks 1907, is known from California. This species was described from a single female labeled "California."

Genus *Banksiola* Martynov 1924

Eight species have been placed in this genus and only a single species, *B. crotchi* Banks 1943 (fig. 10:27*e*) (Lassen County), has been recorded from California.

Family LIMNEPHILIDAE

This family probably contains more species than any other in the Nearctic region. There are about forty-five genera and two hundred and fifty species. At present only sixteen genera and thirty-eight species are known from California. All the larvae build cases (fig. 10:2*c,g*) and are found in a variety of habitats, ranging from temporary spring ponds to clear, cold, swiftly flowing mountain streams. The larvae which build the so-called "log-cabin type" cases belong to this family. Adult males may be distinguished by the maxillary palpi which are never more than three-

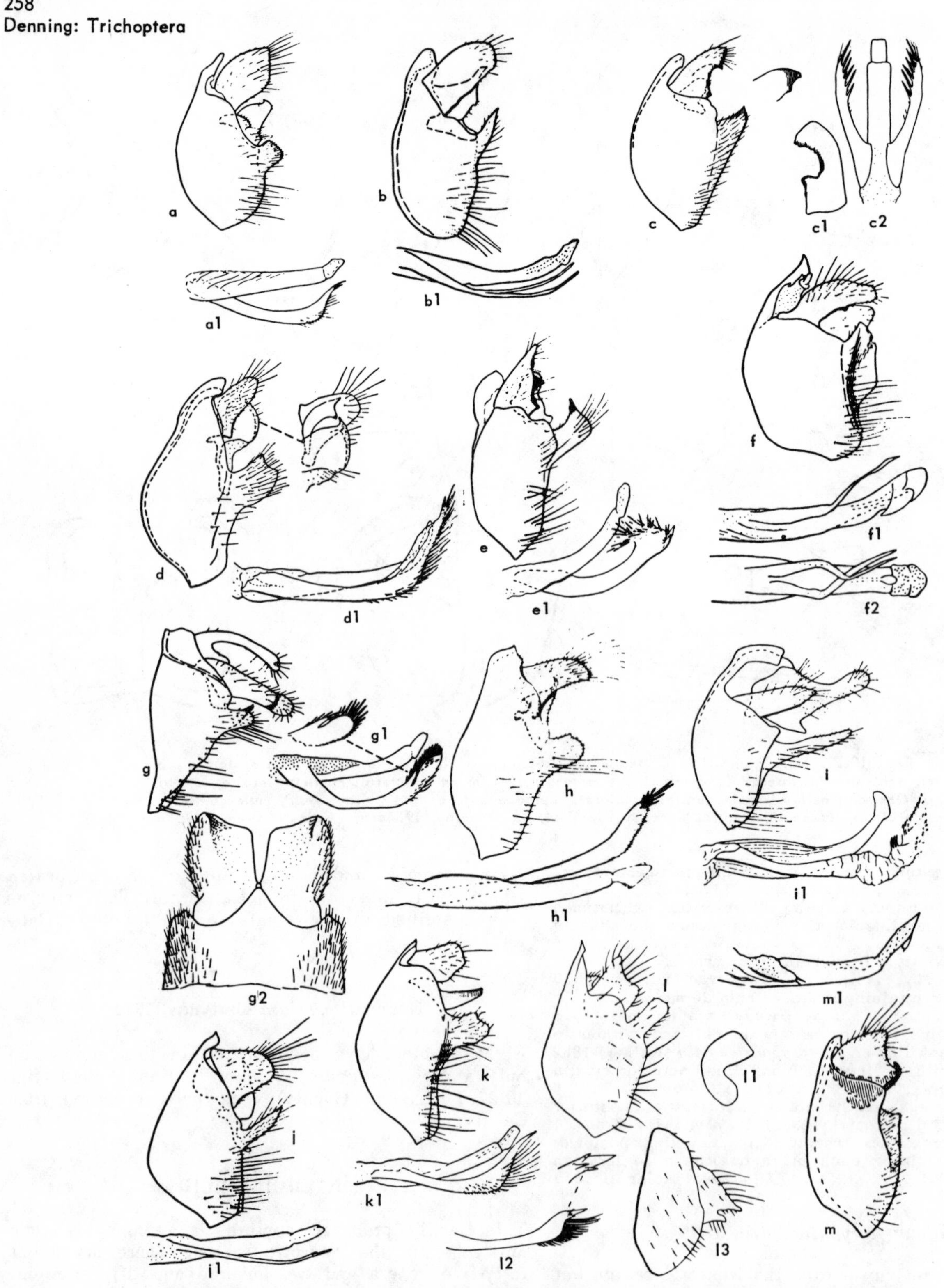

Fig. 10:28. Male and female genitalia of *Limnephilus*. *a*, *aretto*, lateral; a_1, aedeagus, lateral; *b*, *assimilis*, lateral; b_1, aedeagus, lateral; *c*, *acnestus*, lateral; c_1, cercus; c_2, aedeagus; *d*, *frijole*, lateral; d_1, aedeagus, lateral; *e*, *externus*, lateral; e_1, aedeagus, lateral; *f*, *coloradensis*, lateral f_1, aedeagus, lateral; f_2, aedeagus, dorsal; *g*, *nogus*, lateral; g_1, aedeagus, lateral; g_2, female genitalia, ventral; *h*, *lunonus*, lateral; h_1, aedeagus, lateral; *i*, *morrisoni*, lateral; i_1, aedeagus, lateral; *j*, *spinatus*, lateral; j_1, aedeagus, lateral; *k*, *occidentalis*, lateral; k_1, aedeagus, lateral; *l*, *productus*, lateral; l_1, clasper, caudal view; l_2, lateral arm of aedeagus; l_3, female genitalia, lateral; *m*, *secludens*, lateral; m_1, aedeagus, lateral (*a, b, e, f, i, j, k, m*, Ross, 1938, Ross, 1938; *d*, Ross, 1949c; *g*, Ross, 1944; *h*, Ross, 1941; *l*, Denning, 1948a).

segmented. A diagnostic feature of the larvae is the location of the antennae midway between the eye and the base of the mandibles (fig. 10:5*j*).

Key to Genera of Limnephilidae

Known to Occur in California

Adults

1. Spurs 0-2-2 .... *Desmona* Denning 1954
— Spurs 1-2-2, 1-3-3, or 1-3-4 2
2. Maxillary palpi abnormally long in both ♂ and ♀ (fig. 10:25*g*) .... *Chyranda* Ross 1944
— Maxillary palpi normal 3
3. Front wings with Sc ending in a straight, oblique cross vein (fig. 10:26*a*) .... *Apatania* Kolenati 1848
— Front wings with Sc not ending in an oblique cross vein 4
4. M_{1+2} in hind wings undivided .... *Neophylax* McLachlan 1871
— M_{1+2} in hind wings divided into M_1 and M_2 5
5. Vertex covered with close, appressed silky hairs .. .. *Hesperophylax* Banks 1916
— Vertex not covered with silky appressed hairs..... 6
6. Front wings with short rounded stigma, R_2 curved, parallel with R_1 (fig. 10:26*b*) 7
— Front wings with long narrower stigma if well marked, R_2 not parallel with R_1 8
7. Front wings with well-marked longitudinal silvery line (fig. 10:26*b*) .... *Psychoglypha* Ross 1944
— Front wings with no longitudinal silvery line .... *Glyphopsyche* Banks 1904
8. Clasper of male segmented (fig. 10:26*j*) 9
— Clasper of male not segmented (fig. 10:28*c*) 10
9. Discal cell (first R_3) of fore wing short, fork of R_{2+3} in hind wing near wing margin ... *Pedomoecus* Ross 1947
— Discal cell (first R_3) of fore wing long, fork of R_{2+3} in hind wing arises at discal cell .... *Dicosmoecus* McLachlan 1875
10. Mesoscutal warts on mesonotum poorly defined, represented by a linear area of setae (fig. 10:26*c*) 11
— Mesoscutal warts on mesonotum ovate and well defined, (fig. 10:26*f*) 12
11. Ninth tergite broad, usually forming a wide band with remainder of 9th segment (fig. 10:29*a*) .... *Lenarchus* Martynov 1915
— Ninth tergite reduced to a thin bridge (fig. 10:28*a*) . ... *Limnephilus* Leach 1815
12. Last tarsal segment of all legs without black spines, apex of front wing rounded ...... *Drusinus* Betten 1934
— Last tarsal segment of at least 1 pair of legs with 1 or more short black spines 13
13. Hind wings with hooks along costal margin, spurs 1-3-3 .... *Oligophlebodes* Ulmer 1905
— Hind wings with no hooks along costal margin, spurs 1-3-4.. 14
14. Dorsal surface of 8th tergite reticulated (fig. 10:31*c*) .... *Hydatophylax* Wallengren 1891
— Dorsal surface of 8th tergite not reticulated 15
15. Claspers extending caudad to apex of male genitalia (fig. 10:26*h*) .... *Clostoeca* Banks 1943
— Claspers do not extend caudad as far as any other part of male genitalia (fig. 10:29*l*) .... *Ecclisomyia* Banks 1907

Genus *Apatania* Kolenati 1848

About eleven species have been placed in this genus with only a single species, *A. sorex* Ross 1941 (fig. 10:26*d*) (Plumas County), recorded from California. The majority of the species are found in the Arctic region or at high elevations.

Genus *Pedomoecus* Ross 1947

The genus *Pedomoecus* contains but one species known only from California, *P. sierra* Ross 1947 (fig. 10:26*g*) (Mono and Fresno counties). The genus appears to be closely related to *Dicosmoecus*, especially in having segmented claspers, and is a member of the subfamily Dicosmoecinae.

Genus *Dicosmoecus* McLachlan 1875

The genus is represented in the West by several species; all these are large, varying from light brown to blackish in color. The wings are broadly rounded and the surface is granulated. *Dicosmoecus* larvae are found in lakes or large, slowly moving streams.

Two fairly common species are known to occur in California. These may be separated on the basis of the male genitalia. About fifteen species have been placed in the genus, but the species are poorly defined and the genus is in need of revision.

atripes (Hagen) 1875 (fig. 10:26*i*)	Placer County
unicolor (Banks) 1897 (fig. 10:26*j*)	Inyo, Shasta

Genus *Chyranda* Ross 1944

At present only one species, *C. centralis* (Banks) 1900 (fig. 10:26*e*) (Shasta and Lassen counties), is known. The very long maxillary palpi (fig. 10:25*g*) distinguish this genus from others.

Genus *Clostoeca* Banks 1943

The genus *Clostoeca* contains only two species, both limited to the West and both found in California.

disjunctus (Banks) 1914 (fig. 10:26*h*)	San Bernardino, Marin
sperryae Banks 1943. This species is poorly defined.	Siskiyou County

Genus *Desmona* Denning 1954

Only a single species, *D. bethula* Denning 1954 (fig. 10:27*d*) (Plumas and Sierra counties), has been referred to this genus.

Genus *Hesperophylax* Banks 1916

Records of only three of the six Nearctic species are available. The large, often cylindrical cases, constructed of small stones, are quite conspicuous and often occur in abundance in clear, swiftly flowing streams (fig. 10:2*f*). All described species have a longitudinal silvery stripe on the front wings, and the vertex and anal portions of the front wings are covered with appressed silky hair.

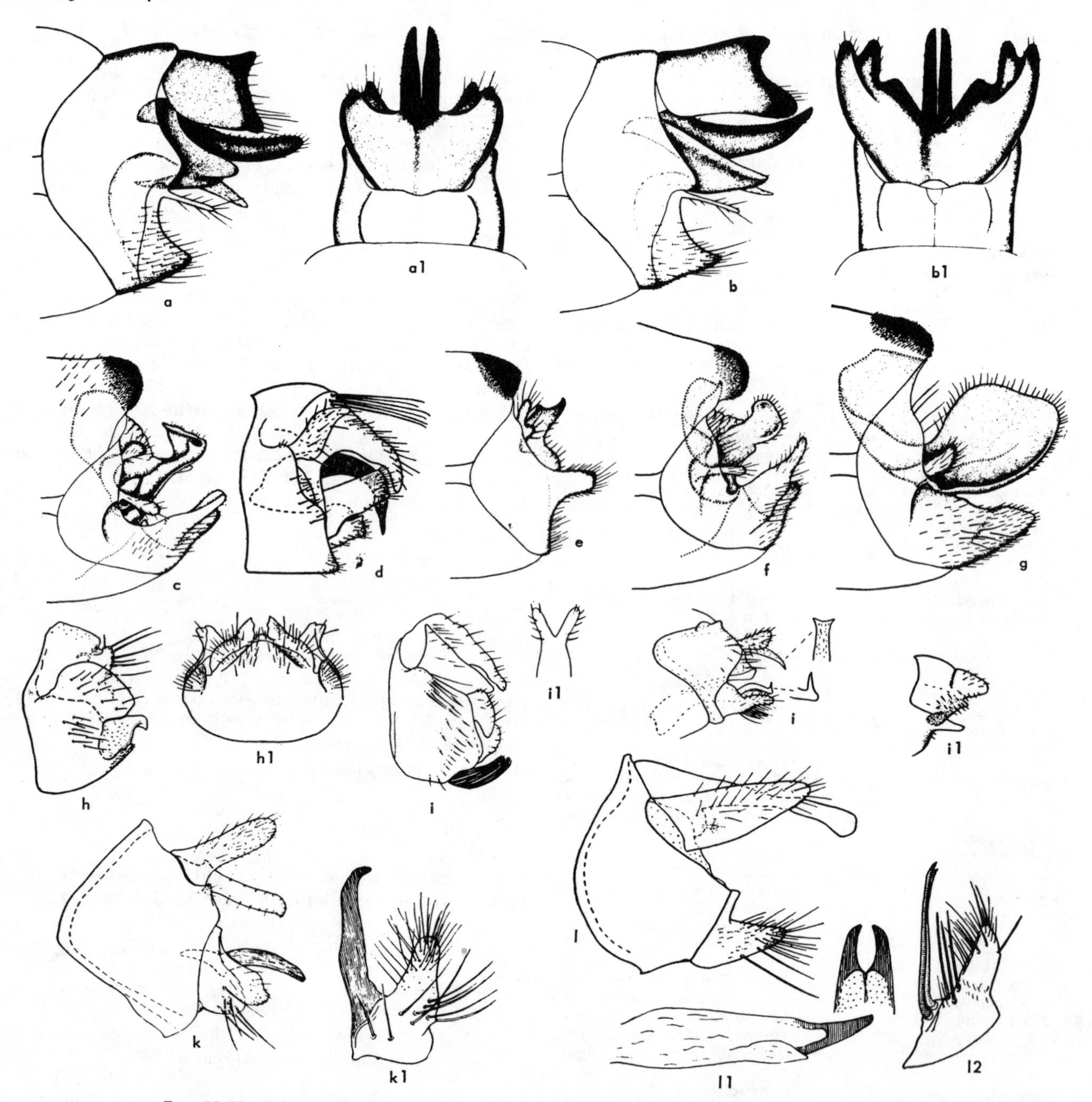

Fig. 10:29. Male and female genitalia. *a*, *Lenarchus gravidus*, lateral; a_1, tenth tergite, dorsal; *b*, *Lenarchus rillus*, lateral; b_1, tenth tergite, dorsal; c, *Psychoglypha avigo*, lateral; *d*, *Neophylax occidentis*, lateral; *e*, *Glyphopsyche irrorata*, lateral; *f*, *Psychoglypha ormiae*, lateral; *g*, *Psychoglypha bella*, lateral; *h*, *Oligophlebodes sierra*, lateral; h_1, ventral view; *i*, *Neophylax rickeri*, lateral; i_1, tenth tergite, dorsal; *j*, *Ecclisomyia bilera*, lateral; j_1, female genitalia, lateral; *k*, *Ecclisomyia conspersa*, lateral; k_1, clasper, ventral; *l*, *Ecclisomyia simulata*, lateral; l_1, aedeagus; l_2, clasper, ventral (*a,b*, Schmid, 1952*a*, *c,e-g*, Schmid, 1952*b*; *d*, Ross, 1938; *h*, Ross, 1944; *i*, Denning, 1948*c*; *j*, Denning, 1951; *k,l*, Ross, 1950*b*).

Key to Adult Males

1. Small specimens, approximately 10-12 mm. in length. (No California records are available but it occurs in Oregon.) .... *minutus* Ling 1938
— Larger specimens, 20 mm. or more in length 2
2. Apex of 10th tergite acute, appearing pointed from both lateral and caudal view; apicoventral corner of cerci produced into a long, narrow finger (fig. 10:27*c*); Mono and San Bernardino counties. .*magnus* Banks 1918
— Apex of 10th tergite rounded from either lateral or caudal view, apicoventral corner of cerci not produced into a narrow finger 3
3. Apicoventral corner of cerci produced into a thick, short rounded process (fig. 10:27*a*); Lassen to San Bernardino County .... *incisus* Banks 1943

— Apicoventral corner of cerci not so produced (fig. 10:27*b*); San Bernardino County .... *occidentalis* (Banks) 1908

Genus *Drusinus* Betten 1934

Except for a doubtful species known only from a female collected in British Columbia, *D. edwardsi* (Banks) 1920 (fig. 10:30*a*) (Shasta to Marin counties) is the only western species. The three other species are eastern in distribution. The immature stages have not been described.

The status of the name *Drusinus* is in doubt. Schmid (1955) considers *Drusinus* a synonym of *Pseudostenophylax* Martynov 1909.

Genus *Limnephilus* Leach 1815

Approximately ninety species have been described in this dominant genus of the family. Only thirteen species are recorded from California; doubtless many more remain to be collected. Members of the genus occur along streams, lakes, or ponds. There is still a difference of opinion regarding the limits of *Limnephilus,* and in the past a wide variety of species have been referred to the genus. Even today a precise definition of the genus is difficult. The larvae (fig. 10:6) of only a few species have been described.

Key to Adult Males[14]

1. Front basitarsus as long or longer than next segment 2
— Front basitarsus not more than 1/2 length of succeeding segment 9
2. Front basitarsus subequal in length to 2nd segment 3
— Front basitarsus usually 1½ times longer than 2nd segment 5
3. Tenth tergite produced into a sharp recurved tip (fig. 10:28*h*); Marin County .... *lunonus* Ross 1941
— Tenth tergite low and without a sharp tip 4
4. Clasper triangular, large, and broad (fig. 10:28*c*); Inyo County .... *acnestus* Ross 1938
— Clasper straight, fingerlike, not particularly large (fig. 10:28*l*); Inyo County ..... *productus* Banks 1914
5. Lateral appendages of aedeagus rodlike, crossed over aedeagus (fig. 10:28*f*); Mono and Tuolumne counties .... *coloradensis* (Banks) 1899
— Lateral appendages of aedeagus absent, or if present not crossed over aedeagus 6
6. Tenth tergite knobbed and large, much larger than cerci which are slender and elongate (fig. 10:28*i*); Placer County .... *morrisoni* Banks 1920
— Tenth tergite smaller, or with apex small 7
7. Tenth tergite slender, long, arched in middle higher than cercus; 8th tergite without apical patch of black setae (fig. 19:28*g*); Marin County .... *nogus* Ross 1944
— Tenth tergite not arched higher than cercus; 8th tergite with a patch of short black or yellow setae 8
8. Posterior edge of cerci flattened and flangelike; clasper with sharp apical point (fig. 10:28*e*); Modoc County .... *externus* Hagen 1861
— Posterior edge of cerci sharp or rounded in cross section, not flanged; clasper without sharp, acute point (fig. 10:28*k*); Lassen County .... *occidentalis* Banks 1908
9. Clasper formed as a narrow sclerite extending along margin of 9th segment (fig. 10:28*m*); Modoc, Placer counties .... *secludens* Banks 1914
— Clasper formed as a triangular or thumblike process . 10
10. Eighth tergite with a mesoapical patch of short dark setae 11
— Eighth tergite simple, lateral arms of aedeagus with basal half much thicker than remainder (fig. 10:28*b*); Inyo County .... *assimilis* Banks 1908
11. Clasper longer than deep, lobes of 10th tergite with long dorsal process, lateral arms of aedeagus filiform (fig. 10:28*j*); Mono County .... *spinatus* Banks 1914
— Clasper shorter than deep, lobes of 10th tergite not produced into a long dorsal process 12
12. Cerci triangular, small and scarcely surpassing clasper in lateral area (fig. 10:28*d*); Stanislaus, Tuolumne, Sacramento counties .... *frijole* Ross 1944
— Cerci more or less rhomboidal, much larger than clasper (fig. 10:28*a*); Tehama and Fresno counties .... *aretto* Ross 1938

[14]Adapted from Ross and Merkley (1952).

Genus *Lenarchus* Martynov 1914

Eleven species have been assigned to the genus, but only two have been definitely recorded from California. The genus is closely related to the preceding one and may be separated from it mainly on genitalic differences.

gravidus (Hagen) 1861 (fig. 10:29*a*)	Madera, Tuolumne, Mariposa
rillus (Milne) 1935 (fig. 10:29*b*)	Mariposa, Sierra

Genus *Psychoglypha* Ross 1944

The seven species are western or subarctic except *alascensis* Banks, which is also recorded from Michigan. At present three species are known to occur in California. Front wings are characteristically colored, the stigma is distinct and usually red, and a silvery streak runs through the center. Both wings are long and narrow (fig. 10:26*b*).

avigo (Ross) 1941 (figs. 10:26*b*, 10:29*c*)	Plumas County
bella (Banks) 1903 (fig. 10:29*g*)	Placer, Marin, Santa Cruz
ormiae (Ross) 1938 (fig. 10:29*f*)	Placer, Marin

Genus *Glyphopsyche* Banks 1904

Only two species have been referred to this genus, one in the Midwest and the other in the West and the Arctic. The single species known from California is *G. irrorata* (Fabricius) 1781 (fig. 10:29*e*) (Marin County).

Genus *Oligophlebodes* Ulmer 1905

All five Nearctic species are western in distribution. As yet, only one species, *O. sierra* Ross 1944 (fig.

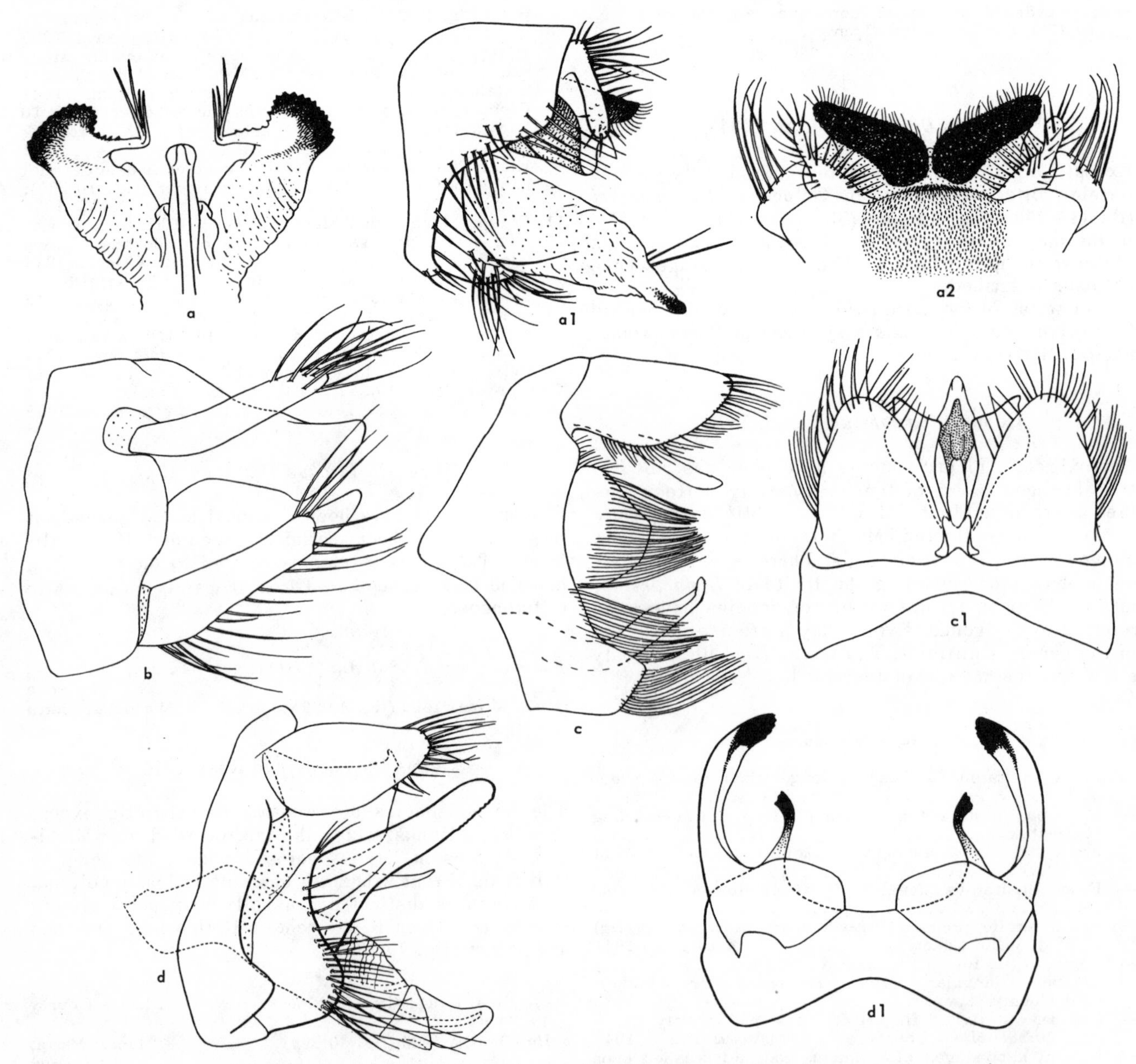

Fig. 10:30. Male genitalia. *a, Drusinus edwardsi*, aedeagus, ventral; a_1, lateral view; a_2, dorsal view; *b, Marilia flexuosa*, lateral; *c, Namamyia plutonis*, lateral; c_1, dorsal view; *d, Nerophilus californicus*, lateral; d_1, claspers, ventral view (Celeste Green, 1955).

10:29*h*) (Shasta and Tuolumne counties), is known to occur in California.

Genus *Neophylax* McLachlan 1871

Seventeen Nearctic species are known to occur in the genus, but only two have been collected in California. The larvae are found in rapidly flowing streams, and their cases are made of sand with ballast stones at the sides (fig. 10:2*j*). The adults are commonly taken in cool places under damp bridges or in the foliage along the sides of streams. At high altitudes the adults emerge late in the fall. The fairly short, triangular wings are characteristic of the genus.

occidentis Banks 1924 (fig. 10:29*d*) — Sierra, Modoc, Placer, Tuolumne
rickeri Milne 1935 (fig. 10:29*i*) — Marin, Contra Costa

Genus *Ecclisomyia* Banks 1907

The genus contains seven species, five of which have been described from western United States and Canada. Three species are known to occur in California.

The larvae are apparently restricted to cold, rapidly flowing streams in mountainous regions.

Key to Adult Males

1. Apex of 10th tergite directed dorsad or ventrad; claspers without mesal spines 2
— Apex of 10th tergite held horizontally; claspers with long, black mesal spines or spurs 3
2. Apex of 10th tergite directed dorsad, aedeagus with a pair of sinuate lateral processes .... *scylla* Milne 1935
— Apex of 10th tergite directed ventrad; aedeagus with a pair of sinuate lateral processes hooked apically (fig. 10:29*j*); Lassen County ...... *bilera* Denning 1951
3. Clasper with a single, stout, mesal seta; aedeagus bearing a pair of stout spines near apex (fig. 10:29*k*) Inyo County .... *conspersa* Banks 1907
— Clasper with 2 or more stout, mesal spines; aedeagus terminating in an ovate apex with a mesal notch 4
4. Claspers short, almost as wide at base as long (fig. 10:29*l*); Placer, Plumas, Lassen, and Tuolumne counties .... *simulata* Banks 1920
— Claspers long, twice as long as width of base .... *maculosa* Banks 1907

Genus *Hydatophylax* Wallengren 1891

Members of this genus are large, and the wings usually have a distinct pattern. The dorsal surface of the eighth tergite is reticulated. Three Nearctic species are known but only one, *H. hesperus* (Banks) 1914 (fig. 10:31*c*), has been reported from "Sutley," California.

Family ODONTOCERIDAE

This is a small family with but five genera, four of which are western and one, *Psilotreta,* apparently restricted to the East. Larvae of only one species, *Psilotreta frontalis* Banks, have been described. The California fauna contains four genera and four species.

Key to Genera

Adults[15]

1. Apex of fore wing rounded 2
— Apex of fore wing oblique, wings narrow .... *Marilia* Müller 1878
2. Cell R_4 not pedicellate in fore and hind wings 3
— Cell R_4 pedicellate in fore and hind wings 4
3. R_1 not fused with R_2 at tip in fore and hind wings ... . *Nerophilus* Banks 1899
— R_1 fused with R_2 at tip in fore and hind wings .... *Namamyia* Banks 1905
4. Cerci long, prominent; 10th tergite bearing a pair of hooks .... *Psilotreta* Banks 1899
— Cerci short, slender; 10th tergite with no hooks .... *Parthina* Denning 1954

Genus *Marilia* Muller 1878

Most of the species of the genus are southern in distribution. Only two species are known from the United States. The one California species is *M. flexuosa* Ulmer 1905 (fig. 10:30*b*) (Monterey and Ventura counties).

[15]Adapted from Betten (1934).

Genus *Namamyia* Banks 1905

The only species in this genus was described from California. It is *N. plutonis* Banks 1905 (fig. 10:30*c*) (Butte, Humboldt, Shasta, and Tuolumne counties).

Genus *Nerophilus* Banks 1899

The only species in the genus is *N. californicus* (Hagen) 1861 (fig. 10:30*d*) (Tehama, Alpine, Tulare, Sonoma, Napa, and Solano counties). It is a fairly large species and is well distributed throughout Oregon and California.

Genus *Parthina* Denning 1954

Only a single species has been placed in this genus. It is *P. linea* Denning 1954 (fig. 10:31*b*) (Tuolumne County).

Family CALAMOCERATIDAE

The family is represented by four genera; two are eastern and two are essentially southwestern and western in distribution. There are only six species in the family. The known California fauna consists of one genus with one species.

Key to the Genera

Adults[16]

1. Cell R_2 present in hind wing; no furrow of scales in fore wing of males 2
— Cell R_2 not present in hind wing; a furrow of scales present in fore wings .... *Notiomyia* Banks 1905
2. R_1 of fore wings not fused with R_2 at tip 3
— R_1 of fore wings fused with R_2 at tip .... *Ganonema* McLachlan 1866
3. Spurs 2-4-3 or 1-4-2; 2nd segment of maxillary palpus short ....*Anisocentropus* McLachlan 1863
— Spurs 2-4-2 or 2-4-4; 2nd segment of maxillary palpus long ....*Heteroplectron* McLachlan 1871

Genus *Heteroplectron* McLachlan 1871

The only species in this genus is *H. californicum* McLachlan 1871 (fig. 10:31*a*) (northern California). It is well distributed throughout the northern part of the state and has been recently taken in Fresno County.

[16]Adapted from Betten (1934).

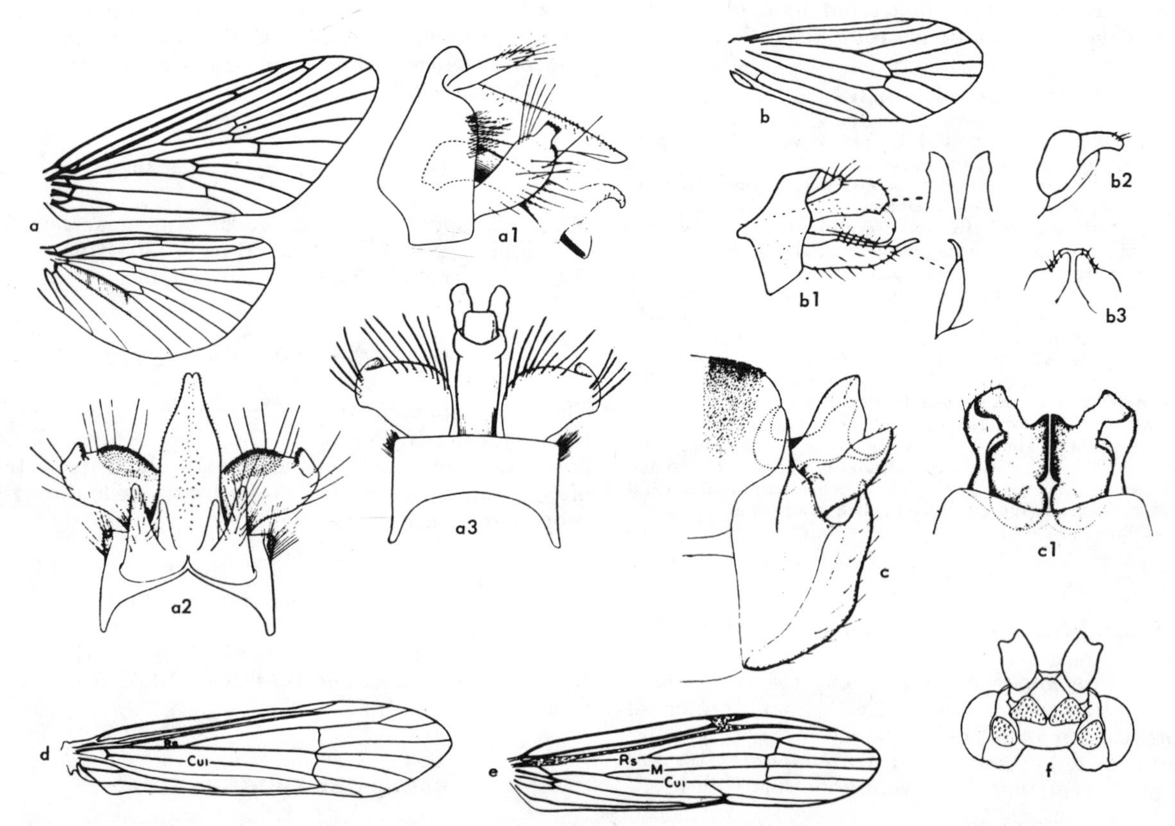

Fig. 10:31. Structures of adult Trichoptera. *a*, *Heteroplectron californicum*, wings of male; a_1, male genitalia, lateral; a_2, male genitalia, dorsal; a_1, ventral; *b*, *Parthina linea*, forewing; b_1, male genitalia, dorsal; b_2, female genitalia, lateral; b_3, female genitalia, dorsal; *c*, *Hydatophylax hesperus*, male genitalia, lateral; c_1, tenth tergite, dorsal; *d*, *Triaenodes injusta*, wing; e. *Oecetis inconspicua*, wing; *f*, *Athripsodes tarsi-punctatus*, head (*a*, Kimmins and Denning, 1951; *b*, Denning 1954*a*, c, Schmid, 1950*b*; *d-f*, Ross, 1944).

Family LEPTOCERIDAE

Four of the seven genera of Leptoceridae and nine species are known to occur in the state. The larvae are found in lakes, ponds, or streams and all build cases (fig. 10:3*c*,*d*). Some members of the family are widely distributed and occur in abundance. The adults are slender and possess long, slender antennae.

Key to the Genera

Adults[17]

1. Front wing with stem of media atrophied (fig. 10:31*d*) .... *Triaenodes* McLachlan 1865
— Front wing with stem of media present 2
2. Media apparently not branched (fig. 10:31*e*) .... *Oecetis* McLachlan 1877
— Media obviously branched 3
3. Head with epicranial stem distinct, lateral sutures absent or indistinct 4
— Head with epicranial stem absent or indistinct, lateral sutures distinct (fig. 10:31*f*) 5
4. Color yellowish to light-brown ... *Setodes* Rambur 1842
— Color dark brown to bluish-black................. .... *Mystacides* Berthold 1827
5. Color whitish, most of Rs of hind wings and its branches atrophied .... *Leptocella* Banks 1899
— Color brownish, Rs and its branches in hind wing present 6
6. Front tibiae with 2 apical spurs .... *Athripsodes* Billberg 1820
— Front tibiae with no apical spurs .... *Leptocerus* Leach 1815

[17]Adapted from Ross (1944).

Larvae[18]

1. Middle legs with claw stout and hook-shaped, tarsus bent; case transparent .... *Leptocerus* Leach 1815
— Middle legs with claw slender, slightly curved, tarsus straight; case seldom transparent 2
2. Maxillary palpi nearly as long as stipes; mandibles long, sharp at apex, the teeth considerably below apex .... *Oecetis* McLachlan 1877
— Maxillary palpi short, about ½ length of stipes; mandibles shorter, blunt at apex, the teeth near or at apex 3
3. Head with a suturelike line paralleling the epicranial arms (fig. 10:5*i*) .... *Athripsodes* Billberg 1820
— Head without a suturelike line in addition to the epicranial arms 4
4. Mesonotum membranous with a pair of sclerotized, narrow, curved, or angled bars (fig. 10:5*i*) .... *Athripsodes* Billberg 1820

[18]From Ross (1944).

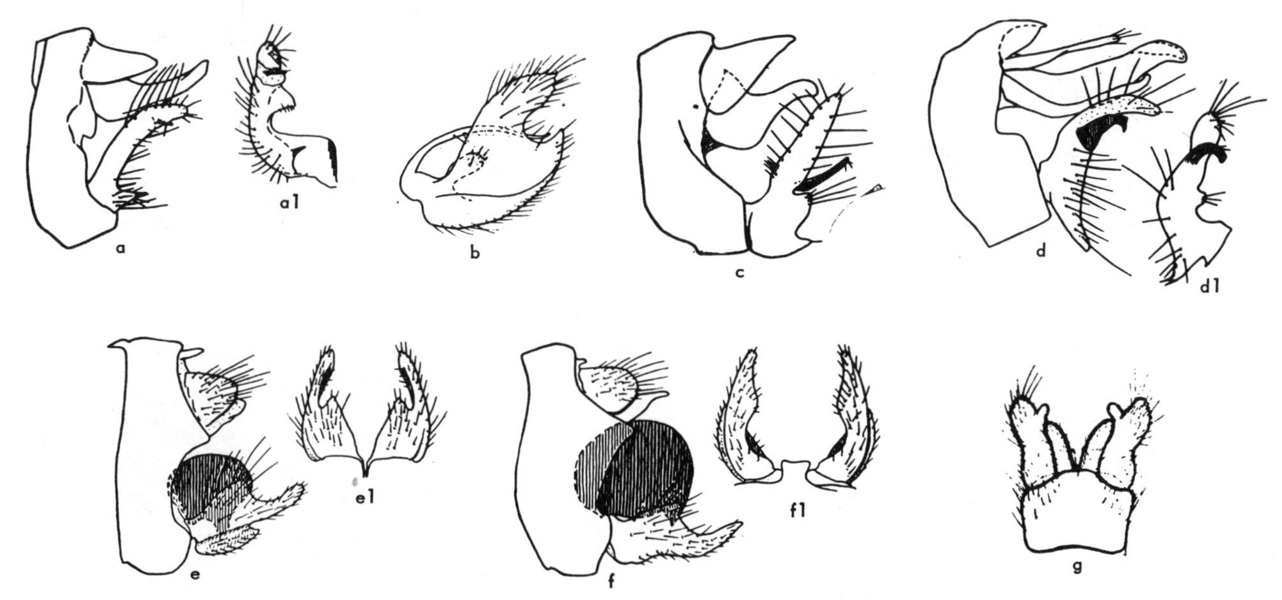

Fig. 10:32. Male genitalia. *a*, *Athripsodes annulicornis*, lateral; a_1, clasper, caudal view; *b*, *Trianodes frontalis*, clasper, lateral; *c*, *Athripsodes transversus*, lateral; *d*, *Athripsodes tarsipunctatus*, lateral; d_1, clasper, caudal view; *e*, *Oecetis ochracea*, lateral; e_1, claspers, ventral; *f*, *Oecetis inconspicua*, lateral; f_1, claspers, ventral; *g*, *Mystacides alafimbriata*, ventral (*a-f*, Ross, 1944; *g*, Ling, 1938, unpublished fig.).

— Mesonotum without such a pair of sclerotized bars .. 5
5. Anal segment developed into a pair of sclerotized, concave plates, with spinose dorsolateral and mesal carinae, and an overhanging ventral flap .... *Setodes* Rambur 1842
— Anal segment convex and without carinae between anal hooks 6
6. Hind tibia entirely sclerotized, without a fracture in middle; abdomen without gills .... *Leptocella* Banks 1899
— Hind tibia with a fracture near middle which appears to divide tibia into 2 segments; abdomen with at least a few gills 7
7. Hind tibia with a regular fringe of long hair .... *Triaenodes* McLachlan 1865
— Hind tibia with only irregularly placed hairs .... *Mystacides* Berthold 1827

Genus *Athripsodes* Billberg 1820

Although there are about thirty species in the genus, only three are known to occur in the state. Several times that number will be eventually collected.

annulicornis (Stephens) 1836 (fig. 10:32*a*) — Placer County
tarsipunctatus (Vorhies) 1909 (fig. 10:32*d*) — Stanislaus
transversus (Hagen) 1861 (fig. 10:32*c*) — Lake County

Genus *Mystacides* Berthold 1827

This is a small genus with only three Nearctic species. Members of the genus may occur in abundance, especially along the shores of some lakes. Larval and pupal cases are cylindrical, slightly tapering, and composed of small grains of sand (fig. 10:2*d*).

Key to Adult Males

1. Apical process of 9th sternite forked (fig. 10:34*b*) 2
— Apical process of 9th sternite not forked (fig. 10:34*c*) simple and wide; eastern half of U.S. .... *longicornis* (Linneaus) 1758
2. Forked apical process of 9th sternite with arms long and slender, only slightly divergent (fig. 10:34*b*), Nevada County .... *sepulchralis* (Walker) 1852
— Forked process of 9th sternite with arms widely divergent (fig. 10:32*g*); northern California and Fresno County .... *alafimbriata* Griffin 1912

Genus *Oecetis* McLachlan 1877

Only three of the eighteen Nearctic species have thus far been recorded from California. Some of the most widely distributed and most abundant caddisflies, such as *O. inconspicua* (Walker), belong to this genus.

disjuncta (Banks) 1920. — Northern California, Fresno County
This species may be a color variant of *avara* (Banks); the genitalia are very similar, that for *avara* as in fig. 10:34*d*.
inconspicua (Walker) 1852 (fig. 10:32*f*) — Widespread in California
ochracea (Curtis) 1825 (fig. 10:32*e*) — Shasta County

Genus *Triaenodes* McLachlan 1865

Only one of the nineteen described Nearctic species is known to occur in California, and that species, *T. frontalis* Banks 1907 (fig. 10:32*b*) (Ventura, Mono,

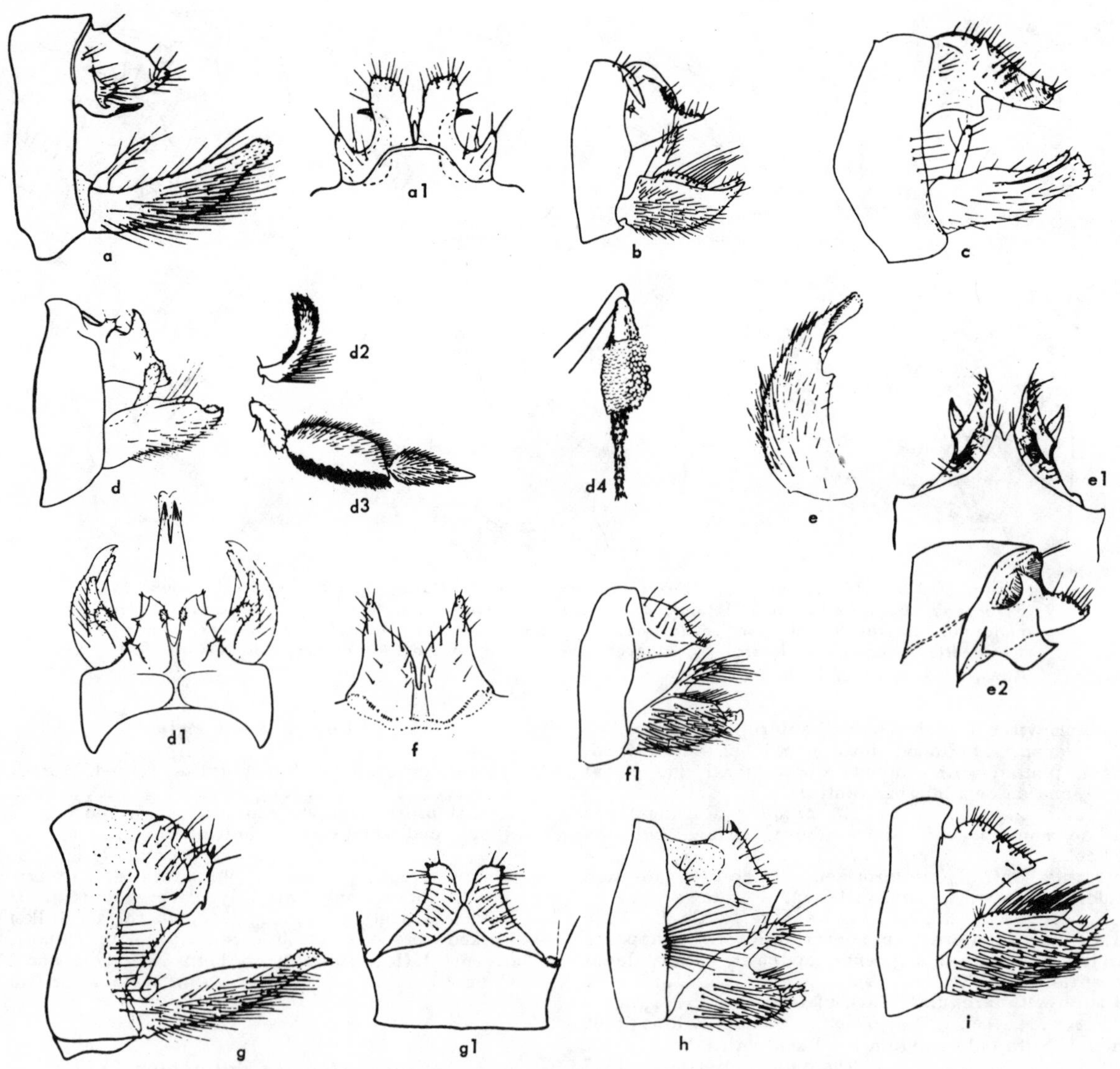

Fig. 10:33. Genitalia and other structures of male *Lepidostoma*. *a*, *cantha*, lateral; a_1, tenth tergite, dorsal; *b*, *cascadensis*, lateral; *c*, *jewetti*, lateral; *d*, *podager*, lateral; d_1, dorsal view; d_2, maxillary palpus; d_3, labial palpus; d_4, fore tibia and tarsus; *e*, *lotor*, clasper, ventral; e_1, tenth tergite, dorsal; e_2, ninth and tenth tergite, lateral; *f*, *strophis*, tenth tergite, dorsal; f_1, genitalia, lateral; *g*, *rayneri*, genitalia, lateral; g_1, dorsal view; *h*, *unicolor*, genitalia, lateral; *i*, *roafi*, genitalia, lateral (*a,g*, Ross, 1941; *b,f*, Ross, 1938; *c,h*, Ross, 1946; *d*-d_3, Kimmins; and Denning, 1951; d_4, *e,f,h,i*, Ross, 1946).

and Monterey counties), is widely distributed in the West. Most of the larvae build cases in a spiral pattern (fig. 10:3*d*).

Family LEPIDOSTOMATIDAE

Two genera and forty-nine species are recognized. One genus, *Lepidostoma,* is widespread in the United States and Canada and contains forty-five species; *Theliopsyche* is apparently restricted to the eastern states and contains four species. One genus and twelve species are recorded from California.

Sexual dimorphism has been developed to such an extent in the genus *Lepidostoma* that a large number of very unusual and bizarre characters are to be found in the males. Some of the more unusual developments are seen in the maxillary palpi, the basal segment of the antennae, and the legs and wings. In some of the furrows and reflexed portions of the wings, scales occur that are somewhat similar to those of the Lepidoptera. This group has exceeded all others in the Trichoptera in development of secondary sexual characters.

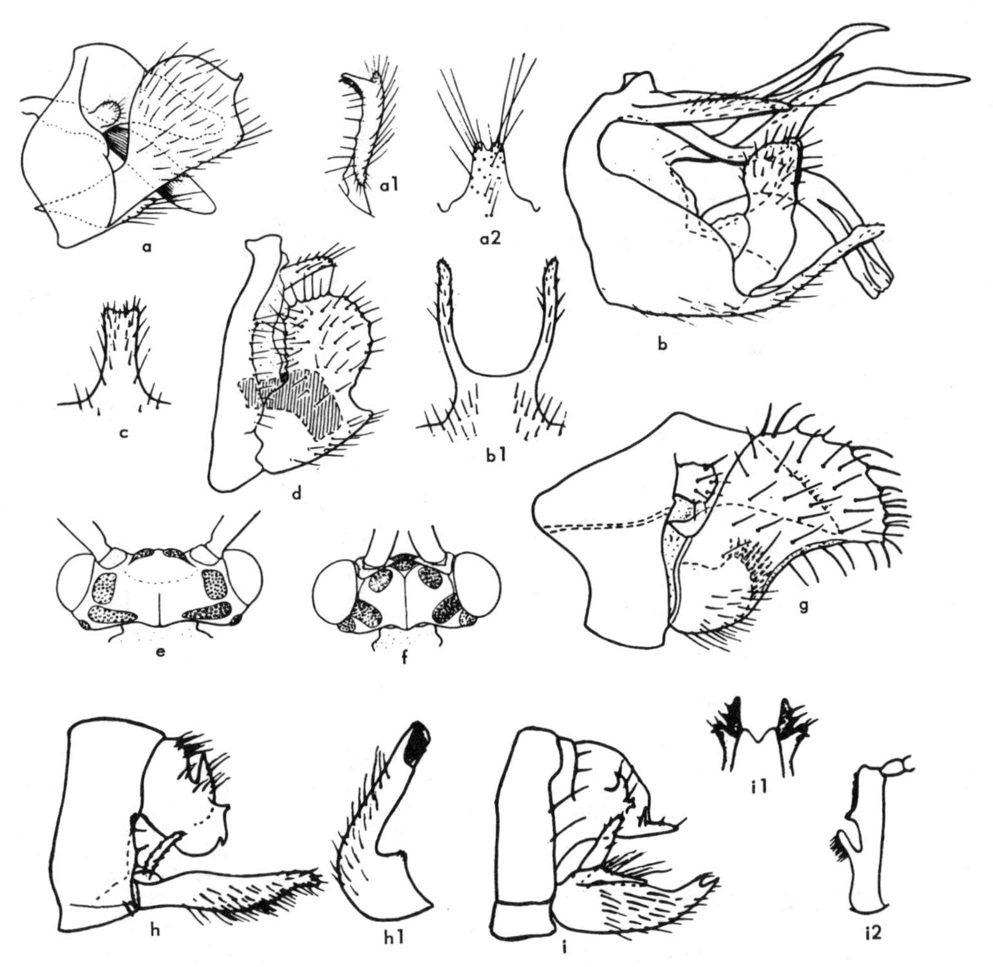

Fig. 10:34. Genitalia and other structures of male Trichoptera. *a*, *Sericostoma griseolum*, genitalia, lateral; a_1, clasper, dorsal; a_2, ventral plate, ventral view; *b*, *Mystacides sepulchralis*, genitalia, lateral; b_1, ninth sternite; *c*, *Mystacides longicornis*, ninth sternite; *d*, *Oecetes avara*, genitalia, lateral; *e*, *Lepidostoma liba*, head; *f*, *Theliopsyche* sp., head; *g*, *Helicopsyche borealis*, genitalia, lateral; *h*, *Lepidostoma errigena*, genitalia, lateral; h_1, clasper, ventral; *i*, *Lepidostoma mira*, genitalia, lateral; i_1, tenth tergite, dorsal; i_2, first antennal segment (*a*, Kimmins and Denning, 1951; *b-g*, Ross, 1944; *h,i*, Denning, 1954*b*).

Keys to Genera[19]

Adults

1. Head with posterior warts fairly wide, triangular or curved (fig. 10:34*e*) ...... *Lepidostoma* Rambur 1842
— Head with posterior warts long, narrow, and straight (fig. 10:34*f*) .... *Theliopsyche* Banks 1911

Genus *Lepidostoma* Rambur 1842

Of the forty-five known Nearctic species only twelve have been recorded from California. Although the larvae superficially resemble some of the Limnephilidae they may be readily separated by the absence of the dorsal spacing hump (fig. 10:6), and by the proximity of the antennae to the eye (fig. 10:5*k*). An example of the larval case is shown in figure 10:2*k*.

[19] Adapted from Ross (1944).

Key to Adult Males[20]

1. Front wing with entire costal cell reflexed, turned back as a flap at least over subcostal cell 2
— Front wing with at most only a small basal portion of the costal cell reflexed 4
2. Tenth tergite having ventrolateral portion forming a broad, flat divergent flange (fig. 10:33*e*); Santa Cruz County .... *lotor* Ross 1946
— Tenth tergite with only a short sharp spur projecting from the ventrolateral portion 3
3. Tenth tergite with a large prominent dorsal lobe (fig. 10:34*h*); San Bernardino, Los Angeles, and San Diego counties .... *errigena* Denning 1954
— Tenth tergite with at most only a short, sharp spur dorsally (fig. 10:33*g*); Mono, Sierra, and Placer counties .... *rayneri* Ross 1941
4. Base of 10th tergite with a pair of heavy ventral spurs, laterad in position; clasper with only a dorsal finger-like process near base (fig. 10:33*a*); Monterey, Contra Costa, Los Angeles counties .... *cantha* Ross 1941

[20] Adapted from Ross (1946).

— Base of 10th tergite without such a pair of spurs; claspers with 2 or 3 processes 5
5. Basal process of clasper very short, barely reaching above clasper (fig. 10:33*i*); Mono, Tuolumne, and Placer counties .... *roafi* (Milne) 1936
— Basal process of clasper long, reaching considerably above clasper 6
6. Lobe of 10th tergite with apical portion long, ovate, and concave; apex of claspers truncate (fig. 10:33*c*); Marin and Plumas counties .... *jewetti* Ross 1946
— Lobe of 10th tergite short, not of above type; claspers tapering toward apex 7
7. Clasper, lateral aspect, with apex acute 8
— Clasper, lateral aspect, with apex fairly blunt, or with a buttonlike ventral lobe 10
8. Tenth tergite with a prominent spur along lateral aspect (fig. 10:34*i*); Tuolumne County .... *mira* Denning 1954
— Tenth tergite with no spur along lateral aspect 9
9. Lobes of 10th tergite with a sharp dorsal projection; front legs scaled (fig. 10:33*d*); northern California .... *podager* (McLachlan) 1871
— Lobes of 10th tergite without a dorsal projection; (fig. 10:33*b*) front legs not scaled; Tulare, Tuolumne, El Dorado, and Madera counties .... *cascadensis* (Milne) 1936
10. Tenth tergite with no dorsal projection, ventral corner attenuated (fig. 10:33*f*); northern California .... *strophis* Ross 1938
— Tenth tergite with dorsal projection, ventral corner not attenuated 11
11. Tenth tergite lobes with a short, dorsal spine, ventral margin incised (fig. 10:33*h*); widespread .... *unicolor* (Banks) 1911
— Tenth tergite lobe with prominent dorsal spine, ventral lobe elongated (fig. 10:35*e*); Del Norte County .... *astanea* Denning 1954

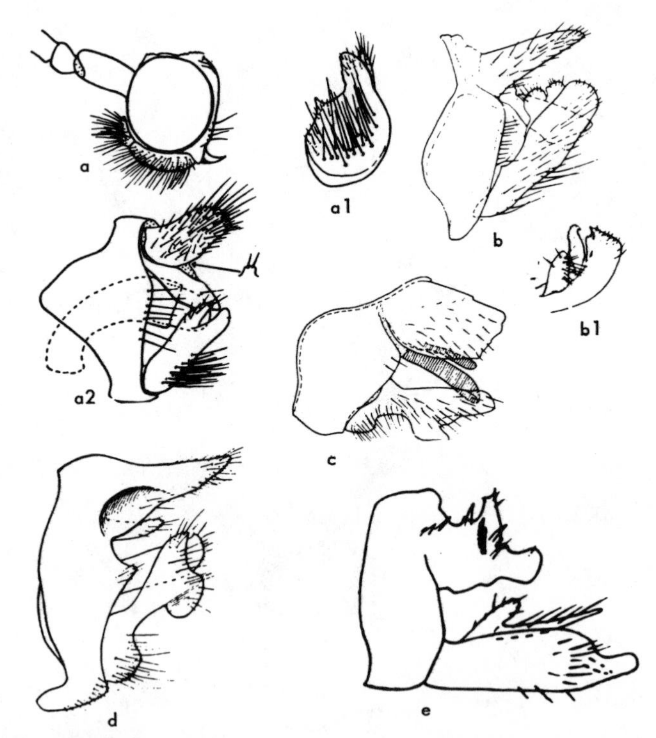

Fig. 10:35. Head and genitalia of male Trichoptera. *a*, *Micrasema aspilus*, head; a_1, clasper; a_2, genitalia, lateral; *b*, *Micrasema onisca*, lateral; b_1, clasper, dorsolateral view; *c*, *Oligoplectrum echo*, genitalia, lateral; *d*, *Brachycentrus americanus*, genitalia, lateral; *e*, *Lepidostoma astanea*, genitalia, lateral (*a*,*c*, Ross, 1938; *b*, Ross, 1947; *d*, Ross, 1938; *e*, Denning, 1954*b*).

Family BRACHYCENTRIDAE

The family is represented in the Nearctic region by three genera and twenty-eight species. All the known genera but only four species have been recorded from California.

Key to the Genera

Adults[21]

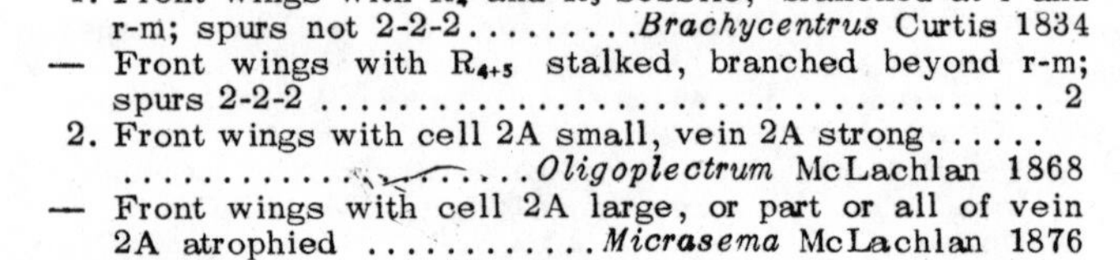

1. Front wings with R_4 and R_5 sessile, branched at r and r-m; spurs not 2-2-2......... *Brachycentrus* Curtis 1834
— Front wings with R_{4+5} stalked, branched beyond r-m; spurs 2-2-2 2
2. Front wings with cell 2A small, vein 2A strong .... *Oligoplectrum* McLachlan 1868
— Front wings with cell 2A large, or part or all of vein 2A atrophied .... *Micrasema* McLachlan 1876

Known Larvae

1. Middle and hind tibiae with an inner, apical spur; mesonotum with sclerites long and narrow; metanotal plates heavily sclerotized .. *Brachycentrus* Curtis 1834
— Middle and hind tibiae without an inner spur; mesonotal sclerites short, wide; metanotal plates lightly sclerotized .... *Micrasema* McLachlan 1876

[21]Adapted from Ross (1947).

Genus *Micrasema* McLachlan 1876

Fourteen Nearctic species are known, but only two have been reported from California. Since most of the species are western in distribution, it is probable that additional species will be collected in the state.

aspilus (Ross) 1938a (fig. 10:35*a*) — widespread
onisca Ross 1947 (fig. 10:35*b*) — Monterey, Fresno, Riverside, and San Diego counties

Genus *Oligoplectrum* McLachlan 1868

Only two North American species have been referred to this genus and one of these has been collected in California, *O. echo* Ross 1947 (fig. 10:35*c*) (Mono and Plumas counties). The larvae are unknown.

Genus *Brachycentrus* Curtis 1834

Twelve Nearctic species are known but only one is recorded from California, *B. americanus* (Banks) 1899 (fig. 10:35*d*) (Mono and Shasta counties). This species is abundant and widely distributed. Emergence, egg

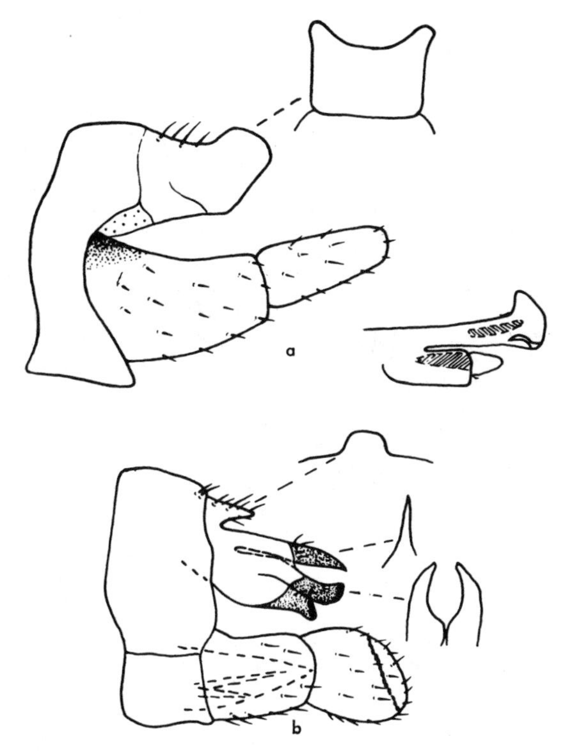

Fig. 10:36. Male genitalia of *Rhyacophila*. *a, chandleri*, lateral; *b, lineata*, lateral (Denning, 1955).

deposition, and immature stages of a closely related species, *fuliginosus*, are described by Denning (1937).

Family SERICOSTOMATIDAE

In North America the family consists of but one genus, *Sericostoma,* containing six species. One species, *S. griseolum* (McLachlan), is quite abundant in certain streams in the arid parts of Arizona and California. The larvae have not been described.

Genus *Sericostoma* Berthold 1827

There are six described species in this genus in the United States. Only one, *S. griseolum* (McLachlan) 1871 (fig. 10:34*a*) (widespread), is known to occur in California.

Family HELICOPSYCHIDAE

There is but one genus in this family, and only four species in the genus. The larvae have distinctive cases that resemble snail shells (fig. 10:3*e*). At times they are very abundant in the slower waters of clear, cool streams.

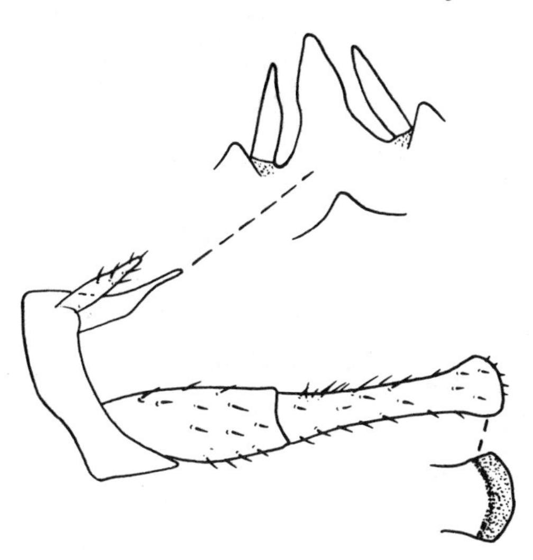

Fig. 10:37. Male genitalia of *Wormaldia pachita*, lateral (Denning, 1955).

Genus *Helicopsyche* Hagen 1866

Only a single species, *H. borealis* (Hagen) 1861 (fig. 10:34*g*) (widespread), is known from California. Another closely related species, *H. mexicana* Banks, should eventually be collected in southern California.

REFERENCES

(A bibliography to 1944 is given in "The Caddis Flies, or Trichoptera of Illinois". Ill. Nat. Hist. Sur. Bull. 23: 1-326, 961 figs., by Herbert H. Ross, 1944. References since that time and additional references are given below.)

BLICKLE, R. L., and MORSE, W. J.
1955. New and Little known Polycentropus (Trichoptera). Bull. Brooklyn Ent. Soc., 50:95-98.

DAVIS, JARED J.
1949. Two new species of caddis flies (Trichoptera) from Washington State. Ann. Ent. Soc. Amer.,42:448-450, 2 figs.

DENNING, DONALD G.
1947*a*. New species and records of North American Hydroptilidae (Trichoptera). Psyche, 54:170-177, 3 figs.
1947*b*. New species and records of Nearctic Hydroptilidae (Trichoptera). Bull. Brooklyn Ent. Soc., 42:145-158, 11 figs.
1948*a*. New and little known species of Nearctic Trichoptera. Psyche, 55:16-27, 7 figs.
1948*b*. Review of the Rhyacophilidae (Trichoptera). Canad. Ent., 80:97-117, 28 figs.
1948*c*. Descriptions of eight new species of Trichoptera. Bull. Brooklyn Ent. Soc., 43:119-129, 11 figs.
1948*d*. New species of Trichoptera. Ann. Ent. Soc. Amer., 41:397-401, 6 figs.
1949*a*. New species of Nearctic caddis flies. Bull. Brooklyn Ent. Soc., 44:37-48, 10 figs.
1949*b*. A new genus and five new species of Trichoptera. Jour. Kansas Ent. Soc., 22:88-93, 5 figs.
1949*c*. New and little known species of caddis flies. Amer. Midl. Nat., 42:112-122, 11 figs.
1950*a*. Records and descriptions of Nearctic caddis flies. Part I. Bull. Brooklyn Ent. Soc., 45:97-104, 7 figs.

1950*b*. Records and descriptions of Nearctic caddis flies. Part II. Jour. Kansas Ent. Soc., 23:115-120, 8 figs.
1951. Records and descriptions of Nearctic caddis flies. Part III. Jour. Kansas Ent. Soc., 24:157-162, 1 pl.
1952. Descriptions of several new species of caddis flies. Canad. Ent., 84:17-22, 11 figs.
1953. A new genus of Limnephilidae (Trichoptera). Pan-Pac. Ent., 29:165-169, 4 figs.
1954*a*. New species of western Trichoptera. Jour. Kansas Ent. Soc., 27:57-64, 10 figs.
1954*b*. New species of *Lepidostoma* (Trichoptera : Lepidostomatidae). Pan-Pac. Ent., 30:187-194, 8 figs.
1956. Several new species of western Trichoptera. Pan-Pac. Ent., 32:73-80, 7 figs.

HICKIN, NORMAN E.
1952. Caddis. London: Methuen and Co. Ltd. 50 pp., illus., 4 color pls.

KIMMINS, D. E., and D. G. DENNING
1951. The McLachlan types of North American Trichoptera in the British Museum. Ann. Ent. Soc. Amer., 44:111-140, 9 pls.

MARTYNOV, ANDREAS B.
1914. Trichoptères de la Sibérie et des regions adjacentes IV. Subf. Limnophilinae. Ann. Mus. Zool. Acad. Sci., St. Petersb., 19:173-285.
1924. Trichoptera. Practical Entomology, 5:67 and 388 pp., illus.

PENNAK, R. W.
1953. Fresh-water invertebrates of the United States. New York: Ronald Press Co. 769 pp., illus.

ROSS, HERBERT H.
1946. A review of the Nearctic Lepidostomatidae (Trichoptera). Ann. Ent. Soc. Amer., 39:265-290, 37 figs.
1947. Descriptions and records of North American Trichoptera with synoptic notes. Trans. Amer. Ent. Soc., 73:125-168, 7 pls.
1948*a*. New Nearctic Rhyacophilidae and Philopotamidae (Trichoptera). Ann. Ent. Soc. Amer., 41:17-26, 8 figs.
1948*b*. New species of Sericostomatid Trichoptera. Proc. Ent. Soc. Wash., 50:151-157, 4 figs.
1949*a*. The caddisfly genus *Neothremma* Banks (Trichoptera, Limnephilidae). Jour. Wash. Acad. Sci., 39:92-93, 2 figs.
1949*b*. A classification for the Nearctic species of *Wormaldia* and *Dolophilodes* (Trichoptera : Philopotamidae). Proc. Ent. Soc. Wash., 51:154-160, 4 figs.
1949*c*. Descriptions of some western Limnephilidae (Trichoptera). Pan-Pac. Ent., 25:119-128, 2 pls.
1950*a*. New species of Nearctic Rhyacophila (Trichoptera, Rhyacophilidae). Jour. Wash. Acad. Sci., 40:260-265, 11 figs.
1950*b*. Synoptic notes on some Nearctic Limnephilid caddisflies (Trichoptera, Limnephilidae). Amer. Midl. Nat., 43:410-429, 22 figs.
1951*a*. The caddisfly genus *Anagapetus* (Trichoptera: Rhyacophilidae). Pan-Pac. Ent., 27:140-142, 4 figs.
1951*b*. The Trichoptera of Lower California. Proc. Calif. Acad. Sci., 27:65-76, 3 figs.
1951*c*. Phylogeny and biogeography of the caddisflies of the genera *Agapetus* and *Electragapetus* (Trichoptera:Rhyacophilidae). Jour. Wash. Acad. Sci., 41: 347-356, 23 figs.
1952. The caddisfly genus *Molannodes* in North America. Ent. News, 63:85-87, 4 figs.

ROSS, HERBERT H., and DON R. MERKLEY
1950. The genus *Tinodes* in North America. Jour. Kansas Ent. Soc., 23:64-67, 3 figs.
1952. An annotated key to the Nearctic males of *Limnephilus* (Trichoptera, Limnephilidae). Amer. Midl. Nat., 47:435-455, 25 figs.

ROSS, HERBERT H., and GEORGE J. SPENCER
1952. A preliminary list of the Trichoptera of British Columbia. Proc. Ent. Soc. B. C., 48:43-51, 6 figs.

SCHMID, FERNAND
1950*a*. Monographie du genre *Grammotaulius* Kol. (Trichoptera, Limnophilidae). Rev. Suisse Zool., 57:317-352, 67 figs.
1950*b*. Le genre Hydatophylax Wall. (Trichop.). Schweiz. Ent. Gesell. Mitt. 23:265-296, 75 figs.
1951. Quelques nouveaux Trichoptères nearctiques. Bul. Inst. Roy. Sci. Nat. Belg., 27:1-16, 28 figs.
1952*a*. Le groupe de *Lenarchus* Mart. (Trichoptera, Limnophilidae). Schweiz. Ent. Gesell. Mitt., 25:157-210, 32 figs.
1952*b*. Le groupe de *Chilostigma* (Trichoptera, Limnophilidae). Arch. f. Hydrobiol., 47:76-163, 197 figs.
1954. Le genre *Asynarchus* McL. (Trichoptera, Limnophilidae). Schweiz. Ent. Gesell. Mitt., 27:57-96, 27 figs.
1955. Contribution a l'etude des Limnophilidae (Trichoptera). Schweiz. Ent. Gesell. Mitt., 28:1-245, 104 figs.

CHAPTER 11

Aquatic Lepidoptera

By W. H. Lange, Jr.
University of California, Davis

The Lepidoptera number more than 100,000 described species and are characteristically terrestrial in their habits. However, there are a few families, including the Cossidae, Cosmopterygidae, Yponomeutidae, Arctiidae, Phalaenidae, Tortricidae, Tineidae, Nepticulidae, Pyralidae, and Sphingidae, containing species capable of aquatic or subaquatic existence either in their immature forms or as adults. The Pyralidae, and particularly members of the subfamily Nymphulinae, have met the challenge of an aquatic existence very successfully. For this reason the present discussion will concern itself primarily with this group. *Acentropus niveus* (Oliv.), in the pyralid subfamily Schoenobiinae, is included in the keys because of its true aquatic existence. It is probably a recent introduction in the United States.

CLASSIFICATION

The recent paper by Lange (1956) is the first attempt to classify our North American genera of aquatic moths on a morphological basis. Before this most investigators placed our species in either *Nymphula* or *Cataclysta (Elophila)*. The check list of McDunnough (1939) lists ten genera in the aquatic pyralids (Nymphulinae) of which *Hydropionea, Geshna, Diathrausta, Stenoides, Piletocera,* and *Eurrhypara* are pyraustine and not nymphuline, leaving the genera *Nymphula, Ambia, Cataclysta,* and *Oxyelophila.* Klima (1937) lists ninety-six genera in the Nymphulinae, but many will be found to belong elsewhere when a complete study is made. Schaus (1924, 1940, and other papers) has named many of the tropical American species in the Nymphulinae.

The classification of the North American Nymphulinae in the past has been based upon the work of many early European students. A few of the more salient references to North American Nymphulinae are those of Grote (1880, 1881), Meyrick (1890, 1895), Hampson (1897, 1906), and Dyar (1906, 1917). Forbes (1923) gives keys to the adults and immature forms of the eastern species.

Our known California species of true aquatic moths belong to the genera *Synclita, Usingeriessa,* and *Paragyractis.* In addition, a single larva of the genus *Parapoynx* has been taken.

BIOLOGY

On the basis of ecology and morphology the North American Nymphulinae can be divided into the Nymphulini, or the typical plant-feeding forms, and the Argyractini, or rock-dwelling forms that feed on algae and diatoms.

The immature stages of many of the plant-feeding species are fairly well known in the United States, but the rock-dwelling forms offer a virgin field for research. Packard (1884) and Hart (1895) were among the first in this country to study our aquatic, plant-feeding caterpillars, followed by Forbes' (1910) studies at Lake Quinsigamond and a study of a case-forming species on *Lemna* (Forbes, 1911)[1]. The investigations of Welch (1914, 1915, 1916, 1919, 1922, 1924) and Welch and Sehon (1928) have added greatly to our knowledge of the biology and respiratory mechanisms of the plant-feeding aquatic moths. Frohne (1938, 1939*a*, 1939*b*) discussed the role of higher aquatic plants in relation to the food cycles of insects and particularly in reference to the food habits of several semiaquatic moths related to those under discussion.

Berg (1949, 1950) greatly increased our knowledge of the various aquatic caterpillars associated with plants of the genus *Potamogeton.* The work of Berg and Frohne was continued in investigations by McGaha (1952, 1954), covering the limnological and biological relations of insects to a number of species of flowering plants. Pennak (1953) gives one of the more recent reviews of aquatic Lepidoptera, placing them all in two genera, *Elophila* and *Nymphula.*

There is considerable variation in the biological

[1] I believe this record applies to *Neocataclysta magnificalis* (Hüb.)

relationships of the various plant-feeding Nymphulinae. In general, however, the eggs are laid on the under sides of the leaves of floating hydrophytes, and the larvae cut out sections of leaves, often forming cases in which they live. Oxygen requirements are met by periodic vibratory movements of the larvae, by cutaneous absorption, or by well-developed tracheal gills. Pupation occurs in a silken cocoon inside the larval case. The adults may emerge under water, but do not usually enter water to deposit their eggs.

Lloyd (1914) was apparently the first investigator in this country to record the immature stages of one of the Argyractini, *Elophila fulicalis* (Clemens) (now in the genus *Parargyractis*), from swift waters in New York, and little biological work has been done since this time.

The rock-dwelling forms are well represented in the West. In the present study the larvae of four species in the genus *Parargyractis* were found in California. The larvae of a typical rock dweller construct silken tents under which they feed on algae and diatoms. Pupation occurs in dome-shaped, feltlike cocoons, which have openings at each end to allow the passage of water. The pupae are found in an inner waterproof silken lining, and just before pupating the larva cuts a semicircular escape slit for the adult to escape from the tough cocoon. The adult females enter the water, using the two hind pairs of legs as oars, and deposit the eggs in groups on rocks, often several feet under water in swift streams. The wings of certain groups, such as the *Parargyractis* spp. are held characteristically—the hind wings, particularly in the males, are tilted at an angle downward from the body.

PARASITES

Both tachinid and ichneumonid parasites have been recorded from aquatic nymphuline moths. Lloyd (1919) found about 50 per cent of the pupae of *Elophila fulicalis* (Clemens) *(Parargyractis)* parasitized by an aquatic tachinid, *Ginglymyia acirostris* Townes. Ichneumonids in the genera *Cryptus, Trichocryptus, Cremastus,* and *Neostricklandia* have been reported from aquatic moths.

I reared an ichneumonid near the genus *Pseuderipternus,* in the Cremastini, from the pupal case of a *Parargyractis* sp., in Colusa County, California.[2]

PRESERVING AND COLLECTING

Adults of the Nymphulinae can be collected at a light often at some distance from water. *Parargyractis* adults often rest upon the cement or wooden supports of bridges over the streams they frequent and can be collected during the day in such locations. They also can be frequently swept from bushes or trees growing along the edges of streams, and it is not unusual to find them concentrating in certain trees back from the water's edge. They frequent the shady and protected sides of large boulders along the edges of streams, and can be flushed with an insect net. They often occur with adult Trichoptera and frequent similar locations. Adults of the plant-frequenting nymphulines can be often collected during the daytime, resting on aquatic plants. Toward evening they become more active and can be collected as they fly from plant to plant.

Adults should be mounted and spread at the time of collecting, but can be relaxed, in certain groups at least, quite satisfactorily. Larvae can be dropped into 70 per cent alcohol, but are best preserved by killing first in hot water, then transferring to alcohol. The rock-dwelling forms are collected in more oxygenated niches in streams, lakes, or springs where the larvae occur under silken sheets, or sometimes in old cocoons. The cocoons are made of a tough, feltlike material, and can be easily detached from the rocks with a pair of forceps. It is difficult to rear adults from detached cocoons, but those about ready to emerge will do so in moist battery jars. Larvae or cocoons should be left on rocks and placed in laboratory aquaria without disturbing them, if best results are to be obtained. Larvae of the plant-feeding forms are usually taken in cases, either floating on the surface or attached to the floating leaves or stems of aquatic plants. They will often complete their life history out of water, and adults will emerge from pupae in cocoons if they are placed in moist rearing chambers.

TAXONOMIC CHARACTERS

The wing venation, wing pattern and color, maxillary and labial palpi, hairs and spines on the legs, the antennae, sclerotized parts of the abdominal segments, male sexual tufts on the abdomen, and the male and female genitalia are all used in the classification of this group of moths. Examination of the genital structures is necessary as color and patterns often vary. The entire abdomen is usually removed and dropped into 10 per cent KOH and heated for ten minutes to several hours depending upon the age and individual differences. It is well at the time of placing in KOH to observe and record the color pattern of the abdomen and the presence or absence of sexual tufts which can be taxonomic characters. Under most conditions the females will require a longer period of KOH treatment than the males. Superficial hairs and extraneous materials are teased out in distilled water. In the females, the eggs and other internal structures can be removed by slitting the sides of the body and teasing them out so as to leave the genital structures. The specimen is then transferred to a solution of approximately 40 per cent glacial acetic acid, 30 per cent 95 per cent alcohol, and 30 per cent water. After fifteen minutes the specimen is transferred to cellosolve to which has been added a few drops of clove oil, left fifteen to twenty minutes, and mounted directly in euparol or balsam under a cover slip. In the male the valva are usually spread apart, and the aedeagus

[2]Determined by Miss L. M. Walkley.

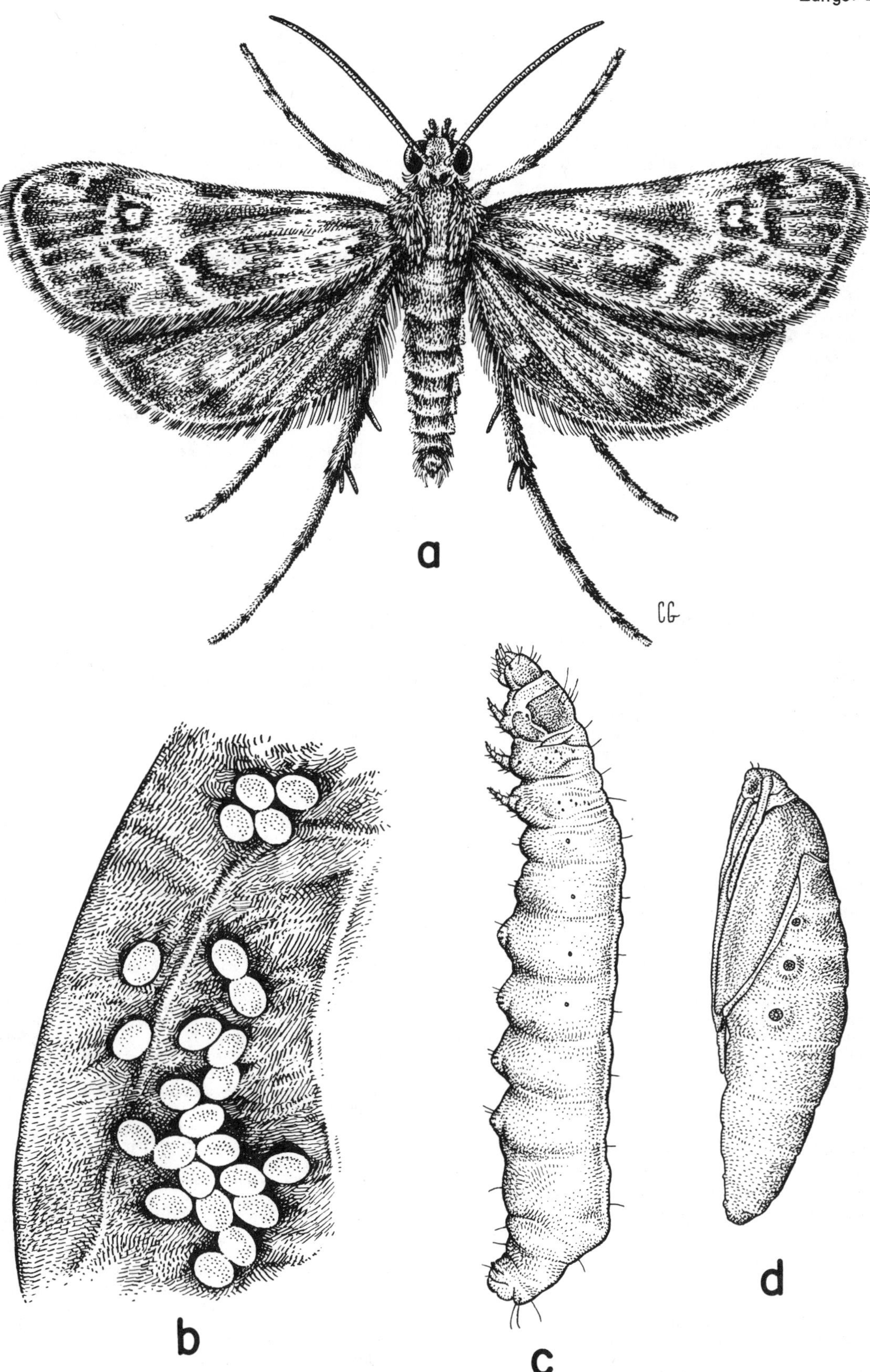

Fig. 11:1. *Synclita occidentalis* Lange. *a*, adult male; *b*, eggs on under surface of leaf of *Echinodorus cordifolius*; *c*, lateral view of mature larva; *d*, pupa.

teased out and mounted to one side. In the females it is usually better to build up the slide by means of broken pieces of cover slip or rings so as not to flatten out and distort the structures to be observed. Staining may be of some help in particular cases.

Wing venation can be examined by removing the wings on one side and placing them on a slide with a drop of cellosolve. Euparol can be added and a permanent mount made which permits the wing venation to be observed and likewise the wing pattern. The wing venation can be easily projected on a bioscope and the details filled in under the microscope.

Some larvae have been prepared by boiling in 10 per cent KOH, splitting the body longitudinally, and preparing by the same methods already discussed. The mandibles can be examined *in situ* or dissected out to one side. Palpi can be examined *in situ* or mounted on slides using the same techniques already discussed.

Family PYRALIDAE

Subfamily NYMPHULINAE (HYDROCAMPINAE)

Proboscis well developed. In fore wings R_2 usually, but not always, stalked with R_{3+4}. R_5 free, or basally stalked with R_{2+3+4}. Maxillary palpi well developed. Labial palpi long, ascending; in many forms, but not all, the 3rd segment is longer than half the median segment. Male genitalia with well-developed uncus; gnathos well developed, arising at posterior extremity of tegumen[3], often, but not always, toothed. Female genitalia with base of ductus bursae often sclerotized; bursa with or without signum. Larvae of known species aquatic, either feeding on aquatic plants, or rock dwellers feeding on algae or diatoms.

The recent paper of Lange (1956) suggests the presence of two tribes to receive the North American members of the subfamily. Until more tropical material is examined, it is impossible to know if this classification will apply on a world-wide basis.

Tribe NYMPHULINI

Nymphulites Duponchel, 1844:201. emen.

Fore wings with vein 1A absent. Hind wings with vein M_2 present; vein 1A present, entire. Middle and hind legs of females lacking tibial swimming hairs. Female genitalia with cornuti absent, or if present, arranged as compact groups of spines.

The larvae of this group are plant feeders, with or without gills, often making tubes or cases of leaves in which they live. The mandibles of the larvae are small and the teeth are arranged in a semicircular fashion for feeding on leaves (fig. 11:7*f*-*g*).

Tribe ARGYRACTINI

Argyractini Lange, 1956.

Fore wings with vein 1A present. Hind wings with vein M_2 absent; vein 1A vestigial, occasionally entire. Middle and hind legs of females with a well-developed row of tibial, swimming hairs. Female genitalia with cornuti absent, or with scattered or spirally arranged thornlike spines, or with small spines arranged in extensive bands.

The larvae of this group are suited to an aquatic habitat, feeding on algae and diatoms on the surface of rocks, often in fast-flowing streams, or in lakes, or springs. The mandibles of the known species are large, flattened, and the teeth arranged in a flat plane. The known larvae have blood gills. The adult females of the known species are capable of entering water and depositing their eggs on rocks, sometimes several feet under water.

Key to Genera of North American Nymphulinae

Based Upon Superficial Characters of the Adults[4]

1. Labial palpi pendant; 2A of fore wings extending only to center of inner margin ..... *Acentropus* Curtis, 1834
— Labial palpi ascending; 2A of fore wings reaching anal angle 2
2. M_2 present in hind wings 3
— M_2 absent in hind wings 13
3. Hind wings strongly incised below apex 4
— Hind wings not incised or with slight indication 5
4. Male with costal glandular swelling on fore wing; median tibial spurs of metathoracic legs very long, attaining apex .... *Undulambia* Lange, 1956
— Males lacking costal glandular swellings; medial tibial spurs short, not attaining apex ... *Contiger* Lange, 1956
5. Males with basal costal fold; M_2 in hind wing stalked with M_3 near outer margin; Florida .... *Oligostigma* Guenée, 1854
— Males lacking basal costal fold; M_2 not stalked with M_3 in hind wing 6
6. Outer margin of hind wings with series of black and metallic spots 7
— Outer margin of hind wings lacking spots 9
7. Black and metallic spots limited to small area on emarginate subapical part of termen of hind wing .... *Oligostigmoides* Lange, 1956
— Black and metallic spots extending the entire length of the termen of the hind wing 8
8. Black spots distinctly separate with pupillate, bluish centers; vein 1A in secondaries complete .... *Neocataclysta* Lange, 1956
— Black spots not distinctly separate; bluish or other metallic spots not pupillate; vein 1A in secondaries vestigial .... *Chrysendeton* Grote, 1881
9. Color of wings fuscous to reddish-brown; small, wing expanse of males usually not exceeding 16 mm. 10
— Light-colored species with wings variously banded, immaculate white, streaked with dark lines, or with ocellate spots; wing expanse of males usually exceeds 16 mm. 11
10. Apex of fore wing pointed; dark fuscous species; 2nd segment of labial palpi triangularly tufted apically .... *Nymphuliella* Lange, 1956
— Apex of fore wing rounded; fuscous, with reddish or brownish markings; 2nd segment of palpi lacking triangular tuftings .... *Synclita* Lederer, 1863
11. Antennae, especially in males and on terminal segments, ringed distally with tufts of scales, giving enlarged appearance to segments; labial palpi long, ascending to center or apex of frons and with scales projecting downward from segments 1 and 2 .... *Parapoynx* Hübner, 1826
— Antennae simple, ciliate, some segments with appressed scales on one side; labial palpi lack downwardly projecting scales from segments 1 and 2 12

[3]In *Eoparargyractis* the tegumen is short and the uncus very long, so that the gnathos appears to arise from the basal junction of the valva.

[4]After Lange (1956).

12. Fore wings with prominent subterminal white band and three large ocellate spots margined with black .... *Nymphula* Schrank, 1802
— Fore wings lacking subterminal white band and 3 ocellate spots .... *Munroessa* Lange, 1956
13. Fore wings falcate apically . .*Oxyelophila* Forbes, 1922
— Fore wings not falcate apically 14
14. Cell in hind wings open ... *Eoparargyractis* Lange, 1956
— Cell in hind wings closed 15
15. Cell in hind wings extending beyond center of hind wing; tornus of fore wing in most species rounded; outer margin of hind wings not emarginate below apex .... *Parargyractis* Lange, 1956
— Cell in hind wings extending to about center or just beyond center of wing; outer margin of hind wings emarginate below apex 16
16. Veins SC and RS branching at about one-half distance to apex of hind wing 17
— Veins Sc and Rs branching just before apex of hind wing 18
17. Tegulae long, particularly in male; black spots on secondaries small, confined chiefly to veins; usually light-colored species of contrasting colors .... *Argyractoides* Lange, 1956
— Tegulae in male short; black spots on secondaries usually large, not confined to veins and inwardly edged with a linear, black line; dark-colored species .... *Usingeriessa* Lange, 1956
18. Fore wing broad, apex broadly rounded .... *Neargyractis* Lange, 1956
— Fore wing slender, apex distinctly pointed .... *Argyractis* Hampson, 1897

Key to Genera of North American Nymphulinae

Based upon Male Genitalia[5]

1. Costa of valva with 4 long, sickle-shaped processes .... *Nymphula* Schrank, 1802
— Costa of valva without sickle-shaped processes ... 2
2. Eighth sternite with a central, posteriorly projecting process 3
— Eighth sternite lacking a posteriorly projecting process 5
3. Central process of 8th sternite wide, triangulate; aedeagus very long, angulate ... *Contiger* Lange, 1956
— Central process of 8th sternite narrowed terminally; aedeagus shorter, not angulate 4
4. Uncus long, slender, fingerlike; cupped process near base of tegumen on each side near base, absent .... *Synclita* Lederer, 1863
— Uncus short, wide at base, constricted at middle, spatulate at tip; cupped process near base of tegumen present .... *Nymphuliella* Lange, 1956
5. Aedeagus with 1 to several large, internal spines or cornuti 6
— Aedeagus with cornuti as groups of small, external, or internal spines, or cornuti lacking 9
6. Saccus of vinculum wider than basal junction of gnathos; cornutus of several large spines; male with sexual tufts on 8th sternite .. *Oxyelophila* Forbes, 1922
— Saccus of vinculum not as wide as basal junction of gnathos 7
7. Gnathos arising near articulation of valva......... .... *Eoparargyractis* Lange, 1956
— Gnathos arising distant from articulation of valva 8
8. Uncus penciliform, long, slender, greatly surpassing gnathos; sexual tufts on 8th sternite absent; cornutus distinct, but not greatly enlarged .... *Chrysendeton* Grote, 1881
— Uncus not penciliform, enlarged apically; sexual tufts in male on 8th sternite and 1st sternite present; cornutus greatly enlarged, recurved .... *Argyractoides* Lange, 1956

[5]After Lange (1956).

9. Gnathos wide apically and with separate nipplelike process; aedeagus longer than valva; base of valva distinctly wider than apical part .... *Acentropus* Curtis, 1834
— Gnathos not wide apically, attenuated, and nipplelike process absent; aedeagus same length or shorter than valva; base of valva slightly narrower or about same width as valva at widest point 10
10. Valva distinctly narrowed apically; aedeagus clavate, with bulbous base; juxta lobes surround aedeagus .... *Undulambia* Lange, 1956
— Valva rectangular in general outline, or expanded apically, not narrowed; aedeagus elongate, inflated basally or terminally but not clavate with bulbous base; juxta lobes not developed 11
11. Sacculus of left valva with thumblike process or small papillalike processes with projecting setae 12
— Sacculus of left valva without processes; tufts of hair may be present at base of valva 13
12. Sacculus of valva with thumblike process set with setae; distal end of sacculus with projecting tuft of hairs; apex of valva broadly falcate .... *Oligostigma* Guenée, 1854
— Sacculus of valva usually with 1 or 2 small setiferous, papillalike processes[6]; sacculus distally lacking tufts of hair; apex of valva evenly rounded or truncate .... *Parapoynx* Hübner, 1826
13. Tuft of setae arising from verrucalike area at basal margin, or near basal margin of valva 14
— Tuft of setae absent 15
14. Tuft of setae located above saccus near basal margin of valva; aedeagus 1/2 length of valva; valva greatly elongated with strong costal area and long, apical spines .... *Neocataclysta* Lange, 1956
— Tuft of setae from base of valva; aedeagus more than 1/2 length of valva; valva enlarged, truncate, or emarginate apically, apical spines short .... *Munroessa* Lange, 1956
15. Valva with apical thumblike process or harpe near costal margin; outer margin of valva truncate .... *Oligostigmoides* Lange, 1956
— Thumblike process on valva absent 16
16. Valva simple, evenly rounded apically; sexual tufts on 1st and 8th sternites absent in male .... *Parargyractis* Lange, 1956
— Valva peaked, or with median, apical, thumblike process; often with apical recurved spines; sexual tufts often present 17
17. Valva bladelike with median, apical, thumblike process and apical recurved spines .... *Neararagyractis* Lange, 1956
— Valva peaked apically 18
18. Juxta with 2 converging lateral arms; spines of gnathos greatly produced .... *Usingeriessa* Lange, 1956
— Juxta lacking converging arms; spines of gnathos reduced .... *Argyractis* Hampson, 1897

Key to Genera of North American Nymphulinae

Based Upon Female Genitalia[7]

1. Ostium small; base of ductus bursae sclerotized only at ostium .... *Acentropus* Curtis, 1834
— Ostium large, opening into a sclerotized part of ductus bursae 2
2. Ductus bursae with spines 3
— Ductus bursae lacking spines 4
3. Ductus with rows of thornlike spines; bursa copulatrix ovate .... *Argyractis* Hampson, 1897
— Ductus with a valvelike group of large spines some of which project into bursa copulatrix; bursa elongate .... *Neargyractis* Lange, 1956
4. Bursa copulatrix with spines or sclerotized plates or bands 5

[6]*P. maculalis* (Clemens) 1860, lacks processes.
[7]After Lange (1956).

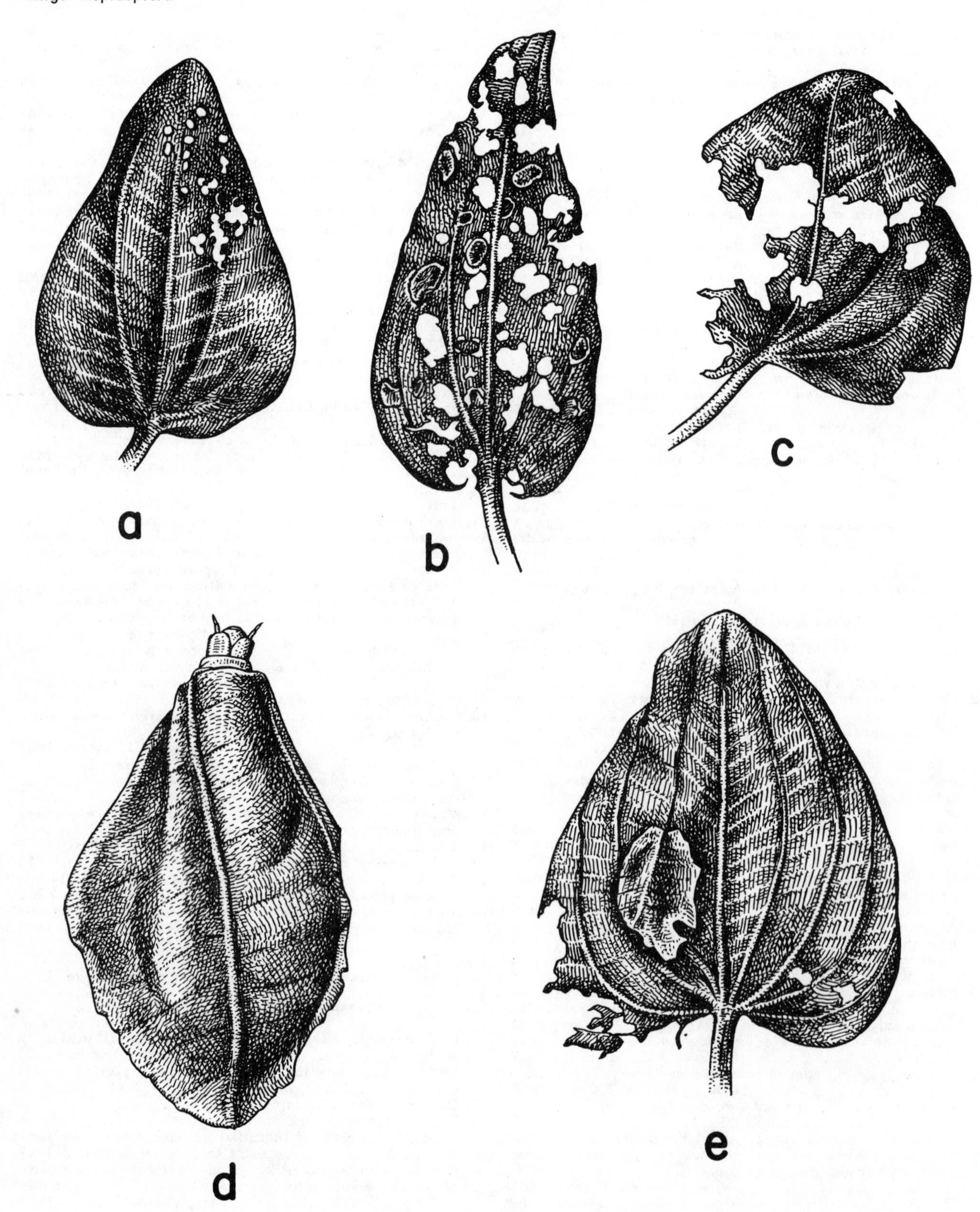

Fig. 11:2. *Synclita occidentalis* Lange associated with *Echinodorus cordifolius*. *a*, young larvae cutting out circular holes; *b*, damage, and larvae feeding under single pieces of leaves; *c*, damage by older larvae; *d*, larval case; *e*, cocoon of two pieces of leaves attached to under surface of floating leaf.

— Bursa copulatrix lacking ornamentations 11
5. Bursa copulatrix with separate, thornlike spines, few to many in number; spines in some species arranged spirally . . . . *Parargyractis* Lange, 1956 in part
— Bursa copulatrix with sclerotized plates, groups of small, dentate or spinulate spines 6
6. Bursa copulatrix with scattered, dentate or spinulate spines . . . . *Usingeriessa* Lange, 1956
— Bursa copulatrix with sclerotized plates or spines arranged in a definite pattern 7
7. Bursa with sclerotized bands or plates, spines vestigial 8
— Bursa with distinct spines arranged in a definite pattern 9
8. Bursa with band in bursa extensive, curved, with ends approximate; collar of ductus bursae long . . . . *Oligostigma* Guenée, 1854
— Bursa of 1 or 2 small bands of closely set spines 9
9. Bursa with 2 groups of closely arranged spines . . . . *Neocataclysta* Lange, 1956
— Bursa with restricted, linear bands composed of very small spines 10
10. Bursa with 1 linear band made up of small spicules . . . . *Oxyelophila* Forbes, 1922
— Bursa composed of 2 linear bands . . . . *Parapoynx* Hübner, 1826 in part
11. Ductus bursae slender, long, may extend to 1st or 2nd abdominal segment in some species; lightly sclerotized and little differentiation of basal part . . . . *Parapoynx* Hübner, 1826 in part
— Ductus bursae shorter, never attaining base of abdomen; base of ductus sclerotized, usually differentiated from distal part 12
12. Bursa copulatrix opaque, surface roughened, reticulated, or with raised elevations or spines 13
— Bursa copulatrix transparent, surface markings absent 17
13. Bursa copulatrix constricted anteriorly with a bowed sclerotization . . . . *Argyractoides* Lange, 1956
— Bursa lacking sclerotized structure 14
14. Surface of bursa finely reticulated; collar section of ductus bursae between base and bursa enlarged, greatly attenuated distally; bursa weakly developed . . . . *Synclita* Lederer, 1863
— Surface of bursa roughened, or with raised elevations collar section of ductus bursae not attenuated distally; bursa saclike, developed 15
15. Bursa with sparsely scattered spines over surface . . . . *Nymphuliella* Lange, 1956
— Surface covered with densely distributed spicules or raised elevations 16
16. Base of ductus bursae wide, joined distally to valvelike spined structure; ovipositor short; anterior apophyses not joined at base . . . . . *Munroessa* Lange, 1956
— Base of ductus bursae elongated; sclerotized collar of bursa in place of spined structure; ovipositor very long; anterior apophyses greatly developed, joined at base . . . . *Chrysendeton* Grote, 1881
17. Ductus bursae with longitudinal striations; little differentiation between basal and terminal parts of ductus bursae . . . . *Undulambia* Lange, 1956
— Ductus bursae lacking longitudinal striations; usually a basal sclerotized part of ductus bursae clearly separated from terminal part 18
18. Ductus bursae very wide; bursa a spheriodal, membranous sac . . . . *Oligostigmoides* Lange, 1956
— Ductus bursae narrow; bursa elongate, or if spheroidal, clearly differentiated from ductus bursae 19
19. Base of ductus bursae well developed, attenuated terminally; surface scobinate or spiculate 20
— Base of ductus bursae more restricted; scobinations or spicules absent 21
20. Surface of ductus bursae scobinate; a distinct valvelike sclerotization between base and terminal part of ductus bursae . . . . *Nymphula* Schrank, 1802
— Surface of ductus bursae spined; valvelike sclerotized plate absent . . . . *Contiger* Lange, 1956
21. Base of seminal duct enlarged and sclerotized; basal part of ductus bursae not distinctly separate from terminal part . . . . *Eoparargyractis* Lange, 1956
— Base of seminal duct not enlarged; basal part of ductus bursae distinctly separate from terminal part . . . . *Parargyractis* Lange, 1956 in part

The morphology of the larval and pupal stages of the Nymphulinae are imperfectly known. For this reason adequate keys to the North American genera cannot be given at this time. Forbes (1923) utilized gill numbers to separate out the eastern species. The investigations of Berg (1950) indicate that gill numbers are difficult to use for taxonomic purposes as the number varies with the instar available and with individuals of the same instar.

Preliminary Key to Known Larvae of North American Genera

1. Mandibles small, teeth arranged in a semicircle; species constructing cases and feeding on floating or partly submerged plants 2
— Mandibles large, flattened, teeth arranged in a flat plane; with simple, blood gills; species living under webs on rocks under water, feeding on algae and diatoms . . . . *Parargyractis* Lange, 1956
2. With branching, tracheal gills; in cases made of excised portions of leaves of aquatic plants . . . . *Parapoynx* Hübner, 1826
— Gills absent 3
3. Case spheroidal, made of *Lemna* leaves . . . . *Neocataclysta* Lange, 1956
— Case oblong, with sharp lateral edge, made of excised portions of leaves of various aquatic plants . . . . *Nymphula* Schrank, 1802; . . . . *Synclita* Lederer, 1863; . . . . *Munroessa* Lange, 1956

Key to California Adults Based on Superficial Characters

1. Outer margin of hind wings lacking a series of black metallic spots; M_2 of hind wings present; 1A of fore wings absent . . . . . *Synclita occidentalis* Lange 1956
— Outer margin of hind wings with a series of black and metallic spots; M_2 of hind wings absent; 1A of fore wings present, vestigial 2
2. Outer margin of hind wings excised below apex; marginal black spots of hind wings distinct, preceded with a narrow, black line . . . . *Usingeriessa brunnildalis* (Dyar) 1906
— Outer margin of hind wings not excised, but evenly rounded; black spots of hind wings not as clearly separated, inner line preceding spots interrupted, if present 3
3. Both wings uniformly irrorated with gray; black spots on outer margin of hind wings reduced; southern California species . . . . *Parargyractis schaefferalis* (Dyar) 1906
— Wings not uniformly irrorated with gray; color of background white or rufous with some scattered gray scales 4
4. Ground color of both wings white; fore wings with ochraceous transverse bands; hind wings with a costodiscal black-outlined mark just basad of white field; southern California species . . . . *Parargyractis kearfottalis* (Dyar) 1906
— Ground color of fore wings reddish-brown; basal ochraceous bands of fore wings absent; costodiscal black

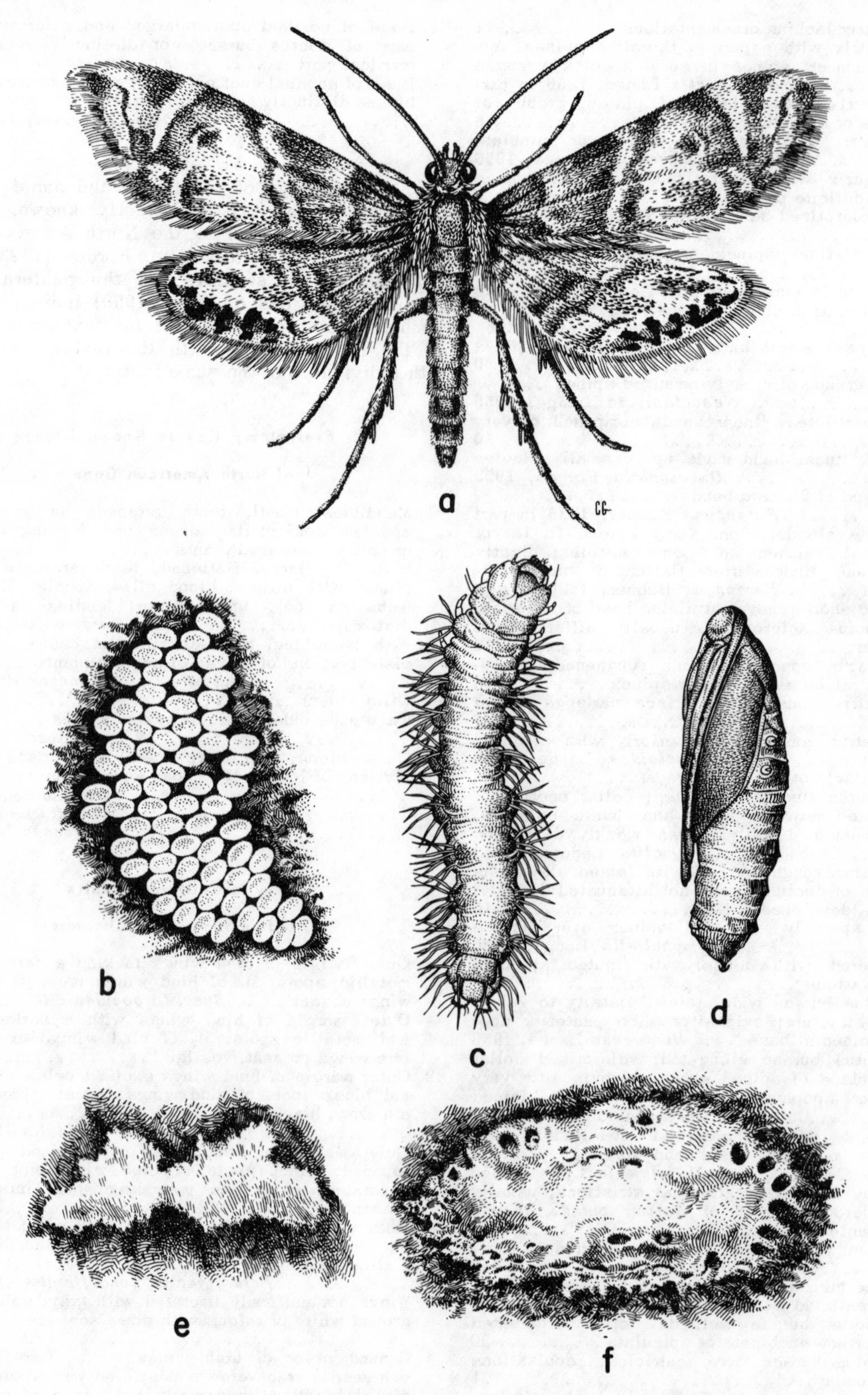

Fig. 11:3. *Parargyractis truckeealis* (Dyar). *a*, adult male; *b*, eggs laid on rock under water; *c*, mature larva; *d*, pupa; *e*, larval web on rock; *f*, cocoon attached to rock.

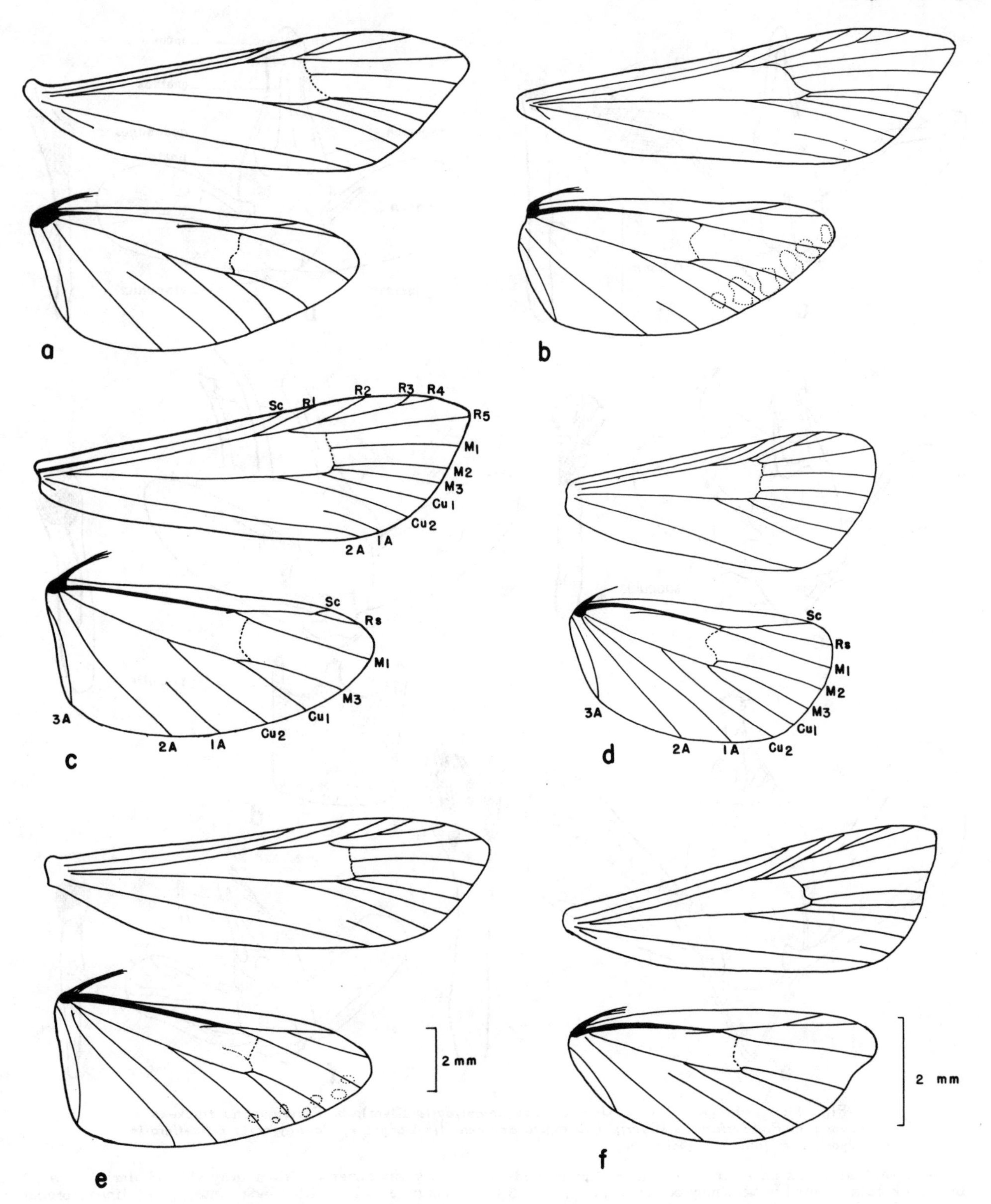

Fig. 11:4. Wing venation. *a, Parargyractis kearfottalis* (Dyar); *b, P. truckeealis* (Dyar); *c, P. jaliscalis* (Schaus); *d, Synclita occidentalis* Lange; *e, Parargyractis schaefferalis* (Dyar); *f, Usingeriessa brunnildalis* (Dyar) (same scale except for e).

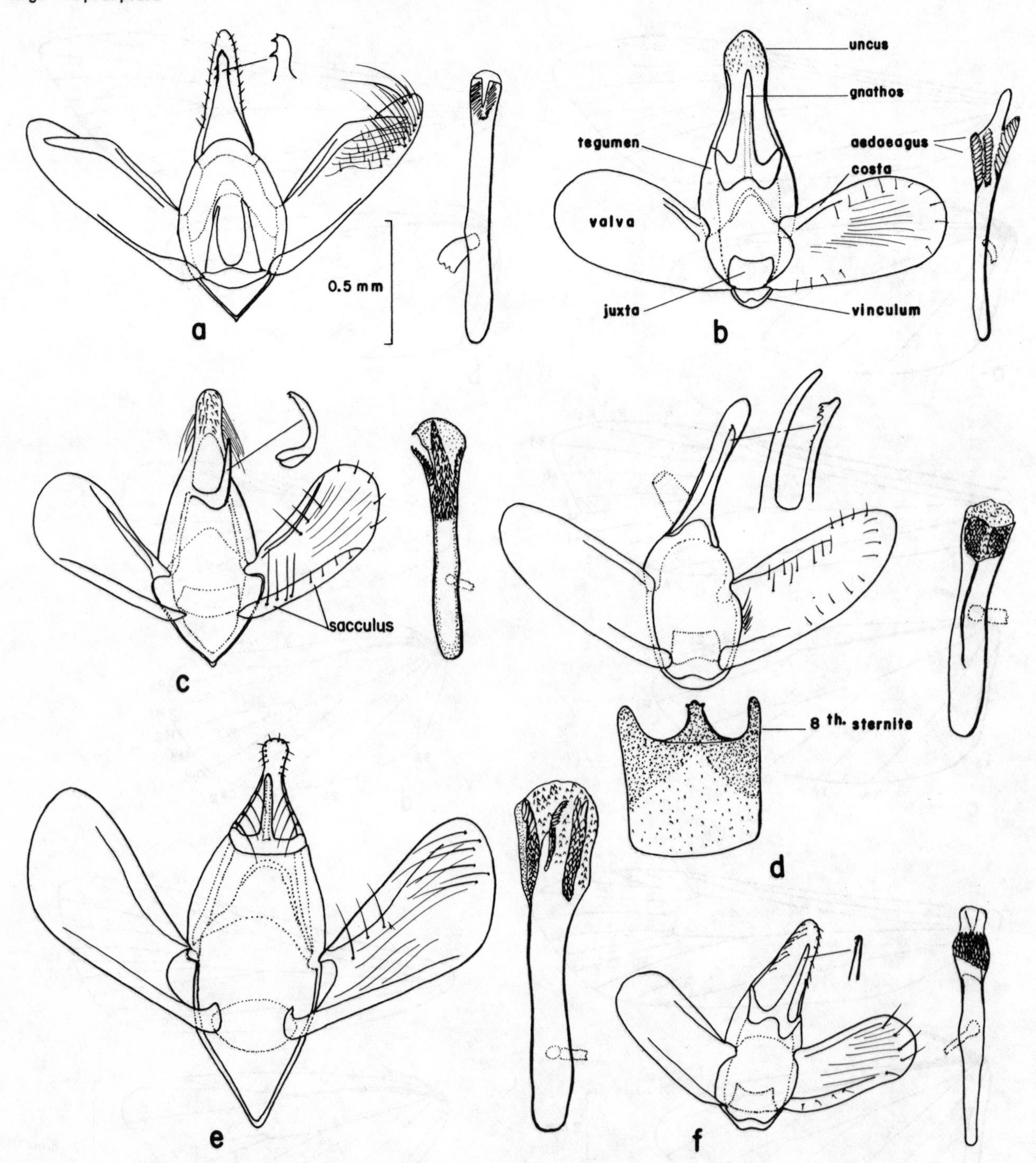

Fig. 11:5. Male genitalia. *a, Usingeriessa brunnildalis* (Dyar); *b, Parargyractis truckeealis* (Dyar); *c, P. kearfottalis* (Dyar); *d, Synclita occidentalis* Lange; *e, Parargyractis schaefferalis* (Dyar); *f, P. jaliscalis* (Schaus).

lines absent although an interrupted black line preceding black dots on hind wings, may occur 5

5. Reddish markings on both wings deeper in color and more distinct; hind wings usually lacking interrupted black lines preceding black dots; subterminal white streak often interrupted below; black irrorations in median space distinct and bright .... *Parargyractis jaliscalis* (Schaus) 1906

— Reddish markings duller in color and less distinct, many specimens with a grayish and darker cast; hind wings usually with black, interrupted lines, preceding black dots; subterminal with streak usually not interrupted below; black irrorations less distinct .... *Parargyractis truckeealis* (Dyar) 1917

Key to California Adults Based on Male Genitalia

1. Eighth sternite with a central sclerotized arm; gnathos

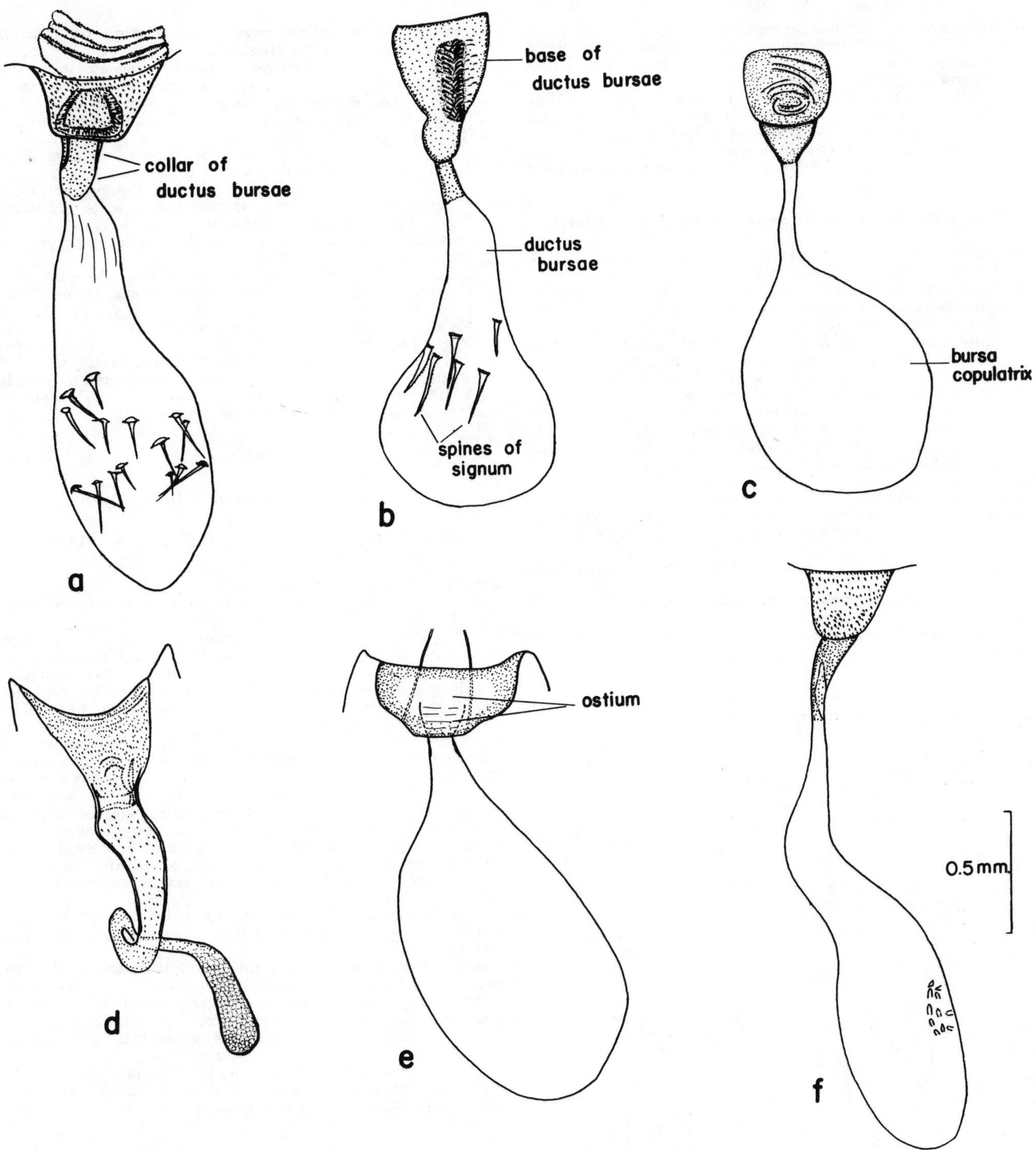

Fig. 11:6. Female genitalia. *a*, *Parargyractis schaefferalis* (Dyar); *b*, *P. kearfottalis* (Dyar); *c*, *P. jaliscalis* (Schaus); *d*, *Synclita occidentalis* Lange; *e*, *Parargyractis truckeealis* (Dyar); *f*, *Usingeriessa brunnildalis* (Dyar).

bladelike and set with numerous teeth .... *Synclita occidentalis* Lange 1956

— Eighth sternite lacking sclerotized central arm; gnathos not bladelike, with or without teeth 2

2. Valva peaked apically; juxta with 2 lateral arms .... *Usingeriessa brunnildalis* (Dyar) 1906

— Valva evenly rounded apically; juxta lacking arms, entire 3

3. Vinculum greatly produced anteriorly 4

— Vinculum not produced anteriorly 5

4. Aedeagus toothed terminally .... *Parargyractis kearfottalis* (Dyar) 1906

— Aedeagus not toothed terminally ....
.... *Parargyractis schaefferalis* (Dyar) 1906
5. Aedeagus set with 3 external groups of spines; valva expanded apically; uncus constricted subapically...
.... *Parargyractis truckeealis* (Dyar) 1917
— Aedeagus with a single closely set group of external spines; valva not expanded apically; uncus not constricted, but evenly attenuated apically ....
.... *Parargyractis jaliscalis* (Schaus) 1906

Key to California Adults Based on Female Genitalia

1. Bursa copulatrix with spines 2
— Bursa copulatrix lacking spines 4
2. Spines in bursa small, spiculate or dentate, more or less vestigial ..*Usingeriessa brunnildalis* (Dyar) 1906
— Spines large, thornlike 3
3. Spines not forming a complete spiral, about 6 in number; base of ductus bursae with group of spines ....
.... *Parargyractis kearfottalis* (Dyar) 1906
— Spines arranged in a spiral fashion, more than 6 in number; base of ductus bursae lacking group of closely set spines ....*Parargyractis schaefferalis* (Dyar) 1906
4. Bursa copulatrix short, elongate; collar of ductus bursae heavily sclerotized ....
.... *Synclita occidentalis* Lange 1956
— Bursae copulatrix large, spheroidal 5
5. Base of ductus bursae purselike, appearing tripartite
.... *Parargyractis truckeealis* (Dyar) 1917
— Base of ductus bursae small, globose ....
.... *Parargyractis jaliscalis* (Schaus) 1906

Key to Known California Larvae

1. Mandibles small, teeth arranged in a semicircle; species constructing cases and feeding on floating or submerged aquatic plants 2
— Mandibles large, flattened, teeth arranged in a flat plane; with simple, blood gills; species living under webs on rocks, feeding on algae and diatoms 3
2. Tracheal gills branched .... *Parapoynx* sp.
— Tracheal gills absent ....
.... *Synclita occidentalis* Lange 1956
3. Head black; gills short, white; Inyo County ....
.... *Parargyractis* sp.
— Head light-colored or brown, not black; gills of mature larvae brown, green or black, long 4
4. Mandibles short, wide; distal margin not greatly narrowed (fig. 11:7*a*) ....
.... *Parargyractis jaliscalis* (Schaus) 1906
— Mandibles greatly elongated, distal margin narrowed (fig. 11:7*d, e*) 5
5. Mandibles greatly enlarged; width at divided tooth approximately 1/5 length; common species (fig. 11:7*e*)
.... *Parargyractis truckeealis* (Dyar) 1917
— Mandibles with width at divided tooth approximately 1/3.3 length; limited distribution in Lake, Colusa, Humboldt, and Plumas counties (fig. 11:7*d*) ....
.... *Parargyractis* sp.

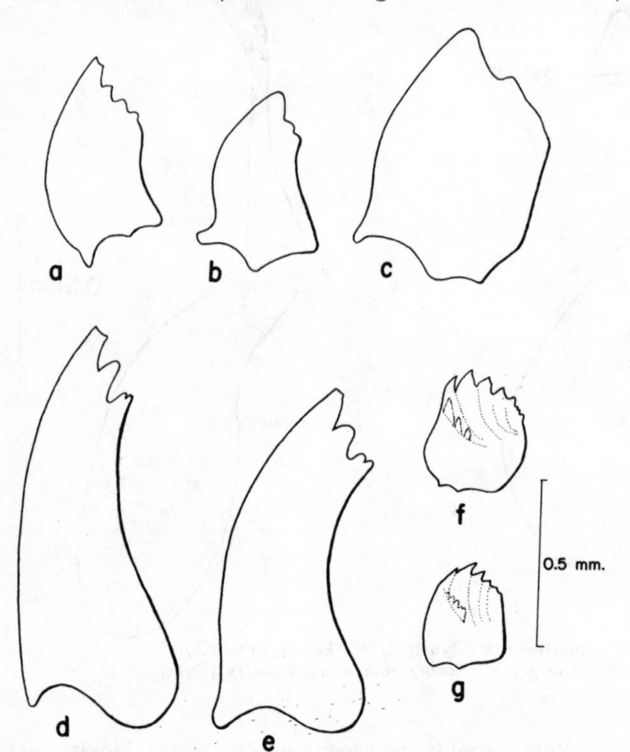

Fig. 11:7. Mandibles of larvae. *a, Parargyractis jaliscalis* (Schaus); *b, P. fulicalis* (Clemens); *c, Parargyractis* sp. (Inyo Co.); *d, Parargyractis* sp. (Little Stony Creek, Colusa Co.); *e, P. truckeealis* (Dyar); *f, Parapoynx* sp. (San Francisco); *g, Synclita occidentalis* Lange.

Synclita occidentalis Lange

(Figs. 11:1; 11:2; 11:4*d*; 11:5*d*; 11:6*d*; 11:7*d*)

Synclita occidentalis Lange, 1956.

A fuscous species with reddish and white markings, closely allied to *S. obliteralis* (Walker).

Male.—Expanse 11-15 mm. Labial palpi fuscous; 2nd segment with projecting scales. Maxillary palpi attaining end of 2nd segment of labial palpi; broadly tufted apically. Ocelli present. Antennae ciliate; tufted terminally. Fore wings fuscous with rufous and white markings; basal dark band, followed by postbasal lighter transverse line and a submedian fuscous band; median area fuscous with a few scattered, white scales, followed by an interrupted transverse white line reaching costa; discal spot white; subapical area reddish with some fuscous scales; submarginal area with some scattered white scales; marginal line banded with alternate light and dark scales. Hind wings fuscous, with reddish area at base and an irregular reddish dash near angle. Legs dark brown; hind tarsi with 2 pairs of subequal stout spurs; tarsal segments dark-banded apically.

Female.—Expanse 15-22 mm. As male except larger and usually lighter in color. Light markings less distinct than male.

Male genitalia.—Valva simple, bladelike; uncus slender, well developed; gnathos well developed, knifelike, with group of large teeth above and several small teeth below; aedeagus stout, with terminal group of spiculate spines. Eighth sternite with characteristic central arm bearing 2 small, stout laterally placed spines.

Female genitalia.—Base of ductus bursae wide, surface minutely spiculate; collar of ductus bursae elongate, well sclerotized; bursa copulatrix reduced, elongate.

Taxonomic position.—There appear to be several species involved in what has previously been called *Nymphula obliteralis* (Walker). This species can be readily identified by the peculiar, three-pronged central process of the eighth sternite in the males. In *obliteralis* this process is spinelike, composed of three slender spines, and in *gurgitalis* Lederer, the central process is greatly enlarged and inflated apically. *S. occidentalis* is, in general, larger than *obliteralis,* and the larger, irregular reddish spots on the hind wings are characteristic. The color varies almost with individuals, particularly the white markings, but the general appearance is much the same. Specimens from Texas and Hawaii seem to be the same as eastern *obliteralis* and are not *occidentalis.*

Distribution.—California and Arizona. California records include the following counties: Alameda, Butte, Colusa, Riverside, San Diego, Stanislaus, and Yolo. It will undoubtedly have a wider distribution when extensive collections are made.

Biology.—Larvae and pupae of this species were first found July 9, 1954, feeding on *Echinodorus cordifolius* growing in a permanent pond at Biggs, Butte County. The life history seems similar to *S. obliteralis* (Walker) as worked out by Williams (1944) and other investigators. The larvae of *occidentalis* are gill-less and feed on a number of aquatic plants growing in permanent ponds and in rice fields. Hosts recorded are: hysop *(Bacopa rotundifolia),* potamogeton *(Potamogeton gramineus),* arrowhead (*Sagittaria* sp.), *Echinodorus cordifolius, Jussiaea californica,* and young cattail *(Typha californica).*

The eggs are disclike, ovate in outline, dirty-white in color, and are laid singly or slightly overlapping on the undersurfaces of aquatic plants near the edges of the leaves. In size they vary from 0.48 to 0.53 mm. wide and 0.56 to 0.71 mm. long, and the surface is roughened. Eggs are laid in groups of from ten to fifty-seven. Larvae upon hatching cut small circular areas out of the leaves and live under single pieces of leaves attached to the undersides of the leaves by a silken web. As the larvae enlarge they cut larger pieces of leaf, forming when they are mature an oblong case made of two pieces of leaf fastened together with silk. The young larvae are bathed in water and apparently breathe cutaneously, whereas it appears that the larger larvae inside the cases are exposed to air, living in a more or less waterproof case from which they reach to feed on plants. The larvae, when ready to pupate, make an inner silken cocoon, fastening the case to the leaves or petioles, sometimes more or less exposed to air and at other times completely submerged.

The small larvae are white, about 1 mm. long and translucent, later turning to a pale green, although the dorsal vessel may show as a darker, dorsal line. The mature larva is robust, with small head and reduced prolegs, attaining a length of about 16 mm. The setae are inconspicuous, and pale-colored. The head and prothoracic shield are often a pale brown in color, glossy, and more or less translucent. The crockets are arranged in a biordinal mesoseries. The spiracles are small, but distinct, and dark-ringed. The pupa is pale brown, 9-16 mm. long by 3-6 mm. wide at the greatest point. Abdominal segments two, three, and four, above, have large, almost stalked spiracles.

There was evidence of at least two and possibly three generations a year of this moth. In rice fields at Biggs, the common host was hysop which in one instance was almost completely destroyed by the larvae of this moth. As rice grows, this weed seems to disappear naturally and the moth maintains itself mostly on *Echinodorus* growing in permanent ponds. On August 5, 1954, moths were collected fairly commonly from five to six P.M. when they could be flushed from rice plants and collected when they alighted two to four inches above the water on rice stems in an area where hysop was being attacked by the larvae. Some adults were collected resting on algae on the surface of the water, and adults were also collected during the day resting on the floating leaves of other plants at Biggs and Maxwell.

No parasites were detected during the period this moth was under observation.

Parargyractis schaefferalis (Dyar)

(Figs. 11:4*e*; 11:5*e*; 11:6*a*)

Elophila schaefferalis Dyar, 1906:95.
Parargyractis schaefferalis, Lange 1956.

A dark gray species with small, marginal dark spots on outer margin of hind wings.

Male.—Expanse 17-20 mm. Labial palpi short, brown, with scattered, white scales. Maxillary palpi short, brown, with apical white scales. Head, thorax, and abdomen with scattered, brown and grayish-white scales. Vertex of epicranium dome-shaped, projecting upward behind antennae. Ocelli absent. Antennae ciliate, with tufts of scales at segments. Fore wings with gray-brown scales on a gray-white background; basal area dark gray; median area lighter, in some specimens outlined with darker-margined lighter, transverse fasciae; a subapical dark streak, followed by a white, apical wedge-shaped mark; cilia dark banded. Hind wings with indication of dark, discal spot; in some specimens a median transverse fascia, dark-outlined on outer side; a submarginal group of 4 black spots, alternating with 4, marginal black spots, located in an orange field, and separated outwardly with metallic, silvery scales.

Female.—Expanse 24-29 mm. As male except larger and dark-margined fasciae more distinct; swimming hairs present on tarsi of middle and hind legs.

Male genitalia.—Valva narrowed at base, expanded bladelike apically; uncus small, spatulate; gnathos short, stout, not reaching end of uncus, set dorsally with small, scattered teeth; vinculum well developed; aedeagus stout, with three groups of apical teeth.

Female genitalia.—Base of ductus bursae large, cuplike, set with numerous small spines; collar sclerotized; ductus bursae with small surface spicules; bursa copulatrix elongate, with a group of about 15 large, thornlike teeth arranged spirally.

Taxonomic position.—The genitalia of this species indicate that it is closest to *P. longipennis* (Hampson). The spiral arrangement of the signa in the female bursa suggests relationships with several other Mexican species.

Distribution.—California and Arizona. In California it has a southern range in Inyo, Los Angeles, and Riverside counties, where it is sparingly taken from May through September.

Biology.—Unknown. It is possible that larvae collected in March at Surprise Canyon, Inyo County, could be either this species or *kearfottalis* Dyar. These larvae are characterized by a black head and small, white gills.

Usingeriessa brunnildalis (Dyar)

(Figs. 11:4*f*; 11:5*a*; 11:6*f*)

Elophila brunnildalis Dyar, 1906:91.
Usingeriessa brunnildalis, Lange 1956.

A dark brown species, with outer margin of hind wings excised below apex and with a marginal row of black dots which are inwardly edged with a black line.

Male.—Expanse 15-20 mm. Labial palpi brown, long, ascending; 3rd segment long, slender, attaining apex. Maxillary palpi brown, slender, reaching end of middle segment of labial palpi. Ocelli absent. Antennae ciliate, inwardly with tufts of parallel scales which become tufted apically. Fore wings dark brown with indistinct lighter lines; basal darker area with indistinct basal transverse lighter line; median area lighter, set off basally with lighter transverse line, which is edged with darker line outwardly, and set off distally with indistinct darker transverse line; a subapical triangular dark patch set off inwardly by an oblique lighter line, and by an outer apical dash; cilia alternately dark and light banded. Hind wings gray-brown; basal darker area followed by an indefinite lighter transverse fascia (distinct in San Diego specimens); a darker discal spot; outer margin with row of about 5 black spots, sometimes indistinct, with silvery scales between, inwardly edged with a black line; cilia dark basally.

Female.—Expanse 15-22 mm. As male except usually larger and antennae not ciliate, but tufted with dark scales.

Male genitalia.—Valva peaked apically, set with long, recurved spines; uncus tongue-shaped, well developed, surpassing gnathos; gnathos stout, with downward projecting teeth on dorsal surface; juxta with 2 lateral sclerotized arms; vinculum well developed; aedeagus long, stout, set with two groups of spiculate spines.

Female genitalia.—Base of ductus bursae sclerotized, numerous spicules; collar of ductus well sclerotized; bursa copulatrix with a group of vestigial spines, surface roughened.

Taxonomic position.—Related to *U. onyxalis* (Hampson) and *U. symphonalis* (Dyar), and an unidentified species from Peru. The San Diego specimens are lighter in color and the markings more distinct. The Placer County specimens tend to be evenly suffused with brown and the marginal spots run together and are indistinct.

Distribution.—California, Arizona, and Texas (April-Oct.). California localities include Placer, Riverside, San Diego, Tulare, and Yolo counties. Only one Texas specimen (Del Rio, Lange slide U-152, female) is associated with this species. The longest series is in the California Academy collection from Bear River, Placer County, California, collected in July.

Biology.—Unknown. The morphological features of this species indicate that the larvae should live on rocks similar to *Parargyractis.*

Parargyractis kearfottalis (Dyar)

(Figs. 11:4*a*; 11:5*c*; 11:6*b*)

Elophila kearfottalis Dyar, 1906:92.
Parargyractis kearfottalis, Lange 1956.

A white species with ochraceous bands on fore wings and marginal black spots on hind wings distinct, and preceded by a costodiscal, subrectangular, black-margined figure.

Male.—Expanse 14-17 mm. Labial palpi white, with a few buff scales. Maxillary palpi short, white. Antennae ciliate, with apical tufts. Head, thorax, and abdomen white, with buff scaling. Fore legs with dark-scaled apical tufts on tibiae; tarsi with dark, apical annulations. Fore wings with white background; a basal group of dark scales; 2 transverse, median bands separated by white area, with dark scales at costa; a postmedian white area with brown irrorations; 2 basally converging, apical white dashes, margined and separated by 3 ochraceous bands, with silvery scales at base and dark scales at costa; marginal, ochraceous band extending almost to tornus, with silvery dash at base on margin; base of cilia dark. Hind wings white; a median ochraceous spot, followed by a silvery dash and an indefinite transverse band extending to margin; a costodiscal, black-outlined, elongate mark enclosing a white field; a marginal group of black spots, separated by silvery scales, margined on inside by an interrupted, black, thin line; cilia dark at base.

Female.—Expanse 17-22 mm. As male, except larger; antennae less ciliate.

Male genitalia.—Valva broad, narrowed basally; uncus stout, curved at tip and with a small spine on dorsal surface; aedeagus with distinct apical tooth and 3 groups of small spines.

Female genitalia.—Base of ductus bursae broad, ostium set with posteriorly projecting spines; collar distinct, sclerotized; signum of 6 or 7 large thornlike spines in bursa.

Taxonomic position.—This species was named by Dyar as a variety of *bifascialis* Robinson from Phoenix, Arizona. It seems to be a distinct species, with quite characteristic genitalia. I have examined a series sent by Dr. Forbes from Phoenix. It is difficult to separate this species by superficial features from some specimens of *bifascialis* lacking the solid costodiscal bar, *daemonalis* Dyar, and *cappsi* Lange from Texas. The spined ostium in the female and the toothed aedeagus of the male are characteristic. This species is lighter in color than the related species, and the interrupted black lines preceding the black, marginal spots are a helpful character.

Distribution.—Widely distributed, but not numerous in collections. The only California material I have seen are specimens from Vidal, San Bernardino County, collected by D. Weedmark, October 5, 1947, and sent by Dr. Munroe from the G. H. and J. L. Sperry collection.

Specimens have been also examined from Arizona, Colorado, Idaho, Nevada, New Mexico, Oregon, and Washington.

Biology.—Nothing is known of the biology.

Parargyractis jaliscalis (Schaus)

(Figs. 11:4*c*; 11:5*f*; 11:6*c*; 11:7*a*)

Cataclysta jaliscalis Schaus, 1906:135.
Parargyractis jaliscalis, Lange 1956.
Elophila satanalis Dyar, 1917:75-76.

A reddish-brown species with rufous, white, and silvery markings, and marginal black spots on hind wings not preceded by a black, interrupted line.

Male.—Expanse 15-17 mm. Labial palpi light brown; 2nd segment with rough scales beneath; 3rd segment long. Maxillary palpi short, brown with buff apical scaling. Antennae ciliate; apically tufted. Head, thorax, and abdomen buff, with scattered darker scales. Fore wings reddish-brown with rufous, white, and silvery markings; basal area dark brown followed by a white, transverse band; a submedian red band, followed by a white fascia; a postmedian dark, reddish brown area on a grayish-white background; 2 converging, apical, white streaks with reddish spot at base and separated by a dark, reddish-brown area; a marginal, reddish band; tornus with oblique, reddish dash with a few silvery scales. Hind wings with basal, reddish, zigzag reddish band, followed by a silvery streak and a white fascia to margin; postmedian area white with grayish-brown irrorations; submarginal black dots on a white background, with silvery and white scales between; marginal dots black, contiguous, separated by a few reddish scales; cilia darker at base.

Female.—Expanse 17-24 mm.; larger than male; antennae less ciliate and tufted.

Male genitalia.—Valva short, wide at base; uncus stout, broadly joined to gnathos, set with apical spines; gnathos well developed, curved at tip and with inconspicuous tooth on dorsal surface; aedeagus narrowed basally; apically with group of closely set spines, preceded by a clear area.

Female genitalia.—Ostium wide above, funnel-shaped ductus bursae not sclerotized; bursa copulatrix lacking signum.

Taxonomic position.—Dyar's *satanalis* from Texas is undoubtedly *jaliscalis* Schaus. This species can be confused on superficial characters with certain individuals of *truckeealis,* but it is more brightly marked than *truckeealis,* and the genitalia will settle doubtful cases.

Distribution.—A widely distributed species extending from Mexico to Redding, Shasta County, California, collected from January to November. California records include: Colusa, Fresno, Glenn, Inyo, Madera, Riverside, San Bernardino, Shasta, Stanislaus, and Yolo counties. I have also examined specimens from Arizona, Mexico, New Mexico, and Texas. In addition to a female sent by Mr. Capps from the National Museum collection determined by Mr. Schaus from Guadalajara, Mexico, I have seen specimens collected by E. Schlinger during July, 1954, from five miles East of Pairita, Canon de Santa Clara, Chihuahua, Mexico, and Nombre de Dios, Durango, Mexico.

Biology.—Larvae of this species were first found at Stony Creek near Hamilton City, Glenn County, on June 5, 1954. The biology is similar to *P. truckeealis* (Dyar), and in certain northern localities it can occur together with *truckeealis.* Mr. D. Abell has found *jaliscalis* associated with both the intermittent and permanent part of Dry Creek, Fresno County, and has supplied me with a number of specimens. Long series of adults have been collected by Dr. John Comstock and Mr. Lloyd Martin from Madera Canyon, Arizona, throughout a period of several years, in an area where the stream is dry during most of the year. At Ladoga, Colusa County, on October 13, 1954, Mr. A. A. Grigarick and I collected larvae of *jaliscalis* and those of an unidentified larva in a stream where only a trickle of water was running and the water was ill smelling and almost stagnant in the pools.

As with *truckeealis,* the females apparently enter the water to deposit their eggs under rocks. At Ladoga, eggs of what are probably this species, were found under rocks in shallow water where a trickle of water was still flowing, but not in the deep, stagnant pools. The flattened, ovate, tan-colored eggs were laid in overlapping groups under the rocks, and varied in width from 0.24 to 0.29 mm. and in length from 0.36 to 0.42 mm. In many localities the rocks were covered with algal growth, much more so than for *truckeealis.* The larvae live under thin, silkenlike webs and apparently feed on the algae and diatoms occurring on the rocks. These larval webs have small openings all around the edge and vary in size as they conform to depressions or grooves in the rocks. The pupal cases are similar to *truckeealis,* although they usually have fewer openings on the ends, and the openings are larger. The pupal case is firmly attached by two reinforced anchorlike attachments on each side; it varies in size, but may be about 12 mm. wide by 18 mm. long, dome-shaped, and attached to bed rock, small rocks, or larger rocks.

The mature larva of *jaliscalis* has characteristic mandibles which separate it easily from related species (fig. 11:7*a*). The larvae usually are lighter in color than *truckeealis,* but can range in color from dirty-white to blackish, with a brown head and prothoracic shield. The blood gills are more numerous and thinner than associated species. The mature larva is about 15 mm. long and 3 mm. wide (without gills). Abdominal segments one to eight have a subspiracular group of four to five filamentous blood gills, and a supraspiracular group with the same number. The bases of the gills in the next to last instar are dark (preserved specimens), a character found in the mature larva of the unknown Stony Creek species. The larvae of *truckeealis* usually have only three subspiracular gills on abdominal segments one to eight, and only one to three supraspiracular gills.

The pupa occurs in an apparently air tight silk-lined inner cocoon. The pupa is brown, 8 mm. long by 3 mm. wide at the greatest point. The cremaster is anchorlike, and two-pronged, stouter than that found in *truckeealis.* Abdominal segments two and three, have well-developed, dorsal, stalked spiracles and abdominal segments four to seven have dorsal and lateral transverse ridges.

Several generations occur each year, judging from the occurrence of larvae throughout the year. In Colusa County larvae were found to overwinter.

Parargyractis truckeealis (Dyar)

(Figs. 11:3; 11:4*b*; 11:5*b*; 11:6*e*; 11:7*e*)

Elophila truckeealis Dyar, 1917:76.
Parargyractis truckeealis, Lange 1956.

A grayish-red to fuscous species with white and fuscous markings, and black spots on hind wings preceded (in most specimens) at apex and tornus by a black line.

Male.—Expanse 11-19 mm. Labial palpi light brown; 2nd and 3rd segments with a few rough scales below. Maxillary palpi broadly tufted apically. Antennae ciliate basally, with a few scales on inside, and with apical tufts. Head, thorax, and abdomen variable in color, buff to dark brown. Fore wings with basal dark area followed by a lighter, transverse line; a dark, submedian band followed by a white, transverse line which is dark edged; a median, gray-irrorated area; a discal dark spot followed by a luteous mark; 2 basally converging, white apical dashes, separated by a dark brown area, with a luteous and silvery mark below; a marginal, luteous line; a luteous dash at tornus; cilia darker at base. Hind wings with reduced basal, dark area, followed by a reddish-brown band; a median, silvery zigzag line, followed by a grayish, irrorated area; a dark, discal spot; a marginal group of black spots in a yellow field, separated by silvery spots and a marginal group of luteous spots; dark spots preceded at apex and tornus with a thin, interrupted, black line; cilia dark at base.

Female.—Expanse 16-22 mm. As male, except larger; antennae less ciliate; swimming hairs on meso- and metathoracic legs.

Male genitalia.—Valva broad, narrowed slightly basally; uncus stout; gnathos strong, curved at tip, with inconspicuous dorsal spines; aedeagus with 3, externally, basally hinged, groups of spines.

Female genitalia.—Base of ductus bursae wide, purselike, appearing 3-partitioned; ostium centrally situated; ductus bursae nonsclerotized; bursa copulatrix lacking signum.

Taxonomic position.—This species is the western counterpart of the eastern *fulicalis* and related to the northern *opulentalis,* judging from the genitalia, but is distinct. *P. fulicalis* (Clemens) extends west as far as Texas, and *opulentalis* (Lederer) is very common in Canada.

Distribution.—This is our commonest species in northern California. I have examined it from British Columbia, California, Idaho, Montana, Nevada, Oregon, Washington, and Wyoming. In California specimens have been examined from the following counties: Butte, Colusa, El Dorado, Glenn, Humboldt, Lake, Mendocino, Napa, Nevada, Placer, Plumas, Sacramento, Shasta, Sierra, Sonoma, and Stanislaus. I found this species quite common on the west shore of Rattlesnake Island in Clear Lake, Lake County, and it is associated with fast-flowing streams. The most southern locality found by me is the Stanislaus River near Oakdale, but more extensive collecting is needed.

Biology.—I collected larvae of this species at Trout Creek, Samuel Springs, Napa County, on April 11, 1954, and since this initial collection, I have found it quite common in California, Oregon, Washington, and Idaho. It has not been collected to date in the more coastal, acid, cold streams flowing into the ocean. Mr. H. P. Chandler collected this species in many localities in connection with his work with the Department of Fish and Game, and I am indebted for the privilege of examining his collections and including some of his observations.

The larvae of this species occur on rocks, often those covered with algae and diatoms, and make silken webs, often in depressions, under which they feed. They also occur occasionally on submerged tree trunks or metal objects in certain situations. Their food consists of algae and diatoms. Often thirty or forty larvae per square foot of rock surface may occur. The larval tents are irregular in outline since they conform to depressions or grooves in rocks. The characteristic cleared areas under the cocoons, which indicate the earlier work of this species, often remain on the rocks after the water level drops. They have a number of small openings around the edge, somewhat like hemstitching, to let in a circulation of water. As the larvae enlarge, the tents are extended to enclose more rock surface. When ready to pupate the larva fills in the space beneath the tent with loose, woollike filaments, then spins several blankets of closely matted filaments about itself which are closely appressed and stuck together. The larva spins an anchor to stick the cocoon to the rock, and also to leave openings to let in water. In order to enable its escape from the cocoon the larva cuts a semicircular "escape hatch" which is rather inconspicuous but is pushed out flaplike when the adults emerge. The larva then spins an inner, apparently air-tight silken cocoon in which it pupates. The inner cocoon apparently contains air, as reported by Lloyd (1917) for *fulicalis.* The felty cocoons vary in different streams from 7-15 mm. wide to 15-20 mm. long.

The mature larvae of *truckeealis* vary in color from blackish to greenish and have dark brown heads and thoracic shields. They are about 15 mm. long and 3 mm. wide (without gills), and the crochets are arranged in a biordinal semicircle. The larvae have characteristic filamentous blood gills, a subspiracular group of three gills on abdominal segments one to six and usually a single supraspiracular gill on each segment. The number of gills varies to a certain extent in different localities and in individuals. The larvae of this species seem to have a wide adaptability to different conditions as they have been collected on small rocks in fast-flowing, shallow riffles, to larger rocks in streams, to bedrock areas of streams in situations where the water is well aerated, to exposed rocks in lakes where the water is aerated.

The pupa is dark brown, and approximately 2.5 mm. wide at the widest point, by 8 mm. long, and with a characteristic, anchorlike cremaster. The enlarged spiracles on abdominal segments two and three and the ridges on the abdominal segments are similar to those of *jaliscalis.* Pupae emerge in the laboratory in six to nine days.

In Rumsey Canyon where the species has been observed throughout the year it is thought that three generations occur. As the water recedes, the old cocoons of earlier generations can be seen, and the fall-collected adults are often very small. Larvae overwinter, and adults disappear during the winter. In the laboratory, adult females were found to crawl down rocks and deposit their eggs under water on the undersides of the rocks. They utilize a plastron type of respiration as the wings are folded back and covered with a silvery sheen. Eggs were found on rocks in many streams, often under several feet of water, and many times in the center of deep holes where it would seem impossible for larvae to crawl. It is assumed that adult females can dive under water and use their swimming legs to reach rocks. There is some indication that the small larvae crawl about, but not to any great extent. Mr. Chandler found larvae on rocks below Pitt 4 Dam, Shasta County, under four to six feet of water, in a part of the stream where no rocks are available for the adults to crawl down, and where the water is very seldom lowered. I found larvae in a fast-flowing section of the North Fork of the American River near Auburn on May 28, 1954, under two feet of water, and in a section of the stream where it was almost impossible to remain on one's feet because of the swiftness of the water.

Eggs are laid in groups of a few, to more than two hundred and fifty, are yellow to tan in color, flattish, elongate-ovate in outline, glued to the surface of the rocks and often slightly overlapping. Eggs vary in size from 0.25 to 0.33 mm. wide, and from 0.40 to 0.45 mm. long, and the surface of the eggs is roughened. Eggs may be laid on top of the rocks under some conditions.

Larvae often live several days out of water. Eggs collected in Rumsey Canyon on April 25, 1954, hatched in an aquarium three days later. The first instar larvae were found to be gill-less, white, with a black head, 1.02 mm. long, and with a width at the head of 0.24 mm. The crochets were found to be in a uniordinal series unlike the biordinal arrangement of later instars.

Larvae probably serve as fish food, as several small larvae of this species were found in the stomachs of a perch, *Hysterocarpus traskii,* collected by A. Cordone at Pitt 4 power house, on the Pitt River, near Big Bend, Shasta County on September 1, 1953.

Unplaced California Larvae

Several larvae have been examined from California for which no adults have been associated.

Parargyractis sp.

(Fig. 11:7*d*)

On October 13, 1954, A. A. Grigarick and I collected larvae and a few pupae of an apparently distinct species at Little Stony Creek, Colusa County. The species seems related to *truckeealis* Dyar, but the large mandibles are quite distinct. No adults were reared, but several parasitized pupae and two damaged ichneumon adults were collected. Larvae were also found in a stream near Ladoga, about three miles from the Little Stony Creek locality, in association with *P. jaliscalis*. The larvae were found under silken tents in pools on fairly large rocks or on bed rock, and unlike other species seen to date, the larva makes a larval chamber near the center of the web, which projects upwards slightly, and from which it can be collected. Other larvae were found in old abandoned cocoons. The pupal cases appeared white, and stiff, owing to impregnation with salts. The creek runs through a serpentine area, and it was assumed the salts were of magnesium origin. On November 17 small- to medium-sized larvae were found, and one adult ichneumon parasite emerged from a pupa soon after it was collected.

I also associate as this species larvae from Bartlett Creek, Lake County, collected September 8, 1946, el. 1,000 ft., by H. P. Chandler; Howell's, Plumas Co., August 14, 1946, by H. P. Chandler, where it occurs together with *truckeealis;* Dyerville, Humboldt Co., August 14, 1946, H. P. Chandler; N. Fork of the Feather River, Plumas Co., el. 3,000 ft., by H. P. Chandler, together with *truckeealis*.

The large mandibles of this species will easily separate it from *truckeealis*. The pupae are similar to *truckeealis,* but the abdominal spiracles are noticeably larger, and more stalked.

Parargyractis sp.

(Fig. 11:7c)

Larvae from Surprise Canyon, Inyo County, collected March 31, 1951, by R. L. Usinger, constitute a distinct species. Both *P. kearfottalis* and *P. schaefferalis* occur in this area, and so this could be either species. The black head, short, white gills, and black mandible with only a single strong tooth are characteristic of this species.

Parapoynx sp.

(Fig. 11:7*f*)

A single larva of this genus, with typical branched gills, was examined. It was collected on *Ceratophyllum* in an aquarium at San Francisco, September 5, 1947, by W. B. Richardson. It is probably an introduced species.

REFERENCES

BERG, C. O.
1949. Limnological relations of insects to plants of the genus *Potamogeton*. Trans. Amer. Micr. Soc., 68: 279-291.
1950. Biology of certain aquatic caterpillars (Pyralididae: *Nymphula* spp.) which feed on *Potamogeton*. Trans. Amer. Micr. Soc., 69:254-266.

DYAR, H. G.
1906. The North American Nymphulinae and Scopariinae. Jour. N.Y. Ent. Soc., 14:77-107.
1917. Notes on North American Nymphulinae (Lepid., Pyralidae). Insec. Inscit. Menst., 5:75-79.

FORBES, W. T. M.
1910. The aquatic caterpillars of Lake Quinsigamond. Psyche, 17:219-227.
1911. Another aquatic caterpillar *(Elophila)*. Psyche, 18:120-21.
1923. The Lepidoptera of New York and neighboring states. Memoir 68. Cornell Agr. Exp. Sta., 1-729 pp. (Nymphulinae p. 574-581).

FROHNE, W. C.
1938. Contribution to knowledge of the limnological role of the higher aquatic plants. Trans. Amer. Micr. Soc., 57:256-268.
1938*a*. Biology of *Chilo forbesellus* Fernald, an hygrophilous crambine moth. Trans. Amer. Micr. Soc., 58:304-326.
1939*b*. Observations on the biology of three semiaquatic lacustrine moths. Trans. Amer. Micr. Soc., 58: 327-348.

GROTE, A. R.
1880. New species of moths. The North Amer. Ent., 1:97-98.
1881. New Pyralidae. Papilio, 1:15-19.

HAMPSON, G. F.
1897. VII. On the classification of two subfamilies of of the family Pyralidae: the Hydrocampinae and Scopiarianae. Trans. Ent. Soc. London, 1897(2)127-240.
1906. LX. Descriptions of new Pyralidae of the subfamilies Hydrocampinae and Scoparianae. Ann. Mag. Nat. Hist., 18:373-393.
HART, C. A.
1895. On the entomology of the Illinois River and adjacent waters. Bull. Illinois Lab. Nat. Hist., 4:164-183.
KLIMA, A.
1937. Pyralididae: Subfam. Scopariinae, Nymphulinae. Lepidopterorum Catalogus pars 84 (edited by F. Bryk) Gravenhage: W. Junk. 1-226 pp.
LANGE, W. H., JR.
1956. A generic revision of the aquatic moths of North America (Lepidoptera: Pyralidae, Nymphulinae.) Wasmann Jour. Biol., 14:59-144.
LLOYD, J. T.
1914. Lepidopterous larvae from rapid streams. Jour. N.Y. Ent. Soc., 22:145-152, 2 pl.
1919. An aquatic dipterous parasite and additional notes on its lepidopterous host. Jour. N.Y. Ent. Soc., 27:263-65, 1 pl.
MEYRICK, E.
1890. XIII. On the classification of the Pyralidina of the European fauna. Trans. Ent. Soc. London. pp. 429-492.
1895. A handbook of British Lepidoptera. London: MacMillan and Co. vi + 843 pp.
McDUNNOUGH, J.
1939. Check list of the Lepidoptera of Canada and the United States of America. Mem. So. Calif. Acad. Sci., 11:1-171.
McGAHA, Y. J.
1952. The limnological relations of insects to certain aquatic flowering plants. Trans. Amer . Micr. Soc., 71:355-381.
1954. Contribution to the biology of some Lepidoptera which feed on certain aquatic flowering plants. Trans. Amer. Micr. Soc., 73:167-177.
PACKARD, A. S.
1884. Habits of an aquatic caterpillar. Amer. Nat., 18:824-826.
PENNAK, R. W.
1953. Fresh-water invertebrates of the United States. New York: Ronald Press Co. ix + 769 pp. (aquatic Lepidoptera p. 585-587).
SCHAUS, W.
1924. New species of Pyralidae of the subfamily Nymphulinae from tropical America. (Lepid.). Proc. Ent. Soc. Wash., 26:93-130.
1940. Insects of Porto Rico and the Virgin Islands — moths of the families Geometridae and Pyralididae. N.Y. Acad. Sci., 12:291-471 (Pyralididae, Nymphulinae pp. 379-387).
WELCH, P. S.
1914. Habits of the larva of *Bellura melanopyga* Grote. Biol. Bull., 27:97-114.
1915. The Lepidoptera of the Douglas Lake region, northern Michigan. Ent. News, 26:115-119.
1916. Contributions to the biology of certain aquatic Lepidoptera. Ann. Ent. Soc. Amer., 9:159-187, 3 pl.
1919. The aquatic adaptations of *Pyrausta penitalis* Grt. Ann. Ent. Soc. Amer., 12:213-26.
1922. The respiratory mechanism in certain aquatic Lepidoptera. Trans. Amer. Micr. Soc., 41:29-50, 2 pl.
1924. Observations on the early larval activities of *Nymphula maculalis* Clemens (Lepidoptera). Ann. Ent. Soc. Amer., 17:395-402.
WELCH, P. S., and G. L. SEHON
1928. The periodic vibratory movements of the larva of *Nymphula maculalis* Clemens (Lepidoptera) and their respiratory significance. Ann. Ent. Soc. Amer., 21:243-58.
WILLIAMS, F. X.
1944. Biological studies in Hawaiian water-loving insects. Part. IV. Lepidoptera or moths and butterflies. Proc. Hawaii. Ent. Soc., 12:180-185, pl. 10-11.

CHAPTER 12

Aquatic Hymenoptera

By K. S. Hagen
University of California, Albany

All the Hymenoptera known to be aquatic are apparently entomophagous. The superfamilies Ichneumonoidea, Chalcidoidea, Proctotrupoidea, and the family Pompilidae include species in which the adults have become adapted to penetrating water surfaces. This adaptation was necessitated by the occurrence beneath the water surface of the host in which they develop. A pompilid species is an exception in that it enters the water to obtain its prey which it then transports from the water to its terrestrial larval cell.

Respiration.—Since larvae of endoparasitic Hymenoptera live surrounded by animal tissues, the microhabitats of aquatic species are quite similar to those of species which infest terrestrial hosts. It is not surprising, therefore, that parasitic larvae of terrestrial and aquatic Hymenoptera do not differ greatly in their adaptations, and that some genera (including both *Trichopria* and *Dacnusa*) contain both terrestrial and aquatic species (Berg, 1949).

Among ectoparasitic insects an outstanding example of a distinct respiratory adaptation in the immature stages is found in the ichneumonoid family Agriotypidae. The two known species of this family are parasitic upon the prepupae and pupae of caddisflies in their cases. *Agriotypus armatus* Walker of Europe (Fisher, 1932) and *A. gracilis* Wtstn. of Japan (Clausen, 1931) attach their eggs to the host body, and the larvae feed externally throughout their period of activity. Although they are immersed in water, the early larval instars have no morphological adaptations for respiration, their oxygen requirements evidently being fulfilled by diffusion from the water that flows through the case (Clausen, 1950).

Clausen (1950) goes on to say that "as soon as the host dies, which is during the last larval stage of the parasite, the flow of water through the case ceases, and the parasite larva must obtain an adequate air supply in some other way. Various early observers had noted that the caddisfly cases containing mature larvae, pupae, or adults of *A. armatus* always possessed, at the anterior end, a silken ribbon about one millimeter in width and several centimeters in length, which was free-floating in the surrounding water. W. Muller (1889, 1891) determined that death of the parasite usually followed the removal of the ribbon, and Henriksen (1922) verified this observation and was first to ascribe a respiratory function to it. The ribbon of *A. gracilis* not only serves the respiratory needs of the full grown larva and pupa but also the adult through its entire winter hibernation period.

"The adult female, which must enter the water for oviposition, is clothed with a dense pubescence that holds an air bubble over the entire body while she is submerged. Both *A. armatus* and *A. gracilis* need to emerge at intervals of about ten to fifteen minutes to renew their air supply. The oxygen content of the enveloping bubble is doubtless appreciably supplemented from the surrounding water" (Clausen, 1950).

Survey of aquatic species.—Many hymenopterous species that parasitize aquatic insects will not be treated in this discussion. These species are omitted because they do not attack their hosts beneath the water surface, but oviposit in or on a host that is not submerged.

In California there are only two species of Hymenoptera recorded as being aquatic. One is a mymarid and the other a braconid. However, as soon as a special search is made, numerous species of aquatic Hymenoptera undoubtedly will be discovered.

The following key treats the recognized Nearctic genera which include species that are either known or suspected of entering the water in the adult stage. The discussion is of necessity based largely on published observations by investigators in the eastern United States.

A Generic Key to the Adults of Nearctic Aquatic Hymenoptera

1. Pronotum extending back to tegulae 2
— Pronotum not extending back to tegulae .... CHALCIDOIDEA ... 10
2. Trochanters 1-segmented 3
— Trochanters 2-segmented .. ICHNEUMONOIDEA ... 5
3. Wing venation well developed, hind wing with several veins, and at least 1 closed cell; large species .....

... VESPOIDEA, POMPILIDAE, *Anoplius* Dufour 1834
— Wing venation more or less reduced; hind wing nearly veinless and with no closed cells; minute species .... PROCTOTRUPOIDEA ... 4
4. Antennae inserted close to clypeus ....
.... SCELIONIDAE, *Tiphodytes* Bradley 1902
— Antennae inserted on middle of face ....
.... DIAPRIIDAE, *Trichopria* Ashmead 1893
5. Fore wing with 2 recurrent nervures; 1st cubital and 1st discoidal cells not separated; all the abdominal segments freely movable .... ICHNEUMONIDAE ... 6
— Fore wing with only 1 or without any recurrent nervure; usually with the 2nd and 3rd abdominal segments immovably united above .... BRACONIDAE ... 7
6. Abdomen more or less compressed; areolet (= a small cell near the center of the wing) more or less triangular or absent ....
.... Ophioninae .. *Cremastus* Gravenhorst 1829
— Abdomen depressed or cylindrical; areolet more or less pentagonal, quadrangular, or occasionally absent .... Cryptinae .. *Apsilops* Förster 1868
7. Mandibles attached normally, their tips opposed and meeting when closed; clypeus emarginate anteriorly, forming with the mandibles a semicircular opening; head posteriorly margined ....
.... Rhogadinae, *Ademon* Haliday 1833
— Mandibles widely separated, not meeting when closed, the tips concave, the teeth curving outward instead of inward .... Dacnusinae ... 8
8. Eyes bare .... *Dacnusa* Haliday 1833
— Eyes hairy 9
9. Stigma short, wide; fore wing with Rs+M and M absent, 1st cubital and 1st discoidal cells confluent ....
.... *Chorebidella* Riegel 1950
— Stigma long; fore wing with vein Rs+M and M present or obsolescent, 1st cubital and 1st discoidal cells separated ....
.. *Chorebus* Haliday 1833 and *Chorebidea* Viereck 1914
10. Tarsi 4- or 5-segmented ....MYMARIDAE.... 11
— Tarsi 3-segmented .... TRICHOGRAMMATIDAE.... 13
11. Phragma plainly projecting into abdomen; abdomen sessile .... *Anagrus* Haliday 1833
— Phragma never projecting beyond petiole; abdomen petiolate or subpetiolate 12
12. Abdomen distinctly petiolate; female antennae 9-segmented .... *Caraphractus* Haliday 1846
— Abdomen subpetiolate; female antennae 10-segmented .... *Patasson* Walker 1846
13. Antennal club solid; marginal vein curved ....
.... *Trichogramma* Westwood 1833
— Antennal club segmented; marginal vein straight.... 14
14. Antennae 7-segmented (1 funicle segment) ....
.... *Prestwichia* Lubbock 1863
— Antennae 8-segmented (2 funicle segments) ....
.... *Hydrophylita* Ghesquière 1946

Family POMPILIDAE

The only pompilid known to be aquatic is *Anoplius (Anoplius) depressipes* Banks 1919. It is a medium-sized, entirely black spider wasp occurring throughout the Alleghanian fauna of northeastern and north central United States. Evans (1949) found that this species had been misidentified in the literature under such names as *Priocnemis flavicornis,* now known as *Priocnemoides fulvicornis* (Cresson) 1867, and as *Anoplius illinoiensis* (Robt.) 1901. The prey of *A. depressipes,* according to Evans, consists of pisaurid spiders belonging to the genus *Dolomedes.*

Needham and Lloyd (1916) described how *A. depressipes,* under the name of *Priocnemis flavicornis,* drags its spider prey over water. The female is unable to carry such a heavy burden in sustained flight, but the buoyancy of the spider body permits the wasp to skim along the water surface with it (Rau, 1934).

Evans (1949) calls attention to an interesting paper by Caudell (1922) which had been overlooked by authors dealing with aquatic Hymenoptera. Caudell describes the adult aquatic habits of *A. depressipes* under the name *A. illinoiensis.* This species crawls beneath a floating leaf or down an edge of a stone when entering the water and has been observed running across the bottom of a stagnant pool.

Superfamily PROCTOTRUPOIDEA

In the superfamily Proctotrupoidea, only two species are recorded as exhibiting aquatic habits, *Tiphodytes gerriphagus* (Marchal) 1900 of the family Scelionidae and *Trichopria columbiana* (Ashmead) 1893, a diapriid.

Family SCELIONIDAE

Marchal (1900) described *Tiphodytes gerriphagus* under the generic name *Limnodytes.* In 1902, Bradley proposed the new name *Tiphodytes* for *Limnodytes* which was preoccupied. Marchal found the insect through his embryological studies of Gerridae. He was made aware of the parasite's presence by the undulating movements of larvae which he could see through the chorion of the gerrid egg. This species is found in Europe and eastern United States, and is a parasite of *Gerris* eggs, reaching them readily by swimming with both the wings and the legs (Matheson and Crosby, 1912). However, Martin (1928) observed: "The adults flying about lily pads or walking over them. After the female landed on a leaf she usually walked straight to the edge of it. When a suitable place was found, she first dipped her head under the surface film of the water and after a few struggles with it, succeeded in pulling herself entirely under water. Clinging to the underside of the pad, she walked about searching for gerrid eggs." Martin observed a second manner of oviposition. Instead of going head first, the female backed into the water and clung to the lily pad with the front pair of legs, while the middle pair hung in the water and the wings as well as the hind legs floated on the surface film of the water. In this position the female was able to parasitize the gerrid eggs. *Trepobates* eggs were also found to be parasitized, and a single parasite emerged from each egg.

The first instar larva of *T. gerriphagus* is "teleaform." It is characterized by a complete lack of segmentation but has the body divided by a sharp constriction into two more or less equal parts. Clausen (1940) draws attention to Marchal's (1900) illustration of a *T. gerriphagus* larva showing distinct tufts upon the summits of a pair of fleshy lobes situated at the lateroventral margins of the anterior part of the abdomen. Martin (1928), dealing with same species, shows the hairs in a transverse row. Perhaps our North American *Tiphodytes* is an undescribed species though the adults resemble each other.

Family DIAPRIIDAE

The other aquatic proctotrupoid species is *Trichopria columbiana* (Ashmead) 1893, recorded in Virginia and Michigan. Berg (1949), who studied the relation of insects to plants of the genus *Potamogeton,* reared *T. columbiana* from four species of ephydrid flies belonging to the genus *Hydrellia* in Michigan. Berg implies that they are good swimmers. From the few known life histories of *Trichopria,* it is suspected that *T. columbiana* is perhaps a gregarious internal parasite of *Hydrellia* puparia and thus would oviposit into the puparium.

Superfamily ICHNEUMONOIDEA

The aquatic species of Ichneumonoidea are relatively larger than any of the aquatic Chalcidoidea or Proctotrupoidea. This is not surprising since all the aquatic species of Ichneumonoidea parasitize either lepidopterous or dipterous larvae, whereas the species of the other two superfamilies except *Trichopria,* attack eggs of various aquatic species.

Family ICHNEUMONIDAE

The majority of the aquatic species of Ichneumonidae are referred to as undescribed species of several different genera. Chagnon (1922) reported an ichneumonid which he observed crawling down a rush beneath the water, and Cushman (1933) discussed aquatic ichneumonids and the probable species of Chagnon's observation, deciding upon *Neostricklandia sericata* Viereck 1925, a species now placed in the genus *Apsilops* (Muesebeck, 1950). Frohne (1939) observed this species and four additional undescribed ichneumonids of similar habits in lakes of northern Michigan. He concluded that *A. sericata* parasitizes the larvae of *Occidentalia comptulatalis* (Hulst) 1886 and *Chilo forbesellus* Fernald 1896. It was observed by Frohne, ovipositing into the galleries of these pyralids in *Scirpus occidentalis* (Watson). Another species of the genus *Apsilops,* under the generic name *Trichocryptus,* was recorded as a possible parasite of *Schoenobius melinellus disperellus* Robison 1870 which bores in spike rush (Frohne, 1939). Berg (1949) reared a *Cremastus* species from *Nymphula icciusalis* Walker 1859, a pyralid which attacks *Potamogeton natans* Linn. in Michigan. Frohne (1939) bred *C. chilonis* Cushman 1935 from numerous cocoons found in the larval galleries of two pyralids, *C. forbesellus* and *O. comptulatalis.* Dr. W. H. Lange recently reared a parasite from an aquatic pyralid, *Parargyractis* sp. close to *truckeealis* (Dyar), which occurs near Colusa, California. The parasite was determined by Miss L. M. Walkley of the United States National Museum as a new genus and species belonging to the tribe Cremastini, and is near the genus *Pseuderipternus* Viereck (1917). The genus *Cremastus* is characterized by having the thorax much stouter in front, or short and thick, not subcylindrical whereas the thorax of *Pseuderipternus* is very long and slender so as to be nearly cylindrical.

Family BRACONIDAE

Like the Ichneumonidae, most of the aquatic braconids have been referred to as undescribed species. Since *Chorebus aquaticus* Muesebeck 1950 was reared from *Hydrellia griseola* var. *scapularis* Loew (Muesebeck, 1950), it is quite possible that the adult enters the water to parasitize the aquatic ephydrid fly. The closely allied genera *Chorebidea, Chorebidella,* and *Dacnusa* contain several undescribed species which likewise are parasites of various species of *Hydrellia* attacking *Potamogeton* in Michigan (Berg, 1949). The type locality of *C. aquaticus* is Sacramento, California. Riegel (1950) described *Chorebidella bergi* from specimens which Berg reared from *Hydrellia cruralis* Cresson on *Potamogeton amplifolius* Tuckerm. in Michigan. The species *Ademon niger* (Ashmead) 1895 is another parasite of various species of *Hydrellia* (Berg, 1949), and is evidently found from Utah to the eastern coast of the United States.

Superfamily CHALCIDOIDEA

Family MYMARIDAE

There are three aquatic mymarids known to occur in the Nearctic region. *Caraphractus cinctus* Walker 1846 (fig. 1) (=*Polynema natans* Lubbock 1863) is found in eastern United States, and parasitizes the eggs of *Notonecta, Dytiscus,* and *Calopteryx virgo.* Four to five larvae were found in each *Notonecta* egg. This species swims with its wings, and has been recorded to have lived submerged in water for twelve hours (Matheson and Crosby, 1912). An *Anagrus* sp., near the European *A. subfuscus* Förster 1847, is recorded as parasitizing gyrinid eggs in Kansas (Hoffmann, 1932) and also was taken from the stomach of a trout in New York. The European species, *A. brocheri* Schultz is unable to swim but moves about in the water by walking upon foliage, plant stems, or other objects (Clausen, 1940).

Patasson gerrisophaga (Doutt) 1949 is the only hymenopteron, other than one braconid species suspected of having aquatic habits, to be recorded from California. Doutt (1949) records the species from Lake Britton, Shasta County, and from El Cerrito, Contra Costa County. It has been reared from eggs of *Gerris* sp., but the possible aquatic habits of this parasite are undetermined.

Family TRICHOGRAMMATIDAE

The trichogrammatid genera which include adults of aquatic habit are *Prestwichia, Hydrophylita,* and possibly *Trichogramma.*

Several aquatic Hymenoptera which were found originally in Europe were later discovered in North

Fig. 12:1. *Caraphractus cinctus* Walker (Matheson and Crosby, 1912).

America. *Prestwichia aquatica* Lubbock 1863 of Europe possibly occurs in the Nearctic region, for Martin (1928) implies that it has been reported in New York. It was placed in the above key in the event that it is established in North America. There are said to be three forms or races of *P. aquatica* (Henriksen, 1922), the typical one having females with well-developed wings and males with rudimentary wings. The second form has wings rudimentary in both sexes. The third form has the wings of the females reduced. This species has been recorded from the eggs of *Ranatra, Nepa, Aphelocheirus, Dytiscus,* and *Pelobius* (Ruschka and Thienemann, 1913). A maximum of fifty individuals was secured from a single egg of the larger species of *Dytiscus* (Clausen, 1940). Enock (1889) claims that only the middle legs are used by the adults for swimming, but Henriksen states that the hind legs serve this purpose. Heymons (1908) states that the adults remained under water for five days.

Hydrophylita aquivolans (Matheson and Crosby) 1912 is known from the eastern United States, and has been recorded by Matheson and Crosby (1912) as parasitizing the eggs of *Ischnura* sp. This species swims by means of its wings.

Hoffmann (1932) exposed eggs of several species of aquatic insects and eggs of a water mite to *Trichogramma minutum* Riley 1871, but obtained progeny only from the mite eggs. There is no distinct indication that *T. minutum* actually entered the water for oviposition, since the adult parasites were found webbed in the scum covering the water mite egg mass. This host record must be considered questionable until supported by further experimentation. If the observation is valid, it would be the first record of any stage of a spider mite to be attacked by an internal parasite. Martin (1928) describes the immature stages of *T. minutum* and its host range.

REFERENCES

BERG, CLIFFORD O.
1949. Limnological relations of insects to plants of the genus *Potamogeton*. Trans. Amer. Micr. Soc., 68:279:291.
BRADLEY, J. CHESTER
1902. A recently discovered genus and species of aquatic hymenoptera. Canad. Ent., 34:179-180.
CAUDELL, A. N.
1922. "A diving wasp." Proc. Ent. Soc. Wash., 24: 125-126.
CHAGNON, G.
1922. A hymenopteron of aquatic habits. Canad. Ent., 65:24.
CLAUSEN, C. P.
1931. Biological observations in Agriotypus. Proc. Ent. Soc. Wash., 33:29-37.
1940. Entomophagous insects. New York: McGraw-Hill, 688 pp.
1950. Respiratory adaptations in the immature stages of parasitic insects. Arthropoda, 1:198-224.
CUSHMAN, R. A.
1933. Aquatic ichneumon flies. Canad. Ent., 65:24.
1935. New Ichneumon-flies. Jour. Wash. Acad. Sci., 25:547-564.
DOUTT, R. L.
1949. A synopsis of North American *Anaphoidea*. Pan-Pac. Ent., 25:155-160.
ENOCK, F.
1898. Notes on the early stages of *Prestwichia aquatica* Lubbock. Ent. Mon. Mag., 34:152-153.
EVANS, HOWARD F.
1949. The strange habits of *Anoplius depressipes* Banks: A mystery solved. Proc. Ent. Soc. Wash., 51:206-208.
FISHER, K.
1932. *Agriotypus armatus* (Walk.) and its relations with its hosts. Proc. Zool. Soc. London, (1932):451-461.
FROHNE, W. C.
1939. Semiaquatic hymenoptera in north Michigan lakes. Trans. Amer. Micr. Soc., 58:228-240.
HENRIKSEN, K. L.
1922. Notes on some aquatic hymenoptera. Ann. Biol. Lacustre, 11:19-37.
HEYMONS, R.
1908. Süsswasser-Hymenopteren aus der Umgebung Berlins. Deut. Ent. Ztschr. (1908):137-150.
HOFFMANN, C. H.
1932. Hymenopterous parasites from the eggs of aquatic and semi aquatic insects. Jour. Kansas Ent. Soc., 5:33-37.
LUBBOCK, J.
1863. On 2 aquatic hymenoptera, 1 of which uses its wings in swimming. Trans. Linn. Soc. Lond. (Zool), 24:135-141.
MARCHAL, P.
1900. Sur un nouvel hyménoptère aquatique, le *Limnodytes gerriphagus* n. gen., n. sp., Ann. Soc. Ent. Fr., 59:171-176.
MARTIN, CHARLES H.
1928. Biological studies of 2 hymenopterous parasites of aquatic insect eggs. Ent. Amer., 8 (n.s.):105-151.
MATHESON, ROBERT, and C. R. CROSBY
1912. Aquatic hymenoptera in America. Ann. Ent. Soc. Amer., 5:65-71.
MUESEBECK, C. F. W.
1950. Two new genera and 3 species of Braconidae. Proc. Ent. Soc. Wash., 52:77-81.
MUESEBECK, C. F. W., K. V. KROMBEIN, H. K. TOWNES
1951. Hymenoptera of America north of Mexico. Synoptic catalog. U.S.D.A. Agr. Monogr. No. 2, 1420 pp.
RIEGEL, GARLAND T.
1950. A new genus and species of Dacnusini. Ent. News, 61:125-130.
RUSCHKA, F., and A. THIENEMANN
1913. Zurkenntnis der Wasser-Hymenopteren. Zeit. Wissensch. Insektenbiol., 9(2):48-52; (3):82-87.

CHAPTER 13

Aquatic Coleoptera

By H. B. Leech and H. P. Chandler[1]
California Academy of Sciences, San Francisco

The Aquatic Coleoptera are commonly termed "water beetles." This is convenient and descriptive, but does not imply close relationship among all the included species. In the Nearctic region alone there are ten families in which both larvae and adults of virtually all species are aquatic, three in which at least one stage is aquatic, two in which the larvae occur in water or in the underwater parts of plants and the adults are usually semiaquatic. These all live in fresh (including mineralized and saline) or land waters, in contrast to examples of five other families (see Carabidae, Staphylinidae, Melyridae, Eurystethidae, and Limnichidae) which are found in the intertidal zone of the Pacific Coast sea beaches.

In addition there are many species which burrow in wet mud and sand or hunt and hide under debris and stones at the water's edge. These forms, and insects which have merely fallen into the water are often caught in collecting true aquatics. Hence, the commoner riparian groups are included in the following keys to families.

Sabrosky (1953) claims that some 277,000 species of beetles were described from the world fauna by the end of 1948. In the sense of the present study (but excluding the terrestrial Hydrophilidae) about 5,000 species are "aquatics." Some three hundred and thirty of these occur in California, more than one-third of them being Dytiscidae. Comparatively little collecting has been done here as yet, and there are undoubtedly many undescribed species of water beetles.

The Nearctic water beetles represent two of the three currently recognized suborders of Coleoptera. The families Amphizoidae, Haliplidae, Dytiscidae, Noteridae, and Gyrinidae belong to the Adephaga; the remainder to the Polyphaga. Adults of the aquatic Adephaga carry an air supply under the elytra, and when returning to the surface to renew it, break the surface film with the tips of the elytra and abdomen; the underside of the body is virtually hairless. The aquatic Polyphaga also carry air between the elytra and abdomen, but in addition almost all of them have the underside of the body covered with a short, dense hydrofuge pubescence. This holds a sheet of air, which is in connection with the elytral reservoirs. When coming up to renew their supply, hydrophiloid beetles break the surface film with an antenna.

The mature larvae of virtually all aquatic beetles leave the water to form their pupal cells. Exceptions include those which pupate in gas-filled cocoons (*Noterus* of Noteridae, *Donacia,* and *Neohaemonia* of Chrysomelidae, *Lissorhoptrus* of Curculionidae); and the genus *Psephenoides* of Psephenidae, which normally has the water actually in contact with the pupal cuticle.

Details of relationships of the various families, methods of respiration, life histories, habitats, and distribution will be discussed separately for each family. All Nearctic families having one or more aquatic or semiaquatic species are included except Chelonariidae, the larvae of which are said to be semiaquatic, but are unknown to us in nature.

Fossil water beetles.—The first undoubted Coleoptera fossils are from the Upper Permian, of about 200 million years ago. Some elytra of this age, from Australia, appear to be of beetles related to the modern family Hydrophilidae. What appear to be dytiscid types occur in the Upper Jurassic, and Hatch refers one Liassic or Lower Jurassic fossil definitely to the Gyrinidae. Specimens which can be referred to modern genera date from about Eocene time, the period preceding the Oligocene of Baltic Amber fame.

Sixty-eight species of fossil aquatic or semiaquatic water beetles have been described from North America: Dytiscidae (25), Gyrinidae (2), Hydrophilidae (25), Psephenidae (3), Chelonariidae (1), Limnichidae (1), Dryopidae (2), Helodidae (1), Chrysomelidae (8). The Hydrophilidae are largely Eocene and Miocene, with a few from the Pleistocene; the Dytiscidae are predominantly Pleistocene, with some from the Miocene; species of the other families are from the Miocene, except the Chrysomelidae *(Donacia)* which are Pleis-

[1]All the keys to larvae, except that for Helodidae, are to be credited to the late Harry P. Chandler. He also wrote the texts for the families Limnichidae, Psephenidae, Dryopidae, Elmidae, Ptilodactylidae, Chrysomelidae, and Curculionidae. The remainder of the article was written by Hugh B. Leech.

tocene (7) and Miocene (1). Several of the Pleistocene dytiscids and hydrophids are from the La Brea, California, tar pits, and have been identified as species still living in the region.

Economic importance.—A few species of water beetles, both larvae and adults, are actually pests of cultivated plants (the rice weevils, *Lissorhoptrus* spp.; Kuschel, 1951), or occasionally attack field crops (*Helophorus* spp. on turnips, cabbages, swedes, and other crucifers). Adults are sometimes troublesome in domestic water supplies (Hurst, 1945); larvae of the larger dytiscids *(Dytiscus, Cybister),* hydrophilids *(Hydrophilus),* and perhaps gyrinids *(Dineutus),* may kill small fish in rearing ponds and presumably in nature (Wilson, 1923); others may attack tadpoles (Williams, 1936). Some species are the secondary hosts of trematodes (see section under Parasites). Adults of the Dryopidae, and both adults and larvae of the Elmidae, Chrysomelidae, and Curculionidae feed on the roots and other parts of aquatic plants, but the total damage they do has never been assessed.

These bad features are offset by the uses to which water beetles are put. Predaceous species of the terrestrial hydrophilid genus *Dactylosternum* have been used in control programs against the sugar-cane beetle borer in the Philippine and Hawaiian islands, and against the banana borer in Jamaica. Both adults and larvae of the Haliplidae, Dytiscidae, and Hydrophilidae (Martin and Uhler, 1939) comprise a regular, and sometimes major part of the food of aquatic birds, as well as of turtles, frogs, toads, salamanders, and fish (Wilson, 1923). Many water beetles fly well, and during the spring and fall dispersal flights there are tremendous numbers in the air, especially during the afternoon and evening; birds and bats must eat numbers of them. In various parts of China large dytiscids, *Cybister* spp., and hydrophilids, *Hydrophilus* spp., are commonly eaten by humans, both as medicine and as confection (Hoffmann, 1947; Bodenheimer, 1951). On the other hand water beetles are suitable as aquarium animals (Lutz, 1930; Boardman, 1939). In addition, certain aquatic beetles have proved to be good subjects for research in physiology (Hodgson, 1953).

Parasites.—Most parasites of water beetles are of academic interest only, but further research may prove some to be of more importance than suspected. Pujatti (1953) reported the metacercaria of a trematode of the genus *Lecithodendrium,* whose primary host was thought to be a bat, from both dytiscids *(Hydaticus, Cybister)* and hydrophilids *(Hydrophilus, Neohydrophilus),* from Bangalore, Mysore, India. Hall (1929) listed the European dytiscid *Ilybius fuliginosus* as the secondary host of *Haplometra cylindracea,* a trematode of frogs, and *Dytiscus marginalis* as containing the cercaria of a trematode. Since water beetles are eaten by fish, reptiles, amphibians, birds, and some mammals, it will be surprising if they do not prove to be involved in the life cycles of other trematodes, as well as of cestodes, nematodes, and acanthocephalids. (For references to nematodes in water beetles, see Todd, 1942, 1944; La Rivers, 1949; Jackson, 1951).

It is not uncommon to find red or reddish-brown egglike things attached to the dorsum of the abdomen of water beetles, under the elytra and wings, and occasionally on the legs or ventral surface. These are immature water mites (Hydracarina), which may occur in several stages, some being active, others sessile. They actually suck the blood of their hosts.

Tiny proctotrupid and chalcid wasps parasitize the eggs of Dytiscidae, and larger wasps parasitize the pupae of Gyrinidae (q.v.). Many small dytiscids and hydrophilids, especially species of *Hydroporus,* may be so heavily coated on the undersurface by Protophyta and Protozoa that parts are obscured. Protozoan parasites occur in the digestive tracts of some water beetles.

A very interesting order of ascomycete fungi, the Laboulbeniales, occurs on insects. Despite the classical series of monographs by Thaxter (1896-1931), knowledge of the species on water beetles is preliminary. R. K. Benjamin writes (1952): "The Laboulbeniales are a large and diverse order of very small Ascomycetes which are known to develop only upon the chitinous integument of living insects and, more rarely, of mites (arachnids). Thaxter (1896, 1908, 1924, 1926, 1931), who studied the Laboulbeniales more extensively than any other student, frequently observed that certain species exhibit a remarkable tendency to grow on very restricted portions of the host integument. Such specificity of position may be characteristic only of certain isolated species or may be a regularly occurring phenomenon within a given genus. In the genus *Chitonomyces* Thaxter which infests aquatic beetles belonging to the families Dytiscidae, Gyrinidae, and Haliplidae, for example, most species, especially those occurring on Gyrinidae, Haliplidae, and the Dytiscid tribe Laccophilini, are so restricted to definite areas of the host surface that identification of the parasites, when the host is known, can be made with remarkable accuracy by noting merely their position upon the integument. Frequently one insect species is subject to infection by a number of different species of Laboulbeniales, often representing several genera. Outstanding examples of multiple infection coupled with position specificity are the parasites of *Orectogyrus specularis* Aube (Gyrinidae) and *Laccophilus maculosus* Germ. (Dytiscidae). *Orectogyrus specularis* is known to be infested by no fewer than sixteen species of *Chitonomyces* (Thaxter, 1926, p. 511-512), all of which are found on certain very restricted areas of the terminal abdominal segment and the neighboring genital lobes. As many as ten of these species of *Chitonomyces* have been found growing on the same host individual. *Laccophilus maculosus* is parasitized by at least twelve known species of *Chitonomyces* which are all very specific for their point of attachment to the host."

Dr. Benjamin has been so kind as to provide the following list of genera attacking water beetles (based on the world fauna):

Haliplidae—*Chitonomyces; Hydraeomyces.*

Dytiscidae—*Chitonomyces.*

Gyrinidae—*Laboulbenia.*

Hydrophilidae—*Autoicomyces* (on *Berosus*); *Ceratomyces* (on *Tropisternus, Hydrochus*); *Rhyncho-*

phoromyces (on *Hydrobius, Sperchopsis, Cymbiodyta, Enochrus*); *Hydrophilomyces* (on *Phaenonotum*); *Eusynaptomyces* (on *Hydrochara*); *Limnaiomyces* (on *Hydrochara*).

Hydraenidae—*Thripomyces* (on *Hydraena*); *Hydrophilomyces,* (on *Ochthebius. H.digitalis* has also been placed in the genus *Misgomyces.*)

Dryopidae—*Cantharomyces* (on *Parnus, Parygrus, Pelonomus*); *Helodiomyces* (on *Parnus*).

Collecting.—Success in collecting water beetles depends largely on knowing what habitats to probe. Some data are given for each family, under "Habitat and distribution"; there is a good general discussion and much information on individual species in Young's *The Water Beetles of Florida* (1954).

With an ordinary kitchen sieve, most people can, without even wetting their feet, obtain more beetles in a couple of hours than they are able to prepare for the cabinet in as many days. For general collecting the following are adequate:

1 strongly made kitchen sieve or soup-strainer, 6½ or 7 inches across, with about 17 wires to the inch.

1 similar sieve with a finer mesh, for smaller beetles.

1 pair of forceps (and a second pair for emergencies). Vials of 80 per cent alcohol. Three useful sizes are 95×25 mm., 70×20 mm., and 55×15 mm., the smallest being reserved for mated pairs of beetles, or for very small species.

Vials, empty and clean, for bringing home live specimens.

A square of white canvas or rubberized cloth, about 4×4 ft.

To work deep water, some form of long-handled water net or small dredge is most helpful. In the swifter parts of rivers and larger streams, a kitchen sieve is too small; a copper-mesh window screen, frame and all, may be used.

It is best to mount the beetles within a day or two after they have been collected. If they must be stored in alcohol for long periods, the alcohol should be changed after the first day or two, since the first lot is diluted by water when the beetles are put in, and dirtied by the grease dissolved out of them. Specimens killed in an ethyl acetate jar should not be left in more than twenty-four hours; they may be put directly into fresh 70 per cent alcohol, or layered in cotton and subsequently relaxed.

The majority of water beetles prefer shallow water, where they hide among aquatic plants and underwater debris near the shore. In general, the more the water is roiled by the collector the higher the percentage of beetles present he will catch. The sieve is worked back and forth just above the bottom, with a turn at the end of each stroke to keep the contents in place, a dozen or so strokes being made through the eddies created. If not too much debris is present, the beetles may be picked out of the net with tweezers and popped into alcohol or into an ethyl acetate killing bottle (Valentine, 1942). If the water contains masses of dead leaves, silt, or algae, dump the contents of the net onto the white sheet, preferably in bright sunshine, and pick out the beetles as they leave the pile. In lieu of a sheet, use a flat rock or bare sloping ground. If there is no time for this, put the entire contents of the sieve into a large tin or a waterproof bag, and examine it at home or run it through a Berlese funnel. Water less than a foot deep usually gives the best results, and mere puddles or wet marshy places should not be overlooked.

In running water, hold the sieve or window screen against the bottom, at a slight downstream angle, and turn over stones and the like, upstream from it, letting the current wash the dislodged insects onto the netting.

Following is an annotated list of some of the habitats to search for water beetles.

Lakes.—The open waters of large lakes are almost barren of water beetles. Rocky shore lines may provide Dytiscidae, Psephenidae, and Elmidae. Shallow weedy inlets and narrow bays, protected from wave action, are good for Hydrophilidae, and if the host plants are present, Chrysomelidae and Curculionidae. Large schools of Gyrinidae occur here, as on ponds and the backwaters of rivers.

Ponds.—Ponds are of various sizes, and it depends on local naming whether a given body of water is a pond or a lake. In general, shallow ponds in open meadows, with abundant plant growth, have the greatest number of beetles. Here all families except the Amphizoidae, Ptilodactylidae, Dryopidae, and Elmidae occur. Woodland ponds tend to have colder and more acid water; they contain fewer beetles, but species which may not be found elsewhere; here look for Helodidae. Ecologically ponds grade insensibly into the quiet backwaters of streams and rivers, and into seepage areas. Most aquatic Chrysomelidae and Curculionidae are pond or lake shore forms.

Streams.—The smallest streams are hardly more than seepage areas; their inhabitants are nocturnal and best collected at night by flashlight. Large rapid streams are ecologically like small rapid rivers. In the backwaters and around the edges of pools, expect Haliplidae, Dytiscidae, Gyrinidae, Hydrophilidae; on and under stones in the faster water, even far out from shore, Dryopidae, Elmidae, Psephenidae, Ptilodactylidae; on waterlogged wood and accumulations of leaves, Amphizoidae, Hydraenidae, Hydrophilidae, Helodidae, Dryopidae, and Elmidae; in sand, stones and silt along the shore, Hydrophilidae, Hydraenidae; in algae, Haliplidae, Hydraenidae, and Hydroscaphidae.

Rivers.—Rivers are essentially large streams, and carry much the same fauna, though the actual species may be different.

Swamps.—Swamps are usually difficult places in which to collect. They provide chiefly Hydrophilidae, Chrysomelidae, and Curculionidae.

Springs.—Cold-water springs usually contain Dytiscidae, Hydrophilidae, Gyrinidae, and Helodidae. Hot springs may have Dytiscidae, Hydrophilidae, Hydraenidae, Hydroscaphidae, and Elmidae.

Tree holes.—Some species of Helodidae live in the water which collects in holes in trees.

Shorelines.—Many small forms, especially Hydraenidae and Hydrophilidae, occur in the wet sand and mud a few inches above the water's edge. They float to the surface when this material is pushed into the water and stirred up. They may also be taken at night with the aid of a light.

Seashore.—By splitting open cracks in the rocks with hammer and chisel, in the intertidal zone of the Pacific Coast, the collector can find examples of the families Carabidae, Staphylinidae, Hydraenidae, Melyridae, and Elacatidae. A species of the family Limnichidae has been taken in the mud of tidal flats, and in clumps of grasses submerged at high tide.

Miscellaneous.—Many species of water beetles fly well, and during the spring and fall dispersal flights (usually in May and early September) large numbers may be taken at lights. Where the evenings are sufficiently warm, light collecting is profitable throughout the summer. During the day the beetles sometimes mistake the shiny tops of cars (see Duncan, 1927) or the glass of greenhouses and building tops (Felt and Chamberlain, 1935) for water, and land with a sound like hail. Knaus (1909) reported an evening fall flight of the big *Hydrophilus triangularis* Say at McPherson, Kansas; thousands of them fairly blackened the sky and produced a buzzing and humming noise. Two non-entomologists fired into them with a shotgun and brought down a number.

Mounting.—In general, beetles less than 6 mm. in length should be mounted on triangular "points"; colorless fingernail polish is a suitable adhesive. Since many species of water beetles are identified by the form of the genitalia (usually of males), it will save hours of work if the organs are extruded, or completely dissected out, as a regular procedure in mounting. This virtually requires the use of a binocular steroscopic microscope. Specimens taken from alcohol or ethyl acetate jars are relaxed enough. By holding them upside down between forefinger and thumb, and probing and prying a little with a pin which has a tiny right-angled hook on the end, it is not hard to extrude the genitalia. In most cases the parts are then clearly visible, but if necessary they may be dissected off and mounted on a point under the insect, or in a tiny vial of glycerin; often they are not bilaterally symmetrical, and not suited for mounting on a microscope slide.

The males of many species of water beetles have one or more segments of the front and middle tarsi dilated and clothed beneath with hairs which may be simple, dilated apically, or formed into adhesion discs. Others obviously differ from the females because of their curiously shaped tarsal claws, tibiae, femora, trochanters, or antennae, bearded or tufted areas on the ventral surface, modified abdominal sternites, or dorsal sexual sculpture. In many cases (unfortunately including a number of small forms), it is necessary to dissect the beetle to be sure of its sex.

REFERENCES

BENJAMIN, R. K.
1952. Sex of host specificity and position specificity of certain species of Laboulbenia on *Bembidion picipes*. Amer. Jour. Bot., 39:125-131, 19 figs.

BOARDMAN, E. T.
1939. Field guide to lower aquarium animals. Cranbrook Inst. of Sci. (Bloomfield Hills, Mich.), Bull. No. 16, 186 pp., 51 figs.

BODENHEIMER, F. S.
1951. Insects as human food, a chapter of the ecology of man. The Hague: W. Junk, 352 pp., illus.

CARPENTER, F. M.
1930. A review of our present knowledge of the geological history of insects. Psyche, 37:15-34, 3 tables.
1947. Early insect life. Psyche, 54:65-85, 9 text figs.
1950. Fossil insects. Gamma Alpha Record, 40:60-68, (1+) 3 text figs.

DARLINGTON, P. J., JR.
1929. Notes on the structure and significance of *Palaeogyrinus*. Psyche, 36:216-219.

DUNCAN, D. K.
1927. An unusual condition found in collecting water beetles in Arizona. Bull. Brooklyn Ent. Soc., 22:143.

FELT, E. P., and K. F. CHAMBERLAIN
1935. The occurrence of insects at some height in the air, especially on the roofs of high buildings. N.Y. St. Mus. Cir. 17, 70 pp., 4 text figs.

HALL, M. C.
1929. Arthropods as intermediate hosts of heminths. Smithson. Misc. Coll., 81:1-77.

HATCH, M. H.
1926. Palaeocoleopterology. Bull. Brooklyn Ent. Soc., 21:137-144.
1926. Tillyard on Permian Coleoptera. Bull. Brooklyn Ent. Soc., 21:193.
1927. A revision of fossil Gyrinidae. Bull. Brooklyn Ent. Soc., 22:89-97, 11 figs.

HODGSON, E. S.
1953. A study of chemoreception in aqueous and gas phases. Biol. Bull., 105:115-127, 3 figs.

HOFFMANN, W. E.
1947. Insects as human food. Proc. Ent. Soc. Wash., 49:233-237.

HURST, W. D.
1945. Predaceous diving beetles in Winnipeg's water supply. Jour. Amer. Wat. Wks. Assoc., 37:1204-1206.

JACKSON, D. J.
1951. Nematodes infesting water beetles. Ent. Mon. Mag. (4), 12:265-268.

KNAUS, W.
1909. Gunning for bugs. Ent. News, 20:364.

KUSCHEL, G.
1951. Revision de *Lissorhoptrus* LeConte y generos vecinos de America. Revista Chilena Ent., 1:23-74, 46 figs.

LA RIVERS, IRA
1949. Entomic nematode literature from 1926 to 1946, exclusive of medical and veterinary titles. Wasmann Collector (San Francisco, Calif.), 7:177-206.

LUTZ, F. E.
1930. Aquatic insect pets. Nat. Hist., (1930):389-401, illus.

MARTIN, A. C., and F. M. UHLER
1939. Food of game ducks in the United States and Canada. U.S. Dept. Agric. Tech. Bull. No. 634. 156 pp., illus.

PUJATTI, D.
1953. Hydrophilidae e Dytiscidae del Sud India con metacercaria del gen. *Lecithodendrium* Looss (Coleoptera et Trematoda). Boll. Soc. Ent. Ital., 83:137-139, 1 text fig.

SABROSKY, C. W.
1953. How many insects are there? System. Zool., 2:31-36.

THAXTER, R.
1896-1931. Contribution toward a monograph of the Laboulbeniaceae. I-V. Mem. Amer. Acad. Arts Sci., 12-16.

VALENTINE, J. M.
1942. On the preparation and preservation of insects, with particular reference to Coleoptera. Smithson. Misc. Coll., 103:1-16, 5 text figs. (Publication No. 3696).

WILLIAMS, F. X.
1936. Two water beetles that lay their eggs in the frothy egg masses of a frog or tree toad. Pan-Pac. Ent., 12:6-7.

Key to Nearctic Families of Aquatic and Semiaquatic Coleoptera[2]

Adults

1. First visible abdominal sternite completely divided by hind coxal cavities (fig. 13:9), hind coxae immovably united to metathorax (if first 2 or 3 sternites are hidden by expanded hind coxal plates (fig. 13:3*a*), see family Haliplidae, following); first 3 visible abdominal sternites connate, sutures partly obliterated (less obviously so in Gyrinidae, which have a dorsal and a ventral pair of eyes, and short, irregular antennae) (fig. 13:23); antennae usually filiform or nearly so. Suborder ADEPHAGA 2
— First visible abdominal sternite extending for its entire breadth posterior to hind coxal cavities, undivided (if both a dorsal and a ventral pair of eyes present, see Gyrinidae in Adephaga, above); hind coxae not immovably united to metathorax. Suborder POLYPHAGA 9
2. Hind coxae not contacting elytral epipleura (fig. 13:1*d*), separated from them by metasternal epimera which contact 1st abdominal sternite as may the metasternal episterna (except in *Trachypachus* of Carabidae, but note the numerous setae, as long as segments are wide, on its antennal segments); at least some parts of body with erect individual setae; typically terrestrial beetles (fig. 13:1*e*), with a few species adapted to the intertidal zone of the sea beach 3
— Hind coxae contacting elytral epipleura and separating metapleura from 1st abdominal sternite (see *Trachypachus,* above); body often with vestiture, but without individualized setae; aquatic beetles 4
3. Scutellum present .... CARABIDAE
— Scutellum absent (fig. 13:1*e*) ...... OMOPHRONIDAE
4. Eyes divided by sides of head, appearing as 4: a dorsal and a ventral pair (fig. 13:23); antennae short, stout, 3rd enlarged and earlike, the following segments more or less closely unified; middle and hind legs greatly modified for swimming, short and flattened, tarsi folding fanwise (fig. 13:22*b*) .... GYRINIDAE
— Eyes 2, antennae elongate, slender; hind legs suitable for crawling or definitely modified for swimming, tarsi often flattened somewhat but never folding fanwise 5
5. Hind coxae expanded into large plates, covering first 2 or 3 abdominal sternites and bases of hind femora (fig. 13:3*b*); legs hardly modified for swimming, hind tarsi slightly flattened and fringed with long hairs (which are usually closely applied to the tarsi in dry specimens); small beetles, 5.5 mm. or less in length (fig. 13:4*a*) .... HALIPLIDAE
— Hind coxae not expanded into large plates, not covering hind femora nor more than 1st abdominal sternite 6
6. Metasternum with a transverse, triangular antecoxal sclerite separated by a well-marked suture (as in fig. 13:3*a*); hind legs not adapted for swimming, tarsi not flattened or fringed with hairs but simple and carabidlike; black or brownish beetles, 11 to 15.5 mm. long (fig. 13:1*a*) .... AMPHIZOIDAE
— Metasternum without a transverse suture; no antecoxal sclerite; hind legs modified for swimming, tarsi flattened, usually fringed with long hairs (which are commonly stuck to tarsi in dry specimens)......... 7
7. Middle of prosternum and its postcoxal process in same plane (fig. 13:8*a*); front and middle pairs of tarsi distinctly 5-segmented, segment 4 approximately as long as 3; scutellum fully exposed, or concealed 8
— Middle of prosternum not in same plane as its process, the latter bent (fig. 13:8*b*); front and middle tarsi 4-segmented, or 5-segmented with 4th very small and almost concealed between lobes of 3rd; scutellum concealed (except in *Celina* spp.) (fig. 13:14*b*). Subfamily Hydroporinae of .... DYTISCIDAE
8. Scutellum covered by bases of elytra and hind margin of pronotum; hind tarsi with 2 slender curved claws of equal length, tarsal segments nearly parallel-sided; front tibiae (except in *Notomicrus* sp., a beetle 1.5 mm. or less in length) with a curved spur or hook at apex (fig. 13:18*c*) .... NOTERIDAE
— Scutellum fully visible (fig. 13:16*e*); or if covered, hind tarsi with a single straight claw, segments lobed behind on outer side (fig. 13:9)..................
.... (in part) DYTISCIDAE
9. Tiny beetles, from about 0.5 to 1 mm. in length; hind coxae usually laminate, tarsi 3-segmented; hind wings fringed with long hairs 10
— Beetles of various sizes, from nearly 1 mm. to 40 mm. in length; at least hind tarsi with more than 3 segments; in very small dorsally glabrous forms, never with hind coxae laminate 12
10. Hind coxae widely separated; antennae 8-segmented, short, compact, without long hairs (fig. 13:42*e*); staphylinidlike beetles with truncate elytra and exposed conical abdomen; aquatic species ....
.... HYDROSCAPHIDAE
— Hind coxae contiguous or almost so; antennae 11-segmented; if elytra somewhat truncate and a few abdominal segments exposed, note antennae; littoral species 11
11. Only 3 visible abdominal sternites; phalacrid-like beetles, convex, polished, glabrous....SPHAERIIDAE
— Abdomen with at least 5 visible abdominal sternites; body form usually rather flattened, dorsally somewhat to very hairy ....PTILIIDAE
12. Antennae short, true segment 6 modified to form a cupule (fig. 13:25*b*), 7-11 (often reduced in number to 3) forming a strong pubescent club, 1-5 (sometimes reduced to 3 or 4) simple and glabrous; maxillary palpi nearly always longer than antennae; head usually with a Y-shaped impressed line on vertex 13
— Antennae not so constructed, usually longer than maxillary palpi; head without Y-shaped impressed line .. 15
13. Antennal club of 5 pubescent segments; abdomen with 6 or 7 visible sternites; transverse suture of head not joined at middle by a posterior median suture; tiny beetles, not more than 2.5 mm. in length (fig. 13:25*e*) .... HYDRAENIDAE
— Antennae with only 3 pubescent segments beyond cupule; 5 visible abdominal sternites, if a 6th is present it is membranous *(Hydrochus)* or more or less retracted under 5th *(Berosus, Laccobius);* transverse suture of head, when present, directed angularly backward at middle, meeting coronal suture to form a Y; size varied, but species approaching 1 mm. in length are convex and rounded 14

[2]If beetles actually found in water do not fit any of the choices offered in the key, it is probable that they do not belong in this habitat at all. Beetles commonly fly, fall, or are blown or knocked into the water, and are washed in by rain.

Many species live close to the water's edge, yet are not truly aquatic. Without prior knowledge of the groups, it is not always easy to recognize them as littoral rather than aquatic species. The following keys to families (adults and larvae) include the common shore dwellers, but only those which are aquatic in some stage are treated in detail below the family level. A partial exception is made for the few forms which occur within the intertidal zone of the California sea shore (see Carabidae, *Ochthebius vandykei* in Hydraenidae, Staphylinidae, Melyridae, Eurystethidae, *Throscinus crotchi* in Limnichidae).

14. All tarsi apparently 4-segmented (1st segment reduced to a minute nodule); small, squat beetles, head usually deflexed, in contact with front coxae, not visible from above .... GEORYSSIDAE
— At least 1 pair of tarsi with more than 4 segments, and other characters not as above .. HYDROPHILIDAE
15. Elytra truncate, strongly abbreviated, wings usually present and capable of being folded up beneath them; more than 2 tergites and usually the greater part of dorsal surface of abdomen exposed, segments entirely chitinized (fig. 13:43*a*) .... STAPHYLINIDAE
— Elytra usually covering entire abdomen or exposing 1 or at most 2 segments; if elytra much abbreviated, the beetles are truly apterous and have extensile membranous vesicles (fig. 13:44*a*) on prothorax and between metathorax and abdomen 16
16. Antennae short, more or less geniculate, last 3 segments forming a compact club with segmentation indistinct; middle and hind coxae widely separated on median line; front tibiae dentate externally; elytra truncate, exposing 2 abdominal tergites which form a pygidium .... HISTERIDAE
— Not having above combination of characters 17
17. Head beyond eyes produced into a distinct beak, on the end of which are the mandibles and trophi .... CURCULIONIDAE
— Head not prolonged into a distinct beak 18
18. All tarsi actually 5-segmented, but segment 4 very small and nearly concealed within lobes of 3 (fig. 13:60*c*); first 3 segments dilated, with adhesive (hairy) pads beneath; beetles of from 5 to 10 mm. in length .... CHRYSOMELIDAE
— Tarsi with 5 or fewer segments, but not fitting above description 19
19. Head with 2 dorsal ocelli, near mid-line and behind an imaginary line across hind margin of compound eyes; antennae filiform, tarsi 5-segmented, elytra polished, glabrous except for lines of long, nearly erect setae .... BRATHINIDAE
— Head lacking ocelli 20
20. Hind coxa with projection covering base of trochanter, this projection sometimes extended laterally along coxa to form a groove into which anterior margin of trochanter and femur fit when retracted 21
— Hind coxa without a projection covering base of trochanter, anterior margin of femur never fitting into a groove in coxa 27
21. Front coxae more or less conically projecting; hind margin of pronotum never crenulate (fig. 13:47*e*) ... . HELODIDAE
— Front coxae variously formed; if projecting, then hind margin of pronotum is crenulate 22
22. Front coxae more or less exerted and projecting, and/or hind margin of pronotum crenulate 23
— Front coxae transverse or rounded 24
23. Mandibles concealed when closed; labrum usually invisible or nearly so from in front; head usually slightly rostrate, because of clypeus projecting from between somewhat approximated antennal insertions (fig. 13:58*c*) .... PSEPHENIDAE
— Mandibles always partly visible, frequently conspicuous; labrium nearly always in full view from in front; head not at all rostrate .... PTILODACTYLIDAE
24. Tarsi 4-segmented, last segment shorter than rest, claws slender; front and middle tibiae broad and spinose along their outer edges; antennae short, much thickened .... HETEROCERIDAE
— Tarsi almost always 5-segmented, last segment usually as long as rest together, claws usually large and strong; front and middle tibiae narrow, not spinose; antennae often filiform 25
25. Middle coxae widely separated, hind coxae nearly or quite contiguous; legs retractile, last segment of tarsi shorter than rest together (fig. 13:57*a*) .... LIMNICHIDAE
— If middle coxae are separated, hind coxae are equally so; legs not retractile, last segment of tarsi usually as long as rest together, claws large 26
26. Antennae very short, 6 or more apical segments forming a close pectinate club; female genitalia without styli, usually asymmetrical (fig. 13:49*a*) .... DRYOPIDAE
— Antennae usually slender, never with apical segments forming a close pectinate club; female genitalia symmetrical, with movable styli (fig. 13:54*c*) .... ELMIDAE
27. Tarsi all 5-segmented; elytra abbreviated, wings completely absent; body with extensile membranous vesicles on prothorax and between metathorax and abdomen (fig. 13:44) .... MELYRIDAE
— Hind tarsi 4-segmented; elytra not abbreviated; body without extensile vesicles 28
28. First 3 visible abdominal sternites connate; eyes transverse and emarginate .... TENEBRIONIDAE
— All visible abdominal sternites similar and movably articulated to each other; eyes rounded, not emarginate 28
29. All coxae widely separated on median line; slow-moving species occurring in cracks of rocks in the intertidal zone .... EURYSTETHIDAE
— Front and middle coxae nearly or quite contiguous on median line; active species, often near margins of streams or alkali lakes .... ANTHICIDAE

Key to Nearctic Families of Aquatic and Semiaquatic Coleoptera

Larvae

1. With legs 2
— Without legs 28
2. Legs apparently 5-segmented, tarsi with 2 movable claws (except Haliplidae which are 1-clawed). Suborder ADEPHAGA 3
— Legs apparently 4-segmented, tarsi united with a single claw. Suborder POLYPHAGA 9
3. Tenth abdominal segment armed with 4 hooks; spiracles absent; lateral gills present on all abdominal segments; cardo very large; (fig. 13:21). GYRINOIDEA .... GYRINIDAE
— Tenth segment not armed with 4 hooks; spiracles usually present; with or without lateral gills; cardo of moderate to small size. CARABOIDEA 4
4. Abdomen 9-10 segmented 5
— Abdomen with 8 visible segments 7
5. Tarsi 1-clawed; abdomen 9-10 segmented (fig. 13:5) .... HALIPLIDAE
— Tarsi 2-clawed; abdomen 9 segmented 6
6. Terminal tarsal setae much shorter than claws; retinaculum single or absent; littoral in part .... CARABIDAE
— Terminal setae much longer than claws; retinaculum bicuspidate; littoral .... OMOPHRONIDAE
7. Legs fossorial, larval form elaterid (wireworm) like (fig. 13:19*a*), mandible with distinct retinaculum, inner margin neither sulcate nor tubular .... NOTERIDAE
— Legs ambulatory or natatory, larval form not elaterid-like; mandible without distinct retinaculum, inner margin either sulcate or tubular 8
8. Larvae flattened, thoracic and abdominal sides greatly expanded into thin lateral plates; cerci 1-segmented, thick, blunt at tips; gular suture median and simple (fig. 13:2) .... AMPHIZOIDAE
— Larvae not with sides greatly extended; gular suture double, at least anteriorly (fig. 13:12*a*, *b*) .... DYTISCIDAE
9. Urogomphi (cerci) segmented or individually movable (often retracted into a terminal breathing pocket in 8th abdominal segment in Hydrophilidae) 10

— Urogomphi solidly united at base, or absent 15
10. Maxillary palpiger free and segmentlike, usually carrying a finger-shaped galea; spiracles biforous; HYDROPHILOIDEA 11
— Maxillary palpiger closely united with stipes, without finger-shaped galea; spiracles annular or absent; ocelli in groups of 5. STAPHYLINOIDEA 12
11. With only 1 or no ocelli on each side of head; 9 complete abdominal segments, 10th small; mostly terrestial but occasionally littoral ....HISTERIDAE
— Ocelli usually in groups of 6; usually with 8 complete abdominal segments (9 in *Helophorus*) (figs. 13:37, 13:38) .... HYDROPHILIDAE
12. Mandible lacking asperate or tubercular molar portion, usually without molar parts; some littoral species. .... STAPHYLINIDAE
— Mandible with molar part usually large, asperate or tubercular 13
13. Tenth abdominal segment provided with a pair of recurved hooks (fig. 13:28) .... HYDRAENIDAE
— Tenth segment lacking terminal hooks 14
14. Spiracles absent; balloonlike appendices on prothorax, 1st and 8th abdominal segments; antenna very short and 2-segmented (fig. 13:42*a-d*) .... HYDROSCAPHIDAE
— Spiracles present; no balloonlike appendices; antennae 3-segmented; mandibles multiserrated .... PTILIIDAE
15. Ninth abdominal segment with a ventral movable operculum closing a caudal chamber 16
— Ninth abdominal segment without operculum 19
16. Caudal chamber with 2 prehensile claws and 3 tufts of retractile gills 17
— Caudal chamber without prehensile claws, with or without 3 tufts of retractile gills 18
17. Form robust with head retracted into 1st thoracic segment but nearly as wide as thorax or abdomen; thorax with pleura sclerotized but sternites absent or membranous, abdomen with pleura set off by sutures on segments 1-4 .... LIMNICHIDAE
— Form usually more slender, head more or less retracted into 1st thoracic segment but thorax widening posteriorly, sternites present or at least venter sclerotized from side to side; abdomen with 6 to 8 pleurae set off by sutures (rarely with sutures obliterated or indistinct) (fig. 13:50) .... ELMIDAE
18. Body form cylindrical in cross section, without tufts or retractile gills; abdominal sterna and pleura greatly reduced on segments 1-5 and absent on segments 6-9, terga forming a complete sclerotized ring, abdominal spiracles on segments 1-7 lateral, on 8 dorsal .... DRYOPIDAE
— Body form greatly flattened with margins extended so as to resemble a suction cup; cloacal chamber containing 3 tufts of fingerlike gills (fig. 13:59) .... (in part) PSEPHENIDAE
19. Four or more of the first 7 abdominal segments bearing 2 ventral tufts of filamentous gills 20
— Abdominal segments without such gills 21
20. Body greatly flattened with margins extended so as to resemble a suction cup; with ventral gills on 4 or more of the first 6 abdominal segments (fig. 13:59) .... (in part) PSEPHENIDAE
— Body form elongate cylindrical; with gills on under side of abdominal segments 1-7 .... (in part) PTILODACTYLIDAE
21. With gills restricted to anal region, more or less in a caudal chamber 22
— Without conspicuous gills 23
22. Body form elongate cylindrical; 9th segment with 5 to 21 mamillae-form "gills" and 2 prehensile curved appendages covered with short, stout spines in a shallow caudal chamber; lateral spiracles on first 8 abdominal segments...(in part) PTILODACTYLIDAE
— Body form more eliptical and flattened; 8th segment with gills in the terminal caudal chamber and a pair of enlarged spiracles on the apex above; other spiracles vestigial or absent, 9th abdominal segment vestigial; 3rd antennal segment multiarticulate, long and tapering (fig. 13:46) .... HELODIDAE
23. Ninth abdominal segment without urogomphi, if solidly chitinized it is similar to the anterior segment and without hornlike projections 24
— Ninth abdominal segment conspicuous with urogomphi fused at base and immovable or more heavily chitinized with various hornlike projections 25
24. Mouth parts directed anteriorly, maxillae large and conspicuous; legs well developed; abdominal segments 1-9 similar in form, each with a lateral spiracle and with a solid tergite covering dorsal half of segment; littoral .... HETEROCERIDAE
— Mouth parts ventral, maxillae normal, distinct gular region or gular suture absent; mandibles either simple or possessing a pseudomola; legs often poorly developed; abdominal segments mostly membranous, or with small, well-separated sclerites, 9th segment much smaller than others; habits: surface feeding or submerged on stocks and roots (fig. 13:60*a,b*) ... . CHRYSOMELIDAE
25. Abdominal spiracles located in disclike scleromes; urogomphi branched with outer prongs curved out and inner prongs strongly curved in; intertidal zone, seashore .(fig. 13:45) .... EURYSTETHIDAE
— Abdominal spiracles not located in disclike scleromes, urogomphi not branched thus; not truly aquatic but some species found in the littoral zone 26
26. Maxillary articulating area either large or indistinct; when it is indistinct, mandible has a mola; without lacinia mandibulae 27
— Maxillary articulating area absent, or very small, or concealed by mentum, not large and cushioned; mandible with lacinia mandibulae but lacking molar part .... MELYRIDAE
27. Mandible with a fleshy, hairy lobe behind base of mola; hypopharyngeal sclerome small, cup-shaped, on top of slightly chitinized dome; 1 ocellus on each side of head .... ANTHICIDAE
— Mandible without such a lobe, hypopharnynx with sclerome at base .... TENEBRIONIDAE
28. Mouth parts terminal, body mostly membranous with pro-, meso-, and metatergum and propleurae and sternum chitinized, 8th abdominal segment more or less truncate, terminated by a flat plate below which is 9th segment bearing 2 large spiracles; *Cercyon* in .... HYDROPHILIDAE
— Mouth parts ventral, body membranous except for protergum; no dorsal plate on 8th abdominal segment ..............................CURCULIONIDAE

Family CARABIDAE

Ground Beetles

This is one of the four largest families of beetles; like the Staphylinidae, it has some 20,000 described species in the world.

Two small, flattened kinds of the genus *Thalassotrechus* have been described from the intertidal zone of the California coast, but they appear to be northern and southern forms of one species, intergrading north of Santa Barbara. They have strongly prognathous mandibles, stout, well-separated legs, but no wings. They and their larvae can readily wedge themselves into cracks in rocks which are submerged at high tide, and presumably can make use of air trapped there. As they are predators they may find their food in such crevices, though they are capable of foraging elsewhere while the tide is out.

There are related genera of harpaline Carabidae

with similar habits in other parts of the world: *Aëpus* from the coasts of Europe and the island of Madeira; *Aepopsis* from England, Spain, Morocco, etc., *Thalassobius* from Chile; *Illaphanus* from the southeast coast of Australia. In *Aëpus* some of the tracheae are expanded into large internal air reservoirs, but the structure of *Thalassotrechus* has not been reported upon.

Key to the California Species of Thalassotrechus

Adults

1. Head, pronotum and elytra rufotestaceous; Santa Barbara County .... *barbarae* (Horn) 1892
— Head and pronotum rufotestaceous, elytra black except narrow marginal bead, and epipleura; Dillon's Beach, Marin County; Moss Beach, San Mateo County; Carmel and Pacific Beach, Monterey County .... *nigripennis* Van Dyke 1918

REFERENCE

VAN DYKE, E. C.
1918. New Inter-tidal rock-dwelling Coleoptera from California. Ent. News, 29:303-308.

Family AMPHIZOIDAE

The amphizoids are sluggish beetles, 11 to 15 mm. long, to be found in many streams and rivers. All belong to the single genus *Amphizoa*. The adults (fig. 13:1*a*) are black to reddish-brown and resemble certain Tenebrionidae; in fact, one kind was described as a new genus and species of that family. The larvae (fig. 13:2) are also aquatic, and at a glance resemble those of the cychrine carabids and some silphids. Both adults and larvae crawl but do not swim. They occur under stones and in trash along the edges of streams and rivers, and in log jams, eddies, and backwaters. Darlington (1929) says: "*Amphizoa* emits an odor which is rather pleasant, at least to the collector, and which Horn compares to that of decaying wood. The beetles also exude a thickish, yellow fluid from the joints, so that they leave a cigarette-like stain on the fingers." However, Edwards (1953) reports that the yellowish fluid is actually "discharged from the anus and is immediately spread over the entire abdominal surface by the threshing about of the hind legs."

Relationships.—The family Amphizoidae belongs to the Adephaga and is of caraboid stock. It shows relationships with the Carabidae and Dytiscidae but is more closely allied to the Old World family Hygrobiidae. Like this last it is distinguished from the Dytiscidae by having a transverse, triangular antecoxal sclerite separated by a well-marked suture. It differs from the Hygrobiidae, but resembles some Dytiscidae, in that the metasternal episterna reach the middle coxal cavities, and it differs from both in that the legs are ambulatory and not specialized for swimming. The wings (fig. 13:1*b*) are venationally like those of the Dytiscidae. The larvae resemble those of some carabids, but have eight instead of nine or ten visible segments; they differ from dytiscid larvae in having nonsuctorial mandibles.

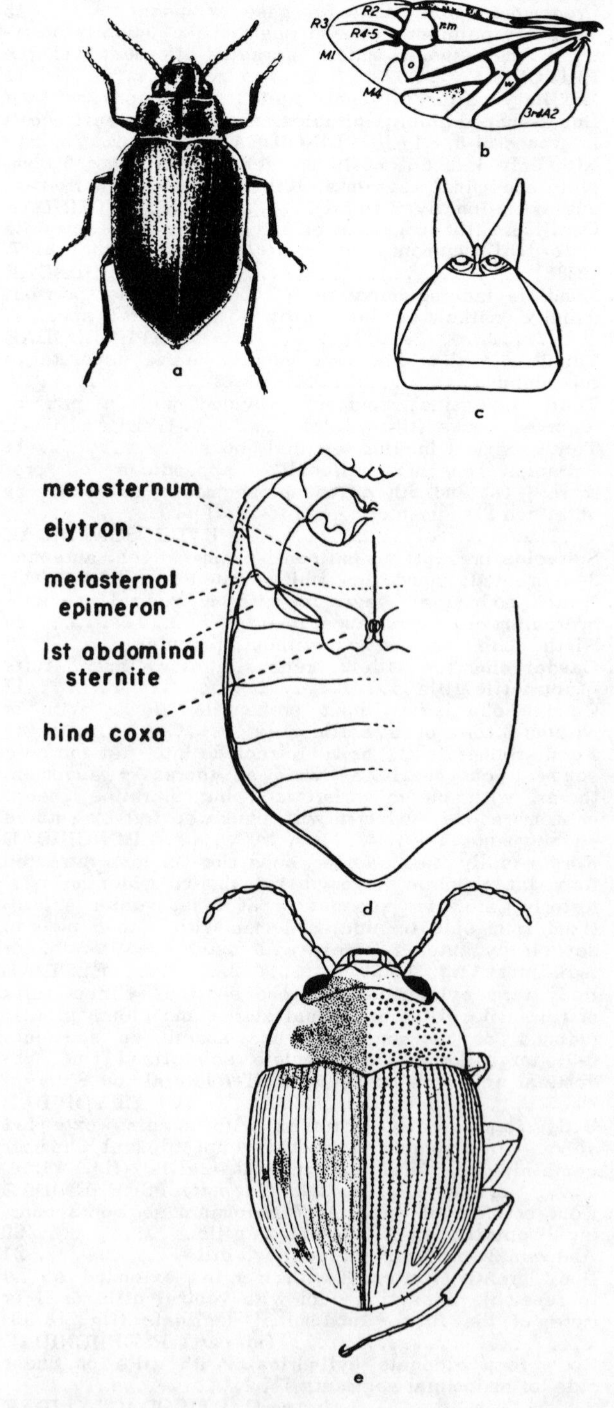

Fig. 13:1. *a*, adult of *Amphizoa insolens*; *b*, wing of *Amphizoa lecontei*; *c*, eighth abdominal tergite of larva of *Amphizoa lecontei*, showing large spiracles; *d*, ventral aspect of *Calosoma semilaeve* to show body parts; *e*, *Omophron tanneri proximum*, adult (*a*, Sharp, 1882; *b*, Edwards, 1950; *c*, Hubbard, 1892; *d*, Essig, 1926; *e*, Chandler, 1941).

Hubbard (1892) thought that both larvae and adults

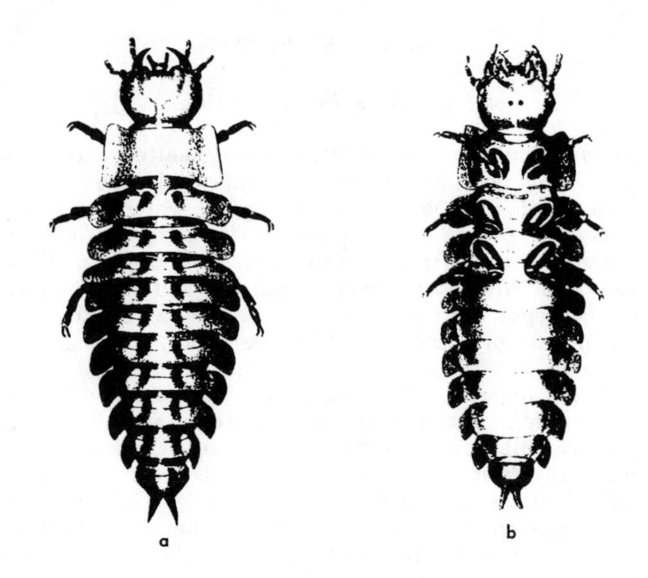

Fig. 13:2. Larvae of *Amphizoa lecontei*. *a*, dorsal aspect; *b*, ventral (Hubbard, 1892).

of *Amphizoa* gave "the impression of a terrestrial bettle with amphibious or semiaquatic habits." Edwards (1951) suggested that the bristles and long silky hairs along the inner edge of the tibiae are vestiges of former swimming hairs, and thus implied an ancestry with a more active aquatic life.

Respiration.—The larvae have the eighth pair of abdominal spiracles very large and close together on the tergite (fig. 13:1*c*). Edwards (1954) reports that when dislodged and floating in relatively quiet water young larvae assume a characteristic pose, with the abdomen horizontal and the eighth pair of spiracles at the surface of the water, but with the thorax and head tucked under the abdomen so that the mandibles lie beneath the abdominal tip. He found that in captivity the older larvae always climbed up some object till they were above the surface of the water, and stayed there till they reëntered it to capture prey. Adults depend on refilling the subelytral chambers at intervals, and carry a bubble of air at the tip of the abdomen when they submerge.

Life history.—Little was known of amphizoid history until Edwards published his data (1954). He found the surprisingly large eggs (2.1 mm. by 1 mm.) at Glacier Park, Montana, in late August. They were fastened loosely in cracks on the undersurface of a piece of floating, weathered driftwood. Larvae nearly ready to emerge were tightly tucked up in the eggs, but upon removal were found to measure a little more than 3.5 mm. in length!

Very young larvae have been taken in November, and mature larvae, 12 to 15 mm. long, in late June, so they must overwinter in the first or second instars.

Pupae are still unknown, but Edwards states that H. P. Chandler found two mature larvae, apparently ready to pupate, in protective cases "lodged in debris-filled crevices between logs, one at a distance of about one and a half feet above the water and the other about four feet above the water level and two feet back from the edge of the stream." This was at the north fork of the Fresno River, Madera County, elevation 4,000 feet, on June 1.

Adults are abundant in middle and late August. Most specimens in collections were taken during the period May to September, inclusive. This is the time when collectors are in the field, and does not prove that the beetles cannot be found in other months. Darlington (1929) records freshly emerged adults of *A. insolens* LeConte at North Bend, Washington, July 30.

Habitat and distribution.—The American species occur in streams, rivers, and sometimes in lakes, from near sea level to an elevation of at least ten thousand feet. They are most numerous at elevations between 1,000 and 3,000 feet.

Adults are often found clinging to water-soaked logs, especially old decaying ones full of cracks and holes, lying well out in the current. With the larva they crawl among stones and gravel at the water's edge, on logs which project out into the water, and particularly on debris washed to the sides of eddies or held by log jams. Darlington records many *A. lecontei* Matthews and a few *A. insolens* from wind-driven logs and trash floating at the outlet of a lake. Such stations are correlated with their food habits: they were formerly presumed to eat chiefly sluggish or drowned insects. Edwards (1954) found that adults of *A. insolens* and *A. lecontei* showed amazing speed in capturing agile plecopteran nymphs, and would touch no other food; all stages of the larvae also fed only on living stonefly nymphs.

Dispersal is presumably by flight, but the only record of flight is by Darlington, who knocked an *A. lecontei* down as it flew over Lake Minnewanka at Banff, Alberta.

Except for *A. davidi* Lucas of eastern Tibet, all the described species are American. They occur in the Rocky Mountains and westward almost to sea level, and from Alaska to southern California. The common *A. insolens* and *A. lecontei* were described in 1853 and 1872 respectively, *A. striata* Van Dyke in 1927, and *A. carinata* Edwards in 1951.

Taxonomic characters.—Three of the four American species have characteristic differences of outline, convexity, and elytral carination, by which they may be recognized at a glance. They show slight structural differences in pronotal widths and lateral crenulations, grooving of the outer rear edge of the front tibiae, median lobes of the male genitalia, and valvifers and paraprocts of the female genitalia. Males may be distinguished by having a single brush of short hairs on the underside of the first segment of the front and middle tarsi; in females this segment is armed with bristles like those of the following segments.

Genus *Amphizoa* LeConte

Only one species has been reported from California. Elytra evenly convex from side to side (fig. 13:1*a*); pronotum at least as broad at middle as at base, with coarse lateral crenulations; groove on outer hind edge of front tibiae faint and confined to distal half; elytra very coarsely rugose; color dull black; length 11 to 14.5 mm.; Alaska to southern California, east to Montana; in California, south at least to San Jacinto Mountains of Riverside County .... *insolens* LeConte 1853

REFERENCES

(For a full bibliography see Edwards, 1951)

EDWARDS, J. G.
1951. Amphizoidae (Coleoptera) of the World. Wasmann Jour. Biol., 8:303-332, 4 pls.
1953. The real source of *Amphizoa* secretions. Coleopt. Bull., 7:4.
1954. Observations on the biology of Amphizoidae. Coleopt. Bull., 8:19-24.
HATCH, M. H.
1953. The beetles of the Pacific Northwest. Part I: Introduction and Adephaga. Univ. Wash. Publ. Biol., 16:vii+1-340, incl., front., 2 text figs., 37 pls.

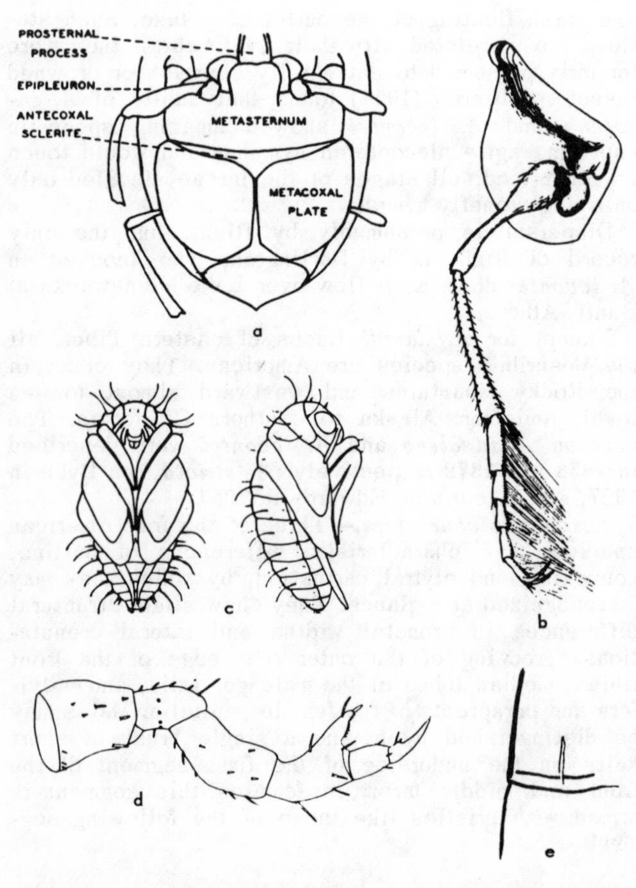

Fig. 13:3. Haliplidae. *a*, *Peltodytes callosus*, part of undersurface; *b*, *Haliplus* sp., underside of hind leg to show coxa, trochanter, and part of femur covered by hind coxal plate; *c*, *Peltodytes lengi*, pupa; *d*, *Peltodytes lengi*, front leg of larva; *e*, *Haliplus longulus*, left side of pronotum to show basal plica (*a*, Leech, 1948; *b*, Matheson, 1912; *c*, *d*, Hickman, 1930; *e*, Wallis, 1933).

Family HALIPLIDAE

Crawling Water Beetles

Though known as the crawling water beetles, haliplids are fairly good swimmers in ponds and lakes. They are found in running and in quiet waters, usually near the shore, and especially in association with algae (*Chara*, *Nitella*, *Spirogyra*, and others). They eat both plant and animal matter, and are in turn preyed upon by fish. Adults and larvae are aquatic, but pupal cells are formed on the shore, in damp soil or sand.

Relationships.—The Haliplidae are included in the Caraboidea, but are at once distinguished by their greatly enlarged platelike hind coxae, which cover at least the first two abdominal sternites and the basal halves of the hind femora (fig. 13:3*a*, *b*). The antennae are inserted more nearly between the eyes than in other Caraboidea (except the Cicindelidae), and the elytra almost always have longitudinal rows of large punctures (fig. 13:4*a*). An origin from terrestrial forms, but independently of other aquatic groups, has been suggested.

Respiration.—The young larvae of haliplids obtain oxygen by cutaneous respiration; after the second instar they have spiracles. Adults must come to the surface of the water for air, and their methods of respiration have been reported by Hickman (1931). He found that a beetle which has a supply of air in both the reservoir under the elytra and its connection, the chambers formed by the expanded hind coxae, is able to float up and break the surface film with the tip of its abdomen. With air supply replenished it dives, and shows a large bubble protruding from the hind coxal cavity. This bubble has both respiratory and hydrostatic functions; a beetle deprived of it cannot float up and break the surface film with its abdomen, but must swim up and crawl out to obtain air. Since the subelytral and hind coxal air reservoirs are connected, a beetle can live if only the latter are allowed contact

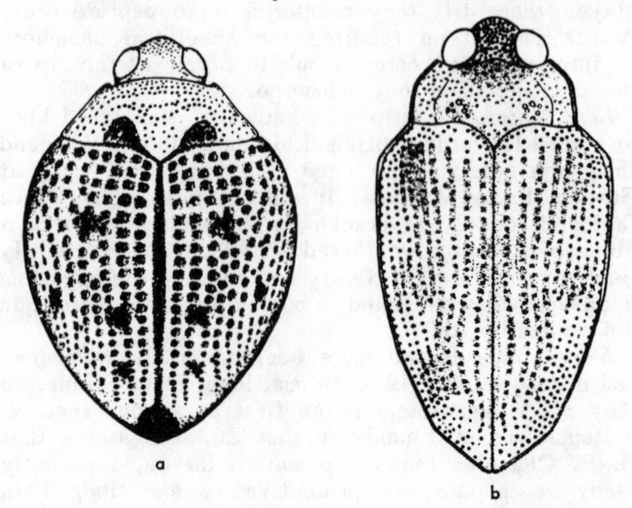

Fig. 13:4. Haliplidae, adults. *a*, *Peltodytes simplex*; *b*, *Brychius hungerfordi* (*a*, Leech, 1948; *b*, Spangler, 1954).

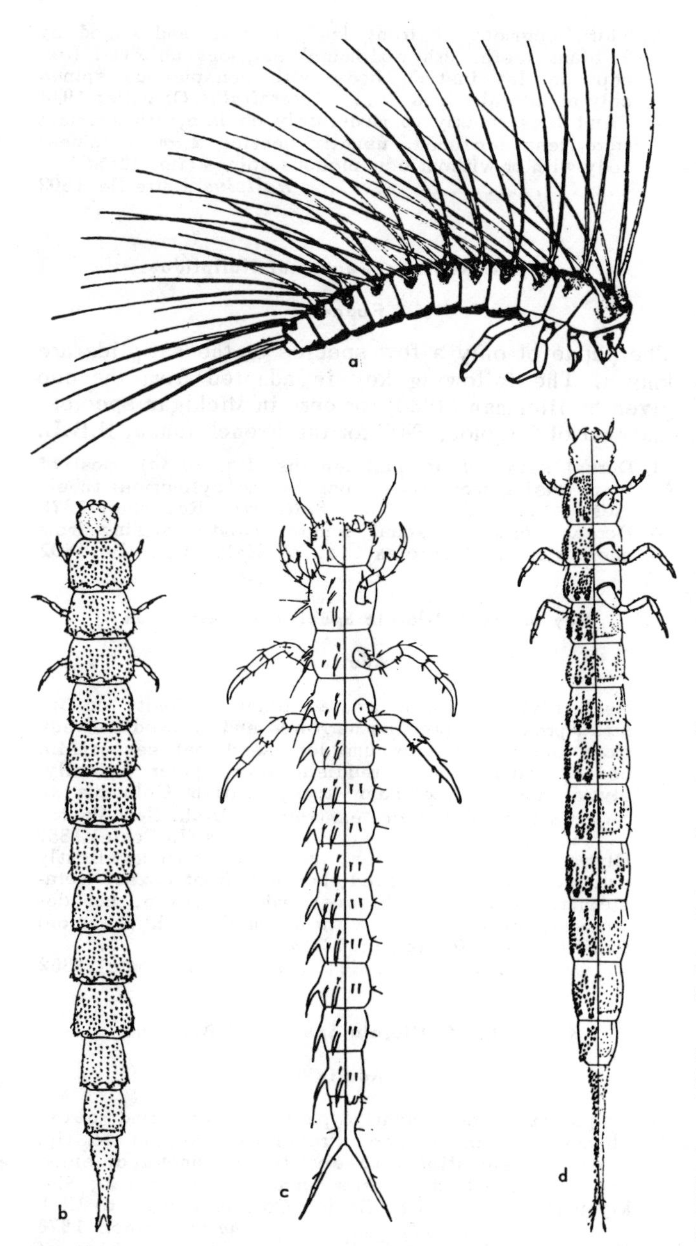

Fig. 13:5. Haliplidae, larvae. *a, Peltodytes edentulus; b, Brychius elevatus; c, Haliplus immaculicollis,* first instar; *d, H. immaculicollis,* third instar (*a,* Wilson, 1923; *b,* Bertrand, 1928; *c, d,* Hickman, 1930).

with fresh air; but it cannot obtain enough oxygen in the water by diffusion from the coxal bubble to sustain life.

Life history.—The biology of haliplids has been treated, in this country, by Matheson (1912), and Hickman (1931). The immature stages of six North American species have been described in detail by Hickman (1930) and Wilson (1923). At least some species overwinter in the larval stage, commonly in damp soil above the water line. Probably all species pass the winter in the adult stage. Some kinds hibernate but others remain active even under a foot or more of ice; in the latter case we do not know how they obtain the regular supplies of fresh air needed for such activity.

Adults of *Peltodytes* spp. lay their yellow eggs on aquatic vegetation (*Elodea, Ceratophyllum,* and filamentous algae). Those of *Haliplus* are said to bite holes in the plant tissue and insert their (white) eggs into the emptied cells. Egg-laying has not been recorded for *Brychius* or *Apteraliplus.* The larvae (fig. 13:5*a-d*) are most commonly found crawling slowly on and among filamentous algae, upon which they feed.

At one time both adults and larvae were thought to be predatory. Hickman and others reared specimens from eggs and found larvae and adults to be vegetarian. The larvae of some species ate filamentous algae *(Spirogyra)* which they grasped with their specialized front legs (fig. 13:3*d*), whereas species not so adapted fed on *Chara* and *Nitella;* they took animal food only when starved. However, F. Balfour-Browne (1940) reported that in addition to taking plant food, his adult specimens fed eagerly and regularly on cut-up aquatic insect larvae, crustaceans, entomostracans, and fish. He quoted Beier as writing that the beetles searched for oligochaete and nematode worms, and even attacked mosquito and chironomid larvae. Balfour-Browne also mentions that he often found remains of chitin in the oesophagus of haliplid beetles.

The mature larva crawls up on the shore, digs into a suitable spot, and forms a pupal chamber, often under a stone or log. The pupa (fig. 13:3*c*) is a typical exarate type.

Habitat and distribution.—The species of *Brychius* are few, but occur in both the Nearctic and Palaearctic regions. Those of *Peltodytes,* more numerous, are in the same regions and also attain the Oriental and Ethiopian. The species of *Haliplus* comprise the bulk of the family, and are found in all the above areas, as well as the Australian and Neotropical regions; the greatest number are in the Holarctic realm. *Apteraliplus* is known only from California and Washington; the wings are reduced and the species is flightless.

Because they are poor swimmers except in still water, haliplids are most numerous in ponds (*Apteraliplus, Haliplus, Peltodytes*), the sheltered parts of lakes and streams or rivers (*Haliplus, Peltodytes*), or in the current of streams and shallow rivers but where stones offer shelter (*Brychius, Peltodytes*). Few species are permanent residents of saline waters, and none become adapted to hot springs.

In North America the species of *Brychius* are of local occurrence and until lately were rare in collections; only two species have been reported from east of the Rocky Mountains. They prefer small streams with gravelly or stony bottoms. *Apteraliplus parvulus* (Roberts) is known from San Mateo County, California (though *Haliplus wallisi* Hatch from the Grand Coulee of Washington is said to be a synonym). In California *A. parvulus,* like certain small species of *Hydroporus* (Dytiscidae), seems to have adapted its life cycle to take advantage of ponds formed during the winter rains in an otherwise dry region. The greater number of species of both *Haliplus* and *Peltodytes* are found east of the Rocky Mountains, whereas in the Pacific Northwest some of the species form races or distinguishable populations in nearly every mountain valley, so that their taxonomy is difficult.

Taxonomic characters.—Examination of the male genitalia is almost imperative in making specific identifications. Color, and the sizes and arrangements of the elytral punctures, are of value in *Peltodytes* and *Haliplus,* but the characters most used are on the ventral surface (prosternum, metasternum, trochanters, hind coxae). Secondary sex characters are found in the claws and tarsal segments of the front and middle legs of the males. The first three segments of the front and middle tarsi of the males are slightly broadened and/or pedunculate, and clothed beneath with a dense pad of short hairs.

Key to Nearctic Genera of Haliplidae

Adults

1. Last segment of palpi cone-shaped, as long as or longer than next to last; hind coxal plates large, only last abdominal sternite completely exposed; elytra with fine sutural striae in at least apical half (fig. 13:4*a*) .... *Peltodytes* Régimbart 1878
— Last segment of palpi subulate, shorter than next to last; hind coxal plates smaller, leaving last 3 abdominal sternites exposed; elytra without fine sutural striae 2
2. Pronotum with sides of basal two-thirds nearly parallel; epipleura broad, extending almost to tips of elytra, which are never truncate; metasternum reaching epipleura (fig. 13:4*b*) ....*Brychius* Thomson 1860
— Pronotum with sides widest at base, convergent anteriorly; epipleura evenly narrowed, usually ending near base of last abdominal sternite, never reaching elytral apices; episternum completely separating metasternum from epipleura 3
3. Median part of prosternum and base of prosternal process forming a plateaulike elevation, at least in part angularly separated from sides of prosternum. .... *Haliplus* Latreille 1802
— Prosternum evenly rounded from side to side .... *Apteraliplus* Chandler 1943

Key to Nearctic Genera of Haliplidae

Larvae

1. Each body segment with 2 or more erect, segmented, hollow spine-tipped filaments, each half as long as body (fig. 13:5*a*); abdomen with 9 segments; ocelli located on a small prominence; front legs chelate, 4th segment produced apically and edged with a solid row of small teeth, so that 5th segment and claw can be closed on it, pincherlike (fig. 13:3*d*) .... *Peltodytes* Régimbart 1878
— Body spines, except in the 1st instar, never stalked or much longer than the length of 1 body segment; 10th abdominal segment produced posteriorly in a forked or unforked horn; front legs, if chelate, with segment less produced and without solid row of small teeth 2
2. Third antennal segment shorter than 2nd; front legs moderately chelate but 3rd instead of 4th segment produced, edged with 2 blunt teeth; 9th abdominal segment not forked, not strongly curved ventrally, body without conspicuous spines (fig. 13:5*b*) .... *Brychius* Thomson 1860
— Third antennal segment 2-3 times as long as 2nd 3
3. Third segment of front leg produced and edged by 2 blunt teeth; 9th abdominal segment unforked (except in 1st instar); body with conspicuous spines only at lateral edges ..... *Apteraliplus* Chandler 1943
— Front legs weakly to moderately chelate, 4th segment more less produced, usually bearing 2 or 3 spines; body with or without conspicuous spines (fig. 13:5*d*) .... *Haliplus* Latreille 1802

Key to Nearctic Genera of Haliplidae

Pupae

The pupae of only a few species of the haliplids are known. The following key is adapted from the one given by Hickman (1930) for certain Michigan species, and that of Guignot (1947) for the French fauna. H.B.L.

1. Dorsal setae of unequal lengths (fig. 13:3*c*), most of them arising from more elongate and cylindrical tubercles .... *Peltodytes* Régimbart 1878
— Dorsal setae of equal length, mostly arising from short, conical tubercles ..... *Haliplus* Latreille 1802

Key to the California Species of Peltodytes

Adults

1. Each elytron with a median black callosity on 3rd stria; prosternal process narrowed and grooved between front coxae; metasternum depressed between middle coxae; hind femora reddish-brown, paler apically; elytra not dentate near apex; British Columbia to Baja California, Mexico, eastward to Utah, New Mexico .... *callosus* (LeConte) 1852
— Elytra without callosities; prosternal process slightly narrowed, but not grooved, between front coxae; metasternum nearly flat between middle coxae, not depressed; each elytron with a small tooth at apical four-fifths; California (fig. 13:4*a*)................. .... *simplex* (LeConte) 1852

Key to the California Species of Brychius

Adults

1. Epipleural sides nearly parallel from hind coxae almost to tip, thence strongly convergent to tip; prosternal elevation sparsely, finely punctured, sides not constricted at anterior edge of front coxae; Siskiyou County, north to British Columbia, west to Utah .... *hornii* Crotch 1873
— Epipleural sides gradually convergent from base of coxae to apex of elytra; prosternal elevation moderately punctate, sides constricted just below anterior edge of front coxae; Humboldt, Mendocino, and Alameda counties ....*pacificus* Carr 1928

Key to the California Species of Haliplus

Adults

1. Pronotum with a basal plica on each side, a little nearer to hind angles than to median line (fig. 13:3*e*) 2
— Pronotum without basal plicae 3
2. Short, broad species; basal pronotal plicae shorter, less than 1/4 of length measured from plical base along plica to anterior pronotal margin; elytra immaculate except for darkened elytral punctures which may cluster to form indefinite spots; Humboldt County,

north to British Columbia, east to Wyoming ....
.... *robertsi* Zimmermann 1924
— More elongate species, ovate; basal pronotal plicae more than 1/4 of length as measured above; elytra with a common postmedian sutural blotch, and each usually with 1 to 3 additional small blackish marks in apical half; Humboldt and Lassen counties, north to British Columbia .... *dorsomaculatus* Zimmermann 1924
3. Prosternal process plainly, usually broadly, margined at sides 4
— Prosternal process not at all margined at sides, or at most feebly so near apex 6
4. Mid-metasternum with a large fovea on each side of middle, just behind an imaginary line joining posterior margins of hind coxae; prosternal process almost parallel-sided; San Diego, San Bernardino, Santa Barbara, and San Mateo counties ....
.... *concolor* LeConte 1852[3]
— Mid-metasternum without a large fovea on each side of middle, though often with a longitudinal impression; elytra almost unicolorous 5
5. Punctures of first 5 rows of elytral striae confused and confluent, those of 6th to 10th distinctly separated; fine punctures of elytral interspaces numerous, crowded and mixed up with those of the striae; "California"; Baja California at U.S. boundary ....
.... *rugosus* Roberts 1913
— Punctures of all elytral striae clearly separated, not confused or confluent; each elytral interval with a row of small deep punctures, sutural row irregularly double, other rows single with some irregularities, not mixed with strial punctures except along basal margin; San Diego County ....
.... *mimeticus* Matheson 1912
6. Central part of prosternum, anterior to coxae, sharply margined laterally and distinguishable from the side parts; metasternum between middle coxae on slightly different plane from part behind coxae, with a central fovea in declivity 7
— Central part not margined laterally, merging into side parts; mid-metasternum on a plane between and behind middle coxae 8
7. Delimiting lines of central part of prosternum extending to anterior margin; Tulare County, Mariposa County
.........................*cylindricus* Roberts 1913
— Delimiting lines disappearing halfway between coxae and front margin of prosternum; Monterey County north to Oregon ....*gracilis* Roberts 1913
8. Subsutural row of interstitial punctures with at least an occasional misplaced puncture, and usually more or less double anteriorly; strial punctures large; British Columbia to Oregon, Utah; California.......
.... *leechi* Wallis 1933
— Subsutural row of interstitial punctures regular and rather sparsely placed basally; strial punctures smaller and of almost uniform size; California to British Columbia to Massachusetts ....
.... *subguttatus* Roberts 1913

Genus Apteraliplus Chandler

A tiny (1.5 - 2.5 mm.) spindle-shaped, humpbacked species; wings reduced to little pads, nonfunctional; San Mateo County; Stanford University campus, Santa Clara County
.... *parvulus* (Roberts) 1913

REFERENCES

BALFOUR-BROWNE, F.
1940. British water beetles. Vol. 1. London: Ray Soc., xx+375 pp., 89 text figs., 5 pls., 72 maps.

[3] *H. tumidus* LeConte 1880, a Texas species, has been recorded from "California" but presumably on the basis of mislabeled specimens. It would run to *concolor* in the key, but the humeral umbone area is asperate, not smooth between the usual punctures.

CHANDLER, H. P.
1943. A new genus of Haliplidae (Coleoptera) from California. Pan-Pac. Ent., 19:154-158, 7 text figs.
GUIGNOT, F.
1947. Coleoptères Hydrocanthares. *In* Faune de France, 48:1-288, 128 text figs.
HATCH, M. H.
1953. The beetles of the Pacific Northwest. Part I: Introduction and Adephaga. Univ. Wash. Publ. Biol., 16:vii+1-340 incl. front., 2 text figs., 37 pls.
HICKMAN, J. R.
1930. Life-histories of Michigan Haliplidae (Coleoptera). Pap. Mich. Acad. Sci., 11:399-424, pls. XLVII-LV.
1931*a*. Respiration of the Haliplidae (Coleoptera). Pap. Mich. Acad. Sci., 13:277-289, 4 text figs., 1 table.
1931*b*. Contribution to the biology of the Haliplidae (Coleoptera). Ann. Ent. Soc. Amer., 24:129-142.
MATHESON, R.
1912. The Haliplidae of America north of Mexico. Jour. N.Y. Ent. Soc., 20:156-193, pls. 10-15.
ROBERTS, C. H.
1913. Critical notes on the species of Haliplidae of America north of Mexico with descriptions of new species. Jour. N.Y. Ent. Soc., 21:91-123.
SPANGLER, P. J.
1954. A new species of water beetle from Michigan (Coleoptera: Haliplidae). Ent. News, 65:113-117, 4 figs.
WALLIS, J. B.
1933. Revision of the North American species (north of Mexico) of the genus *Haliplus,* Latreille. Trans. Roy. Canad. Inst., 19:1-76, 38 text figs.
WILSON, C. B.
1923. Water beetles in relation to pondfish culture, with life histories of those found in fishponds at Fairport, Iowa. Bull. Bur. Fish., 39:231-345, 148 figs.
YOUNG, F. N.
1954. The water beetles of Florida. Univ. Florida Stud., Biol. Sci. Ser. 5:IX+238 pp., 31 figs.

Family DYTISCIDAE

Predaceous Water Beetles

Dytiscids are widespread, numerous in species, locally common, active during the day, and sometimes countless numbers are attracted to lights at night. Except for the Gyrinidae, they are more readily seen than other aquatics, and most people who refer casually to "water beetles" have them in mind.

The adults spend most of their time under water, but must obtain air either by breaking through the surface film or from bubbles attached to aquatic plants. Their hind legs are used like oars, and in articulation, form, and vestiture are specially adapted for swimming. Eggs are laid in wet places just out of the water, or more commonly on or in plants under water. The larvae, like the adults, are predaceous; those of most genera must come to the surface to obtain fresh air. When mature they leave the water to form pupal cells in the damp soil or sand nearby.

Relationships.— With the hind coxae fused to the metathorax and completely dividing the first visible abdominal sternite, the Dytiscidae, with related families of aquatic habits (Amphizoidae, Haliplidae, Hygrobiidae, Noteridae, Gyrinidae), form the section Hydradephaga of the superfamily Caraboidea, suborder Adephaga. The terrestrial families (Cicindelidae, Carabidae) form the section Geodephaga. The aquatic

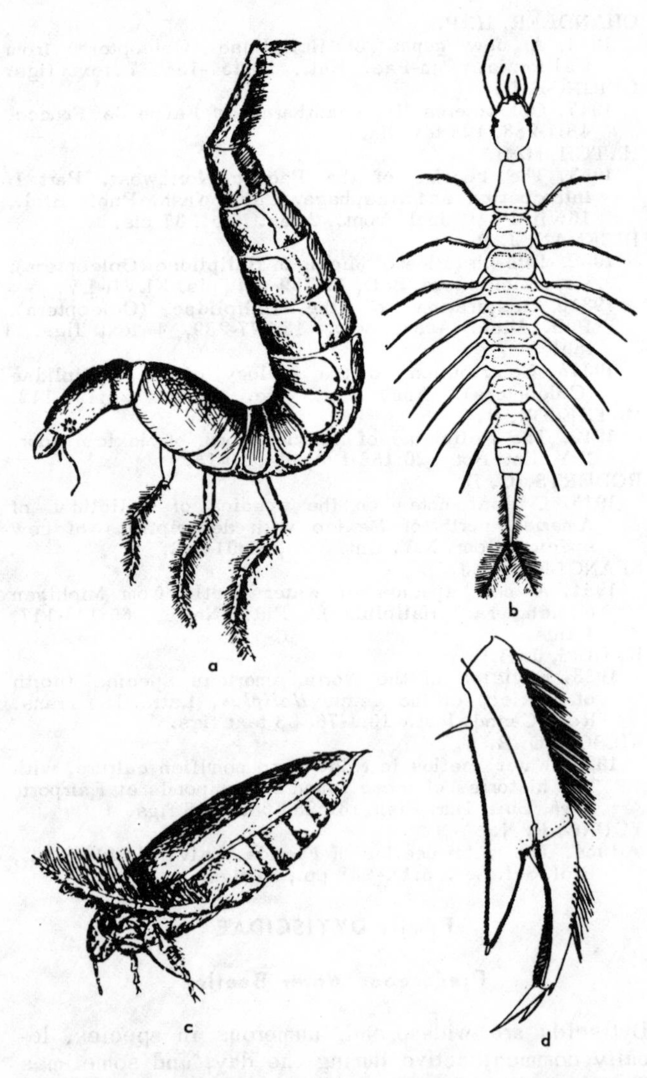

Fig. 13:6. Dytiscidae. *a, Dytiscus* sp., larva taking air at surface of the water; *b, Coptotomus* sp., larva, to show abdominal gills; *c, Acilius* sp., adult renewing air supply at surface of water; *d, Matus bicarinatus,* tibia and tarsus of middle leg of larva showing chela (*a, c,* Wesenberg-Lund, 1912; *b,* Peterson, 1945; *d,* J. Balfour-Browne, 1947).

forms are considered to be derived from terrestrial adephagids which took to the water and became adapted to conditions found there. Representatives of the ancient families Amphizoidae (hind legs not modified for swimming; larvae *Carabus*-like in appearance) and Hygrobiidae (hind legs moved alternately in swimming; mandibles of larvae not channeled) are more primitive than the Dytiscidae; the Hygrobiidae (Great Britain, western and southern Europe, western China, Australia) are probably most like the ancestral group from which the Dytiscidae and others arose. However, we have no proof that all these families had a common ancestor, nor that all the aquatic families result from a single successful venture from land to water.

Respiration.—The Dytiscidae are almost glabrous underneath and lack the ventral film of air, held by hydrofuge pubescence, found in the Hydraenidae, Hydrophilidae, Elmidae and others. Like adults of the other Hydradephaga they have their main reservoir of air under the elytra. To fill this subelytral chamber with fresh air, the beetles must break the surface film with the tip of the abdomen. When undisturbed they often float slowly upward, tail first, since the center of gravity is near the ventral anterior part of the body. But when in danger or perturbed they swim up quickly, do a half somersault, poke the end of the abdomen out (fig. 13:6*c*), and dive almost immediately.

The hairs on the last dorsal segment (pygidium) are kept oiled by unicellular hypodermal glands and are not wetted when the abdominal tip breaks the surface film. In conjunction with the elytral apices this pygidium forms an opening through which stale air can be forced out of the subelytral chamber by an up-and-down movement of the flexible dorsal surface of the abdomen. The large last pair of abdominal spiracles opens to the fresh air at the pygidium, enabling the tracheal system to be filled directly; the other spiracles connect the two lateral longitudinal tracheae and the thoracic air sacs with the subelytral reservoir.

The adults of many species of Dytiscidae hibernate during the winter; because of their inactivity and the low temperature, their oxygen requirements are doubtless very small. Other species remain active, even under ice, and probably obtain air from the layer trapped beneath the ice as it rises from aquatic plants.

The larvae of Dytiscidae have two longitudinal tracheae, but no enlargements or storage areas. As a result, most of them are forced to swim or crawl to the surface at intervals to renew their air supply. This they do by breaking the surface film with the abdominal tip or the hairy caudal cerci, by which they hang suspended (fig. 13:6*a*); only the large caudal spiracles can take in air. According to the literature the other spiracles (except perhaps the mesothoracic pair, through which air may be forced out) are closed and nonfunctional except during each larval molt, and at maturity when the larvae leave the water to pupate.

The larvae of *Coptotomus* spp. have lateral abdominal gills, a pair on each of the first six segments (fig. 13:6*b*). They are able to remain continuously beneath the surface, obtaining their air from the water. The larvae of the Hydroporinae do not usually come to the surface, and are also able to obtain air from the water though they have no gills. The venters of these and some other dytiscid larvae have a network of tracheae near the surface, so they may be able to breathe through the skin.

Life history.—In the north temperate zone there is but one generation a year, though larvae of various sizes, pupae, and teneral adults may all be present at one time in late summer. Most species, if not all, pass the winter as adults; some hibernate, others remain active even under the ice. In certain regions which lack summer rain, such as the coast of northern California, some species of the Hydroporinae can be found only during the winter, at which time they breed in the ephemeral ponds formed by the winter rains. Some species of dytiscids overwinter both as adults and as larvae; adults of certain kinds aestivate in the soil when ponds dry up during the summer.

The kind of ovipositor a female dytiscid has is an indication of where she will lay her eggs. There are three major types: (1) those in which the genital valves are somewhat dorsoventrally flattened, blunt, weak, or short, and not adapted for piercing, but often hairy and doubtless tactile (Hydroporinae; various unassociated genera such as *Colymbetes*); such beetles lay their eggs externally on plants and other underwater objects, or in soft mud; (2) those in which the valves are suited for piercing *(Dytiscus, Cybister)* or sawing *(Ilybius, Laccophilus)* plant tissue; the eggs are usually laid within the stems and leaves of plants growing in the water; and (3) a greatly elongated ovipositor with which eggs are laid in cracks and protected places under loose bark, in moss or among grass roots, sometimes out of the water though still in damp places *(Acilius)*. It should be noted that oviposition habits cannot always be inferred from the form of ovipositor. Some species with nonpiercing genital valves are reported to lay their eggs in plant tissues, in holes first made with the mandibles.

The eggs hatch in about five to eight days, though in *Dytiscus* they may take weeks or even months, depending on the water temperature. There are three larval instars. The larvae are predaceous and cannibalistic; they find their prey—chiefly the larvae and adults of aquatic insects, shrimps, worms, leeches, snails, tadpoles, and even small fish—either by hunting actively, or by waiting in ambush. The jaws are sharp, and channeled; when the prey is caught, the jaw tips pierce through the body wall and a brownish digestive fluid is injected through the channels. This serves both to kill the prey, and to start preoral digestion of the body contents, all but the insoluble parts of which are then sucked back through the mandibles.

What appears at first glance to be a remarkable raptorial development of the front and middle legs of the larva of *Matus bicarinatus* (Say) was reported by J. Balfour-Browne (1947). The serrated inner edge of the tarsus opposes the marginally serrate edge of a prolongation of the tibia (fig. 13:6*d*), forming an obvious chela. By watching living larvae, Balfour-Browne found that they never used the front legs in a raptorial manner, but did use them most successfully to burrow in mud. The tibial projection acted both as a protection to the tarsus, and as a digging scoop. It is thus a fossorial adaptation otherwise unknown in the family Dytiscidae, and quite unlike the short broad fossorial legs of the Noteridae.

Mature larvae crawl up the shore and burrow in a suitably damp spot, forming a pupal chamber in the soil. Some species tunnel into the mud near the water line; others crawl under logs or stones, then dig down; a few build turretlike cells above the soil (see Matheson 1914, fig. 4) in sheltered places. Beetles freshly emerged from the pupal skin usually stay in their pupal cells for a week or more, while the integuments harden and attain full coloration; they break out of the cells about a month or six weeks after having formed them.

Parasites.—Few, if any, parasites have been recorded from the Nearctic Dytiscidae. Both larvae and adults of Old World species are known to suffer from the effects of Protozoa (internal and external forms), hairworms (*Gordius* and allied genera), trematodes, and mites. Species of several genera of proctrotrupid and chalcid Hymenoptera parasitize dytiscid eggs, even those laid in the underwater stems of plants.

Dytiscidae, particularly adults, fall prey to various aquatic animals. There are numerous records of finding them in the crop and stomach contents of reptiles, amphibians, fishes, birds, especially ducks and wading birds, as well as land birds such as swallows, and animals such as raccoons and skunks.

Habitat and distribution.—Dytiscid beetles occur in almost any spot damp enough to be called wet, except the open sea and extremely hot or saline waters. Streams, rivers, ponds, lakes, acid bogs, salt marshes, stagnant pools, all have their characteristic species. Just as these habitats grade into or resemble one another, the shallow weedy bays of lakes and the quiet backwaters of rivers being comparable to ponds, so the species typical of ponds may be often found in rivers or lakes. Some species are restricted to specialized habitats such as hot springs or saline desert pools, so that their populations are small and isolated. Other forms may be remarkably tolerant; for instance *Deronectes striatellus* (LeConte) may be found from sea level to an altitude of 10,000 feet, in clear, cold mountain streams, in ponds choked with decaying sawdust, and in the pools of a drying up stream where the water is turbid and so warm that the fish are dying. Large species, such as those of the genera *Dytiscus* and *Cybister,* tend to be in the larger, though not necessarily deeper, bodies of water, whereas small members of the *vilis* section of the genus *Hydroporus* may be in isolated moss-covered seepages no bigger than one's hand.

Keeping in mind the microhabitats within major habitats, and the overlapping of species into several habitats, one may generalize about Nearctic species as follows. In fast-flowing streams and rivers: *Bidessus, Deronectes, Oreodytes, Agabinus, Agabus.* In slowly flowing waters: the same plus *Laccophilus, Hydrovatus, Desmopachria, Hygrotus, Hydroporus Hydrotrupes, Ilybius, Matus, Copelatus, Rhantus, Colymbetes, Dytiscus, Hydaticus.* In ponds and lakes: *Macrovatellus, Laccophilus, Desmopachria, Bidessus, Celina, Hygrotus, Hydroporus, Deronectes, Agabus, Ilybius, Matus, Copelatus, Coptotomus, Neoscutopterus, Rhantus, Dytiscus, Hydaticus, Acilius, Thermonectus, Graphoderus, Eretes, Cybister.* In saline pools: *Hygrotus, Deronectes.* In hot springs (up to about 40° C.): *Bidessus, Hygrotus.*

Those species which occur in fast water usually crawl among stones or stay in the slack water behind them. In quiet water, weedy shallows are more productive than those without much vegetation; very few species (though sometimes many specimens) are found where there is a dense growth of filamentous algae.

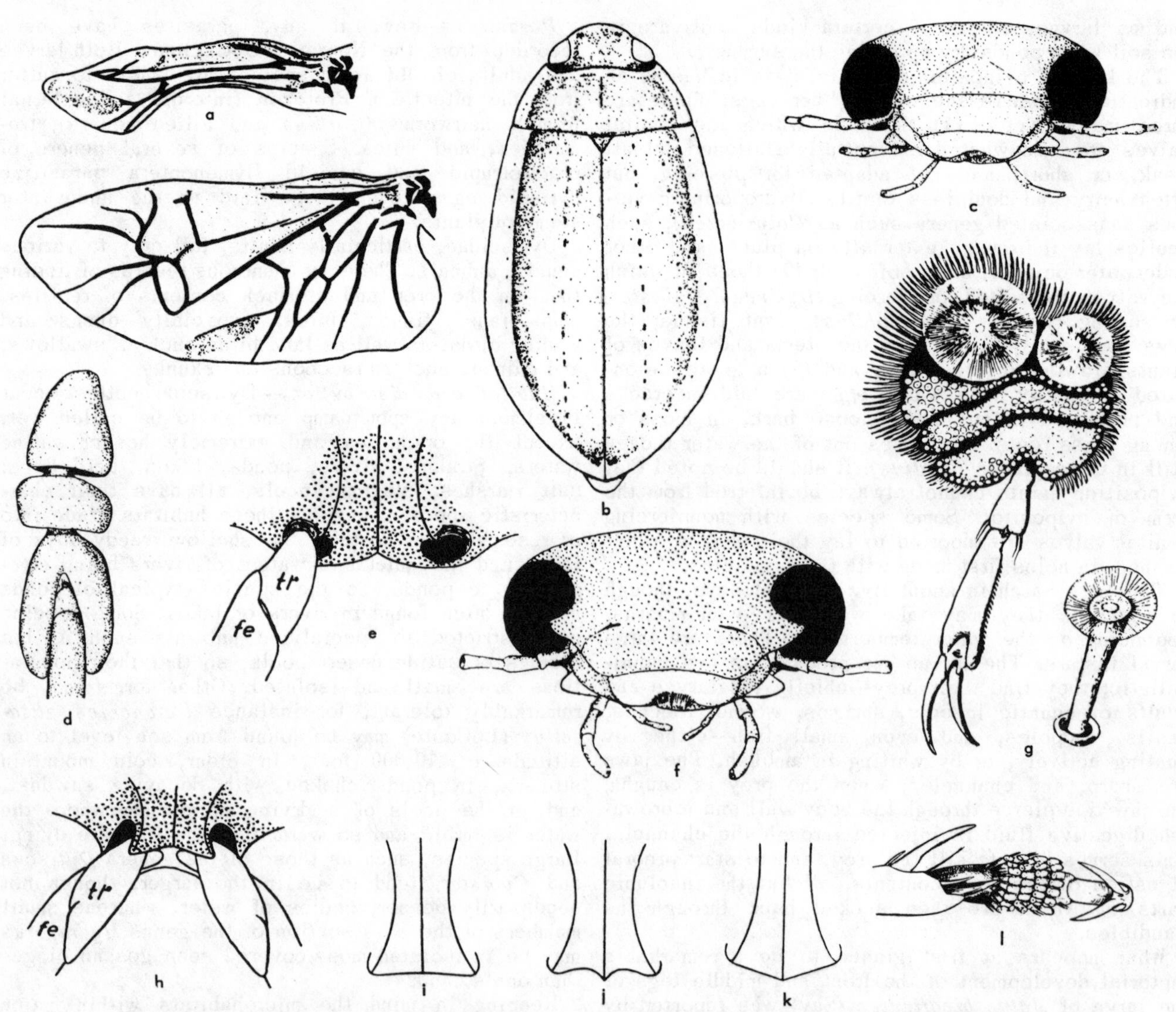

Fig. 13:7. Dytiscidae, structures of adults. *a*, *Agabus bifarius*, reduced and fully developed wings; *b*, *Cybister explanatus*, adult, showing scutellum; *c*, *Dytiscus* sp., head showing non-emarginate eyes; *d*, *Hydroporus* sp., front tarsus showing very small fourth segment; *e*, *Bidessus* sp., showing trochanter (tr) of hind leg of adult free, not covered by hind coxal process; *f*, *Agabus* sp., head of adult to show emarginate eyes; *g*, *Dytiscus* sp., front tarsus of male; *h*, *Hygrotus* sp., showing trochanters (tr) of hind legs covered at base by lobes of hind coxal processes; *i*, *j*, *k*, *Hydroporus* spp., posterior margin of hind coxal process showing variation of this structure within the genus; *l*, *Rhantus* sp., front tarsus of male (*a*, Leech, 1942; *b*, Leech, 1948; *c*, *f*, F. Balfour-Browne, 1950; *d*, *e*, *h*, F. Balfour-Browne, 1940; *g*, Miall, 1903; *i-k*, Guignot, 1950; *l*, Williams, 1936).

Waters which have had but few beetles may become populous for a time during the spring and fall migration flights, usually May and September. Very little is known about the flight habits of our species, or indeed even about which ones can fly. Leech (1942) discussed dimorphism in the wings (fig. 13:7*a*) of *Agabus bifarius* (Kirby), a species which often occurs in ephemeral ponds. As late as 1948. Brinck was able to cite only six species of dytiscids in which the wings were known to be reduced. But in 1950 and 1952 Jackson, working in Scotland, opened up a whole field for study; she found by dissection that many species which had fully developed wings had atrophied flight muscles, and were in fact flightless. However, not all examples of some species, or even all specimens from a single pond, had the same development of flight muscles. In other cases it seemed that the muscles were functional for a short time after the beetles emerged from their pupal cells, then gradually atrophied. It is interesting to compare this with the case of some solitary large queen ants. Richards (1953:137) says that after a nuptial flight and loss of her wings, such a queen "... hunts for a crevice or digs a small cell under a stone. Here she lives without food for nearly a year, being sustained by food reserves derived from the breakdown of her wing-muscles." It will thus be of value to record what species actually fly to lights, or land on the shiny tops of cars, mistaking them for water.

Most species crawl or run easily on dry land; some, especially *Laccophilus* spp. and *Hydrotrupes palpalis* Sharp, can jump almost like flea-beetles.

Dytiscidae occur on all the major land masses and on a fair number of oceanic islands. Many of the genera are widespread; of the eleven listed by J. Balfour-Browne from Oceania, seven occur in California, and of the nineteen cited by him and by Brinck for Manchuria, all but one occur in California. The species of *Matus* are found in the triangular area bounded by southern Ontario, Texas, and Florida; the only other known genus in the tribe, *Batrachomatus,* is confined to the Australian region. One species, *Eretes sticticus* (Linnaeus), is almost cosmopolitan, and has been taken in southern California.

Taxonomic characters.—In many genera specific separations are based on the secondary sex characters of the males. These are most often seen as a broadening of the front and/or middle tarsi, toothed or otherwise modified tarsal claws, smooth and shiny rather than rough or dull elytral and pronotal sculpture, enlarged antennal segments, peculiar forms of the tibiae and femora, or differences in sculpture of the last abdominal sternite. The male genitalia show specific characters within most genera of Laccophilinae, Hydroporinae, and Colymbetinae, but probably not in the Dytiscinae except for the genus *Dytiscus.* The dorsal color pattern is helpful in genera such as *Hydroporus, Hygrotus, Rhantus, Thermonectus,* and *Dytiscus.*

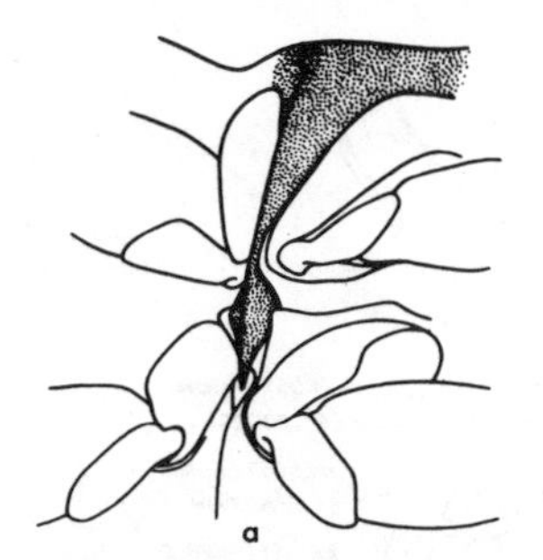

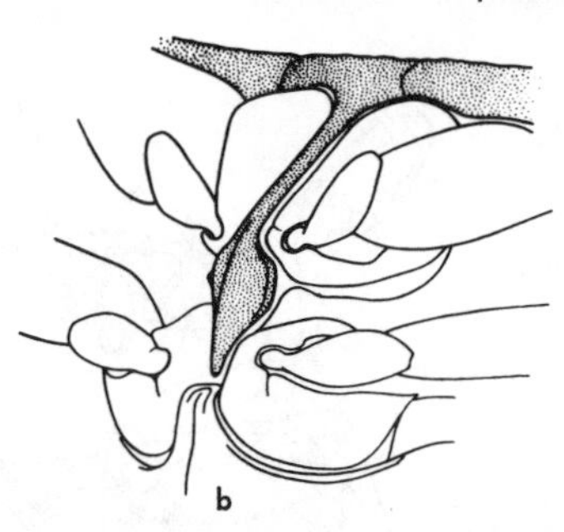

Fig. 13:8. Ventrolateral view of thorax showing relationship of middle of prosternum to its postcoxal process. *a, Laccophilus decipiens; b, Hygrotus unguicularis* (Leech, originals).

Key to the Genera of Dytiscidae of the United States and Canada[4]

Adults

1. Middle of prosternum and its postcoxal process (fig. 13:8*a*) in same plane; front and middle tarsi distinctly 5-segmented, segment 4 approximately as long as 3; scutellum fully exposed (fig. 13:7*b*) or concealed .. 17
— Middle of prosternum not in same plane as its process (fig. 13:8*b*); front and middle tarsi 4-segmented, or 5-segmented with 4th very small and almost concealed between lobes of 3rd (fig. 13:7*d*); scutellum concealed, except in *Celina* spp. which have elytral apices and apex of last abdominal sternite acuminate. Subfamily Hydroporinae 2
2. Scutellum fully visible; apices of elytra and last abdominal sternite produced, acuminate. METHLINI (fig. 13:14*b*) .... *Celina* Aubé 1837-38
— Scutellum covered by pronotum; apices of elytra rounded, subtruncate, or acute 3
3. Episterna of metathorax not reaching middle coxal cavities (fig. 13:9), being excluded by mesepimera (VATELLINI). Prosternal process short, broad, not reaching metasternum, its tip ending at front of the contiguous middle coxae .... *Derovatellus* Sharp 1882
— Episterna of metathorax reaching middle coxal cavities; apex of prosternal process reaching metasternum ... 4
4. Broad apex of hind coxal processes conjointly divided into 3 parts: 2 widely separated narrow lateral lobes and a broad depressed middle region (fig. 13:13*c*) (HYDROVATINI). Small broadly ovate beetles about 2.5 mm. long .... *Hydrovatus* Motschulsky 1855
— Hind coxal processes not divided into 3 parts as described above, but either without lateral lobes, or with these lobes covering bases of trochanters 5
5. Hind coxal processes without lateral lobes, bases of hind trochanters entirely free (fig. 13:7*e*) 6
— Sides of hind coxal processes divergent, more or less produced into lobes which cover bases of hind trochanters. HYDROPORINI (fig. 13:7*h*) 11
6. Hind tibiae straight, of almost uniform width from near base to apex; hind tarsal claws unequal; prosternal process short and broad, or rhomboid; epipleura with a diagonal carina crossing near base; glabrous, ovate, ventrally convex beetles. HYPHYDRINI 7
— Hind tibiae slightly arcuate, narrow at base, gradually widening to apex; hind tarsal claws equal; prosternal process oblong; epipleura without a diagonal carina near base (except in *Brachyvatus).* BIDESSINI 8
7. Middle coxae separated by about width of a middle coxa; prosternal process short and broad, apex obtuse .... *Pachydrus* Sharp 1882
— Middle coxae separated by only 1/2 the width of a middle coxa; prosternal process rhomboid, apex acute (fig. 13:15*d*) .... *Desmopachria* Babington 1841
8. Each elytron with a sharp narrow carina, starting at base opposite pronotal plica and fading out at declivity; pronotum transversely impressed at base .... *Anodocheilus* Babington 1841
-- Elytra without sharp narrow carinae, though often with a short basal groove opposite each pronotal plica; pronotum not transversely impressed at base 9
9. Hind coxal lines strongly sulcate-impressed, parallel posteriorly, converging as they continue forward across mid-metasternum to meet at middle coxae (fig. 13:13*b*); front and middle tarsi clearly 5-segmented .... *Bidessonotus* Régimbart 1895
— Hind coxal lines not continued anteriorly across metasternum; front and middle tarsi apparently 4-segmented 10
10. Epipleura crossed near base by an oblique carina .... *Brachyvatus* Zimmermann 1919
— No oblique carina near base of epipleura (fig. 13:14*c,e*) .... *Bidessus* Sharp 1882
11. Bases of hind femora contacting hind coxal lobes .... *Laccornis* Des Gozis 1914
— Hind femora separated from hind coxal lobes by basal part of trochanters 12

[4]The taxonomy of the Hydroporinae is still unsettled, and can not be solved by study of any one continental fauna. The present key does not recognize all genera proposed for Nearctic species. The *Oreodytes picturatus* (Horn) of this paper is the type of the genus *Deuteronectes* Guignot 1945, and *O. abbreviatus* the type of *Nectoporus* Guignot 1950. *Deronectes quadrimaculatus* (Horn) has sometimes been placed in *Neonectes* Zimmermann, the type of which is *Hydroporus natrix* Sharp 1884, from Japan.

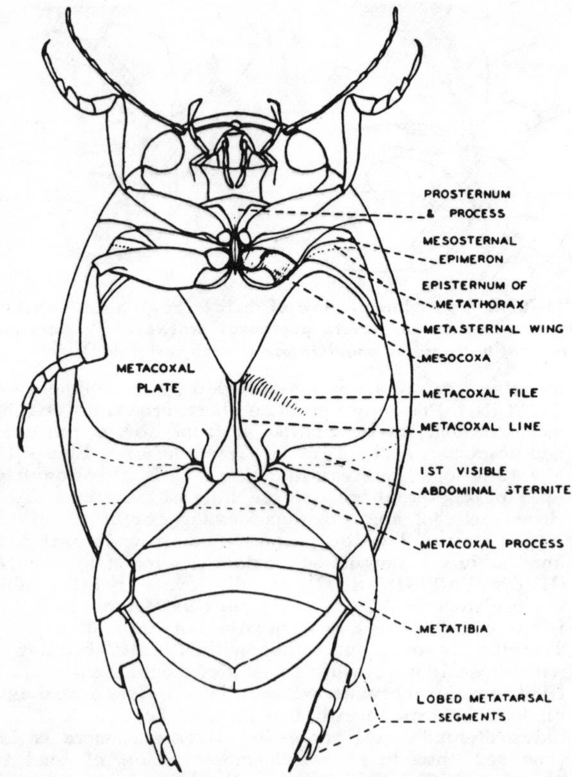

Fig. 13:9. *Laccophilus terminalis*, underside of adult to show episternum of metathorax cut off from middle coxal cavity by mesosternal epimeron (Leech, 1948).

12. A diagonal carina crossing epipleura near base (fig. 13:13*a*); front and middle tarsi 4-segmented .... *Hygrotus* Stephens 1828
— No carina crossing epipleura; front and middle tarsi actually 5-segmented, though 4th is usually very small and hidden between lobes of 3rd 13
13. Hind margin of hind coxal processes (best viewed with head of insect toward observer) together virtually straight across (fig. 13:7*i*) or sinuate and angularly prominent at middle (fig. 13:7*j*), or obtusely angulate (fig. 13:7*k*), but never triangularly incised at middle; median line of processes thus as long as or longer than lateral coxal lines 14
— Hind margins of hind coxal processes slightly, to deeply and more or less triangularly incised at middle, median line thus shorter than lateral coxal lines 16
14. Hind margins of hind coxal processes together either truncate or angularly prominent (fig. 13:7*i, k*)....... .... (in part) *Hydroporus* Clairville 1806
— Hind margins of hind coxal processes conjointly sinuate and somewhat angularly prominent at middle (fig. 13:7*j*) 15
15. Hind angles of pronotum rectangular or obtuse .... (in part) *Hydroporus* Clairville 1806
— Hind angles of pronotum acute .... (in part) *Deronectes* Sharp 1882
16. Pronotum with a longitudinally impressed line or crease on each side, and usually with a shallow transverse impression near base or an impression on each side near base; hind femora with median line of setigerous punctures, otherwise sparsely punctate or nearly smooth; undersurface of body densely finely punctate or shagreened, with scattered or numerous coarser punctures .... *Oreodytes* Seidlitz 1886
— Pronotum without sublateral impressed lines, usually without basal impression; hind femora usually densely punctate over entire surface; undersurface of body densely finely punctate to subgranulated, usually lacking scattered large punctures (figs. 13:15*a, b*; 13:16*a, b*) .... (in part) *Deronectes* Sharp 1882
17. Scutellum covered by hind margin of pronotum, or rarely a small tip visible; hind tarsi each with a single straight claw (fig. 13:9). LACCOPHILINAE . 18
— Scutellum entirely visible (fig. 13:7*b*) 19
18. Spines of hind tibiae notched or bifid at tip; apical 3rd of prosternal process lanceolate, only moderately broad (figs. 13:8*a*, 13:9); larger species, 2.5 to 6.5 mm. long .... *Laccophilus* Leach 1817
— Spines of hind tibiae simple, acute at tip; apical 3rd of prosternal process somewhat diamond-shaped, dilated behind front coxae and with tip acute .... *Laccodytes* Régimbart 1895
19. Eyes emarginate above bases of antennae (fig. 13:7*f*); first 3 segments of front tarsi of male widened and with adhesion discs (fig. 13:7*l*) or not, but never together forming a nearly round plate. COLYMBETINAE 20
— Eyes not emarginate above bases of antennae (fig. 13:7*c*); first 3 segments of front tarsi of male greatly broadened, forming a nearly round (fig. 13:7*g*) or an oval (fig. 13:10*a*) plate with adhesion discs. DYTISCINAE 32
20. Hind femora with a linear group of ciliae near posterior apical angle (fig. 13:10*b*). AGABINI 21
— Hind femora without such a group of ciliae 25
21. Hind coxal processes in form of rounded lobes (fig. 13:10*d*) 22
— Hind coxal processes parallel-sided, lateral margins straight to apices (fig. 13:10*e*) .... *Agabinus* Crotch 1873
22. Hind tarsal claws of equal length; if slightly unequal, then both are very short, only 1/3 length of 5th tarsal segment 23
— Hind tarsal claws obviously unequal, outer one of each pair 2/3 or less length of inner claw (fig. 13:10*f*) 24
23. Labial palpi very short, terminal segments subquadrate (fig. 13:10*g*) .... *Hydrotrupes* Sharp 1882
— Labial palpi approximately as long as maxillary palpi, terminal segments linear, not subquadrate (fig. 13:10*h*) .... *Agabus* Leach 1817
24. Labial palpi with penultimate segments enlarged, triangular in cross section, the faces concave and unequal (fig. 13:10*i*); genital valves of female dorsoventrally flattened, not armed with teeth .... *Carrhydrus* Fall 1922
— Labial palpi with penultimate segments linear, not enlarged and triangular; genital valves of female laterally compressed, sawlike, with series of sharp teeth along upper edge (fig. 13:10*c*).............. .... *Ilybius* Erichson 1832
25. Prosternum with a median longitudinal furrow, from near front margin to apex of prosternal process; first 4 hind tarsal segments distinctly produced at upper (inner) posterior corners. MATINI.............. .... *Matus* Aubé 1836
— Prosternum without a median longitudinal furrow; hind tarsal segments not lobate at upper hind corners 26
26. Hind coxal lines divergent anteriorly, coming so close together posteriorly as almost to touch median line, thence turning outward almost at right angles onto hind coxal processes (fig. 13:10*k*); hind tarsal claws equal; pronotum clearly but narrowly margined laterally. COPELATINI .... *Copelatus* Erichson 1832
— Hind coxal lines never almost touching median line and thence turning outward almost at right angles onto coxal processes; hind tarsal claws equal or not; pronotum margined or not 27
27. Hind claws of same length or virtually so; smaller species, 6 to 9 mm. long 28
— Hind claws obviously unequal, outer ones only from 1/3 to 2/3 length of inner ones; larger species, 9 to 20

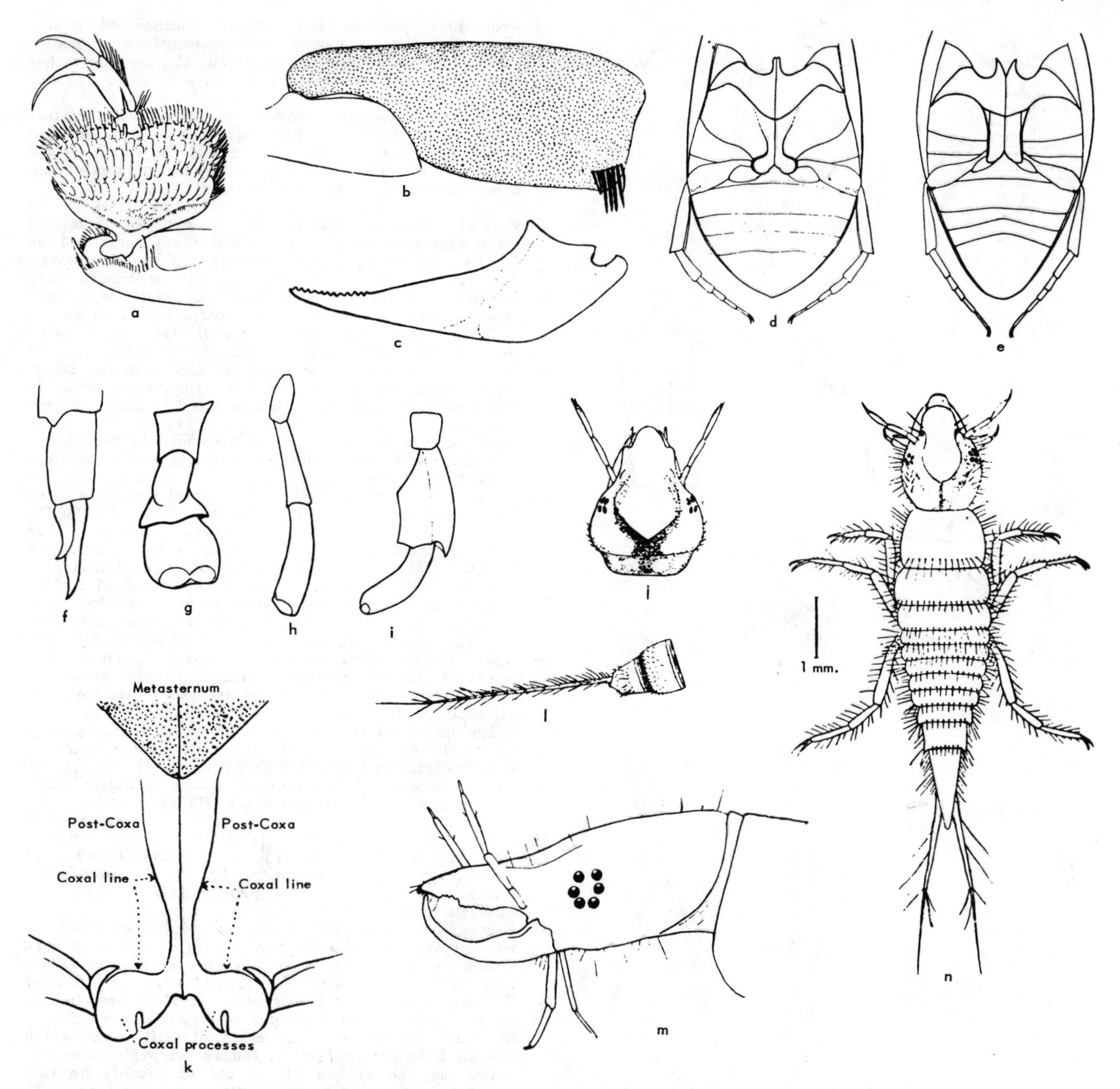

Fig. 13:10. Dytiscidae. *a, Cybister* sp., front tarsus of male; *b, Agabus* sp., trochanter and femur of hind leg of adult, showing linear group of ciliae; *c, Ilybius* sp., lateral view of serrated ovipositor of female; *d, Agabus* sp., ventral surface of adult; *e, Agabus* sp., ventral surface of adult; *f, Ilybius* sp., last two segments and tarsal claws of hind tarsus; *g, Hydrotrupes palpalis*, labial palp of adult; *h, Agabus* sp., labial palp of adult; *i, Carrhydrus crassipes*, labial palp of adult; *j, Deronectes* sp., head of larva to show notch in frontal projection; *k, Copelatus* sp., underside of adult to show coxal lines; *l, Deronectes* sp., tip of abdomen and a cercus to show hairs; *m, Hydroporus* sp., larva, side view of head to show frontal projection; *n, Hydroporus niger*, larva (*a*, Chatanay, 1910; *b,c*, F. Balfour-Browne, 1950; *d-i*, Leech, 1942; *j,l*, Bertrand, 1927; *k*, F. Balfour-Browne, 1935; *m*, Needham and Williamson, 1907; *n*, Wilson, 1923).

mm. long. COLYMBETINI 29

28. Terminal segment of palpi (especially of labial palpi) notched or emarginate at apex; pronotum clearly though narrowly margined laterally. COPTOTOMINI .... *Coptotomus* Say 1834

— Terminal segment of palpi not emarginate at apex; pronotum with an exceedingly fine line along lateral edges, but not margined. AGABETINI .... *Agabetes* Crotch 1873

29. Anterior point of metasternum (between middle coxae) clearly triangularly split to receive tip of prosternal process, the triangular channel usually deep with

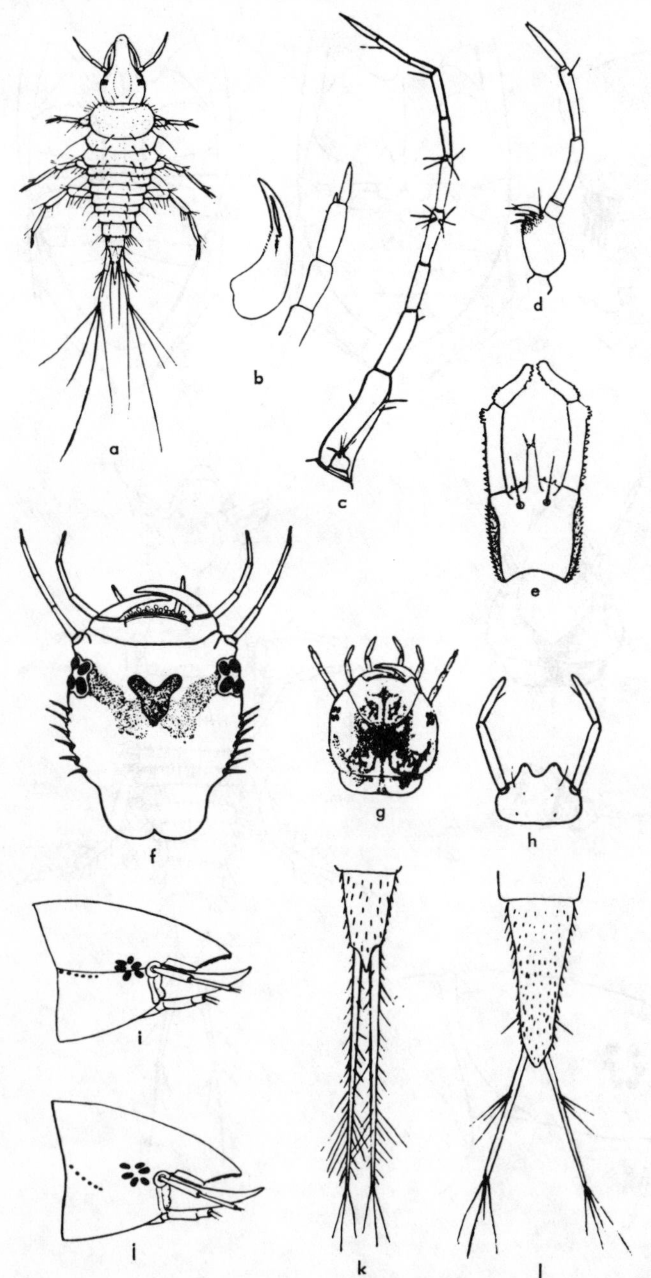

Fig. 13:11. Dytiscidae, larval structures. *a, Hydrovatus confertus; b, Copelatus parvulus,* mandible and antenna; *c, Cybister fimbriolatus,* maxilla; *d, Laccophilus maculosus,* maxilla; *e, Thermonectus basillaris,* ventral view of labium to show ligula; *f, Laccophilus proximus,* dorsal view of head; *g, Ilybius subaeneus,* dorsal view of head; *h, Hydaticus* sp., labium; *i, Ilybius* sp., lateral view of head to show marginal keel and temporal spines; *j, Agabus* sp., lateral view of head; *k, Laccophilus* sp., abdominal tip and cerci; *l, Agabus,* abdominal tip and cerci (*a,b,* Williams, 1936; *c,d,e,f,* Wilson, 1923; *g,h,* Bertrand, 1927; *i,j,* F. Balfour-Browne, 1950; *k,l,* Wesenberg-Lund, 1912).

its apex about on a line with hind margins of middle coxae; pronotum usually margined laterally 30

— Anterior tip of metasternum depressed, with a shallow pit or broad notch to receive tip of prosternal process, never with a sharply outlined triangular excavation; pronotum not margined 31

30. Prosternal process flat; dorsal surface of beetle unusually flat, pronotum widely margined laterally; elytra lightly reticulate throughout, the meshes rather coarse, unequal and irregular in shape .... *Hoperius* Fall 1927

— Prosternal process convex or cariniform; elytral reticulation lightly impressed, meshes of unequal sizes and shapes, but very small (a superimposed secondary reticulation of deeply impressed lines occurs over parts of elytra of some females) ....*Rhantus* Dejean 1833

31. Elytral sculpture consisting of numerous parallel transverse grooves .... *Colymbetes* Clairville 1806

— Elytra coarsely reticulate, without transverse grooves .... *Neoscutopterus* J. Balfour-Browne 1943

32. Inferior spur at apex of hind tibiae dilated, *much* broader than the other large spur; first 3 segments of front tarsi of male forming an oval plate (fig. 13:10*a*); large beetles, 20 to 32 mm. long. CYBISTRINI 38

— Inferior spur not or but little broader than the other; first 3 segments of front tarsi of male forming a nearly round plate (fig. 13:7*g*); medium-sized to large beetles 33

33. Posterior margins of first 4 hind tarsal segments beset with a dense fringe of flat golden cilia; smaller beetles, about 8 to 15 mm. long 34

— Posterior margins of first 4 hind tarsal segments bare; large beetles, about 20 to 38 mm. long. DYTISCINI .... *Dytiscus* Linnaeus 1758

34. Apex of prosternal process sharply pointed, pronotum margined laterally; external edge of each elytron from behind middle to about apical 5th margined with short spines; eyes prominent; upper surface of hind tarsi punctate and with fine appressed hairs. ERETINI .... *Eretes* Laporte 1833

— Apex of prosternal process rounded; pronotum not margined laterally; elytra without spines at sides; upper surface of hind tarsi naked except for marginal cilia... 35

35. Outer margin of metasternal side wings arcuate; outer (shorter) spur at apex of hind tibiae blunt, more or less emarginate. THERMONECTINI 36

— Outer margin of metasternal wings straight; outer spur at apex of hind tibiae acute. HYDATICINI .... *Hydaticus* Leach 1817

36. Elytra densely punctate, and in addition usually fluted and hairy in female .... *Acilius* Leach 1817

— Elytral punctation extremely fine or absent; some females with a superimposed sexual sculpture of elongate grooves, or granulate 37

37. Elytra basically yellowish, uniformly speckled or vermiculate with black; hind margin of middle femora with a series of stiff setae which are only about 1/2 as long as femora are wide ...... *Graphoderus* Sturm 1834

— Elytra black with yellow maculae or transverse bands, or yellow with black spots, or irrorate; hind margin of middle femora with a series of stiff setae which are as long or longer than femora are wide (warning: setae may be broken off in old or roughly handled specimens) .... *Thermonectus* Dejean 1833

38. Apex of hind tarsi of males with 2 claws, of females with a longer outer and a rudimentary inner claw .... *Megadytes* Sharp 1882

— Apex of hind tarsi of males always, of females usually, with only 1 claw .... *Cybister* Curtis 1827

Key to the Known Nearctic Genera of Dytiscidae

Larvae

1. Head with a frontal projection (fig. 13:10*m*); body lacking lateral fringes of swimming hairs, maxillary palpus 3-segmented. HYDROPORINAE 2

— Head without a frontal projection; body with or without swimming hairs; maxillary palpus 4- or more segmented 6

2. Frontal projection with a notch at each side (fig. 13:10*j*) 3
— Frontal projection without notches 4
3. Cerci with only primary hairs, 6-7 in number (fig. 13:10*n*) ....*Hydroporus* Clairville 1806 and .... *Hygrotus* Stephens 1828
— Cerci with additional secondary hairs (fig. 13:10*l*) .... *Oreodytes* Seidlitz 1886 (in part), ... and *Deronectes* Sharp 1882
4. Larva not greatly widened; last abdominal segment long and tapering; cerci with only primary hairs .... *Bidessus* Sharp 1882
— Larva greatly widened in middle; last abdominal segment long or short; cerci long with secondary hairs, or short with primary hairs only 5
5. Last abdominal segment long and tapering; cerci short, arising beneath segment, projecting beyond it, and having primary hairs only (fig. 13:11*a*) .... *Hydrovatus* Motschulsky 1855
— Last abdominal segment short; cerci long, with secondary hairs...........*Oreodytes* Seidlitz 1886 (in part)
6. Maxillary stipes broad, suboval, usually with 1 to 2 strong inner marginal hooks (fig. 13:11*d*).......... 7
— Maxillary stipes very long and slender, with no inner marginal hooks (fig. 13:11*c*)..................... 18
7. Abdominal segments 7 and/or 8 laterally without fringe of long swimming hairs; ligula absent 8
— Segments 7-8 with swimming hairs; ligula present .. 14
8. Fourth (last) segment of antennae less than 2/3 the length of 3rd 9
— Fourth segment more than 2/3 the length of 3rd ....*Rhantus* Dejean 1833 and *Colymbetes* Clairville 1806
9. Mandibles toothed along inner edge, 4th antennal segment double (biramous) (fig. 13:11*b*) .... *Copelatus* Erichson 1832
— Mandibles not toothed, 4th antennal segment not double 10
10. Cerci with numerous secondary hairs (fig. 13:11*k*); 4th antennal segment less than 1/4 as long as 3rd; front constricted just below acute apex, head gradually narrowed posteriorly without occipital suture or shallow groove defining neck area, anterior abdominal spiracles in the membranous area (fig. 13:11*f*) .... *Laccophilus* Leach 1817
— Cerci usually with 7 primary hairs in 2 whorls, 3 near middle, 4 apically (fig. 13:11*l*); 4th antennal segment about 1/2 as long as 3rd, apex of front obtuse, anterior abdominal spiracles on the margin of the tergites ... 11
11. Thorax as wide as long, about equal in length to abdomen, sides of head nearly round without neck area set off by shallow groove .... *Hydrotrupes* Sharp 1882
— Thorax longer than wide, never as long as abdomen, sides of head round or squarish but with neck area set off by occipital suture or shallow groove (fig. 13:11*g*) 12
12. Tibia and tarsus with conspicuous spines confined to apical half, mostly terminal; dorsal margin of middle and hind femora without row of spines; tergites, from 2nd thoracic to 8th abdominal, bearing 3 or more pairs of long conspicuous setae, venter of abdominal segments 4 to 6 with a similar pair of setae .... *Agabinus* Crotch 1873
— Tibia and tarsus with spines not confined to apical half; dorsal margin of middle and hind femora with row of spines; anterior tergites and sternites (see above) usually without conspicuous setae, or if present they are much smaller than those on 8th and 9th segments 13
13. Lateral margin of head more or less compressed or keeled, with temporal spines on a line which would intersect the ocelli or pass just below them (fig. 13:11*i*) .... *Ilybius* Erichson 1832
— Lateral margin of head not keeled, temporal spines on a line which would run well below the eye (figs. 13:11*j*, 13:12*a*) ....*Agabus* Leach 1817

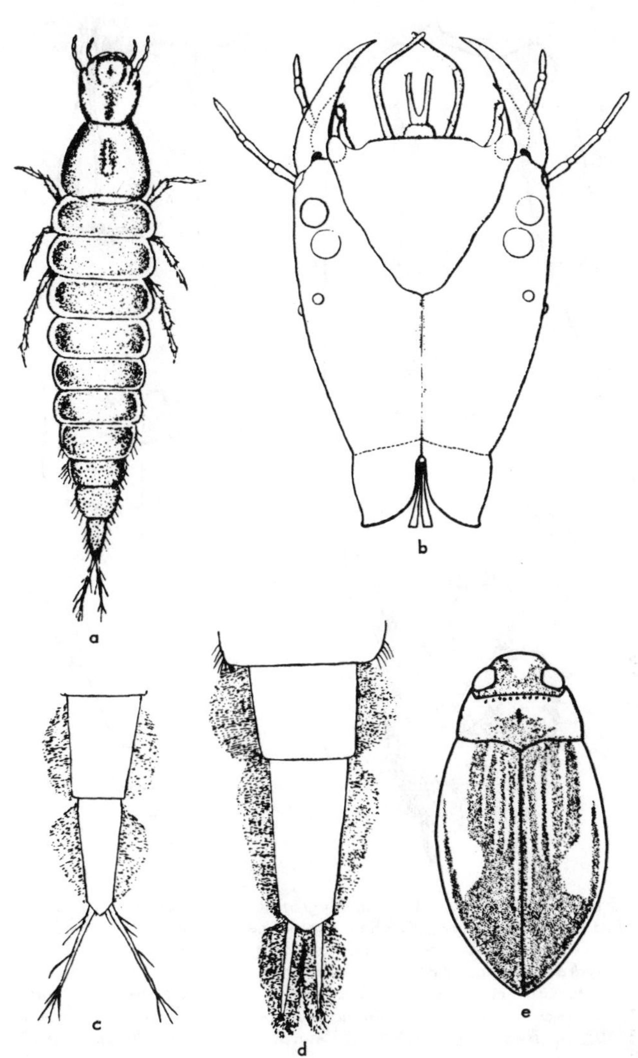

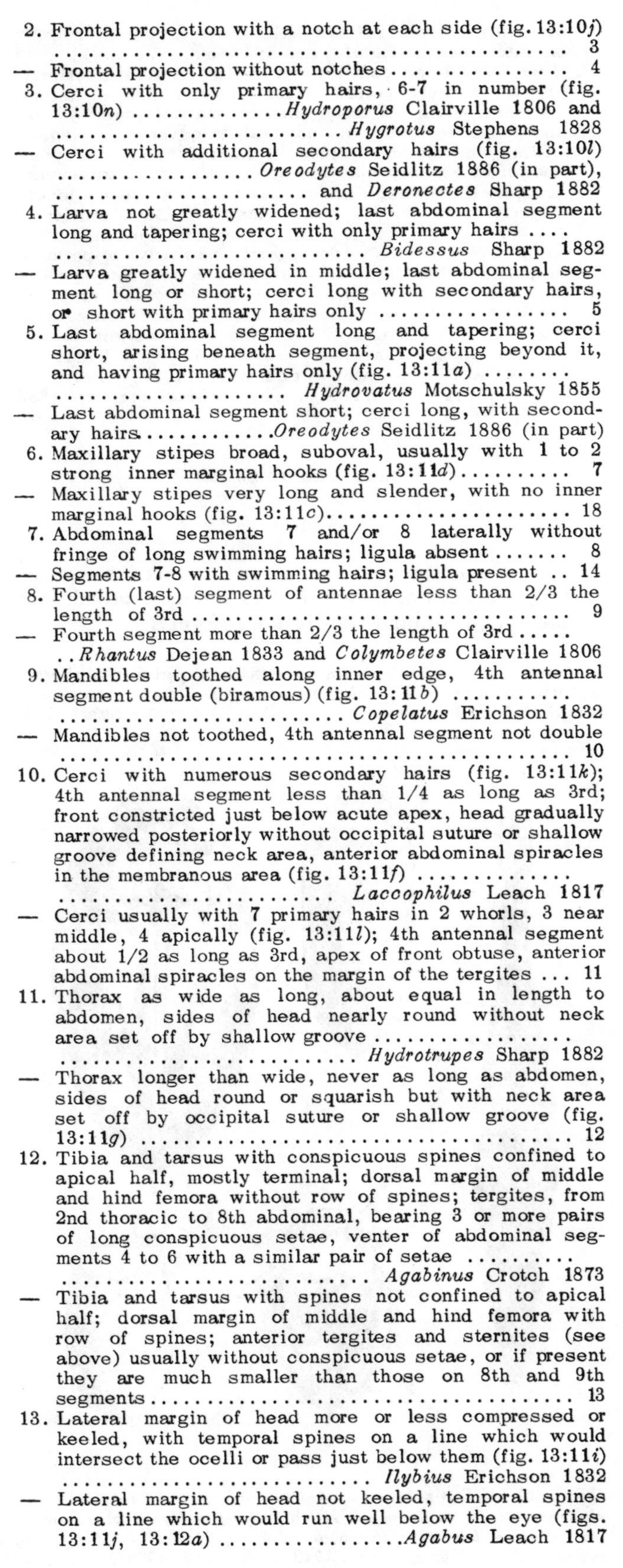

Fig. 13:12. Dytiscidae. *a*, *Agabus tristis*, larva; *b*, *Acilius sulcatus*, head of larva to show bifid ligula; *c*, *Hydaticus* sp., abdominal tip and cerci of larva; *d*, *Dytiscus* sp., abdominal tip and cerci of larva; *e*, *Hygrotus impressopunctatus*, female form *lir..ellus*, dorsal view of adult (*a*, Kincaid, 1900; *b*, Fiori, 1949; *c,d*, Wesenberg-Lund, 1912; *e*, Gellermann, 1928).

14. With a pair of long, lateral gills on the 6 anterior abdominal segments (fig. 13:6*b*) COPTOTOMINI ... . *Coptotomus* Say 1834
— Lacking gills on abdominal segments. THERMONECTINAE 15
15. Ligula very short, armed with 4 spines. ERETINI .... *Eretes* Laporte 1833
— Ligula long, simple or bifid but without 4 spines (fig. 13:11*e*). THERMONECTINI 16
16. Ligula apically bifid (figs. 13:12*b*, 13:14*a*) .... *Acilius* Leach 1817
— Ligula simple 17
17. Ligula not as long as 1st segment of labial palp .... *Thermonectus* Dejean 1833
— Ligula nearly equal to or exceeding length of 1st labial palpal segment .... *Graphoderus* Sturm 1834
18. Head anteriorly dentate (fig. 13:13*d*); ligula long; cerci absent. CYBISTRINI .... *Cybister* Curtis 1827
— Head anteriorly lacking dentation; ligula absent, or low and bilobed; cerci present. DYTISCINI 19

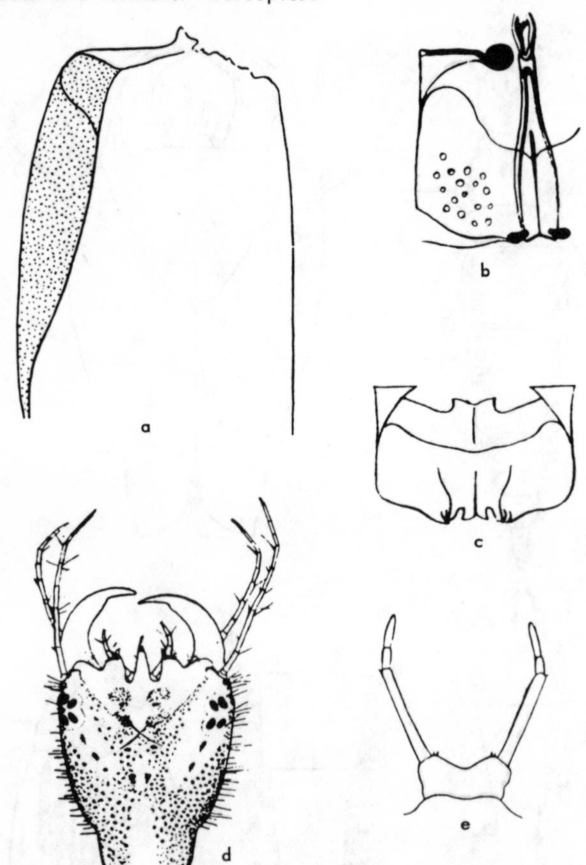

Fig. 13:13. Dytiscidae. *a, Hygrotus* sp., eipileural margin of adult in ventral view, to show carina near base; *b, Bidessonotus* sp., meso- and metasternal area of adult, to show metacoxal lines; *c, Hydrovatus* sp., meso- and metasternal region of adult to show metasternal processes; *d, Cybister fimbriolatus*, head of larva; *e, Dytiscus verticalis*, labium of larva (*a*, F. Balfour-Browne; 1940; *b*, Régimbart, 1895; *c*, Sharp, 1882; *d,e*, Wilson, 1923).

19. Cerci with lateral fringes (fig. 13:12*d*); labium without projecting lobes (fig. 13:13*e*) .... *Dytiscus* Linnaeus 1758
— Cerci lacking lateral fringes (fig. 13:12*c*); labium with 2 projecting lobes (fig. 13:11*h*) .... *Hydaticus* Leach 1817

Key to the California Species of Laccophilus

Adults

1. Elytra predominantly yellowish or brownish, variously maculate or irrorate with brown, never regularly vittate; male with an arcuate "stridulatory file" of impressed nearly parallel lines on each hind coxal plate, starting near apex of metasternum 2
— Brown elytral markings forming regular vittae on at least basal 3rd of elytral disc; hind coxal plates of male without file; Texas, to Riverside and Imperial counties .... *quadrilineatus* Horn 1871
2. Metasternum and hind coxal plates black, abdominal sternites largely piceous; elytra nearly uniformly irrorate with brown, vaguely paler laterally and at apex; Baja California and California to Utah and Texas .... *atristernalis* Crotch 1873
— Metasternum, hind coxal plates and abdominal sternites yellow or brownish-yellow; elytra irrorate with brown, but with sharply defined pale areas at sides, usually on suture behind middle, and at base 3
3. Hind coxal lines parallel and well separated anteriorly; elytra narrowly yellow along anterior 2/3 of lateral margins, each elytron usually marked with yellow spots as follows: a subbasal transverse series, a median lateral spot joined to lateral band, one post-median at suture, a series at apical 3/4, a transverse subapical series; larger species, average length slightly more than 5 mm.; Alaska to Baja California .... *decipiens* LeConte 1852
— Hind coxal lines convergent, narrowly separated, in anterior 3rd (fig. 13:9); elytra predominantly yellow in anterior half laterally and basally, and at apex, and with a dark brown area at lateral 2/3; irrorated sections of elytra sharply outlined with darker brown; smaller species, average length 4.75 mm.; Texas to Lower California, Imperial county .... *terminalis* Sharp 1882

Genus *Macrovatellus*

Adults

A single species may occur in southern or southeastern California, since it has been recorded from the Mexican mainland and Baja California.

Pronotum much narrower than elytral base, giving a characteristic discontinuity of outline; eyes large, prominent; length 5.5 to 6.5 mm. .... *mexicanus* Sharp 1882

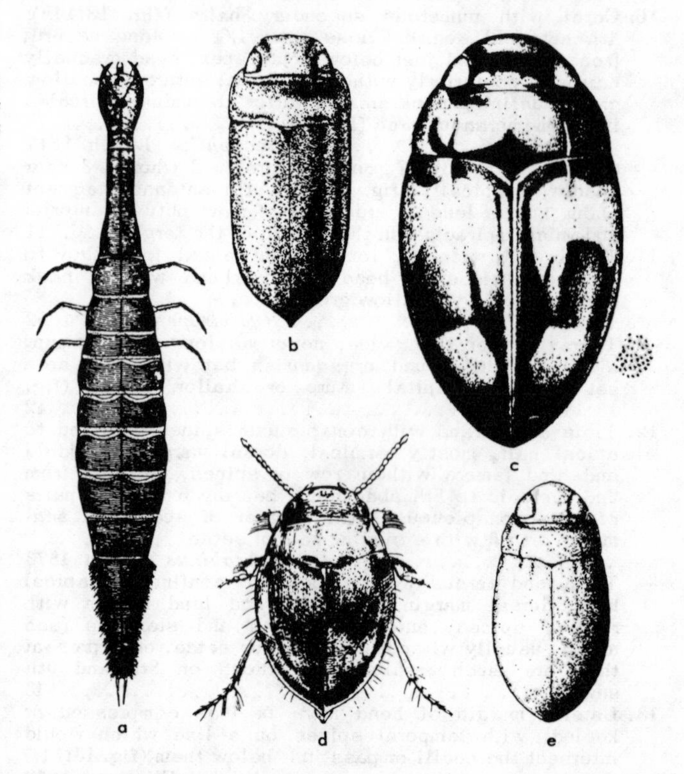

Fig. 13:14. Dytiscidae. *a, Acilius sulcatus*, larva; *b, Celina slossoni*, adult; *c, Bidessus leachi*, adult; *d, Hydrovatus confertus*, adult; *e, Bidessus shermani*, adult (*a*, Fiori, 1949; *b,e*, Leng and Mutchler, 1918; *c*, Leech, 1948; *d*, Williams, 1936).

Genus *Hydrovatus*

Adults

Only one species is recorded from California. Form short, broad, convex (as in fig. 13:14*d*); length 2-2.5 mm.; elytra with dual punctation; an oblique raised line across base of each epipleuron, as in *Hygrotus* and *Desmopachria;* Imperial, Tulare, Fresno, San Joaquin counties .... *brevipes* Sharp 1882

Genus *Celina*

Adults

These are small reddish beetles, 4 to 6 mm. long, of rather narrow and parallel form. The apices of elytra and last abdominal sternite are produced behind, acuminate (fig. 13:14*b*).

Two California specimens, representing two species, have been seen. Both were collected near Blythe, Riverside County, and both may be undescribed. One is the same as the teneral male recorded from Baja California as "*Celina angustata* Aubé?" by Leech (1948).

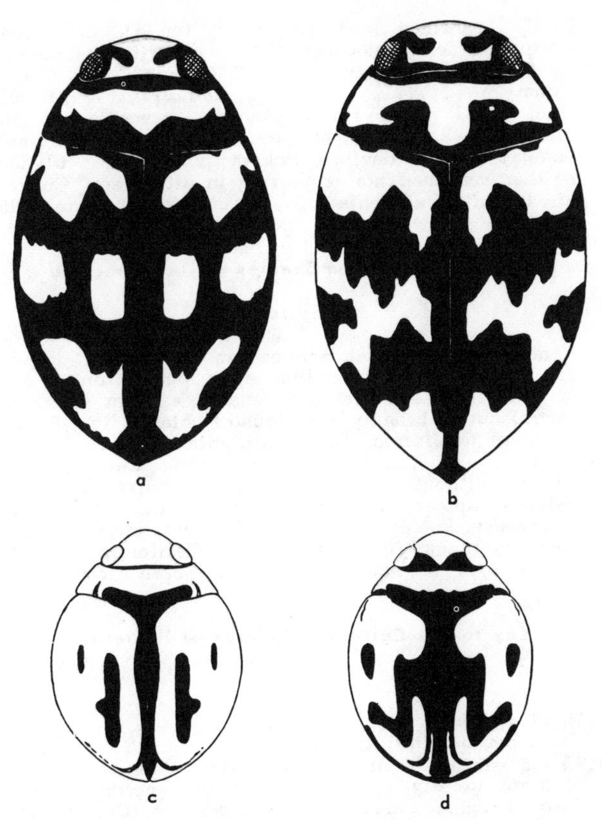

Fig. 13:15. Dytiscidae adults. *a, Deronectes deceptus; b, Deronectes eximius; c, Desmopachria dispersa; d, Desmopachria latissima* (a,b, Leech, original; c,d, Young, 1951).

Key to the California Species of Bidessus[5]

Adults

1. Metasternum, hind coxal plates, and first 1 or 2 abdominal sternites coarsely punctate, punctures obviously coarser than those of dorsal surface; elytra with 4 large pale areas, or sometimes with antemedian pale areas joined laterally and near suture to smaller postmedian ones; clypeus lightly margined anteriorly 2
— Metasternum, and usually hind coxal plates and abdominal sternites finely punctate, punctures finer than those of dorsal surface; elytra unicolorous, vittate, quadrimaculate, or sexmaculate; clypeus not margined anteriorly (except in male of *ornatellus*) 6
2. Each elytron with a slightly oblique rounded longitudinal carina, continuing to beyond middle from basal plica 3
— Each elytron with only the usual short impressed basal plica 4
3. Inner margin of rounded elytral carinae following directly behind basal plicae; sutural margin usually raised, much more finely punctate than elytral disc; Ventura County, north at least through Mariposa and Lake counties .... *plicipennis* (Crotch) 1873
— Inner margin of elytral carinae following from appreciably inward of basal plicae (fig. 13:14*c*); sutural margin differentiated by fine punctation but not usually raised; northwest California, Sonoma County, north to southwest Oregon .... *leachi* Leech 1948
4. Pale spots of elytra large, usually occupying more than 1/2 the elytra; sutural dark areas of elytra, separating anterior spots, together about as wide as a compound eye; basal plicae of pronotum more deeply impressed, sinuate, equal to or shorter than adjacent elytral plicae 5
— Pale spots of elytra smaller, occupying less than 1/2 of elytral area; sutural dark parts separating anterior spots each about as wide as an eye; basal plicae of pronotum lightly impressed, almost straight, equal to or longer than adjacent elytral plicae; Santa Barbara County to San Diego County .... *quadripustulatus* Fall 1917
5. Form almost evenly oval, not much narrower posteriorly; sides of pronotum nearly straight, base very slightly narrower than base of elytra but outline (in dorsal view) not discontinuous; disc of elytra rounded, not flattened; Santa Barbara County, Ventura County .... *pictodes* Sharp 1882
— Form less evenly oval, more parallel-sided, much narrower posteriorly than anteriorly; sides of pronotum arcuate, forming more of a discontinuity in outline with elytral base; disc of elytra flat. (Some specimens show traces of longitudinal carinae and might be run to *leachi*, but are distinguished by their much finer elytral punctation and very broad anterior tip of metasternum; Sonoma, Mendocino, and Lake counties .... sp.
6. Each elytron with a shallow sutural stria, area between it and suture raised and more finely punctured than elytral disc; pale markings on elytra in form of spots or fasciae 7
— Elytra without sutural striae; elytral punctation coarser than that of base of pronotal disc; elytral coloring varying from uniformly piceous, through vittate forms, to almost uniformly luteous; elytra shining between punctures (all males, most females) or alutaceous (some females); Alaska to Argentina ....complex centering around *affinis* (Say) 1823
7. Hind coxal plates coarsely punctate, punctures at least twice as large as those on base of disc of pronotum; elytra coarsely punctate with sparse inconspicuous fine hairs, appearing glabrous; clypeus

[5]Fall (1901) recorded *B. cinctellus* (LeConte) 1852, from California, on the basis of an old specimen labeled only "So. Cal."; no modern examples have been seen, though it may occur in the region of the Colorado River. It will trace to Couplet 3, but does not fit either choice. Examples of "*subtilis*" from about the San Francisco Bay area north may be *B. subplicatus* Hatch 1953; if so, the two appear to integrate in a cline.

finely margined and projecting (male), or neither margined nor projecting (female); British Columbia to California, "Crane Valley" (=Bass Lake, Madera County?) .... *ornatellus* Fall 1917
— Hind coxal plates finely punctate, punctures smaller than those of base of pronotal disc; elytra finely punctate, with obvious golden pubescence; clypeus neither margined nor projecting in either sex; California to Baja California .... *subtilis* (LeConte) 1852

Key to the California Species of Desmopachria

Adults

1. Each elytron with an antemedian longitudinal impression in a marginal reddish spot; punctation of disc of pronotum as coarse and sparse as that of elytra; elytral suture broadly and sinuately black (fig. 13:15*d*); southern California, Baja California .... *latissima* LeConte 1852
— Elytra without antemedian lateral impressions and reddish spots; punctations of pronotal disc much finer and denser than that of elytra; elytral suture narrowly black (fig. 13:15*c*); Baja California, Arizona, Texas .... *dispersa* Crotch 1873

Key to the California Species of Hygrotus

Adults

1. Head with front margined anteriorly 2
— Front not margined anteriorly 5
2. Under surface black; larger, more elongate species with vittate elytra, length 4-4.8 mm.; northern United States .... *masculinus* (Crotch) 1874
— Under surface rufotestaceous; small, convex, ovate species, 2.5–3.2 mm. long 3
3. Form more elongate-oval, not markedly convex; northwest U.S. .... *intermedius* (Fall) 1919
— Form broad and very convex 4
4. Large punctures of basal half of elytral disc rounded, shallow, the bottoms clearly visible; fore tarsi of male dilated; California ....*hydropicus* (LeConte) 1852
— Large punctures of elytral disc rounded or irregular, deep, the bottoms not visible; fore tarsi of male hardly wider than in female; northern U.S. and adjacent Canada .... *sayi* Balfour-Browne 1944
5. Last ventral abdominal sternite with an obtuse tumidity on each side, more prominent in the male; northwest U.S. and adjacent Canada... *tumidiventris* (Fall) 1919
— Last ventral sternite without tumidities 6
6. Dorsum fulvo- to rufotestaceous, elytra with distinct vittiform markings; large (4.5–5.5 mm.), elongate species 7
— Dorsum flavo-, rufo-, or piceo-testaceous or piceous-brown, elytral disc often with broad nubilous markings, which are never distinctly vittate (under surface black) 8
7. Elytra shining, with impressed lines of serial punctures in basal half and intervals with intermixed fine and very coarse punctures; or dull, serial punctures not in impressed lines, and of same size as those of intervals (fig. 13:12*e*) (Holarctic) ....(in part) *impressopunctatus* (Schaller) 1783
— Elytra shining or dull, serial punctures larger than those of interspaces; anterior (inner) protarsal claw of male with broad tooth at base; Nevada, Washington; Lassen County .... *infuscatus* (Sharp) 1882
8. Elytral interspaces with intermixed coarser and finer punctures 9
— Punctation of elytral interspaces nearly uniform, not obviously dual 10
9. Larger species, 4.5–5.5 mm. long; metathoracic episternum shagreened, usually coarsely punctate (Holarctic) ....(=males and some females of) .... *impressopunctatus* (Schaller) 1783
— Smaller species, 3.5–4 mm. long; metathoracic episternum shining, usually impunctate. Sierra of California .... (=males and some females of) .... *nigrescens* (Fall) 1919
10. Front femora and middle tibiae of male modified 11
— Front femora and middle tibiae of male normal, similar to those of female 12
11. Head with median pale spot at base; middle femur robust, curved, lower margin impressed and flattened in apical half for reception of tibia; front femur of male with a transverse impression whose lower margin forms a deep semicircular emargination in ventral edge of femur; lower margin of tibia of male sinuate, widest at basal quarter and apex; middle tibia of female parallel-sided, lower margin suddenly constricted at base; known only from Oakley, Contra Costa County .... *curvipes* (Leech) 1938
— Head without median pale spot at base; middle femur slender, nearly straight and parallel-sided, apex very slightly impressed for reception of tibia; front femur of male with patch of dense short pubescence apically, middle tibia parallel-sided, lower margin produced inwardly at apex; middle tibia of female an elongate triangle, gradually widening from base to apex; southern California, north at least to Oakley, Contra Costa County; Marin County .... *pedalis* (Fall) 1901
12. Larger species, 4.5–5.5 mm. long................. 13
— Smaller species, 3–4 mm. long 14
13. Ventral surface uniformly shagreened between punctures, giving a velvety luster (fig. 13:12*e*) (Holarctic) .... (=female form *lineellus* (Gyllenhal) 1808, of) .... *impressopunctatus* (Schaller) 1783
— Ventral surface polished, microreticulate in part; male with broad subbasal tooth on anterior (inner) front tarsal claw; northwest U.S................. .... *infuscatus* (Sharp) 1882
14. Form narrow, twice as long as wide, or even longer 15
— Form less elongate, varying from very distinctly less than, to nearly twice, as long as wide; anterior claw of fore tarsus of male perceptibly differing from its fellow 17
15. Anterior claw of front tarsus of male broader, thicker, a little shorter and more sharply bent at base than its fellow (if 2nd segment of front tarsus of male is outstandingly the largest, see couplet 17); California .... *collatus* (Fall) 1919
— Anterior claw of front tarsus of male slightly thickened but otherwise like its fellow 16
16. Front tarsi of male narrow, hardly wider than those of female; smaller species about 3 mm. long; Sierran; Mono Lake .... *artus* (Fall) 1919
— Front tarsi of male broader; larger species, 3.7–4.2 mm. long; coastal; Santa Cruz County to Humboldt County ....*sharpi* (Van den Branden) 1885
17. Dorsal surface dull, opaque 18
— Dorsal surface shining 19
18. Head and pronotum black, elytra piceous-brown (=some females); California....(in part) *nigrescens* (Fall) 1919
— Head, pronotum, and elytra brownish-yellow or brownish-ferrugineous, with variable piceous markings; northwest U.S....(=some females of) *obscureplagiatus* (Fall) 1919
— Head, pronotum, and elytra pale flavotestaceous, elytra usually with piceous markings in apical half; broadly oval species; southern California ....(=females of) *fraternus* (LeConte) 1852
19. Front tarsi of male at most only moderately dilated 20
— Front tarsi of male broadly dilated, nearly or quite as wide as the apical width of the tibia........... 21
20. Color above testaceous with a large elytral cloud extending nearly to base; form slightly narrower, size usually a little smaller; California .... *lutescens* (LeConte) 1852
— Color above nearly as in *lutescens,* but elytral cloud usually less extended basally; elytral markings very variable sometimes almost absent; California .... *medialis* (LeConte) 1852

21. Dorsal surface pale flavotestaceous; 2nd segment of front tarsus not outstandingly dilated, only slightly wider than 3rd; broadly oval species; southern California .... (=males of) *fraternus* (LeConte) 1852
— Dorsum brownish-yellow to brownish-ferrugineous; 2nd segment of front tarsus of male dilated, nearly 1/2 again as wide as 3rd; elongate-oval species; northwest Pacific Coast .... (in part) *obscureplagiatus* (Fall) 1919

Key to the California Species of Hydroporus

Adults

1. Hind margin of metacoxal processes conjointly distinctly sinuate, somewhat depressed at median line and there angulate and sloping to meet first visible abdominal sternite; hind coxal process glabrous. Hind trochanters usually elongate, their apices nearly halfway from hind angles of hind coxal processes to apices of femora. Mostly small, dorsally glabrous, reddish or brownish species 2
— Hind margin of metacoxal processes truncate or somewhat rounded, if distinctly angulate medially then median line not depressed and sloping to meet 1st sternite; hind coxal processes usually hairy. Hind trochanters shorter and stouter, clearly less than 1/2 length of distance from hind coxal processes to apices of femora. Mostly larger species, piceous in color, with an obvious elytral vestiture 11
2. Hind coxal lines narrowly separated, curving strongly outward in apical half; hind trochanters shorter, not reaching halfway from hind coxal processes to apices of femora; pronotal side margins broad, about as wide as a median antennal segment; form narrower, size small, 3 mm. or less in length; median lobe of male genitalia simple apically 3
— Hind coxal lines more widely separated, curved slightly outward near apices; hind trochanters longer, pronotal side margins narrower than a median antennal segment; body form broader, size various; median lobe of male genitalia bifid apically 5
3. Size smaller, 2.1–2.4 mm. long; form elongate-oval, more convex, elytra notably coarsely punctate; Mendocino County .... *bidessoides* Leech 1941
— Size larger, 2.5–3 mm.; form more broadly oval, flatter, elytra finely punctate 4
4. Slightly larger species, 3 mm. long; front tarsal claws of male nearly twice as large as those of middle tarsi; southern California .... *barbarae* Fall 1932
— Slightly smaller species, 2.5–2.8 mm.; front tarsal claws of male like those of middle tarsi; California, southern Oregon .... *terminalis* Sharp 1882
5. Area between metacoxal lines nearly flat, smooth, with at most a few tiny punctures; apex of last abdominal sternite of female produced, lobate; California ....(a complex of species centering around .... *rossi* Leech 1941)
— Area between metacoxal lines clearly though irregularly punctured, often somewhat inflated; last abdominal sternite of female evenly rounded at apex 6
6. Form broader, more convex; ventral punctation coarse. The undersurface shining, polished, without slightest trace of alutaceous sculpture between punctures. Sides of first 3 abdominal sternites as coarsely punctate as sides of metasternum, more so than metacoxal plates; next 2 sternites less coarsely but very densely punctate; last sternite with a broad, heavily punctate transverse groove. Length 3.2–3.5 mm.; southern California, north to Sequoia National Park .... *latebrosus* LeConte 1852
— Form narrower and less convex, undersurface with at least in part an alutaceous sculpture on metacoxal plates or abdominal sternites; punctation finer and generally sparser 7
7. Elytra nearly or quite uniform in color, varying, however, from yellowish-testaceous to dark brown or castaneous 8
— Elytra conspicuously and rather abruptly paler at base, coast of central and northern California .... *palliatus* Horn 1883
8. Size large, 4 mm. long; elytra finely punctate; San Francisco Bay region .... *hardyi* Sharp 1882
— Size smaller, never as long as 4 mm., rarely more than 3.5 mm. 9
9. Larger species, 3.0–3.5 mm. long; California....... .... *vilis* LeConte 1852
— Smaller species, 2.4–2.8 mm. long 10
10. Strongly flattened, more nearly parallel-sided; usually rufous beneath, with clearly defined alutaceous sculpture; southwest U.S. .... *belfragei* Sharp 1882
— Rather broadly subovate, evidently convex; piceous-brown beneath, with scarcely detectable alutaceous sculpture; southern California.. *barbarensis* Wallis 1933
11. Size larger, 3.5–6.5 mm. long 12
— Size smaller, rarely as much as 3.5 mm. long. Form narrowly oblong-oval, color black to brownish, elytra usually with a few small yellowish spots near humeri and in apical 3rd; elytral surface alutaceous, punctures not obscured by pubescence; last 4 abdominal sternites sparsely, very finely punctate; Pacific Coast .... *occidentalis* Sharp 1882
12. Size very large, more than 6 mm. long; elytra densely pubescent, punctation very fine; abdominal sternites densely, nearly uniformly punctate. Second segment of front tarsi of male wider than 1st, claws elongate, slender, equal; Pacific Coast..... *fortis* LeConte 1852
— Size smaller, not more than 5.2 mm. long 13
13. Hind coxal plates with coarse, shallow punctures, like those on sides of metasternum and 1st few abdominal sternites 15
— Hind coxal plates impunctate, or at most with small deep punctures 14
14. Hind coxal plates impunctate; last abdominal sternite densely punctate. Antennae less stout, intermediate segments about twice as long as broad; antennae fuscous, paler at base; western U.S. .... *axillaris* LeConte 1853
— Hind coxal plates with sparse, small, deep punctures. Last sternite rather sparsely punctate. Antennae relatively stout, intermediate segments much less than twice as long as broad; antennae and legs almost black throughout; Sierra of California .... *funestus* Fall 1923
15. Lateral margins of elytra, as viewed from the side, rather strongly arcuately ascending at base. Sides of pronotum and elytra, in dorsal view, less continuous in outline than usual. Lateral marginal bead of pronotum very narrow, low, its inner edge indefinite towards base; hind margins of pronotum almost straight from hind angles to scutellar lobe. Elytra of female duller, more finely punctate than that of male 16
— Lateral margins of elytra much less strongly ascending at base; sides of pronotum and elytra with nearly continuous outline, in dorsal view 17
16. Form more robust and less depressed. Hind angles of pronotum more broadly rounded, sides arcuate, discontinuity of outline with elytra more apparent. Second protarsal segment of male asymmetric, more produced on inner (anterior) side of mid-line than on outer; 3rd segment more nearly parallel-sided. Typically a species of ponds and lakes; south central British Columbia to Sierra of California...*tademus* Leech 1949
— Form less robust, more depressed. Hind angles of pronotum more nearly rectangular, sides straighter. Second protarsal segment of male symmetric; 3rd segment not appreciably parallel-sided. A species of streams and rivers; Alaska to northern coast of California .... *mannerheimi* J. Balfour-Browne 1944
17. Larger species, 4.5–5.3 mm. long. Elytra of male shining, of female dull; Pacific Coast .... *subpubescens* LeConte 1852
— Smaller species, 3.7–4.3 mm. long. Elytra shining in both sexes 18

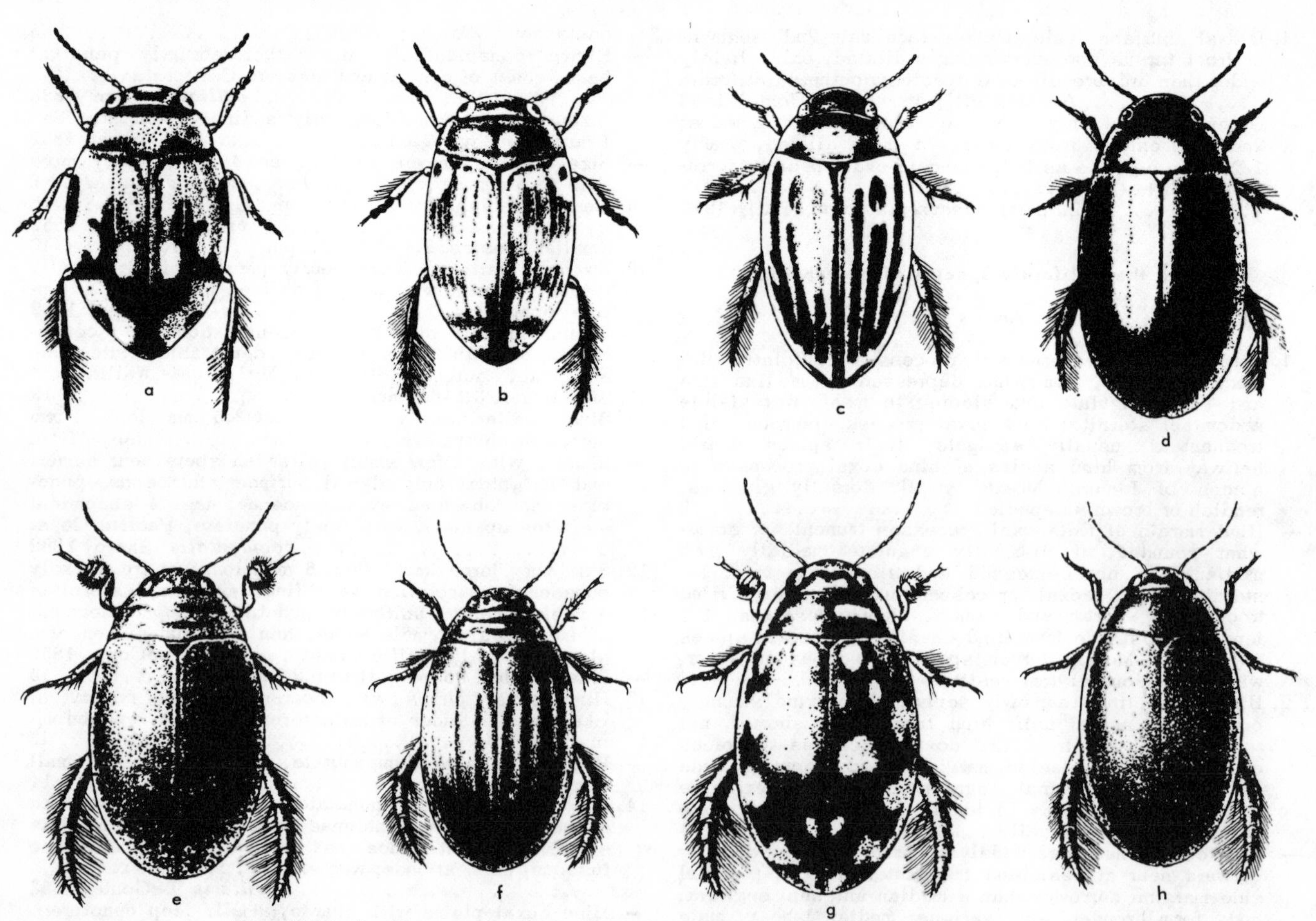

Fig. 13:16. Dytiscidae, adults. *a, Deronectes roffi; b, Deronectes striatellus; c, Agabus disintegratus; d, Agabus regularis; e, Acilius semisulcatus* male; *f, A. semisulcatus* female; *g, Thermonectus marmoratus; h, Thermonectus basillaris* (Sharp, 1882).

18. Anterior (inner) claw of front tarsus of male a little shorter, 2/3 to 3/4 length of outer claw; Sierra of California, north to British Columbia, east to Saskatchewan .... *pervicinus* Fall 1923
— Claws of front tarsi of male equal in length; California .... *hirtellus* LeConte 1852

Key to the California Species of Deronectes

Adults

1. Hind femora densely, evenly punctate throughout 2
— Hind femora usually closely punctate in posterior half, more sparsely punctate or nearly smooth in front of median row of setigerous punctures 10
2. Form more broadly rotundate-oval and strongly convex (less noticeably in *eximius*), mid-coxae more widely separated 3
— Form more elongate and less convex, mid-coxae more narrowly separated 6
3. Each elytron with 3 lines (sutural, discal, subhumeral) of lightly impressed punctures, discal series twice as far from sutural series as latter is from suture. Hind coxal lines slightly to moderately divergent anteriorly. Undersurface usually black 4
— Each elytron with a sutural and more than 1 discal line of punctures. The discal series are less distinctly impressed, especially toward base, and are more or less confused with similar interstitial punctures posteriorly. Hind coxal lines widely divergent anteriorly; if continued, they would pass outside the mesocoxae. Undersurface rufous. Rotundate oval, strongly convex species; elytra black, marked with yellow as follows: a transverse posteriorly indented basal spot, 2 smaller median spots arranged transversely, and 1 or more irregular apical spots (fig. 13:15*a*); California .... *deceptus* (Fall) 1932
4. Form less convex and less rotundate; median area of elytra (i.e., between discal striae) flattened and slightly depressed. Under surface black, very densely, finely punctate and irregularly wrinkled, opaque. Elytra black with yellow fasciae (fig. 13:15*b*), rather than with spots as in *deceptus;* California.......... .... *eximius* (Motschulsky) 1859
— Form broader and more convex, elytra not flattened or depressed medially. The undersurface rufopiceous to black, densely, finely punctate, mid-metasternum more sparsely punctate, shining 5
5. Form more elongate, less obese. Elytra black with a yellow fascia near base, an indistinct median fascia, and vague yellow markings laterally near apex (fig. 13:16*a*). Elytra virtually glabrous; California[6] [*D. addendus* (Crotch), *nec* Fall] .... *roffi* (Clark) 1862
— Form more strongly convex. Elytra yellowish, with ante- and postmedian interrupted black fasciae. Elytra finely pubescent; California[6].. *corpulentus* (Fall) 1923

[6]Said to occur in California, on the basis of some old specimens in the LeConte collection. All modern examples so determined have proved to be misidentified specimens of *deceptus* or *eximius*.

6. Elytra with several distinctly impressed and entire discal striae, 1st scarcely more distant from sutural than latter is from suture (form rather broad as a rule, distinctly less than twice as long as wide) 7
— Elytra with 2 lightly impressed or unimpressed discal series of coarser punctures, 1st twice as distant from sutural series as the latter is from suture 9
7. Outer discal striae nearly as distinct and complete as the inner ones; lateral margin of pronotum not appreciably wider posteriorly; prosternal intercoxal carina not terminating anteriorly in a prominence; front and middle tarsi of male very little wider than those of female; elytral coloration varying from almost entirely yellowish to entirely black (fig. 13:16*b*); smaller species, 3.5–4.5 mm. long; Pacific Coast .. ..*striatellus* (LeConte) 1852
— Outer discal elytral striae feeble and more or less indistinct or incomplete; prosternal intercoxal carina terminating anteriorly in a distinct prominence; front and middle tarsi of male broader than in female 8
8. Side margin of pronotum gradually widening from apex to base, where it is at least 1/2 as wide as a median antennal segment; pronotum with longitudinal rugae at sides and base, surface shiny between the punctures, which are irregular in size and distribution and much larger than those of elytra; elytra usually maculate; larger and more robust species, 5.0–5.5 mm. long; southern California .... *funereus* (Crotch) 1873
— Side margin of pronotum only slightly wider posteriorly; pronotum with few or no rugae, surface often reticulate between the punctures which are evenly distributed and only a little larger than those of elytra; elytra black; smaller species, 4.5–4.9 mm. long; California ....*dolerosus* Leech 1945
9. Claws of front tarsi of male elongate, nearly ½ as long as tarsi; smaller species, 4 mm. long; Oregon, northern California .... *expositus* (Fall) 1923
— Claws of front tarsi of males not appreciably modified, not longer than claw-bearing segment; size larger, 4.25–5 mm.; Holarctic.... *griseostriatus* (Degeer) 1774
10. Elytra maculate, with 2 large yellow blotches at base, 2 smaller ones behind middle, and usually 2 little ones at apices; elytral punctation dual, consisting of fine close punctures, with numerous scattered coarse punctures which become sparser and smaller toward sides and apex of elytra; northern California .... *quadrimaculatus* (Horn) 1883
— Elytral markings vittiform; elytral punctation close, fine, uniform, not dual (except for the usual longitudinal series of larger punctures) 11
11. Outline of pronotum and elytra nearly uniform, pronotum with hind angles well defined, nearly rectangular .... (return to 9)
— Outline of pronotum and elytra discontinuous, pronotum broadly rounded posteriorly at sides, without evident hind angles................................... 12
12. Size larger (average length 4.35 mm.); elytra dull, surface deeply reticulate between punctures which are separated by their own width or less; middle femora of male subangulate on lower (posterior) edge; middle tibiae of males broad in apical third, thence somewhat sinuately narrower toward base on inner margin sharp and bladelike; New Mexico, southeast California.................*coelamboides* (Fall) 1923
— Size smaller (average length 3.55 mm.); elytra moderately shining, surface finely reticulated between punctures which are separated by more than their own widths; middle femora and tibiae of male as in preceding species; Death Valley region .... *panaminti* (Fall) 1923

Key to the California Species of Oreodytes

Adults

1. Hind femora strongly curved at inner end, at attachment to trochanters; elytra yellow, each elytron with suture, 6 narrow vittae, and 2 or 3 lateral spots, black or piceous; head yellow with V-shaped median piceous mark; elytra sparsely, finely punctate; form elongate, depressed; elytra of female toothed externally near apex; length 4.5–5 mm.; Colorado, British Columbia, Alaska, ?California ......*semiclarus* (Fall) 1923
— Anterior margin of hind femora almost straight to base; elytra variously marked, never with 6 narrow vittae; head all yellow, or black adjacent to eyes; elytra of female not toothed 2
2. Form elongate-oval; pronotum much narrower than elytral bases, the outline thus discontinuous; head black adjacent to eyes; a groove paralleling and just within each metacoxal line; elytra yellow or reddish-yellow, with black markings; length 4 mm.; northern California .... *bisulcatus* (Fall) 1923
— Form more broadly oval; pronotum only a trifle narrower than elytral bases; head pale.................... 3
3. Elytra with impressed subsutural series of punctures, traces of discal and lateral longitudinal series, and a transverse line of coarse punctures at base of disc; elytra with broad discal piceous cloud; length about 3.5 mm.; Pacific Coast.......*subrotundus* (Fall) 1923
— Elytra without an impressed subsutural line of punctures 4
4. Elytra virtually or entirely impunctate, except for a discal and subhumeral line of fine punctures; mid-metasternum not differentiated, smooth and evenly, rather finely punctate; elytral markings linear, lines frequently more or less fused. Length 2.7–3 mm.; Pacific Coast.............. *obesus* (LeConte) 1866
— Elytra with numerous irregularly spaced and rather coarse shallow punctures; mid-metasternum differentiated, sulcate and coarsely, irregularly punctate 5
5. Mid-metasternum broadly, shallowly depressed, coarsely, irregularly punctate, subrugose at sides; sides of pronotum nearly straight; in dorsal view, hind margin of pronotum decidedly not at right angles to longitudinal axis of body; lateral margins of elytra, as viewed from side, rather strongly arcuately ascending at base; length 3–3.5 mm.; northern California..... .... *abbreviatus* (Fall) 1923
— Mid-metasternum narrowly, more deeply depressed, sides nearly straight, sharply margined, margins joining anterior ends of hind coxal lines; sides of pronotum evenly arcuate; hind margin of pronotum, except for scutellar lobe, virtually at right angles to longitudinal axis of body; lateral margins of elytra much less strongly ascending at base; length 2.8 to 3 mm.; California, Nevada .... *picturatus* (Horn) 1883

Key to the California Species of Agabus

Adults

1. Vittate species, elytra black with testaceous stripes which are often broader than black areas 2
— Nonvittate species 3
2. Elytral reticulation with meshes of irregular shapes and very unequal sizes, each usually with 1 or 2 small punctures; pronotum rufotestaceous to piceous, paler laterally; elytra pale testaceous, hyaline, each elytron with about 5 black vittae, discal ones often coalescing; northern California .... *lineellus* LeConte 1861
— Elytral meshes irregular in shape, but very small and of fairly uniform size, without small punctures; pronotum rufotestaceous with anterior and posterior discal transverse black band; elytra varying from testaceous with 3 or 4 black vittae on each elytron, to black with outer margin only pale (fig. 13:16*c*); United States .... (in part) *disintegratus* (Crotch) 1873
3. Outline of pronotum and elytra discontinuous, pronotum somewhat constricted basally where it is only 7/8 width of base of elytra; northwest America, Lassen County .... *bjorkmanae* Hatch 1939

— Outline of pronotum and elytra not or very slightly discontinuous, pronotum not constricted basally, at most but slightly narrower than elytral base 4

4. Row of punctures paralleling lower posterior margin of hind tibiae closely set, forming an almost continuous groove from near base almost to apex, or with a less regular double row; metasternal wings narrow (fig. 13:17*b*) 5

— Lower posterior margin of hind tibiae without a row of punctures paralleling margin, or with a single row of distinctly separated punctures; metasternal wings broad (fig. 13:17*c*) or narrow 9

5. Outer 3rd of metasternal wings narrow, tonguelike, nearly parallel-sided, apically sinuate, about as wide as an epipleuron adjacent to 2nd abdominal sternite (fig. 13:17*b*) 6

— Outer 3rd of metasternal wing more nearly triangular, straighter, all but apex much wider than an epipleuron adjacent to 2nd sternite (fig. 13:17*a*) 7

6. Many meshes of elytral reticulation in basal half of disc joined end to end, forming irregular series, but those at sides and apex discrete as usual; a notably convex species, hind tibiae short and broad, elytra usually with a brassy tinge (fig. 13:16*d*); California. .... *regularis* (LeConte) 1852

— Meshes of elytral reticulation often lightly impressed on disc, but of same type throughout; flatter species, elytra black; San Francisco Bay area south to Monterey region ....*brevicollis* LeConte 1857

7. Hind femora strongly obliquely strigate; inner claws of front tarsi of male with small basal tooth; elytral sculpture of female like that of male, or with meshes very elongate and narrow, strongly oblique, the surface opaque; northern North America, Siberia........... .... *tristis* Aubé 1838

— Hind femora punctate; claws not toothed 8

8. Hind femora and trochanters usually with coarse punctures, carination of prosternum continued along prosternal process; form narrower, more convex, elytral sculpture deeply impressed; Mendocino County (common), south to Tulare and Los Angeles counties (rare) ....*ilybiiformis* (Zimmermann) 1928

— Hind femora and trochanters very finely punctate, prosternal process broader and almost flat; elytral sculpture lightly impressed; San Francisco Bay northward .... *pandurus* Leech 1942

9. Prosternal process rather broad, sharply acuminate, only moderately convex; hind tibiae with full row of punctures paralleling lower posterior margin 10

— Prosternal process narrower, varying from moderately convex to angularly or acutely carinate 11

10. Metasternal wings narrower (fig. 13:17*c*), least distance from middle coxae to hind coxal plates less than 1/2 width of the latter, measured along the same line; more elongate, subparallel, rather flat species, elytral reticulation lightly impressed; western U.S. and Canada ....*seriatus intersectus* (Crotch) 1873

— Metasternal wings broader, least distance from middle coxae to hind coxal plates more than 1/2 the width of the latter; oval, broader, more convex species, elytral reticulation more deeply impressed; western North America .... *lugens* LeConte 1852

11. Hind tibiae with a row of punctures paralleling lower posterior margin.............................. 12

— Hind tibiae without such a row of punctures, or with only a few punctures near base or in basal half 15

12. Individual meshes of elytral reticulation greatly elongated; larger species, 10–11 mm. long; Pacific Coast .... *discors* LeConte 1861

— Elytral meshes more or less quadrate 13

13. Pronotum sharply but very narrowly margined at sides; smaller species, 7 mm. long; California to Oregon .... *vandykei* Leech 1942

— Pronotal margin almost as broad as a median antennal segment, rather flat; larger species, 8–10 mm. long 14

14. Hind angles of pronotum nearly rectangular; last abdominal sternite of male with strigosities at each side of middle at an angle to median line of the insect, and more numerous; Pacific Coast .... *walsinghami* (Crotch) 1873

— Hind angles of pronotum obviously obtuse; strigosities of last male sternite nearly parallel with long axis of body, less numerous; Pacific Coast .... *confertus* LeConte 1861

— Hind angles of pronotum less than a right angle; prosternal process elongate, sharply acuminate, on same plane as prosternum; hind femora coarsely punctate (see *ilybiiformis*, couplet 8)

15. Elytral reticulation coarse in both sexes, the individual meshes often about as wide as is a median antennal segment 16

— Elytra finely or minutely reticulate in the male, more coarsely so in the female 17

16. Reddish-brown, metasternum and hind coxal plates blackish; prosternal process glabrous; larger, more robust species, 8.5–9.0 mm. long; Pacific Coast .... *austini* Sharp 1882

— Black, elytra brownish or fuscous; prosternal process finely pubescent; smaller nearly parallel-sided species, 6.75–8.3 mm. long; Pacific Coast .... *strigulosus* (Crotch) 1873

17. Front tarsi of male broadly dilated, inner claws with a median or subapical tooth 18

— Front tarsi of male moderately or narrowly dilated, inner (anterior) claws not toothed 19

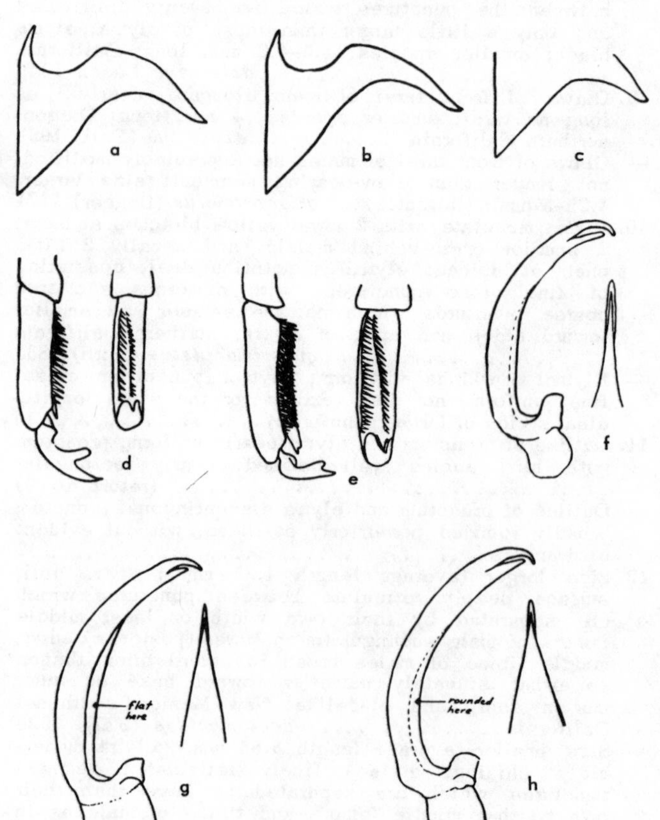

Fig. 13:17. Dytiscidae, adult structures of *Agabus*. *a*, *pandurus*, metasternal wing; *b*, *brevicollis*, metasternal wing; *c*, *seriatus intersectus*, metasternal wing; *d*, *lutosus*, apical segment and claw of front tarsus of male; *e*, *griseipennis*, apical segment and claw of front tarsus of male; *f*, *ancillus*, aedeagus; *g*, *obsoletus*, aedeagus; *h*, *morosus*, aedeagus (Leech, 1942).

18. Tooth of anterior claw of front tarsi of male outstanding, nearly median in position (fig. 13:17*d*); Pacific Coast .... *lutosus* LeConte 1853
— Tooth subapical, not standing out at an angle (fig. 13:17*e*); California to Wyoming .... *griseipennis* LeConte 1859
19. Elytra black, sharply margined with reddish-yellow; pronotum rufotestaceous with an anterior and posterior transverse black band; U.S. .... (in part) *disintegratus* (Crotch) 1873
— Color not as above 20
20. Front tarsi of male feebly dilated, 3rd segment scarcely wider than 4th; epipleurae pale; northwest U.S. and Canada .... *approximatus* Fall 1922
— Front tarsi of male rather strongly dilated, 3rd segment evidently wider than 4th, though less than twice as wide; epipleura in great part black 21
21. Side of pronotum (viewed laterally) inflated anteriorly; lateral margin of pronotum very narrow at base, progressively wider anteriorly; California .... *hoppingi* Leech 1942
— Side of pronotum not inflated anteriorly; lateral margin of nearly uniform width 22
22. Elytral reticulation fine, meshes even, small, everywhere rounded and alike; California .... (sp. near *smithi* Brown 1930)
— Elytral reticulation fine, meshes of at least discal area obviously unequal and of irregular form 23
23. Smaller species (less than 7 mm. long), flatter; apical 3rd of median lobe of male genitalia laterally compressed, nearly parallel-sided in dorsal or ventral view (fig. 13:17*f*); British Columbia to California ... . *ancillus* Fall 1922
— Larger species (7.0–8.5 mm.), more convex; median lobe of male genitalia gradually narrowed from near base to apex, or broad, and nearly parallel-sided to near apex, thence sharply narrowed 24
24. Larger, broader species, averaging 8 mm. long; median lobe of male genitalia flat in basal two-thirds of concave side (fig. 13:17*g*); San Diego to San Francisco, chiefly coastal .... *obsoletus* LeConte 1858
— Smaller, narrower species, averaging 7 mm. long; median lobe of male genitalia rounded in basal two-thirds of concave side (fig. 13:17*h*) 25
25. Prosternal process narrower, lanceolate, on only a slightly different plane from narrow part between front coxae, carina of prosternum often continuing sensibly onto base of it; Colorado, Utah, Nevada, Sierra of California .... *obliteratus* LeConte 1859
— Prosternal process broader basally, appreciably cordate, and on a different plane from narrow part between front coxae, carina of prosternum fading out between coxae; San Francisco northward .... *morosus* LeConte 1852

For another key, see Fall, 1922.

Key to the California Species of Ilybius

Adults

1. Hind tarsal segments of male margined externally above; last sternite of female very shallowly emarginate apically; Pacific Coast .... *quadrimaculatus* Aubé 1838
2. Hind tarsi of male not margined; last sternite of female with a deep triangular emargination; northwest America .... *fraterculus* (LeConte) 1862

Key to the California Species of Agabinus

1. Ovate; elytra finely punctate, punctures more conspicuous than the lightly engraved reticulation; lines of reticulation fine, meshes irregular in size and shape, not at all concave; metasternal wings broad, shortest distance between middle coxa and hind coxal plate 3/4 width of latter, measured along same line; (fig. 13:10*e*); British Columbia to Texas .... *glabrellus* (Motschulsky) 1859
— Broadly oval; elytra very finely punctate, punctures less conspicuous than the reticulation; meshes irregular in size and shape, smaller and more numerous than in *glabrellus,* most of them concave, giving elytra a dimpled appearance; metasternal wings narrow, shortest distance between middle coxa and hind coxal plate less than 2/3 width of latter, measured along same line; British Columbia to California .... *sculpturellus* Zimmermann 1919

Genus *Hydrotrupes* Sharp

Adults

The single described species occurs in northern California and in Oregon. Length 4–4.5 mm., black or slightly aeneous; pronotum margined at sides; elytra finely lightly reticulate, with punctures at intersections; prosternal process almost as broad as long, nearly flat, margined; hind trochanters large, fully 1/2 as long as hind femora .... *palpalis* Sharp 1882

Key to the California Species of Copelatus

Adults

1. Each elytron with 10 discal striae (a shorter submarginal one not being counted); southwest U.S., ?California .... *impressicollis* Sharp 1882
— Each elytron with 8 discal striae; southwest U.S., Imperial County .... *chevrolati renovatus* Guignot 1952

Key to the California Species of Coptotomus

Adults

1. Form elongate-oval, nearly parallel-sided; color testaceous marked with black or piceous. Metasternum broad at narrowest point behind a middle coxa, wider than an epipleuron; each hind coxal process notched at about mid-point of hind margin; pronotum flattened or slightly depressed near sides, margins appearing slightly explanate; northwest North America, Lassen County .... *longulus* LeConte 1852
— Form evenly ovate, widest at middle; color decidedly ferrugineous, marked with rufopiceous. Metasternum behind a middle coxa not as broad as an epipleuron; hind coxal processes notably broad, hind margins not notched; sides of pronotum not at all explanate; Arizona, California, Yuma region .... *difficilis* LeConte 1852

Key to the California Species of Rhantus

Adults

1. Head solidly black between and behind eyes 2
— Head with transverse bilobed yellow or reddish spot between hind margin of eyes 3
2. Metasternum and hind coxal plates piceous or black; pronotum with a median black spot; California...... .... *consimilis* Motschulsky 1859
— Metasternum and hind coxal plates flavous to piceoferrugineous, pronotum usually without a median black spot; north U.S. .... *tostus* LeConte 1866
3. Black above and below, mouth parts and legs slightly paler, antennae and 2 median spots on head rufous;

claws of mesotarsi of male of equal length; larger species, 14–16 mm. long; southwest U.S. .... *atricolor* (Aubé) 1838
— Elytra yellow, finely irrorated with black, or black except at apex and around margins; pronotum yellowish, usually marked with black; ventral surface largely piceous; mesotarsal claws of male unequal; smaller species, 10–13 mm. long 4
4. Claws of front tarsi of male elongate, anterior ones 1/3 larger than their fellows and longer than claw-bearing segment; female with an elongate-oval roughened area on each elytron, from humerus to beyond middle; carina of prosternal process usually black; California .... *anisonychus* (Crotch) 1873
— Claws of front tarsi of male not as long as claw-bearing segment; female without roughened area on elytra; prosternal carina usually pale 5
5. Both claws of front tarsi of male straight, sinuate along lower margin; hind femora pale to almost completely piceous; Nearctic...... *binotatus* (Harris) 1828
— Claws of front tarsi of male short, rather evenly arcuate, anterior ones a little broader medially than their fellows; hind femora pale 6
6. Anterior claws of front tarsi of male more abruptly sinuate along lower edge, broader at middle than at base; aedeagus of male with apical quarter thinner, more strongly twisted to one side; Pacific Coast north of Los Angeles County ...... *hoppingi* (Wallis) 1933
— Anterior claws of front tarsi of male less abruptly sinuate along lower edge, not broader at middle than at base; aedeagus with apical quarter thicker, less strongly twisted, southwest United States, Mexico .... *gutticollis* (Say) 1834

Key to the California Species of Colymbetes

Adults

1. Front tarsi of male densely clothed beneath with glandular hairs, but without enlarged suction cups (palettes); undersurface, including legs, black; Pacific Coast, California? .... *seminiger* LeConte 1862
— Frontal tarsi of male with rounded palettes beneath, at least in part 2
2. Front tarsi of male with 4 rows of similar rounded palettes, but no basal rows of small elongate palettes or of hairs 3
— Front tarsi of male with 3 rows of rounded palettes, and basal rows of elongate palettes or of hair 4
3. Form stout, convex; legs rufopiceous. Sculpture of pronotum consisting of irregular impressed lines of varying lengths, few of which join or enclose spaces; metasternal wing at narrowest point behind middle coxa, fully ½ as wide as hind coxal plate measured along continuation of same line; southern California, San Diego County, San Bernardino County .... *strigatus* LeConte 1852
— Form narrower, flatter; legs testaceous, femora sometimes tinged with rufous or piceous. Pronotal sculpture vermiculate, most of lines joining and enclosing spaces; width of metasternal wing behind middle coxa less than 1/2 that of hind coxal plate measured along same line; northwest north America, Sierra of California .... (subsp. of) *rugipennis* Sharp 1882
4. Protarsi of male with 3 rows of rounded palettes and a basal strip of glandular pubescence 5
— Protarsi of male with 3 rows of rounded palettes and a somewhat double basal row of smaller elongate palettes, each a little less than half as big as the round ones; San Francisco Bay area, east to foothills of Sierra .... (species)
5. Protarsi of male with 3 rows of rounded palettes; apical half of 1st segment with glandular pubescence; elytra of female dull, transverse lines in basal ⅔ broad, very deeply engraved, intervals between being scabrous; Oregon, Humboldt County ...... *inaequalis* Horn 1871
— Apical third of 1st protarsal segment of male clothed beneath with glandular pubescence; elytra of female shining, sculptured as in male; Monterey County, Napa County ....? *exaratus* LeConte 1862

Genus *Eretes* Laporte

Adults

The single species occurring in North America is at once recognizable by its narrow form, pale color, size (14 to 17 mm. long), and the fact that the elytral margins, from behind the middle to about posterior fifth, are set with short flat spines and fine golden hairs; cosmopolitan; in California, at least as far north as northern Fresno County .... *sticticus* (Linnaeus) 1767

Genus *Hydaticus* Leach

Adults

Length 11.5–14.5 mm.; black, head reddish anteriorly, pronotum black only across base discally, elytra usually with indications of pale vittae (especially in female), often pale across base, and pale laterally; elytra extremely finely punctate; northern north America; northern California ? .... *modestus* Sharp 1882

Genus *Acilius* Leach

Adults

Length 13–15 mm.; yellowish-brown dorsally, elytra irrorated with black, pronotum with 2 transverse discal black bands (fig. 13:16*e*); elytra of male rather densely punctate, punctures in basal region somewhat lunate; elytra of female (except in var. *simplex* LeConte) fluted, with 4 broad pubescent grooves, outer 3 reaching to humeral area, inner one to a little in front of middle (fig. 13:16*f*); northern north America, south to San Diego County *(simplex)* in California ...*semisulcatus abbreviatus* Mannerheim 1843

Key to the California Species of Thermonectus

Adults

1. Elytra black, each elytron conspicuously marked with 10 or 11 yellow spots of various sizes, lateral spots often connected along elytral margins (fig. 13:16*g*); length 12–14 mm.; southwest U.S., Central America, San Diego County, Riverside County .... *marmoratus* (Hope) 1832
— Elytra black to rufopiceous, with irregular marginal longitudinal yellowish marks which may be divided by dark lines or specks; subbasal transverse yellow band present or not (fig. 13:16*h*); length 9–11 mm.; southern U.S., Guatemala, San Diego County north to Shasta and Lassen counties ... *basillaris* (Harris) 1829

Genus *Graphoderus* Dejean

Adults

Length 13–15 mm.; pronotum black discally along anterior and posterior margins; elytra yellowish, irrorated with black, extremely finely punctate; northern North America; Lassen County .... *occidentalis* (Horn) 1883

Key to the California Species of Dytiscus

1. Hind coxal processes broadly rounded or bluntly pointed apically 2

— Hind coxal processes spinous apically 3
2. Hind coxal processes broadly rounded; apices of hind trochanters blunt; middle coxae less widely separated, metasternum between middle coxae narrower than greatest width of prosternal process; adhesive pads on 2nd and 3rd middle tarsal segments of male uniformly dense; elytra of female sulcate (fluted) in basal two-thirds; length 24-26 mm.; British Columbia, to Humboldt County . . . . *hatchi* Wallis 1950
— Hind coxal processes bluntly pointed; apices of hind trochanters sharply pointed; metasternum between middle coxae broader than greatest width of prosternal process; adhesive pads on 2nd and 3rd middle tarsal segments of male each divided longitudinally by a nearly median bare space; elytra of female smooth, as in male; length 27–30 mm.; Alaska to Baja California, Alberta to Colorado . . . . . . *marginicollis* LeConte 1845
3. Dorsum of head with a narrow pale margin adjacent to eyes; abdominal sternites pale; form narrower; Palearctic, Alaska to Manitoba; California ? . . . . *anxius* Mannerheim 1843
— Dorsum of head not pale adjacent to eyes, except at front corners; abdominal sternites black, broadly pale along hind margins; form broader; Palearctic; Alaska to New Hampshire, Sierra south at least to Fresno County . . . . *dauricus* Gebler 1832

Key to the California Species of Cybister

Adults

1. Outer posterior angle of hind femur acute, produced posteriorly in both sexes; female with only a single claw on hind tarsus, and without sexual sculpture on elytra; Nevada to Baja California; Butte County to Imperial County . . . . *explanatus* LeConte 1852
— Outer posterior angle of hind femur not produced, not acute; female with rudimentary 2nd claw on hind tarsus and with sexual sculpture of impressed lines on sides of head and pronotum, and elytra except near suture; Texas to Arizona, Imperial County to San Diego County . . . . *ellipticus* LeConte 1852

REFERENCES

BALDUF, W. V.
1935. The bionomics of entomophagous Coleoptera. N.Y.: John Swift. 220 pp., 108 text figs. Planograph.
BALFOUR-BROWNE, F.
1940. British water beetles. Vol. I. London: Ray Soc. XX+375 pp., 89 text figs., 5 pls., 72 text maps.
1950. British water beetles. Vol. II. London: Ray Soc. XX+394 pp., 90 text figs., 1 pl., 56 text maps.
BALFOUR-BROWNE, J.
1944. Remarks on the *Deronectes*-complex. Entomologist, 77:186-189.
1947. On the false-chelate leg of an aquatic beetle larva. Proc. Roy. Ent. Soc. London (A), 22:38-41, 7 text figs.
BERTRAND, H.
1927. Les larves des Dytiscides, Hygrobiides, Haliplides. Université Paris. XV+370 pp., 33 pls., 208 text figs. (3 pp. Errata et addenda interpolated between pp. 358 and 359).
1928. Les larves des Dytiscides, Hygrobiides, Haliplides. Encycl. ent. (A)10:vi+366 pp., 33 pls., 207 text figs. (3 pp. Errata et addenda interpolated between pp. 360 and 361).
BÖVING, A. G.
1913. Studies relating to the anatomy, the biological adaptations and the mechanism of oviposition in the various genera of Dytiscidae. Internationale Revue der gesamten Hydrobiologie und Hydrographie, Biol. Suppl., Ser. V(1912):1-28, pls. I-VI.
BÖVING, A. G., and F. C. CRAIGHEAD
1930. An illustrated synopsis of the principal larval forms of the order Coleoptera. Ent. Amer. 11(n.s.) (1):1-80, (2):81-160, incl. pls. 1-36.
BRINCK, P.
1945. Nomenklatorische und systematische studien über Dytisciden. III. Die classifikation der Cybisterinen. Lunds Universitets Årsskrift N. F. Avd. 2. 41:1-20, 1 text fig.
1948. Coleoptera of Tristan da Cunha. *In* Results of the Norwegian Scientific Expedition to Tristan da Cunha 1937-1938. No. 17:1-122, 26 text figs., 1 pl., map on inside back cover. (Publ. by Det Norske Videnskaps-Akademi, Oslo.)
CROWSON, R. A.
1950. The classification of the families of British Coleoptera. Ent. Mon. Mag., 86:149-160, 9 text figs.
FALL, H. C.
1919. The North American species of *Coelambus*. Mt. Vernon, N.Y.: John D. Sherman, Jr., pp. 1-20.
1922. A revision of the North American species of *Agabus* together with a description of a new genus and species of the tribe Agabini. Mt. Vernon, N.Y.: John D. Sherman, Jr., pp. 1-36.
1923. A revision of the North American species of *Hydroporus* and *Agaporus*. Privately printed. Pp. 1-129.
GUIGNOT, F.
1945. Dix-neuvième note sur les Hydrocanthares. Bull. Soc. nat. Vaucluse, (1943)1945:5-9.
1950. Trente-et-unième note sur les Hydrocanthares. Bull. Mensuel Soc. Linnéenne de Lyon, 19:25-28.
1952. Description de Dytiscides inédits de la collection Régimbart et de quelques autres espèces et variétés nouvelles. Rev. franc. d'Ent., 19:166-171, 1 text fig.
HATCH, M. H.
1929. Studies on Dytiscidae. Bull. Brooklyn Ent. Soc., 23:217-229.
1933. Studies on *Hydroporus*. Bull. Brooklyn Ent. Soc., 28:21-27.
1953. The beetles of the Pacific Northwest. Part I: Introduction and Adephaga. Univ. Wash. Publ. Biol., 16:vii+1-340, incl. front., 2 text figs., 37 pls.
JACKSON, DOROTHY J.
1952. Observations on the capacity for flight of water beetles. Proc. Roy. Ent. Soc. London (A), 27:57-70, 3 text figs.
LA RIVERS, I.
1951. Nevada Dytiscidae (Coleoptera). Amer. Midl. Nat., 45:392-406.
LEECH, HUGH B.
1941. Note on the species of *Agabinus*. Canad. Ent., 73:53
1942*a*. Dimorphism in the flying wings of a species of water beetle, *Agabus bifarius* (Kirby). Ann. Ent. Soc. Amer., 35:76-80, 3 text figs.
1942*b*. New or insufficiently known nearctic species and subspecies of *Agabus*. Canad. Ent., 74:125-136, 1 pl.
1942*c*. Key to the Nearctic genera of water beetles of the tribe Agabini, with some generic synonymy. Ann. Ent. Soc. Amer., 35:355-362, 1 pl.
1945. Three new species of nearctic *Deronectes*. Canad. Ent., 77:105-110, 4 text figs.
1948. Coleoptera: Haliplidae, Dytiscidae, Gyrinidae, Hydrophilidae, Limnebiidae. No. 11 *in* Contributions toward a knowledge of the insect fauna of Lower California. Proc. Calif. Acad. Sci., Ser. 4, 24:375-484, 2 pls.
MATHESON, R.
1914. Life-history of a dytiscid beetle (*Hydroporus septentrionalis* Gyll.). Canad. Ent., 46:37-40, 1 text fig., pl. I.
RICHARDS, O. W.
1953. The social insects. London: MacDonald; New York: Philosophical Library, xiii+219 pp., 12 text figs., 51 photographs on 19 pls.

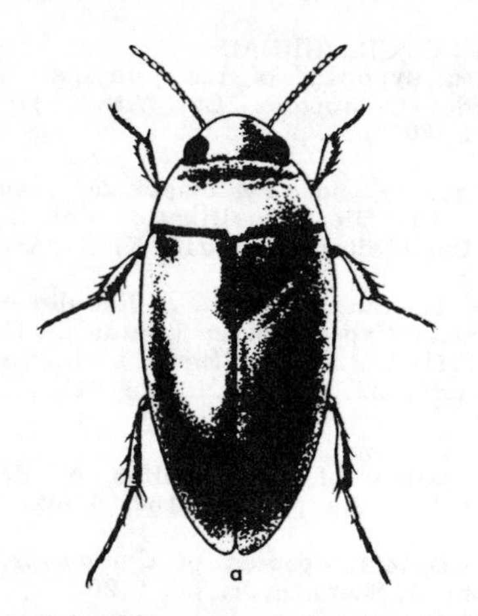

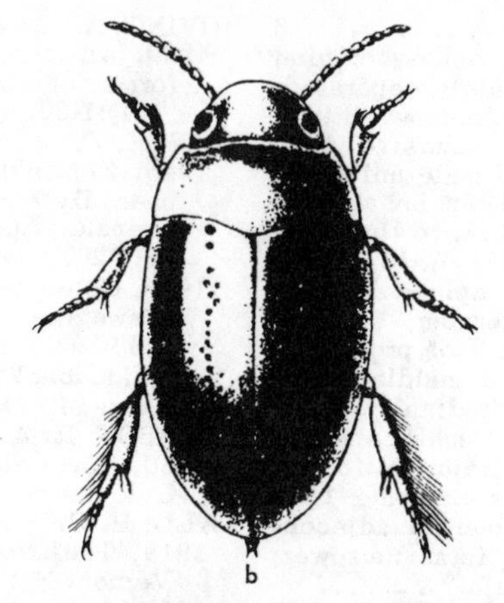

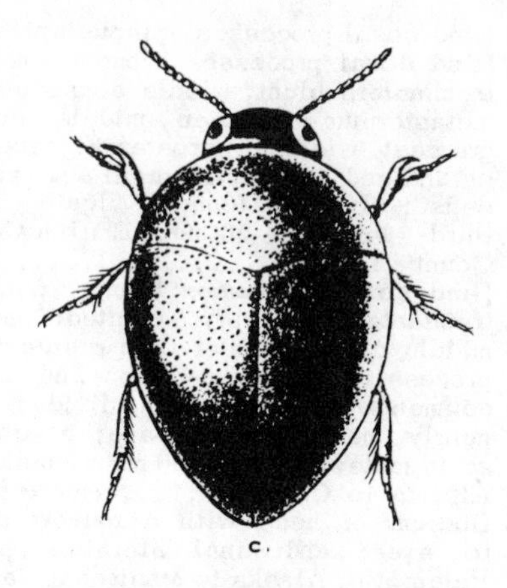

Fig. 13:18. Noteridae, adults. *a, Notomicrus* sp.; *b, Pronoterus* sp.; *c, Colpius inflatus* (Sharp, 1882).

SCHAEFFER, C.
1908. On North American and some Cuban *Copelatus*. Jour. Y.Y. Ent. Soc., 16:16-18.
WALLIS, J. B.
1939*a*. *Hydaticus modestus* Sharp versus *Hydaticus stagnalis* Fab. in North America. Canad. Ent., 71: 126-127.
1939*b*. The genus *Graphoderus* Aubé in North America (north of Mexico). Canad. Ent., 71:128-130.
1939*c*. The genus *Ilybius* Er. in North America. Canad. Ent., 71:192-199.
1950. A new species of *Dytiscus* Linn. Canad. Ent., 82:50-52.
WESENBERG-LUND, C.
1912. Biologische Studien über Dytisciden. International Revue der gesamten Hydrobiologie und Hydrographie, Biol. Suppl., Series V(1912):1-129, 4 text figs., pls. I-IX.
WILLIAMS, F. X.
1936. Biological studies in Hawaiian water-loving insects. Part 8. Coleoptera or beetles. Proc. Hawaii. Ent. Soc. (1935), 9:235-273, incl. pls. II-VI.
WILSON, C. B.
1923. Water beetles in relation to pondfish culture, with life histories of those found in fishponds at Fairport, Iowa. Bull. Bur. Fish., 39:231-345, 148 figs.
YOUNG, F. N.
1951. A new water beetle from Florida, with a key to the species of *Desmopachria* of the United States and Canada. Bull. Brooklyn Ent. Soc., 46:107-112, incl. pl. VI.
YOUNG, F. N.
1954. The water beetles of Florida. Univ. Florida Stud., Biol. Sci. Ser., 5:IX+238 pp., 31 figs.

Family NOTERIDAE

Burrowing Water Beetles

The North American members of this family are small, rarely if ever more than 5.5 mm. long. The adults (fig. 13:20) of most genera are strong and active swimmers. The common name is suggested by the habits of the larvae, which have powerful fossorial legs and can dig their way at an astonishing pace through the wet mud around the roots of aquatic plants. The adults also burrow in underwater debris. The adults are predaceous, but the food of the larvae is unknown, though mandibular structure suggests they may eat both plant and animal matter.

Relationships.—Long recognized as a distinctive group in the family Dytiscidae, accorded subfamily rank by Régimbart in 1878 on adult morphology and by Meinert in 1901 on larval characters, the genera whose larvae were known were given family status by Bertrand (1927) and Böving and Craighead (1931). On adult characters other genera have been included, though the genus *Notomicrus* is discordant, and upon discovery of its larvae may have to be removed from the Noteridae.

The known larvae of Noteridae differ from those of other Hydradephaga in having fossorial legs; according to Meinert, the larvae of Hygrobiidae, Noteridae, and Amphizoidae resemble those of Carabidae in having an open mouth, whereas in dytiscid larvae it is reduced to a narrow slit.

Respiration.—The respiration of adult Noteridae is similar to that of Dytiscidae (q.v.). The larvae of North American species of the family have not been described, but are presumed to resemble those of allied Old World genera in appearance and in their remarkable habits.

The larvae of *Noterus capricornis* Herbst (see Balfour-Browne and Balfour-Browne, 1940) live in the mud around the roots of plants such as *Iris, Alisma, Sparganium,* under one to two feet of water, and are unable to come to the surface for air. They are elateridlike in form (fig. 13:19*a*) with a chitinous point at the tail end, and are presumed to obtain their air supply by piercing plant roots and taking the intercellular air, as do *Donacia* spp. larvae (Chrysomelidae). The air in the plant must reach the tracheal system via the eighth pair of spiracles, which are side by side at the extreme apex of the segment, immediately beneath the chitinous point. When placed in shallow water, the larvae raise the tip of the abdomen through the surface film, and take air in the regular manner.

The mature larvae make their cocoons of bits of vegetable material mixed with mud particles (see fig. 4 in Balfour-Browne). They are attached to the roots of plants growing in the water; the point of attachment

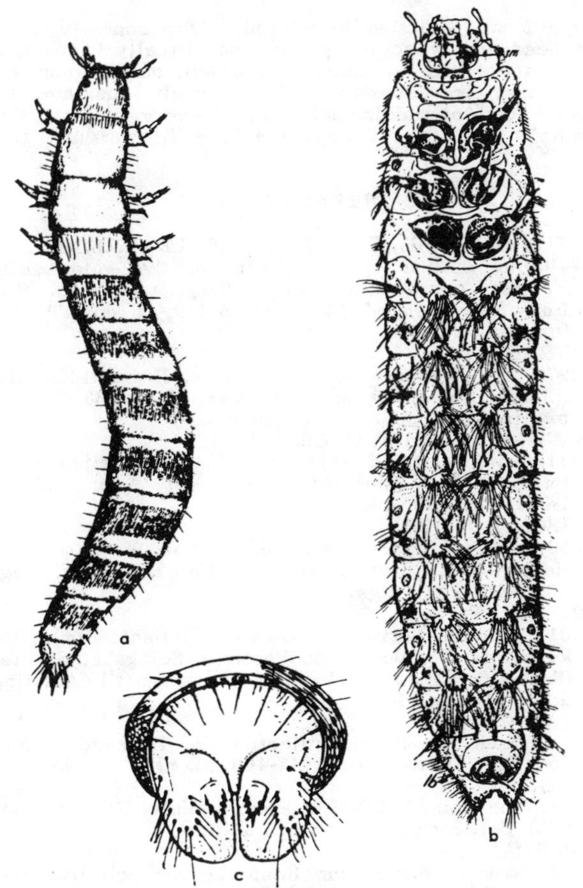

Fig. 13:19. Larvae of Noteridae and Ptilodactylidae. *a*, *Noterus* sp.; *b*, *Stenocolus* sp.; *c*, *Anchytarsus* sp., ninth abdominal segment (*a*, Wesenberg-Lund, 1912; *b*,*c*, Böving and Craighead, 1931).

is over a hole bitten into the root, the cocoons being filled with air which escapes from the root.

Life history.—The eggs are presumed to be laid in summer, on the roots of plants a foot or so under water, or in the mud close by. Freshly emerged adults appear in early August, and the winter is passed in the adult and perhaps also in the larval stage. Cocoons are formed on plant roots under water, as mentioned in the preceding paragraph (see fig. 3 in Balfour-Browne).

Habitat and *distribution.*—The typical streamlined noterines, such as the Old World *Noterus, Hydrocanthus* (Old and New Worlds), the New World *Suphisellus,* and to a lesser extent our *Pronoterus,* are active beetles and strong swimmers. The tiny *Notomicrus* is more suited for crawling, and clinging to underwater objects, whereas the nearly hemispherical *Colpius* (and the very similar Central and South American *Suphis*), with hind legs but little modified for swimming, spends most of its time in the debris at the bottom. All of them occur in weedy ponds and lakes. The species of *Pronoterus* fly to lights at night. Jackson (1950) studied the wings of the two British species of *Noterus.* In *N. capricornis* Herbst, she found that though the wings were fully formed and of normal size, the muscles of flight were absent. In *N. clavicornis* Degeer the wings were either brachypterous, or partly reduced in size. However, in earlier studies in weevils of the genus *Sitona,* she discovered that certain fully winged individuals which were seen to fly in the fall, had the flight muscles degenerate and functionless by spring.

The family is represented in the warmer parts of both hemispheres, and widely distributed therein. *Hydrocanthus* occurs in North, Central, and South America, Africa, Madagascar, Burma, Java, and in adjacent areas; *Suphisellus* of the New World is closely allied to *Canthydrus* of the Old; *Colpius* and *Pronoterus* are Neotropical, reaching Florida in the United States, whereas the aberrant *Notomicrus* has a similar distribution, but extends also to Australia.

Taxonomic characters.—Color pattern is of some value in separating the species of *Suphisellus,* and punctation of the dorsum is useful in most genera. Characters of greatest use are found in the shape of the tibiae, the form of the prosternal and hind coxal process, the punctation of the ventral surface. In some genera the apex of the prosternal process may differ in shape in the sexes of one species, as may the form of the metasternum.

Key to the Genera of Noteridae of the United States[7]

Adults

1. Apex of front tibiae bearing more or less conspicuous curved spurs or hooks (fig. 13:18*c*) 2
— No spurs or hooks on front tibiae; small beetles, rarely exceeding 1.5 mm. in length (fig. 13:18*a*) .... *Notomicrus* Sharp 1882
2. Front tibial spurs strong and conspicuous; hind femora with angular ciliae; prosternal process truncate behind (or if rounded in male, body form very broad, almost hemispherical) 3
— Front tibial spurs weak and inconspicuous; hind femora usually without angular ciliae; prosternal process rounded behind in both sexes (fig. 13:18*b*) .... *Pronoterus* Sharp 1882
3. Laminate inner plates of hind coxae truncate at apex with an arcuate emargination on each side of the depressed middle; hind coxal cavities separated; prosternal process truncate in female, rounded in male (fig. 13:18*c*) .... *Colpius* LeConte 1861
— Laminate inner plates of hind coxae with a broad and deep angular excision at apex, leaving on each side a diverging triangular process (fig. 13:20); hind coxal cavities contiguous 4
4. Apex of prosternal process at least twice its breadth between coxae, not broader than long; last segment of maxillary palpus emarginate at apex; pronotum with lateral marginal lines originating at hind angles but disappearing at about middle; beetles usually less than 3 mm. in length .... *Suphisellus* Crotch 1873
— Apex of prosternal process very broad, at least 2.5 to 3 times its breadth between coxae, broader than long; last segment of maxillary palpus truncate at apex; pronotum with lateral marginal lines the entire length of margins and joining front margin anteriorly; beetles usually between 4 and 5.5 mm. in length .... *Hydrocanthus* Say 1823

Key to the Known Genera of Nearctic Noteridae

Larvae

1. Mandibles slender, narrowly sulcate at tip, simple;

[7]Key supplied by Dr. Frank N. Young.

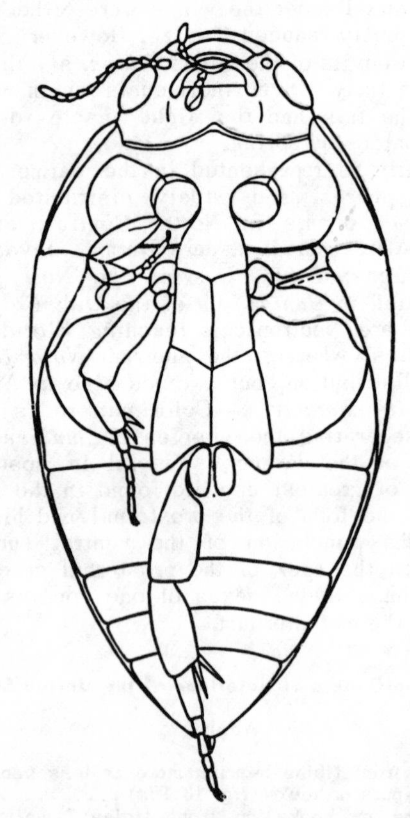

Fig. 13:20. *Hydrocanthus iricolor*, ventral surface of adult (Young, 1954).

3rd antennal segment more than twice as long as 4th .... *Hydrocanthus* Say 1823
— Mandibles stout, bifid at tip; 3rd antennal segment no longer than 4th .... *Suphisellus* Crotch 1873

Genus *Suphisellus* Crotch

Two species, *S. levis* (Fall) and *S. lineatus* (Horn), have been recorded from Baja California. The former is known only by the type from San José del Cabo; the latter has been taken in Texas, and may possibly occur in southeastern California.

Oval, 3 mm. long; pronotum reddish-yellow, piceous or black discally; elytra reddish-yellow, each elytron with a sutural, discal, humeral, and submarginal black vitta .... *lineatus* (Horn) 1871

Genus *Hydrocanthus* Say

In 1928 Zimmermann described *H. similator* from "Nordamerika: Mass.; Californien." No other California specimens are known, so there is a possibility that his example was mislabeled. On the other hand Horn (1894) listed the eastern *H. iricolor* Say from Santa Anita, Baja California, and since Zimmermann describes *H. similator* as like *H. iricolor* in form and size, there may be a species of *Hydrocanthus* in California.

Length 5 mm. Color yellowish-red, elytra scarcely darker than head and pronotum; prosternum virtually impunctate, except sometimes on sides of process; metasternum and hind coxal processes densely, strongly punctate, the former smooth at middle just behind anterior margin ("Californien") .... *similator* Zimmermann 1928

REFERENCES

BALFOUR-BROWNE, F., and J. BALFOUR-BROWNE
1940. An outline of the habits of the water-beetle, *Noterus capricornis* Herbst (Coleopt.). Proc. Roy. Ent. Soc. London (A), 15:105-112, 4 comp'd. text figs., pl. I.

BERTRAND, H.
1927. Les larves des Dytiscides, Hygrobiides, Haliplides. Université de Paris. XV+370 pp., 33 pls., 208 text figs., + 3 pp. errata et addenda.

BÖVING, A. G., and F. C. CRAIGHEAD
1931. An illustrated synopsis of the principal larval forms of the order Coleoptera. Ent. Amer., 11 NS. 1-351, incl. 125 pls.

JACKSON, DOROTHY J.
1950. *Noterus clavicornis* Degeer and *N. capricornis* Herbst (Col., Dytiscidae) in Fife. Ent. Mon. Mag., 86:39-43, 2 text figs.

MEINERT, FR.
1901. Vandkalvelarverne (Larvæ Dytiscidarum). Det kongelige Danske Videnskabernes Selskabs, Skrifter, (Series 6, Naturvidenskabelig og mathematik) Afdeling, 9:341-440, pls. I-VI.

RÉGIMBART, M.
1878. Étude sur la Classification des Dytiscidae. Ann. Soc. Ent. France, (5), 8:447-466, pl. #10.

YOUNG, F. N.
1954. The water beetles of Florida. Univ. Florida Stud., Biol. Sci. Ser., 5:IX+238 pp., 31 figs.

ZIMMERMANN, A.
1928. Neuer Beitrag zur Kenntnis der Schwimmkäfer. Weiner Ent. Zeit., 44:165-187.

Family GYRINIDAE

Whirligig Beetles

Commonly known as whirligig and waltzing beetles and scuttle bugs, the gyrinids (fig. 13:22*a*, *b*, *c*) live up to their names as they actively whirl and gyrate on the surface of the water. They have other vernacular names such as "apple smeller" and "mellow bugs," descriptive of the odors of a milky fluid excreted chiefly in the region of the prothorax. The adults dive readily, and can take flight after climbing out of the water. They are the only aquatic beetles which regularly take advantage of the surface film for support.[8] Although the undersurface is wettable and breaks through the film, the sides of the beetle are supported as is shown by their virtual inability to keep afloat when the surface tension is reduced by the addition of a detergent. The compound eyes are specially adapted for surface life; each eye is divided into two widely separated parts (fig. 13:23), the lower being beneath the surface film as the beetle swims,

[8]The species of Hydraenidae, Spercheidae, and various small Hydrophilidae can break through the surface film from below with their tarsal claws. They thus gain traction and support and can walk along upside down, but are always partly buoyed by a film of air held on their ventral hydrofuge pubescence.

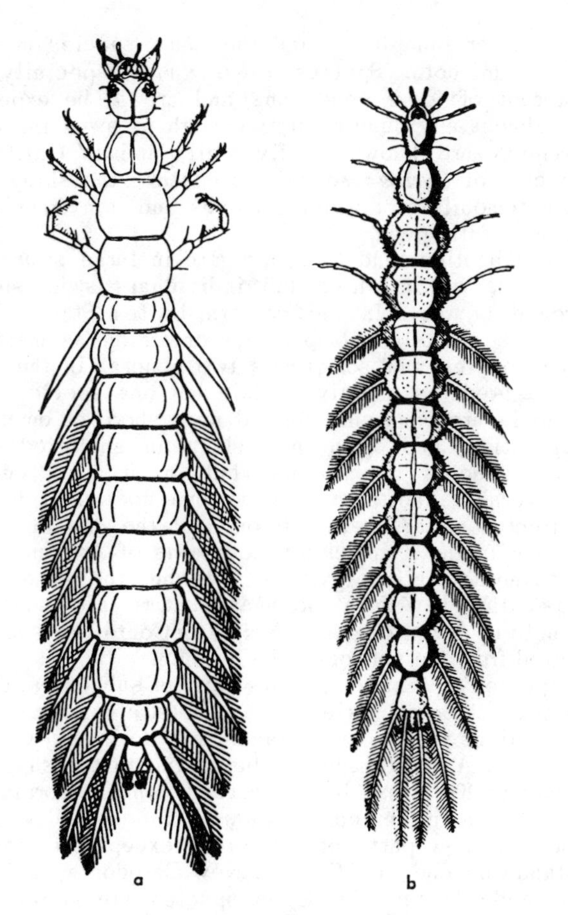

Fig. 13:21. Gyrinidae, larvae. *a*, *Dineutus* sp.; *b*, *Gyrinus* sp. (*a*, Wilson, 1923; *b*, Miall, 1903).

the upper in the air. Each of these parts is structurally best suited for vision in its respective medium.

The larvae (fig. 13:21) are strictly aquatic. They are long, slender, and have abdominal gills.

Relationships.—Much has been written on the relationships of the Gyrinidae, and even among recent authors there are wide differences of opinion. If specialization for their mode of life (divided eyes; short thick antennae, second segment with earlike process; middle and hind legs as fanlike paddles) is not overstressed, the gyrinids clearly belong in the suborder Adephaga and to a caraboid stock. The larvae have two free tarsal claws on each leg, a condition not found outside the Adephaga.

Respiration.—The adults take in air at a groove just inside the posterior end of the lateral margin of each elytron. It is stored in a dorsal reservoir under the elytra, into which the abdominal spiracles open. When a beetle dives, some of this air protrudes as a bubble which glistens like quicksilver, and which is held in place by the vestiture of the last abdominal segment. In warm weather gyrinids go beneath the surface only when frightened, or to lay eggs, and do not stay under long. However, in regions where winter forces them to hibernate they are said to pass that time either buried in the bottom mud and debris or clinging to submerged plants. Their air reservoir presumably suffices during such a period of low metabolic rate.

The larvae (fig. 13:21) are independent of surface air, since they have tracheal gills, a pair on each of the first eight abdominal segments and two pairs on the ninth, by which they obtain air directly from the water. An undulant or trembling motion of the abdomen provides the necessary current and movement of water. The pupae are in mud cells (fig. 12:22*d*) above the water, and breathe air in the regular manner.

Life history.—In the north temperate zone winter is passed in the adult stage, or perhaps in the larval in some forms. Eggs are laid in the spring, attached to surfaces of the submerged parts of plants. They are laid in rows or clusters and sometimes at a considerable depth. Hatching takes place in from one to three weeks. The young larvae drop to the bottom, where they hide and hunt in the debris until nearly full-grown. The larvae are predaceous and cannibalistic, feeding on bloodworms, Odonata nymphs, and even small fish, into which they inject a killing poison through a canal in their slender curved mandibles.

When mature, a larva of *Gyrinus* leaves the water and either crawls up a suitable plant *(Scirpus, Typha)* or onto some object at the water's edge. In the former case it collects what adherent material it can scrape from the surface of the reeds near the water line, mixes in an adhesive substance, and places the mass on its back. When it has enough it crawls up a stem, shifts the load to the end of its abdomen, and while in that position bends back and attaches the mass to to the reed. It then bites a hole in the mass, works its way in, closes the opening, and forms a pupal cell within. A larva which goes to shore may collect mud or sand and carry the load up a standing object, or it may form its pupal cell under some shelter. The larvae of *Dineutus* are said to build their cells around them, a piece at a time, reaching to the ground each time for the next load.

Parasites.—The student is referred to the fine paper by F. G. Butcher, 1933 (Hymenopterous parasites of Gyrinidae with descriptions of new species of *Hemiteles.*" *Annals of the Entomological Society of America,* 26:76-85, incl. 1 pl.). Following is a list of species reared from the cocoons of North American Gyrinidae, and known to be primary parasites:

Parasite	Host
Bathythrix gyrini (Ashmead)	*Gyrinus* sp. (this is the *Hemiteles gyrinophagus* Cushman of Butcher's paper).
Bathythrix pimplae Howard	*Gyrinus* sp.
Hemiteles cheboyganensis Butcher	*Gyrinus* sp.
Hemiteles hungerfordi Cushman	*Gyrinus* sp.
Eriplanus cushmani (Butcher)	*Gyrinus* sp.
Gyrinophagus dineutis (Ashmead)	*Dineutus assimilis, Gyrinus* sp.

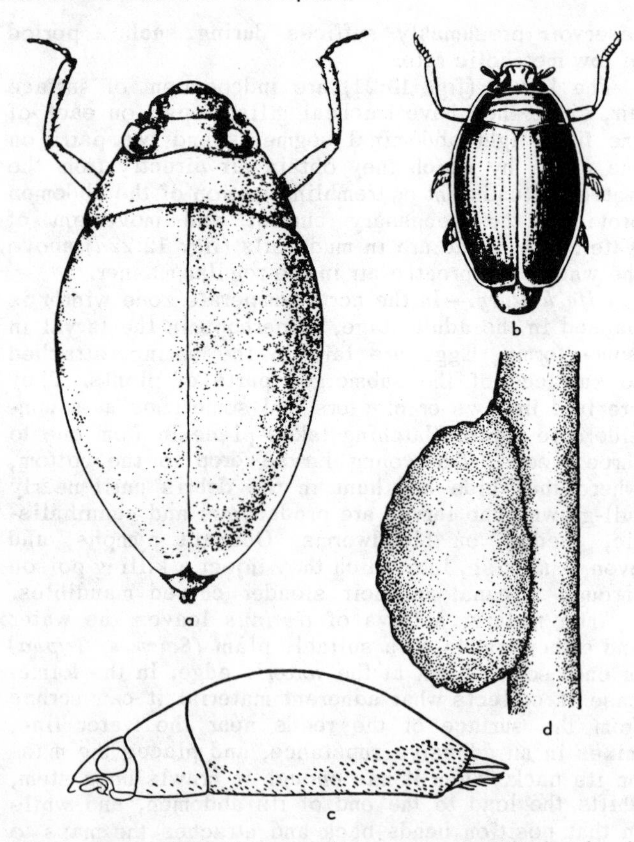

Fig. 13:22. Gyrinidae. *a, Dineutus* sp., adult female; *b, Gyrinus* sp. adult; *c, Gyretes sinuatus*, adult, lateral view; *d, Gyrinus* sp., mud cocoon on reed (*a*, Borror and Delong, 1954; Reitter, 1909; *c*, Young, 1954; *d*, Butcher, 1933).

The listing of *Spilocryptus incertus* Cresson by Dimmock and Knab resulted from an error of host record, perhaps the effect of contaminated cage material.

Of greater interest is the parasitization of *Dineutus* by the larvae of *Brachynus,* a genus of bombardier beetles. Wickham (1893, 1894) was the first to record larvae of *B. janthinipennis* Dejean as external parasites of the pupae of *D. assimilis* (Aubé); Dimmock and Knab (1904) described the *Brachynus* larva and pupa in detail and illustrated them.

Fruiting bodies of a parasitic fungus, *Laboulbenia gyrinidarum* Thaxter of the family Laboulbeniaceae, are often seen attached to the elytra and pronotum of individual gyrinids, especially in the genus *Gyrinus.* The fungi resemble minute, swollen, black bristles, and tend to be near the margins of the dorsal surface, whence they stick out nearly at right angles. They are said not to harm or bother the beetles in any way.

Habitat and distribution.—Adult gyrinids are found on streams, rivers, ponds, and lakes (but only transiently on brackish or saline waters), and rarely far from shore. The species of *Gyretes* prefer and are especially adapted for swift currents such as occur in rivers. Species of *Dineutus* are typically stream inhabitants, whereas those of *Gyrinus* are most common on lakes and ponds. However, quiet pools and eddies in streams and rivers are often comparable to still-water inhabitats, and the same species may be found in both. Species of *Gyrinus* especially are tolerant of local conditions and are to be expected on almost any puddle large enough to swim in. Many gyrinids are known to fly well, and in California species of *Gyrinus* sometimes mistake the shiny tops of automobiles for water surfaces and rain down upon them at full speed.

The beetles tend to congregate in large swarms or schools, within which the individuals swim slowly around or rest. In the eastern United States these schools are often composite, containing as many as five species and sometimes two genera. In the West the aggregates usually contain but one species. The schools are commonly formed near shore or emergent vegetation, and often in a shady or sheltered spot. Most species are diurnal, but in at least certain unfavorable habitats some become nocturnal. I have collected water beetles throughout the day in a tiny stream in the Tumacacori Mountains of Arizona without finding a single gyrinid. Yet when the same water was visited with a Coleman lantern after dusk, a couple of dozen *Gyrinus plicifer* LeConte were quickly netted from the school on the surface.

Species of *Dineutus,* in its various subgenera, occur in the Ethiopian, Oriental, and Australasian regions, as well as in the Fiji Islands and North and Central America. A single species belonging to the subgenus *Cyclinus* Kirby has been recorded from California and is common in Mexico. *Gyrinus* is widespread and occurs in most parts of the world except the Pacific Islands beyond New Guinea, New Caledonia, and New Zealand. At least a dozen species are known from South and Central America, and more than forty from North America. *Gyretes* is a Neotropical genus with many species in South and Central America; two species have been described from California, but one is of doubtful validity.

Taxonomic characters.—The species of gyrinids, especially of *Gyrinus,* are monotonously alike in general appearance, and in our fauna show few color differences. It is often difficult to identify females to species unless they can be associated by locality with males. The latter offer some clear external differences, but for verification the male genitalia must be extruded or dissected out. Fall (1922) has illustrated the male genitalia of most of our species of *Gyrinus,* and Roberts (1895) those of *Dineutus.* The color of the under surface is used to distinguish groups in both genera, but is difficult to apply in immature specimens. In males the front tarsi are broadened, and clothed beneath with dense pads of sucker hairs.

The chief external structural features of value in *Gyrinus* are: size and convexity of the beetle; surface sculpture of the elytra, not always alike in the two sexes; relative sizes of the serial punctures of the elytra, the extent to which they are impressed as striae, and the position of the marginal series; and the form of the anterior margin of the metasternum, of the outer apical angles of the elytra, and of the basal segment of the front tarsi of the male. In *Dineutus* the form of the front tibiae, ciliation and dentition of the front femora, shape of the elytral apices, and size of the insect are important. In *Gyretes* size and form, width of the lateral band of vestiture, and form of the outer apical elytral angle are of most use.

Key to the Nearctic Genera of Gyrinidae

Adults

1. Dorsum glabrous; last abdominal segment rounded, sternite without a median longitudinal line of hairs; scutellum visible or not 2
— Sides of pronotum and elytra pubescent; last abdominal segment elongate, conical, sternite with a median of golden hairs; scutellum not visible (fig. 13:22*c*). ORECTOCHILINAE .... *Gyretes* Brullé 1835
2. Scutellum visible; elytral striae punctate, suture margined; smaller and more convex species, 3 to 8 mm. long (fig. 13:22*b*). GYRININAE .... *Gyrinus* (Geoffroy *in*) Muller 1764
— Scutellum not visible; elytral striae not punctate, suture not margined; larger and flatter species, 9 to 16 mm. long (fig. 13:22*a*). ENHYDRINAE .... *Dineutus* MacLeay 1825

Key to Nearctic Genera of Gyrinidae

Larvae

1. Head subcircular with collum narrow and distinct; mandible falcate without retinaculum (fig. 13:21*a*). ENHYDRINAE .... *Dineutus* MacLeay 1825
— Head elongate (fig. 13:21*b*), with collum about as wide as rest of head and not distinct; mandible with retinaculum 2
2. Nasale without teeth. ORECTOCHILINAE .... *Gyretes* Brullé 1835
— Nasale with 2 to 4 teeth in a transverse row. GYRININAE .... *Gyrinus* (Geoffroy *in*) Müller 1764

Key to the California Species of Gyrinus

Adults

1. Under surface, except legs but including hypomera and epipleura, testaceous to ferrugineous (subgenus *Gyrinus*) 2
— Under surface, except legs, entirely metallic black,

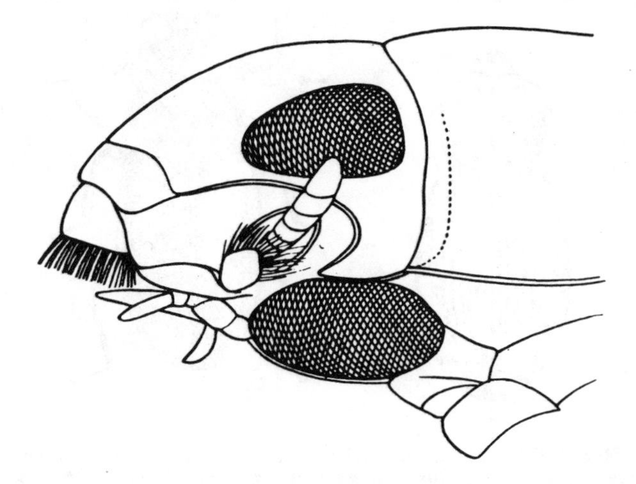

Fig. 13:23. *Gyrinus* sp., lateral view of adult head showing divided eyes (Leech, original).

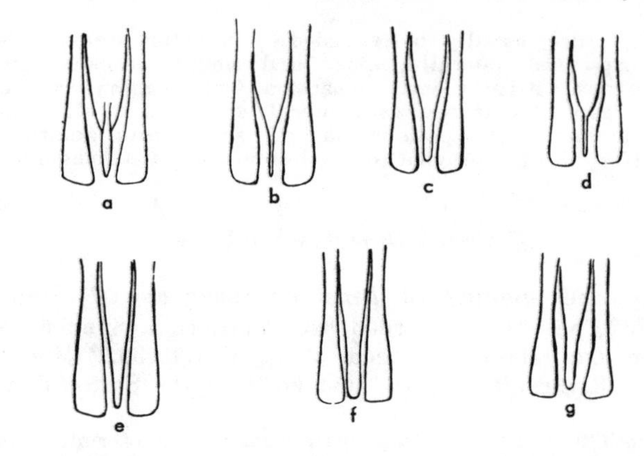

Fig. 13:24. *Gyrinus*, male genitalia. *a*, *bifarius*; *b*, *plicifer*; *c*, *consobrinus*; *d*, *affinis*; *e*, *pleuralis*; *f*, *parcus*; *g*, *picipes* (Fall, 1922).

or sides of ventral segments rarely dull rufous, epipleura often testaceous or rufous (subgenus *Oreogyrinus*) 4
2. Elytra minutely punctulate and very finely alutaceous, more noticeably so in female; median lobe of aedeagus of male very like that of *G. bifarius* Fall (fig. 13:24*a*), constricted at apical third, thence flat, narrow, nearly parallel-sided, tip rounded; British Columbia to California, New Mexico .... *punctellus* Ochs 1949
— Elytra highly polished, either not at all alutaceous or punctulate, or only visibly so under high power . 3
3. Outer apical angle of elytra with an inflated plica across angle close to margin, margin thereby depressed in this area; color beneath nearly uniform, surface without trace of microsculpture in either sex; median lobe of aedeagus of male (fig. 13:24*b*) very narrow, parallel-sided; British Columbia to Baja California, Colorado to Texas .... *plicifer* LeConte 1852
— Outer apical angle of elytra without an inflated fold across angle; under surface normally darker medially than along margins; median lobe of aedeagus (fig. 13:24*c*) much broader, at least 1/3 as wide as a lateral lobe; California to British Columbia, Utah .... *consobrinus* LeConte 1852
4. Under surface metallic black, hypomera and epipleura testaceous or ferrugineous; larger species, 5.5 to 7 mm. long 5
— Under surface entirely metallic black, or virtually so, epipleura normally showing no more than an obscure rufous tint in certain lights; smaller species, 4.25 to 6 mm. long 6
5. Elytra in both sexes thickly covered with very fine, short, oblique striolae; median lobe of aedeagus (fig. 13:24*d*) very slender in about apical third, thence gradually widened to base, carinate medially, flattened apically; northern North America, California .... *affinis* Aubé 1838
— Elytra in both sexes micropunctulate and minutely alutaceous; median lobe of aedeagus (fig. 13:24*e*) gradually decreasing in width from base to narrowly rounded tip, dorsally carinate in more than apical half; California to Colorado, British Columbia and Alberta .... *pleuralis* Fall 1922
6. Form strongly convex, nearly symmetrically arched in profile; pronotum strongly bi-impressed on each side, anterior impressions usually joined across disc, sides and lateral margin finely wrinkled; 11th elytral striae strictly marginal, lateral striae slightly canaliculate basally; elytra of males shining, of females dulled by a sculpture of rounded meshes, except near suture and across base; male genitalia similar to that of typical *parcus* (fig. 13:24*f*); San Diego, Orange counties .... *parcus californicus* Ochs 1949
— Form only moderately convex, greatest convexity

(profile) usually before middle; pronotum weakly bi-impressed laterally, sides and margin smooth; 11th elytral striae clearly separated from marginal groove except at extreme base; lateral striae not at all canaliculate; elytra polished in both sexes; male genitalia (fig. 13:24*g*); Alaska to California . *picipes* Aube 1838

Genus *Dineutus* MacLeay

Only one species of *Dineutus* (subgenus *Cyclinus* Kirby), has been recorded from California. Specimens seen are labeled "Mecca, Cal., VIII.15.34." Mecca is in Riverside County, just north of the Salton Sea.

Length 9 to 10 mm., form oval; front femora of male with a slight angulation and tiny tooth at apical three-fourths on lower anterior margin; apex of median lobe of male genitalia distinctly arched, not flattened, apices of parameres broadly rounded .... *solitarius* Aubé 1838

Genus *Gyretes* Brullé

In 1852 LeConte described *G. sinuatus* from "Ad flumen Colorado." In his redescription of 1868 he added "Abundant in the Colorado River, near Fort Yuma, California." In 1907 Régimbart described *G. californicus* on the basis of a single female labeled only "Californie"; it is not possible to distinguish this from *G. sinuatus* on the basis of the original description.

It is peculiar that no modern specimens from California have been seen.

Elongate-oval, convex, shining; pronotum and elytra with tomentose lateral band, which is narrower basally on elytra than on pronotum, broadening apically and reaching suture at apex in male but not in female (fig. 13:22*c*) .... *sinuatus* LeConte 1852

REFERENCES

DIMMOCK, G., and F. KNAB
1904. Early stages of Carabidae. Springfield (Mass.) Mus. Nat. Hist. Bull., 1:1-56, 4 text figs., pls. 1-4.
FALL, H. C.
1922. The North American species of *Gyrinus*. Trans. Amer. Ent. Soc., 47:269-306, pl. No. 16.
HATCH, M. H.
1930. Records and new species of Coleoptera from Oklahoma and western Arkansas, with subsidiary studies. Publ. Univ. Oklahoma Biol. Surv., 2:15-26, 2 text figs.
1953. The beetles of the Pacific Northwest. Part I: Introduction and Adephaga. Univ. Wash. Publ. Biol., 16:vii+1-340, incl. front. 2 text figs., 37 pls.
LEECH, H. B.
1940. *Dineutus* in California. Pan-Pac. Ent., 16:74.
OCHS, G.
1949. A revision of the Gyrinoidea of Central America. Rev. de Entomologia, 20:253-300.
ROBERTS, C. H.
1895. The species of *Dineutes* of America north of Mexico. Trans. Amer. Ent. Soc., 22:279-288, 2 pls.
WICKHAM, H. F.
1893. Description of the early stages of several North American Coleoptera. Bull. Nat. Hist., Univ. Iowa, 2:330-344, pl. IX.
1894. On some aquatic larvae, with notice of their parasites. Canad. Ent., 26:39-41.
YOUNG, F. N.
1954. The water beetles of Florida. Univ. Florida Stud., Biol. Sci. Ser., 5:ix+238 pp., 31 figs.

Family HYDRAENIDAE

The species of this family often have been included in the Hydrophilidae. All hydraenids are small, most of them between 1 and 2.5 mm. long. Many kinds cling to rocks and waterlogged wood in rapid streams, but others crawl among the pebbles and through the debris at the shore line or tunnel in the damp sand nearby. Some occur in brackish waters, and a few in the intertidal zone. They can only creep or run, since their legs are not modified for swimming. The adults are aquatic and largely vegetarian, the larvae more terrestrial than aquatic, and carnivorous.

Three genera are known from our fauna. *Ochthebius* has many species; most are beautifully sculptured

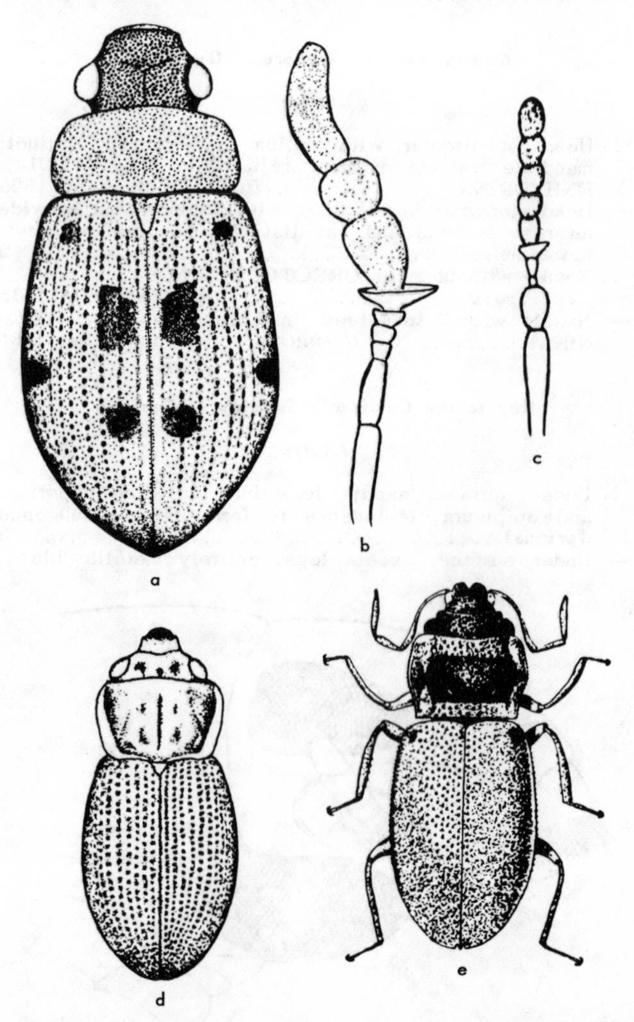

Fig. 13:25. *a*, *Berosus dolerosus*, adult; *b*, *Helochares* sp., antenna; *c*, *Ochthebius* sp., antenna; *d*, *Ochthebius interruptus*, adult; *e*, *Hydraena needhami* (*a,d*, Leech, 1948; *b,c*, Crowson, 1950; *e*, D'Orchymont, 1929).

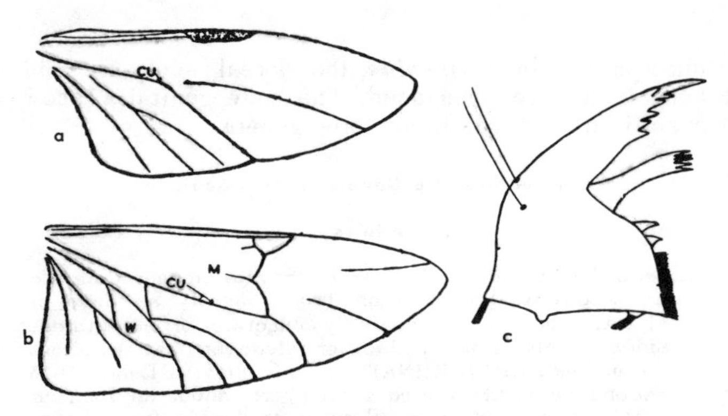

Fig. 13:26. *a*, Staphylinoid wing; *b*, Hydrophiloid wing; *c*, *Ochthebius* sp., mandible of larva (*a,b*, Crowson, 1950; *c*, Richmond, 1920).

and some have metallic colors; almost all have a thin transparent border (fig. 13:25*d*) around the pronotum, which is narrower at base than the elytra. The species of *Hydraena* have greatly elongated maxillary palpi (fig. 13:25*e*) and closely, rather coarsely punctate elytra. Species of *Limnebius* have short palpi, are black or dark brown, smooth, with very finely punctate elytra, and a pronotum as broad at base as the elytra.

Relationships.—In earlier studies the genera *Hydraena* and *Ochthebius* were often associated with *Hydrochus* and *Helophorus* in a family Helophoridae, and *Limnebius* placed in the Hydrophilidae. Later they were placed as the tribes Hydraeninae and Limnebiinae in the Hydrophilidae. In 1931 Böving and Craighead kept them together in a family Limnebiidae, but removed the family from the Hydrophiloidea to the Staphylinoidea, basing their opinion on larval characters and the wing venation of the adults.

The hydraenids are like most hydrophilids (fig. 13:25*a*) in having a transverse (clypeofrontal) suture on the head, but entirely lack the posterior median suture which gives it a Y-shaped form in the Hydrophilidae. They differ also in having six or seven rather than five visible abdominal sternites; the antennae have five pubescent segments beyond the cupule (fig. 13:25*c*) instead of only three (fig. 13:25*b*); there is no distinct bisetose empodium between the tarsal claws.

They resemble the Staphylinoidea in wing venation (the m-cu loop is greatly reduced or absent, compare figs. 13:26*a* and *b*), but Crowson (1950) has suggested that this may be an incidental effect of small body size and not of phylogenetic value. They resemble the Staphylinoidea also in having a ninth abdominal sternite and no basal piece to the aedeagus and dorsal ocelli in a few species, but are like hydrophiloids in the form of the metendosternite, absence of femoral plates on the hind coxae, elytra fully covering the abdomen, and aquatic habits.

Respiration. —The ventral surface of an adult, except the head, legs, and abdominal tip, is covered with short dense hydrofuge pubescence. This unwettable pile holds a pillowlike bubble of air over the lower surface of the beetle while it is under water. To replenish this and the subelytral supply, a beetle crawls to the surface and assumes a nearly horizontal position. It then inclines its body to one side, to bring the angle between the head and prothorax to the surface, and breaks through the surface film with its hydrofuge antennal club. This opens a funnel-shaped passage to the ventral air bubble, and by pulsating its abdominal segments the beetle is able to change the old air for fresh in a short time. Except at this time the antennae are tucked away beside the eyes, and the tactile role they play in terrestrial beetles is taken over by the elongate maxillary palpi.

Hydraenid larvae are holopneustic, that is, all nine pairs of abdominal spiracles are open and functional; hydrophilid larvae (except those of *Helophorus*) are metapneustic, having only the terminal ones open.

Life history.—The eggs are laid in the spring singly, on stones or algae in shallow water, or out of the water but in damp places. Those of *Limnebius* and *Ochthebius* are either naked, or more or less covered with loosely applied silk (fig. 13:27*b*); those of *Hydraena* are protected by a blanketlike covering of closely applied silk (fig. 13:27*a*). They hatch in seven to ten days.

The larvae (fig. 13:28) are holopneustic, as mentioned above, and swim clumsily or not at all. They walk and run quickly, and being predators, spend their time hunting in the damp sand and silt at the waters edge. They mature in about two months and form pupal cells in damp situations. Some species of *Ochthebius* make mud cocoons on the sides of stones. Winter is passed in the adult stage.

Habitat and distribution.—The larvae are adapted as land crawlers and are found at the margins of ponds, lakes, streams, and rivers, especially where filamentous algae are present. Adults of some species tunnel in the damp sand or soil at the waters edge, or make use of tunnels excavated by *Bledius* (Staphylinidae) and *Dyschirius* (Carabidae) and similar littoral forms. Adults of other species cling to stones well out in running water. Species of *Ochthebius* are commonest in and around ponds, lakes, and saline waters; those of *Hydraena* and *Limnebius* along streams and rivers.

Ochthebius bruesi Darlington was described from a hot spring (temperature 38.8°C) south of Beowawe,

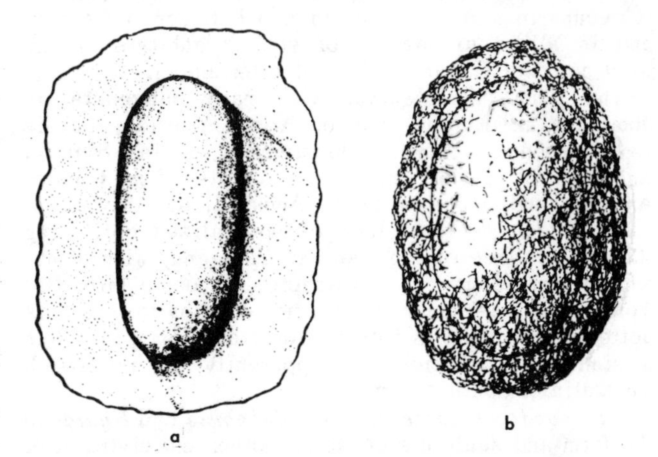

Fig. 13:27. *a*, *Hydraena pennsylvanica*, egg; *b*, *Ochthebius fossatus*, egg (Richmond, 1920).

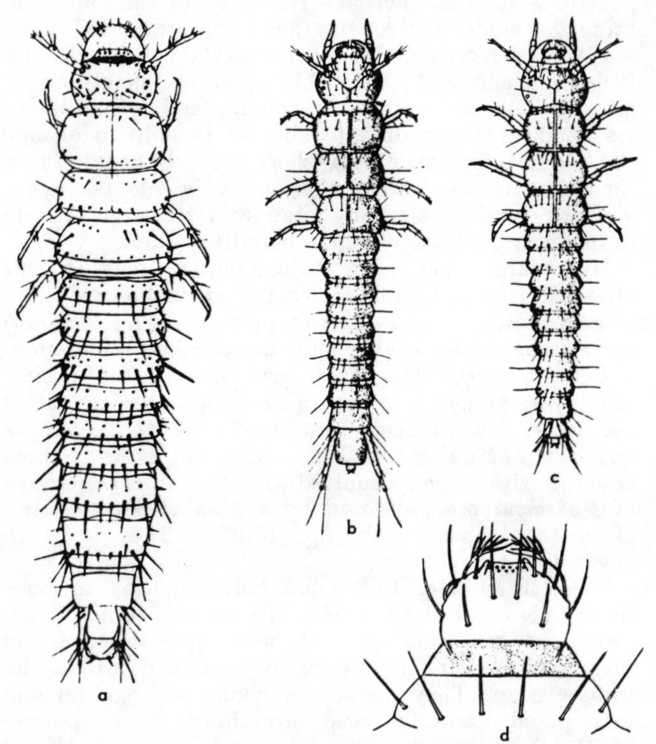

Fig. 13:28. Hydraenidae, larvae. *a, Limnebius truncatellus; b, Hydraena pensylvanica; c, Ochthebius fossatus; d, Hydraena* sp., labrum and clypeus of larva (*a*, D'Orchymont, 1913; *b-d*, Richmond, 1920).

Eureka County, Nevada, and from one at Amedee, Lassen County, California; it has since been reported from other Nevada hot springs (see La Rivers, 1950). A species allied to *O. rectus* LeConte is common in the saline pools at Bad Water, 276 feet below sea level in Death Valley, Inyo County, California, and has been recorded from other waters in the valley (Thorpe, 1931). One apterous species, *O. vandykei* Knisch, is known only from the intertidal zone of the Pacific Coast of North America, from British Columbia south at least to San Luis Obispo County, California. It occurs in narrow cracks in and between the rocks, and is allied to species of similar habitats on the west coast of Europe and the Mediterranean.

The genus *Ochthebius,* with many subgenera, is dominant in Europe and in North America, but is represented in Africa, Madagascar, India, Siberia, and Australia. One species is reported from Central America and one from South America, but the hiatus may be due more to lack of specialized collecting than to absence of the beetles. *Hydraena,* with somewhat fewer subgenera and species, is dominantly Old World and has much the same distribution, but is better represented in Central America. *Limnebius* has a similar distribution, but apparently did not reach Australia or South America.

Taxonomic characters.—In *Ochthebius* and *Hydraena* the form and sculpture of the pronotum and elytra offer readily seen differences; in the former, the shape of the labrum; in the latter the legs often show sexual dimorphism. In *Limnebius* the dorsal sculpture and body form have been used. The male genitalia show specific differences in all three genera.

Key to Nearctic Genera of Hydraenidae

Adults

1. Second segment of hind tarsi elongate, longer than the 3rd segment; pronotum as broad basally as base of elytra, smooth, not coarsely punctate or sculptured, sides evenly arcuate; black or rufescent beetles, about 1 mm. long. LIMNEBIINAE...... *Limnebius* Leach 1815
— Second segment of hind tarsi short, about as long as 3rd; pronotum at base slightly or decidedly narrower than base of elytra, surface uneven, coarsely punctate or with a transparent lateral boarder, sides sinuate or irregular; black, reddish, or aenescent beetles, 1 to 2 mm. long. HYDRAENINAE................ 2
2. Maxillary palpi very long, much longer than antennae; pronotum coarsely, closely punctate, sides without a transparent border (fig. 13:25*e*) .... *Hydraena* Kugelann 1794
— Maxillary palpi shorter than antennae; pronotum variously sculptured, often with deep fossae and grooves, almost always with a transparent border in at least basal half (fig. 13:25*d*)....... *Ochthebius* Leach 1815

Key to Nearctic Genera of Hydraenidae

Larvae

1. Setae on clypeus not placed at anterior margin and 2 median ones distant from each other; lacinia mobilis narrower (fig. 13:26*c*); inner lobe of maxillae not distinctly divided apically, cerci nearly contiguous proximally and divergent (fig. 13:28*c*) .... *Ochthebius* Leach 1815
— Setae on clypeus placed at anterior margin and equidistant; lacinia mobilis broader; inner maxillary lobe distinctly divided apically; cerci widely separated proximally and nearly parallel.................... 2
2. Third antennal segment without inner swellings, segment 2 with a single antennal appendage; a pair of pectinate setae at anterior margin of labrum (fig. 13:28*b,d*) .... *Hydraena* Kugelann 1794
— Third antennal segment with an inner swelling; 2nd with 2 slender antennal appendages; no pectinate setae at anterior labral margin; inner maxillary lobe strongly divided (fig. 13:28*a*) .... *Limnebius* Leach 1815

Key to California Species of Limnebius

Adults

1. Dorsal surface more or less polished; California .... *piceus* (Horn) 1872
— Dorsal surface more or less alutaceous 2
2. Elytral base slightly narrower than prothorax; pronotum and elytra strongly alutaceous; California .... *alutaceus* (Casey) 1886
— Elytral base equal in width to prothorax; pronotum and elytra only feebly alutaceous; California .... *congener* (Casey) 1886

Genus *Hydraena* Kugelann

H. pennsylvanica Kiesenwetter used to be credited to the Pacific Coast, but California specimens were later separated and described as *H. vandykei* d'Orchymont.

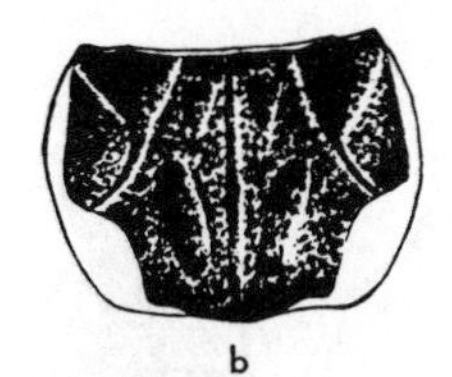
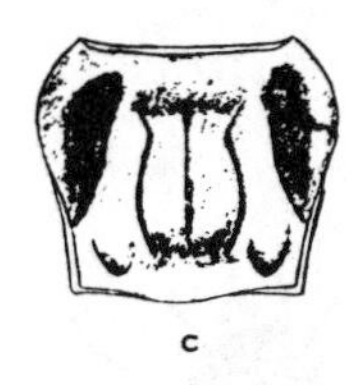
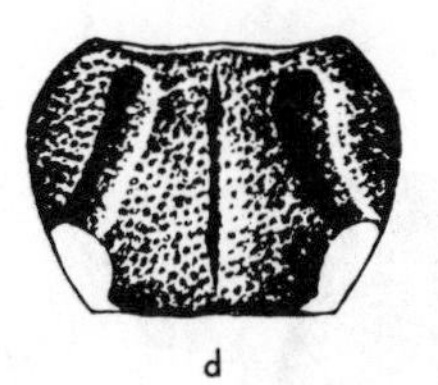
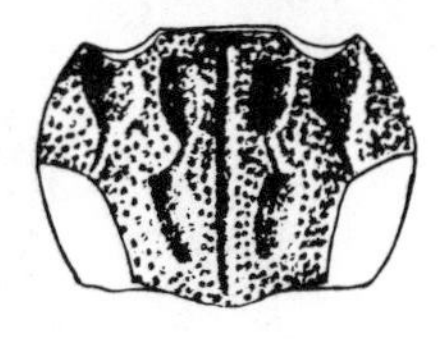

a b c d e

Fig. 13:29. Prothorax, in dorsal view, of *Ochthebius* spp. *a*, *holmbergi*; *b*, *rectus*; *c*, *lineatus*; *d*, *cribricollis*; *e*, *puncticollis* (Horn, 1890).

Pronotum appreciably hexagonal, shallowly longitudinally impressed on each side of disc, evenly and closely punctate; elytra densely punctate, punctures arranged in longitudinal series; length 1.8-2-5 mm.; California .... *vandykei* d'Orchymont 1923

Key to the California Species of Ochthebius

Adults

1. Sides of prothorax at least in part with an evident transparent border (fig. 13:29*b*,*c*) (fresh-water species) 2
— Sides of prothorax without trace of transparent border. Small, narrow, blackish species, found along seacoast in intertidal zone; British Columbia to San Luis Obispo, California; San Francisco Bay ..*vandykei* Knisch 1924
2. Elytral sides explanate, the flange as wide as a tibia; large, broad species, 2 mm. long; Alameda, Santa Clara counties .... *martini* Fall 1919
— Elytral sides not broadly explanate 3
3. Lateral prothoracic margin abruptly narrowing from near front angles, with pronounced angulation at middle; transparent border broad 4
— Lateral margin abruptly narrowed from middle, or deeply notched near hind angles (fig. 13:29*d*,*e*), or gradually sinuately narrowed to base (fig. 13:29*a*) 5
4. Pronotum anteriorly with a discal and a lateral impression on each side of median longitudinal impression, and 1 discal depression on each side of it near base; elytra with well-marked rows of punctures, striate; Yuma, Imperial counties ...... *fossatus* LeConte 1855
— Pronotum with median longitudinal impression, but without discal posterior pair and with only traces of anterior pairs; elytra polished, virtually impunctate Kern, Riverside counties ....*laevipennis* LeConte 1878
5. Prothorax strongly narrowed or excavated behind middle; transparent border broad 6
— Prothorax gradually sinuately narrowed, not excavated, at most a slight projection at posterior one-third; transparent border narrow (fig. 13:29*c*) 10
6. Sides of prothorax nearly parallel anteriorly, transparent border broadly attaining front angles (fig. 13:29*b*); western U.S. .... *rectus* LeConte 1878
— Sides of prothorax distinctly arcuate anteriorly, transparent border narrowly attaining front angles, or beginning at sinuation (fig. 13:29*d*) 7
7. Elytral interspaces very narrow, strongly costate throughout; tiny species, 1–1.2 mm. long, with deeply striate and coarsely punctate elytra; Ventura, Santa Barbara counties .... *costipennis* Fall 1901
— Elytral interspaces flat or slightly convex 8
8. Pronotum with discal foveae on each side of median impression 9
— Pronotum without discal foveae, though coarsely punctured; sides deeply emarginate near hind angles (fig. 13:29*d*); northern U.S.; Santa Barbara County ....*cribricollis* LeConte 1850
9. Each anterior lateral discal fovea of pronotum united to corresponding posterior fovea by a deep groove; pronotum coarsely punctured (fig. 13:29*e*); southwest U.S. .... *puncticollis* LeConte 1852
— Anterior lateral foveae clearly separated from posterior foveae; sides of pronotum abruptly deeply notched at middle, transparent border narrowly attaining front angles; western U.S. .... *discretus* LeConte 1878
10. Median line of pronotum distinct, 2/3 length of prothorax 11
— Median line short or absent 12
11. Lateral thoracic foveae united into a sinuate line on each side of median impression (fig. 13:29*c*); southwest U.S...................... *lineatus* LeConte 1852
— Lateral foveae not united (fig. 13:25*d*); western U.S. .... *interruptus* LeConte 1852
12. Lateral posterior and anterior discal prothoracic foveae deep, united; western U.S. .... *sculptus* LeConte 1878
— Anterior foveae indistinct, posterior foveae broad, shallow (fig. 13:29*a*); Pacific Coast .... *holmbergi* Mannerheim 1853

REFERENCES

BÖVING, A. G., and F. C. CRAIGHEAD
1931. An illustrated synopsis of the principal larval forms of the Order Coleoptera. Ent. Amer., n.s. 11: 1-351, incl. 125 pls.
CASEY, T. L.
1886. Descriptive notices of Northern American Coleoptera. I. Bull. Calif. Acad. Sci., 2:157-264, pl. no. 7 (*Limnocharis (Limnebius)*, pp. 167-171.)
1900. Review of the American Corylophidae, Cryptophagidae, Tritomidae and Dermestidae, with other studies. Jour. N.Y. Ent. Soc., 8:51-172, 4 text figs. (*Limnebius*, pp. 51-53).
CROWSON, R. A.
1950. The classification of the families of British Coleoptera. Ent. Mon. Mag., 86(1032):149-160, (1033): 161-171, 22 text figs.
HORN, G. H.
1890. Notes on the species of *Ochthebius* of Boreal America. Trans. Amer. Ent. Soc., 17:17-26, 1 pl.
HRBÁČEK, JAROSLAV.
1950. On the morphology and function of the antennae of the central European Hydrophilidae (Coleoptera). Trans. Roy. Ent. Soc. London, 101:239-256, 17 text figs. (Note that Hrbáček uses the term Hydrophilidae in the broad sense, to include the Hydraenidae and Spercheidae. Also that his use of the names *Hydrous* and *Hydrophilus* are equivalent to the *Hydrophilus* and *Hydrochara*, respectively, of the present chapter.)
LA RIVERS, I.
1950. The staphylinoid and dascilloid aquatic Coleoptera of the Nevada area. Great Basin Naturalist, 10:66-70.
1951. Erratum page (for above article). Great Basin Naturalist, 11:52.
LEECH, H. B.
1948. Haliplidae, Dytiscidae, Gyrinidae, Hydrophilidae, Limnebiidae. *In* Contributions toward a knowledge of the insect fauna of Lower California. No. 11, Coleoptera. Proc. Calif. Acad. Sci. (4), 24:375-484, pls. 20-21.
D'ORCHYMONT, A.
1913. Contribution à l'étude des larves hydrophilides.

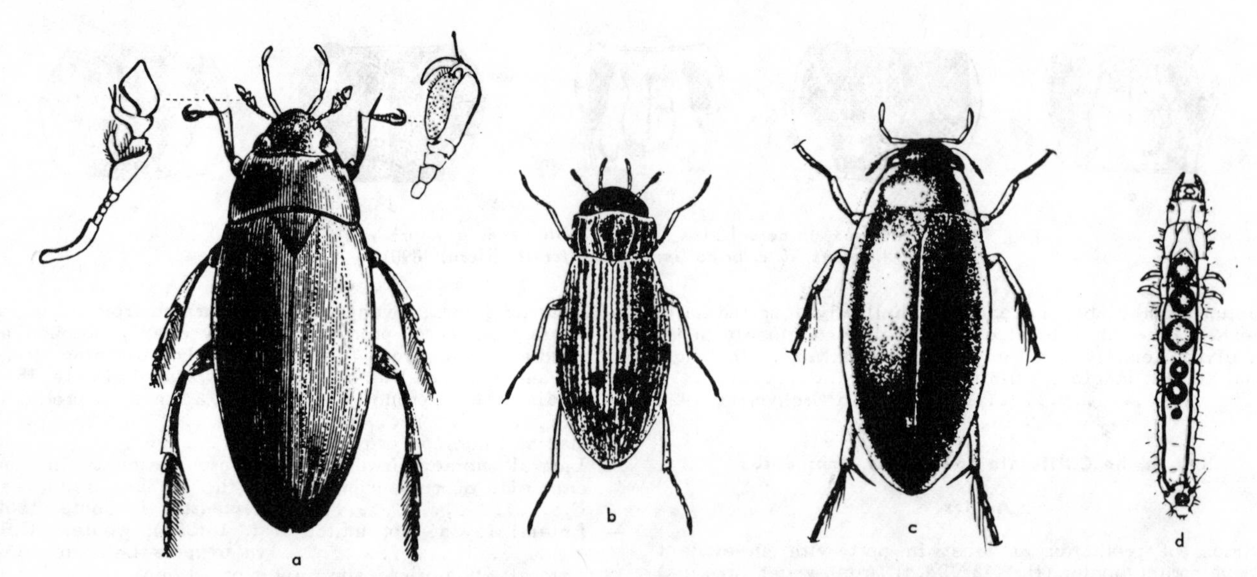

Fig. 13:30. Hydrophilidae. *a*, *Hydrophilus triangularis*, adult; *b*, *Helophorus minutus*, adult; *c*, *Tropisternus lateralis*, adult; *d*, *Enochrus* sp., young larva containing bubbles of air it has swallowed (*a*, Riley, 1881; *b*, Wesenberg-Lund, 1943; *c*, Blatchley, 1910; *d*, Williams, 1936).

Annales de Biologie lacustre, 6:173-214, 23 text figs.
1929. Contribution à l'étude des Palpicornia. VII. Bull. et Ann. Soc. Ent. Belg., 69:79-96, 1 pl.

RICHMOND, E. A.
1920. Studies on the biology of aquatic Hydrophilidae. Bull. Amer. Mus. Nat. Hist., 42:1-94, pls. 1-16 incl.

THORPE, W. H.
1931. Miscellaneous records of insects inhabiting the saline waters of the California desert regions. Pan-Pac. Ent., 7:145-153.

Family HYDROPHILIDAE

Water Scavenger Beetles

The Hydrophilidae and Dytiscidae comprise a majority of the species of water beetles. Since they occur together in the same ponds and streams, we tend to overlook the fact that they belong to different suborders of the Coleoptera, the Dytiscidae to the Adephaga and the Hydrophilidae to the Polyphaga. The hydrophilids are divided into two groups, one aquatic, the other terrestrial. The aquatic species in general differ from the Dytiscidae in being flatter beneath, more convex above, darker in color, and poorer swimmers. The predaceous dytiscids move their hind legs in unison when swimming and obtain a powerful oarlike stroke, but the hydrophilids move theirs alternately after the manner of a walking insect. This is perhaps correlated with food habits, for the largely vegetarian hydrophilids do not have to be such fast swimmers as the predaceous dytiscids. There is also a major difference in their manners of obtaining fresh air at the surface of the water. The Hydrophilidae, which have their largest and most used spiracles on the thorax, break the surface film with their antennae, adjacent to a front corner of the prothorax; the Dytiscidae, with major spiracles on the abdomen, come up tail first, as do the larvae of both families.

The Hydrophilidae belong to a group of beetles called the Clavicornia, because of the club-shaped antennae. The name Palpicornia has also been used, having reference to the palpi as tactile organs when the beetles are under water. A vernacular name, "silver beetles," describes the appearance of their undersurface as they swim, for it is covered by a silvery film of air, held by a pile of fine hairs.

Adults of both aquatic and terrestrial species are largely herbivorous, but will eat dead animal tissue, and a few are predaceous. The larvae of both groups are carnivorous and cannibalistic, except perhaps those of the genus *Berosus,* which reportedly feed only on green algae. (Böving doubts this, because of the mandibular structure.)

The beetles and their larvae are an important source of food for certain ducks and other aquatic birds, as well as for shallow-water fish. The adults are eaten by frogs and toads, and during flight periods are presumably caught by birds and bats, and perhaps by predaceous flies.

Relationships.—The Hydrophilidae (in the broad sense) do not show close relationship with any superfamily of the suborder Polyphaga except the Staphylinoidea, with which they seem to be connected by means of the Spercheidae and Hydraenidae. They differ from the staphylinoid forms in having cantharoid wing venation (the m-cu loop distinct; see fig. 13:26*a* and *b*), short antennae with the true sixth segment modified to form a cupule (fig. 13:25*b*), maxillary palpi which are usually longer than the antennae (fig. 13:36*a*), head usually with a Y-shaped suture, elytra covering the abdomen, and (except Sphaeridiinae) aquatic habits. The genus *Hydrochus* here retained in the Hydrophilidae, forms the family Hydrochidae of some classifications.

Respiration.—Adults come to the surface head first, turn slightly sideways, and break the surface film

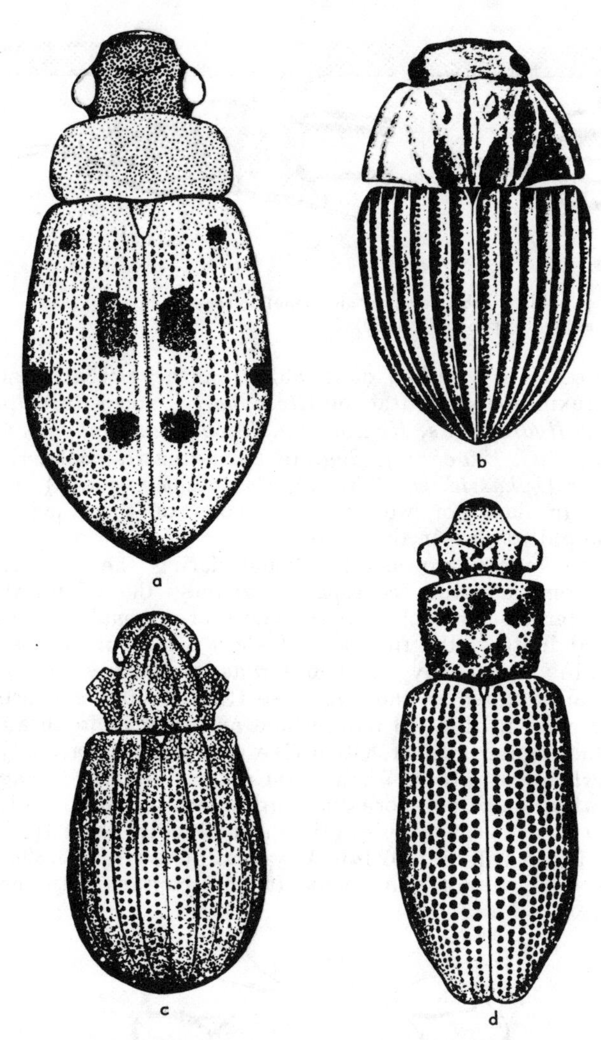

Fig. 13:31. Hydrophilidae, adults. *a, Berosus dolerosus; b, Pemelus costatus; c, Epimetopus thermarum; d, Hydrochus variolatus* (*a,c,d,* Leech, 1948; *b,* Horn, 1890).

with the unwettable hairy club of an antenna. This provides a funnellike point of contact with the extensive bubble held on the undersurface by the hydrofuge hair, the bubble in turn being connected with the subelytral storage chamber. A pumping action of the abdominal segments forces out the old supply and brings in fresh air.

As has been shown by Hŕbăcek (1950), the manner in which contact is made with the air varies among different genera. In all but *Hydrochus* the antenna is raised (fig. 13:33*a*), with the unwettable part of the club in contact with the film of air on the hydrofuge pubescence behind the eye. As the top of the bent antenna reaches the surface (fig. 13:33*b*), the surface film breaks to form a funnel (fig. 13:33*c*), and the antenna is moved to another position. In *Hydrochus* the antenna is held in front of the eye as the beetle surfaces (fig. 13:34*c*).

The species of all genera Hŕbăcek studied, except *Helophorus,* normally turn a little sideways as they reach the surface, and break the film with one antenna. The species of *Helophorus* rise so that the whole back of the head is near the surface, then push up both antennae to form a broad funnel across the back of the head (fig. 13:34*a*).

The length of the funnel is important, for at and above a critical distance of about 1.5 mm., only an air canal, comprising the unwettable antennal segments themselves, is formed. But at less than 1.5 mm. the surface tension is stronger than the hydrostatic pressure, and a real funnel is possible. The length of the unwettable part of the antennal club is also significant. Beetles with long and specially formed club segments, such as *Hydrophilus* (the *Hydrous* of Hŕbăcek's paper) are able to reach fresh air while the body is relatively deeply submerged (fig. 13:34*b*). Some strong swimmers, such as species of *Berosus,* are able to rise forcefully enough to obtain air even when their antennae have been removed. They break the surface film with the unwettable sides of the head, especially the genae, to form a funnel.

The coprophagous Sphaeridiinae, which live in damp or wet places but are rarely aquatic, can break the surface film with the antennae, but not so efficiently as do the truly aquatic forms. In fact, Hŕbăcek found that such coprophagous Scarabaeidae as *Aphodius* and *Onthophagus* readily used their antennae to make air funnels, though the air film on the body could not last more than twelve to twenty-four hours because of the shape of the body and lack of hydrofuge pubescence.

Hydrophilid larvae (except those of *Helophorus* and *Berosus*) are metapneustic, that is, only the posterior pair of spiracles is open and functional; the tracheae leading to these spiracles may be enlarged to form air reservoirs. Larvae of *Helophorus* spp. have nine

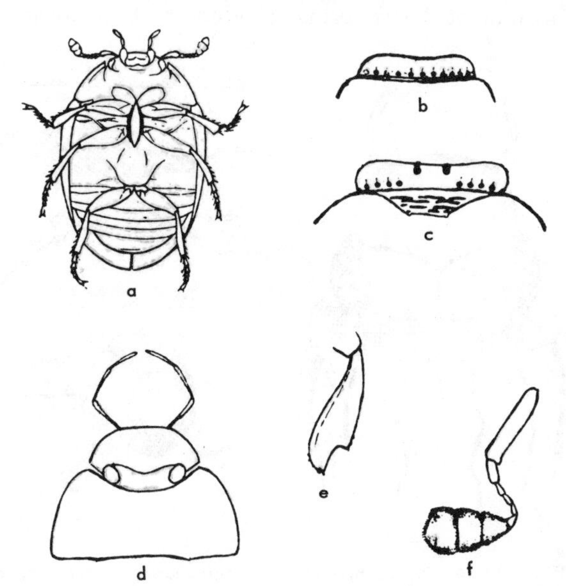

Fig. 13:32. Hydrophilidae. *a, Cercyon* sp., ventral surface of adult; *b, Hydrochara* sp., clypeus and labrum of adult; *c, Neohydrophilus* sp., clypeus and labrum of adult; *d, Helobata* sp., head and prothorax of adult; *e, Megasternum* sp., front tibia of adult; *f, Cercyon* sp., antenna of adult (*a,* Mulsant, 1844; *b,c,* D'Orchymont, 1912; *d-f,* Horn, 1890).

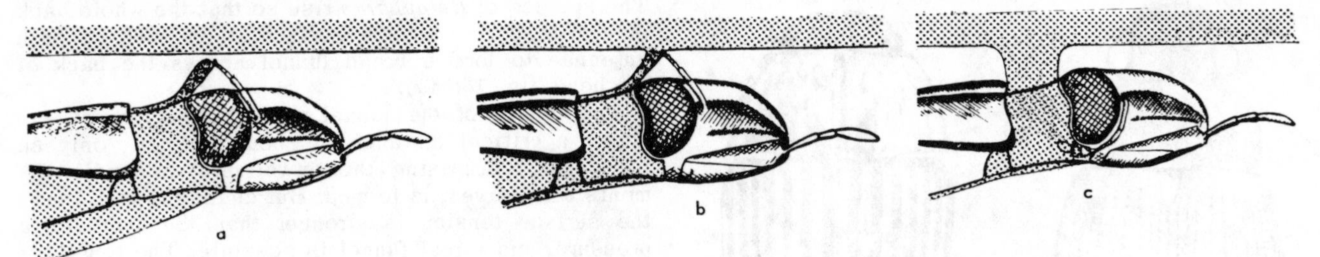

Fig. 13:33. *Helophorus aquaticus*, head and prothorax of adult extablishing an air funnel to renew its air supply; dotted portion represents air (Hrbácek, 1950).

pairs of well-developed biforous spiracles. The larvae of *Berosus* (fig. 13:37*a*) have abdominal gills containing tracheoles, hence need not come to the surface to breathe, and are able to live in deeper water, farther from shore, than are other hydrophilid larvae.

Life history.—Adults of many species of hydrophilids fly well, and like the Dytiscidae have spring dispersal flights, usually in April and May. The life cycle is completed during the summer, and there is commonly a dispersal of the newly emerged adults in August and September. At such times, especially in the spring, the flying beetles may be attracted in thousands to electric and other lights.

The eggs are usually laid in the spring, but some species oviposit throughout the summer. Eggs may be deposited singly (*Hydrochus* spp., some species of *Berosus*), or collectively. The egg mass is enclosed in a nearly transparent bag-shaped case and carried by the female beneath her abdomen in the genera *Helochares* (fig. 13:35*a*), *Epimetopus*, and *Spercheus* of the exotic allied family Spercheidae. In other known examples the eggs are either embedded in a loose web (*Sphaeridium, Cercyon, Cymbiodyta, Paracymus*) or a blanketlike covering (*Coelostoma*), or completely enclosed in a silken case which has a vertical mast or flexible ribbon attached (*Helophorus*, some *Berosus* spp., *Hydrophilus, Hydrochara, Laccobius, Anacaena, Hydrobius, Enochrus*). Eggs of the terrestrial Sphaeridiinae (*Sphaeridium, Cercyon, Coelostoma*, above) are laid in damp or wet places; those of the aquatic hydrophilids are in the water.

The larvae are usually found during the summer but some species are reported to pass the winter in both larval and adult stages. There are normally three larval instars, the first two of short duration. In most species the newly hatched larvae swallow some of the air trapped in the egg case (fig. 13:30*d*) and are thus able to float to the surface and take in fresh air in the regular manner. Otherwise they must laboriously crawl up, or drown. When mature, most aquatic larvae crawl up onto the shore and burrow into the damp soil, where some make well-defined pupal cells (see figs. 14, 15, in Wilson, 1923*a*). A typical pupa (fig. 13:39*a*) has long spines which keep it from contact with the

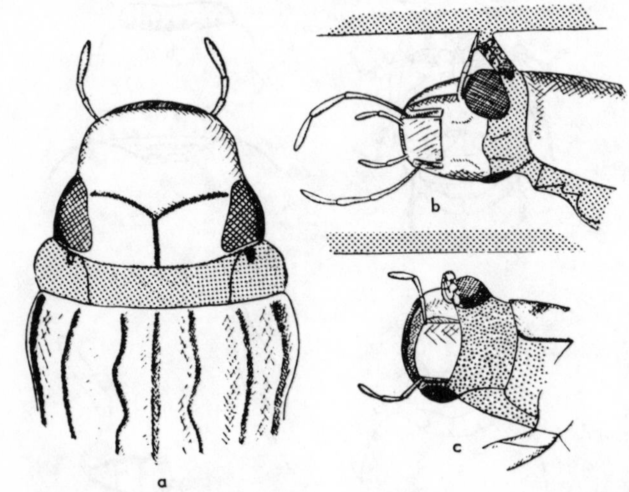

Fig. 13:34. Hydrophilidae. *a*, *Helophorus aquaticus*, adult taking in fresh air at the water surface; dotted area indicates the region of the air funnel; *b*, *Hydrophilus aterrimus*, adult obtaining air with body deeply submerged; dotted area indicates air film; *c*, *Hydrochus elongatus*, head and prothorax of an adult approaching the surface of the water to replenish its air supply; dotted area indicates air film on beetle (Hrbácek, 1950).

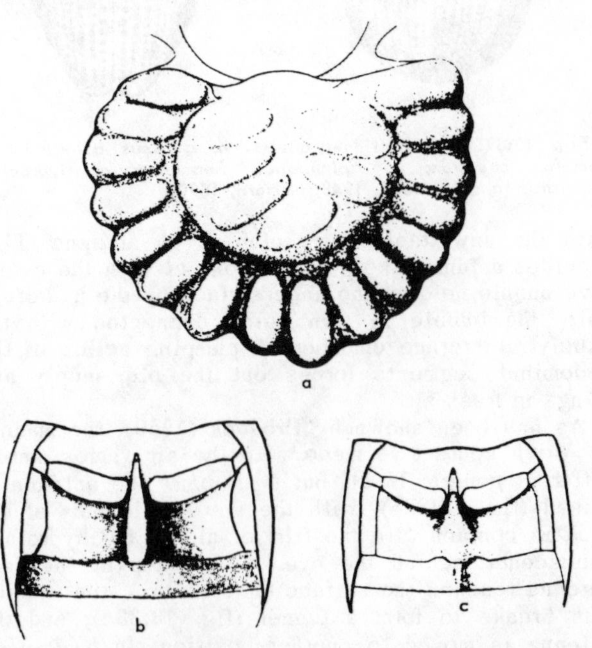

Fig. 13:35. Hydrophilidae. *a*, *Helochares maculicollis*, ventral view of egg case, detached from abdomen of female; *b*, *Phaenonotum* sp., undersurface of adult to show lack of carina on first abdominal sternite; *c*, *Dactylosternum* sp., undersurface of adult to show carina on first abdominal sternite (*a*, Richmond, 1920; *b*,*c*, Horn, 1890).

damp earth, and posterior cerci which enable it to move around. Since there is no covering of hydrofuge hair, the thoracic and abdominal spiracles cannot be kept dry, and the pupa is readily drowned if water enters its cell. Mature larvae of some species of *Enochrus* are said to form pupal cases in the water by tying algae together.

Parasites.—Several species of nematode worms have been recorded as parasites of Old World hydrophilids (adults). Todd (1942) described a new genus and species of nematode, *Zonothrix tropisterna,* from *Tropisternus lateralis nimbatus* (Say) from Nebraska; in 1944 he described two new species, *Pseudonymus brachycercus* and *P. leptocercus* from *Hydrophilus triangularis* (Say) from Nebraska and Louisiana; all worms were in the large intestine. Wilson (1923) records hymenopterous larvae from a *H. triangularis* larva in its pupal cell.

Habitat and distribution.—The larvae of most species are aquatic, or live in wet sand or soil at the waters edge; those of the Sphaeridiinae are in dung, wet rotting vegetable matter (especially cacti), or under dead leaves in damp places near the water. Of the aquatic larvae, all but *Berosus* spp. are found in shallow water, usually among algae and weeds, or in the marginal debris.

None of the New World species has been reported as an agricultural pest, but in England two kinds of *Helophorus,* both larvae and adults, attack turnips, cabbages, swedes, and other cruciferous crops; in Russia a species of *Helophorus* is said to damage grasses. At least three species of predaceous *Dactylosternum* have been used in attempts to control the sugar cane beetle borer in the Philippine Islands and Hawaii, and one against the banana borer in Jamaica.

Adult hydrophilids are commonest in marshy places, in weedy, shallow ponds, or in flooded areas where there is grass growing out of the water. Some species prefer saline or mineralized waters; others occur in running water, especially where algae grow.

The family Hydrophilidae is comparable to the Dytiscidae in number of species, but has been less studied. The dytiscids are more numerous in the colder parts of the world, the hydrophilids in the warmer; this applies especially to the terrestrial Sphaeridiinae. All the subfamilies and a majority of the tribes are represented in the North American and even in the California fauna.

Taxonomic characters.—Body form is usually characteristic, and except the smallest, most North American species can be placed to genus at sight. Color is of specific value in a few genera, and is correlated somewhat with generic recognition (that is, metallic colors are rare except in *Helophorus, Hydrochus,* and *Berosus*), but most hydrophilids are so uniformly unicolorous that a color pattern is rarely available. The punctation and/or striation of the elytra, punctation of the head and pronotum, forms of the elytral apices *(Berosus),* form and punctation of the pro-, meso-, and metasternum (Sphaeridiinae, Hydrophilinae, Hydrobiinae), and extent of hydrofuge pubescence on the ventral surface (Hydrophilinae, Hydrobiinae) are the chief characters of value.

The sexes are often difficult or impossible to separate on external features, but males of some species are distinguished by enlarged or peculiarly shaped front and middle tarsal claws, curved or otherwise modified tibiae, cristate abdominal sternites, and differently punctate mesosternal areas.

Key to the Genera of Hydrophilidae of the United States and Canada

Adults

1. Pronotum with 5 longitudinal grooves (fig. 13:34*a*), or produced anteriorly at middle so as to hide much of head (fig. 13:31*c*) 2
— Pronotum not with 5 longitudinal grooves, not produced anteriorly to hide much of head 3
2. Pronotum with 5 longitudinal grooves; antennae short, 9-segmented, not compact; eyes not divided by a canthus; form more or less elongate, not very convex, elytra not projecting below abdomen. HELOPHORINAE (fig. 13:30*b*) .... *Helophorus* Fabricius 1775
— Pronotum without longitudinal grooves, but produced anteriorly at middle, covering much of head; antennae 9-segmented, club pubescent, compact; eyes divided horizontally, partly or completely, by a canthus; form short and convex, elytra projecting much below abdomen. EPIMETOPINAE (fig. 13:31*c*).......... *Epimetopus* Lacordaire 1854
3. Pronotum conspicuously narrower than elytral bases; scutellum very small, eyes protuberant; maxillary palpi never very long; antennae with no more than 3 segments *before* the cupule. HYDROCHINAE (fig. 13:31*d*) .... *Hydrochus* Leach 1817
— Pronotum not appreciably narrower than elytral bases (except in some Berosini, but there note elongate triangular scutellum); eyes prominent or not; antennae usually with 5 well-developed segments before the cupule 4
4. Basal segment (which may be very small) of hind tarsi shorter than 2nd; antennae shorter, about as long as or shorter than maxillary palpi (fig. 13:36*a*); last glabrous antennal segment asymmetrical, often cuplike, embracing 1st segment of pubescent always triarticulate club; 2nd segment of maxillary palpi not or little thicker than 3rd or 4th; aquatic species 5
— Basal segment of hind tarsi longer than 2nd segment; antennae usually longer than maxillary palpi which

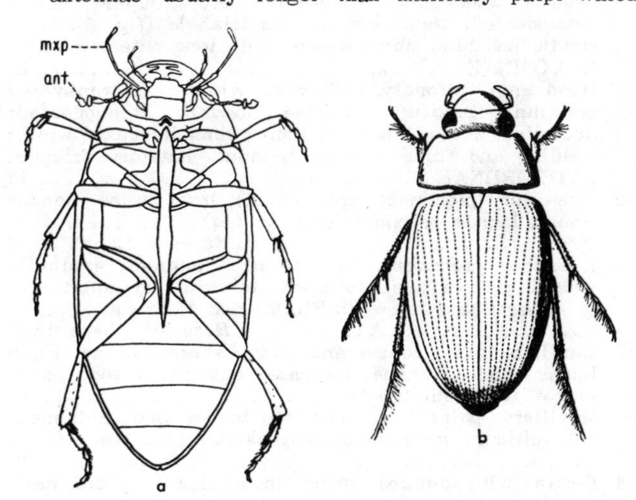

Fig. 13:36. Hydrophilidae, adults. *a, Hydrophilus triangularis; b, Hydrobius fuscipes* (*a*, Borror and Delong, 1954; *b*, Miall, 1903).

are never very long; last glabrous antennal segment obconic, fitted more or less tightly against 1st segment of pubescent club, which may be loose or compact (fig. 13:32*f*); 2nd segment of maxillary palpi (1st is very small) much thicker than 3rd or 4th; species terrestial or in damp places. SPHAERIDIINAE 24

5. Meso- and metasterna with a continuous median longitudinal keel which is prolonged posteriorly into a spine between hind coxae (fig. 13:36*a*). HYDROPHILINAE 6
— Meso- and metasterna without a continuous common keel 10
6. Prosternum sulcate to receive anterior part of mesosternal keel; metasternal keel projecting beyond hind trochanters as a spine (fig. 13:36*a*) 7
— Prosternum carinate, not sulcate; metasternal keel not or hardly reaching beyond bases of hind trochanters 9
7. Larger species, 30 to 45 mm. long 8
— Smaller species, 6 to 15 mm. long (fig. 13:30*c*) .... *Tropisternus* Solier 1834
8. Prosternum bifurcate at middle, so that anterior tip of mesosternal keel could contact head .... *Dibolocelus* Bedel 1892
— Prosternum sulcate at middle behind to receive mesosternal keel, but closed anteriorly, hooded (fig. 13:30*a*) .... *Hydrophilus* Geoffroy 1762
9. Front margin of clypeus simply truncate or arcuate; labrum lacking 2 isolated pores at middle, but with continuous row of setigerous punctures along posterior margin (fig. 13:32*b*); antennal club compact, almost symmetrical .... *Hydrochara* Berthold 1827
— Front margin of clypeus arcuate, and emarginate at middle to expose a preclypeus; labrum with 2 small isolated pores at middle, posterior margin with setigerous punctures at each side only (fig. 13:32*c*); antennae perfoliate, of very asymmetric segments .... *Neohydrophilus* d'Orchymont 1928
10. First 2 abdominal sternites with a common excavation, large and spectacle-shaped, which is normally filled with a hyaline transversely bilobed mass, the latter supported by a fringe of long stiff golden hairs arising from anterior margin of 1st sternite; small beetles (1-2.5 mm. long) with ability to roll up partly ...... CHAETARTHRIINAE .... *Chaetarthria* Stephens 1833
— First 2 abdominal sternites without a broad excavation in common; 1st sternite without a fringe of long stiff golden hairs projecting back from anterior margin ... 11
11. Head markedly deflexed, often with deep transverse groove delimiting a postoccipital region; antennae usually with only 3 segments before cupule, hence 7-segmented; scutellum a long triangle (fig. 13:31*a*); middle and hind tibiae fringed with long natatory hairs. BEROSINAE 12
— Head not strongly deflexed, without a transverse occipital groove; antennae normally 9-segmented; scutellum not or not much longer than its basal width; middle and hind tibiae without natatory fringes. HYDROBIINAE 13
12. Eyes very prominent (protuberant); front tibiae slender linear; labrum prominent (fig. 13:31*a*) .... *Berosus* Leach 1817
— Eyes not prominent; front tibiae triangular, gradually widening from base to apex; labrum very short, not prominent; New Jersey to Florida and Alabama..... .... *Derallus* Sharp 1882
13. Maxillary palpi robust and short, shorter or not much longer than antennae, ultimate segment as long as or longer than penultimate 14
— Maxillary palpi more slender, longer than antennae, with ultimate segment usually shorter than penultimate 20
14. Elytra with sutural striae in at least apical half; usually only 5 abdominal sternites visible; hind tibiae not arcuate; hind trochanters normal, closely applied to femora.. 15
— Elytra with rows of punctures but no sutural striae; 5th abdominal sternite truncate or emarginate, usually exposing 6th; hind tibiae arcuate; hind trochanters large, about 1/3 as long as femora, their apices distinct from femora .... *Laccobius* Erichson 1837
15. Larger species, at least 4.5 mm. long; elytra striate or with pronounced rows of punctures 16
— Smaller species, not more than 3 mm. long; elytra impunctate or confusedly punctate, never striate, at most with punctures subserially arranged 18
16. Segments 2 to 5 of middle and hind tarsi with a fringe of long, fine swimming hairs, which arise from a series of punctures or a narrow groove along upper inner edge of tarsi (hairs often closely stuck along groove in dry or greasy specimens); lateral margins of elytra even .... *Hydrobius* Leach 1815
— Middle and hind tarsi completely without groove and fringe of long swimming hairs along upper inner edge; lateral margins of elytra weakly serrate at least basally 17
17. Form strongly convex, almost hemispherical in profile; clypeus more deeply emarginate, median part nearly truncate; occurs in eastern United States and Canada .... *Sperchopsis* LeConte 1862
— Form oval; clypeus evenly, shallowly, arcuately emarginate; occurs in western United States and Canada .... *Ametor* Semenov 1900
18. Prosternum longitudinally carinate at middle; mesosternum usually with longitudinal median carina behind its anterior transverse Λ-shaped protuberance; 1st abdominal sternite sometimes with fine longitudinal carina at middle............ *Paracymus* Thomson 1867
— Prosternum not carinate; mesosternum simple, or with transverse Λ-shaped or arcuate carina or protuberance, but no median longitudinal carina behind it; 1st abdominal sternite never carinate at middle 19
19. Mesosternum simple, noncarinate, or with a small transverse protuberance before middle coxae .... *Crenitis* Bedel 1881
— Mesosternum with a prominent angularly elevated or dentiform protuberance before middle coxae .... *Anacaena* Thomson 1859
20. All tarsi 5-segmented, though basal segment may be very small and difficult to see 21
— Middle and hind tarsi 4-segmented 22
21. Curved pseudobasal segment (basal segment is very small) of maxillary palpi with convexity to front; mesosternum with projecting longitudinal lamina; elytra confusedly punctate .... *Enochrus* Thomson 1859
— Curved pseudobasal segment of maxillary palpi with convexity to rear; mesosternum at most feebly protuberant .... *Helochares* Mulsant 1844
22. Anterior coxal cavities closed behind; labrum concealed beneath projecting nonemarginate clypeus, which extends around to about middle of each eye, and outward for a distance equal to about width of an eye (fig. 13:32*d*) .... *Helobata* Bergroth 1888
— Anterior coxal cavities open behind; labrum fully exposed, clypeus truncate or emarginate, not extending laterally in front of eyes 23
23. Maxillary palpi long and slender, pseudobasal segment (= 1st long one) at least 2/3 as long as a front tibia; tarsal claws with broad basal tooth in male, less prominently toothed in female ...*Helocombus* Horn 1890
— Maxillary palpi shorter, stouter, first 2 long segments together subequal in length to a front tibia; tarsal claws simple in both sexes .... *Cymbiodyta* Bedel 1881
24. Head not narrowed just in front of eyes, but with expanded margin reaching almost to outer edge of eyes, and hiding bases of antennae (fig. 13:41*a*). SPHAERIDIINI 25
— Head narrowed in front of eyes, outer margin forming an angle with latter; antennal bases visible from above, not concealed under margin of head (fig. 13:41*b*) 27

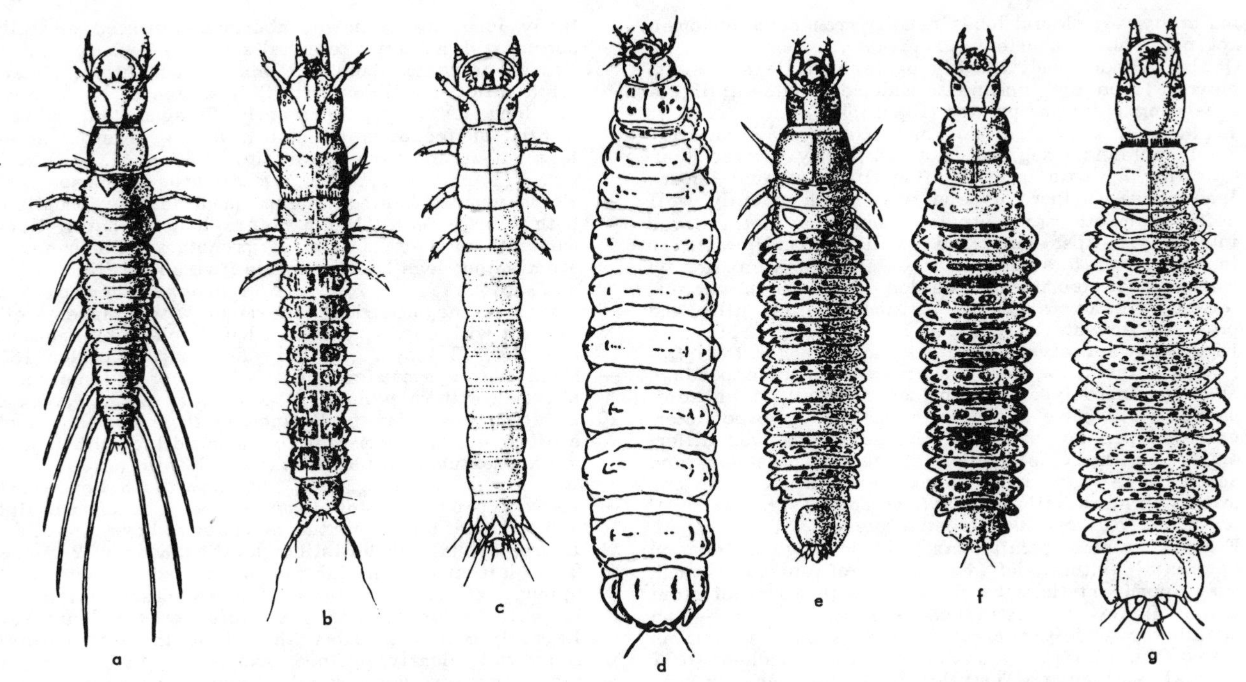

Fig. 13:37. Hydrophilidae, larvae. *a, Berosus peregrinus; b, Helophorus* sp.; *c, Hydrochus* sp.; *d, Cercyon* sp.; *e, Laccobius* sp.; *f, Paracymus* sp.; *g, Anacaena* sp. (*a-c, e-g,* Richmond, 1920; *d,* Böving and Henriksen, 1938).

25. Apex of abdomen not covered by elytra; scutellum an elongate triangle; elytra with sutural striae; 1st abdominal sternite not longitudinally carinate at middle; antennae 8-segmented ..... *Sphaeridium* Fabricius 1775
— Elytra covering apex of abdomen; scutellum triangular, but short; elytra with sutural striae or not; 1st abdominal sternite carinate or not; antennae 9-segmented 26
26. First abdominal sternite longitudinally carinate at middle (fig. 13:35*c*); sutural striae distinct .... *Dactylosternum* Wollaston 1854
— First abdominal sternite not carinate at middle (fig. 13:35*b*); sutural striae absent .... *Phaenonotum* Sharp 1882
27. Elytra with well-formed epipleurae in basal half or more; prosternum carinate at middle, or rarely elevated, flat or tectiform (fig. 13:32*a*). CERCYONINI 28
— Elytra inflexed tightly against sides of body, without distinct epipleurae except at extreme base; prosternal process always elevated at middle to plane of mesosternal protuberance as a flat plate, not carinate. MEGASTERNINI 31
28. Mesosternal elevation appearing to be a continuation of mid-metasternum, the 2 in broad contact; elytra never with strongly costate intervals 29
— Mesosternal elevation discrete, contacting metasternum at a single narrow point or separated by a hiatus 30
29. Mesosternal elevation a broad, flat or slightly concave plate, pentagonal, the anterior end pointed .... *Pelosoma* Mulsant 1844
— Mesosternal elevation narrow, almost carinate along middle, sides steep, anteriorly arrowheadlike .... *Omicrus* Sharp 1879
30. Middle of prosternum differentiated from sides, forming a plate like a low roof, with fine but not acute median longitudinal carina .... *Oosternum* Sharp 1882
— Middle of prosternum carinate but not differentiated from sides as a roof-shaped platelike area. .... *Cercyon* Leach 1817
31. Lateral margins of pronotum nearly evenly arcuate as usual; front tibiae entire, or with outer edge abruptly notched, narrowed at apical two-thirds 32
— Lateral margins of pronotum deeply angulate at middle; front tibiae entire .... *Cryptopleurum* Mulsant 1844
32. Front tibiae abruptly notched and narrowed at apical two-thirds (fig. 13:32*e*); elytra finely striate and punctate, or only punctate ...... *Megasternum* Mulsant 1844
— Front tibiae entire; elytra strongly costate (fig. 13:31*b*) .... *Pemelus* Horn 1890

Key to the Known Genera of Aquatic Nearctic Hydrophilidae

Larvae

1. Nine complete abdominal segments, 10th reduced but distinct; integument noticeably chitinized, spiracles lateral, well developed (fig. 13:37*b*) .... *Helophorus* Fabricius 1775
— Eight complete abdominal segments, 9-10 reduced and forming a stigmatic atrium (except in *Berosus* in which atrium has not developed) 2
2. Antennae with their points of insertion nearer the externofrontal angles than are those of the mandibles (mandible with a terminal seta, inner tooth, and lacinia mobilis); labium and maxillae inserted in a furrow on under side of head; gula well developed and attaining occipital opening, (abdominal segments with well-developed chitinous plates) (fig. 13:37*c*) .... *Hydrochus* Leach 1817
— Antennae with their points of insertion farther from externofrontal angles than are those of mandibles; labium and maxillae inserted at anterior margin of under side of head; gula reduced and not attaining occipital opening 3
3. First 7 abdominal segments with long lateral tracheal gills; segments 9-10 reduced, no stigmatic atrium present (fig. 13:37*a*) .... *Berosus* Leach 1817
— Tracheal gills not nearly so prominent, or else absent; abdominal segments 9-10 reduced, forming a stigmatic atrium 4
4. Ocular areas oval, larger, aggregated but more distant; legs well developed, visible from above except

in *Paracymus*; pleural lobes usually prominent; abdomen not noticeably truncate 5
— Ocular areas round, closely aggregated; legs absent; pleural lobes not prominent; abdomen truncate; ligula exceeding palpi and pointed (fig. 13:37*d*) .... *Cercyon* Leach 1817
5. First antennal segment not distinctly longer than following 2 taken together; fingerlike antennal appendage present; labro-clypeus with teeth usually well defined; mouth parts stouter; mandibles not grooved internally; stipes large and swollen, usually with an inner row of 5 stout setae; externofrontal angles of mentum not prominent, rounded; legs much shorter, femora without fringes of swimming hairs; gills and prostyles absent 6
— First segment of antennae distinctly longer than following 2 taken together; fingerlike antennal appendage absent; labroclypeus with teeth small or absent; mouth parts more slender; mandibles grooved internally; stipes not swollen, with setae arranged differently; externofrontal angles of mentum prominent and acute; legs very long; femora with fringes of long swimming hairs; gills present *(Hydrochara, Tropisternus)* or absent *(Hydrophilus)*; prostyle present 12
6. Frontal sutures parallel and not uniting to form an epicranial suture; left expansion of epistoma much more prominent than the right and with a row of stout setae; ligula absent; reduced sclerites of meso- and metathorax widely separated; tarsi about as long as tibiae (fig. 13:37*e*) .... *Laccobius* Erichson 1837
— Frontal sutures not parallel, and they may or may not unite to form an epicranial suture; lateral expansions of epistoma similar and usually in line with anterior margin of labro-clypeus, ligula present and longer than segment 1 of palpus; sclerites of meso- and metathorax reduced but not so widely separated; tarsi shorter than tibiae 7
7. Antennae shorter and antennal appendage more prominent; epicranial suture absent; legs reduced; abdomen more truncate; cercus with long terminal seta......... 8
— Antennae longer and antennal appendage less prominent; epicranial suture present but usually short; legs fairly long, not reduced; abdomen narrowed caudally; cercus with a shorter terminal seta 9
8. Frons truncate behind; labrum tridentate, lateral teeth bifid; mandibles with 2 inner teeth; ligula about as long as palpi, apparently 2-segmented; anterior margin of pronotum without a fringe of stout setae; legs not visible from above (fig. 13:37*f*).......... .... *Paracymus* Thomson 1867
— Frons rounded behind; labrum quadridentate; mandibles with 3 inner teeth; ligula not as long as palpi, 1-segmented; anterior margin of pronotum with fringe of stout setae; legs barely visible from above (fig. 13:37*g*) .... *Anacaena* Thomson 1859
9. Mandibles asymmetrical, the right with 2 inner teeth, the left with only 1; abdomen with prolegs on segments 3-7 (fig. 13:38*a*) .... *Enochrus* Thomson 1859
— Mandibles symmetrical each with 2 or 3 inner teeth; abdomen without prolegs 10
10. Labroclypeus with 5 distinct teeth, outer left tooth a little distant from rest; each mandible with 3 inner teeth; mentum wider at base (fig. 13:38*b*).......... ....*Hydrobius* Leach 1815
— Labroclypeus with at least 6 teeth; each mandible with 2 inner teeth; mentum narrower at base 11
11. Labroclypeus with 6 distinct teeth, placed in 2 groups, 2 on left and 4 on right; mentum covered with small spines; anterior sclerites of metathorax with caudal projections (fig. 13:38*c*) .... *Helochares* Mulsant 1844
— Labroclypeus with more than 6 teeth, those toward right not clearly defined and with several smaller teeth; mentum with small spines only toward base; anterior sclerites of metathorax without caudal projections, rectangular (fig. 13:38*d*) .. .. *Cymbiodyta* Bedel 1881
12. Head subspherical; labroclypeus without teeth; each mandible with a single inner tooth; ligula not longer than 1st segment of palp; gills absent; pronotum not entirely chitinized (fig. 13:38*e*) .... *Hydrophilus* Geoffroy 1762
— Head subquadrangular, narrowed behind; labroclypeus with inconspicuous teeth; each mandible with more than 1, usually with 2 inner teeth; ligula distinctly

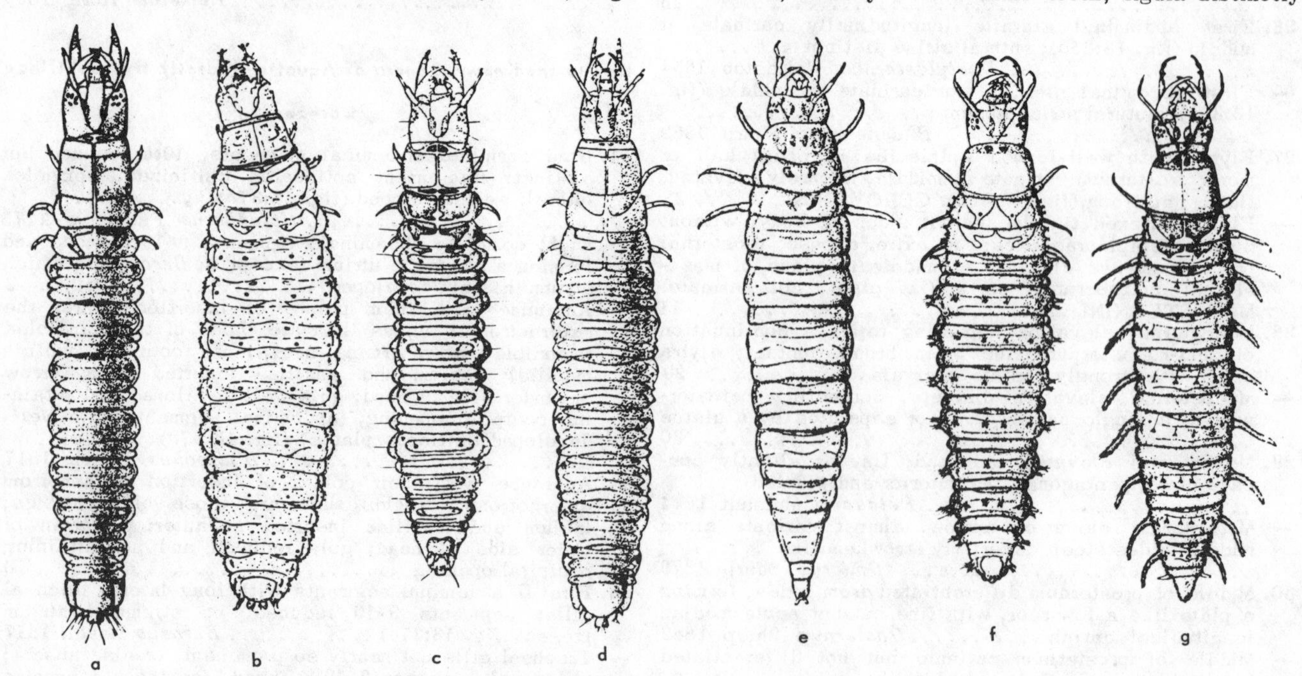

Fig. 13:38. Hydrophilidae, larvae. *a, Enochrus* sp.; *b, Hydrobius fusipes*; *c, Helochares maculicollis*; *d, Cymbiodyta* sp.; *e, Hydrophilus triangularis*; *f, Tropisternus* sp.; *g, Hydrochara* sp. (*a,c-g*, Richmond, *b*, Böving and Henriksen, 1938).

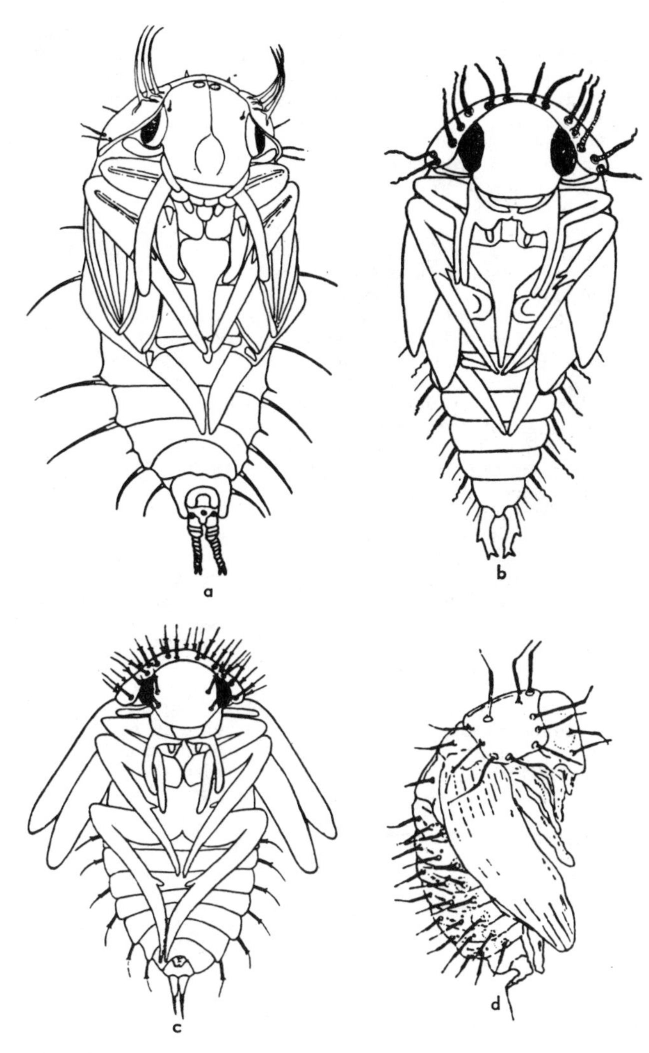

Fig. 13:39. Hydrophilidae, pupae. *a, Hydrophilus triangularis; b, Tropisternus lateralis; c, Berosus* sp.; *d, Cercyon analis* (a-c, Wilson, 1923; *d*, Böving and Henriksen, 1938).

longer than 1st palpal segment; pronotum entirely chitinized; gills present but more or less rudimentary 13

13. Mentum with sides nearly straight; frontoexternal angles very prominent; pleural gills rudimentary but indicated by tubercular projections, each with several terminal setae (fig. 13:38*f*) ... *Tropisternus* Solier 1834

— Mentum convex towards basal half; frontoexternal angles less prominent; pleural gills fairly well developed and pubescent (fig. 13:38*g*)................ .... *Hydrochara* Berthold 1827

Key to Certain Genera of Nearctic Hydrophilidae

Pupae

The pupae of many genera of Hydrophilidae are unknown.

The following key is adapted from one given by Böving and Henriksen (1938), with *Tropisternus* interpolated from the short key in Wilson (1923). A somewhat different key to most of the same genera is given by Richmond (1920). It should be noted that because of taxonomic changes, the *Hydrophilus* and *Hydrochara* of the following key are equivalent to the *Hydrous* and *Hydrophilus*, respectively, of the original keys mentioned above. H.B.L.

1. Only 2 styli present on anterior margin of pronotum, 1 on each side, and a small group of styli present beneath each anterior angle of pronotum .... *Helophorus* Fabricius 1775

— Anterior margin of pronotum differently outfitted, generally with a transverse row of more than 2 styli (*Hydrophilus*, however, with only a small group of styli at each anterior angle) 2

2. Hind wings not visible dorsally; 2nd to 7th abdominal segments each with a transverse row of 4 or 6 styli on the tergite and 1 stylus on each pleurite (fig. 13:39*d*) 3

— Hind wings partly visible dorsally, not entirely covered by posterior edges of fore wings; 2nd to 7th abdominal segments each with a transverse row of 4 or 6 styli on the tergite and 1 stylus on each pleurite 4

3. Cerci of about same length as 9th abdominal segment, placed closely together, slender, with a little dark, hard tip bearing a seta; cerci indistinctly multiarticulate .... *Cercyon* Leach 1817

— Cerci much shorter than 9th abdominal segment, conical and 2-segmented .. *Sphaeridium* Fabricius 1775

4. Cerci long and distally provided with some minute chitinous scales, or slightly bifid (fig. 13:39*b*); either 1 pair of supraorbital and 16 pronotal styli, or with 2 pairs of supraorbital and 32 pronotal styli 5

— Cerci not bifid, nor with scales distally; 2 pairs of supraorbital styli (except in *Hydrobius fuscipes* which has none), and 18–24 pronotal styli 7

5. Styli on dorsal surface of abdomen in transverse rows of 4 on each segment; spiracles partly concealed by lateral tubercles 6

— Styli on dorsal surface of abdomen in transverse rows of 6 on each segment; spiracles not concealed, no tubercles; 2 pairs of supraorbital styli, 32 pronotal styli in all .... *Hydrochara* Berthold 1827

6. Size large, length 25 to 30 mm.· 1 pair of supraorbital styli (fig. 13:39*a*), and 16 pronotal styli in all...... .... *Hydrophilus* Geoffroy 1762

— Size smaller, length 15 mm. or less; no supraorbital styli (fig. 13:39*b*), 26 pronotal styli in all .... *Tropisternus* Solier 1834

7. Tergites of 2nd to 7th abdominal segments each with a transverse row of 6 styli and 1 pleural stylus on each side; 24 pronotal styli 8

— Tergites of 2nd to 7th abdominal segments each with a transverse row of 4 styli and 1 pleural stylus on each side; pronotal styli frequently fewer than 24 ... 9

8. Styli short; terminal seta much longer than stylus itself; anterior margin of pronotum with a row of 10 styli .... *Paracymus* Thomson 1867

— Styli and terminal setae of nearly equal length, or styli longer than terminal setae; anterior margin of pronotum with a row of 6 or 8 styli, the external styli of anterior row in a lateral position .... *Cymbiodyta, Helochares, Enochrus*

9. Supraorbital styli only vestigial; 20 pronotal styli in all; cerci long, finely articulated and tapering (species *fuscipes* L. of genus) .... *Hydrobius* Leach 1815

— With 2 pairs of well-developed supraorbital styli (fig. 13:39*c*) 10

10. Eighth abdominal segment with 4 sharp, upcurved, triangular processes posteriorly; cerci not articulated .... *Laccobius* Erichson 1837

— Eighth abdominal segment without triangular processes posteriorly; cerci articulated distally (fig. 13:39*c*) .... *Berosus* Leach 1817

Genus *Helophorus* Fabricius

It is not possible to give a satisfactory key to the California species of *Helophorus* until the publication of a revision of the North American forms, by the late F. E. Winters and K. F. Chamberlain. Following is a list of the species reported from the state.

H. angustulus Mannerheim 1853 California.
H. fortis LeConte 1866 California.
H. linearis LeConte 1855 Western U.S.
H. obscurus LeConte 1852 California.

Key to the California Species of Hydrochus

Adults

1. Head and pronotum with numerous small, rounded, flat scales, which obscure the punctation (northern U.S. and adjacent Canada) ... *squamifer* LeConte 1855
— Head and pronotum without scales, the punctation clearly defined 2
2. Pronotal foveae deep; strial punctures of elytra large, wider than the interspaces (fig. 13:31*d*); California .... *variolatus* LeConte 1852
— Pronotal foveae shallower; elytral interspaces as wide as striae; California .... *vagus* LeConte 1852

Key to the California Species of Berosus

Adults

1. Elytra emarginate and/or spinous or bispinous apically (subgenus *Enoplurus* Hope); larger species, 6–8 mm. long 2
— Elytra not emarginate, not spinose, the apices acute or a little produced at most (subgenus *Berosus* s. str.); smaller species, 2.75–6 mm. long 3
2. Femora bicolored, basal section (covered with hydrofuge pubescence) black; pronotum coarsely, densely punctate, punctures tending to be subserially arranged, often confluent, surface between either shining or alutaceous; elytral interspaces roughened by coarse punctures with scabrous or almost asperate margins; Washington–Baja California, Arizona .... *punctatissimus* LeConte 1852
— Femora entirely yellow; pronotum with dual punctation, the relatively sparse coarse ones intermixed with numerous much finer ones, surface shining or alutaceous; elytral interspaces smooth, punctures slightly scabrous on some lateral intervals; Kings, Eldorado counties ....? *maculosus* Mannerheim 1853
3. Apical emargination of 5th abdominal sternite rounded at middle or vaguely shallowly emarginate, not dentate; pronotum immaculate; only alternate elytral intervals punctate, if any; southern California to Tres Marias Island, Mexico, in saline pools, lagoons, ocean beaches .... *metalliceps* Sharp 1882
— Apical emargination of 5th abdominal sternite with a single prominent blunt median tooth, or with 2 sharp, often small, teeth 4
4. Hind edge of apical emargination of 5th abdominal sternite with a single glabrous median tooth; small species, 3–3.75 mm. long; strial punctures of basal discal region of elytra very coarse, each elytron with a pale spot on disc just behind middle; California ....? *californicus* Motschulsky 1859
— Hind edge of apical emargination with 2 small glabrous teeth; larger species, more than 4 mm. long 5
5. Crest of mesosternal protuberance with a prominent falcate tooth anteriorly; California to Texas, South Dakota; Butte County, Tulare County .... *styliferus* Horn 1873
— Crest of mesosternal protuberance at most with a low tooth on its anterior part 6
6. Raised area of mid-metasternum trifid posteriorly, median tooth projecting much farther back than lateral ones 7
— Mid-metasternum nearly triangular posteriorly, without obvious lateral teeth; elytra of female either shining as in male, or alutaceous; California to British Columbia to New York .... *fraternus* LeConte 1855
7. Elytra of female shining, as in male; male with basal segment of front tarsi only slightly enlarged, apical half covered with a dense pad of adhesive hairs, tips of which are not obviously dilated; hind edge of hind femora of male margined above and below, angulate at middle; coarsely punctate species, discal striae impressed almost or quite to base; California to British Columbia to South Carolina....... *striatus* (Say) 1823
— Elytra of female dull, alutaceous; male with basal segment of front tarsi enlarged, somewhat triangular, all but extreme base covered by a pad of adhesive hairs, tips of those near apex expanded to form palettes; hind edge of hind femora of male not margined above, not angulate at middle; more finely punctate species, discal striae not or very lightly impressed toward base 8
8. Parameres of aedeagus of male much swollen dorsally in apical third, forming a declivity, apices abruptly narrowed, together triangular in dorsal view, not on same plane as declivity; San Diego north at least to Napa County .... *ingeminatus* d'Orchymont 1946
— Parameres of aedeagus of male forming a gradual declivity from about mid-point to apices, slightly widened at apical third, thence gradually narrowing to apex; Louisiana to Arizona, and southeast corner of California, San Bernardino and Imperial counties . ... *infuscatus* LeConte 1855

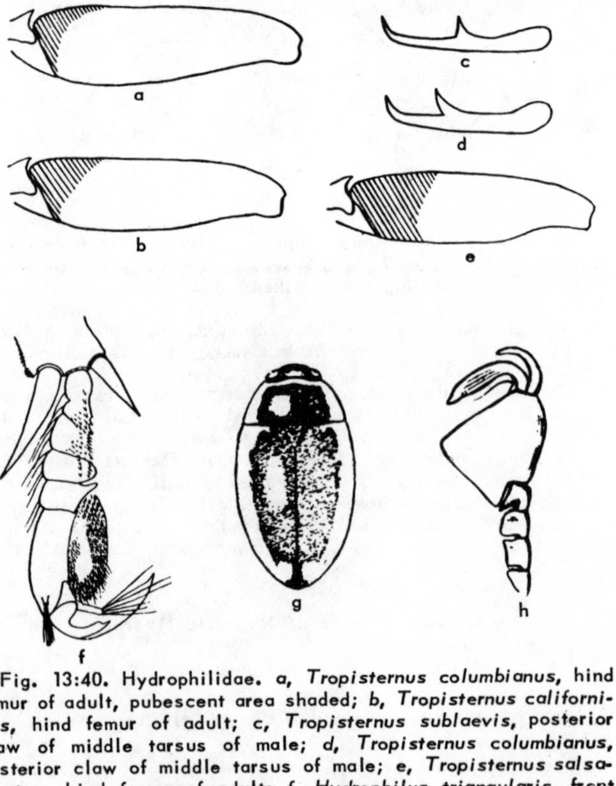

Fig. 13:40. Hydrophilidae. *a*, *Tropisternus columbianus*, hind femur of adult, pubescent area shaded; *b*, *Tropisternus californicus*, hind femur of adult; *c*, *Tropisternus sublaevis*, posterior claw of middle tarsus of male; *d*, *Tropisternus columbianus*, posterior claw of middle tarsus of male; *e*, *Tropisternus salsamentus*, hind femur of adult; *f*, *Hydrophilus triangularis*, front tarsus of male; *g*, *Tropisternus lateralis*, adult; *h*, *Hydrophilus insularis*, front tarsus of male (*a–e*, Leech, 1945; *f*, Wilson, 1923; *g*, Leech, 1948; *h*, Régimbart, 1901).

Key to the California Species of Hydrochara

Adults

1. Meso- and metasternal keel in one plane (in profile view); metasternal section of sternal keel as wide as a hind femur; pronotum unicolorous, color of head, pronotum, elytra, and legs varying from almost black to a beautiful blue-green; coarse elytral punctures 2–3 times as large as those on pronotum; southwest U.S. . . . . *lineata* LeConte 1855
— Metasternal section of sternal keel only as wide as a metatibia, apical two-thirds not on same plane as rest and as mesosternal keel; dorsal surface greenish-black, pronotum and elytra margined with yellow laterally, legs yellow to rufopiceous; coarse elytral punctures subequal to those on pronotum; San Francisco Bay area . . . . *rickseckeri* Horn 1895

Key to the California Species of Hydrophilus

Adults

1. Elytra with a fine spine apically, at suture; 5th segment of protarsi of male broadly, triangularly enlarged (fig. 13:40*h*), the under surface with a series of suckers paralleling each margin, those near the larger claw being stalked, the rest sessile; larger species, 33–45 mm. long; southwest U.S., ? California. . . . . *insularis* Laporte 1840
— Elytra not spinose apically; 5th segment of anterior tarsi of male broadened, but not triangularly so suckers not restricted to its margins and angles (fig. 13:40*f*); smaller species, 30–35 mm. (fig. 13:30*a*); North America . . . . *triangularis* Say 1823

Key to the California Species of Tropisternus

Adults

1. Head, pronotum, and elytra margined with yellow (figs. 13:30*c* and 13:40*g*); North and South America . . . . *lateralis* (Fabricius) 1775
— Pronotum and elytra concolorous 2
2. Pseudobasal segment of maxillary palps short, barely or not reaching eyes, subequal in length to ultimate segment; elytral humeri turned under, not visible from above; northern California to Nevada to southern Oregon . . . . *orvus* Leech 1945
— Pseudobasal segment longer, reaching anterior margin of eye, distinctly longer than ultimate or penultimate; elytral humeri not turned under, clearly visible from above 3
3. Outer margin of epipleura completely without setigerous punctures; epipleura behind hind coxae gradually narrowed, not turned under until just before elytral apex; California . . . . *obscurus* Sharp 1882
— Outer margin of epipleura, from apex almost or quite to base, with a series of setigerous punctures; epipleura behind hind coxae folded in, so that their lower (inner) margin runs along the inner face of the elytra 4
4. Lateral median series of pronotal punctures coarser, 1–2, rarely 4, the anterior ones smaller; body equally obtuse anteriorly and posteriorly, strongly convex in profile; western U.S. . . . . *ellipticus* (LeConte) 1855
— Lateral median series of punctures of pronotum fine, 4–12 in number, forming an oblique line; body not strongly convex in profile 5
5. Basal quarter of hind femora pubescent (fig. 13:40*e*); saline lagoons, coast of southern half of California . . . . *salsamentus* Fall 1901
— Pubescent area of hind femora small, not extending much beyond tips of trochanters 6
6. Pubescent area of hind femora more extensive though still small, clearly reaching to tip of trochanters (fig. 13:40*b*); elytra with dual punctation in apical quarter, the large punctures numerous, irregularly spaced, nearly obliterating the small primary punctures; California . . . . *californicus* (LeConte) 1855
— Pubescent area of hind femora very small, confined to extreme base, barely or not reaching tip of trochanters (fig. 13:40*a*); elytra nearly uniformly punctate (except for the usual coarse, serial punctures) 7
7. Hind femora parallel-sided in basal two-thirds, flat, sparsely and moderately coarsely punctate, in large part testaceous; mesosternal keel very finely punctate in both sexes; posterior claw of middle tarsi of male with the tooth about equidistant from tip to base (fig. 13:40*c*); larger species, less parallel-sided, elytra distinctly greenish; Mono County . . . . .*sublaevis* (LeConte) 1855
— Hind femora not parallel-sided in basal two-thirds, the posterior margin gradually curved (fig. 13:40*a*); hind femora distinctly inflated, more coarsely and numerously punctate, in large part piceous or black; punctation of mesosternal keel fine in female, rather sparse and moderately coarse in male; posterior claw of middle tarsi of male with the tooth larger and more apical (fig. 13:40*d*); smaller species, more parallel-sided, dorsal color blacker with a slight bronze luster; western U.S. . . . . *columbianus* Brown 1931

Key to the California Species of Chaetarthria

Adults

1. Dorsum not black; U.S. . . . . . . *pallida* (LeConte) 1861
— Dorsum black 2
2. Smaller, 1.2–1.5 mm. long; southern California . . . . *minor* Fall 1901
— Larger, 1.6–2.2 mm. long; California . . . . *nigrella* (LeConte) 1861

Key to the California Species of Laccobius

Adults

1. Elytral punctuation in alternate series of fairly regular punctures, some larger and denser, contrasting with others less regular, finer and sparser; Los Angeles County . . . . *insolitus* d'Orchymont 1942
— Punctuation not in evident series of alternately coarser and finer punctures 2
2. Ends of parameres of male genitalia, viewed laterad, with an open-forceps form, the entire aedeagus resembling, in its slenderness, that of *insolitus;* pronotum largely light on lateral borders, this light color ordinarily penetrating deeply in front, less deeply to rear of middle in the obscure median spot, which later is more or less transverse; occasionally this spot is almost reduced to median third of disc; Alberta to Colorado; northern California. . . .*carri* d'Orchymont 1942
— Ends of parameres, from side view, not forcepslike; obscure pronotal spot more transverse, approaching nearer to lateral prothoracic borders which are then darker 3
3. Form more briefly oval, with obscure transverse pronotal spot less broken at sides; elytral punctuation less irregular, more or less arranged in brownish-colored lines; base of head and pronotum more or less distinctly granulated between punctuation; southern Canada, northern U.S. . . . . *agilis* Randall 1838
— Form more elongate or elliptical with elytral punctuation more irregular, less or not at all aligned in colored series 4
4. Obscure pronotal spot more broken laterally, less transverse; elytral punctuation a trifle less dense and a bit stronger, particularly along suture; Oregon, California, Nevada . . . . *ellipticus* LeConte 1855

— Spot transverse, nearly entire on sides; elytral punctuation somewhat less dense and finer along suture; Washington to southern California .... *californicus* d'Orchymont 1942

Genus *Hydrobius* Leach

Adults

Length 6.5 to 8 mm. Longitudinal serial punctures of elytra varying from lightly impressed as striae in apical half only, to deeply impressed as striae from apices to basal quarter; interspaces finely uniformly punctate, with somewhat serial larger punctures on alternate ones (fig. 13:36*b*); Holarctic; northern U.S. and Canada, California .... *fuscipes* (Linnaeus) 1758

Key to the California Species of Ametor

Adults

1. Hind femora with hydrofuge pubescence on slightly more than basal third; elytra with clearly defined longitudinal rows of punctures, becoming striate apically, lateral rows closer together and more coarsely punctured; elytra evenly convex, surface shining or finely scabrous; larger, more strongly convex species, 6–7 mm. long; mountains of northern California to Idaho, British Columbia .... *latus* (Horn) 1873
— Hind femora pubescent adjacent to trochanters and for a short distance along anterior margin; elytra striate, striae punctures usually present in discal area basally, or both punctures and striae effaced except on declivity; elytra subcostate from humeral umbone to apical fourth (that is, along 6th interspace) and at apical fourth of 4th and 5th interspaces, depressed along 7th, 8th, and 9th interspaces, and on 2nd and 3rd interspaces at apical fourth; elytral surface usually strongly scabrous; smaller, flatter species, 4–5.75 mm. long; Alaska to Wyoming to California .... *scabrosus* (Horn) 1873

Key to the California Species of Crenitis

Adults

1. Form oblong, somewhat depressed; mesosternum simple, or with a median protuberance; sides of pronotum nearly straight, or weakly arcuate with a broad but appreciable hind angle; hind tarsi long, slender 2
— Form elliptical, convex; mesosternum with a short transverse groove before mesocoxae, its anterior edge slightly raised; sides of pronotum strongly evenly arcuate, without trace of hind angles, surface shining between punctures; hind tarsi shorter, stouter 4
2. Pronotum piceous with sides pale, or almost uniformly rufotestaceous; front edge of mentum not margined, usually depressed at middle; mesosternum simple ... 3
— Head, pronotum, and elytra unicolorous, black with a slight brassy tinge, surface of pronotum alutaceous at sides between punctures; front edge of mentum strongly margined, not depressed at middle; mesosternum with a low transverse median protuberance; Alaska to California .... *morata* (Horn) 1890
3. Mentum nearly square, almost as long as broad; abdomen dull, finely, densely, uniformly punctate and pubescent; antennae 9-segmented; California to Oregon .... *dissimilis* (Horn) 1873
— Mentum rectangular, 1/2 again as wide as long; abdomen shining, sparsely pubescent, punctures fine and tending to be in transverse series; antennae 8-segmented; California, Arizona, at high altitudes .... *alticola* (Fall) 1924
4. Smaller species, 1.8–2.2 mm. long; form a short oval, narrower behind, somewhat egg-shaped, very convex, almost hemispherical; coarse punctures of lateral rows on elytra more widely spaced in the rows, the rows not impressed to simulate striae; British Columbia to California .... *rufiventris* (Horn) 1873
— Larger species, 2.2–2.75 mm. long; form more elongate-oval, moderately convex; coarse punctures of lateral rows on elytra close-set, the rows evidently impressed, simulating striae; British Columbia to California ... . *seriellus* (Fall) 1924

Key to the California Species of Anacaena

Adults

1. Head entirely black, or with a vague rufopiceous tinge in front of eyes; pronotum varying from piceous except for lateral margins, to rufopiceous with testaceous margins and vague darker rufopiceous markings on disc; body form more narrowly oval, less convex; average size larger, 2–2.7 mm.; Holarctic; Alaska to New York to California, south at least to Los Angeles County .... *limbata* (Fabricius) 1792
— Head black with clear testaceous margin in front of each eye; pronotum testaceous to luteous, with piceous discal markings like a broad H with a blotch on central bar; body form more broadly oval, more convex; average size smaller, 1.75–2.5 mm.; southern California, northward at least to Siskiyou County .... *signaticollis* (Fall) 1924

Key to the California Species of Paracymus

Adults

1. Form short, strongly convex in profile; hind femora finely strigose at least on anterior half; head, pronotum, and elytra strongly punctate, punctures comparable in size to facets of eyes; mesosternal protuberance smaller; antennae 8-segmented; northern America .... *subcupreus* (Say) 1825
— Form more elongate, flatly oval in profile; hind femora polished, not strigate; head and pronotum finely punctate to almost impunctate; mesosternal protuberance prominent, almost laminiform; antennae 7-segmented 2
2. Pronotum and elytra with sharply defined pale margin which is broader behind; saline waters, especially coast of southern California ..... *elegans* (Fall) 1901
— Pronotum and elytra gradually diffusedly paler at sides; hot springs and mineralized waters, Arizona, southern California including Death Valley .... *ellipsis* (Fall) 1910

Key to the California Species of Cymbiodyta

Adults

1. Elytra with sutural striae at least in apical half, but elsewhere at most with serial punctures, the rows of which are not impressed as striae; head black 2
— Elytra with distinct striae, in addition to sutural; head pale in front of eyes; form broad, all rows of punctures entire, including scutellar row; California .... *punctatostriatus* (Horn) 1873
2. Hind femora pubescent in slightly less than basal two-thirds; larger punctures of sublateral group on pronotum hardly half as big as coarse punctures of lateral series of elytra; [illegible]lifornia to British Columbia..... .... *dorsalis* (Motschulsky) 1859
— Hind femora pubescent in basal three-fourths; punctures of sublateral group on pronotum as large as or larger than those of lateral series of elytra 3

3. Elytra with regular series of punctures clearly traceable in apical one-fourth to one-third, and laterally with at least 3 rows of closely spaced regular punctures from apical margin to near umbone; elytra, and pronotum except discally, usually testaceous; California to Oregon .... *imbellis* (LeConte) 1861
— Elytra with lateral rows of serial punctures not regular and clearly defined, but confused with coarse punctures of interspaces; pronotum and elytra piceous or black, paler laterally; California to British Columbia..... .... *pacifica* Leech 1948

Key to the California Species of Enochrus

Adults

1. Fifth abdominal sternite with a small apical emargination, from which projects a differentiated fringe of flat golden ciliae 2
— Fifth abdominal sternite entire apically, not emarginate, without differentiated fringe of flat ciliae 8
2. Last 2 segments of maxillary palpi of equal length, or last longer than penultimate 3
— Last segment of maxillary palpi shorter than penultimate 5
3. Head luteous in front of eyes, elsewhere black; prosternum not carinate; smaller species, 3.5–4 mm. long; pronotum black, sides and parts of anterior and posterior margins luteous; California............. .... *cuspidatus* (LeConte) 1878
— Head entirely black, or vaguely narrowly rufopiceous in front of eyes; prosternum with low poorly defined median longitudinal carina; larger species, 4–5 mm. long 4
4. Head black, pronotum and elytra rufotestaceous; southwest U.S. .... *fucatus* (Horn) 1873
— Head, pronotum, and elytra black, usually only front angles of pronotum pale; Pacific Coast, Arizona .... *carinatus* (LeConte) 1855
5. Larger species, 4.5–5 mm. long; prosternum not carinate; mesosternal protuberance nearly level, its form somewhat obscured by pubescence; California to British Columbia .... *californicus* (Horn) 1890
— Smaller species, 2.75–3.75 mm. long; prosternum carinate; mesosternal protuberance laminiform, acutely prominent anteriorly, not at all obscured by vestiture 6
6. Pronotum piceous on disc 7
— Pronotum entirely testaceous; elytra polished, minutely punctate, except for several more or less evident longitudinal series of coarser punctures; southwest U.S. .... *pectoralis* (LeConte) 1855
7. Elytra extremely finely, sparsely punctate, except for serial punctures which stand out contrastingly; pronotum usually broadly piceous, elytra tinged with piceous; narrower species; San Diego County .... *cristatus* (LeConte) 1855
— Elytra rather densely, moderately coarsely punctate, serial punctures not conspicuous; pronotum usually piceous only at middle of disc, elytra not tinged with piceous; broader species; California to Oregon to Utah .... *obtusiusculus* (Motschulsky) 1859
8. Dorsal surface black or piceous, sides of head and pronotum often paler; California to Oregon .... *conjunctus* (Fall) 1901
— Dorsal surface testaceous, except for piceous markings on head and sometimes on pronotum; pronotum with 4 small black dots as if at corners of a discal quadrangle 9
9. Front margin of clypeus evenly arcuate-emarginate, without trace of a secondary emargination at middle; species occurring in brackish or saline water along seacoast, near lower Colorado River, and in Death Valley 10
— Arcuate front margin of clypeus with a secondary emargination at middle, exposing a preclypeus 11
10. Elytra more coarsely punctate; salt marsh and saline pools, coast of California and up estuary of Colorado River .... *hamiltoni pacificus* Leech 1950
— Elytra more finely punctate; saline waters, Death Valley, California ...... *hamiltoni pyretus* Leech 1950
11. Smaller species, 3.5–4.5 mm. long; hind edge of hind femora simple; Central Valley of California .... *?latiusculus* (Motschulsky) 1859
— Larger species, 5–6 mm. long; hind edge of hind femora of male raised and slightly produced at middle; California to British Columbia ... *diffusus* (LeConte) 1855

Genus *Helochares* Mulsant

Adults

Elytra with 10 rows of coarse punctures, the rows not impressed as striae; front margin of labrum arcuate inwards; mesosternal protuberance distinctly longitudinally carinate; length 4–5.5 mm.; California to Baja California to Texas .... *normatus* (LeConte) 1861

Key to Certain California Species of Cercyon

Adults

Several species of *Cercyon* occur under kelp and other debris on ocean beaches, and a few are found in dead grass and other decomposing vegetation beside streams or lakes. Most of the species are strictly terrestrial, though found only in damp places, especially in rotting vegetable matter and in the dung of animals. Preliminary studies by a specialist at the British Museum have shown that the nomenclature of the species in North America is sadly confused. It is not practicable to give a key to all the species occurring in damp places; the following are found near water.

1. Form broad, rather flat, head held obliquely; head, pronotum, and intervals of elytral disc rather densely, uniformly punctate, surface shining between punctures; prosternum with at most a very low, weak carina on median line; differentiated metasternal area limited to mid-metasternum; larger species 2.5–3.5 mm. long, occurring on sea beaches 2
— Form oval, convex, head held nearly vertically; head, pronotum and elytral intervals sparsely, finely punctate, surface alutaceous; prosternum with a high, sharp, longitudinal median carina; discal elytral striae lightly impressed, closely punctured, lateral series not impressed, consisting only of coarser punctures; differentiated metasternal area limited to mid-metasternum; mesosternal protuberance oval twice as long as wide, flat, punctate; smaller species,

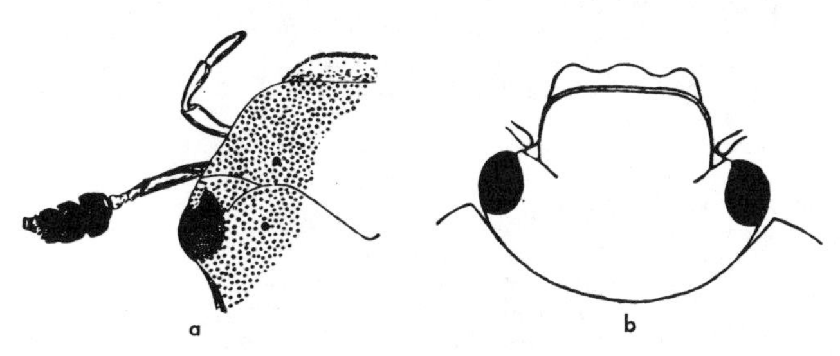

Fig. 13:41. Hydrophilidae. *a*, head of an adult representative of Sphaeridiini to show margin of head hiding base of antenna; *b*, head of an adult representative of Cercyonini (D'Orchymont, 1919).

1.75–2.5 mm. long, occurring near margins of fresh water; California to Washington .... *kulzeri* Knisch 1922

2. Elytral striae deeply impressed from base to apex, only lateral ones visibly punctate; mesosternal protuberance narrowly fusiform, 3 times as long as wide, acute anteriorly, rounded posteriorly, slightly convex, punctate; color variable: pronotum and elytra yellow, to almost entirely black; Alaska to Baja California .... *fimbriatus* Mannerheim 1852

— Striae lightly impressed and finely punctate on disc, evanescent laterally; mesosternal protuberance a narrow, finely punctate carina; color variable, usually with elytra testaceous, pronotum rufotestaceous with median black mark; Alaska to Baja California; Carmel, Monterey County; Palo Verde, Imperial County! .... *luniger* Mannerheim 1853

REFERENCES

BLACKWELDER, R. E.
1931. The Sphaeridiinae of the Pacific Coast. Coleoptera, Hydrophilidae. Pan-Pac. Ent., 8:19-32.

BÖVING, A. G., and F. C. CRAIGHEAD
1931. An illustrated synopsis of the principal larval forms of the Order Coleoptera. Ent. Amer., 11 NS. (1-4):1-351, incl. 125 pls.

BÖVING, A. G., and KAI L. HENRIKSEN
1938. The developmental stages of the Danish Hydrophilidae (Ins., Coleoptera). Vidensk. Medd. fra Dansk naturh. Foren., 102:27-162, 55 text figs., many compound.

FALL, H. C.
1901. List of the Coleoptera of southern California, with notes on habits and distribution and descriptions of new species. Occ. Pap. Calif. Acad. Sci., 8:1-282.
1924a. Some notes on *Cercyon,* with descriptions of 3 new species. Psyche, 31:247-253.
1924b. New species of North American Hydrobiini. Jour. N.Y. Ent. Soc., 32:85-89.
1930. On *Tropisternus sublaevis* Lec. and *T. quadristriatus* Horn. Ent. News, 41:238-240.

HORN, G. H.
1873. Revision of the genera and species of the tribe Hydrobiini. Proc. Amer. Philos. Soc., 13:118-137.
1890. Notes on some Hydrobiini of Boreal America. Trans. Amer. Ent. Soc., 17:237-278, pls. 3-4.

HRBÁČEK, JAROSLAV.
1950. On the morphology and function of the antennae of the central European Hydrophilidae (Coleoptera). Trans. Roy. Ent. Soc. London, 101:239-256, 17 text figs. (Note that Hrbáček uses the term Hydrophilidae in the broad sense, to include the Hydraenidae and Spercheidae. Also that his use of the names *Hydrous* and *Hydrophilus* are equivalent to the *Hydrophilus* and *Hydrochara,* respectively, of the present article.)

LEECH, H. B.
1946. Remarks on some Pacific Coast species of *Tropisternus* (Coleoptera: Hydrophilidae). Canad. Ent., 77:179-184, 5 text figs.
1948. Haliplidae, Dytiscidae, Gyrinidae, Hydrophilidae, Limnebiidae. *In* Contributions toward a knowledge of the insect fauna of Lower California. No. 11, Coleoptera. Proc. Calif. Acad. Sci., (4), 24:375-484, pls. 20-21.
1950. New species and subspecies of Nearctic water beetles. Wasmann Coll., 7:243-256, 6 text figs.

D'ORCHYMONT, A.
1921. Le genre *Tropisternus.* I. (Col. Hydrophilidae). Ann. Soc. Ent. Belg., 61:349-374.
1922. Le genre *Tropisternus.* II. (Col. Hydrophilidae). Ann. Soc. Ent. Belg., 62:11-47, 4 text figs.
1928. Revision des *Neohydrophilus* américains. Bull. & Ann. Soc. Ent. Belg., 68:158-168.
1938. Contribution à l'étude des Palpicornia. XII. Bull. & Ann. Soc. Ent. Belg., 78:426-438.
1942*a*. Revision des *Laccobius* américains. (Coleoptera Hydrophilidae Hydrobiini). Bull. Mus. Roy. Hist. Nat. Belg., 18:1-18, 10 text figs.
1942*b*. Contribution à l'étude de la tribu Hydrobiini Bedel spécialement de sa sous-tribu Hydrobiae. Mém. Mus. Roy. Hist. Nat. Belg., (2), 24:1-68, 4 text figs.
1946. Notes on some American *Berosus* (s. str.). Bull. Mus. Roy. Hist. Nat. Belg., 22:1-20, 11 text figs.

RICHMOND, E. A.
1920. Studies on the biology of the aquatic Hydrophilidae. Bull. Amer. Mus. Nat. Hist., 42:1-94, pls. 1-16.

SHARP, D.
1882. Haliplidae, Dytiscidae, Gyrinidae, Hydrophilidae. *In* Biologia Cent.-Amer., Insecta, Coleoptera, 1(2):1-116, pls. 1-3, 4 partim.
1883. Revision of the species included in the genus *Tropisternus.* (Fam. Hydrophilidae). Trans. Roy. Ent. Soc. London, 1883, (2):91-117.

TODD, A. C.
1942. A new parasitic nematode from a water scavenger beetle. Trans. Amer. Micr. Soc., 61:286-289, 4 text figs.
1944. Two new nematodes from the aquatic beetle *Hydrous triangularis* (Say). Jour. Parasit., 30:269-272, incl. 1 pl.

WILSON, C. B.
1923*a*. Life history of the scavenger water beetle, *Hydrous (Hydrophilus) triangularis,* and its economic relation to fish breeding. Bull. Bur. Fish., 39:9-38, 22 text figs.
1923*b*. Water beetles in relation to pondfish culture, with life histories of those found in fishponds at Fairport, Iowa. Bull. Bur. Fish., 39:231-345, 148 text figs.

WINTERS, F. E.
1926. Notes on the Hydrobiini (Coleoptera - Hydrophilidae) of Boreal America. Pan-Pac. Ent., 3:49-58.
1927. Key to the subtribe Helocharae Orchym. (Coleoptera—Hydrophilidae) of Boreal America. Pan-Pac. Ent., 4:19-29.

Family HYDROSCAPHIDAE

These tiny staphylinidlike beetles, about 1 mm. long, are aquatic and live among filamentous algae. The family was proposed for the one described North American species, *Hydroscapha natans* LeConte; Palaerctic species were added later. *Hydroscapha* is the only genus.

The affinities of *Hydroscapha* are still not clear. Besides serving as the type of the subfamily Hydroscaphinae Böving, it has at various times been included with *Limnebius* (then in the Hydrophilidae), in the Scaphidiidae, and in the Trichopterygidae (now Ptiliidae). In 1950 Crowson included both it and Sphaeriidae in the superfamily Staphylinoidea. In 1954 he suggested that these two families, with Lepiceridae, might form a fourth suborder of the Coleoptera.

Both adults and larvae (fig. 13:42*a-d*) may be found abundantly in suitable streams, especially on algal-covered rocks in the marginal shallows. However, they are remarkably tolerant; La Rivers (1950:68) records finding them on December 29, 1946 in Nevada: "A quite populous colony was located in the icy Amargosa river just south of Beatty in moderately swift, rough water which froze at the banks each night. Individuals were found clinging lethargically to the undersurface of rhyolite stones well-grown with thin algal layers." On March 31, 1953, I found them

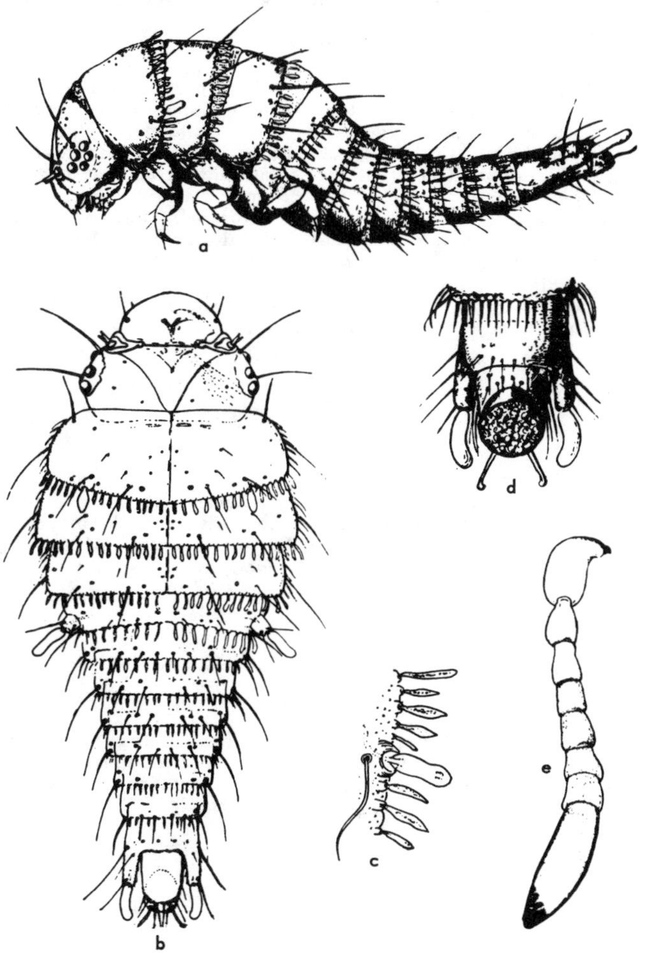

Fig. 13:42. Hydroscaphidae. *a-d, Hydroscapha natans*, larva; *a*, lateral; *b*, dorsal; *c*, detail of prothoracic filament; *d*, ventral view of tip of abdomen; *e, Hydroscapha granulum* (European sp.), antenna of adult (*a-d*, Böving, 1914; *e*, D'Orchymont, 1945).

numerous in algae in a hot spring about five miles north of Beatty, where the temperature was 40°C.[9] Schwarz (1914) reported *H. natans* from water of about 45°C. at Castle Hot Springs, Yavapai County, Arizona.

The egg is large in proportion to the female's abdomen, and only one is developed at a time. It is laid on the algae. The larva has mouth parts suitable for feeding on algae. It is strictly aquatic, yet has exceedingly small spiracles; respiration is presumably by means of the three pairs of curious filaments (fig. 13:42*a-c*), the apical section of which is in each case filled by an enlargement of the trachea.[10] The tip of the abdomen is formed into an adhesive sucker ventrally (fig. 13:42*d*). Adults carry a supply of air under the elytra.

[9]This is the "Hot Springs Station, Nevada" of Brues (1928:144). The property has changed hands in recent years, and has been known as Burrell Hot Spring, Hicks Hot Springs, and (1953) Amargosa Hot Springs.

[10]For a discussion on the action of tracheal gills in some other larvae, see the family Helodidae.

The single Nearctic species has been recorded from the southern parts of California, Nevada, and Arizona.

Genus *Hydroscapha* LeConte 1874

Length (head, prothorax, and elytra) about 1 mm., facies of a tiny reddish-brown tachyporine Staphylinidae; elytral apices truncate, with fringe hairs of folded wings projecting out behind them; abdomen conical, 4 or 5 segments visible behind the short elytra; antennae 8-segmented (fig. 13:42*e*), hind coxae separated by a width equal to 1 of them; Los Angeles, San Bernardino, San Luis Obispo, and Lake counties .... *natans* LeConte 1874

REFERENCES

BÖVING, A. G.
1914. Notes on the larvae of *Hydroscapha* and some other aquatic larvae from Arizona. Proc. Ent. Soc. Wash., 16:169-174, 2 text figs., pls. xvii-xviii.

BRUES, C. T.
1928. Studies on the fauna of hot springs in the western United States and the biology of thermophilous animals. Proc. Amer. Acad. Arts Sci., 63:139-228, 7 text figs. pls. i-vi.

CROWSON, R. A.
1950*a*. The classification of the families of British Coleoptera. Superfamily 3 : Staphylinoidea. Ent. Mon. Mag., 86:274-288, text figs. 23-43.
1950*b*. The classification of the families of British Coleoptera (concluded). Ent. Mon. Mag., 90:57-63, 1 text fig.

LA RIVERS, IRA
1950. The staphylinoid and dascilloid aquatic Coleoptera of the Nevada area. Great Basin Naturalist, 10(1-4): 66-70. (Also an erratum page, *loc. cit.*, 11:52.)

D'ORCHYMONT, A.
1945. Notes sur le genre *Hydroscapha* LeConte (Coleoptera Polyphaga Staphyliniformia). Bull. Mus. Roy. Hist. Nat. Belg., 21:1-16, 8 text figs.

SCHWARZ, E. A.
1914. Aquatic beetles, especially *Hydroscapha*, in hot springs, in Arizona. Proc. Ent. Soc. Wash., 16:163-168.

Family STAPHYLINIDAE

Rove Beetles

This is a family of some 20,000 described species (world), many kinds of which live in grass and debris along the margins of fresh water, or actually burrow in the wet sand and mud.

Adults (fig. 13:43*a, b*), larvae (fig. 13:43*c, d*), and pupae of a few kinds, all belonging to the subfamily Aleocharinae, occur in cracks in rocks below high tide mark. Some of them prowl about at the bottom of tide pools, as much at ease as are the shrimps. They carry a film of air entangled in the hairs of the body, but according to Saunders (1928) cannot stay under water for more than three or four hours. The species of *Emplenota* are winged; the others are wingless or have nonfunctional wings.

All the intertidal species are small, some tiny, and to study them it is essential that they be properly prepared and mounted on slides. Allen (1953) has recorded the English *Trogophloeus arcuatus* Stephens as occurring beneath the surface of fresh water.

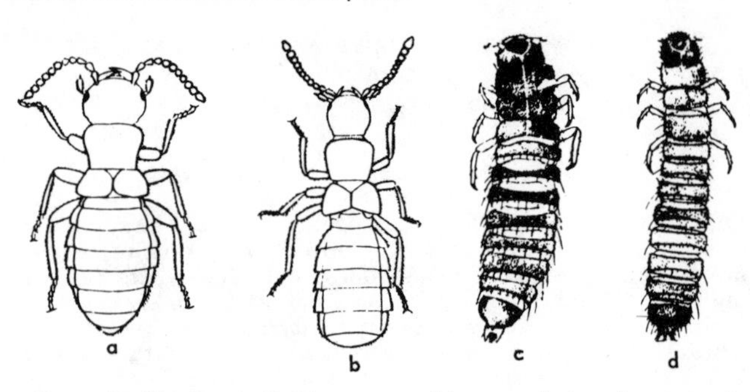

Fig. 13:43. Staphylinidae. *a,c, Liparocephalus brevipennis,* adult and larva, respectively; *b,d, Diaulota densissima,* adult and larva, respectively (Saunders, 1928).

Key to the Genera of Staphylinidae of the Intertidal Zone of the California Coast[11]

Adults

1. Front and middle tarsi 4-segmented, hind tarsi 5-segmented (but see preceding footnote). BOLITOCHARINI 2
— Front tarsi 4-segmented, middle and hind pairs 5-segmented; or all tarsi 5-segmented 5
2. Anterior margin of labrum strongly rounded; eyes hairy (fig. 13:43*b*) .... *Diaulota* Casey 1893
— Anterior margin of labrum truncate; eyes hairy or not 3
3. Eyes not hairy; antennal segments 9 and 10 clearly longer than broad (fig. 13:43*a*) .... *Liparocephalus* Mäklin 1853
— Eyes hairy; antennal segments 9 and 10 obviously broader than long 4
4. Tibiae very short, anterior and middle pairs spinulose externally; 5th abdominal segment longer than 4th .... *Thinusa* Casey 1893
— Tibiae rather slender, clothed with rather coarse pubescence, without trace of spinules; 4th and 5th abdominal segments of equal length .... *Bryobiota* Casey 1893
5. Front tarsi 4-segmented, middle and hind tarsi 5-segmented. ZYRINI .... *Pontomalota* Casey 1885
— All tarsi 5-segmented. ALEOCHARINI .... *Emplenota* Casey 1884

Key to the California Species of Diaulota

Adults

1. Body coloration uniformly dark throughout; length about 2.7 mm. (fig. 13:43*b*); Alaska to California; Moss Beach, San Mateo County .... *densissima* Casey 1893
— At least head and abdominal tip reddish; length about 1.7 mm.; Alaska to California; Moss Beach, San Mateo County .... *brevipes* (Casey) 1893

[11]The key is largely a compilation. The primary divisions are from one by Fenyes (1918), the only one available. Chamberlin and Ferris (1929:142) have shown that the hind tarsi of *Diaulota brevipes* Casey may be 4- or 5-segmented, or with fusions giving intermediate conditions.

After this manuscript was sent to the printer, there appeared "A revision of the Pacific Coast Phytosi with a review of the foreign genera", by Ian Moore. Trans. San Diego Soc. Nat. Hist., 12:103-152, incl. pls. 8-11. March, 1956. It contains new keys to the genera, and species.

Key to the California Species of Liparocephalus

Adults

1. Head conspicuously broader than prothorax and distinctly broader than long (length taken from clypeal suture to extreme posterior border of head), the proportions being as 7:6; head lighter in color than rest of body; length (on slide and somewhat expanded) 6 mm.; Moss Beach, San Mateo County; Mendocino County .... *cordicollis* LeConte 1880
— Head and thorax practically equal in width, length of head equal to width or but very slightly greater; of a uniformly dark color; length (on slide and somewhat expanded) 4.5–5 mm. (fig. 13:43*a*); Alaska to California; Moss Beach, San Mateo County.................. .... *brevipennis* Mäklin 1853

Genus *Thinusa* Casey

Adults

Length about 2.2 mm.: form very slender; abdomen black, head slightly paler, pronotum and elytra dark piceous-brown; legs and antennae reddish-testaceous; Alaska to California; San Francisco, San Diego .... *maritima* (Casey) 1885

Genus *Bryobiota* Casey

Adults

Length about 2.3 mm.; form very slender; anterior parts, legs and antennae pale reddish-testaceous, abdomen black, apex of last segment slightly paler; southern California; San Diego .... *bicolor* (Casey) 1885

Key to the California Species of Pontomalota[12]

Adults

1. Head testaceous; 1st segment of hind tarsi fully 2/3 longer than 2nd; 4th abdominal segment, and sometimes base of 5th, clouded with castaneous color; southern California .... *opaca* (LeConte) 1863
— Head blackish 2
2. First segment of hind tarsi scarcely longer than 2nd; abdomen entirely black; Washington to southern California .... *californica* Casey 1885
— First segment of hind tarsi about 1/3 longer than 2nd; apex of 3rd, 4th, and base of 5th abdominal segments clouded with blackish castaneous; California; Santa Cruz .... *nigriceps* Casey 1885

Key to Some California Species of Emplenota[12]

Adults

1. Basal segment of hind tarsi short, only slightly longer than 2nd; southern California, San Diego County north to Monterey County .... *arenaria* (Casey) 1893
— Basal segment of hind tarsi much longer, equal to next two combined; southern California............ .... *pacifica* (Casey) 1893

REFERENCES

ALLEN, A. A.
1953. A note on the habits of *Trogophloeus arcuatus* Steph. (Col., Staphylinidae). Ent. Mon. Mag., 89:286.

[12]After Casey, 1885. *P. bakeri* Bernhauer 1912, from southern California, is omitted.

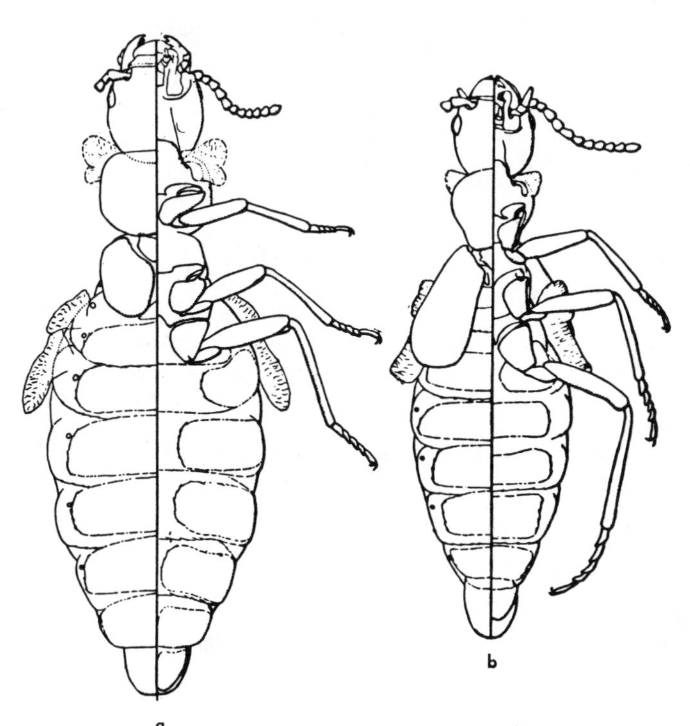

Fig. 13:44. Melyridae, adults, showing dorsal surface on left side and ventral on the right. a. *Endeodes collaris*; b, *Endeodes basalis* (Blackwelder, 1932).

CASEY, T. L.
1885. New genera and species of California Coleoptera. Bull. Calif. Acad. Sci., 1:283-336, 1 pl.
1893. Coleopterological notices V. Ann. N.Y. Acad. Sci., 7:281-606, 1 pl.
CHAMBERLIN, J. C., and G. F. FERRIS
1929. On *Liparocephalus* and allied genera (Coleoptera; Staphylinidae). Pan-Pac. Ent., 5:137-143; 5:153-162, 5 compound text figs.
FENYES, A.
1918. Family Staphylinidae, subfamily Aleocharinae, *In* Genera Insectorum, Coleoptera, Fasc. 173A, pp. 1-110, 15 figs.
1920. Same, Fasc. 173B, pp. 111-414, 97 text figs., 7 colored pls.
SAUNDERS, L. G.
1928. Some marine insects of the Pacific Coast of Canada. Ann. Ent. Soc. Amer., 21:521-545, 9 compound text figs.

Family MELYRIDAE

The species of one genus of this family are restricted to the seashore of western North America, from British Columbia to Baja California. *Endeodes* spp. are small, wingless, yellow, or orange and black predaceous beetles. Blackwelder (1932) records finding *Endeodes* ". . . near high tide mark but definitely in the damp areas wet regularly by high tides. They were all found under rubbish, especially boards." In an unpublished manuscript the late E. C. Van Dyke wrote "Ordinarily they are to be observed running over the rocks and drift wood situated above the high tide line, though they often frequent the inter-tidal area. They are at times covered by the water, for I have on several occasions pried them out from their retreats which had been but recently uncovered."

Endeodes belongs to the subfamily Malachiinae, characterized by extensile membranous vesicles on the prothorax and between the metathorax and abdomen (fig. 13:44). Many species of Malachiinae look like cantharids, and the family Melyridae has been included in the Cantharoidea, but belongs in the Cleroidea. The larvae of *Endeodes* spp. occur with the adults, and resemble clerid larvae.

Key to the Species of Endeodes Lec.

Adults

1. Each elytron at least twice as long as wide 2
— Each elytron not much longer than wide 3
2. Elytra concolorous, ferrugineous .... *blaisdelli* Moore, 1954
— Elytra of 2 colors, ferrugineous basally and black apically .... *basalis* (LeConte) 1852
3. Legs, antennae, and mouth parts darker than thorax 4
— Legs, antennae, and mouth parts pale .... *insularis* Blackwelder 1932
4. Head black................. *collaris* (LeConte) 1852
— Head reddish .... *rugiceps* Blackwelder 1932

REFERENCES

BLACKWELDER, R. E.
1932. The genus *Endeodes* LeConte (Coleoptera, Melyridae). Pan-Pac. Ent., 8:128-136, 3 compound text figs.
MOORE, IAN M.
1954. Notes on *Endeodes* LeConte with a description of a new species from Baja California (Coleoptera: Malachidiidae). Pan-Pac. Ent., 30:195-198.

Family EURYSTETHIDAE

The four known species of *Eurystethes* occur in the intertidal zone of the seashore. Two are found on the coast of California; one (*E. californicus* Motschulsky) ranges from the Aleutian Islands to Vancouver Island, but not to California; the last is known from Robben Island, near the east coast of Sakhalin Island, U.S.S.R., just north of Japan.

The flattened form of these slow-moving beetles and their larvae enables them to live in the crevices of rocks in the intertidal area, commonly in the barnacle zone. There they occur in colonies, and are presumed to feed on mites and other animal organisms. Adults of the California species are 2 to 3 mm. long, black or dark metallic green above, brown below, wingless, with sparse short hairs. The larva and pupa of *E. californica* have been described; the larva (fig. 13:45*a*) has four urogomphi, an up-curved outer pair, and a smaller inward-curving inner pair; the pupa (fig. 13:45*b*) shows the wide separation of the hind coxae, so characteristic of the adults.

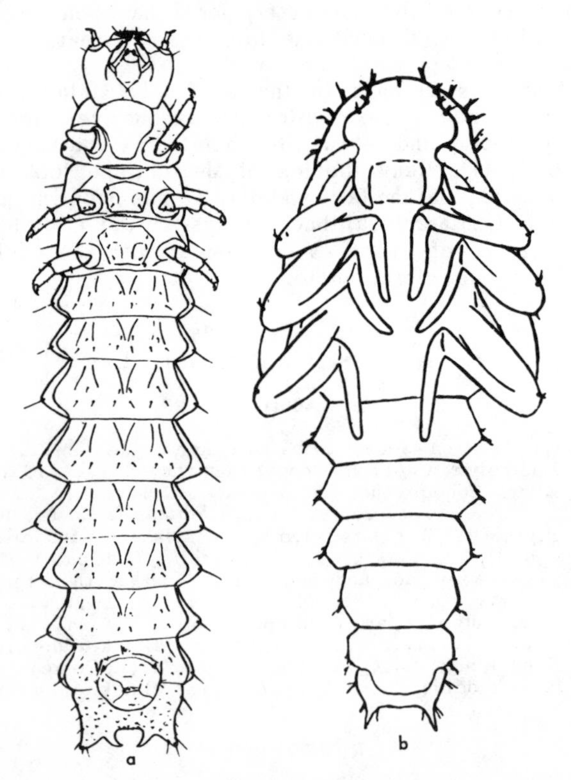

Fig. 13:45. Eurystethidae, *Eurystethes californica*. *a*, ventral view of larva; *b*, pupa, ventral (Wickham, 1904).

Key to the California Species of Eurystethes

1. Elytra striate; head, thorax, and elytra shining, metallic; Mendocino County, south at least to Farallon Islands, San Francisco County .... *fuschsii* (Horn) 1893
— Elytra not striate; head, thorax, and elytra alutaceous and subopaque; Marin County, south to San Mateo County .... *subopacus* Van Dyke 1918

REFERENCES

BLACKWELDER, R. E.
1945. Checklist of the Coleopterous insects of Mexico, Central America, the West Indies, and South America. U.S. Nat. Mus., Bull. 185: 343-550.
CROWSON, R. A.
1953. The classification of the families of British Coleoptera (continued). Ent. Mon. Mag., 89:49-59.
HORN, G. H.
1893. Miscellaneous Coleopterous studies. Trans. Amer. Ent. Soc., 20:136-144, 1 pl.
VAN DYKE, E. C.
1918. New Inter-tidal rock-dwelling Coleoptera from California. Ent. News, 29:303-308.
WICKHAM, H. F.
1904. The metamorphosis of *Aegialites*. Canad. Ent., 36:57-60, pl. 2.

Family HELODIDAE

This family has a world-wide distribution. The North American helodids are small, mostly 3 or 4 mm. in length and rarely exceeding 6 mm. Many of the species are yellowish-brown, some are rufous, some black, and a few are bicolored. The larvae are associated with water or very damp places.

The adults are never strictly aquatic, but those of some species unhesitatingly enter the water to escape from danger. Adults of certain genera, such as *Scirtes*, occur on foliage near the larval habitat, but are not known to enter the water; some, including species of *Cyphon*, may be taken on flowers.

Only three genera, *Elodes*, *Cyphon*, and *Scirtes* have been reported from California, but *Prionocyphon* has been found in western Arizona.

Relationships.—The family Helodidae belongs in the superfamily Dascilloidea, and has often been included as but a division or group within the Dascillidae. On the other hand, even in some fairly recent works the Eucinetidae, and *Eubria* of the Psephenidae, have been treated as subfamilies of Helodidae.

The rather *Lepisma*-like larvae are at once known by their long, multiarticulat antennae (fig. 13:46*c-e*). Such antennae are not found in any other larval Coleoptera, nor indeed, according to Crowson, in the larvae of any other holometabolous insects. However, Kraatz (1918) has shown that the first and second instar larvae of *Scirtes tibialis* Guérin have fewer antennal segments (fig. 13:46*c,d*), and Peyerimhoff (1913) said that those of the European *Prionocyphon serricornis* Müller changed into short and more typically coleopterous antennae when the larvae were exposed to dry environments.

Respiration.—Adults are not known to stay under water for long at a time, and presumably rely on what air they carry down under their elytra. By means of simple experiments with nitrogen-saturated and with oxygen-saturated water, Popham (1954) has shown that the air bubbles and/or film carried down by aquatic insects do function also as physical gills. To benefit the insects these air supplies, exposed to the water, must be connected to the tracheal system. For most species it remains to be found what percentage of the total oxygen requirement is supplied in this manner; the temperature of the environment and activity of the insect are major factors in the appraisal.

The spiracles of the larvae are vestigial or absent, except for an annuliform pair on the eighth abdominal segment. The larvae have to come to the surface periodically to replenish their air supply (fig. 13:48*a*). At least some species have very large tracheal reservoirs; those of *Prionocyphon limbatus* have been illustrated by Good (fig. 13:46*a*). In addition, the larvae have terminal retractile gills[13] (fig. 13:47*a,b*) with which they have been presumed to extract some oxygen from the water.

Benjamin Walsh (*in* Osten Sacken, 1862:117) stated that submerged larvae of *P. discoideus* extruded the gills "vibrating [them] vigorously up and down . . ." Such action suggests that the gills may be of considerable use as such, but Treherne, in controlled

[13]It is remarkable that of the two Nearctic species of *Prionocyphon*, the larvae of *P. discoideus* (Say) have three prominent tassels of bipectinate gills (fig. 13:47*a*), whereas those of *P. limbatus* LeConte have five fingerlike gills resembling those of figure 13:47*b*.

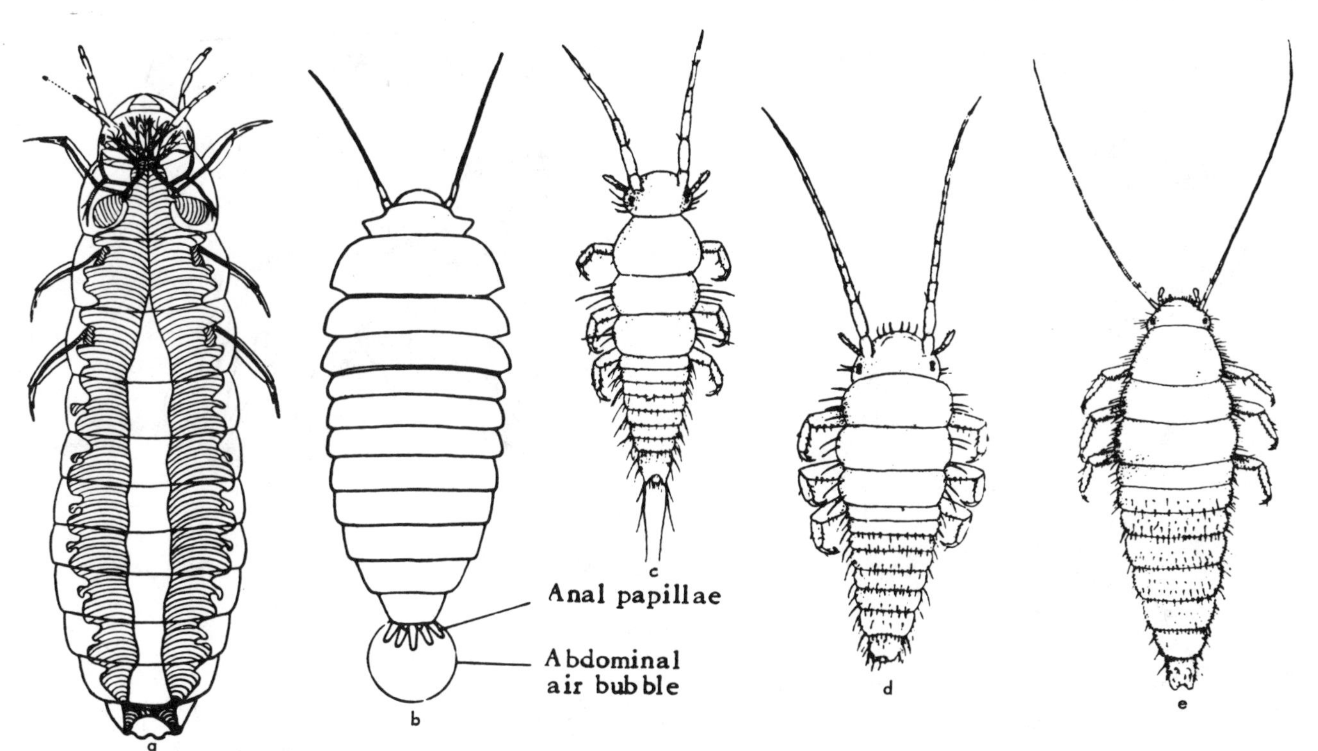

Fig. 13:46. Helodidae, larvae. *a*, *Prionocyphon limbatus*, showing tracheal reservoirs; *b*, *Elodes minuta* (European sp.), to show extruded anal papillae and abdominal air bubble; c-e, *Scirtes tibialis*, first, second, and later instar larvae (*a*, Good, 1924; *b*, Treherne, 1952; c-e, Kraatz, 1918).

experiments with larvae of the European *Elodes minuta* (1952) and of *E. minuta* and *E. marginata* (1954), concluded that the anal papillae do not play an important part in the respiration of the larvae (neither are they of consequence in the various dipterous larvae which have them). Their chief function seems to be to absorb salts in the process of stabilizing the osmotic pressure of the haemolymph.

The larva of *E. minuta* has large tracheal air sacs. By contraction of the abdominal sacs it extrudes a bubble from the last pair of abdominal spiracles. This bubble is held in place chiefly by apical semi-hydrofuge hairs, and is in direct connection with the air in the tracheal system (fig. 13:46*b*). Treherne found that larvae which had the terminal spiracles blocked to prevent formation of an air bubble, and the tip of the abdomen sealed to forestall protrusion of the papillae, could live less than half as long when submerged in well-aerated water as could untreated larvae. The same result was obtained when only the spiracles were plugged. Thus the terminal air bubble was the important respiratory mechanism. Normal larvae were found to consume about 4.50 cc. of oxygen per gram of dry tissue per hour, and the bubble functioned to extract oxygen from water even at relatively high concentrations. It also functioned to eliminate CO_2 at a greater rate than did the body surface; the papillae allowed it to escape much less rapidly, but still more than twice as fast as did the general body surface.

Life history.—So little is known of the life histories of the Nearctic species of Helodidae that it is unsafe to generalize. The eggs are presumably laid in damp places near water; in the case of *Prionocyphon* this may be in water-filled tree holes. The larvae are aquatic; those of only three North American species have been adequately described.

Larvae of *Scirtes tibialis* Guérin were said by Kraatz (1918:393) to eat the leaves of duckweed (*Lemna minor*); but Beerbower (1944:677) found those of *S. orbiculatus* (Fabricius) feeding on minute algae. Larvae of *Prionocyphon limbatus* are reported to eat the broken-down epidermal cells of various dead leaves in ponds. The larvae of *Scirtes grandis* Motschulsky of Ceylon were said by Nowrojee to be predaceous, but later authors doubt this. Those of *S. championi* Picado occur in water in the leaf axils of epiphytic bromeliads in Costa Rica.

The specialized mouth parts of *Elodes hausmanni* Gredler (fig. 13:47*d,g*), and of some other European helodids have been illustrated by Beier. The mandibles with their brushes and the toothed parts of the hypopharynx work with machinelike regularity, gathering and filtering fine detritus from the surface of stones and underwater vegetation. The teeth on the anterior hypopharyngeal comb vary in shape and number, depending on the species and genus of the larva.

Mature larvae of most genera leave the water and form pupal cells in damp soil. Those of *Scirtes tibialis* remain on the undersides of *Lemna* leaves; the pupa is attached by its anterior end and hangs freely (fig. 13:47*c*). Walsh found the pupae of *Prionocyphon dis-*

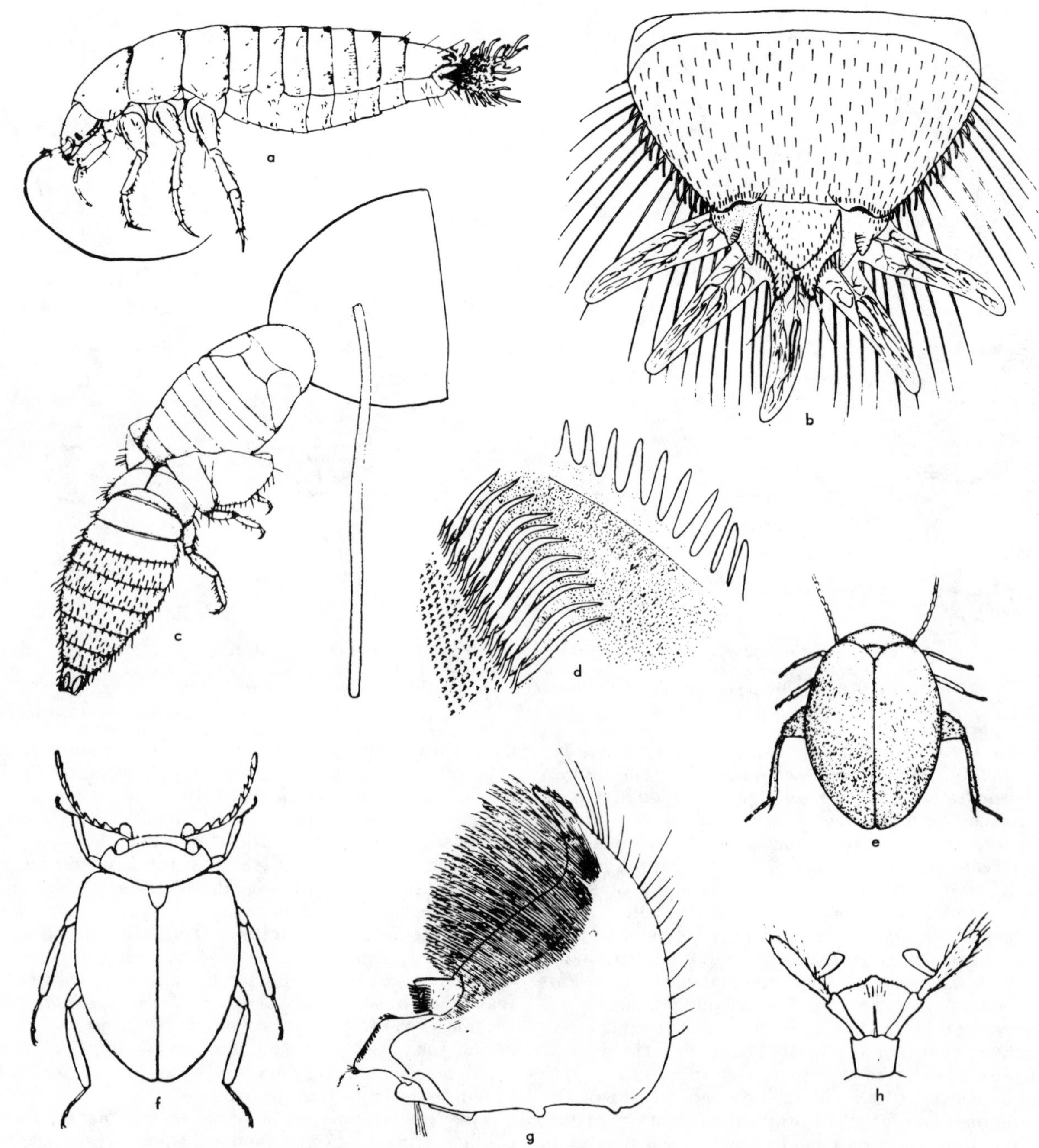

Fig. 13:47. Helodidae. *a, Prionocyphon discoideus*, larva; *b, Elodes hausmanni*, apex of larval abdomen to show retractile gills; *c, Scirtes tibialis*, underside of a *Lemna* leaf showing attached pupa and last larval skin; *d, Elodes hausmanni*, hypopharynx of fourth instar larva; *e, Scirtes tibialis*, adult; *f, Prionocyphon serricornis*, adult (European sp.); *g, Elodes hausmanni*, mandible of fifth instar larva; *h*, labial palpi of an adult helodid (*a*, Böving and Craighead, 1931; *b,d,g*, Beier, 1949; *c,e*, Kraatz, 1918; *f,h*, Portevin, 1931).

coideus (Say) in the damp decaying wood of an oak stump, the lower part of which contained water and numerous larvae of the same species. The pupa of *P. limbatus* Le Conte is remarkable for the four huge spines on the pronotum (fig. 13:48*b*). There appears to be but one generation a year, and some, if not most, species overwinter as larvae.

Habitat and distribution.—Adults of species of *Ora, Scirtes, Cyphon, Microcara,* and *Sarabandus* are usually found on vegetation near the water or in damp places in which the larvae occur. Those of *Prionocyphon* hide under dead leaves or crawl on dead fallen branches in more forested areas. Adults of *Elodes* fly along streams or around springs, or hide

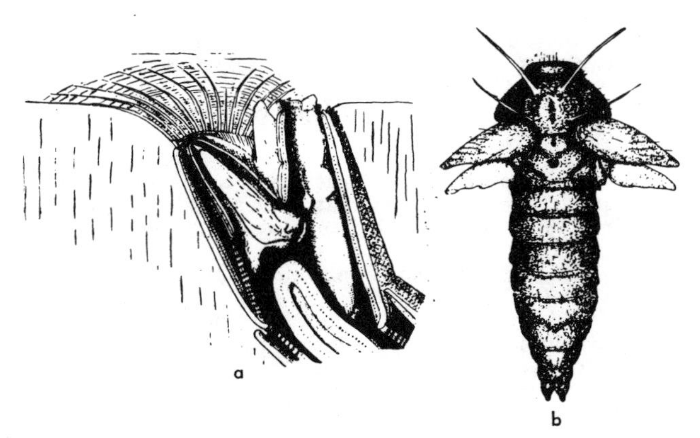

Fig. 13:48. Helodidae. *a, Elodes hausmanni*, semidiagramatic longitudinal section through abdomen of larva replenishing its air supply at the water surface; *b, Prionocyphon limbatus*, dorsal view of pupa (*a*, Beier, 1949; *b*, Good, 1924).

under stones and logs projecting into the water in which the larvae are found.

The only Nearctic adults which are even partly aquatic are those of *Elodes*. Following the original description of *E. aquatica*, Blaisdell wrote, "A colony was discovered on the undersurface of a rather large flat rock, that projected out of and over the surface of the water, at an angle of a few degrees, in a rather swiftly flowing stream. When the rock was lifted up the beetles were disturbed, most of them falling upon the water. As soon as they struck the water, they began to swim briskly and to gyrate as do the Gyrinids; they dove beneath the surface and swam rapidly under the surface, all endeavoring to return to the rock."

Wintersteiner recorded larvae of *Cyphon* sp. from among dead leaves in a pond, and I once found them numerous in wet moss around the edge of a cold spring in wooded hills. R. L. Usinger *(in litt.)* found them under stones in a spring.

Pratt took larvae of *Prionocyphon discoideus* "in the very dirty water in holes in the middle of poplar stumps . . . ," and remarked that the body of a dead one had been eaten out and rendered transparent by the larvae of a heleid fly.

W. E. Snow (unpublished data) has found larvae of species of *Cyphon, Elodes,* and *Prionocyphon* in tree holes in Illinois. A. M. Laessle *(in litt.)* found those of *Cyphon* spp. (det. van Emden) in the rain-water pockets of Jamaican bromeliads. Picado (1913) reared his *Scirtes championi* from larvae found in bromeliads in Costa Rica.

The genera *Elodes, Microcara,* and *Prionocyphon* are Holarctic, with additional species in the Neotropics: *Cyphon* is nearly world-wide, *Ora* is dominately New World but recorded also from India, and *Scirtes* has a host of species in Australia and Africa as well as in the Neotropics.

Key to the Nearctic Genera of Helodidae

Adults

1. Hind femora much like those of front and middle legs, not greatly enlarged for jumping; spurs of hind tibiae small 2
— Hind femora greatly enlarged, broadened, saltatorial; spurs of hind tibiae long, one being at least 1/2 as long as first tarsal segment (fig. 13:47*e*) 6
2. Meso- and metasternal processes in contact between middle coxae, separating them 4
— Mesosternal process short, narrow, not contacting metasternum, middle coxae contiguous in about apical ½ 3
3. First segment of hind tarsi flattened above, finely margined laterally, 2nd with part of hind margin prolonged, hiding basal section of 3rd; labial palpi with 3rd segment arising from side (about mid-point) of 2nd (fig. 13:47*h*) . . . . *Elodes* Latreille 1796
— First segment of hind tarsi rounded above, not laterally margined, 2nd not produced posteriorly, not hiding part of 3rd; labial palpi with 3rd segment arising from apex of 2nd . . . . *Sarabandus* Leech 1955
4. First antennal segment large, fully twice as broad as any of those following, expanded anteriorly; 2nd segment arising from posterior apical angle of 1st and from under a slight margin; 3rd segment very small, 1/2 as long as 2nd; broadly ovate species (fig. 13:47*f*) . . . . *Prionocyphon* Redtenbacher 1858
— Antennae not as above; body shape various 5
5. Labial palpi with 3rd segment arising from side (about mid-point) of 2nd (fig. 13:47*h*) . . . . *Microcara* Thomson 1859
— Labial palpi with 3rd segment arising from end of 2nd . . . . *Cyphon* Paykull 1799
6. Hind coxae meeting along full length of median line, hind margins conjointly forming a subquadrate plate which is not on same plane as intercoxal process of abdomen[14] (fig. 13:47*e*) . . . . *Scirtes* Illiger 1807
— Hind coxae touching each other only anteriorly, arcuately diverging posteriorly where they are about on a plane with the abdominal process which separates them[14] . . . . *Ora* Clark 1864

Key to the Mature Larvae of Certain Genera of Nearctic Helodidae[15]

1. Anterior margin of hypopharynx with a central cone bearing 2 pairs of flat, and usually serrate, spines; head with 1 or 2 ocelli on each side 2
— Cone of hypopharynx with 1 pair of spines; head with 3 ocelli on each side . . . . *Elodes* Latreille 1796
2. Abdominal segments 3 to 6 with a regular series of short, flattened differentiated setae along lateral margins 3
— Sides of abdominal segments with only scattered thin setae, like those of dorsum . . . . . . *Cyphon* Paykull 1799
3. Anterior angles of labrum bent under, inner margin projecting from under transverse front margin (fig. 13:47*a*) . . . . *Prionocyphon* Redtenbacher 1858
— Anterior angles of labrum on same plane as rest, labrum thus simply emarginate (fig. 13:46*c,e*) . . . . *Scirtes* Illiger 1807

Key to the California Species of Elodes[16]

Adults

1. Last visible abdominal sternite of male with a deep, rather narrow V-shaped median apical emargination;

[14]This separation will not hold for all exotic species.

[15]With the aid of unpublished data supplied by Drs. W. E. Snow and F. van Emden. However, so few larval helodids have been described that the characters chosen above may not all hold.

[16]The North American species of this genus are difficult to separate, and have not been seriously studied since Horn's paper

elytra piceous, pronotum piceous discally, luteous to rufotestaceous laterally; Marin County to San Mateo County .... *aquatica* Blaisdell 1940
— Last visible abdominal sternite of male with a broad and rather shallow U-shaped emargination; elytra piceous, luteous or bicolored 2
2. Discal 3rd of pronotum piceous, sides luteous to rufotestaceous; elytra piceous, or piceous near suture and luteous laterally; Del Norte County .... *nunenmacheri* Wolcott 1922
— Pronotum luteous, or with a semicircular darker area showing through from underside of apex, above head; elytra luteous, apices piceous; San Francisco Bay region .... *apicalis* LeConte 1866

REFERENCES

BEERBOWER, FRED V.
1944. Life history of *Scirtes orbiculatus* Fabricius (Coleoptera : Helodidae). Ann. Ent. Soc. Amer., 36: 672-680, 1 pl.

BEIER, M.
1949. Koerperbau und Lebensweise der larve von *Helodes hausmanni* Gredler (Col. Helodidae). Eos, 25: 49-100, 17 text figs.
1952. Bau und funktion der Munderzeuge bei den Helodiden-larven (Col.). Trans. IXth Internat. Congress Ent. (Amsterdam, 1951), 1 (Sect. 1): 135-138, 2 figs.

BLAISDELL, F. E.
1940. A new species of *Helodes* from Marin County, California (Coleoptera : Dascillidae). Ent. News, 51: 190-191.

GOOD, H. G.
1924. Notes on the life history of *Prionocyphon limbatus* Lec. (Helodidae, Coleoptera). Jour. N.Y. Ent. Soc., 32: 79-84, pls. VIII and IX.

HORN, G. H.
1880. Synopsis of the Dascyllidae of the United States. Trans. Amer. Ent. Soc., 8:76-114, 1 pl.

KEMPERS, K. J. W. BERNET
1944. De larven der Helodidae (Cyphonidae). Tijdschrift voor Ent., (1943), 86: 85-91, 38 figs. (Published April 7, 1944).

KRAATZ, W. C.
1918. *Scirtes tibialis,* Guer. (Coleoptera-Dascyllidae), with observations on its life history. Ann. Ent. Soc. Amer., 11: 393-400, pl. XXXV.

LAESSLE, ALBERT M.
(information in letters, 1954).

LEECH, H. B.
1955. A new genus for *Cyphon robustus* LeConte (Coleoptera : Helodidae). Pan-Pac. Ent., 31:34.

LOMBARDI, DINA
1928. Contributo alla conoscenza dello *Scirtes hemisphaericus* L. (Coleoptera-Helodidae). Boll. Lab. Ent. Bologna, 1: 236-258, 11 text figs.

OSTEN SACKEN, R.
1862. Descriptions of some larvae of North American Coleoptera. Proc. Ent. Soc. Phila., 1: 105-130, pl. I, (*Prionocyphon discoideus,* pp. 115-118, no figs.)

PEYERIMHOFF, P. DE
1913. Le double type larvaire de *Prionocyphon serricornis* Müll. (Col. Helodidae). Bull. Soc. Ent. France, (1913), No. 6, pp. 148-151.

PICADO, C.
1913. La larve du genre *Scirtes*. Bull. Soc. zool. Fr., 37: 315-319. ("Paru le 31 Janvier 1913".)

POPHAM, E. J.
1954. A new and simple method of demonstrating the physical gill of aquatic insects. Proc. Roy. Ent. Soc. London (Ser. A), 29: 51-54, 1 text fig.

of 1880. Examples of about a dozen undescribed species from California have been seen. Despite couplet 2 of the above key, color is probably not of value as a primary character in distinguishing species.

PRATT, F. C.
1907. Notes on "punkies." (Ceratopogon spp.). U.S.D.A., Bur. of Ent., Bull. 64: (2+) 23-28, text figs. 3-6.

SNOW, WILLIS E.
1954. (Dr. Snow kindly provided data from his unpublished thesis.)

TREHERNE, J. E.
1952. The respiration of the larva of *Helodes minuta* (Col.). Trans. IXth Internat. Congress Ent. (Amsterdam, 1951), 1: 311-314, 1 text fig., 2 tables.

WALSH, B. D.
1862. (as quoted in Osten Sacken, q.v.)

WINTERSTEINER, F.
1913. (as reported in) Proceedings of the New York Entomological Society. Jour. N.Y. Ent. Soc., 21: 90.

WOLCOTT, A. B.
1922. A new species of *Helodes* (Helodidae, Col.). Bull. Brooklyn Ent. Soc., 17: 94.

Family DRYOPIDAE[17]

Dryopids (fig. 13:49*a*) are usually moderately small (8 mm. or less), dull silver gray, black, or brown beetles found crawling about upon the bottom of streams. They vary from semi- to truly aquatic although none can swim. The claws are usually large and adapted for holding on to the substrate. In most species the head may be retracted into the thorax, so that the mouth parts and most of the antennae are hidden. Thus only the front of the head and part of the eyes are exposed.

The larvae are elateroid. The body is long and cylindrical with the integument smooth. Abdominal tergites form a complete ring except for a narrow ventral groove on the first five segments. Ninth segment with caudal chamber is closed by an operculum but without gills or prehensile hooks.

Respiration.—The adult beetles have an air reservoir beneath the elytra as do most aquatic beetles but they seldom ascend to the surface to replenish their air supply as do those of many other families. Instead they depend largely upon oxygen obtained by diffusion while submerged. Much of the body and legs are covered by hydrofuge tomentum or pile which surrounds them with a blanket of air. Into this blanket oxygen diffuses from the water, and carbon dioxide diffuses outward. Since they usually inhabit well-aerated running streams, it seems that oxygen obtained in this way is sufficient to maintain their unhurried activity. In addition they are frequently observed to pick up small bubbles of oxygen produced by aquatic plants during photosynthesis. They are often found clustered under rough, irregular stones which have trapped small bubbles of air in the crevices on the underside.

None of the known dryopid larvae of North America have caudal gills. All have the usual lateral spiracles

[17] After this manuscript was sent to the printer, Hinton's redefinition of the family appeared (Proc. Zool. Soc. Lond., 125: 565-566. November, 1955). He writes that ". . . a large section of the family (*Sostea* and allied genera) is strictly terrestrial in the adult stage." He states that the larvae of *Dryops* and *Helichus* have been taken in the soil or in decaying wood, and that the spiracles of the mesothorax and first eight abdominal segments are functional in all instars in the known dryopid larvae.

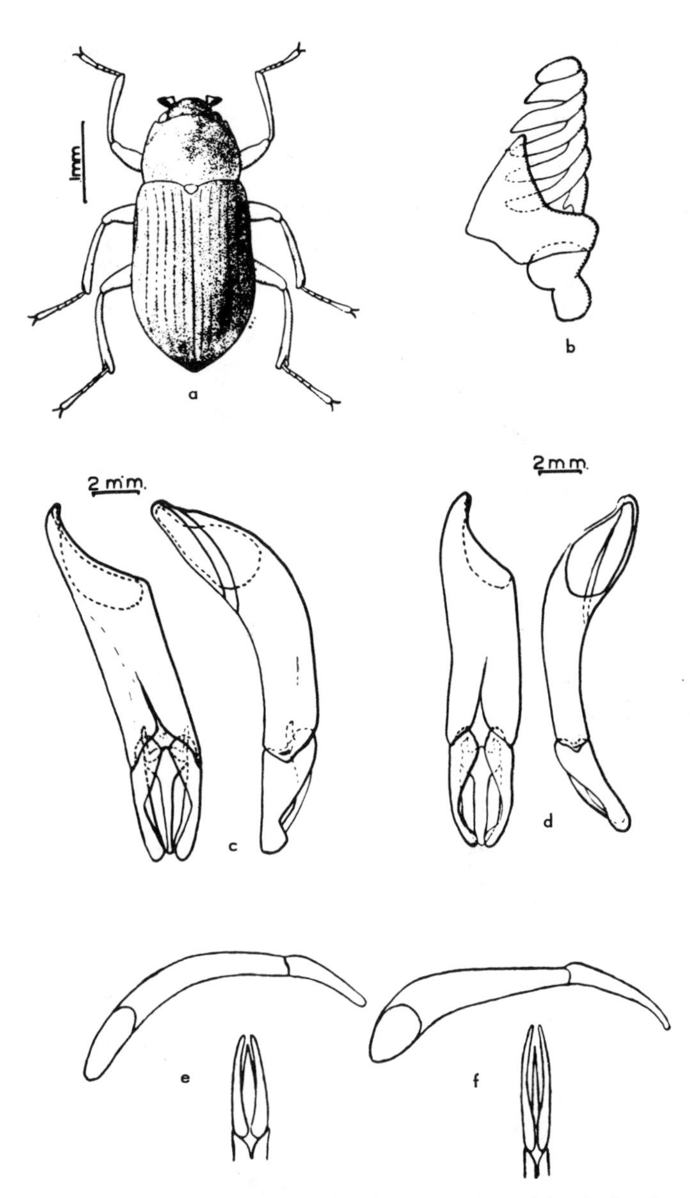

Fig. 13:49. Dryopidae. *a, Helichus suturalis*, adult; *b, Dryops* sp., antenna; *c, Helichus immsi*, dorsal and lateral view of male genitalia; *d, Helichus productus*, dorsal and lateral view of male genitalia; *e, Helichus striatus*, lateral and dorsal view of male genitalia; *f, Helichus suturalis*, lateral and dorsal view of male genitalia (*a,c,d*, Hinton, 1937; *b*, Hinton, 1939; *e,f*, Musgrave, 1935).

on the mesothorax and the first seven abdominal segments and a pair of dorsal spiracles on the eighth, but it is not known if any of these are functional. *Dryops luridus* Erichson from Europe, which is aquatic but has been reported also from damp soil, is entirely surrounded by a film of air when submerged. The integument has a smooth, glassy texture. The habitat of the larva of *D. arizonensis* is not known. The larva of *Pelonomus* is similar in appearance to *D. luridus*. *Helichus* larvae have two large, irregular, raised patches of microscopic hydrofuge pile on each of the segments with lateral spiracles; the spiracle is on the anterior margin of the narrow isthmus between the two patches. The enlarged dorsal pair of spiracles on the eighth abdominal segment are close together and in a position to suggest that they may be used for surface breathing.

Life history.—Little is known of the life history of these beetles. The adults of *Helichus* are abundant in many places, yet only two larvae are known to have been collected in the United States. They were collected in north central California from a very small and a medium-sized creek about a mile apart. Both were taken from swift, turbulent water by turning rocks and agitating the bottom. The shape of the last abdominal segment suggests a burrowing habitat. Larvae of *Pelonomus* were found by Sanderson in a damp flood-plain depression. The pupa of *Dryops luridus* is terrestrial.

The adults feed on vegetation at the bottom of streams, but the larvae are believed to be root feeders. Although the wings are well developed I have never seen the adults in flight. They are widely distributed in desert areas indicating dispersal by this method. I found several adults of two species in a large open-top steel water-storage tank for a stock-watering trough. The water was supplied from a well by a windmill driven pump. It was about fifteen miles from this tank to the nearest stream.

Habitat and distribution.—The Dryopidae are found mostly in the tropical and warmer temperate regions of the world with an occasional species extending into the cooler regions. *Dryops* is found on every continent except Australia. *Pelonomus* is from southeastern United States. *Helichus* is found in Mexico and most of the United States, with *H. lithophilus* in the east and *H. foveatus* in the west extending into Canada. In California it is found in moderately to rapidly flowing creeks or rivers with rocky or gravel bottoms, from sea level up to an elevation of 6,000 feet. *H. suturalis* ranges from Mexico to northern California and southern Utah.

Taxonomic characters.—The placement and form of the antennae serve to separate the Nearctic genera but cannot be relied on in other parts of the world. In *Helichus* pubescence is the most used character since it covers most of the body, masking many other possible characters. It should always be used with discretion since the specimen may assume quite a different appearance if it is rubbed off or soiled with grease or dirt. If one is not familiar with the group, it is best to check the male genitalia, especially with *H. productus, H. immsi,* and several related species found in Arizona, New Mexico, Texas, and Mexico, for which no dependable distinguishing characters have been found. Size and slight differences in shape are of limited usefulness and difficult to describe. There is a tendency in *H. foveatus* for alternate interstrial spaces to be more convex. Several species and subspecies have been based largely on the presence or lack of this character, but study of specimens from many localities indicates that it is of minor importance. *H. suturalis* is also very variable and is reported by Hinton to have the number of antennal segments sometimes reduced owing to the fusing together of the terminal segments of the pectinate club.

Key to Nearctic Genera of Dryopidae

Adults

1. Second segment of antennae produced into an earlike process (fig. 13:49*b*) 2
— Second antennal segment not thus produced; southeast U.S. .... *Pelonomus* Erichson 1847
2. Antennae approximate; thorax with sharp-edged longitudinal line on each side .... *Dryops* Olivier 1791
.... *Dryops arizonensis* Schaeffer 1905; Arizona
— Antennae widely separated, thorax without longitudinal line; northern America .... *Helichus* Erichson 1847

Larvae

1. Abdominal segments 1-7 with patches of raised hydrofuge pile above and below each lateral spiracle, 9th abdominal segment with dorsal surface flattened, ventral operculum with 2 clawlike tubercles on posterior margin .... *Helichus* Erichson 1847
— Lateral spiracles without differentiated or raised areas, 9th abdominal segment with dorsal surface strongly convex, operculum without tubercles 2
2. Body integument with a reticulate pattern of small punctures; gular sutures obliterated, with 2 pairs of setae near where the sutures would be. ....
.... *Pelonomus* Erichson 1847
— Body integument without reticulate pattern of small punctures; gular sutures present, only 1 pair of setae on gula near suture .... *Dryops* Olivier 1791

Key to the California Species of Helichus Erichson

Adults

1. Abdominal sternites uniformly densely pubescent ... 2
— Pubescence of last sternite not matching that of preceding sternites 3
2. No dependable external characters; male genitalia (fig. 13:49*c*) with paramere, as viewed from side, 1/2 as wide as long, clasping aedeagus below from apex past the middle; Utah, New Mexico, central and southern California in valley and foothill streams ....
.... *immsi* Hinton 1937
— Paramere (fig. 13:49*d*) as viewed from side much narrower, 1/2 as wide as long, clasping aedeagus only at the tip; central and southern California in valley and foothill streams .... *productus* LeConte 1852
3. Thorax foveate each side behind middle; male genitalia as in fig. 13:49*e*; British Columbia, Arizona, and California in streams up to 6,000 feet.........
.... *foveatus* LeConte 1852
— Thorax without such indentations (fig. 13:49*f*); Mexico, Utah, and California in streams at lower elevations .... *suturalis* LeConte 1852

REFERENCES

HINTON, H. E.
1937. *Helichus immsi*, sp. n. and notes on other North American species of the genus. Ann. Ent. Soc. Amer., 30:317-23, 1 text fig., 1 pl.
1939. An inquiry into the natural classification of the Dryopoidea, based partly on a study of their internal anatomy (Col.). Trans. Roy. Ent. Soc. London, 89: 133-184, 105 text figs., 1 pl.
MUSGRAVE, P. N.
1935. A synopsis of the genus *Helichus* Erichson in the United States and Canada, with description of new species. Proc. Ent. Soc. Wash., 37:137-145, incl. pl. No. 17.

Family ELMIDAE

Riffle Beetles

The larvae and most of the adults of this family are truly aquatic but are unable to swim. They are found crawling about on the bottom of the stream or cling-

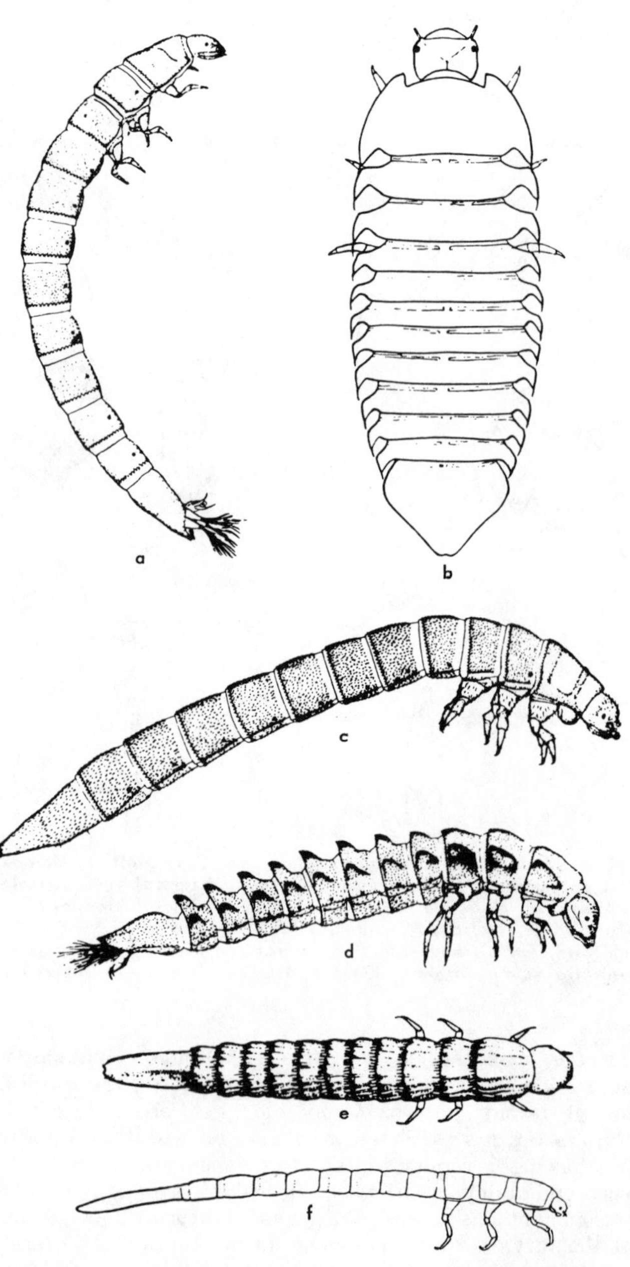

Fig. 13:50. Elmidae, larvae. *a*, *Zaitzevia* sp.; *b*, *Phanocerus clavicornis*; *c*, *Narpus* sp.; *d*, *Promoresia* sp.; *e*, *Heterelmis* sp.; *f*, *Dubiraphia* sp. (*a,c,d*, West, 1929; *b,e*, Hinton, 1940; *f*, Sanderson, 1954).

ing to vegetation. Adult elmids are small, compact beetles with an average size of only 2-3 mm. The color of the dorsal surface may be black, gray, brown, or red, with the ventral surface clothed with pile or tomentum so that it usually has a silver gray color. Like the Dryopidae, they have the last tarsal segment and claws greatly enlarged for anchoring themselves to the substrate, and the head may be retracted into the prothorax, concealing the mouth parts. The larvae are elongate, cylindrical, hemicylindrical or subtriangular in cross section, with the head conspicuously narrower than the thorax and the abdomen tapering posteriorly. The last abdominal segment has a ventral caudal chamber, closed by an operculum, into which can be retracted two prehensile hooks and three tufts of filamentous gills.

Relationships.—The Elmidae are closely related to both the Dryopidae and the Limnichidae. General body form closely resembles many of the dryopids, but the antennae are never pectinate. There are also numerous differences in the internal anatomy. The larvae are quite different. Adults of the Limnichidae do not closely resemble elmids, but the larvae are quite similar.

Respiration.—The method of respiration of adults is similar to that discussed in the section on Dryopidae. All elmid larval instars except the first have nine pairs of more or less evident spiracles situated on the mesothorax and first eight abdominal segments. These are reported by Susskind not to be connected with the tracheae until the last larval instar. Respiration in the water is apparently accomplished entirely by means of the three tufts of retractile gills situated in the caudal chamber. The number and length of these gill filaments will vary with the different species. Certain species which live in fast, cold water have been noted to have shorter or fewer gills than normal. The larva of *Dubiraphia* which has the last abdominal segment much longer than in the other genera, also has the caudal filaments two to three times as long as in most genera. This may explain why *Dubiraphia* is one of the few genera to inhabit lakes and slow moving rivers. Larvae have been observed to expand and contract their filaments rhythmically when the oxygen balance of the water is reduced.

Life history.—The female genitalia are not adapted for inserting the eggs into plant tissue as are those of Dryopidae. It has been suggested that a life cycle may take two years in this group because adults and larvae of the same species were taken at the same time. However, in California as many as five size groups of larvae, as well as adults, have been taken by processing stream screenings through a Berlese funnel. When mature the larvae crawl out of the water and pupate under stones on the bank. The few species that have been reared usually emerged as adults in about two weeks. Both larvae and adults feed upon algae, moss, and other vegetative matter in the water, including the roots of larger plants.

Habitat and distribution.—Elmids are found in streams of all types but may be rare in those with seasonal flow, heavy sediment load, mud or sand bottoms, low gradient, or low oxygen content. Only a few genera are found in lakes, ponds, or slow moving streams. The most common in California is *Dubiraphia*. *Microcylloepus* may be found along rocky shore lines where wave action aerates the water. *Microcylloepus moapus*, *M. thermarum*, and *Stenelmis calida* are all restricted to warm springs in Nevada. *Heterlimnius koebelei* Martin has been collected at altitudes up to 8,000 feet in California. Adults of *Lara* are not aquatic, but are found under logs and trash at the water's edge. They are more active than those of other genera and will quickly hide if exposed, even scrambling over the surface of the water to reach shore. Most elmids are very lethargic and will play dead for some time after being exposed. *Cleptelmis* may not start to emerge from masses of moss placed in a Berlese funnel for twelve to fifteen hours.

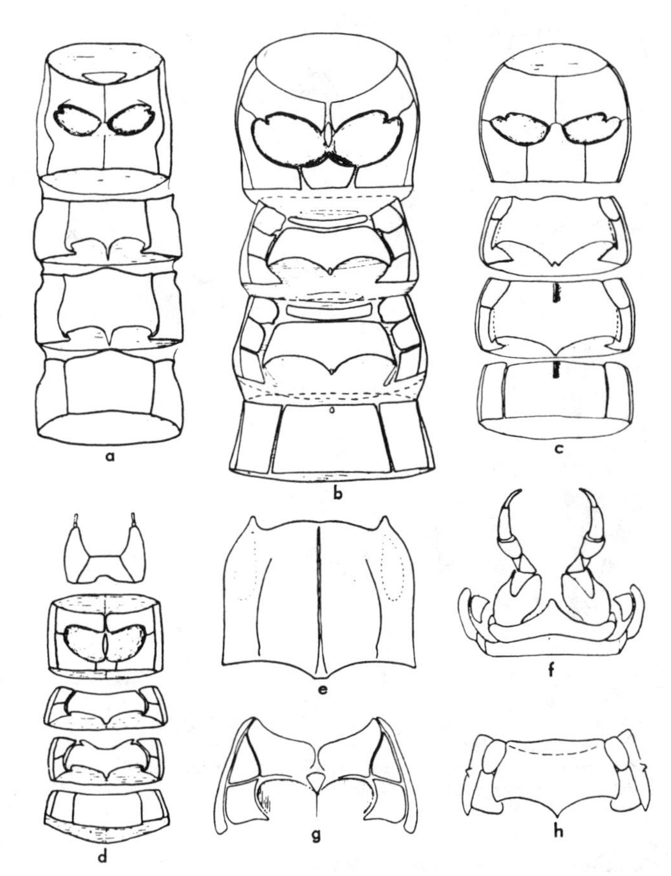

Fig. 13:51. Elmidae. *a-d*, ventral view of thorax and first abdominal segment; *a*, *Cylloepus* sp.; *b*, *Heterelmis* sp.; *c*, *Elsianus* sp. (second abdominal segment shown here); *d*, *Microcylloepus* sp.; *e*, *Ampumixis dispar*, adult pronotum; *f*, *Optioservus* sp., larval mesothorax, ventral; *g*, *Heterlimnius* sp., larval prothorax, ventral; *h*, *Ordobrevia* sp., larval mesothorax, ventral (*a-d*, Hinton, 1940; *e,h*, Sanderson, 1953; *f,g*, Sanderson, 1954).

Elmids are found throughout the tropical and temperate regions of the world. In North America they extend from Mexico through the United States into southern Canada. I have seen larvae from as far north as Great Slave Lake. In general, the Nearctic genera may be placed in six distribution groups. The genera

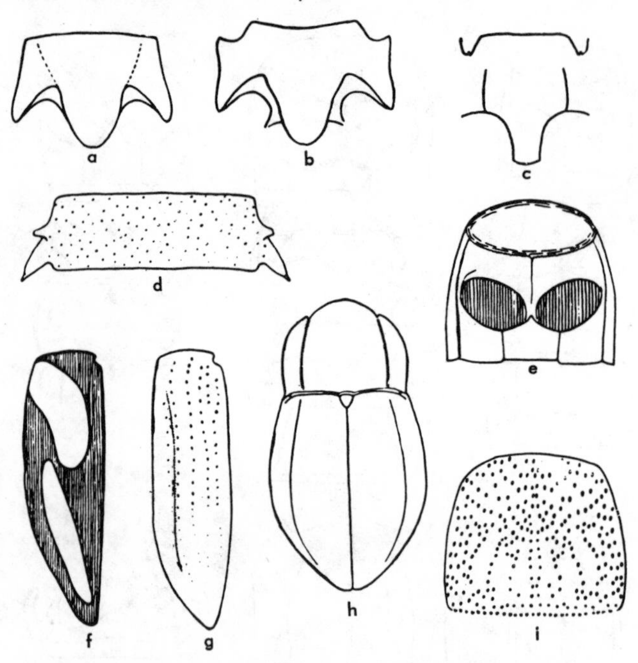

Fig. 13:52. Elmidae. *a, Dubiraphia* sp., adult prosternum; *b, Gonielmis dietrichi*, adult prosternum; *c, Microcylloepus pusillus*, adult prosternum; *d, Ancyronyx* sp., third larval abdominal tergite; *e, Stenelmis* sp., larval prothorax, ventral; *f, Gonielmis dietrichi*, left elytra; *g, Ordobrevia* sp., left elytra; *h, Limnius* sp., adult; *i, Microcylloepus* sp., larval pronotum (*a,b,d-i*, Sanderson, 1953; *c*, Hinton, 1935).

Phanocerus, Cylloepus, Neoelmis, Elsianus, Heterelmis, and *Hexacylloepus* are mainly Mexican in distribution, extending only into the southern states of the United States. *Dubiraphia, Optioservus,* and *Microcylloepus* are found throughout the United States, *Optioservus* ranging well into Canada and *Microcylloepus* into Mexico. *Gonielmis* is restricted to the southeastern United States. *Stenelmis, Limnius, Macronychus,* and *Ancyronyx* are in the eastern part of the continent. *Stenelmis,* which contains the largest number of species in the United States, has only one species west of the Rocky Mountains. It is significant that this species, *S. calida,* is taken in warm springs in Nevada. *Cleptelmis, Heterlimnius, Narpus,* and *Zaitzevia* are taken throughout the western United States and into Canada. *Lara, Ordobrevia, Ampumixis, Atractelmis,* and *Rhizelmis* are found along the Pacific Coast.

Taxonomic characters.—Variations in the epipleura and the locking devices for holding it firmly against the ventral abdominal segments to form a rigid airtight chamber beneath the elytra, serve as useful characters in breaking the genera down into related groups. *Lara,* like the Dryopidae and Limnichidae, has the epipleura broadly reaching the medial suture at the apex of the elytra, but the epipleura is twisted under and applied to the inner side of the elytra near the apex. In the Elmini the epipleura is always tapered to a point and reaches the apex only in group two, containing *Ordobrevia* and *Stenelmis* (fig. 13:53*a*). The epipleura is shortened and usually gradually tapered to its apex near the end of the fourth or middle of the fifth abdominal segment, in the next composite group (fig. 13:53*e*). This contains *Ampumixis, Atractelmis, Cleptelmis,* and *Rhizelmis,* which are closely related and characterized by the three-segmented maxillary palpi. Also in this group are *Macronychus, Ancyronyx, Narpus,* and *Dubiraphia,* which are not closely related to each other or to any other North American genera. Closely related *Promoresia, Optioservus,* and *Heterlimnius,* together with the more distantly related *Zaitzevia* and *Limnius,* form a group; they have the posterior lateral angle of the fourth abdominal segment produced into a prominent tooth which is usually bent vertically to fit into a slightly widened area near the apex of the shortened epipleura (fig. 13:53*b*). The fifth group contains *Elsianus, Neoelmis, Hexacylloepus,* and *Cylloepus;* each has a tooth at the middle of the fifth abdominal segment, fitting into a conspicuously widened area near the apex of the shortened epipleura (fig. 13:53*d*). Groups two and five have a number of characters in common including the clasping of the epipleura by the basal half of the fifth abdominal segment, indicating close ancestral relationship.

The number of segments in the maxillary palpi (3 or 4) is a useful character but difficult to see, especially if the head is retracted. The shape and sculpture of the pronotum, and the sublateral carinae of the pronotum and elytra, characterize many of the genera. Two unrelated genera, *Macronychus* and *Zaitzevia,* have the antennal segments reduced in number and the last segment enlarged. *Heterlimnius* has the antennae ten-segmented in one species and eleven in the other. In a few cases it has been necessary to use the distribution of the patches of tomentum on the body or legs.

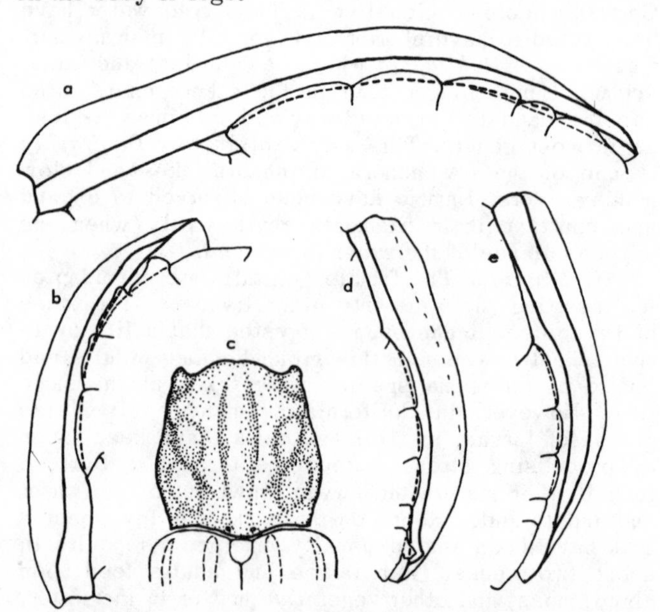

Fig. 13:53. Elmidae. *a, Stenelmis calida*, epipleura and abdominal segments; *b, Optiservus quadrimaculata*, epipleura and abdominal segments; *c, Stenelmis* sp., adult thorax, dorsal; *d, Microcylloepus* sp., epipleura and abdominal segments; *e, Cleptelmis addenda*, epipleura and abdominal segments (Chandler, original).

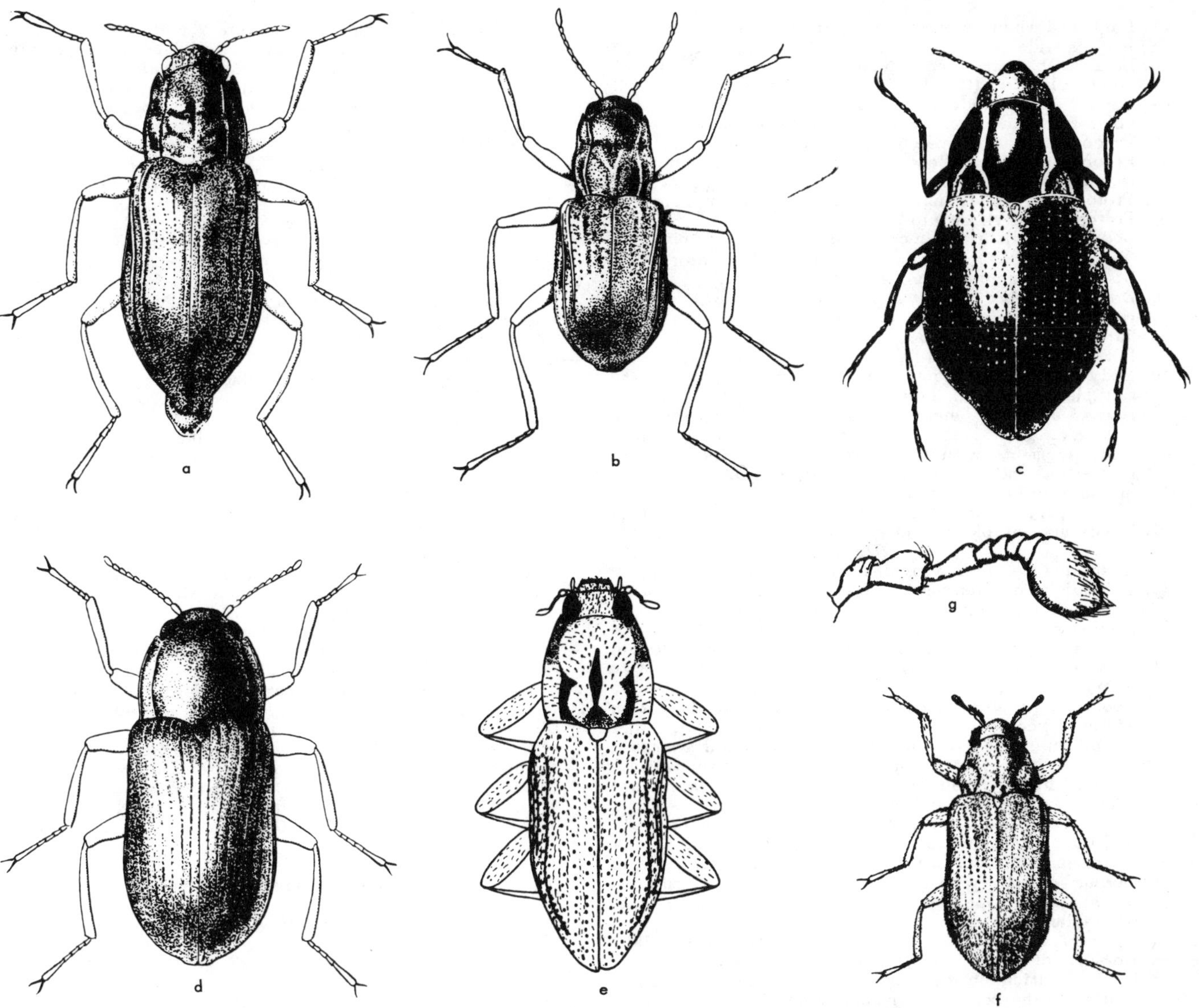

Fig. 13:54. Elmidae, adults. *a*, *Heterelmis longula*; *b*, *Microcylloepus* sp.; *c*, *Cleptelmis* sp.; *d*, *Elsianus* sp.; *e*, *Zaitzevia parvulus*; *f*, *Phanocerus clavicornis*; *g*, *Zaitzevia* sp.; adult antenna (*a*,*b*,*d*,*f*, Hinton, 1940; *c*, Hinton, 1935; *e*, Hatch, 1938; *g*, Hinton, 1939).

The larvae are distinguished mainly by the number and arrangement of the sternites and pleurae of the thorax and the number of abdominal pleurae. The latter is more variable. *Fuselmis* has eight, whereas closely related *Cleptelmis* has seven for one species and six for the other. *Microcylloepus* which has two meso- and metapleurae in the younger instars was found to have three in the last instars. This may occur in other genera. Body shape and the shape of the ninth abdominal segment are also helpful. The marginal spines are characteristic but have been little used here because a compound spine may appear spatulate at lower magnification.

Key to Known Genera of Nearctic Elmidae

Larvae

1. Abdominal segments 1-8 with pleura 2

— Abdominal segments 1-7 or 1-6 with pleura. ELMINI in part 6

2. Lateral margins of thorax and abdomen expanded, sometimes bearing dense rows of erect spatulate spines. LARINI 3

— Body hemicylindrical or subtriangular, lateral margins not expanded, without conspicuous erect spines. ELMINI in part 4

3. Sides expanded and body greatly flattened, resembling an elongate suction cup, procoxal cavities closed behind (fig. 13:50*b*) .... *Phanocerus* Sharp 1882

— Sides moderately expanded with coarse compound spines along edge; dorsum with ridge on either side, more pronounced posteriorly, so that last segment is square-sided and flat-topped; procoxal cavities open behind; larger species, up to 16 mm. long .... *Lara* LeConte 1852

4. Body very elongate, last abdominal segment 6 times longer than wide, operculum only 1/3 as long, tubercles and spines small, inconspicuous; entire body with pubescent appearance (fig. 13:50*f*) .... *Dubiraphia* Sanderson 1954

— Last abdominal segment never more than 4 times as long as wide 5
5. Head with short erect spines, body more cylindrical, color yellowish (fig. 13:50*c*) ..... *Narpus* Casey 1893
— Head without erect spines, body more hemicylindrical, color dusky .... *Rhizelmis* Chandler 1954
6. Procoxal cavities open behind (fig. 13:51*g*) 7
— Procoxal cavities closed behind by a definite sclerite or sclerites 16
7. Protopleura divided into 3 parts (fig. 13:51*g*) 8
— Protopleura divided into 2 parts 13
8. Body hemicylindrical in cross section (fig. 13:50*a*) . ... *Zaitzevia* Champion 1923
— Body subtriangular in cross section 9
9. Dorsum of body with a medial and lateral pair of prominences or humps on each body segment (fig. 13:50*d*) 10
— Dorsum without humps 11
10. Body segments without lateral prominences; meso- and metapleura divided into 2 parts; 2 longitudinal dark marks on each thoracic tergite; southeastern U.S. .... *Gonielmis* Sanderson 1954
— Body segments with lateral prominences (fig. 13:50*d*); meso- and metapleura in 1 piece; no dark marks on thoracic tergites; eastern U.S. .... *Promoresia* Sanderson 1954
11. Meso- and metapleura 1 piece (fig. 13:51*f*)......... .... *Optioservus* Sanderson 1954
— Meso- and metapleura divided into 2 parts 12
12. Tubercles on dorsum separated by less than their own widths, uniform in distribution except along posterior margin of segments where they are nearly touching, with broad, flat spines less than length of tubercles; anterior mesopleurae about 1/2 the size of posterior .... *Heterlimnius* Hinton 1935
— Tubercles sparse, separated by more than their own widths except double row along medial line, large smooth areas on pronotum, and 9th abdominal segment; marginal tubercles separated by their own widths, spines longer than tubercles; anterior mesopleural sclerite equal to posterior .... *Limnius* Erichson 1847
13. Abdomen with pleura on segments 1 to 7........... 14
— Abdomen with pleura on segments 1 to 6........... 15
14. Dorsum of body with dense medial and lateral patches of spatulate spines, making the body appear subtriangular in cross section ....*Ampumixis* Sanderson 1954
— Dorsum of body evenly rounded from side to side, hemicylindrical in cross section, with spatulate spines limited to the posterior fringe of each segment..... .... *Cleptelmis* Sanderson 1954
15. Form more slender and cylindrical, abdominal segments twice as wide as long, 9th abdominal segment with convex sides, tip roundly and shallowly emarginate dorsally without spines .... *Cleptelmis* Sanderson 1954
— Form more robust and hemicylindrical, abdominal segments 3 times as wide as long, 9th abdominal segment with conical sides, tip deeply emarginate dorsally, so as to form 2 lateral spines .... *Macronychus* Müller 1806
16. Lateral margins of abdominal segments 1 to 8 produced posteriorly into spinelike processes (fig. 13:52*d*); body form wider and less convex .... *Ancyronyx* Erichson 1847
— Lateral margins of segments not produced, body shape hemicylindrical to cylindrical 17
17. Protopleura extending to medial line both anterior and posterior to coxal cavities, sternum surrounded and reduced to a triangular sclerite between posterior ½ of coxal cavities, meso- and metapleura undivided, with only sternopleural suture present (fig. 13:51*a*) .... *Cylloepus* Erichson 1847
— Prosternum separating the pleura posteriorly, meso- and metapleura divided into 2 or 3 parts 18
18. Meso- and metapleura divided into 3 parts 19
— Meso- and metapleura divided into 2 parts (fig. 13:51*h*) 20
19. Dorsal half of all segments except prothorax and 9th abdominal segment with tubercles bearing erect spatulate spines, arranged in 10 prominent longitudinal rows; 9th abdominal segment strongly triangular with a medial and lateral rows of tubercles (figs. 13:50*e*, 13:51*b*) .... *Heterelmis* Sharp 1882
— Only the medial part of the dorsal half of the body showing any tendency for tubercles to be arranged in rows (fig. 13:52*i*); tubercles without erect spatulate spines; 9th abdominal segment rounded (last instar of) .... *Microcylloepus* Hinton 1935
20. Procoxal cavities not closely approaching lateral margin, separated from each other by a small, unattached sclerite or by the projection of the sternum forward to unite with pleura 21
— Procoxal cavities narrowly separated from lateral margin, separated from each other by propleural intercoxal process projecting posteriorly under the sternal intercoxal process (fig. 13:52*e*) 23
21. Procoxae separated by a small, elongate, unattached sclerite lying between procoxal cavities, not covered by pleura anteriorly or sternum posteriorly (fig. 13:51*d*); (first instars of) .... *Microcylloepus* Hinton 1935
— Procoxal cavities usually separated by the extension of the sternum forward to meet the pleura; if sclerite between coxae is present it is reduced in size and partly covered by the sternum 22
22. Anterior margin of head on each side with a large, conspicuous tooth (fig. 13:51*c*) .... *Elsianus* Sharp 1882
— Anterior margin of head without a distinct tooth on each side .... *Neoelmis* Musgrave 1935
23. Propleura divided into 2 parts by a short suture from coxal cavity to margin, also with a medial suture in front of coxae; 1st antennal segment more than 1/2 as long as 2nd .... *Stenelmis* Dufour 1835
— Propleura with above sutures obliterated, only sternal pleural sutures faintly visible; 1st antennal segment less than 1/2 as long as 2nd.................... .... *Ordobrevia* Sanderson 1953

Key to Nearctic Genera of Elmidae

Adults

1. Adults terrestrial, rarely entering the water; body with pubescence but without tomentum; procoxae transverse, trochantin exposed. LARINI 2
— Adults aquatic; body beneath with tomentum; front coxae rounded, trochantin concealed. ELMINI 3
2. Antennae clubbed; pronotum with basal sublateral sulcus connecting with margin at middle; elytra without accessory striae .... *Phanocerus* Sharp 1882

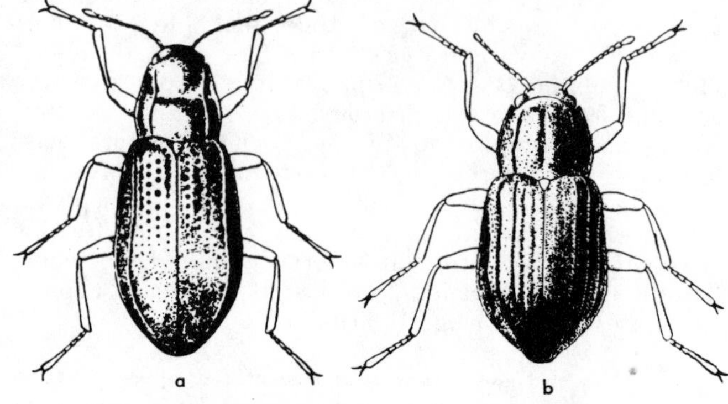

Fig. 13:55. Elmidae, adults. *a*, *Neoelmis longula*; *b*, *Hexacylloepus apicalis* (Hinton, 1940).

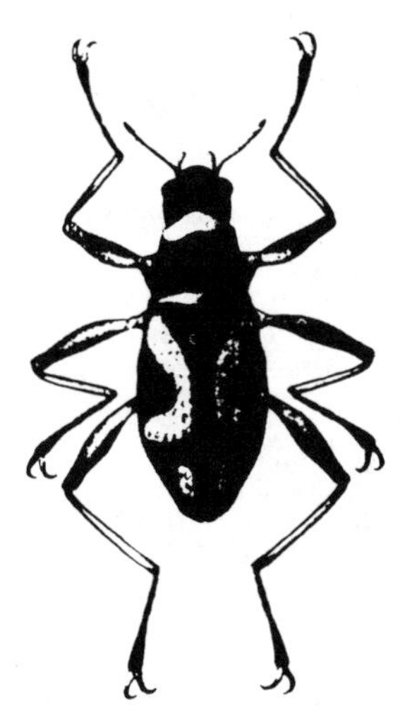

Fig. 13:56. *Ancyronyx variegatus*, adult (Garman, 1912).

Phanocerus clavicornis Sharp 1882; fig. 18; 2–3 mm.; brown; Texas, Mexico, Central America.

— Antennae not clubbed; pronotum without sublateral sulcus; each elytron with an accessory stria near suture below the scutellum; length 6–7 mm.; black; western .... *Lara* LeConte 1852

3. Coxae very widely separated and of about the same size, located more on the side of body than on the nearly flat venter; metacoxae almost globular; posterior margin of prosternal process almost as wide as head 4

— Coxae less widely separated, especially the procoxae; metacoxae distinctly larger and transverse; posterior margin of prosternal process less than 1/2 the width of the head 5

4. Pronotum with 2 oblique, transverse depressions at anterior 3rd, without sublateral carinae; antennae filiform, with 11 segments, not clubbed; tibia without patches of tomentum; elytra without sublateral carinae (fig. 13:56) .... *Ancyronyx* Erichson 1847

   Ancyronyx variegatus (Germar) 1824; 2.3–2.6 mm.; black, maculate; eastern U.S.

— Pronotum without transverse depressions, but with basal sublateral carinae; antennae 7-segmented, last segment enlarged; elytra with tomentum extending up to sublateral carinae; tarsi with 1 or 2 patches of tomentum .... *Macronychus* Müller 1806

   Macronychus glabratus Say 1825; 2.8–3.4 mm.; black; eastern U.S.

5. Posterior lateral corner of 4th or middle of lateral margin of 5th abdominal sternites not produced into a prominent upturned tooth; half of epipleura of elytra usually uniformly and narrowly tapered to fine point, (fig. 13:53*a, e*) 6

— Posterior lateral corner of 4th or middle of lateral margin of 5th abdominal segment produced into a prominent tooth, which is bent upward to clasp epipleura (less conspicuously bent in *Zaitzevia* which has 8-segmented antennae); epipleura slightly to conspicuously widened to receive tooth, usually bluntly narrowing to apex posterior to tooth (fig. 13:53*b,d*). . 13

6. Epipleura extending to apex of elytra (fig. 13:53*a*); protibia without a patch of tomentum 7

— Epipleura usually ending near base or middle of 5th abdominal segment (fig. 13:53*e*); protibia with a patch of tomentum on anterior side 8

7. First medial elytral stria short, about 1/6 the length of 2nd (fig. 13:52*g*); granules of head and legs elongate .... *Ordobrevia* Sanderson 1953

   Ordobrevia nubifera (Fall) 1901; 2.2–2.4 mm.; black, maculate; California to Washington, in foothill streams

— First elytral stria as long as 2nd; granules of head and legs round (fig. 13:53*c*); mostly eastern with 1 species from Nevada .... *Stenelmis* Dufour 1835

   Stenelmis calida Chandler 1949; 3–3.4 mm.; brown, gray pile; Nevada, in warm springs

8. Pronotum smooth, without sublateral carina; maxillary palpi 3- or 4-segmented 9

— Pronotum with sublateral carina extending from base to or nearly to anterior edge (fig. 13:51*e*); maxillary palpi 3-segmented 10

9. Three-segmented maxillary palpi; larger species, length 3–4 mm.; markings, if present, transverse; western .... *Narpus* Casey 1893

— Four-segmented maxillary palpi; smaller species 2–3.3 mm. long; markings, if present, longitudinal (fig. 13:52*a*); U.S. .... *Dubiraphia* Sanderson 1954

10. Size larger, 2.5–2.6 mm. long; prosternum projecting under head; epipleura extending to middle of 5th abdominal segment .... *Rhizelmis* Chandler 1954

   Rhizelmis nigra Chandler 1954; 2.5 mm.; black; California in shaded streams at 2,000 to 5,000 feet

— Size smaller, 1.8–2.2 mm. long; prosternum not projecting under head; epipleura ending at base of 5th abdominal segment 11

11. Pronotum with sides parallel basally, strongly convergent anteriorly (fig. 13:51*e*), lateral margins densely punctured, nearly obliterating short, lateral branch of the carinae running to posterior margin; anteriorly the area between margin and carinae depressed; elytra broad, very convex or dome-shaped .... *Ampumixis* Sanderson 1954

   Ampumixis dispar (Fall) 1925; 1.8–2.2 mm.; black, elytra black to red; California, Oregon; in rapid clear steams, up to 6,000 feet

— Sides of pronotum not parallel, converging anteriorly from base, anterior marginal area not depressed, elytra less convex 12

12. Pronotum with shallow, transverse, basal impression connecting the carinae, sides nearly straight, carinae gradually less distinct anteriorly to apical 4th, not forked posteriorly, marginal area densely punctate; spindle-shaped species ..... *Atractelmis* Chandler 1954

   Atractelmis wawona Chandler 1954; 2 mm.; black, elytra with humeral and subapical maculae; rare in rapid streams, 2,000 to 5,000 feet elevation in California

— Sides of pronotum convex, carinae with posterior lateral branch reaching posterior margin, marginal area not so densely punctate, without transverse impression; sides of elytra more parallel (fig. 13:54*c*).......... .... *Cleptelmis* Sanderson 1954

13. Fourth abdominal sternite with prominent tooth on posterior angle, 5th abdominal segment not clasping epipleura (fig. 13:53*b*) 14

— Fifth abdominal sternite with basal half of lateral margin strongly clasping epipleura, middle of segment with a prominent tooth which is bent vertically to fit into widened area of epipleura (fig. 13:53*d*) 19

14. Antennae with 8 segments, last segment enlarged (fig. 13:54*g*); pronotum with median, longitudinal groove, elytra with 3 sublateral carinae .... *Zaitzevia* Champion 1923

   Zaitzevia parvula Horn 1870, fig. 30; 2 mm.; black, with rufous to black elytra; western U.S. and Canada; in streams up to 6,000 feet in California

— Antennae 10- or 11-segmented; posterior lateral corners of 4th abdominal sternite with prominent tooth sharply bent upward to clasp the epipleura 15

15. Pronotum with steplike sublateral carinae extending from base to anterior margin (fig. 13:52*h*); elytra with 3 sublateral carinae; size small, 1.4 mm. long .... *Limnius* Erichson 1847

   Limnius latiusculus (LeConte) 1866, 1.4 mm.; black; eastern U.S.

— Pronotum with sublateral carinae short or absent, elytra without sublateral carinae 16

16. Pronotum without distinct basal carinae, general form spindle-shaped .... *Gonielmis* Sanderson 1954
    Gonielmis dietrichi (Musgrave) 1933; 2.3 mm.; black, diagonal elytral markings (fig. 13:52*f*); southeastern U.S.
— Pronotum with short sublateral carinae, extending from base to basal 4th or halfway to anterior margin 17
17. Tarsi stout, nearly as long as tibia, distal half of last segment as thick as tibia, claws very large, curved more than 90°; last abdominal segment with side nearly straight, rounded only near apex; 9th antennal segment little wider than 8th, club poorly differentiated; lateral and posterior margins of prothorax smooth; eastern U.S. and Canada .... *Promoresia* Sanderson 1954
— Tarsi less developed, never as thick as tibia, claws shorter and more slender, curved about 90°; last abdominal segment usually rounded; last 3 antennal segments more definitely enlarged; lateral margins of prothorax slightly serrate, posterior margin with many small closely placed teeth 18
18. Claws curved about 90°, last abdominal segment usually evenly rounded; antennae 11-segmented with last 3 segments only slightly enlarged, about 1½ times as wide as 8th segment; body form less convex with sides of elytra near base nearly straight; U.S. and Canada .... *Optioservus* Sanderson 1954
— Claws smaller, less curved (about 70°), last abdominal segment with apex slightly truncate or emarginate; antennae with 10 or 11 segments, last 3 more differentiated, about twice as wide as segments next to them; form variable but usually moderately to strongly convex along medial line and sides of elytra; western .... *Heterlimnius* Hinton 1935
19. Second elytral striae short, only about 1/5 as long as 1st; larger species, 3.5 to 6 mm.; color black (fig. 13:54*d*); southwestern .... *Elsianus* Sharp 1882
    Elsianus moestus (Horn) 1870; 4.7–4.9 mm.; black; Arizona, may extend into southeastern California
— Second elytral striae long, equal to 1st; smaller species, 1.4–3 mm............................ 20
20. Each elytron with 1 sublateral carina; pronotum smoothly convex except for sublateral carinae and transverse impression at anterior two-fifths, no other depressions or humps (fig. 13:55*a*).................. .... *Neoelmis* Musgrave 1935
    Neoelmis caesa (LeConte) 1874; 1.6 mm.; testaceous; Texas
— Each elytron with 2 lateral carinae (rarely 1 in *Microcylloepus*), pronotum either with other impressions or humps, or transverse depression is not located at anterior two-fifths 21
21. Pronotal hypomera with a belt of tomentum extending from coxae to lateral margin (fig. 13:57*b*).......... .... *Hexacylloepus* Hinton 1940
    Hexacylloepus ferrugineus (Horn) 1870; 1.7 mm.; testaceous; Texas, Oklahoma
— Hypomera with or without tomentum, but if present it never reaches lateral margin 22
22. Pronotum without a pronounced medial longitudinal depression or groove, usually with a transverse depression at middle; hypomera with tomentum near coxae; prosternal process broad, without parallel sides (fig. 13:54*a*) .... *Heterelmis* Sharp 1882
    Heterelmis glaber Horn 1870; 1.9–2.2 mm.; dark brown; San Pedro River, Arizona, Indian Springs, Nye County, Nevada; probably will be taken in southeastern California
— Pronotum with a pronounced medial longitudinal depression or groove, transverse depression, if present, at anterior two-fifths; prosternal process with sides nearly parallel for at least 1/2 its length (fig. 13:54*b*) 23
23. Pronotum with a more or less distinct transverse depression at anterior two-fifths; hypomera and epipleura without tomentum; 1.4–2.4 mm. long (fig. 13:52*c*); U.S. and Mexico .... *Microcylloepus* Hinton 1935
— Pronotum without transverse depression, at anterior two-fifths; epipleura without, hypomera usually without tomentum; U.S. and Mexico .... *Cylloepus* Erichson 1847
    Cylloepus parkeri Sanderson 1953; 2.7 mm.; black with rufous maculae; Bloody Basin, Yavapai County, Arizona; may occur in southeastern California

Key to California Species of Lara

Adults

1. Pronotum with hind angles acute but scarcely more prominent than middle lobes; elytral pubescence uniform; length 6.5 mm.; black; Washington to middle California, along margins of rapid clear streams up to 6,000 feet .... *gehringi* Darlington 1929
— Pronotum with hind angles acute and prominent; alternate elytral intervals with pubescence less decumbent, so that elytra appear dark with sericeous lines; length 7.5 mm.; black; British Columbia to southern California, along rapid, clear streams up to 5,000 feet .... *avara* LeConte 1852

Key to California Species of Narpus

Adults

1. Color black; form narrower, 3 times as long as wide; pronotum less convex, sides nearly straight, anterior width 4/5 of posterior width; length 3–4 mm.; California coast range, in rapid, clear streams .... *angustus* Casey 1893
— Elytra black to red, usually with black band across middle; form wider, twice as long as wide; pronotum more convex, sides curving inward anteriorly, anterior width 2/3 posterior; length 3.2–4.2 mm.; western U.S. and Canada, in rapid, clear streams .... *concolor* LeConte 1881

Key to California Species of Dubiraphia

Adults

1. Elytra of a uniform dark brown, occasionally with a faint yellowish humeral or subapical spot; length 1.8–2.5 mm.; known only from Clear Lake area, Lake County, where it is abundant along rocky sections of the shore line .... *brunnescens* (Fall) 1925
— Elytra piceous with humeral and subapical light yellow markings, sometimes united into a discal vitta nearly reaching apex; 2.1–2.3 mm.; from vegetation in slow, moving section of Russian River, and a steep foothill stream Tehama County ..... *giulianii* Van Dyke 1949
  (Both of these may prove to be synonyms or subspecies of *D. vittata* Melsheimer, which is common throughout the eastern U.S., with one record from Utah.)

Key to California Species and Subspecies of Microcylloepus

Adults

1. Size smaller (1.4–1.6 mm.) and narrower (.5–.6 mm.), pronotum longer than wide, wing reduced, not reaching tip of elytra; from warm springs in Nevada 2
— Size larger (1.8–2 mm.) and wider (.8–.9 mm.); pronotum wider than long; described from eastern U.S. but with several western subspecies or forms, including *similis* (Horn) 1870, and *foveatus* LeConte 1874 .... *pusillus* LeConte 1852

2. Elytra with 1 sublateral carina; form with nearly parallel sides, elytra only slightly wider than pronotum; sculpturing of pronotum reduced; size smaller (1.4×.5 mm.); from warm springs in northwest corner of Nevada .... *thermarum* (Darlington) 1928
— Elytra with 2 sublateral carina; sides less parallel, elytra distinctly wider than pronotum; size larger (1.6×.5 mm.); from warm spring in southeastern Nevada .... *moapus* La Rivers 1949

Key to California Species of Cleptelmis

Adults

1. Elytra black with humeral and subapical red spot; length 1.8–2.2 mm.; British Columbia, Montana, Utah, northern California, in rapid streams on roots and moss .... *ornata* (Schaeffer) 1911
— Elytra black without spots except humeral angle paler, never distinctly red; length 1.7–2.1 mm.; California to South Dakota in rapid streams up to 6,000 feet elevation, on roots and moss .... *addenda* (Fall) 1907

Key to California Species of Heterlimnius

Adults

1. Antennae 11-segmented; length 2–2.5 mm.; black; British Columbia to northern California, in rapid streams .... *koebelei* Martin 1927
— Antennae 10-segmented; length 2.6–2.9 mm.; brown, black, sometimes base of elytra rufous; British Columbia, California, from rapid streams in northern Sierra Nevada above 6,000 feet . . .*corpulentus* (LeConte) 1874

Key to California Species of Optioservus

Adults

1. Elytra black, without trace of pale area at the humeral angles; length 2–2.5 mm.; southern California and Arizona .... *divergens* (LeConte) 1874
— Elytra never entirely black 2
2. Dull colored, elytra dull black to rufous, spots less conspicuous, elytra densely micropunctate on intervals, hairs on elytra and pronotum more conspicuous giving a grizzled appearance; length 2.2–2.5 mm.; west central California .... *canus* Chandler 1953
— Elytra shining black, microsculpture and hairs not conspicuous, not giving a grizzled appearance; humeral and subapical red spots larger and more distinct 3
3. Form broader, less parallel-sided, elytra noticeably wider than prothorax, elytral striae slightly impressed, humeral spot larger usually reaching 2nd striae; length 1.8–2.2 mm.; Sierra Nevada and northeastern California up to 6,000 feet .... *quadrimaculatus* (Horn) 1870
— Form more elongate, elytra scarcely wider than prothorax, sides nearly parallel, striae not impressed; humeral spot reaching only to 3rd striae; north Coastal California, Oregon .... *seriatus* (LeConte) 1874

REFERENCES

HATCH, M. H.
1938. Two new species of Helmidae from a warm spring in Montana. Ent. News, 49:16-19, 2 text figs.

HINTON, H. E.
1935. Notes on the Dryopoidea (Col.). Stylops, 4:169-179, 7 text figs.
1939. An inquiry into the natural classification of the Dryopoidea, based partly on a study of their internal anatomy. Trans. Roy. Ent. Soc. London, 89:133-184, 105 text figs., 1 pl.
1940. A monographic revision of the Mexican water beetles of the family Elmidae. Novit. Zool., 42:19-396, 401 text figs.

SANDERSON, M. W.
1938. A monographic revision of the North American species of *Stenelmis* (Dryopidae: Coleoptera). Univ. Kansas Sci. Bull., 25:635-717, incl. pls. No. lxxx, lxxxi.
1953. A revision of the Nearctic genera of Elmidae (Coleoptera) (Part I). Jour. Kansas Ent. Soc., 26: 148-163, incl. 3 pls.
1954. A revision of the Nearctic Genera of Elmidae (Coleoptera) (Part II). Jour. Kansas Ent. Soc., 27:1-13, incl. 1 pl.

WEST, L. S.
1929. A preliminary study of larval structures in the Dryopidae. Ann. Ent. Soc. Amer., 22:691-727, incl. 6 pls.

Family LIMNICHIDAE

These small beetles are found along the margins of streams. They are clothed with dense, short hairs, each hair arising from a puncture (fig. 13:57*a*). In some species these punctures are enlarged, giving a honeycomb appearance. Occasionally the body has a checkerboard metallic luster. The Limnichinae are

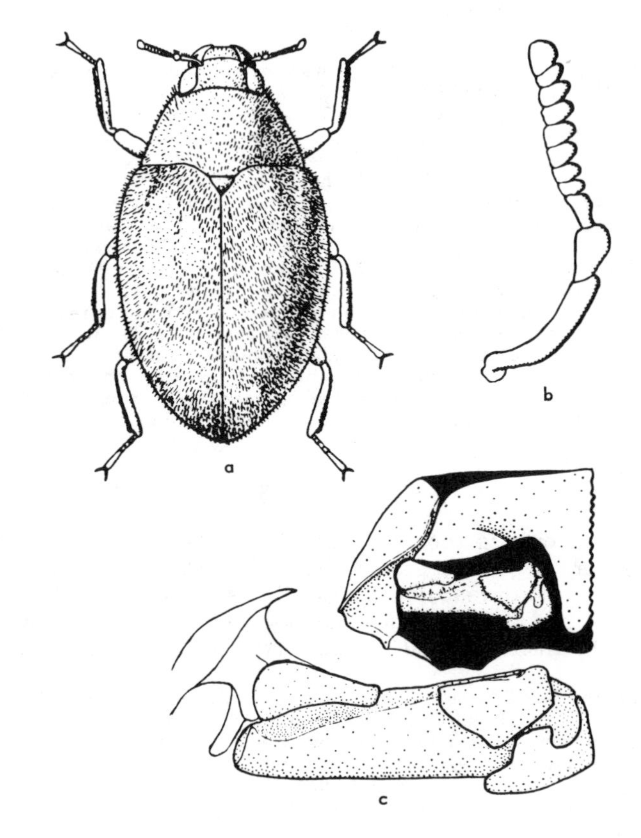

Fig. 13:57. Limnichidae. *a*, *Lutrochus gigas*, adult (Peruvian sp.); *b*, *L. gigas*, antenna; *c*, *L. erichsonianus*, side view of prosternum and hypomera of adult, hypomeral ridge showing at left (Mexican sp.) (Hinton, 1939).

broadly oval and very convex, often with the appendages retractile and fitting into grooves in the body. Some species of *Limnichus* retract the head into the thorax so that even the eyes are concealed, leaving only the two basal segments of the antennae exposed.

Relationships.—The adults have many characters in common with the Dryopidae in which *Throscinus* and *Lutrochus* were included for many years. The larvae more closely resemble those of the Elmidae.

Respiration.—The adults are riparian, but the dense hairs on some species suggests that they may be able to submerge with a film of air about their bodies. The larva of *Lutrochus* has three tufts of retractile, filamentous gills in a caudal chamber similar to those of the elmids. This would indicate that it is truly aquatic.

Habitat and distribution.—The adults of *Limnichus* have been taken with Omophronidae and Georyssidae along the margins of streams on wet sand or loam. They may be flushed from their hiding places by throwing water on the bank. Fall (1901:104) reports *Limnichus tenuicornis* (Casey) from a wet, springy hillside. *Throscinus* was reported by Fall from mud flats near San Diego which are covered at high tide.

Key to the Nearctic Genera of Limnichidae

Adults

1. Hypomera with a transverse or oblique ridge (fig. 13:57*c*) LIMNICHINAE 2
— Hypomera without ridges; Texas to southern California CEPHALOBYRRHINAE .... *Throscinus* LeConte 1874
2. Antennae 11-segmented, segments 1-2 large, 3-11 pectinate (fig. 13:57*b*); head less retractile, mouth parts and eyes exposed; 1st abdominal sternite not grooved to receive femora and tibiae; eastern U.S. .... *Lutrochus* Erichson 1847
— Antennae 10-segmented; 1st abdominal sternite with grooves into which hind femora and tibiae fit 3
3. Pronotum without cavities on dorsal surface for reception of antennae in repose; antennae slender, without distinct club .... *Limnichus* Latreille 1829
— Pronotum with polished sublateral cavities on dorsal surface to receive antennae in repose; antennae with distinct club .... *Physemus* LeConte 1854

Genus *Throscinus*

One species is recorded from California, but the very similar *T. politus* Casey of the Gulf Coast of Texas and the northwest coast of Mexico should be watched for. It has the third antennal segment longer than the fourth, almost as long as the fourth and fifth together.

Length 2-2.5 mm., form elongate-oval, nearly parallel-sided. Dorsal surface black, polished punctate, hairy; elytral punctures subserially arranged on disc, becoming minute and sparse at sides and apically; elytral vestiture dual: of sparse longer, thinner, semierect hairs, and dense recumbent, short, broad hairs, the latter less dense at middle of disc and forming a "whorl" toward sides just before middle. Undersurface rufopiceous, opaque except between prosternal carinae; antennal segments 3 and 4 subequal; Coronado, San Diego County, on mud flats covered at high tide; Naples, Los Angeles County, among shoots of a beach plant (J. O. Martin) .... *crotchi* LeConte 1874

Key to the California Species of Limnichus

Adults

1. Eyes very convex and prominent, rather coarsely faceted and conspicuously visible from above, sides of front above them not acute or cariniform 2
— Eyes vertical in plane and more or less finely faceted, edges of front of head above them acute and projecting, usually almost concealing them from dorsal view ... 3
2. Elytral punctures large, very deep, so densely crowded as to be regularly polygonal with very fine separating walls, resembling a honeycomb; southwest U.S.; Tulare, Calaveras, and Yolo counties .... *nebulosus* LeConte 1879
— Elytral punctures almost as large and very deep, but impressed, circular, isolated, separated by ½ to fully once their own diameters; Mono, Santa Clara, Riverside, San Diego counties .... *perforatus* (Casey) 1890
3. Prosternum not sulcate; Humboldt, San Bernardino counties .... *tenuicornis* (Casey) 1890
Prosternum sulcate 4
4. Elytral punctures minute, feeble, very scattered sometimes almost obsolete; southwest U.S.; Riverside County .... *perpolitus* (Casey) 1890
— Elytral punctures always readily visible, separated by slightly more than, to evidently less than, twice their own diameters 5
5. Elytral vestiture simple, of subrecumbent, robust, and aciculate hairs 6
— Elytral vestiture dual, of very small short confusedly matted and densely placed hairs near the surface, and finer, longer, sparser suberect hairs interspersed; southwestern U.S.; San Diego and Riverside counties .... *naviculatus* (Casey) 1890
6. Smaller species, less than 2 mm. long; elytral pubescence sparser, punctures very shallow and separated by twice their own widths; California .... *californicus* LeConte 1879
— Larger species, more than 2 mm. long; elytral pubescence dense, punctures impressed, separated by 1½ times their own widths; southwestern U.S.; Shasta County to Tulare and Sonoma counties to Santa Clara County .... *analis* LeConte 1879

Genus *Physemus*

Adults

A single North American species. Length 0.8-1 mm. Form elliptical, strongly convex; color blackish above, reddish below. Elytra shining, finely punctate, with recumbent pubescence; southwestern U.S.; Imperial, Los Angeles counties .... *minutus* LeConte 1854

REFERENCES

CASEY, T. L.
1889-1890. Coleopterological notices. I. With an appendix of the termitophilous Staphylinidae of Panama. Ann. N.Y. Acad. Sci., 5:39-198.
1912. Descriptive catalogue of the American Byrrhidae. Memoirs on the Coleoptera, 3:1-69.
FALL, H. C.
1901. List of the Coleoptera of southern California, with notes on habits and distribution and descriptions of new species. Occ. Pap. Calif. Acad. Sci., 8:1-282.
HINTON, H. E.
1939. An inquiry into the natural classification of the Dryopoidea, based partly on a study of their internal anatomy (Col.). Trans. Roy. Ent. Soc. London, 89: 133-184, 105 text figs., 1 pl.

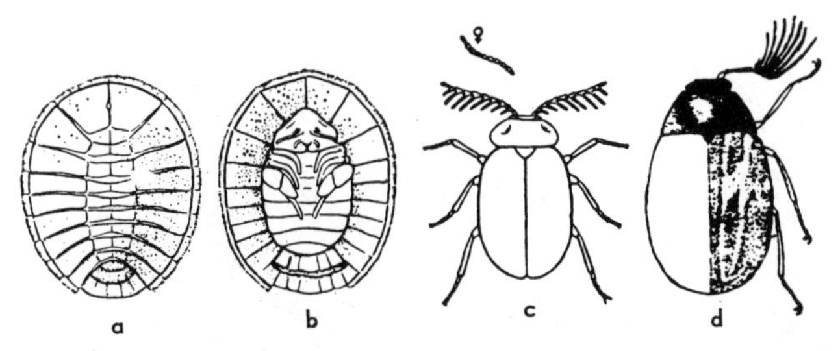

Fig. 13:58. Psephenidae. *a,b,c,* larva, pupa and adult of *Eubrianax edwardsi; d, Acneus quadrimaculatus,* adult (a-c, Essig, 1926; *d,* Horn, 1881).

Family PSEPHENIDAE

Water Pennies

These oval, depressed, moderately small (6 mm.), brown to black beetles (fig. 13:58) are occasionally seen on the rocks along the margins of streams. The adults are terrestrial but may enter the water to lay their eggs. The aquatic larvae are more familiar to those who sample stream life. They are commonly called water pennies because of their broad flat oval shape. They are found clinging to stones in slow to moderately rapid streams.

Relationships.—The genera *Psephenus* and *Eubrianax* are quite closely related, although they were placed in different families for many years. *Acneus* and *Ectoparia* in a general way resemble *Psephenus* and *Eubrianax* both as larvae and adults but are much more distantly related to them. The presence of the trochantin in the adult, and the location of the gills in the caudal chamber closed by an operculum in the larva, are strong indications that they should be placed in a separate family. The larva of *Ectoparia* has been described in a number of publications as that of *Helichus* (Dryopidae).

Respiration.—When the adults enter the water to deposit their eggs, they are surrounded by a film of air supported by the dense body hairs.

The larvae of *Psephenus* and *Eubrianax* have two tufts of gill filaments below various of the abdominal segments one to six. Spiracles are on the dorsal side of the body near the base of the pleura of the mesothorax and the first eight abdominal segments. In *Psephenus* the pleura of the eighth segment is reduced or lacking, and the spiracles are on the lateral posterior margin of the terga.

The pupae are found on stones just above the water line and have no gills.

Larvae of *Acneus* and *Ectoparia* have three tufts of retractile gill filaments which can be drawn into the caudal chamber. The caudal chamber is closed by an operculum. The only functional spiracles are at the end of the long, curved, lateral projections of the pleura of the eighth abdominal segment. What appear to be scars of spiracles can be seen on the dorsal side of all the abdominal segments in *Acneus.*

Life history.—The adults apparently deposit their eggs beneath the water on the face of stones which they can reach by crawling down from the surface. They are occasionally thus exposed when turning stones of which only the tops protrude from the water in midstream. The pupa of *Eubrianax* retains all except the last three segments of the larval skin as a shieldlike cover. The body is light colored at first, except the last three segments which closely resemble and replace those of the larval shield. *Psephenus* pupate farther above the water surface. The pupae are of uniform color with the last segments not replacing those of the shield.

Adults of *Acneus* are terrestrial and the larvae are aquatic. The pupae are not known, but a related genus *Psephenoides* has one of the few truly aquatic pupae in the Coleoptera. It is oval and flattened like the larvae with tufts of fingerlike gills along the margin of the abdominal segments.

Habitat and distribution.—This family is usually found in clear, moderate to rapid streams with gravel or rocky bottoms. The adults of *Acneus* are taken along small, rapid, low elevation streams frequently near waterfalls; the known larvae (all collected by me) were taken in rapid sections of a stream, in pools of quiet water protected from any current by large boulders. It would appear that they require well-aerated water, and protection from erosion or silting.

The adults of most of the genera are occasionally taken flying along the streams they inhabit. *Psephenus* is found in the northeastern United States and from California to Idaho on the west coast. In California they are taken mostly at elevations below 4,000 feet. *Eubrianax* is found throughout California up to 6,000 feet. *Ectoparia* is restricted to the northeastern and central United States. *Acneus* is found in California and Oregon up to 4,000 feet.

Key to Nearctic Genera of Psephenidae

Adults

1. Posterior margin of prothorax smooth 2
— Posterior margin of prothorax finely beaded or crenulate 3
2. Head hidden beneath the expanded pronotum; base of claws with a menbranous appendage nearly reaching the tip of claw .... *Eubrianax* Kiesenwetter 1874
Eubrianax edwardsi (LeConte) 1874 (fig. 13:58*a, c*), California, in streams
— Head visible from above; base of claws without such an appendage; northeastern and west coast of U.S. .... *Psephenus* Haldeman 1853
3. Prosternum of moderate width, not depressed between the coxae; claws toothed at base; antennae simple; northeastern and north central U.S. ....*Ectoparia* LeConte 1853
— Prosternum narrow, depressed between the coxae; claws slightly broader at base in female or toothed in male; antennae of male flabellate; Washington to California .... *Acneus* Horn 1880

Key to Nearctic Genera of Psephenidae

Larvae

1. Ninth abdominal segment without a ventral operculum, abdominal segments fitting tightly together to the outer margin (fig. 13:59*a*), 4 or more segments with tufts of ventral filamentous gills 2

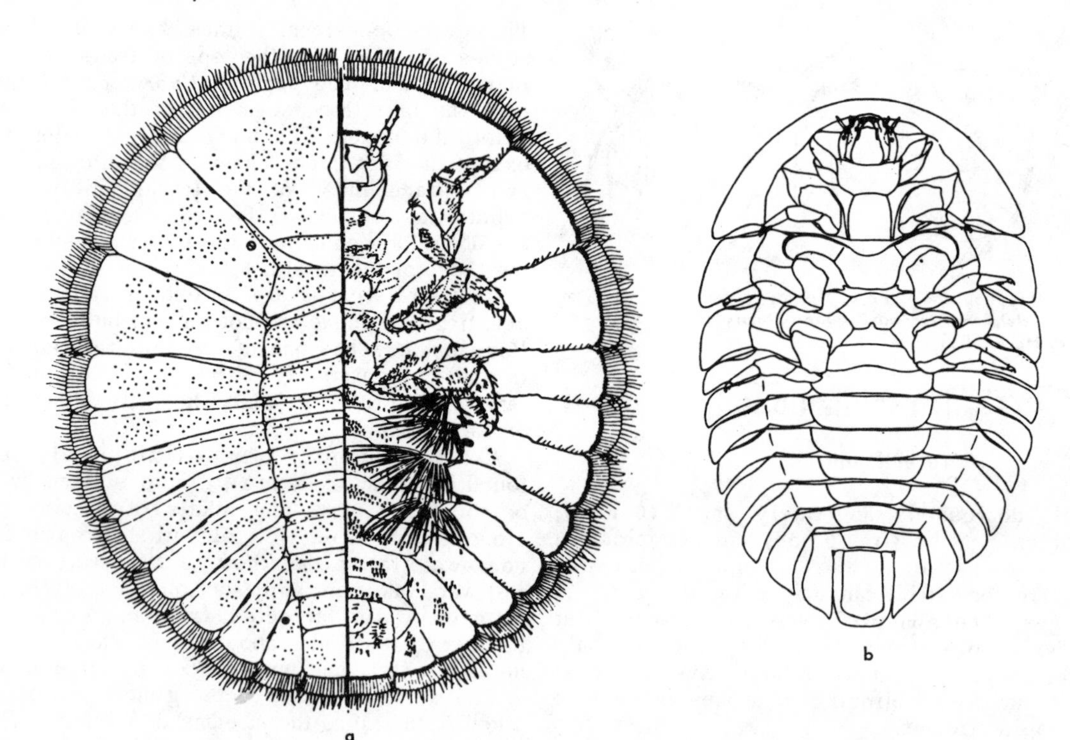

Fig. 13:59. Psephenidae. *a*, *Eubrianax edwardsi*, larva, left side, dorsal and right side, ventral view; *b*, *Ectopria* sp., larva, ventral (*a*, Blackwelder, 1930; *b*, West, 1929).

— Ninth abdominal segment with a ventral operculum closing a caudal chamber containing 3 tufts of retractile filamentous gills (fig. 13:59*b*), no other abdominal gills; extensions of abdominal pleurae separated from each other 3

2. Eighth and 9th abdominal segments without lateral expansions; gills on segments 1-6 or 2-6 .... *Psephenus* Haldeman 1853

— Eighth abdominal segment with lateral expansions (fig. 13:59*a*); gills on segments 1-4 .... *Eubrianax* Kiesenwetter 1874

3. Ninth abdominal segment expanded, with hind margin bilobed .... *Acneus* Horn 1880

— Ninth abdominal segment quadrate (fig. 13:59*b*), posterior margin straight and little wider than anterior margin .... *Ectoparia* LeConte 1853

Key to Western Species of Psephenus

Adults [18]

1. Head impression longitudinally divided; elytra pale at base; common throughout California along streams .... *haldemani* Horn 1870

— Head impression not longitudinally divided; elytra uniform in coloration, except very narrowly on the basal margin 2

2. Pronotum velvety black 3

— Pronotum dull and opaque, not velvety 4

3. Pronotal sides evenly and feebly arcuate from base to apex; pronotal punctures rather strong and dense anteriorly, becoming finer and sparse behind; California along streams .... *veluticollis* Casey 1893

— Pronotal sides straight or feebly sinuate in middle two-fourths, anteriorly arcuately continuous with the rounded apical angles, apex arcuate; labrum impressed at middle; pronotal punctures scarcely recognizable (female); California along streams .... *calaveras* Blaisdell 1923

4. Pronotal sides broadly rounded and subparallel toward base, more convergent and nearly straight anteriorly, apex truncate; elytra with impressed lines, disc feebly elevated along the suture (male); southern California along streams .... *falli* Casey 1893

— Pronotal sides broadly arcuate, somewhat broadly feebly sinuate at middle, apex arcuate; elytra without impressed lines, not elevated along suture in male; Idaho along streams .... *lanei* Blaisdell 1923

[18] Based on Blaisdell's key (1923). It is possible that couplet 2 will prove to be a sexual character.

Key to Western Species of Acneus Horn

Adults

1. Elytra black, with 4 pale spots which do not reach the elytral suture (fig. 13:58*d*); California, Oregon along swift, rocky streams ...... *quadrimaculatus* Horn 1880

— Elytra predominantly pale with pale area reaching suture, each elytron with 7 black spots; Oregon along swift, rocky streams .... *oregonensis* Fender 1951

REFERENCES

BLAISDELL, F. E.
1923. Two new species of *Psephenus* Hald., with a note on *Narpus angustus* Casey. Ent. News, 34:234-238.

FENDER, K. M.
1951. A new species of *Acneus*. Proc. Ent. Soc. Wash., 51:271-272.

Family PTILODACTYLIDAE

These oval to elongate beetles are brown to black in color and are usually found along the margins of streams. The larvae of some are aquatic but those of others feed upon the roots of grasses or are found in leaf mold. The larvae are of the elongate cylindrical burrowing type. The aquatic forms have the last abdominal segment flattened above and may have gills situated about the anus or on the ventral side of the abdominal segments.

Relationships.—There is considerable confusion as to which genera should be included in this family. Originally it was placed as a subfamily of *Helodidae* and included *Lachnodactyla* and *Ptilodactyla*. On the basis of larval characters, Böving (1931) made it a family and included *Eurypogon* and *Anchytarsus,* formerly placed in *Dascillidae*. He also figured an unassociated larva from China as belonging in this family. Larvae almost identical to that figured have been collected in California, and on the basis of distribution and size have been associated with *Stenocolus*. Another California larva close to *Anchytarsus* on the basis of size has been associated with *Anchycteis*.

Respiration.—Only the larvae of certain genera are known to be truly aquatic. Whether the adults ever enter the water is not known. The larvae of all genera have the usual spiracles on the mesothorax and the first eight abdominal segments. In addition, two types of gills are found on the aquatic forms. Larvae of *Anchytarsus* and *Anchycteis* have fingerlike gills in the anal region on the ventral side of the ninth abdominal segment. There are only five such gills in *Anchycteis,* which would seem insufficient to maintain them below the surface for any length of time. However, the integument has the shining gloss seen in the larva of *Dryops,* which is reported to carry a film of air about the entire body. The *Stenocolus* larva has two tufts each composed of eleven to twelve fingerlike gills on the underside of abdominal segments one to seven.

Habitat and distribution.—Adults of *Stenocolus* are found along streams entering the Sacramento and San Joaquin valleys of California, up to an elevation of 4,000 feet. Associated larvae have been taken in about the same area in streams varying from small creeks to rivers. They burrow in the substratum of the stream, feeding upon the roots of trees and other vegetation. Larvae associated with *Anchycteis* have been taken in northern California in spring areas and small- to medium-sized rapidly flowing creeks. Presumably they have approximately the same habits as *Stenocolus*.

Key to the Nearctic Genera of Ptilodactylidae

Adults

1. Mandibles prominent, acutely margined above (margin often hidden by pubescence), rectangularly flexed at tip, head not retracted; thorax acutely margined .... *Stenocolus* LeConte 1853
   Stenocolus scutellaris LeConte 1853; central California along streams.

— Mandibles not prominent, arcuate at tip, not acutely margined above, head strongly deflexed; tarsi slender 2

2. Antennae slender; middle coxae not more widely separated than the anterior pair; thorax obtusely margined, prosternum moderately long before the coxae .... *Anchytarsus* Guerin 1843
   Anchytarsus bicolor (Melsh.) 1846; New York to Georgia

   Antennae serrate (pectinate in male), moderately long; middle coxae twice as widely separated as the anterior pair; margin of thorax very obtusely rounded, prosternum short in front of the coxae...... *Anchycteis* Horn 1880
   Anchycteis velutina Horn 1880; northern California, in springs and rapid streams.

Larvae

1. Two tufts of gill filaments on underside of abdominal segments 1-7, 9th segment without prehensile hooklike appendages; submentum not divided (fig. 13:19*b*) .... *Stenocolus* LeConte 1853

— Anal area of 9th abdominal segment with short fingerlike gills and 2 prehensile curved appendages covered with short stout spine; submentum divided longitudinally into 3 parts 2

2. With only 5 fingerlike gills, 3 anal and 1 on outer side of each prehensile appendage; dorsal flattened apex of 9th abdominal segment with small raised projection .... *Anchycteis* Horn 1880

— With 21 fingerlike gills; apex of 9th abdominal segment without projection (fig. 13:19*c*).................. .... *Anchytarsus* Guérin 1843

REFERENCES

BÖVING, A. G., and F. C. CRAIGHEAD
1931. An illustrated synopsis of the principal larval forms of the order Coleoptera. Ent. Amer., 11:1-351, 125 pls.

LECONTE, J. L., and G. H. HORN
1883. Classification of the Coleoptera of North America. Smithson. Misc. Coll., No. 507, xxxviii+568 pp., 70 text figs.

Family CHRYSOMELIDAE

This large family contains one subfamily, Donaciinae, and one species in the subfamily Galerucinae which are closely associated with aquatic plants. The grublike larvae of Donaciinae (which contains about sixty species) are truly aquatic, and the adults of some species may make short excursions under water. The larvae and adults of *Galerucella nymphaeae* feed only on the upper surface of floating leaves. The Chrysomelidae are characterized by their tarsi which have the fourth segment reduced and hidden between the lobes of the third. They are usually separated from the Cerambycidae by their oval form and shorter antennae, though the Donaciinae resemble the Cerambycidae to some extent in this respect. For this reason they are sometimes called the longhorn leaf beetles. *Galerucella* is a typical chrysomelid.

Relationships.—The Donaciinae have developed their aquatic adaptations independently of the rest of the Chrysomelidae, and for this and other reasons are sometimes placed in a separate family Donaciidae.

Respiration.—The larvae of *Donacia* have the dorsally placed spiracles of the eighth abdominal

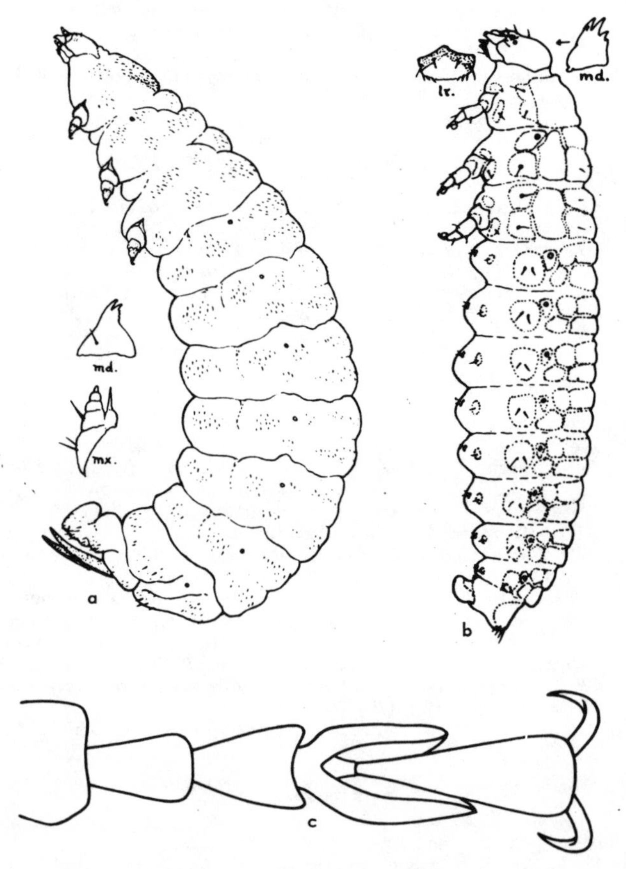

Fig. 13:60. Chrysomelidae, larvae. *a*, *Donacia* sp., lateral; *b*, *Galerucella nymphaeae*, lateral; *c*, *Donacia hirticollis*, adult hind tarsus (*a*,*b*, Peterson, 1951; *c*, Leech, original).

segment biforous and projecting like a pair of spurs (fig. 13:60*a*). The posterior part of the abdomen is bent beneath the body, so that the spiracles are ventral in relation to the rest of the body. They pierce the stems of aquatic plants and obtain the intercellular air in them. The cocoons are attached to the bases or stems of the same plants and the pupae obtain their oxygen in much the same manner.

Life history.—According to MaGaha (1952) mating takes place on the surface of the leaves or in the flowers of the aquatic plant on which *Donacia* feed. The females of some species cut holes in the floating leaves through which they reach the ovipositor to deposit their eggs in concentric circles on the underside of the leaf. Others deposit their eggs under the edge of the leaf. *D. piscatrix* crawls down the flower peduncle to deposit its eggs about 5 cm. below the surface of the water. The larvae feed on the submerged roots, stems, and petioles. The cocoons are constructed on the same part of the plant on which the larvae are found. They are spun from silk produced by glands in the mouth and are free of water inside when completed. The insect may remain in its cocoons for ten months or more. The pupa transforms to the adult long before it is time for the adult to emerge. In due course the end of the cocoon is broken off, and the beetle emerges carrying enough air on the ventral pubescence and under the elytra to last it until it reaches the surface.

All stages of *Galerucella nymphaeae* occur on the upper surface of the leaves of *Nuphar* or *Nymphaea*. Each female may lay from thirty-six to one hundred and fifteen eggs over a ten to twenty day period. These are deposited in masses of six to twelve each, usually at daily intervals. The larvae cannot swim.

Habitat and distribution.—The species of *Donacia* have been observed to feed on a wide variety of aquatic plants, but many of the forms are quite specific. The adults sometimes feed on other plants if the leaves of the preferred plant have not yet reached the surface of the water, but will quickly desert them when the preferred plant becomes available. Some adults feed on pollen. Among the plants used are *Nuphar, Nymphaea, Myriophyllum, Sagittaria, Sparganium,* and *Potamogeton*.

Key to the Nearctic Genera of Aquatic Chrysomelidae

Adults

1. Prothorax with a distinct, thin, lateral margin. GALERUCINAE .... *Galerucella* Crotch 1873
   Galerucella nymphaeae (Linnaeus) 1758, U.S.
— Prothorax without such a margin. DONACIINAE 2
2. Tarsi dilated, spongy beneath; 5th segment of tarsi subequal to or shorter than 2nd and 3rd together .... *Donacia* Fabricius 1775
— Tarsi not dilated, narrow, glabrous; 5th segment of tarsi distinctly longer than 2nd and 3rd together .... *Neohaemonia* Latreille 1829
   Neohaemonia nigricornis (Kirby) 1837, Canada and U.S., east of Rocky Mts.

Key to the Nearctic Genera of Aquatic Chrysomelidae

Larvae

1. Dorsal surface of 8th abdominal segment without a pair of long spines; abdominal prolegs present (fig. 13:60*b*). GALERUCINAE .....*Galerucella* Crotch 1873
— Dorsal surface of segment 8 with a pair of pointed spines; abdominal prolegs absent. DONACIINAE 2
2. Color white or cream; dorsum of 7th abdominal segment always rounded transversely and dorsal to posterior in position (fig. 13:60*a*) ..... *Donacia* Fabricius 1775
— Color green; half of dorsum of 7th abdominal segment posterior and other half flattened and ventral; usually found on *Potamogeton natans* .... *Neohaemonia* Latreille 1829

Key to the California Species of Donacia

Adults

1. Sutural bead of elytra approximate to sutural margin throughout entire length. Subgenus *Donacia* 2
— Sutural bead and sutural margin of elytra divergent at apical 6th. Subgenus *Plateumaris* 6
2. Entire dorsum pubescent, elytral epipleura not limited dorsally by an elevated ridge; California .... *pubescens* LeConte 1867
— Elytra glabrous, epipleura limited dorsally by a distinct elevated ridge 3
3. Pronotum finely pubescent; California .... *hirticollis* Kirby 1837
— Pronotum glabrous 4
4. Hind femora extending to elytral apices, pronotal

punctuation fine and sparse; dorsum metallic green and/or cupreous, strial punctuation coarse, basal triangulate excavation deep; Oregon, California .....
.... *proxima californica* LeConte 1861
— Hind femora never extending to elytral apices, pronotal punctuation coarse 5
5. Elytra with transverse rugae, median line of pronotum usually absent, head only slightly constricted behind eyes, eyes large and moderately prominent; fine and dense strigate-rugous sculpturing on elytra, coarse transverse rugae sparse, antennae stouter, hind femora less clavate; California ....
.... *subtilis magistrigata* Mead 1938
— Elytra without transverse rugae, median line of pronotum always present, head strongly constricted behind eyes, eyes small and prominent; hind femur less clavate with a very small obscure tooth, apical 3rd of elytra curved ventrad; California ....
.... *distincta occidentalis* Mead 1938
6. Hind femora entirely metallic, last abdominal tergite deeply emarginate; California ....
.... *emarginata* Kirby 1837
— Hind femora bicolored or entirely rufous; or if metallic, last abdominal tergite truncate 7
7. Prosternal episterna coarsely striate; California
.... *longicollis* Schaeffer 1925
— Prosternal episterna punctate-rugose, or at most only posterior half finely strigate 8
8. Second and 3rd segments of antennae small, equal, or subequal, pronotum densely and coarsely punctate, hind femur smaller at base than apex; western U.S.
.... *pusilla pyritosa* LeConte 1857
— Third antennal segment a 3rd longer than 2nd and equal or subequal to 4th, pronotum alutaceous and usually finely punctate, hind femur as wide or wider at base than at apex; U.S. ....
.... *germari* Mannerheim 1843

REFERENCES

HOFFMAN, C. E.
1940. Morphology of the immature stages of some northern Michigan Donaciini (Chrysomelidae, Coleoptera). Pap. Mich. Acad. Sci., 25:243-290, 101 figs.
McGAHA, Y. J.
1952. Limnological relations of insects to certain aquatic flowering plants. Trans. Amer. Micr. Soc., 71:355-381.
MacGILLIVRAY, A. D.
1903. Aquatic Chrysomelidae and a table of the families of coleopterous larvae. Part 5 *in* Aquatic insects in New York State. N.Y. State Mus. Bull. 68, pp. 288-331, pls. 20-31.
MEAD, A. R.
1938. New subspecies and notes on *Donacia* with key to the species of the Pacific States (Coleoptera, Chrysomelidae). Pan-Pac. Ent., 14:113-120, 2 text figs.
SCHAEFFER, CHARLES
1925. Revision of the New World species of the tribe Donaciini of the coleopterous family Chrysomelidae. Brooklyn Mus. Sci. Bull., 3:45-165, 1 pl. (No. 5).
SZEHESSY, V.
1941. Die zu Gattung *Haemonia* Lats. gestellten Arten aus America. Mitteil. Münschner Ent. Ges., 31:148-154.

Family CURCULIONIDAE

Weevils

This large family is characterized by having the head produced into a long snout on the end of which are the mouth parts. Both adults and larvae feed on plants and have invaded almost every environment where the higher plants are found. Even the aquatic plants are not free from their attacks. Only a fraction of a per cent of the species in the family have been associated with aquatic plants; however, this is still a sizable number (seventy to one hundred species) because of the extent of the family. The degree to which these are actually aquatic is known for only a few species. Some species live and feed only on the emergent parts of aquatic plants. Still others burrow through the submerged stems or roots but do not actually come in contact with the water, the burrow being filled with air.

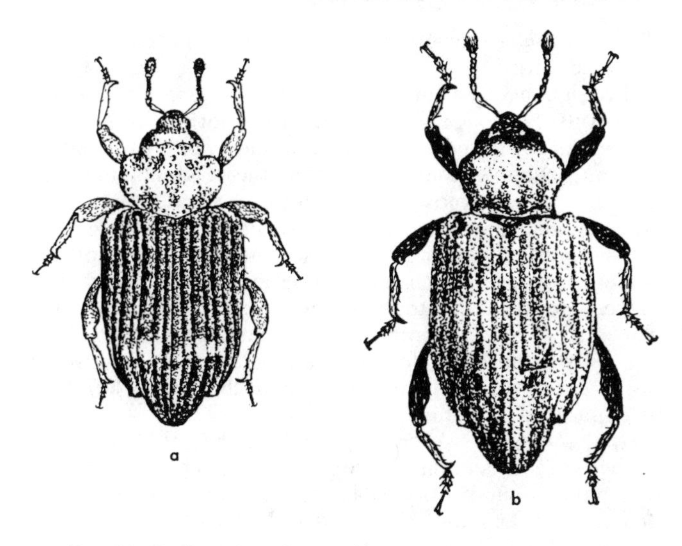

Fig. 13:61. Curculionidae adults. *a*, *Pnigodes setosus*; *b*, *Bagous tingi* (Tanner, 1943).

Relationships.—There are several sections in this family which are associated with the aquatic environment. Some of these are in different subfamilies and have developed their aquatic adaptations quite independently of each other. They are only remotely related to other aquatic families.

Respiration.—Most of the larvae will live for some time if placed under water; but their inability or even lack of inclination to try to swim, crawl, or burrow into plant tissue on which they rest would indicate that in nature they would soon perish if separated from their burrows or natural habitat.

The larvae of *Lissorhoptrus simplex* (Say) have the second to seventh abdominal segments with the spiracles hook-shaped and situated dorsally so they can be inserted into the plant tissue above them as they work along between the leaf sheath and the stem. Air is thus obtained from the plant tissue. The larva exudes a watertight cocoon which is attached to a rice stem and is connected to the air-filled interior of the stem by a hole.

Larvae believed to be those of *Bagous* and of an unassociated genus did not have any such special adaptations. The latter, taken from aquatic vegetation, had normal spiracles. Larvae of *Bagous* have the abdomen truncate behind, with a pair of enlarged spiracles on the truncate surface and the spiracles along the sides of the abdomen reduced or absent.

Thus, *Bagous* can utilize the air-filled burrows behind it for respiration.

Presumably the adults are able to store air beneath the elytra. *Bagous americanus* is reported by McGaha to spend long periods on the underside of floating leaves. The extent to which the Calendrinae are aquatic is not known, but I took several species by sweeping among aquatic vegetation which was mostly submerged. This prompted me to place two of the species on rush stems in a container half filled with water, so that the tops of the stems were above the surface. These specimens remained submerged more than half the time for a weeıi or more. They would walk up or down the stems at will but were helpless if separated from them.

Life history.—According to McGaha, the adults of *Bagous* feed and may even mate below the surface. He never found more than six to eight eggs in the females at any one time and only half of these were mature. *B. longirostris* may insert its eggs into the petioles of *Nymphaea tuberosa* as much as six feet below the surface. The larvae burrow through the petioles where they also pupate in an air-filled chamber. *B. americanus* mines the leaves of *Nymphae odorata* for the first three larval instars and then enters the pedioles where it pupates at their junction with the leaves.

Habitat and distribution.—Aquatic curculionids are usually closely associated with their host plants. My first contact with this group was while collecting on the shore of Utah Lake in a shallow, temporary pool in which vegetation was starting to grow. After collecting the first three specimens of the type series *B. chandleri* while sweeping the bottom, a closer look was taken to see where they were coming from. They were found walking about on the bottom of this shallow pool.

Tanner and McGaha record the following plant associations: *Lissorhoptrus simplex* is a serious pest of cultivated rice; *Brachybamus electus* on *Sagittaria; Endalus limatulus* on *Scirpus; Tanysphyrus lemnae* on *Lemna; Onychylis nigrirostris* on *Sagittaria, Nymphaea* and adults on the flowers of *Pontederia; Anchodemus angustus* on *Sagittaria; Lixellus filiformis* on *Scirpus, Ptilimnium,* and *Lepidum*. Additional records for *Bagous* are *Carex, Eleocharis, Cyporus,* and *Potamogeton. Listronotus appendiculatus* and *Hyperodes* sp. are recorded by McGaha from *Sagittaria,* but no mention is made as to whether this group is aquatic.

Taxonomic characters.—Böving and Craighead set up the subfamily Lissorhoptrinae for *Lissorhoptrus simplex,* based largely on the specialized respiratory structures of the larvae. *Lissorhoptrus* has usually been placed with *Bagous* in the Hydronomi. The respiratory structures of the latter are not at all like those of *Lissorhoptrus.*

Key to North American Genera of Curculionidae

Presumed to be Aquatic

Adults

1. Abdomen of male with an apparent extra segment at tip, pygidium and anal segment separated by a suture; club of antennae usually annulated, sensitive, and not shining. CURCULIONINAE: ERIRHININI, HYDRONOMI 2
— Abdomen similar in both sexes, pygidium not divided by a suture as the anal segment of male is at least partly free and retractile; club of antennae usually with its basal segment enlarged or shining, or both, without or with indistinct sutures. CALENDRINAE (Not further treated here).
2. Metasternum as long as 1st sternite, covered with dense, varnishlike, waterproof coating of scales 3
— Metasternum 1/3 length of 1st sternite; front coxae narrowly separated .... *Phycoetes* LeConte 1876
Phycoetes testaceus LeConte 1876, So. Calif.
3. Beak very short and broad, not longer than head; tarsi narrow, 3rd segment deeply emarginate .... *Stenopelmus* Schoenherr 1845
Stenopelmus rufinasus Gyllenhal 1835, Florida to southern California.
— Beak cylindrical, much longer than head 4
4. Third segment of hind tarsi emarginate or bilobed 5
— Third segment of hind tarsi simple; legs long and slender 10
5. Beak curved; antennal funicle of 6 segments, the 2nd short; 3rd segment of tarsi broad, deeply bilobed, last segment short 6
— Beak straight; 2nd segment of funicle long, as is also last segment of tarsi 9
6. Tarsi with a single claw.... *Brachybamus* Germar 1836
Brachybamus electus Germar 1836, Massachusetts, Indiana, Florida
— Tarsi with 2 claws 7
7. Last segment of tarsi broad, claws well separated 8
— Last segment of tarsi narrow, projecting beyond lobes of 3rd, claws slender; eastern .... *Onychylis* LeConte 1876
8. Elytra but slightly if at all wider than thorax; length usually 2 mm. or more .... *Endalus* Laporte 1840
— Elytra much wider than thorax; length less than 1.5 mm. .... *Tanysphyrus* Germar 1817
9. Front and middle tibiae serrate on inner side; 3rd tarsal segment narrow, slightly emarginate; funicle of 6 segments .... *Lixellus* LeConte 1876
Lixellus filiformis LeConte 1876, Ontario, Alberta, Oregon
— Tibiae not serrate within; 3rd tarsal segment broad, deeptly bilobed; funicle of 7 segments; eastern .... *Anchodemus* LeConte 1876
10. Club of antennae partly smooth and shining, funicle of 6 segments; prosternum not excavated .... *Lissorhoptrus* LeConte 1876
Lissorhoptrus simplex (Say) 1831; eastern.
— Club entirely pubescent and sensitive, funicle of 7 segments; prosternum broadly and deeply excavated in front of coxae 11
11. Pronotum feebly constricted in front (fig. 13:61*b*) .... *Bagous* Germar 1817
— Pronotum very strongly constricted and tubulate in front (fig. 13:61*a*) .... *Pnigodes* LeConte 1876
Pnigodes setosus LeConte 1876; U.S.

Key to California Species of Endalus

Adults

1. Last segment of tarsi slightly prominent, claws moderately large, elytra wider than prothorax; U.S. .... *limatulus* (Gyllenhal) 1836
— Last segment of tarsi not prominent; body oval; U.S. .... *ovalis* LeConte 1876

Key to California Species of Bagous

Adults

1. Fourth tarsal segment long, claws divergent, elytra uniform in color 2
— Fourth tarsal segment not so long, claws less divergent, elytra with color spots; U.S. .... *restrictus* LeConte 1876
2. Elytra uniform in color, beak short; southern California .... *californicus* LeConte 1876
— Elytra with some mixing of scales of different colors, (fig. 13:61*b*); San Mateo County .... *tingi* Tanner 1943

REFERENCES

McGAHA, Y. J.
1952. Limnological relations of insects to certain aquatic flowering plants. Trans. Amer. Micr. Soc., 71:355-382.
TANNER, V. M.
1943. Study of the subtribe Hydronomi with description of new species. Great Basin Nat., 4:1-38, incl. 12 text figs., 1 pl.

CHAPTER 14

Aquatic Diptera

By Willis W. Wirth and Alan Stone
Entomology Research Branch
United States Department of Agriculture

The Diptera are a large and diverse group of insects which we know commonly under the names of flies, gnats, midges, and mosquitoes. According to Peterson (1951) fully half of the species live in water, utilizing nearly all possible habitats where water may be present. Among the aquatic species are such important economic pests as mosquitoes, punkies, buffalo gnats, and horse flies, all of which have bloodsucking habits and many of which are vectors of such important diseases as malaria, dengue, filariasis, yellow fever, and encephalitis. On the other hand aquatic Diptera such as midges, crane flies, and brine flies often occur in sufficient quantities to be very important food items of game and food fish and aquatic game birds.

As denoted by the name Diptera, this order is recognized by the presence of only one pair of wings, the hind pair being reduced to slender, club-shaped balancing organs called halteres. The mouth parts are adapted for sucking liquid food and in many families form a piercing apparatus, at least in the female sex. Other structures are quite subject to modification within the general evolutionary pattern of the order.

Metamorphosis is complete in the Diptera, the process entailing the usual egg, larval, pupal, and adult stages, although in some families there may occur at times a telescoping of from one to three of these stages (e.g., larviparous Sarcophagidae, Pupipara, paedogenetic Itonididae). The larvae never possess true legs, although several pairs of prolegs may be present; the body form is extremely diverse, ranging from a simple fusiform type with well-developed head capsule in the primitive groups to the typical, maggotlike larva of the muscoid families. Nearly all types of spiracular arrangement are found in the order, but the aquatic species are most commonly amphipneustic, metapneustic, or apneustic. The pupae of the primitive families are usually free and active, but in the progressively higher families there is a tendency for the pupa to be enclosed in the last larval skin, which forms a protective, highly sclerotized, barrel-shaped puparium.

The variety of habitats occupied by the immature stages of aquatic Diptera is perhaps greater than that of any other order. Aquatic Diptera are known from the intertidal zone of exposed, wave-swept rocks along seacoasts (Tendipedidae; Clunioninae), from thermal waters issuing from springs or geysers at temperatures up to 120° F (Stratiomyidae), from natural seeps of crude petroleum (Ephydridae: *Helaeomyia petrolei*), and from saturated brine pools *(Ephydra cinerea)*. More normal habitats include the usual standing and running waters where dipterous larvae play their various roles as herbivores, scavengers, and predators.

We would like to acknowledge again our indebtedness to O. A. Johannsen of Cornell University whose great work on aquatic Diptera provided the basis for our treatment of this group, and to C. P. Alexander of the University of Massachusetts who supplied the basic information in the first edition for the tipulid section. We have had valuable assistance from published papers from which we have borrowed freely, manuscript keys which we were permitted to use, and general review and criticism from the following specialists to whom we extend our sincere thanks: Larry W. Quate (Psychodidae), S. B. Freeborn and R. M. Bohart (Culicidae), H. K. Townes (Tendipedidae), M. T. James (Stratiomyidae), W. W. Middlekauff (Tabanidae) and C. W. Sabrosky (Muscidae, Sarcophagidae, and Tetanoceridae). The superb work, *Die Larvenformen der Dipteren,* which has just been completed by Willi Hennig (1948-1952) has been of extraordinary help in our search of the literature and in our adaptation of generic keys to larvae and pupae.

It might well be admitted that this study is almost entirely a compilation of keys, descriptions, and figures from many sources, although in great part from the specialists we have named above. For any errors or omissions that we have made in interpreting or copying their work we assume full responsibility. The only families in which our work is in any sense original are the Simuliidae and Tabanidae-Tabaninae (Stone), and Heleidae, Tendipedidae-Clunioninae, Ephydridae, and Canaceidae (Wirth).

To go into any great detail in the special terminology of characters used in the keys and diagnoses is impossible here. We hope that reference to the figures will be sufficient to use the keys in most cases. Torre-Bueno's (1937) glossary is almost indispensable and should be consulted freely. If available to the student, the following references should be used to supplement our keys: Curran (1934) and Brues and Melander (1954) for keys and illustrations to the adults; Malloch (1917), Johannsen (1934-1938), Hennig (1948-1952), Chu (1949), and Peterson (1951) for keys and illustrations to the immature stages. Crampton's (1942) review in the Diptera of Connecticut of the external morphology of adult Diptera is extremely helpful. We have followed (fig. 14:1,2) the Tillyard modification of the Comstock-Needham system of designating the wing venation, outlined in the same volume by Friend (1942). References to the special literature of each family will be found at the end of each family section; however, to conserve space when such references are to general Diptera works, the reader is referred to the end of this section for the complete citation.

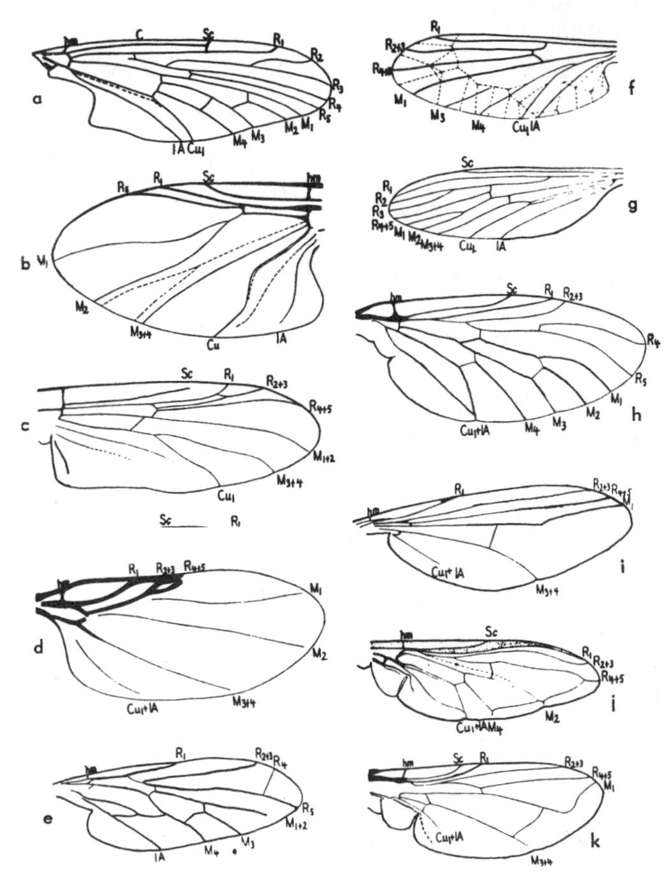

Fig. 14:1. Wings of Diptera. Venation labeled according to Tillyard modification of Comstock-Needham system: A, anal; C, costa; Cu, cubitus; M, media; R, radius; Sc, subcosta; hm, humeral. *a*, Tanyderidae, *Protoplasa;* *b*, Simuliidae, *Simulium;* *c*, Tendipedidae, *Procladius;* *d*, Phoridae, *Dohrniphora;* *e*, Empididae, *Empis;* *f*, Blephariceridae, *Blepharicera;* *g*, Culicidae, *Aedes;* *h*, Rhagionidae, *Chrysopilus;* *i*, Dolichopodidae, *Dolichopus;* *j*, Syrphidae, *Syrphus;* *k*, Muscidae, *Musca* (Friend, 1942).

Keys to the Families of Aquatic Diptera

Adults

1. Antenna usually longer than the thorax (figs. 14:3*a*, 14:16*b*, 14:36), consisting of a flagellum of 6 or more similar free segments, with 2 enlarged basal segments 2
— Antenna usually 3-segmented, 3rd with a style (fig. 14:44*g*) or arista (figs. 14:2*e,f*; 14:54*i*; 14:63*a*) or with more or less distinct annulations (fig. 14:44*d,e*) 13
2. Mesonotum with a V-shaped transverse suture beginning on each side in front of root of wings, the pointed middle part close to the scutellum (fig. 14:3*a*); female with a conical ovipositor (TIPULOIDEA) 3
— Mesonotum without a V-shaped suture; if a transverse suture present it is interrupted in middle and there are less than 9 veins meeting wing margin (figs. 14:15*a,b*; 14:19*i,j*; 14:28*a*; 14:29*i*) 5
3. Only 1 anal vein reaches wing margin (fig. 14:1*a*) 4
— Two anal veins reach the wing margin (fig. 14:3*a*) .... TIPULIDAE
4. Second and 3rd veins each with 2 branches (radius 5-branched) (fig. 14:1*a*) .... TANYDERIDAE
— Second and 3rd veins with only 3 branches reaching the wing margin (radius 4-branched) (fig. 14:8*f*) .... LIRIOPEIDAE
5. Wing membrane with a secondary venation due to creases from folding of wing in the pupa (fig. 14:1*f*) 6
— Wings without an extensive secondary venation 7
6. Wings large, with fine hairs, true veins almost absent but an elaborate fanlike development of secondary folds present; male antenna exceedingly long, 6-segmented (fig. 14:12) .... DEUTEROPHLEBIIDAE
— Wings with well-developed primary veins, the secondary folds forming a delicate network like spider webbing (fig. 14:1*f*) .... BLEPHARICERIDAE
7. Costa continuing around wing margin, although often weaker along hind margin of wing 8
— Costa ending at or near the wing tip 11
8. Seven longitudinal veins reaching wing margin (fig. 14:43*a*) .... THAUMALEIDAE
— At least 9 veins reaching wing margin (figs. 14:16*c*; 14:18*b*) 9
9. Veins and body lacking scales; subcosta ending in costa at or beyond middle of wing (fig. 14:18*b*) .... DIXIDAE
— Veins, including hind margin of wing, very hairy or scaly; body and legs hairy or scaly (fig. 14:9*a*) 10
10. Wings short, broadly ovate or pointed, cross veins absent except near base of wing (fig. 14:9*a*); wings held rooflike over body when at rest.... PSYCHODIDAE
— Wings longer and narrow, not held rooflike over the body (fig. 14:15*a,b*) .... CULICIDAE
11. Wings with anterior veins thick, others very weak, 5th longitudinal vein (mediocubitus) forked at the base, not petiolate (fig. 14:19*b*) .... SIMULIIDAE
— Anterior veins not greatly thicker than other veins, 5th vein usually forked near middle of wing and petiolate 12
12. Fourth vein (anterior media) not forked; mouth parts not fitted for piercing; postscutellum usually with median longitudinal furrow or keel ... TENDIPEDIDAE
— Fourth vein forked; mouth parts fitted for piercing; postscutellum rounded, without median furrow or keel .... HELEIDAE
13. Tarsi with 3 nearly equal pads under tarsal claws (empodium developed pulvilliform, fig. 14:2*d*) 14
— Tarsi with 2 pads under claws (empodium hairlike or absent, fig. 14:2*c*); 3rd antennal segment never annulated 16

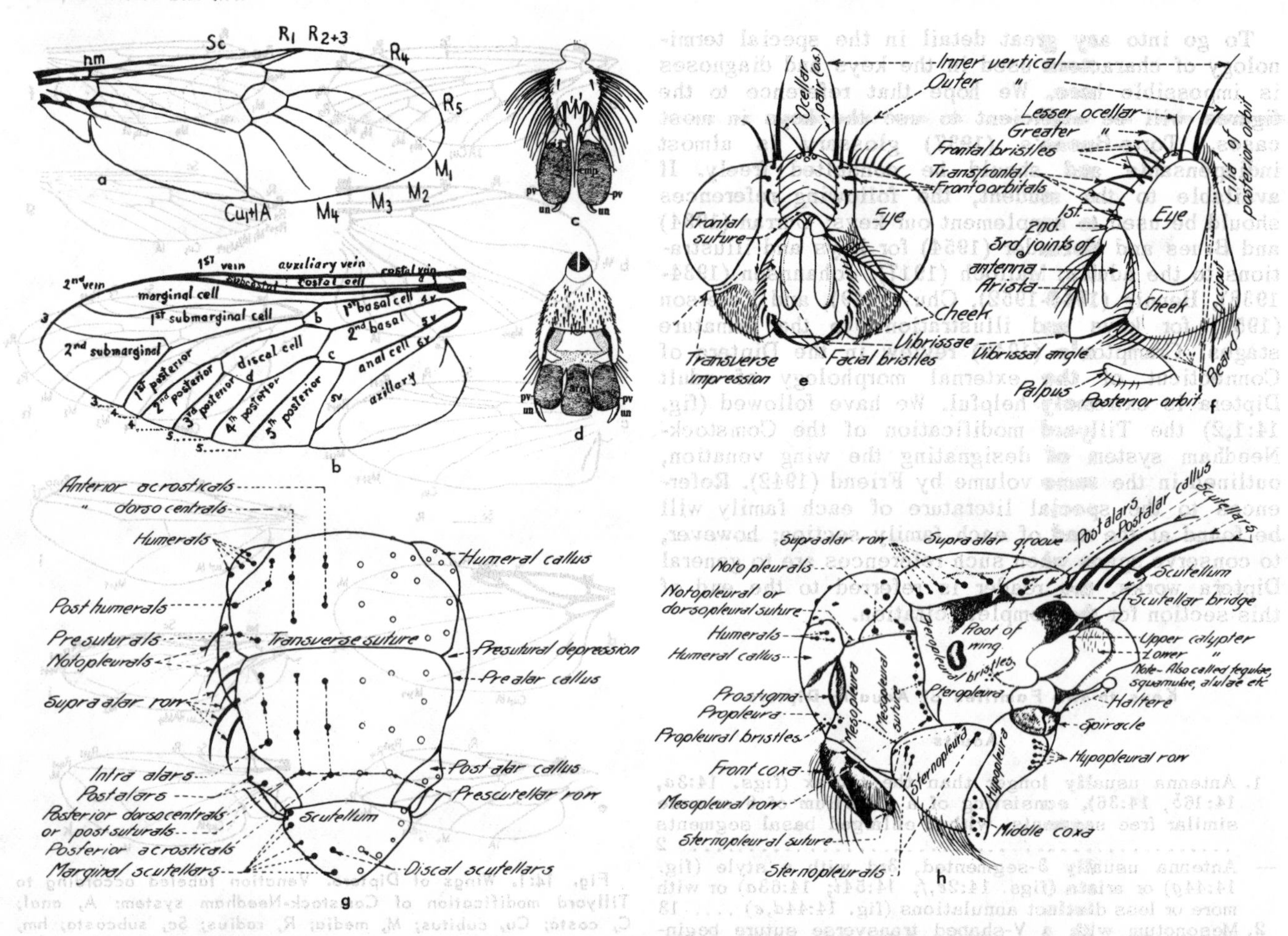

Fig. 14:2. Adult Diptera. *a*, Stratiomyidae, *Ptecticus*, veins labeled according to Tillyard System (see explanation of fig. 14:1); *b*, Tabanidae, *Tabanus*, veins and cells labeled according to Loew and Williston System; *c*, Asilidae, *Mallophora*, last tarsal segment; ventral view (emp, empodium; pv, pulvillus; un, ungue or claw); *d*, Tabanidae, *Tabanus*, same (aro, arolium or pulvilliform empodium); *e*, Calliphoridae, *Calliphora*, head, anterior view; *f*, Same, head, lateral view; *g*, Same, thorax, dorsal view; *h*, Same, thorax, lateral view (*a*, Friend, 1942; *b*, Williston, 1908; *c*,*d*, Crampton, 1942; *e*-*h*, Walton, 1909).

14. Third antennal segment compact, not annulated, bearing an elongate arista .... RHAGIONIDAE
— Third antennal segment annulated into 3-8 apparent segments, rarely bearing an elongate arista (figs. 14:44*d, e, g*; 14:47*d, e*) 15
15. Tibial spurs absent; costa ending before wing tips; squamae small or vestigial .... STRATIOMYIDAE
— Spurs present at least on mid-tibia; costa continuing around hind margin of wing; squamae large and conspicuous .... TABANIDAE
16. Radial veins stout, running into costa near middle of wing; medial veins weak and extending obliquely across wing; cross veins and basal cells absent (fig. 14:1*d*) .... PHORIDAE
— Radial veins not stronger than others; medial veins not oblique 17
17. Anal cell closed just before wing margin, therefore short-petiolate; spurious vein running obliquely between 3rd (R_5) and 4th (M_1) veins and crossing the anterior cross vein (r-m) (fig. 14:1*j*) .... SYRPHIDAE
— Anal cell short, transverse, oblique or convex apically, or rarely entirely absent 18
18. Frontal suture or lunule entirely absent; front uniformly sclerotized; no alula 19
— Frontal suture well developed as a horseshoe-shaped groove over the antennae, continuing down to separate the center of the face from the sides; frontal lunule a crescentic sclerite between antennae and frontal suture; mesofrons nearly always differentiated from the frontal orbits (fig. 14:2*e*) 20
19. Anterior cross vein (r-m) situated at or before the basal 4th of the wing; 2nd basal and discal cells always united; 3rd vein never furcate (fig. 14:1*i*) .... DOLICHOPODIDAE
— Anterior cross vein (r-m) situated far beyond the basal 4th of the wing; 2nd basal and discal cells usually separated; 3rd vein usually furcate (fig. 14:1*e*) .... EMPIDIDAE
20. Second antennal segment without a longitudinal seam on its upper outer edge; thorax without a complete transverse suture before the wings; squamae small (ACALYPTRATAE) 21
— Second antennal segment with a longitudinal seam along upper outer edge which extends quite to the base; thorax with a complete transverse suture before the wings (fig. 14:2*g, h*); squamae large (fig. 14:2*h*) (CALYPTRATAE) 24
21. Costa fractured just before end of subcosta; subcosta usually incomplete, not attaining costa 22

— Costa entire, subcosta complete, ending in the costa; legs with dorsal preapical bristle ...... SCIOMYZIDAE

22. Costa also broken at humeral cross vein; 2nd basal and anal cells not formed (fig. 14:54*j*)... EPHYDRIDAE

— Costa not broken at humeral cross vein; 2nd basal and anal cells complete (fig. 14:1*k*)................. 23

23. Vibrissae present at the angle formed by the facial ridges and oral margin; preapical tibial bristles present; transverse mesonotal suture nearly complete ... . SCOPEUMATIDAE (part)

— Vibrissae absent, only peristomial hairs or setae present; preapical tibial bristles absent; transverse suture not developed .... CANACEIDAE

24. Hypopleuron with a vertical row of bristles (fig. 14:2*h*); 4th vein (M_1) curving or bending forward, narrowing the apical cell at the wing margin (fig. 14:1*k*) .... SARCOPHAGIDAE

— Hypopleuron with short fine hairs or bare; 4th vein nearly straight, the apical cell not narrowed (fig. 14:63*a*) 25

25. Eyes of male not widely separated; cruciate frontal bristles present; lower calypter longer than the upper (fig. 14:2*h*); transverse mesonotal suture complete . ... MUSCIDAE

— Eyes widely separated in both sexes; no cruciate frontal bristles; lower calypter not longer than the upper; transverse suture incomplete .... SCOPEUMATIDAE (part)

Larvae

1. Head capsule complete (figs. 14:8*a*; 14:16*e*; 14:35*d*), or the posterior part with deep longitudinal incisions (fig. 14:4*h*); mandibles opposed, moving in a horizontal or oblique plane (NEMATOCERA) 3

— Head capsule incomplete or lacking (figs. 14:49*e*; 14:54*a*; 14:63*l*); mandibles or mouth hooks parallel, moving in a vertical plane, or if their motion is obliquely inward, then the head is not sharply differentiated from the 1st thoracic segment (BRACHYCERA) 2

2. Mandibles normally sickle-shaped, not protruding much beyond apices of the well-developed maxillae (fig. 14:49*p*); maxillary palpi distinct; dorsal part of head capsule more or less developed (fig. 14:49*s*, *t*); antennae well developed, situated on the sclerotized dorsal plate; pharyngeal skeleton loosely articulated; slender metacephalic and tentorial rods present, produced into thorax (fig. 14:49*e*) (ORTHORRHAPHA) 15

— Mandibles short and hooklike, usually capable of protrusion much beyond the poorly developed maxillae; palpi rarely visible; no head capsule; antennae when present situated on a membranous surface; pharyngeal skeleton with posteriorly directed branches fused anteriorly, forming a compact body (figs. 14:54*a*; 14:56*e*; 14:63*k*, *l*) (CYCLORRHAPHA)................... 20

3. Head capsule complete, not retractable within the thorax (fig. 14:8*a*, *l*) 4

— Head capsule incomplete posteriorly (fig. 14:4*h*) and more or less retractable within the thorax (figs. 14:4*n*, *o*; 14:5*e*) .... TIPULIDAE

4. Head, thorax, and 1st abdominal segment fused into a compound body division; ventral side with a median row of sucker discs by which larva attaches itself to the surface of rocks of mountain stream beds (fig. 14:11*a*, *b*, *d-g*, *i*, *j*) .... BLEPHARICERIDAE

— Head not fused with thorax and 1st abdominal segment (figs. 14:16*e*; 14:19*a*; 14:22*j*) 5

5. Body flattened, thorax with 3 distinct segments; abdomen with 7 pairs of large lateral pseudopods, each supplied with several rows of concentric minute hooks; antenna long, 2-branched (fig. 14:13*e*) .... DEUTEROPHLEBIIDAE

— With other characters 6

6. Prolegs (pseudopods) absent (fig. 14:15*c*) 7

— Prolegs present either at one or both ends of body or on intermediate abdominal segments (figs. 14:19*a*; 14:28*l*) 9

7. The 3 thoracic segments fused, forming a more or less enlarged section of the body distinctly thicker than the abdomen (figs. 14:14; 14:16*e*; 14:17*f*) .... CULICIDAE

— Thorax and abdomen about equal in diameter (figs. 14:9*c*; 14:18*c*, *d*, *f*, *j*, *k*) 8

8. Thorax and abdomen with a secondary segmentation, usually 2 annuli on each thoracic and 1st abdominal segment and 3 for each of abdominal segments 2-7, some or all of the annuli with narrow, transverse, dorsal plates; respiratory openings on prothorax and anal segment (fig. 14:9*c*) .... PSYCHODIDAE

— Body without distinct secondary annuli or narrow dorsal plates; spiracles absent.......................... .... HELEIDAE (part)

9. Prolegs present on intermediate body segments (fig. 14:8*h*; 14:18*c*, *f*) 10

— Prolegs confined to anterior or posterior ends of body, or both 11

10. Abdominal segments 1 and 2 each with a pair of ventral prolegs, each provided with many small hooks; posterior end of body with 2 pairs of fringed processes and a terminal, bristly lobe behind the posterior spiracles (fig. 14:18*d*, *j*) .... DIXIDAE

— Abdominal segments 1, 2, and 3 each with ventral prolegs, each with 1 or more claws; posterior abdominal segments retractile, terminating in a long slender respiratory tube (fig. 14:8*e*, *h*) .... LIRIOPEIDAE

11. Prolegs present on prothorax only; posterior part of abdomen swollen and armed caudally with an adhesive disc for attachment to rocks or objects in running streams; maxillae usually with a fan of long hairs (fig. 14:19*a*) .... SIMULIIDAE

— Prolegs present on prothorax and posterior end of body, or at posterior end only; body not club-shaped; caudal adhesive disc absent; maxillae without long fanlike hairs 12

12. Prothoracic and caudal spiracles present 13

— Prothoracic and caudal spiracles absent, no functional spiracles 14

13. Prothoracic prolegs absent; long, paired, posterior prolegs present; caudal end of body with 6 long filaments; caudal spiracles paired (fig. 14:6) .... TANYDERIDAE

— Prothoracic and posterior prolegs present and unpaired; caudal end of body with a single spiracle between 2 short processes (fig. 14:43*b*) ...... THAUMALEIDAE

14. Both anterior and posterior prolegs normally present, well developed and provided with claws (figs. 14:22*j*; 14:28*l*), or if posterior pair is reduced, the anterior pair is at least represented by patches of spinules (fig. 14:28*g*); body hairs not strongly developed .... TENDIPEDIDAE

— If both anterior and posterior prolegs are present, the body segments are provided with strong spines or bristles (fig. 14:39*ff*); or if body is smooth, anterior proleg may be absent and posterior proleg represented by a few ventral claws (fig. 14:38*t*) .... HELEIDAE (part)

15. Free part of head not retractile; prolegs wanting; body more or less depressed; cleft of respiratory chamber transverse; integument with calcium carbonate crystals; pupa enclosed in last larval skin (fig. 14:45) .... STRATIOMYIDAE

— Free part of head more or less retractile; prolegs present, or, if absent, the body cylindrical and nearly circular in cross section; pupa free 16

16. Apex of abdomen with 1 or 2 pairs of setigerous or ciliated processes; abdominal segments each with a pair of strong prolegs bearing claws (fig. 14:51) ... 17

— Either the caudal processes or the prolegs wanting 18

17. Upper divergent pair of caudal processes as long as anterior prolegs, each limb bearing 3 setae longer

than itself *(Roederiodes)* .... EMPIDIDAE (part)
— Caudal processes longer than prolegs, ciliated when larva is fully grown; dorsal head plate pyriform; each antenna with 3 terminal hairs; hypopharynx not V- or Y-shaped *(Atherix)* .... RHAGIONIDAE
18. Body cylindrical, circular in cross section, tapering at both ends; abdomen finely, longitudinally striate, with a girdle of prolegs or pseudopods on each segment; posterior respiratory organs close together, situated in a vertical cleft (fig. 14:48*a, b, g*) ...... TABANIDAE
— Abdomen not distinctly finely striate, with a girdle of prolegs or pseudopods on each segment; hypopharynx V- or Y-shaped, the branches directed posteriorly 19
19. Caudal end terminating in a spiracular pit surrounded by several pointed lobes (fig. 14:49*b, e, i, l, n, v, y*) . ... DOLICHOPODIDAE
— Caudal end rounded with the spiracles borne on the surface or on distinct raised processes (fig. 14:51) .... EMPIDIDAE (part)
20. Pharyngeal skeleton with the "hypopharynx" produced in the form of a median tooth; prothoracic spiracles reduced, simple (fig. 14:52*b*) .... PHORIDAE
— "Hypopharynx" not so produced; prothoracic spiracles each with several to many openings, rarely absent 21
21. Mouth hooks vestigial or wanting in the aquatic species; posterior spiracles close together at the extremity of a tube which may be either very short, or more or less elongate and in part retractile; integument transversely wrinkled (fig. 14:53) .... SYRPHIDAE
— Mouth hooks present; spiracles in well-separated discs (figs. 14:56*l*; 14:63*k*); if at the tip of a retractile tube (fig. 14:59), the mouth hooks are palmate 22
22. Posterior spiracular plates ending in a needlelike spine (fig. 14:56*d*); anterior spiracles absent; mouth hooks serrate below (fig. 14:56*e*)............... .... EPHYDRIDAE (part)
— Posterior spiracular plates (PS) not ending in a needlelike spine (figs. 14:56*l*; 14:63*d*); anterior spiracles (AS) present (figs. 14:60*d*; 14:63*d, f, l, v*) 23
23. Posterior spiracles borne at the end of a more or less elongated, sometimes branched or retractile air tube (figs. 14:56*l*; 14:58*c*; 14:59*a, c*; 14:60*d*; 14:63*e, g, k*) 24
— Posterior spiracles borne flush or on short sclerotized plates (figs. 14:62*a,d*; 14:63*d*; 14:64*a-c*) or in a cavity on the caudally truncated or rounded posterior segment 26
24. Spiracles borne on a distinctly lobed disc or tube; mouth hooks joined together behind the mouth opening (fig. 14:61*a-c*) .... SCIOMYZIDAE
— Spiracular disc or tube without fleshy marginal lobes (figs. 14:57*c, d*; 14:60*d*; 14:63*u*); mouth hooks not joined together behind the mouth opening 25
25. Mouth hooks slender and pointed, without ventral teeth (fig. 14:63*f, k, r*) .... MUSCIDAE (part)
— Mouth hooks palmate or digitate below (fig. 14:58*a*) .... EPHYDRIDAE (part)
CANACEIDAE
26. Posterior spiracles borne deep inside a cuplike, expanded chamber, the elongated slits arranged vertically (fig. 14:64*a-c*) .... SARCOPHAGIDAE
— Posterior spiracles not inside a cuplike chamber, the slits short and radially arranged (figs. 14:62; 14:63*d, n, p*) .... MUSCIDAE (part)
SCOPEUMATIDAE

Pupae

1. Pupa free, not completely covered by the last larval skin, though it may be covered by a cocoon (figs. 14:7*a*; 14:16*h*; 14:19*n, o*; 14:27*g*; 14:49*u, w*) 2
— Pupa remaining wholly within the last larval skin, the puparium, which is heavily sclerotized and in many cases is shortened to form an ellipsoidal or egg-shaped body (figs. 14:52*c*; 14:53*c*; 14:63*c*) (BRACHYCERA, part) 19
2. Antennal sacs elongated, lying over the compound eyes and extending to, or beyond, the bases of the wing sheaths; prothoracic respiratory organs in most cases conspicuous (figs. 14:7*a*; 14:16*h*; 14:22*i*) (NEMATOCERA) 3
— Antennal sacs short, directed posteriorly and laterally, not lying over the eyes; prothoracic respiratory organs usually rudimentary or lacking (figs. 14:48*f*; 14:49*c, u*) (BRACHYCERA-ORTHORRHAPHA) 17
3. Pupa usually more or less enclosed by a fibrous cocoon (figs. 14:19*n, p*; 14:27*g*) 4
— Pupa without cocoon, though sometimes in a silken tube 5
4. Cocoon cone-shaped or slipperlike (fig. 14:19*p*), from open end of which the respiratory organs of pupa project; respiratory organs consisting of 4-60 or more coarse filaments (fig. 14:19*o*) .... SIMULIIDAE
— Cocoon conical or cylindrical, with several filaments on margin; respiratory organ of pupa a slender, unbranched filament .... TENDIPEDIDAE (part)
5. Body of pupa more or less convex, in most cases hard shelled, attached limpetlike to rocks at stream bottom (figs. 14:10*a*; 14:11*h*; 14:13*a, b, f, g*) 6
— Body of pupa not so attached and not limpetlike 8
6. Body flattened, shield-shaped; respiratory organs simple, subcylindrical, or ovoid (fig. 14:10*a*) *(Maruina)* .... PSYCHODIDAE (part)
— Body more convex 7
7. Respiratory organs usually consisting of 4 simple lamellae; antennae not twice-coiled; ventral adhesive discs median in position (fig. 14:11*h*) .... BLEPHARICERIDAE
— Respiratory organs consisting of 3 or 4 crooked projections; antennae of male excessively long, each forming a double coil on the venter; 3 pairs of lateral adhesive discs (fig. 14:13*a, b, f, g*) .... DEUTEROPHLEBIIDAE
8. Leg sheaths straight, projecting beyond ends of wing sheaths, in most cases far beyond; caudal end not paddlelike; usually more than 8 mm. long; pupation usually in sand or mud (figs. 14:4*i*; 14:7*a*; 14:8*d*) 9
— Leg sheaths often curved or folded, projecting little if any beyond ends of wing sheaths (figs. 14:9*b*; 14:18*g*); caudal end often terminating in a paddle (figs. 14:15*f*; 14:16*h*; 14:22*h, i*); usually less than 8 mm. long; pupation usually in open water 11
9. One of the prothoracic respiratory organs longer than the body, the other short (fig. 14:8*c,d,h*) .... LIRIOPEIDAE
— Prothoracic respiratory organs less than ¾ of body length, nearly always much shorter and subequal in length (figs. 14:4*d, i, k, l, m*; 14:7) 10
10. Head with a high bispinose crest (fig. 14:7); respiratory organs small, smooth, and gradually narrowed to a terminal point (fig. 14:7*b*) .... TANYDERIDAE
— Without the above combination of characters .... TIPULIDAE
11. Abdominal segments angular when viewed from above, owing to the acute posterolateral angles of the tergites; thorax corrugated into numerous ridges and hollows; spiracles present on all abdominal segments except the 1st and last .... THAUMALEIDAE
— With other characters 12
12. Wing sheaths ending about mid-length of pupa; leg sheaths short, straight, and superimposed (fig. 14:9*b*) .... PSYCHODIDAE (part)
— Wing sheaths ending distinctly before mid-length of pupa when body and leg sheaths are straight, or body arched and leg sheaths undulatory 13
13. Caudal end of pupa with 2 paddles, each with a midrib (figs. 14:15*h*; 14:16*h*) .... CULICIDAE
— Paddles, if present, without a midrib (fig. 14:22*h*) 14
14. Swimming paddles consisting of 2 long pointed lobes fused basally, not movable, without long hairs and spines; spiracle situated beyond middle of prothoracic

respiratory organ; pupa normally lying on its side nearly motionless in an arched position (fig. 14:18*g*) .... DIXIDAE

— Swimming paddles absent (fig. 14:41*b-d*), or if present, provided with hairs or spines (figs. 14:22*h*; 14:31*j*), or pupa differing in other characters 15

15. Prothoracic respiratory organs without open stigmata, or composed of numerous filaments (figs. 14:31*j*; 14:35*c*), or absent .... TENDIPEDIDAE (part)

— Prothoracic respiratory organs with open stigmata (figs. 14:22*b*; 14:39*l-q, u, v, hh*) 16

16. Anal segment with a 2-lobed swimming paddle .... TENDIPEDIDAE (part)

— Anal segment ending in a pair of pointed processes (figs. 14:41*b-d*; 14:42*o-t*) .... HELEIDAE

17. Pupa robust, short, with greatly elongated prothoracic respiratory organs (fig. 14:49*c, j, m, q, u*).......... .... DOLICHOPODIDAE

— Pupa more elongate, prothoracic respiratory organs not so conspicuous 18

18. Prothorax with a large aperture mesad of, and connected with, the spiracle (fig. 14:48*f*) .... TABANIDAE

— Prothorax without such an aperture ...... EMPIDIDAE

19. Puparium within last larval skin, which is unchanged in shape; larval head capsule distinct (fig. 14:45) .... STRATIOMYIDAE

— Puparium without distinct head, the larval skin modified in shape owing to the changed form of pupa within (figs. 14:52*c*; 14:53*c, d*; 14:58*b*; 14:59*b*; 14:60*c*; 14:63*c, h, j, q, s*) (CYCLORRHAPHA) go to couplet 20 of larval key for characters of spiracles or pharyngeal skeleton which remain on the surface of the puparium.

GENERAL DIPTERA REFERENCES

BRUES, C. T., and A. L. MELANDER
1954. Classification of insects. Rev. ed., Cambridge, Mass. 917 pp.

CHU, H. F.
1949. How to know the immature insects. Dubuque, Iowa. 234 pp.

CRAMPTON, G. C.
1942. The external morphology of the Diptera. *In* Insects of Conn. Pt. VI. Diptera, fasc. 1. Bull. Conn. Geol. Nat. Hist. Surv., 64, pp. 10-165.

CRISP, G., and L. LLOYD
1954. The community of insects in a patch of woodland mud. Trans. Roy. Ent. Soc. London, 105:269-314, 3 pls.

CURRAN, C. H.
1934. The families and genera of North American Diptera. New York. 512 pp.
1942. Key to families. *In* Insects of Conn. Pt. VI. Diptera, fasc. 1. Bull. Conn. Geol. Nat. Hist. Surv., 64 pp. 175-182.

FRIEND, R. B.
1942. Taxonomy. Wing venation. *In* Insects of Conn. Pt. VI. Diptera, fasc. 1. Bull. Conn. Geol. Nat. Hist. Surv., 64, pp. 166-174.

HENDEL, F.
1928. Die Tierwelt Deutschlands. 11. Teil. Zweiflugler order Diptera II. Allgemeiner Teil. Jena. 135 pp.

HENNIG, W.
1948-1952. Die Larvenformen der Dipteren. 1 Teil, 186 pp. 2 Teil, 458 pp. 3 Teil, 628 pp. Berlin.

JOHANNSEN, O. A.
1934. Aquatic Diptera. Part I. Nemocera, exclusive of Chironomidae and Ceratopogonidae. Mem. 164, Cornell Univ. Agric. Exp. Sta., 71 pp., 24 pls.
1935. Aquatic Diptera. Part II. Orthorrhapha-Brachycera and Cyclorrhapha. Mem. 177, Cornell Univ. Agric. Exp. Sta., 62 pp., 12 pls.
1937. Aquatic Diptera. Part III. Chironomidae: subfamilies Tanypodinae, Diamesinae, and Orthocladiinae. Mem. 205, Cornell Univ. Agric. Exp. Sta., 84 pp., 18 pls.
1938. Aquatic Diptera. Part IV. Chironomidae: subfamily Chironominae. Mem. 210, Cornell Univ. Agric. Exp. Sta., pp. 1-56, pls. 1-9.

MALLOCH, J. R.
1917. A preliminary classification of Diptera, exclusive of Pupipara, based upon larval and pupal characters, with keys to imagines in certain families. Part I. Bull. Illinois Lab. Nat. Hist., 12:161-409, 29 pls.

NEEDHAM, J. G., and C. BETTEN
1901. Aquatic insects in the Adirondacks. Bull. N.Y. St. Mus., 47:383-612.

PETERSON, A.
1951. Larvae of insects. Coleoptera, Diptera, Neuroptera, Siphonaptera, Mecoptera, Trichoptera. Part II. 416 pp. Columbus, Ohio.

TORRE-BUENO, J. R. DE LA
1937. A glossary of entomology. 336 pp., 9 pls., Brooklyn, New York.

WALTON, W. R.
1909. An illustrated glossary of chaetotaxy and anatomical terms used in describing Diptera. Ent. News, 20:307-319, 4 pls.

Family TIPULIDAE

The family of true crane flies is very large, and the members are rather diverse in size and form, distinguished, in addition to the general crane fly facies, (fig. 14:3*a*), by the well-developed V-suture, lack of ocelli, and presence of two anal veins reaching the wing margin. The legs break off very readily between the trochanter and femur and are used very awkwardly

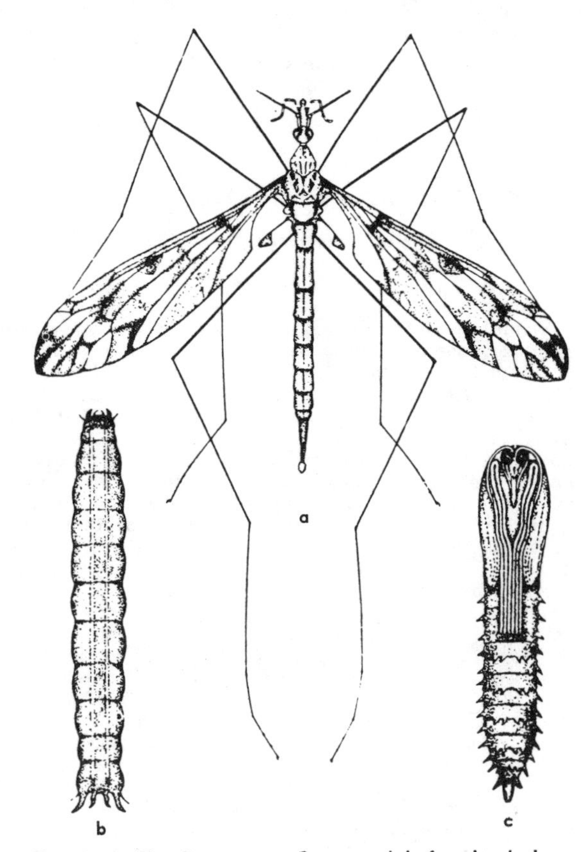

Fig. 14:3. *Tipula trivittata* Say. *a*, adult female; *b*, larva; *c*, pupa (Alexander, *in* Curran, 1934).

in walking. The variation in size is remarkable, from the 2 mm. wing length of *Dasymolophilus* to the gigantic *Holorusia* with wings 40 mm. long. Several genera have the wings more or less reduced, including the snow fly, *Chionea*, with wings practically absent. Species of this genus are commonly seen crawling around on the snow in winter, but the larvae are terrestrial in leafmold. Larvae (fig. 14:3*b*) of some species of *Tipula* known as "leather jackets" often seriously damage meadows and pastures in California by feeding on the roots of the plants. As these larvae come to the surface of the ground at night they can be destroyed by poison baits, and sometimes they are feasted upon by large flocks of birds. The most aquatic genera are *Antocha* (fig. 14:4*f*), whose larvae build silken cases, and *Hexatoma* (fig. 14:4*j*). The other species, as far as known, occur at the margins of streams and ponds, either strictly aquatic or in saturated earth, saturated moss cushions, or algal films, on dripping cliff faces, and in similar habitats. The immature stages of very few of the California species have been described so that the keys to the immature stages that follow, adapted from Johannsen (1934), may be quite unsatisfactory for California. The literature is very scattered, and there is no comprehensive work on the western species. Some of the more important papers dealing with North American species are listed at the end of this section.

Two nomenclatorial changes should be noted. We use the Meigen 1800 names; hence the genus currently listed as *Erioptera* is changed to *Polymeda*. However, by a prior type designation than that usually accepted (see Stone, 1941, p. 413) *Erioptera* Meigen replaces *Molophilus* Curtis.

Keys to the Aquatic or Semiaquatic Genera of North American Tipulidae

Adults[1]

1. Terminal segment of maxillary palpus elongate, whiplashlike; nasus usually distinct; antenna usually with 13 segments; wing with Sc_1 usually atrophied; vein Cu_1 constricted at m-cu, the latter usually at or close to fork of M_{3+4}; body size usually large (TIPULINAE) 2
— Terminal segment of maxillary palpus short; no distinct nasus; antenna usually with 14 or 16 segments; wing with Sc_1 present, its extreme tip atrophied in some Cylindrotominae; vein Cu_1 straight, not constricted at m-cu, the latter placed far before the fork of M_{3+4} usually at or close to fork of M; body size usually small or medium 6
2. Legs unusually long and filiform; wing with vein R_{1+2} atrophied and with Sc_2 ending in Sc close to origin of R_s .... *Dolichopeza* Curtis
— Legs of normal stoutness for the family; wing with vein R_{1+2} preserved; when atrophied with Sc of moderate length, Sc_1 atrophied before fork of Rs and Sc_2 ending at or near mid-length of Rs (exception, some species of *Longurio*); cell second A of normal width 3
3. Wing with vein R_3 bent strongly caudad before end, thence angularly deflected cephalad, cell R_3 thus being much constricted at or near mid-length .... *Holorusia* Loew
— Wing with vein R_3 straight or only gently arcuated throughout its length, not constricting the cell 4
4. Flagellar segments without verticils, the lower face of individual segments produced to give the organ a serrate appearance; terminal flagellar segment abruptly more slender .... *Prionocera* Loew
— Flagellar segments verticillate, simple or nearly so 5
5. Abdomen of both sexes greatly elongated, somewhat resembling that of a dragonfly; verticils of outer flagellar segments very long and conspicuous; valves (cerci) of ovipositor with smooth margins .... *Longurio* Loew
— Abdomen not so elongated or valves with serrate margins; antennal verticils of moderate length only .... *Tipula* Linnaeus
6. Wing with tip of R_{1+2} atrophied, giving the appearance of a long fusion back from margin of veins R_1 and anterior branch of Rs; free tip of Sc_2 preserved (CYLINDROTOMINAE) 7
— Wing sometimes with tip of R_{1+2} atrophied but not giving the appearance of a long fusion backward from margin of veins R_1 and anterior branch of Rs; free tip of Sc_2 preserved in many species of the tribe Limoniini, lacking in other tribes (LIMONIINAE) 8
7. Head and intervals of mesonotal praescutum with numerous deep punctures; a deep median groove on praescutum .... *Triogma* Schiner
— Head and intervals of mesonotal praescutum smooth; no median praescutal groove .... *Phalacrocera* Schiner
8. Eyes hairy; wing with vein Sc_1 very long, Sc_2 lying basad of origin of Rs 9
— Eyes glabrous; wing with Sc_1 short or of moderate length; when long Sc_2 lying distad of origin of Rs; where Sc_2 lies basad of origin of Rs the entire vein Sc is shortened 10
9. Antenna usually with 16 segments; size large, wing usually more than 10 mm. .... *Pedicia* Latirelle
— Antenna usually with 13 or 14 segments; size small, wing usually less than 8 mm. ... *Dicranota* Zetterstedt
10. Wing with free tip of Sc_2 often present; veins R_4 and R_5 fused to margin, only 2 branches of Rs being present; Antenna usually with 14 or 16 segments 30
— Wing with free tip of Sc_2 atrophied; veins R_4 and R_5 separate, the former usually transferred to the upper branch, R_{2+3}, to form a distinct element R_{2+3+4}; usually with 3 branches of Rs present; antenna usually with 16 segments 11
11. Tibial spurs present 12
— Tibial spurs lacking 21
12. Antenna with not more than 12 segments .... *Hexatoma* Latreille
— Antenna with more than 14 segments 13
13. Apical cells of wing with macrotrichia 14
— Cells of wing without macrotrichia (excepting in stigmal area) 17
14. A supernumerary cross vein in cell M .... *Limnophila* (in part)
— No supernumerary cross vein in cell M 15
15. Cell R_3 of wing sessile, subsessile or short-petiolate; R_{2+3+4} lacking or much shorter than m-cu 16
— Cell R_3 of wing long-petiolate, R_{2+3+4} being as long as or longer tham m-cu .... *Paradelphomyia* Alexander
16. Wing with macrotrichia abundant, involving the cells basad of cord .... *Ulomorpha* Osten Sacken
— Wing with sparse macrotrichia in cells beyond cord only .... *Limnophila* (in part)
17. Wing with vein R_2 lacking; m-cu at outer end of cell 1st M_2 .... *Phyllolabis* Osten Sacken
— Wing with vein R_2 present; m-cu at or before ⅔ the length of cell first M_2 when the latter is present 18
18. Wing with m-cu at or close to fork of M; anterior arculus lacking .... *Dactylolabis* Osten Sacken
— Wing with m-cu beyond the fork of M, at from ⅓ to about ½ the length of cell first M_2; where close to fork of M the arculus complete 19

[1]Adapted from Alexander, in Curran, 1934.

19. Wing with Sc relatively short, Sc_1 ending before the level of the fork of Rs; antennae with long conspicuous verticils .... *Pilaria* Sintenis
— Wing with Sc longer, Sc_1 ending opposite or beyond the fork of Rs; antennal verticils not unusually long 20
20. Head strongly narrowed and prolonged behind; radial and medial veins beyond cord long and sinuous; vein R_3 extending generally parallel to vein R_4, not diverging markedly at tips; vein second A strongly curved to margin .... *Pseudolimnophila* Alexander
— Head broad, not conspicuously narrowed behind; radial and medial veins beyond cord more nearly straight; vein R_3 diverging strongly from vein R_4, cell R_3 conspicuously widened at margin; vein second A not curved strongly to margin .... *Limnophila* Macquart (in part)
21. Wing with cell M_1 present .... *Crypteria* Bergroth
— Wing with cell M_1 lacking 22
22. Two branches of Rs reach the wing margin .... *Gonomyia* Meigen (in part)
— Three branches of Rs reach the wing margin 23
23. Wing with cell R_3 short, vein R_3 shorter than the petiole of cell R_3 24
— Wing with cell R_3 deep, vein R_3 longer than the petiole of cell R_3 27
24. Wing with vein R_2 present 25
— Wing with vein R_2 lacking 26
25. Wing with Rs long and straight, exceeding the distal section of M_{1+2}; tuberculate pits on cephalic part of praescutum; trochanters elongate .... *Rhabdomastix* Skuse (in part)
— Wing with Rs shorter, less than the distal section of M_{1+2}; tuberculate pits removed from cephalic margin of praescutum; trochanters short .... *Polymeda* Meigen (in part)
26. Wing with Sc long, Sc_1 extending to near opposite or beyond mid-length of Rs; m-cu at or beyond fork of M .... *Rhabdomastix* Skuse (in part)
— Wing with Sc short, not extending to beyond mid-length of Rs; if sc is relatively long, m-cu lies more than its own length before the fork of M .... *Gonomyia* Meigen
27. Wings with distinct macrotrichia in outer cells 28
— Wings with outer cells glabrous 29
28. Wing with Rs shortened, its union with R_{2+3+4} forming an angle, so cell R_1 is nearly equilateral in outline .... *Cryptolabis* Osten Sacken
— Wing with Rs long, normal in position, cell R_1 elongate .... *Ormosia* Rondani
29. Wing with Rs ending in cell R_3 there being no element R_{2+3+4} .... *Erioptera* Meigen
— Wing with Rs ending in cell R_4, cell R_3 being petiolate by a distinct element R_{2+3+4} .... *Polymeda* Meigen (in part)
30. Wing with vein R_2 lacking 31
— Wing with vein R_2 present 32
31. Rostrum short and inconspicuous; Rs long and straight, running close to R_1 and in alignment with R_{2+3}; r-m distinct .... *Elliptera* Schiner (in part)
— Rostrum of moderate length, about equal in length to remainder of head; Rs short, gently arcuated, not in alignment with R_{2+3}; r-m often shortened or obliterated by approximation of adjoining veins .... *Helius* Lepeletier
32. Wing with Rs long and straight; antenna 16-segmented 33
— Wing with Rs shorter, and more arcuated; antenna 14-segmented .... *Limonia* Meigen
33. Anal angle of wing very prominent, almost square; Rs long, diverging at an acute angle from R_1, ending approximately between the branches of Rs or in alignment with R_{4+5} .... *Antocha* Osten Sacken
— Anal angle of wing normally rounded; Rs long, lying very close to R_1 and nearly parallel to it, its end in alignment with R_{2+3}; basal section of R_{2+3} short and arcuated, diverging from the end of Rs at nearly a right angle .... *Elliptera* Schiner (in part)

Larvae [2]

1. Body cylindrical, if more or less depressed then with long spines or leaflike projections (figs. 14:3*b*; 14:4*c, f, j, n, o*) 2
— Body much depressed, ventral surface flattened, almost leechlike; hairy .... *Dactylolabis*
2. Body provided with elongated spines or leaflike projections 3
— Body without distinct spines 4
3. Body appendages very long and filiform (fig. 14:4*e*) .... *Phalacrocera*
— Body appendages short, leaflike or tuberculate, with 3 or 4 teeth on anterior convex side of some of the appendages .... *Triogma*
4. Spiracular disc surrounded by 6 or 8 lobes (fig. 14:4*a, b*) 24
— Spiracular disc not as above 5
5. Spiracular disc with 2 long ventral lobes (fig. 14:4*f*) 6
— Spiracular disc not as above 8
6. Spiracles lacking or vestigial (fig. 14:4*f*); mentum not completely divided medially; hypopharynx a rigid double comb; larvae in silken cases .... *Antocha*
— Spiracles large, exposed (fig. 14:4*o*); mentum completely divided medially; hypopharynx labriform 7
7. Abdomen without prolegs but with raised welts on segments 4 to 7, these covered with microscopic scurf (fig. 14:4*o*) .... *Pedicia*
— Abdomen with conspicuous cylindrical prolegs on segments 3 to 7, these with circlets of conspicuous hooklets around their ends .... *Dicranota*
8. Head capsule massive, complete, posterior incisions usually narrow (fig. 14:5*e*) 9
— Head capsule of 4 or 6 slender rods, posterior incisions deep 17
9. Mentum completely divided, a toothed plate on each side; abdominal segments without distinct creeping welts 10
— Mentum, if present, and sclerotized, not completely divided (fig. 14:5*c*); abdominal segments with basal creeping welts 15
10. Spiracular disc with 4 lobes (fig. 14:4*g*)........... 11
— Spiracular disc squarely truncated, surrounded by 5 lobes 12
11. Each mental plate 4-toothed; hypopharynx labriform .... *Paradelphomyia*
— Each mental plate with 7 or 8 teeth; hypopharynx a comblike ring .... *Pseudolimnophila*
12. Spiracular disc ending in 5 flattened black plates which are finely toothed along their margins .... *Ormosia* (in part)
— Spiracular disc not as noted above 13
13. Ventral plates of head capsule expanded and with 5 to 8 teeth at anterior end; color green; spiracular disc reduced, almost unmarked .... *Polymeda* (in part)
— Ventral plates not toothed as described above 14
14. Marks of all the lobes of spiracular plates solidly black .... *Polymeda* (in part)
.... *Ormosia* (in part)
— All paired lobes dark within, with a median light line .... *Gonomyia*
.... *Polymeda* (in part)
.... *Ormosia* (in part)
15. Body cylindrical; spiracular disc with 4 or 5 lobes (fig. 14:5*f*), or in some cases nearly obsolete 16
— Body depressed; spiracular disc with 4 lobes; abdominal segments 3 to 9 with dorsal and ventral creeping welts, the latter naked .... *Elliptera*
16. With dorsal and ventral welts on abdominal segments (fig. 14:4*n*); mentum with more than 5 teeth (fig. 14:5*c*) .... *Limonia*
— With 6 ventral welts only; body with long dark pubescence .... *Helius*
17. Blades of maxillae not produced beyond the retractile

[2] Adapted from Johannsen, 1934.

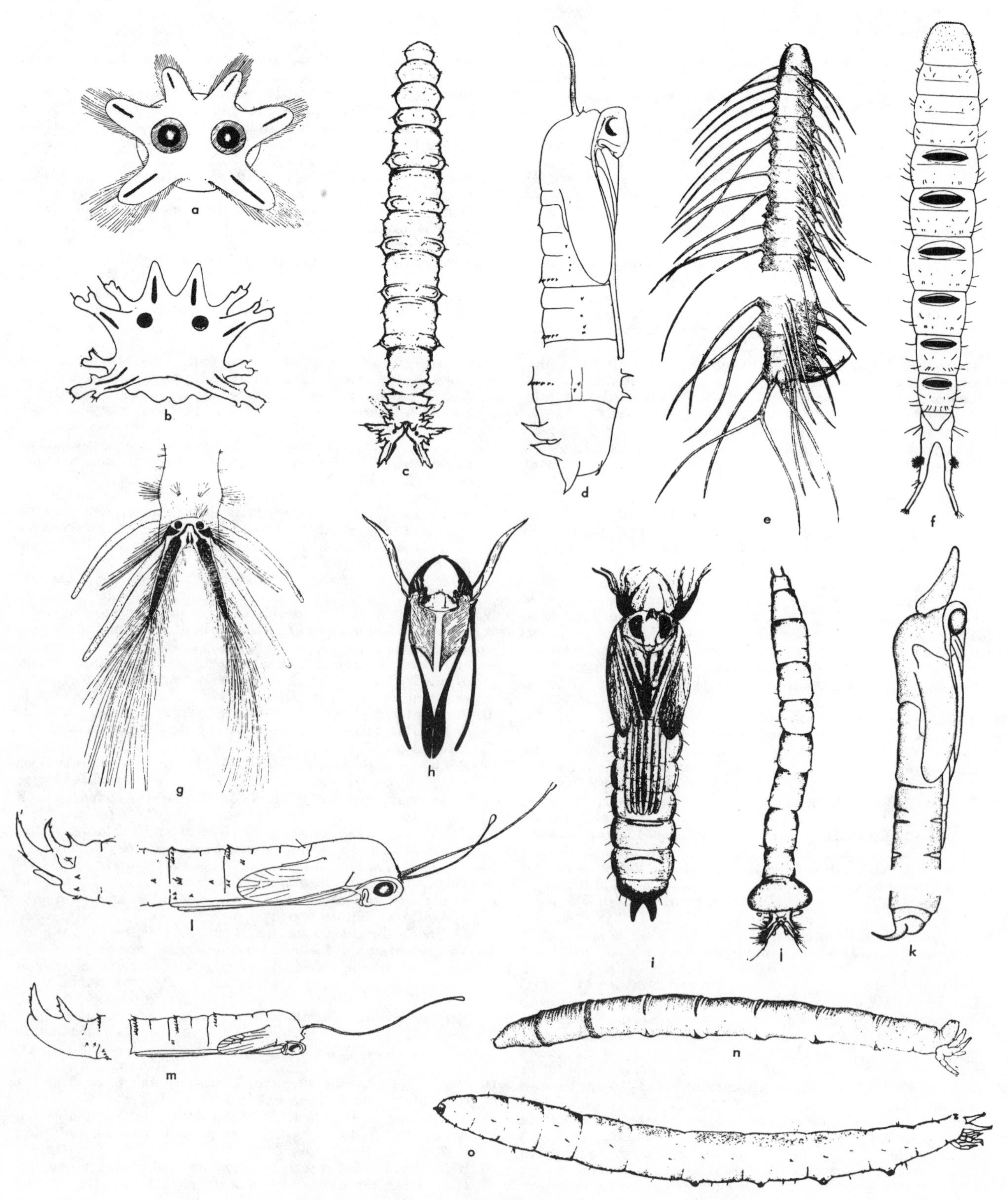

Fig. 14:4. Tipulidae. *a,d, Holorusia rubiginosa* Loew: *a*, spiracular disc of larva; *d*, pupa. *b,c, Tipula abdominalis* Say: *b*, spiracular disc of larva; c, larva. e, *Phalacrocera replicata* (L.), larva; *f,i, Antocha saxicola* O.S.: *f*, larva; *i*, pupa. *g, Pseudolimnophila luteipennis* (O.S.), spiracular disc of larva; *h, Hexatoma spinosa* (O.S.), head capsule of larva; *i, Hexatoma fultonensis* (Alex.), larva; *k, Limonia canadensis* (Westw.), pupa; *l, Tipula illustris* Doane, pupa; *m, Longurio testaceus* Loew, pupa; *n, Limonia rostrata* (Say), larva; o, *Pedicia albivitta* Walk., larva (e,*j*,*k*,n,o, Johannsen, 1934; others from Johannsen, after Alexander 1920).

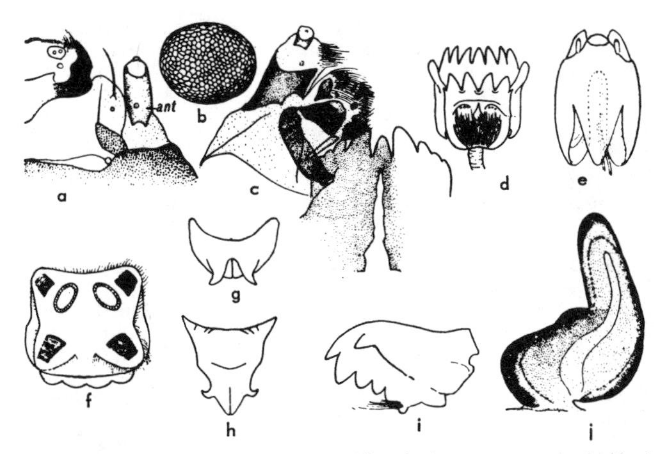

Fig. 14:5. *Limonia signipennis* (Coq.). Larva: *a*, right half of labrum and antenna; *c*, right maxilla and labium; *d*, hypopharynx; *e*, head capsule; *f*, spiracular disc; *i*, mandible. Egg: *b*. Pupa: *g*, last segment, male; *h*, last segment, female; *j*, respiratory horn (Saunders, 1928).

head; form long and slender...... (back to couplet 12)
— Blades of maxillae produced as elongate appendages with tips projecting beyond the retractile head (fig. 14:4*h*); body stout 18
18. Mental region a sclerotized narrow transverse bar, finely striate; mandibles not hinged; dorsal plate of head capsule not spatulate ...... *Limnophila* (in part)
— Mental plate not sclerotized 19
19. Mandibles hinged; maxillae and labrum densely hairy; dorsal plate of head capsule united into a spatula .. 20
— Mandibles not hinged; maxillae and labrum not densely hairy; dorsal plate of head capsule separated 21
20. Length less than 12 mm.; basal tooth of mandibular plate nearly ½ length of blade .... *Ulomorpha*
— Length more than 14 mm.; basal tooth of mandibular plate about ⅓ length of blade .... *Pilaria*
21. Coloration of body deep saturated orange-yellow; spiracular disc with ventral lobes not darkly lined, bearing at tips a few very long hairs .... *Hexatoma* (in part)
— Coloration of body pale yellow, whitish, or greenish; spiracular disc with lobes lined with dark brown or black, or disc much reduced 22
22. Grown larva 14 to 15 mm. long, 1 to 1.3 mm. in diameter (fig. 14:4*j*) .... *Hexatoma* (in part)
— Grown larva more than 15 mm. long, 1.6 mm. in diameter 23
23. Mandible with 2 teeth about mid-length .... *Hexatoma* (in part)
— Mandible sickle-shaped .... *Limnophila* (in part)
24. Anal gills pinnately branched .... *Longurio*
— Anal gills not pinnately branched 25
25. Antepenultimate segment of abdomen with strong lateral tubercle .... *Dolichopeza*
— Antepenultimate segment without such a tubercle ... 26
26. Lobes surrounding spiracular disc elongate, digitiform, fringed with long hairs .... *Prionocera*
— Lobes surrounding disc less elongate 27
27. Length of full grown larva more than 50 mm.; spiracular disc with 6 moderately long lobes fringed with long hairs (fig. 14:4*a*); mandibles small, with a dorsal and a ventral tooth .... *Holorusia*
— Smaller species; if large the lobes surrounding disc bifid (fig. 14:4*b*); mandible with 2 or 3 ventral teeth .... *Tipula*

Pupae[3]

1. Maxillary palpi more or less recurved at tips, if straight the pupae very large (30 mm. or more in length) and in *Longurio* the breathing horns very unequal; the longer one more than ½ the length of the body; species usually more than 12 mm. in length; basal abdominal segments each armed with a transverse row of teeth or projections before posterior margin 2
— Maxillary palpi not recurved at tips 6
2. Dorsal abdominal segments each with 2 slender spines before margin; abdominal spines long and branched .... *Triogma*
— Not as above 3
3. Pronotal breathing horns very long and slender, the longer one nearly, if not quite ½ the length of the body 4
— Pronotal breathing horns subequal and not very long; maxillary palpi recurved at tips 5
4. Length 40 mm.; longer breathing horn 18 mm. long (fig. 14:4*m*); maxillary palpi not recurved at tips .... *Longurio*
— Length 20 mm.; longer breathing horn 9 to 10 mm. long; maxillary palpi recurved at tips .... *Prionocera*
5. Mesonotum with a large, roughly triangular, reticulated area on each side of median line; dorsum of cauda with 4 lobes .... *Dolichopeza*
— Mesonotum unarmed or with 4 to 6 lobes; dorsum of cauda with 6 or rarely 4 lobes (fig. 14:4*d*, *l*).. *Holorusia* .... *Tipula*
6. Last larval skin adhering to posterior end; color bright green; mesonotum unarmed; basal abdominal segments unarmed with teeth or spinose projections before posterior margins .... *Phalacrocera*
— Not as above 7
7. Pronotal breathing horns 8-branched (fig. 14:4*i*) .... *Antocha*
— Breathing horns simple, unbranched (fig. 14:4*k*) 8
8. Rostral sheath elongated (fig. 14:4*k*) .... *Limonia* (in part)
— Rostral sheath not elongated 9
9. Dorsal spiracles on 8th abdominal segment large and functional; pronotal breathing horns slender, cylindrical; abdominal tergites with transverse rows of small spines .... *Helius*
— Dorsal spiracles on 8th abdominal segment small or lacking 10
10. Abdominal pleurites with circular areas set with numerous microscopic spinules; pronotal breathing horns short, usually truncated at tips which are margined with breathing pores .... *Pedicia* .... *Dicranota*
— Abdominal pleurites not as above, if with spines these are large and few in number; breathing horns long and cylindrical 11
11. Abdominal segments with broad transverse bands or welts on basal rings of 3rd to 7th segments 12
— Abdominal segments with basal rings unarmed as above, posterior rings before margin with a transverse row of spines or stiff setae 13
12. Pronotal breathing horns large, earlike, contiguous basally; in silken cocoons .... *Elliptera*
— Pronotal breathing horns not contiguous basally; cephalic crest small or lacking; a pair of small spiracles on dorsum of 8th abdominal segment .. *Limonia* (in part)
13. A distinct crest on mesonotal prescutum armed with tubercles, spines, or setae; usually less than 9 mm. long .... *Gonomyia*, *Ormosia*, *Polymeda*, *Erioptera*
— No distinct crest on mesonotal prescutum (scutellum armed in some *Eriocera*); usually more than 10 mm. long 14
14. Less than 6 mm. in length; abdominal armature weak, lacking on 7th segment............ *Paradelphomyia*
— Larger size; abdominal armature spinose 15
15. Abdominal segments with large protuberant spiracles, those on the 2nd segment very large .... *Dactylolabis*
— Abdominal segments without conspicuous protuberant spiracles 16
16. Pronotal breathing horns elongate, split into 2 flaps at tip 17
— Breathing horns not split into 2 flaps at tip 19
17. Abdominal segments with 5 or 6 rows of setiferous tubercles .... *Pseudolimnophila*

[3]Adapted from Johannsen, 1934.

— Abdominal segments without such rows of tubercles 18
18. Abdominal segments with tubercles or spines only near posterior margins of segments .... *Ulomorpha*
— Abdominal segments each with 3 or 4 pairs of blunt naked tubercles .... *Pilaria*
19. Head and thorax without spines or tubercles; lateral abdominal spiracles small .... *Limnophila*
— Head and thorax often with tubercles on antennal scape, labrum, or mesonotal scutellum; lateral abdominal spiracles large, functional .... *Hexatoma*

California Species of the Genera Containing Aquatic or Semiaquatic Species

This list is undoubtedly quite incomplete since there are probably many scattered records that have not been seen as well as many specimens standing in collections that have not been reported in the literature.

Subfamily TIPULINAE

Holorusia grandis (Bergroth) 1888. Humboldt to Los Angeles County
(= *rubiginosa* Loew 1863).
Tipula acuta Doane 1901. Sacramento, San Mateo, Los Angeles Counties
Tipula acutipleura Doane 1912. San Diego County
Tipula aequalis Doane 1901.
Tipula aitkeniana Alexander 1944. Santa Clara County
Tipula aspersa Doane 1912. Monterey County
Tipula atrisumma Doane 1912. San Diego County
Tipula awanachi Alexander 1947. San Diego County
Tipula beatula Osten Sacken 1877. Sonoma to San Diego
Tipula bernardinensis Alexander 1946. Riverside County
Tipula bifalcata Doane 1912. San Diego County
Tipula biproducta Alexander 1947. San Diego County
Tipula bituberculata Doane 1901. Monterey to San Diego
Tipula biunca Doane 1912. Southern California
Tipula boregoensis Alexander 1946. San Diego County
Tipula cahuilla Alexander 1946. San Luis Obispo, San Diego counties
Tipula calaveras Alexander 1944. Alameda County
Tipula californica (Doane) 1908. San Mateo County
(= *xanthomela* Dietz 1918).
Tipula capistrano Alexander 1946. San Francisco to San Diego County
Tipula carunculata Alexander 1945. Mono County
Tipula cazieri Alexander 1942. Los Angeles County
Tipula circularis Alexander 1947. Alameda County
Tipula cladacantha Alexander 1945. Mariposa County
Tipula cognata Doane 1901. Sierra, Mariposa
Tipula comstockiana Alexander 1947. San Diego County
Tipula contatrix Alexander 1944. Nevada County
Tipula cylindrata Doane 1912. San Diego County
Tipula degeneri Alexander 1944. Tulare County
Tipula derbyi Doane 1912. San Mateo County
Tipula desertorum Alexander 1946. San Diego County
Tipula diacanthophora Alexander 1945. Tuolumne County
Tipula doaneiana Alexander 1919. San Mateo County
(= *californica* Doane 1912).
Tipula dorsimacula shasta Alexander 1919. Shasta County
Tipula downesi Alexander 1944. Mariposa County
Tipula evidens Alexander 1920. Fresno County
Tipula fallax Loew 1863. San Mateo County
Tipula flavocauda Doane 1912. San Diego County
(= *buenoi* Alexander 1946).
Tipula flavomarginata Doane 1912. San Diego County
Tipula fragmentata Dietz 1919. San Mateo County
Tipula fulvolineata Doane 1912. San Mateo County
(= *graphica* Doane 1901).
Tipula furialis Alexander 1946. Inyo, San Diego
Tipula gothicana Alexander 1943.
Tipula graciae Alexander 1947. San Bernardino County
Tipula graminivora Alexander 1921. Solano County
Tipula grata Loew 1863. Mono County
Tipula hastingsae Alexander 1951. Contra Costa, San Diego
Tipula illustris Doane 1901.
Tipula inusitata Alexander 1949. San Mateo County
Tipula inyoensis Alexander 1946. Inyo County
Tipula jacintoensis Alexander 1946. Riverside County
Tipula lacteipes Alexander 1943. Tuolumne County
Tipula lamellata Doane 1908. Mono County
Tipula leechi Alexander 1938.
Tipula linsdalei Alexander 1951. Monterey County
Tipula lucida Doane 1901.
Tipula lygropis Alexander 1920. Santa Cruz I.
Tipula macracantha Alexander 1946. Mohave County
Tipula macrophallus (Dietz) 1918. Contra Costa to Los Angeles counties
Tipula madera Doane 1911. San Mateo County
Tipula magnifolia Alexander 1948.
Tipula marina Doane 1912. San Mateo County
Tipula mariposa Alexander 1946. Mariposa, Madera
Tipula mayedai Alexander 1946. Inyo, San Diego
Tipula megalabiata Alexander 1915.
Tipula megalabiata referta Alexander 1947. San Diego County
Tipula megalodonta Alexander 1946. Los Angeles, Riverside, San Diego
Tipula megatergata Alexander 1920. Los Angeles County
Tipula mesotergata Alexander 1930. Los Angeles County
Tipula micheneri Alexander 1944. Riverside County
Tipula miwok Alexander 1945. Mariposa County
Tipula modoc Alexander 1944. Modoc County
Tipula mohavensis Alexander 1946. Mohave County
Tipula mono Alexander 1945. Mono County
Tipula mutica Dietz 1919. Sonoma County
Tipula newcomeri Doane 1911. Placer County
Tipula occidentalis Doane 1912. San Diego County
Tipula opisthocera Dietz 1919. Sonoma County
Tipula ovalis Alexander 1951. Monterey County
Tipula pacifica Doane 1912. Placer County
Tipula palmarum Alexander 1947. San Diego County
Tipula pellucida Doane 1912.
(= *pyramis* Doane 1912).
Tipula perfidiosa Alexander 1945. Kern County
Tipula planicornis Doane 1912. San Diego County
Tipula plutonis Alexander 1919. Tulare County
Tipula praecisa Loew 1872.
Tipula pubera Loew 1864. San Francisco, Mono, Monterey counties
Tipula quaylei Doane 1909. Sutter County
Tipula repulsa Alexander 1943. Solano, Mariposa
Tipula sanctae-luciae Alexander 1951. Monterey County
Tipula sayleriana Alexander 1946. San Diego County
Tipula sequoiarum Alexander 1945. Tulare County
Tipula sequoicola Alexander 1947. Tulare County
Tipula shastensis Alexander 1944. Shasta County
Tipula silvestra Doane 1909. Monterey County
Tipula simplex Doane 1901. San Mateo, Mariposa
Tipula spernax Osten Sacken 1877. Humboldt, Placer, Monterey counties
Tipula spinerecta Alexander 1947. Kern County
Tipula sternata Doane 1912. San Mateo County
Tipula subapache Alexander 1947. San Bernardino County
Tipula subtilis Doane 1901
Tipula supplicata Alexander 1944. Riverside County
Tipula sweetae Alexander 1930. Los Angeles County
Tipula tenaya Alexander 1946. Mono County
Tipula timberlakei Alexander 1947. Riverside County
Tipula tingi Alexander 1939. Sacramento to San Diego
Tipula trichophora Alexander 1920. Santa Cruz, Mariposa
Tipula tristis Doane 1901. San Mateo County
Tipula truculenta Alexander 1943. Monterey, San Diego
Tipula trypetophora Dietz 1919. San Mateo County
Tipula tuberculata Doane 1901.
Tipula ungulata Doane 1912. San Diego County
Tipula vestigipennis Doane 1908. San Francisco County
Tipula vitabilis Alexander 1947. San Diego County

Tipula vittatipennis Doane 1912.
Tipula williamsii Doane 1909. San Francisco County
Tipula yosemite Alexander 1945. Mariposa County
Tipula zelootype Alexander 1946. Imperial County

Subfamily LIMONIINAE

Antocha monticola Alexander 1917. Placer County
Crypteria americana Alexander 1917.
Cryptolabis bidenticulata Alexander 1948. San Bernardino County
Cryptolabis brachyphallus Alexander 1950. San Diego County
Cryptolabis mixta Alexander 1948. Mariposa County
Cryptolabis retrorsa Alexander 1950. Del Norte County
Dactylolabis damula (Osten Sacken) 1877. "Crofton's"
Dactylolabis imitata Alexander 1945. Plumas, Marin, San Mateo counties
Dactylolabis parviloba Alexander 1944. Nevada County
Dicranoptycha laevis Alexander 1947. San Diego County
Dicranoptycha melampygia Alexander 1950. Humboldt County
Dicranoptycha occidentalis Alexander 1927. San Diego County
Dicranoptycha spinosissima Alexander 1950. Shasta County
Dicranoptycha stenophallus Alexander 1950. Del Norte County
Dicranota cazieriana Alexander 1944. Alameda County
Dicranota debilis (Williston) 1893.
Dicranota maculata (Doane) 1900. Humboldt County
Dicranota nuptialis Alexander 1947. Mariposa County
Dicranota polymeroides (Alexander) 1914. Humboldt County
Dicranota reducta (Alexander) 1921.
Dicranota tehama Alexander 1950. Shasta County
Dicranota vanduzeei (Alexander) 1930. Mt. St. Helena
Elliptera astigmatica Alexander 1912. San Mateo County
Elliptera clausa Osten Sacken 1877. Mariposa County
Erioptera bispinosa (Alexander) 1919. Alameda County
Erioptera fenderi (Alexander) 1952. Shasta County
Erioptera millardi (Alexander) 1944. San Diego County
Erioptera nitida (Coquillett) 1905. Humboldt County
Erioptera palomarica (Alexander) 1947. San Diego County
Erioptera perflaveola (Alexander) 1918. Humboldt County
Erioptera sackeniana (Alexander) 1926. Marin County
Erioptera sequoiae (Alexander) 1952. Tulare County
Erioptera squamosa (Alexander) 1919. San Diego County
Gonomyia aciculifera Alexander 1919. Contra Costa County
Gonomyia californica Alexander 1916. Humboldt County
Gonomyia cinerea (Doane) 1900. Humboldt County
Gonomyia coloradica Alexander 1920.
Gonomyia flavibasis Alexander 1916. Monterey County
(= *tuberculata* Alexander 1925).
Gonomyia hesperia Alexander 1926. Riverside County
Gonomyia lindseyi Alexander 1944. Modoc County
Gonomyia percomplexa Alexander 1945.
Gonomyia pelingi Alexander 1946.
Gonomyia poliocephala (Alexander) 1924. Plumas County
Gonomyia proserpina Alexander 1943.
Gonomyia spinifer Alexander 1918. Contra Costa County
Gonomyia virgata Doane 1900. Humboldt County
Hexatoma albihirta (Alexander) 1912.
Hexatoma azrael Alexander 1943. Nevada County
Hexatoma brevipila (Alexander) 1918. Humboldt County
Hexatoma californica (Osten Sacken) 1877. Marin County
Hexatoma mariposa Alexander 1943. Mariposa County
Hexatoma palomarensis Alexander 1947. San Diego County
Hexatoma parva (Doane) 1900. Humboldt, El Dorado, San Mateo counties
(= *obscura* Williston 1893, *austera* Doane 1900).
Hexatoma saturata (Alexander) 1919. San Diego County
Hexatoma sculleni Alexander 1943.
Limnophila amabilis Alexander 1950. Shasta County
Limnophila antennata Coquillett 1905.
Limnophila apiculata Alexander 1919. San Diego County
Limnophila barberi Alexander 1916. Humboldt County
Limnophila claggi Alexander 1930.
Limnophila cressoni Alexander 1917. Marin County
Limnophila edentata Alexander 1919. San Diego County
Limnophila freeborni Alexander 1943. El Dorado County
Limnophila hepatica Alexander 1919. Humboldt County
Limnophila modoc Alexander 1944. Modoc County
Limnophila nitidithorax Alexander 1918. Alameda County
Limnophila nupta Alexander 1947. Mariposa County
Limnophila occidens Alexander 1924.
Limnophila sequoiarum Alexander 1943. Tulare County
Limnophila strepens Alexander 1916. Marin County
Limnophila subaptera Alexander 1917. Tulare County
Limonia adjecta (Doane) 1908. San Mateo County
Limonia anteapicalis Alexander 1946. Inyo County
Limonia bistigma (Coquillett) 1905 Del Norte County
(= *tributaria* Alexander 1943).
Limonia californica (Osten Sacken) 1861
Limonia canadensis (Westwood) 1835
Limonia catalinae Alexander 1944. Catalina Island
Limonia citrina (Doane) 1900. San Mateo County
Limonia distans (Osten Sacken) 1869.
(= *cervina* Doane 1908).
Limonia diversa (Osten Sacken) 1859.
Limonia halterata (Osten Sacken) 1869.
Limonia humidicola (Osten Sacken) 1859.
Limonia libertoides (Alexander) 1912. Marin County
Limonia linsdalei Alexander 1943. Monterey County
Limonia longipennis (Schummel) 1829.
Limonia maculata (Meigen) 1819.
Limonia particeps Doane 1908. Monterey County
Limonia rhipidoides (Alexander) 1918. Alameda County
Limonia sciophila (Osten Sacken) 1877. Sonoma, Marin, San Francisco counties
Limonia signipennis (Coquillett) 1905. Humboldt, San Mateo, Monterey
(= *marmorata* Osten Sacken 1861).
Limonia simulans concinna (Williston) 1893.
Limonia stigmata Doane 1900. San Mateo County
Limonia venusta (Bergroth) 1888.
(= *duplicata* Doane 1900; *negligens* Alexander 1927).
Limonia viridicans (Doane) 1908. San Mateo County
Lipsothrix shasta Alexander 1946. Shasta County
Ormosia burneyensis Alexander 1950. Shasta County
Ormosia cornuta (Doane) 1908. San Mateo County
Ormosia heptacantha Alexander 1948. Humboldt County
Ormosia legata Alexander 1948. Fresno County
Ormosia leptorhabda Alexander 1943. Del Norte County
Ormosia manicata (Doane) 1900. Humboldt, Plumas
Ormosia modica Dietz 1916. Sonoma County
(= *stylifer* Alexander 1919).
Ormosia pernodosa Alexander 1950. Placer County
Ormosia profunda Alexander 1943. Del Norte County
Ormosia sequoiarum Alexander 1944. Tulare County
Ormosia taeniocera Dietz 1916. Sonoma County
Ormosia tahoensis Alexander 1950. Placer County
Ormosia tricornis Alexander 1948. Humboldt County
Pedicia actaeon Alexander 1947. Humboldt County
Pedicia ampla truncata Alexander 1941. Riverside County
Pedicia aperta (Coquillett) 1905.
Pedicia bidentifera Alexander 1950. Shasta County
Pedicia obtusa Osten Sacken 1877. Siskiyou, Marin
Pedicia parvicellula Alexander 1938. Humboldt County
Pedicia septentrionalis (Bergroth) 1888.
Pedicia simplicistyla (Alexander) 1930. Marin County
Pedicia subaptera (Alexander) 1917.
Pedicia subobtusa Alexander 1949. Placer County
Phyllolabis claviger Osten Sacken 1877. "Crofton's Retreat"
Phyllolabis encausta Osten Sacken 1877. Marin County
Phyllolabis flavida Alexander 1918. San Diego County
Phyllolabis hirtiloba Alexander 1947. Mariposa County
Phyllolabis meridionalis Alexander 1945. San Diego County
Phyllolabis myriosticta Alexander 1945. San Diego County
Phyllolabis sequoiensis Alexander 1945. Tulare County
Polymeda alicia (Alexander) 1914. Humboldt County
Polymeda bipartita (Osten Sacken) 1877. San Francisco County

Polymeda bisulca (Alexander) 1948. Mariposa County
Polymeda burra (Alexander) 1924. Placer County
Polymeda cana (Walker) 1848. San Mateo, Mono, Riverside
Polymeda cinctipennis (Alexander) 1918. Los Angeles County
Polymeda crickmeri (Alexander) 1948. San Bernardino County
Polymeda dulcis (Osten Sacken) 1877. Del Norte, Placer, Mono counties
Polymeda dyari (Alexander) 1924. Plumas County
Polymeda laticeps (Alexander) 1916. Humboldt County
Polymeda melanderi (Alexander) 1944. Mariposa County
Polymeda melanderiana (Alexander) 1944.
Polymeda pilipes var. *anomala* (Osten Sacken) 1861. San Mateo, Monterey, San Diego counties
Polymeda rainieria (Alexander) 1943. Mariposa County
Polymeda sparsa (Alexander) 1919. Alameda County
Polymeda tripartita (Osten Sacken) 1877. Marin, San Francisco counties
Pseudolimnophila luteipennis (Osten Sacken) 1859.
Rhabdomastix californiensis Alexander 1921. Monterey County
Rhabdomastix fasciger Alexander 1920. Santa Cruz County
Rhabdomastix trichophora Alexander 1943.
Tasiocera subnuda (Alexander) 1926. Alameda County
Ulomorpha nigronitida Alexander 1920. Fresno County
Ulomorpha quinquecellula Alexander 1920. Fresno County
Ulomorpha vanduzeei Alexander 1920. Fresno County

REFERENCES

ALEXANDER, C. P.
1919. The crane-flies of New York. Part I. Distribution and taxonomy of the adult flies. Mem. Cornell Univ. Agric. Exp. Sta., 25:767-993.
1920. The crane-flies of New York. Part II. Biology and phylogeny. Mem. Cornell Univ. Agric. Exp. Sta., 38: 691-1133.
1934. Tipuloidea, *In* Curran, The families and genera of North American Diptera. pp. 33-58.
1942. Family Tipulidae. Guide to the insects of Connecticut, VI. The Diptera or true flies of Connecticut, fasc., 1:196-486.
1944-1947. Undescribed species of Tipulidae from the western United States I. Pan-Pac. Ent., 20:9-97; II. 21:91-97; III. 23:91-96.
1945-1947. New or little known crane flies from California. I. Bull. So. Calif. Acad. Sci., 44:33-45; II. 45:1-16; III. 46:35-50.
CURRAN, C. H.
1934. *See* Diptera references.
DIETZ, W. G.
1913. A synopsis of the described North American species of the dipterous genus *Tipula* L. Ann. Ent. Soc. Amer., 6:461-484.
DOANE, R. W.
1900. New North American Tipulidae. Jour. N.Y. Ent. Soc., 8:182-198.
1901. Descriptions of new Tipulidae. Jour. N.Y. Ent. Soc., 9:97-127.
1912. New western *Tipula*. Ann. Ent. Soc. Amer., 5: 41-61.
JOHANNSEN, O. A.
1934. *See* Diptera references.
OSTEN SACKEN, C. R.
1869. The North American Tipulidae. Mongr. N. Amer. Diptera. IV. Smithson. Misc. Coll., 8, 345 pp.
SAUNDERS, L. G.
1928. Some marine insects of the Pacific Coast of Canada. Ann. Ent. Soc. Amer., 21:521-545.
STONE, ALAN
1941. The generic names of Meigen 1800 and their proper application. Ann. Ent. Soc. Amer., 34:404-418.

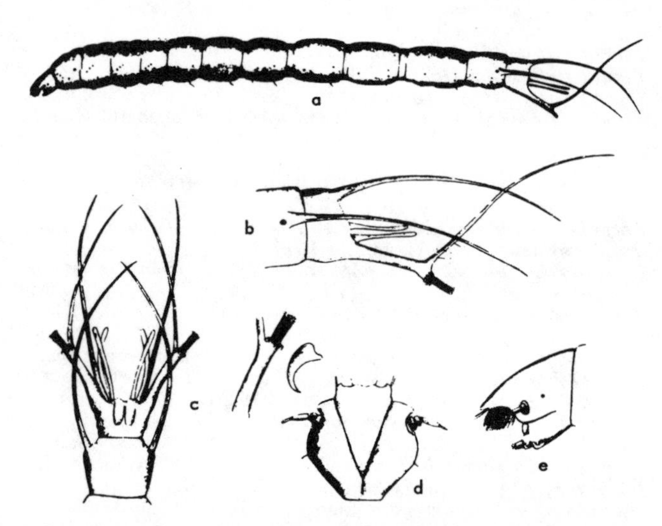

Fig. 14:6. *Protoplasa fitchii* O.S., larva. *a*, lateral view; *b*, posterior end, lateral view; *c*, dorsal view; *d*, head capsule, dorsal view; *e*, lateral view (Alexander, 1930).

Family TANYDERIDAE

The only species of this rare family of primitive crane flies whose immature stages have been discovered is *Protoplasa fitchii* O. S. of eastern North America. They were found in wet sandy soil at the margins of major streams. It is quite possible that the larvae of the two known western species will also be found in an aquatic environment so the family is

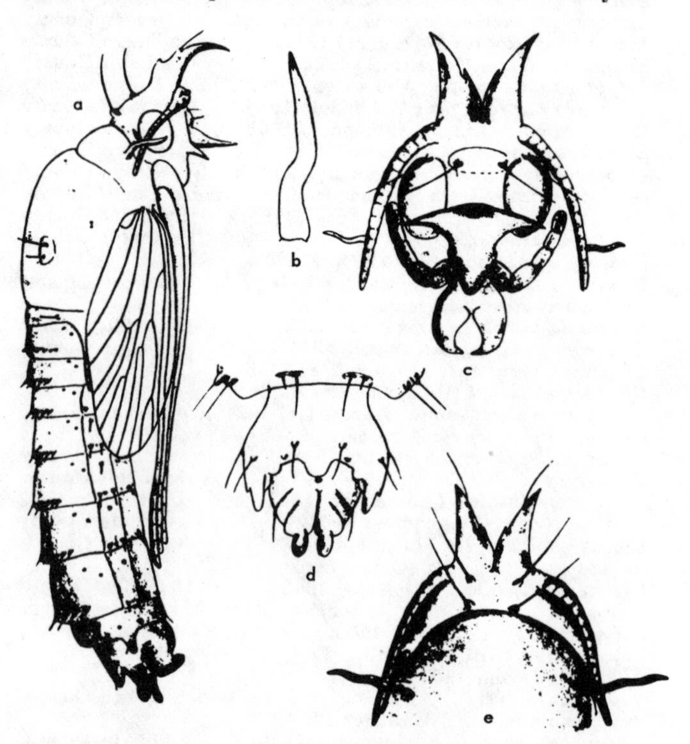

Fig. 14:7. *Protoplasa fitchii* O.S., pupa. *a*, lateral view; *b*, pronotal breathing horn; *c*, head, ventral aspect; *d*, genital sheath, female; *e*, head, dorsal view (Alexander, 1930).

included here, and figures of the larvae and pupae of *Protoplasa* are included (figs. 14:6-14:7).

Key to North American Genera of Tanyderidae

1. A supernumerary cross vein in cell M of the wing (fig. 14:1*a*) .... *Protoplasa* Osten Sacken
— Wings without supernumerary cross veins .... *Protanyderus* Handlirsch

California Species of Tanyderidae

Protanyderus vanduzeei (Alexander) 1918. Contra Costa County
Protanyderus vipio (Osten Sacken) 1877. San Mateo Creek, San Francisco

REFERENCES

ALEXANDER, C. P.
1930. Observations on the dipterous family Tanyderidae. Proc. Linn. Soc. N.S.W., 55:221-230, 2 pls., 1 fig.

CRAMPTON, G. C.
1930. A comparison of the more important structural details of the larva of the archaic Tanyderid Dipteron *Protoplasa fitchii*, with other Holometabola, from the standpoint of phylogeny. Bull. Brooklyn Ent. Soc., 35:235-258, 4 pls.

WILLIAMS, INEZ
1933. The external morphology of the primitive Tanyderid Dipteron *Protoplasa fitchii* O. S., with notes on the other Tanyderidae. Jour. N.Y. Ent. Soc., 41:1-36, 7 pls.

Family LIRIOPEIDAE (=PTYCHOPTERIDAE)

False Crane Flies

Members of this family differ from the true crane flies (Tipulidae) and the winter crane flies (Melusinidae) in having only one anal vein (fig. 14:8*f*), and they differ from the primitive crane flies (Tanyderidae) in having the radial vein only four-branched. The conspicuous and interesting members of the subfamily

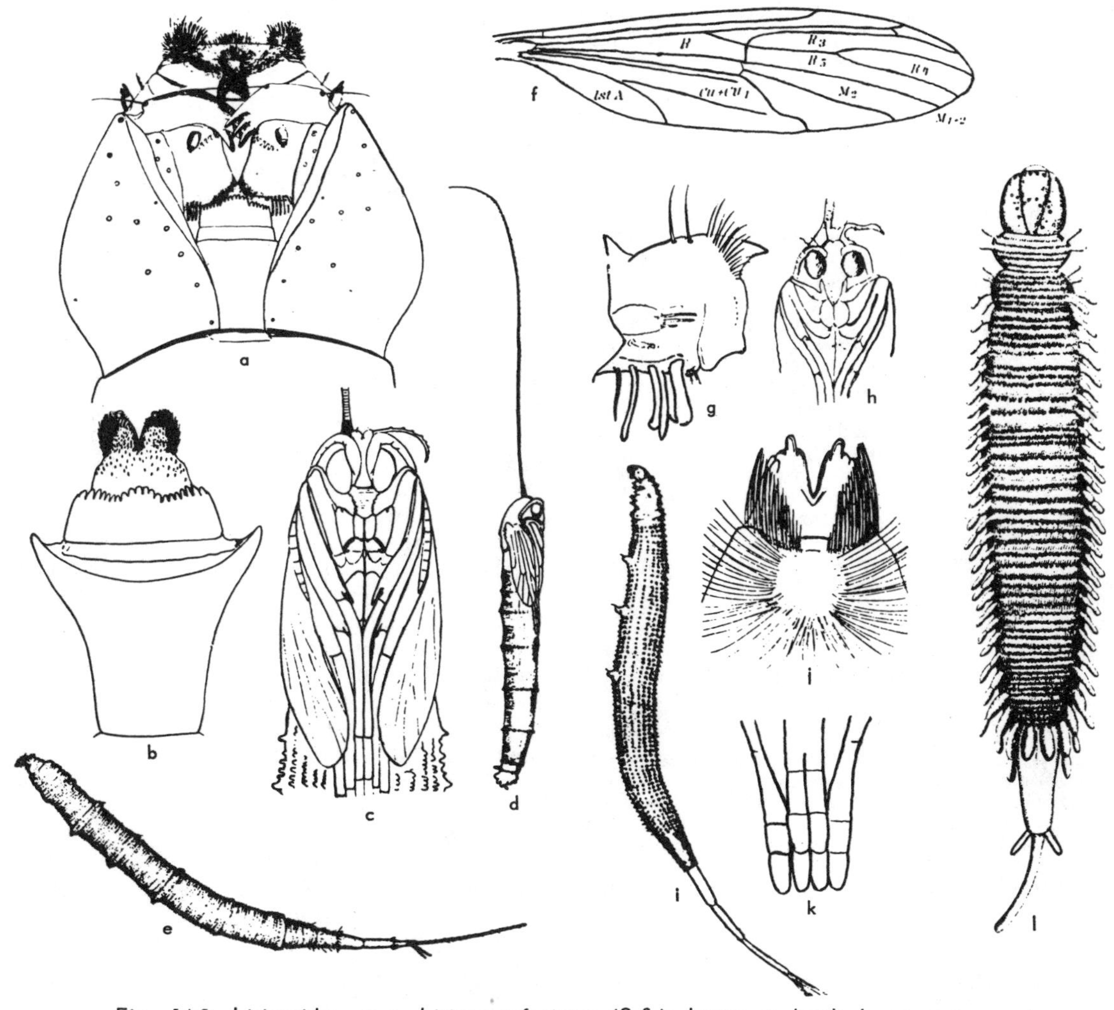

Fig. 14:8. Liriopeidae. *a-e*, *Liriope rufocincta* (O.S.): Larva: *a*, head; *b*, prementum, and submentum; *e*, lateral view. Pupa: *c*, ventral view; *d*, lateral view. *f*, *Bittacomorpha occidentalis* Aldr., wing; *g-k*, *Bittacomorpha clavipes* (F.): Larva: *g*, mandible; *i*, lateral view; *j*, labium. Pupa: *h*, ventral view; *k*, leg cases. *l*, *Bittacomorphella jonesi* (Johns.), larva (*i*, Johannsen, 1934; *a-e,g,h,j,k*, from Johannsen after Alexander, 1920; *f,l*, Alexander, 1927).

Bittacomorphinae are known as phantom midges because of their appearance in flight. Their long black- and white-banded legs are outstretched like the spokes of a wheel to catch the breeze, and they float through the air with very little wing action. Species of the genus *Liriope* resemble large fungus gnats.

The immature stages (fig. 14:8) occur in the saturated soil, rich in decaying vegetable matter, of swamps and pond margins. The larvae possess a long caudal breathing tube, whereas the pupae have a single, very long respiratory horn from the pronotum.

All three of the known genera are found in California.

Keys to the Genera of Liriopeidae

Adults

1. Antennae 16-segmented; wings with cell M_1 present .... *Liriope* Meigen
— Antennae 20-segmented; wings with cell M_1 absent 2
2. Wings with macrotrichia in distal ends of radial and medial cells; basitarsi not dilated .... *Bittacomorphella* Alexander
— Wings without macrotrichia in cells; basitarsi conspicuously dilated (fig. 14:8*f*)..... *Bittacomorpha* Westwood

Larvae

1. Mentum with outer margin finely serrated; mandibles with 3 large outer teeth (fig. 14:8*a*); pseudopods small; coloration yellow or brown............*Liriope* Meigen
— Mentum bilobed, not toothed; mandibles with a single large outer tooth (fig. 14:8*g*); pseudopods prominent, each with a conspicuous curved claw; coloration rusty-red or black 2
2. Size small (length less than 20 mm.) (fig. 14:8*l*); coloration black, breathing tube light yellow, entirely retractile; body covered with very long projections which are encased in a black horny substance; mandibles with an inner comb of teeth .... *Bittacomorphella* Alexander
— Size larger (length more than 40 mm.) (fig. 14:8*i*); coloration rusty-red; body tapering to the long slender, partly retractile breathing tube; body covered with transverse rows of shorter stellate tubercles; mandibles without an inner comb of teeth .... *Bittacomorpha* Westwood

Pupae

1. All tarsi lying parallel (fig. 14:8*c*); wing pads with M branched........ *Liriope* Meigen
— Fore tarsi lying above middle tarsi (fig. 14:8*k*); wing pads with M unbranched 2
2. Size small (length, excluding breathing horn, less than 12 mm.); right breathing horn small, degenerate; abdominal tubercles weak, tipped with several strong setae .... *Bittacomorphella* Alexander
— Size larger (length, excluding breathing horn, more than 14 mm.); right breathing horn elongate, filiform, longer than the body; abdominal tubercles strong, elongate, crowned by a circlet of 4 or 5 spines and tipped with a setiferous papilla .... *Bittacomorpha* Westwood

California Species of Liriopeidae

Subfamily BITTACOMORPHINAE

Genus *Bittacomorpha* Westwood

occidentalis Aldrich 1895. — Humboldt County

Genus *Bittacomorphella* Alexander

sackenii (von Roeder) 1890. — Mono, Humboldt

Subfamily LIRIOPEINAE

Genus *Liriope* Meigen

lenis (Osten Sacken) 1877. — Mariposa, Mono, Monterey
minor (Alexander) 1920. — Monterey County
monoensis (Alexander) 1947. — Mono County
sculleni (Alexander) 1943.

At least six other species have been described from the West, but have not been recorded from California.

REFERENCES

ALEXANDER, C. P.
1920. *See* Diptera references.
1927. *In* Wytsman, Diptera, Fam. Ptychopteridae. Genera Insectorum, fasc. 188, 12 pp.
JOHANNSEN, O. A.
1934. *See* Diptera references.

Family PSYCHODIDAE

Adult psychodids (fig. 14:9*a*) are small, hairy, moth-like species commonly known as "mothflies." They are found about moist places, often in considerable numbers. Some *Psychoda* and *Telmatoscopus* species breed in drainpipes in houses, from which they may emerge in swarms. The larvae in such situations are able to withstand soap and hot water. Some members of the family are biting pests, and one genus, *Phlebotomus,* is the vector of certain diseases prevalent in the tropics and subtropics.

Our treatment of the immature stages of aquatic Psychodidae is drawn very largely from an unpublished manuscript by Dr. Larry W. Quate, which he has been kind enough to let us use. The immature stages of but a few species of North American Psychodidae are known. Malloch (1917) and Johannsen (1934) have given the most complete accounts available. For the most complete treatment of the immature stages of the psychodids, we must turn to Europe and the publications of Satchell (1947*a,b,* 1948, 1949), who has worked extensively on the British fauna and has described the immature stages and the biology of the majority of the British species.

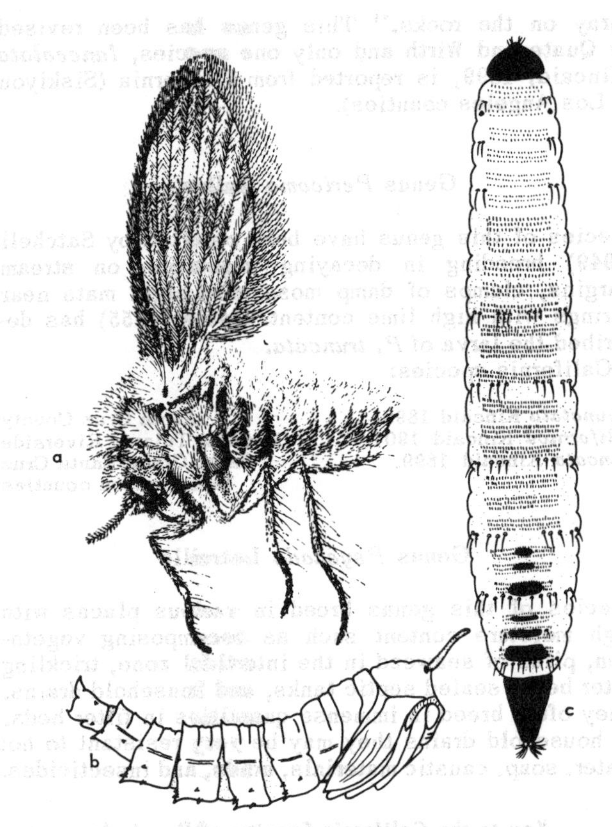

Fig. 14:9. Psychodidae. *a*, *Pericoma* sp., female; *b-c*, *Psychoda* sp.: *b*, pupa; *c*, larva (*a*, Smart, 1948; *b,c*, Quate 1955).

The immature stages of the Psychodidae have the following characteristics:

Egg.—Pale or dark brown in color; oval; usually without prominent reticulations or other conspicuous markings.

Larva (fig. 14:9*c*).—Elongate cylindrical, except *Maruina* (fig. 14:10*c*) which is onisciform, flattened ventrally, with eight ventral suckers (fig. 14:10*b*). Head and mouth parts complete, head nonretractile, except *Sycorax;* mandibles toothed, opposed, usually with long setae, leaflike structures, and serrate bristles; epicranial suture U-shaped, conspicuous; antenna small, domelike or composed of several small rods, or long and three-segmented. Thoracic and abdominal segments usually divided secondarily into annuli, ordinarily two for each thoracic and first abdominal segment and three for abdominal segments two to seven, some or all annuli with tergal plates. Respiration amphipneustic, anterior spiracles borne on apex of domelike protuberance on thorax; posterior spiracle at apex of siphon or located dorsally on terminal abdominal segment.

Pupa.—Body cylindrical (fig. 14:9*b*) or flattened ventrally (fig. 14:10*a*); head, legs, and wing distinct, closely applied to body; tips of legs not extending beyond apices of wing covers; respiratory horn present on prothorax, usually elongate and slender, exterior surface with double row of pits extending from tip to base or near base, pits connected to inner chamber by lateral channels. Abdominal segments with variously shaped spines and marginal fringes. Terminal segment elongate, roughly rectangular in lateral outline.

The larvae of psychodids are easily recognized because the body segments are secondarily divided

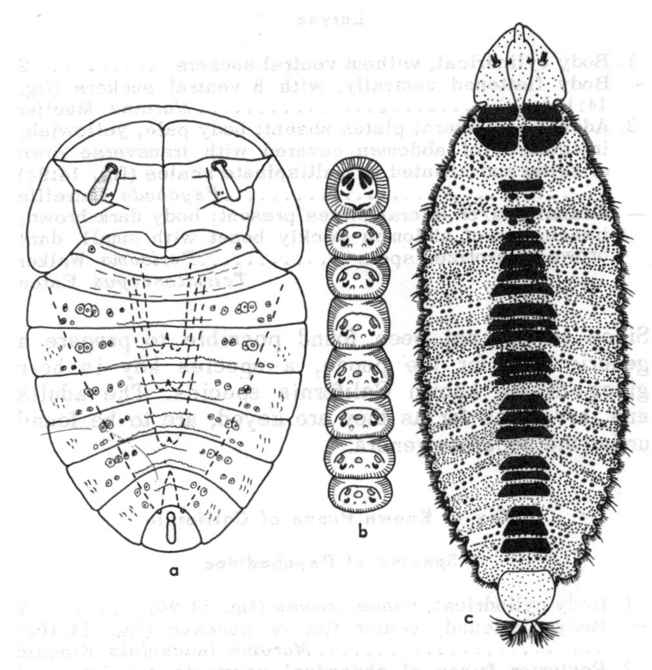

Fig. 14:10. *Maruina* sp. *a*, pupa; *b*, ventral suckers of larva; *c*, larva (Quate 1955).

into two or three annuli. The commoner species generally bear a tergal plate on each annulus. The pupae are less easily separated from other Nematocera, and at present generic pupal characters are not available.

With the exception of *Maruina, Phlebotomus,* and possibly *Trichomyia,* North American psychodids are semiaquatic, living under very moist or semiliquid conditions or rarely, in open water. Common habitats include moist, decaying vegetable matter of many types, heavily saturated mud and sand at stream margins, and moss and algae floating on still or slow-moving streams or growing in water-splash areas. The pupae can withstand drier conditions, although they do not appear to prefer these, but they have been found intermixed with larvae in very moist situations.

Keys to the North American Genera of Aquatic Psychodidae

Adults

1. Only 1 longitudinal vein between the 2 forked veins .... *Maruina* Mueller
— Two longitudinal veins between the 2 forked veins (fig. 14:9*a*) 2
2. Antenna with 14 to 16 segments, the last ones diminutive and often united to each other. *Psychoda* Latreille
— Antenna with 15 to 16 segments, the last ones not diminutive or united to each other 3
3. Flagellar segments barrel-shaped or fusiform without very pronounced basal bulb .... *Pericoma* Walker
— Flagellar segments with notable basal bulb, often more bulging on one side, and with a distinct distal neck .... *Telmatoscopus* Eaton

Larvae

1. Body cylindrical, without ventral suckers 2
— Body flattened ventrally, with 8 ventral suckers (fig. 14:10*b,c*) .... *Maruina* Mueller
2. Adanal and lateral plates absent; body pale, yellowish; integument of abdomen covered with transverse rows of setae and ciliated or multispinate scales (fig. 14:9*c*) .... *Psychoda* Latreille
— Adanal and/or lateral plates present; body dark brown; integument of abdomen thickly beset with small, dark brown sclerotized spots .... *Pericoma* Walker *Telmatoscopus* Eaton

Since it has not been found possible to prepare a generic key for the pupae, a species key is here given to the known California species. The adults and larvae, as far as they are keyed, are to be found under the separate genera.

Key to Known Pupae of California

Species of Psychodidae

1. Body cylindrical, venter convex (fig. 14:9*b*) 2
— Body flattened, venter flat or concave (fig. 14:10*a*) .... *Maruina lanceolata* Kincaid
2. Posterior fringe of abdominal segments consisting of dense rows of setae, several setae arising from most bases 3
— Posterior fringe of abdominal segments consisting of sparse rows of setae or bristles, each arising from single base, or without fringe of setae 4
3. Respiratory horn with noticeable darkened area near center .... *Psychoda sigma* Kincaid
— Respiratory horn without noticeable darkened area near center .... *Psychoda severini* Tonnoir
4. Pits of respiratory horn confined to apical third, with single pit near base of horn; pits absent on central part of horn .... *Telmatoscopus furcatus* Kincaid
— Pits of respiratory horn extending in irregular double row from apex to at least 1/3 distance from base of horn 3
5. Respiratory horn with large darkened area enclosing pits; large subapical feathery, conelike protuberance on either side of mid-line of posterior margin of sternites, twice as large as other marginal elements, terminating in single, broad bristle and spur .... *Pericoma truncata* Kincaid
— Respiratory horn unicolorous, without darkened area surrounding pits; no marginal elements of body with feather base 6
6. Large elements of marginal fringe of abdominal segments consisting of bristle arising from broad base; bristles as long as base; base with 1 or 2 apical spurs; second abdominal tergite with 2 pairs of setae in center of segment arising from broad base .... *Psychoda cinerea* Banks
— Large elements of marginal fringe of abdominal segments consisting of bristle arising from large, cone-shaped base; bristle much shorter than base; base without apical spurs; second abdominal tergite without setae in center of segment or 1 small pair, arising from small, cone-shaped bases .... *Psychoda alternata* Say

Genus *Maruina* Mueller

According to Quate and Wirth (1951), "The larvae and pupae of this genus are found adhering to rocks in rather swift-flowing streams, and although usually not below the water line, they are kept wet by water spray on the rocks." This genus has been revised by Quate and Wirth and only one species, *lanceolata* (Kincaid) 1899, is reported from California (Siskiyou to Los Angeles counties).

Genus *Pericoma* Walker

Species of this genus have been reported by Satchell (1949) breeding in decaying vegetation on stream margins, clumps of damp moss, and algal mats near springs with high lime content. Quate (1955) has described the larva of *P. truncata*.

California species:

bipunctata Kincaid 1899.	Santa Cruz County
californica Kincaid 1901.	Madera, Tulare, Riverside
truncata Kincaid 1899.	Santa Clara, Santa Cruz Los Angeles counties

Genus *Psychoda* Latreille

Species of this genus breed in various places with high moisture content such as decomposing vegetation, piles of seaweed in the intertidal zone, trickling filter beds, sealed septic tanks, and household drains. They often breed in immense quantities in filter beds. In household drains they may be very resistant to hot water, soap, caustic materials, acids, and insecticides.

Key to the California Species of Psychoda

Adults

1. Antenna of 16 segments; San Diego County.......... .... *cinerea* Banks 1894
— Antenna of 14 or 15 segments 2
2. Antenna of 15 segments 3
— Antenna of 14 segments 4
3. Segment 14 distinctly separated from 13 and about equal in size to 15; tip of labium with 2 long and 2 short teeth, outer surface with 2 labial spines; no small patches of dark hairs at the wing margin; Santa Clara, El Dorado, and Los Angeles counties .... *phalaenoides* (L.) 1758
— Segment 14 intimately united to 13; 15 smaller than either 13 or 14; tip of labium with 1 short and 4 long teeth; tips of several veins of wing with small patches of dark hairs; Siskiyou, Fresno, and Riverside counties .... *alternata* Say 1824
4. Wing without any markings; tip of labium with 4 labial spines; ventral plate not narrow near its base, lobes with rounded ends, their sides almost parallel; Humboldt County .... *severini* Tonnoir 1922
— Wing with a black band; tip of labium with 2 labial spines; ventral plate narrowing near its base, lobes with pointed ends, their sides rather divergent; San Francisco County............... *sigma* Kincaid 1899

Larvae

1. Anterior body segments without plates, not more than 9 plates in all (fig. 14:9*c*) 2
— All body segments with 2 or 3 tergal plates, 26 plates in all 3
2. Tergal plates on abdominal segment 5 absent or but faintly present and much smaller than plates on segments 6 and 7........................ *alternata* Say
— Tergal plates on abdominal segment 5 always present,

as large as plates on 6 and 7 .... *phalaenoides* (L.)
3. Siphonal sensillum with long seta adjacent to base; anterior tergal plates of abdominal segment 1 much smaller than posterior plate............ *cinerea* Banks
— Siphonal sensillum without long seta near base; anterior tergal plate of abdominal segment 1 only slightly smaller than posterior plates........ *severini* Tonnoir

The larva of *sigma* is not known.

Genus *Telmatoscopus* Eaton

Species of this genus have a biology very similar to that of *Psychoda* although they seem to prefer a somewhat wetter environment. The larva is not known of either of the California species.

Key to Adults of California Species of Telmatoscopus

1. Wings heavily pigmented, giving a mottled appearance; legs annulated: Monterey and San Diego counties .... *furcatus* (Kincaid) 1899
— Wings not heavily pigmented; legs not annulated; El Dorado and San Diego counties..... *niger* (Banks) 1894

REFERENCES

JOHANNSEN, O. A.
1934. *See* Diptera references.
MALLOCH, J. R.
1917. *See* Diptera references.
QUATE, L. W., and W. W. WIRTH
1951. A taxonomic revision of the genus *Maruina* (Diptera: Psychodidae). Wasmann Jour. Biol., 9:151-166.
1955. A revision of the Psychodidae (Diptera) in America North of Mexico. Univ. Calif. Publ. Entom., 10:103-273.
SATCHELL, G. H.
1947*a*. The larvae of the British species of *Psychoda* (Diptera: Psychodidae). Parasitology, 38:51-69.
1947*b*. The ecology of the British species of *Psychoda* (Diptera: Psychodidae). Ann. Appl. Biol., 34:611-621.
1948. The respiratory horns of *Psychoda* pupae (Diptera, Psychodidae) Parasitology, 39:43-52.
1949. The early stages of the British species of *Pericoma* Walker (Diptera, Psychodidae). Trans. Royal Ent. Soc. London, 100:411-447.
SMART, JOHN
1948. A handbook for the identification of insects of medical importance. Psychodidae, pp. 35-38. British Museum.

Family BLEPHARICERIDAE

The members of this family (fig. 14:11), commonly known as "net-winged midges" (fig. 14:11*c*), are slender, rather delicate flies of medium size, usually having a network of delicate creases in the wing owing to the folding in the pupal case. The adults are generally found resting on streamside rocks or foliage near swift streams. The larvae may be found crawling on rocks in the rushing water of mountain or hill streams, living either directly in the water or in the perpetual spray of waterfalls. The larvae appear to feed on the algal scum on the surface of the rocks. Pupation takes place in the same locality, the pupae being attached firmly to the rocks. It has been suggested, with some evidence, that the females dive under water during oviposition.

Although a good bit of work has been done on the immature stages of Blephariceridae by Kellogg (1903, 1907), Kitakami (1931), and Johannsen (1934), the results have been very confusing so far as generic characters are concerned, and not enough western species have been positively associated with adults to permit the preparation of species keys. We attempted to make a thorough study of the Nearctic species, but found that though the larvae and pupae could be determined to genera in Kitakami's keys, the adults associated with them did not fall into the expected genera and the larvae did not always agree with the pupae generically. For this reason we are not presenting any keys to the immature stages.

Key to the North American Genera of Blephariceridae

Adults

1. Second basal cell open apically (fig. 14:1*f*) .... *Blepharicera* Macquart
— Second basal cell closed apically (fig. 14:11*c*) 2
2. Vein R_{2+3} fused and continuing to wing margin .... *Philorus* Kellogg
— Vein R_{2+3} branched, the posterior branch either simulating a cross vein near the base of R_1 (fig. 14:11*c*), or elongate 3
3. Pleura pilose ....*Bibiocephala* Osten Sacken
— Pleura bare .... *Agathon* Roeder

California Species of Blephariceridae

Agathon doanei Kellogg 1903. Santa Clara, Santa Cruz, Monterey counties
Agathon canadensis (Garrett) 1922. Siskiyou to San Bernardino county
Bibiocephala comstocki Kellogg 1903. Santa Clara County
Bibiocephala grandis Osten Sacken 1874. Siskiyou County
Blepharicera jordani Kellogg 1903. San Mateo, Riverside
Blepharicera osten sackeni Kellogg 1903. Humboldt, Shasta, Napa counties
Philorus ancillus (Osten Sacken) 1878.
Philorus aylmeri Garrett 1923. Modoc to Tulare
Philorus cheaini Garrett 1925. Sierra County
Philorus markii Garrett 1925. Alpine County
Philorus sequoiarum Alexander 1952. Tulare County
Philorus yosemite (Osten Sacken) 1877. Humboldt, Mariposa, Tulare
Mariposa, Tulare counties

REFERENCES

JOHANNSEN, O. A.
1934. *See* Diptera references.
KELLOGG, V. L.
1903. The net-winged midges (Blepharoceridae) of North America. Proc. Calif. Acad. Sci., 3:187-226.
1907. Family Blephariceridae. *In* Wytsman, Genera Insectorum fasc. 56, 15 pp.
KITAKAMI, SHIRO
1931. The Blephariceridae of Japan, Mem. Coll. Sci., Kyoto Imp. Univ. (B)6:53-108.
WALLEY, G. S.
1927. Review of the Canadian species of the dipterous family Blepharoceridae. Canad. Ent., 59:112-116.

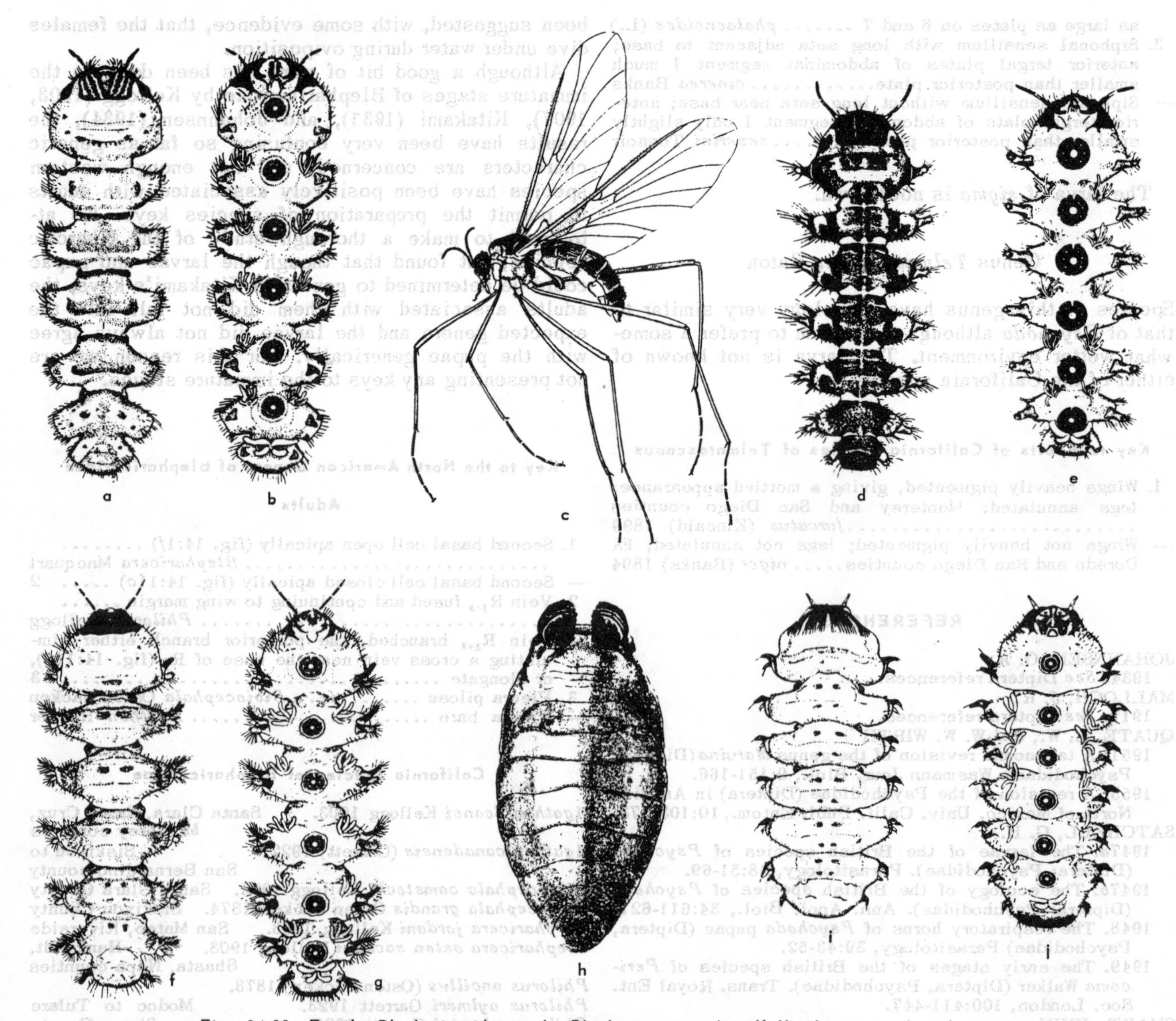

Fig. 14:11. Family Blephariceridae. *a,b, Blepharicera jordani* Kell., larva: *a,* dorsal view; *b,* ventral. *c,i,j, Agathon elegantula* (v. Roed.): *c,* adult; *i,* larva, dorsal view; *j,* same, ventral. *d,e, Blepharicera osten-sackeni* Kell., larva: *d,* dorsal view; *e,* ventral. *f,g,h, Bibiocephala comstocki* Kell.: Larva: *f,* dorsal view, *g,* ventral; *h,* pupa (Kellogg, 1903).

Family DEUTEROPHLEBIIDAE

This family contains a single genus with five described species, two of which occur in the western United States. The North American species were described by Pennak (1945) and Wirth (1951), and Pennak (1951) summarizes what is known about the family. Since the species generally occur at high altitudes, the name "mountain midges" has been given to the family. The adults are rarely collected. *Deuterophlebia shasta* Wirth (fig. 14:13) has been found only in Shasta County, California, and *D. coloradensis* Pennak (fig. 14:12) is possibly confined to the Rocky Mountain region.

The larvae and pupae are found in cold mountain streams on the upper surface of smooth rocks, either in the riffle or splash area or where the water flow is very shallow and swift.

REFERENCES

PENNAK, R. W.
1945. Notes on mountain midges (Deuterophlebiidae) with a description of the immature stages of a new species from Colorado. Amer. Mus. Nov. No. 1276, pp. 1-10.
1951. Description of the imago of the mountain midge *Deuterophlebia coloradensis* Pennak (Diptera, Deuterophlebiidae). Amer. Mus. Nov. No. 1534, pp. 1-11.
WIRTH, W. W.
1951. A new mountain midge from California (Diptera, Deuterophlebiidae). Pan.-Pac. Ent., 27:49-57.

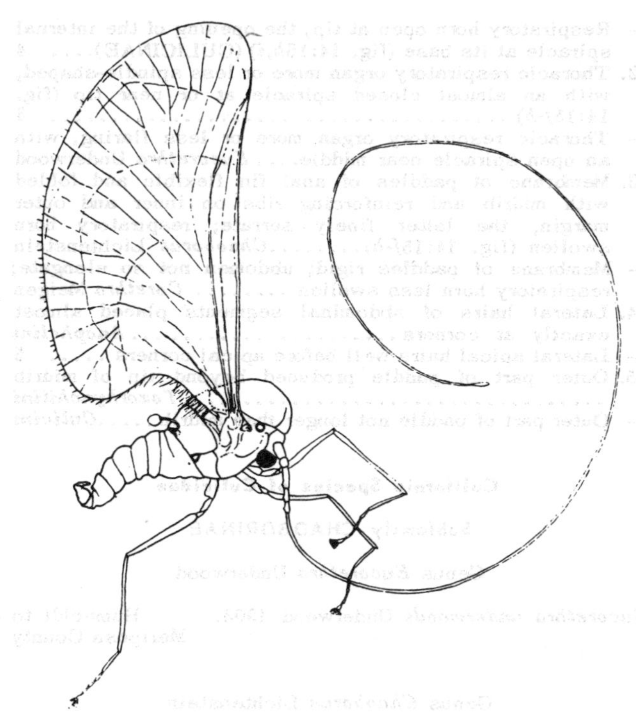

Fig. 14:12. *Deuterophlebia coloradensis* Pennak (Pennak, 1951).

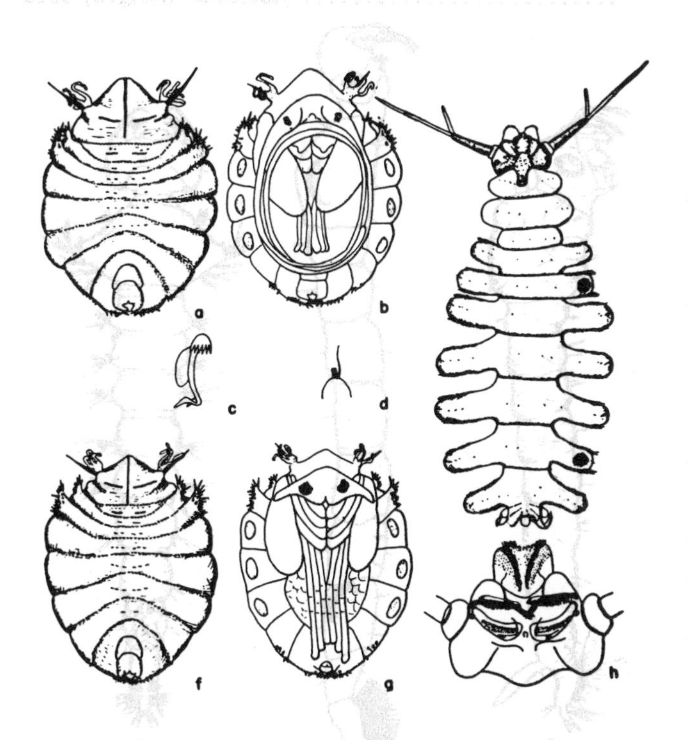

Fig. 14:13. *Deuterophlebia shasta* Wirth. *a*, male pupa, dorsal view; *b*, same, ventral; *c*, detail of hook of larval proleg; *d*, detail of tip of eighth abdominal segment of larva; *e*, mature larva, dorsal view (hooks of second and sixth prolegs, right side, inverted); *f*, female pupa, dorsal view; *g*, same, ventral; *h*, larval head, ventral view, showing mouth parts (Wirth, 1951).

Family CULICIDAE

Mosquitoes as a group need no introduction. They occur wherever man may go, from the arctic to the tropics. As a nuisance they have stimulated strong invectives and have prompted the formation of many mosquito abatement projects throughout the country. But contrary to popular opinion, not all mosquitoes bite. The males of all species are innocuous, and both sexes of the Chaoborinae are nonbiters. Of greater consequence is the role of mosquitoes in the transmission of disease. Malaria, yellow fever, dengue, encephalitis, filariasis, and other mosquito-borne diseases have plagued the human race for centuries.

Inspired largely by wartime epidemics, knowledge of mosquitoes probably surpasses that of any other group of insects. In California the only really confusing group of mosquitoes left to study is the *Aedes* group of the high mountain snow pools. The California species of Culicinae have been recently revised by Freeborn and Bohart (1951). Mosquito larvae are the familiar "wrigglers" which play an important part in the aquatic community where they form a staple item in the diet of fish and many aquatic insects.

Keys to the North American Genera of Culicidae

Adults

1. Proboscis not elongate, extending but little beyond clypeus; wings with scales (when present) confined mostly to fringe (CHAOBORINAE) 2
— Proboscis elongate, extending far beyond clypeus; wings with scales on veins and margin (CULICINAE) 4
2. Anal vein ends before the fork of the fifth vein .... *Eucorethra* Underwood
— Anal vein ends beyond the fork of the fifth vein 3
3. First tarsal segment much shorter than the second .. .. *Corethra* Meigen
— First tarsal segment longer than the second (fig. 14:15*a*)............ *Chaoborus* Lichtenstein
4. Proboscis rigid, the basal half stout, the apical half more slender and bent sharply backward .... *Toxorhynchites* Theobald
— Proboscis not rigid, of nearly uniform thickness or the tip swollen 5
5. Scutellum rounded posteriorly and with marginal hairs arranged in an unbroken line; abdomen without scales; females with palpi about as long as proboscis (fig. 14:16*b*)..................... *Anopheles* Meigen
— Scutellum trilobed posteriorly and with bristles in 3 groups; abdomen with evident scales; females with short palpi 6
6. Postnotum with a tuft of setae; squamae without a fringe of hairs .... *Wyeomyia* Theobald
— Postnotum without a tuft of setae; squamae with a fringe of hairs 7
7. Second marginal cell (bounded by branches of first fork vein) less than half as long as its petiole .... *Uranotaenia* Lynch
— Second marginal cell at least as long as its petiole 8
8. Fourth fore tarsal segment about as long as wide; mesonotum marked with longitudinal white lines .... *Orthopodomyia* Theobald
— Fourth fore tarsal segment much longer than wide ... 9
9. First segment of hind tarsus with a median pale ring; wing scales mixed dark and white 10

— First segment of hind tarsus without a median pale ring 11
10. Spiracular and postspiracular bristles present (fig. 14:16*j*); abdominal tergites with many pale apical scales, frequently forming a triangle .... *Psorophora* Robineau Desvoidy
— Spiracular and postspiracular bristles absent; abdominal tergites with laterobasal pale spots .... *Mansonia* Blanchard
11. Postspiracular bristles present; spiracular bristles absent; female abdomen pointed at tip ....*Aedes* Meigen
— Postspiracular bristles absent; female abdomen blunt at tip 12
12. Spiracular bristles present, yellowish; mostly large mosquitoes ....*Culiseta* Felt
— Spiracular bristles absent; medium to small mosquitoes .... *Culex* Meigen

Larvae

1. Antennae prehensile, with long and strong apical spines (fig. 14:14) (CHAOBORINAE) 2
— Antennae not prehensile and lacking the strong apical spines (fig. 14:17*f*) (CULICINAE) 4
2. Eighth abdominal segment with an elongate dorsal air tube or respiratory siphon (fig. 14:14*a*). *Corethra* Meigen
— Eighth abdominal segment without an elongate air tube or respiratory siphon 3
3. Air sacs present in thorax and 7th abdominal segment (fig. 14:14*b*) .... *Chaoborus* Lichtenstein
— Air sacs absent; 8th abdominal segment with a well-developed spiracular disc with prominent valves (fig. 14:14*c*) ....*Eucorethra* Underwood
4. Eighth abdominal segment without a respiratory siphon (fig. 14:16*e*) ....*Anopheles* Meigen
— Eighth abdominal segment with a respiratory siphon (fig. 14:17*f*) 5
5. Mouth brushes prehensile, each composed of 10 stout rods ....*Toxorhynchites* Theobald
— Mouth brushes not or rarely prehensile, composed of 30 or more hairs 6
6. Siphon without a pecten (figs. 14:17*d,e*) 7
— Siphon with a pecten (figs. 14:17*a-c*) 9
7. Anal segment without a median ventral brush but with a pair of ventrolateral tufts only ....*Wyeomyia* Theobald
— Anal segment with a median ventral brush (fig. 14:17) 8
8. Distal half of siphon adapted for piercing underwater plant tissue (fig. 14:17*d*).........*Mansonia* Blanchard
— Siphon not adapted for piercing plant tissue (fig.14:17*e*) ....*Orthopodomyia* Theobald
9. At least lower head hair single and coarse, spinelike; head longer than wide ....*Uranotaenia* Lynch
— Both head hairs slender, branched or unbranched; head wider than long (fig. 14:17*f*) 10
10. Siphon with a pair of large basoventral hair tufts (fig. 14:17*b*) ....*Culiseta* Felt
— Siphon without a pair of large basoventral hair tufts 11
11. Siphon with 3 or more pairs of ventrally located hair tufts (inserted below the lateral line) ..*Culex* Linnaeus
— Siphon with 1 or rarely 2 pairs of ventrally located hair tufts 12
12. Tufts of ventral brush not inserted in saddle (fig. 14:17*f*) .... *Aedes* Meigen
— Tufts of ventral brush inserted in saddle which completely rings anal segment (fig. 14:17*c*) .... *Psorophora* Robineau Desvoidy

Pupae

1. Respiratory horn either almost closed apically or with the spiracular opening near its middle (fig. 14:15*f-h*) (CHAOBORINAE) 2
— Respiratory horn open at tip, the opening of the internal spiracle at its base (fig. 14:16*h,i*) (CULICINAE).... 4
2. Thoracic respiratory organ more or less spindle-shaped, with an almost closed spiracle at or near tip (fig. 14:15*f-h*) 3
— Thoracic respiratory organ more or less flaring, with an open spiracle near middle.... *Eucorethra* Underwood
3. Membrane of paddles of anal fin flexible and folded with midrib and reinforcing ribs on inner and outer margin, the latter finely serrate; respiratory horn swollen (fig. 14:15*f-h*) ....*Chaoborus* Lichtenstein
— Membrane of paddles rigid; abdomen not so elongate; respiratory horn less swollen .... *Corethra* Meigen
4. Lateral hairs of abdominal segments placed almost exactly at corners ....*Anophelini*
— Lateral apical hairs well before apical corners 5
5. Outer part of paddle produced beyond tip of midrib ....*Toxorhynchitini*
— Outer part of paddle not longer than midrib .....*Culicini*

California Species of Culicidae

Subfamily CHAOBORINAE

Genus *Eucorethra* Underwood

Eucorethra underwoodi Underwood 1903. Humboldt to Mariposa County

Genus *Chaoborus* Lichtenstein

Key to Adults of California Species

1. Wings unspotted; Modoc County .... *flavicans* (Meigen) 1818

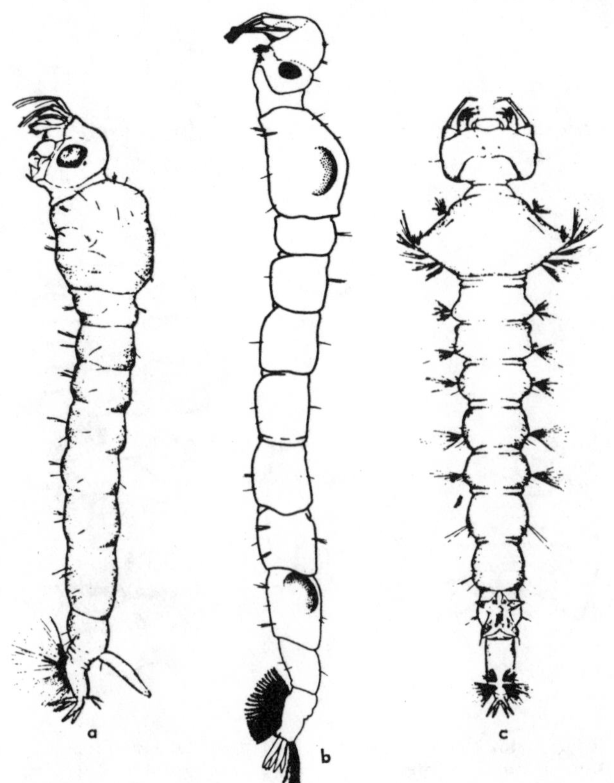

Fig. 14:14. Chaoborinae larvae. *a*, *Corethra cinctipes* Coq.; *b*, *Chaoborus* sp.; *c*, *Ecorethra underwoodi* Und. (*a,c*, Matheson, 1944; *b*, Peterson, 1951).

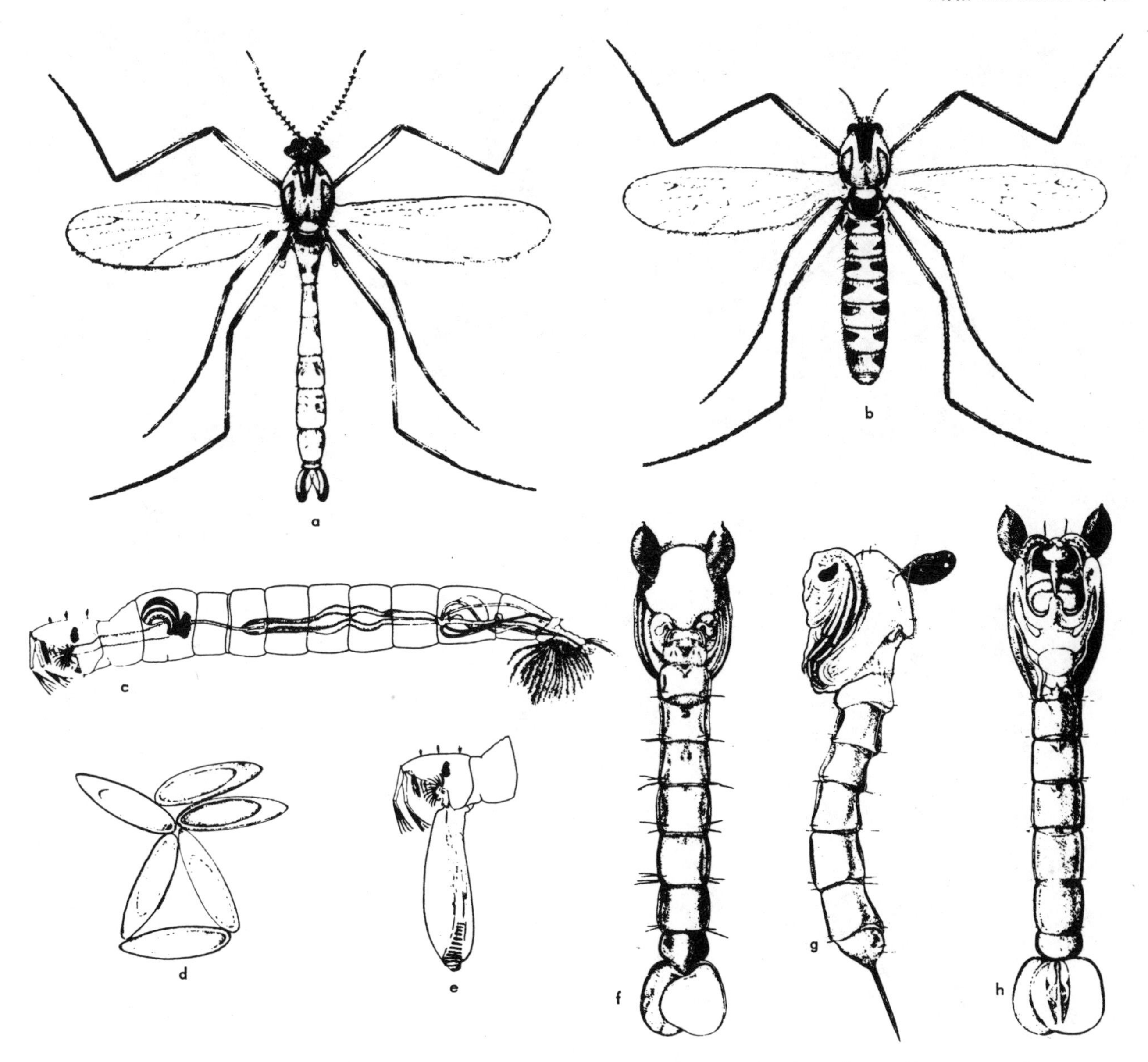

Fig. 14:15. *Chaoborus astictopus* Dyar and Shannon. *a*, male; *b*, female; *c*, larva; *d*, eggs; *e*, eversible pharyngeal sac of larva; *f-h*, pupa (Herms, 1937).

— Wings distinctly spotted 2
2. Large species, the wing 5 mm. long; costa dark scaled with 3 paler areas; Humboldt to San Luis Obispo counties .... *nyblaei* Zetterstedt 1838
— Smaller species, the wing 3 mm. or less; costa mostly pale scaled with a few small, dark spots 3
3. Femora and tibiae with many distinct dark spots; sides of scutum anteriorly with dark spots; Imperial County .... *punctipennis* (Say) 1823
— Femora and tibiae not darkened except near joints; (fig. 14:15*b*) sides of scutum anteriorly without dark spots; Siskiyou to Ventura counties .... *astictopus* Dyar and Shannon 1924

Subfamily CULICINAE[4]

Key to the California Species of Anopheles

Females

1. Wing scales unicolorous but clumped to produce dark spots 2
— Wing with patches of pale scales 3

[4]The Keys that follow are taken with slight modifications from Freeborn and Bohart, 1951.

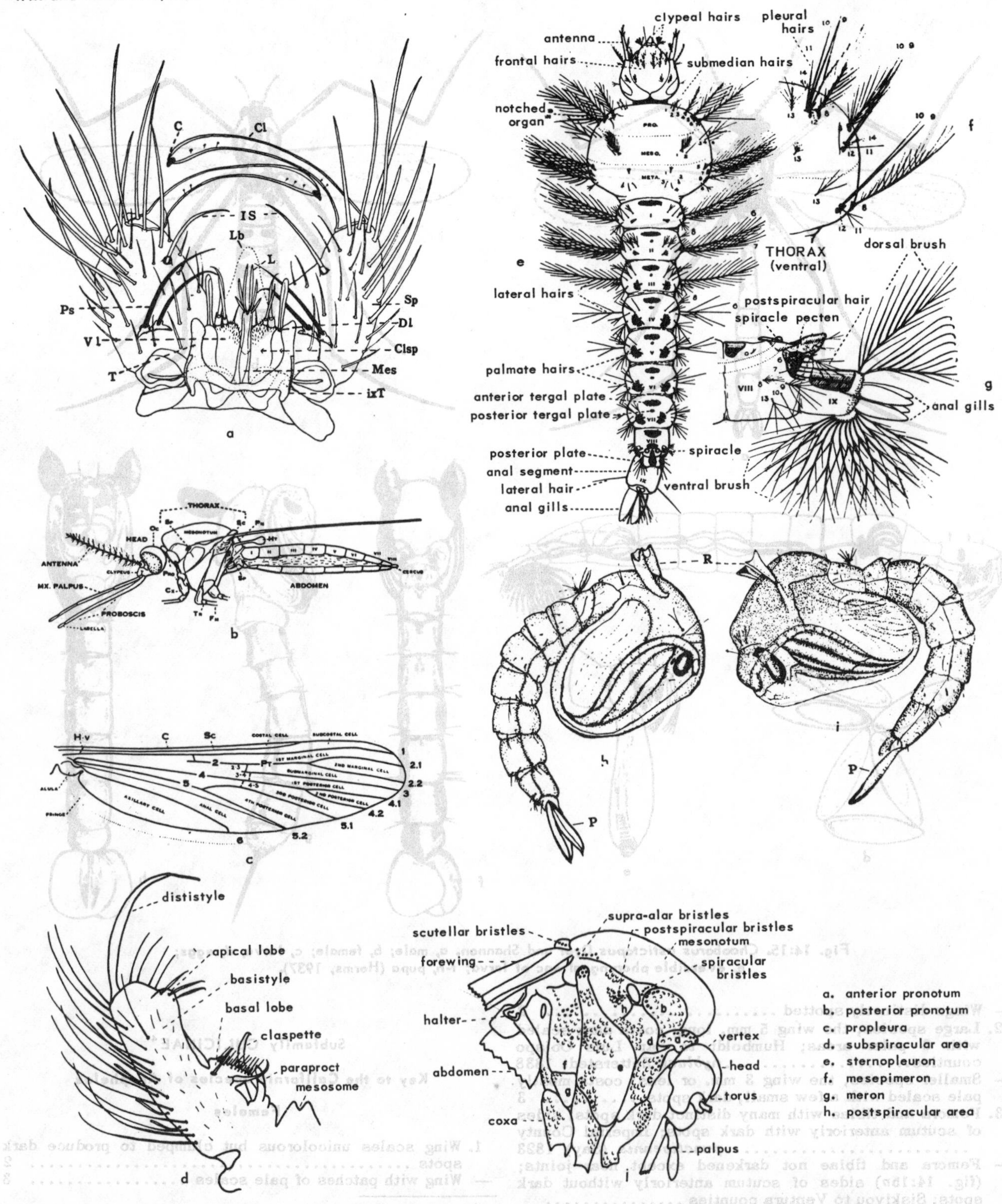

Fig. 14:16. Culicinae. *a,e,f,g, Anopheles quadrimaculatus* Say: *a*, male genitalia, dorsal view (C, claw of dististyle; Cl, dististyle; Clsp, claspette, Dl, dorsal lobe of claspette, IS, internal spines, L, leaflets of mesosome, Lb, anal lobes, Mes, mesosome; Ps, parabasal spines; Sp, basistyle; T, lobe of ninth tergite; ixT, ninth tergite; Vl, ventral lobe); *e*, larva; *f*, thorax of larva; *g*, abdominal segment 8 and 9 of larva. *b,c, Anopheles* sp.: *b*, adult female (Cx, coxa; Fm, femure; Ht, halter; Mx, maxillary; Oc, occiput; Pn, postnotum; Pr, prothorax, Sc, scutellum; Sp, spiracle; Tr, trochanter); *c*, wing venation. *d, Aedes dorsalis* (Mg.), male genitalia; *h, Aedes cinereus* Mg., pupa (R, respiratory trumpet; P, paddle); *i, Anopheles punctipennis* (Say), pupa; *j*, thorax of adult mosquito (*a,h,i*, Matheson, 1944; *b,c,e-g*, Ross and Roberts, 1943; *d,j*, Freeborn and Bohart, 1951).

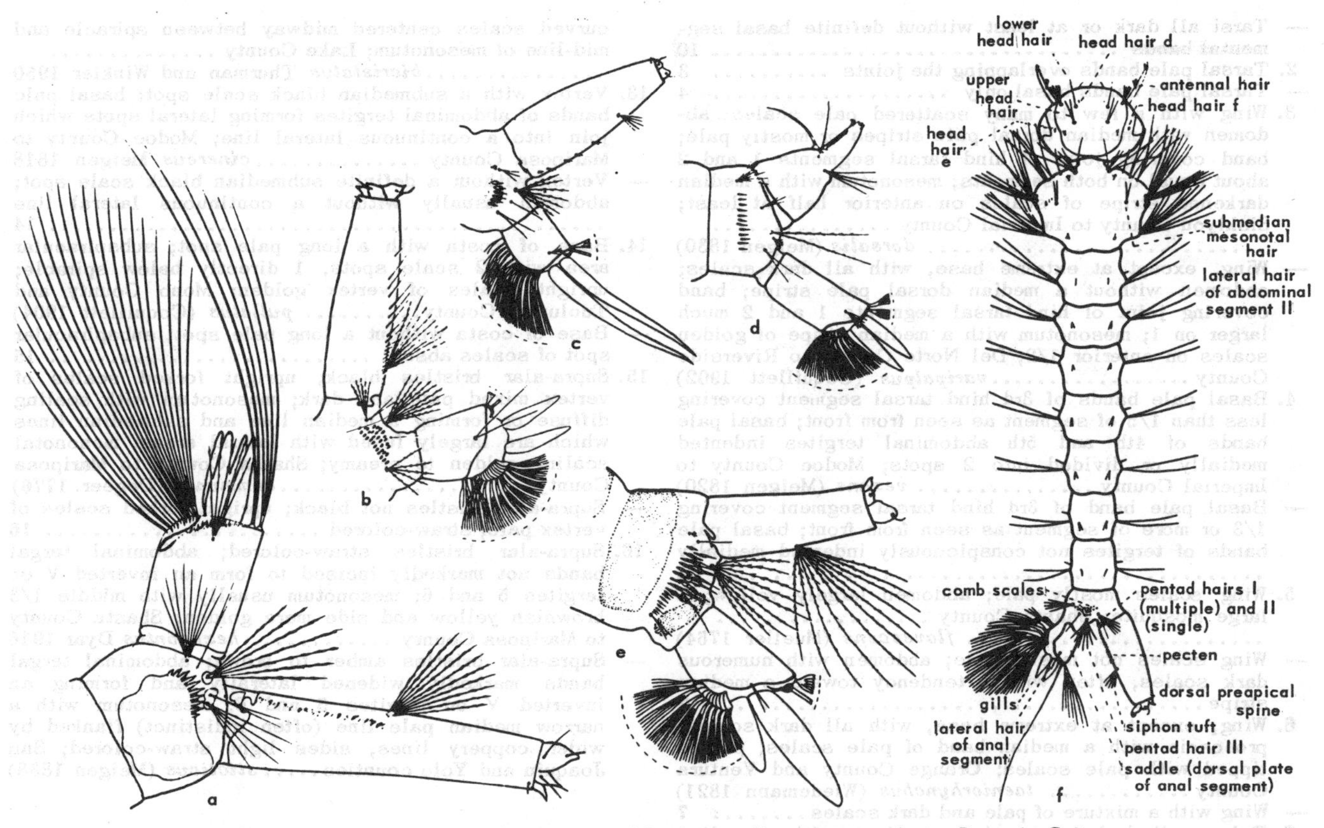

Fig. 14:17. Culicinae, terminal segments of larvae. *a, Uranotaenia anhydor* Dyar; *b, Culiseta inornata* (Will.); *c, Psorophora confinnis* (L.A.); *d, Mansonia perturbans* (Walk.); *e, Orthopodomyia californica* Bohart; *f, Aedes dorsalis* (Mg.); dorsal view of larva (*a*, Yamaguti and LaCasse, 1951; *b-f*, Freeborn and Bohart, 1951).

2. Fringe scales of wing uniformly dark; Del Norte County to San Diego County ....*freeborni* Aitken 1939
— Fringe scales at wing tip whitish to bronzy; coastal from Humboldt to Orange counties ....*occidentalis* Dyar and Knab 1906
3. Palpus banded; Siskiyou County to San Diego County *pseudopunctipennis franciscanus* McCracken 1904
— Palpus entirely dark scaled; Central Valley and north coast ....*punctipennis* (Say) 1823

Male genitalia

1. Claspette with inconspicuous bristles on subapical lobe and 2 long slender ones on apical lobe; 9th tergite without well-developed lobes....................... ....*pseudopunctipennis franciscanus* McCracken
— Claspette with 2 stout setae on subapical lobe and 1 on apical lobe which is usually associated with a smaller internal seta and a very small one between; 9th tergite with fingerlike lobes .... *punctipennis* (Say)
freeborni Aitken
occidentalis Dyar and Knab

Larvae

1. Inner clypeal hairs widely spaced, outer clypeals bare; inner antennal surface with small stout spicules ... .*pseudopunctipennis franciscanus* McCracken
— Inner clypeal hairs separated by less than width of basal tubercle; outer clypeals densely branched; inner antennal surface with rather slender bristles 2

2. Antepalmate hair (hair 2) of 4th abdominal segment usually single...........*occidentalis* Dyar and Knab
— Antepalmate hair (hair 2) of 4th abdominal segment usually double or triple .... *freeborni* Aitken
punctipennis (Say)

Genus *Uranotaenia* Lynch Arribalzaga

Uranotaenia anhydor Dyar 1907. San Diego County

Genus *Orthopodomyia* Theobald

Orthopodomyia californica Bohart 1950. Lake to Riverside

Genus *Psorophora* Robineau-Desvoidy

Psorophora confinnis (Lynch Arribalzaga 1891.) Imperial, Riverside County

Genus *Mansonia* Blanchard

Mansonia perturbans (Walker 1856.) Yolo to Riverside

Keys to the California Species of Aedes

Females

1. Tarsi with pale bands on some segments 2

— Tarsi all dark or at least without definite basal segmental bands 10
2. Tarsal pale bands overlapping the joints 3
— Tarsal pale bands basal only 4
3. Wing with a few to many scattered pale scales: abdomen with median dorsal pale stripes or mostly pale; band covering joint of hind tarsal segments 1 and 2 about equal on both segments; mesonotum with a median darkened stripe of scales on anterior half at least; Siskiyou County to Imperial County .... *dorsalis* (Meigen 1830)
— Wing, except at extreme base, with all dark scales; abdomen without a median dorsal pale stripe; band covering joint of hind tarsal segments 1 and 2 much larger on 1; mesonotum with a median stripe of golden scales on anterior 1/2; Del Norte County to Riverside County .... *varipalpus* (Coquillett 1902)
4. Basal pale bands of 3rd hind tarsal segment covering less than 1/5 of segment as seen from front; basal pale bands of 4th and 5th abdominal tergites indented medially or divided into 2 spots; Modoc County to Imperial County .... *vexans* (Meigen 1820)
— Basal pale band of 3rd hind tarsal segment covering 1/3 or more of segment as seen from front; basal pale bands of tergites not conspicuously indented medially 5
5. Wing scales mostly pale; abdomen largely yellow; a large mosquito; Shasta County .... *flavescens* (Mueller 1764)
— Wing scales not mostly pale; abdomen with numerous dark scales, often with a tendency toward a median stripe 6
6. Wing, except at extreme base, with all dark scales; proboscis with a median band of pale scales; palpus tipped with pale scales; Orange County and Ventura County .... *taeniorhynchus* (Wiedemann 1821)
— Wing with a mixture of pale and dark scales 7
7. Palpus all dark; proboscis sometimes with a median patch or band of pale scales; Siskiyou to Kern counties .... *nigromaculis* (Ludlow 1907)
— Palpus with some pale scales; proboscis dark or with mixed pale and dark scales 8
8. Pale wing scales confined almost entirely to front part of wing; torus with a few inconspicuous pale scales on inner surface; Modoc County to Los Angeles County .... *increpitus* Dyar 1916
— Pale wing scales scattered generally over wing; torus with a conspicuous inner patch of pale scales...... 9
9. Hind wing margin at base of fringe bordered with mixed black and white scales; wing scales mostly very broad; Sonoma to Riverside counties .... *squamiger* (Coquillett 1902)
— Hind wing margin at base of fringe bordered with minute unicolorous scales; wing scales narrow to moderately broad; Shasta County to El Dorado County .... *fitchii* (Felt and Young 1904)
10. Wing with scattered to numerous whitish scales along costa, subcosta, and radius 11
— Scales of costa, subcosta, and radius dark except for basal spots which are about as long or shorter than halter 13
11. Palpus with all dark scales; subspiracular area of pleuron with a lower scale patch only; abdominal tergite 2 with pale band usually narrowed at middle or divided; Shasta County to Mariposa County; high Sierra .... *ventrovittis* Dyar 1916
— Palpus with numerous pale scales, especially toward base; subspiracular area with 2 scale spots, 1 directly below spiracle; abdominal tergite 2 with pale band broadened at middle 12
12. Palpus not much longer than twice greatest diameter of torus; mesonotum with slender or only slightly broadened scales between spiracle and mid-line of mesonotum; Shasta County to Mariposa County.......... .... *cataphylla* Dyar 1916
— Palpus much longer than twice greatest diameter of torus; mesonotum with a large patch of pale broadened curved scales centered midway between spiracle and mid-line of mesonotum; Lake County .... *bicristatus* Thurman and Winkler 1950
13. Vertex with a submedian black scale spot; basal pale bands of abdominal tergites forming lateral spots which join into a continuous lateral line; Modoc County to Mariposa County .... *cinereus* Meigen 1818
— Vertex without a definite submedian black scale spot; abdomen usually without a continuous lateral line 14
14. Base of costa with a long pale spot; subspiracular area with 2 scale spots, 1 directly below spiracle; upright scales of vertex golden; Mono County and Tuolumne County.......... *pullatus* (Coquillett 1904)
— Base of costa without a long pale spot; subspiracular spot of scales absent 15
15. Supra-alar bristles black; upright forked scales of vertex mixed pale and dark; mesonotum with scaling diffuse or forming a median line and submedian lines which are largely fused with lateral areas; mesonotal scaling golden to creamy; Shasta County to Mariposa County .... *communis* (Degeer 1776)
— Supra-alar bristles not black; upright forked scales of vertex pale, straw-colored 16
16. Supra-alar bristles straw-colored; abdominal tergal bands not markedly incised to form an inverted V on tergites 5 and 6; mesonotum usually with middle 1/3 brownish yellow and side more golden; Shasta County to Mariposa County .... *hexodontus* Dyar 1916
— Supra-alar bristles amber to brown; abdominal tergal bands markedly widened laterally and forming an inverted V on tergites 5 and 6; mesonotum with a narrow median pale line (often indistinct) flanked by wider coppery lines, sides light straw-colored; San Joaquin and Yolo counties..... *sticticus* (Meigen 1838)

Male Genitalia

1. Claspette absent, or fused to basistyle and without a filament 2
— Claspette present, attached at base only (fig. 14:16*d*) 3
2. Dististyle unequally bifurcate near base, longer branch equally bifurcate at apex .... *cinereus* Meigen
— Dististyle broad, flattened, not bifurcate; apex blunt with tooth borne at inner apical four-fifths .... *vexans* (Meigen)
3. Filament (distal part) of claspette short, not bladelike .... *bicristatus* Thurman and Winkler
— Filament of claspette bladelike 4
4. Basal lobe of basistyle with bristles which are nearly equal in size or grading from small to large, but without a single bristle distinguished by its greater size..... 5
— Basal lobe of basistyle with a single stout bristle or spine distinguished by its greater size, sometimes with smaller spines in addition to slender bristles (fig. 14:16*d*) 8
5. Apical lobe of basistyle present; claspette filament with a sharp angle on convex side..... *increpitus* Dyar
— Apical lobe of basistyle undeveloped 6
6. Claspette filament with a retrorse projection .... *taeniorhynchus* (Wiedemann)
— Claspette filament without a retrorse projection 7
7. Basal lobe of basistyle a dense clump of long, stout, apically curved bristles of which longer ones are about 4 times as long as tooth of dististyle .... *varipalpus* (Coquillett)
— Basal lobe of basistyle a small clump of slender nearly straight bristles of which longer ones are about 2.5 to 3 times as long as tooth of dististyle .... *nigromaculis* (Ludlow)
8. Basal lobe of basistyle with 1 or 2 unusually stout bristles in addition to long stout one (fig. 14:16*d*)... 9
— Basal lobe of basistyle without differentiated bristles other than a long stout one 10

9. Basal lobe of basistyle with 1 short stout bristle, a long curved one and a dense group of slender bristles: apical lobe of basistyle with a group of bladelike setae; claspette stem not sharply angled near middle (fig. 14:16*d*) .... *dorsalis* (Meigen)
— Basal lobe of basistyle with 2 stout inner bristles close together and a long outer spine with a few inconspicuous bristles in between; apical lobe of basistyle with bristles of varying length; claspette stem sharply angled near middle...*pullatus* (Coquillett)
10. Claspette filament with a sharp angle near base of concave edge 11
— Claspette filament without sharp angle near base of concave edge 12
11. Claspette filament only a little wider at basal 1/3 than apical diameter of claspette stem .... *fitchii* (Felt and Young)
— Claspette filament expanded at basal 1/3 to about twice the apical diameter of claspette stem .... *sticticus* (Meigen) *ventrovittis* Dyar
12. Claspette filament with a median sharp angle or retrorse point on convex side 13
— Claspette filament rather evenly convex at middle of one side 14
13. Claspette filament with expansion forming an acute angle; basal lobe of basistyle well separated from apical lobe .... *squamiger* (Coquillett)
— Claspette filament with expansion forming an obtuse angle; basal lobe of basistyle reaching almost to apical lobe .... *flavescens* (Mueller)
14. Claspette filament markedly broadened at basal 1/3 ....*cataphylla* Dyar
— Claspette filament slender, not markedly broadened at basal 1/3 15
15. Apical lobe of basistyle with numerous short, curved, broadened setae; basal lobe of basistyle with stout spine broadly curved from near middle; claspette distinctly shorter than dististyle .... *hexodontus* Dyar
— Apical lobe of basistyle with bristles only; basal lobe of basistyle with stout spine bent or curved near apex; claspette about as long as dististyle .... *communis* (Degeer)

Larvae

1. Pecten terminating in 1 or more rather widely separated teeth 2
— Pecten teeth rather evenly spaced 8
2. Siphon with detached pecten teeth distad of insertion of siphon tuft; upper and lower head hairs almost invariably single 3
— Siphon with pecten row ending before insertion of siphon tuft................................... 4
3. Siphon with only a single pair of tufts; dorsal preapical spine straight; Sierran meadow species, usually in full sun .... *cataphylla* Dyar
— Siphon with a small tuft above pecten row basad of main tuft; also a single or double hair inserted dorsally and subapically on siphon; dorsal preapical spine curled at end; Coast Range meadow species .... *bicristatus* Thurman and Winkler
4. Comb of at least 20 scales; lateral hair of anal segment more than 1/2 as long as saddle; typical comb tooth with a strong central thorn and weak side teeth; flooded meadows .... *flavescens* (Mueller)
— Comb of less than 20 scales 5
5. Pentad hair 2 located obliquely ventral to hair 1 and about on a line between 3 and 1; upper and lower head hairs not both single 6
— Pentad hair 2 directly posterior or dorsoposterior to 1; upper and lower head hairs usually single, rarely double 7
6. Lower head hair with 4 to 8 branches; upper with 5 to 9; Sierran meadow or pine needle pools in partial or dense shade .... *cinereus* Meigen
— Lower head hair single to triple, upper with 2 to 4 branches; river overflow pools or sometimes in irrigated pastures............................ *vexans* (Meigen)
7. Anal segment completely ringed by saddle; dorsal preapical spine of siphon about as long as middle pecten tooth; pecten extending about 3/4 of siphon tube length; flooded meadows and irrigated pastures .... *nigromaculis* (Ludlow)
— Anal segment not completely ringed by saddle; dorsal preapical spine of siphon inconspicuous; pecten extending not more than 2/3 of siphon tube length; Sierran meadows ....*ventrovittis* Dyar
8. Anal segment completely ringed by saddle 9
— Anal segment not completely ringed by saddle (fig. 14:17*f*) 10
9. Comb of many apically fringed scales in a patch; upper and lower head hairs single; basal diameter of siphon greater than dorsal length of saddle; coastal marshes of southern California..... *taeniorhynchus* (Wiedemann)
— Comb usually with 5 to 7 thornlike scales in a row; upper head hair single to triple; lower single or double; basal diameter of siphon about equal to dorsal length of saddle; Sierran pools in shade or sun .... *hexodontus* Dyar
10. Lateral hair of anal segment as long as or longer than saddle, sometimes with 2 or more branches......... 11
— Lateral hair of anal segment shorter than saddle and usually single...................................... 13
11. Pentad hair 2 located obliquely ventral to hair 1 and about on a line between 3 and 1; typical comb scale slender and with a row of bristles across rounded or truncate apex; gills large, usually about as long as siphon; hair *d* large and multiple; tree holes........ .... *varipalpus* (Coquillett)
— Pentad hair 2 located directly posterior or dorsoposterior to 1; gills much shorter than siphon; head hair *d* minute 12
12. Siphon tube about 4 times as long as its greatest diameter; terminal pecten tooth almost as long as apical diameter of siphon tube; upper head hair with 3 or more branches (usually 4), lower with 2 or more (usually 3); Sierran meadow or tule-filled pools............ ....*fitchii* (Felt and Young)
— Siphon tube less than 3 times as long as its greatest diameter; terminal pecten tooth about 1/2 as long as apical diameter of siphon tube; upper and lower head hairs single to triple (upper usually double or triple, lower usually double); brackish coastal marshes .... *squamiger* (Coquillett)
13. Anal segment about 7/8 ringed by saddle; lower head hairs single or double, upper usually with 2 or more branches; shaded river overflow pools........... ....*sticticus* (Meigen)
— Anal segment 1/2 to 4/7 ringed by saddle.......... 14
14. Upper head hair with 5 or more branches, lower with 3 or more; Sierran meadows.......*pullatus* (Coquillett)
— Upper head hair with fewer than 5 branches, lower with fewer than 3 15
15. Dorsal microsetae toward apex of saddle well developed, more than twice as long as those toward base of saddle and longer than diameter of setal ring at base of lateral hair of anal segment; head hairs with 2 to 4 branches or sometimes all single (especially in northeast California); siphon tuft inserted before middle of siphon tube; mostly in flooded grassy pools, common in the Sierra, less so at lower elevations down to sea level..............................*increpitus* Dyar
— Dorsal microsetae toward apex of saddle weakly developed, less than twice as long as those toward base of saddle and about as long as diameter of the setal ring of lateral hair 16
16. Dorsal preapical spine of siphon inserted less than its length from apical margin of siphon tube; siphon tuft often inserted at or beyond middle of inner surface of siphon tube; gills usually short but sometimes longer than anal segment; upper and lower head hairs usually single, sometimes double or upper rarely triple; irri-

gated pastures, brackish marshes, flooded meadows, mostly in full sun (fig. 14:17*f*) ..... *dorsalis* (Meigen)
— Dorsal preapical spine of siphon inserted at least its length from apical margin of siphon tube; siphon tuft inserted before middle of inner surface of siphon tube; gills longer than anal segment, sometimes twice as long; upper and lower head hairs almost invariably single; shaded Sierran pools, usually with pine needles .... *communis* (Degeer)

Keys to the California Species of Culiseta

Females

1. Hind tarsus with pale basal bands; wing with distinct scale spots 2
— Hind tarsus unbanded; wing without distinct scale spots 3
2. Hind tarsal bands narrow, that of segment 2 covering about 1/10 of segment; cross veins unscaled; Del Norte County to San Diego County...*incidens* (Thomson 1868)
— Hind tarsal bands broad, that of segment 2 covering 1/4 to 1/3 of segment; cross veins scaled; Humboldt to San Diego counties .... *maccrackenae* Dyar and Knab 1906
3. Wing speckled with pale scales, costa with pale scales near base; mid-femur speckled with pale scales in front; dorsum of abdomen brownish, with yellowish basal bands which may cover most of segment; Siskiyou to Imperial counties ....*inornata* (Williston 1893)
— Wing dark scaled, costa dark near base; mid-femur dark scaled in front; dorsum of abdomen black with whitish basal bands; Shasta, Mono, and Mariposa counties............................*impatiens* (Walker 1848)

Male Genitalia

1. Dististyle about 8 times as long as its median diameter; lobes of 9th tergite with many short stout spines ....*inornata* (Williston)
— Dististyle more than 10 times as long as its median diameter; lobes of 9th tergite with many slender bristles 2
2. Basal lobe of basistyle with only a single, rather slender, differentiated bristle; a long row of short spines across apex of 8th segment ventrally ....*impatiens* (Walker)
— Basal lobe of basistyle with 2 or 3 stout spines 3
3. Eighth segment ventrally with a median apical clump of 3 to 8 toothlike spines...............*incidens* (Thomson)
— Eighth segment ventrally with a single median apical toothlike spine .... *maccrackenae* Dyar and Knab

Larvae

1. Saddle with a conspicuous group of dorsal bristles at apex; submedian mesonotal hair strongly developed, with 3 or more branches and about as long as antenna .... *maccrackenae* Dyar and Knab
— Saddle without dorsal bristles at apex; submedian mesonotal hair (hair 1) weak, single, considerably shorter than antenna 2
2. Upper head hair similar to lower head hair in length and number of branches .... *impatiens* (Walker)
— Upper head hair shorter than lower and with more branches 3
3. Lateral hairs of anal segment about as long as or longer than saddle (fig. 14:17*b*); head hair *d* more strongly developed than head hair *e* and often with 3 or more branches.......................*inornata* (Williston)
— Lateral hair of anal segment considerably shorter than saddle; head hair *d* about equal in size to hair *e* and usually double .... *incidens* (Thomson)

Keys to the California Species of Culex

Females

1. Tarsi with pale bands 2
— Tarsi unbanded 5
2. Proboscis completely encircled by a white band near the middle 3
— Proboscis dark above 4
3. Outer surfaces of femora and tibiae with white scales arranged in a narrow line or in a row of spots; venter of abdominal segments with an inverted black V; mesonotom typically with a line of white scales around front margin back to angle above spiracle, in supra-alar area, and on either side of prescutellar area; Del Norte to Imperial counties ....*tarsalis* Coquillett 1896
— Outer surfaces of femora and tibiae without white lines or rows of spots, venter of abdominal segments with median black spots; mesonotum without definite white lines; Siskiyou County to San Diego County .... *stigmatosoma* Dyar 1907
4. Proboscis with a definite median pale area beneath; tarsal bands distinct and whitish; Shasta to San Diego counties .... *thriambus* Dyar 1921
— Proboscis without a definite median pale area beneath; tarsal bands rather indistinct and with a brownish yellow tint; San Luis Obispo, Orange, and San Diego counties........................*restuans* Theobald 1901
5. Abdominal segments 2 to 4 all dark as seen from above 6
— Abdominal segments 2 to 4 with bands or lateral spots as seen from above 7
6. Vertex with a row of broad appressed scales forming a narrow border back of eyes; San Diego County..... .... *anips* Dyar 1916
— Vertex without a row of broad appressed scales behind eyes; Monterey County .... *reevesi* Wirth 1948
7. Pale markings of abdominal tergites apical 8
— Pale markings of abdominal tergites basal 10
8. Ventral pale stripe of hind femur ending shortly before apex; palpus about 3 times as long as flagellar segment 4 as seen in lateral view; Trinity County to San Diego County .... *apicalis* Adams 1903
— Ventral pale stripe of hind femur complete; palpus about 2 times as long as flagellar segment 4 as seen in lateral view.. 9
9. Pale scales of vertex ashy white; abdominal segments 5 of unengorged dried specimens about 1.3 times as broad as long; Humboldt County to Mariposa County .... *territans* Walker 1856
— Pale scales of vertex with a yellowish tinge; abdominal segment 5 of unengorged dried specimens 1.5 to 1.7 times as broad as long; Trinity to San Diego counties ....*boharti* Brookman and Reeves 1950
10. Mesonotal integument brownish red; bands on abdominal tergites 2 to 5 narrow and with indistinct outlines; Colusa County to Riverside County................ ....*erythrothorax* Dyar 1907
— Mesonotal integument brown or dark brown, not conspicuously reddish; abdominal tergites 2 to 5 with rather broad and definite bands 11
11. Bands of abdominal tergites 3 and 4 connected with lateral spots; mesonotal integument usually reddish brown; Colusa County to San Diego County .... *pipiens* L. 1758
— Bands of abdominal tergites not connected with lateral spots; mesonotal integument usually dark brown; San Joaquin to Imperial counties .*quinquefasciatus* Say 1823

Male Genitalia

1. Paraproct (10th sternite) crowned with a single row of teeth of which the outer are peglike, blunt 2
— Paraproct crowned with a clump of many bristles 6
2. Subapical lobe of basistyle deeply divided, bearing several hooked rods and a broad leaf; dististyle tipped

with a curved horn and crested with a clump of fine bristles .... *anips* Dyar
— Subapical lobe of basistyle not divided, not bearing a leaf, dististyle not crested 3
3. Mesosome lobes not bridged, broad and toothed apically .... *apicalis* Adams
— Mesosome lobes connected by a sclerotized bridge 4
4. Mesosome lobes broad apically but without teeth; dististyle strongly narrowed on apical 1/3 ... *reevesi* Wirth
— Mesosome lobes toothed apically; dististyle narrowing rather gradually toward apex 5
5. Mesosome lobes strongly narrowed and heavily sclerotized apically, bridge stout; dististyle membrane not divided into lobes toward apex .... *boharti* Reeves and Brookman
— Mesosome lobes broad and not unusually sclerotized apically, bridge narrow; dististyle membrane divided into lobes toward apex............... *territans* Walker
6. Paraproct with basal arm at most weakly developed 7
— Paraproct with a well-developed curving basal arm 8
7. Median process of mesosome stout toward apex and usually blunt or slanted apically, ventral cornu proportionately smaller than in *quinquefasciatus* .... *pipiens* Linnaeus
— Median process of mesosome slender toward apex, pointed, straight, or curved near tip; ventral cornu very large .... *quinquefasciatus* Say
8. Median process of mesosome with several stout teeth bordered by a group of close-set slender ones on inner side of ventral cornu........................... 9
— Median process of mesosome not separated from ventral cornu by a group of close-set slender teeth 10
9. Subapical lobe of basistyle (starting basally) with 3 rods (basal one pointed), a short and slender hooked spine, a leaf, and a long slender spine .... *stigmatosoma* Dyar
— Subapical lobe of basistyle with 3 rods (basal one blunt), no hooked spine, a leaf, and a slender spine .... *thriambus* Dyar
10. Median process with about 5 teeth, 1 of which is very long and slender, matching long external process; subapical lobe of basistyle with 3 rods (basal 2 stout), a spine (slightly hooked), a narrow leaf, and a slender bristle .... *tarsalis* Coquillett
— Median process without several teeth of which one is long and matches external process; subapical lobe of basistyle with 3 rods, slender and hooked club, broad leaf, and bristle 11
11. Median process of mesosome with 7 to 12 short and stout teeth; external process large and very stout in basal 3/4; basal process well developed, thumblike .... *erythrothorax* Dyar
— Median process of mesosome with a single tooth surpassed by stout external process; basal process weakly developed .... *restuans* Theobald

Larvae

1. Siphon hairs mostly single 2
— Siphon hairs double or multiple 3
2. Antennal tuft inserted at apical 1/3 of shaft; lateral abdominal hairs of segments 3 and 4 double .... *thriambus* Dyar
— Antennal tuft inserted slightly before middle of shaft; lateral abdominal hairs of segments 3 and 4 single .... *restuans* Theobald
3. Lower head hair with 3 or more branches, upper with 4 or more branches........................... 4
— Lower head hair single or double, upper single to triple 7
4. Siphon with ventral tufts only...... *tarsalis* Coquillett
— Siphon with subapical tuft lateral or sublateral 5
5. Dorsal microsetae toward apex of saddle conspicuously enlarged as compared with those at dorsal middle; lateral hairs of abdominal segments 3 and 4 usually triple .... *stigmatosoma* Dyar
— Dorsal microsetae toward apex of saddle about as large as those at middle; lateral hairs of abdominal segments 3 and 4 usually double 6
6. Siphon tube 6-7 times as long as its basal diameter and bearing 5 or 6 pairs of tufts .... *erythrothorax* Dyar
— Siphon tube a little more than 4 *(pipiens)* to a little less than 4 *(quinquefasciatus)* times as long as its basal diameter and bearing 4 pairs of tufts........ .... *pipiens* Linnaeus
quinquefasciatus Say
7. Siphon with 2 dorsal tufts; lateral hair of anal segment small and with more than 4 branches; upper head hair short and double, lower long and single..... *anips* Dyar
— Siphon without dorsal tufts; lateral hair of anal segment with fewer than 5 branches................. 8
8. Siphon tube 7-9 times as long as its basal diameter which is about 2 times the apical diameter; upper head hair double or triple, lower double or rarely single; siphon tufts relatively short.......... *apicalis* Adams
— Siphon tube slightly to considerably less than 7 times its basal diameter which is distinctly less than 2 times the apical diameter; siphon tufts often 1/3 to 1/2 as long as siphon 9
9. Lower head hair double, upper triple..... *reevesi* Wirth
— Lower head hair single, upper usually double....... 10
10. Abdominal segment 4 much paler than 3 or 5; spicules near dorsal apex of saddle relatively slender; upper head hair double or rarely triple .... *boharti* Reeves and Brookman
— Abdominal segments rather evenly pigmented; spicules near dorsal apex of saddle becoming coarse; upper head hair double or rarely single........... *territans* Walker

REFERENCES

AITKEN, T, H. G.
1945. Studies in the Anopheline complex of western North America. Univ. Calif. Publ. Entom., 7:273-364.
COOK, E. F.
1956. The Nearctic Chaoborinae. Univ. Minn. Agric. Exp. Sta., Tech. Bull. 218, pp. 1-102.
DEONIER, C. C.
1943. Biology of the immature stages of the Clear Lake gnat (Diptera, Culicidae). Ann. Ent. Soc. Amer., 36: 383-388.
DYAR, H. G.
1928. The mosquitoes of the Americas. Publ. Carnegie Instn., 387:1-616.
DYAR, H. G., and R. C. SHANNON
1924. The American Chaoborinae. Insec. Inscit. menst., 12:201-216.
FREEBORN, S. B., and R. M. BOHART
1951. The mosquitoes of California. Bull. Calif. Insect. Surv. 1:25-78.
HERMS, W. B.
1937. The Clear Lake Gnat. Univ. Calif. Bull. 607, 22 pp.
MATHESON, R.
1944. Handbook of the mosquitoes of North America, 2d ed, 314 pp.
PETERSON, A.
1951. *See* Diptera references.
ROSS, EDWARD S., and H. RADCLYFFE ROBERTS
1943. Mosquito Atlas, Part I. Ent. Soc. Amer., 44 pp.
YAMAGUTI, S., and W. J. LA CASSE
1951. Mosquito fauna of North America, Parts I-V.

Family DIXIDAE

These small, delicate flies (fig. 14:18*a*) are sometimes included in the Culicidae, from which they differ in

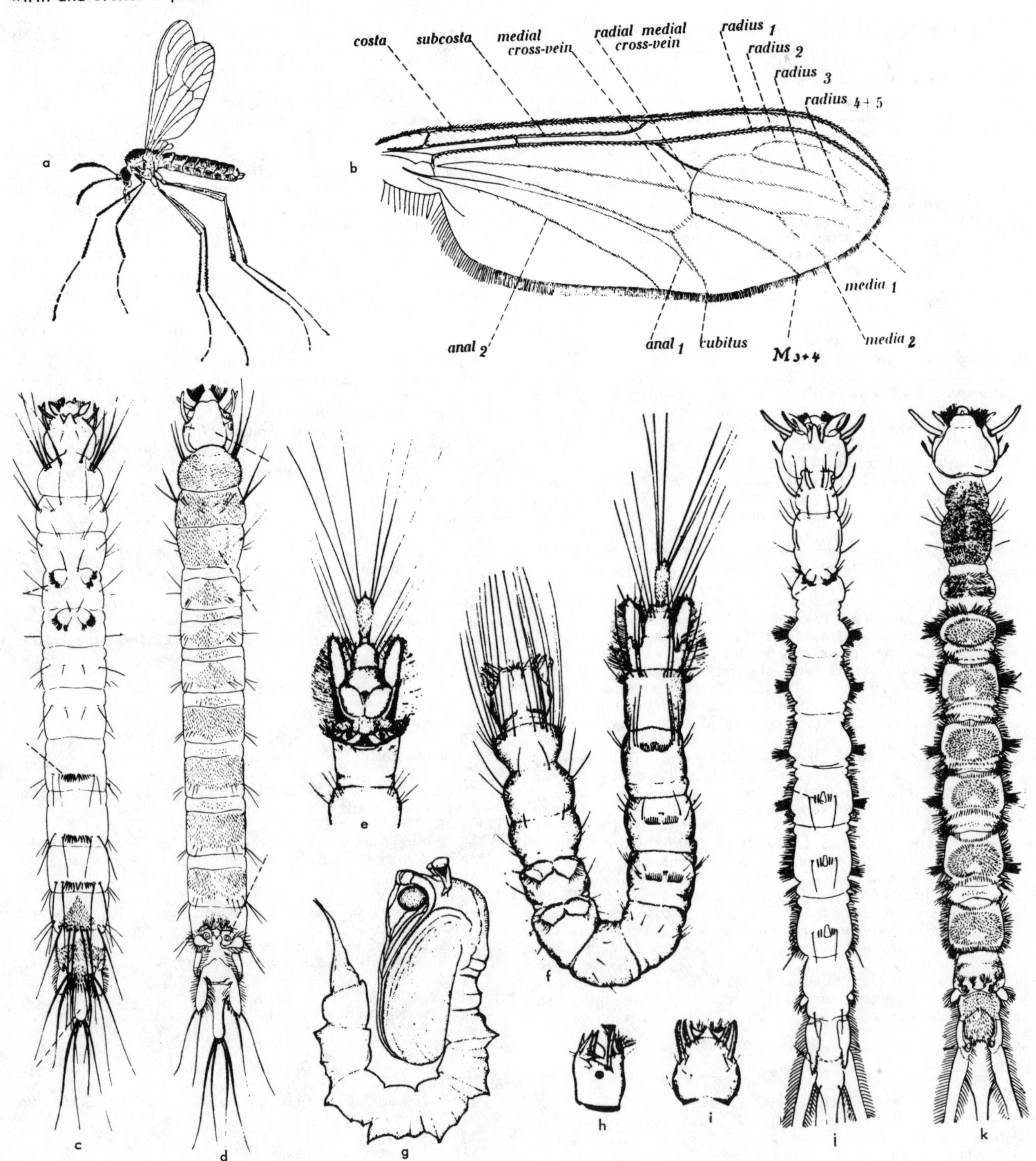

Fig. 14:18. Dixidae. *a*, *Dixa* sp., female; *b*, *Dixa brevis* Garr., wing; *c,d*, *Dixa californica* Joh. larva: c, ventral view; *d*, dorsal. *e-i*, *Dixa aliciae* Joh.: *e*, caudal end of larva, dorsal view; *f*, larva, ventral; *g*, pupa; *h*, head of larva, lateral view; *i*, same, dorsal. *j,k*, *Dixa chalonensis* Now., larva: *j*, ventral view; *k*, dorsal (*a*, Kellogg, 1905; *b-d*, *j,k*, Nowell, 1951; *e-i*, Johannsen, 1934).

their lack of scales and the curvature of the apical wing veins. The larvae resemble those of anopheline mosquitoes, except that the thoracic segments are not fused into a broadened body, one or two pairs of prolegs are present on the abdomen, and the caudal appendages are quite different.

The adult flies are usually found resting on the shaded side of rocks a few inches above the surface of streams or ponds, always facing upward. It is not known whether there is any feeding in the adult stage. The dixids do very little flying during the day, but in the evening the males may form small mating swarms, which are apparently oriented opposite vegetation overhanging the water. Mating can occur in flight or on the rocks bordering the stream. The eggs are laid in a mass of jelly on a solid substratum above the water line.

The larvae always occur along the downstream margin of rocks or floating branches. They may be divided into at least two groups on the basis of their habits. Those of the subgenus *Meringodixa* lie on the surface of the water with their bodies at right angles to the substratum, the caudal end being in the meniscus and the head directed away from the substratum. The larvae of *Dixa* and *Paradixa* assume the typical inverted-U position at or just above the water surface with the head and caudal end in the water. The latter group can crawl about on the rocks by sliding on a film of water pushed ahead of them. The larval food of the Dixidae is believed to consist of microscopic organisms in the surface film. The pupae (fig. 14:18*g*) are found attached by their sides to the substratum as much as two inches above the water.

Recently Nowell (1951) published a large study of the family in western North America in which he recognized three subfamilies and eight genera. The larvae of the California species can easily be divided into three morphological groups, but the adults of *Dixa* and *Meringodixa* are difficult to separate. Therefore it does not seem advisable to recognize them as full genera at this time. The species are all very similar in appearance, and many of them can be separated only by a study of the male genitalia. The California species are not well enough known for keys to be presented, particularly since, in a period of two years Johannsen (1923, 1924), Dyar and Shannon (1924), and Garrett (1924) published papers on western species, and considerable synonymy is involved.

Keys to California Subgenera of Dixa

Larvae

1. One pair of abdominal prolegs (fig. 14:18*j*) .... *Meringodixa* Nowell 1951
— Two pairs of abdominal prolegs (fig. 14:18*c*) 2
2. Dorsum of abdomen bare or nearly so (fig. 14:18*d*) .... *Paradixa* Tonnoir 1924
— Dorsum of abdomen with rosettes of hairs on segments 2 to 7.......... *Dixa* Meigen 1818

Adults

1. Sternopleura bare; 1st flagellar segment of antenna cylindrical.......... *Paradixa*
— Sternopleura with at least a few hairs; 1st flagellar segment of antenna fusiform 2
2. First flagellar segment of antenna 6 times as long as broad .... *Meringodixa*
— First flagellar segment of antenna 4 times as long as broad .... *Dixa*

California Species of Dixa

Species	Distribution
Dixa (Dixa) hegemonica Dyar and Shannon 1924.	Humboldt County
Dixa (Dixa) modesta Johannsen 1903.	Riverside County
Dixa (Dixa) rhathyme Dyar and Shannon 1924.	Lake, Santa Clara, Ventura counties
Dixa (Dixa) xavia Dyar and Shannon 1924.	Lake County to Santa Clara County
Dixa (Meringodixa) chalonensis Nowell 1951.	San Benito County
Dixa (Paradixa) aliciae Johannsen 1924.	Santa Clara and Merced counties
Dixa (Paradixa) californica Johannsen 1923.	Humboldt to Monterey and Tulare counties
Dixa (Paradixa) somnolenta Dyar and Shannon 1924.	Tulare County

REFERENCES

DYAR, H. G. and R. C. SHANNON
1924. Some new species of American *Dixa* Meigen. Insec. Inscit. Menst., 12:193-201.
GARRETT, C. B. D.
1924. New American Dixidae. Privately published at Cranbrook, B.C.
JOHANNSEN, O. A.
1923. North American Dixidae. Psyche, 30:52-58.
1924. A new species of *Dixa* from California. Psyche, 31:45-46.
1934. *See* Diptera references.
KELLOGG, VERNON L.
1905. American Insects.
NOWELL, W. R.
1951. The dipterous family Dixidae in western North America (Insecta: Diptera). Microentomology, 16:187-270, figs. 74-88.

Family SIMULIIDAE

Black flies, sometimes known as buffalo gnats or turkey gnats, are small, usually dark, compactly built flies (fig. 14:19*i,j*) with a rounded back and short, broad wings, with the heavy veins concentrated near the anterior margin of the wing. The antennae are nine- to eleven-segmented, rather short and stout with no long hairs. The females of most of the species are bloodsucking, and certain species attack man readily, but others prefer domestic animals or birds. They feed during the daylight hours and sometimes into the night, and adults, particularly the males, are collected readily at light traps. Many species will fly considerable distances from the breeding areas.

The larvae (fig. 14:19*a*) are found in flowing water on stones, vegetation, or other objects, usually in

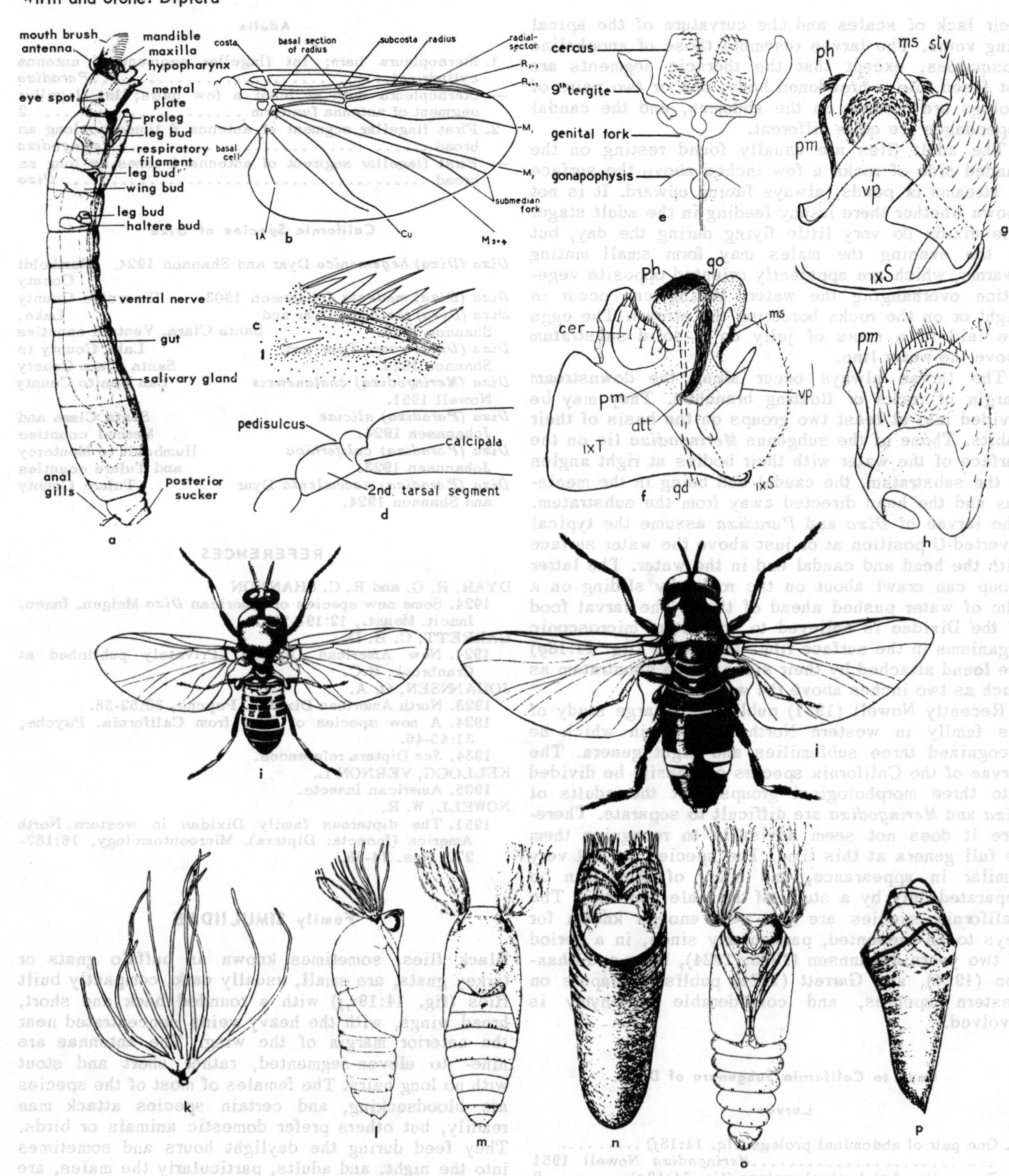

Fig. 14:19. Simuliidae. *a,c-e, Simulium ornatum* Mg.: *a,* larva; *c,* detail of anterior wing margin; *d,* part of hind tarsus; *e,* female genitalia. *b, Prosimulium hirtipes* (Fries), wing; *f-h, Simulium salopiense* Edw., male genitalia: *f,* lateral view with near basistyle removed, other indicated by a dotted line; *g,* ventral view; *h,* inner view of left basistyle (att., point of attachment of paramere to basistyle; cer., cercus; gd., genital duct; go., genital opening; ms., median sclerite; ph., parameral hooks; pm., paramere; sty., dististyle; vp., ventral plate; ixS., ninth sternite; ixT, ninth tergite). *i-p, Simulium arcticum* Mall.: *i,* female; *j,* male (not same scale); *k,* respiratory organ of pupa; *l,* pupa, lateral view; *m,* same, dorsal; *n,* cocoon, dorsal view; *o,* pupa, ventral view; *p,* cocoon, lateral view (*a-e,* Smart, 1944; *f-h,* Freeman, 1950; *i-p,* Cameron, 1922).

the swiftest part of the stream. Some are found in the larger streams, often in immense masses, making a mosslike covering over rocks, but many others are found in the more rapidly flowing parts of very small water courses. They attach by means of a posterior sucker, with their heads pointed downstream, and they move about by a looping movement involving this posterior sucker and a median prothoracic leg, and also by means of silken threads which they can let out, permitting them to move rapidly without being swept completely away. In all but the far northern genus *Gymnopais*, the larva possesses a pair of large mouth brushes which are spread out to strain the water for food particles. The body is soft and sacklike, widest near the posterior third, and when fully grown, the darkened histoblasts of the developing respiratory organs of the pupa can be seen at the sides of the thorax.

The larva spins a cocoon (fig. 14:19*n,p*), attaching it to the surface on which it rests. In *Prosimulium* and some *Cnephia* species this is loose and irregular and may cover the respiratory organs anteriorly. In most other species, however, this cocoon is a well-defined wall-pocket or boot-shaped structure, often possessing characters of value in specific identification. The pupa lies in the cocoon with the variously formed thoracic respiratory organs partly protruding from the anterior downstream opening of the cocoon. The adult emerges in an air bubble if the cocoon is still under water at the time of emergence, and the adults can fly immediately.

The adults may occur in huge numbers and since the reaction to the bite is often very severe, they can cause serious losses to livestock, wild mammals, and birds. Some species carry onchocerciasis, a filarial disease of man in the tropics, and others transmit blood diseases of birds.

Keys to the North American Genera of Simuliidae

Adults

1. A bulla behind eye laterally; scutum with stout erect hairs but no fine recumbent hairs ..... *Gymnopais* Stone
— No bulla behind eye; scutum usually with fine recumbent hairs but never with stout, erect ones 2
2. Macrotrichia of anterior wing veins hairlike, not intermingled with spiniform ones 3
— Macrotrichia of anterior wing veins mixed hairlike and spiniform (fig. 14:19*c*) 4
3. Vein R_1 joining costa at about middle of wing .... *Parasimulium* Malloch
— Vein R_1 joining costa well beyond middle of wing (fig. 14:19*b*) .... *Prosimulium* Roubaud
4. Second hind tarsal segment without a distinct notch (pedisulcus) dorsally near base; vein R with hairs dorsally .... *Cnephia* Enderlein
— Second hind tarsal segment with a distinct pedisulcus (fig. 14:19*d*); vein R usually without hairs dorsally .... *Simulium* Latreille

Pupae

1. Dorsum of abdomen with no hooks; sternites 4-6 each with about 10 hooks in more than one transverse row; almost no cocoon.................. *Gymnopais* Stone
— Dorsum of abdomen with hooks on some of the segments; if sternites 4-6 have more than 4 hooks these are in a single transverse row 2
2. Strong terminal hooks present; cocoon irregular without any clearly defined anterior margin; respiratory filaments if fewer than 14 not arising from more than 2 main trunks; if more than 14 not arising from a rounded knob on a short petiole .... *Prosimulium* Roubaud
— Terminal hooks usually very weak or absent and cocoon well formed and with a clearly defined anterior margin; if not, the respiratory organ of about 11 filaments arising at a considerable distance from the base from 2 main trunks, or the filaments more numerous, arising from a rounded knob on a short petiole 3
3. Either terminal hooks well developed and cocoon poorly formed or cocoon with a broad anteroventral collar so that it is boot-shaped and tergites 5-7 each have an anterior row of small spines .... *Cnephia* Enderlein
— Terminal hooks weak or absent; cocoon well developed; either cocoon is boot-shaped with a series of loops forming the anterior margin (fig. 14:19*p*) and tergites 5-7 are free of spines anteriorly, or the cocoon is wall-pocked-shaped.............. *Simulium* Latreille

The pupa of *Parasimulium* is not known.

California Species of Simuliidae

Genus *Parasimulium* Malloch

This is an unusual genus known only from a single male specimen collected at Bair's Ranch, Humboldt County, and described by Malloch (1914) as *Parasimulium furcatum*. Discovery of the immature stages would be of very great interest.

Genus *Prosimulium* Roubaud

Species of this genus appear to have only one emergence in a season, and most of them appear rather early in the spring or summer. The pupae are often found in dense masses on sticks and stones. There is no evidence that any of the California species are annoying as pests of man, as is *Prosimulium hirtipes* in the North and East. There are at least four undescribed species in California in addition to those listed here. These will not run to any of the species in the pupal key here given, but cannot be separated from all those given in the adult keys.

Keys to California Species of Prosimulium

Females

1. Claws bifid; that is, with the basal swelling produced into a strong, acute tooth; mesoscutum predominately reddish brown; Calaveras County ....*onychodactylum* Dyar and Shannon 1927
— Claws simple, with not more than a basal swelling or with a very minute basal tooth adjacent to the basal swelling.. 2
2. Integument orange; Plumas and Marin counties to Tulare County .... *fulvum* (Coquillett) 1902
— Integument dark brown to black 3
3. Antenna with 9 segments; Plumas County.......... .... *novum* Dyar and Shannon 1927
— Antenna with 11 segments 4

4. Stem vein with pale yellow hairs; Del Norte and Modoc counties to San Luis Obispo County ....
.... *exigens* Dyar and Shannon 1927
Plumas, El Dorado and Mariposa counties..........
............................*hirtipes* (Fries) 1824
Nevada County...................*travisi* Stone 1952
— Stem vein with dark hairs (Trinity, Shasta, and Sierra counties) ....*dicum* Dyar and Shannon 1927

Males

1. Integument orange................*fulvum* (Coquillett)
— Integument dark brown to black.................. 2
2. Antenna with 11 segments 3
— Antenna with 9 segments (presumably) ....
..........................*novum* Dyar and Shannon
3. Hind femora, at least, yellow ....
..................*onychodactylum* Dyar and Shannon
travisi Stone
— Hind femora brown or blackish.................... 4
4. Ventral plate with a compressed median keel; parameres large, flattened, sclerotized plates, bluntly rounded apically.................. *exigens* Dyar and Shannon
— Ventral plate centrally with a down-curving lip, but not forming a compressed median keel; parameres much smaller, somewhat tapering distally...............
..........................*dicum* Dyar and Shannon
hirtipes (Fries)

Pupae

1. Respiratory organ consisting of 2 stout, divergent, distinctly ringed tubes, on a short petiole, from which slender filaments arise ....
..................*onychodactylum* Dyar and Shannon
— Respiratory organ not so formed 2
2. Respiratory filaments 16 3
— Respiratory filaments more than 40 4
3. Filaments of respiratory organ closely clumped; dorsum of thorax rugose .... *travisi* Stone
— Filaments of respiratory organ with the main trunks divergent; dorsum of thorax not rugose ....
...............................*fulvum* (Coquillett)
hirtipes (Fries)
4. Thorax strongly rugose ......*dicum* Dyar and Shannon
— Thorax not rugose.........*exigens* Dyar and Shannon

The pupa of *Prosimulium novum* is not known.

Genus *Cnephia* Enderlein

Species of this genus have a biology rather similar to those of *Prosimulium*, there apparently being only one emergence in the year in the early summer. Two of the species will sometimes be found annoying to man, although *osborni* was collected from a grouse.

Keys to the California Species of Cnephia

Females

1. Tarsal claws with a distinct tooth or strong basal projection 2
— Tarsal claws simple; Plumas, Placer, and El Dorado counties .... *mutata* (Malloch) 1914
2. Radial sector with a short but distinct anterior branch 3
— Radial sector unbranched; Plumas, Placer, and El Dorado counties ...... *minus* (Dyar and Shannon) 1927
3. Wing length from humeral cross vein 4 mm.; basal tooth of claw long, more than 1/2 length of claw; palpal sense organ 0.3 length of segment; Plumas and Sierra counties..................... *stewarti* Coleman 1953
— Wing length from humeral cross vein less than 3.5 mm.; basal tooth of claw not more than half as long as claw; palpal sense organ 0.5 length of segment ....
.................*osborni* (Stains and Knowlton) 1943

Males

1. Radial sector with a short but distinct anterior branch .... *stewarti* Coleman (presumably also *osborni* Stains and Knowlton)
— Radial sector unbranched 2
2. Dististyle with 2 rather blunt terminal teeth; calcipala large, rounded apically .... *mutata* (Malloch)
— Dististyle with only a single terminal tooth; calcipala smaller, the tip more acute .. *minus* (Dyar and Shannon)

Pupae

Pupae of only two of the California species are known. The pupa of *mutata* has usually twelve respiratory filaments arising irregularly from two main trunks, and terminal hooks are present on the abdomen. That of *stewarti* has six swollen respiratory filaments in three pairs, one pair pointing forward, the other two somewhat petiolate and pointing backward.

Genus *Simulium* Latreille

Of the nineteen species here included in the genus from California, four (*aureum, canonicolum, latipes*, and *pugetense*) belong in the subgenus *Eusimulium;* two (*argus* and *vittatum*) to the subgenus *Neosimulium;* four (*bivittatum, griseum, trivittatum*, and *venator*) to the subgenus *Lanea;* one (*virgatum*) to the subgenus *Dyarella;* one (*canadense*) to the subgenus *Hearlea;* and the remaining seven belong to the typical subgenus. There are at least three undescribed species from California not included in the following keys. Since they are not all known in both sexes and the pupa, where the unknown sex or pupa will run in the following keys is not known, so caution should be used when attempting to determine species from these keys.

There is a rather wide variety of habitats and habits within the genus, and much remains to be learned about seasonal and geographical distribution and feeding habits. Species of the subgenus *Lanea* rather typically inhabit desert regions.

Keys to California Species of Simulium

Females

1. Vein R with hair dorsally......................... 2
— Vein R without hair dorsally 4
2. Postscutellum with a patch of yellow appressed scales; scape and pedicel of antenna pale brown; Siskiyou and Lassen counties: Contra Costa and Mono counties to Los Angeles and San Bernardino counties.........
.... *aureum* Fries 1824
— Postscutellum without scales; antenna entirely dark 3
3. Stem vein with dark hair which may appear coppery in reflected light; Plumas County....*latipes* Meigen 1804

Humboldt and Modoc counties to San Mateo and Mariposa counties .... *pugetense* Dyar and Shannon 1927
— Stem vein with light yellow hair; Placer and San Joaquin counties to Inyo County .... *canonicolum* Dyar and Shannon 1927
4. Claws simple 5
— Claws with a basal projection or subbasal tooth 13
5. Frons and terminal abdominal tergites shining black or brown 6
— Frons and terminal abdominal tergites distinctly pollinose 8
6. Fore coxae darkened, brown; Inyo and Ventura counties to San Diego County... *jacumbae* Dyar and Shannon 1927
— Fore coxae clear yellow 7
7. Hairs on stem vein and base of costa pale; Plumas County to El Dorado County .... *venustum* Say 1823
— Hairs on stem vein and base of costa dark; Plumas County to Monterey County..................... .... *tuberosum* (Lundstroem) 1911
8. Mesonotum with 1 or more distinct, rather broad, straight stripes; if 1 it may be rather diffuse and not reach scutellum 9
— Mesonotum unstriped or the stripes very narrow and some of them not straight 11
9. A single median orange brown to black stripe; Mono and Inyo counties ..... *venator* Dyar and Shannon 1927
— Seven alternating stripes of contrasting color 10
10. Darker stripes of scutum orange; lateral dark spots of dorsum of abdomen absent or indistinct on most of the segments, never as prominent as the median dark sclerites; ventral projections of paraprocts not long enough to cross tips when in normal position (Monterey and Fresno counties to Riverside County) .... *bivittatum* Malloch 1914
— Darker stripes of scutum dark brown to blackish; dorsum of abdomen with pronounced dark lateral spots on several of the segments, nearly as dark as the median sclerites; ventral projections of paraprocts distinctly crossing each other when in normal position (Butte County to Santa Barbara and San Bernardino counties) .... *trivittatum* Malloch 1914
11. Predominantly ash-gray and black species, the thorax with a pattern of blackish stripes on a gray ground; coxae gray 12
— Predominantly yellow or yellowish gray; thorax with almost no pattern; fore coxae clear yellow; San Bernardino, Riverside, and Imperial counties .... *griseum* Coquillett 1898
12. Arm of genital fork with a somewhat darkened external process; the internal process smaller and paler and removed from one on other side; Modoc County to Monterey and San Diego counties .... *vittatum* Zetterstedt 1838
— Arm of genital fork with no external process; the internal process large, pale, rather close to one of other side; Siskiyou County to San Diego County .... *argus* Williston 1893
13. Claws each with a strong basal projection; Butte County .... *meridionale* Riley 1887
— Claws each with a small subbasal tooth 14
14. Fore coxae dark; Plumas County to Riverside County .... *piperi* Dyar and Shannon 1927
— Fore coxae yellow 15
15. Frons shining or subshining 16
— Frons opaque, covered with pollen 17
16. Mesoscutum with 3 lines, the median one straight and slender, the lateral ones curved and somewhat wider; Mariposa and Inyo counties ...... *hunteri* Malloch 1914
— Mesoscutum not lined (fig. 14:19*i*); Siskiyou County to Santa Barbara County .... *arcticum* Malloch 1914
17. Anterior gonapophyses elongate, their inner margins subparallel; hairs of paraproct long and stout; Siskiyou County to Riverside County ... *virgatum* Coquillett 1902
— Anterior gonapophyses short, their inner margins concave; hairs of paraproct short and rather slender; Siskiyou County to San Diego County .... *canadense* Hearle 1932

Males

1. Radial vein with hair dorsally 2
— Radial vein bare dorsally 4
2. Postnotum with appressed yellow scales; ventral plate very narrow and acutely pointed medially, the basal arms strongly divergent .... *aureum* Fries
— Postnotum without scales; ventral plate broad 3
3. Dististyle, when viewed from end, showing a flattened, shiny, triangular area, 1 corner of this forming an inner lobe .... *latipes* Meigen .... *pugetense* (Dyar and Shannon)
— Dististyle tapering evenly to a point, not truncate in this manner .... *canonicolum* (Dyar and Shannon)
4. Dististyle short and stout with 3 or more teeth 5
— Dististyle with only 1 or 2 teeth, or no teeth, often longer 6
5. White submedian area of scutum usually extending back to white prescutellar area; dististyle subquadrate, an obtuse rounded angle between the lateral and apical margins; teeth small and set close together .... *argus* Williston
— Pale submedian areas of scutum fading out before reaching prescutellar area; dististyle subtriangular, the apicolateral margin a continuous curve; teeth larger and set farther apart .... *vittatum* Zetterstedt
6. The white submedian areas of scutum visible in an anterior view extending back to white prescutellar area .... *trivittatum* Malloch
— These white submedian areas, if present, not reaching white or denuded prescutellar area although the dark lines of this aspect may be white when viewed posteriorly 7
7. Dististyle flat, quadrangular, with a distal internal angle more or less prolonged toward the median line; dististyle smaller than the basistyle 8
— Dististyle more or less cylindrical, or if flattened distinctly longer than wide 10
8. Median area of scutum broadly orange except for anterior part .... *venator* Dyar and Shannon
— Median area of scutum not orange 9
9. Thorax mostly velvety black; white submedian marks extending back from anterior margin of scutum about 1/3 of distance to scutellum .... *bivittatum* Malloch
— Thorax gray; white submedian marks on front of scutum not prolonged posteriorly .... *griseum* Coquillett
10. Dististyle with lateral angles which give a sinuous appearance, not more than 3 times as long as wide and without basal projection; ventral plate rather broad with a strong, narrow median projection nearly 1/2 as long as dististyle, and lateral, rounded rectangles .... *virgatum* Coquillett
— Dististyle with lateral margins more regular and/or ventral plate not so formed 11
11. Dististyle more than 4 times as long as wide, narrowed at basal 1/3; without basal process or pronounced angle; ventral plate semicircular in shape with a median notch .... *canadense* Hearle
— Dististyle not more than 3 times as long as wide; not narrowed at basal 1/3; if longer, then a basal process present; ventral plate of various shapes but not semicircular 12
12. Dististyle with a stout spine, sclerotized lobe, or distinct tubercle at base internally................ 13
— Dististyle without a stout spine or distinct tubercle at the base 16
13. Base of dististyle with a stout spine or horny projection internally 14
— Base of dististyle with a rounded lobe bearing short, stout, spinelike hairs internally .... *tuberosum* (Lundstroem)
14. Basal prongs of ventral plate with short lateral projections; apex of adminiculum hyaline, the sides set off by a notch, hairy .... *piperi* Dyar and Shannon
— Basal prongs of ventral plate without lateral projections; if apex of ventral plate is smooth and pale it is much longer and narrower 15

15. Ventral plate with a prolonged hyaline tip .... *hunteri* Malloch
— Ventral plate not prolonged apically, unicolorous; diststyle slender; basal tooth at extreme base, extending basad .... *jacumbae* Dyar and Shannon
16. Ventral plate more or less compressed, with denticles on margin 17
— Ventral plate broadly rounded without denticles on margin .... *meridionale* Riley
17. Ventral plate narrow, in the shape of an inverted Y, with ventral process or keel .... *arcticum* Malloch
— Ventral plate broader, tooth-shaped, without ventral process .... *venustum* Say

Pupae

1. Respiratory organ consisting of a large ringed club and 2 smooth, curved basal projections .... *canadense* Hearle
— Respiratory organ consisting of slender branched or unbranched filaments 2
2. Respiratory filaments 4 3
— Respiratory filaments more than 4 5
3. Anterior margin of cocoon with a rather long median projection .... *latipes* Meigen
— Anterior margin of cocoon nearly or quite straight 4
4. Dorsal respiratory filament strongly divergent from other 3 .... *aureum* Fries
— Dorsal respiratory filament not strongly divergent from other 3 .... *pugetense* Dyar and Shannon
5. Respiratory filaments 6 6
— Respiratory filaments more than 6 7
6. Filaments all arising rather close to base (front of cocoon open ventrally with no transverse band below opening) .... *venustum* Say .... *tuberosum* (Lundstroem)
— At least 2 filaments arising at a considerable distance from base .... *trivittatum* Malloch
7. Front of cocoon with a basal collar at distinct angle to surface so that the cocoon is boot-shaped (fig. 14:19*p*) 8
— Front of cocoon with a narrow collar, little raised above the surface, or the sides not touching antero-ventrally, so the cocoon is wall-pocket-shaped 9
8. Respiratory filaments 8, closely clumped ....*virgatum* Coquillett
— Respiratory filaments 12, more divergent (fig. 14:19*k*) .... *arcticum* Malloch
9. Cocoon with a median projection from the anterior margin .... *piperi* Dyar and Shannon
— Cocoon with no long median projection from anterior margin 10
10. Respiratory filaments 8 .... *bivittatum* Malloch
— Respiratory filaments 10 or more 11
11. Respiratory filaments 10.............. *argus* Williston
— Respiratory filaments 16 or more..................... 12
12. Respiratory filaments 16 .... *vittatum* Zetterstedt
— Respiratory filaments more than 16 13
13. Respiratory filaments 22-26......... *meridionale* Riley
— Respiratory organ of more than 100 fine filaments ... . *hunteri* Malloch

The pupae of *canonicolum, griseum, venator,* and *jacumbae* are not certainly known.

REFERENCES

CAMERON, A. C.
1922. The morphology and biology of a Canadian cattle-infesting black fly, *Simulium simile* Mall. (Diptera, Simuliidae). Bull. Canad. Dept. Agric., 5:3-26.
DYAR, H. G., and R. C. SHANNON
1927. The North American two-winged flies of the family Simuliidae. Proc. U.S. Nat. Mus. 69:1-54.
FREEMAN, PAUL
1950. The external genitalia of male Simuliidae. Ann. Trop. Med. Parasit., 44:146-152.
HEARLE, E.
1932. The blackflies of British Columbia (Simuliidae, Diptera). Proc. Ent. Soc. B.C., 29:5-19.
MALLOCH, J. R.
1914. American black flies or buffalo gnats. U.S.D.A. Bur. Ent. Tech. Ser., 26:1-70.
SMART, JOHN
1944. The British Simuliidae, with keys to the species in the adult, pupal and larval stages. Freshwater Biol. Assoc. Brit. Empire, Sci. Bull. 9:1-57.
STAINS, G. S., and G. F. KNOWLTON
1943. A taxonomic and distributional study of Simuliidae of western United States. Ann. Ent. Soc. Amer., 36:259-280.
STONE, A.
1952. The Simuliidae of Alaska. Proc. Ent. Soc. Wash., 54:69-96.
VARGAS, L., A. MARTINEZ PALACIOS, and A. DIAZ NAJERA
1946. Simulidos de Mexico. Rev. Inst. Salub. y Enferm. Trop., 7:101-192, 25 pls.

Family TENDIPEDIDAE (=CHIRONOMIDAE)

The "midges" or "nonbiting midges" of the family Tendipedidae form one of the most important single groups in the ecology of aquatic environments. Midge larvae are most abundant in the shallow water areas of lakes, ponds, and streams favored by a heavy growth of aquatic plants. However, in such areas they are preyed upon more heavily by larger insects and by fish so that more adult midges may actually emerge from the deeper regions. Sandy bottoms are also much less productive than soft mucky bottoms on which most of the bottom-dwelling species are dependent for building their tubes. Nearly all students of fish food habits agree on the importance of midge larvae, pupae, and adults in the diet of fresh water fish (Townes, 1938); these stages are eaten by the young of many species of fish and may be almost the exclusive food of adult bass, trout, and whitefish. The aquatic immature stages are also eaten by many shore and water birds, especially those which are bottom feeders. Although the adults cannot bite, they may emerge in such numbers that they become distinct pests around lake and stream-side resorts.

Adult midges are readily distinguished by their long antennae which are usually plumose in the males (fig. 14:29*i*), shorter in the females (fig. 14:29*j*); by the absence of a "V-suture" on the mesonotum, but by the presence of a distinct median longitudinal furrow or keel on the postscutellum (figs. 14 28*a,k;* 14:29*i*); by the wing with the costa not continuing around the wing margin past the wing tip, the media not forked, but the mediocubitus forked at about the middle of the wing; and by the reduced mouth parts.

The larvae are distinguished by a complete head capsule which is not retractable within the thorax; by mandibles opposed; by anterior and posterior, often long and slender, prolegs being nearly always present (fig. 14:22*j*), on the prothorax and last abdominal segment; and by the complete absence of functional spiracles. The pupae are free, although sometimes

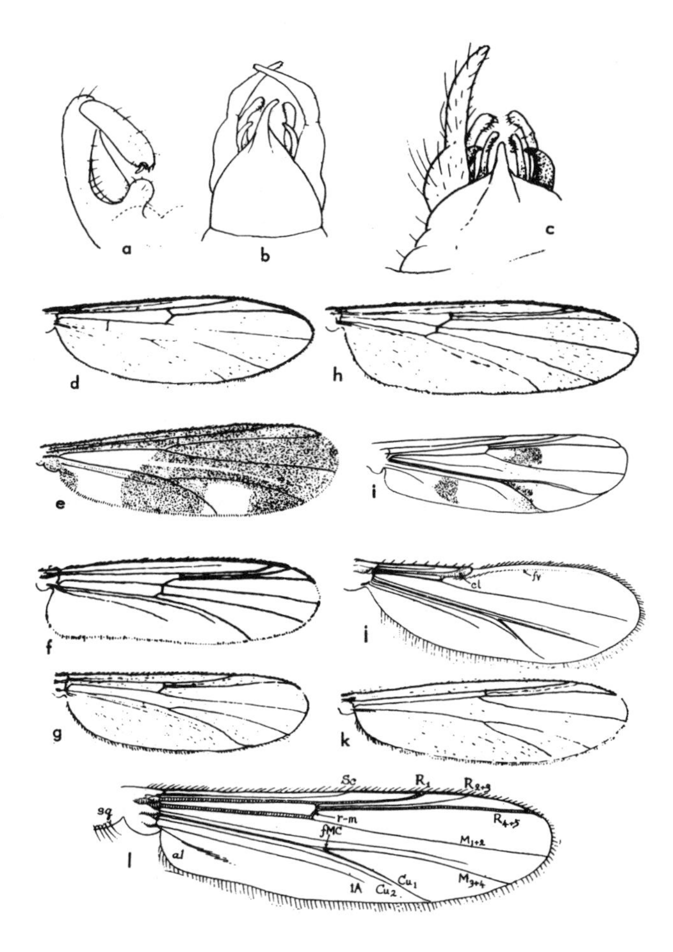

Fig. 14:20. Tendipedidae adults. a-c, male genitalia: a, *Cricotopus;* b, *Tendipes;* c, *Calopsectra;* d-l, Wings: d, *Boreochlus;* e, *Chasmatonotus;* f, *Diamesa;* g, *Metriocnemus;* h, *Pentaneura;* i, *Polypedilum;* j, *Corynoneura* (cl, clavus; fv, false vein); k, *Calopsectra;* l, *Cricotopus* (al, anal lobe; fMC, mediocubital fork; sq, squama) (a-i,k, Johannsen, 1952; j,l, Tolinaga, 1936).

with the last larval skin loosely attached (fig. 14:35*c*), sometimes formed in a silken cocoon or gelatinous case (fig. 14:27*g*); the leg sheaths are free from the body and do not project far beyond the tips of the wing cases; usually there is a functional prothoracic respiratory organ; the body is rarely strongly arched (fig. 14:22*i*), and ends caudally in a pair of pointed (fig. 14:22*h*) or oval processes or in a two-lobed, unribbed, swimming paddle which often bears a fringe of hairs or slender filaments (fig. 14:34*k*). The eggs are usually enclosed in a gelatinous string which may in turn be deposited in a gelatinous mass whose structure may vary with the species in its spiral arrangement, convolutions, and the like (figs. 14:28*c,e;* 14:29*h;* 14:35*i*).

The life history may vary considerably; some species breeding in warm water may have a number of generations a year, whereas the same species breeding in cold lake bottoms may require a year for emergence; still other species may have but one generation a year. Feeding habits differ widely; most of the Pelopiinae and some Tendipedini are predaceous, chiefly on other tendipedid larva, but in general other species are detritus or plankton feeders. Walshe (1951) has reported in detail the habits of several genera which habitually spin a salivary net (fig. 14:26*o*) to catch plankton and detritus, devouring the net and its contents at regular intervals. She (1950) has also investigated the function of hemoglobin in the respiration of bloodworms, concluding that it may enable the larvae to continue active feeding under relatively anaerobic conditions. It may act in oxygen transport at very low oxygen concentrations, thereby enabling continued respiratory irrigation, and it may greatly increase the rate of recovery from periods of oxygen lack, making such recovery possible even under adverse respiratory conditions.

The larvae pupate in their larval tunnels or cases and come to the surface of the water only slightly before the imago emerges. Advantage may be taken of this fact to place floating conical traps at the water surface over the habitat in which the larvae breed, to capture emerging adults together with their pupal skins. Thence by bottom-dredging larvae and pupae with larval skins attached, a complete, correlated series of specimens can be obtained by which each species can be identified from its respective habitat and its comparative abundance determined.

Although the importance of tendipedids in freshwater biology has long been recognized in Europe, and a fairly complete system of classification of all stages has been worked out by Edwards (1929), Coe (1950), Chernovsky (1948), Thienemann (1908 to date) and his students Lenz (1936), Pagast (1947), Strenzke (1950), and Zavrel (1947), our knowledge for the rest of the world is woefully lacking. In North America the classic descriptive work of Malloch (1915) is sadly out of date, although Johnnsen (1952) has almost singlehandedly fought to keep our nomenclature in line with that of European workers and has given us, in his aquatic memoirs (1937), a remarkable set of papers on our known immature stages in North America. Only the North American adults of the tribe Tendipedini have been given modern revision (by Townes, 1945) although Hauber, Walley, and Roback have published some smaller papers on limited groups.

Keys to the North American Genera of Tendipedidae

Adults

1. Posterior cross vein m-cu present (fig. 14:20*h*) 2
— Posterior cross vein m-cu absent (fig. 14:20*g*) 4
2. Second branch of radius present and forked (figs. 14:1*c*; 14:20*h*), or absent altogether leaving but 2 radial branches (fig. 14:20*d*) 3
— Second branch of radius present and simple, not forked (fig. 14:20*f*) (DIAMESINAE) 14
3. Second branch of radius forked (fig. 14:20*h*), or if absent the anterior and posterior branches of radius are almost in contact (PELOPIINAE) 6
— Second branch of radius entirely absent, the anterior and posterior branches of radius well separated (fig. 14:20*d*) (PODONOMINAE) 11
4. Front basitarsus shorter than front tibia (fig. 14:29*i*); male dististyle folded inward (fig. 14:20*a*); middle and hind tibial spurs not modified into combs 5

— Front basitarsus longer than front tibia (fig. 14:28*b*); male dististyle directed rigidly backward (fig. 14:20*b,c*); combs present on middle and hind tibiae (fig. 14:21*g-n*) (TENDIPEDINAE) 32
5. Pronotum not or slightly notched anteriorly on median line; anepisternal suture well developed; male antennae normally plumose (HYDROBAENINAE) 18
— Pronotum widely divided into lateral lobes; anepisternal suture short or absent (fig. 14:30*j*); male antennae almost bare, never plumose (fig. 14:30*b*) (CLUNIONINAE) 27
6. Costa not produced or but very slightly produced beyond tip of the posterior radial branch (fig. 14:20*h*); wing hairy; female antenna with 12 or rarely 13 segments; 4th tarsal segment linear......... *Pentaneura* Philippi
— Costa distinctly produced beyond tip of posterior radial branch (fig. 14:1*c*); female antennae with 13 to 15 segments 7
7. Fourth tarsal segment linear; wing either hairy or bare 8
— Fourth tarsal segment cordiform (bilobed above), shorter than 1/5; wing bare 10
8. Mediocubital fork sessile; antennae of both sexes 15-segmented; wings more or less hairy, at least at tip .... *Anatopynia* Johannsen
— Mediocubital fork with petiole 9
9. Petiole of mediocubital fork measured from the cross vein not 1/3 as long as posterior branch; wing hairy; acrostichal hairs absent; wings spotted. *Pelopia* Meigen
— Petiole of mediocubital fork at least 1/2 as long as posterior branch (fig. 14:1*c*); wing hairy or bare; acrostichal hairs present .... *Procladius* Skuse
10. Petiole of mediocubital fork measured from the cross vein 1/4 or more as long as the posterior branch; minute acrostichal hairs present on median line of thorax .... *Clinotanypus* Kieffer
— Petiole of fork less than 1/6 as long as posterior branch or fork wholly sessile; no acrostichals ....*Coelotanypus* Kieffer
11. Large pulvilli present; empodium much-branched; mediocubitus with long petiole; eyes with short dense pile; antennae 15-segmented; hind tibiae without comb ... . *Trichotanypus* Kieffer
— Pulvilli absent 12
12. Mediocubital fork petiolate (fig. 14:20*d*); eyes bare; antennae 14-segmented; hind tibia with a single pubescent spur; outer spur absent, with a loose comb of about 6 bristles; empodium of moderate length...... ....*Boreochlus* Edwards
— Mediocubital fork sessile 13
13. Hind tibia without a definite comb on inner side at tip, both spurs long; 1st branch of radius long and slender throughout in both sexes; female antennae elongate, 15-segmented.........*Lasiodiamesa* Kieffer
— Hind tibial comb present, 1 spur short; 1st radial branch shorter, more or less swollen at tip in female; female antenna short with 8-12 segments ....*Podonomus* Philippi
14. Mediocubital fork just beyond posterior cross vein; eyes bare; 4th tarsal segment cylindrical ....*Prodiamesa* Kieffer
— Mediocubital fork just before posterior cross vein (fig. 14:20*f*) 15
15. Fourth tarsal segment more or less cordiform, shorter or at least not longer than the 5th 16
— Fourth tarsal segment cylindrical, longer than the 5th ....*Syndiamesa* Kieffer
16. Eyes finely pubescent; male antennae plumose .... *Diamesa* Meigen
— Eyes bare 17
17. Legs annulate with light and dark; male antennae not plumose ....*Heptagyia* Philippi
— Legs not annulate .... *Psilodiamesa* Kieffer
18. Radius fused with costa and not reaching 3/4 its length; a false vein running close to the anterior margin of the wing (fig. 14:20*j*) ..*Corynoneura* Winnertz
— Radius free; no false vein (fig. 14:20*l*) 19
19. Wing membrane with macrotrichiae, at least at tip (fig. 14:20*g*) 20
— Wing membrane bare 23
20. Cross vein r-m very long and oblique, appearing as the base of the posterior branch of the radius; male dististyles forked; small pulvilli present................ 21
— Cross vein r-m short (fig. 14:20*l*); male dististyles not forked 22
21. Mesonotum conically produced over the head in front; wings spotted.............. *Eurycnemus* van der Wulp
— Mesonotum not produced; wing unmarked. *Brillia* Kieffer
22. Pulvilli absent; wing hairs decumbent .... *Metriocnemus* van der Wulp
— Pulvilli present but small; wing hairs suberect, restricted to wing tip .... *Hydrobaenus* Fries (part)
23. Mesonotum with a longitudinal fissure; wings black with white markings in most species (fig. 14:20*e*) ....*Chasmatonotus* Loew
— Mesonotum without distinct longitudinal fissure..... 24
24. Palpi porrect, 3-segmented......*Symbiocladius* Kieffer
— Palpi flexible, 4-segmented 25
25. Fourth tarsal segment of at least hind leg cordiform, shorter than 5th .... *Cardiocladius* Kieffer
— Fourth tarsal segment linear 26
26. Dorsocentral hairs minute and decumbent; tibiae often with pale rings; terminalia often white; eyes densely pubescent ....*Cricotopus* van der Wulp
— Dorsocentral hairs stronger and suberect; tibiae not banded; eyes usually bare ....*Hydrobaenus* Fries (part)
27. Hind tarsi with the 2nd segment not longer than the 3rd; all tarsi with the 4th segment cylindrical and simple, the 5th simple and never trilobed; antennae 4- to 11-segmented, often with sexual dimorphism; eyes usually hairy; male genitalia moderate to large; female abdomen rounded caudad 28
— Hind tarsi with the 2nd segment longer than 3rd; all tarsi with the 4th segment cordiform, the 5th simple or deeply trilobed at tip; antennae 7-segmented in both sexes; eyes bare; male genitalia small; female abdomen tapered 30
28. Second hind tarsal segment much shorter than 3rd, the 5th segment slightly bilobed; wings fully developed (male) (fig. 14:29*b*) or absent (female) (fig. 14:29*a*); antenna usually 11-segmented (male) or 7-segmented (female); Florida Coasts .... *Clunio* Haliday
— Second hind tarsal segment subequal to 3rd, the 5th segment simple; wings straplike or practically absent, similar in the two sexes; antennae 4- to 7-segmented 29
29. Wings straplike, reaching to 4th segment of abdomen, halteres present; palpi long, 3- or 4-segmented; antennae 6-segmented............... *Eretmoptera* Kellogg
— Wings vestigial, not reaching to abdomen, halteres absent; palpi short, 1-segmented; antennae 7-segmented (fig. 14:30*a,b*) .... *Tethymyia* Wirth
30. Fifth tarsal segment simple or slightly bilobed; Florida coasts ....*Thalassomya* Schiner
— Fifth tarsal segment deeply trilobed at tip (fig. 14:30*k*) 31
31. Legs unmodified; hairs of legs weak (fig. 14:30*j*); California, New York, Florida ...... *Telmatogeton* Schiner
— Front legs of male modified, the femora swollen with an angular projection near apex which interlocks with a basal projection of the tibia; hairs of legs strong, sometimes flattened as appressed scales (fig. 14:30*i*); Pacific Coast ....*Paraclunio* Kieffer
32. Wing membrane with macrotrichia at least toward the apex; squama without a fringe of hairs; r-m cross vein nearly longitudinal in position, apparently continuous with posterior branch of radius (fig. 14:20*k*) (CALOPSECTRINI) ....*Calopsectra* Kieffer
— Wing membrane without macrotrichia, or if macrotrichia are present, then squama with a marginal fringe of long hairs; r-m cross vein distinctly oblique to direction of posterior branch of radius (TENDIPEDINI) 33

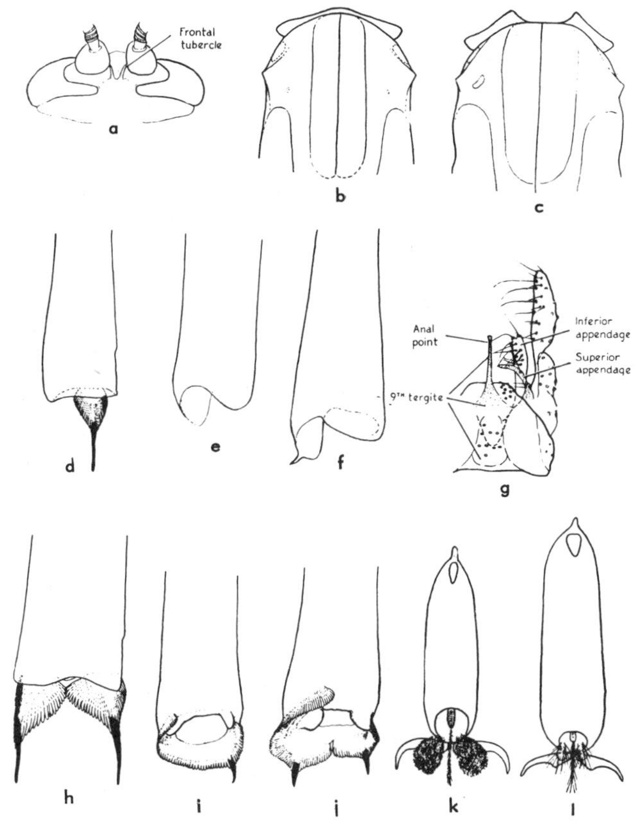

Fig. 14:21. Tendipedinae adults. *a, Tendipes plumosus* (L.), dorsal view of head; *b,c,* anterior end of thorax, dorsal view, showing shape of pronotum: *b, Cryptochironomus; c, Glyptotendipes. d-f,* Apex of left front tibia showing armature: *d, Pseudochironomus; e, Tendipes; f, Polypedilum. g, Polypedilum,* male genitalia, dorsal view, with parts labeled; *h-j,* apex of left hind tibia, showing combs and spines: *h, Pseudochironomus; i, Tanytarsus (Stictochironomus); j, Tendipes. k,l,* Last tarsal segment, ventral view, showing pulvilli: *k, Tendipes; l, Paratendipes* (Townes, 1945).

33. Flagellum of male antenna with 11 segments; middle tibia usually with 2 spines on the combs (1 on each comb) (fig. 14:21*h*); apex of front tibia on inner side with a low rounded scale which is not distinctly projecting (fig. 14:21*f*) 34
— Flagellum of male with 13 segments; middle tibia with a single spine on the combs (fig. 14:21*i*), or, if 2 spines are present, then the apex of the front tibia with a distinctly projecting subtriangular spine or scale (fig. 14:21*a-e*) 38
34. Pronotum, as seen from above, completely interrupted in the middle, the 2 lateral halves distinctly separated (fig. 14:21*c*) 35
— Pronotum not interrupted in the middle, though usually distinctly notched (fig. 14:21*b*) 36
35. Median notch of pronotum narrow, not so broad as deep; superior appendage of male genitalia broad and with numerous bristles; anal point of genitalia very broad .... *Xenochironomus* Kieffer
— Median notch of pronotum broadly V-shaped, broader than deep, and often about 0.3 times as broad as mesonotum (fig. 14:21*c*), superior appendage of male genitalia horn-shaped, with not more than 1 bristle beyond the base; anal point of genitalia narrow .... *Glyptotendipes* Kieffer
36. Inferior appendage of male genitalia large, longer than the superior appendage and with many bristles (fig. 14:20*b*); basistyle and dististyle not ankylosed; frons usually with a pair of tubercles at its center (fig. 14:21*a*) .... *Tendipes* Meigen
— Inferior appendage of male genitalia absent or very small, usually smaller than the superior appendage, with few or no bristles; basistyle and dististyle ankylosed; frons usually without tubercles 37
37. Inferior appendage of male genitalia with 1 or more apical bristles; frons often with a pair of tubercles near its center .... *Cryptochironomus* Kieffer
— Inferior appendage of male genitalia absent, or, when present, without bristles (though with numerous microtrichia); frons without tubercles ... *Harnischia* Kieffer
38. Pulvilli vestigial, not visible except in slide mounts (fig. 14:21*l*)................................. 39
— Pulvilli well developed, about 1/2 as long as the claws (fig. 14:21*k*) 41
39. Fore tibia with an apical spine which with its broader base is about 1.2 times as long as the tibial diameter; apical segment of male flagellum less than 1/2 as long as the combined length of remaining segments; squama without a fringe of hairs .... *Kribioxenus* Kieffer
— Fore tibia with an apical spine which with its broader base is not more than 0.6 times as long as the tibial diameter, or without a spine; apical segment of male flagellum longer than 0.7 times the combined length of remaining segments; squama often with a fringe of hairs 40
40. Fore tibia with an apical spine; middle tibia usually with 2 spines on the combs (one on each comb); squama often with a marginal fringe of hairs; male genitalia with a ventral pair of appendages. *Paratendipes* Kieffer
— Fore tibia without an apical spine; middle tibia with a single spine on the combs; squama without a marginal fringe of hairs; male genitalia without a ventral pair of appendages .... *Apedilum* Townes
41. Squama without a marginal fringe of hairs .... *Lauterborniella* Bause
— Squama with a marginal fringe of long hairs 42
42. Both combs of middle and hind tibiae triangular in shape, widely separated at the base, and each with an apical spine (fig. 14:21*h*); apex of front tibia with a black ventral triangular spur (fig. 14:21*d*); pronotum wide and with a broad deep notch at its center..... .... *Pseudochironomus* Malloch
— Both combs of middle and hind tibiae neither triangular in shape nor widely separated (fig. 14:21*i,j*); apex of front tibia without a spur, but usually with a spine or scale on the inner side (fig. 14:21*f*) 43
43. Mesonotum without a median longitudinal row of hairs; apex of fore tibia truncate, without a scale or spine; fore femur of male with a patch of backward-directed hairs at its apical 4th on the inner side .... *Microtendipes* Kieffer
— Mesonotum with a double median longitudinal row of hairs; apex of fore tibia with a rounded or mucronate scale (fig. 14:21*f*); fore femur of male without a patch of backward-directed hairs 44
44. Mesonotum extending far forward beyond the pronotum; middle tibia with a spine on each comb .... *Stenochironomus* Kieffer
— Mesonotum extending moderately or not at all beyond the pronotum; middle tibia with a spine on the inner comb but rarely on the outer comb (fig. 14:21*i*) 45
45. Pronotum in the middle projecting forward from the most anterior point on the mesonotum, slightly broader at the middle than at the side; hind tibia with a spine on the outer comb, the inner comb unarmed .... *Omisus* Townes
— Pronotum not or scarcely reaching as far forward as the most anterior point on the mesonotum, distinctly narrower at the middle than at the side 46
46. Base of 8th tergite and sternite of male triangularly produced; male dististyle with long bristles on the inner side (fig. 14:21*g*); fore tibia of both sexes with a pointed scale ending in a conspicuous spine (fig. 14:21*f*); outer comb of hind tibia with a spine, the

inner comb unarmed (fig. 14:21*i*); pulvilli deeply bifid .... *Polypedilum* Kieffer
— Base of 8th tergite and sternite of male subtruncate; male diststyle with short bristles on the inner side, some of which are microscopically forked near the apex; fore tibia of both sexes with a round-ended scale which has a minute apical spine or is unarmed; usually both combs of hind tibia with a spine; pulvilli entire .... *Tanytarsus* van der Wulp

Larvae

1. Antennae not retractile; epipharynx present (fig. 14:27*h*); frontoclypeus narrow behind (fig. 14:35*d*); labium[5] simple (figs. 14:27*i*, 14:35*g*) 2
— Antennae retractile; epipharynx absent; frontoclypeus broadly rounded behind; labium complex (fig. 14:23*b*) (PELOPIINAE)............................... 6
2. Tormae (premandibles) well developed (fig. 14:27*h*) 3
— Tormae rudimentary (PODONOMINAE)............ 11
3. Paralabial plates usually absent (fig. 14:27*i*) or if present are not radially striated (fig. 14:25*b*,*d*) 4
— Paralabial plates present and radially striated (figs. 14:31*g*, 14:32*e*) (TENDIPEDINAE) 25
4. Third antennal segment usually annulated (figs. 14:26*i*,*l*; 14:29*f*), or if not, paralabial plates are present but not radially striated (DIAMESINAE) 13
— Third antennal segment not annulated (fig. 14:27*b*,*k*,*n*); paralabial plates absent 5
5. Restricted to intertidal rocks on seacoasts; preanal papillae absent, instead 1-3 single hairs; anal gills absent; basal bristle on labium absent; posterior prolegs short, only twice as long as broad (fig. 14:28*l*) (CLUNIONINAE) 15
— Not restricted to intertidal rocks; preanal papillae usually well developed and bearing more than 3 hairs; anal gills and posterior prolegs usually well developed (fig. 14:27*f*) (HYDROBAENINAE) 18
6. Body slender, segments with a few scattered bristles (fig. 14:22*j*); head about 0.45-0.66 times as broad as long; labrum with numerous clavate bristles on anterior margin (fig. 14:22*d*); labium without paralabial comb, but with labial vesicles (fig. 14:22*c*); both pairs of anal gills close to anal opening ... *Pentaneura* Philippi
— Body stout, segments with a longitudinal hair fringe on each side; head broader, about 0.66-1.0 times as broad as long; labrum with 6 sense vesicles on anterior margin (fig. 14:23*a*); anal gills short, triangular, the ventral pair remote from anus, on base or prolegs ... 7
7. Labium without paralabial comb, but instead with a row of separate chitin points on each side (fig. 14:23*q*); head about 1.5 times as long as broad; antennae 1/2 to 3/4 as long as head 8
— Labium with paralabial comb (figs. 14:23*b*,*l*; 14:24*d*); head short, never more than 1/3 longer than broad; antennae 1/4 to 1/3 as long as head 9
8. Antennae very long, about 3/4 as long as head; mandibles hooklike .... *Clinotanypus* Kieffer
— Antennae 1/2 as long as head; mandibles curved (fig. 14:23*p*) .... *Coelotanypus* Kieffer
9. Larva with 6 anal gills; lingua ("glossa") of hypopharynx with 5 subequal yellow teeth ..*Pelopia* Meigen
— Larva with 4 anal gills 10
10. Superlinguae ("paraglossae") of hypopharynx scalelike, with several teeth on outer margin; lingua ("glossa") with 5 teeth (fig. 14:24*f*) .... *Procladius* Skuse
— Superlinguae of hypopharynx slender, with forked apex, or if with toothed margin the lingua with 4 subequal yellow teeth (fig. 14:23*c*) .... *Anatopynia* Johannsen
11. Labrum broad, the epistomal suture very weakly curved cephalad; labral chaetae developed as short, fine hairs (fig. 14:26*d*) 12
— Labrum narrow, the epistomal suture strongly curved cephalad; labral chaetae developed as 2 groups of strongly curved, distally serrate bristles; 3 bristles in epipharyngeal comb .... *Podonomus* Philippi
Boreochlus Edwards
12. Tubercles bearing the labral bristles scarcely twice as high as broad (fig. 14:26*d*); preanal papillae as long as posterior prolegs, each with 13 bristles (fig. 14:26*c*); head with 2 eyespots on each side; annulate part of antennae a little longer than the nonannulate 2nd segment (fig. 14:26*i*)............. *Lasiodiamesa* Kieffer
— Tubercles bearing the labral bristles about 5 times as high as broad; preanal papillae much shorter than the posterior prolegs, each with 6 bristles; head with 3 eyespots on each side; annulate part of antennae about 1/2 as long as the non-annulate 2nd segment (fig. 14:26*l*) .... *Trichotanypus* Kieffer
13. Head capsule with warts or tubercles (fig. 14:29*c*,*d*); abdominal segments with numerous small stellate hairs (fig. 14:29*o*); posterior prolegs with sucker discs (fig. 14:29*o*) .... *Heptagyia* Philippi
— Head and abdomen without such processes 14
14. Paralabial plate with an arched margin, with or without fringe of hairs, on each side of labium (fig. 14:25*b*,*d*); 2 eyespots on each side of head....*Prodiamesa* Kieffer
— Paralabial plates absent .... *Diamesa* Meigen
Psilodiamesa Kieffer
Syndiamesa Kieffer
15. Antennae 5-segmented (fig. 14:30*f*); tormae (premandibles) with lateral teeth (fig. 14:30*e*); maxillae broader than long, the palpi segmented; labial plate with a lateral pair of subbasal setae (fig. 14:30*h*) 16
— Antennae 4-segmented; tormae (premandibles) with distal teeth of apex entire; maxillae not broader than long, the palpi nonsegmented; labial plate without setae 17
16. Color of integument bluish or purplish; 2nd antennal segment much shorter than the proximal segment; labrum with a median sclerotized epipharyngeal bridge .... *Tethymyia* Wirth
— Color of integument greenish or whitish; 2nd antennal segment about as long as proximal segment; labrum without a median sclerotized epipharyngeal bridge .... *Clunio* Haliday
17. Frontoclypeal suture absent; torma with broad flattened lobe with an appressed hyaline frayed veil .... *Thalassomya* Schiner
— Frontoclypeal suture present; torma with 3 blunt distal teeth, without hyaline veil....... *Telmatogeton* Schiner
Paraclunio Kieffer
18. Antennae at least 1/2 as long as head, 2nd segment slightly bent and darkened; posterior prolegs with a basal ventral spur; 1st thoracic segment elongated necklike, the 2nd and 3rd segments not separated (fig. 14:27*e*) .... *Corynoneura* Winnertz
— Antennae not 1/2 as long as head, 2nd segment not bent; spur on posterior prolegs absent; 1st thoracic segment not elongated necklike (fig. 14:27*f*) 19
19. Larva living in symbiotic relationship on body or under wing covers of mayflies or stoneflies (fig. 14:31*m*) 20
— Larva not symbiotic with mayflies or stoneflies..... 21
20. Living under wing covers of mayflies; head unusually small .... *Symbiocladius* Kieffer
— Head not unusually small .....*Hydrobaenus* Fries (part)
21. Bristle-bearing preanal papillae of last segment at least twice as long as broad (fig. 14:27*f*) 22
— Bristle-bearing preanal papillae shorter or entirely absent (fig. 14:28*g*) 23
22. Basal antennal segment slightly but distinctly bent ....*Brillia* Kieffer
— Basal antennal segment straight (fig. 14:27*k*) .... *Metriocnemus* van der Wulp
23. Labial plate with 11 teeth, the middle one broadly truncated, hypopharynx with distally projecting hairs; labrum with simple spines but without pectinate or toothed plates; anterior prolegs fused, terminating in

[5]More correctly called the hypostome in recent morphological papers.

2 crowns of claws; lives in running water .... *Cardiocladius* Kieffer
— Not as above 24
24. Convex side of mandibles with several transverse wrinkles (fig. 14:27*o*) ....*Cricotopus* van der Wulp
— Convex side of mandibles at most with 1 or 2 furrows .... *Hydrobaenus* Fries (part)
25. Antennae 5-segmented, more or less elongate, somewhat curved, mounted on a prominent tubercle (figs. 14:28*d*; 14:32*a*); abdominal segments 2-6 each with a bifid plumose bristle on each lateroposterior angle; claws of posterior prolegs arranged in the form of a horseshoe .... *Calopsectra* Kieffer
— Antennae 5- or 6-segmented, shorter, usually straight, not mounted on tubercles (fig. 14:35*d*); abdominal segments without lateral plumose bristles (fig. 14:31*l*) 26
26. Antennae with 6 segments, a Lauterborn organ on or near apex of 2nd segment and another on apex of 3rd segment (fig. 14:33*a,i*); labial plate usually with even number of teeth (fig. 14:33*c,i*) 27
— Antennae with 5 segments (figs. 14:31*a,e*; 14:32*i,n*) 31
27. Two to 4 middle teeth of labial plate pale, the others dark (fig. 14:33*c,i*)............................ 28
— All labial teeth dark 29
28. Two middle teeth of labial plate pale (fig. 14:33*c*) .... *Microtendipes* Kieffer
— Four middle teeth pale (fig. 14:33*i*)............. ....*Paratendipes* Kieffer
29. Striations of paralabial plates scarcely visible; middle pair of teeth projecting beyond laterals (fig. 14:32*r*) ....*Stenochironomus* Kieffer
— Striations of paralabials distinct 30
30. Middle pair of teeth slightly shorter than 1st laterals .... *Tanytarsus*
— Middle pair of teeth longer than first laterals .... *Lauterborniella* Bause
31. Labial plate with a broad toothless middle part, flanked by 2 obliquely placed rows of darker lateral teeth (fig. 14:34*l*); maxillary palpus 3-4 times as long as broad; antennal sensorium at distal 3rd of basal segment (fig. 14:34*j*) .... *Cryptochironomus* Kieffer (part)
— Labium a toothed plate (figs. 14:31*b,g*; 14:35*a,g,h*); maxillary palpus only about twice as long as broad; antennal sensorium on proximal or middle 3rd of basal segment (figs. 14:32*i,n*; 14:34*a,g*)................ 32
32. Ventral gills present on 8th abdominal segment (figs. 14:31*l*; 14:35*j*) ... *Tendipes* Meigen (in strictest sense)
— Ventral gills absent on 8th abdominal segment...... 33
33. Labial plate with an even number of teeth (figs. 14:31*b*; 14:32*s*; 14:34*b,d*) ..... *Cryptochironomus* Kieffer (part)
Tanytarsus van der Wulp
Polypedilum Kieffer
— Labial plate with an odd number of teeth (figs. 14:31*g*, 14:32*j*; 14:35*a,g,h*) 34
34. Paralabial plates nearly touching on median line, each more than 4 times as broad as long, as in *Calopsectra* (fig. 14:32*j*); labial plate with 9 teeth; preanal papillae nearly 3 times as long as broad; anal gills short and plump (fig. 14:32*o*)........*Pseudochironomus* Malloch
— Paralabial plates broadly separated along median line, or not so broad as above (figs. 14:31*g*; 14:35*a*); preanal papillae shorter (fig. 14:35*j*).. *Tendipes* Meigen (part)
Glyptotendipes Kieffer
Harnischia Kieffer

Pupae

1. Cephalothorax swollen and differentiated from the slender abdomen; tubercles, spines, or bristles absent from anterior margin of thorax (fig. 14:22*i*); prothoracic respiratory organs with open stigmata (figs. 14:22*b,e*; 14:24*a*); anal segment often with a 2-lobed swimming paddle or caudal fin (fig. 14:22*h*) (PELOPIINAE) ... 5
— Cephalothorax not swollen and strongly differentiated from the abdomen (figs. 14:27*g*; 14:28*f,h*; 14:39*j*); a pair of tubercles, spines, or bristles present on the anterior margin of the thorax 2
2. Prothoracic respiratory organs usually without open stigmata (figs. 14:25*c*; 14:32*g,l*) 3
— Prothoracic respiratory organs with open stigmata (fig. 14:26*b,m*) (PODONOMINAE) 11
3. Prothoracic respiratory organs a tuft of numerous filaments (fig. 14:31*j*) or if simple (fig. 14:33*b,h*) the hind corner of 8th abdominal segment with a strong spine or comb (fig. 14:33*e,g*); anal segment usually a cleft swimming paddle with a fringe of slender filaments (fig. 14:32*h,p*) (TENDIPEDINAE) 28
— Prothoracic respiratory organ simple or absent; hind corner of 8th abdominal segment without spine or comb, anal segment at most with 3-5 strong bristles at or near the apex....................................... 4
4. Hind margins of abdominal segments with a row of strong, almost perpendicular spines or bristles (fig. (fig. 14:25*a*) (DIAMESINAE).................... 13
— Hind margins of abdominal segments without a row of strong erect spines or bristles, sometimes with small warts, or posteriorly directed points or spines (HYDROBAENINAE, CLUNIONINAE).................... 17
5. Lobes of caudal fin well developed, the outer margin bare or beset with small prickles, the 2 filaments near middle of margin (fig. 14:22*h*)..... *Pentaneura* Philippi
— Lobes of caudal fin not well developed (fig. 14:23*o*), or if so, the outer margin hairy or distinctly dentate, the 2 filaments proximad of the middle (fig. 14:23*f*) 6
6. Lobes of caudal fin not well developed; prothoracic respiratory organ brown, short ovoid, with a single small opening; no distinct mark in middle of 1st abdominal segment .... *Pelopia* Meigen
— Lobes of caudal fin well developed 7
7. Abdominal segments dorsally with strong bristles set on tubercles; lobes of caudal fin pointed (fig. 14:23*f*) ....*Anatopynia* Johannsen (part)
— Abdominal segments dorsally without strong bristles set on tubercles.............................. 8
8. Lobes of caudal fin with the outer and hind margins beset with numerous small prickles or dentate 9
— Lobes of caudal fin fringed with long, fine hairs 10
9. Prothoracic respiratory organs clavate, about 2.3 times as long as broad; spiracular disc large, transversely oval, the chamber alveolate (fig. 14:23*e*); venter and inner margins of lobes of caudal fins finely dentate (fig. 14:23*f*) ....*Anatopynia* Johannsen (part)
— Prothoracic respiratory organs fusiform, the chamber broadened distally forming a funnel under the circular spiracular disc (fig. 14:24*a*); margin of lobes of caudal fin with numerous prickles (fig. 14:24*b*).......... .... *Procladius* Skuse
10. Prothoracic respiratory organ flattened, a little more than twice as long as broad; lateral margin of 7th abdominal segment with 6 rather short, stiff bristles .... *Clinotanypus* Kieffer
— Prothoracic respiratory organs more than 3 times as long as broad; lateral margin of 7th abdominal segment with 8 long, lanceolate filaments . *Coelotanypus* Kieffer
11. Posterior margins of abdominal segments with spinose crossbands and with lateral spines; caudal segment at most with 2 strong distal spines or long bristles (fig. 14:26*j*) ...... *Boreochlus* Edwards, *Podonomus* Philippi
— Posterior margins of abdominal segments without spinose crossbands or lateral spines; caudal segment with 2 long, slender appendages or broad, bilobed swimming paddles 12
12. Posterior margin of 8th abdominal segment without emargination; 9th segment with 2 long caudal points (fig. 14:26*a*); prothoracic respiratory organ obconical (fig. 14:26*b*) .... *Lasiodiamesa* Kieffer
— Posterior margin of 8th abdominal segment deeply emarginate mesad; 9th segment with roundly pointed caudal lobes (fig. 14:26*k*); prothoracic respiratory organ spindle-shaped (fig. 14:26*m*) .....*Trichotanypus* Kieffer

13. Lateral lobes of anal segment without a fringe of spines or bristles (fig. 14:25*a*) 14
— Lateral lobes of anal segment with a fringe of spines or bristles (fig. 14:25*g*)........................ 15
14. Abdominal segments uniformly shagreened; anal segment on each side with 3 short stiff bristles hooked at apex; pupa 3-4 mm. long, in a gelatinous semiellipsoid case (fig. 14:29*g*)...............*Heptagyia* Philippi
— Pupa not in a gelatinous case; anal bristles long (fig. 14:25*a*); length usually more than 4 mm. .... *Diamesa* Meigen
15. Several branched hairs near lateral margins of abdominal segments; lobes of anal segment each with 3 bristles near apex........................... 16
— No lateral branched hairs on abdominal segments; anal segment with 4 or 5 apical bristles (fig. 14:25*g*) ....*Prodiamesa* Kieffer
16. Two or 3 pairs of darker "chitin-spots" regularly arranged on each segment; a fringe of very fine hairlike points uniformly spaced on lateral margin of lobes of 9th segment, in addition to the apical bristles .... *Syndiamesa* Kieffer
— Without paired dark "chitin-spots"; lateral lobes of 9th segment with a dense fringe of fine hairs .... *Psilodiamesa* Kieffer
17. Last abdominal segment obliquely truncated, forming an elliptical highly sclerotized disc (fig. 14:28*j*); prothoracic respiratory organs present (fig. 14:28*i*) (marine species)............................. 18
— Last abdominal segment not forming an oblique, highly sclerotized, flattened disc..................... 19
18. Prothoracic respiratory organs wedge-shaped, the point directed forward and downward, the spiracle near the base (fig. 14:28*i*) .... *Paraclunio* Kieffer
Telmatogeton Schiner
— Prothoracic respiratory organ more or less tubular, with terminal spiracle .... *Thalassomya* Schiner
19. Minute species, 3 mm. or less in length, attached to aquatic plants and enclosed in a gelatinous case; prothoracic respiratory organs lacking; abdominal segments each with 4 hairs or filaments on each side; tergites shagreened; caudal lobes each margined with swim hairs (fig. 14:27*d*) ....*Corynoneura* Winnertz
— Larger species, not as above.................... 20
20. Symbiotic on mayflies or stoneflies 21
— Not symbiotic on mayflies or stoneflies............ 22
21. In a web under wing covers of mayfly nymphs; respiratory organs absent .... *Symbiocladius* Kieffer
— Respiratory organs ovoid or elongate and tapering ....*Hydrobaenus* Fries (part)
22. Posterior margins of segments with closely set rounded warts; respiratory organs present or lacking (figs. 14:27*g*, 14:29*k*) ....*Metriocnemus* van der Wulp
— Posterior margins of segments with posteriorly directed points or spines (fig. 14:28*f*) 23
23. Respiratory organs broadened distally, with spinules, apex more or less emarginate or if not, posterior margin of 2nd tergite without a row of hooks; tergites shagreened; caudal lobes margined with swim hairs and each terminating in 3 more or less curved bristles ....*Brillia* Kieffer
— Not as above 24
24. Prothoracic respiratory organs absent; posterior margins of tergites 1-8 with a row of sharp, closely set, caudal-projecting spines; tergites 2-6 each with transverse patches of coarse spinules; length of pupa 4-5 mm. .... *Cardiocladius* Kieffer
— Tergites with smaller spinules, or otherwise differing 25
25. Intermediate tergites with shagreen covering a considerable part, leaving anterior and lateral margins free, or arranged in wide transverse bands; prothoracic respiratory organs cylindrical, tubular, with or without spinules, rarely clavate and rarely absent; caudal lobes each with 3 closely set, curved, apical bristles but without a fringe of swim hairs (fig. 14:27*l*) ....*Cricotopus* van der Wulp
— Shagreen distributed differently or respiratory organs or armature of caudal lobe otherwise 26
26. Exclusively marine species 27
— Principally fresh-water or terrestrial species, although a few marine forms are known .*Hydrobaenus* Fries (part)
27. Abdominal sternites with rows of spinules on posterior margins; sheaths of gonostyles with sharp-pointed apices ....*Clunio* Haliday
— Abdominal sternites without rows of spinules on posterior margins; sheaths of gonostyles without sharp-pointed apices (fig. 14:30*d,g*) .... *Tethymyia* Wirth
28. Prothoracic respiratory organs simple, seldom absent (fig. 14:32*g*) ....*Calopsectra* Kieffer
— Prothoracic respiratory organs composed of at least 2 branches, usually a tuft of numerous filaments 29
29. Lateroposterior angles of 8th segment without a comb or a spur (fig. 14:34*f*); cephalic tubercles present (fig. 14:34*e*) .... *Cryptochironomus* Kieffer
— Lateroposterior angles of 8th segment with either a comb or a spur 30
30. Prothoracic respiratory organs each with 2-12 branches (figs. 14:32*l*; 14:33*b,h*) 31
— Prothoracic respiratory organs each with numerous (more than 20) filaments (figs. 14:31*j*; 14:35*c*)...... 36
31. Each lateroposterior angle of 8th abdominal segment with a strong spur bearing some spinules (fig. 14:31*d*); prothoracic respiratory organs each with 4-8 branches ....*Polypedilum* Kieffer
— Lateroposterior angles of 8th segment with a comb; prothoracic respiratory organs each with 4, 7, or 11 branches 32
32. Prothoracic respiratory organs each with 4-12 branches; cephalic tubercles present (fig. 14:33*f*) 33
— Prothoracic respiratory organs each with 2 or 4 branches; cephalic tubercles absent....................... 35
33. Prothoracic respiratory organ 11-branched, one branch much larger than the others (fig. 14:33*h*); lateral margin of 8th abdominal segment with 4 filaments; combs of 8th segment each with 5-6 strong teeth and some finer spinules (fig. 14:33*g*) .... *Paratendipes* Kieffer
— Prothoracic respiratory organ 4- or 7-branched, the largest branch bearing apically a number of acute papillae (fig. 14:33*b*); lateral margins of 8th segment each with 4 or 5 filaments; combs of 8th segment each with 3-6 teeth (fig. 14:33*e*) 34
34. Prothoracic respiratory organ with 7 branches; 8th segment with 5 filaments on each lateral margin .. ..*Microtendipes* Kieffer
— Prothoracic respiratory organ with 4 branches; 8th segment with 4 filaments on each lateral margin ....*Stenochironomus* Kieffer
35. Prothoracic respiratory organ with 2 branches, both smooth (fig. 14:32*l*)........*Pseudochironomus* Malloch
— Prothoracic respiratory organ with 2 or 4 branches, one of them covered with short, pointed papillae or evaginations .... *Lauterborniella* Bause
36. Lateroposterior angles of 8th abdominal segment each with a comb of 2, 3, or more teeth or spines (figs. 14:34*c,m*; 14:35*e*)............................ 37
— Lateroposterior angles of 8th segment with a simple or compound spur (fig. 14:35*b,f*)........*Tendipes* Meigen
37. Some tergites of abdomen with a spiked macelike process (fig. 14:31*i*)...........*Glyptotendipes* Kieffer
— Tergites without such a process ....*Tanytarsus* van der Wulp
Harnischia Kieffer

California Species of Tendipedidae

Subfamily PELOPIINAE (=TANYPODINAE)

The larvae of the Pelopiinae are predaceous, feeding largely on other tendipedid larvae and often found in

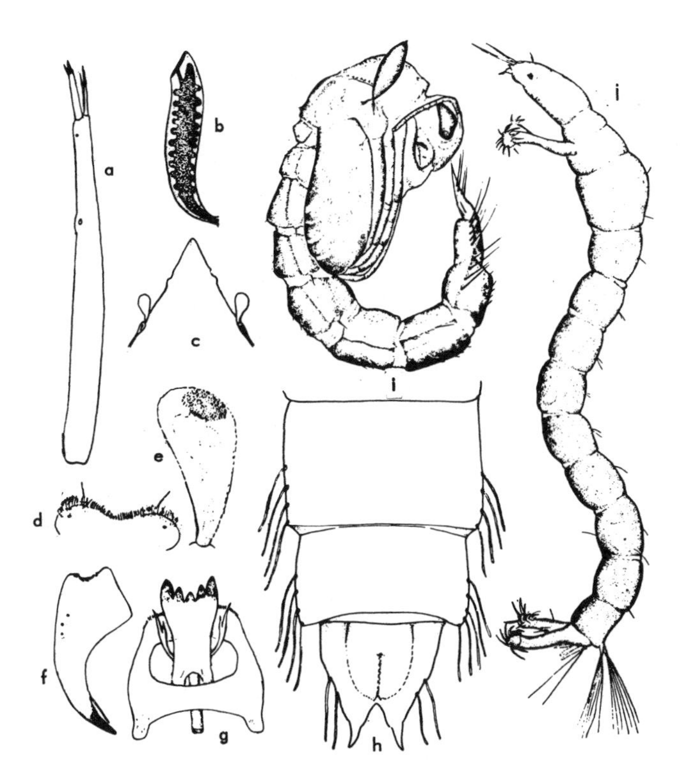

Fig. 14:22. Pelopiinae. *a,c-h*, *Pentaneura flavifrons* (Joh.). Larva: *a*, antenna; *c*, labium; *d*, labrum; *f*, mandible; *g*, hypopharynx with lingua (l), superlinguae (s) and pecten (p). Pupa: *e*, respiratory organ; *h*, caudal segments, ventral view. *b*, *P. decolorata* (Mall.), pupa, respiratory organ. *i,j*, *P. monilis* (L.): *i*, pupa, habitus; *j*, larva, habitus (Johannsen, 1937).

the cases made by their prey, although they build none themselves. They are especially abundant in permanent streams, ponds, and lakes. The larvae (fig. 14:22*j*) are distinguished by their retractile antennae; labrum without tormae or epipharyngeal appendages but with small rods or hairs (fig. 14:22*d*) or with six marginal sense vesicles (fig. 14:23*a*); the mandibles sickle-shaped (fig. 14:22*f*) or hooked, moving in a horizontal plane; the hypopharynx greatly specialized and consisting of strong supporting frame or suspensoria laterally, with a comb or pecten of numerous teeth and in the center with a four- to seven-toothed lingua (glossa), at the sides of which are the superlinguae (paraglossae) (fig. 14:23*m*); the labium consisting of a colorless labial plate having on each side a labial vesicle (fig. 14:22*c*) or a paralabial comb (fig. 14:23*b*); the body segments with a few scattered hairs on each side or a dense fringe of delicate hairs; the pair of preanal papillae on the dorsum of the ninth abdominal segment cylindrical and bearing from seven to more than twenty bristles in a tuft; anterior and posterior prolegs well developed, often very elongate (fig. 14:22*j*), bearing apical claws; and four or six anal gills present. The pupae are active and superficially resemble mosquito pupae (fig. 14:22*i*); the prothoracic respiratory organs are ovate (fig. 14:23*i*) or trumpet-shaped (fig. 14:22*b,e*) with a reticulate or finely spinous surface; the lateral margins of abdominal segments seven and eight usually bear four or more slender filaments (fig. 14:22*h*); the last segment bears a two-lobed caudal fin with a variable vestiture of hairs, spines, or teeth. Zavrel (1921) and Lenz (1936) have outlined the classification of the immature stages of the European fauna, and Johannsen (1937) has treated the known North American species.

Genus *Pentaneura* Philippi (= *Ablabesmyia* auct.)

(Figs. 14:20*h*; 14:22)

The North American species have been revised by Johannsen (1946). The larvae are quite slender with

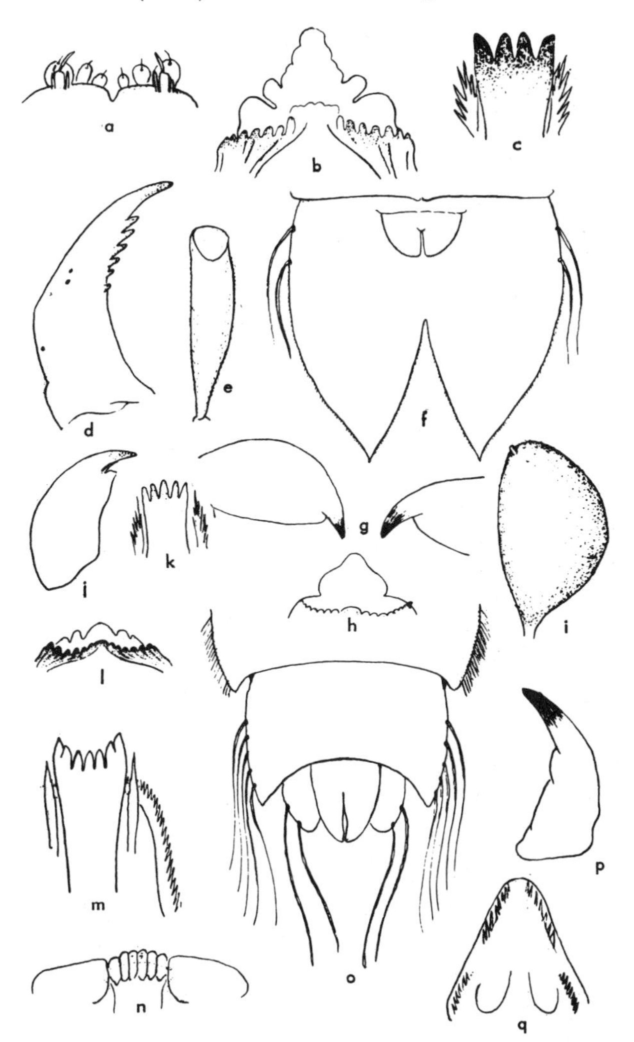

Fig. 14:23. Pelopiinae. *a-f*, *Anatopynia dyari* (Coq.). Larva: *a*, labrum; *b*, labium and paralabials; *c*, lingua; *d*, mandible. Pupa: *e*, respiratory organ; *f*, caudal fin, ventral view. *g-i*, *Pelopia punctipennis* (Mg.). Larva: *g*, mandibles; *h*, labium and paralabials. Pupa: *i*, respiratory organ. *j-l,o*, *P. stellata* (Coq.). Larva: *j*, mandible; *k*, lingua and superlinguae; *l*, labium and paralabials. Pupa: *o*, caudal end. *m,n,p,q*, *Coelotanypus concinnus* (Coq.), larva: *m*, lingua superlingua and pecten; *n*, labrum; *p*, mandible; *q*, labium (Johannsen, 1937).

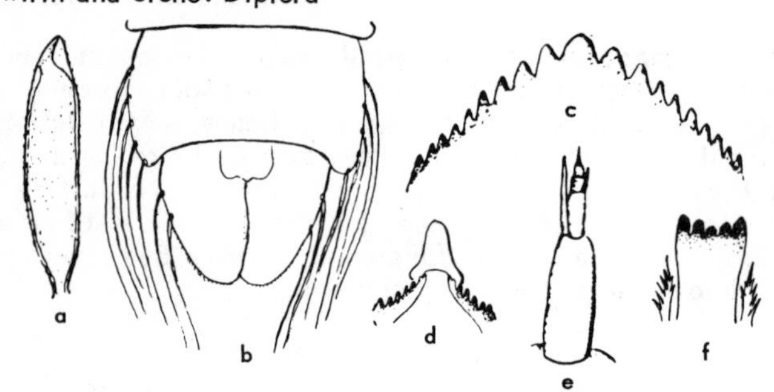

Fig. 14:24. Pelopiinae and Diamesinae. *a,b,d,f, Procladius adumbratus* Joh. Pupa: *a*, respiratory organ; *b*, caudal end, ventral view. Larva: *d*, labium and paralabials; *f*, lingua and superlinguae. *c,e, Diamesa nivoriunda* Fitch, larva: *c*, labial plate; *e*, antenna (Johannsen, 1937).

swollen thorax, long prolegs, and long tapering head.
California species:

carnea (Fabricius) 1805.	Lake Tahoe
flavifrons (Johannsen) 1905.	Mono, San Diego counties Lake Tahoe
monilis (Linnaeus) 1758.	Widespread
pilosellus (Loew) 1866.	Humboldt County

Genus *Anatopynia* Johannsen

(Figure 14:23*a-f*)

algens (Coquillett) 1902.	Humboldt County
guttularis (Coquillett) 1902.	Northern and central California
marginellus (Malloch) 1915.	Santa Clara County
venusta (Coquillett) 1902.	Shasta, Santa Clara

Genus *Pelopia* Meigen

(= *Tanypus* Meigen, *Protenthes* Johannsen)

(Fig. 14:23*g-o*)

punctipennis (Meigen) 1818.	Monterey County

Genus *Procladius* Skuse

(Figs. 14:1*c*; 14:24*a,b,d,f*)

The larvae of this genus are very common predators of the profundal zone of northern lakes.
California species:

bellus (Loew) 1866.	Lassen County, Lake Tahoe
culiciformis (Linnaeus) 1767.	Lake Tahoe

Genera *Clinotanypus* Kieffer and *Coelotanypus* Kieffer

These genera are very closely related in adult and larval morphology, the adults being robust and hump-backed in appearance, with cordiform fourth tarsal segments and bare wings. Neither genus has been reported from California but they undoubtedly occur there.

Subfamily PODONOMINAE

(Figs. 14:20*d*; 14:26*a-n*)

This subfamily, which only recently has been separated from the Pelopiinae, is apparently restricted to the colder regions of the world. The immature stages are known to breed in cold mountain streams and lakes and in bog pools. We have reared an undetermined species of *Boreochlus* from a sphagnum bog near Washington, D.C., apparently the first North American breeding record. The larvae and pupae of the Podonominae have been described and figured by Thienemann (1937). None of the four Nearctic genera have been reported yet from California.

Subfamily DIAMESINAE

(Figs. 14:20*f*; 14:24*c,e*; 14:25; 14:27*a*; 14:29*c-g, q*)

This group resembles the Podonominae as well as the Hydrobaeninae in its greater abundance in regions of colder climate. The immature stages are characteristic of mountain brooks or other swift streams, where they live at the splash line or on smooth rocky bottoms in the shallow water of rapids (hygropetric environment). Saunders (1928, 1930) has described and figured the immature stages of *Heptagyia*, and Johannsen (1937) has included descriptions of that genus as well as the other four known Nearctic genera in his memoir. The serious student contemplating work on this group will find the recent revision by Pagast (1947) indispensable. No species have yet been reported from California although many await description or discovery.

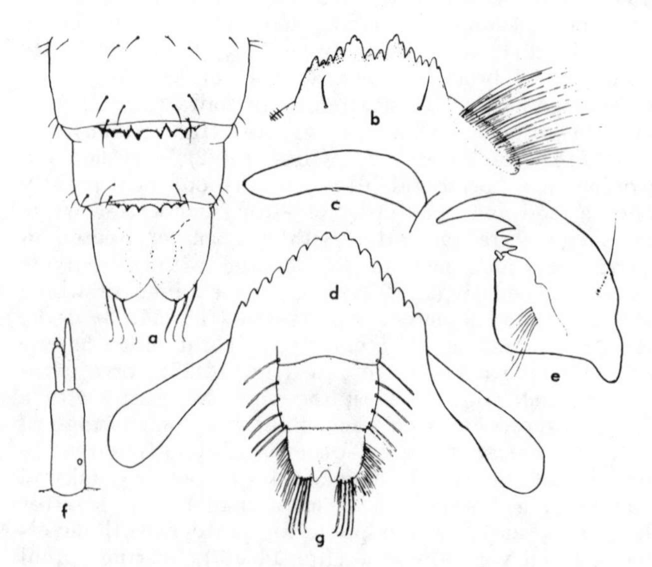

Fig. 14:25. Diamesinae. *a, Diamesa nivoriunda* Fitch, caudal end of pupa, dorsal view; *b,g, Prodiamesa olivacea* (Mg.): *b*, labial plate and paralabials of larva; *g*, caudal end of pupa. *c-f, Prodiamesa bathyphila* (K.): *c*, respiratory organ of pupa; *d*, labial plate and paralabials of larva; *e*, larval mandible; *f*, larval antenna (Johannsen, 1937).

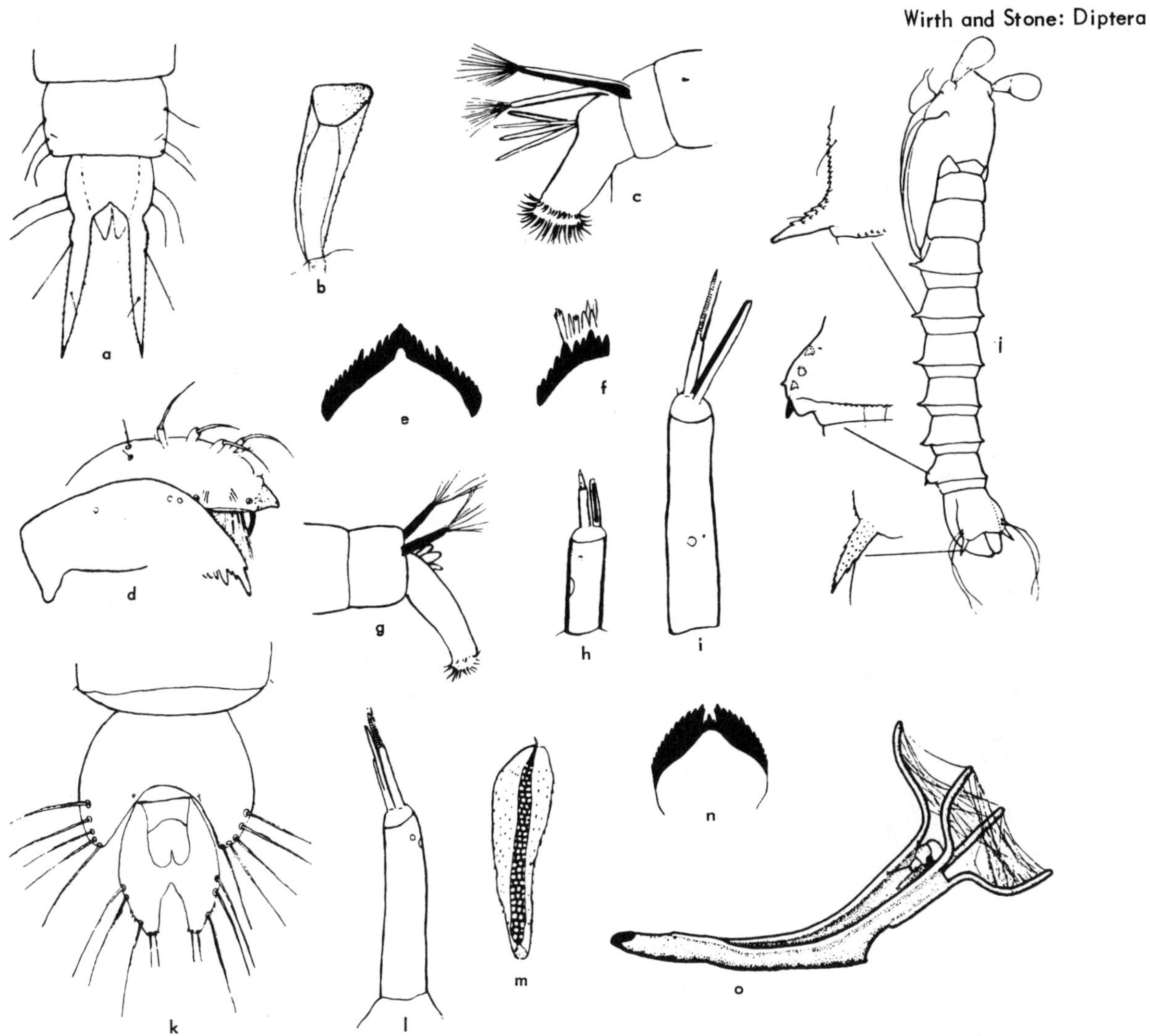

Fig. 14:26. Podonominae and Tendipedinae. *a-e,i, Lasiodiamesa*, Pupa: *a*, caudal segments; *b*, respiratory organ. Larva: *c*, caudal segments; *d*, labrum and mandible; *e*, labial plate; *i*, antenna. *f-h,j, Podonomus*, larva: *f*, labial plate, left half; *g*, caudal segments; *h*, antenna; *j*, pupa. *k-n; Trichotanypus*, Pupa: *k*, caudal segments; *m*, respiratory organ. Larva: *l*, antenna; *n*, labial plate. *o, Calopsectra (Rheotanytarsus) rivulorum* (K.), larva and case showing salivary net (*a-n*, Thienemann, 1937; *o*, Walshe, 1951).

Subfamily CLUNIONINAE

California, with its long, rugged coastline, has been exceptionally well provided with its share of the world's "marine midges." These remarkable insects probably form a highly evolved and specialized offshoot of the Hydrobaeninae, from which in many cases the immature stages cannot well be distinguished. This subfamily is restricted, with the exception of some secondarily adapted mountain-stream Hawaiian species, to the zone of algae-covered rocks between tide levels on seacoasts. The larvae form silken cases on the rocks and between the bases of the algae filaments on which they feed. They pupate in reinforced larval cases, and several genera have evolved a remarkable, highly sclerotized, obliquely flattened, caudal disc on the pupa, probably to offset the high-water pressures set up in the tunnels by wave action. This subfamily has been revised by Wirth (1949), including intensive field collections and study in California where the following species have been found.

Genus *Eretmoptera* Kellogg

browni Kellogg 1900. Point Lobos, Monterey County

Genus *Tethymyia* Wirth

(Fig. 14:30*a-h*)

aptena Wirth 1949. Mendocino to Monterey County

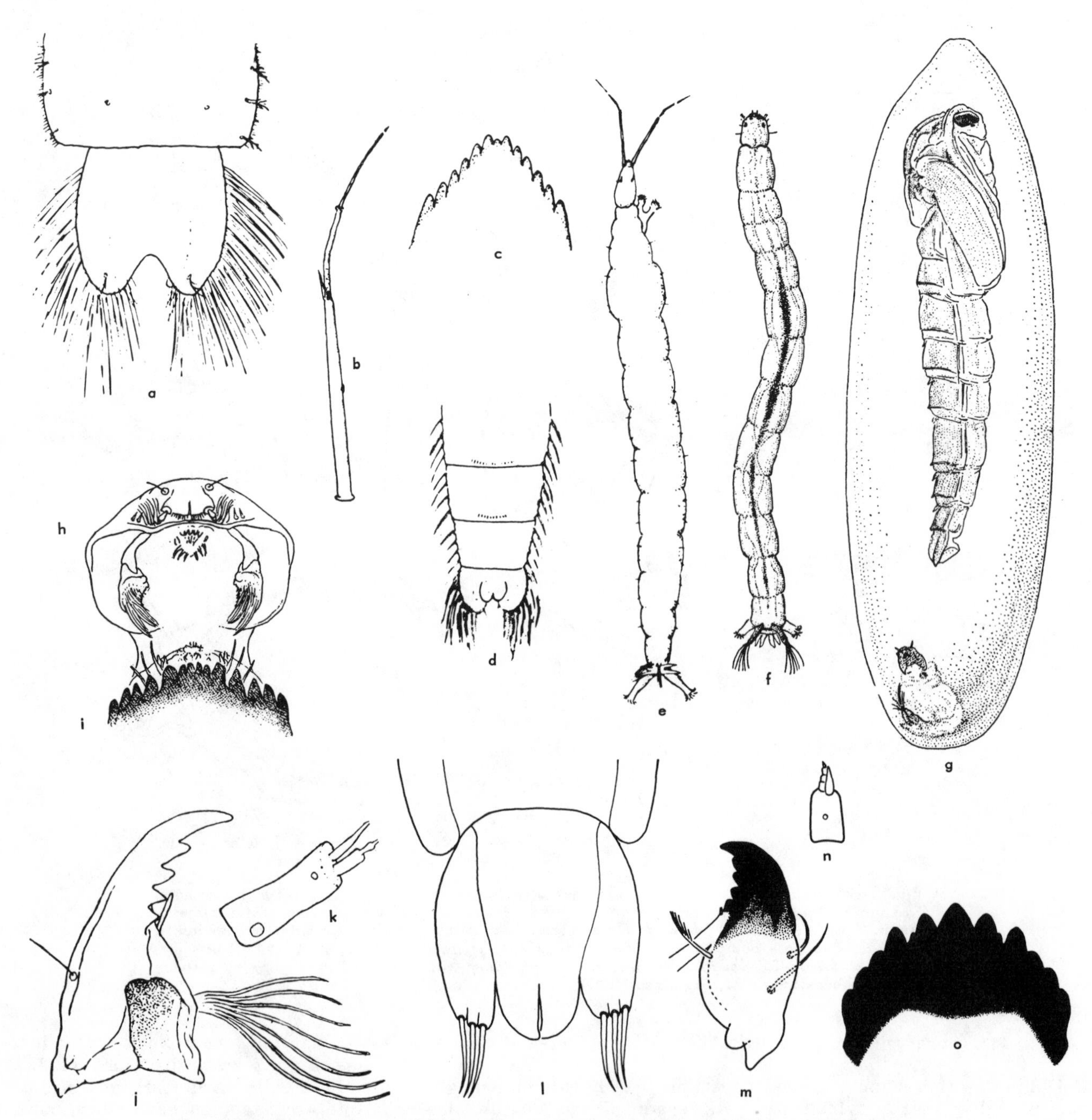

Fig. 14:27. Hydrobaeninae and Diamesinae. *a*, *Prodiamesa bathyphila* (K.), caudal end of pupa; *b-e*, *Corynoneura scutellata* Winn. Larva: *b*, antenna; *c*, labial plate; *e*, habitus; *d*, caudal segments of pupa. *f-k*, *Metriocnemus knabi* Coq., larva: *f*, habitus; *h*, labrum; *i*, labial plate; *j*, mandible; *k*, antenna; *g*, pupa, with larval skin in case. *l-o*, *Cricotopus elegans* Joh. Pupa: *l*, caudal end. Larva: *m*, mandible; *n*, antenna; o, labial plate (*a-k*, Johannsen, 1937; *l-o*, Berg, 1950).

Genus *Telmatogeton* Schiner

(Fig. 14:28*h-l*; 14:30*j,k*)

macswaini Wirth 1949. Mendocino County

Genus *Paraclunio* Kieffer

alaskensis (Coquillett) 1900. Del Norte to San Diego
trilobatus Kieffer 1911. Del Norte to Monterey County
(fig. 14:30*i*)

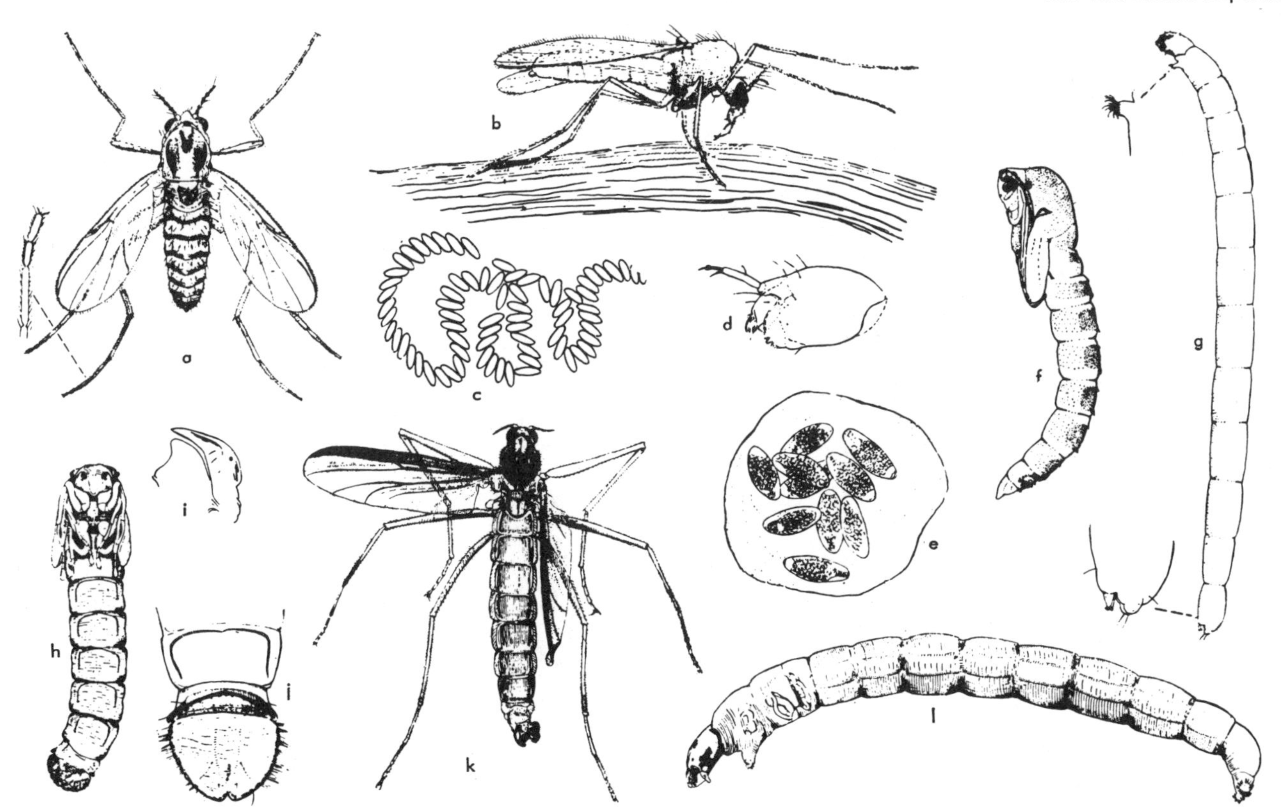

Fig. 14:28. Tendipedinae, Hydrobaeninae, and Clunioninae. *a,e-g, Hydrobaenus: a,* female; *e,* globule of eggs; *f,* pupa; *g,* larva. *b-d, Calopsectra: b,* female; *c,* string of eggs; *d,* larval head. *h-l, Telmatogeton,* Pupa: *h,* ventral view; *i,* respiratory organ; *j,* caudal shield. Male: *k,* habitus. Larva: *l,* lateral view (Williams, 1944).

Subfamily HYDROBAENINAE

(Figs. 14:20*a,j,l*; 14:27*b-o*; 14:28*a,e-g*; 14:29*h-p*; 14:31*m*)

This subfamily is so large and so little known except in Europe that it is practically hopeless to attempt any sort of discussion of it. The recently published key by Thienemann (1944), however, as well as the recent papers by Strenzke on the terrestrial forms (1950, particularly), will attract much needed attention to these tiny midges in this country. Habits are as diverse as the species and genera are numerous. Some species (Saunders, 1928) breed together with the Clunioninae on algae-covered, intertidal rocks along seacoasts where their swarms are often apparent in contrast to the nonswarming scampering movements of the true marine midges. The species of *Cricotopus* (figs. 14:20*a,l*; 14:27*l-o*), *Corynoneura* (fig. 14:27*b-e*), *Metriocnemus,* and *Brillia* are more often strictly aquatic, breeding in a wide variety of ponds and streams. Some *Cricotopus* (Berg, 1950) mine the leaves of aquatic plants, and several species are economic pests of rice (Risbec, 1951). Two undescribed California species of *Cricotopus* have recently been found living in pads of the alga *Nostoc,* probably in close biological association. Two species of *Metriocnemus* breed in the water-filled cups of pitcher plants, *M. knabi* Coquillett (fig. 14:27*f-k*) in the eastern *Sarracenia* and *M. edwardsi* Jones (fig. 14:29*h-p*) (1916) in *Darlingtonia,* the California pitcher plant. The genus *Cardiocladius* is restricted to rapidly flowing water, whereas species of *Symbiocladius* (fig. 14:31*m*) live as commensals or parasites under the wing pads of mayfly nymphs of the genus *Rithrogena* (Claassen, 1922; Codreanu, 1939; Roback, 1951). The genus *Hydrobaenus* (fig. 14:28*a,e-g*) as known in this country is a dumping ground for most of the remaining members of the subfamily including many aquatic species, but especially for the subaquatic and terrestrial species which breed in damp and wet ground along stream margins, swampy areas, mossy banks, bogs, or in almost any moist situation. Many of the more terrestrial species show a reduction of the posterior prolegs or anal gills of the larva or the prothoracic respiratory organs of the pupa.

Subfamily TENDIPEDINAE

Students of stream and lake ecology and fish production are most concerned with the larvae of the larger species of this subfamily, which are commonly spoken of as bloodworms. With the exception of a few species of *Calopsectra* most of the members of the Tendipedinae live in sluggish or quiet water, some in lakes at great depths.

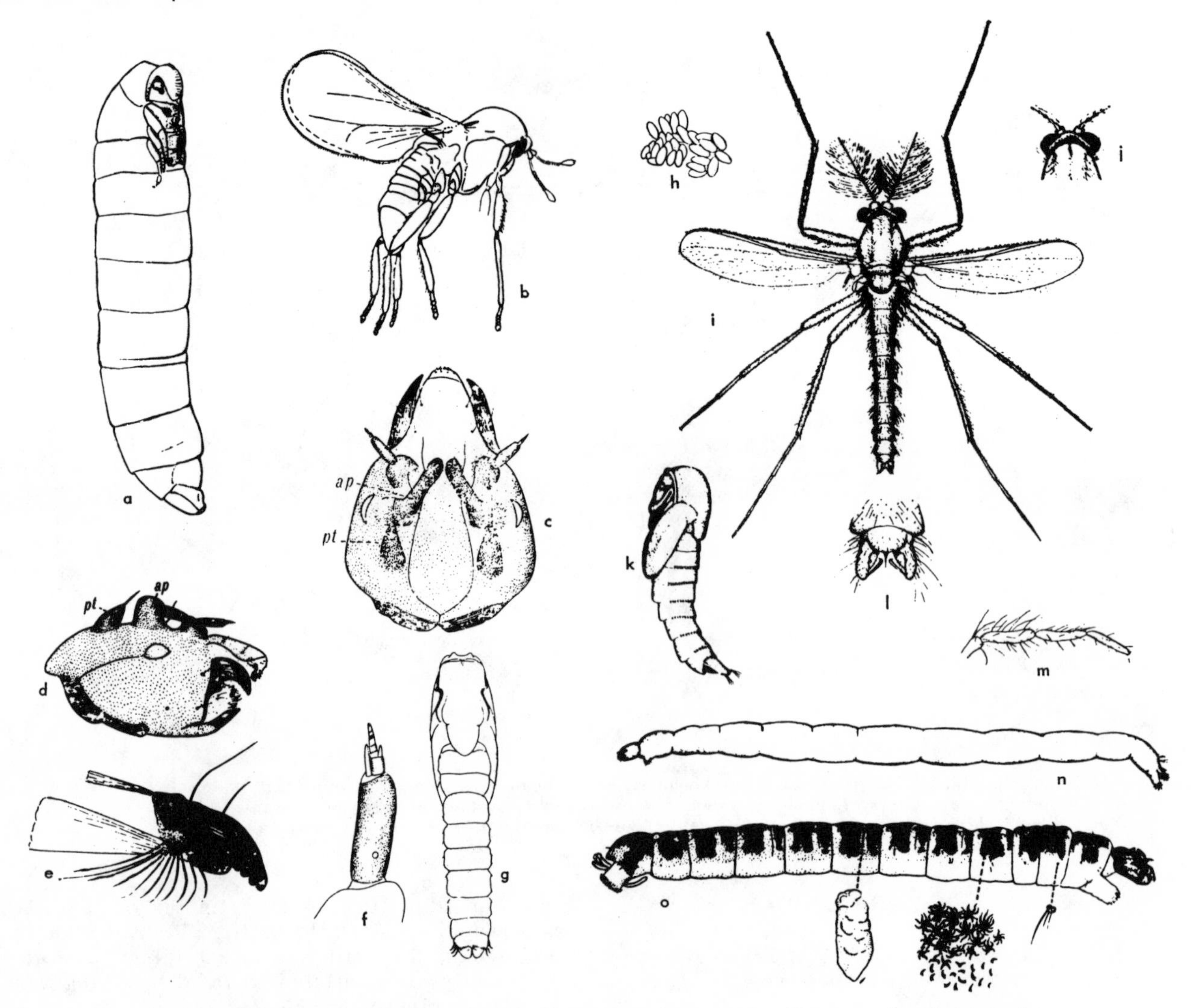

Fig. 14:29. Clunioninae, Hydrobaeninae, and Diamesinae. *a,b*, *Clunio*: *a*, female; *b*, male. c-g,o, *Heptagyia lurida* (Garrett), larva: *c*, head, dorsal view; *d*, head, lateral view; *e*, mandible; *f*, antenna; *o*, larva, lateral view; *g*, dorsal view of pupa. *h-n*, *Metriocnemus edwardsi* Jones: *h*, egg mass; *i*, male, habitus; *j*, female head; *k*, pupa; *l*, male genitalia, dorsal view; *m*, female palpus; *n*, larva, lateral view (*a,b*, Williams, 1944; c-g,o, Saunders, 1928; *h-n*, Jones, 1916).

In the subgenus *Tendipes* (Andersen, 1949), the genus *Calopsectra* (Johannsen, 1937), and in several other genera the classification of the immature stages has to a great extent proceeded independently of that of the imagoes to the great confusion of present students. Only by the careful rearing of many species and the comparison of these correlated specimens with the types and descriptions of the older species will it be possible to merge the two systems and produce order.

Tribe CALOPSECTRINI (= TANYTARSINI)

Genus *Calopsectra* Kieffer (= *Tanytarsus* auct.)

(Figs. 14:20*c,k*; 14:26*o*; 14:28*b-d*; 14:32*a-h*)

The larvae of this genus have long five-segmented antennae, of which the basal segment is long and curved and borne on a distinct tubercle or prominence of the head (fig. 14:32*a*); the paralabial plates are usually very long and narrow and nearly touching on the mid-line; (fig. 14:32*e*), and most of the abdominal segments bear bifid plumose bristles posterolaterally. The pupae have simple, usually slender, prothoracic respiratory organs (fig. 14:32*g*) (as in the Hydrobaeninae) in contrast to the branched or tufted organs of the Tendipedini. A number of species build *Hydra*-like tubes of silt with the number of arms differing for different species, the arms used to support the salivary net which catches food particles (fig. 14:26*o*). Two groups of species (*Zavrelia* and *Stempellina*) build portable cases of sand grains very like those of leptocerid caddisflies (Walshe, 1951).

No California records of this genus are available.

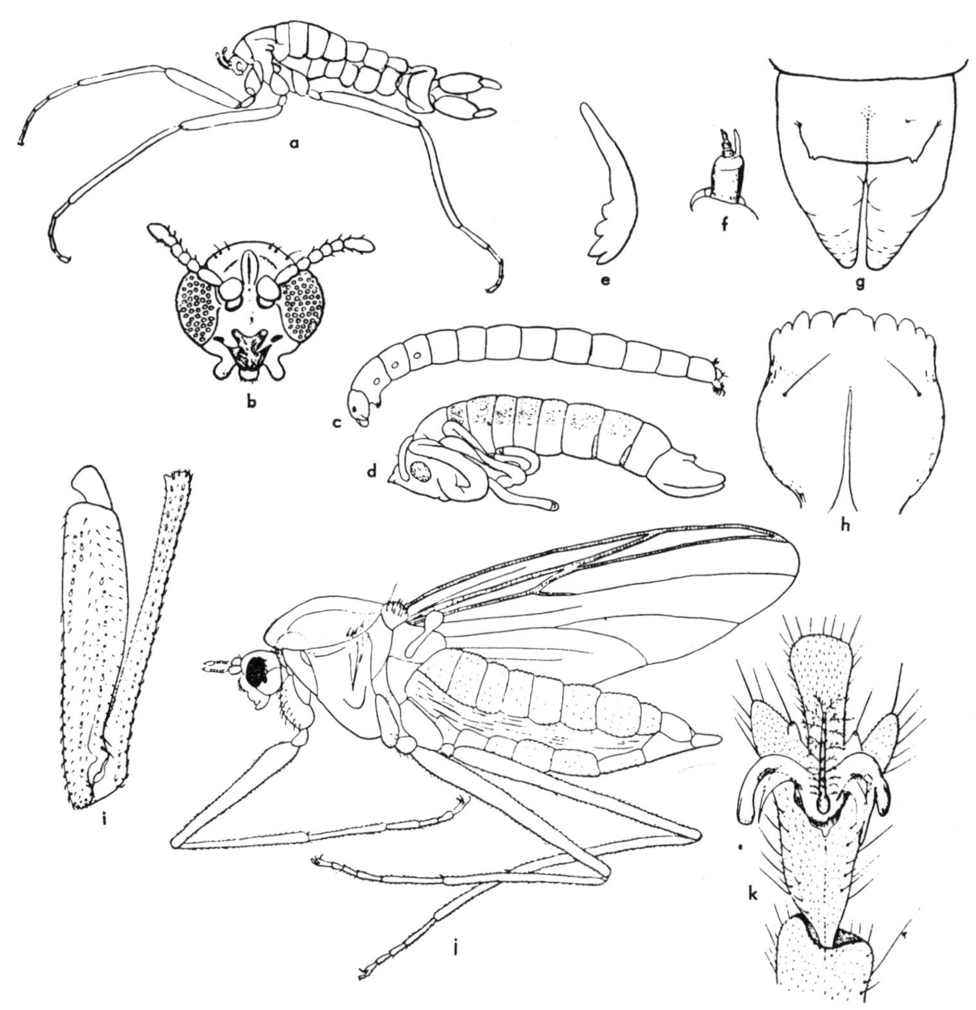

Fig. 14:30. Clunioninae. *a-h*, *Tethymyia aptena* Wirth: *a*, male, lateral view; *b*, male head, anterior view; *c*, larva, lateral view; *e*, larval torma; *f*, larval antenna; *h*, larval labial plate; *d*, male pupa in lateral view; *g*, last segment of pupa in dorsal view. *i*, *Paraclunio trilobatus* K., front femur and tibia; *j,k*, *Telmatogeton macswaini* Wirth: *j*, female; lateral view; *k*, front tarsus of male, fifth segment (Wirth, 1949).

Tribe TENDIPEDINI (=CHIRONOMINI)

In striking contrast to the unorganized state of our knowledge of the bulk of the North American tendipedids, the Tendipedini have been thoroughly revised in a recent publication by Townes (1945). Because of their larger size and economic importance in stream and lake biology, the members of the genus *Tendipes* (=*Chironomus*) have become much better known than the numerous species of smaller greenish tendipedines, but the great variability of characters of both adult and immature stages makes their classification extremely difficult. Members of this tribe are most numerous in the weedy coves of larger rivers and bays of lakes, although smaller streams and ponds also produce numerous species. Most of the species breed in beds of aquatic vegetation or in bottom mud of water less than ten feet deep, and comparatively few species occur in the great profundal depths.

Genus *Pseudochironomus* Malloch

(Figs. 14:21*d,h*; 14:32*i-l,o,p*)

This and following genera through *Stenochironomus* form a natural group which is well set off from the remaining "*Chironomus*" of authors, and has been termed the "Chironomariae connectentes" by European workers. They share in the larva, with few exceptions, the characters of six-segmented antennae (figs. 14:32*i,m*; 14:33*a,j*) and an even number of labial teeth (figs. 14:32*r,s*; 14:33*c,i*); in the pupa, the prothoracic respiratory organs usually with two to twelve branches (fig. 14:32*l*; 14:33*b,h*); the adults with one long spur on the hind tibia, or if with two short spurs, the wings are banded or the tarsi lack pulvilli (fig. 14:21*l*); the male with fourteen antennal segments.

The genus *Pseudochironomus* is set off in an anomalous position from the other members of this group; the larva by the nearly contiguous paralabial plates

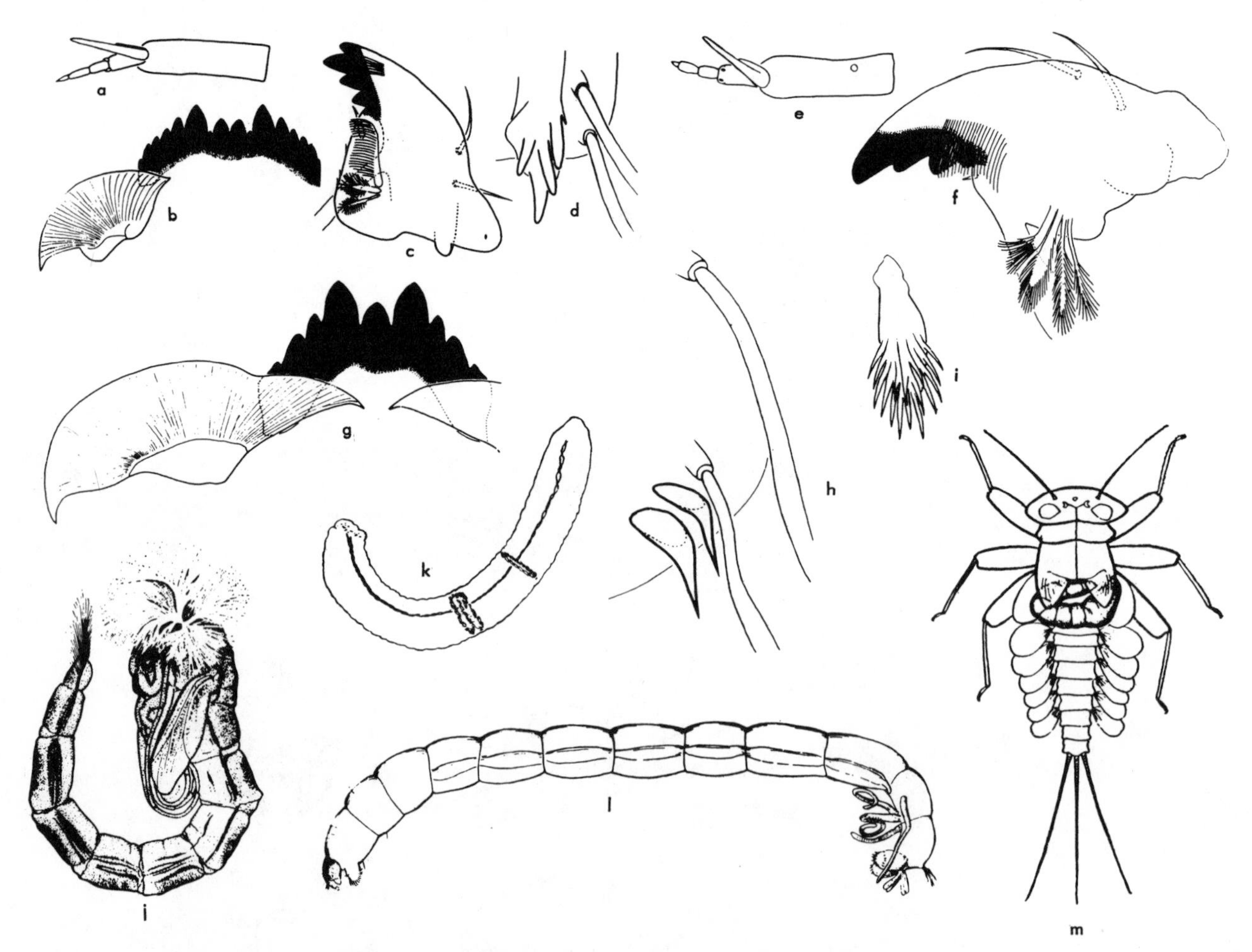

Fig. 14:31. Tendipedini and Hydrobaeninae. *a-d*, *Polypedilum ophioides* Townes, larva: *a*, antenna; *b*, labial plate and paralabials; *c*, mandible; *d*, spur of eighth segment of pupa. *e-i*, *Glyptotendipes dreisbachi* Townes. Larva: *e*, antenna; *f*, mandible; *g*, labial plate and paralabials. Pupa: *h*, comb of eighth segment; *i*, macelike process of sixth segment. *j-l*, *Tendipes: j*, pupa; *k*, egg mass; *l* larva. *m*, *Symbiocladius equitans* Claasen: mayfly nymph with larva under wing pads (*a-i*, Berg, 1950; *j-l*, Williams, 1944; *m*, Claasen, 1922).

(fig. 14:32*j*) as in *Calopsectra,* the elongated preanal papillae (fig. 14:32*o*) as in the Pelopiinae and some Hydrobaeninae, and the labial plate with only nine teeth. The pupae have only two-branched prothoracic respiratory organs (fig. 14:32*l*) and lack cephalic tubercles. No species have been reported from California.

Genus *Lauterborniella* Bause

Larvae of this genus are unique among the Tendipedini in their habit of building a movable case of sand grains resembling those of certain caddisflies, agreein this habit with certain *Calopsectra* species. The larvae have six-segmented antennae, an even pair of labial teeth of which the middle pair are longer than the first laterals, and the pupae have two (subgenus *Lauterborniella*) or four (subgenus *Zavreliella*) branches on the prothoracic respiratory organs, one branch of which bears minute, pointed papillae. Species of the subgenus *Zavreliella* have the front femora clubbed apically and the wings spotted. No California species have yet been reported.

Genus *Microtendipes* Kieffer

(Fig. 14:33*b-e,j*)

The larvae of this genus differ from others in the group with six-segmented antennae by having the two middle teeth of the labial plate pale (fig. 14:33*c*). The pupae have seven-branched prothoracic respiratory organs.

One California species:

caducus Townes 1945. Placer County

Genus *Paratendipes* Kieffer

(Figs. 14:21*l*; 14:33*a,f-i*)

The larvae of *Paratendipes* differ from those of related

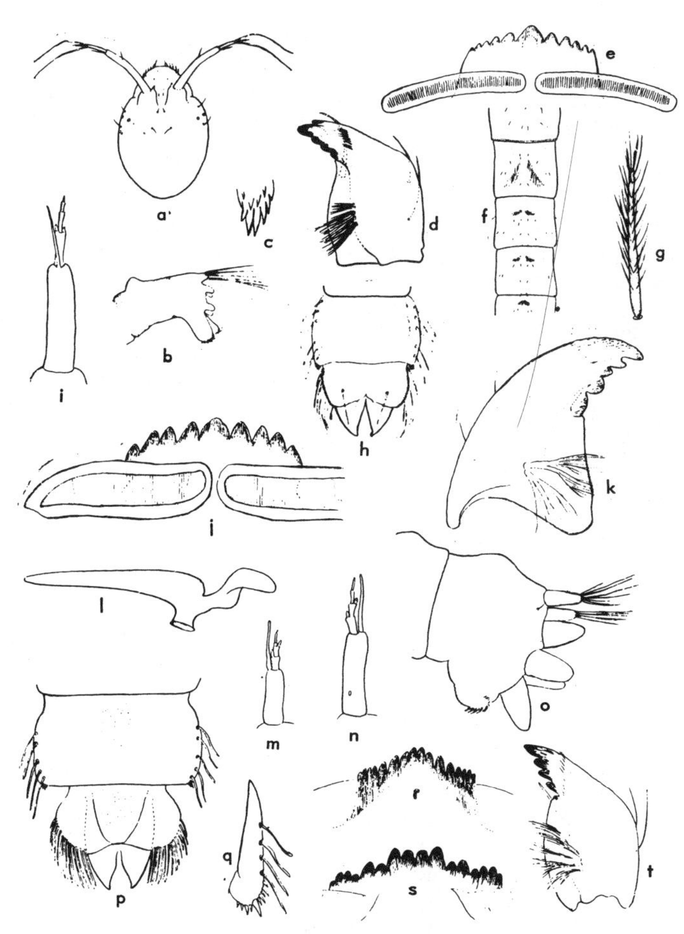

Fig. 14:32. Tendipedinae. *a-h, Calopsectra dives* (Joh.). Larva: *a*, head; *b*, caudal segments; *d*, mandible; *e*, labial plate and paralabials. Pupa: *c*, comb of eighth segment; *f*, second to sixth tergites; *g*, respiratory organ; *h*, eighth and ninth segments *i-l, o,p, Pseudochironomus richardsoni* Mall. Larva: *i*, antenna; *j*, labial plate and paralabials; *k*, mandible; *o*, caudal segments. Pupa: *l*, respiratory organ; *p*, caudal segments. *m,r, Stenochironomus*, larva: *m*, antenna; *r*, labial plate. *n,q,s,t, Tanytarsus (Stictochironomus)*, larva: *n*, antenna; *s*, labial plate; *t*, mandible; *q*, lateral fin of eighth segment of pupa (Johannsen, 1937).

genera in having the four middle teeth of the labial plate pale (fig. 14:33*i*); the pupae have about twelve branches in the prothoracic respiratory organ (fig. 14:33*h*), one of which is much larger than the others.

One California species:

albimanus (Meigen) 1818. Plumas, Placer counties

Genus *Polypedilum* Kieffer

(Figs. 14:20*i*; 14:21*f,g*; 14:31*a-d*)

Larvae of this genus differ from others of the "Chironomariae connectentes" in having five-segmented antennae (fig. 14:31*a*) as in the *Tendipes* group, but the labial plate has an even number of teeth (16), usually with the middle pair the longest (fig. 14:31*b*); the ring organ of the antenna situated at the proximal fifth. The widely separated paralabial plates and the anal gills constricted near the middle are also distinctive of the larva. The pupae have prothoracic respiratory organs with four to eight branches, and the cephalic tubercles may be wartlike, each apically with an area of minute pointed papillae (subgenus *Pentapedilum*), or low and conical without apical bristles *(Polypedilum)*. The wings are usually distinctly maculate (fig. 14:20*i*). Adults of the subgenus *Pentapedilum* resemble those of *Calopsectra* in having hairy wings.

The habits of this large and widespread genus, although incompletely known, are diverse; of particular interest are the species which mine the leaves and stems of aquatic plants. Here again the species of the subgenus *Pentapedilum* seem to use their burrows for the purpose of supporting the salivary net on which they collect their plankton food, whereas species of the subgenus *Polypedilum* actually feed on the leaf tissue (Berg, 1950; Walshe, 1951).

California species:

aviceps Townes 1945.	Santa Clara County
digitifer Townes 1945.	Kern County
isocerus Townes 1945.	Placer, Mono
laetum (Meigen) 1818.	Placer County
pedatum excelsius Townes 1945.	Plumas, Del Norte
scalaenum (Schrank) 1803. [= *needhami* (Johannsen)]	Laguna Canyon

Genus *Stenochironomus* Kieffer

In this genus the larvae have six-segmented antennae (fig. 14:32*m*), sixteen dark teeth in the labial plate, the middle pair projecting beyond the others, and the paralabials are well developed but the striations are very weak (fig. 14:32*m*). The pupae have four branches on the prothoracic respiratory organs. The wings are

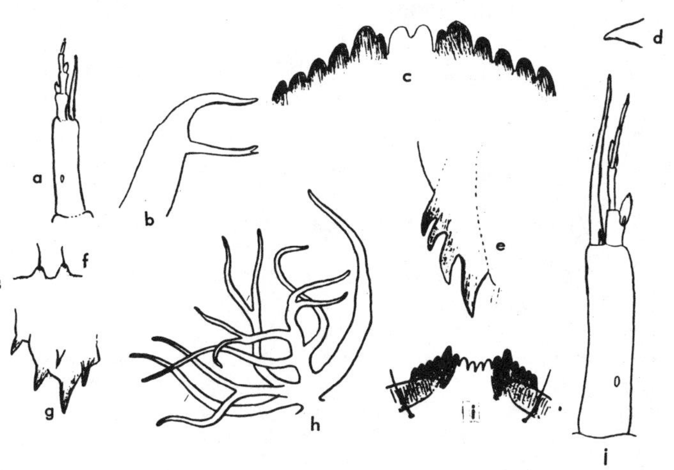

Fig. 14:33. Tendipedinae. *a,f-i, Paratendipes albimanus* (Mg.). Larva: *a*, antenna; *i*, labial plate. Pupa: *f*, cephalic tubercles; *g*, comb of eighth segment; *h*, respiratory organ. *b-e,j, Microtendipes pedellus* (DeGeer). Pupa: *b*, tip of branch of respiratory organ; *d*, cephalic tubercle; *e*, comb of eighth segment. Larva: *c*, labial plate; *j*, antenna (Johannsen, 1937).

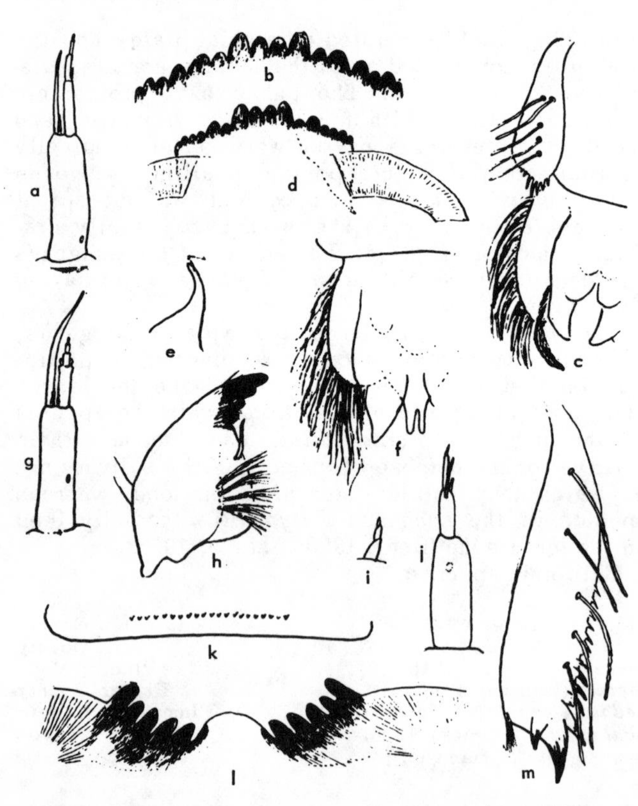

Fig. 14:34. Tendipedinae. *a-c*, *Tanytarsus (Endochironomus) nigricans* (Joh.). Larva: *a*, antenna; *b*, labial plate. Pupa: *c*, fin and lobe of eighth and ninth segments. *d,g,h,m*, *Tanytarsus (Tribelos) jucundus* var. *dimorphus* (Mall.). Larva: *d*, labial plate and paralabials; *g*, antenna; *h*, mandible. Pupa: *m*, fin of eighth segment. *e,f,i-l*, *Cryptochironomus psittacinus* (Mg.). Pupa: *e*, cephalic tubercles; *f*, anal lobe; *k*, posterior margin of fourth tergite. Larva: *i*, labral papilla; *j*, antenna; *l*, labial plate (Johannsen, 1937).

often conspicuously banded. One species is a common, leaf-mining pest of water lilies in Japan (Tokunaga and Kuroda, 1935).

California species:

colei (Malloch) 1919.	Del Norte County
taeniapennis (Coquillett) 1901.	Placer County

Genus *Tanytarsus* van der Wulp

(Fig. 14:21*i*; 14:32*n,q,s,t*; 14:34*a-d,g,h,m*)

This genus is not to be confused with *Calopsectra* Kieffer, which until Townes' (1945) paper was generally known as *Tanytarsus*. This genus includes as subgenera those species which have gone under the generic names of *Endochironomus* Kieffer and *Stictochironomus* Kieffer; the genera *Lauterbornia* Kieffer, *Phaenopsectra* Kieffer, *Sergentia* Kieffer, and *Lenzia* Kieffer are synonyms of *Tanytarsus (Tanytarsus)*.

The pupae of *Tanytarsus* have prothoracic respiratory organs consisting of a tuft of numerous white filaments as in *Tendipes*, but differ from that genus in having a comb of several teeth (figs. 14:32*q*; 14:34*c,m*) rather than a simple or compound spur on the hind corners of the eighth abdominal segment, and they differ from *Glyptotendipes* in the absence of a spike macelike process on the abdominal tergites. The larvae of the subgenus *Stictochironomus* have six-segmented antennae (fig. 14:32*n*) and an even number of labial teeth (fig. 14:32*s*), whereas the other subgenera have five-segmented antennae (fig. 14:34*a,g*) and an odd or even (fig. 14:34*b,d*) number of labial teeth, not being clearly differentiated from other genera. *T. (E.) nigricans* (Johannsen) (fig. 14:34*a-c*) is also a net-spinning plankton feeder, spinning its tubes within rolled or folded leaves of aquatic plants.

California species:

albescens Townes 1945.	Plumas County
dyari Townes 1945.	Humboldt, Placer counties
flavipes (Meigen) 1818.	Humboldt County
nigricans (Johannsen) 1905.	Del Norte, Santa Clara
profusus Townes 1945.	Central California
quagga Townes 1945.	Placer County

Genus *Xenochironomus* Kieffer

The only species whose immature stages are known is the Holarctic *X. xenolabis* (Kieffer), which lives as inquilines in sponges. The antennae of the larva are five-segmented and the labial plate has an odd number of teeth, the paralabials are broad and long, not widely separated, and the anterior margin of the labrum has two pairs of styles. The pupae have neither a comb nor a spur on the hind corners of the eighth abdominal segment and segments two to five each bear a number of dark brown spots, each with a curved black spinule on the dorsum. This genus has not yet been reported from California although five species are known in North America.

Genus *Cryptochironomus* Kieffer

(Figs. 14:21*b*; 14:34*e,f,i-l*)

The larvae of most of the species of this genus are predaceous, feeding on oligochaetes and midge larvae, and differ strikingly from those of other Tendipedini. The labial plate (fig. 14:34*l*) consists of a broad toothless middle part flanked by two obliquely placed rows of darker lateral teeth; the maxillary palpi are three to four times as long as broad; the antennal blade is attached to the second segment; and the sensorium is on the distal third of the first segment (fig. 14:34*j*). The prothoracic respiratory organs of the pupae consist of a tuft of numerous filaments as in *Tendipes*, but typically there is neither a comb nor a spur on the hind corners of the eighth abdominal segment (fig. 14:34*f*).

California species:

digitatus (Malloch) 1915.	Placer County
fulvus (Johannsen) 1905.	Lake County

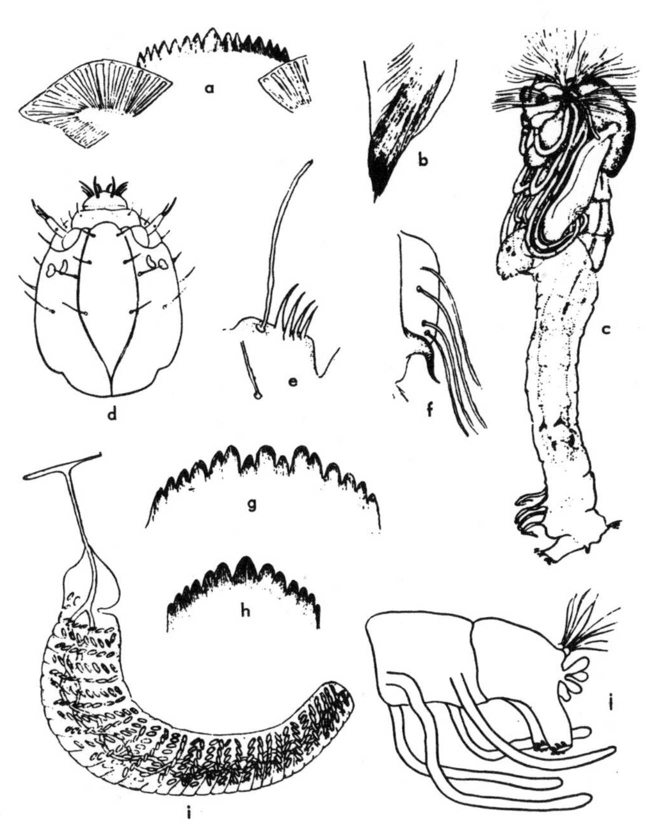

Fig. 14:35. Tendipedinae. *a,e, Harnischia abortiva* (Mall.): *a,* labial plate and paralabials of larva; *e,* comb of eighth segment of pupa. *b-d,g,i,j, Tendipes (Tendipes) riparius* (Mg.). Pupa: *b,* spur of eighth segment; *c,* pupa emerging from larval exuviae. Larva: *d,* head, dorsal view; *g,* labial plate; *i,* egg mass; *j,* caudal segments. *f,h, Tendipes (Limnochironomus) modestus* (Say): *f,* spur of eighth segment of pupa; *h,* labial plate of larva (Johannsen, 1937).

Genus *Harnischia* Kieffer

The larvae of *Harnischia* very closely resemble those of *Cryptochironomus,* differing principally in that the median part of the labial plate (fig. 14:35*a*) is more nearly the same dark color as the lateral, toothed parts, and the antennal blade is situated on the basal part rather than the distal end of the second segment. In the pupae the prothoracic respiratory organs consist of a tuft of whitish filaments; the hind corners of the eighth abdominal segment may lack either a comb or spur (subgenus *Harnischia*), or there may be a comb of two or three spines (fig. 14:35*e*) in which case there may be a transverse row of colorless, slender spines on the second sternite (subgenus *Parachironomus*) or on sternites one to three (subgenus *Cladopelma*). No records are available.

Genus *Glyptotendipes* Kieffer

(Figs. 14:21*c*; 14:31*e-i*)

The larvae of *Glyptotendipes* cannot be readily separated from those of *Tendipes,* but the pupae are easily distinguished by the presence of a spiked, macelike process (fig. 14:31*i*) on tergites two to six or (in the subgenus *Phytochironomus*) on tergites three to six. The larvae of some species mine in stems or leaves of aquatic plants where they are net-spinning plankton feeders. An undetermined species from Concord, Contra Costa County, is the only California record available.

Genus *Tendipes* Meigen 1800

(=*Chironomus* Meigen 1803)

(Figs. 14:20*b*; 14:21*a,e,j,k*; 14:31*j-l*; 14:35*b-d,f-j*)

As restricted by Townes (1945), *Tendipes* contains only the species of the subgenera *Limnochironomus* Kieffer, *Kiefferulus* Goetghebuer, *Einfeldia* Kieffer, *Chaetolabis* Townes, and *Tendipes s. str.* The larvae of this genus are characterized by five-segmented antennae, the labial plate with an odd number of teeth (fig. 14:35*g,h*), and short preanal papillae, and in the subgenus *Tendipes* there are often ventral gills present on the eighth abdominal segment (figs. 14:31*l*; 14:35*j*). The pupae have prothoracic respiratory organs consisting of a tuft of numerous whitish filaments (figs. 14:31*j*; 14:35*c*), the cephalic tubercles are acute, the posterior angles of the eighth abdominal segment may bear a comb (subgenus *Kiefferulus*) or a simple spur (fig. 14:35*b*) (subgenus *Limnochironomus*) or a compound spur (fig. 14:35*f*) (subgenus *Tendipes*).

The widespread large species *T. plumosus* (Linnaeus) is probably the most important organism in the profundal zone of eutrophic lakes, where it may become as abundant as 2,500 larvae per square meter in summer, the smaller larvae (hatched in the current season) preferring depths less than five meters while the large larvae (hatched the year before) move to the greater depths. In these large lakes there is one complete generation a year with the peak of adult emergence in May (Townes, 1938). Walshe (1951) states that *plumosus* is exceptional among the mud-dwelling *Tendipes* larvae in its habit of spinning a salivary net across the lumen of its mud tube, of irrigating the tube with undulations of its body, and eating the net with its entrapped phytoplankton and organic detritus. In shallower oxygen-deficient ponds and streams, *T. decorus* (Johannsen) may become the most important species, developing up to five generations a year in warmer waters.

California species:

anthracinus (Zetterstedt) 1860.	Plumas County
atrella Townes 1945.	Placer County
californicus (Johannsen) 1905.	Los Angeles County
decorus (Johannsen) 1905.	Widespread in California
nervosus (Staeger) 1839.	Placer County
plumosus (Linnaeus) 1758.	Southern California
utahensis (Malloch) 1915.	Los Angeles County

REFERENCES

ANDERSON, F. S.
1949. On the subgenus *Chironomus*. Vidensk. Medd fra Dansk Naturh. For., 111:1-65.
BERG, C. O.
1950. Biology of certain Chironomidae reared from *Potamogeton*. Ecol. Monogr., 20:83-101, 2 pls.
CHERNOVSKY, A. A.
1949. Detector of larvae of midges of the family Tendipedidae. Akad. Nauk U.S.S.R. Insect Detector no. 31, 186 pp. [in Russian].
CLAASEN, P. W.
1922. The larva of a chironomid (*Trissocladius equitans* n. sp.) which is parasitic upon a May-fly nymph (*Rithrogena* sp.). Univ. Kansas Sci. Bull., 14:395-405, 3 pls.
CODREANU, R.
1939. Recherches biologiques sur un chironomide, *Symbiocladius rhithrogenae* (Zavr.). Ectoparasite "cancérigène" des Ephémères torrenticoles. Arch. de Zool. Expt. et Gen., 81:1-283.
COE, R. L.
1950. Chironomidae. *In* Handbooks for the identification of British insects. 9:121-206. Royal Ent. Soc. London.
EDWARDS, F. W.
1929. British non-biting midges (Diptera, Chironomidae). Trans. Ent. Soc. Lond., 77:279-430, 3 pls.
JOHANNSEN, O. A.
1905. Aquatic Nematocerous Diptera II. Chironomidae, Bull. N.Y. St. Mus., 86:76-331, pls. 16-37.
1937. *See* Diptera references.
1946. Revision of the North American species of the genus *Pentaneura* (Tendipedidae: Chironomidae, Diptera). Jour. N.Y. Ent. Soc., 54:267-289, 1 pl.
JOHANNSEN, O. A., and H. K. TOWNES, JR.
1952. Tendipedidae (Chironomidae). *In* Guide to the insects of Connecticut. Part VI. Diptera. Fifth Fascicle. Midges and gnats pp. 1-148, 22 pls.
JONES, F. M.
1916. Two insect associates of the California pitcher plant, *Darlingtonia californica* (Dipt.). Ent. News, 27:385-392, 2 pls.
LENZ, F.
1921. Die Metamorphose der *Chironomus*-Gruppe. Morphologie der Puppen und Larven. Deutsch. Ent. Zeitschr., 1921:1-15 (separate).
1936. Die Metamorphose der Pelopiinae. *In* Lindner, Die Fliegen der palaarktischen Region. Lieferung 100, pp. 51-78.
1941. Die Jugenstadien der Sectio Chironomariae (Tendipedini) connectentes (Subf. Chironominae = Tendipedinae). Zusammenfassung und Revision. Arch. f. Hydrobiol., 38:1-69.
MALLOCH, J. R.
1915. The Chironomidae, or midges, of Illinois, with particular reference to the species occurring in the Illinois River. Bull. Illinois Lab. Nat. Hist., 10: 275-543, 23 pls.
PAGAST, F.
1947. Systematik und Verbreitung der um de Gattung *Diamesa* gruppierten Chironomiden. Arch. f. Hydrobiol. 41:435-596.
RISBEC, J.
1951. Les Dipteres nuisibles au riz de Camargue au debut de son developpement. Rev. de Path. Veg. et d'Ent. Agr. de France, 30:211-227.
ROBACK, S. S.
1953. Savannah River Tendipedid larvae (Diptera: Tendipedidae (=Chironomidae)). Proc. Acad. Nat. Sci. Phila., 105:91-132.
SAUNDERS, L. G.
1928. Some marine insects of the Pacific Coast of Canada. Ann. Ent. Soc. Amer., 21:521-545.
1929. The early stages of *Diamesa (Psilodiamesa) lurida* Garrett (Diptera, Chironomidae). Canad. Ent., 60: 261-264.
1930. The larvae of the genus *Heptagia*, with description of a new species (Diptera, Chironomidae). Ent. Mon. Mag., 66:209-214.
STRENZKE, K.
1950. Systematik, Morphologie und Okologie der terrestrischen Chironomiden. Arch f. Hydrobiol. Suppl., 18:207-414, 2 pls.
THIENEMANN, A.
1937. Podonominae, eine neue Unterfamilie der Chironomiden. (Chironomiden aus Lappland I.). Int. Rev. ges. Hydrobiol. und Hydrogr., 35:65-112, 3 pls.
1944. Bestimmungstabellen für die bis jetzt bekannten Larven und Puppen der Orthocladiinen. Arch. f. Hydrobiol., 39:551-664.
1954. *Chironomus*. Leben, Verbreitung und wirtschaftliche Bedeutung der Chironomiden. Die Binnengewässer, Band XX, 834 pp.
TOKUNAGA, M.
1936. Japanese *Cricotopus* and *Corynoneura* species (Chironomidae, Diptera). Tenthredo, 1:9-52.
TOKUNAGA, M., and M. KURODA
1935. Stenochironomid flies from Japan (Diptera), with a description of a new species. Trans. Kansai Ent. Soc., no. 6:1-6.
TOWNES, H. K., JR.
1938. VI. Studies on the food organisms of fish. *In* Biol. Surv. Allegheny and Chemung Watersheds, suppl. 27th Ann. Rept. N.Y. St. Conserv. Dept. pp. 162-175.
1945. The Nearctic species of Tendipedini [Diptera, Tendipedidae (=Chironomidae)]. Amer. Midl. Nat., 34:1-206, 261 figs.
WALSHE, B. M.
1950. The function of haemoglobin in *Chironomus plumosus* under natural conditions. Jour. Exp. Biol., 27: 73-95.
1951. The feeding habits of certain chironomid larvae (subfamily Tendipedinae). Proc. Zool. Soc. Lond., 121:63-79.
WIRTH, W. W.
1949. A revision of the clunionine midges with descriptions of a new genus and four new species (Diptera: Tendipedidae). Univ. Calif. Publ. Entom., 8:151-182.
ZAVREL, J., and A. THIENEMANN
1921. Die Metamorphose der Tanypinen. II. Arch. f. Hydrobiol. Suppl., 2:655-784.

Family HELEIDAE (=CERATOPOGONIDAE)

The biting midges, often called sand-flies, punkies, or "no-see-ums," are blood-sucking or predaceous relatives of the true midges or Tendipedidae, from which they can be separated by the presence of a well-developed proboscis, wings nearly always superimposed over the back when at rest, the media nearly always forked, and the postscutellum gently rounded without median furrow (fig. 14:36). Species of the genera *Culicoides, Lasiohelea,* and *Leptoconops* are bloodsuckers. Most of the other genera are predaceous on other small, soft-bodied insects, but special habits have been developed by some species of *Atrichopogon* and *Forcipomyia,* which suck blood from the wing veins or bodies of larger insects, and by the genus *Pterobosca* in which the tarsi are remarkably modified for attachment to the wings of dragonflies. In the Orient one species of *Culicoides* habitually takes secondhand vertebrate blood from the abdomens of engorged mosquitoes.

The immature stages of Heleidae are mostly aquatic, their habitats ranging from fresh to salt water. The larvae of most genera are elongate, eellike, and

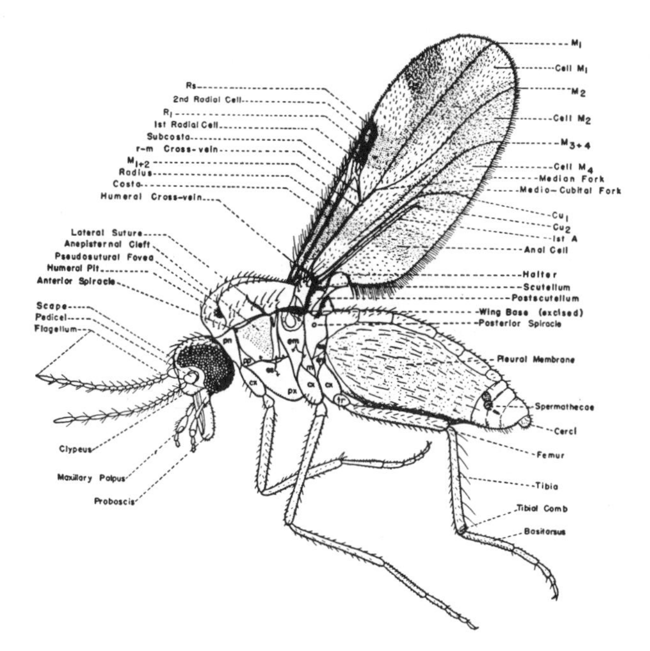

Fig. 14:36. *Culicoides palmerae* James, lateral view of female, left wing and right legs removed, with parts labeled (Wirth, 1952).

legless, and live as predators and scavengers in sand or mud, freely in the water and in algal and moss mats in the water and along the margins of watercourses. Accumulations of water in tree holes or in the bracts of plants are specially favored but limited habitats. Some of the terrestrial species are stouter and setose, with well-developed prolegs, and breed under moist decaying tree bark, in rich damp soil, or in compost piles (figs. 14:39*ff*; 14:41*a*). The pupae (fig. 14:41*b-d*) are typically conical in shape, with spinose abdomens bearing a pair of pointed apicolateral processes, which in the aquatic species usually aid in working the pupa to just above the water margin when the imago is about to emerge.

Most of the references to the taxonomy and biology of the biting midges can be found in the paper by Wirth (1952) on California Heleidae, from which most of the following keys and figures were taken. Special mention should be made, however, of Malloch's (1915) paper on the Illinois species, Johannsen's (1943) review of the genera and checklist of North American species and his (1952) synopsis of the New England species, Mayer's (1934) extensive treatment of the immature stages of the European species, and Thomsen's (1937) on the North American larvae and pupae.

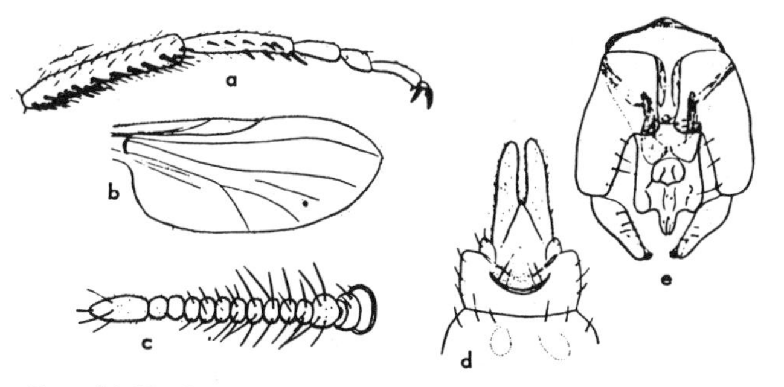

Fig. 14:37. *Leptoconops*. *a-d*, *L. freeborni* Wirth, female: *a*, hind tarsus; *b*, wing; *c*, antenna; *d*, apex of abdomen, ventral view. *e*, *L. kerteszi* K., male genitalia (Wirth, 1952).

Keys to the North American Genera of Heleidae

Adults

1. Cross vein r-m absent (fig. 14:37*b*); anterior media not forked; antennae of female with 12-14 segments (fig. 14:37*c*) (LEPTOCONOPINAE) .... *Leptoconops* Skuse
— Cross vein r-m present; anterior media with 2 branches, the posterior branch occasionally interrupted at base or rarely absent; antennae of female 15-segmented (fig. 14:36) 2
2. Empodium well developed, at least in the female (FORCIPOMYIINAE) 3
— Empodium small or vestigial 5
3. Wing with microtrichia large, macrotrichia erect and sparse when present; costa to about distal third of wing; 2nd radial cell longer than 1st (fig. 14:38*k*) .... *Atrichopogon* Kieffer
— Wing with microtrichia minute, macrotrichia abundant, long and sloping, often scalelike; costa usually to about middle of wing (fig. 14:38*l*) 4
4. Second radial cell longer than 1st, very narrow, ending well beyond middle of wing; posterior wing fringe hairlike (not California) .... *Lasiohelea* Kieffer
— Second radial cell short, not much longer than 1st; posterior wing fringe usually lanceolate .... *Forcipomyia* Meigen
5. First radial cell nearly or completely obliterated; second obliterated or square-ended, usually ending at or before middle of wing; eyes very short pubescent (fig. 14:38*g*) (DASYHELEINAE) ..... *Dasyhelea* Kieffer
— One or both radial cells well developed, the 2nd not markedly square ended, ending past middle of wing; eyes bare (HELEINAE) 6
6. Media petiolate, forking distad of level of r-m cross vein (fig. 14:38*h*) 7
— Media sessile, forking at or proximad of level of cross vein (fig. 14:39*h*) (STENOXENINI) 13
7. Claws of both sexes small, equal, and simple; macrotrichia usually abundant; 2 more or less equal radial cells; humeral pits prominent (fig. 14:36) (CULICOIDINI) .... *Culicoides* Latreille
— Claws of female large, equal or unequal; macrotrichia absent or scanty; 1 or 2 radial cells, the 2nd often larger than 1st; humeral pits small when present 8
8. One radial cell; claws equal; femora unarmed (not California .... *Parabezzia* Malloch
— Two anterior radial cells, if the 1st is obsolete the claws are unequal 9
9. Microtrichia of wing absent, membrane more or less milky white; 2 anterior radial cells more or less equal (HELEINI) 10
— Microtrichia present; 2nd radial cell longer than 1st (STILOBEZZIINI) 11
10. Claws of hind leg of female equal; wing usually unmarked; macrotrichia scanty or absent (fig. 14:38*r*) .... *Helea* Meigen
— Claws of hind leg of female unequal; wing with 2 to 12, small, black dots or streaks; macrotrichia numerous (fig. 14:38*h*)............... *Alluaudomyia* Kieffer
11. Claws of female unequal on all legs; 4th tarsal segments bilobed .... *Stilobezzia* Kieffer
— Claws on fore and midlegs of female equal; 4th tarsal segments cylindrical or cordate 12
12. Hind femora greatly swollen and armed with numerous spines; 4th tarsal segments cordate... *Serromyia* Meigen
— Hind femora without spines, moderately swollen; 4th tarsal segments cylindrical .... *Monohelea* Kieffer
13. Two radial cells (fig. 14:39*a,e*)................... 14
— One radial cell (fig. 14:39*f,h*) 22

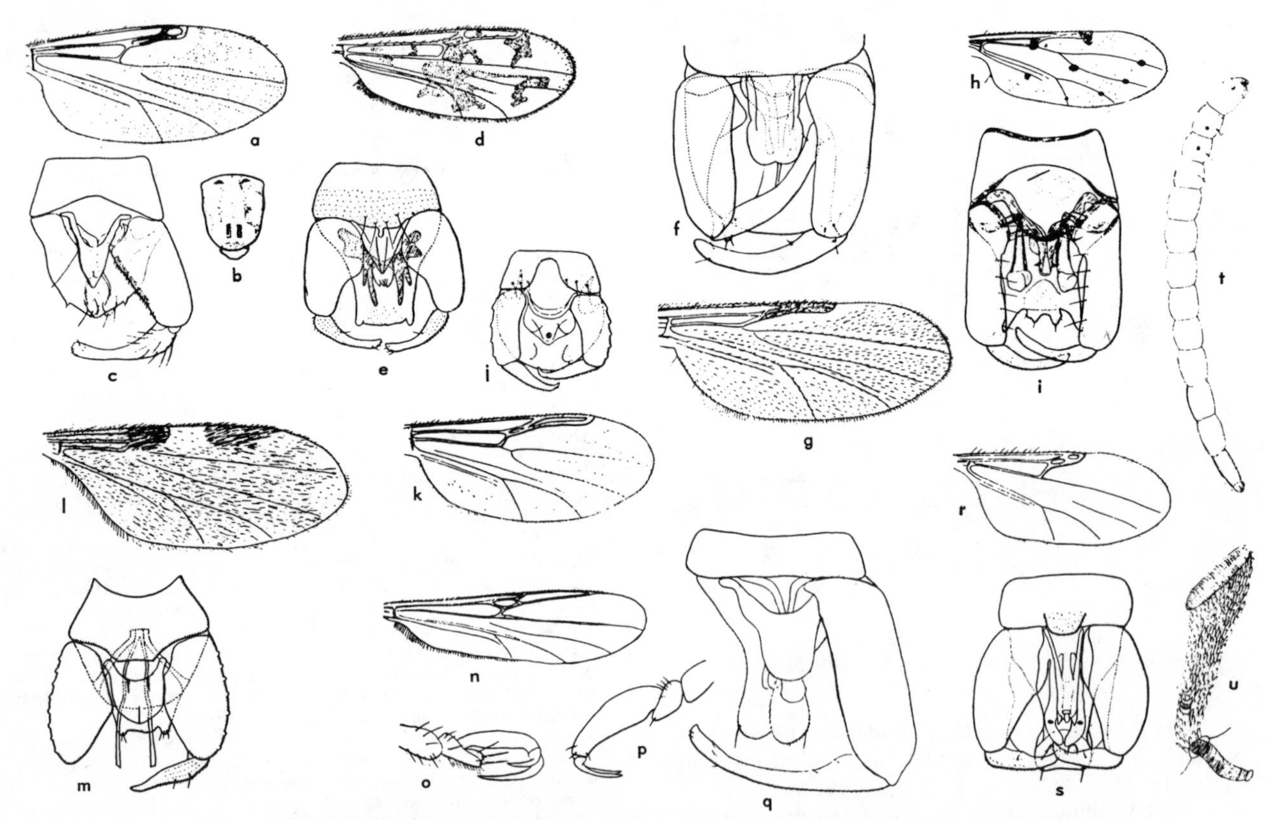

Fig. 14:38. Adult Heleidae. a-c, *Culicoides cockerellii* (Coq.): *a*, wing; *b*, mesonotum; *c*, male genitalia, right gonostyles removed. *d,e*, *Monohelea maculipennis* (Coq.): *d*, wing; *e*, male genitalia. *f,g*, *Dasyhelea cincta* (Coq.): *f*, male genitalia; *g*, wing. *h,i*, *Alluaudomyia bella* (Coq.): *h*, wing; *i*, male genitalia. *j,k*, *Atrichopogon occidentalis* Wirth: *j*, male genitalia; *k*, wing. *l,m*, *Forcipomyia macswaini* Wirth: *l*, wing; *m*, male genitalia. *n-q*, *Clinohelea usingeri* Wirth: *n*, wing; *o*, female hind tarsus, fourth and fifth segments; *p*, same, fore tarsus; *q*, male genitalia, right gonostyles removed. *r,s*, *Helea stigmalis* (Coq.): *r*, wing; *s*, male genitalia. *t*, *Dasyhelea grisea* (Coq.), larva; *u*, *Alluaudomyia bella* (Coq.), respiratory organ of pupa (a-s, Wirth, 1952; *t,u*, Thomsen, 1937).

14. Fifth tarsal segments armed below with straight, blunt, black spines (batonnets) (fig. 14:39*c*) 15
— Fifth tarsal segments without ventral batonnets (fig. 14:39*d*), sometimes with sharp, tapering spines 18
15. Claws of female unequal at least on mid and hind legs; not California .... *Dicrohelea* Kieffer
— Claws of female equal......................... 16
16. Femora armed below with short, stout, black spines .... *Sphaeromias* Stephens
— Femora unarmed 17
17. Wing broad, especially at base, anal lobe expanded; not California................. *Jenkinshelea* Macfie
— Wing narrow, anal lobe small.. *Johannsenomyia* Malloch
18. Femora armed below with strong spines; 5th tarsal segments not inflated 20
— Femora unarmed; 5th segment inflated on fore tarsi (fig. 14:38p) 19
19. Mid and hind legs of female with claws unequal or a single claw with basal barb, 4th tarsal segment deeply bilobed and armed with spines (fig. 14:38*o*); costa not produced beyond apex of 2nd anterior radial cell (fig. 14:38*n*); pleura with transverse silvery band........ .... *Clinohelea* Kieffer
— Claws equal on all legs; 4th tarsal segment cordiform; costa produced past apex of 2nd anterior radial cell .... *Neurohelea* Kieffer
20. Fore femora greatly swollen, the fore tibiae arched; apical 3rd of hind tibiae swollen; hind tarsi of female with a single, long basally barbed claw; wings fasciate; not California .... *Heteromyia* Say
— Fore femora not swollen or, if moderately so, fore tibiae not arched 21
21. Femora and tibiae all armed with long, sharp scattered spines; not California .... *Echinohelea* Macfie
— Fore femora, and often mid and hind pairs, armed ventrally with short, stout, black spines .... *Palpomyia* Meigen
22. Fifth tarsal segments of female with long, blunt, black spines (batonnets) below (fig. 14:39*i*); costa in female extending nearly to tip of wing (fig. 14:39*h*); femora unarmed .... *Probezzia* Kieffer
— Fifth tarsal segments unarmed.................... 23
23. Cross vein r-m very short, basal sections of radius and media very close together; M_2 strongly elbowed at base in female; costa extending to wing tip; femora unarmed; not California .... *Stenoxenus* Coquillett
— Cross vein r-m longer, basal radial cell not unusually narrow; costa to about 3/4 of wing length (fig. 14:39*f*) .... *Bezzia* Kieffer

Larvae

1. All body segments with short spines; prothoracic prolegs present (figs. 14:39*ff*; 14:41*a*) (FORCIPOMYIINAE) 2
— No spines on body segments; prothoracic prolegs absent (fig. 14:38*t*) 3

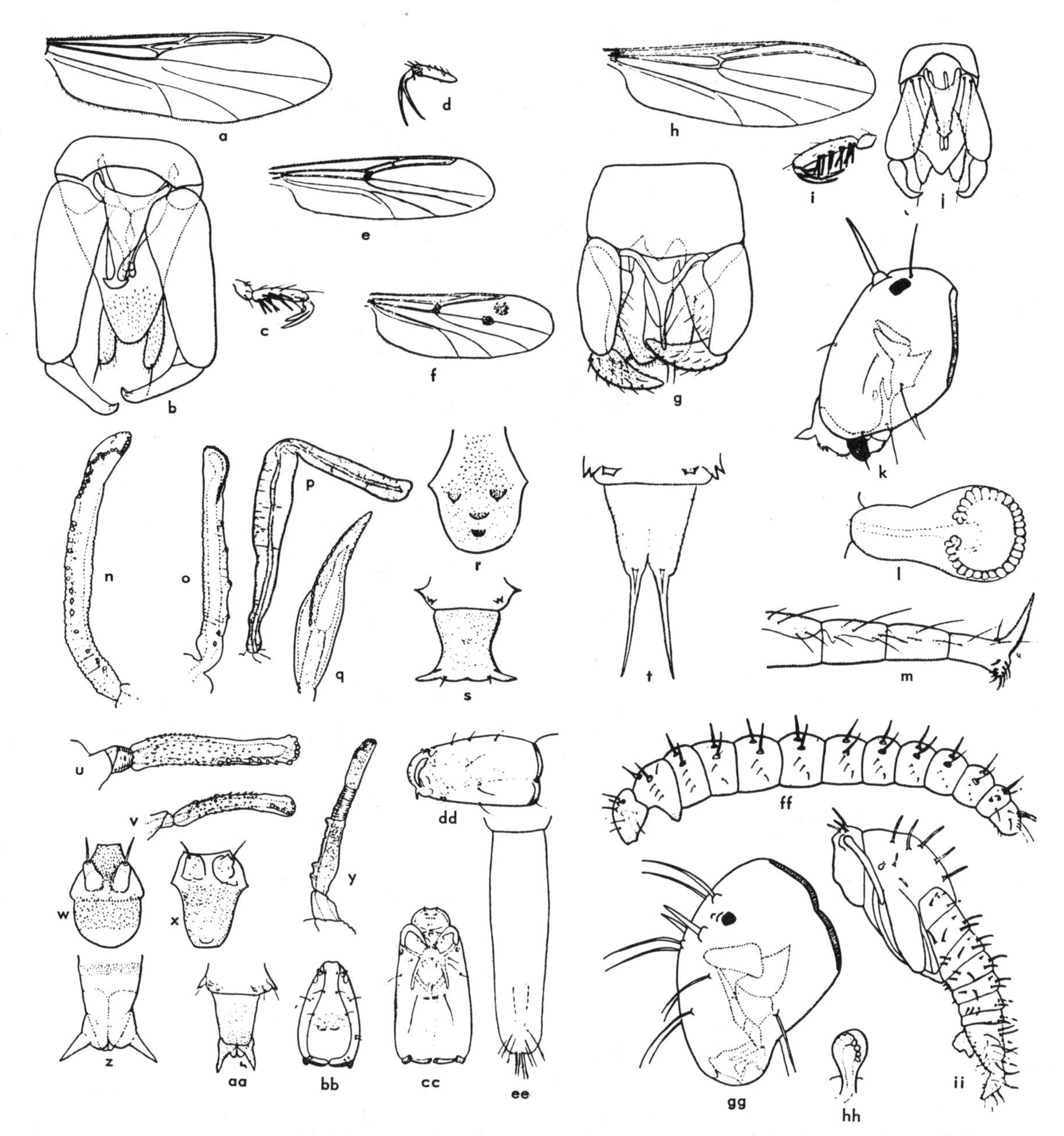

Fig. 14:39. Heleidae. a-c, *Johannsenomyia sybleae* Wirth: a, wing; b, male genitalia; c, fifth tarsal segment of female. d,e, *Palpomyia kernensis* Wirth: d, fifth tarsal segment of female; e, wing. f,g, *Bezzia punctipennis* (Will.): f, wing; g, male genitalia. h-j, *Probezzia flavonigra* (Coq.); h, wing; i, fifth tarsal segment of female; j, male genitalia. k-m, *Forcipomyia calcarata* (Coq.), larva: k, head, lateral view; m, caudal segments, lateral view; l, respiratory organ of pupa. n-t, *Dasyhelea* spp. Respiratory organ of pupa: n, *johannseni* (Mall.); o, *mutabilis* (Coq.); p, sp., *longipalpus* group; q, *pollinosa* Wirth. Operculum of pupa: r, *mutabilis*. Last segments of pupae: s, sp., *longipalpis* group; t, *pollinosa* Wirth. u-ee, *Culicoides*, immature stages. Respiratory organs of pupa: u, *luteovenus* R. and H.; v, *unicolor* (Coq.); y, *variipennis* (Coq.). Operculum of pupa: w, *variipennis*; x, *luteovenus*. Last segment of pupa: z, *variipennis*; aa, *unicolor*. Head of larva: bb, *variipennis*; cc,dd, *unicolor*. Last segment of larva: ee, *unicolor*. ff-ii, *Forcipomyia cinctipes* (Coq.), Larva: ff, lateral view; gg, head. Pupa: hh, respiratory organ; ii, pupa with larval exuviae attached (Wirth, 1952).

2. Body flattened, oval in cross section; lateral processes of body at least as long as segments (fig. 14:41*a*) .... *Atrichopogon* Kieffer
— Body circular in cross section, or if flattened slightly, segments with processes less than 1/2 as long as segment (fig. 14:39*k,m,ff-hh*)......... *Forcipomyia* Meigen
3. Head well developed, head capsule well sclerotized (figs. 14:38*t*, 14:39*bb-dd*; 14:42*i,j*) 4
— Head capsule not sclerotized; internal system of heavily sclerotized rods and levers (fig. 14:40*a*) (LEPTOCONOPINAE) .... *Leptoconops* Skuse
4. Last body segment with retractile proleg bearing 10 to

12 hooks; habits clambering (fig. 14:38*t*) (DASYHELEINAE) .... *Dasyhelea* Kieffer
— Last body segment without proleg or hooks, with fine hairs only; habits swimming (fig. 14:39*ee*; 14:42*h*) (HELEINAE) 5
5. Head short and thick, not more than 1.5 times as long as broad (fig. 14:39*dd*); body segments each slightly longer than head 6
— Head long, more than twice as long as broad (fig. 14:42*i,j*); body segments long and slender 7
6. Head pear-shaped; body segments wider than head; length more than 5 mm. .... *Palpomyia* Meigen (part), *Sphaeromias* Curtis
— Head oval; body segments not wider than head; length 4-5 mm. .... *Culicoides* Latreille
7. Anal hairs as long as, or longer than, last segment .... *Alluaudomyia* Kieffer
— Anal hairs usually shorter than last segment (fig. 14:42*h*) .... *Palpomyia* Meigen (part, *Bezzia* Kieffer, *Johannsenomyia* Malloch

Pupae

1. Pupa with larval exuviae attached to last 3 segments (fig. 14:39*ii*); respiratory organ short, knoblike (fig. 14:39*l,hh*) (FORCIPOMYIINAE) 2
— Pupa free from larval exuviae; respiratory organ elongated (fig. 14:39*n-q,u,v*) 3
2. Abdomen with branched or setaceous projections on first 5 segments (fig. 14:41*b*)..... *Atrichopogon* Kieffer
— Abdomen with spines or stumplike projections on all but last segment (fig. 14:39*ii*)..... *Forcipomyia* Meigen
3. Anal segment with pair of setigerous protuberances in addition to apicolateral processes (fig. 14:39*s*) (DASYHELEINAE) .... *Dasyhelea* Kieffer
— Anal segment without additional setigerous protuberances (fig. 14:42*o-t*) 4
4. Respiratory organ ending in short, dark, ovate, or barrel-shaped structure with about 10 spiracles (fig. 14:40*b,d*) (LEPTOCONOPINAE)....*Leptoconops* Skuse
— Respiratory organ funnel-shaped, clavate, or tubular (fig. 14:42*a-f,k*) (HELEINAE) 5
5. Respiratory organ funnel-shaped, entirely covered with scales; with spiracles in pairs (fig. 14:38*u*) .... *Alluaudomyia* Kieffer
— Respiratory organ clavate or tubular, with spiracles not in pairs 6
6. Respiratory organ tubular (fig. 14:39*u,v,y*) 7
— Respiratory organ clavate or spoon-shaped 8
7. Respiratory organ with constriction near base; operculum without spines ....*Stilobezzia* Kieffer
— Respiratory organ with constriction at basal 4th or 3rd, operculum with spines (fig. 14:39*w,x*) .... *Culicoides* Latreille
8. Operculum with 1 pair of setae (fig. 14:42*g*) .... *Palpomyia* Meigen
— Operculum with 2 to 4 pairs of setae (fig. 14:42*l-n*) .... *Bezzia* Kieffer

California Species of Heleidae

Subfamily LEPTOCONOPINAE

Genus *Leptoconops* Skuse

(Figs. 14:37*a-e*; 14:40)

Species of *Leptoconops* are most prevalent in desert and semidesert regions the world over. The shiny black females with milky white wings are vicious biters, preferring to attack in the open in the bright

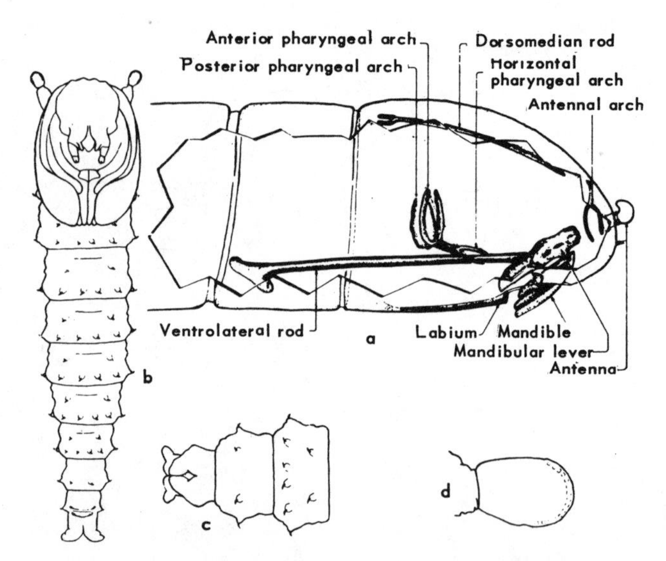

Fig. 14:40. *Leptoconops.* *a,c,d, L. torrens* (Towns.). Larva: *a*, cutaway view of head and two thoracic segments. Pupa: *c*, caudal segments; *d*, respiratory organ. *b, L. kerteszi* K., pupa, ventral view (Smith and Loew, 1948).

sunlight. The biology of the "Bodega gnat" and the "Valley black gnat," our two commonest American species, has been described in detail by Smith and Lowe (1948). The orange-colored larvae of *L. kerteszi* Kieffer, the Bodega gnat, seem to prefer sandy soil near the water line in tidal marshes or on saline lake margins of the western states, whereas those of *L. torrens* (Townsend), the valley black gnat, have been found in wet sinks of the finely compacted clay-adobe soil regions of the western Sacramento Valley and Santa Clara Valley. Development is slow, requiring a year for a generation of *kerteszi* and at least two years for *torrens*.

Keys to the California Species of Leptoconops

Adults

1. Frons with numerous long hairs; basitarsi with 15 to 20 short, stout, dark spines (fig. 14:37*a-d*); body and legs with vestiture of dense, long hairs; coastal marshes in Ventura County; Monterey County; San Nicolas Island .... *freeborni* Wirth 1952
— Frons bare or with a pair of short hairs between eyes; basitarsi with no more than 8 slender spines; body and legs with sparse hairs, mostly short 2
2. Antenna of female 14-segmented (flagellum 12-segmented); basitarsi with no strong differentiated ventral spines; 9th tergite of male with 2 long, widely spaced, apicolateral processes; west side of Central Valley; Santa Clara Valley; Riverside County .... *torrens* (Townsend) 1893
— Antenna of female 13-segmented (flagellum 11-segmented); basitarsi with 2 to 8 spines; 9th tergite of male with a caudomesal pair of short processes (fig. 14:37*e*); coastal salt marshes and salt and alkaline lakes throughout California ...... *kerteszi* Kieffer 1908

Larvae

1. Mature larva orange colored; body with apparently 23

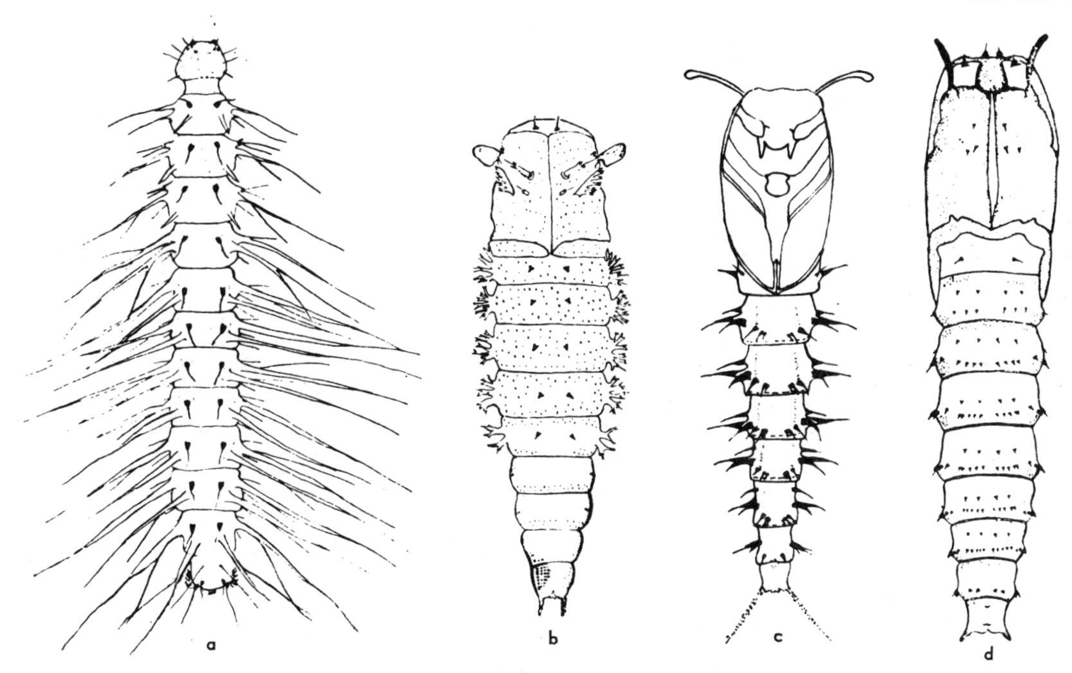

Fig. 14:41. Heleidae, immature stages. *a,b, Atrichopogon peregrinus* (Joh.): *a*, larva; *b*, pupa. *c, Bezzia glabra* (Coq.), pupa. *d, Stilobezzia bulla* Thomsen, pupa (Thomsen, 1937).

segments; eyespots present; dorsolateral rods of pharyngeal apparatus present .... *kerteszi* Kieffer

— Mature larva whitish; body with apparently 21 segments; eyespots absent; dorsolateral rods of pharyngeal apparatus absent (fig. 14:40*a*) ..... *torrens* (Townsend)

Subfamily FORCIPOMYIINAE

Genus *Atrichopogon* Kieffer

The species of this genus are mostly aquatic, but one very common North American species, *levis* (Coquillett), breeds in soil at the roots of lawn grass. The biology and immature stages of several aquatic North American species were discussed by Thomsen (1937), and one of the best biological papers on this genus was recently published by Nielsen (1951). The immature stages commonly occur on wet stones in or beside small streams, on floating logs, in blanket algae in ponds, or on the sides of tanks or concrete pools. The larvae have flattened bodies with prominent, segmental, dorsal, and lateral processes (fig. 14:41*a*). Larval and pupal characters (fig. 14:41*b*) of the few known species are good, but many groups of adults can be separated only by quantitative characters, and many of the species keyed out below are probably complexes. Several species suck the blood of blister beetles (Meloidae).

Keys to Adults of the California Species of Atrichopogon

Females

1. Proboscis about 1.4 times as long as height of head; clypeus triangularly produced between antennae; large species (length 2.0 mm.); widespread throughout California .... *fusculus* (Coquillett) 1901

— Proboscis not longer than height of head; clypeus with base truncate or not strongly produced between antennae; size smaller (1-1.6 mm.) 2

2. Scutellum yellow with 4 long, black, marginal bristles; humeri usually extensively yellow; antennae long, distal segments about 3 times as long as broad; wings nearly bare; widespread .... *levis* (Coquillett) 1901

— Scutellum dark brown; mesonotum brown or black... 3

3. Scutellum with 2 bristles; wing nearly bare; small, polished black species ; Humboldt and Siskiyou counties .... *minutus* (Meigen) 1830

— Scutellum with 4 bristles 4

4. Mesonotum usually with a sublateral pair of translucent brown lines from ends of scutellum to humeral depressions; mesonotal setulae in rows; widespread .... *websteri* (Coquillett) 1901

— Mesonotum without translucent lines, with dense scattered setae 5

5. Wing nearly bare, macrotrichia confined to tips of cells R_5 and M_1; 7th abdominal sternite with median, forked, spinous appendage; Humboldt, Siskiyou, Modoc, Tulare, and Monterey counties; Transition Zone ... . *arcticus* (Coquillett) 1900

— Wing with extensive macrotrichia at least to cell M_2; 7th sternite unmodified 6

6. Macrotrichia only to cell M_2; mesonotum shining; distal antennal segments each twice as long as broad; eyes pubescent; Transition Zone of northern Coast Ranges and Sierra Nevada .... *transversus* Wirth 1952

— Macrotrichia to middle of anal cell (fig. 14:38*k*); distal antennal segments 3 times as long as broad; eyes bare; northern and central California .... *occidentalis* Wirth 1952

Male Genitalia

1. Ninth sternite deeply excavated in middle 3rd of posterior margin, with marginal hairs not extending entirely across sternite (fig. 14:38*j*) 2

— Ninth sternite transverse or shallowly and broadly

excavated, with marginal row of hairs entirely across sternite 3
2. Ninth sternite with 3 hairs on each caudolateral margin; caudomedian lobe of aedeagus rounded (fig. 14:38*j*) .... *occidentalis* Wirth
— Ninth sternite with compact group of 3 to 5 hairs in row in middle of posterior margin; median point of aedegus sharp .... *websteri* (Coquillett)
3. Parameres expanded broad and platelike and almost meeting mesad, anterior arm reduced; aedeagus quadrate with semidetached, dorsoposterior sclerite bearing 2 small distal spines .... *arcticus* (Coquillett)
— Parameres hooklike with distinct anterior arm; aedeagus without spinose, detached, dorsoposterior sclerite 4
4. Aedeagus with broad, caplike apex 5
— Aedeagus with apex rounded or with sharp median point 6
5. Ninth sternite about twice as broad as long, caudal margin transverse; diststyles with very slender tips; parameres expanded caudad ...... *fusculus* (Coquillett)
— Ninth sternite more than 3 times as broad as long, with broad, shallow, mesal excavation; diststyles tapering to moderate tips; parameres not expanded mesad .... *minutus* (Meigen)
6. Aedeagus with caudomedian rounded lobe; 9th sternite transverse, with about 20 hairs...... *transversus* Wirth
— Aedeagus with distinct sharp caudomedian point; 9th sternite with broad, shallow, excavation with about 10 hairs .... *levis* (Coquillett)

Genus *Forcipomyia* Meigen

(Figs. 14:38*l,m*; 14:39*k-m,ff-ii*)

Saunders (1924) has published our best account of the immature stages of *Forcipomyia*. Most of the species are terrestrial, the gregarious larvae breeding under tree bark, in manure or moist decaying vegetable matter, in ant nests, and the like. However, there are many subaquatic species frequenting damp moss and algae, and in tropical America they often breed in the water inside the leaf bases of bromeliads and other plants. One such species was imported to Hawaii in pineapple plants. *F. calcarata* (Coquillett) larvae (fig. 14:39*k,m*) were found along the rocky margin of a warm mineral spring at Alum Rock Park in Santa Clara County (Wirth, 1952). Females of *eques* (Johannsen) and related species suck the blood from the wing veins of lacewing flies, butterflies, and other insects. The closely related genus *Pterobosca,* which is found in the southeastern states and more commonly throughout the tropics, has modified tarsi for attachment to the wings of dragonflies.

Keys to Adults of the California Species of Forcipomyia

Females

1. Tarsal ratio (length of 1st segment of hind tarsus divided by length of 2nd) less than 1.2; 2nd radial cell not elongate (subgenus *Forcipomyia*)............ 2
— Tarsal ratio greater than 1.2; 2nd radial cell elongate 14
2. Tarsal ratio 0.4–0.9; 3rd palpal segment swollen to tip; tibiae without lanceolate scales 3
— Tarsal ratio 0.75–1.2; 3rd palpal segment swollen at base only; tibiae with or without lanceolate scales 5
3. Abdominal segments without apical yellow bands: legs with long yellow bristles, femora and bases of tibiae brown; wing with dark stigma of long, shaggy black hairs; tarsal ratio 0.5–0.6; San Joaquin Valley...... .... *brookmani* Wirth 1952
— Abdominal segments with apical yellow bands; legs uniform brown, with long black bristles; wing without dark costal spot; tarsal ratio 0.67–0.9; larvae under tree bark 4
4. Tarsal ratio 0.9; southern California .... *texana texana* (Long) 1902
— Tarsal ratio 0.67; northern California .... *texana simulata* Walley 1932
5. Mid and hind tibiae each with an external row of suberect, lanceolate scales 6
— Mid and hind tibiae without lanceolate scales 8
6. Front tibia without external row of lanceolate scales; widespread in California ..*bipuncatata* (Linnaeus) 1767
— Front tibia with an external row of lanceolate scales 7
7. Shining black species with prominent, long, black bristles; wing about 2.0 mm.; transition in northern California .... *cilipes* (Coquillett) 1900
— Dull brown species without prominent bristles; wing about 1.5 mm.; widespread. *squamipes* (Coquillett) 1902
8. Mesonotum shining black (tarsi whitish, wings, extensively pale; tarsal ratio 1.2); widespread; larvae in manure, etc.............. *brevipennis* (Macquart) 1826
— Mesonotum dull brown to yellow 9
9. Wing with at least 2 light marginal spots........... 10
— Wing with not more than 1 light costal spot......... 11
10. Wing with 6 pale marginal spots (fig. 14:38*l*); legs unbanded; central California; larvae in compost piles .... *macswaini* Wirth 1952
— Wing with 2 pale costal spots; legs with broad yellow bands; widespread; larvae under pine bark .... *cinctipes* (Coquillett) 1905
11. Halteres with dark knobs; unmarked, dull brown species; tarsal ratio 0.5; widespread................ .... *occidentalis* Wirth 1952
— Halteres entirely whitish 12
12. Mesonotum dull golden yellow with lighter median band; abdomen banded; Kern County...*quatei* Wirth 1952
— Mesonotum uniform dull brown 13
13. Wings with definite yellow costal spot; northern and central California .... *townesi* Wirth 1952
— Wings with uniform brown hairs; Riverside County .... *hurdi* Wirth 1952
14. Tarsal ratio 1.5–2.0; (subgenus *Euforcipomyia*); southern and central California ..*calcarata* (Coquillett) 1905
var. *sonora* Wirth 1952
— Tarsal ratio 2.0 or more (subgenus *Thyridomyia*).... 15
15. Last 5 antennal segments moniliform; tarsal ratio 2.9; northern and central California .... *monilicornis* (Coquillett) 1905
— Last 5 antennal segments each about twice as long as broad; tarsal ratio 2.2; southern and central California .... *colemani* Wirth 1952

Male Genitalia

1. Parameres present and well developed; generally elongate (fig. 14:38*m*)........................... 2
— Parameres absent or in form of short hyaline plates (subgenus *Euforcipomyia*)...................... 16
2. Parameres short and stout, not connected mesad at bases, arising from an expanded triangular plate extending cephalad of root of basistyles; 9th sternite with deep caudomesal excavation or window (subgenus *Thyridomyia*)................................ 15
— Parameres usually nearly as long as basistyles (fig. 14:38*m*), connected mesad on posterior margins (subgenus *Forcipomyia*) 3
3. Parameres fused at bases for distance greater than 4 times basal width of free parts 4

— Parameres not or but very shortly fused at bases (fig. 14:38*m*) 10
4. Parameres fused 0.6 of distance to tip, free parts pressed together halfway, abruptly narrowed to well-separated fingerlike apices on distal 0.2; aedeagus narrow, posterior margin with lateral teeth .. *hurdi* Wirth
— Parameres fused less than halfway to tips 5
5. Apices of parameres expanded, anterior margin of fused part with deep mesal cleft 6
— Apices of parameres tapering to filiform tips 7
6. Apices of parameres blunt and greatly surpassing tip of aedeagus caudad; anterior margin of fused part with deep mesal cleft .... *texana texana* (Long)
— Apices of parameres sharp, subequal to, and appressed to, tip of aedeagus; anterior margin of fused part truncate .... *texana simulata* Walley
7. Fused part of parameres about 1/4 of total length 8
— Fused part of parameres 1/3 to 1/2 of total length .. 9
8. Aedeagus with pair of low teeth on lateral margins; fused part of parameres broader than long; dististyles arcuate; 9th sternite with bilobate caudal margin ... .*cilipes* (Coquillett)
— Aedeagus with lateral margins entire; fused part of parameres longer than broad; dististyles sinuate; 9th sternite rounded caudad................ *quatei* Wirth
9. Aedeagus with apex pointed; fused part of parameres about 1/3 of total length.......*brevipennis* (Macquart)
— Aedeagus with rounded apex; fused part of parameres about 1/2 of total length ....*brookmani* Wirth
10. Parameres filiform on distal 1/3 or more 11
— Parameres stout distad 13
11. Parameres fused at base for distance about equal to basal width.................................. 12
— Parameres at least slightly separated at bases .... *bipunctata* (Linnaeus)
12. Basistyles with distinct patch of dark spines on mesal margin; aedeagus more than 1.2 times as long as broad .... *townesi* Wirth
— Basistyles without patch of spines on mesal margin; aedeagus as broad as long .....*squamipes* (Coquillett)
13. Parameres fused at bases 14
— Parameres well separated at bases by about 1/4 their total length, straight and rodlike, of uniform thickness to tips; aedeagus more than 1.5 times as long as broad .... *cinctipes* (Coquillett)
14. Parameres longer than basistyle; thickened from bases to tips; aedeagus shield-shaped (fig. 14:38*m*)....... .... *macswaini* Wirth
— Parameres about 0.75 times as long as basistyles, rather heavy, but apices tapered; aedeagus heavily sclerotized, with strong, expanded sublateral, caudal lobes, deeply notched mesad both on fore and hind margins .... *occidentalis* Wirth
15. Parameres straight and clublike, with expanded caudal apices; 9th tergite broadly rounded, reaching nearly to apices of basistyles .... *colemani* Wirth
— Parameres sinuate, sharp and incurved toward aedeagus distad; 9th tergite short, not reaching more than 1/2 the length of basistyles ...... *monilicornis* (Coquillett)
16. Aedeagus about as long as broad, bell-shaped in outline, anterior basal arch extending more than halfway to tip; dististyle slender, more than 5 times as long as greatest breadth .... *calcarata* (Coquillett)
— Aedeagus more than 1.5 times as broad as long, in outline shaped like a derby hat, anterior arch shallow, about 1/3 of total length; dististyle very broad, about 3 times as long as greatest breadth... var. *sonora* Wirth

Subfamily DASYHELEINAE

Genus *Dasyhelea* Keiffer

(Figs. 14:38*f,g,t*; 14:39*n-t*)

This genus is more aquatic than the preceding ones, and favorite habitats include blanket algae of ponds or slow streams, algae on dripping banks, and the fermenting sap of tree ulcers. The larvae cannot swim, but travel by crawling over the substrate and will drown if submerged completely.

Keys to the California Species of Dasyhelea

Females

1. Posterior margins of abdominal tergites light yellow 2
— Abdominal tergites uniformly dark................. 6
2. Abdominal pleura with small black streaks 3
— Abdominal pleura with large dark patches on segments 3-6 or all light 5
3. Mesonotum with compact central tuft of long, scalelike bristles; Kern, Imperial, and Riverside counties .... *cincta* (Coquillett) 1901
— Mesonotum without central tuft of long, scalelike bristles 4
4. Anterior veins of wing brownish, intermediate antennal segments elongate; Kern and Inyo counties......... .... *brookmani* Wirth 1952
— Anterior veins of wing whitish, intermediate antennal segments moniliform; Colorado River ....*pallens* Wirth 1952
5. Mesonotum grayish-green pruinose with 3 contrasting darker, narrow, longitudinal vittae; abdominal pleura 3-6 extensively black; southern and central California .... *grisea* (Coquillett) 1901
— Mesonotum uniformly brassy pollinose; abdominal pleura uniformly pale; Imperial, Kern, and San Luis Obispo counties ....*pollinosa* Wirth 1952
6. Halteres with knobs entirely brown to black; mesonotum velvety black; scutellum black; Modoc, Tulare, and San Luis Obispo counties.............. *atrata* Wirth 1952
— Halteres with knobs yellow or white, at least above 7
7. Mesonotum shining 8
— Mesonotum pruinose or quite dull 9
8. Mesonotum extensively yellowish on sides and humeri; 3 broad, brown vittae in middle; Kern and Merced counties .... *ancora* (Coquillett) 1902
— Mesonotum uniformly black, humeri narrowly yellow; southern California .... *tristyla* Wirth 1952
9. Mesonotum extensively whitish or silvery pruinose; pattern in lines or spots 10
— Mesonotum dull or velvety black or both, with no pruinose markings 13
10. Humeri with large quadrate area yellow; silvery pruinose mesonotal pattern extensive; Imperial and Riverside counties.................. *festiva* Wirth 1952
— Humeri black or very narrowly yellow 11
11. Large species (more than 2 mm.); many long bristlelike hairs; mesonotum with 3, narrow, longitudinal, setigerous, pruinose gray vittae; Humboldt County .... *tenebrosa* (Coquillett) 1905

— Smaller species (1.5 mm. or less); hairs not prominent; mesonotum with different pattern 12
12. Scutellum brown in middle; mesonotal pruinescence broken into spots and patches; abdomen uniformly black; halteres with knobs black below; Lake, San Benito, and Riverside counties .... *pritchardi* Wirth 1952
— Scutellum usually entirely yellow; mesonotum with velvet-black and silvery pruinose vittae; abdomen yellow below; halteres with knobs entirely white; widespread in California .... *mutabilis* (Coquillett) 1901
13. Halteres with knob dark except flat end yellowish; northern and central California .... *johannseni* (Malloch) 1915
— Halteres with knob entirely whitish (separable only by male genitalia) central and southern California .... *thomsenae* Wirth 1952
Modoc, Mono, and Tulare counties . *bifurcata* Wirth 1952
Santa Clara, Santa Barbara, and San Luis Obispo counties.................. *sanctaemariae* Wirth 1952

Male Genitalia

1. Parameres symmetrical (fig. 14:38*f*) 2
— Parameres asymmetrical 7
2. Hyaline envelope over ventral face of aedeagus; apicolateral processes of 9th tergite short (fig. 14:38*f*) 3
— No hyaline envelope over ventral side of aedeagus; apicolateral processes more than twice as long as broad.......................... *johannseni* (Malloch)
3. Hyaline envelope not covering distal half of posterior apodemes of aedeagus; these long, with obliquely truncated, pointed apices .... *pollinosa* Wirth
— Hyaline envelope completely covering aedeagus, including posterior apodemes; these with rounded apices 4
4. Hyaline envelope over aedeagus about 1/2 again as long as broad; apicolateral processes with slender tips (fig. 14:38*f*) .... *cincta* (Coquillett)
— Hyaline envelope nearly as broad as long; apicolateral processes with short rounded lobes 5
5. Posterior margin of hyaline envelope with irregular serrations............................ *pallens* Wirth
— Posterior margin of hyaline envelope smoothly rounded 6
6. Dististyles short and stout, about 3/4 as long as basistyles; anterior arch of aedeagus 3/4 of distance to posterior margin .... *brookmani* Wirth
— Dististyles long and slender, nearly as long as basistyles; anterior arch of aedeagus about 1/2 of distance to posterior margin .... *tenebrosa* (Coquillett)
7. Dististyles cleft nearly to base in 3 unequal teeth; apicolateral processes a pair of large, stout, triangular lobes with closely approximated bases ... *tristyla* Wirth
— Dististyles simple; apicolateral processes small and widely separated 8
8. Aedeagus much broader than long, with 2 lateral pairs of posterior lobes or apodemes 9
— Aedeagus about as long as broad, with a pair of slender, submedian apodemes and often a median posterior lobe 10
9. Apicolateral processes long and slender; posterior sclerite of parameres short, broad, with blunt apex .... *atrata* Wirth
— Apicolateral processes broader than long; posterior sclerite of parameres long, abruptly bent ventrad past tip of aedeagus, with slender recurved, pointed tip 1/2 as long as basal part .... *pritchardi* Wirth
10. Inner margin of basistyles with prominent, curved, sclerotized hook; dististyles stout with rounded tips; posterior sclerite of parameres stout, sinuate, 1/2 again as long as basistyles, with slender tip .... *grisea* (Coquillett)
— Inner margin of basistyles smooth or with low triangular prominence; dististyles with pointed tips 11
11. Ninth sternite produced mesally on posterior margin past level of basal 1/2 of aedeagus 12
— Ninth sternite not produced mesally past anterior arch of aedeagus 14
12. Posterior extension of 9th sternite less than 1/3 of width of aedeagus in narrowest part 13
— Posterior extension of 9th sternite broadly rounded, filling space between base of basistyles; apicolateral processes long, with constricted bases .... *sanctaemariae* Wirth
13. Ninth sternite with fine mesal point at level of basal bridge of aedeagus, articulating with a heavy mesal sclerite of aedeagus................ *thomsenae* Wirth
— Ninth sternite with Y-shaped posterior bifurcation extending past tip of aedeagal apodemes, with arms twice as broad as narrowest part; aedeagus without mesal sclerite .... *bifurcata* Wirth
14. Posterior sclerite of parameres very stout, distal 1/3 abruptly recurved ventrolaterad.......... *festiva* Wirth
— Posterior sclerite of parameres slender and nearly straight 15
15. Apicolateral processes long and slender; aedeagus with rounded ventroposterior lobe, mesal sclerite absent .... *ancora* (Coquillett)
— Apicolateral processes as broad as long; aedeagus with slender mesal sclerite as long as submedian apodemes .... *mutabilis* (Coquillett)

Pupae

1. Respiratory organs very long, much flattened and twisted, entirely made up of very narrow annuli, each annulus with a pair of minute spiracles on lateral margins (fig. 14:39*p*) (*longipalpus* group) .... sp. 1, Wirth 1952
— Respiratory organs clavate or wedge-shaped, not greatly flattened, annuli if present are broader and not extending more than 3/4 way to apices 2
2. Respiratory organs wedge-shaped, with sharp-pointed apices (fig. 14:39*q*); apicolateral processes of abdomen very long and pointed, not divergent (fig. 14:39*t*).... .... *pollinosa* Wirth
— Respiratory organs clavate, the apices broadly rounded; apicolateral processes bluntly pointed and divergent 3
3. Respiratory organs about 4 times as long as broad, not annulate but ventral sides granulose, spiracles in a double row .... *grisea* (Coquillett)
— Respiratory organs about 8 times as long as broad, broad proximal parts annulate, spiracles in a single row 4
4. About 7 widely spaced spiracles on annulate part of respiratory organ (fig. 14:39*o*).... *mutabilis* (Coquillett)
— About 14 scattered spiracles on annulate part of respiratory organ (fig. 14:39*n*) .... *johannseni* (Malloch)

Subfamily HELEINAE

Tribe CULICOIDINI

Genus *Culicoides* Latreille

(Figs. 14:36; 14:38*a-c*; 14:39*u-ee*)

The California species of this genus which have a bad reputation as biters are *obsoletus* (Meigen) of the northern mountains, *tristriatulus* Hoffman of the northern California coasts ranging to Alaska, and *reevesi* Wirth of the hilly parts of the southern San Joaquin Valley. Other species have been reported elsewhere, but not in California, as biting man, and it is likely that wild hosts, particularly birds, form their most important source of food. The immature stages are aquatic, the larvae often occuring in tremendous concentrations along stream, lake, or pond margins, in tree holes, or in polluted puddles and ditches. Foote and Pratt (1954) have recently published a well-illustrated revision of the species of the eastern United States. Hill (1947), Williams (1951), and Kettle and Lawson (1952) have contributed valuable papers on the biology of *Culicoides*.

Key to the California Species of Culicoides

Females

1. Wing uniformly colored, without light or dark markings; small, dark pruinose species 2
— Wing with at least 1 dark and usually 2 or more light spots 5
2. Mesonotum uniformly colored, without definite pattern 3
— Mesonotum with distinct pattern 4
3. Mesonotum polished black; Mono, Plumas, and San counties .... *monoensis* Wirth 1952 (part)
— Mesonotum pruinose brown ..... *brookmani* Wirth 1952
4. Mesonotal pattern largely bluish pruinose, an irregular, median, dark line and 2 very irregular, broken lateral, dark patches; southern and central California....... .... *hieroglyphicus* Malloch 1915
— Mesonotum mostly darker, the light pruinose markings largely confined to 2 narrow submedian stripes; northern and central California ....*jamesi* Fox 1946
coast Range................. *tenuistylus* Wirth 1952
northern and central California .... *unicolor* (Coquillett) 1905
5. Second anterior radial cell wholly or mainly included in a light spot (fig. 14:38*a*) 6
— Second anterior radial cell mainly included in a very dark spot 10
6. Mesonotum uniformly colored, without definite pattern 7
— Mesonotum with distinct pattern, the dark markings large and distinct 9
7. Wing macrotrichia scanty, at tip of wing only; smaller species; northern counties and Sierra Nevada....... .... *obsoletus* (Meigen) 1818
— Wing covered almost entirely with macrotrichia; larger species 8
8. Third palpal segment slender; mesonotum with 3 distinct, very narrow, dark striae; larger brown species, less hairy; eyes separated by width of at least 1 facet; Humboldt and Marin counties .*tristriatulus* Hoffman 1925
— Third palpal segment short and swollen; mesonotal striae very indistinct; smaller grayish, hairy subspecies; Imperial and San Bernardino counties...... .... *cockerellii saltonensis* Wirth 1952
9. Mesonotal pattern typically with 2 oblique, lateral, dark-brown spots (fig. 14:38*b*); large, slender subspecies; northern California................... .... *cockerelli cockerellii* (Coquillett) 1901
— Mesonotal pattern typically with front and sides dark, the prescutellar area pruinose gray with angularly twice-stepped anterior margin; shorter, robust species; widespread in California; breeds in tree holes .... *luteovenus* Root and Hoffman 1937
10. Wing with only anterior radial cells dark, remainder unicolorous gray; mesonotum shining black; small black species .... *monoensis* Wirth (part)
— Wing with 2 or more light spots in addition to dark anterior radial cells; mesonotum not shining black 11
11. Mesonotum without distinct pattern................ 12
— Mesonotum with distinct color pattern 16
12. Wing with only 1 or 2 light spots, on r-m cross vein and just behind 2nd anterior radial cell 13
— Wing with light areas on posterior margin in addition to 2 light costal spots 14
13. Wing with 2 light spots, at cross vein and just beyond 2nd anterior radial cell; color differences in wing itself; macrotrichia scattered and abundant; Shasta County .... *unicolor* (Coquillett) (part)
..............................*usingeri* Wirth 1952
— Wing with 1 light spot just beyond 2nd anterior radial cell owing to absence of macrotrichia; these arranged in rows in cells M_1 and M_2; Ventura, Riverside, and San Diego counties ....
...............*copiosus* Root and Hoffman 1937 (part)
14. Wing almost devoid of macrotrichia, a few at wing tip and along ends of veins; small species, 1 mm. or less in length; Kern and Tulare counties ....
..............................*reevesi* Wirth 1952
— Wing with abundant macrotrichia; larger species, more than 1.5 mm. in length 15
15. Mesonotum dark brown, contrasting strongly with the yellowish abdomen; Mohave Desert ..*mohave* Wirth 1952
— Mesonotum and abdomen grayish brown, not contrasting; breeds in tree holes .... *unicolor* (Coquillett) (part)
widespread ....*utahensis* Fox 1946
Shasta County ....*palmerae* James 1943
16. Mesonotum marked with many small dark spots, each with a hair arising from center; widespread in California ....*variipennis* (Coquillett) 1901
— Mesonotum marked with a few large dark stripes, usually more or less cross connected................. 17
17. Wing with 3 light spots along anterior margin between 2nd anterior radial cell and tip of vein M_1; Mono, Kern, and Ventura counties........ *stellifer* (Coquillett) 1901
— Wing with only 2 light spots between 2nd anterior radial cell and tip of vein M_1 18
18. Hind tibia with small subbasal light band, the tip dark 19
— Hind tibia with small subbasal and broad subapical light bands; cell R_5 of wing with light spots at base and extreme apex; widespread ....
.... *haematopotus* (Malloch) 1915
19. Second light spot in cell R_5 double; median dark stripe of mesonotum connected with lateral dark stripes; widespread ....*baueri* Hoffman 1925
— Second light spot in cell R_5 single; median dark stripe of mesonotum not connected with lateral dark stripes 20
20 .Larger species; wing pattern includes light spots in base of cell M_1 and middle of cell M_2; widespread ....*crepuscularis* (Malloch) 1915
— Smaller species; wing pattern includes light spots on base of vein M_1 and middle of vein M_2 ....
.................*copiosus* Root and Hoffman (part)

Male Genitalia

1. Ninth tergite rounded at tip, apicolateral processes absent or very small, not surpassing apex of tergite (fig. 14:38*c*) 2
— Ninth tergite of various shapes, apicolateral processes well developed, surpassing apex of tergite 5
2. Inner side of basistyle not spinose; apicolateral processes of 9th tergite absent, apex of 9th tergite blunt; ventral root of basistyle much longer than dorsal root ....*obsoletus* (Meigen)
— Inner side of basistyle with patch of stout spines; apicolateral processes present but small; ventral and dorsal roots subequal (fig. 14:38*c*) 3
3. Tip of aedeagus with a distinctly elongated, slender point arising from the broad body . *tristriatulus* Hoffman
.... *luteovenus* Root and Hoffman

— Tip of aedeagus not elongated and slender 4
.... *cockerellii cockerellii* (Coquillett)
4. Tip of aedeagus broadly rounded (fig. 14:38*c*) ......
— Tip of aedeagus distinctly angular ....
.... *cockerellii saltonensis* Wirth
5. Parameres fused in a broad flattened plate 6
— Parameres separated their entire length........... 10
6. Aedeagus with bifid tip; parameres fused at base, with tips separate; ventral root of basistyle much reduced .... *variipennis* (Coquillett)
— Aedeagus with single tip or fused tip; parameres fused at apex, with more or less prominent basal arch; ventral root of basistyle well developed but partly internal, often with rugose membrane connecting with parameres 7
7. Ninth sternite with distinct medioposterior lobes; aedeagus with basal arms separate, fused only at extreme tips................ *hieroglyphicus* Malloch
— Ninth sternite without medioposterior projections; aedeagus fused on distal 1/2 8
8. Ninth sternite with distinct median posterior cleft; dististyles with greatly expanded bell-shaped apices .... *jamesi* Fox
— Ninth sternite without median cleft; dististyles with very slender, pointed apices 9
9. Apicolateral processes of 9th tergite as long as distance between their bases; parameres about twice as long as broad; aedeagus broadly rounded on distal 1/2 .... *tenuistylus* Wirth
— Apicolateral processes of 9th tergite not 1/2 as long as distance between their bases; parameres about as broad as long; aedeagus conical on distal 1/2 ......
.... *brookmani* Wirth
10. Parameres with lateral fringe of spines at apices; ventral root of basistyle in shape of boat hook; anterior arch of aedeagus 2/3 of total length, basal arms narrow and distinctly curved, posterior lateral margin often with posterior processes 11
— Parameres with apices simple or bearing terminal spines; anterior arch of aedeagus less than 1/2 of total length, basal arms stout, not greatly curved, posterior lateral margin without lateral processes 15
11. Posterior margin of arch of aedeagus with distinct posterior projections 12
— Posterior margin of arch of aedeagus without posterior projections 13
12. Parameres with lobelike projection on stem before tip, tip broadly expanded and flattened, as wide as length of fringing spines, aedeagus with truncate apex.....
.... *haematopotus* Malloch
— Parameres with slender stem without lobelike swelling, tip not 1/2 as wide as length of spines, aedeagus with rounded apex .... *baueri* Hoffman
13. Parameres slender and sinuate, without lobelike swelling.......................... *unicolor* (Coquillett)
— Parameres with lobelike projection on stem before tip 14
14. Anterior arch of aedeagus broadly rounded, apex truncate; parameres with lobe much swollen, pouchlike, apices slender, not flattened, lateral spines short .... *mohave* Wirth
— Anterior arch of aedeagus acutely notched on median line, apex rounded; parameres with lobe small, flattened, apices rather broad and flattened with long lateral spines .... *stellifer* (Coquillett)
15. Parameres with stems subparallel, extreme apices flaring in several spinose points ....*usingeri* Wirth
— Parameres without spinose apices 16
16. Aedeagus conical with sides straight, apex distinctly notched with sublateral points .... *monoensis* Wirth
— Aedeagus with apex truncate or rounded 17
17. Ninth tergite with apicolateral processes subparallel, more than 4 times as long as broad; aedeagus broad with flaring apex 18
— Ninth tergite with apicolateral processes triangular, shorter; aedeagus with apex narrow, tapering, or roundly pointed .... *crepuscularis* Malloch
18. Dististyle abruptly bent and narrowed near base, the latter bulbous; parameres with swollen, crooked bases, distal parts flattened and broad.................. 19
— Dististyle gently curved and tapering from base; aedeagus with distal part conically tapering, apex only slightly flaring if at all; parameres with bases abruptly bent, not expanded, distal part very slender .... *copiosus* Root and Hoffman
19. Aedeagus with distal 3rd subparallel with broader flaring apex .... *utahensis* Fox
— Aedeagus broad with a pair of prominent apicolateral projections at distal 3rd, the part beyond abruptly narrowed to a narrow bell-shaped tip ...*palmerae* James

Tribe HELEINI

Genus *Alluaudomyia* Kieffer

(Fig. 14:38*h,i,u*)

Little is known of the habits of this genus. Thomsen (1937) found the larvae of *A. bella* (Coquillett) in algae, dead leaves, and mud from swamps and ponds in New York, and larvae of *A. needhami* Thomsen in blanket algae in ponds. The larvae are predaceous. The North American species have been revised by Wirth (1952), and Williams (1953) has reported on the bionomics of two Georgia species.

Key to Adults of the California Species of Alluaudomyia

1. Wing with marginal black spots on the membrane midway between the veins; Tulare County ....
.... *stictipennis* Wirth 1952
— Wing with marginal black spots on the veins........ 2
2. Wing with 8 well-defined, rounded, black spots (fig. 14:38*h*); hind tibia with 3 light rings; Merced County .... *bella* (Coquillett) 1902
— Wing with 10, more or less, black spots, those on posterior part of wing reduced to lines along veins; hind tibia with 2 light rings; Riverside County.....
.... *needhami* Thomsen 1935

Key to Adults of the California Species of Helea

1. Claws of all legs long and similar; anterior radial cells well formed; wing without microtrichia, appearing white 2
— Claws of fore and midlegs long, of hind legs 1/2 as long; 1st anterior radial cell obsolescent, 2nd very small; wing with microtrichia, appearing grayish hyaline; Tulare and Shasta counties ....
.... *longipennis* Wirth 1952
2. M_2 complete or only narrowly interrupted at base (subgenus *Helea*)................................ 3
— M_2 with at least basal 1/2 missing (subgenus *Isohelea*) 4
3. Anterior veins deeply infuscated; wing 2.6 times as long as broad; humeral pits small and narrow; Lassen and Mono counties............. *mallochi* (Cole) 1921
— Anterior veins white; wing broad and rounded, about 2.2 times as long as broad; humeral pits large and

rounded; Modoc and Mono counties .... *culicoidithorax* (Hoffman) 1926
4. Mesonotum with marked bluish-gray pruinescence; wings intensely whitish, the surface with ridges; anterior radial cells only slightly infuscated; Siskyou and Mono counties .... *pruinosa* Wirth 1952
— Mesonotum shining black; wings milky white; anterior radial cells deeply infuscated, forming a black stigma (fig. 14:38*r,s*); Modoc, Mono, Nevada, and Humboldt counties................ *stigmalis* (Coquillett) 1902

Tribe STILOBEZZIINI

Genus *Stilobezzia* Kieffer

(Fig. 14:41*d*)

According to Thomsen (1937) the larvae of *Stilobezzia* are commonly found in mud and algae along pond margins, and because of their curved bodies and slow, crawling motion they closely resemble the larvae of *Dasyhelea*. The food habits were not stated. The North American species have been revised by Wirth (1953).

Key to Adults of the California Species of Stilobezzia

1. Wings without macrotrichia (subgenus *Stilobezzia*), small, pruinose brown species with distinct gray pruinose mesonotal pattern; El Dorado and San Luis Obispo counties .... *pruinosa* Wirth 1952
— Macrotrichia present on distal part of wing (subgenus *Neostilobezzia*). Larger dull yellowish-brown species; mesonotum with 3 indistinctly darker vittae; Tulare County .... *fuscula* Wirth 1952

Genus *Serromyia* Meigen

Strenzke (1950) found the larvae of *S. femorata* (Meigen) in damp moss in Germany. There is only one known California species.

barberi Wirth 1952. Humboldt, San Luis Obispo

Genus *Monohelea* Kieffer

(Fig. 14:38*d,e*)

The American species have been revised by Wirth (1953). Their habits are not known. Strenzke (1950) found larvae of *M. calcarata* Goetghebuer in moss in a spring.

maculipennis (Coquillett) 1905. Monterey County

Tribe STENOXENINI

Genus *Neurohelea* Kieffer

nigra Wirth 1952. Mono, Tulare counties

Genus *Clinohelea* Kieffer

usingeri Wirth 1952. (Fig. 14:38*n-q*) San Luis Obispo County

Genus *Johannsenomyia* Malloch

Wirth (1952) described the pupa of *J. sybleae* Wirth (fig. 14:42*k,t*) which was taken from a cattail-choked pond near the beach. Malloch (1914) also described the pupa of *J. flavidula* (Malloch) taken from the Illinois River. The pupae are characterized by extremely short prothoracic respiratory organs.

Key to Adults of the California Species of Johannsenomyia

1. Halteres and pedicel of antennae white to yellowish (fig. 14:39*a-c*); northern and central California...... .... *sybleae* Wirth 1952
— Halteres and pedicel of antennae brown to black.... 2
2. Hind tibia yellowish on basal 2/3; anterior wing veins yellowish; Modoc County .....*halteralis* (Malloch) 1915
— Hind tibia all black; anterior wing veins brown 3
3. Femora and tibiae extensively yellow; size 2.5 mm.; northern and central California................... .... *caudellii* (Coquillett) 1905
— Femora and tibiae entirely black; size 2.0 mm.; Imperial and San Bernardino counties .... *pullata* Wirth 1952

Genus *Sphaeromias* Curtis

Species of this predaceous genus are among the largest in the family. Thomsen (1937) described the larva and pupa of *S. longipennis* (Loew). The larvae of this genus, together with those of *Palpomyia* and *Bezzia*, are predaceous, and are often present in large numbers in blanket algae in ponds and shallow lakes.

Key to Adults of the California Species of Sphaeromias

1. Femora, especially on fore legs, with apices yellowish, broad brown bands in middle; legs very spiny; large species more than 4 mm.; antennae long, flagellar segments annulated; Modoc County .... *longipennis* (Loew) 1861
— Femora with apices brown, bases broadly yellowish; legs moderately spiny; medium-size or small species less than 3.5 mm.; antennae long or short, segments not annulated 2
2. Antennae shorter than height of head including proboscis; abdominal tergites uniformly pale yellow; size 2.5-3.5 mm.; Colorado River..... *brevicornis* Wirth 1952
— Antennae nearly twice as long as height of head; at least some abdominal tergites brown, with narrow apical light border; size less than 2.0 mm.; Kern, Mono, San Bernardino, and Imperial counties............ .... *minor* Wirth 1952

Genus *Palpomyia* Meigen

(Fig. 14:39*d,e*; 14:42*a,f,g,o,r*)

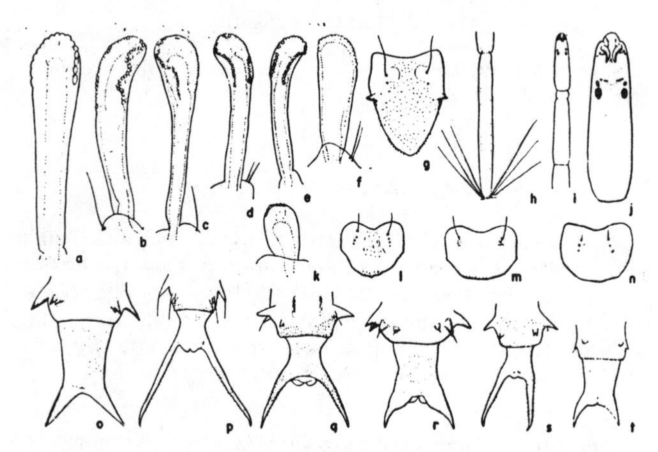

Fig. 14:42. Heleidae, immature stages. *a,g,o, Palpomyia flavipes* (Mg.), pupa: *a*, respiratory organ; *g*, operculum; *o*, caudal segment. *b,l,q, Bezzia varicolor* (Coq.), pupa: *b*, respiratory organ; *l*, operculum; *q*, caudal segment. *c,h-j,m,p, B. glabra* (Coq.). Pupa: *c*, respiratory organ; *m*, operculum; *p*, caudal segment. Larva: *h*, last abdominal segment; *i*, head and prothoracic segment; *j*, head, ventral view. *d,e,n,s, B. biannulata* Wirth, pupa: *d,e*, respiratory organs; *n*, operculum; *s*, caudal segment. *f,r, Palpomyia essigi* Wirth, pupa: *f*, respiratory organ; *r*, caudal segment. *k,t, Johannsenomyia sybleae* Wirth, pupa: *k*, respiratory organ; *t*, caudal segment (Wirth, 1952).

This is a large genus with a broad range of breeding habits. In general, the larvae are aquatic and most often occur in blanket algae or in the mucky or sandy margins or bottoms of lakes, ponds, and streams. Some species breed in the profundal zone of deep lakes. The adults are predaceous with strong, spinose legs.

Keys to Adults of the California Species of Palpomyia

Females

1. Only fore femora armed below with spines 2
— All 3 pairs of femora armed below with 1 or more stout, black spines 6
2. Fore femora slightly swollen, with 3 to 7 ventral spines 3
— Fore femora markedly swollen, with 15 to 20 spines; halteres yellow 5
3. Halteres yellow; fore femora with 3 or 4 spines; northern and central California ..... *aldrichi* (Malloch) 1915
— Halteres brown or black; fore femora with 5 to 7 spines 4
4. Legs predominantly yellow; mesonotum nearly bare; northern and central California .. *flavipes* (Meigen) 1818
— Legs predominantly black; mesonotum with dense long pubescence and intermixed fine black hairs; northern and central California .... *nigripes* (Meigen) 1830
5. Mesonotum shining reddish with 3 prominent blackish fasciae; Humboldt County ...... *trifasciata* Wirth 1952
— Mesonotum unicolorous shining black; northern and central California .... *armatipes* Wirth 1952
6. Mesonotum grayish pollinose with brown fasciae or spots; tarsal claws as long as, or longer than, 5th segment (fig. 14:39*d*); halteres yellowish; abdomen yellow or with light bands 7
— Mesonotum subshining reddish brown with 3 black fasciae; 5th tarsal segment with ventral spines; claws 1/2 as long as segment; halteres dark brown; abdomen brown; Central Valley and Colorado River Valley .... *essigi* Wirth 1952
7. Wing with small, dark brown spot over cross vein (fig. 14:39*e*); pedicel of antenna and abdomen yellow; length 2.5 mm.; Kern County .... *kernensis* Wirth 1952
— Wing without brown spot at cross vein; pedicel black; abdomen brown with grayish bands at apices of segments; length 3.0 mm.; widespread in California .... *linsleyi* Wirth 1952

Male Genitalia

1. Aedeagus greatly reduced; parameres with narrow median anterior arm, distal part greatly expanded, oval; basistyles with setose basal lobe ... *nigripes* (Meigen)
— Aedeagus normal, with caplike hyaline apex; parameres with lateral anterior arms; posterior part slenderer; lobe on basistyle past middle or absent 2
2. Basistyle with triangular or thumblike setigerous lobe at or past middle on inner ventral side 3
— Basistyle without ventromesal setigerous lobe 4
3. Lobe of basistyle thumblike, nearly 1/2 as long as basistyle; aedeagus with bluntly rounded cap on apex; parameres separated on distal 1/2, with slender, bare apices .... *aldrichi* (Malloch)
— Lobe of basistyle short and triangular; aedeagus with pointed triangular cap on apex; parameres with free part tonguelike, pubescent .... *armatipes* Wirth
4. Parameres separated or cleft to base, basal arms simple and curved; aedeagus conical with ventral face entire 5
— Parameres with median part fused in a straight or dorsally bent slender rod with slightly enlarged apex, basal arms short, stout, and bilobed; aedeagus with ventral face incompletely closed..... *flavipes* (Meigen)
5. Genitalia much longer than broad, 9th tergite twice as long as broad; parameres entirely separate, slender, with capitate, bent, finely hairy tips .... *essigi* Wirth
— Genitalia much broader than long, 9th tergite as broad as long; parameres fused at base, closely contiguous, swollen in middle with slightly tapered tips .... *linsleyi* Wirth

Males of *trifasciata* and *kernensis* unknown.

Genus *Bezzia* Kieffer

(Figs. 14:39*f,g*; 14:42*b-e, h-j, l-n, p-q, s*)

The discussion for the genus *Palpomyia* also applies for the most part to this large genus. To date no adequate characters have been found for separating larvae or pupae of these two genera. Descriptions of larvae and pupae of several North American species have been presented by Thomsen (1937) and Wirth (1952), the latter dealing with California species.

Key to Adults of the California Species of Bezzia

1. At least fore femora armed below with black spines 2
— All femora unarmed ventrally 10
2. Wings with large, subapical, black spots (fig. 14:39*f*); Inyo County .... *punctipennis* (Williston) 1896
— Wings unspotted 3
3. Halteres yellowish 4

— Halteres brown or black 5
4. Scutellum brown; hind tibia yellow with broad subbasal and narrow apical brown rings; mesonotum uniformly grayish-brown pruinose; large species (3.9 mm); Alameda and Tulare counties .. *varicolor* (Coquillett) 1902
— Scutellum yellow; hind tibia brown with middle 3rd yellow; mesonotum with prominent black vittae; small species (2.5 mm.); widespread in California .... *setulosa* (Loew) 1861
5. Mesonotum uniformly dull brown; hind femora spinose 6
— Mesonotum shining black or dull black with silvery pollinose areas; hind femora with or without spines 7
6. Antennae with flagellar segments yellow at bases; legs yellow with indistinct brown bands; Humboldt, San Mateo, and San Luis Obispo counties .... *sordida* Wirth 1952
— Antennae and legs uniformly dark brown; Modoc County .... *modocensis* Wirth 1952
7. Mesonotum with anterior corners silvery pollinose and with broad median dull-black stripe; Mariposa and Los Angeles counties.......... *pulverea* (Coquillett) 1901
— Mesonotum shining black, at most with 2 short, narrow, submedian, silvery lines......................... 8
8. Mesonotum with pair of narrow submedian white lines on anterior 1/2; Mono and Plumas counties .... *bilineata* Wirth 1952
— Mesonotum uniformly polished black............... 9
9. Femora and tibiae entirely black; San Luis Obispo County .... *flavitarsis* Malloch 1914
— Femora and tibiae of fore legs broadly yellow-banded; Kern County .... *expolita* (Coquillett) 1901
10. Halteres with knob yellow or white 11
— Halteres with knob brown or black 14
11. Hind tarsal claws of female very large and unequal; costa produced past tip of radial cell to 0.9 of wing length; mesonotum uniformly grayish, granular pruinose; Monterey County .... *granulosa* Wirth 1952
— Hind tarsal claws of female small and equal; costa to 0.8 of wing length, not produced past tip of radial cell; mesonotum otherwise 12
12. Mesonotum uniform jet black; fore tibia yellow; only apex narrowly dark; Kern, Riverside, and Imperial counties .... *coloradensis* Wirth 1952
— Mesonotum pollinose gray with darker vittae; fore tibiae annulate 13
13. Fore tibia with single narrow brown ring in middle, absent on mid and hind tibiae; mesonotum pearly gray with broad, median, anterior, reddish-brown vitta; San Joaquin Valley.............. *glabra* (Coquillett) 1902
— All tibiae with broad submedian and apical brown bands; mesonotum dark pollinose gray with submedian anterior lines and posterior lateral spots brown; Modoc, San Benito, and San Luis Obispo counties .... *biannulata* Wirth 1952
14. Thorax shining black with short, submedian, anterior pair of silvery lines on mesonotum; fore- and midlegs dark brown with narrow yellow rings; widespread.... .... *bivittata* (Coquillett) 1905
— Thorax uniformly dull, reddish brown with dense, long pubescence; legs yellowish to brown; widespread .... *opaca* (Loew) 1861

Genus *Probezzia* Kieffer

According to Mayer (1934), the pupae of this genus cannot be separated from those of *Johannsenomyia*. The North American species have been revised by Wirth (1951). There is only one known California species.

flavonigra (Coquillett) 1905. (fig. 14:39*h-j*). — Great Basin and Central Valley

REFERENCES

FOOTE, R. H., and H. D. PRATT
1954. The *Culicoides* of the eastern United States. Publ. Hlth. Monogr. no. 18:1-53.

HILL, M. A.
1947. The life cycle and habits of *Culicoides impunctatus* Geotghebuer and *Culicoides obsoletus* Meigen, together with some observations on the life cycle of *Culicoides odibilis* Austen, *Culicoides pallidicornis* Kieffer, *Culicoides cubitalis* Edwards, and *Culicoides chiropterus* Meigen. Ann. Trop. Med. Parasit., 41:55-115.

JOHANNSEN, O. A.
1943. A generic synopsis of the Ceratopogonidae (Heleidae) of the Americas, a bibliography, and a list of the North American species. Ann. Ent. Soc. Amer., 36: 763-791.
1952. Heleidae (= Ceratopogonidae) *In* Guide to the Insects of Connecticut Part VI, The Diptera or true flies. Fifth Fasc: Midges and gnats. Bull. Conn. Geol. Nat. Hist. Surv., 80:149-175.

KETTLE, D. S., and J. W. H. LAWSON
1952. The early stages of British biting midges, *Culicoides* Latreille (Diptera: Ceratopogonidae) and allied genera. Bull. Ent. Res., 43:421-467, 6 pls.

MALLOCH, J. R.
1914. Notes on North American Diptera. Bull. Illinois Lab. Nat. Hist., 10:213-243, 3 pls.
1915. The Chironomidae or midges of Illinois. Bull. Illinois Lab. Nat. Hist., 10:275-543, 23 pls.

MAYER, K.
1934. Die Metamorphose der Ceratopogonidae (Dipt.). Ein Beitrag zur Morphologie, Systematik, Oekologie und Biologie der Jugenstadien dieser Dipterenfamilie. Arch. f. Naturges. (Leipzig), 3:205-288.

NIELSEN, A.
1951. Contributions to the metamorphosis and biology of the genus *Atrichopogon* Kieffer (Diptera, Ceratopogonidae), with remarks on the evolution and taxonomy of the genus. Kong. Danske Vidensk. Selskab Biol. Skr., 6:1-95, 2 pls.

SAUNDERS, L. G.
1924. On the life history and the anatomy of the early stages of *Forcipomyia* (Diptera), Nemat., Ceratopogonidae). Parasitology, 16:164-213, 3 pls.

SMITH, L. M., and H. LOWE
1948. The black gnats of California. Hilgardia, 18: 157-183.

STRENZKE, K.
1950. Systematik, Morphologie und Okologie der terrestrischen Chironomiden. Arch. f. Hydrobiol. Suppl.-Bd. 18:207-414, 2 pls.

THOMSEN, L.
1937. Aquatic Diptera. Part V. Ceratopogonidae. Mem. Cornell Univ. Agric. Exp. Sta., 210:57-80, 9 pls.

WILLIAMS, R. W.
1951. Observations on the bionomics of *Culicoides tristriatulus* Hoffman with notes on *C. alaskensis* Wirth and other species at Valdez, Alaska, summer 1949 (Diptera, Heleidae). Ann. Ent. Soc. Amer., 44:173-440.
1953. Notes on the bionomics of the *Alluaudomyia* of Baker County, Georgia. I. Observations on breeding habitats of *bella* and *needhami*. Proc. Ent. Soc. Wash., 55:283-285.

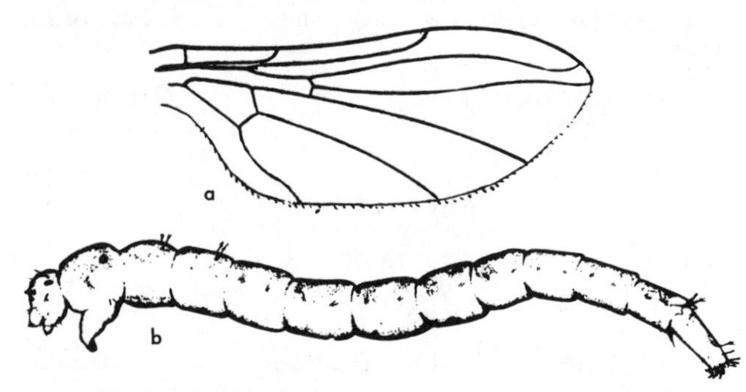

Fig. 14:43. *a, Thaumalea* sp., wing; *b, Thaumalea americana* Bezzi, larva (*a,* Curran, 1934; *b,* Johannsen, 1934).

WIRTH, W. W.
1951. The genus *Probezzia* in North America. Proc. Ent. Soc. Wash., 53:25-34.
1952. The Heleidae of California. Univ. Calif. Publ. Entom., 9:95-266.
1952. The genus *Alluaudomyia* Kieffer in North America (Diptera, Heleidae). Ann. Ent. Soc. Amer., 45:423-434.
1953. Biting midges of the heleid genus *Stilobezzia* in North America. Proc. U.S. Nat. Mus., 103:57-85.
1953. American biting midges of the heleid genus *Monohelea.* Proc. U.S. Nat. Mus., 103:135-154.

Family THAUMALEIDAE

Only two of the four described genera are known to occur in America. The genus *Thaumalea* has been collected in Humboldt County, but since the specimens are females their identity cannot be established. It is possible that they are *T. fusca* (Garrett 1925). Johannsen (1934) has described the larva and pupa of *Thaumalea,* and Dyar and Shannon (1924) reviewed the North American species.

"The adults are found along the edges of streams, particularly those with mossy banks, and are not common in collections. They are small flies, under 6 mm. in length, and the wings bend sharply near the base in death, folding downward as in the Psychodidae.

"The larvae (fig. 14:43*b*), which resemble those of the Chironomidae, are found in small brooks and streams where the clear water flows very thinly over the rocks, so that the back if the larva is always exposed above the surface. They feed on detritus and diatoms and move about in search of food. The pupae are found in the bottom of the stream between stones, etc." (Curran, 1934).

Key to the North American Genera of Thaumaleidae

1. Sc_1 ending in costa (fig. 14:43*a*); no macrotrichia on the wing membrane .... *Thaumalea* Ruthe
— Sc_1 obsolete apically, ending free, weaker than Sc_2; wing membrane with fine macrotrichia in addition to the microtrichia .... *Trichothaumalea* Edwards

References

CURRAN, C. H.
1934. *See* Diptera references.
DYAR, H. G. and R. C. SHANNON
1924. The American species of Thaumaleidae (Orphnephilidae). Jour. Wash. Acad. Sci., 14:432-434.
EDWARDS, F. W.
1929. A revision of the Thaumaleidae. Zool. Anz., 82:121-142.
JOHANNSEN, O. A.
1934. *See* Diptera references.

Family STRATIOMYIDAE

The soldier flies are small to moderately large flies, often of bright coloration, which habitually frequent flowers. They are characterized by a general absence of bristles, usually possessing instead a soft pubescence over the body; the wing veins are crowded

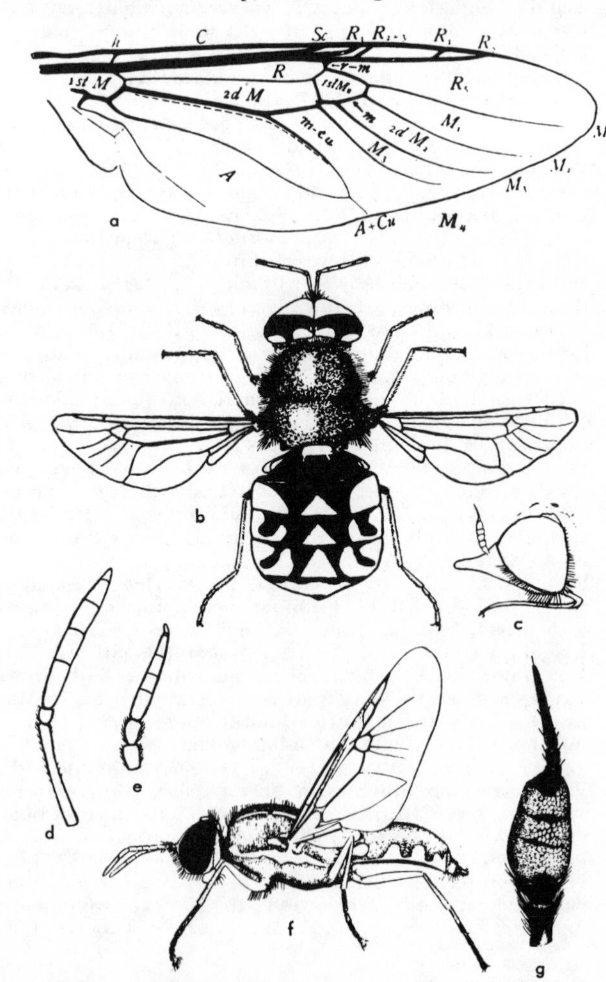

Fig. 14:44. Stratiomyidae, adults. *a, Stratiomys,* wing venation; *b, S. maculosa* Loew, male; *c, Nemotelus,* head; *d, Stratiomys,* antenna; *e, Eulalia,* antenna; *f, Euparyphus lagunae* Cole, female; *g, Adoxomyia subulatala* (Loew), antenna (*a,* Comstock, 1935; *b,* Essig, 1926; *c,* Williston, 1908; *d,e,* James, 1936; *f,* Cole, 1912; *g,* James, 1943).

anteriorly (fig. 14:44*a*), with the posterior veins weak, the discal cell of characteristic shape, and the antennae (fig. 14:44*d,e,g*) of exceedingly variable form. The eggs of some aquatic species are deposited on plants over the water. The larvae (fig. 14:45*a-e*) are elongate, more or less flattened dorsoventrally, with the head elongate and tapering, much narrower than the body and capable of being retracted halfway within the thorax. The body surface is characteristically shagreened, the thickening consisting of a heavy deposit of calcium carbonate secreted by the integument. Respiration is accomplished by a pair of anterior spiracles situated on tubercles on the prothorax and by a posterior pair hidden in a transverse, slitlike chamber which opens on the dorsum or the apex of the last segment. In the aquatic species this slit is fringed with long, plumose hairs which serve as a float when the tip of the abdomen is at the water surface, or which support an air bubble when the larva submerges. The aquatic larvae are predominantly vegetarians and feed on algae, decaying vegetable matter, and some small microörganisms. Pupation takes place within the last larval skin, forming a puparium, with essentially the same characters as the mature larva. Descriptions and figures of aquatic stratiomyid larvae may be found in the papers by Malloch (1917) and Johannsen (1922, 1935).

Nearly all the known aquatic stratiomyids belong in the subfamilies Stratiomyinae and Adoxomyiinae, which so far as known are entirely aquatic. For this reason, keys to all the North American genera of these two subfamilies are included.

Keys to the North American Genera of Aquatic Stratiomyidae

Adults

1. Abdomen with 7 visible segments (BERIDINAE), or antenna terminating in a hairlike arista (GEOSARGINAE, PACHYGASTRINAE), or with elbowed distal segment longer than proximal flagellar segments combined (HERMETIINAE) .... Nonaquatic
— Abdomen with not more than 5 clearly defined segments (fig. 14:44*b*); antenna never terminating in a hairlike arista or with elbowed distal segment longer than proximal flagellar segments combined 2
2. All posterior veins arising from the discal cell (fig. 14:44*f*) (ADOXOMYIINAE) 3
— The fourth posterior vein arises from the 2nd basal cell (fig. 14:44*a*) (STRATIOMYINAE) (key adapted from James, 1936) 9
3. Scutellum without spines 4
— Scutellum with spines 5
4. Face conically produced (fig. 14:44*c*) ....*Nemotelus* Geoffroy
— Face receding below; Florida.. *Euryneurasoma* Johnson
5. Antenna short, with a subterminal arista; eastern United States .....*Hermione* Meigen (= *Oxycera* Meigen)
— Antenna more or less elongate 6
6. Antennal style not differentiated 7
— Antennal style distinctly differentiated 8
7. Antenna with 8 distinct segments, the 2nd not longer than the 1st (fig. 14:44*f*) .....*Euparyphus* Gerstaecker
— Antenna with 3 segments, the 3rd annulate (not California) .... *Scoliopelta* Williston
8. Eyes hairy; 3rd vein branched .... *Adoxomyia* Kertész
— Eyes bare; 3rd vein simple; Arizona and New Mexico ....*Aochletus* Osten Sacken
9. Antenna 10-segmented, those beyond the 2nd more or less fused into a flagellum 10
— Antenna 7- or 8-segmented, those beyond the 2nd more or less fused into a flagellum (fig. 14:44*d,e*) 11
10. Tenth antennal segment forming a style; scutellar spines small, preapical, their tips barely reaching the apex of the scutellum; Texas to Panama.......... ....*Dicyphoma* James
— Tenth antennal segment not stylelike; scutellar spines strong, approximate, equal; Neotropical, one species to Florida .... *Cyphomyia* Wiedemann
11. Face produced conically downward; abdomen slender; usually unicolorous .... *Myxosargus* Brauer
— Face sometimes protuberant, but not conically produced below the oral margin; abdomen usually with pale tegumentary markings (fig. 14:44*b*) 12
12. Flagellum of antenna 5-segmented, the apical segments never forming a style; ratio of 1st to 2nd segment usually 2:1 or greater (fig. 14:44*d*) 13
— Flagellum 6-segmented, the 5th segment short and ringlike, the 6th set at an angle to the rest of the flagellum, segments 5 and 6 forming a definite style; ratio of 1st to 2nd segment less than 2:1 (fig. 14:44*e*) 15
13. Face produced; proboscis elongate; scutellum semielliptical, the spines located at the apex on the median 3rd 14
— Face receding; proboscis short; scutellum trapezoidal, the spines located on the outer corners (fig. 14:44*b*) ....*Stratiomys* Geoffroy
14. Flagellum flattened; head wider than thorax; eyes angular in profile; spines of scutellum strong; wings infumated in the region of the strong veins; not California .... *Hoplitimyia* James
— Flagellum terete; head no wider than the thorax; eyes rounded in profile; spines of scutellum weak or evanescent; wings hyaline or uniformly infuscated .... *Labostigmina* Enderlein
15. Spines of scutellum present, strong and situated within the median 3rd, close together; vein r-m missing or if present, at least 1 of the branches of the media is strongly abbreviated or absent 16
— Spines of scutellum absent; branches of media well developed; vein r-m present; not California ....*Anoplodonta* James
16. Cross vein r-m absent; media with 3 branches, all extending almost to posterior wing margin ....*Hedriodiscus* Enderlein
— Cross vein r-m present; at least the 3rd branch of the media much abbreviated............. *Eulalia* Meigen

Larvae and Puparia

1. Posterior spiracular chamber not margined with long, soft, plumose or pubescent hairs (BERIDINAE, GEOSARGINAE, PACHYGASTRINAE, and HERMETIINAE) .... Not aquatic
— Posterior spiracular chamber margined with long, soft, plumose or pubescent hairs 2
2. Antennae placed dorsally on head remote from margin; last abdominal segment not over twice as long as its basal width (fig. 14:45*a-c*) (ADOXOMYIINAE) 3
— Antennae placed at lateroanterior angles of head; last abdominal segment very much longer than broad and tapering (fig. 14:45*d,e*) (STRATIOMYINAE) 5
3. Posterior margin of ventral side of next to last segment with a pair of stout, curved spines ....*Euparyphus* Gerstaecker
— Posterior margin of ventral side of next to last segment without curved, hooklike spines 4
4. Posterior spiracular chamber located on dorsal side of last segment, the fringe of hairs rather short; last segment emarginate and armed with 4 or 6 long, marginal hairs (fig. 14:45*b*) ....*Nemotelus* Geoffroy

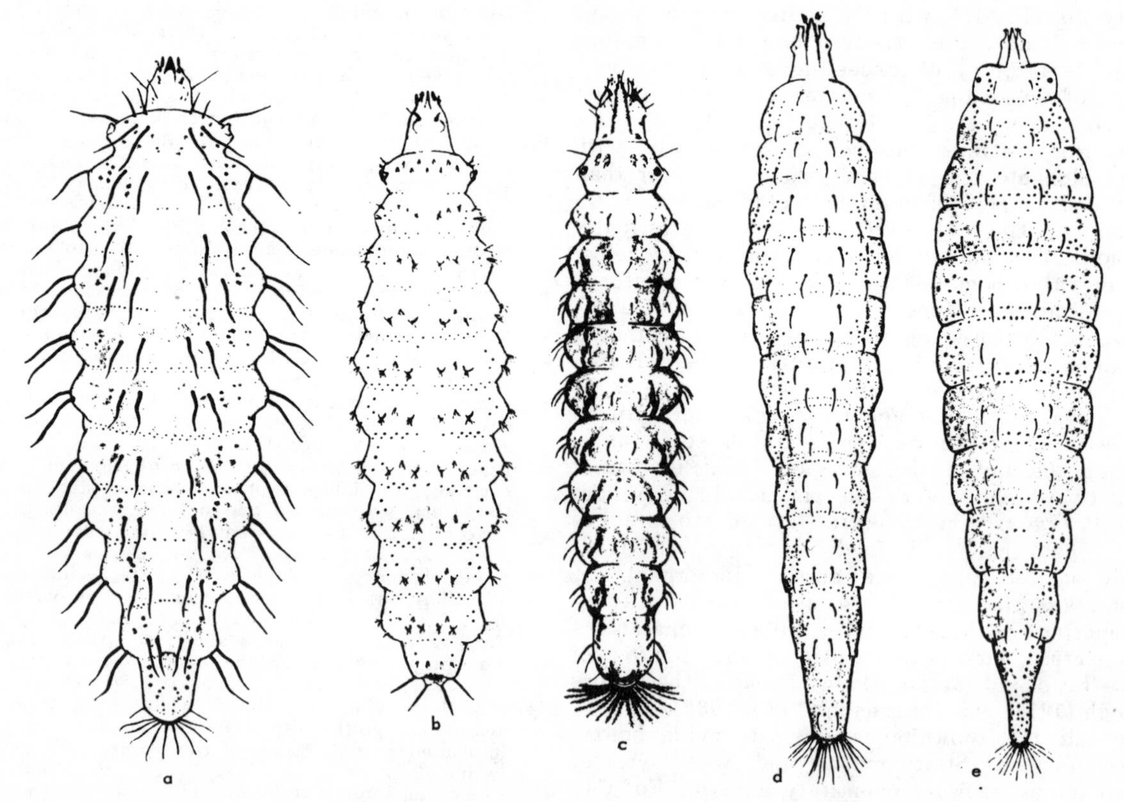

Fig. 14:45. Stratiomyidae, larvae. a, *Hermione*; b, *Nemotelus*; c, *Euparyphus*; d, *Eulalia cincta* (Oliv.); e, *Stratiomys* (Johannsen, 1935).

— Posterior spiracular chamber situated at apex of last segment (fig. 14:45*a*), with fringe of long hairs; last segment without long marginal hairs .. *Hermione* Meigen

5. Last segment very elongate, more than 3 times as long as basal width (fig. 14:45*e*); caudal margin of next to last segment without ventral hooks; antenna about 3 times as long as its diameter ..... *Stratiomys* Geoffroy

— Last segment shorter, at most 3 times as long as basal width (fig. 14:45*d*); caudal margin of next to last segment with 2-4 strong, curved, ventral hooks; antenna about 6 times as long as its diameter .... *Hedriodiscus* Enderlein, *Eulalia* Meigen

The immature stages of the remaining genera named in the key to adults are not known.

California Species of Aquatic Stratiomyidae

Subfamily ADOXOMYIINAE

Genus *Nemotelus* Geoffroy

(Figs. 14:44*c*; 14:45*b*)

Although the larvae of none of the North American species are known, those of several European species have been described (see Johannsen, 1935). Brues (1928) figured an unnamed species of this genus which he took in a hot spring in Mount Lassen National Park. There are numerous North American species which badly need revision as Melander's (1903) key is out of date.

California species:

albomarginatus James 1936. — California
arator Melander 1903. — Santa Clara, San Diego
canadensis Loew 1863. — Santa Clara, Los Angeles
glaber Loew 1872. — Siskiyou, Los Angeles
lambda James 1933. — California
rufoabdominalis Cole 1923. — Riverside County
trinotatus Melander 1903. — California
tristis Bigot 1887. — Santa Clara County
unicolor Loew 1863. — California

Genus *Euparyphus* Gerstaecker

The genus *Euparyphus* reaches its greatest development in the southwestern United States, with the genus *Hermione* replacing it in large part in the eastern states and in Europe. The larvae (fig. 14:45*c*) are usually found in wet moss and algae in and beside small streams, although Brues (1928) has found the larvae of an undetermined species in a hot spring at Mount Lassen National Park. Lenz (1923) has described the larvae of several European species as *Oxycera* [=*Hermione*] which on the basis of both larval and adult characters probably belong to *Euparyphus*. Johannsen (1922) has described larvae of several eastern species of both genera.

California species:

amplus Coquillett 1902. — San Diego County
apicalis Coquillett 1902. — Siskiyou, Orange
crotchii (Osten Sacken) 1877. — Northern and Central Calif.
crucigerous Coquillett 1902. — Modoc, Plumas
flaviventris James 1936. — Modoc County; Lake Tahoe
lagunae Cole 1912. (fig. 14:44*f*) — Orange County

major Hine 1901.	Lake, Santa Clara
mariposa James 1939.	Mariposa County
niger Bigot 1879.	California
pardalinus James 1936.	Inyo County
tahoensis Coquillett 1902.	Lake Tahoe

Genus *Adoxomyia* Kertész (*Clitellaria* of authors)

It is probable that this genus may have aquatic habits, but the only information available on the breeding habits consists of two specimens in the National Museum, of *claripennis* James from Pima County, Arizona, which were reared by R. E. Ryckman from decomposing cactus. The following key to the California species is adapted from a recent revision of the genus by James (1943).

Key to Adults of the California Species of Adoxomyia

1. Legs, including tarsi, wholly black, or black except for the knees 2
— Tarsi pale in large part, and contrasting with the black tibiae; California .... *argentata* (Williston) 1885
2. Antennal flagellum red, at least on the basal segments (fig. 14:44*g*) 3
— Antennal flagellum entirely black; northwestern and central California .... *rustica* (Osten Sacken) 1877
3. Pleura predominantly black-pilose; mesonotum of male with abundant, erect black pile; venter black-pilose (reddish-brown, in certain lights) on the basal segments; northern and central California .... *lata* (Loew) 1872
— Thorax entirely pale-pilose; pile of mesonotum, in both sexes, appressed; venter entirely pale-pilose; southern California .... *appressa appressa* James 1935

Subfamily STRATIOMYINAE

Genus *Myxosargus* Brauer

According to James (1942) who revised this group, *Myxosargus* belongs in a distinct tribe of the Stratiomyinae together with several other small genera, the habits of none of which are known. It is entirely possible that they may not be aquatic.

California species:

knowltoni Curran 1929.	Inyo, Los Angeles

Genus *Stratiomys* Geoffroy

The larvae of *Stratiomys* are among the largest of the family and are recognized by their fusiform shape and the long, tapering caudal air tube (fig. 14:45*e*). There are no transverse rows of long body hairs or ventral hooks on the next to the last segment. They may breed in a wide variety of aquatic habitats from ponds and streams to saline pools and hot springs. Needham and Christenson found *S. melastoma* Loew breeding in shallow, slow-flowing water in Logan River, Utah. Brues (1928) described the larvae of two unnamed species of *Stratiomys* from hot springs in Mount Lassen National Park. Other descriptions of larvae of this genus may be found in Malloch (1917) and Johannsen (1922). The following key to the California species has been adapted from James and Steyskal (1952):

Key to Adults of the California Species of Stratiomys

1. Antenna short, the 1st segment normally not more than 3 times as long as the 2nd, the flagellum at least 2 times as long as the 1st segment; abdomen flattened, the venter either entirely yellow or with black basal bands or spots 2
— Antenna longer, the 1st segment normally 4 or more times as long as the 2nd, the flagellum less than 2 times as long as the 1st segment; venter partly or wholly black 3
2. Abdominal sternites each with a black basal band; male with eyes usually distinctly hairy; both sexes with femora, except extreme apices, black, only in rare cases brown, but still darker than the tibiae; Sierra Nevada .... *laticeps* Loew 1866
— Abdominal sternites wholly pale or with narrow basal spots or bands; eyes in both sexes bare; female with femora yellow, at base, apex in part black, male with femora black except apical 4th or more; California .... *currani* James 1932
3. Fourth abdominal tergite with an oblique black mark interrupting lateral yellow spots on each side (fig. 14:44*b*); eyes hairy in both sexes; widespread in California .... *maculosa* Loew 1866
— Fourth abdominal tergite not so colored; eyes bare in both sexes 4
4. Posterior orbit broader below than above and broadly yellow at least in its lower part; in the female yellow throughout or at most black in its upper 4th, bare except for short, inconspicuous hairs, and broad throughout, its minimum breadth at least equal to length of 2nd antennal segment; in the male distinctly broadened in the lower part; facial orbits broadly yellow (except in some high altitude males of *S. barbata*); sternites with posterior yellow part approximately equally developed on all segments, or at most only slightly more so on the anterior segments 5
— Posterior orbit not much broader in lower part, narrow throughout in the male, relatively so in the female, its minimum breadth usually distinctly less than length of 2nd antennal segment; yellow color of posterior orbit usually narrow or lacking; if posterior orbit is relatively wide or somewhat broadened and yellow on lower half or more, it is clothed in part with conspicuous tomentum and the posterior yellow portion of the sternites is definitely more strongly developed on the anterior than on the posterior segments 6
5. Each sternite usually at least half black, the black part forming a parallel-sided anterior transverse band; posterior pale part of 4th tergite usually interrupted medianly; pale marking of 5th tergite consisting of a median triangle which is widest posteriorly; femora, except knees, black; face of male usually largely black, usually with black pile; California .... *barbata* Loew 1865
— Sternites usually predominantly yellow, each with a black anterior transverse band which is broadest in the middle; posterior pale part of 4th tergite continuous, rarely narrowly interrupted, but when so without an isolated median triangle; pale marking of 5th tergite a conspicuous median pentagonal marking which is usually distinctly broader anteriorly; face pale, oral margin black; Sierra Nevada .....*melastoma* Loew 1865
6. Predominantly black species, the abdominal tergites wholly black except occasionally traces of yellow markings on the median line of the 5th segment, and laterally or at the anterior corners of the 2nd segment, the ventrites also black except for narrow posterior

margins which are only rarely extended; mesonotum with thick gray pile; Sierra Nevada .*nevadae* Bigot 1887
— Abdominal tergites with conspicuous markings on at least the 2nd segment, if without, ventrites with considerable yellow laterally 7
7. Abdomen no broader than thorax, rather strongly convex in transverse section; mesonotal pile cinereous; 3rd tergite without transverse apical markings; California .... *griseata* Curran 1923
— Abdomen distinctly broader than thorax, only moderately convex; mesonotal pile yellowish; 3rd tergite sometimes with transverse apical or even broad lateral markings; Sierra Nevada ....*discaloides* Curran 1923

Genus *Labostigmina* Enderlein

similis (Johnson) 1895. Lake Tahoe

Genus *Hedriodiscus* Enderlein

The larva of the eastern *H. vertebrata* (Say) which breeds in small streams cannot be distinguished from those of *Eulalia*.

Key to Adults of the California Species of Hedriodiscus

1. Face prominent below the base of the antennae and marked with a black spot on each side in ♀; branches of the media weak; length 10 mm. or less; Imperial County .... *trivittatus* (Say) 1829
— Face receding to the oral margin and in the ♀ wholly pale; branches of media strong; length 12 mm. or more 2
2. Mesonotum broadly margined with yellow, the disc with 2 yellow spots; scutellum entirely yellow; widespread in California............*truquii truquii* (Bellardi) 1861
— Yellow markings of mesonotum greatly reduced 3
3. Base of scutellum rarely and but narrowly black; pile of scutellum pale; southern California ....*truquii innotata* (Curran) 1927
— Base of scutellum black; pile of scutellum black; Imperial County .... *currani* James 1932

Genus *Eulalia* Meigen (=*Odontomyia* Meigen)

The larvae resemble those of *Stratiomys,* but the antennae are longer, the last segment of the body is shorter, and there are one or two pairs of stout hooks on the ventroposterior margin of the next to the last segment and in some species also on the preceding one (fig. 14:45*d*). The larvae breed in small sluggish streams or ponds, especially those with muddy bottoms, and feed on algae, small microörganisms, and decaying vegetable matter. Brues (1928) records five unnamed species from western hot springs. The following key to adults is adapted from James (1936*c*) and the one on larvae from Johannsen (1922).

Keys to the Adults of the California Species of Eulalia

Males (eyes contiguous)

1. Eyes clothed with conspicuous pile; southern and central California ....*hirtocculata* (James) 1936
— Eyes bare 2
2. Femora black, the apices often pale 3
— Femora wholly yellow, at most reddish or brownish . 5
3. The pale markings on the 2nd abdominal segment reaching toward and attaining, or nearly attaining, the base of the segment; San Joaquin Valley . *colei* (James) 1936
— The pale markings on the 2nd abdominal segment not encroaching upon the basal half of the segment, usually narrower and linear 4
4. Face unusually broad, the oral margin pale; vein R_4 distinctly present; southern and central California ....*pilosa* (Day) 1882
— Face of moderate breadth; the oral margin black; vein R_4 usually absent; widespread in California .... *hoodiana* (Bigot) 1887
5. Face above the depression wholly yellow, the frontal triangle sometimes black 6
— Face above the depression and the frontal triangle wholly black 9
6. Lateral margins of mesonotum wholly green or yellow; Sonoma County .... *cincta* (Olivier) 1811
— Lateral margins of mesonotum black, the posterior calli sometimes pale 7
7. Inner ocular orbits strongly bowed; the width of the face across the oral margin greater than the distance from the upper angle of the oral margin to the base of the antennae; Riverside County.... *tumida* (Banks) 1926
— Inner ocular orbits almost straight; the width of the face across the oral margin about equal to the distance from the upper angle of the oral margin to the base of the antennae 8
8. Abdominal band as wide on the 2nd and 3rd segments as on the 4th; in some cases the pattern on the 2nd and 3rd segments is broken, so that the width must include the separated marginal spots; southern and central California .... *arcuata* (Loew) 1872
— Abdominal bands definitely narrower on the 2nd and 3rd segments than on the 4th, and without separated marginal spots; southern California .... *alticola* (James) 1932
9. Oral margin yellow 10
— Oral margin black; Kern County...*pubescens* (Day) 1882
10. Face retreating below the antennae; California ....*americana* (Day) 1882
— Face with a keeled prominence below the base of the antennae; Santa Clara County .. *inaequalis* (Loew) 1865

Females (eyes widely separated)

1. Eyes clothed with conspicuous pile .... *hirtocculata* (James)
— Eyes bare 2
2. Mesonotum with lateral yellow markings extending at least for a short distance in front of the suture .... *cincta* (Olivier)
— Mesonotum wholly black in front of the suture, the humeri sometimes yellow 3
3. Posterior femora wholly, or at least half, black..... 4
— Posterior femora yellow or at most reddish......... 5
4. Vein R_4 present 6
— Vein R_4 absent .... *hoodiana* (Bigot)
5. Front wholly black; head unusually wide...*pilosa* (Day)
— Front with yellow markings along the midfrontal suture and to each side of the ocellar triangle; head of normal width ....*colei* (James)
6. Head chiefly black, more or less marked with yellow; dorsum of thorax wholly black, at most the posterior calli yellow 7
— Head yellow, marked with black; sides of mesonotum yellow at least along the wing bases; humeri yellow 8
7. Abdomen black in ground color, with pale transverse linear markings at the posterior margins of the segments .... *pubescens* (Day)
— Abdomen pale in ground color, with a dorsal black stripe which may, however, be almost the width of the abdomen ....*americana* (Day)
8. Face with median black or brown vitta which widens

out just above the oral margin ......*inaequalis* (Loew)
— Face wholly green or yellow, or marked with black other than along the midfacial line 9
9. The black markings on the 3rd abdominal segment semielliptical, broadest at the middle of the segment .... *alticola* (James)
— The black marking of the 3rd abdominal segment with a biarcuate posterior margin, as broad or nearly as broad laterally as medially 10
10. Face in lateral profile extending forward as far as width of the eye; abdominal bands never contiguous .... *tumida* (Banks)
— Face in lateral profile extending forward much less than width of the eye; abdominal bands often contiguous .... *arcuata* (Loew)

REFERENCES

BRUES, C. T.
1928. *See* Diptera references.
CURRAN, C. H.
1927. Synopsis of the Canadian Stratiomyidae (Diptera). Trans. Roy. Soc. Can. Sec. V, 1927, pp. 191-228.
JAMES, M. T.
1936*a*. A proposed classification of the Nearctic Stratiomyinae (Diptera: Stratiomyidae). Trans. Amer. Ent. Soc., 62:31-36.
1936*b*. The Stratiomyidae of Colorado and Utah. Jour. Kansas Ent. Soc., 9:33-48.
1936*c*. The genus *Odontomyia* in America North of Mexico (Diptera, Stratiomyidae). Ann. Ent. Soc. Amer., 29: 517-550.
1942. A review of the Myxosargini (Diptera, Stratiomyidae). Pan-Pac. Ent., 18:49-60.
1943. A revision of the Nearctic species of *Adoxomyia* (Diptera, Stratiomyidae). Proc. Ent. Soc. Wash., 45:163-171.
JAMES, M. T., and G. C. STEYSKAL
1952. A review of the Nearctic Stratiomyini (Diptera, Stratiomyidae). Ann. Ent. Soc. Amer., 45:385-412.
JOHANNSEN, O. A.
1922. Stratiomyiid larvae and puparia of the North Eastern States. Jour. N.Y. Ent. Soc., 30:141-153, 2 pls.
1935. *See* Diptera references.
LENZ, F.
1923. Stratiomyidenlarven aus Zuellen. Arch. Naturgesch. Abt. A, 89:39-62.
MALLOCH, J. R.
1917. *See* Diptera references.
MELANDER, A. L.
1903. A review of the North American species of *Nemotelus*. Psyche, 10:171-183, 1 pl.

Family RHAGIONIDAE

The larvae of *Atherix* are the only known aquatics of the Rhagionidae, or snipe flies. "The eggs of *A. variegata*, the only American species, are laid in masses on twigs over streams into which the young larvae fall immediately after hatching; the female dies, clinging to the egg mass, after deposition. Other females then lay eggs on this mass until a ball of considerable size is formed, composed of egg masses and dead females." (Johannsen, 1935). The larva (fig. 14:46*a,b*) was described by Greene (1926). *A. variegata* Walker 1848, has been recorded from California.

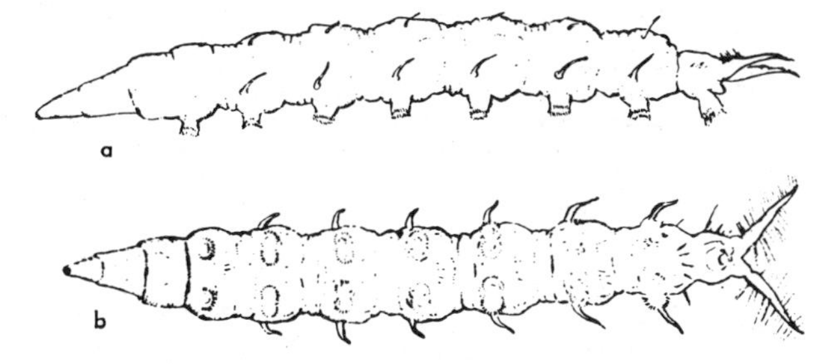

Fig. 14:46. Rhagionidae. *a,b*, *Atherix variegata* Walk., larva, lateral and dorsal views (Johannsen, 1935).

REFERENCES

GREENE, C. T.
1926. Descriptions of larvae and pupae of two-winged flies belonging to the family Leptidae. Proc. U.S. Nat. Mus. 70:1-20.
JOHANNSEN, O. A.
1935. *See* Diptera references.
LEONARD, M. D.
1930. A revision of the dipterous family Rhagionidae (Leptidae) in the United States. Mem. Amer. Ent. Soc. 7:1-181.

Family TABANIDAE

The horse- and deer-flies are familiar objects to everyone. The latter are persistent biters of hunters and campers as well as the larger warm-blooded vertebrates of the forest. Some species have been incriminated as vectors of certain diseases affecting man. Only the females suck blood, and many males are nectar and pollen feeders. It is often difficult to determine the association of the sexes because of the marked dissimilarity between males and females.

The larvae are all predaceous on soft-bodied invertebrates, and most of them are aquatic or semiaquatic.

Brennan (1935) and Stone (1938) reviewed the two

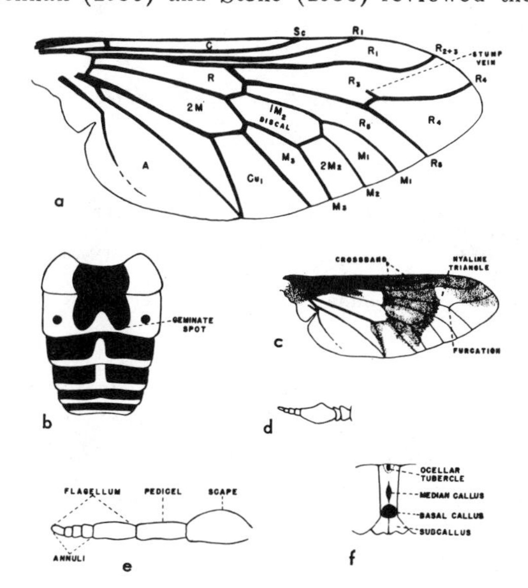

Fig. 14:47. Tabanidae. *a,d,f*, *Tabanus* sp.: *a*, wing venation; *d*, antenna; *f*, frons of female. *b*, *Chrysops surda* O.S., abdomen of female; *c*, *Chrysops coloradensis* Bigot, wing; *e*, *Chrysops coquilletti* Hine, antenna (Middlekauff, 1950).

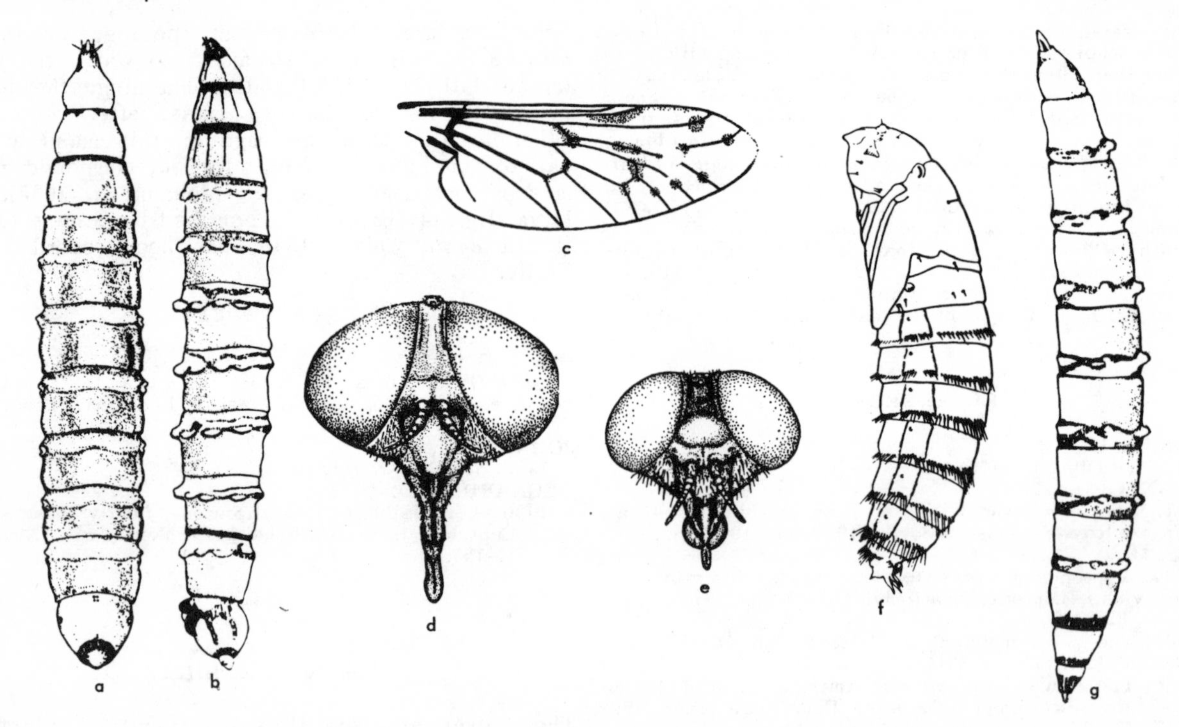

Fig. 14:48. Tabanidae. *a, Chrysozona americana* (O.S.), larva; *b, Atylotus incisuralis* (Macq.), larva; *c, Silvius quadrivittatus* (Say), wing; *d, Pilimas californica* (Bigot), head of female; *e, Apatolestes comastes* Will., head of female; *f, Tabanus tetricus* Mar., pupa; *g, Chrysops discalis* Will., larva (*c,d,e*, Brennan, 1935; *a,b,f,g*, Cameron, 1926).

North American subfamilies, and Philip (1947) published a catalogue of the Nearctic species with a key to the genera. The early literature on the larvae and pupae was reviewed by Marchand (1920), and later papers on the biology and immature stages are by Stone (1930), Webb and Wells (1924), and Cameron (1926). The California species have been treated by Middlekauff (1950).

Key to Nearctic Genera of Tabanidae[6]

Adults

1. Hind tibiae with 2 apical, usually strong spurs (weakest in the rare *Merycomyia*); subepaulets bare, uninflated (PANGONIINAE) 2
— Hind tibiae without apical spurs (TABANINAE)..... 12
2. Antennae 9-10 segmented (flagellum with at least 7 divisions, the basal one not greatly differentiated) (PANGONIINI) 3
— Antennae with not more than 7 segments (flagellum with at most 5 divisions, the basal one considerably elongated) (fig. 14:47*e*); proboscis seldom exceeding height of head 8
3. Eyes of female with upper angles sharply acute; wings with a wide, sharply marked, irregular brown pattern .... *Goniops* Aldrich
— Eyes above at most rectangulate; wings hyaline or diffusely tinted 4
4. Proboscis hardly 1/2 the height of the head; palpi usually shorter than the antennae and swollen basally (fig. 14:48*e*) 5
— Proboscis about equal to height of head (fig. 14:48*d*); palpi subequal to length of the antennae, 1/2 or less that of the proboscis, not swollen basally 6

[6]Adapted from Philip, 1947.

5. Eyes bare .... *Apatolestes* Williston
— Eyes and body abundantly pilose..... *Brennania* Philip
6. Cell R_5 closed and petiolate ..... *Esenbeckia* Rondani
— Cell R_5 open at wing margin 7
7. Eyes hairy (short in female); base of vein R_4 with short spur .... *Pilimas* Brennan
— Eyes bare; R_4 often without spur ... *Stonemyia* Brennan
8. Flagellum with 5 divisions (fig. 14:47*a*); usually slender, often picture-winged flies (CHRYSOPINI) 9
— Third antennal segment with 3 divisions (MERYCOMYIINI) .... *Merycomyia* Hine
9. Eyes in life (or relaxed) "freckled" with small rounded spots; pedicel of antenna either about 1/2 the length of the scape or distinctly longer than the flagellum; wings hyaline or with isolated clouds on cross veins (fig. 14:48*c*) 10
— Eyes in life (or relaxed) with large angular spots; scape and pedicel usually subequal, the flagellum as long or longer than either; wing usually with irregular, extensive dark pattern 11
10. Flagellum of antenna much shorter than either of the basal segments and shaped like a bowling pin with a bent neck; pedicel nearly as long as scape .... *Assipala* Philip
— Flagellum at least longer than pedicel, the latter only about 1/2 as long as scape .... *Silvius* Meigen
11. Abdomen globose, much wider than thorax, wings evenly fumose, a spur at base of vein R_4 .... *Neochrysops* Walton
— Abdomen slender; wings mostly with irregular dark patterns, no spurs (except adventitiously) (fig. 14:47*c*) .... *Chrysops* Meigen
12. Subepaulet (basicosta) bare and flat in Nearctic species 13
— Subepaulet with hairs like those on the costal vein; often inflated 17
13. Labellae sclerotized, smooth and shining 14
— Labellae fleshy, shrunken and distorted when dry (DIACHLORINI) 15

14. Body compact, uniformly dull greenish or yellow; females without frontal callosities, otherwise head, wing, and leg characters not unusual (CHLOROTABANINI) .... *Chlorotabanus* Lutz
— Bodies more slender, predominantly black; subcallus, antennal scape and tibiae swollen; vein R_4 curved abruptly forward apically (BOLBODIMYIINI) .... *Bolbodimyia* Bigot
15. Fore tibia swollen, more thickened than the respective femur or other tibiae; wing with a fumose apical spot and no spur vein on base of vein R_4 .... *Diachlorus* Osten Sacken
— Fore tibia normal; wing hyaline or the cross veins clouded, and spur present on base of R_4 16
16. Body pruinose without distinct pattern (in Nearctic species); female with very wide front and no ocelligerous tubercle; male with occipital tubercle suppressed .... *Aegialomyia* Philip
— Body bicolored or with definite pattern; female with normal front and distinct ocelligerous tubercle; male with prominent tubercle in the occipital notch .... *Stenotabanus* Lutz
17. Flagellum of antenna with 4 divisions; scape considerably longer than thick; eyes in life (or relaxed) and wings with irregular bands and spots (CHRYSOZONINI) .... *Chrysozona* Meigen
— Flagellum usually with 5 divisions (rarely fewer by apical fusion and then wings hyaline); scape scarcely longer than thick (fig. 14:47*d*); eyes in life banded or unicolorous (TABANINI) 18
18. Plate of 3rd antennal segment with a hooklike dorsobasal extension; eyes sparsely hairy 19
— Plate rarely with dorsobasal directed tooth, and if so, eyes bare 20
19. Subcallus above antennae with erect black hairs laterally .... *Agkistrocerus* Philip
— Subcallus pollinose, no hair ..... *Hamatabanus* Philip
20. Flagellum modified with 2 to 3 terminal divisions more or less fused and indistinct; small flies, 11 mm. and less 21
— Flagellum normal, 4 distinct divisions; small to large flies 22
21. Very small flies, length less than 9.5 or rarely 10 mm.; frontal callosity of female missing or rudimentary; eyes ostensibly bare, unbanded in life .... *Microtabanus* Fairchild
— Length 9.5 to 11 mm.; frontal callosity distinct; eyes bare or very sparsely pilose, banded in life (or relaxed) .... *Glaucops* Szilady
22. Vertex of female with a distinct denuded ocelligerous tubercle; of male with an elevated, anteriorly shining tubercle; in doubtful cases eyes are densely pilose 23
— Vertex without such tubercles; if raised in male then completely pollinose; eyes bare, or sparsely pilose, or in certain males densely pilose 24
23. Eyes ostensibly bare; front of female narrow, the callosity long and linear; upper eye facets of male strongly differentiated and enlarged; scutellum pallid .... *Leucotabanus* Lutz
— Eyes usually distinctly pilose; front of female variable, but the basal callosity correspondingly broad; eye facets of male relatively undifferentiated; scutellum usually dark ..... *Tabanus* (subg. *Hybomitra* Enderlein)
24. Eyes yellow or brown, distinctly pilose, particularly in males; front of female without callosities, or they are reduced to small isolated bare spots; a single diagonal, dark band on eyes, often showing even in dried specimens .... *Atylotus* Osten Sacken
— Eyes blackish when dried (certain males only lower area of small facets), bare or if sparsely hairy, with more than 1 purple band in life and the abdomen may have a uniform pale, middorsal stripe; front of female with a broad basal callosity 25
25. Antennae including flagellar divisions, and palpi with unusually erect hairs; palpi blunt and stout, the proboscis relatively quite small; postocular rim (between eye margins and occipital hairs) unusually wide .... *Anacimas* Enderlein
— Not with this combination of characters; hairs of flagellar divisions very inconspicuous 26
26. Body and wings entirely deep brown to black (abdomen may have pale lateral vestiture); frontal and facial callosities unusually protuberant and shining; eyes bare, in life with the upper of 4 purple bands strongly decurved and widened outwardly .... *Whitneyomyia* Bequaert
— Body and wings varicolored (when entirely black, more than 15 mm. long); eyes bare or sparsely hairy .... *Tabanus (s. str.)* Linnaeus

Larvae

1. Body club-shaped, the thoracic segments slender, the abdominal segments robust (terrestrial) .... *Goniops* Aldrich
— Body tapering anteriorly and posteriorly, not club-shaped 2
2. Small, usually less than 20 mm. when mature; 2nd and 3rd antennal segments subequal, or the distal one the longer; tracheal trunks (seen in living larvae) unswollen, sinuous; usually a protrusile stigmatal spine; body almost entirely striated........ *Chrysops* Meigen
— Large, or very large, usually more than 20 mm. when mature; antenna with the terminal segment much the shortest and smallest; tracheal trunks usually swollen; almost never with a protrusile spine at tip of siphon; ventral and dorsal areas often smooth, especially on thorax 3
3. Anal segment rounded, the siphon short, hardly exsertile; dark ornamentation confined to anal and 3 thoracic segments (fig. 14:48*a*)............*Chrysozona* Meigen
— Anal segment usually tapering, the siphon more or less elongate when extended; dark markings, when present, usually on other abdominal segments as well, or confined to a prothoracic collar only. .*Tabanus* Linnaeus

Pupae

1. One spine on each side of median line on dorsum of each abdominal segment much stronger than the others in the series .... *Goniops* Aldrich
— Spines of each series either of an almost uniform strength or at least no 2 spines conspicuously stronger than the others 2
2. Dorsolateral and lateral combs absent on caudal segment; bristles on frontal tubercles double; antenna about as long as basal width, surpassing adjacent line of dehiscence (best seen after emergence) .... *Chrysops* Meigen
— Dorsolateral or lateral combs present; frontal bristles single; antenna long or short 3
3. Antenna elongate, the tip surpassing adjacent line of dehiscence ....*Chrysozona* Meigen
— Antenna short, only about as long as basal width, the tip not attaining adjacent line of dehiscence .... *Tabanus* Linnaeus

The California Species of Tabanidae[7]

Key to the California Species of Apatolestes

1. Females; eyes not contiguous 2
— Males; eyes contiguous 9
2. Front basally inflated, shining black or dark brown, pollinose only at the vertex around the ocelli (fig. 14:48*e*) 3
— Front either normally pollinose, or with a bare restricted area, not with a calluslike swelling above the subcallus 5

[7]The following keys, with slight alteration, are from Middlekauff, 1950.

3. Body entirely subshining black, the wings smoky; Santa Barbara to Riverside counties. . *ater* Brennan 1935
— Body grayish to dull blackish, only the costal cell sometimes infuscated 4
4. Costal cell hyaline; palpi with pale hairs; Lassen to San Diego counties. *comastes comastes* Williston 1885
— Costal cell infuscated; palpi usually with some black hairs; San Bernardino, Riverside, and San Diego counties .... *comastes willistoni* Brennan 1935
5. Scutellum yellowish 6
— Scutellum blackish or cinereous 7
6. Flagellum black; front entirely pollinose when not irregularly worn; Riverside and Imperial counties .... *parkeri* Philip 1941
— Flagellum mostly yellow; front with an expansive bare area; Fresno and Tulare counties .... *albipilosus* Brennan 1935
7. Front very wide, height and basal width subequal; a narrow mesal, bare brown streak about 2/3 its height; cheeks unusually swollen; Riverside County .... *colei* Philip 1941
— Front narrower, about 1:2 or 2:5, the bare area more extensive 8
8. Body predominantly gray; wings subhyaline; midfrontal bare area widely separated from eye margins; Contra Costa and Los Angeles counties. . *similis* Brennan 1935
— Body predominantly brownish or darker; wings fumose, bare area almost touching eye margins; Mariposa and Riverside counties .... *hinei* Brennan 1935
9. Wings hyaline with or without infuscated costal cells or isolated clouds; pleural hairs, at least, largely whitish 10
— Wings smoky, or palpal and pleural hairs black 13
10. Costal cell fumose, distinctly darker than remainder of wing .... *comastes willistoni* Brennan 1935
— Costal cell completely hyaline 11
11. Palpi truncated apically, short and stubby; hairs sparse .... *similis* Brennan 1935
— Palpi elongated, more slender; sometimes obscured by long dense pile 12
12. Size 14 mm. or more; palpal and facial vestiture dense .... *colei* Philip
— Size less than 14 mm.; these hairs not especially dense .... *comastes comastes* Williston 1885
13. Shining black species .... *ater* Brennan 1935
— Dull black species with pale abdominal incisures .... *hinei* Brennan 1935

Apatolestes rossi Philip 1949 from Lake County is not included in this key.

Genus *Brennania* Philip

Brennania hera (Osten Sacken) 1877 San Francisco, San Mateo, Monterey

Key to the California Species of Pilimas

1. Front subparallel, not divergent below; antennae, palpi, and legs reddish including their vestiture; San Bernardino County.................... *beameri* Philip 1942
— Front divergent below (fig. 14:48*d*); apical 1/2 of flagellum black; palpi with black hairs; legs fuscous with dark vestiture; Siskiyou to Riverside counties .... *californica* (Bigot) 1892
(= *Pangonia dives* Williston 1887 and *Silvius jonesi* Cresson 1919)

Key to the California Species of Stonemyia

1. Large robust species (15 mm.), black; the mesonotum of female clothed with pale yellow pile; wing dark brown; a patch of pale yellow hair on middle of posterior margin of 2nd tergite; Madera County .... (= *albomacula* Stone 1940) *velutina* (Bigot) 1892
— Body length usually less than 15 mm.; body yellowish with or without black markings 2
2. Proboscis long, at least equaling width of head; abdomen orange with black markings; legs black; Sierra County to Mariposa County .... *tranquilla fera* (Williston) 1887
— Proboscis shorter than width of head; abdomen and legs golden yellow 3
3. Entire thorax dark, the 1st tergite and terminal tergites and sternites predominantly so; spots on segments 2–4 variable (male); Madera and Mariposa counties .... *ruficornis* (Bigot) 1892
— Integument of entire body and legs concolorous yellowish (male); Shasta County to Riverside County .... *abaureus* Philip 1941

Key to California Species of Silvius

1. Females; eyes not contiguous 2
— Males; eyes contiguous 6
2. Wings immaculate; orange species; Lassen County to Riverside County .... *gigantulus* (Loew) 1872
— Wings maculate; gray species 3
3. Costal cell hyaline 4
— Costal cell infuscated.......................... 5
4. Basal callus about 1/2 width of front, broadly separated from eyes; veins of radial sector with subapical spots (fig. 14:48*c*); Riverside and San Diego counties .... *quadrivittatus* (Say) 1823
— Basal callus distinctly more than 1/2 width of front, narrowly separated from eyes; veins of radial sector without subapical spots; Trinity County to San Diego County .... *notatus* (Bigot) 1892
5. Scape robust; flagellum about as long as scape; Humboldt County.................. *philipi* Pechuman 1938
— Scape slender; flagellum about twice as long as scape .... *sayi* Brennan 1935
6. Wings immaculate; orange species .... *gigantulus* (Loew) 1872
— Wings maculate; gray species 7
7. Abdomen predominantly yellowish; subapical spots on veins of radial sector; not densely haired species; venter of abdomen with 3 rows of spots .... *quadrivittatus* (Say) 1823
— Abdomen predominantly dark; no subapical spots on veins of radial sector; thorax and base of abdomen with long white hair; venter of abdomen mostly dark or with a broad dark median band .... *notatus* (Bigot) 1892

Key to California Species of Chrysops

1. Females; eyes not contiguous 2
— Males; eyes contiguous.......................... 25
2. Apex of wing beyond crossband hyaline, sometimes just a trace of a cloud confined to cell R_1 3
— Apex of wing beyond crossband distinctly infuscated 6
3. Femora banded in middle with reddish brown; abdomen greenish yellow with black markings; Butte County to Merced County.............. *hirsuticallus* Philip 1941
— Femora black; abdomen otherwise 4
4. Pile of pleura golden yellow or orange; basal tergites laterally orange; Shasta, Plumas, and Sierra counties .... *excitans* Walker 1848
— Pile of pleura whitish to gray; abdomen black with or without grayish pile 5
5. Base of cell Cu_1 with a small hyaline spot .... *carbonaria* Walker 1848
— Base of cell Cu_1 infuscated; large obscure gray middorsal triangles more likely to be present; Humboldt and El Dorado counties ...... *mitis* Osten Sacken 1875
6. Basal callus orange or orange with lateral and upper margins black.......................... 7
— Basal callus entirely black 14

7. Hyaline triangle attains posterior margin of wing (fig. 14:47*c*) 8
— Hyaline triangle enclosed, not attaining posterior margin of wing; Los Angeles County ....
.... *fulvaster* Osten Sacken 1877
8. Knob of halter brown or black; scape not incrassate; Plumas County to San Bernardino County ....
.... *coloradensis* Bigot 1892
— Knob of halter yellow; scape incrassate 9
9. First segment of flagellum distinctly longer than the sum of the 4 apical segments; outer margin of crossband sinuous with toothlike projections in cells R_3 and M_1 .... *virgulata* Bellardi 1859
— First segment of flagellum not longer than the sum of the 4 apical segments; outer margin of crossband evenly convex or with no projection in R_3 10
10. Apical spot of wing exceeds vein R_5; Madera County to San Diego County....... *clavicornis* Brennan 1935
— Apical spot of wing not exceeding vein R_5, as a rule occupying about half of cell R_4 11
11. Second tergite typically with double geminate black spot; outer margin of crossband sinuous; Ventura and Los Angeles counties.......... *robusta* Brennan 1935
— Second tergite with single geminate spot; outer margin of crossband straight 12
12. Scape much enlarged, widest diameter almost twice that of pedicel (fig. 14:47*e*); Shasta County to San Diego County................ *coquillettii* Hine 1904
— Scape enlarged, but widest diameter not much more than pedicel 13
13. Tergites caudad from 3 with black lateral spots; Contra Costa County .....*pachycera hungerfordi* Brennan 1935
— Tergites caudad from 3 without black lateral spots; San Diego County . *pachycera pachycera* Williston 1887
14. Cell 2M infuscated over at least half its length; tergites with no yellow on mid-line 15
— Cell 2M hyaline, although there may be some infuscation at base and apex; tergites with yellow markings on mid-line and posterior margins 17
15. Dorsum of abdomen with orange spot on anterolateral margin; Lassen County to Calaveras County ....
.... *noctifera noctifera* Osten Sacken 1877
— Dorsum of abdomen black 16
16. Crossband abbreviated in cell M_3 and with a distinct toothlike projection from the crossband in cell R_3; Humboldt County to Marin County ....
.... *noctifera pertinax* Williston 1887
— Crossband practically complete to hind margin of wing and without a toothlike projection; Marin County to San Luis Obispo County .... *pechumani* Philip 1941
17. Radial cell hyaline, although possibly with a small amount of infuscation at base and apex 18
— Radial cell infuscated over more than 1/2 its area .. 19
18. First tergite black, posterolateral margins may be gray
.... *aestuans* van der Wulp 1867
— First tergite yellow with brown quadrate spot on anteromedian margin; Modoc County to Inyo County.......
.... *bishoppi* Brennan 1935
19. Discal cell (first M_2) infuscated 20
— Discal cell hyaline, although possibly infuscated around borders; Del Norte County to Inyo County ...
.... *discalis* Williston 1880
20. Second tergite with a small black anteromedian quadrate or reniform spot, the remainder of the segment yellow; face black with 3 yellow pollinose stripes; Shasta, Plumas, and Sierra counties ..... *asbestos* Philip 1949
— Second tergite otherwise marked; face with some integumental yellow 21
21. Radial cell with an infuscated band for its entire length, sometimes dilute toward outer two-thirds; a narrow hyaline line on posterior border, widest apically; fore coxae black 22
— Radial cell with apical 3rd hyaline; fore coxae yellow
.... *furcata* Walker 1848
22. Second tergite with an isolated black spot laterad to geminate spot (fig. 14:47*b*) 23
— Sides of 2nd tergite entirely yellow or the black spot enlarged and connected with the median figure 24
23. Apical spot beyond crossband narrow, not wider than the basal width of cell M_1; yellow stripe on frontoclypeus narrow; Siskiyou County to Tulare County ..
.... *surda* Osten Sacken 1877
— Apical spot beyond crossband wide, about twice the basal width of cell M_1; yellow spot on frontoclypeus wider than lateral brown lobes (Humboldt to Tulare counties) ...... *proclivis proclivis* Osten Sacken 1877
24. Tergite 3 with 4, more or less distinct black spots; yellow spot on frontoclypeus wider than lateral brown lobes; Plumas County to Tulare County ....
.... *proclivis imfurcata* Philip 1935
— Tergite 3 black except for narrow yellow stripe near incisure; yellow spot on frontoclypeus reduced to a very narrow line; Fresno County.................
.... *proclivis picea* Philip 1935
25. Apex of the wing beyond the crossband hyaline 26
— Apex of the wing beyond the crossband more or less infuscated 29
26. Base of cell Cu_1 typically with a hyaline spot .....
.... *carbonaria* Walker 1848
— Base of cell Cu_1 infuscated 27
27. Tergites with some yellowish pubescence, especially near the apex and forming median triangles ....
.... *excitans* Walker 1848
— Tergites black pubescent 28
28. Mid-tibia black............. *mitis* Osten Sacken 1875
— Mid-tibia reddish brown....... *hirsuticallus* Philip 1941
29. Apical spot approximately the same width for its entire length; including only the apex of cell R_4 30
— Apical spot not as above, variable, usually including nearly 1/2 of cell R_4, often extending considerably beyond middle of cell 34
30. Wholly black species; frontoclypeus black; apical spot practically separated from crossband ....
.... *noctifera pertinax* Williston 1887
— Not wholly black species; frontoclypeus yellow; apical spot distinctly united with the crossband 31
31. Frontoclypeus and oral margins of genae yellow 32
— Frontoclypeus and oral margins of genae not entirely yellow, usually with a black spot on each side 33
32. Wing picture dilute .... *bishoppi* Brennan 1935
— Wing picture saturate ..... *aestuans* van der Wulp 1867
33. Frontoclypeus black with a narrow yellow mid-streak
.... *surda* Osten Sacken 1877
— Frontoclypeus yellow with a black spot on each side
.... *proclivis proclivis* Osten Sacken 1877
34. Antenna incrassate (distinctly swollen) 35
— Antennae not incrassate 41
35. Annulate part of flagellum much shorter than scape
.... *virgulata* Bellardi 1859
— Annulate part of flagellum not shorter than scape, usually longer 36
36. Apical spot of the wing extending at least into cell R_5
.... 37
— Apical spot of the wing not extending beyond vein R_5, usually not even attaining this vein 38
37. Antenna yellowish, the scape distinctly bottle-shaped; second tergite yellow with a median broad black geminate figure and a black spot on each side .....
.... *clavicornis* Brennan 1935
— Antenna black, scape not bottle-shaped; 2nd tergite not patterned as above ... *fulvaster* Osten Sacken 1877
38. Second tergite with a double geminate black figure
.... *robusta* Brennan
— Second tergite without a double geminate black figure
.... 39
39. Second tergite with lateral black spots ....
.... *pachycera hungerfordi* Brennan 1935
— Second tergite without lateral black spots 40
40. Cell 2M infuscated subequally with R; crossband sometimes fenestrated; tergites 3 and 4 rarely with a black spot on each side...*pachycera pachycera* Williston 1887
— Cell 2M infuscated about 1/2 that of R; crossband

never fenestrate; tergites 3 and 4 always with a black spot on each side. . . . . *coquillettii* Hine 1904
41. Discal cell hyaline . . . . *discalis* Williston 1904
— Discal cell infuscated 42
42. Body jet black . . . .*pechumani* Philip 1941
— Body not entirely black. 43
43. Tergite 2 with a double geminate spot . . . .
. . . . *coloradensis* Bigot 1892
— Tergite 2 with a small black anteromedian quadrate spot . . . . *asbestos* Philip 1949

Genus *Chrysozona* Meigen

Chrysozona americana (Osten Sacken) 1875.

Genus *Glaucops* Szilady

Glaucops fratellus (Williston) 1887. El Dorado County

Genus *Atylotus* Osten Sacken

Atylotus incisuralis (Macquart 1847). Siskiyou to Tulare (= *Tabanus insuetus* Osten Sacken 1877)

Genus *Tabanus* Linnaeus

Certain hairy eyed species with an ocelligerous tubercle have been placed in the genus *Hybomitra* Enderlein by Philip, and Middlekauff adopted this procedure, but it appears to be more properly considered as no more than a subgenus.

Key to the California Species of *Tabanus*

1. Vertex of female with a distinct denuded ocelligerous tubercle; of male with an elevated, anteriorly shining tubercle (in doubtful cases eyes are densely pilose) *(Hybomitra)* 2
— Vertex without such tubercle; if raised in male then completely pollinose (eyes bare, or sparsely pilose, or in certain males densely pilose) *(Tabanus)* 22
2. Females, eyes not contiguous 3
— Males, eyes contiguous 14
3. Entirely black; including pubescence and pollinosity; Modoc County to Santa Cruz County . . . .
. . . . *procyon* (Osten Sacken) 1877
— Not entirely black 4
4. Abdomen broadly orange brown laterally, tergites 2 and 3 usually unicolorous orange to margins; the median black area usually constricted on tergite 3; sublateral pale spots not conspicuous 5
— Abdomen not broadly orange laterally; if paler laterally, the median dark area is rather broad and not constricted on tergite 3; sublateral pale spots, if present, distinct
. . . . 10
5. Hind tibial fringe well developed, mostly golden orange; length usually 18 mm. or more; prescutal lobe yellowish; Mendocino, San Mateo, and Santa Cruz counties . . . .
. . . . *californica* Marten 1882
— Hind tibial fringe weak, or, if well developed, not golden orange, frequently black; prescutal lobe reddish; usually less than 18 mm. 6
6. Antenna black, rarely with a little orange on 2nd and base of 3rd segments; 3rd segment narrow, with obtuse dorsal angle and little if any dorsal excision; vein R_4 often with a stump 7
— Not with this combination of characters 9
7. Frons slightly less than 3 times as high as wide, the sides nearly parallel. . . . . *philipi* Stone 1938
— Frons about 4 times as high as width at base, usually somewhat divergent above 8
8. Venter of abdomen with mostly yellowish hair, although there may be some scattered black hairs medianly; Del Norte County to Inyo County
. . . .*sonomensis sonomensis* Osten Sacken 1877
— Venter of abdomen almost uniformly clothed with fine black hair, the yellowish hair usually sparse except along incisures; Del Norte County to Inyo County
. . . . *sonomensis phaenops* Osten Sacken 1877
9. Subcallus pollinose; Plumas County to Mono County
. . . . *haemaphorus* Marten 1882
— Subcallus denuded; Del Norte County to Madera County
. . . . *captonis* Marten 1882
10. Subcallus denuded, shining 11
— Subcallus pollinose 13
11. Prescutal lobe black (Trinity County to Kern County)
. . . . *rhombicus* Osten Sacken 1876
— Prescutal lobe reddish brown 12
12. Stump vein usually present; length 15 mm. or more; front usually more than 3½ times as high as width at base; Shasta, Plumas, and Mono counties . . . .
. . . .*tetricus tetricus* Marten 1883
— Stump vein usually absent; length 14 mm. or less; front about 3 times as high as width at base; Modoc County to Alpine County . . . . .*melanorhinus* Bigot 1892
13. Stump vein usually present; pollen of head gray; Shasta, Plumas, and Inyo counties . *tetricus hirtulus* Bigot 1892
— Stump vein rarely present; pollen around face usually faintly tinged with yellow; Lassen, Tuolumne, and Mono counties. . . . .*opacus* Coquillett 1904
14. Entirely black, including palpus *procyon* Osten Sacken
— Not entirely black 15
15. Abdomen broadly orange brown laterally, the median black area usually constricted on tergite 3; sublateral pale spots not conspicuous 16
— Abdomen not broadly orange brown laterally; if paler laterally, the median dark area rather broad and not constricted on tergite 3; sublateral pale spots, if present, distinct 19
16. Antennae red with black annuli . . . . . *californica* Marten
— Antennae with black more extensive 17
17. Venter with black hair rather sparse, and, at least sublaterally a considerable amount of yellowish hair
. . . . 18
— Venter with only black hair . . . .
. . . . *sonomensis phaenops* Osten Sacken
18. Second sternite with a large black median spot; 2nd palpal segment stout, but apex not acute . . . .
. . . . *sonomensis sonomensis* Osten Sacken
— Second sternite with scarcely any black medianly; 2nd palpal segment stout, with apex acute . . . .
. . . . *haemaphorus* Marten
19. Prescutal lobe black (subcallus denuded) . . . .
. . . . *rhombicus* Osten Sacken
— Prescutal lobe pale, reddish orange 20
20. Subcallus denuded. . . . .*melanorhinus* Bigot
— Subcallus pollinose 21
21. Stump vein present . . . . *tetricus* Marten
— Stump vein absent. . . . . *opacus* Coquillett
22. Females; eyes not contiguous 23
— Males; eyes contiguous 31
23. Abdominal tergites with both median and sublateral pale spots, those of median row may be contiguous but do not form a parallel-sided stripe 24
— Abdominal tergites without sublateral pale spots, or median row of spots contiguous forming a parallel-sided stripe 25
24. Eyes sparsely pilose (high power); furcation without a black spot; Modoc County to Inyo County . . . .
. . . . *laticeps* Hine 1904
— Eyes bare; furcation with a black spot (Sierra County to Los Angeles County. . . . . *monoensis* Hine 1924
25. Abdomen unicolorous black or rarely with a median row of small whitish triangles 26
— Median dorsal row of spots contiguous, forming a more or less distinct parallel-sided stripe 28

26. Furcation without a distinct dark spot, although entire wing may be infuscated; mesonotum and fore tibia black 27
— Furcation usually with a distinct dark spot; mesonotum covered with creamy hair over a dark reddish background; fore tibia bicolored; Del Norte County to Imperial County .... *punctifer* Osten Sacken 1876
27. At least tergite 1 with an inconspicuous tuft of white hairs beneath the scutellum; frontal callosity subquadrate, the median callus narrow; El Dorado and Santa Clara counties .... *kesseli* Philip 1950
— None of tergites with median tufts of white hairs; frontal callosity merging broadly with median callus above; Humboldt, Sonoma, and Santa Cruz counties .... *aegrotus* Osten Sacken 1877
28. Base of vein R_4 with stump vein; Sierra, Inyo, and San Luis Obispo counties .... *productus* Hine 1904
Base of vein R_4 without stump vein 29
29. Eyes pilose; median spot a triangle on tergite 2; Mono County .... *stonei* Philip 1941
— Eyes bare; median spot on tergite 2 rectangular 30
30. Outer hind tibial fringe mostly black; body markings contrasting; front usually evenly convergent; Trinity County to Imperial County .... *lineola scutellaris* Walker 1850
— Outer hind tibial fringe mostly white; body color faded in appearance, the middorsal stripe not especially paler than the sublateral ones; frontal margins rather bowed in the middle; Riverside and Imperial counties .... *vittiger nippontucki* Philip 1942
31. Eyes pilose 32
— Eyes bare, or only sparsely pilose 33
32. Abdomen with a median row of pale spots forming a parallel-sided stripe; upper eye facets greatly enlarged .... *vittiger nippontucki* Philip
— Abdomen without a parallel-sided stripe; 4 black spots on abdominal tergite 2; upper eye facets not greatly enlarged.......................... *laticeps* Hine
33. Abdomen unicolorous, black (rarely a median row of small whitish-haired triangles) 34
— Abdomen not unicolorous 36
34. Mesonotum black; no spot at furcation 35
— Mesonotum bordered by a band of whitish pile; spot at furcation .... *punctifer* Osten Sacken
35. Eyes with distinct upper area of enlarged facets; a tuft of white hairs beneath scutellum on tergite 1 .... *kesseli* Philip
— Eyes with upper facets but little differentiated; abdomen entirely black-haired ..... *aegrotus* Osten Sacken
36. Abdomen with a median row of pale spots forming a parallel-sided stripe 37
— Abdomen not so colored; furcation with a dilute infuscated spot .... *monoensis* Hine
37. Stump vein present; antennae mostly black......... .... *productus* Hine
— Stump vein absent; antennae pale .... *lineola scutellaris* Walker

REFERENCES

BRENNAN, J. M.
1935. The Pangoniinae of Nearctic America (Tabanidae, Diptera). Univ. Kansas Sci. Bull., 22:249-401.
CAMERON, A. E.
1926. Bionomics of the Tabanidae (Diptera) of the Canadian Prairie. Bull. Ent. Res., 17:1-42.
MARCHAND, W.
1920. The early stages of Tabanidae (horse-flies). Monogr. Rockefeller Inst. Med. Res., 13:1-203.
MIDDLEKAUFF, W. W.
1950. The horse flies and deer flies of California. Bull. Calif. Insect Surv., 1:1-24.
PHILIP, C. B.
1947. A catalog of the blood-sucking fly family Tabanidae (horseflies and deerflies) of the Nearctic region north of Mexico. Amer. Midl. Nat., 37:257-324.
STONE, ALAN
1930. The bionomics of some Tabanidae (Diptera). Ann. Ent. Soc. Amer., 23:261-304.
1938. The horseflies of the subfamily Tabaninae of the Nearctic region. U.S.D.A. Misc. Publ., 305:1-171.
WEBB, J. L., and H. W. WELLS
1924. Horseflies: biologies and relation to western agriculture. U.S.D.A. Bull. 1218, 35 pp.

Family DOLICHOPODIDAE

The small, usually metallic blue or green "long-legged flies" of the family Dolichopodidae are seen everywhere, especially where patches of sunlight play on damp, shaded, heavily vegetated areas. The family is a large one, and the biology is in general so little known that one cannot yet generalize about the aquatic groups. For this reason only those few genera whose larvae are known to be aquatic are included here, and more complete keys to adults can be found in the list of references cited. The adults share with those of the Empididae the curious habit of mating dances, for which the males very often have striking, specific ornamentations of the antennae, legs, or wings. Sexual dimorphism is often great, and the females of many genera are difficult, if not impossible to separate. The larvae as well as the adults of this family are predaceous, and larval habitats range all the way from mud and water (*Dolichopus, Argyra, Campsicnemus,* and *Hydrophorus*), moist sandy beaches (*Tachytrechus* and *Hypocharassus*), dry sandy beaches (*Asyndetus*), intertidal rocks (*Aphrosylus, Hygrocele uthus, Hydrophorus,* and *Cymatotopus*), tree cavities (*Systenus*), and beneath tree bark (*Medetera*), to mines in the stems of grasses and sedges (*Thrypticus* and *Oligochaetus*).

The larvae of this family can be distinguished by the small, retractile, semimembranous head with a triangular dorsal plate and long, slender, dorsal and tentorial rods extending far back into the body (fig. 14:49*p, s, t*); minute anterior spiracles (fig. 14:49*n*), the posterior spiracles small (fig. 14:49*g, k, o, x*) and borne on the concave caudal end of the body between several unequal pairs of short lobes (fig. 14:49*b, e, g, i, n, y*). The pupae (fig. 14:49*c, f, j, m, q, u*) are usually formed within a cocoon made of pieces of the larval medium cemented together with a silken lining (fig. 14:49*f, w*). The elongate thoracic respiratory organs project far in front of the head through the wall of the cocoon (fig. 14:49*a, w*); paired cephalic spines may be present; the leg sacs extend beyond the wing cases; and the abdomen bears transverse bands of spinules. Since the immature stages of very few genera have been sufficiently characterized, the construction of a key is still not practical. Smith (1952) has reviewed what is known of the biology of the immature stages.

Key to Adults of the North American Genera of Aquatic Dolichopodidae

1. Posterior basitarsus with stout bristles above (fig.

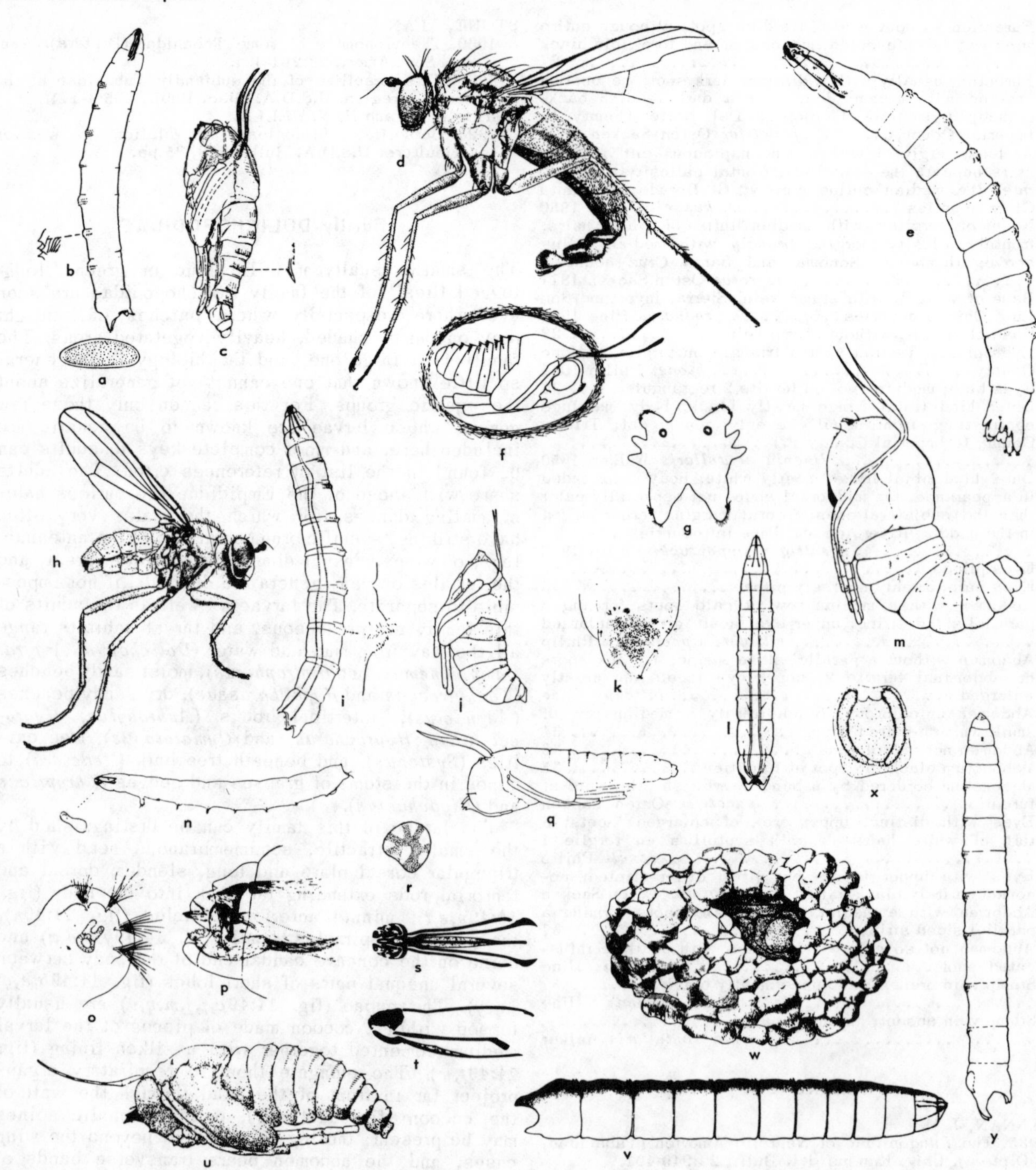

Fig. 14:49. Dolichopodidae. *a-c*, *Campsicnemus fumipennis* Parent: *a*, egg; *b*, larva; *c*, pupa. *d*, *Dolichopus bolsteri* V.D., adult male; *e*, *Aphrosylus*, larva; *f,g*, *Aphrosylus praedator* Whr.: *f*, pupa; *g*, spiracular disc of larva. *h-k*, *Hydrophorus pacifica* V.D.: *h*, adult; *i*, larva; *j*, pupa; *k*, spiracular disc of larva. *l,m*, *Argyra albicans* Loew: *l*, larva; *m*, pupa. *n-q*, *Systenus albimanus* Wirth; *n*, larva; *o*, right dorsal lobe of caudal segment of larva; *p*, head of larva, lateral view; *q*, pupa. *r-w*, *Hydrophorus agalma* Wheeler, larva: *r*, spiracular disc; *s*, dorsal, and *t*, lateral view of cephalopharyngeal skeleton; *v*, lateral view. Pupa, *u*. Cocoon, *w*. *x,y*, *Dolichopus ramifer* Loew, larva: *x*, posterior spiracle; *y*, lateral view (*a-c*, Williams, 1938; *d*, Cole and Aldrich, 1921; *e*, Wheeler, 1897; *f,g*, Saunders, 1928; *h-k*, Williams, 1939; *l,m*, Marchand, 1918; *n-q*, Wirth, 1952; *r-w*, Greene, 1923; *x,y*, Peterson, 1951).

14;49*d*) .... *Dolichopus* Latreille
— Posterior basitarsus without bristles above (fig. 14:49*h*) 2
2. First antennal segment hairy above 3
— First antennal segment bare above 5
3. Arista apical.. 4
— Arista dorsal (fig. 14:49*d,h*); eyes very hairy and vertically elongated; face narrow in middle, broad below, and reaching the lower corners of the eyes .... *Tachytrechus* Loew
4. Third antennal segment subtriangular or triangular; wing broad at base; body with silvery white dust .... *Argyra* Macquart
— Third antennal segment either furcate or with a strong angular projection basally; mesonotal bristles reduced; face broad, palpi large........... *Hypocharassus* Mik
5. Third antennal segment long and pointed at least in the male with an apical arista; male genitalia long and free 6
— Third antennal segment shorter, never pointed at tip and arista usually dorsal or subapical; male genitalia short and imbedded 7
6. Hind cross vein distant from margin of wing; palpi appressed; genitalia of male pedunculated .... *Systenus* Loew
— Hind cross vein situated close to wing margin; palpi disengaged and hanging downward; male genitalia not pedunculated .... *Aphrosylus* Walker
7. Hind cross vein equal to or longer than last section of 5th vein (fig. 14:49*h*); fore femur and tibia usually with short, stout ventral spines 8
— Hind cross vein much shorter than last section of 5th vein; midleg of male usually greatly modified and distorted .... *Campsicnemus* Walker
8. Mouth parts with outer lobe of labella greatly enlarged in the form of a movable sclerotized mandiblelike structure .... *Melanderia* Aldrich
— Outer lobe of labella not modified..*Hydrophorus* Meigen

California Species of Aquatic Dolichopodidae

Subfamily DOLICHOPODINAE

Genus *Dolichopus* Latreille

(Fig. 14:49*d,x,y*)

This genus includes most of the commoner, large, strong-bodied, bristly members of the family. Little is known of their breeding habits, although Johannsen (1935) reported that Needham took an adult of *D. scoparius* Loew in a tent trap over a brook in the Adirondacks, and Peterson (1951) figured the larva of *D. ramifer* Loew reared from decayed vegetation. Nielsen, *et al.* (1954) figured the larva of *D. plumipes* Scopoli from soil in a variety of wet and dry situations in Iceland. Our knowledge of the biology, distribution, and taxonomy of the genus is admirably presented in the monograph by Van Duzee, Cole, and Aldrich (1921). The following key to the California males is adapted from a later revision of their keys by Van Duzee and Curran (1934). The females are not nearly so well known and are more difficult to separate.

Key to Males of the California Species of Dolichopus

1. At least 1 pair of femora mostly black or the femora with blackish stripes below 2
— Femora yellow; at most the tips of the posterior pair black 4
2. Lower orbital cilia pale 3
— Lower orbital cilia wholly black; rarely a few pale hairs between the black ones.................... 8
3. Middle tibia black 14
— Middle tibia yellow 23
4. Lower orbital cilia pale 5
— Lower orbital cilia black, at most 1 or 2 pale ones below; northern and central California .... *bruesi* Van Duzee 1921
5. Squamal cilia black, the short hairs often yellowish 6
— Squamal cilia pale 25
6. Posterior tibia yellowish, at most slightly brownish at the apex 7
— Posterior tibia with black apex 27
7. Posterior tarsus wholly black 37
— Posterior tarsus with the 1st segment at least 1/2 yellow 43
8. Anterior tibia black or blackish.................. 9
— Anterior tibia yellow or yellowish; Humboldt County .... *andersoni* Curran 1924
9. Last segment of front tarsus compressed........... 10
— Front tarsus plain 12
10. Fifth segment of front tarsus somewhat obcordate, 2nd and 3rd usually yellowish; Shasta and Mariposa counties .... *manicula* Van Duzee 1921
— Front tarsus wholly black, the 5th segment not notched at tip 11
11. First antennal segment yellow below; 5th segment of front tarsus cut off rather straight at tip; Sierra Nevada .... *acricola* Van Duzee 1921
— Antenna wholly black; 5th segment of front tarsus extended a little at upper corner; northern and central California .... *corax* Osten Sacken 1877
12. Genital lamellae wholly black, or brown with a black border; Mono and San Benito counties .... *barbaricus* Van Duzee 1921
— Genital lamellae whitish or yellowish with a black border... 13
13. Posterior femur ciliated below; central and southern California ....*paluster* Melander and Brues 1900
— Posterior femur without cilia below; Placer County .... *monticola* Van Duzee 1921
14. Cilia of the squamae pale 15
— Cilia of the squamae black........................ 18
15. Costa considerably enlarged before the tip of 1st vein; Mariposa and Modoc counties .. *viridis* Van Duzee 1921
— Costa not or with a small enlargement at tip of 1st vein 16
16. Genital lamellae large, blackish, rather pointed at tip; Mono County .... *nigricauda* Van Duzee 1921
— Genital lamellae whitish with a black border, at most a little brownish 17
17. All segments of middle tarsus slightly compressed; Inyo County ....*squamosus* Van Duzee 1921
— Middle tarsus plain, normal; Tulare County....... .... *formosus* Van Duzee 1921 (part)
18. Hind femur ciliated on lower surface 19
— Hind femur without cilia below 20
19. Middle and hind femora each with 2 preapical bristles, placed one before the other; Humboldt County .... *aedequatus* Van Duzee 1921
— Middle and hind femora each with 1 preapical bristle; central and southern California................. .... *myosota* Osten Sacken 1887 (part)
20. First segment of posterior tarsus with 10-12 large bristles; Sierra Nevada.. *multisetosus* Van Duzee 1921
— First segment of posterior tarsus with 2-6 large bristles 21

21. Genital lamellae somewhat triangular in outline ....*myosota* Osten Sacken (part)
— Genital lamellae oval in outline 22
22. Genital lamellae with long, blackish hairs along the inner side and 2 stout, angulated bristles; Tulare County .... *nigrimanus* Van Duzee 1921
— Genital lamellae fringed with short, delicate hairs .. ..*formosus* Van Duzee (part)
23. Costa greatly enlarged before tip of 1st vein; middle tarsus very long and slender, each segment enlarged at tip; face extending below the eyes; San Diego County .... *appendiculatus* Van Duzee 1921
— Not as above 24
24. Hind tibia distinctly thickened, mostly black; central California ....*californicus* Van Duzee 1921
— Hind tibia rather slender, yellow with a black tip; California ... *amnicola* Melander and Brues 1900 (part)
25. Second abdominal segment with a tuft of long yellow hairs on the sides and a shorter and small tuft on 3rd segment; central and southern California Coast Ranges ....*afflictus* Osten Sacken 1877
— Abdomen without such tufts of hairs.............. 26
26. First antennal segment long, thick, densely hairy; California .... *crenatus* Osten Sacken 1877
— First antennal segment normal, enlargement at tip of 1st vein as long as cross vein; northern and central California ....*idahoensis* Aldrich 1893
27. First antennal segment black, at least on upper edge 28
— First antennal segment wholly yellow 34
28. Hind femur blackened at apex, at least above; Placer County ....*affluens* Van Duzee 1921
— Hind femur not at all blackened apically 29
29. Front purple, blue, or violet; California ....*ramifer* Loew 1861
— Front green or bronze in color 30
30. Hind femur ciliated on lower inner edge 31
— Hind femur without cilia 32
31. Fifth segment of front tarsus notched at tip so as to form 2 nearly equal lobes; northern California ....*obcordatus* Aldrich 1893
— Fifth segment of front tarsus with its tip divided into very unequal lobes; central California ....*pollex* Osten Sacken 1877
32. Anterior coxa wholly black........................ .... *amnicola* Melander and Brues 1900 (part)
— Anterior coxa yellowish in front, with a black or green stripe on the outer surface 33
33. Last 4 segments of front tarsus a little compressed, black; Humboldt County .. *aldrichii* Wheeler 1897 (part)
— All tarsi plain, normal; Alameda and San Diego counties .... *penicillatus* Van Duzee 1921 (part)
34. Arista enlarged at tip; wing with a lobe at tip of 6th vein 35
— Arista plain 36
35. Genital lamellae large; hind margin of wing with a deep sinus before the lobe at 6th vein; Humboldt and Tulare counties .... *hastatus* Loew 1864 (part)
— Genital lamellae small; hind margin of wing normal, except for the lobe at the tip of the 6th vein; Placer County .... *comptus* Van Duzee 1921 (part)
36. Basal segment of middle tarsus laterally fringed; Monterey County........ *plumipes* (Scopoli) 1763 (part)
— Middle tarsus plain; Fresno County ....*uxorcula* Van Duzee 1921
37. First antennal segment wholly yellow 38
— Antenna black, 1st segment may be yellow with upper edge narrowly black.............................. 41
38. Arista enlarged at tip; wing with a conspicuous lobe at tip of 6th vein .... (see couplet 35)
— Arista plain.................................... 39
39. Front tarsus ornamented........................... 40
— Front tarsus plain; basal segment of mid tarsus fringed with black hairs on both sides .... *plumipes* (Scopoli) 1763 (part)
40. Wing with a projecting lobe at tip of 6th vein; Mono County .... *completus* Van Duzee 1921 (part)
— Wing without or with scarcely a trace of such a lobe Humboldt County .... *sufflavus* Van Duzee 1921
41. Fifth segment of front tarsus distinctly but only a little enlarged and but little longer than 4th; California .... *coquilletti* Aldrich 1893 (part)
— Enlargement of 5th segment very conspicuous 42
42. Second segment of the anterior tarsus shorter than the 1st; Sacramento County ...... *walkeri* Van Duzee 1921
— Second segment of the anterior tarsus as long as or longer than the 1st; Inyo and Yuba counties ....*jugalis* Tucker 1911
43. Antenna black, 1st and 2nd segments may be yellow below 44
— First antennal segment wholly yellow 51
44. Antenna wholly black; central and southern California ....*bakeri* Cole 1912
— First antennal segment yellow below 45
45. Front tarsus ornamented 46
— Front tarsus plain 48
46. First and 2nd segments of front tarsus of equal length; central California...*canaliculatus* Thomson 1868 (part)
— Second segment of front tarsus not much more than 2/3 as long as 1st.................................. 47
47. Fifth segment of front tarsus only a little enlarged .... *coquilletti* Aldrich 1893 (part)
— Fifth segment of front tarsus much compressed and widened; Los Angeles and San Diego counties...... .... *talus* Van Duzee 1921
48. Hind femur with a single preapical bristle.......... .... *penicillatus* Van Duzee 1921 (part)
— The usual preapical bristles ending a row of bristles of increasing length; outer posterior edge of front coxa with a green or blackish streak, sometimes the front coxa wholly green 49
49. Last 4 segments of middle tarsus compressed ....*aldrichii* Wheeler 1897 (part)
— Middle tarsus plain as usual 50
50. First antennal segment normal in size and with short hairs; Sonoma County....... *cavatus* Van Duzee 1921
— First antennal segment long, with long bushy hair; central and southern California.................. .... *consanguineus* Wheeler 1899
51. Front tarsus ornamented 52
— Front tarsus plain......... *hastatus* Loew 1864 (part)
52. Hind femur ciliated below........................ 53
— Hind femur without cilia 57
53. Hind coxa infuscated on basal 1/2, at least with a large blackish spot at base on outer surface 54
— Hind coxa wholly yellow, or nearly so 56
54. First 3 segments of front tarsus of nearly equal length, 3rd being a little the longest .... *canaliculatus* Thomson 1868 (part)
— Second segment of front tarsus longer than 1st, 3rd very much shorter 55
55. Genital lamellae truncate at apex, with a short but rather acute point at upper corner; Humboldt and Shasta counties ....*grandis* Aldrich 1893
— Genital lamellae with a long acute point in the center of apical margin; Monterey County .... *superbus* Van Duzee 1921 (part)
56. Costa with a small, short enlargement at tip of 1st vein; 3rd and 4th veins parallel at their tips; Shasta County .... *cuprinus* Wiedemann 1830
— Costa considerably enlarged at tip of 1st vein, tapering to its tip; 3rd and 4th veins approaching each other toward their tips; Inyo County ....*subcostatus* Van Duzee 1930

57. Front tarsus with 2nd segment nearly 1/4 longer than 1st 58
— Front tarsus with 2nd segment as long as, or shorter than 1st 59
58. Genital lamellae scarcely twice as long as wide; northern and central California .. *tenuipes* Aldrich 1894
— Genital lamellae 4 times as long as wide .... *superbus* Van Duzee 1921 (part)
59. First and 2nd segments of front tarsus of nearly equal length .... *canaliculatus* Thomson 1868 (part)
— Second segment of front tarsus distinctly shorter than 1st 60
60. Genital lamellae with a deep incision on upper edge; Santa Clara and San Diego counties .... *duplicatus* Aldrich 1893
— Genital lamellae without an incision, normal 61
61. Wing with a small lobe at tip of 6th vein; costa not enlarged at tip of 1st vein; middle basitarsus with a large bristle above; hind basitarsus only a little yellow at base .... *completus* Van Duzee 1921 (part)
— Wing without a lobe at tip of 6th vein; costa with a long tapering enlargement at tip of 1st vein; middle basitarsus without a bristle; hind basitarsus yellow, narrowly black at tip; Humboldt County .... *occidentalis* Aldrich 1893

Genus *Tachytrechus* Walker

This genus was named for the habit the adults have, of running on the sand or mud, or flying low along the margins of lakes or the seashore. The only account of the immature stages is Lundbeck's (1912) description of the pupa of the European *T. insignis* (Stannius) which was taken from the sand on a lake margin. The following key is adapted from one by Harmston and Knowlton (1940) for the Nearctic species.

Key to Males of the California Species of Tachytrechus

1. Antennal arista normal, pointed, without terminal lamella 6
— Arista long, with a terminal lamella 2
2. Femora entirely yellow; central and southern California .... *auratus* (Aldrich) 1896
— Femora yellow with black markings or entirely black 3
3. Front femur broad at base, usually with black marking on outer surface 4
— Front femur not broadened at base, entirely yellow 5
4. Front femur with a large jet black spot covering nearly the whole of outer surface; lamella of arista scarcely whitish at base; Monterey, Mono, and Inyo counties .... *olympiae* (Aldrich) 1896
— Front femur scarcely infuscated on outer surface; lamella of arista large with apical 1/2 black, basal 1/2 white except for slight narrow infuscation near the stem; Lake Tahoe .... *tahoensis* Harmston and Knowlton 1940
5. Hind femur black at base below; bristles on anterior surface of fore tibia longer than the row of bristles on outer posterior surface, the latter row of bristles of decreasing length, those at base much longer than those near distal end; Humboldt County and Sierra Nevada .... *sanus* Osten Sacken 1877
— Hind femur slightly or not at all blackened at base below; bristles of anterior surface of fore tibia not longer than the row of bristles on outer posterior surface, the latter row of bristles of nearly equal length throughout; Tulare County.. *spinitarsis* Van Duzee 1924
6. Legs wholly black; hind tibia with 2 flattened bristles on posterior side near middle; California .... *angustipennis* Loew 1861
— All femora blackish-green except for apical 4th; tibiae yellow; middle tarsus with 2nd joint flattened; central and southern California..... *granditarsus* Greene 1922

Subfamily HYDROPHORINAE

Genus *Hydrophorus* Fallén

(Fig. 14:49*h-j, r-w*)

Flies of the genus *Hydrophorus* are characteristic skaters over ponds and near the shores of lakes and marine estuaries. The adults (fig. 14:49*h*) are characterized by dull pruinose bodies, long legs with the fore ones raptorial, and long wings with the discal cell especially lengthened. According to Williams (1939) who published a remarkably detailed account of the biology of the Hawaiian *H. pacificus* Van Duzee, the adults are fond of bloodworms, the larvae of Tendipedidae, which they pull from the mud of wet shores. The immature stages of *H. agalma* Wheeler (fig. 14:49*r-w*) have been nicely figured by Greene (1923) from specimens collected on the shores of Lake Winnipeg, Manitoba. The following key is adapted from one to the Nearctic *Hydrophorus* by Van Duzee (1926).

Key to Adults of the California Species of Hydrophorus

1. Knobs of halteres wholly yellow 2
— Knobs of halteres infuscated, at least on outer surface 12
2. Postvertical bristles in a row of 4 or more on each side 3
— Postverticals only 2, as usual 5
3. Scutellum with 2 or more pairs of marginal bristles 4
— Scutellum with only 1 pair of marginal bristles; California .... *gratiosus* Aldrich 1911
4. Wing with distinct whitish spots or clouds in the cells and at base; Lake County ...... *plumbeus* Aldrich 1911
— Wing without whitish spots, grayish hyaline as usual; Alameda and San Diego counties .... *argentatus* Van Duzee 1918
5. Black propleural bristles above fore coxa absent 6
— With 1 black propleural bristle above each fore coxa 7
6. Dorsocentrals minute, white; fore femur of male with a deep notch near the tip, its tibia bent; San Bernardino County .... *canescens* Wheeler 1896
— Dorsocentrals black; San Bernardino County .... *flavipennis* Van Duzee 1926
7. Face opaque with pollen, the ground color not showing through 8
— Face showing more or less metallic color through the pollen 9
8. Face ochre-yellow; wings with very conspicuous spots on the cross vein and last section of 4th vein; veins blackish, scarcely paler at base; Tioga Pass .... *alter* Parent 1934

— Pollen of face white; wings without spots on the veins; veins broadly yellow at root of wing, sometimes mostly yellow; California............*praecox* (Lehman) 1822
9. Tip of fore tibia in both sexes with an acute angle produced toward the femur; fore coxa with from 1-3 black bristles on upper outer corner; central and southern California..............*philombreus* Wheeler 1890
— Tip of fore tibia not, or but little, angulated........ 10
10. Wing with a cloud on the cross vein; female with a row of longer bristles near the tip of posterior surface, at the lower edge of fore femur; Alameda, Sonora, and San Luis Obispo counties .... *breviseta* Thomson 1868
— Wing without a cloud on the cross vein; female with a double row of spines on under side of fore femur 11
11. Pleura with thin brownish or yellowish pollen; length 4.5 mm.; both rows of spines on lower surface of fore femur of about equal length; Ventura and Riverside counties.................. *magdalenae* Wheeler 1899
— Pleura with white pollen; length 3 mm.; fore femur with a row of 5 spines on lower posterior edge which are much longer than those in anterior row; California .... *sodalis* Wheeler 1899
12. Wing with a brown spot on the cross vein and another on the middle of last section of 4th vein; Del Norte County..........................*phoca* Aldrich 1911
— Wing without spots on the veins; Del Norte to San Francisco counties .... *innotatus* Loew 1864

Genus *Hypocharassus* Mik

Smith (1952) described and figured larvae, pupae, and cocoons of *Hypocharassus pruinosus* (Wheeler) from the sandy bars and beaches of the bays at Cape Cod, Massachusetts. They are quite similar to those of *Hydrophorus*. The two known species of this genus are known only from the Atlantic coastal beaches of the United States.

Genus *Melanderia* Aldrich

The only two known species are both found along California sea beaches. The genus is characteristic because of the extreme development of the labella forming a pair of sclerotized, mandiblelike lobes which are used in seizing the prey. The habits are similar to those of the closely related *Hydrophorus*.

Key to Adults of the California Species of Melanderia

1. Front and face purple; 3rd and 4th veins parallel and far apart, ending in the apex of wing; front femur of male with a close bunch of 6-8 bristles on a slight protuberance on inner side near base; Humboldt County .... *mandibulata* Aldrich 1922
— Front and face green; 3rd and 4th veins close together, the 3rd curving back near wing tip; front femur of male without such tuft on inner side of front femur; Santa Barbara to San Diego counties .... *curvipes* Van Duzee 1918

Subfamily APHROSYLINAE

Genus *Aphrosylus* Walker

(Fig. 14:49*e-g*)

The North American species of this Holarctic genus have been considered by some authors as generically or subgenerically distinct under the name *Paraphrosylus* Becker. All the known species are confined to algae-covered intertidal rocks along sea coasts, where the adults prey on adults and larvae of midges and other small Diptera or on small annelids. Both Wheeler (1897) and Saunders (1928) have good accounts including excellent figures of the immature stages. The larvae are unusual in having nine caudal lobes (fig. 14:49*e,g*) rather than the usual four, but the supernumerary dorsomedian lobe and small lateral pairs lack the usual hair fringe possessed by the two large dorso- and ventrolateral pairs.

California species:

californicus Harmston 1952.	Orange County
direptor Wheeler 1897.	Humboldt, San Diego
grassator Wheeler 1897.	Monterey County
nigripennis (Van Duzee) 1924.	Humboldt County
praedator Wheeler 1897.	Humboldt, Monterey
wirthi Harmston 1952.	San Francisco, San Mateo

Genus *Cymatotopus* Kertesz

This genus replaces *Aphrosylus* in the Australasian region. It is merely mentioned here to call attention to the beautiful figures and fine description by Williams (1939) of all stages of the Hawaiian *C. acrosticalis* Parent.

Subfamily RHAPHIINAE

Genus *Systenus* Loew

(Fig. 14:49*n-q*)

Although this genus is at most only semiaquatic it is included here because of the general limnological interest in the study of tree-hole insects. Wirth (1952) has recently described several North American species with figures of the immature stages. The larvae live exclusively in the wet and decaying wood pulp of tree cavities where they prey on other dipterous larvae such as *Dasyhelea* which breed in the same medium. There are four American species known only from the eastern states.

Subfamily DIAPHORINAE

Genus *Argyra* Meigen

(Fig. 14:49*l,m*)

Marchand (1918) described and figured the immature stages of *Argyra albicans* Loew which he found in mud on the edge of a pond. The following adult key is an adaptation of one to the North American species by Van Duzee (1925).

Key to Males of the California Species of Argyra

1. Abdomen without yellow on the sides, or very nearly so 2
— Abdomen with distinct yellow ground color on some of the segments 8
2. Anterior coxa wholly black, or nearly so 3
— Anterior coxa yellow, at least on apical 1/2 7
3. Hind basitarsus with long bristles 4
— Hind basitarsus with only the usual short hairs 5
4. All femora black with their tips narrowly yellow; Del Norte and Humboldt counties ....
......................*nigriventris* Van Duzee 1925
— All femora yellow, except apical 1/2 of posterior pair; Mono County .....*argentiventris* Van Duzee 1925 (part)
5. Middle femur widened below near basal 3rd, so as to form an obtuse angle; its tibia with a brush of hairs near the middle; Los Angeles and San Diego counties
.... *femoralis* Van Duzee 1925
— Middle femur nearly evenly rounded or straight below
.... 6
6. Face and front velvety black; California ....
............................*nigripes* Loew 1864
— Face white or grayish white; front metallic green with more or less gray pollen; Humboldt to San Mateo counties .... *barbipes* Van Duzee 1925
7. Hind basitarsus with long bristles ....
...............*argentiventris* Van Duzee 1925 (part)
— Hind basitarsus with only the usual short hair; California .... *cylindrica* Loew 1864
8. One pair of femora more than 1/2 black or green 9
— All femora yellow, tips of posterior pair may be broadly black 10
9. All femora black or green, their tips may be yellow; Del Norte County .... *albiventris* Loew 1864
— One or 2 pairs of femora largely yellow; Alameda County .... *splendida* Van Duzee 1925
10. Fore coxa almost wholly black; Alameda and Los Angeles counties......... *californica* Van Duzee 1925
— Fore coxa yellow, sometimes considerably blackened at base.......... *argentiventris* Van Duzee 1925 (part)

Subfamily CAMPSICNEMINAE

Genus *Campsicnemus* Haliday

(Fig. 14:49*a-c*)

The males of many species of *Campsicnemus* have remarkable secondary sexual modifications, usually on the middle tibiae, from which the genus takes its name. The adults habitually run or skate on the surface of still water, or they frequent damp vegetation. Williams (1939) goes into remarkable detail on the habits of the numerous Hawaiian species, each of which seems to prefer a special little niche in the stream- or pond-side environment. He was able to rear one aquatic species through its life cycle on mud in the laboratory, and made a more complete larval study of another species which breeds in decaying banana stems. Nielsen, *et al.* (1954) described and figured the larva of *C. armatus* Zett. which lives in wet soil of bogs and meadows in Iceland. The following key is taken from Harmston and Knowlton's (1942) key to the North American species.

Key to Males of the California Species of Campsicnemus

1. All femora black, though tips may be narrowly yellow; northern and central California ....
..........................*nigripes* Van Duzee 1917
— At least 1 pair of femora 1/2 yellow 2
2. Middle tibia greatly thickened and deformed, at least near middle 3
— Middle tibia not thickened; Del Norte to San Mateo counties..................... *degener* Wheeler 1899
3. Middle tibia deeply excavated near apex, bearing in hollow of the excavation a conspicuous peduncled process; Modoc County ..... *philoctetes* Wheeler 1899
— Middle tibia without such a peduncled process near the apex; Lake Tahoe .... *claudicans* Loew 1864

REFERENCES

GREENE, C. T.
1923. The immature stages of *Hydrophorus agalma* Wheeler (Diptera). Proc. Ent. Soc. Wash., 25:66-69.
HARMSTON, F. C., and G. F. KNOWLTON
1940. *Tachytrechus* studies (Dolichopodidae, Diptera). Canad. Ent., 72:111-115.
1942. The Dipterous genus *Campsicnemus* in North America. Bull. Brooklyn Ent. Soc., 37:10-17.
JOHANNSEN, O. A.
1935. *See* Diptera references.
LUNDBECK, W.
1912. Diptera Danica. Part IV Dolichopodidae. Copenhagen. 407 pp.
MARCHAND, W.
1918. The larval stages of *Argyra albicans* Lw. (Diptera, Dolichopodidae). Ent. News, 29:216-220.
NIELSEN, P., O. RINGDAHL, and S. L. TUXEN
1954. The Zoology of Iceland. Vol. III, part 48a. Diptera, 1:1-189.
PETERSON, A.
1951. *See* Diptera references.
SAUNDERS, L. G.
1928. *See* Diptera references.
SMITH, M. E.
1952. Immature stages of the marine fly, *Hypocharassus pruinosus* Wh., with a review of the biology of immature Dolichopodidae. Amer. Midl. Nat., 48:421-432.
VAN DUZEE, M. C.
1925. A revision of the North American species of the genus *Argyra* Macquart, two-winged flies of the family Dolichopodidae. Proc. U.S. Nat. Mus., 66:1-43, 1 pl.
1926. A table of the North American species of *Hydrophorus* with the description of a new form (Diptera). Pan-Pac. Ent., 3:4-11.
VAN DUZEE, M. C., and C. H. CURRAN
1934. Key to the males of Nearctic *Dolichopus* Latreille (Diptera). Amer. Mus. Novit. no. 683, 26 pp.

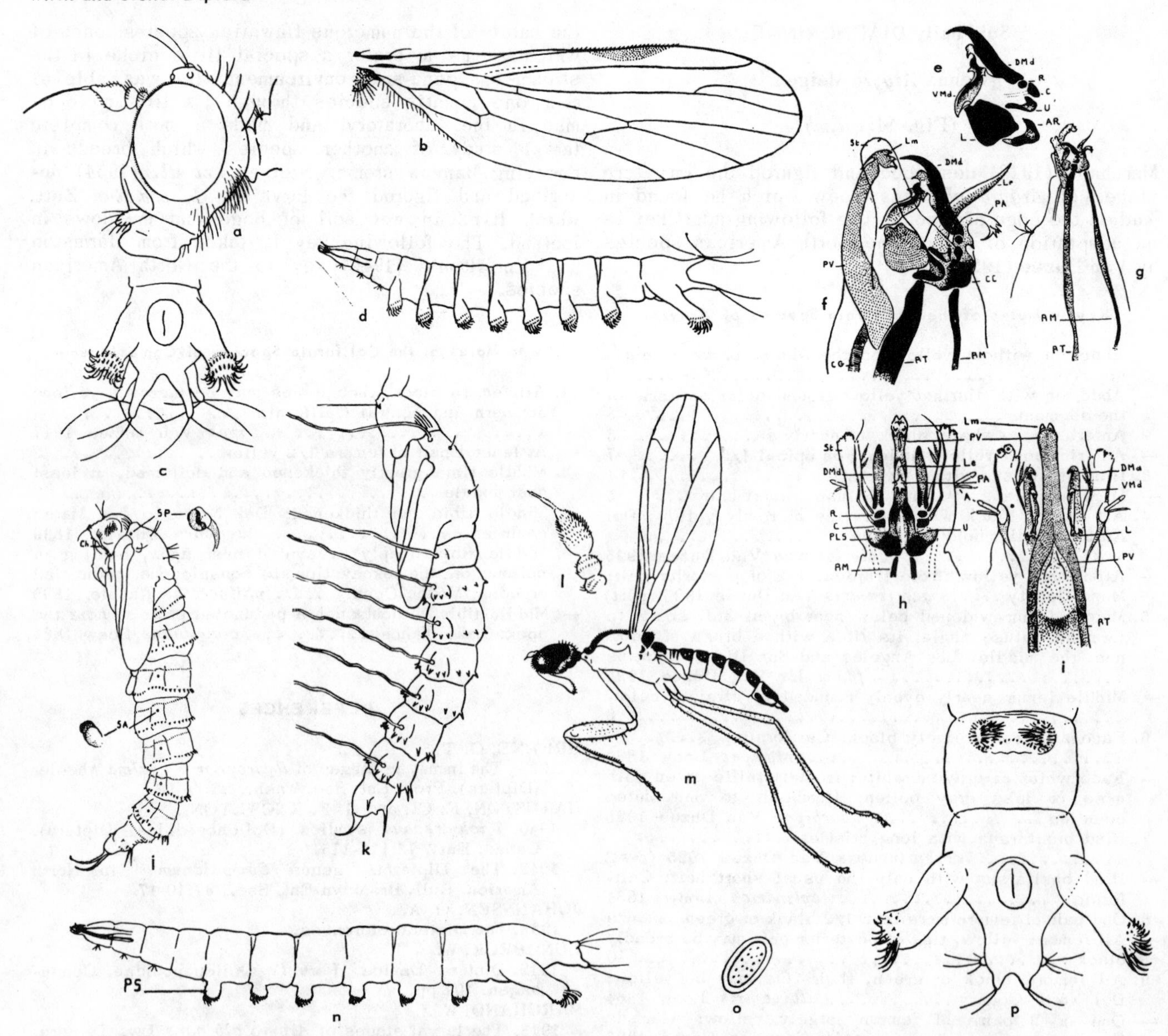

Fig. 14:50. Empididae. *a-j, Wiedemannia (Roederella) ouedorum* Vaillant: *a,* head of adult; *b,* wing; *c,* last body segment of larva, ventral view; *d,* larva, lateral view; *e-i,* sclerotized parts of head (*e-g,* lateral; *h,* ventral; and *i,* dorsal views); *j,* pupa. Abbreviations: (A, antenna; AR, articular sclerite; C, cardo; CC, cephalic capsule; CG, canal of the salivary gland; DMd, dorsal arm of mandible; L, laminar sclerite; Le, labrum; Lm, labium; Pa, anterior projection of cephalic capsule; PSL, lateral superficial projection of cephalic capsule; Ps, pseudopod; Pv, hypopharynx; R, rotule or ball-and-socket sclerite; RM, metacephalic rods; RT, tentorial rods; Sa, abdominal spiracle; Sp, prothoracic spiracle; St, stipes of the maxilla; U, U-shaped sclerite; VMd, ventral arm of mandible); *k-p, Hemerodromia seguyi* Vaillant: *k,* pupa; *l,* adult antenna; *m,* female, lateral view; *n,* larva, lateral view; *p,* larva, posterior end of body, ventral view; *o,* egg (Vaillant, 1952, 1953).

VAN DUZEE, M. C., F. R. COLE, and J. M. ALDRICH
1921. The dipterous genus *Dolichopus* Latreille in North America. Bull. U.S. Nat. Mus. 116, 304 pp., 16 pls.

WHEELER, W. M.
1897. A genus of maritime Dolichopodidae new to America. Proc. Calif. Acad. Sci., 1:145-152, 1 pl.

WILLIAMS, F. X.
1939. Biological studies in Hawaiian water-loving insects. Part III. Diptera or flies. B. Asteiidae, Syrphidae, and Dolichopodidae. Proc. Hawaii. Ent. Soc., 10:281-315.

WIRTH, W. W.
1952. Three new species of *Systenus* (Diptera, Dolichopodidae), with a description of the immature stages from tree cavities. Proc. Ent. Soc. Wash., 54:236-244.

Family EMPIDIDAE

(Figs. 14:50, 14:51)

Adult empidids or "dance flies" frequent moist situations and all are predaceous on smaller animals. Some species live along seashores, feeding upon small invertebrates at the water's edge. The unique mating habits of the adults have been described by many observers. During their mating swarms or dances, the males of some species offer food to the females, and still other males build a characteristic silvery balloon to lure the female.

The immature stages of most groups, where known, breed in damp earth, decaying wood, or vegetation, usually in wooded areas, or under bark of trees. In the subfamilies Clinocerinae and Hemerodromiinae, however, all the known larvae are aquatic. In fact all the known aquatic empidids belong here. The larvae of this family are predators and resemble those of the Dolichopodidae in many characteristics. They differ mainly in the shape of the last body segment which is rounded caudad in the terrestrial forms, or has tapered processes in the aquatic larvae, (figs. 14:50*d,n*; 14:51*b,d*), rather than concave and peripherally lobed. The aquatic larvae also possess well-developed ventral pseudopods similar to those of *Atherix*. The pupae (figs. 14:50*j,k*; 14:51*a,c*) lack the elongate prothoracic spiracular organs, the body is provided with rows of well developed spines, and in some genera there are several pairs of very long, ventrolateral hairs or filaments.

Keys to the North American Genera of Aquatic Empididae

Adults

1. Thorax elongate, not highly arched, mesopleura obliquely longer than vertically high; front coxa always longer than posterior ones (fig. 14:50*m*); anal angle of wing not projecting; 2nd basal cell closed apically (fig. 14:50*b*) 2
— Thorax not elongate, the mesopleura distinctly higher vertically than obliquely long; front coxa not elongate; anal angle of wing more or less projecting (EMPIDINAE, OCYDROMIINAE, and HYBOTINAE); or the discal cell united with the second basal cell and all veins extending straight to wing margin (TACHYDROMIINAE) .... Not aquatic
2. Anterior pair of legs far from the middle pair, raptorial (fig. 14:50*m*) (HEMERODROMIINAE) 3
— Anterior legs not distant from the middle pair and not raptorial (CLINOCERINAE) 9
3. Antennal style shorter than 3rd segment (fig. 14:50*l*); thorax without discal macrochaetae, metapleura bare; 3rd vein normally forked; anal cell when present broader toward apex, the anal cross vein usually straight (fig. 14:50*m*) 4
— Antennal arista more than twice as long as 3rd antennal segment; thorax with some discal macrochaetae; metapleura with some fine setulae; 3rd vein simple; anal cell with parallel sides and more or less rounding cross vein, subequal to the small second basal cell 8
4. Humeral cross vein absent, subcosta fused with costa close to base of wing; 1st vein short, ending before middle of wing; eyes nearly or quite contiguous on face (fig. 14:50*m*) ....*Hemerodromia* Meigen
— Humeral cross vein present, subcosta separate from costa; 1st vein ending at or beyond middle of wing, anal cell complete or at least anal vein weak; eyes separated on face.. 5
5. Discal cell complete ....*Chelifera* Macquart
— Discal cell incomplete, fused with either 2nd basal or 3rd posterior cell 6
6. Second posterior cell petiolate; front femora more or less spinose; 1st vein ending beyond middle of wing, 2nd vein not shortened 7
— Second posterior cell sessile; front femora weak and not strongly spinose, front tibia pubescent; 1st vein ending at middle of wing, 2nd vein usually short.... .... *Neoplasta* Coquillett
7. Discal cell fused with 3rd posterior cell with anterior cross vein beyond end of 2nd basal cell; style microscopic .... *Thanategia* Melander
— Discal cell fused with 2nd basal cell, with anterior cross vein much before end of cell; style 1/3 as long as 3rd antennal segment .... *Metachela* Coquillett
8. Discal cell complete, emitting 3 veins apically; anal cross vein rounding the anal cell ..*Chelipoda* Macquart
— Discal cell open, the posterior cross vein absent and the 4th vein acutely forked; anal cell truncate, slightly shorter than 2nd basal cell....*Phyllodromia* Zetterstedt
9. Third antennal segment remarkably lengthened, straplike, and without evident style 10
— Third antennal segment not remarkably long (fig. 14:50*a*) 11
10. Antenna inserted above the middle of the head; a row of acrostichals present.......... *Niphogenia* Melander
— Antenna inserted below the middle of the head; acrostichals absent.................*Ceratempis* Melander
11. Antenna inserted at the middle of the head, the 3rd segment with a short style which terminates in a bristlelike segment; face not constricted from the cheeks by a suture; 3rd vein not furcate ....*Boreodroma* Coquillett
— Antenna inserted above the middle of the head, the 3rd segment usually with a long arista; oral margin of the cheeks with a more or less distinct incision or suture extending toward the eyes (fig. 14:50*a*) 12
12. Third vein simple 13
— Third vein branched (fig. 14:50*b*) 14
13. Anal cross vein greatly reflexed, nearly parallel with axis of wing; costa with setulae; 1st vein ending before middle of wing; front femora with flexor thornlike spines .... *Oreothalia* Melander
— Anal cross vein perpendicular to the axis of the wing; costa without setulae; 1st vein ending beyond the middle of wing; femora not spinose .... *Heleodromia* Haliday
14. Fourth vein arising near base of anal cell, the 2nd basal cell as long as anal cell; subcosta obsolete apically......................*Proclinopyga* Melander
— Fourth vein arising near the end of the basal 3rd of anal cell, the 2nd basal cell therefore shortened; subcosta ending in the costa (fig. 14:50*b*)............. 15
15. Proboscis as long as the head, slender, pointed, rigid, without labella; cheeks rostriform............... ....*Roederiodes* Coquillett
— Proboscis usually short, thick, fleshy, not inflexed; cheeks usually narrow (fig. 14:50*a*) 16
16. Head articulated to the thorax high up on the occiput, extending obliquely forward; legs yellowish........ ....*Dolichocephala* Macquart
— Head articulated nearer the center of the occiput, hanging vertically (fig. 14:50*a*); legs usually wholly or in part blackish................................. 17
17. Face separated from the lower occiput by an incision on the cheeks that extends to the eye; face not descending beneath the eyes; no acrostichals .... *Clinocera* Meigen

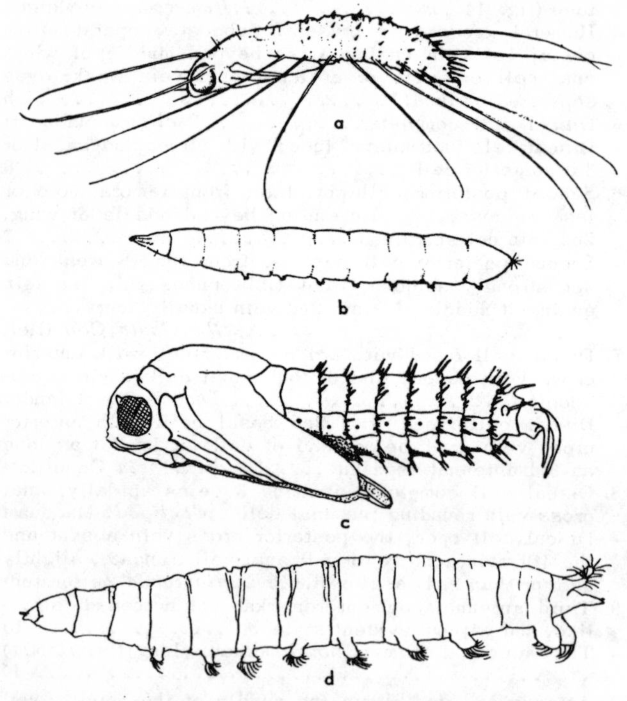

Fig. 14:51. Empididae, immature stages. *a, Hemerodromia rogatoris* Coq. (?), pupa; *b, Chelifera precatoria* (Fallén), larva; *c,d, Clinocera stagnalis* Haliday: *c*, pupa; *d*, larva (*a,b*, Johannsen, 1935; *c,d*, Nielsen *et al.*, 1954).

— Face fusing with lower occiput and extending more or less beneath the level of the eyes, its front margin deeply emarginate or impressed and carinate, no incision between face and occiput; cheeks broader; acrostichals usually abundant (fig. 14:50*a*) .... *Wiedemannia* Zetterstedt

California Species of Aquatic Empididae

Subfamily HEMERODROMIINAE

The following keys and distribution records have been taken from the recent revision of the subfamily by Melander (1947).

Genus *Hemerodromia* Meigen

Johannsen (1935) has described and figured the pupa (fig. 14:51*a*) of (?) *H. rogatoris* Coquillett which was found on the rocky bottom of small brooks near Ithaca, New York. Vaillant (1953) described and illustrated in excellent detail all stages of *H. seguyi* Vaillant (fig. 14:50*k-p*) the larvae of which prey upon larvae of *Simulium* in swift rocky streams of Algeria.

Key to Adults of the California Species of Hemerodromia

1. Front femur without distinct basal tubercle beneath, though with basal spine; San Diego and Humboldt counties .... *empiformis* (Say) var. *brevifrons* Melander 1947
— Front femur with prominent spine-tipped tubercle at base beneath; California ..... *rogatoris* Coquillett 1896

Genus *Chelifera* Macquart

Brocher (1909) has described and figured the immature stages of the European *C. precatoria* (Fallén) larva, (fig. 14:51*b*) which breeds in mud in small streams. The abdominal spiracles of the pupa are drawn out into long threads which are supposed to be tracheal gills. None of the eleven North American species are yet known from California.

Genus *Metachela* Coquillett

Melander records one of the two North American species, *M. collusor* Melander 1902, from California. The immature stages are not known.

Key to Adults of the California Species of Neoplasta Coquillett

1. Dorsal valve of male pygidium large and more or less hemispherical, ventral part of pygidium globular, not compressed or carinate, the dehiscence closed over at base; humeri and most of propleura yellowish 2
— Dorsal valve of male pygidium inverted triangular, ventral part of pygidium usually compressed and carinate, its dehiscence open at base; humeri black, abdomen of male piceous, of female dull yellowish; California .... *hebes* Melander 1947
2. Dorsal valve of pygidium longer than lateral valve, sometimes very large; tumescence of middle femur of male pronounced, confined to basal 5th, its setae usually close-set; Mariposa, Humboldt, and Santa Cruz counties..... *scapularis* var. *megorchis* Melander 1947
— Dorsal valve of pygidium scarcely larger than lateral valve; tumescence of middle femur of male extending along basal 4th or 3rd, not pronounced, its setae evenly spaced; California.. *scapularis scapularis* (Loew) 1862

Subfamily CLINOCERINAE

Genus *Boreodromia* Coquillett

maculipes (Bigot 1887) California

Key to Males of the California Species of Proclinopyga Melander

1. Under side of middle femur bearing 3 pale setae at base; middle tibia swollen near middle and there bearing many flexor setulae; middle basitarsus as long as tibia; 6th abdominal segment with many long black bristles; Woodside, California .... *amplectens* Melander 1927
— Under side of middle femur more extensively setose, the setae evenly distributed and about 12 in number; middle tibia not deformed and as long as basal 2

segments of mid-tarsus; 6th segment of abdomen not densely setose; Siskiyou County........................ *monogramma* Melander 1927

Genus *Roederiodes* Coquillett

Needham and Betten (1901) described the immature stages of *R. juncta* Coquillett which were taken in a creek rapids in the Adirondacks. The larvae bears a remarkable resemblance to that of the rhagionid *Atherix* with two pairs of setigerous processes at the tip of the abdomen and with well developed ventral prolegs.

Genus *Clinocera* Meigen

Bischoff (1924) recorded finding larvae of *Clinocera* in swift streams among moss and described the larva. Nielsen, *et al.* (1954) described the immature stages of *Clinocera stagnalis* Haliday (fig. 14:51*c,d*) from running water and small ponds in Iceland.

Key to Adults of the California Species of Clinocera

1. Discal cell longer than the 2nd posterior; thorax not vittate; stigma faint; Santa Clara County........................ *brunnipennis* Melander 1927
— Discal cell equal to or shorter than the 2nd posterior; thorax bivittate; stigma more or less evident; Mono County.....................*prasinata* Melander 1927

Genus *Wiedemannia* Zetterstedt

Johannsen (1935) has described the pupa of *W. (Philolutra) simplex* (Loew) from a stream in Yellowstone National Park. Evidently the long abdominal filaments of the pupae of Hemerodromiinae are absent in this genus and in *Roederiodes*. Vaillant (1951) described and excellently illustrated all stages of *W. (Roederella) oucdorum* Vaillant (fig. 14:50*a-j*) which prey on larvae of *Simulium* in Algeria. Wirth (unpublished) has observed adults of this genus dragging simuliid larvae from the water's edge in swift streams of the Sierra Nevada in California. There are several undescribed species from California in the United States National Museum collection.

REFERENCES

BISCHOFF, W.
1924. *See* Diptera references.
BROCHER, F.
1909-1911. Métamorphoses de l'*Hemerodromia praecatoria* Fall. Ann. Biol. Lacustre, 4:44-45.
JOHANNSEN, O. A.
1935. *See* Diptera references.
MELANDER, A. L.
1927. Family Empididae. Genera Insectorum, Fasc. 185, 434 pp., 8 pls.
1947. Synopsis of the Hemerodromiinae (Diptera, Empididae). Jour. N.Y. Ent. Soc., 55:237-273.
NEEDHAM, J. G., and C. BETTEN
1901. *See* Diptera references.
NIELSEN, P., O. RINGDAHL, and S. L. TUXEN
1954. The Zoology of Iceland. Vol. III, part 48a, Diptera 1:1-189.
VAILLANT, F.
1951. Un empidide destructeur de simulies. Bull Soc. Zool. Fr., 76:371-379.
1953. *Hemerodromia Seguyi,* nouvel empidide d'algérie destructeur de simulies. Hydrobiologia, 5:180-188.

Family PHORIDAE

The "hump-backed flies" of the family Phoridae have a few aquatic members although no North American species are known to be strictly aquatic. Most phorids live in decaying animal or vegetable matter, in fleshy fungi, or as parasites or predators on a number of insects or other invertebrates. The larvae of *Diploneura cornuta* (Bigot) have been reported (as *Dohrniphora venusta* Coquillett) from pitchers of *Sarracenia flava* by Jones (1918) who described and figured all stages (fig. 14:52*a-c*). More strictly aquatic phorids have been reported by Schmitz from pitchers of *Nepenthes* in the Dutch East Indies.

REFERENCES

JONES, F. M.
1918. *Dohrniphora venusta* Coquillett (Dipt.) in *Sarracenia flava.* Ent. News, 29:299-302, 1 pl.

Family SYRPHIDAE

(Fig. 14:53)

The "flower flies" are small to large, conspicuous, brightly colored insects (fig. 14:53*h*) which frequent flowers and as a group are second in importance only to bees as pollinators. In this family, mimicry is often very pronounced, with many diverse genera bearing close superficial resemblances to various bees and wasps. The larvae have widely divergent habits which place them roughly in four types: (1) *Microdon*-type, hemispherical, unsegmented larvae

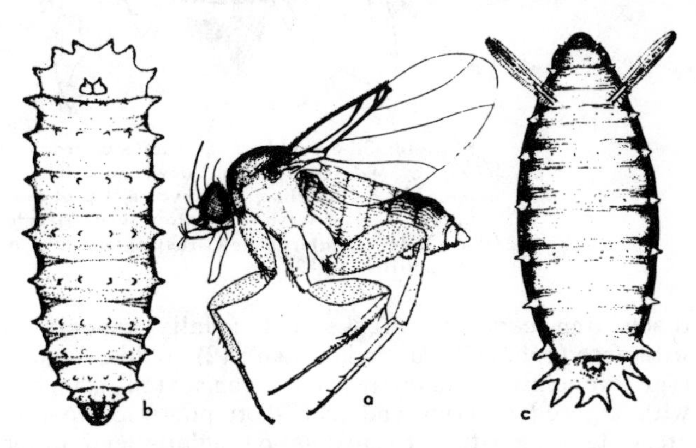

Fig. 14:52. *Diploneura cornuta* (Bigot). *a*, adult; *b*, larva, dorsal view; *c*, puparium, dorsal view (Jones, 1918).

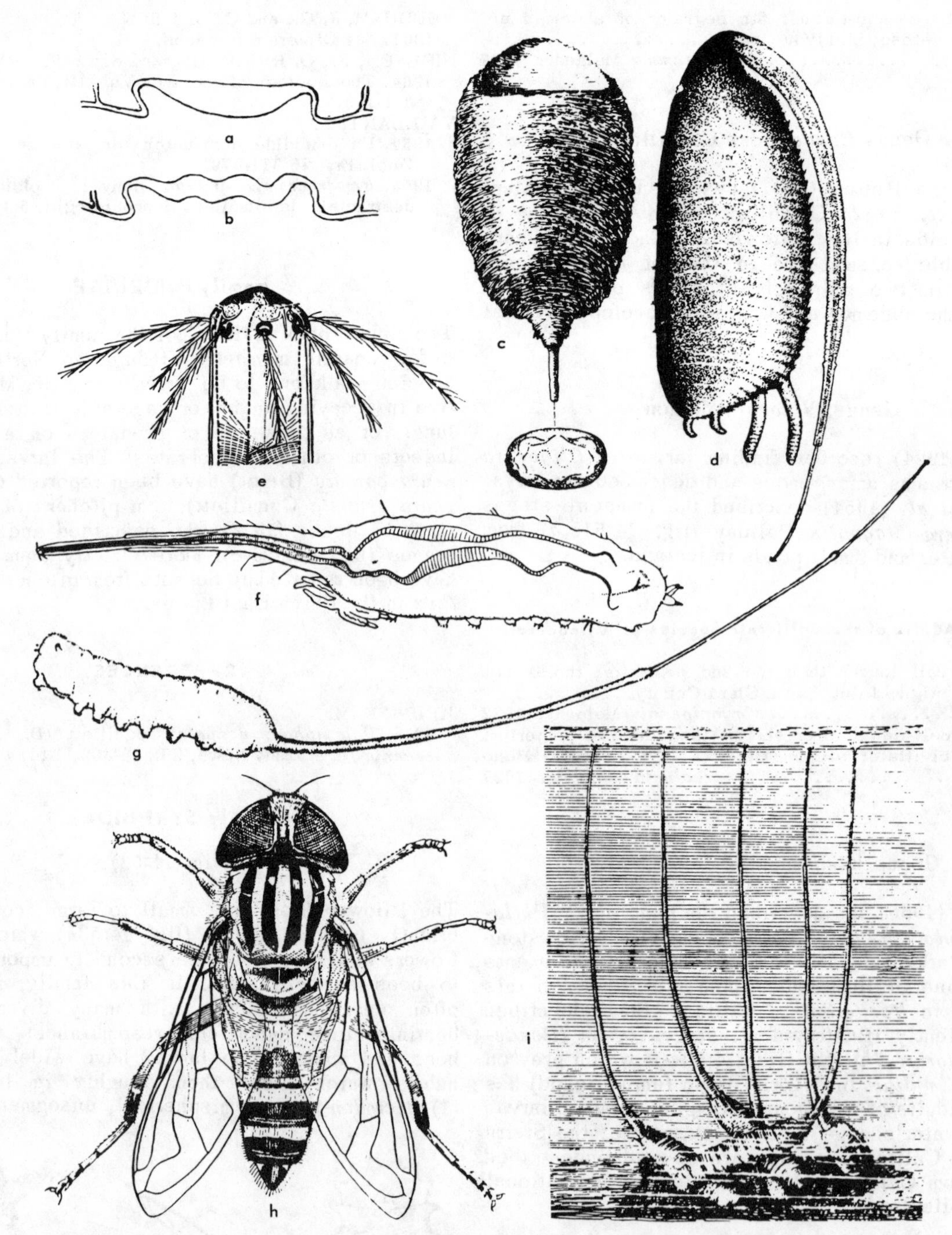

Fig. 14:53. Syrphidae. *a, Tubifera*, tracheal trunk of larva; *b, Helophilus*, tracheal trunk of larva; *c, Myiolepta nigra* Loew, puparium, and posterior spiracle enlarged; *d-f,h,i, Tubifera arvorum* (F.): Puparium, *d.* Larva: *e*, tip of respiratory tube showing spiracular disc, lateral view; *f*, lateral view of larva respiratory tube retracted; *i*, larval habitus. Adult, *h. g, Tubifera bastardii* (Macq.), larva, lateral view, respiratory tube extended (*a,b,g*, Johannsen, 1935; *c*, Greene, 1923; *d-f,h,i*, Williams, 1939).

resembling certain mollusks and usually associated with ants (subfamily Microdontinae); (2) aphidophagous type, leechlike, subcylindrical, segmented maggots with tapered anterior end and short posterior respiratory tube, mostly predaceous on aphids and other soft-bodied Homoptera (subfamily Syrphinae); (3) short-tailed type, cylindrical, grublike larvae with short posterior air tube, often with subapical, lateral parapods, mostly feeding saprophytically on decomposing plant material; and (4) long-tailed type, the "rat-tailed maggots" which resemble the last except for the long, filamentous air tube, and which usually live in more aquatic, filthy environments (fig. 14:53*i*). The Syrphidae with strictly aquatic larvae belong to the long-tailed type comprising the subfamilies Sericomyiinae and Eristalinae. A few Cheilosiinae with short-tailed larvae have aquatic habits, but the majority of this subfamily, together with the Xylotinae and Volucellinae, prefer drier environments under bark, in decaying wood and plant material, in tree holes, or in refuse. Metcalf (1913, 1916) and Heiss (1938) have published excellent general discussions of syrphid biologies with emphasis on the nonaquatic species, and Johannsen (1935) has constructed a key and given descriptions of the known aquatic larvae. The following keys are only to the genera known to be aquatic.

Keys to the North American Genera of Aquatic Syrphidae

Adults

1. Posterior femora lacking a patch of black setulae on their anterior bases; 3rd vein at most moderately curved into apical cell 2
— All femora with such setulae; 3rd vein strongly curved into apical cell (fig. 14:53*h*) (ERISTALINAE) 4
2. Anterior cross vein situated well before middle of discal cell, or the mesonotum with bristles; arista bare or slightly pubescent (CHEILOSIINAE) 3
— Anterior cross vein situated at or beyond the middle of discal cell; thorax rarely with short spines; arista plumose (SERICOMYIINAE) .... *Sericomyia* Meigen
3. Face and frons rugose; disc of abdomen opaque black, the sides shining .... *Chrysogaster* Meigen
— Face and frons smooth; disc of abdomen either wholly shining or with shining spots or bands .... *Myiolepta* Newman
4. Marginal cell closed and petiolate (fig. 14:53*h*) .... *Tubifera* Meigen
— Marginal cell open 5
5. Large, robust species; thorax thickly yellow or orange pilose; hind femur swollen and arcuate . *Mallota* Meigen
— Smaller and more slender species; hind femur not swollen and arcuate .... *Helophilus* Meigen

Larvae and Puparia

1. Respiratory tube shorter than body when extended (fig. 14:53*c*); last 1-3 abdominal segments each with a ciliate, pointed lobe on each side .. *Myiolepta* Newman *Chrysogaster* Meigen
— Caudal respiratory organ when extended, much longer than the body (fig. 14:53*j*) (*Sericomyia* also keys out here)... 2
2. The 2 longitudinal tracheal trunks of larva undulating (fig. 14:53*b*) .... *Helophilus* Meigen
— Tracheal trunks of larva straight (fig. 14:53*a*) 3
3. Larva distinctly flattened posteriorly with distinct papillae bearing coarse hairs .... *Mallota* Meigen
— Larva not flattened, without papillae bearing hairs .... *Tubifefa* Meigen

California Species of Aquatic Syrphidae

Subfamily CHEILOSIINAE

Genus *Chrysogaster* Meigen

Johannsen (1935) found puparia of *C. pulchella* Williston in water near the shore of a pond in New York. The following key is adapted from Shannon (1916).

Key to Adults of the California Species of Chrysogaster

1. Lower posterior edge of scutellum with a single row of downward projecting pile; Humboldt County .... *chilosioides* (Shannon) 1922
— Lower posterior edge of scutellum without downward projecting pile 2
2. At least the first 2 tarsal segments yellow or yellowish red; antennae elongate; mesonotum with coppery vittae; central and southern California. . .*bellula* Williston 1882
— Legs entirely dark 3
3. Third antennal segment approximately 3 times as long as broad....................................... 4
— Third antennal segment less elongate 5
4. Pile on frons and ocellar triangle black; San Francisco and Marin counties .... *nigrovittata* (Loew) 1876
— Pile on frons and ocellar triangle whitish; Alameda and Sonoma counties .... (♂) *stigmata* Williston 1882
5. Squamae and halteres darkened 6
— Squamae and halteres whitish 7
6. Dark, greenish black species; pile on ocellar triangle and frons rather long and black; California .... *pacifica* Shannon 1916
— Dark, steel-blue species with very short whitish pile on frons and ocellar triangle; Reno, Nevada .... *unicolor* Shannon 1916
7. The transverse pollinose band below antenna reduced almost to 2 spots at the eye margins; frons black; arista a little shorter than the antennae; Alameda and Sonoma counties................. (♀) *stigmata* Williston 1882
— The pollinose facial band complete and distinct; frons greenish metallic 8
8. The vittae on the mesonotum very faint, paler than the rest of dorsum; arista a little longer than the antenna; rather large, robust species; Monterey County .... *robusta* Shannon 1916
— The vittae on mesonotum coppery colored; arista a little shorter than antenna; small species; Lake Tahoe; Mono County .... *parva* Shannon 1916

Genus *Myiolepta* Newman

Greene (1923) described the puparium (fig. 14:53*c*) of *M. nigra* Loew which he found in a hole in a tulip tree. Immature stages of several European species have been described from decaying wood of hollow poplars.

californica Shannon 1923. Mendocino County
lunulata Bigot 1884 Sierra Nevada
(= *varipes* auct. non Loew).

Subfamily SERICOMYIINAE

Genus *Sericomyia* Meigen

Bloomfield (1897) recorded the immature stages of the European *S. borealis* Fallén from a pool in a peat bog. Curran (1934) keys twelve North American species, none of which have been recorded from California.

Subfamily ERISTALINAE

Genus *Tubifera* Meigen 1800

[=*Eristalis* Latreille 1804, *Elophilus* Meigen 1803]

Johannsen (1935) has published descriptions of seven North American species of rat-tailed maggots of this genus. Excellent figures of the immature stages of *T. tenax* (L.) and *aeneus* (F.) also have been published by Metcalf (1913), and an exceptionally fine account of the Australasian *T. arvorum* (F.) in Hawaii was given by Williams (1939) (fig. 14:53*d-f, h, i*).

Key to Adults of the California Species of Tubifera

1. Eyes bare, reddish brown, with more or less confluent dark brown spots; scutellum concolorous with mesonotum; California. . . . . *aenea* (Scopoli) 1763
— Eyes not spotted, hairy; scutellum more or less yellowish 2
2. Mesonotum with a transverse metallic fascia situated between the scutellum and the suture; San Bernardino County . . . . *alhambra* (Hull) 1925
— Mesonotum without a prescutellar band 3
3. Pile of eyes confined to a vertical stripe; large, honeybeelike species, moderately pilose; widespread . . . . *tenax* (Linnaeus) 1758
— Pile of eyes not confined to a vertical stripe 4
4. Third abdominal segment of males with conspicuous, lateral, yellow spots; moderately pilose species; mesonotum with a pair of anterior, longitudinal sublateral, pollinose bands; California . . . . *occidentalis* (Williston) 1882
— Third abdominal segment of males dark except for narrow basal and apical bands; thinly pilose species; mesonotum without pollinose bands 5
5. Arista with long pubescence on basal 1/2; California . . . . *hirta* (Loew) 1865
Arista entirely bare; widespread . . *latifrons* (Loew) 1865

Genus *Mallota* Meigen

Johannsen (1935) gives descriptions of the immature stages of the common eastern species *M. posticata* (Fabricius) and *M. cimbiciformis* (Fallén) which breed in tree holes. Apparently they are not reported from California.

Genus *Helophilus* Meigen

The larva of *Helophilus latifrons* Loew, which was described and figured by Jones (1922), is similar to that of *Tubifera* except that the two longitudinal tracheal trunks are undulating (fig. 14:53*b*).

California species:

fasciatus Walker 1849.	Widespread
latifrons Loew 1863.	Widespread

REFERENCES

BLOOMFIELD, E. N.
1897. Habits of *Sericomyia borealis* Fln. Ent. Mon. Mag., 33:222-223.
CURRAN, C. H.
1934. Notes on the Syrphidae in the Slosson collection of Diptera. Amer. Mus. Novit. no. 724, 7 pp.
GREENE, C. T.
1923. A contribution to the biology of N. A. Diptera. Proc. Ent. Soc. Wash., 25:82-89, 2 pls.
HEISS, E. M.
1938. A classification of the larvae and puparia of the Syrphidae of Illinois exclusive of aquatic forms. Illinois Biol. Monogr., 16:1-142, 17 pls.
JOHANNSEN, O. A.
1935. *See* Diptera references.
JONES, C. R.
1922. A contribution to our knowledge of the Syrphidae of Colorado. Colorado Agr. Coll. Agr. Exp. Sta. Bull. 269:1-72, 8 pls.
METCALF, C. L.
1913. The Syrphidae of Ohio. Ohio Biol. Surv. Bull. 1:1-123, 11 pls.
1916. Syrphidae of Maine. Maine Agr. Exp. Sta. Bull. 253:193-264, 9 pls.
SHANNON, R. C.
1916. Notes on some genera of Syrphidae with descriptions of new species. Proc. Ent. Soc. Washington, 18:101-113.
WILLIAMS, F. X.
1939. Biological Studies in Hawaiian water-loving insects. Part III. Diptera or flies. B. Asteiidae, Syrphidae and Dolichopodidae. Proc. Hawaii. Ent. Soc., 10: 281-315, 8 pls.

Family EPHYDRIDAE

Shore Flies and Brine Flies

Ephydrid adults are small to minute in size, usually dull in color with characteristic prominent face and gaping mouth. They inhabit moist environments where they are commonly seen walking about on the surface of the mud or water, lapping up or scraping off the minute particles of algae or diatoms which make up much of their food. The ephydrids are especially prominent around inland salt and alkaline ponds and marshes and thermal and mineral springs.

Most of the larvae are aquatic or semiaquatic, and some of them live in or on the leaves and stems of aquatic plants. The peculiar petroleum and saline species have already been mentioned under the general discussion of Diptera. The larvae of these saline and alkaline pond inhabitants are occasionally so multitudinous that they occur as large balls of larvae floating through the water, or so densely packed in shallow waters about pond and lake margins that nothing else can be seen.

Cresson (1942-1949) has admirably brought up to date the adult taxonomy including keys and the known distribution of the North American ephydrids with the exception of the Ephydrinae and a part of the Parydrinae which have currently been revised by Sturtevant and Wheeler (1954). Hennig (1943) collected all the information known about the immature stages. The following keys are adapted from the preceding references.

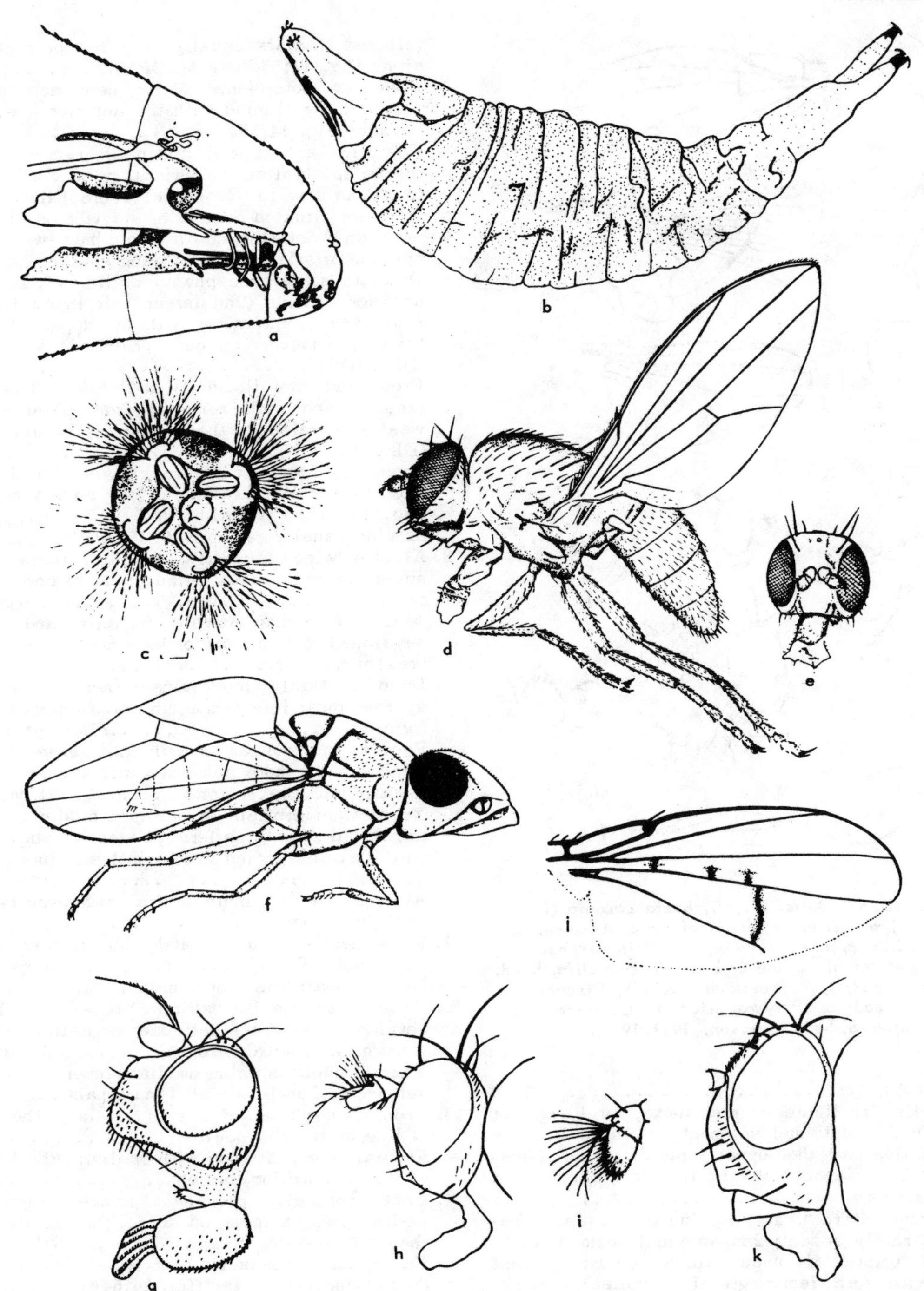

Fig. 14:54. Ephydridae. a-c, *Scatella stagnalis* (Fall.), larva: a, anterior segments with pharyngeal skeleton; c, posterior spiracle; b, puparium. d,e, *Atissa litoralis* Cole, adult: d, lateral view; e, head, anterior view. f, *Lipochaeta slossonae* Coq., adult; g, *Neoscatella oahuensis* (Will.), head; h, *Notiphila virgata* Coq., adult, head; i-k, *Paralimna punctipennis* (Wied.), adult: i, antenna; j, wing; k, head (a-c, Tuxen, 1944; d-f, Cole, 1912; g, Wirth, 1948; h, Cresson, 1917; i-k, Cresson, 1916).

Keys to the North American Genera of Ephydridae

Adults

1. Median facial area setulose, the facial series of bristles converging above; mouth opening large and cavernous, its margin often ciliate with long bristles (fig. 14:54*g*); presutural dorsocentral bristles present and well developed (EPHYDRINAE) 49
— Median facial area bare, the facial bristles in more or less vertical series, parallel with the orbits (figs. 14:54*e*, 14:55*a*,*d*,*f-m*); presutural dorsocentrals present

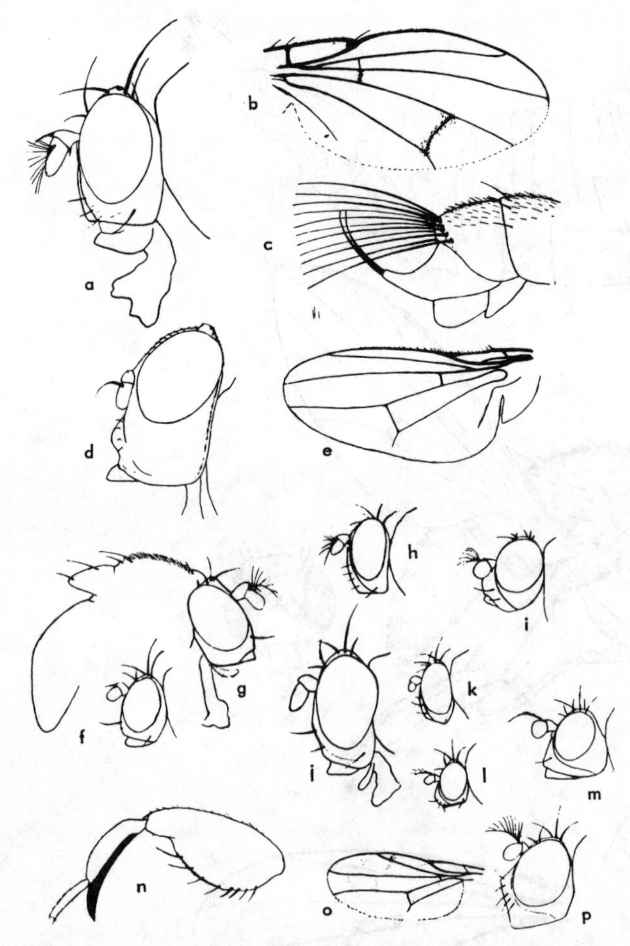

Fig. 14:55. Ephydridae adults. a-c, *Dichaeta caudata* (Fall.): a, head; b, wing; c, posterior segments of male abdomen. d,e, *Gymnopa tibialis* (Cr.): d, head; e, wing. f, *Psilopa*, head; g, *Plagiopsis*, head and dorsal outline of body; h, *Hydrellia*, head; i, *Pseudohecamede*, head; j, *Typopsilopa*, head; k, *Discocerina*, head; l, *Ptilomyia*, head; m, *Parydra*, head; n, *Ochthera*, front leg; o,p. *Zeros*: o, wing; p, head (Cresson, 1917, 1918).

or absent 2

2. Frontal bristles reclinate or proclinate, rarely absent (fig. 14:55*d*); 2nd antennal segment bearing 1 or more spinous bristles at the upper apical corner (figs. 14:54*h,i*; 14:55*a*) (except absent in *Glenanthe*, which also has arista bare) 3
— Frontal bristles lateroclinate, curving over eyes, sometimes small, rarely absent; 2nd antennal segment lacking spinous bristle at upper apical corner (except *Ochthera*, with fore femur greatly swollen) (PARYDRINAE) 41
3. Mesonotum without discal seriate macrochaetae, at most with a pair of interalars and a pair of prescutellar acrostichals (figs. 14:54*d*; 14:55*g*) (PSILOPINAE) 4
— Mesonotum with well-developed seriate macrochaetae, 2 or 3 pairs of true dorsocentrals present (NOTIPHILINAE) 31
4. Shining to opaque species with setation usually well developed; postbucca generally setulose and its posterior margin if sharp is not marginate; facial area usually not sculptured 5
— Shining black species with setation much reduced, postbucca bare and its posterior margin sharp and reflexed; facials usually in pitted or rugulose depressions (fig. 14:55*d*) (GYMNOPINI) 8
5. Posterior notopleural at or near notopleural suture, horizontally aligned with the anterior one; aristal hairs straight (fig. 14:55*g*) 6
— Posterior notopleural distinctly removed dorsad from the notopleural suture; aristal hairs curved (fig. 14:55*i*) (or arista bare in *Glenanthe*) (ATISSINI) 10
6. Ocellars situated caudad of anterior ocellus, sometimes between posterior ocelli; antennae with upper distal spinous bristle; mostly shining species 7
— Ocellars situated cephalad of line of, or in line with, anterior ocellus (2nd larger pair in ocellar triangle in *Paratissa*); antennae without upper distal spinous bristle; mostly opaque pruinose species (DISCOCERININI) 17
7. Frons and face distinctly sculptured; 3 or more facials present and well removed from parafacies; setation weakly developed, the mesonotal setulae not seriated (DISCOMYZINI) 23
— Frons and face at most weakly sculptured mesad; rarely more than 2 facials present and usually near parafacies (fig. 14:55*f,g*); setation well developed, the mesonotal setulae usually seriated (PSILOPINI) 24
8. Alula of wings broad (fig. 14:55*e*); frontals and ocellars absent or minute (fig. 14:55*d*); arista bare; 1 notopleural present ....*Gymnopa* Fallén
— Alula of wings narrow; frontals and ocellars well developed; arista more or less pectinate; 2 notopleurals present.. 9
9. Face medianly tuberculose; acrostichal hairs in 6 sparse rows; fore femur without flexor comb (not California) ....*Parathyroglossa* Hendel
— Face evenly convex or with transverse gibbosity; acrostichal hairs more numerous and nonseriate; fore femur usually with flexor comb ....*Athyroglossa* Loew
10. No distinct orbitals or ocellars evident, the front with small hairs only; arista pubsecent above and below; tiny whitish species with bristles reduced to fine hairs ....*Asmeringa* Becker
— At least the reclinate orbital and usually the ocellars well developed 11
11. Eyes pear-shaped, strongly constricted below; arista bare; not California.............. *Glenanthe* Haliday
— Eyes normal; arista pectinate 12
12. Face with median shining tubercle; cheeks broad, setulose; mesofrons setulose; scutellum with more than 4 marginals; not California ....*Hecamede* Haliday
— Face without shining median tubercle; mesofrons not setulose; scutellum with 4 marginals 13
13. Frontals cephalad of line of ocellars, the latter caudad of line of anterior ocellus 14
— Frontals about aligned with ocellars, the latter cephalad of line of anterior ocellus........................ 16
14. Face conically prominent above epistoma, with a median pair of upcurved setae; 2nd costal section less than 1.5 times as long as 3rd .....*Ptilomyia* Coquillett
— Face most prominent near epistoma, without median pair of upcurved setae (fig. 14:54*e*) 15
15. Ocellars minute; a pair of larger divergent interfrontals anterior to anterior ocellus ....*Pelignellus* Sturtevant and Wheeler
— Ocellars strong, no interfrontals present anterior to anterior ocellus ....*Atissa* Haliday
16. Proboscis slender, geniculate, with lanceolate labellum .... *Pseudohecamede* Hendel
— Proboscis normal, with fleshy labellum........... ....*Allotrichoma* Becker
17. Face with a secondary series of dorsolaterally inclined setae laterad of primaries; eyes pilose; notopleura setulose....................*Polytrichophora* Cresson
— Face with secondary series absent or suggested only by mesally inclined setulae 18

18. Notopleura setulose; anterior notopleural removed caudad closer to posterior notopleural 19
— Notopleura and parafacies bare; anterior notopleural situated in anterior angle of notopleuron, about equidistant from humeral and posterior notopleural...... 20
19. Densely cinereous, opaque species; parafacies and cheeks broadly developed .... *Hydrochasma* Hendel
— Shining to dark opaque species; parafacies and cheeks rather narrow (fig. 14:55*k*)......*Discocerina* Macquart
20. Facial series of 2 primaries 21
— Facial series of at least 3 primaries, upper facial on distinct, shining papilla; opaque, densely cinereous species ....*Hecamedoides* Hendel
21. Mesofrons with a pair of strong proclinate bristles in front of anterior ocellars; 2 pairs of proclinate and 2 pairs of lateroclinate frontoörbitals present; Florida .... *Paratissa* Coquillett
— Mesofrons bare; 1 pair of proclinate and 1 pair of reclinate frontoörbitals present 22
22. Mostly opaque cinereous species; facialia with upcurved seta at lower extremity; not California .... *Diclasiopa* Hendel
— Mostly shining species; facialia without upcurved seta .... *Ditrichophora* Cresson
23. Second vein closely approximate to costa beyond end of 1st; posterior cross vein with sharp angle; wings immaculate ....*Clanoneurum* Becker
— Second vein and posterior cross vein normal; wings maculate; not California .... *Discomyza* Meigen
24. Antennae normal, 1st segment at most slightly exserted, 2nd segment trigonal with distinct lobe and spine 25
— Antennae elongate; 1st segment exserted, 2nd segment conical, broader apicad, without superior lobe but with hairlike apical spine; 3rd segment elongate, more or less cylindrical; ocellars approximate and caudad, almost between posterior ocelli ...*Ceropsilopa* Cresson
25. Prescutellar acrostichals present 26
— Prescutellars absent; 3rd and 4th tergites large and subquadrate 27
26. Abdomen ovate with margins revolute; shining, weakly granulose species ....*Trimerinoides* Cresson
— Abdomen slender, parallel-sided, with margin sharp; subopaque, granulose species; not California....... ....*Trimerina* Macquart
27. Frontals well developed; ocellars cephalad of posterior ocelli; halteres white 28
— Frontals weak, hairlike; ocellars small, set between posterior ocelli (fig. 14:55*g*); halteres usually dark; wings yellowish with dark bases ..*Plagiopsis* Cresson
28. Face with 3 or more closely set facials, lower part flattened and wrinkled; supra-alar bristle well developed not California .... *Rhysophora* Cresson
— Face with 1 or 2 facials; supra-alar absent or minute 29
29. Face with 1 strong facial (fig. 14:55*f*); not transversely tumid in middle but sometimes weakly carinate longitudinally 30
— Face with 2 strong facials, and transversely carinate or tumid in middle...............*Helaeomyia* Cresson
30. Fore basitarsi black and thickened; middle of face slightly carinate with fine transverse wrinkles, facial at or above midfacial profile .....*Leptopsilopa* Cresson
— Fore basitarsi normal; middle of face smooth and convex, facials below midfacial profile ... *Psilopa* Fallén
31. Posterior notopleural removed from ventral margin of notopleuron 32
— Posterior notopleural in normal position, close to ventral margin of notopleuron 36
32. Face broad and greatly swollen, with broad mouth opening (fig. 14:55*p*); wings uniformly spotted (fig. 14:55*o*) (ILYTHEINI) 33
— Face narrow or if broad is concave; wings not uniformly spotted (HYDRININI)........................... 34
33. Second vein long, subparallel to costa; 2nd costal section over twice as long as 3rd ..... *Ilythea* Haliday
— Second vein short, mostly straight to costa; 2nd costal section not twice as long as 3rd (fig. 14:55*o*); not California ....*Zeros* Cresson
34. Face narrow, not concave and if prominent the epistoma is receding 35
— Face broad, concave, with prominent epistoma; arista long haired; not California........*Lemnaphila* Cresson
35. Arista bare; presutural dorsocentral present........ ....*Philygria* Stenhammar
— Arista short- to long-haired; presutural dorsocentral absent ....*Nostima* Coquillett
36. Ocellars stronger than postocellars which are always weak; eyes bare; humerals and supra-alars strong 37
— Ocellars seldom as strong as postocellars, usually much weaker; eyes pilose; humerals and supra-alars usually weak, hairlike (fig. 14:55*h*) (HYDRELLIINI) ....*Hydrellia* Robineau-Desvoidy
37. Mid-tibia with 1-4 erect extensors; dull species (NOTIPHILINI) 38
— Mid-tibia without erect extensors; shining black species (fig. 14:55*j*) (TYPOPSILOPINI)... *Typopsilopa* Cresson
38. Costa extending to 4th vein (fig. 14:54*j*); antennal spine weak (fig. 14:54*i*) 39
— Costa not reaching to tip of 4th vein (fig. 14:55*b*), rarely beyond 3rd; antennal spine strong (figs. 14:54*h*; 14:55*a*) 40
39. Small cinereous species (2-2.5 mm.) without mesonotal setulae; sternopleural absent .... *Oedenops* Becker
— Larger dark species with well-developed mesonotal setulae and strong sternopleural.......*Paralimna* Loew
40. Facial bristles large and situated close to parafacies; frontoörbitals very strong (fig. 14:55*a*); dark brown to black species .... *Dichaeta* Meigen
— Facials hairlike and well removed from parafacies; frontoörbitals weak (fig. 14:54*h*); ochraceous to cinereous species.......................*Notiphila* Fallén
41. Eyes hairy; antennae small, inserted far apart in cavities, arista atrophied (fig. 14:54*f*) (LIPOCHAETINI) .... *Lipochaeta* Coquillett
— Eyes bare; antennae normal, arista well developed .. 42
42. Mouth opening large and gaping, clypeus transversely elongate and exposed (PARYDRINI) 43
— Mouth opening small to moderately large; clypeus not prominent (exposed but tonguelike in *Ochthera* and *Pelina*) (HYADININI) 44
43. Costa extending to 3rd vein; arista with long dorsal rays (fig. 14:54*j*)...............*Brachydeutera* Loew
— Costa extending to 4th vein; arista bare (fig. 14:55*m*) ....*Parydra* Stenhammar
44. Fore legs raptorial, their femora greatly enlarged, their tibiae ending in a spur (fig. 14:55*n*)..*Ochthera* Latreille
— Fore legs normal, femora slender 45
45. Arista with long hairs; head trigonal in cephalic aspect, face subconically or hemispherically gibbous above; abdomen subglobose, lateral margins revolute, 4th tergite not lengthened; not California .... *Gastrops* Williston
— Arista bare or pubescent; head not trigonal 46
46. Wing with 2nd vein long, 2nd costal section about 3 times as long as 3rd; face flat or carinate, not medianly prominent; 3rd antennal segment rounded at apex above ....*Pelina* Haliday
— Wing with 2nd vein short, 2nd costal section less than twice as long as 3rd; face with low conical median prominence; 3rd antennal segment usually angulate at apex above 47
47. Abdomen not convex above, 4th tergite not lengthened, never longer than 5th .... *Hyadina* Haliday

— Abdomen convex above, with 4th tergite elongated, longer than the 5th. 48
48. Costa not extending beyond apex of 4th vein; lateral lobes of abdominal tergites sharply reflexed against venter; not California . . . . *Axysta* Haliday
— Costa extending beyond apex of 4th vein; lateral lobes of abdominal tergites revolute . . . . . *Lytogaster* Becker
49. Tarsal claws long and straight; pulvilli absent (EPHYDRINI) 50
— Tarsal claws short and curved; pulvilli well developed (SCATELLINI) 54
50. Frontoörbitals reduced to 1 pair; no true dorsocentrals only the pair of interalars present; face with a cluster of large bristles in middle; northeastern states . . . . *Cirrula* Cresson
— Two or 3 pairs of strong frontoörbitals and 3-5 pairs of true dorsocentrals 51
51. Two pairs of strong frontoörbitals 52
— Three pairs of strong frontoörbitals 53
52. Third antennal segment bearing a long hair on outer surface; frontal orbits shining . . . . . . *Setacera* Cresson
— Third antennal segment without a long outer hair; frontal orbits not differentiated from mesofrons . . . . *Dimecoenia* Cresson
53. Five pairs of true dorsocentrals, 2 pairs before the suture; cruciate interfrontals situated along lateral margins of mesofrons . . . . *Hydropyrus* Cresson
— Four pairs of true dorsocentrals, only 1 pair before the suture; cruciate interfrontals located well toward middle of mesofrons . . . . *Ephydra* Fallén
54. Arista with long abundant rays on the basal half above 55
— Arista naked or minutely pubescent on its whole length (fig. 14:54*g*); 2 or 3 pairs of dorsocentrals 57
55. Two pairs of true dorsocentrals; not California . . . . *Philotelma* Becker
— Three or 4 pairs of true dorsocentrals 56
56. Humerals reduced; head much longer than high in profile, the cheeks narrow . . . . . *Coenia* Robineau-Desvoidy
— Humerals well developed; head about as high as long in profile, the cheeks broader; facial hump more prominent . . . . *Paracoenia* Cresson
57. Costa extending to apex of 4th vein 58
— Costa ending at or just beyond apex of 3rd vein; wing and mesonotum usually with dark and light pattern . . . . *Scatophila* Becker
58. Only 1 pair of divergent frontoörbitals; wing with pattern of many irregular light and dark spots . . . . *Limnellia* Malloch
— Two pairs of divergent frontoörbitals 59
59. One postsutural pair of true dorsocentrals, the antesutural pair absent; 1 pair of strong acrostichals; genal bristles strong . . . . *Scatella* Robineau-Desvoidy
— Two or more pairs of true dorsocentrals, or if 1 pair, there are no strong acrostichals 60
60. Only 1 pair of strong acrostichal bristles, located at the suture, none present postsuturally . . . . *Neoscatella* Malloch
— A row of short acrostichals complete postsuturally to base of scutellum 61
61. Wing with hyaline spots; genal bristle well developed . . . . *Parascatella* Cresson
— Wing without hyaline spots; genal bristle undeveloped . . . . *Lamproscatella* Hendel

Larvae

1. Each posterior spiracular plate ending in a sharp-pointed spur (fig. 14:56*a,b,d*), the spiracles with 3 slitlike openings; anterior spiracles absent; miners in aquatic plants (NOTIPHILINAE) 2
— Posterior spiracles not on spur-shaped processes (figs. 14:56*l*; 14:57*a*; 14:58*c*; 14:59*a,c*), each spiracle with 3 or 4 openings (figs. 14:54*c*; 14:56*i*; 14:57*d*); anterior spiracles present (figs. 14:56*h*; 14:57*b*; 14:59*a*); habits not mining 3
2. Pharyngeal skeleton with hypostomal sclerites and anterodorsal bridge fused in a slender, common sclerite, the posterior cornua rodlike; mouth hooks usually fused . . . . *Hydrellia* Robineau-Desvoidy
— Pharyngeal skeleton with separate, small, hypostomal sclerites, a well-differentiated anterodorsal bridge and broad posterior cornua; paired mouth hooks present (fig. 14:56*e*) . . . . *Notiphila* Fallén
3. Posterior spiracles each with 3 openings, the margin of the spiracular plate with short, unbranched hairs (fig. 14:56*i*); anterior spiracles in an invaginated pocket (figs. 14:56*h*; 14:58*c*) (PARYDRINAE, part) . . . . *Brachydeutera* Loew
— Posterior spiracles each with 4 openings, with 4 palmately branched hairs between (figs. 14:54*c*; 14:57*g*); anterior spiracles not as above 4
4. Posterior respiratory tube short, the basal unbranched part as short as the 2 distal branched parts, these not longer than broad (PSILOPINAE) . . *Helaeomyia* Cresson
— Posterior respiratory tube and its branches much longer (EPHYDRINAE) 5
5. Eight pairs of abdominal pseudopods bearing strong claws present, the pair on the 8th segment very strongly developed (fig. 14:59*a,c*) (EPHYDRINI) . . . . *Ephydra* Fallén
Setacera Cresson
Dimecoenia Cresson
— Pseudopods absent or weakly developed without strong claws, the pair on 8th segment not larger than the others (fig. 14:56*l*) 6
6. Mouth hooks not toothed, anterior spiracles with openings sessile (PARYDRINAE, part) . . *Ochthera* Latreille
— Mouth hooks toothed (fig. 14:54*a*), anterior spiracles with openings each on a fingerlike process (SCATELLINI) 7
7. Air tube and its branches more than 1/2 as long as body of larva . . . . *Coenia* Robineau-Desvoidy
— Air tube and its branches shorter (fig. 14:56*l*) . . . . *Scatella* Robineau-Desvoidy
Neoscatella Malloch

California Species of Ephydridae

Subfamily PSILOPINAE

Tribe GYMNOPINI

Key to Adults of the California Species of Gymnopa

1. Face with 2 small, separate, depressed, pruinose white spots on each side between antennae and eye margins; California . . . . *bidentata* Cresson 1926
— Face with 1 large pruinose white spot on each side between antennae and eye margins (fig. 14:55*d,e*); California . . . . *tibialis* (Cresson) 1916

Genus *Parathyroglossa* Hendel

ordinata (Becker) 1896. Humboldt, Napa

Key to Adults of the California Species of Athyroglossa

1. All tibiae black; El Dorado, Kern, and Shasta counties . . . . *glabra* (Meigen) 1830

— Only fore tibia black 2
2. Cheeks narrow, about 1/5 height of head; face narrow, evenly convex, not prominent in profile; southern California .... *glaphyropus* Loew 1878
— Cheeks broader; face broader, about 1/3 width of head, subgibbously prominent in profile; southern and central California .... *melanderi* (Cresson) 1922

Tribe ATISSINI

Genus *Glenanthe* Haliday

litorea Cresson 1925. Imperial County

Genus *Hecamede* Haliday

Bohart and Gressitt (1951) figured the puparium of *Hecamede persimilis* Hendel, which they found abundant on the beaches on Guam and breeding in the sand under a human carcass. Not known to occur in California.

Genus *Pelignellus* Sturtevant and Wheeler

subnudus Sturtevant and Wheeler 1954. Orange County

Genus *Asmeringa* Becker

lindsleyi Sturtevant and Wheeler 1954. San Bernardino County

Key to Adults of the California Species of Atissa

1. Mesonotal setulae in 6 series between the intra-alars; dark brown species (fig. 14:54*d,e*); southern and central California .... *litoralis* (Cole) 1912
— Mesonotal setulae in 4 series; whitish pruinose species; antennae and legs yellow; southern California .... *pygmaea* (Haliday) 1833

Key to Adults of the California Species of Ptilomyia

(Fig. 14:55l)

1. Prominent bristles present at about sutural level in both dorsocentral and median acrostichal rows; abdomen shining black; Kern County .....*enigma* Coquillett 1900
— Sutural bristles absent in dorsocentral and acrostichal series; abdomen pollinose 2
2. Cheeks distinctly broader than width of 3rd antennal segment; face uniformly grayish white pollinose; San Bernardino and Inyo counties .... *occidentalis* Sturtevant and Wheeler 1954
— Cheeks narrower than width of 3rd antennal segment; face grayish to brownish 3
3. Facial series with but 1 ventrally inclined seta; 2nd costal section subequal to 3rd; prescutellars about as far from each other as from scutellum; Imperial, Riverside, and San Bernardino counties .... *pleuriseta* (Cresson) 1942
— Facial series with 2 such setae; 2nd costal section distinctly longer than 3rd; prescutellars usually closer to scutellum then they are to each other; Inyo, Mono, San Bernardino, and San Mateo counties .... *alkalinella* (Cresson) 1942

Genus *Pseudohecamede* Hendel

(Fig. 14:55*i*)

abdominalis (Williston) 1896. Riverside County

Genus *Allotrichoma* Becker

Bohart and Gressitt (1951) found a species of *Allotrichoma*, apparently *livens* Cresson, breeding abundantly in pig droppings in muddy pens on Guam and figured the puparium.

Key to Adults of the California Species of Allotrichoma

1. Clypeus shining black, contrasting with the pruinose face; California .... *yosemite* Cresson 1926
— Clypeus pollinose, opaque, not contrasting with the face 2
2. Male with genital appendages rather spatulate and ciliate on their full length from bases; southern and central California ....*lateralis* (Loew) 1860
— Male with genital appendages more or less setiform, with or without dilated tips 3
3. Genital appendages of male with 3 short bristles at tips; Riverside County .... *trispina* Becker 1896
— Genital appendages of male with about 10 long bristles at tips; California .... *simplex* (Loew) 1861

Tribe DISCOCERININI

Key to Adults of the California Species of Polytrichophora

1. Fore femur with an anteroflexor comb of minute spinules and a few setae on postflexor margin; Alameda, Marin, and Orange counties......... *conciliata* Cresson 1924
— Fore femur without anteroflexor comb, but with a series of several stout postflexor spines; southern and central California .... *orbitalis* (Loew) 1861

Genus *Hecamedoides* Hendel

glaucella (Stenhammar) 1843. Kern, Mono

Key to Adults of the California Species of Ditrichophora

1. Frons with 2 orbitals; frons and mesonotum brown pollinose 2
— Frons with 1 orbital............................. 3
2. Tarsi yellowish; Del Norte and San Mateo counties .... *cana* Cresson 1940
— Tarsi black; Shasta County ..... *lugubris* Cresson 1940
3. Shining black species; notopleura especially polished; legs yellowish; face pale yellow to gray pollinose; California .... *argyrostoma* (Cresson) 1916
— Mesonotum and frons dull pollinose black; legs black 4

4. Mesonotum opaque; 2nd costal section much longer than 3rd; Mono, Placer, Plumas, and San Bernardino counties .... *occidentalis* Cresson 1942
— Mesonotum shining.. 5
5. Face cinereous, broad and convex; 2nd costal section about twice as long as 3rd; Los Angeles and Santa Cruz counties .... *valens* Cresson 1942
— Face black or sparingly dark pollinose, prominent at upper facial; 2nd costal section not much longer than 3rd; California .... *atrata* Cresson 1940

Key to Adults of the California Species of Hydrochasma

1. Mouth opening very large; head subglobose, very broad; fore femoral comb of numerous, minute spinules on anterior margin; Los Angeles, Riverside, and San Diego counties ....*capax* Cresson 1938
— Mouth opening not unusually large; head more flattened in profile; fore femoral comb of several spinules on posterior margin; Imperial, Los Angeles, and Sonoma counties ....*leucoprocta* (Loew) 1861

Genus *Discocerina* Macquart

(Fig. 14:55*k*)

According to Grünberg (1910), the European *Discocerina plumosa* (Fallén) breeds in algae and moss in the forest.

Key to Adults of the California Species of Discocerina

1. Parafacies with series of down-curved setulae 2
— Parafacies bare 3
2. Parafacies contrastingly paler in ground color, the sparser vestiture augmenting this contrast; hind trochanters of male without long caudal seta; California .... *obscurella* (Fallén) 1813
— Parafacies not noticeably differing from medifacies either in ground color or vestiture; hind trochanters of male with a long caudal seta; California .... *trochanterata* Cresson 1939
3. Facial series of only 2 equally long bristles; southern and central California........ *nadineae* (Cresson) 1925
— Facial series of 3 or more equally long bristles; Kern, Riverside and San Diego counties .... *turgidula* Cresson 1940

Tribe DISCOMYZINI

Genus *Clanoneurum* Becker

The larvae of *Clanoneurum menozzii* Séguy mine in leaves of beet in Europe (Hennig, 1943).

americanum Cresson 1940. San Diego County

Genus *Discomyza* Meigen

The larvae of *Discomyza incurva* (Fallén) and *maculipennis* (Wiedemann) have been bred from dead mollusks and carrion. The puparium of the latter has been described and figured by Bohart and Gressitt (1951).

Tribe PSILOPINI

Key to Adults of the California Species of Ceropsilopa

1. Femora mostly black; San Diego County .... *dispar* Cresson 1922
— Femora and tibiae yellow 2
2. Face shining, narrow and wrinkled; southern and central California ....*coquilletti* Cresson 1922
— Face entirely opaque, cinereous; Orange County .. .. *staffordi* Cresson 1925

Genus *Trimerinoides* Cresson

adfinis (Cresson) 1922. Kern County

Genus *Trimerina* Macquart

The larvae of the European *Trimerina madizans* (Fallén) are parasitic in eggs of spicers (Hennig, 1943).

Genus *Psilopa* Fallén

(Fig. 14:55*f*)

The larvae of *Psilopa leucostoma* (Meigen) have been reported as mining in species of *Chenopodium* in Europe (Hennig, 1943).

Key to Adults of the California Species of Psilopa

1. Legs except mid- and hind-coxae yellow; face of male silvery white; Riverside and San Mateo counties ....*leucostoma* (Meigen) 1830
— Coxae and femora black 2
2. Antennae and fore tibia more or less yellow; face flattened, the single pair of strong facials very low; wing hyaline; California .... *compta* (Meigen) 1830
— Antennae and fore tibia black; face more or less convex, 2 pairs of facials usually present and comparatively high on face; wing usually with ends of 3rd and 4th veins infuscated; central and southern California ....*olga* Cresson 1942

Key to Adults of the California Species of Leptopsilopa

1. Mid- and hind-legs entirely yellow; Douglas, Arizona .... *atrimana* (Loew) 1878
— All femora mostly black; mid- and hind-tibiae and their tarsi pale; California........*varipes* (Coquillett) 1900

Genus *Helaeomyia* Cresson

(Fig. 14:57*d,f,g*)

The larvae of *Helaeomyia petrolei* (Coquillett), the

famous "petroleum fly," breed in pools of crude petroleum and waste oil in California and Cuba. The biology and description of the immature stages have been reported in detail by Crawford (1912) and Thorpe (1931).

petrolei (Coquillett) 1900. Los Angeles County

Subfamily NOTIPHILINAE

Tribe ILYTHEINI

Key to Adults of the California Species of *Ilythea*

1. Coxae and apices of tarsi and rest of legs pale; Ventura County .... *flaviceps* Cresson 1916
— Coxae and femora black; at most 2 fuscous spots in distal 1/2 of submarginal cell 2
2. Tibiae paler than femora, yellow to ferruginous, at most infuscated medianly; vestiture of face usually more or less cinereous; Los Angeles, Riverside, and San Diego counties .... *caniceps* Cresson 1918
— Tibiae black, at most narrowly pale apically; vestiture of face except foveae, brown; Humboldt to Santa Clara counties .... *spilota* (Curtis) 1832

Tribe PHILYGRIINI

Key to Adults of the California Species of *Philygria*

1. Wings with fuscous spots on the longitudinal veins, no dark crossbands in 1st posterior cell; Humboldt County .... *opposita* Loew 1861
— Wings with fuscous spots only on the cross veins 2
2. Abdomen cinereous except apical segment of male; southern and central California .... *debilis* Loew 1861
— Abdomen almost entirely shining; San Bernardino County .... *nigrescens* (Cresson) 1930

Key to Adults of the California Species of *Nostima*

1. Last section of 4th vein 4 times as long as preceding section between cross veins; acrostichal area not at all shining; Central Valley ....*picta* (Fallén) 1813
— Last section of 4th vein at least 5 times as long as preceding section; acrostichal area somewhat shining; Riverside and Los Angeles counties ....
scutellaris occidentalis Sturtevant and Wheeler 1954

Tribe HYDRELLIINI

Genus *Hydrellia* Robineau-Desvoidy

(Figs. 14:55*h*; 14:56*a,c,d*)

Species of *Hydrellia* are leaf miners in many species of aquatic and subaquatic plants, especially *Potamogeton* and its relatives (Berg, 1950). *H. griseola* (Fallén), one of the most common and widespread species, mines in Gramineae, including barley and rice, and the European *H. nasturtii* Collin in water cress. The tiny Hawaiian species *H. williamsi* Cresson mines in duckweed *(Lemna)*, having much the same habits as *Lemnaphila scotlandi* Cresson, a related form of the eastern United States.

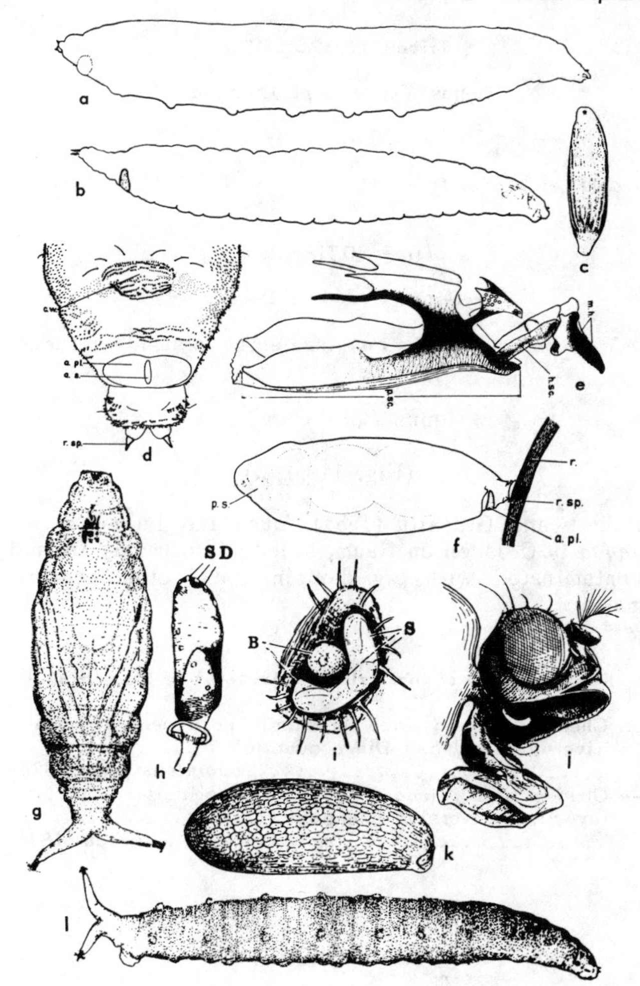

Fig. 14:56. Ephydridae. *a,c,d, Hydrellia*: *a*, larva; *c*, egg; *d*, posterior segments of larva (a. pl., anal plate; a.s., anal slit; c.w., creeping welt; r.sp., respiratory spine). *b,e,f, Notiphila loewi* Cress.: *b*, larva; *e*, pharyngeal skeleton (h.sc., hypopharyngeal sclerite; m.h., mouth hook; p.sc., pharyngeal sclerite); *f*, puparium, attached to root of aquatic plant (a.pl., anal plate; p.s., pharyngeal skeleton; r, root of plant; r.sp., respiratory spine). *g,k,l, Neoscatella*: *g*, puparium; *k*, egg; *l*, larva. *h-j, Brachydeutera hebes* Cress., larva: *h*, anterior spiracle (s.d., spiracular digits); *i*, posterior spiracle (*b*, button or scar from previous molt; *s*, spiracular openings); *j*, head of adult (*a-f*, Berg, 1950; *g-l*, Williams, 1938).

Key to Adults of the California Species of *Hydrellia*

1. Palpi black; mesonotum and pleura black 3
— Palpi yellow; facials in single series; mesonotum brownish, the pleura grayish; California; mostly lowlands .... *griseola* (Fallén) 1813 2
2. Face brown to black ...... var. *obscuriceps* Loew 1862
— Face yellow to golden ..... var. *scapularis* Loew 1862
— Face whitish............... var. *hypoleuca* Loew 1862
3. Facials in a single series; face narrow and carinate southern and central California. . .*tibialis* Cresson 1917
— Facials in a double series; face broad and flat; California; Transition Zone ...... *proclinata* Cresson 1915

Tribe TYPOPSILOPINI

Genus *Typopsilopa* Cresson

(Fig. 14:55*j*)

atra (Loew) 1862. California

Tribe NOTIPHILINI

Genus *Oedenops* Becker

nuda (Coquillett) 1907. Los Angeles, Riverside, San Diego

Genus *Paralimna* Loew

(Fig. 14:54*i-k*)

Bohart and Gressitt (1951) found the larvae of *P. aequalis* Cresson on Guam, breeding commonly in mud contaminated with pig droppings and other organic matter.

Key to Adults of the California Species of Paralimna

1. Cheeks narrow; face brownish pollinose; Imperial, Riverside, and San Diego counties .... *decipiens* Loew 1878
— Cheeks broad; face pale grayish pollinose; Imperial, Inyo, and Riverside counties .... *multipunctata* Williston 1896

Genus *Dichaeta* Meigen

(Fig. 14:55*a-c*)

atriventris Cresson 1915. California

Genus *Notiphila* Fallén

(Figs. 14:54*h*; 14:56*b,e,f*)

According to Hennig (1943), larvae and pupae of most European species live in the mud in the bottom of streams, ponds, and lakes, with the sharp pointed posterior respiratory spine (fig. 14:56*f*) attached to the roots of such plants as *Typha, Nymphaea,* and *Glyceria.* Berg (1950) reared *Notiphila loewi* Cresson from roots of *Potamogeton* in Michigan (fig. 14:56*b,e,f*).

Key to Adults of the California Species of Notiphila

1. Palpi yellow, at least at the apices 2
— Palpi black 3
2. Third antennal segment and tibiae yellow; mesonotum not vittate; mid tibia with 3 extensor bristles; southern California .... *erythrocera* Loew 1878
— Third antennal segment and tibiae black; mesonotum vittate; mid-tibia with 4 extensor bristles; southern and central California........ *pulchrifrons* Loew 1872
3. Ventral hair tuft at base of hind tarsus including at least 1 blackened spine; face golden 4
— Ventral hair tuft at base of hind tarsus entirely pale 5
4. Black spine in hind tarsal tuft of male at least 1/3 as long as basitarsus; California..*macrochaeta* Loew 1878
— Black spine in hind tarsal tuft in both sexes scarcely longer than the pale hairs; California............. .... *atrisetis* Cresson 1917
5. Third antennal segment black 6
— Third antennal segment pale at least in part; Inyo County .... *sicca* Cresson 1940
6. Fore tarsus black; dark fascia on tergite 4 shining, often covering segment except for median stripe; California .... *occidentalis* Cresson 1917
— Fore tarsus yellowish; dark fascia on tergite 4 brown and very little shining; southern and central California .... *olivacea* Cresson 1917

Subfamily PARYDRINAE

Tribe LIPOCHAETINI

Genus *Lipochaeta* Coquillett

slossonae Coquillett 1896. (fig. 14:54*f*) Seacoasts of southern and central California

Tribe PARYDRINI

Genus *Brachydeutera* Loew

Johannsen (1935) has described the active, free-swimming larva of *B. argentata* (Walker), the only North American species, from fish hatchery ponds in New York (fig. 14:58*a-c*). Williams (1938) has given a very detailed account of the biology of *B. hebes* Cresson in Hawaii, including excellent figures of all stages (fig. 14:56*h-j*). The adults of *Brachydeutera* are our most proficient "water-skaters," keeping close to the water surface on their short flights. The larvae feed as scavengers on old plant material but take quantities of filamentous green algae. The puparia (fig. 14:58*b*) are very characteristic with long, divergent, apically dilated, thoracic respiratory horns, and float high on the water surface or are cast upon the shore.

California species:
argentata (Walker) 1856. California

Genus *Parydra* Stenhammar

Nielsen, *et al.* (1954) have figured the larvae of *P. pusilla* Meigen from aquatic habitats in Iceland. The larvae resemble those of *Brachydeutera.*

Key to Adults of the California Species of Parydra

(Figs. 14:55*m*; 14:57*a-c*)

1. Clypeus not visible in profile view; wings hyaline, with numerous definite brown spots (subg. *Callinapaea* Sturtevant and Wheeler); Alameda and Los Angeles

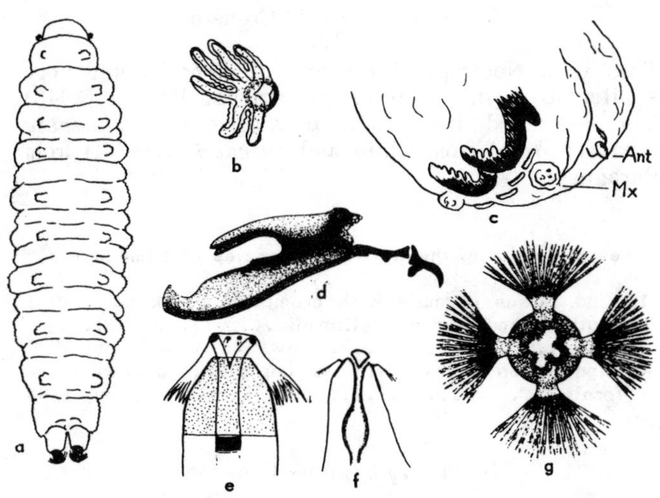

Fig. 14:57. Ephydridae larvae. a-c, *Parydra pusilla* (Mg.): a, dorsal view; b, anterior spiracle; c, oblique view of head region showing mouth hooks (ant, antenna; mx, maxillary palpus). d,f,g, *Helaeomyia petrolei* (Coq.), larva: d, pharyngeal skeleton; f, diagram of posterior spiracle in longitudinal section; g, posterior spiracle. e, *Ephydra riparia* Fall., posterior spiracle as in f (a-c, Nielsen, *et al.*, 1954; d-g, Thorpe, 1930).

counties .... *aldrichi* Sturtevant and Wheeler 1954
— Clypeus visible in profile, gradually tapering in width, narrowing posteriorly; wings hyaline, sometimes with dark spots at apices of veins or on cross veins or dark with pale spots 2
2. Head bristles small and thin, ocellars less than 1/2 as long as the distance between their bases to anterior margin of front; 2nd costal section more than twice as long as 3rd (subgenus *Parydra* Stenhammar) 3
— Head bristles larger, ocellars more than 1/2 as long as distance from their bases to anterior margin of front; 2nd costal section less than twice as long as 3rd (subgenus *Chaetoapnaea* Hendel) 5
3. Postscutellum bare and shining in middle; apical tubercles small and globose; northern California .... *bituberculata nitida* Cresson 1915
— Postscutellum entirely opaque pruinose 4
4. Tibiae mostly reddish; apical scutellar tubercles small and globose; southern California.............. .... *tibialis* Cresson 1916
— Tibiae entirely black; apical scutellar tubercles large, about twice as long as broad; northern California .... *incommoda* Cresson 1930
5. Face strongly convex, a prominent median hump high on face, the face vertical below 6
— Face straight or oblique, prominent at epistoma; frontal bristles long 7

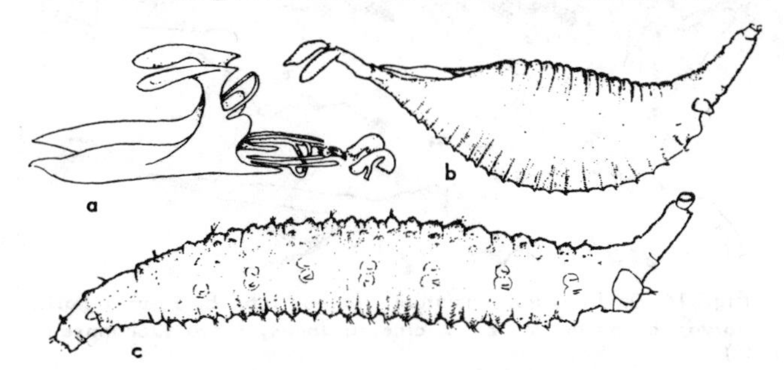

Fig. 14:58. *Brachydeutera argentata* (Walk.). a, larval pharyngeal skeleton; c, larva, lateral view; b, puparium, lateral view (Johannsen, 1935).

6. Second costal section more than 1.5 times as long as 3rd; wing dark with extensive clouds over cross veins; California, common .... *aurata* Jones 1906
— Second costal section scarcely longer than 3rd; wing clear with light clouds over cross veins; California .... *breviceps vicina* Cresson 1940
7. Marginal cell broader than submarginal at the anterior cross vein; head noticeably broader than high; tarsi pale; Humboldt County .... *borealis* (Cresson) 1949
— Marginal cell not noticeably broad 8
8. Second vein acutely entering costa, rarely with an appendage; northern California.... *paullula* Loew 1862
— Second vein long, paralleling costa and more or less abruptly entering it, generally with an apical cloud and appendage 9
9. Wings much wrinkled, with 4 or 5 small, round white spots at the cross veins; prominent, small, round white spot at mesal sutural angles of mesonotum; scutellum elongate, very convex and acute apicad; Humboldt and San Mateo counties .... *varia* Loew 1863
— Wings not conspicuously wrinkled; white spots or areas at cross veins large, not conspicuous 10
10. Face opaque brown; halteres and tarsi pale; common in California .... *appendiculata* Loew 1878
— Face more or less shining, not opaque, overcast with brown or white 11
11. Halteres and tarsi pale; scutellum rather narrow apicad and generally with a small apical tubercle; California .... *socia* (Cresson) 1934
— Halteres and tarsi brown; scutellum broadly rounded at apex; Mono County .... *halteralis* (Cresson) 1930

Tribe HYADININI

Genus *Ochthera* Latreille

Both adults and larvae of *Ochthera* are predaceous on smaller insects. The adults are very common along marshy stream margins and are easily recognized by their strikingly enlarged fore femora and spurred fore tibiae which are held mantislike (fig. 14:55*n*). The larvae of *Ochthera pilimana* Becker from Java which were described by Hennig (1943) have untoothed mandibles fitting their predaceous habits, unique among this family.

Key to Adults of the California Species of Ochthera

1. Face broad, as broad at level of tubercle as height, uniformly brassy; palpi and fore tarsi black; fore tibial spur 1/3 as long as tibia; California .... *mantis* (Degeer) 1776
— Face narrow, twice as high as broad; palpi and fore tarsi yellowish; fore tibial spur 1/2 as long as tibia 2
2. Face with median, shining, black spot above the tubercle; femora black; all tarsi reddish yellow; San Diego to San Bernardino counties............... .... *cuprilineata* Wheeler 1896
— Face uniformly grayish; fore- and mid-femora yellowish brown; hind tarsi black; Riverside County .... *melanderi* Cresson 1944

Key to Adults of the California Species of Pelina

1. Face higher than broad; head not twice as broad as

high; abdomen granulose; mesonotum smooth; mesopleura microrugulose or granulose on anteroventral part; Great Basin region of California .... *compar* Cresson 1934
— Face and head very broad, the former as broad as high, the latter nearly twice as broad as high; mesonotum strongly granulose 2
2. Abdomen irregularly, longitudinally rugulose; California, very common .... *truncatula* Loew 1878
— Abdomen transversely rugulose, strongest on 2nd and 3rd tergites; all tergites smooth laterally; Nevada County .... *canadensis* Cresson 1934

Key to Adults of the California Species of Hyadina

1. Scutellum with sides conspicuously opaque black; mesonotum subshining black; California .... *binotata* (Cresson) 1926
— Scutellum and mesonotum uniformly densely brownish pollinose; mesonotum with grayish spots and stripes; southern California .... *pruinosa* (Cresson) 1926

Key to Adults of the California Species of Lytogaster

1. Frons and mesonotum smooth, at most minutely punctate; scutellum granulose; 3rd antennal segment 1/2 again as long as broad; Los Angeles County..... .... *flavipes* Sturtevant and Wheeler 1954
— Frons, mesonotum, and scutellum rugulose and granulose; 3rd antennal segment as broad as long; California .... *gravida* (Loew) 1863

Subfamily EPHYDRINAE

Tribe EPHYDRINI

Genus *Setacera* Cresson

Johannsen (1935) described the immature stages of *S. atrovirens* (Loew) from a brine pool near Ithaca, New York, and of *S. needhami* Johannsen from Laguna Beach, California. The immature stages are not separable from those of *Ephydra*.

Key to Adults of the California Species of Setacera

1. Males with tuft of long hairs at apex of at least hind tibia 2
— Males without distal hair tufts on mid- or hind-tibia; Placer County ....*aldrichi* Cresson 1935
2. Males with hair tufts on mid- and hind-tibiae 3
— Males with hair tuft at apex of mid-tibia only; Los Angeles and San Diego counties...*durani* Cresson 1935
3. Middle femur of male with flexor cluster of long hairs on basal 1/3; southern and central California ....*needhami* Johannsen 1935
— Middle femur of male without basal cluster of long hairs; southern and central California ....*pacifica* (Cresson) 1926

Genus *Dimecoenia* Cresson

This is a Neotropical relative of *Ephydra* and very similar to it in the immature stages. Hennig (1943) has described the immature stages of *D. caesia* (van der Wulp) from Chile and *zurcheri* (Hendel) from Paraguay.

Key to Adults of the California Species of Dimecoenia

1. Hind tarsus of male with prominent black hair tufts; southern and central California .... *austrina* (Coquillett) 1900
— Hind tarsus of male without hair tufts; southern California ....*spinosa* (Loew) 1864

Genus *Hydropyrus* Cresson

This genus contains only the species *hians* (Say), found in very alkaline lakes from Nebraska to the West Coast. At Mono Lake, California, this species becomes very abundant, and in historical times the puparia which were washed ashore in enormous windrows were collected and eaten by the Pah-Ute Indians who called this food "koo-chah-bee." Aldrich (1912) has given an excellent and very detailed account of the biology and immature stages of this species.

hians (Say) 1830. Borax Lake, Lake County; Owens Lake, Inyo County; Mono Lake, Mono County; Trona, San Bernardino County

Genus *Ephydra* Fallén

Species of *Ephydra* are highly specialized and exclusively aquatic, being commonly known as "brine flies." Habitats range from fresh-water ponds and lakes to the highly saline and alkaline ponds and sinks of desert and semidesert regions (Aldrich, 1912).

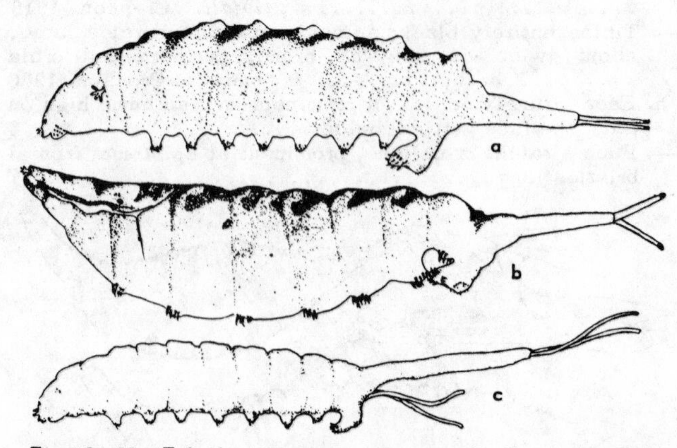

Fig. 14:59. *Ephydra*, immature stages. *a,b*, *E. riparia* Fall.: *a*, larva; *b*, puparium. *c*, *E, cinerea* Jones, larva (Johannsen, 1935).

The commonest and widespread Holarctic species, *riparia* Fallén, breeds in water ranging preferably from fresh to brackish. The variety *macellaria* Egger predominates in the southern part of the range of *riparia*. *E. cinerea* Jones prefers denser waters, up to saturation of salts, and is common in brine ponds of salt works of the West Coast as well as in saline lakes where it becomes extremely abundant. There are several North American species which breed locally in hot springs and geyser effluents. The larvae (figs. 14:57*e*; 14:59*a,c*) of *Ephydra* are found in the masses of blue-green algae on which they feed, and also in the upper layers of the algae-covered bottom ooze in shallow water. The puparia (fig. 14:59*b*) are attached by means of the hooklike last ventral pseudopod of the larva to sticks or debris, when available, but in deeper waters they usually float in large masses on the surface.

Key to Adults of the California Species of Ephydra

1. Pollinose vestiture of the body white or gray 2
— Pollinose vestiture of the body brassy to greenish 3
2. Tibiae dark, contrasting with the bright yellow tarsi; scutellum polished; hind cross vein oblique to the costa; Inyo County .... *auripes* Aldrich 1912
— Tibiae yellowish, concolorous with the tarsi; scutellum pollinose; hind cross vein perpendicular to the costa; southern and central California ..... *cinerea* Jones 1906
3. Legs mostly dark greenish; California ....*riparia* Fallén 1813
— Legs mostly yellowish (pale southern variety) .... *riparia* var. *macellaria* Egger 1862

Genus *Coenia* Robineau-Desvoidy

The larvae of *Coenia fumosa* Stenhammar, a European salt-water species, have been described by Beyer (1939).

turbida Curran 1927. Inyo, San Diego

Key to Adults of the California Species of Paracoenia

1. Acrostichal setulae nonseriate; hind cross vein sinuate; scutellum distinctly convex; central and southern California .... *bisetosa* (Coquillett) 1902
— Acrostichal setulae in 2 distinct series; hind cross vein straight; scutellum flattened; Sierra Nevada .... *platypelta* Cresson 1935

Tribe SCATELLINI

Genus *Scatophila* Becker

Bolwig (1940) has given a fine discussion, including detailed figures of the immature stages of *Scatophila unicornis* Czerny, which he found living on wet stones and wet flowerpots in greenhouses in Denmark. The larvae feed on diatoms and protect themselves by covering their bodies with sand grains and pellets of feces.

Key to Adults of the California Species of Scatophila[8]

1. Marginal cell with 2 or more distinct white spots; male face with a snoutlike protuberance medianly below; Los Angeles and Santa Barbara counties, in greenhouse) .... *unicornis* Czerny 1900
— Marginal cell unspotted, or with a single weak basal white area, or wings very pale and not appearing spotted 2
2. Large quadrate median white spot of submarginal cell with a prominent dark central area; Los Angeles and Riverside counties ....*pulchra* Sturtevant and Wheeler 1954
— Median white area of submarginal cell without a darker center 3
3. The larger facial bristle on outer lower corner turned upward and outward, *Scatella*-like; middle femora of male armed with a row of short spines along inner side; widespread .... *despecta* (Haliday) 1839
— Lower facial not pointed upward and outward; middle femora of male lacking a row of spines 4
4. Body color mostly whitish to grayish white, usually with some darker markings, the disc of scutellum largely to entirely whitish 5
— Body color tan, brown, or black, usually with some lighter markings, the scutellar disc largely dark but often with basal and sometimes with apical gray areas 8
5. Male face flat or concave in middle, with 2nd antennal segment thickly haired along inner side; female face with a pair of diverging setae on upper carina *(disjuncta)* or not *(arenaria)* 6
— Male face prominently convex medianly, the 2nd antennal segment with sparse hairs; female face without a pair of diverging setae on upper carina 7
6. Male face flat to slightly concave medianly but not membranous, the 2-3 stronger hairs on margin at each side rather pale and weak; nearly uniformly whitish species, the abdomen mostly whitish dusted; female face silvery gray, the cheek width greater than width of 3rd antennal segment; Mono county .... *arenaria* Cresson 1935
— Male face concave, membranous medianly behind oral margin, with 2-3 stout black bristles on each side which are stronger than those of *arenaria;* less whitish, especially on abdomen; female face with a pair of diverging setae on upper carina; central California ....*disjuncta* Cresson 1935
7. Male face evenly arched in middle, thickly haired, especially in 2 large lateral clusters in which 3-5 are stronger; wing spots distinct, including apical spot of submarginal cell; female face gray with some brownish discoloration; Kern County to Riverside and Orange counties........*bipiliaris* Sturtevant and Wheeler 1954
— Male face protuberant, the median part forming a broad cone which involves the oral margin, the cone surrounded by clustered hairs except at margin; wing spots rather indistinct; female face silvery gray, the cheek width about equaling width of 3rd antennal segment; Humboldt County .... *conifera* Sturtevant and Wheeler 1954
8. Acrostichal hairs of uniform size, in 2 regular rows reaching nearly to scutellum; tannish species with mesonotal pattern not or badly evident; reclinate orbitals far forward, 3-4 times as far from inner verticals as the latter are from the outer verticals; southern and central California .... *ordinaria* Sturtevant and Wheeler 1954

[8]From Sturtevant and Wheeler, 1954.

— Acrostichals usually irregular in size and position, rarely forming 2 regular rows to scutellum; reclinate orbital not so far forward; mesonotal pattern usually evident 9

9. With a stout pair of presutural acrostichal bristles, as long as the ocellars or nearly so; small dark species with gray mesonotal pattern, highly spotted wings, black legs, strong head bristles; faces of the 2 sexes similar; Los Angeles and Santa Barbara counties ....*exilis* Cresson 1935

— Without presutural acrostichal bristles though 1-2 pairs of hairs in this region are often enlarged; faces of the sexes unlike, that of the male concave, flattened, protuberant, incised, tuberculate, or otherwise modified from the female type 10

10. Abdomen black, noticeably shining, especially on apical tergites, the basal tergites with at most thin pollen.. 11

— Abdomen brown to black, rather heavily pollinose, rarely a bit glossy but never appearing truly shining 12

11. Cheek about as wide as width of 3rd antennal segment; frontal triangle scarcely differentiated in color from the anterior nonorbital areas; mesonotum usually with an olive tint; palpi pale, tarsi usually partly pale; Los Angeles and San Francisco counties ....*viridella* Sturtevant and Wheeler 1954

— Cheeks narrower than width of 3rd antennal segment; frontal triangle and orbits dull brown, leaving 2 black triangular areas anteriorly on front; palpi and tarsi mostly dark; male face sunken in middle, usually membranous to a noticeable degree, the oral margin protuberant; southern California ....*rubribrunnea* Sturtevant and Wheeler 1954

12. Wing base yellow, the blade with conspicuous pattern; scutellum with distinct basal gray triangle; male face flat to concave, the margin protruding, golden yellow; Los Angeles County ....*hesperia* Sturtevant and Wheeler 1954

— Wing base not strongly yellowed; scutellum with indistinct basal gray area or none; face more brown than yellow, that of male protruding at margin, with an elongate median depression bordered (especially anteriorly) by dense curly hairs; female face with some glossiness; Alameda County ....*variofacialis* Sturtevant and Wheeler 1954

Key to Adults of the California Species of Limnellia

1. First posterior cell with 3, small, round, isolated, dark spots not contiguous with 3rd vein and similar to the 2 spots in submarginal cell; central California .... *sejuncta* (Loew) 1863

— First posterior cell with 3, prominent, quadrate, dark spots contiguous on anterior side with 3rd vein; submarginal cell usually without small, round, isolated, dark spots; central California ...*quadrata* (Fallén) 1813

Genus *Scatella* Robineau-Desvoidy

The biology and immature stages of *S. thermarum* Collin, a species very closely related to our *stagnalis* (Fallén) (fig. 14:54*a-c*), have been well described by Tuxen (1944), who found it breeding in large numbers in the hot springs of Iceland. Brauns (1939) and Hennig (1943) have given notes on the European species *subguttata* (Meigen). The adults feed on algae at the water surface, rasping off particles with their proboscces. The larvae are strictly aquatic, crawling through and feeding on the growths of algae and diatoms of their preferred habitats near the surface. The boat-shaped puparia (fig. 14:54*b*) are found in the same places as the larvae.

Key to Adults of the California Species of Scatella

1. Face silvery white; legs, pleura and sides of mesonotum and abdomen silvery gray; wings hyaline, markings very faint; California ......*paludum* (Meigen) 1830

— Face gray or brown; legs, pleura, sides of mesonotum and abdomen dark brown to blackish............. 2

2. Face gray 3

— Face brown 4

3. Scutellum with strong basal bristle in addition to the usual weak downcurved seta near apex; California ....*troi* Cresson 1933

— Scutellum without strong basal bristle; California ....*laxa* Cresson 1933

4. Body entirely dull brownish pollinose; wing spots indistinct; male with basal section of costa swollen about 4 times normal thickness; Los Angeles County ....*obsoleta* Loew 1861

— Mesofrons and mesonotum usually subshining; wing spots distinct; male costa normal................. 5

5. Face tending vertical, the hump very prominent; light spot in submarginal cell very prominent, quadrate, entirely filling cell; size larger; California ....*stagnalis* var. *pentastigma* (Thomson) 1868

— Face more sloping toward epistoma; the hump smaller; light spot in submarginal cell rounded and not filling cell; smaller species; California .... *stagnalis* (Fallén) 1813

Genus *Neoscatella* Malloch

(Figs. 14:54*g*; 14:56*g,k,l*)

The biology and immature stages of several Hawaiian species of *Neoscatella*, which have been admirably described and figured by Williams (1938), are essentially the same as those of *Scatella*.

setosa (Coquillett) 1900. California

Key to Adults of the California Species of Parascatella

1. Mesonotum and scutellum dull; southern California ....*triseta* (Coquillett) 1902

— Mesonotum and scutellum shining 2

2. Face white, with dark spot on the median hump; wing spots indistinct; intermediate frontoörbital seta at least 1/2 as long as the 2 frontoörbital bristles; central California .... *marinensis* Cresson 1935

— Face entirely cinereous to brown; wing spots rather distinct; intermediate frontoörbital seta minute; coastal transition of California........*melanderi* Cresson 1935

Key to Adults of the California Species of Lamproscatella

1. Face and pleura white 2

— Face and pleura brownish....................... 4

2. Disc of mesonotum silvery gray pollinose 3
— Disc of mesonotum brownish 4
3. Mesofrons shining; Inyo and San Bernardino counties .... *salinaria* Sturtevant and Wheeler 1954
— Mesofrons dull; Great Basin of California .... *nivosa* Cresson 1935
4. Bristles fine, acrostichal setulae minute; California .... *dichaeta* (Loew) 1860
— Bristles stout, acrostichal setulae strong; Inyo and Lassen counties .... *cephalotes* Cresson 1935
5. Mesofrons shining; California seacoast ....*quadrisetosa* (Becker) 1896
— Mesofrons dull; California..... *sibilans* (Haliday) 1833

REFERENCES

ALDRICH, J. M.
1912. The biology of some western species of the dipterous genus *Ephydra*. Jour. N.Y. Ent. Soc., 20:77-99.

BERG, C. O.
1950. *Hydrellia* (Ephydridae) and some other acalyptrate Diptera reared from *Potamogeton*. Ann. Ent. Soc. Amer., 43:374-398.

BEYER, A.
1939. Morphologische, Ökologische und physiologische Studien an den Larven der Fliegen: *Ephydra riparia* Fallén, *E. micans* Haliday und *Cänia fumosa* Stenhammar. Kieler Meereforschungen, 3:265-320.

BOHART, G. E., and J. E. GRESSITT
1951. Filth-inhabiting flies of Guam. B. P. Bishop Mus. Bull. 204, 152 pp., 17 pls.

BOLWIG, N.
1940. The description of *Scatophila unicornis* Czerny 1900 (Ephydridae, Diptera). Proc. Roy. Ent. Soc. London (B), 9:129-137.

BRAUNS, A.
1939. Zur Biologie der meerestrandfliege *Scatella subguttata* Meig. (Familie Ephydridae: Diptera). Zool. Anz., 126:273-285.

CRAWFORD, D. O.
1912. The petroleum fly in California, *Psilopa petrolei* Coq. Pomona Coll. Jour. Ent., 4:687-697.

CRESSON, E. T., JR.
Synopses of North American Ephydridae.
1942. (Diptera). I. Psilopinae. Trans. Amer. Ent. Soc., 68:101-128;
1944. II. Hydrelliini, Hydrinini and Ilytheini. 70:159-180;
1946. III. Notiphilini. 72:227-240;
1949. IV. Napaeinae. 74:225-260.

GRÜNBERG, K.
1910. Diptera, *In:* Brauer, A. Die Süsswasserfauna Deutschlands, Heft 2 a, Jena.

HENNIG, W.
1943. Übersicht über die bisher bekannten Metamorphosesstadien der Ephydriden. Neubeschreibungen nach dem Material der Deutschen Limnologischen Sundaexpedition (Diptera: Ephydridae). Arb. Morph. Tax. Ent. Berlin-Dahlem, 10:105-138.

JOHANNSEN, O. A.
135. *See* Diptera references.

NIELSEN, P., O. RINGDAHL, and S. L. TUXEN
1954. The zoology of Iceland. Vol. III, part 48*a*, Diptera 1:1-189.

THORPE, W. H.
1931. The biology of the petroleum fly. Science, 73: 101-103.

TUXEN, S. L.
1944. The hot springs, their animal communities and their zoogeographical significance. The Zoology of Iceland. 1(11):1-206, 7 pls.

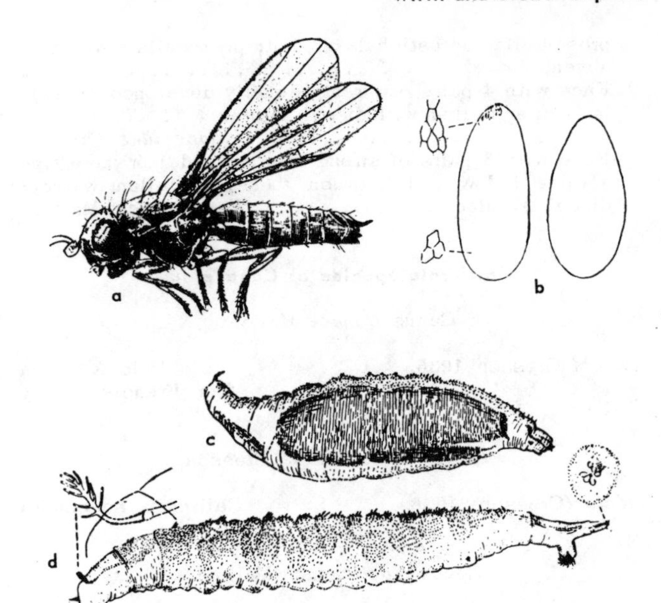

Fig. 14:60. Canaceidae. *a, Procanace nigroviridis* Cress., adult female; *b-d, Canaceoides nudata* (Cress.): *b*, egg; *c*, puparium; *d*, larva (Williams, 1938).

WILLIAMS, F. X.
1938. Biological studies in water-loving insects. Part III. Diptera or flies. A. Ephydridae and Anthomyiidae. Proc. Hawaii. Ent. Soc., 10:85-119.

Family CANACEIDAE

This small family of "beach flies" is closely related in appearance and habits to the preceding family, the Ephydridae. However, canaceids are almost exclusively intertidal in habit, only two fresh-water species having developed in the torrential mountain streams of Hawaii and the East Indies (Wirth, 1951). The adults (fig. 14:60*a*) are quite ephydridlike, especially in the structure of the head, with the large mouth and proboscis and exposed clypeus, but with important differences in wing venation and abdominal structure. The larvae (fig. 14:60*d*) differ from the ephydrids principally in the unbranched posterior air tube with three openings to each posterior spiracle and the presence of one well-developed, spinose, anal proleg. Williams (1938) has given us practically the only detailed description of the immature stages and biology of the Canaceidae in an important, well-illustrated paper on the Hawaiian fresh-water *Procanace nigroviridis* Cresson and on *Canaceoides nudata* (Cresson) (fig. 14:69*b-d*), which also occurs along the United States Pacific Coast.

Key to the North American Genera of Canaceidae

Adults

1. Mesofrons with at least 2 pairs of long interfrontal bristles; prescutellar acrostichals present; postocellars strong .... *Canace* Haliday
— Mesofrons with 1 pair of bristles, just outside ocellars;

prescutellar acrostichals absent; postocellars weak or absent 2
2. Face with 4 pairs of strong, equally developed bristles in line; scutellum with discal setulae ....
.... *Canaceoides* Cresson
— Face with 3 pairs of strong bristles and 1 or more fine setulae below and between these; scutellum without discal setulae .... *Nocticanace* Malloch

California Species of Canaceidae

Genus *Canace* Haliday

aldrichi Cresson 1936. Palo Alto, on San Francisco Bay

Genus *Canaceoides* Cresson

nudata (Cresson) 1926. California seacoasts

Genus *Nocticanace* Malloch

arnaudi Wirth 1954. Marin to Orange counties

REFERENCES

WHEELER, M. R.
1952. The dipterous family Canaceidae in the United States. Ent. News, 63:89-94.
WILLIAMS, F. X.
1938. Biological studies in water-loving insects. Part III. Diptera or flies. A. Ephydridae and Anthomyiidae. Proc. Hawaii. Ent. Soc., 10:85-119.
WIRTH, W. W.
1951. A revision of the dipterous family Canaceidae. Occ. Paps. B. P. Bishop Mus., 20:245-275.
1954. A new intertidal fly from California, with notes on the genus *Nocticanace* Malloch (Diptera; Canaceidae). Pan-Pac. Ent., 30:59-62.

Family SCIOMYZIDAE (=TETANOCERIDAE)

Adult "marsh flies" are frequenters of stream banks, pond shores, and marshy swales. They are easily recognized by their characteristic head shape with porrect antennae and bright colors, often with pictured wings. The known larvae (fig. 14:61*a,c*) are slender and more or less cylindrical, tapering toward the ends; the posterior end has a short, tapered air tube bearing the posterior spiracles surrounded by a number of oval or triangular lobes. The body segments are each provided with a ring of short rounded warts or tubercles. Anterior respiratory organs are small and retractile, with few branches; the posterior spiracles each have three spiracular openings and are usually provided with four palmate tufts of float hairs (fig. 14:61*b*). The puparia are yellowish to brownish, short oval, and usually curved upward at one or both ends, with the caudal lobes of the larva much shrunken in the puparium.

Berg (1953) recently presented convincing evidence that the larvae of Sciomyzidae feed only as predators on snails. Larvae of ten genera have been reared from snails, mostly aquatic species. The puparia found in these snails show extraordinary adaptation in form to fit the curvature of the host shell. Cresson (1920) and Melander (1920) have presented concurrent revisions of the American species of the family; the following keys are taken largely from the latter. The family is presently being studied intensively by Steyskal (taxonomy) and Berg (biology).

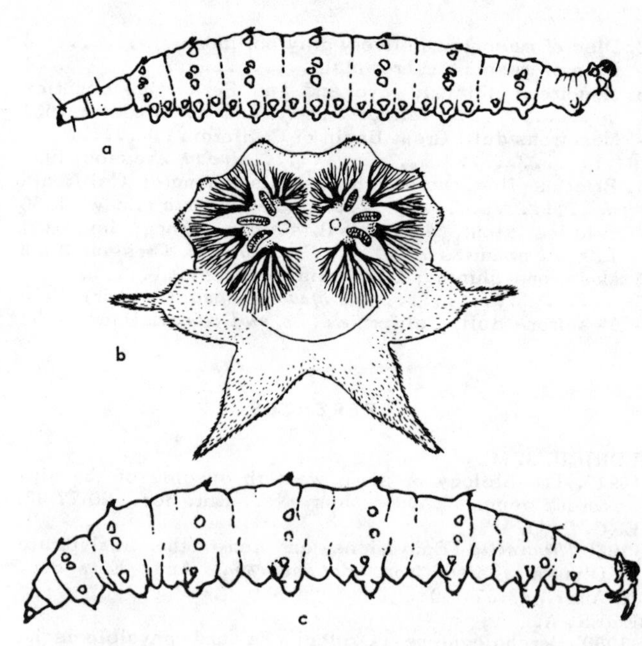

Fig. 14:61. Sciomyzidae. *a*, *Dictya*, larva, lateral view; *b,c*, *Sepedon*, larva: *b*, caudal view of spiracular plate; *c*, lateral view of larva (Peterson, 1951).

Keys to the North American Genera of Sciomyzidae

Adults

1. Propleural bristle present; 2nd antennal segment short (SCIOMYZINAE) 2
— Propleural bristle absent; 2nd antennal segment elongate (TETANOCERINAE) 5
2. Anterior tibia with 2 approximate preapical bristles dorsally 3
— Anterior tibia with 1 dorsal preapical bristle 4
3. Arista densely short white haired; not California
.... *Oidematops* Cresson
— Arista with long, black rays; not California ....
.... *Sciomyza* Fallén
4. Front polished black; cheeks narrow; not California
.... *Pteromicra* Lioy
— Front dull; cheeks moderately broad ....
.... *Pherbellia* Robineau-Desvoidy
5. Scutellum with 1 pair of bristles 6
— Scutellum with 2 pairs of bristles 7
6. Second antennal segment longer than the 3rd ....
.... *Sepedon* Latreille
— Second antennal segment shorter than the 3rd; not California .... *Hemitelopteryx* Cresson
7. Third antennal segment oval, 3 times as long as the 2nd; front distinctly narrowed anteriorly 8
— Third antennal segment rarely oval, usually flattened or concave above, the 2nd segment at least 1/2 as long as the 3rd; sides of front nearly parallel....... 9

8. Posterior tibiae with one preapical dorsal bristle; not California .... *Renocera* Hendel
— Posterior tibiae with 2 preapical dorsal bristles .... *Antichaeta* Haliday
9. Mesopleura and pteropleura with 1 or more bristles 10
— Mesopleura and pteropleura at most with short hairs 11
10. One sternopleural bristle; 2 frontoörbitals .... *Hoplodictya* Cresson
— No sternopleural bristle; 1 frontoörbital. . *Dictya* Meigen
11. Arista almost bare; hind cross vein strongly bent, S-like 12
— Arista pubescent or plumose; hind cross vein sinuous or arcuate 13
12. One pair of dorsocentral bristles; no bristles on callosity beneath calypters; fore femur with weak bristles above; not California............. *Hedroneura* Hendel
— Two pairs of dorsocentrals; bristles present on callosity beneath calypters; fore femur bristly above; not California .... *Elgiva* Meigen
13. Two pairs of dorsocentral bristles; edges of 2nd antennal segment nearly parallel; 1st vein ending almost opposite anterior cross vein..................... 14
— Three pairs of dorsocentrals; 2nd antennal segment obconical; 1st vein ending far before anterior cross vein; not California.......... *Poecilographa* Melander
14. Mesopleura, pteropleura, and callosity beneath calypters bare; only the sternopleura setulose 15
— Mesopleura, pteropleura, sternopleura, and callosity beneath calypters setulose 17
15. Lunule exposed; wings brown, with rounded clear spots 16
— Lunule more or less covered; wings nearly uniformly colored, the cross veins clouded, sometimes with short transverse marks, but no round clear spots ....*Tetanocera* Latreille
16. Second antennal segment slender, much longer than the 3rd; not California ....*Dictyomyia* Cresson
— Second antennal segment quadrate and broad, scarcely longer than the 3rd; not California ..*Euthycera* Latreille
17. Arista loosely black-plumose; interfrontal depression not polished; not California ......*Trypetoptera* Hendel
— Arista closely white-pubescent or short-plumose; interfrontal stripe shining ...... *Limnia* Robineau-Desvoidy

Known Larvae and Puparia

1. Apex of caudal respiratory organ with well-developed palmate hairs (fig. 14:61*b*) 2
— Apex of caudal respiratory organ without well-developed palmate hairs; caudal end of puparium scarcely curved upward; small species ....*Poecilographa* Melander
2. Anterior respiratory organs of larva each with 5 or 6 openings; caudal spiracular disc with 3 pairs of small lobes above and 2 larger pairs below; caudal end of robust puparium sharply curved upward ....*Sepedon* Latreille
— Anterior respiratory organs each with about 8 openings; caudal spiracular disc otherwise; puparium more slender 3
3. Caudal spiracular disc with 8 triangular lobes; anal plate about twice as wide as long; caudal end of puparium short .... *Tetanocera* Latreille
— Caudal spiracular disc with 8 broadly rounded lobes, the ventral lobes larger; anal plate about as wide as long; caudal end of puparium elongate... *Dictya* Meigen

California Species of Sciomyzidae

Subfamily SCIOMYZINAE

Genus *Pherbellia* Robineau-Desvoidy

(= *Melina* Robineau-Desvoidy)

fuscipes (Macquart) 1835.	California
grisescens (Meigen) 1830.	Los Angeles, San Diego, and San Mateo
nana (Fallén) 1820.	Central and southern California
obtusa (Fallén) 1820.	California
pubera (Loew) 1862.	Alameda, Santa Clara
vitalis (Cresson) 1930.	Alameda, Marin

Subfamily TETANOCERINAE

Genus *Sepedon* Latreille

The larvae and puparia of the eastern species, *S. fuscipennis* Loew, have been described and figured by Needham (Needham and Betten, 1901) and Johannsen (1935), and the puparium of *S. armipes* Loew, which also occurs in California, was described by the latter author. Puparia of both species were taken at pond margins under overhanging vegetation. Peterson (1951) has also figured the larva of *Sepedon* (fig. 14:61*a*, *b*). The adult classification of the American species of this genus has been recently revised by Steyskal (1950) from which the following key is adapted.

Key to Adults of the California Species of Sepedon

1. Robust species about 8 mm. long; 2nd antennal segment compressed, less than 3 times as long as wide (widespread in California; syn.: *pacifica* Cresson) ....*praemiosa* Giglio-Tos 1893
— Smaller species, usually less than 6 mm. long; 2nd antennal segment slender, not compressed 2
2. Hind femur of male with a constriction near middle of ventral surface, basad of which there are 2 processes, 1 of which may be bifid; hind tibia distinctly more curved in distal 3rd; abdomen brown with little more than a trace of bluish reflection; oral margin usually raised, rectangular in profile 3
— Hind femur simple in both sexes; hind tibia evenly arcuate; abdomen frequently almost black with bluish reflections; oral margin usually low, the angle with face acute; Shasta County ...... *borealis* Steyskal 1950
3. Frons with no more than a trace of parafrontal black spots; basimedian femoral prong bifid; southern California ....*bifida* Steyskal 1950
— Frons with distinct velvety black, parafrontal spots; basimedian femoral prong not bifid; Modoc and Mono counties ....*armipes* Loew 1859

Genus *Antichaeta* Haliday

robiginosa Melander 1920.	Mono County

Genus *Hoplodictya* Cresson

spinicornis (Loew) 1865.	Los Angeles, Santa Clara, Alpine

Genus *Dictya* Meigen

The immature stages of *Dictya pictipes* Loew have been described and figured by Needham (in Needham and Betten, 1901). They were taken under floating vegetation at the edge of a pond in New York. Peterson (1951) also figured the larva of *Dictya* (fig. 14:61*c*). The species can be separated only on the basis of the male genitalia.

lobifera Curran 1932. — Yosemite
umbrarum (Linnaeus) 1758. — Central California
umbroides Curran 1932. — Central California

Genus *Poecilographa* Melander

Johannsen (1935) has described the puparium of the only American species, *P. decora* (Loew), which was found in a bog in the woods.

Genus *Tetanocera* Latreille

Johannsen (1935) has published a description of the immature stages of *T. ferruginea* Fallén. He took puparia, from which this species was reared, floating on the water surface under overhanging vegetation on a pond margin. Nielsen *et al.* (1954) figured the larvae of *S. robusta* Loew from aquatic environments in Iceland.

California species:

obtusifibula Melander 1920. — Santa Clara County
plebeia Loew 1862. — Humboldt County, Lake Tahoe
vicina Macquart 1843. — Widespread in California

Genus *Limnia* Robineau-Desvoidy

pubescens (Day) 1881. — Shasta County
saratogensis saratogensis (Fitch) 1856. — El Dorado, Lake
saratogensis severa Cresson 1920. — Shasta County

REFERENCES

BERG, C. O.
1953. Sciomyzid larvae (Diptera) that feed on snails. Jour. Parasit., 39:630-636.

CRESSON, E. T., JR.
1920. A revision of the Nearctic Sciomyzidae (Diptera, Acalyptratae). Trans. Amer. Ent. Soc., 46:27-89, 3 pls.

JOHANNSEN, O. A.
1935. *See* Diptera references.

MELANDER, A. L.
1920. Review of the Nearctic Tetanoceridae. Ann. Ent. Soc. Amer., 13:305-322, 1 pl.

NIELSEN, P., O. RINGDAHL, and S. J. TUXEN
1954. The zoology of Iceland. Vol. III, part 48*a*. Diptera, 1:1-189.

PETERSON, A.
1951. *See* Diptera references.

STEYSKAL, G. C.
1950. The genus *Sepedon* Latreille in the Americas (Diptera: Sciomyzidae). Wasmann Jour. Biol., 8: 271-297.

Family SCOPEUMATIDAE

(=CORDILURIDAE or SCATOPHAGIDAE)

Most of the Scopeumatidae or "dung flies" live in dung and decaying vegetation, although a few species are known to be leaf miners and stem borers. At least two genera have been taken boring in underwater stems of aquatic plants and should be considered here. Johannsen (1935) records a third genus, *Acanthocnema,* which was taken in a tent trap over a brook, and it may also be aquatic.

The larvae (fig. 14:62*c*) are whitish and fusiform, largest in diameter before the middle and gradually tapering caudad, the last segment somewhat pointed; the mouth hooks are blunt, broad at the base, and not serrate (fig. 14:62*b*); the anterior respiratory organs (fig. 14:62*e*) are cribriform or fan-shaped and each bears about 35-40 small openings. The puparium is rather straight dorsally, with the pointed caudal end bent upward, the ventral side arched; the integument is longitudinally and transversely wrinkled, with a number of conical papillae between the spiracular disc and the anal plate.

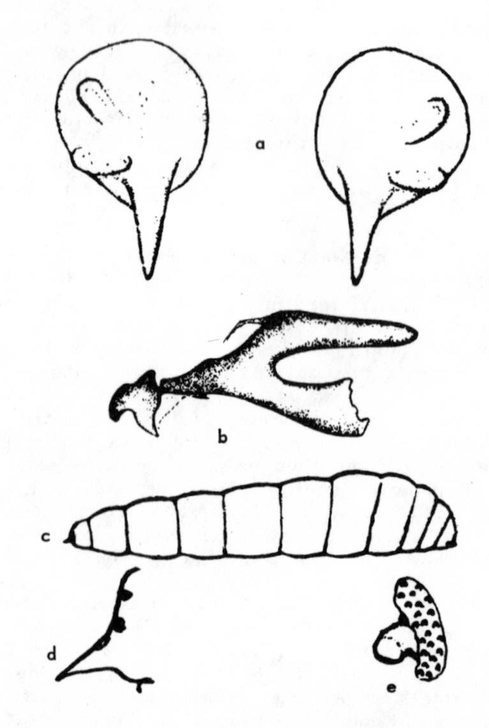

Fig. 14:62. *Hydromyza confluens* Loew, larva. *a*, caudal view of spiracular plates; *b*, pharyngeal skeleton; *c*, lateral view; *d*, spiracular plate, lateral view; *e*, anterior spiracle (*a,b*, Johannsen, 1935; *c-e*, Needham, 1907).

Keys to the Aquatic Genera of North American Scopeumatidae

Adults

1. Frontal bristles or hair long 2
— Frontal bristles extremely short, the front nearly bare .... *Hydromyza* Fallén
2. Front tibia without a short, rectangular apical spine below .... *Cordilura* Fallén
— Front tibia with a short, stout, rectangular spine at apex of ventral surface......... *Acanthocnema* Becker

Larvae and Puparia

1. Pharyngeal skeleton with dorsal processes tapering and extending further caudad than the ventral processes (fig. 14:62*b*); posterior spiracular discs brown, separated from each other by a distance of less than their diameter, the lower spiracular process greatly elongated (fig. 14:62*a,d*) .... *Hydromyza* Fallén
— Pharyngeal skeleton with dorsal processes stout and extending no farther caudad than the ventral processes; posterior spiracular discs yellow, separated from each other by a distance equal to or greater than their diameter, the 3 spiracular processes of each disc subequal and digitiform .... *Cordilura* Fallén

California Species of Aquatic Scopeumatidae

Genus *Hydromyza* Fallén

There is but one North American species, *H. confluens* Loew, which is northeastern and does not occur in California. Studies by Needham (1908) and Welch (1914, 1917) show that this species normally mines the petioles of the yellow water lily in New York and Michigan. Berg (1950) has also reared it from the roots of *Potamogeton* in Michigan.

Genus *Cordilura* Fallén

Frohne (1939) has published very interesting details of the life history of *Cordilura latifrons* Loew in Michigan. This species is a common stem borer of bulrushes (*Scirpus* spp.), attacking the bottom 15 cm. of the culms.

California species:

adrogans Cresson 1918.	Sonoma County
amans Cresson 1918.	Marin County
beringensis Malloch 1923.	Del Norte, Humboldt
luteola Malloch 1924.	Humboldt County
masonina Curran 1931.	Fresno, Inyo

REFERENCES

BERG, C. O.
1950. *Hyurellia* (Ephydridae) and some other acalyptrate Diptera reared from *Potamogeton*. Ann. Ent. Soc. Amer., 43:374-398, 4 pls.
FROHNE, W. C.
1939. Biology of certain subaquatic flies reared from emergent water plants. Pap. Mich. Acad. Sci., Arts and Lttrs., 24:139-147.
JOHANNSEN, O. A.
1935. *See* Diptera references.
NEEDHAM, J. G.
1908. Notes on the aquatic insects of Walnut Lake. *In* A biological survey of Walnut Lake, Michigan, by Thomas C. Hankinson. Mich. Geol. Surv. Rep. 1907: 252-271.
WELCH, P. S.
1914. Observations on the life history and habits of *Hydromyza confluens* Loew (Diptera). Ann. Ent. Soc. Amer., 7:135-147.
1917. Further studies on *Hydromyza confluens* Loew, (Diptera). Ann. Ent. Soc. Amer., 10:35-45, 1 pl.

Family MUSCIDAE (including ANTHOMYIIDAE)

This family contains such well-known pests as the house fly, stable fly, horn fly, and the root maggots whose larvae are saprophagous or phytophagous. The aquatic muscids are predaceous and fall within scattered groups of that section of the family which has been known as the Anthomyiidae. The muscids may be separated from the other suborders and families of aquatic Diptera by the presence of a laterodorsal longitudinal seam on the second antennal segment, the thorax normally with a complete transverse suture on the dorsum, the lower calypter usually large, the postscutellum not prominent, and the hypopleura bare or with only weak hair. The larvae are usually smooth, tapering, and maggotlike (fig. 14:63*d,e,k,m*); the head possesses an internal pharyngeal skeleton (fig. 14:63*f,l*) with well-developed pharyngeal sclerite, short robust hypostomal sclerite, and long slender mouth hooks often appressed together, the posterior spiracles flush with the body surface or on short elevations, the spiracular plates with well-developed button and peritreme, and the slits short and radially arranged (fig. 14:63*u*). The puparia (fig. 14:63*c,h,j,q,s*) are usually ovoid and reddish brown in color.

One North American genus which does not occur in California has known aquatic larvae; *Mydaeina obscura* Malloch from the Canadian Arctic was collected in ponds where it is truly aquatic. Malloch (1919) described and figured the larva and puparium (fig. 14:63*q,r,t*); the latter is very slender with a remarkable subapical constriction forming a caudal float chamber to keep the posterior spiracles in contact with the water surface (fig. 14:63*q*). Other aquatic genera occurring in California are described below.

Key to the Aquatic Genera of North American Muscidae

Adults

1. Anal (6th) vein extending to the wing margin .... *Hydrophoria* Robineau-Desvoidy
— Anal (6th) vein not extending to the wing margin 2
2. Palpi distinctly dilated at apices; parafacies and pteropleura haired .... *Lispe* Latreille

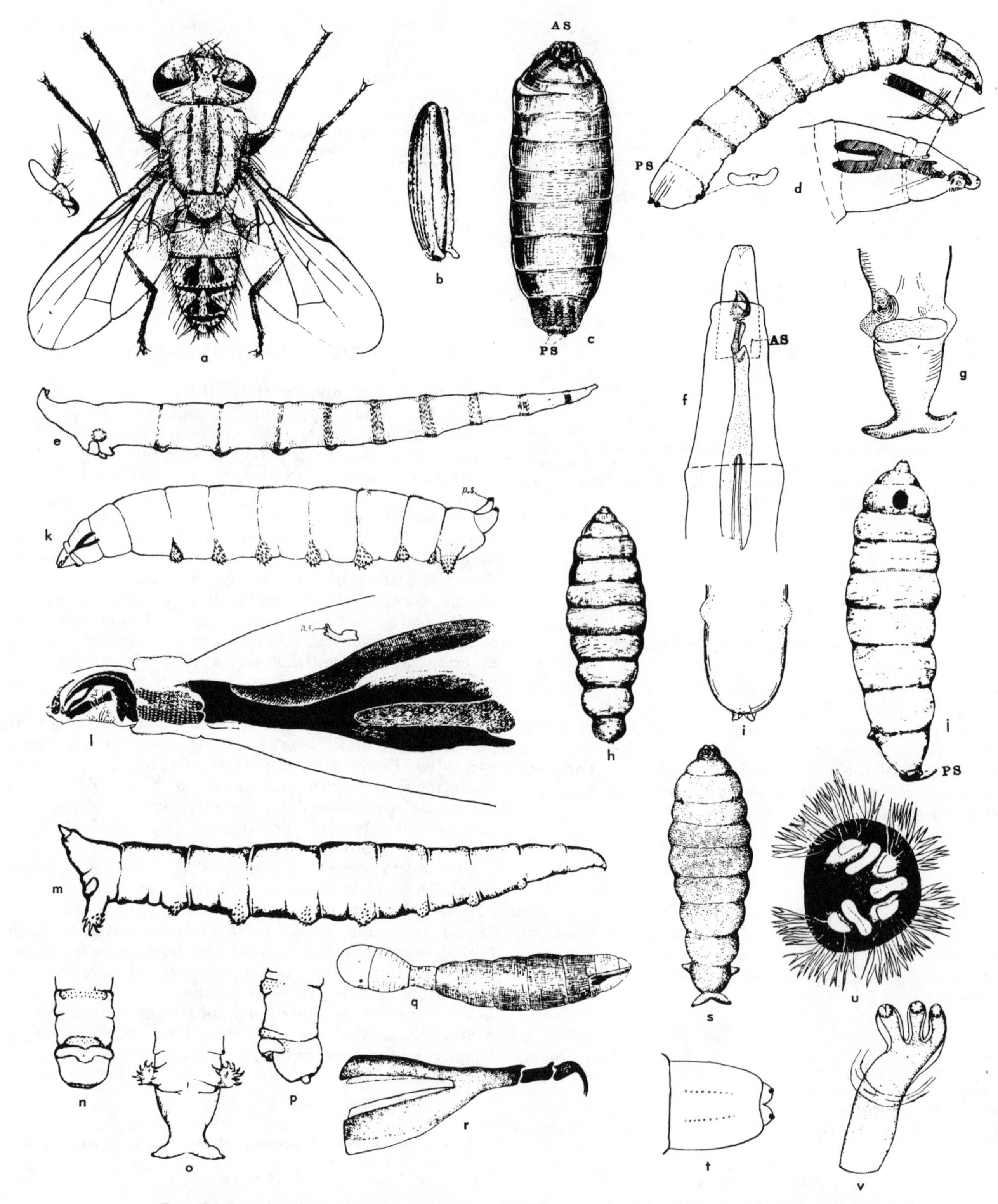

Fig. 14:63. Muscidae. *a-d, Lispe metatarsalis* Thoms.: *a,* adult male; dorsal view, antenna shown separately; *b,* egg; *c,* puparium; *d,* larva. *e-g,j, Lispocephala fusca* Mall., larva: e, lateral view; *f,* pharyngeal skeleton; *g,* posterior segments, ventral view, *j,* puparium. *h,i, L. kaalae* Williams: *h,* puparium; *i,* larva, dorsal view of posterior segments. *k,l,s,u,v, Limnophora exsurda* Pand., larva: *k,* lateral view; *l,* pharyngeal skeleton; *u,* posterior spiracle; *v,* anterior spiracle, *s,* puparium. *m,o, Limnophora aequifrons* Stein, larva: *m,* lateral view; o, posterior segments, ventral view. *n,p, L. discreta* Stein, larva: *n,* ventral view; *p,* lateral view of posterior segments. *q,r,t, Mydaeina obscura* Mall.: Puparium, *q.* Larva: *r,* pharyngeal skeleton; *t,* posterior segments (*a-j,* Williams, 1938; *k,l,s,u,v,* Tate, 1939; *m-p,* Johannsen, 1935; *q,r,t,* Malloch, 1919).

— Palpi not dilated at apices; parafacies and pteropleura bare 3
3. Sternopleural bristles 3 in number, situated in nearly an equilateral triangle .... *Lispocephala* Pokorny
— Sternopleurals not forming a nearly equilateral triangle, if only 3 are present the lower 1 is decidedly farther from the anterior 4
4. Pre-alar bristle present ... *Phaonia* Robineau-Desvoidy
— Pre-alar bristle lacking ....
.... *Limnophora* Robineau-Desvoidy

California Species of Aquatic Muscidae

Subfamily ANTHOMYIINAE

Genus *Hydrophoria* Robineau-Desvoidy

Malloch (1919) states that the larva of *H. arctica* Malloch from the Canadian Arctic is aquatic.

California species:

divisa (Meigen) 1826.	Del Norte, Siskiyou
seticauda Malloch 1919.	Lake Tahoe and San Bernardino Mountains

Subfamily LISPINAE

Genus *Lispe* Latreille (=*Lispa* auct.)

Johannsen (1935) has described the puparium of *Lispe uliginosa* Fallén which was found in the muck of a pond in New York. Williams (1938) described and figured all stages of the Hawaiian *L. metatarsalis* Thomson (fig. 14:63*a-d*), which breeds at the margins of ponds and sluggish streams and is predaceous on midges and other small arthropods. Vaillant (1953) admirably described the immature stages of *L. consanguinea* Loew from North Africa. The larva (fig. 14:63*d*) is rather stout for a muscid, with three-lobed anterior spiracles and blunt posterior end bearing the short, blackish, posterior spiracular tubercles.

California species:

nasoni Stein 1897.	Southern and central California
polita Coquillett 1904.	Lake Tahoe
probohemica Speiser 1914.	Lake Elsinore
salina Aldrich 1913.	Lake County
sordida Aldrich 1913.	Lake County
tentaculata (Degeer) 1776.	Widespread in California

Subfamily COENOSIINAE

Genus *Lispocephala* Pokorny

We are again indebted to Williams (1938) for an excellent account of the habits and immature stages of *Lispocephala,* this time for two of the numerous Hawaiian species, *L. fusca* Malloch (fig. 14:63*e-g, j*) and *L. kaalae* Williams (fig. 14:63*h,i*). These species usually frequent the margins of swifter mountain streams, preying on smaller flies. The larvae (fig. 14:63*e*) are quite slender, with the last body segment tapering to a pair of processes bearing the minute posterior spiracles. These processes are quite long and conical in *fusca* (fig. 14:63*g*) but are short and inconspicuous in *kaalae* (fig. 14:63*i*).

California species:

alma (Meigen) 1826.	Los Angeles County
erythrocera (Robineau-Desvoidy) 1830.	Santa Clara, Plumas, Lake Tahoe
setipes Malloch 1935.	Lake Tahoe

Subfamily PHAONIINAE

Genus *Phaonia* Robineau-Desvoidy

The immature stages and habits of *P. mirabilis* Ringdahl, which breeds in tree holes in Europe where it preys on mosquito larvae, have been discussed by Keilin (1917) and Tate (1935). The larvae apparently molt within the egg and are hatched with the third instar structures well developed, probably as an adaptation of its carnivorous habits. The main tracheae of the larva possess several large bladderlike swellings which probably serve as air reservoirs when the larvae are completely submerged.

California species:

caerulescens (Stein) 1897.	Central and southern California
errans (Meigen) 1826. var. *proxima* (van der Wulp) 1869.	Santa Clara County
fuscicauda Malloch 1918.	Alameda County
limbinervis Stein 1918.	Santa Clara County
luteva (Walker) 1849. [= *varipes* (Coquillett) 1902].	Santa Clara County
nigricauda Malloch 1918.	Humboldt, Santa Cruz
pallida (Stein) 1918.	Mono, Los Angeles
parviceps Malloch 1918. [= *caesia* Stein 1920].	Central and southern California
quieta Stein 1918.	"California"

Genus *Limnophora* Robineau-Desvoidy

To the descriptions of the habits and immature stages of *L. discreta* Stein, presented by Marchand (1923), Johannsen (1935) added notes for that species (fig. 14:63*n,p*) and also *aequifrons* Stein (fig. 14:63*m,o*) and *torreyae* Johannsen, all collected in New York. The larvae (fig. 14:63*k,m*) are rather stout, but with the anterior end tapering to a slender point and the last segment short tapering. There are specific differences in the relative development of the diverging posterior spiracular processes and the ventral creeping welts or prolegs (fig. 14:63*k,m,p*). The puparium (fig. 14:63*s*) is distinctive, with a dark anterior knob formed by the head and first two thoracic segments, the remaining segments forming definite annular constrictions and the larval hind prolegs and posterior respiratory processes remaining as conical lobes. Tate's (1939) account of the early stages of the European *L. exsurda* Pandellé contains superb figures (fig. 14:63*k,l,s,u,v*).

California species:

acuticornis Malloch 1920.	San Bernardino Mountains
aequifrons Stein 1897.	Widespread in California
arcuata Stein 1897.	Central and southern California
bisetosa var. *pruinella* Huckett 1932.	San Bernardino Mountains
discreta Stein 1897.	Los Angeles, Plumas, Lake Tahoe
femorata (Malloch) 1913.	Los Angeles County
magnipunctata Malloch 1919.	Fresno, Tulare
narona (Walker) 1849.	Widespread in California
surda (Zetterstedt) 1845.	Marin County

REFERENCES

ALDRICH, J. M.
1913. The North American species of *Lispa.* Jour. N.Y. Ent. Soc., 21:131-146.
HUCKETT, H. C.
1932. The North American species of the genus *Limnophora,* with descriptions of new species. Jour. N.Y. Ent. Soc., 40:25-76.
1944. A revision of the North American species belonging to the genus *Hydrophoria* Robineau-Desvoidy (Diptera: Muscidae). Ann. Ent. Soc. Amer., 37:261-297, 6 pls.
JOHANNSEN, O. A.
1935. *See* Diptera references.
KEILIN, D.
1917. Recherces sur les anthomyides à larves carnivore. Parasitology, 9:325-450.
MALLOCH, J. R.
1923. Flies of the anthomyiid genus *Phaonia* and related genera known to occur in North America. Trans. Amer. Ent. Soc., 48:227-282, 3 pls.
1919. The Diptera collected by the Canadian Expedition, 1913-1918. (Excluding the Tipulidae and Culicidae). Rep. Canad. Arctic Exped. 1913-1918. 3:34*c*-90*c*, 3 pls.
MARCHAND, W.
1923. The larval stages of *Limnophora discreta* Stein (Diptera, Anthomyidae). Bull. Brooklyn Ent. Soc., 18:58-62.
TATE, P.
1935. The larva of *Phaonia mirabilis* Ringdahl, predatory on mosquito larvae (Diptera, Anthomyidae). Parasitology, 27:556-560.
1939. The early stages of *Limnophora exsurda* Pand. (Diptera, Anthomyidae). Parasitology, 31:479-485.
VAILLANT, F.
1953. Les premiers stades de *Lispa consanguinea* Loew (Muscidae Anthomyinae). Inst. Rech. Sahariennes Univ. d'Alger. Miss. Sc. au Tassili des Ajjer (1949) I. 9 pp.
WILLIAMS, F. X.
1938. Biological studies in Hawaiian water-loving insects. Part III. Diptera or flies. A. Ephydridae and Anthomyiidae. Proc. Hawaii. Ent. Soc., 10:85-119, 9 pls.

Family SARCOPHAGIDAE

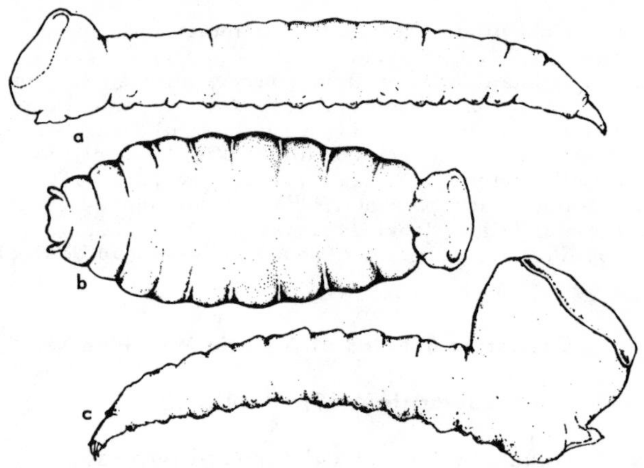

The "flesh flies" regularly breed in excrement or in carrion or are parasitic on other arthropods, but a few species are known to have aquatic habits. They have been taken in pitchers of *Sarracenia* spp., the eastern pitcher plants of North America, feeding on the remains of insects trapped inside. Aldrich (1916) lists six species of *Sarcophaga* which have been reared from *Sarracenia: S. celarata* Aldrich, *fletcheri* Aldrich, *jonesi* Aldrich, *rileyi* Aldrich, *sarraceniae* Riley, and a variety of *utilis* Aldrich. Riley's original account (1874) of *sarraceniae* from South Carolina included descriptions and figures of the immature stages which resemble those of the nonaquatic species. Johannsen's (1935) description and figures (fig. 14:64*a-c*) of "*Sarcophaga dux sarracenioides* (?)" from New York, on the other hand, show important specializations of this species for aquatic life. The posterior spiracles with their long vertical slits are borne at the bottom of a large, thin-walled, cuplike structure which serves as a float, rather than at the bottom of the usual smaller cavity. No sarcophagids have been reported from *Darlingtonia,* the California pitcher plant.

a
b
c

Fig. 14:64. Sarcophagidae. *a-c, Sarcophaga dux* ?: *a*, second stage larva; *c*, third stage larva; *b*, puparium (Johannsen, 1935).

REFERENCES

ALDRICH, J. M.
1916. *Sarcophaga* and allies in North America. 301 pp. 16 pls. Lafayette, Indiana: Thomas Say Foundation.
JOHANNSEN, O. A.
1935. *See* Diptera references.
RILEY, C. V.
1874. Descriptions and natural history of two insects which brave the dangers of *Sarracenia variolaris.* Trans. St. Louis Acad. Sci., 3:235-240.

Glossary

Largely because of uncertainty and confusion concerning anatomical homologies within the Insecta, many of the terms used in entomology cannot be defined so as to have universal application throughout the Class. For this reason such terms used herein have been defined for the purposes of this work only. Many of the terms used in the keys are adequately defined in the introductory parts of the various chapters and others are explained by illustrations. The following list is an attempt to define all other terms with which some readers might not be familiar. For definitions of terms not included here the reader is referred to general works on the group to which the term applies, to textbooks of general entomology, or to J. R. de la Torre-Bueno, *A Glossary of Entomology,* Brooklyn Entomological Society, 1937. Most latin terms are given in the singular. Plural forms are as follows: for words ending in us—i (*stylus, styli*); a—ae (*fossa, fossae*); um—a (*notum, nota*); on—a (*elytron, elytra*); is—es *(penis, penes).*

Acetabulum. The cavity into which an appendage is articulated; specifically, a coxal cavity.

Aciculate. Appearing as if superficially scratched with a needle.

Acuminate. Tapering to a long point.

Aedeagus. In male insects, the intromittent organ or penis.

Aeneous. Bright brassy or golden green color.

Alula. In certain Diptera, the basal lobe along the posterior margin of the wing, also called the axillary lobe.

Alutaceous. Rather pale leather brown; covered with minute cracks like the human skin.

Alveolate. Furnished with cells or alveoli; deeply pitted.

Amphipneustic. In insect larvae, characterized by having only the first thoracic and the last one or two pairs of abdominal spiracles open and functional.

Anal area. The posterior part of a wing supported by the anal veins.

Anal hooks. In Trichoptera larvae, the hooklike claws borne on the anal prolegs.

Anepisternal suture. In certain Diptera, a suture dividing the episternum (=mesopleuron of fig. 14:2*h*) into an upper part (anepisternum) and lower part (katepisternum).

Ankylosed. Grown together at a joint.

Annuliform. In the form of rings or ringlike segments.

Arculus. In some insect wings, especially in Odonata, a transverse vein between the radius and cubitus formed by the oblique basal part of the media and a true cross vein.

Asperate. Roughened.

Atrium. Any chamber at the entrance of a body opening; a chamber just within the spiracle into which the spiracular opening leads, and before the occluding membrane of the trachea.

Auricles. Somewhat earlike outgrowths from the sides of the second abdominal segment of some male Anisoptera.

Basal space. In Odonate wings, the clear space or cell bounded distally by the arculus; also called the basilar space and median space.

Basitarsus. The proximal or basal segment of the tarsus.

Bicuspidate. Having two cusps or points.

Bifid. Cleft or divided into two parts; forked.

Biforous. Of spiracles, having two openings.

Bilamellate. Having or divided into two lamellae or plates.

Bothriothrichia. In Collembola, very fine sensory hairs arising from cuplike pits in the integument.

Bucculae. In Hemiptera, elevated plates or ridges on the under side of the head, on each side of the rostrum.

Bulla. A blister or blisterlike structure; in mayflies, one or more weak spots which may be present in some of the principal wing veins, appearing as weakly sclerotized dilations of the veins.

Callosity. A thick swollen lump, harder than its surrounding area.

Calypteres. In certain Diptera, the two basal lobes at the posterior margin of the wing, termed individually the proximal or lower calypter which is joined to the thorax, and the distal or upper calypter which lies between the latter and the axillary lobe in the outstretched wing.

Canaliculate. Channeled; longitudinally grooved, with a deeper concave line in the middle.

Canthus. The integumentary process more or less completely dividing the eyes of some insects into an upper and lower half.

Carina. An elevated ridge or keel, not necessarily high or acute.

Castaneous. Bright red-brown; chestnut brown.

Caudal. Of or pertaining to the tail or anal end of the body.

Cerci. Paired appendages of the eleventh abdominal segment, often slender, filamentous, and segmented.

Chaetotaxy. The arrangement and nomenclature of the setae or bristles on any part of the exoskeleton of an insect.

Chelate. Bearing a chela, or resembling the nipperlike grasping organ of a crab; in insects, having the femur enlarged and often grooved to receive the curved tibia which may be folded back against it, thus forming an efficient grasping organ.

Chitinized. To have formed, deposited, or filled in with chitin and often erroneously used as a synonym of sclerotized, q.v.

Cinereous. Ash-colored; gray.

Clavate. Clubbed; thickening gradually toward the tip.

Club. In insect antennae, the terminal segments when more or less enlarged to form a knob or head of a club.

Collum. The neck or collar; in Coleoptera, the posterior narrow part of the head or even the prothorax.

Connate. United at the base or along the whole length.

Concolorous. Of a uniform color.

Connexivum. The prominent abdominal margin of some Hemiptera at the junction of the dorsal and ventral plates.

Cord. In some insect wings, a line of transverse joinings, composed of cross veins and bases of principal forks, extending from the stigma backward to the cubital vein.

Cordate. Cordiform, heart-shaped; triangular, with the corners of the base rounded, but not necessarily notched at the middle of the base.

Coriaceous. Like leather; thick, tough, and somewhat rigid.

Cornutus. In genitalia a slender, heavily sclerotized spine which often occurs, usually several at a time, in the ejaculatory duct.

Costate. Furnished with costae or longitudinal raised ribs.

Costiform. In the form of costae or raised ribs.

Crenate. Having the margin notched or scalloped so as to form rounded teeth.

Crenulation. One of a series of rounded projections along a margin.

Cribriform. With perforations like those of a sieve.

Cristate. With a prominent carina or crest on the upper surface; crested.

Cruciate. Shaped like a cross; applied to bristles when they cross in direction.

Cultriform. Sharp-edged and pointed; shaped like a pruning knife with a curved tip.

Cupreous. Coppery; metallic copper-red color.

Cuticle. The material secreted onto or deposited on the outer surface of the epidermal cells and solidifying there to form the exoskeleton.

Declivate. Declivous, sloping gradually downward.

Decumbent. Bending downward; bending down at tip from an upright base.

Decurrent. Applied to wing veins, closely attached to and running along another vein.

Decurved. Bowed or curved downward.

Dehiscence. The splitting of the pupal integument, usually along certain lines of dehiscence, in the emergence of adult insects.

Denticulations. Small teeth or toothlike processes.

Dentiform. Formed or appearing like a tooth.

Depressed. Flattened down.

Dichromatism. The possession of two color varieties or forms.

Disk. The central upper surface of any part.

Distal. Situated away from the point of origin or attachment, as an appendage.

Elytron. One of the anterior leathery or sclerotized wings of beetles, serving as a covering to the hind wing.

Elytroid. *See* operculate.

Emarginate. Notched; with an obtuse, rounded, or quadrate section cut from the margin.

Embolar groove. The deep submarginal gutter on the embolium which may be extended beyond the nodal furrow. Its surface is pruinose.

Entire. With an even, unbroken margin.

Entomophagous. Feeding upon insects.

Epipleura. The deflexed or inflexed parts of the elytra, immediately beneath the edge.

Epistoma. The oral margin and an indefinite space immediately contiguous thereto.

Explanate. Applied to a margin, spread out and flattened.

Extensors. In Diptera, the extensor row of bristles on the upper surface of the femur or tibia.

Falcate. Sickle-shaped; convexly curved.

Fascia. A transverse band or broad line.

Fenestrated. With transparent or perforated windowlike spots or areas.

Ferrugineous. Rusty red-brown.

Fibrillar. Fiber- or threadlike.

Fibrillate. Formed or consisting of fibers or appearing so.

Filiform. Threadlike; slender and of nearly uniform diameter.

Fimbriate. Fringed with hairs of irregular length.

Flabellate. Fan-shaped; with long thin processes lying flat on each other like the folds of a fan.

Flavescent. Yellowish.

Flavotestaceous. Light yellow-brown.

Flavous. Pure, clear yellow.

Flexor comb. In Diptera, a row of bristles along the lower surface of the femur.

Forceps. In mayflies, the movable appendages of the ninth abdominal segment of the male arising from the outer terminal angles of the styliger plate and functioning as clasping organs; also called styli.
Forceps base. *See* styliger plate.
Fossa. A deep pit or furrow.
Fossorial. Formed for or with the habit of digging or burrowing.
Fovea. A deep depression with well-marked sides.
Foveate. With deep depressions or foveae.
Frontoclypeus. The combined front and clypeus when the suture between them is obsolete.
Fulvotestaceous. Brownish yellow.
Fumose. Smoky.
Funicle. In chalcidoid Hymenoptera, the portion of the antenna between the club and the ring segments, the latter being the three small segments adjoining the pedicel.
Fuscous. Dark brownish gray.
Fusiform. Spindle-shaped; broad at the middle and narrowing toward the ends.

Galea-lacinea. In mayfly naiads, the compound maxillary structure known also as the blade of the maxilla borne on the stipes and derived from the fused galea and lacinia.
Ganglionic area. In certain mayfly naiads, the ganglia of the ventral nerve cord are pigmented and may be seen as dark areas beneath the translucent integument.
Geminate. Arranged in pairs composed of two similar parts; doubled; twinned.
Genital hooks. In Plecoptera males, a pair of processes arising from the subanal plates of the eleventh segment and curving forward.
Gibbosity. A protuberance or swelling.
Glabrous. Smooth, hairless and without punctures or structures.
Glossa. One of the two median terminal lobes of the labium.

Hamulus. In some Trichoptera, one of a series of small hooks along the costal margin of the wing.
Hemelytral commissure. In some Hemiptera, a median furrow or imaginary line behind the scutellum formed where the hemelytra meet along the clavus when the wings are folded.
Hyaline. Transparent, glassy.
Hydrofuge. Water repellent; unwettable.
Hypognathus. Having the cranium corresponding in position to the body segments with the mouth parts directed downward.
Hypomeron. In Coleoptera, the inflexed edge of the pronotum (pronotal hypomeron) and the raised lower margin of the epipleura (elytral hypomeron).
Hypoocular. Lying beneath the compound eye.

Immaculate. Without spots or marks.
Incrassate. Thickened; rather suddenly swollen at some point, especially near the tip.
Intercalary vein. An added or extra vein; in mayflies, any of the longitudinal wing veins intercalated between the main veins, as between the forks of R , M, and so on, and often detached basally.
Irrorate. Freckled; covered with minute spots.

Joining. See tail joinings.

Labial mask. The prehensile labium of odonate nymphs.
Lacinia mandibulae. A fleshy or membranous process from the interior face of the mandible.
Lamellate. Sheet or leaflike; composed of or covered with laminae or thin sheets.
Lamelliform. Comprised of or resembling leaves, blades, or lamellae.
Laminate. Formed of thin flat layers or leaves.
Laminiform. Having the appearance or made up of laminae or layers.
Lanceolate. Lance- or spear-shaped; oblong and tapering to a point.
Lateroclinate. Directed laterally or toward the side.
Ligula. The apicocentral sclerite of the labium, borne upon the distal margin of the prementum, usually single but sometimes paired; often used synonymously with "glossa" and "tongue."
Littoral. Pertaining to or inhabiting the shore region of a body of water; in a marine situation, corresponding more or less with the intertidal zone.
Lunate. Crescent-shaped.
Luteous. Yellow, generally orangish or reddish in tint.

Macelike. Resembling a mace, a clublike weapon with a spiked head.
Macrotrichia. The larger microscopic hairs on the surface of the wing.
Maculate. Spotted or marked with figures of any shape, of a color differing from the ground color.
Mammilose. With nipplelike protuberances or processes.
Median Planate. *See* figure 4:41.
Medifacies. In Diptera, the median part of the face, that is, the part between the facial ridges and separated from the eye on each side by the parafacies, q.v.
Meson. The imaginary middle plane dividing a body into right and left halves.
Mesonotal praescutum. In Diptera the region of the mesonotum lying anterior to the transverse suture.
Mesoscutum. The scutum or dorsal surface of the mesothorax.
Mesostigmal lamina. A pair of small transverse plates at the anterior end of the thoracic dorsum, in female Zygoptera forming part of the coupling mechanism used in pairing, being engaged by the inferior appendages of the male.
Metepimeron. In Odonata, the posterior pleural sclerite of the thorax delimited anteriorly from the spiracle-bearing metepisternum by the second lateral suture.
Micropylar. Of or pertaining to the micropyle or aperture in the insect egg through which sperm enter.
Moniliform. Beaded like a necklace.
Mucronate. Having pointed processes; terminating in a sharp point or mucro.

Nasale. An anterior median projection from the frons,

formed either by a fusion of frons, clypeus, and labrum or sometimes by frons and clypeus alone.
Nasus. In some Tipulidae, a noselike point at the apex of the rostrum.
Natatory. Pertaining to, adapted for, or characterized by swimming.
Nervures. The veinlike structures supporting the wing; veins.
Notogaean. Of or pertaining to a proposed zoölogical realm including the Australian, Polynesian, and Hawaiian regions.
Notum. The dorsal sclerotization of a body segment, especially in the thorax.
Nubilous. Cloudy, diffuse, indistinct.

Obconic. In the form of a reversed cone, with the apex as a base and the base apical.
Obsolete. Almost or entirely absent; indistinct; not fully developed.
Occipital. Of or pertaining to the occiput or back part of the head.
Ocellate. Provided with ocelli or simple eyes; having round spots ringed with another color.
Omphalium. In the hemipterous family Gerridae, the elevated median gland opening in the metasternum posteriorly.
Onisciform. Shaped like a sow bug or wood louse of the genus *Oniscus*.
Operculum. In mayflies, one of the pairs of abdominal gills (usually the first, second or fourth) which has become enlarged and thickened so as to form a protective cover for the other gills.
Operculate. Having an operculum, q.v.; elytroid.

Palletes. The disclike structure composed of three tarsal segments on the anterior legs of male Dytiscidae.
Palpiger. A sclerite bearing a palpus; specifically, the palpus-bearing structure of the labial mentum.
Parafacies. In Diptera, the part of the face between the facial ridges and the eyes.
Paraglossae. The lateral terminal lobes of the labium.
Palmately. Shaped like a palm, or a hand with the fingers extended.
Pectinate. Comblike; applied to structures such as antennae with even processes like the teeth of a comb.
Pedicellate. Supported by a pedicel or stalk; stalked or on a stalk or stem.
Penis. In mayflies, one of a pair of tubular processes of the ninth abdominal segment arising dorsal to the styliger plate and carrying the opening of the sperm ducts at their tips.
Perfoliate. Divided into leaflike plates; applied to antennae with disclike expansions connected by a stalk passing nearly through their centers.
Petiolate. Having a petiole, q.v.; stalked; placed upon a stem.
Petiole. A stem or stalk; the slender segment between the thorax and abdomen in many Hymenoptera and certain Diptera.
Piceous. Pitchy black; black with a reddish tinge.
Pile. Thick, fine, short, erect hair giving a surface appearance like velvet.
Pinnately. Having branches or parts arranged on each side of a central axis.
Planate. *See* figure 4:41.
Plastron. In aquatic insects, an air film on the surface of the body, by means of which the insect respires when submerged.
Plica. A fold or wrinkle; a longitudinal pleat of the wing.
Porrect. Extending forward horizontally.
Postocular. Behind the eyes.
Postscutellum. In Diptera the convex area immediately behind or beneath the scutellum.
Prescutal lobe. In Diptera, a callus situated anterior to the wing base, one on each side of the mesonotum.
Proclinate. Directed forward; applied to bristles.
Produced. Drawn out; prolonged.
Prognathous. Having the frontal surface of the cranium horizontal with the mouth parts directed forward.
Proleg. A fleshy appendage of the thorax or abdomen of some immature insects, usually bearing numerous small hooks or crotchets, and functioning as a leg.
Prostyle. In some beetle larvae, one of a pair of lateral, worm-shaped appendages of the anal segment.
Proximal. Situated toward the point of origin or attachment, as of an appendage.
Pruinose. Covered with a frostlike bloom or powdery secretion.
Pseudocelli. Sense organs, resembling ocelli in appearance, distributed over the body in certain Collembola and Protura.
Pseudomola. A structure resembling the mola or roughened grinding surface near the base of the mesal surface of a mandible.
Pterostigma. A thickened opaque spot on the costal margin of a wing near its middle or at the end of the radius.
Pubescence. Short, fine, soft, erect hair or down.
Punctulate. With small punctures or pitlike depressions.
Pupillate. With an eyelike center.
Pygidium. In Curculionidae, the last visible dorsal sclerite of the abdomen.
Pyriform. Pear-shaped.

Radial planate. *See* figure 4:41.
Radial space. The space or cell immediately behind the radial vein.
Raptorial. Adapted for seizing prey; predaceous.
Rastrate. Covered with nearly parallel, longitudinal scratches.
Reclinate. Directed backward.
Reticulate. Covered with a network of lines.
Retinaculum. In Coleoptera, the middle toothlike process of the larval mandible.
Retuse. Having an obtuse or rounded apex with a shallow notch.
Revolute. Spirally rolled backward.
Riparian. Of, pertaining to, or frequenting rivers or their shores.

Rostrate. Having a rostrum or snoutlike projection of the head bearing the mouth parts.
Rostrum. A snoutlike prolongation of the head bearing the mouth parts at the tip, or formed partly by the mouth parts.
Rufescent. Somewhat reddish.
Rufotestaceous. Reddish, yellowish brown.
Rufous. Pale red.
Ruga. A wrinkle.
Rugose. Wrinkled.
Rugulose. Minutely wrinkled; roughened but not rastrate.

Saltatorial. Characterized by or adapted for leaping.
Scabrous. With a rough surface; irregularly and roughly wrinkled.
Sclerotized. Of the integument, hardened and usually darkened in definite areas by the deposition or formation of other substances than chitin in the cuticle.
Sculpturing. The markings or pattern of impression or elevation on an elytron or other body surface.
Scurfy. Having a scaly surface.
Scutellum. In Coleoptera and Hemiptera, the triangular piece between the bases of the elytra or hemelytra.
Scutum. Often the entire mesonotum anterior to the scutellum but sometimes restricted to only that part behind the transverse suture.
Sericeous. Silky; with short, thick, silky down giving a silky sheen.
Serrate. With notched edges like the teeth of a saw.
Sessile. Attached by the base or without any stemlike support or petiole; having the abdomen closely attached for nearly its full width to the thorax.
Setate. Having or bearing setae.
Setiferous. Setigerous, set with or bearing setae.
Setiform. Seta-shaped; hairlike.
Shagreened. Covered with a close-set roughness, like the rough-surfaced horse leather termed shagreen.
Sigmoid. Shaped like the letter S.
Sinuate. Distinctly wavy.
Spatulate. Having a broad rounded end and a narrow attenuate base.
Spinneret. A part of the apparatus by means of which silk is spun, usually a hollow bristle or spine.
Spinulose. Set with little spines.
Squama. Calypteres, q.v.
Sternite. The ventral sclerotized part of a segment.
Stimuli. Small spines.
Stria. In general, any fine longitudinal impressed line; in Coleoptera, a longitudinal depressed line or furrow, frequently punctured, extending from the base to the apex of the elytra.
Striga. A narrow transverse line or slender streak, either on the surface or impressed.
Strigate. Having strigae; applied to a surface on which the strigae are impressed as in the elytra of some beetles or to an ornamentation composed of short fine lines.
Strigose. Rough with rigid bristles.
Striole. A rudimentary or poorly impressed stria.
Stylus. *See* forceps.
Styliger plate. In mayflies, a broad, well-articulated sclerite at the outer terminal angles of which arise the clasping organs (forceps or styli) and thought to be derived from the fusion of the coxites of the primitive appendages of the ninth segment; also called forceps base.
Subanal lobes. In Plecoptera adults, a pair of plates belonging to the eleventh segment, situated beneath the anus and bearing the cerci.
Subcallus. In Tabanidae, the large callus or swelling on the face below the level of the lower inner angles of the eyes and beneath the basal callus which lies at or above the lower inner angles of the eyes.
Subepaulet. In Tabanidae, a small scalelike sclerite at the base of the costa.
Subequal. Similar, but not equal in size, form, or other characteristic.
Subgenital plate. In Plecoptera females, the posterior prolongation of the eighth abdominal sternite.
Subrugose. Slightly or weakly wrinkled.
Subulate. Awl-shaped; linear at the base, attenuate at the tip.
Sulcate. Deeply furrowed or grooved.
Sulcation. A furrow or groove.
Supra-anal process. In Plecoptera males, a median process developed from the supra-anal lobe or dorsal plate of the eleventh abdominal segment.
Synthlipsis. Interocular space.

Tails. A general term designating the long terminal abdominal appendages of mayflies, comprised of two lateral filaments (the cerci) which are always present, and a median filament (the telofilum) which may be well developed, vestigial or absent.
Tail joinings. In Ephemeroptera, the weakly sclerotized or feebly pigmented intersegmental areas of the tails.
Tectiform. Rooflike; sloping from a median ridge.
Tegula. A sclerite or lobelike structure at the base of the fore wing, especially well developed in some Hymenoptera and Lepidoptera.
Teneral. The condition of the imago after emergence from the pupa when its cuticle is not entirely hardened or fully of the mature color.
Terete. Cylindrical or nearly so.
Termen. The outer margin of a wing, between the apex and the hind or anal angle.
Tibiotarsus. The terminal leg segment composed of the fused tibia and tarsus, found in some insects.
Titillator. In Plecoptera males, the median process formed by the processes of the basal lobes.
Tomentum. A form of pubescence composed of matted, woolly hair.
Tornus. In Lepidoptera, the junction of the termen and inner margin of the wing; hind or anal angle of the wing.
Tracheal gills. Filiform or more or less lamellate organs well supplied with tracheae and tracheoles present in many aquatic larvae and some aquatic pupae.
Trochantin. In Coleoptera, a structure often present on the outer side of, and sometimes movable on, the

coxa; also the small sclerite connecting the coxa with the sternum in Dytiscidae.
Truncate. Cut off squarely at the tip.
Tubulate. Formed into or like a tube.
Tumidity. A swelling.
Turbinate. Top-shaped; nearly conical.

Umbone. An embossed elevated knob situated on the humeral angle of elytra.
Urogomphus. A paired fixed or movable process found on the terminal segments of certain larvae.

Ventrite. The ventral surface of one of the body segments.
Vermiculate. Wormlike; with tortuous markings resembling the tracks of a worm.
Verruca. In lepidopterous larvae, somewhat elevated parts of the cuticle bearing tufts of setae.
Verticils. One of the whorls of long, fine hair arranged symetrically on the antennal segments in certain Diptera.
Vestigial. Of an organ or structure, degenerate or imperfectly developed, having little or no utility but which in an earlier evolutionary stage of the organism performed a useful function.
Vestiture. The general surface covering of insects, as scales, hairs, and the like.
Vitreous. Glassy.
Vitta. A longitudinal stripe.

Index